Solid

Rectangular Solid
Surface Area $S = 2\ell w + 2wh + 2\ell h$
Volume $V = \ell wh$

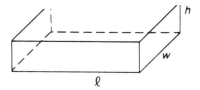

Right Circular Cylinder
Surface Area $S = 2\pi rh + 2\pi r^2$
Volume $V = \pi r^2 h$

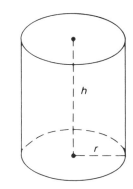

Right Circular Cone
Surface Area $S = \pi r\ell + \pi r^2$
Volume $V = \frac{1}{3}\pi r^2 h$

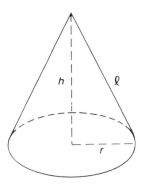

Sphere
Surface Area $S = 4\pi r^2$
Volume $V = \frac{4}{3}\pi r^3$

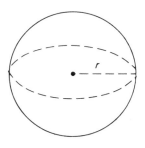

Right Pyramid
Volume $V = \frac{1}{3}bh$
(b is the area of the base)

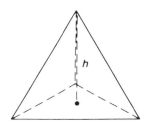

ALGEBRA

FOR COLLEGE STUDENTS

THIRD EDITION

ALGEBRA
FOR COLLEGE STUDENTS

Terry H. Wesner
Henry Ford Community College

Harry L. Nustad

WCB **Wm. C. Brown Publishers**

Book Team

Editor *Earl McPeek*
Developmental Editor *Theresa Grutz*
Production Editor *Eugenia M. Collins*
Designer *K. Wayne Harms*
Photo Editor *Carrie Burger*
Art *Joseph P. O'Connell*

Wm. C. Brown Publishers
A Division of Wm. C. Brown Communications, Inc.

Vice President and General Manager *George Bergquist*
National Sales Manager *Vincent R. Di Blasi*
Assistant Vice President, Editor-in-Chief *Edward G. Jaffe*
Marketing Manager *Elizabeth Robbins*
Advertising Manager *Amy Schmitz*
Managing Editor, Production *Colleen A. Yonda*
Manager of Visuals and Design *Faye M. Schilling*

Design Manager *Jac Tilton*
Art Manager *Janice Roerig*
Photo Manager *Shirley Charley*
Production Editorial Manager *Ann Fuerste*
Publishing Services Manager *Karen J. Slaght*
Permissions/Records Manager *Connie Allendorf*

Wm. C. Brown Communications, Inc.

Chairman Emeritus *Wm. C. Brown*
Chairman and Chief Executive Officer *Mark C. Falb*
President and Chief Operating Officer *G. Franklin Lewis*
Corporate Vice President, Operations *Beverly Kolz*
Corporate Vice President, President of WCB Manufacturing *Roger Meyer*

Cover photo © David Muench 1991

Photo credits: Chapter openers: 1: © Bob Daemmrich/The Image Works; 2: © Comstock/Comstock, Inc.; 3: © Superstock; 4: © Comstock/Comstock, Inc.; 5: © Bob Daemmrich/Stock Boston; 6: © Bob Coyle; 7: © Mike Greenlar/The Image Works; 8: © Dave Schaefer/The Picture Cube; 9: © Wayne Floyd/Unicorn Stock Photos; 10: © Tom McHugh/Photo Researchers Inc.; 11: © Michael J. Howell/The Picture Cube; 12: © Bill Horsman/Stock Boston; 13: © Bob Coyle; 14: © Ellis Herwig/Stock Boston; 15: © Comstock/Comstock, Inc.

Library of Congress Catalog Card Number: 91–55593

ISBN 0-697-07654-7

Printed in the United States of America by Wm. C. Brown Communications, Inc., 2460 Kerper Boulevard, Dubuque, IA 52001

10 9 8 7 6 5 4 3 2

To Mary Ann

Terry

To Dene

Harry

Contents

Chapter 1 ■ Basic Concepts and Properties

Chapter 2 ■ First-Degree Equations and Inequalities

Chapter 3 ■ Exponents and Polynomials

Chapter 4 ■ Rational Expressions

Chapter 5 ■ Exponents, Roots, and Radicals

Chapter 6 ■ Quadratic Equations and Inequalities

Chapter 7 ■ Linear Equations and Inequalities in Two Variables

Chapter 8 ■ Functions

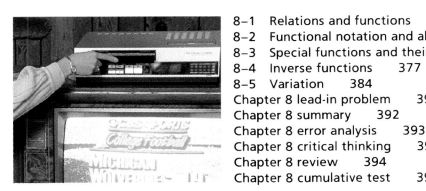

Chapter 9 ■ Conic Sections

Chapter 10 ■ Polynomial and Rational Functions

Chapter 11 ■ Exponential and Logarithmic Functions

Chapter 12 ■ Systems of Linear Equations

Chapter 13 ■ Matrices and Determinants

Chapter 14 ■ Sequences, Series, and Mathematical Induction

Chapter 15 ■ Counting Techniques, Probability, and the Binomial Theorem

20 Point Learning System

Your students will count on Terry Wesner and Harry Nustad's integrated learning system. It is the product of over 50 years of combined teaching experience and has been developed with the help of feedback from users—both professors and students—through various texts and editions by this author team. The authors have fine-tuned and enhanced their learning system for this third edition of *Algebra for College Students*. A full-color design makes an already superb learning system even better. The pedagogical color scheme is used consistently throughout, providing a road map to guide students through the key points of each section. Much more than just adding visual appeal, the color in this text is an integral part of the learning system. Let's take a look at examples of the 20 points that make up the learning system.

1. Chapter Lead-in Problem and Solution
2. Explanations
3. Examples
4. Quick Checks
5. Procedure Boxes
6. Definitions
7. Concepts
8. Notes
9. Problem Solving
10. Mastery Points
11. Section Exercises
12. Quick-Reference Examples
13. Trial Problems
14. Core Exercise Problems
15. Section Review Exercises
16. Chapter Summary
17. Error Analysis
18. Critical Thinking
19. Chapter Review Exercises
20. Cumulative Test

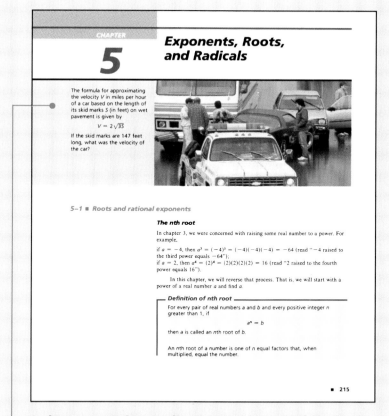

A **chapter-opening application** problem with full-color photo poses a problem that students will learn to solve as they progress through the chapter. Its step-by-step solution is shown before the chapter summary.

Procedure boxes clearly state step-by-step processes for working problems.

Examples include arrows that visually guide students through steps needed to solve the problem. A detailed explanation to the right of each step ensures student understanding of the correct solution method.

Quick checks parallel the development of examples within the text and allow students to immediately test their understanding of the material being studied. The quick check problem is worked step-by-step within the exercise set. The student is able to line-by-line check his or her own solution and use this solution as a quick reference while doing the problems within the exercise set. Notice that these quick check problems become the quick-reference examples on page 44.

Definitions are stated precisely in easy-to-understand terms.

Section 1–6 Sums and Differences of Polynomials **41**

Using the distributive property, the expression can be written

$$4x + 5x = (4 + 5)x = 9x$$

In this expression, $4x$ and $5x$ are terms that we wish to add. Terms are separated by the operations of addition and subtraction.

Simplifying expressions, as in this example, is *combining like terms*. **Like terms** are terms that may differ only in their numerical coefficients. For two or more terms to be called like terms, the variable factors of the terms along with their respective exponents must be identical. However the numerical coefficients may be different.

1. $5x^2y^3$, $-3x^2y^3$, and x^2y^3 are like terms because they differ only in their numerical coefficients.
2. $5a^3b^2$ and $5a^2b^3$ both contain the same variables but are not like terms because the exponents of the respective variables are not the same.

Combining like terms

1. Identify the like terms.
2. If necessary, use the commutative and associative properties to group together the like terms.
3. Combine the numerical coefficients of the like terms and multiply that by the variable factor.
4. Remember that y is the same as $1 \cdot y$ and $-y$ is the same as $-1 \cdot y$.

Note The process of addition or subtraction is performed only with the numerical coefficients, the variable factors remain unchanged.

■ **Example 1–6 A**

Perform the indicated addition or subtraction.

1. $3z - 7z + z - 6 + 15$ Identify like terms
  $= (3 - 7 + 1)z + (-6 + 15)$ Associative and distributive properties
  $= -3z + 9$ Combine numerical coefficients, combine numbers

2. $3a^2b + 5a^2b + 2a^2b = (3 + 5 + 2)a^2b = 10a^2b$
  Like terms

3. $x^3y + 4xy^2 - 3x^3y + 2xy^2$ Identify any like terms
  Like terms
  $= [1 + (-3)]x^3y + (4 + 2)xy^2$ Commutative, associative, and distributive properties
  $= -2x^3y + 6xy^2$ Combine numerical coefficients

Note The numerical coefficient of a term includes the sign that precedes it. Therefore we consider any addition and subtraction of terms as a sum of terms in which the sign that precedes the term is taken as the sign of the numerical coefficient.

▶ **Quick check** Perform the indicated addition or subtraction.
$5xy^3 - 3xy^3 + 7xy^3$ ■

Exponents and Polynomials

The amount A of a radioactive substance remaining after time t can be found using the formula $A = A_0(0.5)^{t/n}$, where A_0 represents the original amount of radioactive material and n is the half-life given in the same units of time as t. If the half-life of radioactive carbon 14 is 5,770 years, how much radioactive carbon 14 will remain after 11,540 years if we start with 100 grams?

3–1 ■ **Properties of exponents**

Exponential form

In chapter 1, we discussed exponents as related to real numbers. Since variables are symbols for real numbers, we shall now apply the properties of exponents to them. The expression a^3 (read "a to the third power") is called the **exponential form** of the product

$$a \cdot a \cdot a$$

We call a the **base** and 3 the **exponent**.

Exponent

Exponential form ── $4^3 = 4 \cdot 4 \cdot 4 = 64$ Standard form

Base   3 factors of 4

Definition of exponents

$$a^n = a \cdot a \cdot a \cdots a$$

n factors of a

where n is a positive integer.

Concept
The exponent tells us how many times the base is used as a factor in an indicated product.

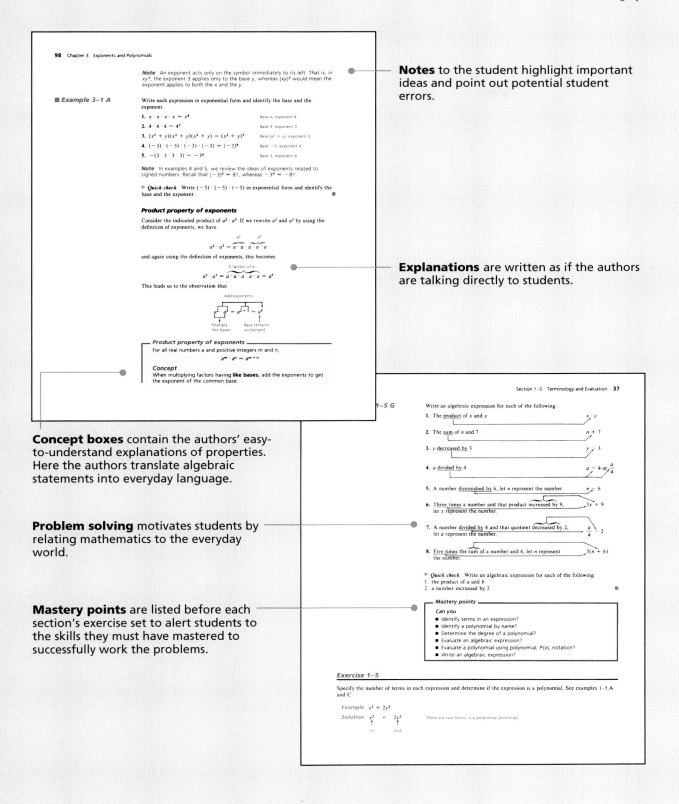

98 Chapter 3 Exponents and Polynomials

Note An exponent acts only on the symbol immediately to its left. That is, in xy^3, the exponent 3 applies only to the base y, whereas $(xy)^3$ would mean the exponent applies to both the x and the y.

■ *Example 3–1 A* Write each expression in exponential form and identify the base and the exponent.

1. $x \cdot x \cdot x \cdot x = x^4$ Base x, exponent 4
2. $4 \cdot 4 \cdot 4 = 4^3$ Base 4, exponent 3
3. $(x^2 + y)(x^2 + y)(x^2 + y) = (x^2 + y)^3$ Base $(x^2 + y)$, exponent 3
4. $(-3) \cdot (-3) \cdot (-3) \cdot (-3) = (-3)^4$ Base -3, exponent 4
5. $-(3 \cdot 3 \cdot 3 \cdot 3) = -3^4$ Base 3, exponent 4

Note In examples 4 and 5, we review the ideas of exponents related to signed numbers. Recall that $(-3)^4 = 81$, whereas $-3^4 = -81$.

▶ *Quick check* Write $(-5) \cdot (-5) \cdot (-5)$ in exponential form and identify the base and the exponent. ■

Product property of exponents

Consider the indicated product of $a^2 \cdot a^3$. If we rewrite a^2 and a^3 by using the definition of exponents, we have

$$a^2 \cdot a^3 = \overbrace{a \cdot a}^{a^2} \cdot \overbrace{a \cdot a \cdot a}^{a^3}$$

and again using the definition of exponents, this becomes

$$a^2 \cdot a^3 = \overbrace{a \cdot a \cdot a \cdot a \cdot a}^{5 \text{ factors of } a} = a^5$$

This leads us to the observation that

Add exponents

$$a^2 \cdot a^3 = a^{2+3} = a^5$$

Multiply like bases   Base remains unchanged

Product property of exponents
For all real numbers a and positive integers m and n,
$$a^m \cdot a^n = a^{m+n}$$

Concept
When multiplying factors having **like bases,** add the exponents to get the exponent of the common base.

1–5 G Write an algebraic expression for each of the following.

1. The product of x and y $x \cdot y$
2. The sum of n and 7 $n + 7$
3. y decreased by 3 $y - 3$
4. a divided by 4 $a \div 4$ or $\dfrac{a}{4}$
5. A number diminished by 6, let n represent the number. $n - 6$
6. Three times a number and that product increased by 9, let x represent the number. $3x + 9$
7. A number divided by 4 and that quotient decreased by 2, let a represent the number. $\dfrac{a}{4} - 2$
8. Five times the sum of a number and 6, let n represent the number. $5(n + 6)$

▶ *Quick check* Write an algebraic expression for each of the following:
1. the product of a and b
2. a number increased by 2 ■

Mastery points

Can you
■ Identify terms in an expression?
■ Identify a polynomial by name?
■ Determine the degree of a polynomial?
■ Evaluate an algebraic expression?
■ Evaluate a polynomial using polynomial, $P(x)$, notation?
■ Write an algebraic expression?

Exercise 1–5

Specify the number of terms in each expression and determine if the expression is a polynomial. See examples 1–5 A and C.

Example $x^2 + 2y^2$

Solution $x^2 + 2y^2$ There are two terms, is a polynomial (binomial)
    ↑   ↑
    1st   2nd

Notes to the student highlight important ideas and point out potential student errors.

Explanations are written as if the authors are talking directly to students.

Concept boxes contain the authors' easy-to-understand explanations of properties. Here the authors translate algebraic statements into everyday language.

Problem solving motivates students by relating mathematics to the everyday world.

Mastery points are listed before each section's exercise set to alert students to the skills they must have mastered to successfully work the problems.

Exercise sets feature both algebraic and word problems that give students ample opportunity to practice their skills.

Fully worked-out quick-reference examples (quick check problems) are included for students to use as a line-by-line check of their work or as an example.

Core exercise problems address the major ideas of the section. The problem numbers for these exercises appear in green type for easy identification.

Trial exercise problems are located in the exercise sets and are denoted with a box around the problem number indicating that the solution is completely worked out in the answer appendix. The problem can be used as an example or line-by-line check of the problem.

---

44   Chapter 1   Basic Concepts and Properties

> **Mastery points**
>
> **Can you**
> ■ Identify like terms?
> ■ Perform addition and subtraction of polynomials?
> ■ Remove grouping symbols?

Exercise 1–6

Perform the indicated addition or subtraction. See example 1–6 A.

Example $5xy^3 - 3xy^3 + 7xy^3$ Determine if we have like terms
Solution $= (5 - 3 + 7)xy^3$ Distributive property
$= 9xy^3$ Combine numerical coefficients

1. $4x + 6x^2 - 2x^2 + 3x - 5x^2 + x$
2. $5y^2 - 12y + 6y^5 - 4y^2 + y$
3. $-a^2 + a - 5a^3 + 2a^2 + 4a$
4. $2x^2y - 4xy^2 + 3x^2y$
5. $5x^2y - 3xy + 6xy - x^2y$
6. $6a^2b - 2ab^2 + 3a^2b - 4a^2b^2$

Remove all grouping symbols and perform the indicated addition or subtraction. See example 1–6 B.

Example $(5x^2 + 2x - 4) - (3x^2 - 7x + 9)$
Solution $= 5x^2 + 2x - 4 - 3x^2 + 7x - 9$ Change the sign of each term in the second parentheses
$= (5x^2 - 3x^2) + (2x + 7x) + (-4 - 9)$ Commutative and associative properties
$= 2x^2 + 9x - 13$ Combine like terms

7. $(3x^2y - 2xy^2 + 9xy) + (2xy^2 - 4x^2y + 3xy)$
8. $(5ab^2 - 2a^2b^2 + 3a) - (4a^2b^2 + 3ab^2)$
9. $(5a^2 - b^2) - (4a^2 + 3b^2) - (a^2 - 7b^2)$
10. $(7x^3 - 2x^2y + 4xy^2 - 6y^3) - (4x^3 - 3x^2y - 2xy^2 - y^3)$
11. $(12x - 24yz) + (46yz - 16x - 26z)$
12. $(19x + 3y) - (-22x - 6y)$
13. $-(5a^2b - 6ab + 16c) + (7ab - 4a^2b)$
14. $(6ab + 11b^2c) - (11ab - 6bc)$
15. $-(14x - 31y) - (14x - 6y + 3z)$
16. $(x^2 - 3x + z) - (x^2 + 5x - 8) + (3x^2 - 4x)$
17. $(2x^2 - 7x + 3) + (5x^2 - 6) - (4 - 3x^2)$
18. $(7x^2 - 2xy + y^2) + (3x^2 - 6y^2 + 3xy) - (4y^2 - 6x^2 - xy)$
19. $(5xy - y^2) - (3yz + 2xy) + (3y^2 - 4xy)$
20. $(5ab + 2b^2) - (4a^2 - 3b^2) + (2ab - 7b^2)$

Perform the following addition and subtraction in column form. See example 1–6 B.

21. $(3x^2 + 4xy - 5y^2)$
$+ (x^2 - 7xy + 3y^2)$
22. $(4a^2 + 2ab + 3b^2)$
$+ (a^2 - 5ab + b^2)$
23. $(3x^2y - 2x^2y^2 + 3xy^2)$
$- (2x^2y + xy - 5xy^2)$
24. $(4st^2 + 3st - 2s^2t)$
$- (2st^2 + 3st - 2s^2t)$
25. $(-6a^2b^2 + 3ab - 4)$
$- (-4a^2b^2 + 2ab - 7)$
26. $(-9x^2y^3 - 4xy + 13)$
$- (-11x^2y^3 + 5xy - 6)$

Section 5–7   C

55. $\dfrac{5-2i}{5-i}$
56. $\dfrac{4-5i}{2+i}$
57. $\dfrac{3-i}{3+i}$
58. $\dfrac{5-i}{5+i}$
59. $\dfrac{2+5i}{3-2i}$
60. $\dfrac{4+3i}{3-i}$
61. $\dfrac{6+3i}{3+4i}$
62. $\dfrac{5-\sqrt{-4}}{3+\sqrt{-9}}$
63. $\dfrac{2+\sqrt{-16}}{4-\sqrt{-1}}$
64. $\dfrac{7-\sqrt{-25}}{3+\sqrt{-36}}$

Solve the following word problems.

65. The impedance of an electrical circuit is the measure of the total opposition to the flow of an electric current. The impedance Z in a series RCL circuit is given by

$Z = R + i(X_L - X_C)$

Determine the impedance if $R = 30$ ohms, $X_L = 16$ ohms, and $X_C = 40$ ohms.

66. Use the formula in exercise 65 to find Z if $R = 28$ ohms, $X_L = 16$ ohms, and $X_C = 38$ ohms.

67. The impedance Z in an AC circuit is given by the formula

$Z = \dfrac{V}{I}$

where V is the voltage and I is the current. Find Z when $V = 0.3 + 1.2i$ and $I = 2.1i$. Round all values to three decimal places.

68. Use the formula in exercise 67 to find Z if $V = 2.2 - 0.3i$ and $I = -1.1i$. Round all values to three decimal places.

69. The total impedance Z_T of an AC circuit containing impedances Z_1 and Z_2 in parallel is given by the formula

$Z_T = \dfrac{Z_1 Z_2}{Z_1 + Z_2}$

Find Z_T when $Z_1 = 3 - i$ and $Z_2 = 2 + i$.

70. Use the formula in exercise 69 to find Z_T if $Z_1 = 4 - i$ and $Z_2 = 3 + i$.

71. If three resistors in an AC circuit are connected in parallel, the total impedance Z_T is given by the formula

$Z_T = \dfrac{Z_1 Z_2 Z_3}{Z_1 Z_2 + Z_1 Z_3 + Z_2 Z_3}$

Find Z_T when $Z_1 = 3 - i$, $Z_2 = 3 + i$, and $Z_3 = 2i$.

72. Use the formula in exercise 71 to find Z_T if $Z_1 = 2 - i$, $Z_2 = 2 + i$, and $Z_3 = 2i$.

73. For what values of x does the expression $\sqrt{5-x}$ represent a real number?

74. For what values of x does the expression $\sqrt{x+4}$ represent a real number?

75. For what values of x does the expression $\sqrt{x+11}$ represent an imaginary number?

76. For what values of x does the expression $\sqrt{8-x}$ represent an imaginary number?

Review exercises

Factor completely. See section 3–8.

1. $x^2 - 12x + 36$
2. $3x^2 + 9x$
3. $x^2 - 16$
4. $9x^2 - 36$
5. $x^2 - 7x + 10$
6. $2x^2 - x - 3$
7. $5x^2 - 14x - 3$
8. $6x^2 - 23x - 4$

Review exercises at the end of each section help students prepare for the following section and keep in touch with previous material.

254 Chapter 5 Exponents, Roots, and Radicals

Chapter 5 lead-in problem

The formula for approximating the velocity V in miles per hour of a car based on the length of its skid marks S (in feet) on wet pavement is given by

$$V = 2\sqrt{3S}$$

If the skid marks are 147 feet long, what was the velocity of the car?

Solution

$V = 2\sqrt{3S}$	Original formula
$V = 2\sqrt{3(\ )}$	Formula ready for substitution
$V = 2\sqrt{3(147)}$	Substitute
$V = 2\sqrt{441}$	Simplify under radical
$V = 2 \cdot 21$	Simplify radical
$V = 42$	Multiply

Hence, the velocity of the car was 42 miles per hour.

Chapter 5 summary

1. $a^{1/n} = \sqrt[n]{a}$, whenever the principal nth root of a is a real number.
2. $a^{m/n} = (\sqrt[n]{a})^m = \sqrt[n]{a^m}$, if the principal nth root of a is a real number.
3. For all nonnegative real numbers a and b and positive integer n greater than 1.

$$\sqrt[n]{ab} = \sqrt[n]{a}\sqrt[n]{b}$$

$$\sqrt[n]{\frac{a}{b}} = \frac{\sqrt[n]{a}}{\sqrt[n]{b}}, \quad b \neq 0$$

and

$$\sqrt[nk]{a^{mk}} = \sqrt[n]{a^m}$$

4. We eliminate radicals from the denominator of a fraction by **rationalizing the denominator**.
5. We can only add or subtract *like* radicals.
6. **Conjugate factors** are used to rationalize the denominator of a fraction when the denominator has two terms where one or both terms contain a square root.
7. We define $i = \sqrt{-1}$, so that $i^2 = -1$.
8. A complex number is any number that can be written in the form $a + bi$, where a and b are real numbers and i represents $\sqrt{-1}$.

Chapter 5 error analysis

1. Principal nth root
   Example: $\sqrt{81} = 9$ or -9
   Correct answer: 9
   What error was made? (*see page 216*)
2. Principal nth roots
   Example: $\sqrt[3]{-8}$ does not exist in the real numbers.
   Correct answer: -2
   What error was made? (*see page 216*)
3. Rational number exponents
   Example: $(32)^{-1/5} = -(32)^{1/5} = -2$
   Correct answer: $\frac{1}{2}$
   What error was made? (*see page 221*)
4. Operations with rational exponents
   Example: $2^{1/3} \cdot 2^{1/3} = 2^{1/9}$
   Correct answer: $2^{2/3}$
   What error was made? (*see page 224*)
5. Product of radicals
   Example: $\sqrt[3]{3} \cdot \sqrt[3]{2} = \sqrt[3]{5}$
   Correct answer: $\sqrt[3]{6}$
   What error was made? (*see page 228*)
6. Reducing the index
   Example: $\sqrt[6]{x^3y} = \sqrt{xy}$
   Correct answer: $\sqrt[6]{x^3y}$
   What error was made? (*see page 229*)
7. Rationalizing the denominator
   Example: $\frac{1}{\sqrt[3]{x}} = \frac{\sqrt[3]{x}}{x}$
   Correct answer: $\frac{\sqrt[3]{x^2}}{x}$
   What error was made? (*see page 234*)
8. Sum of radicals
   Example: $\sqrt{5} + \sqrt{5} = \sqrt{10}$
   Correct answer: $2\sqrt{5}$
   What error was made? (*see page 238*)
9. Radical of a sum
   Example: $\sqrt{16 + 9} = \sqrt{16} + \sqrt{9} = 4 + 3 = 7$
   Correct answer: 5
   What error was made? (*see page 238*)
10. Multiplying radicals
    Example: $\sqrt{6}(\sqrt{6} - 3) = 3$
    Correct answer: $6 - 3\sqrt{6}$
    What error was made? (*see page 242*)

Chapter 5 Review **255**

Chapter 5 critical thinking

Choose a two-digit number whose one's digit is 5 (such as 35). You can find the square of this number if you multiply the ten's digit by one more than the ten's digits and place 25 after this product. For example:

$$35^2$$

3	4	=	12	25

Ten's digit   One more than the ten's digit   Product   Place 25 after the product

Therefore, $35^2 = 1225$
Why is this true?

Chapter 5 review

Assume that all variables represent positive real numbers and no denominator is equal to zero.

[5–1]
Rewrite the following in radical notation and use table 5–1 to simplify.

1. $36^{1/2}$ 2. $16^{-3/4}$ 3. $(-27)^{2/3}$

[5–2]
Perform the indicated operations and simplify. Leave the answer with all exponents positive.

4. $a^{2/3} \cdot a^{1/4}$ 5. $(c^{3/4})^{1/3}$ 6. $(27x^3)^{2/3}$ 7. $(b^{-2/3})^{-3}$

8. $\frac{a^{1/2}}{a^{1/3}}$ 9. $(8x^{12}y^4)^{2/3}$ 10. $\frac{x^3x^{2/3}}{x^{1/3}}$ 11. $\frac{a^{2/3}}{a^{-1/3}}$

[5–3]
Simplify the following.

12. $\sqrt{12}$ 13. $\sqrt{10}\sqrt{15}$ 14. $\sqrt[3]{x^4}\sqrt[3]{x^3}$
15. $\sqrt[3]{6ab^2}\sqrt[3]{4a^2b^4}$ 16. $\sqrt[6]{a^2}$ 17. $\sqrt[3]{8a^2b^4}$
18. If the hypotenuse of a right triangle is 10 inches long and one of the legs is 6 inches long, find the length of the other leg.

[5–4]
Simplify the following expressions leaving no radicals in the denominator.

19. $\sqrt{\frac{49}{64}}$ 20. $\sqrt[3]{\frac{16}{27}}$ 21. $\sqrt[4]{\frac{16x^3y^3}{z^4}}$ 22. $\sqrt{\frac{1}{8}}$ 23. $\frac{6}{\sqrt{2}}$

24. $\sqrt[3]{\frac{8}{25}}$ 25. $\sqrt[3]{\frac{x^3}{y}}$ 26. $\sqrt[3]{\frac{a^2}{b}}$ 27. $\frac{x}{\sqrt[3]{x^2}}$ 28. $\sqrt[3]{\frac{a}{bc^2}}$

29. $\frac{a}{\sqrt[3]{ab^2}}$ 30. $\sqrt{\frac{3x}{y^2}}\sqrt{\frac{xy}{3}}$

[5–5]
Perform the indicated operations and simplify.

31. $4\sqrt{3} - 3\sqrt{3} + 7\sqrt{3}$ 32. $\sqrt{18} + \sqrt{50}$ 33. $\sqrt{8a} + 9\sqrt{18a}$

34. $5x^2\sqrt{xy} - 2x\sqrt{x^3y}$ 35. $\frac{5}{\sqrt{6}} - \frac{1}{\sqrt{3}}$ 36. $\frac{2}{\sqrt{ab}} - \frac{1}{\sqrt{a}}$

The completely worked-out solution for the **chapter-opening word problem** appears at the end of the chapter prior to the chapter summary.

A **chapter summary** synthesizes important concepts.

Error analysis provides a group of problems where a common error has been made. The student is asked to correct the mistake. A page reference is provided so that the student can refer to examples and notes relative to the given problem.

Critical thinking provides special problems that the student must analyze and use their mathematical skills to solve. A series of hints are given in the Instructors Manual.

Chapter review exercises feature problems to help students determine if they need further work on a particular section. The problems are keyed to refer students back to the section from which they were drawn.

256   Chapter 5 Exponents, Roots, and Radicals

[5–6]

Simplify the following expressions leaving no radicals in the denominator.

37. $\sqrt{2}(\sqrt{6} - \sqrt{10})$ **38.** $2\sqrt{a}(\sqrt{ab} + 2\sqrt{a})$ **39.** $(5 - \sqrt{3})^2$

40. $(\sqrt{10} - \sqrt{7})(\sqrt{10} + \sqrt{7})$ **41.** $(2\sqrt{a} + 3\sqrt{b})(2\sqrt{a} - 3\sqrt{b})$ **42.** $(3\sqrt{x} + y)^2$

43. $\dfrac{1}{\sqrt{6} + 2}$ **44.** $\dfrac{10}{4 + \sqrt{6}}$ **45.** $\dfrac{\sqrt{3}}{\sqrt{6} - \sqrt{3}}$

46. $\dfrac{a^2b}{a\sqrt{a} - \sqrt{ab}}$

[5–7]

Simplify the following.

47. $\sqrt{-49}$ **48.** $\sqrt{-28}$ **49.** $(2i)^2$ **50.** $(\sqrt{7}i)^2$ **51.** $\sqrt{-3}\sqrt{-12}$

52. $(\sqrt{-3})^2$ **53.** $\sqrt{-2}\sqrt{-3}$ **54.** i^{37}

Simplify the following and leave the answer in standard form.

55. $(4 + 2i) + (3 + 5i)$ **56.** $(2 - \sqrt{-36}) - (3 + \sqrt{-25})$

57. $(2 - \sqrt{-9})(3 + \sqrt{-16})$ **58.** $(2 + 5i)^2$

59. $\dfrac{3 + 4i}{i}$ **60.** $\dfrac{7 - 6i}{\sqrt{-9}}$ **61.** $\dfrac{4 - i}{2 + i}$ **62.** $\dfrac{9 - \sqrt{-4}}{7 - \sqrt{-9}}$

Chapter 5 cumulative test

Factor completely.

[3–5]   **1.** $a^2 - 7a - 8$ **[3–4]**   **2.** $4x^2 - 3x$ **[3–7]**   **3.** $9x^2 - 36$

[3–6]   **4.** $2x^2 + 11x + 12$ **[3–6]**   **5.** $3a^2 - 11a - 20$ **[3–6]**   **6.** $6x^2 + 17x + 12$

[1–5]   **7.** Evaluate the expression $b^2 - 4ac$ at
(a) $a = 1$, $b = 4$, and $c = -3$;
(b) $a = 2$, $b = -4$, and $c = 3$.

Find the solution set.

[2–1]   **8.** $3(2x - 1) + 4 = x + 3$ **[2–5]**   **9.** $5x + 7 > 2x - 4$ **[2–2]**   **10.** $3x - 2y = 4(x + y)$

[2–4]   **11.** $|3x - 1| = 4$ **[2–6]**   **12.** $|2x + 3| > 8$ **[2–1]**   **13.** $x(x + 1) - (x + 3)^2 = 4$

[2–6]   **14.** $|1 - 4x| \le 7$

Simplify the following and leave in standard form. Assume that all variables represent positive real numbers.

[5–3]   **15.** $\sqrt[3]{64a^{18}b^7}$ **[5–7]**   **16.** $(3 - 4i)(2 + i)$ **[5–3]**   **17.** $\sqrt{48}$ **[5–2]**   **18.** $a^{1/3} \cdot a^{1/4}$

[5–3]   **19.** $\sqrt[3]{8a^4b^2}$ **[3–1]**   **20.** $(2a^2b^4c)^3$ **[5–7]**   **21.** $\sqrt{-18}$ **[5–6]**   **22.** $\dfrac{4}{\sqrt{10} + \sqrt{6}}$

[5–7]   **23.** $\dfrac{1 - 3i}{2 + 3i}$ **[5–2]**   **24.** $(4a^4)^{1/2}$ **[3–3]**   **25.** $\dfrac{3x^{-2}y^4}{9x^{-5}y^2}$ **[5–4]**   **26.** $\sqrt[3]{\dfrac{a^2}{4bc^3}}$

Chapter 5 Cumulative Test   **257**

Solve the following word problems.

[2–3]   **27.** When the length of a side of a square is increased by 4 inches, the area is increased by 72 square inches. Find the original length of a side.

[2–3]   **28.** A metallurgist wishes to form 1,000 kg of an alloy that is 62% copper. This alloy is to be obtained by fusing some alloy that is 80% copper and some that is 50% copper. How many kilograms of each alloy must be used?

[5–1]   **29.** In the theory of ballistics, the ballistic limit v of a material is approximated by the formula
$$v = kT^{6/5}$$
where T is the thickness of a sheet of material and k is a constant that is dependent on the material being used. Compute the ballistic limit if $k = 24,000$ and $T = 0.03125$.

[5–1]   **30.** Use the formula in exercise 29 to find v if $k = 25,000$ and $T = 0.07776$.

Cumulative tests emphasize the "building-block" nature of mathematics and help students retain knowledge and skills from previous chapters.

Preface

*A*lgebra for College Students is designed to be used as an intermediate level text for students who have had some prior exposure to beginning algebra in either high school or college. In this third edition, we have maintained our philosophy of explaining the why's of algebra, rather than simply expecting students to imitate examples. Sections are presented in such a way that, as topics progress, students realize they are actually extending properties they have already learned.

Problem-solving orientation The emphasis on problem solving begins in chapter 1 with word problems that have simple arithmetic solutions. The student also learns to change word phrases into algebraic expressions. In chapter 2 and throughout the rest of the text, the student is shown how to form and solve equations from word problems. **Diagrams** are used to show how the words are translated into mathematical symbols. **Tables** are provided to illustrate how several different word phrases become the same mathematical expression.

Critical thinking To encourage students to approach problems creatively in mathematics and the real world, we have included a critical thinking exercise in each chapter.

Error analysis Students can effectively increase their level of understanding of mathematical concepts by evaluating problems illustrating some of the most common mathematical errors. This strengthens the students' understanding of the concept and provides extra practice restating the concept in their own words.

Readability We have attempted to make the text as readable and accessible to students as possible by presenting the material in a manner similar to that which the instructor might use in the classroom.

Applications We have tried to provide a cross section of applications, mainly in the exercises. These are provided to help answer the perennial question "Why am I studying this stuff?" and to make the learning process itself more interesting. In particular, we have tried to show that algebra has become more important than ever in this age of the digital computer. Most ideas are supported by real-life applications relative to that concept.

Functional use of color In this third edition, color has been used to guide students through the text and clearly show the hierarchy of the text's elements. The effective use of color for each particular text element groups similar kinds of elements and helps students understand the relative importance of the elements.

- **Green** is reserved for the core ideas and core exercise problems presented in each chapter; it is used to highlight procedures, properties, definitions, notes, mastery points, and core exercise problems.
- **Blue** is used to emphasize explanations within the examples and exposition.
- **Red** is used to highlight extra practice opportunities for students within the development of each topic.

Highlights of the learning aids

Examples Examples present all aspects of the material being studied with a step-by-step development showing how the problem is worked. Examples have short phrase statements in **blue** type next to most steps stating exactly what has been done. The student is able to develop a clear understanding of how a problem is worked without having to guess what went on in a particular step.

Quick check exercises These exercises after a set of examples are designed to involve the student with the material while studying it. Quick check exercises directly parallel the development and examples in the text. As each new idea or procedure is illustrated with a set of examples, the student is asked to work a similar problem. A **red** triangle identifies each quick check exercise. Quick check exercises are worked step-by-step as quick-reference examples within the exercise set.

Procedure boxes Procedure boxes clearly state a step-by-step summary of the process by which types of problems are to be worked. **Green** has been consistently used for all procedure boxes to emphasize their importance to students.

Concept boxes Concept boxes include properties, theorems, or definitions along with an explanation in easy-to-understand language. **Green** is used to outline each concept box, emphasizing its importance.

Notes Notes to the student highlight important ideas and point out potential errors that students might make. The notes are printed in **green** type to attract the student's attention and to emphasize their importance.

Mastery points Mastery points are listed before each exercise set. In essence, they are objectives for that section. They are specifically placed in this location to alert students to the particular skills they must know to successfully work the problems. We have found the objectives have more meaning for students after they have completed the section. The **green** outlined box is used to draw the students' attention to the mastery points before they begin the exercise set and to mark the mastery points as covering part of the main ideas of the section.

Exercise sets Exercise sets provide abundant opportunities for students to check their understanding of the concepts being presented. The problems in the exercise sets are carefully paired and graded by level of difficulty to guide the students easily from straightforward computations to more challenging, multi-step problems.

Green type problem numbers identify the core exercise problems in each exercise set.

The directions for each group of problems refer the student to a specific group of parallel examples. After each set of directions is a **quick-reference example**. This example is a specifically chosen quick check exercise from the section and is worked and explained step-by-step. The **red** shading over each quick-reference example tells students it is related to the quick check exercises. Students can use this as a line-by-line check of their solution if they worked the problem while studying the material, or as an example they can refer to while working the exercise set.

Review exercises At the end of each section is a group of review problems. These exercises help reinforce the skills necessary for success in the following section. Answers are provided for all the review exercises.

Trial exercise problems Trial exercise problems appear throughout each exercise set and are denoted by a box around the problem number. This indicates that the solution is shown in its entirety in the answer appendix.

Chapter summaries End-of-chapter summaries synthesize the important ideas of each chapter.

Error analysis At the end of each chapter is a group of problems in which a typical error has been made. The student is asked to find and correct the mistake. If the student cannot find the error, a page reference is given which directs the student back to a specific note or group of examples that focus on this problem. Error analysis helps students increase their ability to find errors when checking their solutions and encourages them to practice restating the important ideas of the chapter in their own words.

Critical thinking Following error analysis is a special problem that requires the student to analyze a problem and use their mathematical skills to answer it. The *Instructor's Resource Manual* contains a series of hints that can be used to guide the student through the analysis of the problem. The *Instructor's Resource Manual* also discusses various ways that critical thinking can be integrated into your course.

Chapter review A chapter review is placed at the end of each chapter. This problem set follows the same organization as the chapter. Each problem is keyed to the section from which it was drawn. Answers to all review problems are provided in the appendix.

Cumulative tests Cumulative tests give students the opportunity to work problems that are drawn from the chapter and from preceding chapters. If students need to review, they can use the section references to review the concept.

Answers Answers are given for all odd-numbered section exercise problems. The answers to all problems in the chapter reviews and cumulative tests are provided in the appendix.

New to this edition

Content

1. Sums and differences of polynomials has been moved from chapter 3 in the second edition to chapter 1 in the third edition. This allows the student to concentrate on adding and subtracting like terms before introducing properties of exponents in chapter 3.

2. Solving absolute value equations and inequalities has been separated in this third edition. Solving absolute value equations is placed after solving word problems and solving absolute value inequalities comes after solving linear inequalities. This separation of absolute value equations from the absolute value inequalities limits possibilities for confusion among students. The separate coverage of absolute value inequalities also reinforces many of the skills developed with absolute value equations.

3. Factoring in chapter 3 has been streamlined. It now includes a greater use of diagrams to illustrate the concepts.

4. Synthetic division moves from the appendix to section 4–6. The remainder theorem and the factor theorem are developed in conjunction with the work with synthetic division.

5. A separate section dealing with operations with rational exponents has been added to chapter 5. This follows a graded approach that is more accessible to students.

6. Functions are given greater emphasis. They are now developed in chapter 8, rather than in chapter 10 of the second edition. This accentuates the importance of functions in mathematics. Functions are examined further in the following chapters.

7. With the introduction of relations and functions in chapter 8, quadratic equations of the form $y = ax^2 + bx + c$ are treated as quadratic functions and defined by $f(x) = ax^2 + bx + c$ in section 9–1.

8. A new section has been added to chapter 9. In section 9–2, "Quadratic relations—more about parabolas," students learn to solve and graph parabolas whose equations are of the form $x = ay^2 + by + c$ $(a \neq 0)$.

9. Chapter 10 is a new chapter, "Polynomial and Rational Functions"; Descartes' rule of signs and bounds of zeros, rational zeros of a polynomial function, irrational zeros of a polynomial function, graphing polynomial functions, and graphing rational functions. This coverage reinforces and expands on the basic coverage of functions in chapter 8.

10. Systems of nonlinear equations and systems of linear equations are both developed in chapter 12. This combining of these two types of systems is a natural extension of students' previous studies of systems of equations.

Section 12–2 now provides real-world examples and exercises applying linear systems of two variables. A new section 12–5 involving systems of linear inequalities and linear programming develops concepts which will be useful to students in business degree programs.

11. Chapter 13 treats matrices and determinants separate from the study of systems of linear equations in two and three variables of chapter 12. We study the matrix and determinant methods of solving systems of linear equations in more depth.

12. Section 14–6 introduces mathematical induction as a method of proving relationships. We feel this gives the student an important introduction to proof in preparation for further mathematical study.

13. Counting techniques and probability are introduced in the new chapter 15 to better prepare students for future mathematics courses or studies in business or statistics.

   The study of the binomial theorem has been moved from section 12–6 of the second edition to section 15–5 of the third edition. We feel this expansion is better treated here after the introduction of factorials during the study of counting techniques since factorials are used in the expansion of a binomial.

Features

1. Each chapter is introduced with an application problem (and accompanying related photo) that can be solved using the procedures studied in the ensuing chapter. The application problem is worked out in detail at the end of the chapter, just prior to the chapter summary for that chapter.

2. Quick check problems have been added within the textual material to give students immediate practice with each new concept.

3. Quick-reference problems are worked out in the exercise sets to serve as further examples for the student.

4. Greater use of arrows to point up important steps taken in the development of an example.

5. Examples have step-by-step development showing how the problem is worked with short phrase state-ments next to most steps stating exactly what has been done.

6. *All* step-by-step procedures outlined for the major concepts are placed in boxes for emphasis.

7. Core exercise problems have been identified with green problem numbers in the exercise sets to highlight the appropriate problems for a basic assignment.

8. Review exercises at the end of each section have been added to help prepare the student for the work of the following section.

9. Error Analysis in each chapter helps students find errors and apply concepts in their own terms.

10. Critical Thinking activities in each chapter help students learn to address multi-step complex problems.

11. Color is used to clearly show which elements are related and to highlight the important concepts for the students.

For the instructor

The **Instructor's Resource Manual** has been expanded to include all critical thinking exercises from the text (with hints and solutions), a guide to the supplements that accompany *Algebra for College Students,* Third Edition, and reproducible quizzes, multiple chapter tests, and extension problems. Also included are a complete listing of all mastery points and suggested course schedules based on the mastery points. The final section of the *Instructor's Resource Manual* contains answers to the reproducible materials.

The **Instructor's Solutions Manual** contains completely worked-out solutions to all of the exercises in the textbook.

The **Educator's Notebook** is designed to assist you in formatting and presenting the concepts of *Algebra for College Students,* Third Edition to your students. Reproducible transparency masters are provided for each section of the textbook.

The **Test Item File/Quiz Item File** is a printed version of the computerized *TestPak* and *QuizPak* that allows you to choose test items based on chapter, section, or objective. The objectives are taken directly from the mastery points in *Algebra for College Students*, Third Edition. The items in the *Test Item File* and *Quiz Item File* are different from those in the prepared tests in the *Instructor's Manual.* Hence, you will have even more items to choose from for your tests.

WCB TestPak 3.0, our computerized testing service, provides you with a call-in/mail-in testing program and the complete *Test Item File* on diskette for use with IBM® PC, Apple®, or Macintosh® computers. *WCB TestPak* requires no programming experience. Tests can be generated randomly, by selecting specific test items or mastery points/objectives. In addition, new test items can be added and existing test items can be edited.

WCB GradePak, also a part of *TestPak 3.0*, is a computerized grade management system for instructors. This program allows you to track students' performance on examinations and assignments. It will compute each student's percentage and corresponding letter grade, as well as the class average. Printouts can be made utilizing both text and graphics.

WCB TestPak 3.0 disks and the WCB call-in service are available free to instructors adopting *Algebra for College Students*, Third Edition.

WCB QuizPak can be used to give your students on-line practice with the topics of elementary algebra. You can choose multiple-choice and true-false items from the Quiz Item File, edit items, or add your own items. Students' on-line test results are graded and scores then recorded in a GradePak file.

For the student

The **Student's Solutions Manual** contains overviews of every chapter of the text, chapter self-tests with solutions, and solutions to all proficiency checks, every other odd-numbered section exercise, and odd-numbered chapter review exercise problems. It is available for student purchase.

On the **Videotapes**, the instructor introduces a concept, provides detailed explanations of example problems that illustrate the concept, including applications, and concludes with a summary. All of the major topics are carefully reinforced by the comprehensive Wesner and Nustad Video series. The tapes are available free to qualified adopters.

The **Audiotapes** have been developed specifically to accompany *Algebra for College Students,* Third Edition. They begin with a complete synopsis of the section, followed by clear discussions of examples with warning and hints where appropriate. Exercises are solved for each section of the text. Students are directed to turn off the tape and solve a specific problem and turn the tape on again for a complete explanation of the correct solution.

The concepts and skills developed in *Algebra for College Students*, Third Edition are reinforced through the interactive **Tutorial Practice Software.** Students practice solving section-referenced problems generated by the computer and review the major topics of elementary algebra. Step-by-step solutions with explanations guide students to mastery of the major concepts and skills of elementary algebra.

WCB QuizPak, a part of *TestPak 3.0*, provides students with true/false and matching questions from the *Quiz Item File* for each chapter in the text. Using this portion of the program will help your students prepare for examinations. Items in *QuizPak* are similar in level and coverage of concepts as the *TestPak* items. Also included with the *WCB QuizPak* is an on-line testing option that allows professors to prepare tests for students to take using the computer. The computer will automatically grade the test and update the gradebook file.

Acknowledgments

We wish to express our sincere thanks for the many comments and suggestions given to us during the preparation of the first edition. In particular, we wish to thank Lynne Hensel, William Lakey, and Douglas Nance for their excellent effort in reviewing each stage of the first edition and supplying us with numerous valuable comments, suggestions, and constructive criticisms.

For their help in typing the manuscript we thank Amy Miyazaki and Debbie Miyazaki, and a special thank you goes to Lisa Miyazaki for her superb aid in preparing the manuscript and for working all of the problems.

Because of his invaluable help and advice in the area of marketing, our sincere thanks go to Harold Elliott.

The authors would like to acknowledge the contribution of Philip Mahler, who introduced to them the idea of using the tabular format to list all possible combinations of factoring in factoring trinomials. Mr. Mahler was also responsible for the idea of using the sign of the product "mn" as an operation in the second column of the table. The chief virtue of this method is that it is algorithmic. The authors have modified the method slightly by listing the greater factor first.

Pauline Chow and Jean Shutters of Harrisburg Area Community College deserve special thanks for their careful preparation of the *Instructor's Solutions Manual*.

Throughout the development, writing, and production of this text, two people have been of such great value that we are truly indebted to them for their excellent work on our behalf. We wish to express our utmost thanks to Suresh Ailawadi and Eugenia M. Collins.

We would like to thank the following reviewers of the third edition of *Algebra for College Students:*

Helen W. Baril
Quinnipiac College

Donald W. Bellairs
Grossmont College

Lou Cleveland
Chipola Junior College

Tom Cochran
Belleville Area College

Thomas E. Covington
Northwestern State University of Louisiana

Gregory J. Davis
University of Wisconsin-Green Bay

JoAnn Everett
Chipola Junior College

Marjorie K. Gross
Carteret Community College

Virginia E. Hanks
Western Kentucky University

Martha C. Jordan
Okaloosa-Walton Community College

Judy Kasabian
El Camino College

Herbert F. Kramer
Longview Community College

Vince McGarry
Austin Community College

Fauline J. Mathis
Chipola Junior College

Patricia B. Mathis
Chipola Junior College

Marilyn Morrison
Volunteer State Community College

Linda J. Padillo
Joliet Junior College

Gus Pekara
Oklahoma City Community College

Bonnie Smith
Chipola Junior College

Gerry C. Vidrine
Louisiana State University

Deborah A. Vrooman
Coastal Carolina College of the University of South Carolina.

In addition, we would like to thank the reviewers of previous editions, whose comments have positively influenced this edition.

Dr. Joseph Altinger
Youngstown State University

Daniel Anderson
University of Iowa

Dorothy Batta
Wilber Wright College

Charles M. Beals
Hartnell College

Philip Beckman
Black Hawk College

Donald W. Bellairs
Grossmont College

James Blackburn
Tulsa Junior College

Nancy Bray
San Diego Mesa College

Ann S. Bretscher
University of Georgia

Dan Burns
Sierra College

Helen Burrier
Kirkwood Community College

Richard A. Butterworth
Massasoit Community College

Vern Byer
University of Maine-Farmington

William Chatfield
University of Wisconsin-Platteville

Al H. Chew
Central Arizona College

Deann Christianson
University of the Pacific

Duane Deal
Ball State University

Mark Dugopolski
Southeastern LA University

Ray Fartch
Eastern Oklahoma State College

Fred W. Fischer
North Seattle Community College

Richard L. Francis
Southeast Missouri State University

Kenneth O. Gamon
Central Washington University

Margaret J. Greene
Florida Junior College

Pamela Hager
College of the Sequoias

Shelby Hawthorne
Thomas Nelson Community College

Lynne Hensel
Henry Ford Community College

Joyce Huntington
Walla Walla Community

College

James Johnson
Modesto Junior College

Glen Just
Mount Saint Clare College

Chris Kolaczewski
University of Akron

William Lakey
Central Michigan University

Carolyn Likins
Millikin University

Karla Martin
Middle Tennessee State University

Raymond A. Maruca
Delaware County Community College

Alfred W. Milligan
Western New Mexico University

Ronald Milne
Goshen College

Jesse Moore
John A. Logan College

Douglas W. Nance
Central Michigan University

Ardash Ozsogomonyan
College of San Mateo

Sue Phillips
University of Nevada, Las Vegas

Thomas Radin
San Joaquin Delta College

George C. Ragland
St. Louis Community College at Florissant Valley

J. Doug Richey
Henderson County Junior College

Beverly Ridenhour
Utah State University

Kenneth Ross
Broward Community College

Grayson Sallez
University of North Carolina-Greensboro

William Schneider
Fairmont St. College

Annalee Scorsone
Lexington Community College

John Snyder
Sinclair Community College

John N. Strange
Hinds Junior College

John Taylor
Hillsborough Community College

Christina Vertullo
Marist College

Gerry Vidrine
Louisiana State University

Daniel Wachter
College of the Desert

George Wales
Ferris State College

Robert Wenger
University of Wisconsin-Green Bay

Richard Werner
Santa Rosa Junior College

Peter Williams
University of Maine

Judy Willoughby
Maples Community College

Jerry Wisnieski
Des Moines Area Community College

Dick Wong
College of Lake County

Donald Zalewski
Northern Michigan University

Ben F. Zirkle
Virginia Western Community College

Richard Zucker
Saddleback College North Campus

Finally, we are grateful to our "book team," for without them there would be no book. In particular, we would like to express our sincere thanks to Earl McPeek, Gene Collins, Theresa Grutz, K. Wayne Harms, and Carrie Burger.

Study Tips

When you work to your full capacity, you can hope to attain the knowledge and skills that will enable you to create your future and control your destiny. If you do not, you will have your future thrust upon you by others.

A Nation at Risk *

There are certain study skills that you as an algebra student need to have, or develop, to assure your success in this course. In addition to the following items listed, acquaint yourself with the text by reading the preface material that precedes these study tips. Then—

1. For every hour spent in class, plan to spend at least two hours studying outside class.

2. Before going to class, read the material to be covered. This will help you more easily understand the instructor's presentation.

3. Take time to become familiar with the learning aids in your textbook. This will allow you to get the maximum benefit from them. In *Algebra for College Students,* Third Edition color has been used to tie related features together.

   • **Green** is used for the core concepts, ideas, and exercises. Be sure you understand everything highlighted with green.

   • **Blue** indicates additional explanation and greater detail.

   • **Red** is used to identify quick checks and quick-reference examples, which give you greater opportunity to check your understanding of each problem type.

4. Review the material related to each exercise set *before* attempting to work the problems. Be sure you understand the underlying concepts in the worked-out examples and the reason for each step.

5. Carefully read the instructions to the exercise set. Look at the examples and determine what is being asked. Remember, these same instructions will most likely appear on tests.

6. When working the exercise set, take your time, think about what you are doing in each step, and ask yourself why you are performing that step. As you become more confident, increase your speed to better prepare yourself for test situations.

7. When working the exercise sets, compare examples to see in what ways they are alike and in what ways they are different. Problems often *look* similar but are not.

If you do not know how to begin a problem, or you get partway through and are unable to proceed, (a) look back through your notes or (b) look for an exercise you can do that has the answer given and try to analyze the similarities. If doing these things does not work, put the problem aside. Often getting away from it for a time will "open the door" when you try it again. Finally, if you need to, consult your instructor and show him/her the work you have done.

The fact that you will be "using tomorrow what you are doing today" makes it imperative that you learn each concept as you go along. Most concepts, especially the ones that give you the most difficulty, need constant review.

The practice of checking your work will aid you in two ways:

1. It will develop confidence, knowing you have done the problem correctly.

2. It will help you discover your errors on an exam that might otherwise have gone undetected had you not checked your work.

When checking your work, use a different method from the one you used to solve the problem. If the same procedure is used, a tendency to make the same mistake exists. Develop methods for checking your work as you do the practice exercises. This checking then becomes automatic when taking a test.

The following hints will aid you in preparing for an exam:

1. Begin studying and reviewing a number of days prior to the exam. This will enable you to contact your instructor for help if you need it. "All-night" sessions the night before the exam seldom (if ever) yield good results.

*The National Commission on Excellence in Education. *A Nation at Risk.* Washington, D.C.: U.S. Government Printing Office, 1983.

2. Take periodic breaks—10 to 15 minutes for each hour of study. Study for no longer than four hours at a time.

3. Work to develop understanding as well as skills. Memorization is seldom useful in an algebra course, so concentrate on understanding the methods and concepts. However do not ignore skill development, since doing so can often lead to what students call "stupid mistakes."

Prior to taking an exam, use the exercise sets, chapter reviews, and/or *Student's Solutions Manual* to make out a practice test, determine where your errors lie, and retake the test to be sure that you have corrected the mistakes. Allot the same amount of time you will be allowed on test day.

When taking the algebra exam you should:

1. Look over the exam to locate the easiest problems.

2. Work these problems first.

3. Work the more difficult and time-consuming problems next. Remember, when stuck on a problem, go on to other problems and return to those giving you difficulty *only after* completing all that you can.

4. Use what time remains to check your answers or to rework those problems that you found most difficult.

Don't panic should you "draw a blank." Avoid thoughts of failure. Should you feel this happening, relax and try to clear your mind. Search out the problems you feel most confident about and begin again. Should you be unable to complete the exam, be sure to check the problems that you have completed. Always be aware of the time remaining. Do not hurry and do not be intimidated by other students completing the exam early.

One final bit of advice. Show your work neatly. Develop this habit when working on your practice problems. There is a close correlation between neatly laid-out work and the correct answer. Your instructor will appreciate this and be more inclined to give you more credit if the answer is wrong.

1

Basic Concepts and Properties

Theresa is conducting a chemistry experiment involving yellow phosphorus. The lab manual states that yellow phosphorus ignites at 34 degrees Celsius. What will the temperature be in degrees Fahrenheit? The formula for changing the temperature measured in degrees Celsius to degrees Fahrenheit is

$$F = 1.8\ C + 32$$

1–1 ■ Sets and real numbers

Set symbolism

We begin our study with a very simple, but important, mathematical concept—the idea of a **set.*** *A set is any collection of objects or things.* We want the sets that we deal with to be *well defined;* that is, given any object, we can determine whether the object is in a given set. For example, the set of old people is not well defined because the meaning of old is not clear. Whereas the set of people whose ages are greater than seventy years is a well-defined set. In mathematics, the idea of a set is used primarily to denote a group of numbers or the set of answers to a problem.

Any one of the things that make up a set is called a **member** or an **element** of that set. One way of writing a set is by listing the elements, separating them by commas, and including this listing within a pair of braces, { }. This way of representing a set is called the **listing** or **roster method.**

■ **Example 1–1 A**

1. Using set notation, write the days of the week that begin with the letter *T*.

   {Tuesday, Thursday}

---

*Georg Cantor (1845–1918) is credited with the development of the ideas of set theory. He described a set as a grouping together of single objects into a whole.

2. Using set notation, write the digits that make up the telephone number for information, 555–1212.

{5,1,2}

Note When we form a set, the elements within the set are never repeated and they can appear in any order.

▶ *Quick check* Using set notation, write the letters of the word "book." ∎

We use capital letters A,B,C,D, and so on, to represent a set. When we wish to show that an element belongs to a particular set, we use the symbol ϵ, which is read "is an element of" or "is a member of." Consider the set $A = \{1,4,5\}$, which is read "the set A whose elements are 1, 4, and 5." If we want to say that 4 is an element of the set A, we write $4 \epsilon A$.

A slash mark, /, is used in mathematics to negate a given symbol. Therefore if ϵ means "is an element of," then $\notin$ means "is *not* an element of." To express the fact that 2 is not an element of the set A, we write $2 \notin A$.

■ *Example 1–1 B*

Using mathematical symbols, write the following statements.

1. 5 is an element of the set C. $5 \epsilon C$

2. 4 is not an element of the set B. $4 \notin B$

▶ *Quick check* Write "7 is an element of the set A" using mathematical symbols. ∎

Subsets

Suppose that P is the set of people in a class and W is the set of women in the same class. It is obvious that the members of W are also members of P. We say that W is a **subset** of P and use the symbol $\subseteq$ to indicate "is a subset of."

— **Definition of A ⊆ B** ————————————————————

The set A is a subset of the set B, written $A \subseteq B$, if every element of A is also an element of B.

■ *Example 1–1 C*

Given the sets $A = \{1,2,3\}$, $B = \{1,2,3,4,5\}$, $C = \{2,3,5\}$, and $D = \{2,3,1\}$, determine if the following statements are true or false.

1. $A \subseteq B$. True, since all the elements in A are also in B.

2. $A \nsubseteq C$. True, since not all of the elements in A are in C. 1 is an element of A but is not an element of C.

Note $A \nsubseteq C$ is read "the set A is not a subset of the set C" and implies that there is at least one element in the set A that is not in the set C.

3. $C \subseteq D$. False, since not all of the elements in C are in D. 5 is an element of C but is not an element of D.

4. $A \subseteq D$. True, since all of the elements in A are in D. ∎

In example 4, since the order of the elements within the set does not change the set, we can conclude that the sets A and D are the same set. Therefore we observe that *the definition of subsets allows a set to be a subset of itself.* That is, for any set A, $A \subseteq A$. We will now use the definition of subsets to define when two sets are equal.

Definition of A = B

The set A is said to be equal to the set B, written $A = B$, if and only if $A \subseteq B$ and $B \subseteq A$.

Suppose that we wanted to form the set of the months of the year that begin with the letter X. Since there are no months that begin with the letter X, this set has no elements and is called the **empty set** or the **null set. A set that contains no elements is called the empty set or the null set and is denoted by the symbol ∅.**

By the definition of subsets, a set A is *not* a subset of a set B if there is at least one element in A that is not in B. The empty set, by definition, has no elements; therefore it can contain no elements that are not in another given set. We must then conclude that *the empty set is a subset of every set.*

Union and intersection of sets

If we wished to form the set of all the students who have blue eyes *or* blond hair, we would be combining the set of students with blue eyes with the set of students with blond hair. This new set that we have formed is the **union** of the two original sets.

Definition of A ∪ B

The union of the sets A and B, written $A \cup B$, is the set of all elements that are in A or in B or in both A and B.

■ *Example 1–1 D*

Given the sets $A = \{2,3,5\}$, $B = \{1,3,7\}$, and $C = \{2,4,6\}$, form the following sets.

1. $A \cup B$
A union B consists of those elements that appear in A, $\{2,3,5\}$, or those in B, $\{1,3,7\}$, or those in both A and B, $\{3\}$. Therefore

$A \cup B = \{1,2,3,5,7\}$

2. $B \cup C = \{1,3,7\} \cup \{2,4,6\} = \{1,2,3,4,6,7\}$ ■

If we wished to form the set of all the students who have blue eyes *and* blond hair, we would be forming a new set of students that possess the characteristic of having *both* blue eyes and blond hair. This new set that we have formed is the **intersection** of the two original sets.

Definition of A ∩ B

The intersection of the sets A and B, written $A \cap B$, is the set of only those elements that are in both A and B.

■ *Example 1–1 E*

Given the sets $A = \{1,2,3\}$, $B = \{2,3,4,5\}$, $C = \{4,5\}$, and $D = \{2,3,6,7\}$, form the following sets.

1. $A \cap B$

$$A \cap B = \{1,2,3\} \cap \{2,3,4,5\} = \{2,3\}$$

A intersection B consists of only those elements that appear in *both A and B*

2. $A \cap C = \{1,2,3\} \cap \{4,5\}$

There are no elements common to both A and C

$$A \cap C = \emptyset$$

The intersection is the empty set

3. $A \cup (B \cap C)$

$$= \{1,2,3\} \cup (\{2,3,4,5\} \cap \{4,5\})$$

Operations enclosed within a grouping symbol must be performed first

$$= \{1,2,3\} \cup \{4,5\}$$

Perform the intersection of B and C

$$= \{1,2,3,4,5\}$$

Perform the union of $B \cap C$ with A   ■

When the intersection of two sets is the empty set, as in example 2, we say that the two sets are **disjoint**. That is, *the sets* A *and* C *are disjoint if and only if* A ∩ C = ∅.

The set of real numbers

Our study of algebra will be primarily concerned with the set of **real numbers** and its properties. We shall now review the set of real numbers and its major subsets.

In each of the previous examples, we could determine the exact number of elements in a set. This type of set is called a **finite** set. Each of the major sets of numbers that make up the set of real numbers has an unlimited number of elements. This is called an **infinite** set.

The set of real numbers is made up of the following sets of numbers.

I. The set of **natural numbers,** or counting numbers, is defined by

$$\{1,2,3,4, \cdots\}$$

and denoted by N.

Note   The three dots tell us to continue this pattern indefinitely. That is, there is no last natural number.

II. The set of **whole numbers** is defined by

$$\{0,1,2,3,4, \cdots\}$$

and denoted by W.

III. The set of **integers** is defined by

$$\{\cdot\ \cdot\ \cdot,\ -2,-1,0,1,2,\ \cdot\ \cdot\ \cdot\}$$

and denoted by J.

To express certain sets of numbers, we need to define the mathematical concept of a **variable. A variable is a symbol (generally a lowercase letter) that acts as a placeholder for an unspecified number.** When we use a variable to represent a set of numbers, the set of numbers is called its **replacement set** or **domain.** For example,

$$\{x \mid x \text{ is a natural number less than 6}\}$$

is read "the set of all elements x such that x is a natural number less than 6." This notation defines the set $\{1,2,3,4,5\}$. The bar is read "such that" and the notation is called **set-builder notation.** In general, set-builder notation has the pattern shown in figure 1–1.

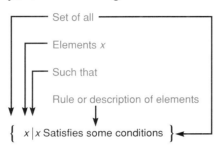

Figure 1–1

■ *Example 1–1 F*

List the elements of the following sets.

1. $\{x \mid x \text{ is an integer between } -4 \text{ and } 2\}$
$= \{-3,-2,-1,0,1\}$

The word "between" indicates not to include -4 and 2

2. $\{x \mid x \text{ is a whole number greater than } 10\}$
$= \{11,12,13,14, \cdot\ \cdot\ \cdot\}$

The words "greater than" indicate that we start with the first whole number after 10

Note Since this is an infinite set, we set up a pattern for the numbers and place three dots after the last number to indicate that this pattern continues indefinitely.

3. $\{x \mid x \text{ is a natural number less than } 1\}$

Since there are no natural numbers less than 1, the set is empty, $\emptyset$. ■

IV. The set of **rational numbers** is defined by

$$\left\{ \frac{p}{q} \middle| p \text{ and } q \in J, q \neq 0 \right\}$$

and denoted by Q. The set-builder notation symbolizes that the set of rational numbers, Q, will consist of all the numbers that can be represented by the quotient of two integers where the denominator is not zero.

A second way that the set of rational numbers can be defined is

$\{x|$ the decimal representation of x is either terminating or repeating$\}$

Examples of terminating and repeating decimals are

$$\frac{1}{2} = 0.5, \quad \frac{1}{3} = 0.\overline{3}, \quad -\frac{1}{6} = -0.1\overline{6}, \quad -\frac{5}{4} = -1.25$$

where a bar placed over a number or group of numbers indicates that the number(s) repeat indefinitely.

V. The set of **irrational numbers** is defined by

$\{x|$ the decimal representation of x is nonterminating and nonrepeating$\}$

and denoted by H. Examples of irrational numbers are

$$\sqrt{3}, \quad -\sqrt{5}, \quad \pi, \quad \frac{\sqrt{2}}{2}$$

The decimal representation of an irrational number will never terminate. We cannot find a repeating pattern of digits, no matter how many digits we write past the decimal point.

Note By using a calculator, the previous numbers can be represented by the following approximations to three decimal places:

$$\sqrt{3} \approx 1.732, \quad -\sqrt{5} \approx -2.236, \quad \pi \approx 3.142, \quad \frac{\sqrt{2}}{2} \approx 0.707$$

The symbol $\approx$ is read "is approximately equal to." π is the distance around a circle (circumference) divided by the distance across the circle through the center (diameter). Common approximations for π are 3.14 and $\frac{22}{7}$.

VI. The set of **real numbers** is defined by

$\{x|x \in Q \text{ or } x \in H\}$

and denoted by R.

Note Whenever we encounter a problem and a specific replacement set or domain for the variable is not indicated, it will be understood that we are dealing with the set of real numbers.

All of the sets that we have considered thus far are subsets of the set of real numbers. Figure 1–2 shows this relationship.

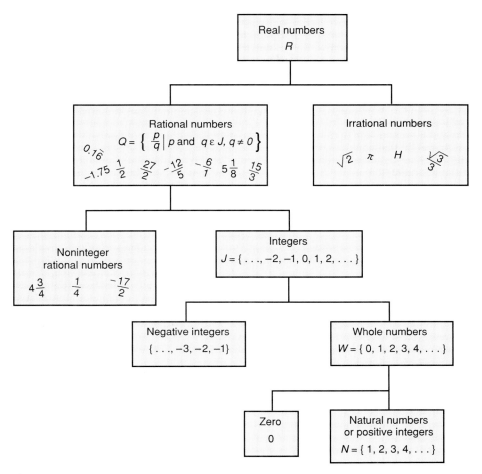

Figure 1–2

The real number line

To visualize the set of real numbers, we use a figure called a **number line.** Any real number can be located on the number line (see figure 1–3).

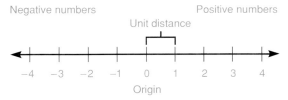

Figure 1–3

The number associated with each point on the number line is called the **coordinate** of the point. The point associated with each number is called the **graph** of that number. In figure 1–4, the numbers -2, $-\dfrac{1}{3}$, $\dfrac{3}{4}$, $\sqrt{3}$, and 3 are the coordinates of the points indicated on the line by solid circles. These solid circles are the graphs of the numbers -2, $-\dfrac{1}{2}$, $\dfrac{3}{4}$, $\sqrt{3}$, and 3.

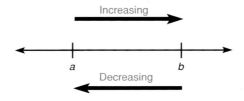

Figure 1–4

Note The coordinate $\sqrt{3}$ represents an irrational number. To graph this point, we can use a decimal approximation from a calculator, $\sqrt{3} \approx 1.732$. The word "number" used in this book means "a real number."

Order and absolute value

The direction we move on the number line is also important. If we move to the right, we move in a positive direction and our numbers **increase.** If we move to the left, we move in a negative direction and our numbers **decrease**.

If we choose arbitrary points on the number line and represent them by a and b, where a and b represent some *unspecified* numbers, we observe that there is an **order** relationship between a and b as in figure 1–5.

Figure 1–5

When the point associated with a is to the left of the point associated with b, we say that a is **less than** b, which in symbols is $a < b$. We might also say that b is **greater than** a, which in symbols is $b > a$. The symbols $<$ (is less than) and $>$ (is greater than) are inequality symbols called **strict inequalities** and denote an **order relationship** between numbers.

■ *Example 1–1 G*

Graph the following pairs of numbers and insert the correct inequality symbol between the numbers to get a true statement.

1. 3 and 5

The graph would be

Since 3 lies to the left of 5, we say that 3 is less than 5, or symbolically $3 < 5$.

Note $3 < 5$, read "3 is less than 5," can also be stated as $5 > 3$, read "5 is greater than 3." No matter which inequality symbol we use, the inequality symbol always points to the *lesser* number.

2. -90 and -100

The graph would be

Since -90 lies to the right of -100, we say that -90 is greater than -100, or symbolically $-90 > -100$. ▪

There are two other inequality symbols, called **weak inequalities,** which are **is less than or equal to,** $\leq$, and **is greater than or equal to,** $\geq$. The weak inequality symbol $\leq$, is less than or equal to, combines the relationship of is less than ($<$) with the relationship of equality ($=$). The weak inequality symbol $\geq$, is greater than or equal to, combines the relationship of is greater than ($>$) with the relationship of equality ($=$).

As we study the number line, we see a very useful property called **symmetry.** The numbers are symmetrical with respect to the origin. That is, if we go two units to the right of 0, we come to the number 2, and if we go two units to the left of 0, we come to the *opposite* of 2, which is -2. This idea of how far a given number is from the origin is called the **absolute value** of that number. *The absolute value of a number is the undirected distance that the number is from the origin.* The symbol for absolute value is $|\quad|$

■ Example 1–1 H

Find the indicated absolute values.

1. $|2| = 2$

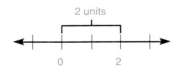

2. $|-3| = 3$

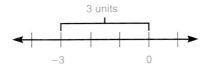

▪

Note   The absolute value of a number is *never* negative. That is, for every $x \in R, |x| \geq 0$.

When we wish to determine the absolute value of a positive number or zero, $x \geq 0$, the absolute value of the number is simply the original number, x. When we wish to determine the absolute value of a negative number, $x < 0$, we find that the absolute value of a negative number is the opposite of the original number. We symbolically write this as $-x$, read "the opposite of x." We can now state the definition of absolute value symbolically.

Definition of $|x|$

$$|x| = \begin{cases} x \text{ if } x \geq 0 \\ -x \text{ if } x < 0 \end{cases}$$

Since x represents a negative number ($x < 0$), the absolute value of x is the opposite of x (or $-x$) and will be a positive number

---

■ **Example 1–1 I**

Find the indicated absolute values.

1. $|4| = 4$ **2.** $|-9| = -(-9) = 9$ **3.** $-|-7| = -(7) = -7$

Note   The negative sign in front of the absolute value in example 3 indicates that we want the opposite of the absolute value. Therefore, the answer is -7.   ■

Mastery points

Can you

■ List the elements of a set?
■ Use set symbolism?
■ Determine when a set is a subset of another set?
■ Determine when sets are equal?
■ Perform the operations of union and intersection on sets?
■ Use set-builder notation?
■ Graph a number?
■ Determine the coordinate of a point?
■ Determine which of two real numbers is greater?
■ Find the absolute value of a number?
■ Use mathematical symbols?

Exercise 1–1

Write each set by listing the elements. See example 1–1 A.

Example   Using set notation, write the letters of the word "book."

Solution $\{b,o,k\}$ The letters are placed within braces in any order, separated by commas, and we do not repeat any of the elements.

1. The days of the week that begin with the letter S **2.** The months of the year that begin with the letter A
3. The even integers between 9 and 15   **4.** The months of the year with less than 28 days
5. The months of the year that begin with the letter C **6.** The days of the week that begin with the letter A

Use mathematical symbols to write the following statements. See examples 1–1 B and C.

Example   7 is an element of the set A.

Solution $7 \in A$ ϵ is the symbol for "is an element of"

7. A is a subset of D.   **8.** B is not a subset of C.
9. The null set is a subset of B.   **10.** The null set is not an element of C.

11. The set whose elements are 6, 7, 8   **12.** The set whose elements are 1, 4, 6

Given the sets $A = \{2,4,6\}$, $B = \{1,3,5\}$, $C = \{1,2,3,4,5,6\}$, $D = \{3,1,5\}$, and $E = \{1,2,3,4,5,6,7,8\}$, determine if the following statements are true or false. See example 1-1 C.

13. $2 \in A$	**14.** $1 \in B$	**15.** $A \subseteq C$	**16.** $B \subseteq E$	**17.** $A \nsubseteq E$
18. $C \subseteq E$	**19.** $B = D$	**20.** $A = D$	**21.** $\emptyset \subseteq A$	**22.** $A \subseteq D$

Given the sets $A = \{1,2,3,4\}$, $B = \{2,3,4,5,6\}$, $C = \{2,3,4\}$, and $D = \{7,8,9\}$, form the following sets. See examples 1-1 D and E.

23. $A \cup C$	**24.** $B \cup C$	**25.** $A \cup D$	**26.** $B \cup D$	**27.** $A \cap D$
28. $C \cap D$	**29.** $(C \cup B) \cap A$	**30.** $(D \cap C) \cup B$	**31.** $A \cup (C \cap D)$	**32.** $(B \cap C) \cup A$

Graph the following sets of numbers on a number line. See example 1-1 G.

33. $\{-3,-1,2,4,5\}$ **34.** $\{-4,-3,-2,1,3\}$ **35.** $\left\{-\dfrac{7}{4},0,\dfrac{5}{4},3,5\right\}$

36. $\{-\sqrt{3},0,\sqrt{2},3,6\}$ **37.** $\{-5,-\sqrt{5},\sqrt{3},3,5\}$ **38.** $\{-4,-2,0,\sqrt{5},\sqrt{7}\}$

Write the value of the following numbers. See examples 1-1 H and I.

39. $|-3|$ **40.** $|-5|$ **41.** $|0|$ **42.** $\left|\dfrac{1}{4}\right|$ **43.** $-|-2|$

44. $-|-7|$ **45.** $-|4|$ **46.** $-|5|$

Replace the comma with the proper strict inequality symbol, $<$ or $>$, between the following numbers to get a true statement. See example 1-1 G.

47. 4, 9   **48.** $-4, -8$ **49.** $-3, -11$ **50.** $-12, -5$ **51.** $-15, -10$

52. $0, -5$ **53.** $-4, 0$

Use mathematical symbols to write each of the following statements.

Example 0 is not an element of the set of natural numbers.

Solution $0 \notin N$ The / mark through the $\in$ symbol means "is **not** an element of the set"

54. The set of whole numbers is not a subset of the set of natural numbers.

55. The set of irrational numbers intersected with the set of rational numbers is the null set.

56. The union of the set of rational numbers with the set of irrational numbers is the set of real numbers.

57. $\{8\}$ is a subset of the set of natural numbers.

58. 8 is not a subset of the set of natural numbers.

59. $\{8\}$ is not an element of the set of natural numbers.

60. 8 is an element of the set of natural numbers.

Determine from the given information whether the answer to each of the following statements is yes or no. The sets A, B, and C are general sets and are not the same as the sets in the previous exercises.

Example If $A \subseteq B$ and $2 \in A$, must 2 be an element of B?

Solution Yes. From the definition of subsets, every element in A must be in B. Therefore if $2 \in A$ and $A \subseteq B$, then $2 \in B$.

61. If $A \subseteq B$ and $3 \in A$, must 3 be an element of B?

62. If $A \subseteq B$ and $7 \in B$, must 7 be an element of A?

63. If $A \nsubseteq B$ and $6 \in A$, does this mean that $6 \notin B$?

64. If $A \nsubseteq B$ and $1 \notin A$, does this mean that 1 must be an element of B?

65. If $A \subseteq B$ and $B \subseteq C$ and $4 \in A$, must 4 be an element of C?

66. If $A \subseteq B$ and $B \subseteq A$, does this mean that $A = B$?

67. If $A \subseteq B$ and $B \subseteq C$, must A be a subset of C?

68. If $A \subseteq B$ and $B \subseteq C$ and $5 \in A$, must 5 be an element of C?

69. If $A \subseteq B$ and $B \subseteq A$ and $8 \notin A$, must 8 be an element of B?

70. If $A \not\subseteq B$, can $B \subseteq A$?

71. Is $\emptyset \subseteq \emptyset$?

72. Is $\emptyset \subseteq B$?

73. Does $\emptyset \cap A = \emptyset$?

74. Does $\emptyset \cup A = A$?

75. Does $A \cap A = A$?

76. Does $A \cup A = A$?

1–2 ■ *Operations with real numbers*

Addition

In this section, we will review the rules for addition and subtraction of real numbers. We use the minus sign ($-$) to indicate a negative number and the plus sign ($+$) to indicate a positive number. A **real number** consists of two parts: its **absolute value** and its **sign** (see figure 1–6).

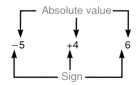

Figure 1–6

We can summarize the rules for adding real numbers.

> ### *Addition of real numbers*
> 1. If the signs are the same, we add their absolute values. The answer is given their common sign.
> 2. If the signs are different, we subtract the lesser absolute value from the greater absolute value. The answer is given the sign of the number with the greater absolute value.

■ *Example 1–2 A*

Add the following.

1. $(+3) + (+6) = (+9)$

         —— Absolute value
         —— Common sign

2. $(-4) + (-9) = (-13)$

         —— Absolute value
         —— Common sign

3. $(-8) + (+5) = (-3)$

         —— Absolute value
         —— Sign of the number with the greater absolute value

4. $(-6) + (+10) = (+4)$

- Absolute value
- Sign of the number with the greater absolute value ■

Subtraction

We are already familiar with the operation of subtraction in problems such as $12 - 8 = 4$. If we were to add (-8) instead of subtracting $(+8)$, we observe that the results are the same, that is

Opposite of $+8$

$$(+12) - (+8) = (+12) + (-8) = (+4)$$

Change to addition

This leads us to the following definition of subtraction.

Definition of subtraction

For any two real numbers a and b, the difference of a and b is
$$a - b = a + (-b)$$

Concept

a minus b is equal to a plus the opposite of b.

Our steps to carry out the subtraction are as follows:

Subtraction of real numbers

Step 1 We change the operation from subtraction to addition.
Step 2 We change the sign of the number that follows the subtraction symbol.
Step 3 We perform the addition using our rules for adding real numbers.

■ *Example 1–2 B*

Subtract the following.

		Stays as it was	*Change to addition*	*Changes to the opposite*	*Add using the rules for addition*
1. $(+7) - (+4)$	$=$	$(+7)$	$+$	(-4)	$=$ ⟶ 3
2. $(-11) - (+3)$	$=$	(-11)	$+$	(-3)	$=$ ⟶ -14
3. $(-14) - (-10)$	$=$	(-14)	$+$	$(+10)$	$=$ ⟶ -4
4. $(-2) - (-12)$	$=$	(-2)	$+$	$(+12)$	$=$ ⟶ 10

▶ *Quick check* $(-6) - (-14)$ ■

When a series of numbers involving addition and subtraction are written horizontally, we can mentally change the operation of subtraction to addition and perform the operations in order from left to right as they occur. For example, consider

$$10 - (-2) + 7 + (-4) - 8$$

Changing the indicated subtractions to additions, we obtain the indicated sum.

$$10 + (+2) + 7 + (-4) + (-8) = 7$$

Many times, part of a problem will have a group of numbers enclosed within grouping symbols such as parentheses ( ), brackets [ ], or braces { }. If a quantity is enclosed within a grouping symbol, we treat the quantity within as a single number. Thus given

$$14 - (5 - 7) + (4 + 8) - (6 - 2)$$

we perform the operations within the parentheses first to get

$$14 - (-2) + (12) - (4)$$

and changing the subtractions to additions

$$14 + (+2) + (12) + (-4)$$

and adding from left to right

$$= 16 + (12) + (-4)$$
$$= 28 + (-4)$$
$$= 24$$

■ Example 1–2 C

Perform the indicated operations.

1. $(-14) - (-8) + (-7) - (4)$
   Change the subtractions to additions.

   $$= (-14) + (+8) + (-7) + (-4)$$

   Perform the addition from left to right.

   $$= (-6) + (-7) + (-4)$$
   $$= (-13) + (-4)$$
   $$= -17$$

2. $12 - (4 - 7) + (8 - 3) + (5 - 9)$
   $= 12 - (-3) + (5) + (-4)$ Simplify within parentheses
   $= 12 + (+3) + (5) + (-4)$ Change to addition
   $= 15 + (5) + (-4)$ Add from left to right
   $= 20 + (-4)$
   $= 16$

Multiplication

We are already familiar with the fact that the product of two positive numbers is positive. We can see this by considering multiplication to be repeated addition. For example, $3 \cdot 4$ means the sum of the three 4s, that is,

$$3 \cdot 4 = 4 + 4 + 4 = 12$$

The number 12 is called the **product** of 3 and 4, furthermore 3 and 4 are called **factors** of 12. *The numbers or variables in an indicated multiplication are referred to as the factors of the product.*

We can summarize the rules of multiplication of real numbers.

> **Multiplication of real numbers**
>
> 1. When we multiply two numbers having the same sign, the product will be positive.
> 2. When we multiply two numbers having different signs, the product will be negative.

■ **Example 1–2 D**

Multiply the following.

1. $(+4)(+8) = +32$ **2.** $(-6)(-3) = +18$

3. $(-4)(+5) = -20$ **4.** $(+2)(-8) = -16$

▶ *Quick check* $(-5)(-7)$ ■

Exponential notation

Consider the indicated products

$$4 \cdot 4 \cdot 4 = 64$$

and

$$3 \cdot 3 \cdot 3 \cdot 3 = 81$$

A more convenient way of writing $4 \cdot 4 \cdot 4$ is 4^3, which is read "4 to the third power" or "4 cubed." We call the number 4 the **base** of the expression and the number 3 to the upper right the **exponent.**
    Thus

$$\text{Expanded form} \longrightarrow 4 \cdot 4 \cdot 4 = \underset{\text{Exponential form}}{4^3} = 64 \longleftarrow \text{Standard form}$$

with labels: Base, Exponent

In like fashion,

$$3 \cdot 3 \cdot 3 \cdot 3$$

may be written 3^4, where 3 is the base and 4 is the exponent, and the expression is read "3 to the fourth power." Then

$$3 \cdot 3 \cdot 3 \cdot 3 = 3^4 = 81$$

Notice that *the exponent tells how many times the base is used as a factor in an indicated product.* We call this form of a product the **exponential form.** That is, the exponential form of the product $3 \cdot 3 \cdot 3 \cdot 3$ is 3^4.
    We will now relate the idea of exponents to signed numbers. Consider the following examples.

■ **Example 1–2 E**

Perform the indicated multiplication.

1. $2^4 = 2 \cdot 2 \cdot 2 \cdot 2 = 16$ The exponent, 4, indicates how many times the base, 2, is used as a factor

2. $(-2)^4 = (-2)(-2)(-2)(-2) = +16$ Even number of negative factors

3. $(-2)^3 = (-2)(-2)(-2) = -8$ Odd number of negative factors   ■

Remember that when we have a negative number, we place it inside parentheses. With this idea in mind, we can see that there is a definite difference between $(-2)^4$ and -2^4. In the first case, the parentheses denote that this is a negative number to a power, $(-2)^4 = (-2)(-2)(-2)(-2) = +16$. In the second case, since there are no parentheses around the number, we understand that this is *not* (-2) to a power, but rather the opposite of what we get for an answer when we find 2^4 as follows:

$$-2^4 = -(2^4) = -(2 \cdot 2 \cdot 2 \cdot 2) = -(16) = -16$$

■ *Example 1–2 F*

Perform the indicated multiplication.

1. $(-3)^3 = (-3)(-3)(-3) = -27$

2. $-3^3 = -(3 \cdot 3 \cdot 3) = -27$ This is the opposite of 3^3, $-(3^3)$

3. $(-3)^4 = (-3)(-3)(-3)(-3) = +81$

4. $-3^4 = -(3 \cdot 3 \cdot 3 \cdot 3) = -81$ This is the opposite of 3^4, $-(3^4)$ ■

Division

We studied subtraction of signed numbers by defining subtraction in terms of addition. We shall use a similar approach for division of signed numbers by defining division in terms of multiplication.

Definition of division

The quotient of any two real numbers a and b ($b \neq 0$) is the unique real number q such that $a = bq$. In symbols,

$$\frac{a}{b} = q \text{ if and only if } a = bq$$

Note In the definition, a is called the dividend, b is the divisor, and q is the quotient.

We can summarize our rules for multiplication and division of real numbers.

Multiplication and division of real numbers

1. When we multiply or divide two numbers having the same sign, our answer will be positive.
2. When we multiply or divide two numbers having different signs, our answer will be negative.

■ *Example 1–2 G*

Perform the indicated division.

1. $\dfrac{(-14)}{(-7)} = +2$

2. $\dfrac{(-24)}{(+8)} = -3$

3. $\dfrac{(+36)}{(-9)} = -4$

4. $\dfrac{(-2)(-8)}{(-4)} = \dfrac{(+16)}{(-4)} = -4$

5. $\dfrac{(+6)(-3)}{(-9)} = \dfrac{(-18)}{(-9)} = +2$

■

Division involving zero

In section 1–1, we defined a rational number to be any real number that can be expressed as a quotient of two integers where the divisor is not zero. The number zero, 0, is the only number that we cannot use as a divisor. To see why we exclude zero as a divisor, recall that we check a division problem by multiplying the divisor times the quotient to get the dividend. If we apply this idea in connection with zero as a divisor, we observe the following. Suppose there were a number q, such that $3 \div 0 = q$. Then $q \cdot 0$ would have to be equal to 3 for our answer to check, but this product is zero regardless of the value of q. Therefore, we cannot find an answer for this problem and we say that the answer is *undefined*. If we try to divide zero by zero and again call our answer q, we have $0 \div 0 = q$, and when we check our work, $0 \cdot q = 0$, we see that any value for q will work. In this situation, we say our answer is *indeterminate*. We therefore decide that **division by zero is not allowed.**

It is important to note that, although division by zero is not allowed, this does not extend to the division of zero by some other number. We can see that $0 \div (-4) = 0$ since $(-4) \cdot 0 = 0$. Thus *the quotient of zero divided by any number other than zero is always 0.*

Division involving zero

If a is any real number except 0,	example
$\dfrac{0}{a} = 0$	$\dfrac{0}{4} = 0$
$\dfrac{a}{0}$ is undefined	$\dfrac{4}{0}$ is undefined
$\dfrac{0}{0}$ is indeterminate	

Problem solving

To solve the following word problems, we will represent gains by positive real numbers and losses by negative real numbers.

■ *Example 1–2 H*

Choose a variable to represent the unknown quantity and find its value.

1. A board that is 8 feet long is joined with a board that is 5 feet long. What is the total length of the board?

Let x equal the total length of the board. To find the total length we must *add* the individual lengths. Thus

$x = 8 + 5 = 13$

The total length of the board is 13 feet.

2. On a given winter's day in Chicago, the temperature was 14° in the afternoon. By 9 P.M. the temperature was −4°. How many degrees did the temperature fall from afternoon to 9 P.M.?

Let x = the number of degrees fall in temperature. We must find the difference between 14° and −4°. Thus

$$x = 14 - (-4) = 14 + 4 = 18$$

There was an 18° fall in temperature.

3. If \$13.58 is spent on 14 audio tapes, how much did each tape cost?

Let c = the cost of each tape. We must divide \$13.58 by 14. Thus

$$c = \frac{13.58}{14} = 0.97$$

Each audio tape costs 97¢.   ∎

Mastery points

Can you
- Add real numbers?
- Subtract real numbers?
- Perform addition and subtraction in order from left to right?
- Treat a quantity within a grouping symbol as a single number?
- Multiply real numbers?
- Divide real numbers?
- Use exponents?
- Apply the rules of division involving zero?
- Solve word problems?

Exercise 1–2

Find each sum or difference. See examples 1–2 A, B, and C.

Example $(-6) - (-14)$

Solution $= (-6) + (+14)$ Change the subtraction to addition, change −14 to +14
$\quad\quad = 8$ Add using the rules for addition

1. $(+6) - (+4)$
2. $(+7) - (-5)$
3. $(-8) - (+6)$
4. $(+7) - (+12)$
5. $(+10) + (-6) + (-3)$
6. $(-5) - (-1)$
7. $(+10) - (-8)$
8. $(-12) + 0$
9. $(-7) - 0$
10. $(+16) + (-7)$
11. $(+9) + (-13)$
12. $0 + (-4)$
13. $(-20) - 0$
14. $6 - 9 + 11 - 8$
15. $-9 - 8 - 7 + 12$
16. $-10 - 12 + 18 + 4$
17. $9 - 4 + 6 - 5 + 7$
18. $(8 - 3) - 6 + (7 + 5)$
19. $-6 - 5 + 8 - (12 - 6)$
20. $14 - [(-4) - (-6)] + (7 - 9)$
21. $[(-6) - 5] + 4 - (16 - 6)$
22. $[4 - (-8)] - 6 - (2 - 11)$

Section 1-2 Operations with Real Numbers **19**

Perform the indicated operations if possible. See examples 1–2 D, E, F, G, and H.

Example $(-5)(-7)$

Solution $= 35$ Signs are the same: positive product

23. $(-2)(-5)$

24. $(-6)(-8)$

25. $(+4)(-7)$

26. $(+5)(-6)$

27. $(-9)(+3)$

28. $(-11)(+4)$

29. $(-3)(-1)(-8)$

30. $(-7)(-2)(-6)$

31. $(+4)(-2)(-6)$

32. $(+5)(-3)(+2)$

33. $(-6)(+10)(+4)$

34. $(+7)(-3)(-2)$

35. $(-4)(-5)(-1)(-3)$

36. $(-9)(+2)(0)(-4)$

37. $(-8)(-10)(+6)(0)$

38. -5^2

39. -2^6

40. $(-4)^2$

41. $(-6)^2$

42. $(-3)^3$

43. $(-4)^3$

44. -5^3

45. -2^3

46. -6^3

47. $\dfrac{(-20)}{(-10)}$

48. $\dfrac{(-32)}{(-8)}$

49. $\dfrac{(+32)}{(-8)}$

50. $\dfrac{(-22)}{(-11)}$

51. $\dfrac{(-21)}{(-7)}$

52. $\dfrac{(-34)}{(-17)}$

53. $\dfrac{(-27)}{(+3)}$

54. $\dfrac{0}{(-8)}$

55. $\dfrac{0}{0}$

56. $\dfrac{(-4)(-3)}{(-6)}$

57. $\dfrac{(-20)(+2)}{(-4)}$

58. $\dfrac{(+18)(+2)}{(-6)}$

59. $\dfrac{(-6)(0)}{(-2)}$

60. $\dfrac{(-8)(0)}{(-2)(+4)}$

61. $\dfrac{(-18)(+12)}{(-9)(+6)}$

62. $\dfrac{(-8)(-6)}{(-2)(0)}$

63. $\dfrac{(-4)(-3)}{(0)(-2)}$

64. $\dfrac{(-6)^2}{(-2)(-2)}$

65. $\dfrac{-10^2}{(-5)(-5)}$

66. $\dfrac{(-12)(0)}{(0)(-3)}$

67. Harry owes Kay and Bill $153 and $56, respectively, and Donna owes Harry $121. In terms of positive and negative symbols, how does Harry stand monetarily?

68. If Helen has $46 just after paying off a debt of $37, how much money did she have before paying off the debt?

69. The temperatures on January 15 in Chicago for the last six years are $14°$, $-6°$, $-4°$, $19°$, $7°$, and $-12°$. What is the average temperature on January 15 for the last six years? (*Hint:* The average is found by adding all of the values and then dividing that sum by the number of values.)

70. The temperatures on February 1 in Ogema, Wisconsin, for the last eight years are $-20°$, $-11°$, $0°$, $-6°$, $-9°$, $-11°$, $4°$, and $-3°$. What is the average temperature on February 1 for the last eight years? (Refer to exercise 69.)

71. Eric Dickerson carried the ball six times as follows: 14-yard gain, 4-yard gain, 6-yard gain, 8-yard loss, 10-yard gain, and 4-yard gain. Represent the gains as positive integers and the losses as negative integers and find his average for the six carries. (Refer to exercise 69.)

72. Over a six-day period, the price of a particular stock suffered losses of $5 the first two days and $4 the last four days. If the stock originally sold for $88, what was its price after the six-day period?

73. Jeff acquires a debt of $10 each day for seven days. If we represent a $10 debt by (-10), write a statement of the change in his assets after seven days. What is the change?

74. Eight days ago Joyce had $48 more in assets than she has today. Let eight days ago be represented by (-8). Write a statement for her daily change in assets if the change is the same each day. What is this daily change?

75. The temperature on a given day in Anchorage, Alaska, was $-10°$ F. The temperature went down $14°$ F. What was the final temperature that day?

76. A TWA flight is flying at an altitude of 27,000 feet. It suddenly hits an air pocket and drops 1,800 feet. What is its new altitude?

77. Mary Ann has $125 in her checking account. She deposits $28, $14, and $17. She then writes checks for $52 and $64. What is her final balance in the checking account?

78. An auditorium contains 68 rows of seats. If each row contains 40 seats, how many people can be seated in the auditorium?

79. In a classroom, there are 4 rows of desks. If each row contains 8 desks, how many students will the classroom hold?

80. Tom received money on his birthday from four different people. He received $10, $6, $8, and $5. How much money did Tom receive for his birthday?

81. Nancy sold 12 glasses of lemonade at her corner stand. If she charged 20 cents a glass, how much did she make in sales?

82. Lisa was born in 1963. How old will she be in the year 2000?

83. A chemist now has 125 ml (milliliters) of acid and she needs 415 ml of the acid. How much more is needed?

84. Beth has $175 and wants to buy a television for $485. How much will she owe?

85. The top of Mt. Everest is 29,028 feet above sea level and the top of Mt. McKinley is 20,320 feet above sea level. How much higher is the top of Mt. Everest than the top of Mt. McKinley?

86. Death Valley is 282 feet below sea level (-282). What is the difference in the altitude between Mt. McKinley and Death Valley? (See exercise 85.)

87. Mrs. Spencer paid $25.20 for 14 watermelons for her fruit market. How much did each watermelon cost her?

88. A carpenter wishes to cut a 12-foot board into 4 pieces that are all the same length. Find the length of each piece.

89. A man drove 374 miles and used 11 gallons of gasoline. How many miles did he drive on each gallon of gasoline?

90. A grocer averages selling 68 gallons of milk each day. How many gallons of milk does he sell in two weeks? (Assume the grocer is open seven days per week.)

91. Rob's blood pressure was 127 over 82. If the first number rose by 6 and the second number dropped by 4, what is the new reading?

92. If Martha drove 252 miles in six hours, how many miles did she travel each hour (in miles per hour) if she drove at a constant speed?

93. Alice can work 62 math problems per hour. How many hours would it take her to work 1,674 problems?

94. Al can type 2,436 words in 29 minutes. How many words can he type per minute?

95. The barometric pressure rose 8 mb (millibars) and then dropped 12 mb. Later that day the pressure dropped another 4 mb and then rose 10 mb. What was the gain or loss in barometric pressure that day?

1–3 ■ *Properties of real numbers*

In mathematics, we begin the development of a set of properties of numbers by making certain assumptions. These assumptions are called **axioms** and are formal statements about numbers that we assume to be true. Such assumptions may arise from the observation of a number of instances of a specific situation or may just be statements that we assume are always valid. This appears to give us freedom to introduce as many such assumptions as we wish, but it is desirable that all such axioms lead to useful consequences. The fewer axioms that we use, the more powerful will be our system.

The first of our assumptions has to do with equality. An **equality** is a mathematical statement that two symbols, or groups of symbols, are names for the same number. For example, $2 + 3$ and $4 + 1$ are different symbols for the number 5, and we can state this by the equality

$$2 + 3 = 4 + 1$$

We shall assume that real numbers have the following properties of equality.

> **Equality properties of real numbers**
>
> For all real numbers a, b, and c
>
> **Reflexive property of equality**
> $a = a$
>
> **Symmetric property of equality**
> If $a = b$, then $b = a$
>
> **Transitive property of equality**
> If $a = b$ and $b = c$, then $a = c$
>
> **Substitution property of equality**
> If $a = b$, then a may be replaced by b or b may be replaced by a in any statement without changing the truth or falsity of the statement.

■ Example 1–3 A

The following statements are examples of the properties of equality.

1. $10 = 10$ Reflexive property

2. If $3 = x$, then $x = 3$ Symmetric property

3. If $m = n$ and $n = 5$, then $m = 5$ Transitive property

4. If $x = 2$ and $x + 1 = 3$, then $2 + 1 = 3$ Substitution property   ■

In section 1–1, we discussed inequalities and the idea of order on the real number line. We will now state two properties of inequality of real numbers.

> **Inequality properties of real numbers**
>
> For all real numbers a, b, and c
>
> **Trichotomy property**
> Exactly one of the following is true.
> $a < b$, $a = b$,   or $a > b$
>
> **Transitive property of inequality**
> If $a < b$ and $b < c$, then $a < c$

■ Example 1–3 B

The following statements are examples of the properties of inequality.

1. $a \not< b$. The statement is read "a is not less than b." From the trichotomy property, we know that if a is not less than b, then a could be equal to b or a could be greater than b. We write this mathematically as $a \geq b$.

2. If $a < b$ and $b < 3$, then $a < 3$.   Transitive property   ■

The following properties of real numbers govern the operations of addition and multiplication.

Properties of real numbers
For all real numbers *a*, *b*, and *c*

Closure property of addition
$a + b \in R$

Closure property of multiplication
$ab \in R$

Commutative property of addition
$a + b = b + a$

Commutative property of multiplication
$ab = ba$

Associative property of addition
$(a + b) + c = a + (b + c)$

Associative property of multiplication
$(ab)c = a(bc)$

Identity property of addition
There is a unique real number 0 such that
$a + 0 = 0 + a = a$ Zero is called the additive identity.

Identity property of multiplication
There is a unique real number 1 such that
$a \cdot 1 = 1 \cdot a = a$ One is called the multiplicative identity.

Additive inverse property
For every real number *a*, there is a unique real number $-a$ (read "the opposite of *a*" or "the negative of *a*" or "the additive inverse of *a*") such that
$a + (-a) = 0$ and $(-a) + a = 0$

Multiplicative inverse property (reciprocal)
For every real number *a*, $a \neq 0$, there is a unique real number $\frac{1}{a}$ such that

$$a \cdot \frac{1}{a} = 1 \text{ and } \frac{1}{a} \cdot a = 1$$

Distributive property of multiplication over addition
$a(b + c) = ab + ac$

■ **Example 1–3 C**

The following statements are applications of the properties of real numbers.

1. $5 + 4 = 4 + 5$ Commutative property of addition
     $9 = 9$

2. $3 \cdot 6 = 6 \cdot 3$ Commutative property of multiplication
     $18 = 18$

3. $(2 + 3) + 4 = 2 + (3 + 4)$ Associative property of addition
     $5 + 4 = 2 + 7$
     $9 = 9$

4. $(2 \cdot 3)4 = 2(3 \cdot 4)$ Associative property of multiplication
$6 \cdot 4 = 2 \cdot 12$
$24 = 24$

5. $5 + 0 = 5$ Identity property of addition

6. $6 \cdot 1 = 6$ Identity property of multiplication

7. $(-7) + 7 = 0$ Additive inverse property

8. $6 \cdot \dfrac{1}{6} = 1$ Multiplicative inverse property

9. $2(3 + 4) = (2 \cdot 3) + (2 \cdot 4)$ Distributive property
$2(7) = 6 + 8$
$14 = 14$

10. $3(8 - 10) = 3[8 + (-10)]$ Distributive property
$3(-2) = 3 \cdot 8 + 3(-10)$
$-6 = 24 + (-30)$
$-6 = -6$

11. $(-8) + (-4) \in R$ Closure property of addition
$-12 \in R$

12. $(-10) \cdot 3 \in R$ Closure property of multiplication
$-30 \in R$

There are many other properties of real numbers that we have not listed. These other properties, called **theorems,** are implied or can be shown to be true from the previous properties and definitions. A theorem is a property that we can prove is true. We will now state several theorems that follow logically from the properties already listed.

Addition property of equality _____

For all real numbers a, b, and c, if $a = b$, then
$$a + c = b + c$$

Concept
We can add the same quantity to both members (also called sides) of an equality without altering the equality.

To prove that this theorem is true for all real numbers, we must start with the given information $a = b$, and, by applying the properties of real numbers, show that the desired result, $a + c = b + c$, is true.

Steps	*Reasons*
1. a, b, and c are real numbers	Given
2. $a + c$ is a real number	Closure property of addition
3. $a + c = a + c$	Reflexive property of equality
4. $a = b$	Given
5. $a + c = b + c$	Substitution property of equality

To discuss the steps in the previous proof, we start with the given information that a, b, and c are real numbers and because of the closure property of addition, $a + c$ must be a real number. We use the reflexive property of equality to write the equality $a + c = a + c$. Finally, since it was given that $a = b$, the substitution property of equality allows us to replace a with b in the right member and finish the proof.

The proofs of the following theorems are provided in the exercise set, where you will be asked to supply the reasons.

Multiplication property of equality

For all real numbers a, b, and c, if $a = b$, then

$$a \cdot c = b \cdot c$$

Concept
We can multiply both members of an equality by the same number without altering the equality.

Zero factor property

For any real number a,

$$a \cdot 0 = 0 \text{ and } 0 \cdot a = 0$$

Concept
The product of zero and any number is zero.

Double-negative property

For any real number a,

$$-(-a) = a$$

Concept
The opposite of the opposite of any real number is the given number.

■ *Example 1–3 D*

The following statements are applications of the previous theorems.

1. If $a = 3$, then $a + 4 = 3 + 4$ Addition property of equality
$$a + 4 = 7$$

2. If $a = 5$, then $4 \cdot a = 4 \cdot 5$ Multiplication property of equality
$$4a = 20$$

3. $4 \cdot 0 = 0$ Zero factor property

4. $-(-4) = 4$ Double-negative property   ■

Mastery points

Can you
- Apply the equality properties for real numbers?
- Apply the inequality properties for real numbers?
- Apply the properties of real numbers?

Exercise 1–3

Apply the indicated property to write a new expression that is equal to the given expression. Assume that all variables represent real numbers. See example 1–3 C.

Examples   Commutative, $5 + y$ Distributive, $3(x + y)$

Solutions $y + 5$ <small>Change the order</small> $3x + 3y$ <small>Distribute multiplication</small>

1. Commutative, $4x + 3y$ 2. Distributive, $4(a - b)$ 3. Associative, $(4 + 2) + 6$
4. Identity, $4 + 0$ 5. Commutative, $5a$ 6. Distributive, $12(3 - y)$
7. Associative, $(2x)y$ 8. Identity, $5 \cdot 1$ 9. Inverse, $(-4) + 4$

10. Inverse, $4 \cdot \left(\dfrac{1}{4}\right)$

Replace each question mark with the correct symbol to make the given statement an application of the given property.

Example   If $b = 4$ and $b + 3 = c$, then $? + 3 = c$; substitution property of equality.

Solution $4 + 3 = c$ <small>Replace b with 4</small>

11. $x + 5 = ?$; reflexive property of equality.   12. Either $x < 0$, $x = 0$, or $x > ?$; trichotomy property.
13. If $2a - 3 = x$, then $x = ?$; symmetric property of equality.   14. If $a = b$ and $b = 5$, then $a = ?$; transitive property of equality.
15. If $a = 7$ and $b + 5 = a$, then $b + 5 = ?$; substitution property of equality.   16. If $5 < x$ and $x < y$, then $? < y$; transitive property of inequality.

Rewrite each relation without the slash mark to obtain an equivalent statement. See example 1–3 B.

17. $a \not< 5$ 18. $3 \not\leq b$ 19. $a \not\geq b$ 20. $b \not> c$
21. $x \neq 5$ 22. $4 \not< y$ 23. $a \not\leq 6$ 24. $b \not\geq 4$

Identify which property is being used. See example 1–3 C.

Examples $5 + a = a + 5$ $(-7) + 7 = 0$

Solutions   Commutative property of addition   Additive inverse property

25. $6 + 7$ is a real number.   26. $8(a + b) = 8a + 8b$ 27. $8 \cdot 1 = 8$

28. $5(x - y) = (x - y)5$ 29. $\dfrac{1}{4} \cdot 4 = 1$ 30. $(-4) + 4 = 0$

31. $0 + 9 = 9$ 32. $(ab)(cd) = (cd)(ab)$ 33. $(6 + 7) + 8 = 6 + (7 + 8)$
34. $(6 \cdot 7)8 = 6(7 \cdot 8)$ 35. $(2x)y = 2(xy)$ 36. $3 + (2 + b) = (2 + b) + 3$

37. $14 + 0 = 14$ 38. $\left(\dfrac{2}{3}\right)\left(\dfrac{3}{2}\right) = 1$ 39. $x + 6 = 6 + x$

40. $4 \cdot 5$ is a real number.   41. $(5 + 4) + y = 5 + (4 + y)$ 42. $x + (-x) = 0$
43. $(a + b) + 0 = (a + b)$ 44. $5(2 + b) = 10 + 5b$

Fill in the missing values. Assume that all variables represent nonzero real numbers.

	Number	Additive inverse	Multiplicative inverse
Example		-4	
Solution	4		$\dfrac{1}{4}$
45.	6		
46.	10		
47.		5	
48.		2	
49.	x		
50.	y		
51.			$\dfrac{1}{7}$
52.			$\dfrac{1}{-5}$
53.		-3	
54.		-8	

Give a reason for each step in the following proofs.

55. Prove: For all real numbers a, b, and c, if $a = b$, then $a \cdot c = b \cdot c$.

a, b, and c are real numbers   a. _____

$a \cdot c$ is a real number   b. _____

$a \cdot c = a \cdot c$ c. _____

$a = b$ d. _____

$a \cdot c = b \cdot c$ e. _____

56. Prove: For any real number a, $a \cdot 0 = 0 \cdot a = 0$.

a is a real number   a. _____

$a = a \cdot 1$ b. _____

$a = a(1 + 0)$ c. _____

$a = a \cdot 1 + a \cdot 0$ d. _____

$a = a + a \cdot 0$ e. _____

$0 = a \cdot 0$ f. _____

$a \cdot 0 = 0$ g. _____

$0 \cdot a = 0$ h. _____

57. Prove: For any real number a, $-(-a) = a$.

a is a real number   a. _____

$-a + a = 0$ b. _____

$-a + [-(-a)] = 0$ c. _____

$a = -(-a)$ d. _____

1–4 ■ *Order of operations*

When we perform several different arithmetic operations within an expression or a formula, we need a standard order in which the operations will be performed. Consider the following example.

If a woman deposits $1,000 in a savings account at 6% simple interest, the amount (A) in her account after 1 year can be found by evaluating

$$A = 1,000 + 1,000(0.06)$$

More than one answer is possible depending on the order in which we perform the operations. For example, if we first add the 1,000 and 1,000 and then multiply by (0.06), we have

$$1,000 + 1,000(0.06) = 2,000(0.06) = 120$$

However if we multiply the 1,000 and (0.06) and to this product add the first 1,000 we have

$$1,000 + 1,000(0.06) = 1,000 + 60 = 1,060*$$

To standardize the answer, we agree to the following order of operations, or priorities:

Order of operations, or priorities

1. **Groups**   Perform any operations within a grouping symbol such as ( ) parentheses, [ ] brackets, { } braces, | | absolute value, or in the numerator or the denominator of a fraction.
2. **Exponents**   Perform operations indicated by exponents.
3. **Multiply and divide**   Perform multiplication and division in order from left to right.
4. **Add and subtract**   Perform addition and subtraction in order from left to right.

Note
a. Within a grouping symbol, the order of operations will still apply.
b. If there are several grouping symbols intermixed, remove them by starting with the innermost one and working outward.

Note   Within any priority, every operation is of equal importance (that is, multiplication is no more important than division in priority 3). Therefore when performing a particular priority, we start at the left and proceed to the right, doing only those operations within the priority that we are performing, as we come to them, without skipping around.

■ *Example 1–4 A*

Perform the indicated operations in the proper order and simplify.

1. $9 + 12 \cdot 3 \div 2 = 9 + 36 \div 2$ Priority 3, multiply
$= 9 + 18$ Priority 3, divide
$= 27$ Priority 4, add

2. $(19 - 1) \div 2 + 3 \cdot 4 = 18 \div 2 + 3 \cdot 4$ Priority 1, parentheses
$= 9 + 12$ Priority 3, divide and multiply
$= 21$ Priority 4, add

*This is the correct answer.

3. $\dfrac{1}{4} + \dfrac{3}{4} \div \dfrac{5}{8} = \dfrac{1}{4} + \dfrac{3}{\overset{}{\underset{1}{4}}} \cdot \dfrac{\overset{2}{8}}{5}$ Invert and multiply

$\quad\quad\quad\quad\quad = \dfrac{1}{4} + \dfrac{6}{5}$ Priority 3, multiply

$\quad\quad\quad\quad\quad = \dfrac{5}{20} + \dfrac{24}{20}$ Least common denominator

$\quad\quad\quad\quad\quad = \dfrac{5 + 24}{20}$ Priority 4, add in the numerator

$\quad\quad\quad\quad\quad = \dfrac{29}{20} \text{ or } 1\dfrac{9}{20}$

4. $2^2 \cdot 5 - 6 \cdot 4 = 4 \cdot 5 - 6 \cdot 4$ Priority 2, exponent
$\quad\quad\quad\quad\quad = 20 - 24$ Priority 3, multiply
$\quad\quad\quad\quad\quad = -4$ Priority 4, subtract

5. $\dfrac{3}{4} - \dfrac{\overset{1}{3}}{\underset{1}{4}} \cdot \dfrac{\overset{1}{4}}{\underset{3}{9}} = \dfrac{3}{4} - \dfrac{1}{3}$ Priority 3, multiply

$\quad\quad\quad\quad\quad = \dfrac{9}{12} - \dfrac{4}{12}$ Least common denominator

$\quad\quad\quad\quad\quad = \dfrac{9 - 4}{12}$ Priority 4, subtract in the numerator

$\quad\quad\quad\quad\quad = \dfrac{5}{12}$

6. $(8.21 + 12.43) \div 2.4 - (6.1)(4.7)$
$\quad = (20.64) \div 2.4 - (6.1)(4.7)$ Priority 1, parentheses
$\quad = 8.6 - 28.67$ Priority 3, divide and multiply
$\quad = -20.07$ Priority 4, subtract

7. $\left(\dfrac{1}{3} + \dfrac{5}{8}\right) \div \dfrac{5}{6} = \left(\dfrac{8}{24} + \dfrac{15}{24}\right) \div \dfrac{5}{6}$ Least common denominator

$\quad\quad\quad\quad\quad = \dfrac{8 + 15}{24} \div \dfrac{5}{6}$ Priority 1, add the fractions

$\quad\quad\quad\quad\quad = \dfrac{23}{24} \div \dfrac{5}{6}$ Add the numerators

$\quad\quad\quad\quad\quad = \dfrac{23}{\underset{4}{24}} \cdot \dfrac{\overset{1}{6}}{5}$ Invert and multiply

$\quad\quad\quad\quad\quad = \dfrac{23}{20} \text{ or } 1\dfrac{3}{20}$ Priority 3

8. $(9.3)^2 - 3(2.1)(4.6)$
$\quad = 86.49 - 3(2.1)(4.6)$ Priority 2, exponent
$\quad = 86.49 - 28.98$ Priority 3, multiply
$\quad = 57.51$ Priority 4, subtract

9. $\dfrac{3(2+4)}{11-8} - \dfrac{4+6}{2} = \dfrac{3(6)}{11-8} - \dfrac{4+6}{2}$ Priority 1, parentheses

$= \dfrac{18}{3} - \dfrac{10}{2}$ Priority 1, numerators and denominators

$= 6 - 5$ Priority 3, divide

$= 1$ Priority 4, subtract

10. $5[4 + 3(10 - 12)]$ We first simplify within the grouping symbol, starting with the innermost one

$= 5[4 + 3(-2)]$ Parentheses

$= 5[4 + (-6)]$ Multiply inside the brackets

$= 5[-2]$ Add inside the brackets

$= -10$ Multiply

▶ *Quick check* $6 + 5(7 - 3) - 2^2$ ◼

Problem solving

Solve the following word problems using the order of operations.

◼ *Example 1–4 B*

Choose a letter to represent the unknown quantity and find its value by performing the indicated operations.

1. Mrs. Miyazaki purchased 8 cans of oil at $1.10 per can and 2 air filters at $3.75 each. What was her total bill?

   Let x = Mrs. Miyazaki's total bill. 8 cans at $1.10 per can cost 8 · $1.10 and 2 air filters at $3.75 each cost 2 · $3.75. The total bill is given by

   $x = 8 \cdot 1.10 + 2 \cdot 3.75$
   $= 8.80 + 7.50 = 16.30$

   Mrs. Miyazaki's total bill was $16.30

2. A woman worked a 40-hour week at $8 per hour. If she worked 6 hours overtime at time and a half, how much money did she receive for the 46 hours of work?

   Let W = the woman's total wages for the week. 40 hours at $8 per hour is 40 · $8 and 6 hours at time and a half is $6 \cdot 1\frac{1}{2} \cdot$ $8. Thus

   $W = 40 \cdot 8 + 6 \cdot 1\frac{1}{2} \cdot 8$

   $= 320 + 72$
   $= 392$

   The woman's total wages for the week was $392.   ◼

┌─ **Mastery points** ─────────────────────────────────┐

Can you
◼ Simplify expressions according to the order of operations?

└──┘

Exercise 1–4

Perform the indicated operations and simplify. See example 1–4 A.

Example $6 + 5(7 - 3) - 2^2$

Solution $= 6 + 5(4) - 2^2$ We first simplify within the grouping symbol
$= 6 + 5(4) - 4$ Perform the indicated power
$= 6 + 20 - 4$ Carry out the multiplication
$= 26 - 4$ Perform the addition
$= 22$ Subtract

1. $(-3) \cdot 4 + 14$

2. $3 \cdot 4 + (-8)$

3. $5 \cdot 6 - (-7)$

4. $6 - (-4)(3)$

5. $12 - (-5)(-4)$

6. $\dfrac{8 + 4}{6} + 5$

7. $\dfrac{18 - 6}{2} + 7 \cdot 5$

8. $3 - 12 \cdot \left(\dfrac{1}{4}\right)$

9. $5 - 15 \cdot \left(\dfrac{1}{3}\right)$

10. $8 + 0(5 - 7)$

11. $10 - 0(8 - 4)$

12. $8 + 6 - 4 \cdot 3 + 12$

13. $10 + 10 \div 10 - 10$

14. $14 \div 7 - 2 + 8 \cdot 4$

15. $24 \div 6 + 6 - 4 \cdot 3$

16. $9 - 18 \div 2 + 7 - 3^2$

17. $14 + 27 \div 3 \cdot 4 - 5^2$

18. $4 - (8 - 6)^2 + 3 \cdot 5$

19. $(4 - 9)^2 \cdot 2 + 7 - 4^2$

20. $12 + 3 \cdot 16 \div 4^2 - 5$

21. $18 - 3^2 \cdot 4 \div 6 + 7$

22. $8 - (12 - 9)^2 \cdot 2 + 5$

23. $26.4 - (3.7)(4.6)$

24. $\dfrac{2}{3} \div \left(\dfrac{5}{6} - \dfrac{4}{9}\right)$

25. $\dfrac{3}{8} - \dfrac{1}{2} \div \dfrac{3}{4}$

26. $\dfrac{1}{8} + \dfrac{7}{12} \cdot \dfrac{9}{14}$

27. $(14.13 + 11.4) \div 3.7 - (3.6) \cdot (4.9)$

28. $(4.6 + 3.1) \cdot (2.7) - (5.4) \cdot (7.3)$

29. $(4.7)^2 \cdot 5 - (14.64) \div (6.1)$

30. $(2.4)^2 + 5(1.9)^2 - 12.6$

31. $\left(\dfrac{11}{12} - \dfrac{5}{6}\right) \div \left(\dfrac{1}{3} + \dfrac{3}{8}\right)$

32. $\left(\dfrac{1}{4} - \dfrac{1}{6}\right) \cdot \left(\dfrac{1}{8} + \dfrac{3}{8}\right)$

33. $5[20 - 3(4 - 6) + 5]$

34. $9 + 2[5 - 3(8 - 3) + 5]$

35. $12 + [14 - 5(7 - 10) + 4]$

36. $(9 - 7)[15 - 4(3 - 6) + 9]$

37. $\dfrac{8(6 - 4)}{2} - \dfrac{27}{-3}$

38. $\dfrac{3(9 - 4)}{5} - \dfrac{10}{-2}$

39. $\dfrac{5 \cdot 6 - 4}{13} - \dfrac{(-18)}{6}$

40. $\dfrac{8 - 3 \cdot 4 + 10}{6 - 3} + \dfrac{(-18) + 6}{-4}$

41. $\left[\dfrac{10 + (-2)}{2(-1)}\right]\left[\dfrac{(-10)(-4)}{(-2)}\right]$

42. $\left[\dfrac{(-11) + (-7)}{(-9)}\right]\left[\dfrac{(-5)(-12)}{10}\right]$

43. $\dfrac{4(3 + 2) - 4^2 + 11}{(-5)(-3)}$

44. $\dfrac{(8 - 4)^2 + (-2)(-3)}{10 - (-4)(3)}$

Perform the indicated operations and simplify. See example 1–4 A.

45. The area of a trapezoid whose bases are 12 meters and 8 meters and whose height is 10 meters is

$$\frac{1}{2} \cdot 10(12 + 8)$$

Find the area in square meters.

46. The surface area of a flat ring whose inside radius is 3 centimeters and whose outside radius is 5 centimeters is approximately

$$\frac{22}{7} \cdot 5^2 - \frac{22}{7} \cdot 3^2$$

Find the area in square centimeters.

47. The water pressure exerted on 1 square foot (144 square inches) of surface area of a diving suit 50 feet below the surface of the water is given by

$$144 [14.7 + 50(0.444)]$$

Perform the indicated operations to find this pressure in pounds per square inch.

48. The surface area of a ring section whose inside diameter is 19 inches and whose outside diameter is 26 inches is approximately

$$\frac{22}{7} \cdot \frac{26 + 19}{2} \cdot \frac{26 - 19}{2}$$

Find the surface area in square inches.

49. The surface area of a rectangular solid whose length is 8 feet, width is 5 feet, and height is 7 feet is given by

$$2 \cdot 8 \cdot 7 + 2 \cdot 8 \cdot 5 + 2 \cdot 5 \cdot 7$$

Find the surface area in square feet.

50. To convert 59 degrees Fahrenheit to degrees Celsius, we use

$$\frac{5}{9} (59 - 32)$$

Find the temperature in degrees Celsius.

51. To convert 25 degrees Celsius to degrees Fahrenheit, we use

$$\left(\frac{9}{5}\right) 25 + 32$$

Find the temperature in degrees Fahrenheit.

Choose a letter to represent the unknown quantity and solve. See example 1–4 B.

52. Norbert makes $6 per hour for a 40-hour week and receives time and a half for every hour he works over 40 hours a week. How much did he earn if he worked 47 hours in one week?

53. Royetta has $120 in her checking account. She deposits three twenty-dollar checks and then writes a check for $87. What is her balance?

54. A man purchased a case of beans (12 cans) at 53¢ per can, 4 pounds of apples at 59¢ per pound, and 3 quarts of milk at 69¢ per quart. What was his total bill (a) in cents, (b) in dollars and cents?

55. A car dealer sold 18 cars at $8,200 each and 4 trucks at $7,100 each. What is the total sale of all the cars and trucks?

56. Forest Road School has 3 first-grade classes, each with 27 students; 2 second-grade classes, each with 31 students; and 2 third-grade classes, each with 29 students. How many students are there all together in first, second, and third grade?

57. While away at school for nine months, a student spends $140 per month on housing, $95 per month on food, and $48 per month on miscellaneous expenses. How much money did the student spend on these expenses during the nine months?

58. A woman wants to carpet two rooms. One room is 4 yards by 5 yards and the second room is 4 yards by 7 yards. How many square yards of carpet are needed? If the carpet costs $8 per square yard and the pad under the carpet costs $2 per square yard, how much will the carpet and pad for these rooms cost?

59. Nancy, Alice, and Jane are typists in an office. If Nancy can type 90 words per minute, Alice can type 85 words per minute, and Jane can type 70 words per minute, how many words can they type together in 20 minutes?

60. Susan charges $5 to wash a car, $11 to mow a lawn, and $27 to wax a car. If in one week, she washed six cars, mowed three lawns, and waxed two cars, how much did she earn that week?

1–5 ■ Terminology and evaluation

Terminology

Before we study the different types of algebraic operations, we must first define a few concepts. *An **algebraic expression** is any meaningful collection of variables, constants, grouping symbols, and symbols of operations.* Examples of algebraic expressions are

$$3x^2y, \quad \frac{ab}{c}, \quad \pi r^2, \quad \frac{x^2 + 2x - 1}{x^2 + 1}, \quad 2x^2 - x + 3, \quad \sqrt{x^2 + y^2}, \quad x, \quad -4$$

In an algebraic expression, *a **term** is any constant, variable, or indicated product, quotient, or root* of constants and variables.* Terms in an algebraic expression are separated by the operations of addition and subtraction.

Note *A constant is a symbol that does not change its value.* In the expression πr^2, π is a symbol that represents only one value (the circumference of a circle divided by its diameter equals 3.14159 · · ·) and therefore is a constant.

■ *Example 1–5 A*

Identify the number of terms in the algebraic expressions.

The plus and minus signs separate the algebraic expression into three terms

1. $3x^2 - 2x + 1$

    1st   2nd   3rd

There are three terms

2. $3x^2 - \dfrac{y^2 - 3z^3}{5x}$

    1st   2nd

There are two terms since the fraction bar acts as a grouping symbol

▶ *Quick check* Identify the number of terms in $x^2 + 2y^2$.   ■

In the expression $3xy$, *each factor or group of factors is called the **coefficient** of the remaining factors.* That is, 3 is the coefficient of xy, x is the coefficient of $3y$, $3x$ is the coefficient of y, and so on. The 3 is called the **numerical coefficient,** and it tells us how many xy's we have in the expression.

Since we most often talk about the numerical coefficient of a term, we will eliminate the word "numerical" and just say "coefficient." It will be understood that we are referring to the numerical coefficient. If no numerical coefficient appears in a term, the coefficient is understood to be 1.

■ *Example 1–5 B*

Determine the numerical coefficient of each term in the algebraic expression $3x^2 - 2y + z$.

3 is the coefficient of x^2, -2 is the coefficient of y, and 1 is understood to be the coefficient of z.

Note The coefficient of a term in an algebraic expression includes the sign that precedes it. In our example, the coefficient of y is -2 and not 2.

*Roots will be discussed in chapter 5.

▶ *Quick check*   Determine the numerical coefficient of each term in $3a^3 - 4a^2 + 2a$. ∎

A special kind of algebraic expression is a *polynomial*. The following are characteristics of a polynomial.

1. It has real number coefficients.
2. All variables in a polynomial are raised only to natural number exponents.
3. The operations performed by the variables are limited to addition, subtraction, and multiplication.

A polynomial that contains just one term is called a *monomial;* one that contains two terms is called a *binomial;* and a polynomial that contains three terms is called a *trinomial.* Any polynomial that contains more than one term is called a *multinomial*, but no special names will be given to polynomials that contain more than three terms.

Note   We should simplify any expression before identifying it. Also, in an expression, the combining of all the constant terms is understood to be a single term. For example, $x + 3 + \pi$ is thought of as $x + (3 + \pi)$ and is a binomial.

■ *Example 1–5 C*

Determine if the algebraic expressions are polynomials.

1. x^2, $5x$, 7, and $-4x^2y^3$ are monomials.

2. $2x - 4$, $x^2 + y^2$, and $27x^3 - y^3$ are binomials.

3. $4x^2 + 3x - 2$ and $6x^2y^2 - 4xy + 3y^2$ are trinomials.

4. $8x^3 - 4x^2 + 3x - 2$ has no special name and is referred to as a polynomial of four terms.

5. $\dfrac{x}{y + z}$ and $5\sqrt{x} + y$ are not polynomials since they contain variables in the denominator or under a radical symbol. ∎

We also identify different types of polynomials by the degree of the polynomial. *The **degree** of a polynomial in one variable is the greatest exponent of that variable in any one term.*

■ *Example 1–5 D*

Determine the degree of the polynomial.

1. $5x^3$ Third degree because the exponent of x is 3

2. $x^4 - 2x^3 + 3x - 5$ Fourth degree because the greatest exponent of x in any one term is 4

Note   In example 2, the polynomial has been arranged in *descending powers* of the variable. This is the form that we will use when we write polynomials in one variable.

▶ *Quick check*   Identify $2x^2 - x + 5$ as a monomial, binomial, or trinomial and determine the degree. ∎

Evaluating an algebraic expression

An extremely important process in algebra is that of calculating the numerical value of an expression when we are given specific replacement values for the variables. This process is called **evaluation.** To carry out this evaluation, we need the following **property of substitution.**

┌── Property of substitution ──────────────────────

If $a = b$, then a may be replaced by b or b may be replaced by a in any expression without altering the value of the expression.

Concept

When two quantities are equal, we can replace one quantity with the other quantity without altering the value of the expression.

We frequently need to evaluate algebraic expressions. For example, the area A of a rectangle is found by multiplying the length ℓ times the width w, $A = \ell w$. If the length is 8 feet and the width is 4 feet, substituting 8 for the length, ℓ, and 4 for the width, w, the expression becomes

$$A = (8)(4) = 32$$

and the area is 32 square feet. We substituted the respective values for the length and the width into the expression and carried out the indicated arithmetic.

Note When replacing variables with the numbers they represent, it is a good procedure to put each of the numbers inside parentheses.

■ *Example 1–5 E*

Evaluate the following expressions for the given real number replacement for the variable or variables.

1. $x^2 + 3x - 4$, when $x = 3$

$= (\ \ )^2 + 3(\ \ ) - 4$	Expression without the x
$= (3)^2 + 3(3) - 4$	Substitute 3 in place of each x and use the order of operations
$= 9 + 3(3) - 4$	Exponents
$= 9 + 9 - 4$	Multiply
$= 14$	Add and subtract

Therefore the expression $x^2 + 3x - 4$ evaluated for $x = 3$ is 14.

2. If we know the temperature in degrees Fahrenheit (F), the temperature in degrees Celsius (C) can be found by the formula $C = \dfrac{5}{9}(F - 32)$. Find the temperature in degrees Celsius if the temperature is 77 degrees Fahrenheit.

$C = \dfrac{5}{9}(F - 32)$	Original formula
$C = \dfrac{5}{9}[(\ \ ) - 32]$	Formula ready for substitution
$C = \dfrac{5}{9}[(77) - 32]$	Substitute
$C = \dfrac{5}{9}[45]$	Order of operations
$C = 25$	

Hence 77 degrees Fahrenheit is equivalent to 25 degrees Celsius.

3. If a company sells n_1 shirts at d_1 dollars each and n_2 ties at d_2 dollars each, the total revenue R from the shirts and ties would be expressed as $R = n_1d_1 + n_2d_2$. Find R if $n_1 = 20$, $d_1 = 26$, $n_2 = 15$, and $d_2 = 8$.

Note In this formula, **subscripts** are used to denote two different values for the number of articles sold (n_1 and n_2) and the price per article (d_1 and d_2). These are read "n sub-one," "n sub-two," "d sub-one," and "d sub-two."

$R = n_1d_1 + n_2d_2$	Original formula
$R = (\ )(\ ) + (\ )(\ )$	Formula ready for substitution
$R = (20)(26) + (15)(8)$	Substitute
$R = 520 + 120$	Order of operations
$R = 640$	

Therefore the total revenue from the sale of shirts and ties is $640.

▶ *Quick check* Find the temperature C for $C = \dfrac{5}{9}(F - 32)$ when $F = 95$. ■

Polynomial notation

Polynomials can be denoted by a single symbol such as P, Q, R, or by P_1, P_2, P_3, and so on. If we are dealing with a polynomial in one variable, the symbol for the polynomial can be used in conjunction with the variable in naming the polynomial. For example, $P(x)$ (read "P of x" or "P at x") could be used to denote the polynomial $x^2 + 3x - 4$ from number 1 of example 1-5 E. Therefore we can write $P(x) = x^2 + 3x - 4$.

The symbol inside the parentheses denotes the variable in the polynomial. We can use this notation to indicate a specific value at which we want to evaluate the polynomial. In example 1-5 E, we evaluated $x^2 + 3x - 4$ at 3. We could have represented this as $P(3)$, which means the polynomial P evaluated at 3, not P times 3.

■ Example 1-5 F

If $P(x) = x^2 - x + 6$ and $Q(y) = y + 2$, find the indicated values.

1. $Q(-5)$

$Q(\ ) = (\ ) + 2$	Polynomial ready for substitution
$Q(-5) = (-5) + 2$	Substitute -5 for y in $Q(y)$
$\quad\quad = -3$	

2. $P(-3)$

$P(\ ) = (\ )^2 - (\ ) + 6$	Polynomial ready for substitution
$P(-3) = (-3)^2 - (-3) + 6$	Substitute -3 for x in $P(x)$
$\quad\quad = 9 - (-3) + 6$	
$\quad\quad = 18$	

3. $P(-4) \cdot Q(-6)$

Substitute -4 for x in $P(x)$ and -6 for y in $Q(y)$.

$= [(-4)^2 - (-4) + 6] \cdot [(-6) + 2]$	Substitute
$= [16 - (-4) + 6] \cdot [-4]$	Order of operations
$= [26] \cdot [-4]$	$P(-4) = 26$ and $Q(-6) = -4$
$= -104$	Multiply

Note $P(-4)$ and $Q(-6)$ represent specific real numbers and we must evaluate each of them at the indicated value before we can perform the indicated multiplication.

▶ **Quick check** If $P(x) = x^2 - x + 6$, find $P(3)$. ■

Algebraic notation

Many problems that we encounter are stated verbally. These need to be translated into algebraic expressions. While there is no standard procedure for changing a verbal phrase into an algebraic expression, these guidelines should be of use.

1. Read the problem carefully, determining useful prior knowledge. Note what information is given and what information we are asked to find.
2. Let some letter represent one of the unknowns. Then express any other unknowns in terms of it.
3. Use the given conditions in the problem and the unknowns from step 2 to write an algebraic expression.

When translating verbal phrases into algebraic expressions, we look for phrases that involve the basic operations of addition, subtraction, multiplication, and division. Table 1–1 shows some examples of phrases that we commonly encounter. We let x represent the unknown number.

■ **Table 1–1**

Phrase	Algebraic expression
Addition	
6 more than a number	
the sum of a number and 6	
6 plus a number	$x + 6$
a number increased by 6	
6 added to a number	
Subtraction	
6 less than a number	
a number diminished by 6	
the difference of a number and 6	
a number minus 6	$x - 6$
a number less 6	
a number decreased by 6	
6 subtracted from a number	
a number reduced by 6	
Multiplication	
a number multiplied by 6	
6 times a number	$6x$
the product of a number and 6	
Division	
a number divided by 6	
the quotient of a number and 6	$\dfrac{x}{6}$
$\dfrac{1}{6}$ of a number	

■ *Example 1–5 G*

Write an algebraic expression for each of the following.

1. The <u>product</u> of x and y $x \cdot y$

2. The <u>sum</u> of n and 7 $n + 7$

3. y <u>decreased by</u> 3 $y - 3$

4. a <u>divided by</u> 4 $a \div 4$ or $\dfrac{a}{4}$

5. A number <u>diminished by</u> 6, let n represent the number. $n - 6$

6. <u>Three times</u> a number and that product <u>increased by</u> 9, let x represent the number. $3x + 9$

7. A number <u>divided by</u> 4 and that quotient <u>decreased by</u> 2, let a represent the number. $\dfrac{a}{4} - 2$

8. <u>Five times</u> the <u>sum</u> of a number and 6, let n represent the number. $5(n + 6)$

▶ *Quick check*   Write an algebraic expression for each of the following:
1. the product of a and b
2. a number increased by 2

___ *Mastery points* ___

Can you
■ Identify terms in an expression?
■ Identify a polynomial by name?
■ Determine the degree of a polynomial?
■ Evaluate an algebraic expression?
■ Evaluate a polynomial using polynomial, $P(x)$, notation?
■ Write an algebraic expression?

Exercise 1–5

Specify the number of terms in each expression and determine if the expression is a polynomial. See examples 1–5 A and C.

Example $x^2 + 2y^2$

Solution x^2 $+$ $2y^2$ There are two terms, is a polynomial (binomial)
          ↑   ↑
         1st   2nd

1. $5x^2 - 2x + 3$ 2. $\dfrac{7x}{2}$ 3. $3xy - \dfrac{3y}{5} + 7x$ **4.** $\dfrac{4a^2 - b^2}{10}$

5. $\sqrt{3x^2 - 2x + 1}$ 6. $\dfrac{x^2 + y^2}{z^2} - a^2 + 2ab$

Determine the numerical coefficient of each term in the following expressions. See example 1–5 B.

Example $3a^3 - 4a^2 + 2a$

Solution The coefficient of a^3 is 3, the coefficient of a^2 is -4, and the coefficient of a is 2.

7. $4x^2 - 7x + y$ **8.** $3x^3 - 2x^2 + x$ 9. $5y^3 + y^2 - 7y$ 10. $4z^4 - z^3 + z^2 + 6z$

Identify each polynomial as a monomial, binomial, or trinomial and determine the degree of the polynomial. See examples 1–5 C and D.

Example $2x^2 - x + 5$

Solution There are three terms, so it is a trinomial, and second degree because the greatest exponent of x in any one term is 2.

11. $5a^3 - 4a^2 + 3$ 12. $4a^2 - a$ 13. $2x^4 - 7x^2$ 14. $6x^2 - 2x + 1$ 15. $4y^5$

Evaluate the following expressions if $a = -3$, $b = 2$, $c = -2$, and $d = 4$. See example 1–5 E.

16. $2(a + 3b)$ 17. $2a - 3b - (2c + d)$ 18. $b - 3(a - 2d)$

19. $(4a + 3b)(c - 2d)$ 20. $5ab^2 + 3cd$ 21. $2a^2 - 3a + 4$

22. $c^2 - d^2$ 23. $a^2 - 4c^2$ **24.** $(3a - 2b) - (2a + b)(c + d)$

25. $a^3b - c^3d$

Evaluate the following polynomials for the specific values of the variable. See example 1–5 F.

Example $P(x) = x^2 - x + 6$, find $P(3)$.

Solution $= (3)^2 - (3) + 6$ Substitute 3 for x, order of operations
$= 9 - 3 + 6$ Exponents
$= 12$ Subtract and add

26. $P(x) = x^2 + x$, $P(3)$, $P(-2)$, $P(0)$ 27. $Q(x) = x^2 - 3x + 4$, $Q(2)$, $Q(-3)$, $Q(0)$

28. $R(x) = 2x^2 - x - 1$, $R(-2)$, $R(1)$, $R(0)$ 29. $P(y) = y^3 - 1$, $P(1)$, $P(-2)$, $P(0)$

30. $Q(z) = z^2 - 5z + 6$, $Q(3)$, $Q(2)$, $Q(0)$

If $P(x) = x^2 + 2x + 1$, $Q(x) = 2x - 1$, and $R(x) = x^2 + x - 6$, find the indicated values. See example 1–5 F.

Example $R[Q(2)]$

Solution $= R[2(2) - 1]$ Substitute 2 for x in $Q(x)$
$= R[4 - 1]$ Multiply
$= R(3)$ Subtract
$= (3)^2 + (3) - 6$ Substitute 3 for x in $R(x)$
$= 9 + 3 - 6$ Exponents
$= 6$ Add and subtract

Note In this example, we had to evaluate the inner polynomial, $Q(x)$, first so that we could determine the value for which we would evaluate the outer polynomial, $R(x)$.

31. $P(2) + Q(-1) - R(2)$ **32.** $P(0) \cdot Q(5)$ **33.** $P(4) - R(0)$

34. $P(-1) \cdot Q(-1) + R(-1)$ **35.** $Q(4) \cdot R(-1)$ **36.** $P[Q(2)]$

37. $P[Q(-1)]$ **38.** $R[P(1)]$ **39.** $Q[R(-2)]$

Evaluate the following formulas. See example 1–5 E.

Example Find the temperature C for $C = \dfrac{5}{9}(F - 32)$ when $F = 95$.

Solution $C = \dfrac{5}{9}[(\ \ ) - 32]$ Formula ready for substitution

$C = \dfrac{5}{9}[(95) - 32]$ Substitute

$C = \dfrac{5}{9}[63]$ Order of operations

$C = 35$

40. $V = e^3, e = 5$

41. $F = ma, m = 12$ and $a = 5$

42. $I = prt; p = 2{,}000; r = 0.09;$ and $t = 3$

43. $V = k + gt, k = 22, g = 11,$ and $t = 3$

44. $H = \dfrac{D^2 N}{2}, D = 4$ and $N = 6$

45. $\ell = a + (n - 1)d, a = 7, n = 20,$ and $d = 4$

46. $V_2 = V_1 + at, V_1 = 80, a = 4,$ and $t = 6$

47. $S = \dfrac{1}{2}gt^2, g = 32$ and $t = 3$

48. $S = V_0 t + \dfrac{1}{2}at^2, V_0 = 22, t = 4,$ and $a = 32$

49. $A = \dfrac{n_1 P_1 + n_2 P_2}{n_1 + n_2}, n_1 = 80, P_1 = 3, n_2 = 110,$ and $P_2 = 5$

50. $R_x = \dfrac{R_1 R_3}{R_2}, R_1 = 8, R_2 = 5,$ and $R_3 = 15$

51. $V_2 = \dfrac{V_1 P_1}{P_2}, V_1 = 14, P_1 = 30,$ and $P_2 = 35$

52. $V_1 = \dfrac{V_2 P_2}{P_1}, V_2 = 12, P_1 = 22,$ and $P_2 = 33$

53. $V_2 = \dfrac{V_1 T_2}{T_1}, V_1 = 14, T_1 = 18,$ and $T_2 = 27$

Evaluate the following formulas. See example 1–5 E.

54. The volume V of a rectangular solid is found by multiplying length ℓ times width w times height h, $V = \ell w h$. Find the volume in cubic feet if $\ell = 12$ feet, $w = 4$ feet, and $h = 5$ feet.

55. The perimeter P of a rectangle is found by the formula $P = 2\ell + 2w$, where ℓ is the length of the rectangle and w is the width. Find the perimeter of the rectangle in meters if $\ell = 12$ meters and $w = 7$ meters.

56. It is necessary to drag a box 600 feet across a level lot in 3 minutes. The force required to pull the box is 2,000 pounds. What is the horsepower (h) needed to do this if h is defined by $h = \dfrac{L \cdot W}{33{,}000 \cdot t}$, where $L =$ distance to be moved, $W =$ force exerted, and $t =$ time in minutes required to move the box through L?

57. A pulley 12 inches in diameter that is running at 320 revolutions per minute (rpm) is connected by a belt to a pulley 9 inches in diameter. How many revolutions per minute will the smaller pulley make if $s = \dfrac{SD}{d}$, where $s =$ speed of smaller pulley, $d =$ diameter of smaller pulley, $S =$ speed of larger pulley, and $D =$ diameter of larger pulley?

58. In a gear system, the velocity V of the driving gear is defined by $V = \dfrac{vn}{N}$, where v = velocity of the follower gear, n = number of teeth of the follower gear, and N = number of teeth of the driving gear. Find V when v = 90 rpm, n = 30 teeth, N = 65 teeth.

59. A formula in electricity is $I = \dfrac{E}{R}$, where I represents the current measured in amperes in a certain part of a circuit, E is the potential difference in voltage across that part of the circuit, and R is the resistance in ohms of that part of the circuit. Find I in amperes if E = 110 volts and R = 44 ohms.

Write an algebraic expression for each of the following. See example 1–5 G.

Examples The *product* of a and b	A *number increased by* 2
Solutions $a \cdot b$	Let n represent the number; hence $n + 2$

60. The sum of x and 8

61. 3 times x

62. 5 less than n

63. 8 more than x

64. x increased by 5 and that sum divided by 6

65. 4 times the sum of x and 7

66. n decreased by 8

67. x decreased by 4

68. a divided by 3 and that quotient increased by 4

69. A number decreased by 15

70. A number added to 6

71. 8 times a number and that product increased by 14

72. A number divided by 5 and that quotient decreased by 2

73. 4 times the sum of a number and 3

74. A number decreased by 6 and that difference divided by 9

75. $\dfrac{1}{4}$ of a number

76. $\dfrac{4}{5}$ of a number

77. One-half of a number

78. One-third of a number

1–6 ■ *Sums and differences of polynomials*

In section 1–5, we learned by the process of evaluation that a polynomial is a symbol representing a real number. Therefore the ideas and properties that apply to operations with real numbers also apply to polynomials. Let us now examine the operations of addition and subtraction of polynomials.

An application of the distributive property

Using the symmetric property of equality and the commutative property, we can write the distributive property of multiplication over addition as

$$ax + bx = (a + b)x$$

Consider the expression

$$4x + 5x$$

Using the distributive property, the expression can be written

$$4x + 5x = (4 + 5)x = 9x$$

In this expression, $4x$ and $5x$ are terms that we wish to add. Terms are separated by the operations of addition and subtraction.

Simplifying expressions, as in this example, is *combining like terms*. **Like terms** are terms that may differ only in their numerical coefficients. For two or more terms to be called like terms, the variable factors of the terms along with their respective exponents must be identical. However the numerical coefficients may be different.

1. $5x^2y^3$, $-3x^2y^3$, and x^2y^3 are like terms because they differ only in their numerical coefficients.

2. $5a^3b^2$ and $5a^2b^3$ both contain the same variables but are not like terms because the exponents of the respective variables are not the same.

> ### Combining like terms
> 1. Identify the like terms.
> 2. If necessary, use the commutative and associative properties to group together the like terms.
> 3. Combine the numerical coefficients of the like terms and multiply that by the variable factor.
> 4. Remember that y is the same as $1 \cdot y$ and $-y$ is the same as $-1 \cdot y$.

Note The process of addition or subtraction is performed only with the numerical coefficients, the variable factors remain unchanged.

■ Example 1–6 A

Perform the indicated addition or subtraction.

1. $3z - 7z + z - 6 + 15$ ⟶ Identify like terms
$= (3 - 7 + 1)z + (-6 + 15)$ ⟶ Associative and distributive properties
$= -3z + 9$ ⟶ Combine numerical coefficients, combine numbers

2. $3a^2b + 5a^2b + 2a^2b = (3 + 5 + 2)a^2b = 10a^2b$

3. $x^3y + 4xy^2 - 3x^3y + 2xy^2$ ⟶ Identify any like terms

(Like terms: x^3y and $-3x^3y$; $4xy^2$ and $2xy^2$)

$= [1 + (-3)]x^3y + (4 + 2)xy^2$ ⟶ Commutative, associative, and distributive properties
$= -2x^3y + 6xy^2$ ⟶ Combine numerical coefficients

Note The numerical coefficient of a term includes the sign that precedes it. Therefore we consider any addition and subtraction of terms as a sum of terms in which the sign that precedes the term is taken as the sign of the numerical coefficient.

▶ *Quick check* Perform the indicated addition or subtraction.
$5xy^3 - 3xy^3 + 7xy^3$ ■

Removing grouping symbols

We learned in chapter 1 that any quantity enclosed within a grouping symbol is treated as a single number. Now we are going to use the distributive property to remove grouping symbols such as ( ), [ ], and { }. Consider the following examples:

1. The quantity $(2x + y)$ can be written $1 \cdot (2x + y)$ because if there is no numerical coefficient, then 1 is understood to be the coefficient. Applying the distributive property,

$$1(2x + y) = 1 \cdot 2x + 1 \cdot y = 2x + y$$

2. The quantity $+ (2x + y)$ can be written $+1 \cdot (2x + y)$, giving

$$+1 \cdot (2x + y) = (+1) \cdot 2x + (+1) \cdot y = 2x + y$$

3. The quantity $-(2x + y)$ can be written $-1 \cdot (2x + y)$, giving

$$-1 \cdot (2x + y) = (-1) \cdot 2x + (-1) \cdot y = -2x - y$$

Removing grouping symbols

1. If a grouping symbol is preceded by no symbol or by a "+" sign, the grouping symbol can be dropped and the enclosed terms remain unchanged.
2. If a grouping symbol is preceded by a "−" sign, when the grouping symbol is dropped, we change the sign of each enclosed term.

■ *Example 1–6 B*

Remove all grouping symbols and perform the indicated addition or subtraction.

1. $(3x^2 + 2xy - 3y^2) + (x^2 - 5xy + 2y^2)$ Remove grouping symbols

$$= 3x^2 + 2xy - 3y^2 + x^2 - 5xy + 2y^2$$ Enclosed terms remain unchanged

$$= (3x^2 + x^2) + (2xy - 5xy) + (-3y^2 + 2y^2)$$ Associative and commutative properties

$$= 4x^2 - 3xy - y^2$$ Combine like terms

Note We can carry out the addition of these same two polynomials by lining them up in columns such that the like terms appear in the same column.

$$
\begin{array}{r}
(3x^2 + 2xy - 3y^2) \\
+ (\ x^2 - 5xy + 2y^2)
\end{array}
\qquad
\begin{array}{r}
3x^2 + 2xy - 3y^2 \\
x^2 - 5xy + 2y^2 \\
\hline
4x^2 - 3xy - y^2
\end{array}
$$

2. $(2x^2 - 7x + 6) - (x^2 - 5x + 9)$ Remove grouping symbols

$$= 2x^2 - 7x + 6 - x^2 + 5x - 9$$ Change the sign of each term in the second parentheses

$$= (2x^2 - x^2) + (-7x + 5x) + (6 - 9)$$ Commutative and associative properties

$$= x^2 - 2x - 3$$ Combine like terms

Note This example may also be worked in the column form, but remember that when we remove the second pair of parentheses, we must change the sign of each enclosed term and then combine the like terms.

$$
\begin{array}{ll}
(2x^2 - 7x + 6) & 2x^2 - 7x + 6 \\
-(\ x^2 - 5x + 9) & \underline{-x^2 + 5x - 9} \\
& x^2 - 2x - 3
\end{array}
$$

3. $(5a^2 + 3a^2b - 3b^2) - (2a^2 - 3ab^2 - 4b^2)$ Remove grouping symbols

$= 5a^2 + 3a^2b - 3b^2 - 2a^2 + 3ab^2 + 4b^2$ Change the sign of each term in the second parentheses

$= (5a^2 - 2a^2) + 3a^2b + 3ab^2 + (-3b^2 + 4b^2)$ Group like terms

$= 3a^2 + 3a^2b + 3ab^2 + b^2$ Combine like terms

By the column method, we have

$$
\begin{array}{ll}
(5a^2 + 3a^2b - 3b^2) & 5a^2 + 3a^2b - 3b^2 \\
-(2a^2 - 3ab^2 - 4b^2) \quad \rightarrow & -\ 2a^2 + 3ab^2 + 4b^2
\end{array}
$$

$$
\begin{array}{l}
5a^2 + 3a^2b \qquad\quad - 3b^2 \\
\underline{-\ 2a^2 \qquad\quad + 3ab^2 + 4b^2} \\
3a^2 + 3a^2b + 3ab^2 + b^2
\end{array}
$$

Rearrange so that like terms appear in the same column

Note In future examples, we will mentally regroup the like terms and use the horizontal method to add or subtract polynomials.

▶ **Quick check** Remove all grouping symbols and perform the indicated addition or subtraction. $(5x^2 + 2x - 4) - (3x^2 - 7x + 9)$ ■

There are many situations in which there will be grouping symbols within grouping symbols. In these cases, *it is usually easier to remove the innermost grouping symbol first.*

■ *Example 1–6 C*   Remove all grouping symbols and perform the indicated addition or subtraction.

1. $2a^2 - [a + (a^2 - 3a)] = 2a^2 - [a + a^2 - 3a]$ Remove parentheses first

$= 2a^2 - [a^2 - 2a]$ Combine like terms within brackets

$= 2a^2 - a^2 + 2a$ Remove brackets

$= a^2 + 2a$ Combine like terms

After removing the parentheses, we combine the like terms before removing the brackets. *Simplify within a group whenever possible.*

2. $3x - \{6x - [y - (3x + 2y)] + 2x\}$

$= 3x - \{6x - [y - 3x - 2y] + 2x\}$ Remove innermost grouping symbol

$= 3x - \{6x - [-y - 3x] + 2x\}$ Combine like terms within brackets

$= 3x - \{6x + y + 3x + 2x\}$ Remove brackets

$= 3x - \{11x + y\}$ Combine like terms within braces

$= 3x - 11x - y$ Remove braces

$= -8x - y$ Combine like terms

▶ **Quick check** Remove all grouping symbols and perform the indicated addition or subtraction. $(5x^2 - 3x + 2) - [2x^2 - (x^2 - 7)]$ ■

Mastery points

Can you

- Identify like terms?
- Perform addition and subtraction of polynomials?
- Remove grouping symbols?

Exercise 1–6

Perform the indicated addition or subtraction. See example 1–6 A.

Example $5xy^3 - 3xy^3 + 7xy^3$ Determine if we have like terms

Solution $= (5 - 3 + 7)xy^3$ Distributive property

$= 9xy^3$ Combine numerical coefficients

1. $4x + 6x^2 - 2x^2 + 3x - 5x^2 + x$

2. $5y^2 - 12y + 6y^5 - 4y^2 + y$

3. $-a^2 + a - 5a^3 + 2a^2 + 4a$

4. $2x^2y - 4xy^2 + 3x^2y$

5. $5x^2y - 3xy + 6xy - x^2y$

6. $6a^2b - 2ab^2 + 3a^2b - 4a^2b^2$

Remove all grouping symbols and perform the indicated addition or subtraction. See example 1–6 B.

Example $(5x^2 + 2x - 4) - (3x^2 - 7x + 9)$

Solution $= 5x^2 + 2x - 4 - 3x^2 + 7x - 9$ Change the sign of each term in the second parentheses

$= (5x^2 - 3x^2) + (2x + 7x) + (-4 - 9)$ Commutative and associative properties

$= 2x^2 + 9x - 13$ Combine like terms

7. $(3x^2y - 2xy^2 + 9xy) + (2xy^2 - 4x^2y + 3xy)$

8. $(5ab^2 - 2a^2b^2 + 3a) - (4a^2b^2 + 3ab^2)$

9. $(5a^2 - b^2) - (4a^2 + 3b^2) - (a^2 - 7b^2)$

10. $(7x^3 - 2x^2y + 4xy^2 - 6y^3) - (4x^3 - 3x^2y - 2xy^2 - y^3)$

11. $(12x - 24yz) + (46yz - 16x - 26z)$

12. $(19x + 3y) - (-22x - 6y)$

13. $-(5a^2b - 6ab + 16c) + (7ab - 4a^2b)$

14. $(6ab + 11b^2c) - (11ab - 6bc)$

15. $-(14x - 31y) - (14x - 6y + 3z)$

16. $(x^2 - 3x + z) - (x^2 + 5x - 8) + (3x^2 - 4x)$

17. $(2x^2 - 7x + 3) + (5x^2 - 6) - (4 - 3x^2)$

18. $(7x^2 - 2xy + y^2) + (3x^2 - 6y^2 + 3xy) - (4y^2 - 6x^2 - xy)$

19. $(5xy - y^2) - (3yz + 2xy) + (3y^2 - 4xy)$

20. $(5ab + 2b^2) - (4a^2 - 3b^2) + (2ab - 7b^2)$

Perform the following addition and subtraction in column form. See example 1–6 B.

21. $(3x^2 + 4xy - 5y^2)$
$+ \ (\ x^2 - 7xy + 3y^2)$

22. $(4a^2 + 2ab + 3b^2)$
$+ \ (\ a^2 - 5ab + \ b^2)$

23. $(3x^2y - 2x^2y^2 + 3xy^2)$
$- \ (2x^2y + \ xy - 5xy^2)$

24. $(4st^2 + 3st - 2s^2t)$
$- \ (2st^2 + 3st - 2s^2t)$

25. $(-6a^2b^2 + 3ab - 4)$
$- \ (-4a^2b^2 + 2ab - 7)$

26. $(- \ 9x^2y^2 - 4xy + 13)$
$- \ (-11x^2y^2 + 5xy - 6)$

Set up the following problems and perform the indicated addition and subtraction.

Example   Subtract $5a^2 - 2a + 4$ from $6a^2 - 7a + 3$.

Solution $(6a^2 - 7a + 3) - (5a^2 - 2a + 4)$ Write the algebraic expression
$= 6a^2 - 7a + 3 - 5a^2 + 2a - 4$ Change the sign of each term in the second parentheses
$= a^2 - 5a - 1$ Combine like terms

27. Subtract $3x^2 - 2x + 1$ from $5x^2 + 3x - 7$.

28. Subtract $2a^2 - 7a + 3$ from $a^2 - 2a + 4$.

29. Subtract $5a^2 - 6a + 7$ from $5a^2 + 2a - 3$.

30. Subtract $-3t^2 + 4t + 5$ from $8t^2 - 11t + 12$.

31. Subtract $-2x^2 + 3x - 7$ from $7x^2 + 9x - 4$.

32. From $6t^2 - 7t + 14$, subtract $8t^2 - 11t + 6$.

33. From $-7y^2 + 3y + 11$, subtract $2y^2 - 13y + 3$.

34. From $6z^2 + 5z - 4$, subtract $-4z^2 + 3z - 9$.

35. From $4a^2 + 7a - 12$, subtract $-8a^2 + 7a - 5$.

36. Subtract $2x^2 - 9x + 4$ from the sum of $6x^2 + 3x$ and $5x^2 - 4x + 2$.

37. Subtract $a^2 - 7a + 11$ from the sum of $3a^2 - 4$ and $-5a^2 + 6a$.

38. Subtract $-4t^2 - 3t + 11$ from the sum of $8t^2 - 6$ and $-5t^2 + 11t + 5$.

39. Subtract $6xy - y^2$ from the sum of $5x^2 + 3xy$ and $2x^2 - 7xy - y^2$.

40. From the sum of $8a^2 + 11$ and $2a^2 - 7a + 6$, subtract $4a^2 - a - 1$.

41. From the sum of $5x^2 - 11x - 7$ and $-2x^2 + 3x - 4$, subtract $3x^2 - 7x + 5$.

42. From the sum of $-3t^2 + 2t - 6$ and $-5t^2 - 4t - 8$, subtract $-8t^2 + 2t + 4$.

43. From the sum of $-t^2 - 7t + 1$ and $-3t^2 + 4t - 8$, subtract $-4t^2 - 6t - 5$.

Remove all grouping symbols and perform the indicated addition or subtraction. See example 1–6 C.

Example $(5x^2 - 3x + 2) - [2x^2 - (x^2 - 7)]$

Solution $= 5x^2 - 3x + 2 - [2x^2 - x^2 + 7]$ Remove both sets of parentheses
$= 5x^2 - 3x + 2 - [x^2 + 7]$ Combine like terms within brackets
$= 5x^2 - 3x + 2 - x^2 - 7$ Remove brackets
$= 4x^2 - 3x - 5$ Combine like terms

44. $5a - [3a - (2a - 3)]$

45. $a - 4 + [3a - (2a + 1)]$

46. $4x + [3x - (x + y)]$

47. $7x - [4x + 3y + (x - 2y)]$

48. $3x - [x - y - (7x - 3y)]$

49. $-(3x - 2y) - (x + y) - [2x - y + (3x - 4y)]$

50. $a - [b + (2a - 4b)] + [5a - 3b]$

51. $(2x - 7y) - \{3x - [4y - (2x + 5y)]\}$

52. $6x^2 - [5x + 3x^2 - (4x^2 - 7x)]$

53. $a - \{a - [a - (2a - b) + 3a]\}$

54. $-\{5x - 3y + [2x - (5x - 7y)] + 4y\}$

55. $(3x + x^2) - \{4x^2 - 3x + [2x^2 - x - (5x^2 + 4x)]\}$

56. $2x + [x - (x^2 - y)] - [2x - (x^2 + y)]$

57. $5a - [6a - (2b - 3c)] - (4b - 3a)$

58. $5a - \{(ab + 4c) + 8b - [7a - (3b - 5c)] + 2a\}$

59. $5x^2 - [3x - (2x^2 + x)] - \{x - [3x^2 - (2x^2 + 3x) - x]\}$

60. $-[-3a^2 + (4a - 3) + 5a] - \{7a^2 + [4a - (a^2 - 3) + 5a^2]\}$

61. $-\{4x^2 - (5x + 3) - 2x\} + \{3x^2 - [7x^2 + (2x - 4) - 5x]\}$

62. $4x^2y - \{3xy^2 - [7x^2y^2 + 3xy^2 - (11x^2y + 2x^2y^2)] - 5x^2y\}$

63. $(9x^2y^2 - 4xy^2) - \{8x^2y^2 - [2xy^2 - (-3x^2y + 7x^2y^2)]\}$

If $P(x) = 2x^2 - x + 3$, $Q(x) = 5x - 4$, and $R(x) = 3x^2 + 4x - 7$, express each of the following in terms of x and perform the indicated operations. See example 1–5 F.

Example $[P(x) - Q(x)] + R(x)$

Solution
$$\begin{aligned}
&= [(2x^2 - x + 3) - (5x - 4)] + (3x^2 + 4x - 7) &&\text{Substitute} \\
&= [2x^2 - x + 3 - 5x + 4] + 3x^2 + 4x - 7 &&\text{Remove parentheses} \\
&= [2x^2 - 6x + 7] + 3x^2 + 4x - 7 &&\text{Combine like terms} \\
&= 2x^2 - 6x + 7 + 3x^2 + 4x - 7 &&\text{Remove brackets} \\
&= 5x^2 - 2x &&\text{Combine like terms}
\end{aligned}$$

64. $P(x) + Q(x) + R(x)$ **65.** $P(x) - Q(x) + R(x)$ **66.** $P(x) - [Q(x) + R(x)]$

67. $Q(x) - [P(x) + R(x)]$ **68.** $R(x) - [Q(x) - P(x)]$ **69.** $[Q(x) - R(x)] - P(x)$

70. $-R(x) + [Q(x) - P(x)]$ **71.** $-P(x) + [Q(x) - R(x)]$

Chapter 1 lead-in problem

Theresa is conducting a chemistry experiment involving yellow phosphorus. The lab manual states that yellow phosphorus ignites at 34 degrees Celsius. What will the temperature be in degrees Fahrenheit? The formula for changing the temperature measured in degrees Celsius to degrees Fahrenheit is

$$F = 1.8C + 32$$

Solution

$$\begin{aligned}
F &= 1.8C + 32 &&\text{Original expression} \\
F &= 1.8(\ \ ) + 32 &&\text{Formula ready for substitution} \\
F &= 1.8(34) + 32 &&\text{Substitute} \\
F &= 61.2 + 32 &&\text{Multiply} \\
F &= 93.2 &&\text{Add}
\end{aligned}$$

The temperature is 93.2 degrees Fahrenheit.

Chapter 1 summary

1. A **set** is any collection of objects or things.
2. The set A is a **subset** of the set B, written $A \subseteq B$, if every element of A is also an element of B.
3. The set A is said to be **equal** to the set B, written $A = B$, if and only if $A \subseteq B$ and $B \subseteq A$.
4. A set that contains no elements is called the **empty set** or the **null set** and is denoted by the symbol $\emptyset$.
5. The **union** of the sets A and B, written $A \cup B$, is the set of all elements that are in A or in B or in both A and B.
6. The **intersection** of the sets A and B, written $A \cap B$, is the set of only those elements that are in both A and B.
7. The sets A and B are **disjoint** if and only if $A \cap B = \emptyset$.
8. A **variable** is a symbol (generally a lowercase letter) that acts as a placeholder for an unspecified number.

9. The **absolute value** of a number is the undirected distance that the number is from the origin. In symbols
$$|x| = \begin{cases} x & \text{if } x \geq 0 \\ -x & \text{if } x < 0 \end{cases}$$
10. The numbers or variables in an indicated multiplication are referred to as the **factors** of the product.
11. The **exponent** tells how many times the base is used as a factor in an indicated product.
12. **Division by zero is not allowed.**
13. The quotient of *zero* divided by any number other than zero is always zero.
14. **Order of operations**
    1. **Groups** Perform any operations within a grouping symbol such as ( ) parentheses, [ ] brackets, { } braces, | | absolute value, or in the numerator or the denominator of a fraction.

2. **Exponents** Perform operations indicated by exponents.

3. **Multiplication and division** Perform multiplication and division in order from left to right.

4. **Addition and subtraction** Perform addition and subtraction in order from left to right.

15. An **algebraic expression** is any meaningful collection of variables, constants, grouping symbols, and symbols of operations.

16. A **term** is any constant, variable, or indicated product, quotient, or root of constants and variables.

17. **Like terms** are terms that may differ only in their numerical coefficients.

18. Each factor or group of factors is called the **coefficient** of the remaining factors. In the expression $5x$, 5 is called the **numerical coefficient.**

19. A **polynomial** is an algebraic expression that consists of one or more terms, has real number coefficients, all variables are raised only to natural number exponents, and the operations involving variables are limited to addition, subtraction, and multiplication.

20. The **degree** of a polynomial in one variable is the greatest exponent of that variable in any one term.

Chapter 1 error analysis

1. Sets
   Example: Write the letters of the word "Mississippi" as a set $\{m,i,s,s,i,s,s,i,p,p,i\}$
   Correct answer: $\{M,i,s,p\}$
   What error was made? (*see page 1*)

2. Intersection of sets
   Example: $\{1,2,3,4\} \cap \{3,5,6,7\} = \emptyset$
   Correct answer: $\{1,2,3,4\} \cap \{3,5,6,7\} = \{3\}$
   What error was made? (*see page 3*)

3. Sets of numbers and set-builder notation
   Example: $\{x|x$ is a whole number less than $5\} = \{1,2,3,4\}$
   Correct answer: $\{x|x$ is a whole number less than $5\}$ $= \{0,1,2,3,4\}$
   What error was made? (*see page 5*)

4. Use of "$\approx$" (is approximately equal to)
   Example: $3\frac{3}{4} \approx 3.75$
   Correct answer: $3\frac{3}{4} = 3.75$
   What error was made? (*see page 6*)

5. Absolute value of a real number
   Example: $-|-5| = 5$
   Correct answer: $-|-5| = -5$
   What error was made? (*see page 10*)

6. Exponential form
   Example: $-3^2 = 9$
   Correct answer: $-3^2 = -(3 \cdot 3) = -9$
   What error was made? (*see page 16*)

7. Division involving zero
   Example: $\dfrac{-5}{0} = 0$
   Correct answer: $\dfrac{-5}{0}$ is undefined while $\dfrac{0}{-5} = 0$.
   What error was made? (*see page 17*)

8. Order of operations
   Example: $10 \div 2 - 6 \cdot 2 + 3 = 1$
   Correct answer: -4
   What error was made? (*see page 28*)

9. Coefficients of variables
   Example: The coefficients of the expression $4x^3 - 5x^2 - 3x$ are 4, 5, and 3.
   Correct answer: The coefficients are 4, -5, and -3.
   What error was made? (*see page 32*)

10. Names of polynomials
    Example: The polynomial $x - (2 + \sqrt{3})$ is a trinomial.
    Correct answer: $x - (2 + \sqrt{3})$ is a binomial.
    What error was made? (*see page 33*)

Chapter 1 critical thinking

If you add any three consecutive even integers, the sum will be a multiple of 3. Why is this true?

Chapter 1 review

[1–1]
Write each set by listing the elements.

1. The letters in the word *college*

2. The first 3 months of the year

3. The months that begin with the letter *L*

Use mathematical symbols to write the following.

4. B is a subset of D.

5. The null set is a subset of A.

6. 5 is an element of C.

7. B is not a subset of C.

Use the following sets to determine if exercises 8–10 are true or false and to form the indicated sets in exercises 11–13.
$A = \{1,2,3,4\}$, $B = \{2,3,4,5,6\}$, $C = \{5,6,7,8\}$, $D = \{3,4\}$.

8. $4 \in A$

9. $D \subseteq B$

10. $A \subseteq B$

11. $A \cup B$

12. $A \cap B$

13. $A \cap C$

Replace the comma with the proper strict inequality symbol, $<$ or $>$, between the following numbers to get a true statement.

14. $0, 4$

15. $-3, -5$

16. $-3, 0$

[1–2]

Find each sum or difference.

17. $(-5) + (-8)$

18. $0 - (-8)$

19. $8 - 11$

20. $0 + (-2) - 4 + (-3)$

21. $(-4) - (-10)$

22. $6 - 8 + 7 - 11 - 4$

Perform the indicated operations if possible.

23. $\dfrac{(-35)}{(-7)}$

24. $\dfrac{(-18)(-2)}{(-6)}$

25. $\dfrac{(-24)}{(+8)}$

26. $\dfrac{(-8)(0)}{(-5)(-4)}$

27. -2^4

28. $\dfrac{(-14)(+12)}{(-4)(+7)}$

29. $\dfrac{(+30)}{(-6)}$

[1–3]

Identify which property is being used in the problem.

30. $(-6) + 6 = 0$

31. $5(a + b) = 5a + 5b$

32. $\left(\dfrac{3}{4}\right)\left(\dfrac{4}{3}\right) = 1$

33. $(3a)b = 3(ab)$

[1–4]

Perform the indicated operations and simplify.

34. $12 - 3 \cdot 4 - 6 + 8$

35. $10 - 16 \div 4 + 7 - 3^2$

36. $26 \div 13 + 8 - 2 \cdot 7$

37. $8 - (4 - 7)^2 + 5 \cdot 6$

38. $15 - (18 - 11)^2 \cdot 2 + 12$

39. $10 - 2[18 - 2(4 - 6) + 11] - 16$

40. $\dfrac{5(4 - 9)}{3} - \dfrac{(-24)}{8}$

41. $\left[\dfrac{(-12) + (-9)}{(-7)}\right]\left[\dfrac{(-8)(-6)}{12}\right]$

42. $\dfrac{5}{8} + \dfrac{1}{2} \div \dfrac{3}{4}$

43. $(2.6) \cdot (5.4) \div (1.8) + 5.7$

44. $(18.47 + 26.89) \div 5.6 + (1.3)^2$

[1–5]

Specify the number of terms in each expression and determine if the expression is a polynomial.

45. $5x^3 - 2x^2 + 7x - 4$

46. $\dfrac{5x^2 - 2x}{4}$

Evaluate the following expressions if $a = -2$, $b = 5$, $c = 3$, and $d = -4$.

47. $2a^2b - cd^2$ **48.** $(b - 2c)^2$ **49.** $(a - 2b)(3c + d)$ **50.** $(ac)^2 - (bd)^2$

51. If $P(x) = x^2 - x - 1$,
find (a) $P(-2)$, (b) $P(0)$, (c) $P(3)$, (d) $P(2) \cdot P(-3)$.

Evaluate the following formulas using the given values.

52. $A = \dfrac{1}{2}bh$, $b = 20$ and $h = 7$ **53.** $A = \dfrac{1}{2}h(b_1 + b_2)$, $h = 10$, $b_1 = 6$, and $b_2 = 9$

54. $R_t = \dfrac{R_1 R_2}{R_1 + R_2}$, $R_1 = 4$ and $R_2 = 6$

Write an algebraic expression for each of the following.

55. The sum of y and 4   **56.** 12 less than a

57. 7 times the sum of a and 6   **58.** x times 5 and that product decreased by 3

59. 5 times the sum of a number and 12   **60.** A number increased by 4 and that sum divided by 8

[1–6]

Perform the indicated addition or subtraction.

61. $7y^2 - 6y + 4y^2 - 3y + 2y^2$ **62.** $7x^2y + 2xy^2 - 3xy^2 + 7x^2y + 5x^2y^2$

63. $3x + 5x + 8 + 9$ **64.** $7a + a - 4a - 7 + 3$

65. $3x - 5x + 7x - 6 + 4$ **66.** $12c + 8c - 9c + 8 - 14$

Remove all grouping symbols and perform the indicated addition or subtraction.

67. $(3x^2 - 2x + 7) - (x^2 + 3x - 4)$ **68.** $7a - [2a - (3a + 4)]$

69. $2x^2 - [4x + 5x^2 - (3x^2 - 6x)]$ **70.** $(2a^2 - b) - \{5a^2 - (a^2 - 2b)\}$

Chapter 1 test

Use the following sets to determine if exercises 1–3 are true or false and to form the indicated sets in exercises 4–6. $A = \{2,4,6\}$, $B = \{4,6,8\}$, and $C = \{1,2,3,4\}$.

1. $A \subseteq C$ **2.** $4 \in B$ **3.** $B \not\subseteq C$ **4.** $A \cup B$ **5.** $A \cap B$ **6.** $A \cap C$

Perform the indicated operations if possible.

7. $-6 - 8$ **8.** $\dfrac{-18}{3}$ **9.** $(-3)(-3)$

10. $(2x - 1) - (3x + 4)$ **11.** -3^2 **12.** $(5)(0)(-2)(3)$

13. $3a + 5a - a$ **14.** $25 - (18 - 3 \cdot 4) - 3^2$ **15.** $15 \div 3 - 2 \cdot 4$

16. $(-2)^4$ **17.** $8 - (-4) - 7 + 3$ **18.** $\dfrac{3}{4} + \dfrac{1}{2} \cdot \dfrac{2}{3} - \dfrac{2}{3}$

19. $\dfrac{-4 + 4}{4}$ **20.** $5a + b - a + 4b$ **21.** If $P(x) = 2x^2 - 3x + 1$, find $P(-2)$.

Evaluate the following expressions if $a = -3$, $b = 2$, and $c = 4$.

22. $a^2 - b^2$ **23.** $ab - ac$

Write an algebraic expression for each of the following.

24. A number diminished by 3   **25.** A number increased by 5 and that sum divided by 8

2

First-Degree Equations and Inequalities

Earl invests a total of $10,000. He invests part in Collins Feline Fanciers that pays a 9% dividend per year and the rest in Grutz Shipyards that pays an 8% dividend per year. If Earl receives $870 per year from his investments, how much did he invest with each company?

2–1 ■ Solving equations

Terminology

An **equation** is a statement of equality. If two expressions represent the same number, we place an equality sign, $=$, between them to form an equation. We use the following example to show the parts of an equation.

$2x + 9$	$=$	$7 - 4x$
Left member of the equation	Equality sign	Right member of the equation

A **mathematical statement** is a mathematical sentence that can be labeled true or false. $2 + 5 = 7$ is a true statement, and $3 + 4 = 8$ is a false statement. An equation that is true for some values of the variable and false for other values of the variable is called a **conditional equation.** The equation $x + 1 = 8$ is a conditional equation since it is true only when $x = 7$ and false otherwise.

A replacement value for the variable that forms a true statement is called a **root,** or a **solution,** of the equation. We say that a solution of the given equation *satisfies* that equation. The set of all those values for the variable that causes the equation to be a true statement is called the **solution set** of the equation. In the equation $x + 1 = 8$, the solution set is $\{7\}$.

If the equation is true for every permissible value of the variable, it is called an **identical equation,** or **identity.** For example,

$$4(a - 2) = 4a - 8$$

is true for any real number replacement for a and is thus called an identity.

Properties such as

$$a \cdot (b \cdot c) = (a \cdot b) \cdot c \quad \text{and} \quad a + b = b + a$$

are further examples of identities. We will be concerned only with conditional equations in this chapter.

In this chapter, we are concerned with **first-degree conditional equations,** also called **linear equations.** In a first-degree conditional equation in one variable, the exponent of the unknown is 1 and the solution set will contain at most one root.

If we wish to solve an equation such as

$$5(2x - 1) = 7x + 10$$

we go through a series of steps whereby we form equations that are **equivalent,** *having identical solution sets,* to the original equation. *Our goal is to form equivalent equations until we isolate the unknown in one member of the equation and our equation is in the form* $x = n$, where *n* is some real number. The following are equivalent equations whose solution set is $\{5\}$.*

$$\begin{aligned}
5(2x - 1) &= 7x + 10 \\
10x - 5 &= 7x + 10 \\
3x - 5 &= 10 \\
3x &= 15 \\
x &= 5
\end{aligned}$$

Since an equation is a statement of equality between the two members of the equation, identical quantities added to or subtracted from each member will produce an equivalent equation. This is called the **addition property of equality.**

Addition property of equality ────────────────────────

For any algebraic expressions A, B, and C, if A = B, then

$$A + C = B + C$$

Concept
The same expression can be added to each member of an equation and the result will be an equivalent equation.

In chapter 1, we defined subtraction in terms of addition, therefore *we can use the addition property of equality to* **subtract** *the same expression from both members of an equation.*

■ *Example 2–1 A*

Find the solution set.

1.
$$\begin{aligned}
4x - 2 &= 3x + 7 \\
4x - 3x - 2 &= 3x - 3x + 7 \qquad \text{Subtract 3x from both members} \\
x - 2 &= 7 \\
x - 2 + 2 &= 7 + 2 \qquad \text{Add 2 to both members} \\
x &= 9 \qquad \text{Solution}
\end{aligned}$$

The solution set is $\{9\}$.

*The formation of these equivalent equations is achieved by using the addition and multiplication properties of equality.

2.
$$2(3x - 1) = 5x + 2x - 3$$

$6x - 2 = 7x - 3$	Simplify
$6x - 6x - 2 = 7x - 6x - 3$	Subtract $6x$ from both members
$-2 = x - 3$	
$-2 + 3 = x - 3 + 3$	Add 3 to both members
$1 = x$	Solution

The solution set is $\{1\}$.   ■

Multiplication property of equality

The addition property of equality along with the properties of real numbers is sufficient to solve many of the equations that we encounter. However, they are not sufficient to solve equations such as

$$4x = 12 \quad \text{or} \quad \frac{3}{4}x = 15$$

Recall that we want our equation to be of the form $x = n$. This means that the coefficient of x must be 1. To achieve this, we need the **multiplication property of equality.**

> ### Multiplication property of equality
>
> For any algebraic expressions A, B, and C, where $C \neq 0$, if $A = B$, then
> $$A \cdot C = B \cdot C$$
>
> **Concept**
> An equivalent equation is obtained when we multiply both members of an equation by the same nonzero expression.

In chapter 1, we defined division in terms of multiplication, therefore *we can use the multiplication property of equality to* **divide** *both members of an equation by the same nonzero expression.*

■ *Example 2–1 B*

Find the solution set.

1. $4x = 12$

$\dfrac{4x}{4} = \dfrac{12}{4}$	Divide both members by 4
$x = 3$	Solution

The solution set is $\{3\}$.

Note We could have multiplied by the reciprocal of 4 to solve the equation. That is,

$$4x = 12$$

$\dfrac{1}{4} \cdot 4x = \dfrac{1}{4} \cdot 12$	Multiply both members by $\dfrac{1}{4}$
$x = 3$	

We should be familiar with the idea that to divide by a number is the same as to multiply by the reciprocal of that number.

2. $\dfrac{3}{4}x = 15$

$\dfrac{4}{3} \cdot \dfrac{3}{4}x = \dfrac{4}{3} \cdot 15$ Multiply both members by the reciprocal of the coefficient of x

$x = 20$ Solution

The solution set is $\{20\}$.

Recall that when we divide by a fraction, we invert and multiply. Therefore if the coefficient of the unknown is a fraction, we will multiply both members of the equation by the reciprocal of the coefficient.

3. $2.6x = 10.4$

$\dfrac{2.6x}{2.6} = \dfrac{10.4}{2.6}$ Divide both members by 2.6

$x = 4$ Solution

The solution set is $\{4\}$.   ■

Using the given theorems and the properties of real numbers, there are four basic steps for solving a linear equation. We shall now apply these to the equation

$$5(2x - 1) = 7x + 10$$

___ *Solving a linear equation* _____

Step 1 *Simplify the equation.* Perform all indicated addition, subtraction, multiplication, and division. Remove all grouping symbols. In our example, step 1 would be to carry out the indicated multiplication in the left member.

$$5(2x - 1) = 7x + 10$$
$$10x - 5 = 7x + 10$$

Step 2 *Use the addition property of equality to form an equivalent equation where all the terms involving the unknown are in one member of the equation.* By subtracting $7x$ from *both* members of the equation, we have

$$10x - 5 = 7x + 10$$
$$10x - 7x - 5 = 7x - 7x + 10$$
$$3x - 5 = 10$$

Step 3 *Use the addition property of equality to form an equivalent equation where all the terms not involving the unknown are in the other member of the equation.* Adding 5 to *both* members of the equation, we have

$$3x - 5 = 10$$
$$3x - 5 + 5 = 10 + 5$$
$$3x = 15$$

Step 4 *Use the multiplication property of equality to form an equivalent equation where the coefficient of the unknown is 1.* That is, $x = n$. Dividing *both* members of the equation by 3, we have

$$3x = 15$$
$$\dfrac{3x}{3} = \dfrac{15}{3}$$
$$x = 5$$

The solution set is $\{5\}$.

To check our solution, we substitute the solution in place of the unknown in the original equation. If we get a true statement, we say the solution "satisfies" the equation.

In our example, $5(2x - 1) = 7x + 10$, we found that $x = 5$. Substituting 5 in place of x in the original equation, we have

$5[2(5) - 1] = 7(5) + 10$	Substitute
$5[10 - 1] = 35 + 10$	Order of operations
$5[9] = 45$	
$45 = 45$ True	Solution checks

■ **Example 2–1 C**

Find the solution set.

1. $-2(y + 3) + 3(2y - 1) = 10$

$-2y - 6 + 6y - 3 = 10$	Carry out the multiplication
$4y - 9 = 10$	Combine like terms
$4y - 9 + 9 = 10 + 9$	Add 9 to both members
$4y = 19$	
$\dfrac{4y}{4} = \dfrac{19}{4}$	Divide both members by 4
$y = \dfrac{19}{4}$	Solution

Check:

$$-2\left[\left(\frac{19}{4}\right) + 3\right] + 3\left[2\left(\frac{19}{4}\right) - 1\right] = 10 \quad \text{Substitute}$$

$$-2\left[\frac{19}{4} + \frac{12}{4}\right] + 3\left[\frac{19}{2} - \frac{2}{2}\right] = 10 \quad \text{Order of operations}$$

$$-2\left[\frac{31}{4}\right] + 3\left[\frac{17}{2}\right] = 10$$

$$-\frac{31}{2} + \frac{51}{2} = 10$$

$$\frac{20}{2} = 10$$

$$10 = 10 \quad \text{True Solution checks}$$

The solution set is $\left\{\dfrac{19}{4}\right\}$.

At this point, we will no longer show the check of our solution, but we should realize that a check of our work is an important final step.

The following equations contain several fractions. When this occurs, it is usually easier to **clear the equation of all fractions.** We do this by multiplying both members by the least common denominator (LCD) of all the fractions. Clearing all the fractions is considered a means of simplifying the equation and will be done as a first step when necessary. Equations containing fractions will be studied more completely in chapter 4.

2.

$$\frac{1}{2}x + 3 = \frac{2}{3}$$

$$6\left(\frac{1}{2}x + 3\right) = 6 \cdot \frac{2}{3}$$ Multiply both members by the least common denominator, 6

$$3x + 18 = 4$$

$$3x + 18 - 18 = 4 - 18$$ Subtract 18 from both members

$$3x = -14$$

$$\frac{3x}{3} = -\frac{14}{3}$$ Divide both members by 3

$$x = -\frac{14}{3}$$ Solution

The solution set is $\left\{-\frac{14}{3}\right\}$.

3.

$$\frac{3}{4}x - \frac{1}{2} = \frac{1}{3}x + 2$$

$$12\left(\frac{3}{4}x - \frac{1}{2}\right) = 12\left(\frac{1}{3}x + 2\right)$$ Multiply both members by the least common denominator, 12

$$9x - 6 = 4x + 24$$

$$9x - 4x - 6 = 4x - 4x + 24$$ Subtract $4x$ from both members

$$5x - 6 = 24$$

$$5x - 6 + 6 = 24 + 6$$ Add 6 to both members

$$5x = 30$$

$$\frac{5x}{5} = \frac{30}{5}$$ Divide both members by 5

$$x = 6$$ Solution

The solution set is $\{6\}$.

4.

$$3.18z + 3.526 = 2(0.73z - 2.709)$$

$$3.18z + 3.526 = 1.46z - 5.418$$ Carry out the multiplication

$$3.18z - 1.46z + 3.526 = 1.46z - 1.46z - 5.418$$ Subtract $1.46z$ from both members

$$1.72z + 3.526 = -5.418$$

$$1.72z + 3.526 - 3.526 = -5.418 - 3.526$$ Subtract 3.526 from both members

$$1.72z = -8.944$$

$$\frac{1.72z}{1.72} = \frac{-8.944}{1.72}$$ Divide both members by 1.72

$$z = -5.2$$ Solution

The solution set is $\{-5.2\}$.

5. $5(x - 4) - 2x = 3x + 7$

$$5x - 20 - 2x = 3x + 7$$ Carry out the multiplication

$$3x - 20 = 3x + 7$$ Combine like terms

$$3x - 3x - 20 = 3x - 3x + 7$$ Subtract $3x$ from both members

$$-20 = 7$$ False

The statement $-20 = 7$ is false and this means that there is *no solution* to the equation. When an equation has no solution, it is called a **contradiction** and its solution set is $\emptyset$.

▶ *Quick check*   Find the solution set. $4(5x - 2) + 7 = 5(3x + 1)$ ■

Problem solving

The following sets of word problems are designed to help us interpret verbal statements and write expressions for them in algebraic symbols. For each problem, we will write an algebraic expression that changes the words into mathematical symbols. These will not be equations. We will use our experience from these problems to help us translate word problems into equations in section 2–3.

■ *Example 2–1 D*

Write an algebraic phrase for each of the following verbal statements.

1. Phil can type 75 words per minute. How many words can he type in m minutes?

   If Phil can type 75 words in one minute, then we multiply

   $$75 \cdot m = 75m$$

   to obtain the number of words he can type in m minutes.

2. If Debbie has d dollars in her savings account and on successive days she deposits $55 and then withdraws $25 to make a purchase, write an expression for the balance in her savings account.

   We *add* the deposits and *subtract* the withdrawals. Thus

   $$d + 55 - 25 = d + 30$$

   represents the balance in dollars in Debbie's savings account after the two transactions.

3. A woman paid d dollars for 20 pounds of ground beef. How much did the beef cost her per pound?

   The price per pound is found by dividing the total cost by the number of pounds. Thus the price in dollars per pound of the beef is represented by

   $$\frac{d}{20}$$

   ■

Mastery points

Can you
- Apply the addition property of equality?
- Apply the multiplication property of equality?
- Solve linear equations?
- Determine when a linear equation has no solution?
- Check your solutions?
- Write an algebraic expression for a verbal statement?

Exercise 2–1

Find the solution set of the following linear equations. See examples 2–1 A, B, and C.

Example $4(5x - 2) + 7 = 5(3x + 1)$

Solution

$20x - 8 + 7 = 15x + 5$	Distributive property
$20x - 1 = 15x + 5$	Combine like terms
$20x - 15x - 1 = 15x - 15x + 5$	Subtract 15x
$5x - 1 = 5$	Combine like terms

$$5x - 1 + 1 = 5 + 1 \qquad \text{Add 1}$$
$$5x = 6 \qquad \text{Combine like terms}$$
$$\frac{5x}{5} = \frac{6}{5} \qquad \text{Divide by 5}$$
$$x = \frac{6}{5} \qquad \text{Solution}$$

The solution set is $\left\{ \dfrac{6}{5} \right\}$.

1. $5x = 15$ **2.** $7x = 28$ **3.** $4y = 10$ **4.** $6x = 27$

5. $a + 5 = 11$ **6.** $x + 4 = -3$ **7.** $z - 6 = -3$ **8.** $x - 9 = 8$

9. $\dfrac{x}{3} = 4$ **10.** $\dfrac{y}{2} = 11$ **11.** $\dfrac{3a}{4} = 8$ **12.** $4.1x = 15.17$

13. $1.8y = 21.6$ **14.** $0.7a = 11.2$ **15.** $\dfrac{2y}{3} = 9$ **16.** $3b - 1 = 8$

17. $5y - 2 = 13$ **18.** $4x + 5 = 5$ **19.** $6x + 4 = -2$ **20.** $9a - 3 = -3$

21. $7x - 4x + 3 = 8$ **22.** $2a + 3a - 7 = 5$

23. $3(y + 1) = 4$ **24.** $2(2z - 3) = 7$

25. $4(3b - 1) + 2b = 11$ **26.** $3(2x + 1) = 7x - 3x + 4$

27. $5(2 - x) + 1 = 3x - 4 + x$ **28.** $7x + 3 - 2x = 4(3 - 2x) + 5$

29. $2(3a - 1) = 3(2a + 5)$ **30.** $5(x - 4) + 12 = 3x + 2x - 6$

31. $\dfrac{1}{3}x + 2 = \dfrac{5}{6}$ **32.** $\dfrac{1}{2}x - 1 = \dfrac{1}{3}x + 3$

33. $\dfrac{3}{4}x + 3 = \dfrac{5}{8}x + 4$ **34.** $\dfrac{3}{8}x + \dfrac{1}{4} = \dfrac{1}{4}x + 3$

35. $\dfrac{5}{12}x + 2 = \dfrac{2}{3}x - 4$ **36.** $\dfrac{7}{12}x - 4 = \dfrac{5}{6}x - 5$

37. $-4(y - 3) + 2(3y - 5) = 11$ **38.** $-2(a + 3) + 5(a - 1) = -4$

39. $3(2y + 4) - 3y = 6(y + 1)$ **40.** $3(2x + 5) - x = 4(x - 3) + 7$

41. $3[2a - (a + 7)] = 10$ **42.** $4[a - (3a - 4)] = a + 5$

43. $-2[3x - (x - 5)] = 3x - 4$ **44.** $5.6z - 22.15 = 24.33$

45. $9.3y - 27.9 + 4.6y = 55.5$ **46.** $7.6a + 18.4 - 3.2a = 66.8$

47. $6.8x + 5.7 = 4.3x - 15.3$ **48.** $2.6(x - 6.3) = 8.9x - 81.9$

49. $6.7 - 4.1(x + 1) = 1.5x - 42.2$ **50.** $3(2x - 1) = 6x + 7$

51. $4a + 3a - 7 = 2(3a + 1) + a$ **52.** $6(z + 3) - 2z = 2(2z + 1)$

Solve the equations for the specified variable. See examples 2–1 A, B, and C.

53. The surface area S of a rectangular solid of length ℓ, width w, and altitude h is given by $S = 2(\ell w + \ell h + hw)$. Find ℓ when $S = 236$, $w = 5$, and $h = 6$.

54. To convert Celsius temperature to Fahrenheit, we use $F = \dfrac{9}{5} C + 32$. Find C when $F = 77$.

55. The amount A of a principal P invested for t years at simple interest with rate r percent per year is given by $A = P(1 + rt)$. Solve for t if $A = 4{,}080$; $P = 3{,}000$; and $r = 9\%$.

Write an algebraic expression for the following verbal statements. See example 2–1 D.

56. Rita can enter 80 words per minute on a word processor. How many words can she enter in m minutes?

57. Express the cost in cents of c cans of oil if each can costs $1.15.

58. A 10-pound bag of dog food costs d dollars. How much does the dog food cost per pound?

59. It costs Pat $18 to rent a posthole digger for h hours. What did it cost him per hour to rent the posthole digger?

60. Megan has n nickels, d dimes, and q quarters in her purse. Express in cents the amount of money she has in her purse. (*Hint:* n nickels are represented in cents by $5n$.)

61. Roger has h half dollars, q quarters, d dimes, and n nickels. Express in cents the amount of money Roger has.

62. Colleen is y years old now. Express her age (a) 21 years from now, (b) 7 years from now.

63. Richard is 3 years old. If Dick is n times as old as Richard, express Dick's age. Express Dick's age 8 years ago.

64. Edna's savings account has a current balance of $457. If she makes a withdrawal of a dollars and then makes a deposit of b dollars, express her new balance.

65. Dale has a balance of d dollars in his checking account. He makes a deposit of $464 and then writes 5 checks for m dollars each. Express his new balance in dollars in terms of d and m.

66. Marty has c cents, all in quarters. Write an expression for the number of quarters Marty has.

67. If w represents a whole number, write an expression for the next greater whole number.

68. If j represents an even integer, write an expression for the next greater even integer.

69. If j represents an odd integer, write an expression for the next greater odd integer.

70. If Joan is f feet and t inches tall, how tall is Joan in inches?

71. Sue earns $500 more than twice the amount Jon earns in a year. If Jon earns d dollars in a year, write an expression for Sue's annual salary.

72. Lisa's annual salary is $1,000 more than n times Bonnie's annual salary. If Bonnie earns $9,000 per year, express Lisa's annual salary.

73. Express the total cost in cents of purchasing c cans of soda pop at 59¢ per can and b loaves of bread at $1.15 per loaf.

74. A gallon of primer coat costs $9.95 and a gallon of latex-base paint costs $12.99. Express the cost of p gallons of primer and q gallons of latex-base paint.

75. Norm can type w words per minute and Rich can type 11 words per minute more than Norm. Write an expression for how many words Rich can type in 20 minutes.

Review exercises

Evaluate the following formulas. See section 1–5.

1. $F = ma$, $m = 24$ and $a = 11$

2. $I = prt$, $p = 5{,}000$; $r = 0.06$; and $t = 2$

3. $V_2 = V_1 + at$, $V_1 = 60$, $a = 32$, and $t = 5$

4. $A = \dfrac{1}{2}h(b_1 + b_2)$, $h = 6.2$, $b_1 = 4.5$, and $b_2 = 8$

5. $S = \dfrac{1}{2}gt^2$, $g = 32$ and $t = 4$

6. $\ell = a + (n - 1)d$, $a = 10$, $n = 15$, and $d = 2$

2–2 ■ *Formulas and literal equations*

In mathematics, a **literal equation** is an equation that contains more than one variable. A **formula** is a mathematical equation that states the relationship between two or more physical conditions.

When we repeatedly use a formula to determine the value of the same variable, it is convenient to solve the equation for that variable in terms of the remaining variables and constants. In the case of the relationship between Celsius and Fahrenheit temperature, it is useful to have one formula for Celsius in terms of Fahrenheit and another formula for Fahrenheit in terms of Celsius. Using our procedure for solving linear equations, we will now solve the Celsius formula for Fahrenheit F.

$$C = \frac{5}{9}(F - 32)$$

$$9C = 9 \cdot \frac{5}{9}(F - 32) \qquad \text{Multiply both members by the least common denominator, 9}$$

$$9C = 5(F - 32)$$

$$9C = 5F - 160 \qquad \text{Distributive property}$$

$$9C + 160 = 5F \qquad \text{Add 160 to both members.}$$

$$\frac{1}{5}(9C + 160) = \frac{1}{5} \cdot 5F \qquad \text{Multiply both members by } \frac{1}{5}$$

$$\frac{9}{5}C + 32 = F \qquad \text{Distributive property}$$

The formula for finding the temperature in degrees Fahrenheit, given the temperature in degrees Celsius, is

$$F = \frac{9}{5}C + 32$$

The two formulas

$$C = \frac{5}{9}(F - 32) \text{ and } F = \frac{9}{5}C + 32$$

may not look the same, but they express the same relationship between C and F. The first formula is solved for C in terms of F, and the second formula is solved for F in terms of C.

The following steps are a restatement of the procedure for solving linear equations. We will now apply these steps to solve formulas and literal equations for the specific variable.

Solving a literal equation or formula

Step 1 Simplify the equation.

Step 2 Obtain all terms with the variable for which we are solving in one member of the equation.

Step 3 Obtain all terms that do not have the variable for which we are solving in the other member of the equation.

Step 4 Determine the coefficient of the variable for which we are solving, and then divide both members of the equation by that coefficient.

■ *Example 2–2 A*

Solve for the specified variable.

1. Solve the literal equation $8x + 4 = 5x + y$, for x.

$$8x + 4 = 5x + y$$
$$8x - 5x + 4 = 5x - 5x + y \qquad \text{Subtract } 5x \text{ from both members}$$
$$3x + 4 = y$$
$$3x + 4 - 4 = y - 4 \qquad \text{Subtract 4 from both members}$$
$$3x = y - 4$$
$$\frac{3x}{3} = \frac{y - 4}{3} \qquad \text{Divide both members by 3}$$
$$x = \frac{y - 4}{3}$$

2. The formula for the perimeter of a rectangle is $P = 2\ell + 2w$. Solve for ℓ.

$$P = 2\ell + 2w$$
$$P - 2w = 2\ell + 2w - 2w \qquad \text{Subtract } 2w \text{ from both members}$$
$$P - 2w = 2\ell$$
$$\frac{P - 2w}{2} = \frac{2\ell}{2} \qquad \text{Divide both members by 2}$$
$$\ell = \frac{P - 2w}{2} \qquad \text{Symmetric property}$$

3. Solve the literal equation $2(3x - y) = 3(x + 3y) + 2$, for x.

$$2(3x - y) = 3(x + 3y) + 2$$
$$6x - 2y = 3x + 9y + 2 \qquad \text{Simplify each member}$$
$$6x - 3x - 2y = 3x - 3x + 9y + 2 \qquad \text{Subtract } 3x \text{ from both members}$$
$$3x - 2y = 9y + 2$$
$$3x - 2y + 2y = 9y + 2y + 2 \qquad \text{Add } 2y \text{ to both members}$$
$$3x = 11y + 2$$
$$\frac{3x}{3} = \frac{11y + 2}{3} \qquad \text{Divide both members by 3}$$
$$x = \frac{11y + 2}{3}$$

4. The area of a trapezoid is given by $A = \dfrac{1}{2}h(b_1 + b_2)$. Solve for b_1.

$$A = \frac{1}{2}h(b_1 + b_2)$$

$$2A = 2 \cdot \frac{1}{2}h(b_1 + b_2) \qquad \text{Multiply both members by 2}$$

$$2A = h(b_1 + b_2)$$

$$2A = b_1h + b_2h \qquad \text{Distributive property}$$

$$2A - b_2h = b_1h \qquad \text{Subtract } b_2h \text{ from both members}$$

$$\frac{2A - b_2h}{h} = b_1 \qquad \text{Divide both members by } h$$

$$b_1 = \frac{2A - b_2h}{h} \qquad \text{Symmetric property}$$

Note Although we have not stated any restrictions on the variables, it is understood that the values that the variables can take on must be such that *no denominator is ever zero*. That is: examples 1, 2, and 3 have no restrictions; example 4, $h \neq 0$.

▶ *Quick check* Solve $P = 2\ell + 2w$, for w. ■

Whether we are solving a linear equation or a literal equation, the procedure is the same.

Linear equation	*Literal equation*	*Solve for* **a**
$5(a + 1) = 2a + 7$	$5(a + b) = 2a + 7b$	Original equation
$5a + 5 = 2a + 7$	$5a + 5b = 2a + 7b$	Simplify (distributive property)
$3a + 5 = 7$	$3a + 5b = 7b$	All a's in one member
$3a = 2$	$3a = 2b$	Terms not containing a in other member
$a = \dfrac{2}{3}$	$a = \dfrac{2b}{3}$	Divide by the coefficient

In the linear equation, we have a solution for a, and in the literal equation, we have solved for a in terms of b.

--- **Mastery points** ---

Can you
- Solve formulas and literal equations for the specified variable in terms of the other variables?

Exercise 2–2

Find the value of the variable whose replacement value is not given.

Example $I = prt$. Solve for p if $I = 320$, $r = 0.08$, and $t = 2$.

Solution

$$I = prt$$
$$320 = p(0.08)(2) \qquad \text{Substitute}$$
$$320 = p(0.16) \qquad \text{Simplify}$$
$$\frac{320}{0.16} = \frac{p(0.16)}{0.16} \qquad \text{Divide both members by (0.16)}$$
$$2{,}000 = p \qquad \text{Simplify}$$
$$p = 2{,}000 \qquad \text{Symmetric property}$$

1. $W = I^2R$; $I = 6$, $W = 324$

2. $L = \dfrac{V}{WH}$; $W = 12$, $H = 6$, $L = 20$

3. $A = P + Pr$; $A = 3{,}240$; $r = (0.08)$

4. $V = k + gt$; $k = 22$, $t = 3$, $V = 55$

5. $\ell = a + (n - 1)d$; $a = 7$, $d = 4$, $\ell = 83$

6. $A = \dfrac{1}{2}h(b_1 + b_2)$; $A = 54$, $b_2 = 10$, $h = 6$

7. $a = \dfrac{V_2 - V_1}{t}$; $a = 4$, $V_2 = 83$, $t = 8$

8. $T = \dfrac{R - R_0}{aR_0}$; $T = 2$, $R_0 = 4$, $a = 3$

Solve the following formulas or literal equations for the specified variable. Assume that no denominator is equal to zero. See example 2–2 A.

Example   The formula for the perimeter of a rectangle is $P = 2\ell + 2w$. Solve for w.

Solution

$$P = 2\ell + 2w$$
$$P - 2\ell = 2\ell + 2w - 2\ell \qquad \text{Subtract } 2\ell \text{ from both members}$$
$$P - 2\ell = 2w \qquad \text{Combine like terms}$$
$$\frac{P - 2\ell}{2} = \frac{2w}{2} \qquad \text{Divide both members by 2}$$
$$w = \frac{P - 2\ell}{2} \qquad \text{Symmetric property}$$

9. $I = prt$; t

10. $E = IR$; R

11. $E = mc^2$; m

12. $V = \ell wh$; h

13. $F = ma$; m

14. $K = PV$; V

15. $A = bh$; b

16. $A = bh$; h

17. $W = I^2R$; R

18. $V = \ell wh$; w

19. $V = k + gt$; k

20. $P = 2\ell + 2w$; w

21. $V = k + gt$; g

22. $A = P + Pr$; r

23. $D = dq + R$; q

24. $m = -p(\ell - x)$; x

25. $m = -p(\ell - x)$; ℓ

26. $R = W - b(2c + b)$; c

27. $R = W - b(2c + b)$; W

28. $2S = 2Vt - gt^2$; V

29. $V = r^2(a - b)$; a

30. $S = \dfrac{n}{2}[2a + (n - 1)d]$; a

31. $S = \dfrac{n}{2}[2a + (n - 1)d]$; d

32. $V = \dfrac{1}{3}\pi h^2(3R - h)$; R

33. $2S = 2Vt - gt^2$; g

34. $P = n(P_2 - P_1) - c$; P_1

35. $\ell = a + (n - 1)d$; d

36. $2x + 3y = 12$; y

37. $2x + 3y = 12$; x

38. $2x - y = 5x + 6y$; x

39. $2x - y = 5x + 6y$; y

40. $3(4x - y) = 2x + y + 6$; y

41. $3(4x - y) = 2(x + y + 3)$; x

42. $ax + 4 = by - 3$; x **43.** $ay + 7 - x = 3x + 4$; y **44.** $a(x + 2) = by$; x

45. $b(y - 4) = a(x + 3)$; y

Solve the following formulas or literal equations for the specified variable. Assume that no denominator is equal to zero. See example 2–2 A.

46. The distance s that a body projected downward with an initial velocity of v falls in t seconds because of the force of gravity is given by

$$s = \frac{1}{2}gt^2 + vt.$$ Solve for g.

47. Solve the formula in exercise 46 for v.

48. The net profit P on sales of n identical cars is given by $P = n(P_2 - P_1) - c$, where P_2 is the selling price, P_1 is the cost to the dealer, and c is the operating expense. Solve for P_2.

49. Solve the formula in exercise 48 for P_1.

50. The perimeter of an isosceles triangle with base b and sides s is given by $P = 2s + b$. Solve for s.

Review exercises

Write an algebraic expression for each of the following. See section 1–5.

1. The product of x and 3

2. 6 times the sum of a and 7

3. y decreased by 2 and that difference divided by 4

4. A number multiplied by 5

5. A number diminished by 12

6. A number divided by 8 and that quotient decreased by 9

2–3 ■ Word problems

Many problems that we encounter are written or stated verbally. We need to translate these word problems into equations that we can solve algebraically. When translating word problems into equations, we should look for phrases involving the basic operations of addition, subtraction, multiplication, and division. Table 1–1 in chapter 1 showed some examples of phrases that are commonly encountered.

We now combine our ability to write an expression and our ability to solve an equation and apply them for solving a word problem. While there is no standard procedure for solving a word problem, the following guidelines should be useful.

1. Read the problem carefully. Determine useful prior knowledge and note what information is given and what information we are asked to find.
2. Whenever possible, draw a picture or use a diagram to represent the information in the problem.
3. Let some letter represent one of the unknowns, then express other unknowns in terms of it.
4. Use the given conditions in the problem and the unknowns from step 3 to write an algebraic equation.
5. Solve the equation for the unknown. Relate this answer to any other unknowns in the problem.
6. Check the results in the original statement of the problem.

■ *Example 2–3 A*   Number problems

Write an equation for the problem and solve for the unknown quantities.

1. One number is 18 more than a second number. If their sum is 62, find the two numbers.

   If we knew the value of the second number, then the first number would be 18 more. Therefore let x be the second number (second number $= x$). Then the first number is $x + 18$. Since these two numbers add up to 62, we write the equation as

first number	sum	second number	is	62
$(x + 18)$	$+$	x	$=$	62

   $$(x + 18) + x = 62 \qquad \text{Equation}$$
   $$x + 18 + x = 62 \qquad \text{Remove parentheses}$$
   $$2x + 18 = 62 \qquad \text{Combine like terms}$$
   $$2x = 44 \qquad \text{Subtract 18 from both members}$$
   $$x = 22 \qquad \text{Divide both members by 2}$$
   $$\text{and } x + 18 = (22) + 18 = 40 \qquad \text{Substitute to determine the other number}$$

   Hence the second number is 22 and the first number is 40.

   To check our answers, we must determine whether they satisfy the conditions stated in the original problem. Since 40 is 18 more than 22 and since the sum of 22 and 40 is 62, we know that our answers are correct. We will not show the checks of the following problems, but we should realize that a check of our work is an important final step.

   ▶ *Quick check*   One number is 9 times a second number and their sum is 120. Find the numbers.

2. If a number is divided by 4 and this quotient is increased by 6, the result is 13. Find the number.

   Let x be the number. Then the number divided by 4 is $\dfrac{x}{4}$ and that quotient increased by 6 is $\dfrac{x}{4} + 6$. Since this equals 13, we have

number divided by 4	increased	by 6	is	13
$\dfrac{x}{4}$	$+$	6	$=$	13

   Solving for x,

   $$\frac{x}{4} + 6 = 13 \qquad \text{Equation}$$

   $$\frac{x}{4} = 7 \qquad \text{Subtract 6 from both members}$$

   $$x = 28 \qquad \text{Multiply both members by 4}$$

   Hence the number is 28.

3. The sum of three numbers is 63. The first number is twice the second number, and the third number is three times the first number. Find the three numbers.

We must know the value of the second number to find the first number, and we must know the value of the first number to find the third number. Therefore everything depends on the second number. If x equals the second number, we have

second number $= x$
first number $= 2x$ (twice the second)
third number $= 3(2x) = 6x$ (three times the first).

Since their sum is 63, we have

first number	sum	second number	sum	third number	is	63
$2x$	$+$	x	$+$	$6x$	$=$	63

Solving for x,

$$2x + x + 6x = 63 \qquad \text{Equation}$$
$$9x = 63 \qquad \text{Combine like terms}$$
$$x = 7 \qquad \text{Divide by 9}$$

Hence, $2x = 2(7) = 14$ and $6x = 6(7) = 42$. Therefore the second number (x) is 7, the first number ($2x$) is 14, and the third number ($6x$) is 42.

▶ *Quick check* The sum of three consecutive odd integers is 51. Find the integers.

Interest problem

4. Phil had $20,000, part of which he invested at 8% interest and the rest at 6%. If his total income from the two investments for one year was $1,460, how much did he invest at each rate?

To solve this problem, we must understand how interest is computed. If we invested $5,000 at 8% interest, after one year we would have 8% of $5,000 or $(0.08)(\$5,000) = \$400$ interest. Thus the amount of interest earned in one year is the product of the rate times the principal (the amount invested).

In the given problem, we have two principals and two rates, and our formula will involve the sum of the two earned interests, which is equal to $1,460.

Let x represent the amount of money invested at 8%. Since this amount and the amount invested at 6% total $20,000, we can describe the 6% principal as the remainder after the amount x has been invested at 8%, that is, $20,000 - x$ is invested at 6% interest. We can use a table to summarize the information.

	Investment earning 8%	Investment earning 6%	Total
Amount invested	x	$20,000 - x$	20,000
Interest received	$(0.08)x$	$(0.06)(20,000 - x)$	1,460

We get the equation for the problem from the bottom row of the table.

amount of interest at 8%	total	amount of interest at 6%	was	total interest
$(0.08)x$	$+$	$(0.06)(20{,}000 - x)$	$=$	$1{,}460$

Solving for x,

$$
\begin{aligned}
(0.08)x + (0.06)(20{,}000 - x) &= 1{,}460 && \text{Equation} \\
(0.08)x + 1{,}200 - (0.06)x &= 1{,}460 && \text{Distributive property} \\
(0.02)x + 1{,}200 &= 1{,}460 && \text{Combine like terms} \\
(0.02)x &= 260 && \text{Subtract 1,200} \\
x &= 13{,}000 && \text{Divide by 0.02}
\end{aligned}
$$

and $20{,}000 - x = 20{,}000 - (13{,}000) = 7{,}000$ Substitute to get other investment

Hence the amount invested at 8% (x) was $13,000 and the amount invested at 6% $(20{,}000 - x)$ was $7,000.

Note When we know the sum of two numbers, we let x equal one number, then the sum minus x (sum $- x$) will be the other number.

▶ *Quick check* Lynne made two investments totaling $25,000. She made an 18% profit on one investment, but she took an 11% loss on the other investment. If her net gain was $2,180, how much was invested at each rate?

Geometry problem

5. The length of a rectangle is 9 feet more than its width. If the perimeter of the rectangle is 58 feet, find the dimensions.

We need the prior knowledge that the perimeter of a rectangle (distance around) is found by the formula $P = 2\ell + 2w$. If we let x represent the width of the rectangle, then the length would be 9 feet longer, or $x + 9$. We can now substitute into the formula.

$$
\begin{aligned}
P &= 2\ell + 2w \\
58 &= 2(x + 9) + 2(x)
\end{aligned}
$$

Solving for x,

$$
\begin{aligned}
58 &= 2(x + 9) + 2(x) && \text{Equation} \\
58 &= 2x + 18 + 2x && \text{Distributive property} \\
58 &= 4x + 18 && \text{Combine like terms} \\
40 &= 4x && \text{Subtract 18} \\
10 &= x && \text{Divide by 4}
\end{aligned}
$$

and $x + 9 = (10) + 9 = 19$ Substitute to get other dimension

Hence the width (x) is 10 feet and the length ($x + 9$) is 19 feet.

▶ *Quick check* The length of a rectangle is 1 inch less than three times the width. Find the dimensions if the perimeter is 70 inches.

```
┌─ Mastery points ──────────────────────────────────────────┐
│                                                            │
│  Can you                                                   │
│    ■ Translate word problems into equations?               │
│    ■ Solve for the unknown quantities?                     │
│                                                            │
└────────────────────────────────────────────────────────────┘
```

Exercise 2–3

Write an equation for the problem and solve for the unknown quantities. See example 2–3 A–1, 2, and 3.

Number problems

Example   One number is 9 times a second number and their sum is 120. Find the numbers.

Solution   Let x represent the second number, then the first number is 9 times the second number, or $9x$. Since their sum is 120, the equation is

first number	sum	second number	is	120
$9x$	$+$	x	$=$	120

Solving for x,

$$9x + x = 120 \qquad \text{Equation}$$
$$10x = 120 \qquad \text{Combine like terms}$$
$$x = 12 \qquad \text{Divide by 10}$$
$$\text{and } 9x = 9(12) = 108 \qquad \text{Substitute to get other number}$$

Therefore the second number (x) is 12 and the first number $(9x)$ is 108.

1. One number is 8 more than a second number. If their sum is 88, find the two numbers.

2. One whole number is 6 times a second whole number and their sum is 63. Find the numbers.

3. One natural number is 8 times another natural number and their sum is 54. Find the natural numbers.

4. One number is 28 more than a second number. If their sum is 62, find the two numbers.

5. One number is 5 less than another number. If their sum is 47, find the two numbers.

6. The difference of two numbers is 17. Find the numbers if their sum is 85.

7. If three times a number is increased by 11 and the result is 65, what is the number?

8. Nine times a number is decreased by 4, leaving 122. What is the number?

9. If a number is divided by 4 and that result is then increased by 6, the answer is 27. Find the number.

10. One-half of a number minus one-third of the number is 8. Find the number.

11. If a number is decreased by 14 and that result is then divided by 5, the answer is -8. Find the number.

12. One-third of a number is 12 less than one-half of the number. Find the number.

13. What number added to its double gives 51?

14. Find a number such that twice the sum of that number and 7 is 38.

15. Find two numbers whose sum is 81 and whose difference is 35.

16. One number is 11 more than twice a second number. If their sum is 53, what are the numbers?

17. Six times a number, increased by 10, gives 88. Find the number.

18. One number is seven times another. If their difference is 28, what are the numbers?

19. The sum of the number of teeth on two gears is 64 and their difference is 12. How many teeth are on each gear?

20. Two gears have a total of 83 teeth. One gear has 15 less teeth than the other. How many teeth are on each gear?

21. Two electrical voltages have a total of 126 volts (V). If one voltage is 32 V more than the other, find the voltages.

22. The sum of two voltages is 85 and their difference is 32. Find the voltages.

23. The sum of two resistances in a series is 24 ohms and their difference is 14 ohms. How many ohms are in each resistor?

24. One resistor exceeds another resistor by 25 ohms and their sum is 67 ohms. How many ohms are in each resistor?

Example The sum of three consecutive odd integers is 51. Find the integers.

Solution First we shall examine consecutive odd integers in general. Consider the list 1, 3, 5, 7 or 215, 217, 219, 221. We observe that the next odd integer on either list is found by adding 2 to the previous integer. Therefore if we let x be the first (least) of the three consecutive odd integers, then $x + 2$ would be the second and $(x + 2) + 2 = x + 4$ must be the third. Since their sum is 51, the equation would be

first odd integer	sum	second odd integer	sum	third odd integer	is	51
x	$+$	$(x + 2)$	$+$	$x + 4$	$=$	51

Solving for x,

$$x + x + 2 + x + 4 = 51 \qquad \text{Remove grouping symbols}$$
$$3x + 6 = 51 \qquad \text{Combine like terms}$$
$$3x = 45 \qquad \text{Subtract 6}$$
$$x = 15 \qquad \text{Divide by 3}$$

Therefore $x + 2 = (15) + 2 = 17$ and $x + 4 = (15) + 4 = 19$

Hence the first consecutive odd integer (x) is 15, the second $(x + 2)$ is 17, and the third $(x + 4)$ is 19.

25. The sum of three consecutive integers is 69. Find the integers.

26. The sum of three consecutive integers is 93. Find the integers.

27. The sum of three consecutive even integers is 48. Find the integers.

28. The sum of three consecutive even integers is −48. Find the integers.

29. The sum of three consecutive odd integers is −63. Find the integers.

30. The sum of three consecutive odd integers is 87. Find the integers.

31. One number is 27 more than another. The smaller number is one-fourth of the larger number. Find the numbers.

32. A number plus one-half of the number plus one-third of the number equals 33. Find the number.

33. A number is decreased by 7 and twice this result is 34. What is the number?

34. The sum of three numbers is 100. The second number is three times the first number and the third number is 6 less than the first number. Find the three numbers.

35. One number is 11 more than another number. Find the two numbers if three times the larger number exceeds four times the smaller number by 4.

36. One number is 8 more than another number. Find the two numbers if two times the larger number is 11 less than five times the smaller number.

37. If the first of two consecutive integers is multiplied by 3, this product is 20 more than the sum of the two integers. Find the integers.

38. Four times the first of three consecutive integers is 1 less than three times the sum of the second and third. Find the integers.

39. Five times the first of three consecutive even integers is 4 less than twice the sum of the second and third. Find the integers.

- Addition property of equality- $A+C = b+C$
- Multiplication Property of equality = $A \cdot C = b \cdot C$
- Divide or multiply by a negative # flip inequality.

$$P \times R \times T = I$$

%	P	R	T	I
	x	.0		

* answer can't be negative in absolute value.

* get absolute value by itself

40. One-fourth of the middle integer of three consecutive even integers is 24 less than one-half of the sum of the other two integers. Find the three integers.

41. The sum of three numbers is 49. The second number is three times the first number and the third number is 6 less than the first number. Find the three numbers.

42. The sum of three numbers is 38; the second number is twice the first number and the third number is 2 more than the first number. Find the three numbers.

43. One number is 7 more than another number. Find the two numbers if three times the larger number exceeds four times the smaller number by 13.

Interest problems

See example 2–3 A–4.

Example   Lynne made two investments totaling $25,000. She made an 18% profit on one investment, but she took an 11% loss on the other investment. If her net gain was $2,180, how much was invested at each rate?

Solution   Let x be the amount invested at 18% profit, then the amount invested at 11% loss was what was left over from the $25,000 or $25,000 - x$. Her profit of 18% on the one investment is denoted by $(0.18)x$ and her loss of 11% on the other investment is $-(0.11)(25,000 - x)$; the loss is denoted as a negative amount. We can use a table to summarize the information in the problem.

	Investment earning 18%	Investment losing 11%	Total
Amount invested	x	$25,000 - x$	25,000
Profit or loss	$(0.18)x$	$-(0.11)(25,000 - x)$	2,180

We get the equation for the problem from the bottom row of the table.

18% profit	net	11% loss	was	2,180
$(0.18)x$	$-$	$(0.11)(25,000 - x)$	$=$	$2,180$

Solving for x,

$$(0.18)x - (0.11)(25,000 - x) = 2,180 \qquad \text{Equation}$$
$$(0.18)x - 2,750 + (0.11)x = 2,180 \qquad \text{Distributive property}$$
$$(0.29)x - 2,750 = 2,180 \qquad \text{Combine like terms}$$
$$(0.29)x = 4,930 \qquad \text{Add 2,750}$$
$$x = 17,000 \qquad \text{Divide by 0.29}$$
$$\text{and } 25,000 - x = 25,000 - (17,000) = 8,000 \qquad \text{Substitute to get other investment}$$

So the amount invested at an 18% profit (x) was $17,000 and the amount invested at an 11% loss ($25,000 - x$) was $8,000.

44. Robert had $37,000, part of which he invested at 8% interest and the rest at 6%. If his total income was $2,600 from the two investments, how much did he invest at each rate?

45. Terry has $15,000. He invests part of this money at 8% and the rest at 6%. His income for one year from these investments totals $1,100. How much is invested at each rate?

46. Tammy had $6,000. She invested part of her money at $7\frac{1}{2}$% interest and the rest at 9%. If her income from the two investments was $511.50, how much did she invest at each rate?

47. Harry invested $26,000, part at 10% and the rest at 12%. If his income for one year from these investments totals $2,860, how much was invested at each rate?

48. Jill invests a total of $12,000, part at 6% and part at 10%. If her total income for one year is $956, how much is invested at each rate?

49. Margot invests a total of $12,000, part at 10% and part at 12%. Her total income for one year from the investments totals $1,340. How much is invested at each rate?

50. Alanzo has $16,875, part of which he invests at 10% interest and the rest at 8%. If his income from each investment is the same, how much did he invest at each rate?

51. Andrew invested a total of $18,000, part at 5% and part at 9%. If his income for one year from the 9% investment was $100 less than his income from the 5% investment, how much was invested at each rate?

52. Peter has invested $5,000 at an 8% rate. How much more must he invest at 10% to make the total income for one year from both sources a 9% rate?

53. Sherri has $4,000 invested at 7% and is going to invest an additional amount at 11% so that her total investment will make 9%. How much does she need to invest at 11% to achieve this?

54. Jeremy has $6,000 invested at 6%; how much must he invest at 10% to realize a net return of 9%?

55. Dick has $26,000, part of which he invests at 10% interest and the rest at 14%. If his income for one year from the 14% investment is $760 more than that from the 10% investment, how much is invested at each rate?

56. Susan had $19,000, part of which she invested at 9% interest and the rest at 7%. If her income from the 7% investment was $40 more than that from the 9% investment, how much did she invest at each rate?

57. Nina made two investments totaling $18,000. She made a 14% profit on one investment, but she took a 9% loss on the other investment. If her net gain was $680, how much was each investment?

58. Donald made two investments totaling $17,500. One investment made him a 13% profit, but on the other investment he took a 9% loss. If his net loss was $475, how much was each investment?

59. Jim made two investments totaling $34,000. One investment made him a 12% profit, but on the other investment he took a 21% loss. If his net loss was $870, how much was each investment?

Geometry problems

See example 2–3 A–5.

Example The length of a rectangle is 1 inch less than three times the width. Find the dimensions if the perimeter is 70 inches.

Solution If x represents the width of the rectangle, then by multiplying the width by three ($3x$) and subtracting 1, ($3x - 1$), represents the length of the rectangle. Using the formula for the perimeter of a rectangle, $P = 2\ell + 2w$, and the fact that the perimeter P is 70, we can write the equation

$P = 2\ell + 2w$	Formula for perimeter
$70 = 2(3x - 1) + 2x$	Substitute

Solving for x,

$70 = 6x - 2 + 2x$	Equation
$70 = 8x - 2$	Combine like terms
$72 = 8x$	Add 2
$9 = x$	Divide by 8
and $3x - 1 = 3(9) - 1 = 27 - 1 = 26$	Substitute

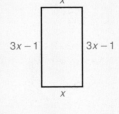

Therefore the width of the rectangle (x) is 9 inches and the length ($3x - 1$) is 26 inches.

60. The length of a rectangle is 9 feet more than its width. The perimeter of the rectangle is 90 feet. Find the dimensions.

61. The length of a rectangle is 5 feet more than its width. If the perimeter is 102 feet, find the length and width.

62. The width of a rectangle is 3 feet less than its length. The perimeter of the rectangle is 110 feet. Find the dimensions.

63. The width of a rectangle is $\frac{1}{3}$ of its length. If the perimeter is 128 feet, find the dimensions.

64. The width of a rectangle is 3 meters less than the length. If the perimeter of the rectangle is 126 meters, find the dimensions of the rectangle.

65. One side of a triangle is twice as long as the second side and the third side is 4 less than three times the second side. If the perimeter is 38 centimeters, find the lengths of the sides.

66. One side of a triangle is three times as long as the second side and the third side is 1 more than two times the second side. If the perimeter is 37 meters, find the lengths of the sides.

Mixture problems

Example   What quantities of 65% pure silver and 45% pure silver must be mixed together to give 100 grams of 50% pure silver?

Solution   To solve this mixture problem, we need to understand that the amount of silver in any given mixture is found by multiplying the percent of silver in the mixture times the amount of mixture. If x represents the amount of 65% pure silver in the final mixture, then $100 - x$ will represent the amount of 45% pure silver. The following table summarizes the information in the problem.

	65% silver mixture	45% silver mixture	50% silver mixture
Number of grams	x	$100 - x$	100
Amount of silver	$(0.65)x$	$(0.45)(100 - x)$	$(0.50)(100)$

We get the equation for the problem from the bottom row of the table.

amount of silver in 65% pure	mixed together	amount of silver in 45% pure	to give	100 grams of 50% pure silver
$(0.65)x$	$+$	$(0.45)(100 - x)$	$=$	$(0.50)(100)$

Solving for x,

$$(0.65)x + (0.45)(100 - x) = (0.50)(100) \qquad \text{Equation}$$
$$(0.65)x + 45 - (0.45)x = 50 \qquad \text{Multiply}$$
$$(0.20)x + 45 = 50 \qquad \text{Combine like terms}$$
$$(0.20)x = 5 \qquad \text{Subtract 45}$$
$$x = 25 \qquad \text{Divide by 0.20}$$
$$\text{and } 100 - x = 100 - (25) = 75 \qquad \text{Substitute to get other amount}$$

So the amount of 65% pure silver (x) is 25 grams and the amount of 45% pure silver ($100 - x$) is 75 grams.

67. An auto mechanic has two bottles of battery acid solutions. One contains 10% acid and the other 4% acid. How many cubic centimeters of each solution must be used to make 120 cm³ of a solution that is 6% acid?

68. A metallurgist wishes to form 2,000 kg of an alloy that is 80% copper. This alloy is to be obtained by fusing some alloy that is 68% copper and some alloy that is 83% copper. How many kilograms of each alloy must be used?

69. If a jeweler wishes to form 12 ounces of 75% pure gold from substances that are 60% and 80% pure gold, how much of each substance must be mixed together to produce this?

70. A chemist wishes to make 1,000 liters of a 3.5% acid solution by mixing a 2.5% solution with a 25% solution. How many liters of each solution is necessary?

71. A pharmacist wishes to fill a total of 200 3-grain and 2-grain capsules using 500 grains of a certain drug. How many capsules of each kind does he fill?

72. A solution that is 38% silver nitrate is to be mixed with a solution that is 3% silver nitrate to obtain 100 centiliters of solution that is 5% silver nitrate. How many centiliters of each solution should be used in the mixture?

73. A druggist has two solutions, one 60% hydrogen peroxide and the other 30% hydrogen peroxide. How many liters of each should she mix to obtain 30 liters of a solution that is 40% hydrogen peroxide?

Review exercises

Write the values of the following numbers. See section 1–1.

1. $|-12|$ **2.** $|0|$ **3.** $\left|-\dfrac{3}{4}\right|$ **4.** $|6|$

Find the solution set. See section 2–1.

5. $3x - 6 = 14$ **6.** $2x + 5 = -7$ **7.** $4 - 3x = -11$ **8.** $3x - 2 = 4x + 5$

2–4 ■ Equations involving absolute value

Absolute value equations

In chapter 1, we defined the absolute value of a number x to be

$$|x| = \begin{cases} x, & \text{if } x \geq 0 \\ -x, & \text{if } x < 0 \end{cases}$$

The absolute value of a number represents the undirected distance from that number to the origin on the number line, that is, the distance from x to 0.
    Consider the equation

$$|x| = 2$$

The right member, 2, denotes the distance that the graph of x is located from the origin. The equation directs us to find all numbers that are 2 units from the origin.

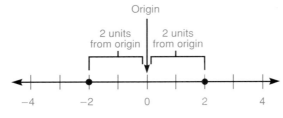

Figure 2–1

    We see from figure 2–1 that 2 and -2 are both 2 units from the origin and, therefore, satisfy the equation $|x| = 2$. The solution set is then given as $\{-2,2\}$.

$|x| = a$

For any real number x and $a \geq 0$,

$$|x| = a \text{ is equivalent to } x = a \text{ or } x = -a$$

Concept

An equation of the form $|x| = a$, called an **absolute value equation,** is equivalent to the equations $x = a$ or $x = -a$.

Note Recall that $|x| \geq 0$. We realize from this that a in our generalizations must be nonnegative, $a \geq 0$. This means that in an equation of this type, there is no solution if a is negative. For example, if $|x| = -2$, then the solution set is $\emptyset$.

Solving absolute value equations

1. Isolate the absolute value in one member of the equation.
2. Write the two equivalent equations.
3. Solve each equation.
4. Check each answer in the original absolute value equation.

Solutions for the absolute value equation $|x| = a$

If a is positive, the solutions are $x = a$ or $x = -a$	If a is zero, the solution is $x = 0$	If a is negative, there is no solution.

■ *Example 2–4 A*

Find the solution set of the following absolute value equations.

1. $|x| = 5$

   $x = 5$ or $x = -5$ Write the two equivalent equations

The solution set is $\{-5,5\}$.

To check the solutions of an absolute value equation, simply substitute the solutions into the original equation. In example 1, this would be:

For 5   For -5

$|5| = 5$ $|-5| = 5$

$5 = 5$ True $5 = 5$ True

2. $|x| + 4 = 10$

   $|x| = 6$ **Isolate** the absolute value by subtracting 4

   $x = 6$ or $x = -6$ Write the two equivalent equations

The solution set is $\{-6,6\}$.

3. $|x + 5| = 8$

   $x + 5 = 8$ or $x + 5 = -8$ Write the two equivalent equations

   $x = 3$ $x = -13$ Subtract 5 from both members

The solution set is $\{-13,3\}$.

4. $|3a - 2| = 7$

$\quad 3a - 2 = 7 \quad$ or $\quad 3a - 2 = -7 \qquad$ Write the two equivalent equations

$\qquad 3a = 9 \qquad\qquad 3a = -5 \qquad$ Add 2 to both members of both equations

$\qquad\quad a = 3 \qquad\qquad\quad a = -\dfrac{5}{3} \qquad$ Divide both members of both equations by 3

The solution set is $\left\{ -\dfrac{5}{3}, 3 \right\}$.

5. $|4x - 3| + 7 = 5$

$\quad\; |4x - 3| = -2 \qquad\qquad\qquad$ Isolate the absolute value in one member

The solution set is $\emptyset$. $\qquad\qquad\qquad$ Since the absolute value of any quantity cannot be negative, the solution set is empty

6. $|2a - 3| = |a + 4|$

The solution set to this equation will be those values that satisfy the condition that either $2a - 3$ and $a + 4$ are equal to each other or are opposites of each other. Hence we have the following equations.

$\qquad\qquad$ Equal to each other $\qquad\qquad$ Opposites of each other

$$2a - 3 = a + 4 \quad \text{or} \quad 2a - 3 = -(a + 4)$$

Solving each of the equations, we have

$2a - 3 = a + 4 \qquad\qquad 2a - 3 = -(a + 4)$

$\quad a - 3 = 4 \qquad$ Subtract a $\qquad 2a - 3 = -a - 4 \qquad$ Remove parentheses

$\qquad\quad a = 7 \qquad$ Add 3 $\qquad\qquad 3a - 3 = -4 \qquad$ Add a

$\qquad\qquad\qquad\qquad\qquad\qquad\qquad 3a = -1 \qquad$ Add 3

$\qquad\qquad\qquad\qquad\qquad\qquad\qquad\; a = -\dfrac{1}{3} \qquad$ Divide by 3

$\left\{ -\dfrac{1}{3}, 7 \right\} \qquad$ Solution set

7. $|2x + 3| = |2x - 4|$

We form the two equations equivalent to the absolute value equation.

$\qquad\qquad$ Equal to each other $\qquad\qquad$ Opposites of each other

$$2x + 3 = 2x - 4 \quad \text{or} \quad 2x + 3 = -(2x - 4)$$

Solving each of the equations, we have

$2x + 3 = 2x - 4 \qquad\qquad\qquad 2x + 3 = -(2x - 4)$

$\qquad\qquad\qquad$ Subtract 2x $\quad 2x + 3 = -2x + 4 \qquad$ Remove parentheses

$\qquad\quad 3 = -4 \qquad$ False $\qquad\quad 4x + 3 = 4 \qquad$ Add 2x

$\qquad\qquad\qquad\qquad\qquad\qquad\qquad 4x = 1 \qquad$ Subtract 3

$\qquad\qquad\qquad\qquad\qquad\qquad\qquad\; x = \dfrac{1}{4} \qquad$ Divide by 4

Since the first equation is a false statement ($3 = -4$), the first equation has no solution. The solution set is only the answer to the second equation and is given by $\left\{ \dfrac{1}{4} \right\}$.

Note   In examples 6 and 7, the solution set of the absolute value equation is the same whether we take the opposite of the first expression or the opposite of the second expression. That is, our solution set would be the same if we had solved $-(2a - 3) = a + 4$ instead of $2a - 3 = -(a + 4)$, or $-(2x + 3) = 2x - 4$ instead of $2x + 3 = -(2x - 4)$.

▶ **Quick check**   Find the solution set. $|4b + 3| = 5$ ■

Mastery points

Can you
- Solve absolute value equations?

Exercise 2-4

Determine whether or not the given value is a solution of the absolute value equation. Use $|2x + 7| = 11$ for exercises 1–4 and $|3x - 1| = 10$ for exercises 5–8.

Example   Is -9 a solution of the absolute value equation $|2x + 7| = 11$?

Solution $|2(-9) + 7| = 11$ Substitute
$|-18 + 7| = 11$
$|-11| = 11$ True

Therefore -9 is a solution of $|2x + 7| = 11$.

1. 2   **2.** -2 **3.** $\dfrac{1}{2}$ **4.** $\dfrac{5}{4}$ **5.** 3   **6.** -3 **7.** $\dfrac{11}{3}$ **8.** $\dfrac{2}{3}$

Find the solution set of the following absolute value equations. See example 2–4 A.

Example $|4b + 3| = 5$

Solution $4b + 3 = 5$ or $4b + 3 = -5$ Write the equivalent equations
$4b = 2$ $4b = -8$ Subtract 3 from all members
$b = \dfrac{1}{2}$ $b = -2$ Divide all members by 4

The solution set is $\left\{ -2, \dfrac{1}{2} \right\}$.

9. $|x| = 9$ **10.** $|x| = 12$ **11.** $|a| = 4$ **12.** $|b| = 5$

13. $|b| + 2 = 6$ **14.** $|x| + 4 = 10$ **15.** $|x| - 5 = 7$ **16.** $|y| - 8 = 2$

17. $|x + 4| = 6$ **18.** $|x + 3| = 6$ **19.** $|a - 3| = 2$ **20.** $|b - 7| = 11$

21. $|3x - 4| = 8$ **22.** $|2x - 1| = 9$ **23.** $|2a + 7| = 9$ **24.** $|3x + 2| = 11$

25. $|5x - 3| = -4$ **26.** $|x + 2| = -1$ **27.** $|4a + 8| + 10 = 3$ **28.** $|3b + 1| + 8 = 5$

29. $|4b - 3| + 2 = 8$ **30.** $|2x + 5| + 3 = 10$ **31.** $|5a + 2| - 7 = 4$ **32.** $|3y - 2| + 4 = 11$

33. $|2.1x - 6.3| = 8.4$ **34.** $|3.2x - 6.4| = 9.6$ **35.** $|1.8x - 10.8| = 5.4$ **36.** $|1.7a - 5.1| = 13.6$

37. $\left|\frac{1}{2}x + 5\right| = 7$ **38.** $\left|\frac{1}{3}x - 2\right| = 13$ **39.** $\left|\frac{3}{4}a - 2\right| = 6$ **40.** $\left|\frac{2}{3}x + 1\right| = 10$

41. $\left|\frac{2}{3}b - 6\right| = 4$ **42.** $\left|\frac{2}{5}x - 3\right| = 17$ **43.** $|5 - 3x| - 4 = 8$ **44.** $|2 - x| = 8$

45. $|3 - 4a| + 2 = 5$ **46.** $|1 - 2x| = 21$ **47.** $|3x - 7| = |5x + 3|$ **48.** $|2x + 1| = |x - 3|$

49. $|2a + 5| = |6a + 7|$ **50.** $|x + 5| = |3x - 4|$ **51.** $|3 - 2a| = |4a + 6|$ **52.** $|1 - 3b| = |2b + 3|$

53. $|3y - 4| = |3y + 8|$ **54.** $|3x - 2| = |3x + 4|$ **55.** $|4b - 3| = |4b + 7|$ **56.** $|2y + 5| = |2y - 7|$

Write an absolute value equation for the following statements and solve for the unknown.

Example   The absolute value of twice a number is 10.

Solution   If we let x represent the unknown number, then twice the number would be $2x$. The absolute value equation would be

$$|2x| = 10 \qquad \text{Equation}$$
$$2x = 10 \quad \text{or} \quad 2x = -10 \qquad \text{Write the two equivalent equations}$$
$$x = 5 \qquad\quad x = -5 \qquad \text{Divide both members of both equations by 2}$$

The solution set is $\{-5, 5\}$.

57. The absolute value of a number is 6.

58. The absolute value of a number is 9.

59. The absolute value of 3 times a number is 12.

60. The absolute value of 5 times a number is 30.

61. If a number is diminished by 8, the absolute value of the result is 14.

62. If a number is diminished by 6, the absolute value of the result is 12.

63. If a number is increased by 7, the absolute value of the result is 15.

64. Twice a number is increased by 3. The absolute value of the result is 11.

65. Three times a number is diminished by 4. The absolute value of the result is 8.

66. One-half of a number is added to 3. The absolute value of the result is 4.

67. One-third of a number is diminished by 7. The absolute value of the result is 14.

68. One-half of a number is diminished by 5. The absolute value of the result is 11.

Review exercises

Write an algebraic expression for each of the following. See section 1–5.

1. The sum of x and 2

2. 6 less than y

3. a diminished by 4

4. A number divided by 5

5. $\frac{1}{3}$ of a number

6. A number decreased by 2 and that difference divided by 8

2–5 ■ *Linear inequalities*

Representing the solution set of a linear inequality

When we replace the equals sign in a linear equation with one of the inequality symbols ($<$ is less than, $>$ is greater than, $\leq$ is less than or equal to, $\geq$ is greater than or equal to), we form a **linear inequality.** The following are examples of linear inequalities.

$$5x + 3 \leq 4, \quad 2(3x - 1) > 5x - 7, \quad 5x - 2x + 1 < x - 3$$

A major difference between a linear equation and a linear inequality is the solution set. The solution set of a conditional linear equation has at most one solution, whereas the solution set of a conditional linear inequality usually consists of an unlimited number of solutions. Consider the inequality $3x \geq 9$. We see by inspection that if we substitute 3, $3\frac{1}{3}$, 4, or 5 for x, the inequality would be true. In fact, we see that if we substitute any number greater than or equal to 3, that is ($x \geq 3$), the inequality would be true. This demonstrates that our solution set has an unlimited number of solutions. We would state it in set-builder notation as $\{x|x \geq 3\}$, which is read "the set of all elements x such that x is greater than or equal to 3."

There are other ways to indicate the solution set of an inequality. One way is to graph the solution set. To graph the solution set $\{x|x \geq 3\}$, we simply draw a number line (as we did in chapter 1), place a *solid circle* (dot) at 3 on the number line, and draw an arrow extending from the circle to the right, as in figure 2–2.

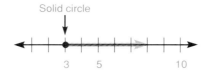

Figure 2–2

The solid circle at 3 represents the fact that 3 is included in the solution set.

■ *Example 2–5 A*

Represent the following solution sets graphically.

1. $\{x|x < -1\}$

Here x is representing all real numbers less than -1, but not -1 itself. To denote the fact that x cannot equal -1, we put a *hollow circle* at -1.

2. $\{x | x \geq 2\}$

The greater than or equal to symbol, $\geq$, indicates that 2 is included in the solution set, so we place a solid circle at 2.

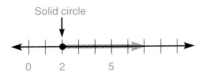

3. $\{x | -2 \leq x < 1\}$

The statement $-2 \leq x < 1$ is called a **compound inequality** and is read "-2 is less than or equal to x and x is less than 1." We place a solid circle at -2 to denote that -2 is included, and we place a hollow circle at 1 to show that 1 is not included. We then draw a line segment between the two circles.

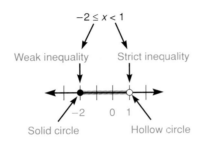

Note When we graph inequalities, a strict inequality, $<$ or $>$, is represented by a hollow circle at the number. A weak inequality, $\leq$ or $\geq$, is represented by a solid circle at the number.

▶ *Quick check* Represent the solution set graphically. $\{x | -3 < x \leq 2\}$ ∎

Interval notation

Another way to represent the solution set of an inequality is to use **interval notation.** In interval notation we use a parenthesis, (or), to represent that the endpoint is *not* part of the interval and a bracket, [or], to represent that the endpoint is part of the interval.

Consider the set $\{x | -3 < x \leq 4\}$. To write this in interval notation, we first write down the endpoints, writing the lesser one first, and separate them by a comma.

$$-3,4$$

If the endpoint is indicated by a strict inequality symbol, $<$ or $>$, place a parenthesis next to that endpoint. If the endpoint is indicated by a weak inequality symbol, $\leq$ or $\geq$, place a bracket next to that endpoint. So,

Strict inequality—parenthesis

$$\{x | -3 < x \leq 4\} \text{ is equivalent to } (-3,4]$$

Weak inequality—bracket

■ *Example 2–5 B*

Represent the following sets with interval notation.

1. $\{x|-2 \le x \le 3\}$

   Since the endpoints are included, we use brackets in interval notation.

   $[-2,3]$

2. $\{x|0 < x < 6\}$

   Since the endpoints are not included, we use parentheses in interval notation.

   $(0,6)$

3. $\{x|x \ge 1\}$

   To indicate that our solution set continues indefinitely, we use $+\infty$, read "positive infinity." We place a parenthesis next to the $+\infty$ symbol because there is no endpoint to be contained.

   $[1,+\infty)$

4. $\{x|x < -3\}$

   Interval $(-\infty,-3)$
   The symbol $-\infty$ is read "negative infinity."

Note In interval notation, the lesser number must come first. In example 4, $(-3,-\infty)$ would be incorrect.

▶ *Quick check* Represent the set $\{x|-3 < x \le 2\}$ with interval notation. ■

Solving a linear inequality

The properties that we will use to solve a linear inequality are similar to those that we used to solve linear equations.

┌─ *Addition property of inequality* ─────────────

For any algebraic expressions A, B, and C, if $A < B$, then

$$A + C < B + C$$

┌─ *Multiplication property of inequality* ─────────────

For any algebraic expressions A, B, and C, if $A < B$, then
1. if $C > 0$ (positive), then

$$A \cdot C < B \cdot C$$

2. if $C < 0$ (negative), then

$$A \cdot C > B \cdot C$$

___ *Concept* _____

1. The same expression can be added to or subtracted from *both* members of an inequality and will not change the direction of the inequality symbol. We can multiply or divide *both* members of the inequality by the same positive expression and still maintain the direction of the inequality symbol.
2. We can multiply or divide *both* members of an inequality by the same *negative* expression provided that we *reverse* the direction of the inequality symbol.

Note The two properties are stated in terms of the is less than ($<$) symbol. These properties also apply for any of the other inequality symbols ($>$, $\leq$, or $\geq$).

Just as with equations, we can use the addition property to subtract the same expression from both members of an inequality. The multiplication property allows us to divide both members of an inequality by the same nonzero expression.

To demonstrate these operations, consider the inequality $8 < 12$.

1. If we add or subtract 4 in each member, we still have a true statement.

$$8 < 12 \qquad \text{or} \qquad 8 < 12 \qquad \text{Original true statement}$$
$$8 + 4 < 12 + 4 \qquad 8 - 4 < 12 - 4 \qquad \text{Add or subtract 4}$$
$$12 < 16 \qquad 4 < 8 \qquad \text{New true statement}$$

2. If we multiply or divide by 4 in each member, we still have a true statement.

$$8 < 12 \qquad \text{or} \qquad 8 < 12 \qquad \text{Original true statement}$$
$$8 \cdot 4 < 12 \cdot 4 \qquad \frac{8}{4} < \frac{12}{4} \qquad \text{Multiply or divide by 4}$$
$$32 < 48 \qquad 2 < 3 \qquad \text{New true statement}$$

3. But if we multiply or divide by -4 in each member, we must reverse the direction of the inequality to have a true statement.

$$8 < 12 \qquad \text{or} \qquad 8 < 12 \qquad \text{Original true statement}$$
$$8(-4) > 12(-4) \qquad \frac{8}{-4} > \frac{12}{-4} \qquad \text{Multiply or divide by } -4 \text{ and reverse direction of the inequality symbol}$$
$$-32 > -48 \qquad -2 > -3 \qquad \text{New true statement}$$

Note When we reverse the direction of the inequality symbol, we say that we **reversed the sense or order** of the inequality.

To summarize our operations, we see that, with one exception, they are the same as the operations for linear equations. **Whenever we multiply or divide both members of an inequality by a negative number, we must reverse the direction of the inequality symbol.**

We shall now solve a linear inequality. The procedure for solving a linear inequality is the same four steps that we used to solve a linear equation.

Consider the inequality

$$3(4x - 1) \geq 5x + 3x + 5$$

Solving linear inequalities

Step 1 *We simplify the inequality by carrying out the indicated multiplication in the left member and the addition in the right member.*

$$3(4x - 1) \geq 5x + 3x + 5$$
$$12x - 3 \geq 8x + 5 \qquad \text{Multiply and combine like terms}$$

Step 2 *We want all terms containing the unknown, x, in one member of the inequality.* Therefore we subtract $8x$ from both members of the inequality.

$$12x - 3 \geq 8x + 5$$
$$12x - 8x - 3 \geq 8x - 8x + 5 \qquad \text{Subtract } 8x \text{ from both members}$$
$$4x - 3 \geq 5 \qquad \text{Combine like terms}$$

Note A negative coefficient of the unknown can be avoided if we form equivalent inequalities where the unknown appears only in the member of the inequality that has the greater coefficient of the unknown.

Step 3 *We want all terms not involving the unknown in the other member of the inequality.* Therefore we add 3 to both members of the inequality.

$$4x - 3 \geq 5$$
$$4x - 3 + 3 \geq 5 + 3 \qquad \text{Add 3 to both members}$$
$$4x \geq 8$$

Step 4 *We form an equivalent inequality where the coefficient of the unknown is 1.* Hence we divide both members of the inequality by 4.

$$4x \geq 8$$
$$\frac{4x}{4} \geq \frac{8}{4} \qquad \text{Divide both members by 4}$$
$$x \geq 2$$

The solution set is $\{x \mid x \geq 2\}$.

Note In step 4, we must be careful to observe whether we are multiplying or dividing by a positive or negative number so that we will form the correct inequality.

■ *Example 2–5 C*

Find the solution set of the following linear inequalities. Leave the answer in (1) set-builder notation, (2) graphical notation, and (3) interval notation.

1. $2(1 - 2x) \geq 4(2 - 3x) + 2x$
$\qquad 2 - 4x \geq 8 - 12x + 2x \qquad$ Distributive property
$\qquad 2 - 4x \geq 8 - 10x \qquad$ Combine like terms
$\qquad 6x + 2 \geq 8 \qquad$ Add 10x to both members
$\qquad 6x \geq 6 \qquad$ Subtract 2 from both members
$\qquad x \geq 1 \qquad$ Divide both members by 6

The solution set is $\{x \mid x \geq 1\}$

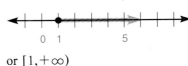

or $[1, +\infty)$

2. $4(3 - 2x) + 3x > 2$

$12 - 8x + 3x > 2$	Distributive property
$12 - 5x > 2$	Combine like terms
$-5x > -10$	Subtract 12 from both members
$\dfrac{-5x}{-5} < \dfrac{-10}{-5}$	Divide both members by -5 and REVERSE THE DIRECTION OF THE INEQUALITY SYMBOL
$x < 2$	Simplify

The solution set is $\{x \mid x < 2\}$

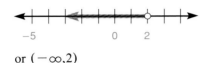

or $(-\infty, 2)$

3. $-5 \leq 2x - 1 < 3$

When solving a compound inequality, the solution must be such that the unknown appears only in the middle member of the inequality. We can still use all of our properties, if we apply them to all *three* members. We must reverse the direction of *all* the inequality symbols when multiplying or dividing by a negative number.

$-5 \leq 2x - 1 < 3$	
$-5 + 1 \leq 2x - 1 + 1 < 3 + 1$	Add 1 to all three members
$-4 \leq 2x < 4$	Simplify
$\dfrac{-4}{2} \leq \dfrac{2x}{2} < \dfrac{4}{2}$	Divide all three members by 2
$-2 \leq x < 2$	Simplify

The solution set is $\{x \mid -2 \leq x < 2\}$

or $[-2, 2)$

4.

$-12 < 8 - 4x \leq 16$	
$-12 - 8 < 8 - 8 - 4x \leq 16 - 8$	Subtract 8 from all three members
$-20 < -4x \leq 8$	Simplify
$\dfrac{-20}{-4} > \dfrac{-4x}{-4} \geq \dfrac{8}{-4}$	Divide all three members by -4 and REVERSE THE DIRECTION OF ALL INEQUALITY SYMBOLS
$5 > x \geq -2$	Simplify

The solution set is $\{x \mid -2 \leq x < 5\}$

or $[-2, 5)$

Note   When we state the answer in set-builder notation, it is customary to have compound inequalities stated with less than inequality symbols. This makes changing from one form of notation to another easier.

▶ *Quick check*   Find the solution set of $3(2x + 1) \geq x + 2x + 7$. Give the answer in set-builder notation, interval notation, and graphical notation. ■

Problem solving

We are now ready to combine our ability to write an expression and our ability to solve an inequality and apply them to solve verbal problems. The guidelines for solving a linear inequality are the same as those for solving a linear equation in section 2–3. The following table shows a number of different ways an inequality symbol could be written with words.

Symbol	$<$	$\leq$	$>$	$\geq$
In words	is less than is fewer than	is at most is no more than is no greater than is less than or equal to	is greater than is more than exceeds	is at least is no less than is no fewer than is greater than or equal to

■ *Example 2–5 D*

Solve the following word problems.

1. Five times a number is added to 6 and the result is no more than 41. Find all numbers that satisfy this condition.

Let x represent the number. Then

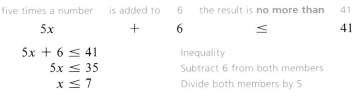

$$5x + 6 \leq 41 \quad \text{Inequality}$$
$$5x \leq 35 \quad \text{Subtract 6 from both members}$$
$$x \leq 7 \quad \text{Divide both members by 5}$$

The solution set is $\{x \mid x \leq 7\}$.

2. Three times a number plus 12 is at least -6 but less than 18. Find all numbers that satisfy these conditions.

Let x represent the number. Then

-6 **is at least**	three times a number	plus	12	**is less than**	18
$-6 \leq$	$3x$	$+$	12	$<$	18

$$-6 \leq 3x + 12 < 18 \quad \text{Compound inequality}$$
$$-18 \leq 3x < 6 \quad \text{Subtract 12 from all three members}$$
$$-6 \leq x < 2 \quad \text{Divide all three members by 3}$$

The solution set is $\{x \mid -6 \leq x < 2\}$.

▶ **Quick check** Two times a number added to 7 is greater than 19. Find all numbers that satisfy this condition. ■

___ **Mastery points** ___

Can you
- Solve linear inequalities and compound inequalities?
- Represent the solution set of a linear inequality in set-builder notation, graphical notation, or interval notation?
- Solve word problems involving linear inequalities?

Exercise 2–5

Represent the following solution sets both graphically and with interval notation. See examples 2–5 A and B.

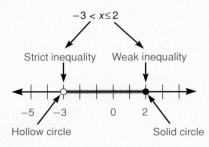

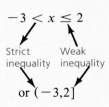

Example $\{x|-3 < x \leq 2\}$

Solution

1. $\{x|-3 \leq x \leq 1\}$ 2. $\{x|0 < x < 4\}$ 3. $\{x|-5 \leq x < -1\}$ 4. $\{x|x > -3\}$
5. $\{x|x \geq 4\}$ **6.** $\{x|x < 2\}$ 7. $\{x|x \leq -1\}$ 8. $\{x|x \leq 0\}$

Find the solution set of the following inequalities. Leave the answer in both set-builder notation and interval notation. See example 2–5 C.

Example $3(2x + 1) \geq x + 2x + 7$

Solution $6x + 3 \geq 3x + 7$ Distributive property and combine like terms
$3x + 3 \geq 7$ Subtract 3x from both members
$3x \geq 4$ Subtract 3 from both members
$x \geq \dfrac{4}{3}$ Divide both members by 3

The solution set is $\left\{x|x \geq \dfrac{4}{3}\right\}$ or $\left[\dfrac{4}{3}, +\infty\right)$.

9. $2x > 18$ 10. $3x < 12$ 11. $\dfrac{2}{3}x \geq 8$

12. $\dfrac{3}{4}x \geq 9$ **13.** $-4x \leq 20$ 14. $-3x > 27$

15. $3x + 2x < x + 6$ 16. $6x - 2x > 5x - 3$ 17. $2x + (4x - 1) > 6 - x$
18. $3(2x + 1) < 9$ 19. $4(2x - 5) \leq 10x + 7$ **20.** $4 - 2(3x + 1) > 8x - 12$
21. $7 - 3(5x - 4) \leq 12 - 9x$ 22. $2(x - 4) - 14 \leq 3(5 - 3x)$ 23. $2(4x + 3) \geq 5 - 4(x - 1)$
24. $6(2x - 3) \leq 9(x - 1) - 4$ 25. $4(2 - x) + 7 > 3x - 3(x - 1)$
26. $4x - 4(x + 2) < 3(2 - x)$ 27. $14 \geq 5(1 - x) + 2(x + 3)$
28. $7 < 4(1 - 2x) + 3(x - 4)$ 29. $8.2x - 3.6x + 7.1 \geq 1.8x + 23.9$
30. $2.1(x - 6) < 0.4x + 7.8$ 31. $4.3(x - 2) \geq 3.1x - 25.4$
32. $7.3(3 - x) < 4.9(2 - x) - 4.7$ 33. $12.6(3x - 2) + 8.9 \leq 8.9(2x - 4) - 0.7$
34. $-2 < 3x + 1 < 4$ 35. $-3 < 3x - 5 < 4$ 36. $-2 \leq 4x + 2 \leq 8$
37. $0 \leq 7x - 2 \leq 6$ 38. $-5 < 4x + 5 \leq 5$ 39. $-4 \leq 3x + 6 \leq 6$

40. $2 \le 1 - x \le 6$ **41.** $3 < 5 - 2x < 7$ **42.** $-2 < 4 - 3x \le 0$

43. $-5 \le 8 - 2x \le 0$ **44.** $0 \le 1 - 4x < 7$ **45.** $-4 \le 3 - 2x \le -1$

46. $-7 \le 4 - 2x \le -4$ **47.** $0 \le 2 - 3x \le 4$ **48.** $4 < 4 - 3x \le 6$

Write an inequality to represent the following statements. See example 2–5 D.

Example A student's score must be below 60 to fail the examination.

Solution Let x = the student's score. Then the inequality would be $x < 60$.

49. A student's score must be at least 90 to receive an A on the exam.

50. A student must score at least 75 on the final exam to pass the course.

51. The temperature today will not get above 42.

52. The temperature today will be at least 80.

53. A salesperson needs to sell at least 10 new cars to make a bonus.

54. The temperature today will range from a low of 18 to a high of 41.

55. On a partly sunny day, there will be at least 96 minutes of sunlight but at most 384 minutes of sunlight.

56. The selling price P must be more than the cost c but less than twice the cost.

57. The selling price P must be at least one and one-half times the cost c but at most three times the cost.

Write an inequality using the given information and solve. See example 2–5 D.

Example Two times a number added to 7 is greater than 19. Find all numbers that satisfy this condition.

Solution Let x = the number. Then the inequality would be

$2x + 7 > 19$ Inequality

$2x > 12$ Subtract 7 from both members

$x > 6$ Divide both members by 2

The solution set is $\{x | x > 6\}$.

58. When 4 is added to three times a number, the result is at least 12. Find all numbers that satisfy this condition.

59. When 6 is subtracted from five times a number, the result is less than 17. Find all numbers that satisfy this condition.

60. Four times a number plus 6 is at least 21. Find all numbers that satisfy this condition.

61. If one-half of a number is added to 16, the result is greater than 24. Find all numbers that satisfy this condition.

62. Three times a number is subtracted from 11 and this result is less than 6. Find all numbers that satisfy this condition.

63. Two times a number is subtracted from 19 and this result is at most 8. Find all numbers that satisfy this condition.

64. A student has scores of 7, 10, and 8 on three quizzes. What must she score on the fourth quiz to have an average of 8 or higher?

65. A student has scores of 66, 71, and 84 on three exams. If an average of 75 is required to pass the course, what is the minimum score he must have on the fourth test to pass?

66. Two times a number minus 6 is greater than 4 but less than 19. Find all numbers that satisfy these conditions.

67. Three times a number plus 2 is greater than 12 but less than 23. Find all numbers that satisfy these conditions.

68. The perimeter of a rectangle must be less than 100 feet. If the length is known to be 30 feet, find all numbers that the width could be. (*Note:* The width of a real rectangle must be a positive number.)

69. The perimeter of a square must be greater than 16 inches but less than 84 inches. Find all values of a side that satisfy these conditions. (*Hint:* The perimeter of a square is given by $P = 4s$, where s represents the length of a side.)

Review exercises

Write the value of the following numbers. See section 1–1.

1. $|-21|$ **2.** $|8|$ **3.** $-|-2|$

Write an algebraic expression for each of the following. See section 1–5.

4. 2 times y **5.** 6 more than a

6. $\frac{1}{4}$ of a number **7.** A number decreased by 12

8. A number increased by 7

2–6 ■ *Inequalities involving absolute value*

Absolute value inequalities of the form $|x| < a$

If the equals sign, $=$, in an absolute value equation is replaced with an inequality symbol, $<$, $>$, $\leq$, or $\geq$, the absolute value equation becomes an **absolute value inequality.** Consider the absolute value inequality

$$|x| < 2$$

This inequality states that the distance between x and the origin is less than 2 units.

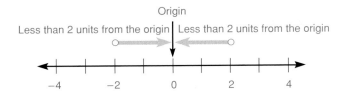

Figure 2–3

We see from figure 2–3 that all numbers between -2 and 2 satisfy the inequality $|x| < 2$. The solution set can be given in any one of the forms that we studied in section 2–5. That is, in set-builder notation, the solution set would be $\{x | -2 < x < 2\}$, in interval notation $(-2,2)$, and in graphical notation, as shown in figure 2–4.

Figure 2–4

We can generalize this observation in the following property.

> **$|x| < a$**
>
> For any real number x and $a > 0$,
> $$|x| < a \text{ is equivalent to } -a < x < a$$
>
> **Concept**
> Given $|x| < a$, then x will be any real number between the opposite of a and a.

Note The property is stated in terms of the strict inequality $<$, but it is still true if we replace the strict inequality symbol with the weak inequality symbol $\leq$. That is, $|x| \leq a$ is equivalent to $-a \leq x \leq a$.

> **Solving absolute value inequalities of the type $|x| < a$**
>
> 1. Isolate the absolute value in one member of the inequality.
> 2. Write the equivalent inequality (a three-member compound inequality).
> 3. Solve the inequality.

> **Solutions for the absolute value inequality $|x| < a$**
>
> If a is positive, the inequality is
> $$-a < x < a$$
> If a is zero or negative, there is no solution.*

■ **Example 2–6 A**

Find the solution set of the following absolute value inequalities and leave the answer in (1) set-builder notation, (2) interval notation, and (3) graphical notation.

1. $|x| - 3 < -2$
   $|x| < 1$ **Isolate** the absolute value by adding 3
   $-1 < x < 1$ Write the equivalent compound inequality

   The solution set is $\{x \mid -1 < x < 1\}$, or $(-1,1)$, or

2. $|x + 3| \leq 6$
   $-6 \leq x + 3 \leq 6$ Write the equivalent compound inequality
   $-9 \leq x \leq 3$ Subtract 3 from all three members

   The solution set is $\{x \mid -9 \leq x \leq 3\}$, or $[-9,3]$, or

*If a is negative or zero, then there is no solution since $|x|$ is always greater than or equal to zero and cannot be less than zero or any negative value that a is representing.

3.
$$|3x - 4| < 5$$
$$-5 < 3x - 4 < 5 \quad \text{Write the equivalent compound inequality}$$
$$-1 < 3x < 9 \quad \text{Add 4 to all three members}$$
$$-\frac{1}{3} < x < 3 \quad \text{Divide all three members by 3}$$

The solution set is $\left\{x \,\middle|\, -\frac{1}{3} < x < 3\right\}$, or $\left(-\frac{1}{3}, 3\right)$, or

4.
$$|5 - 2x| \leq 3$$
$$-3 \leq 5 - 2x \leq 3 \quad \text{Write the equivalent compound inequality}$$
$$-8 \leq -2x \leq -2 \quad \text{Subtract 5 from all three members}$$
$$\frac{-8}{-2} \geq \frac{-2x}{-2} \geq \frac{-2}{-2} \quad \text{Divide by } -2 \text{ and } \textbf{REVERSE the direction of all the inequality symbols}$$
$$4 \geq x \geq 1 \quad \text{Simplify}$$

The solution set is $\{x | 1 \leq x \leq 4\}$, or $[1,4]$, or

5. $|7x - 4| < -5$

The solution set is $\emptyset$. Since the absolute value cannot be less than zero, there is no solution ∎

Absolute value inequalities of the form $|x| > a$

We have only considered those absolute value inequalities that involve "is less than," $<$, or "is less than or equal to," $\leq$. Consider the absolute value inequality

$$|x| > 2$$

This inequality states that the distance between x and the origin is more than 2 units.

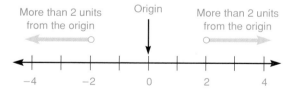

Figure 2–5

We see from figure 2–5 that all numbers greater than 2 or less than -2 satisfy the inequality $|x| > 2$. The solution set is the union of the two intervals, $x > 2$ or $x < -2$, and is given by

$$\{x | x < -2 \text{ or } x > 2\} \text{ or } (-\infty, -2) \cup (2, +\infty)$$

or as shown in figure 2–6.

Figure 2-6

We can generalize this observation in the following property.

$|x| > a$ ────────────────────────

For any real number x and $a > 0$,

$$|x| > a \text{ is equivalent to } x < -a \text{ or } x > a$$

Concept
Given $|x| > a$, then x will be any real number less than the opposite of a or greater than a.

Note The property is stated in terms of the strict inequality $>$, but it is still true if we replace the strict inequality symbol with the weak inequality symbol. That is, $|x| \geq a$ is equivalent to $x \leq -a$ or $x \geq a$.

Solving absolute value inequalities of the form $|x| > a$ ─────

1. Isolate the absolute value in one member of the inequality.
2. Write the equivalent inequalities (a pair of inequalities connected with an ''or'').
3. Solve the inequalities.

Solutions for the absolute value inequality $|x| > a$ ──────

If a is positive, the inequalities are

$$x < -a \text{ or } x > a$$

If a is negative, then the solution set is the set of all real numbers.*

$$R$$

■ *Example 2-6 B*

Find the solution set of the following absolute value inequalities and leave the answer in (1) set-builder notation, (2) interval notation, and (3) graphical notation.

1. $|x + 2| > 2$

$$x + 2 < -2 \quad \text{or} \quad x + 2 > 2 \qquad \text{Write the equivalent inequalities}$$
$$x < -4 \qquad\qquad x > 0 \qquad \text{Subtract 2 from both members of both inequalities}$$

The solution set is $\{x \mid x < -4 \text{ or } x > 0\}$, or $(-\infty, -4) \cup (0, +\infty)$, or

*Any real number will be a solution since the absolute value of any real number is greater than or equal to zero and zero is greater than any negative value that a is representing.

2. $|3x - 4| \geq 2$

$3x - 4 \leq -2$ or $3x - 4 \geq 2$ Write the equivalent inequalities

     $3x \leq 2$ $3x \geq 6$ Add 4 to both members of both inequalities

     $x \leq \dfrac{2}{3}$ $x \geq 2$ Divide all members by 3

The solution set is $\left\{ x \middle| x \leq \dfrac{2}{3} \text{ or } x \geq 2 \right\}$, or $\left(-\infty, \dfrac{2}{3} \right] \cup [2, +\infty)$, or

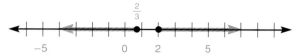

3. $|4x + 3| - 4 > 7$

     $|4x + 3| > 11$ Isolate the absolute value by adding 4

$4x + 3 < -11$ or $4x + 3 > 11$ Write the equivalent inequalities

    $4x < -14$ $4x > 8$ Subtract 3 from all members

     $x < -\dfrac{7}{2}$ $x > 2$ Divide all members by 4

The solution set is $\left\{ x \middle| x < -\dfrac{7}{2} \text{ or } x > 2 \right\}$, or $\left(-\infty, -\dfrac{7}{2} \right) \cup (2, +\infty)$, or

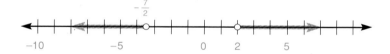

4. $|4x - 5| + 8 \geq 3$

     $|4x - 5| \geq -5$ Isolate the absolute value by subtracting 8

The solution set is $\{ x | x \in R \}$,   Since this is true for all values of x, the solution
or $(-\infty, +\infty)$, or   set is all real numbers

Note A common error in solving absolute value inequalities is to apply the wrong procedure. Once you have the absolute value isolated in one member of the inequality, identify the type of inequality symbol and apply the appropriate property.

▶ *Quick check* Find the solution set of $|5x + 4| - 6 \geq -3$. Give the answer in set-builder notation, interval notation, and graphical notation.   ■

Mastery points

Can you
- Solve absolute value inequalities?

Exercise 2–6

Determine whether or not the given value is an element of the solution set of the absolute value inequality. Use $|3x + 2| \geq 5$ for exercises 1–4 and $|2x - 3| < 7$ for exercises 5–8.

Example   Is 6 an element of the solution set of the absolute value inequality $|2x - 3| < 7$?

Solution $|2(6) - 3| < 7$ Substitute

$\quad\quad\quad |12 - 3| < 7$

$\quad\quad\quad\quad\quad |9| < 7$

$\quad\quad\quad\quad\quad\quad 9 < 7$ (False)

Therefore 6 is not an element of the solution set of $|2x - 3| < 7$.

1. 0   **2.** -1 **3.** 1   **4.** -3 **5.** 0   **6.** 4   **7.** 5   **8.** 10

Solve the following absolute value inequalities. For exercises 9–18, leave the answer in set-builder notation, interval notation, and graphical notation. For exercises 19–47, leave the answer in both set-builder and interval notation. See examples 2–6 A and B.

Example $|5x + 4| - 6 \geq -3$

Solution $|5x + 4| \geq 3$ **Isolate** the absolute value by adding 6

$\quad 5x + 4 \leq -3$ or $5x + 4 \geq 3$ Write the equivalent inequalities

$\quad\quad\quad 5x \leq -7$ $5x \geq -1$ Subtract 4 from all members

$\quad\quad\quad\quad x \leq -\dfrac{7}{5}$ $x \geq -\dfrac{1}{5}$ Divide all members by 5

The solution set is $\left\{ x \mid x \leq -\dfrac{7}{5} \text{ or } x \geq -\dfrac{1}{5} \right\}$, or $\left(-\infty, -\dfrac{7}{5} \right] \cup \left[-\dfrac{1}{5}, +\infty \right)$, or

Graphical notation

9. $|x| < 4$ **10.** $|x| \leq 3$ **11.** $|x| \geq 2$

12. $|x| > 4$ **13.** $|x| - 2 \leq 1$ **14.** $|x| - 1 < 2$

15. $|x| + 3 > 8$ **16.** $|x| + 5 \geq 7$ **17.** $|3x - 4| < 6$

18. $|2x + 5| \leq 4$ **19.** $|5x - 3| \geq 7$ **20.** $|4x - 5| > 9$

21. $|3x - 4| \leq 13$ **22.** $|2x + 7| > 11$ **23.** $|5x + 7| < 12$

24. $|1 - 2x| \leq 5$ **25.** $|4 - 3x| \leq 13$ **26.** $|6 - 3x| < 12$

27. $|5 - 2x| < 15$ **28.** $|4x - 9| < 0$ **29.** $|3x + 6| \leq -1$

30. $|5 - 8x| < -3$ **31.** $|4x - 6| \geq -2$ **32.** $|3x + 7| > -9$

33. $|7x - 4| + 5 < 7$ **34.** $|3x - 11| + 6 < 9$ **35.** $|1 - 3x| - 4 \geq 3$

36. $|2 - 7x| - 3 \geq 4$ **37.** $|3x - 4| + 5 < 4$ **38.** $|7x - 8| + 6 \leq 3$

39. $|4x - 9| + 6 \geq 4$ **40.** $4 + |3x + 1| > 6$ **41.** $8 + |5x - 3| > 10$

42. $4 + |3 - 2x| \geq 7$ **43.** $|4.8x - 18.42| > 11.34$ **44.** $|3.2x - 12.2| > 13.4$

45. $|8.2x - 6.15| \leq 10.25$ **46.** $|10.8x - 10.8| \leq 5.4$ **47.** $|2.1x - 6.3| < 8.4$

Write an absolute value equation or inequality for the following statements and solve for the unknown. Leave the solution of inequalities in both set-builder and interval notation.

Example   The absolute value of a number is more than 3.

Solution   If we let x represent the unknown number, then the absolute value inequality would be $|x| > 3$.

$$x < -3 \text{ or } x > 3 \qquad \text{Write the equivalent inequalities}$$

The solution set is $\{x | x < -3 \text{ or } x > 3\}$ or $(-\infty, -3) \cup (3, +\infty)$.

48. The absolute value of a number is equal to 8.

49. The absolute value of a number is equal to 12.

50. The absolute value of twice a number is equal to 10.

51. The absolute value of $\frac{1}{2}$ a number is equal to 7.

52. If 6 is subtracted from three times a number, the absolute value of the result is equal to 11.

53. If 3 is added to four times a number, the absolute value of the result is equal to 19.

54. The absolute value of a number is less than 4.

55. The absolute value of a number is at most 6.

56. The absolute value of a number is at least 8.

57. The absolute value of twice a number is less than 14.

58. If three times a number is decreased by 4, the absolute value of the result is at least 11.

59. If twice a number is increased by 5, the absolute value of the result is more than 15.

60. If $\frac{1}{2}$ of a number is diminished by 6, the absolute value of the result is at most 14.

61. If $\frac{1}{4}$ of a number is decreased by 8, the absolute value of the result is less than 12.

Review exercises

Perform the indicated operations. See section 1–2.

  1. -4^2 **2.** $(-4)^2$ **3.** -2^4 **4.** $(-2)^4$

Write an algebraic expression for each of the following. See section 1–5.

  5. x raised to the fifth power

6. A number cubed

  7. A number squared

8. The product of x and y

Chapter 2 lead-in problem

Earl invests a total of $10,000. He invests part in Collins Feline Fanciers that pays a 9% dividend per year and the rest in Grutz Shipyards that pays an 8% dividend per year. If Earl receives $870 per year from his investments, how much did he invest with each company?

Solution

Let x represent the amount invested at 9%, then $10,000 - x$ will represent the amount invested at 8%. We can use a table to summarize the information in the problem.

	Investment earning 9%	Investment earning 8%	Total
Amount invested	x	$10,000 - x$	10,000
Dividend received	$(0.09)x$	$(0.08)(10,000 - x)$	870

We get the equation for the problem from the bottom row of the table.

amount of dividend at 9%   total   amount of dividend at 8%   was   total dividend
$$(0.09)x \quad + \quad (0.08)(10,000 - x) \quad = \quad 870$$

Solving for x,

$(0.09)x + (0.08)(10,000 - x) = 870$	Equation
$(0.09)x + 800 - (0.08)x = 870$	Distributive property
$(0.01)x + 800 = 870$	Combine like terms
$(0.01)x = 70$	Subtract 800
$x = 7,000$	Divide by 0.01
and $10,000 - x = 10,000 - 7,000 = 3,000$	Substitute to get other investment

Hence the amount invested at 9% in Collins Feline Fanciers (x) was $7,000 and the amount invested at 8% in Grutz Shipyards ($10,000 - x$) was $3,000.

Chapter 2 summary

1. A **mathematical statement** is a sentence that can be labeled true or false.
2. An **equation** is a statement of equality.
3. A replacement value for the variable that forms a true statement (satisfies that equation) is called a **root,** or **solution,** of that equation.
4. The set of all those values for the variable that causes the equation to be a true statement is called the **solution set** of the equation.

5. An equation that is true for some values of the variable and false for other values of the variable is called a **conditional equation.**
6. An equation that is true for every permissible value of the variable is called an **identical equation,** or **identity.**
7. In a **first-degree conditional equation** in one variable, also called a **linear equation,** the exponent of the unknown is 1 and the solution set will contain at most one root.

8. The **addition property of equality** enables us to add to or subtract the same amount from each member of an equation and the result will be an equivalent equation.
9. The **multiplication property of equality** enables us to multiply or divide both members of an equation by the same nonzero number and the result will be an equivalent equation.
10. Whenever we multiply or divide both members of an inequality by a negative number, we must **reverse the direction of the inequality symbol.**

11. In **interval notation** we use a parenthesis, (or), to represent that the endpoint is *not* part of the interval and a bracket, [or], to represent that the endpoint is part of the interval.
12. If $|x| = a$ and $a \geq 0$, then $x = a$ or $x = -a$.
13. If $|x| < a$ and $a > 0$, then $-a < x < a$.
14. If $|x| > a$ and $a > 0$, then $x < -a$ or $x > a$.

Chapter 2 error analysis

1. Solving fractional equations
   Example: $\dfrac{3}{4}x + 2 = \dfrac{1}{2}x$

   $$4 \cdot \dfrac{3}{4}x + 2 = 4 \cdot \dfrac{1}{2}x$$
   $$3x + 2 = 2x$$
   $$x = -2 \qquad \{-2\}$$
   Correct answer: $\{-8\}$
   What error was made? (*see page 55*)

2. Solving literal equations
   Example: Solve $3x + 2y - 1 = 5x + 4y$ for x.
   $$3x + 2y - 1 = 5x + 4y$$
   $$3x - 5x = 4y - 2y + 1$$
   $$-2x = 2y + 1$$
   $$x = -y + 1$$

   Correct answer: $x = -y - \dfrac{1}{2}$

   What error was made? (*see page 60*)

3. Solving an absolute value equation
   Example: The solution set of the equation $|x| = -2$ is $\{2\}$.
   Correct answer: $\emptyset$
   What error was made? (*see page 73*)

4. Solving an absolute value equation
   Example: Find the solution set of
   $$|2x - 1| = 9$$
   $$2x - 1 = 9$$
   $$2x = 10$$
   $$x = 5 \qquad \{5\}$$
   Correct answer: The solution set is $\{-4, 5\}$.
   What error was made? (*see page 73*)

5. Solution sets of linear inequalities
   Example: The graph of $\{x | x \geq 2\}$ is

   What error was made? (*see page 77*)

6. Multiplying an inequality by a negative number
   Example: If $-2 \leq -\dfrac{1}{3}x < 4$, then

   $$-3 \cdot -2 \leq -3 \cdot -\dfrac{1}{3}x < -3 \cdot 4$$

   Correct answer: $-3 \cdot -2 \geq -3 \cdot -\dfrac{1}{3}x > -3 \cdot 4$
   What error was made? (*see page 80*)

7. Solving linear inequalities
   Example: Find the solution set of
   $$5 - 4x \geq 9$$
   $$-4x \geq 4$$
   $$x \geq -1 \qquad \{x | x \geq -1\}$$
   Correct answer: $\{x | x \leq -1\}$
   What error was made? (*see page 82*)

8. Solving absolute value inequalities
   Example: Find the solution set of
   $$|x - 2| < 5$$
   $$x - 2 < 5$$
   $$x < 7$$
   $$\{x | x < 7\}$$
   Correct answer: The solution set is $\{x | -3 < x < 7\}$.
   What error was made? (*see page 87*)

9. Solving absolute value inequalities
   Example: Find the solution set of $|x + 6| \geq 5$.
   If $|x + 6| \geq 5$, then $-5 \leq x + 6 \leq 5$
   $$-11 \leq x \leq -1$$
   $\{x | -11 \leq x \leq -1\}$.
   Correct answer: $\{x | x \leq -11 \text{ or } x \geq -1\}$
   What error was made? (*see page 89*)

10. Removing grouping symbols
    Example:
    $$(3a^2 + 2a - 1) + [4a^2 - (a + 1)]$$
    $$= 3a^2 + 2a - 1 + 4a^2 - a + 1$$
    $$= 7a^2 + a$$
    Correct answer:
    $$(3a^2 + 2a - 1) + [4a^2 - (a + 1)] = 7a^2 + a - 2$$
    What error was made? (*see page 42*)

Chapter 2 critical thinking

Add any two consecutive integers. Add 7 to that sum and divide the result by 2. If you subtract the original number from this quotient, the result will always be 4. Why is this true?

Chapter 2 review

[2–1]

Find the solution set of the following linear equations.

1. $4x = 32$

2. $x + 11 = 17$

3. $\dfrac{a}{7} = 4$

4. $\dfrac{5b}{2} = 10$

5. $7a - 2 = 13$

6. $2(3z - 4) = 12$

7. $4(3 - 2x) + 3 = 4x - 5$

8. $4(3a + 2) - 2a = 5(a - 1)$

9. $7.8a - 16.9 = 4.3a + 14.6$

[2–2]

Solve the following formulas or literal equations for the specified variable. Assume that no denominator is equal to zero.

10. $v = \ell wh;\ w$

11. $v = k + gt;\ t$

12. $D = dq + R;\ d$

13. $v = r^2(a - b);\ b$

14. $2s = 2vt - gt^2;\ v$

15. $\ell = a + (n - 1)d;\ n$

16. $3x - y = 5x - 4y;\ x$

[2–3]

Write an equation for the problem and solve for the unknown quantities.

17. If three times a number is increased by 15 and the answer is 51, what is the number?

18. One-third of a number is 6 less than one-half of the number. Find the number.

19. The sum of three numbers is 27. The second number is three times the first and the third number is 6 less than the first. Find the three numbers.

20. The length of a rectangle is 8 feet more than twice the width. The perimeter is 82 feet. Find the dimensions.

21. Mary Ann has $24,000, part of which she invests at 10% interest and the rest at 8%. If her income for one year from the two investments was $2,220, how much did she invest at each rate?

22. A solution that is 42% hydrochloric acid is to be mixed with a solution that is 12% hydrochloric acid to obtain 100 centiliters of solution that is 24% hydrochloric acid. How many centiliters of each solution should be used in the mixture?

[2–4]

Find the solution set of the following absolute value equations.

23. $|x| - 4 = 11$

24. $|3a + 5| = 12$

25. $|7 - 2x| = 10$

26. $|4c - 6| + 12 = 18$

27. $|3a + 4| = |2a - 3|$

28. $|4y + 6| = |2y - 5|$

[2–5]

Find the solution set of the following linear inequalities. Leave the answer in both set-builder notation and interval notation.

29. $5x \leq 30$

30. $\dfrac{3}{4}x > 12$

31. $-2x < 9$

32. $2(3x - 4) \leq 1 - 2x$

33. $10 - 2(3x - 4) > 9 - 12x$

34. $5(2x - 3) \leq 7(x + 1) + 3$

35. $-4 < 2x + 3 < 5$

36. $0 \leq 5x + 4 \leq 4$

37. $5 < 3 - x < 8$

38. $-6 \leq 4 - 3x < 2$

39. $6 < 6 - 4x \leq 10$

Write an inequality using the given information and solve.

40. When 5 is subtracted from four times a number, the result is at least 19. Find all numbers that satisfy this condition.

41. Three times a number plus 7 is greater than 22 but less than 34. Find all numbers that satisfy these conditions.

Solve the following absolute value inequalities. Leave the answer in both set-builder and interval notation.

42. $|x| - 4 \geq 6$

43. $|2x + 5| < 6$

44. $|5x - 1| \leq 7$

45. $|4x + 7| > 9$

46. $|1 - 3x| \geq 5$

47. $|3 - 4x| < 12$

48. $|5x + 1| > -3$

49. $6 + |1 - 2x| < 10$

50. $|4x + 5| - 5 \geq 8$

51. $|6x + 3| < -4$

Chapter 2 cumulative test

Determine if the following statements are true or false.

[1–1] **1.** $\frac{1}{2} \in J$

[1–1] **2.** $5 \subseteq R$

[1–1] **3.** $0 \in W$

[1–1] **4.** $Q \cup H = R$

[1–1] **5.** $|-4| < |-6|$

[1–1] **6.** $\{4\} \subseteq \{1,2,3,4\}$

[1–1] **7.** $Q \cap H = \emptyset$

Perform the indicated operations if possible and simplify.

[1–2] **8.** $(-4)(+3)(-2)$

[1–2] **9.** $(-8) - (-7)$

[1–2] **10.** $(-6)(-2)$

[1–2] **11.** $\dfrac{(-5) + 5}{(-2)}$

[1–2] **12.** $25 - (5 - 11) + 6$

[1–2] **13.** $\dfrac{(-4)(-9)}{(2)(-3)}$

[1–2] **14.** -7^2

Identify which property of real numbers is being used.

[1–3] **15.** $a(b + c) = (b + c)a$

[1–3] **16.** $(xy)z = z(xy)$

Form the following sets.

[1–1] **17.** $\{1,2,3\} \cup \{2,3,4,5\}$

[1–1] **18.** $\{8,10,11\} \cup \{4,6,9\}$

[1–1] **19.** $\{10,11,12,13\} \cap \{10,12\}$

Solve the following literal equations for the specified variable and find the solution set for the linear equations and linear inequalities. Assume that no denominator is equal to zero.

[2–1] **20.** $6x + 5 = 2x + 12$

[2–2] **21.** $M = -P(\ell - x); P$

[2–2] **22.** $P = n(P_2 - P_1) - c; P_2$

[2–5] **23.** $-3x \leq 15$

[2–6] **24.** $|4x + 6| \geq 5$

[2–1] **25.** $5(2 - 3x) + 4 = 2x - 7$

[2–5] **26.** $7(3 - 4x) + 6 \leq 6(2 - 4x)$

[2–4] **27.** $|2a + 3| = |3a + 2|$

[2–1] **28.** $6(3a - 5) - 4a = 7(a - 2)$

[2–1] **29.** $3(2x - 4) + 6 = 6x + 8$

[2–6] **30.** $|4 - 3x| \leq 7$

[2–3] **31.** The sum of three numbers is 72. The first number is twice the second number and the third number is three times the first number. Find the three numbers.

[2–3] **32.** Harold has $40,000, part of which he invests at 11% and the rest at 8%. If his income for one year from the 11% investment is $1,740 more than that from the 8% investment, how much did he invest at each rate?

3

Exponents and Polynomials

The amount A of a radioactive substance remaining after time t can be found using the formula $A = A_0(0.5)^{t/n}$, where A_0 represents the original amount of radioactive material and n is the half-life given in the same units of time as t. If the half-life of radioactive carbon 14 is 5,770 years, how much radioactive carbon 14 will remain after 11,540 years if we start with 100 grams?

3–1 ■ Properties of exponents

Exponential form

In chapter 1, we discussed exponents as related to real numbers. Since variables are symbols for real numbers, we shall now apply the properties of exponents to them. The expression a^3 (read "a to the third power") is called the **exponential form** of the product

$$a \cdot a \cdot a$$

We call a the **base** and 3 the **exponent**.

Exponent

Exponential form ⟶ $4^3 = 4 \cdot 4 \cdot 4 = 64$ Standard form

Base   3 factors of 4

┌─ **Definition of exponents** ──────────

$$a^n = \underbrace{a \cdot a \cdot a \cdots a}_{n \text{ factors of } a}$$

where n is a positive integer.

Concept

The exponent tells us how many times the base is used as a factor in an indicated product.

Note An exponent acts only on the symbol immediately to its left. That is, in xy^3, the exponent 3 applies only to the base y, whereas $(xy)^3$ would mean the exponent applies to both the x and the y.

■ *Example 3–1 A*

Write each expression in exponential form and identify the base and the exponent.

1. $x \cdot x \cdot x \cdot x = x^4$ Base x, exponent 4

2. $4 \cdot 4 \cdot 4 = 4^3$ Base 4, exponent 3

3. $(x^2 + y)(x^2 + y)(x^2 + y) = (x^2 + y)^3$ Base $(x^2 + y)$, exponent 3

4. $(-3) \cdot (-3) \cdot (-3) \cdot (-3) = (-3)^4$ Base -3, exponent 4

5. $-(3 \cdot 3 \cdot 3 \cdot 3) = -3^4$ Base 3, exponent 4

Note In examples 4 and 5, we review the ideas of exponents related to signed numbers. Recall that $(-3)^4 = 81$, whereas $-3^4 = -81$.

▶ *Quick check* Write $(-5) \cdot (-5) \cdot (-5)$ in exponential form and identify the base and the exponent.   ■

Product property of exponents

Consider the indicated product of $a^2 \cdot a^3$. If we rewrite a^2 and a^3 by using the definition of exponents, we have

$$a^2 \cdot a^3 = \overbrace{a \cdot a}^{a^2} \cdot \overbrace{a \cdot a \cdot a}^{a^3}$$

and again using the definition of exponents, this becomes

$$a^2 \cdot a^3 = \overbrace{a \cdot a \cdot a \cdot a \cdot a}^{\text{5 factors of } a} = a^5$$

This leads us to the observation that

$$\underset{\substack{\uparrow \\ \text{Multiply} \\ \text{like bases}}}{a^2 \cdot a^3} = a^{2+3} = \underset{\substack{\uparrow \\ \text{Base remains} \\ \text{unchanged}}}{a^5}$$

Add exponents

_____ **Product property of exponents** _____

For all real numbers a and positive integers m and n,

$$a^m \cdot a^n = a^{m+n}$$

Concept

When multiplying factors having **like bases,** add the exponents to get the exponent of the common base.

■ *Example 3–1 B*

Find each of the following products.

1. $y^4 \cdot y^5 = y^{4+5} = y^9$

2. $3^2 \cdot 3^4 = 3^{2+4} = 3^6 = 729$

Note   When we multiply expressions of the same base, we add the exponents; we do not multiply the bases.

$$3^2 \cdot 3^4 \neq 9^6$$

3. $x \cdot x^7 \cdot x^3 = x^{1+7+3} = x^{11}$

Note   If there is no visible exponent associated with a numeral or a variable, the exponent is understood to be 1.

4. $x^{3n} \cdot x^{2n} = x^{3n+2n}$ Multiply like bases by adding the exponents

$\qquad\qquad = x^{5n}$

▶ *Quick check*   Multiply $a^2 \cdot a \cdot a^5$ ■

Power of a power property of exponents

A second property of exponents can be derived by applying the definition of exponents and the product property of exponents. Consider the expression $(a^4)^3$.

$$(a^4)^3 = \underbrace{a^4\, a^4\, a^4}_{\substack{\text{3 factors}\\ \text{of } a^4}} = a^{\overbrace{4+4+4}^{\substack{\text{Adding}\\ \text{exponent 4}\\ \text{three times}}}} = a^{12}$$

From arithmetic, we know that multiplication is repeated addition of the same number. Therefore adding the exponent 4 three times is the same as $3 \cdot 4$. Thus

$$(a^4)^3 = a^{4 \,\cdot\, 3} = a^{12}$$

Power of a power ↓   Multiply exponents ↓

─── *Power of a power property of exponents* ───────

For all real numbers a and positive integers m and n,

$$(a^m)^n = a^{mn}$$

Concept
A power of a power is found by multiplying the exponents.

■ *Example 3–1 C*

Perform the indicated operations.

1. $(x^3)^5 = x^{3\,\cdot\,5} = x^{15}$ **2.** $(2^3)^4 = 2^{3\,\cdot\,4} = 2^{12} = 4{,}096$

3. $(a^{2n})^{3n} = a^{2n\,\cdot\,3n}$ Power of a power, multiply the exponents

$\qquad\qquad = a^{6n^2}$ ■

Group of factors to a power property of exponents

A third property of exponents can be derived using the definition of exponents and the commutative and associative properties of multiplication. Observe that

$$(ab)^3 = \underbrace{ab \cdot ab \cdot ab}_{\text{3 factors of } ab}$$

$$= \underbrace{a \cdot a \cdot a}_{\substack{\text{3 factors} \\ \text{of } a}} \cdot \underbrace{b \cdot b \cdot b}_{\substack{\text{3 factors} \\ \text{of } b}}$$

$$= a^3 b^3$$

Group of factors to a power property of exponents

For all real numbers a and b and positive integers n,

$$(ab)^n = a^n b^n$$

Concept

When a group of factors is raised to a power, we will raise each of the factors in the group to this power.

■ *Example 3–1 D*

Perform the indicated operations.

1. $(xy)^5 = x^5 y^5$

Group of factors   Raise each factor   Standard form
to a power   to the power

2. $(2ab)^4 = 2^4 a^4 b^4 = 16a^4 b^4$

Note A common error is to forget to raise the numerical coefficient to the appropriate power. In example 2, 2 is raised to the fourth power.

3. $(2 + 3)^3 = (5)^3 = 125$

Note The quantity $(2 + 3)^3 \neq 2^3 + 3^3$ because 2 and 3 are *terms*, not factors, as the property requires. If we consider $(2 + 3)$ to be a single number, then by the definition of exponents we have

$$(2 + 3)^3 = (2 + 3)(2 + 3)(2 + 3)$$

or, in general,

$$(a + b)^3 = (a + b)(a + b)(a + b)$$

We will discuss the method of multiplying this product in section 3–3. ■

The following examples illustrate some problems in which more than one property is used within the problem.

■ *Example 3–1 E*

Perform the indicated operations.

1. $(2x^3y^4)^3$

$\quad = 2^3(x^3)^3(y^4)^3$

Each factor in the group is raised to the third power

$\quad = 8x^9y^{12}$

$2^3 = 8$ and power of a power, multiply exponents

2. $(3a^4b^6)^2$

$\quad = 3^2(a^4)^2(b^6)^2$

Each factor in the group is raised to the second power

$\quad = 9a^8b^{12}$

$3^2 = 9$ and power of a power, multiply exponents

3. $(-3a^2)(2ab^3)(-4a^3b^5)$

$\quad = [(-3)(2)(-4)]\,(a^2aa^3)(b^3b^5)$

Multiply like bases using the commutative and associative properties

$\quad = 24a^6b^8$

4. $(3x^2y^5)^2(-2x^4y)^3$

$\quad = 3^2(x^2)^2(y^5)^2(-2)^3(x^4)^3y^3$

Each factor in each group is raised to the power outside

$\quad = 9x^4y^{10} \cdot (-8)x^{12}y^3$

Standard form for numbers, power of a power for variables

$\quad = [9 \cdot (-8)](x^4x^{12})(y^{10}y^3)$

Multiply like bases

$\quad = -72x^{16}y^{13}$

5. $(3x^2)^2x^3 + (2x^3)^2x$

$\quad = 3^2(x^2)^2x^3 + 2^2(x^3)^2x$

Group of factors to a power

$\quad = 9x^4x^3 + 4x^6x$

Power of a power

$\quad = 9x^7 + 4x^7$

Multiplication of like bases, add exponents

$\quad = 13x^7$

Add like terms

6. $(5a^3)^2a^2 + (3a^4)^2a$

$\quad = 5^2(a^3)^2a^2 + 3^2(a^4)^2a$

Group of factors to a power

$\quad = 25a^6a^2 + 9a^8a$

Power of a power

$\quad = 25a^8 + 9a^9$

Multiplication of like bases, add exponents

Note   In example 5, we were able to carry out the addition because we had like terms. In example 6, the addition was not performed because a^8 and a^9 are not like terms.

▶ *Quick check*   Perform the indicated operations. $(5x^4y)^2$ ■

┌─ **Mastery points** ─────────────────────────────────────

   Can you
   ■ Write a product in exponential form?
   ■ Use the product property of exponents?
   ■ Raise a power to a power?
   ■ Raise a group of factors to a power?

Exercise 3–1

Write each expression in exponential form and identify the base and the exponent. See example 3–1 A.

Example $(-5) \cdot (-5) \cdot (-5)$

Solution $= (-5)^3$ Base -5, exponent 3

1. $(-2)(-2)(-2)(-2)$ 2. $5 \cdot 5 \cdot 5 \cdot 5 \cdot 5 \cdot 5$ 3. $x \cdot x \cdot x \cdot x \cdot x$
4. $y \cdot y \cdot y$ 5. $(2x)(2x)(2x)(2x)$ 6. $(ab^2)(ab^2)(ab^2)$
7. $(x^2 + 3y)(x^2 + 3y)(x^2 + 3y)$ 8. $(2a^2 - b)(2a^2 - b)$ 9. $-(2 \cdot 2)$
10. $-(2 \cdot 2 \cdot 2 \cdot 2)$

Perform the indicated operations. See examples 3–1 B, C, D, and E.

Examples $a^2 \cdot a \cdot a^5$ $(5x^4y)^2$

Solutions $= a^{2+1+5}$ Multiply like bases $= 5^2(x^4)^2y^2$ Each factor in the group is raised to the second power
$= a^8$ Add exponents $= 25x^8y^2$ $5^2 = 25$ and power of a power, multiply exponents

11. $a^5 \cdot a^4$ 12. $x^3 \cdot x^9$ 13. $y \cdot y^2$ 14. $b \cdot b^4$ 15. $(-2)^3(-2)^3$
16. $(-3)(-3)^3$ 17. $(-2)(-2)^3$ 18. $(-2)(-2^2)$ 19. $(-2^2)(3^2)$ 20. $(-3^2)(2^2)$
21. $x^2 \cdot x \cdot x^5$ 22. $y^4 \cdot y^2 \cdot y$ 23. $(x^2y^2)(x^5y^3)$ 24. $(a^5b)(a^2b^3)$ 25. $(2ab^2)(3a^2)$
26. $(-2a^2)(4a^5)$ 27. $(-3x^3)(-2x^2)$ 28. $(5ab^2)(2a^5b^4)$ 29. $(8x^2y^5)(3xy^4)$ 30. $(-6a^3b^7)(4ab^4)$
31. $(2^2)^3$ 32. $(-2^2)^3$ 33. $(-3^2)^3$ 34. $(-3^3)^2$ 35. $(-2^3)^2$
36. $(x^4)^7$ 37. $(a^2)^6$ 38. $(a^3b^6)^4$ 39. $(x^2yz^3)^5$ 40. $(5R^2S^5)^2$
41. $(-7s^4t^2)^2$ 42. $(-3a^9b^7)^3$ 43. $(-x^9y^{12}z^8)^4$ 44. $(3x^2y)^2(2xy^3)$ 45. $(a^2b^3)^4(ab^5)$
46. $(2a^3b^2)(a^5b^3)^4$ 47. $(x^2y^6)^2(x^3y^4)^3$ 48. $(-a^2b)(a^5b^2)^3$
49. $(x^4y^5)^2(-x^2y^8)$ 50. $(-2a^2b^4)^3(-3ab^5)^2$ 51. $(-5x^4y^2)^2(-3x^2y^7)$
52. $(2a^2)^2a^3 + (3a)^3a^4$ 53. $(3x^3)^2x^3 + 2x^5(2x^2)^2$ 54. $(5b^2)^22b^7 - (3b^4)^25b^3$
55. $(4a^5)^22a^3 - (3a^4)^34a$ 56. $(-2x^2)^33x^4 + (5x^5)^2$ 57. $(3x^5)^2 - (2x^2)^3x^2$
58. $(x^4)^2x^3 + (x^3)^3x$ 59. $(2a^3)^33a^4 + (3a^5)^22a$ 60. $(b^5)^34b^3 - (2b^3)^4b^2$
61. $(x^4)^33x^5 + (2x^2)^53x^4$ 62. $x^{5n} \cdot x^{4n}$ 63. $a^{2b} \cdot a^{7b}$
64. $x^{6n} \cdot x^n$ 65. $a^{5b} \cdot a^{4b}$ 66. $x^{2n+1} \cdot x^{n+4}$ 67. $a^{2b+5} \cdot a^{3b-2}$ 68. $R^{3S} \cdot R^{2S+3}$
69. $x^{y-4} \cdot x^{2y+9}$ 70. $(a^{3n})^{4n}$ 71. $(x^{3y})^{5y}$ 72. $(R^{2S})^S$

Solve the following word problems.

73. The amount A accumulated in a savings account earning 12% interest per year compounded monthly is given by $A = P(1.01)^n$, where P represents the amount of deposit and n is the number of months the money is left on deposit. Find A, if $P = 5,000$ and $n = 12$.

74. Find A in exercise 73, if $P = 2,000$ and $n = 6$.

75. The amount A of a radioactive substance remaining after time t can be found using the formula

$A = A_0(0.5)^{t/n}$, where A_0 represents the original amount of radioactive material and n is the half-life given in the same units of time as t. If the half-life of nitrogen 13 is 10 minutes, how much radioactive nitrogen will remain after 20 minutes if we start with 10 grams?

76. If the half-life of uranium 229 is 58 minutes, how much radioactive uranium will remain after 174 minutes if we start with 80 ounces? (Refer to exercise 75.)

Review exercises

Perform the indicated multiplication. See section 1–2.

1. $4 \cdot (-6)$ **2.** -3^2 **3.** $4 \cdot 6 \cdot 0 \cdot 3$ **4.** $(-5)^2$

Perform the indicated addition and subtraction. See section 1–6.

5. $6ab + 3ab$ **6.** $a^2 - 5a + 3a - 15$ **7.** $x^2 - 3x + 3x - 9$ **8.** $3x^2 - y^2 - 2x^2 + 3y^2$

3–2 ■ Products of polynomials

Extended distributive property

In chapter 1, we stated the distributive property as

$$a(b + c) = ab + ac$$

Many problems have more than two terms inside the grouping symbol. We will now extend the distributive property to more than two terms and will use subscripts to state the *extended distributive property of multiplication over addition.*

Extended distributive property

$$a(b_1 + b_2 + \cdots + b_n) = ab_1 + ab_2 + \cdots + ab_n$$

Concept
When we multiply a multinomial by a monomial, we multiply each term of the multinomial by the monomial.

■ **Example 3–2 A**

Perform the indicated multiplication.

1. $2x^3(3x^2 - 5x + 7)$
We multiply the monomial $2x^3$ times each term in the trinomial to get

$$(2x^3)(3x^2) + (2x^3)(-5x) + (2x^3)(7)$$

In each indicated product, we multiply the coefficient and add the exponents of the like bases to get

$$2x^3(3x^2 - 5x + 7) = 6x^5 - 10x^4 + 14x^3$$

2. $4a^2b^3(2a^2 - 3ab + 4b^2) = (4a^2b^3)(2a^2) + (4a^2b^3)(-3ab) + (4a^2b^3)(4b^2)$
$$= 8a^4b^3 - 12a^3b^4 + 16a^2b^5$$

▶ *Quick check*   Perform the indicated multiplication. $3b^2(2b^5 - b + 4)$ ■

Multiplication of multinomials

The product of two binomials will require the use of the distributive property several times. That is, in the product

$$(x + y)(2x + y)$$

we consider $(x + y)$ as a single factor and apply the distributive property.

$$(x + y)(2x + y) = (x + y) \cdot 2x + (x + y) \cdot y$$

We now apply the distributive property again.

$$(x + y) \cdot 2x + (x + y) \cdot y = x \cdot 2x + y \cdot 2x + x \cdot y + y \cdot y$$
$$= 2x^2 + 2xy + xy + y^2$$

The last step in the problem is to combine like terms, if there are any.

$$2x^2 + (2xy + xy) + y^2 = 2x^2 + 3xy + y^2$$

Notice that in this product each term of the first factor is multiplied by each term of the second factor. We can generalize our procedure as follows:

Multiplying two multinomials _____

When multiplying two multinomials, we multiply each term of the first multinomial by each term of the second multinomial and then combine like terms.

■ **Example 3–2 B**

Perform the indicated multiplication and simplify.

1. $(a + 4)(a + 1)$

When we perform the multiplication of two binomials, the four products that are obtained can be seen more clearly by using arrows to indicate the multiplication that is being carried out.

$$(a + 4)(a + 1) = a \cdot a + a \cdot 1 + 4 \cdot a + 4 \cdot 1 \qquad \text{Distribute multiplication}$$

$$= a^2 + a + 4a + 4 \qquad \text{Multiply the monomials}$$
$$= a^2 + 5a + 4 \qquad \text{Combine like terms}$$

Note We have drawn arrows to indicate the multiplication that is being carried out. This is a convenient way for us to indicate the multiplication to be performed.

2. $(2x + 3)(5x - 2) = 10x^2 - 4x + 15x - 6 \qquad \text{Distribute and multiply}$

$$= 10x^2 + 11x - 6 \qquad \text{Combine like terms}$$

Note A word that is useful for remembering the multiplication to be performed when multiplying two binomials is **FOIL**. Foil is an abbreviation signifying **F**irst times first, **O**uter times outer, **I**nner times inner, and **L**ast times last.

3. $(3a - 2b)(2a - 5b) = 6a^2 - 15ab - 4ab + 10b^2 \qquad \text{Distribute and multiply}$

$$= 6a^2 - 19ab + 10b^2 \qquad \text{Combine like terms}$$

▶ *Quick check*   Perform the indicated multiplication and simplify.
$(3x - 1)(2x + 3)$ ■

In arithmetic, we multiply numbers stated vertically. We can use this same procedure to multiply two multinomials. To perform the example 3 multiplication vertically, we would proceed as follows:

$$2a - 5b$$
$$3a - 2b$$

Multiply $-2b$ and $2a - 5b$.

$$2a - 5b$$
$$3a - 2b$$
$$-4ab + 10b^2$$

Multiply $3a$ and $2a - 5b$. Line up any like terms in the same columns.

$$2a - 5b$$
$$3a - 2b$$
$$-4ab + 10b^2$$
$$6a^2 - 15ab$$

Add like terms.

$$2a - 5b$$
$$3a - 2b$$
$$-4ab + 10b^2$$
$$6a^2 - 15ab$$
$$\overline{6a^2 - 19ab + 10b^2}$$

Special products

Three special products appear so often that we should be able to write the answer without computation. Consider the product

$$(a + b)^2 = (a + b)(a + b)$$

which becomes

$$a^2 + ab + ab + b^2$$

and combining like terms, we get

$$a^2 + 2ab + b^2$$

This is called the **square of a binomial** and has certain characteristics. Inspection shows us that in

$$(a + b)^2 = a^2 + 2ab + b^2$$

the three terms of the product can be obtained in the following manner:

─ *Square of a binomial* ────────────────

1. The first term of the product is the *square of the first term* of the binomial $[(a)^2 = a^2]$.
2. The second term of the product is *two times the product of the two terms of the binomial* $[2(a \cdot b) = 2ab]$.
3. The third term of the product is the *square of the second term* of the binomial $[(b)^2 = b^2]$.

Our second special product is

$$(a - b)^2 = (a - b)(a - b)$$

which becomes

$$a^2 - ab - ab + b^2$$

and simplifies to

$$a^2 - 2ab + b^2$$

This also is called the square of a binomial. We can apply the previous procedure to this special product if we take the sign of the operation ($+$ or $-$) as the sign of the coefficient.

$$(a - b)^2 = [a + (-b)]^2 = (a)^2 + 2[a \cdot (-b)] + (-b)^2$$
$$= a^2 - 2ab + b^2$$

─ *Square of a binomial* ────────────────

For real numbers a and b,

$$(a + b)^2 = a^2 + 2ab + b^2 \quad \text{and}$$
$$(a - b)^2 = a^2 - 2ab + b^2$$

Concept

$$(\text{1st term} + \text{2nd term})^2 =$$
$$(\text{1st term})^2 + 2(\text{1st term} \cdot \text{2nd term}) + (\text{2nd term})^2$$

■ *Example 3–2 C*

Perform the indicated multiplication and simplify.

1. $(x + 3)^2$

   The first term of the binomial is x and the second term of the binomial is 3. Substituting into the special product property, we have

   $$(x + 3)^2 = (x)^2 + 2(x \cdot 3) + (3)^2 \qquad \text{Special product property}$$

   $ \uparrow \qquad \uparrow \ \ \uparrow \qquad \ \uparrow$
   $ \text{1st} \quad \text{1st 2nd} \quad \text{2nd}$
   $ \text{term} \quad \text{term term} \quad \text{term}$

   $$= x^2 + 6x + 9 \qquad \text{Multiply monomials}$$

   Note $(x + 3)^2 = x^2 + 6x + 9$, *not* $x^2 + 9$. This is a common error.

2. $(2a + b)^2 = (2a)^2 + 2(2a \cdot b) + (b)^2$ Special product property
   $$= 4a^2 + 4ab + b^2 \qquad \text{Multiply monomials}$$

3. $(2x - y)^2$

The first term of the binomial is $2x$ and the second term of the binomial is $-y$. Substituting into the formula, we have

$$(2x - y)^2 = (2x)^2 + 2[(2x)(-y)] + (-y)^2 \qquad \text{Special product property}$$

↑   ↑   ↑   ↑

1st   1st   2nd   2nd
term   term   term   term

$$= 4x^2 - 4xy + y^2 \qquad \text{Multiply monomials}$$

4. $(4a - 3b)^2 = (4a)^2 + 2[(4a)(-3b)] + (-3b)^2$ Special product property
$$= 16a^2 - 24ab + 9b^2 \qquad \text{Multiply monomials}$$

▶ *Quick check*   Perform the indicated multiplication and simplify. $(x + 3y)^2$ ■

The third special product is obtained by multiplying the sum and difference of the same two terms. Consider the product

$$(a + b)(a - b) = a^2 - ab + ab - b^2$$
$$= a^2 - b^2$$

This special product is called the **difference of two squares.**

Difference of two squares

In general, for real numbers a and b,

$$(a + b)(a - b) = a^2 - b^2$$

Concept
The product is obtained by first squaring the first term of the factors and then subtracting the square of the second term of the factors.

$$(\text{1st term} + \text{2nd term})(\text{1st term} - \text{2nd term}) =$$
$$(\text{1st term})^2 - (\text{2nd term})^2$$

Note   We can consider b or $-b$ the second term since the square of each is b^2. That is,

$$(b)^2 = (-b)^2 = b^2$$

■ *Example 3–2 D*

Perform the indicated multiplication and simplify.

1. $(x + 5)(x - 5)$

The first term is x and the second term is 5. Substituting into the special product property, we have

$$(x + 5)(x - 5) = (x)^2 - (5)^2 \qquad \text{Special product property}$$

↑   ↑

1st   2nd
term   term

$$= x^2 - 25 \qquad \text{Standard form}$$

2. $(2a + b)(2a - b) = (2a)^2 - (b)^2$ Special product property
$$= 4a^2 - b^2 \qquad \text{Standard form} \qquad ■$$

In all the examples we have considered, whether they were special products or not, a single procedure is sufficient. When multiplying two multinomials, **we multiply each term of the first multinomial by each term of the second multinomial and then combine like terms.**

■ *Example 3–2 E*

Perform the indicated operations and simplify.

1. $(a + 3)(2a^2 - a + 4)$

$$
\begin{aligned}
&= (a)(2a^2) + (a)(-a) + (a)(4) + (3)(2a^2) \qquad \text{Distribute multiplication} \\
&\quad + (3)(-a) + (3)(4) \\
&= 2a^3 - a^2 + 4a + 6a^2 - 3a + 12 \qquad \text{Multiply monomials} \\
&= 2a^3 + 5a^2 + a + 12 \qquad \text{Combine like terms}
\end{aligned}
$$

2. $(x + y)(x + 2y)(x - y)$

When there are three quantities to be multiplied, we apply the associative property to multiply two of them together first and take that product times the third.

$$
\begin{aligned}
&[(x + y)(x + 2y)](x - y) \\
&= [x^2 + 2xy + xy + 2y^2](x - y) \qquad \text{Multiply the first two groups} \\
&= [x^2 + 3xy + 2y^2](x - y) \qquad \text{Combine like terms in brackets} \\
&= x^3 - x^2y + 3x^2y - 3xy^2 + 2xy^2 - 2y^3 \qquad \text{Distribute multiplication} \\
&= x^3 + 2x^2y - xy^2 - 2y^3 \qquad \text{Combine like terms}
\end{aligned}
$$

Alternate If we had observed in example 2 that the first and last quantities were a special product, we could have used the commutative and associative properties to multiply them together first.

$$
\begin{aligned}
&[(x + y)(x - y)](x + 2y) \\
&= [x^2 - y^2](x + 2y) \qquad \text{Multiply the first two groups (special product)} \\
&= x^3 + 2x^2y - xy^2 - 2y^3 \qquad \text{Distribute multiplication}
\end{aligned}
$$

3. $(a + 2b)^3$

$$
\begin{aligned}
&= (a + 2b)(a + 2b)(a + 2b) \qquad \text{Write in expanded form} \\
&= [(a + 2b)(a + 2b)](a + 2b) \qquad \text{Multiply the first two groups} \\
&= [(a)^2 + 2(a)(2b) + (2b)^2](a + 2b) \qquad \text{Special product, square of a binomial} \\
&= [a^2 + 4ab + 4b^2](a + 2b) \qquad \text{Simplify within brackets} \\
&= a^3 + 2a^2b + 4a^2b + 8ab^2 + 4ab^2 + 8b^3 \qquad \text{Distribute multiplication} \\
&= a^3 + 6a^2b + 12ab^2 + 8b^3 \qquad \text{Combine like terms}
\end{aligned}
$$

4. $(2a - 5)^2 - (a + 3)(a - 4)$

We must take care of all multiplication before we can perform the indicated subtraction.

$$
\begin{aligned}
&(2a - 5)^2 - (a + 3)(a - 4) \\
&= [(2a)^2 + 2(2a)(-5) + (-5)^2] - [a^2 - 4a + 3a - 12] \\
&= [4a^2 - 20a + 25] - [a^2 - a - 12] \\
&= 4a^2 - 20a + 25 - a^2 + a + 12 \qquad \text{Remove brackets} \\
&= 3a^2 - 19a + 37 \qquad \text{Combine like terms}
\end{aligned}
$$

Note In example 4, it is necessary to place grouping symbols around the products since the order of operations requires that multiplication be done before subtraction. In the second line, if we did not use parentheses around the product, the line would be $4a^2 - 20a + 25 - a^2 - a - 12$, and we would have subtracted only the first term of the product and not the entire product as was indicated.

5. $-2\{3x + 2[x - (3x + 1)]\}$

$= -2\{3x + 2[x - 3x - 1]\}$ Remove parentheses

$= -2\{3x + 2[-2x - 1]\}$ Combine like terms

$= -2\{3x - 4x - 2\}$ Distributive property

$= -2\{-x - 2\}$ Combine like terms

$= 2x + 4$ Distributive property ■

___ *Mastery points* ___

Can you
- Multiply a monomial by a multinomial?
- Multiply two multinomials?
- Multiply two multinomials vertically?
- Find the square of a binomial and the difference of two squares using the special products?

Exercise 3–2

Perform the indicated multiplication and simplify. See example 3–2 A.

Example $3b^2(2b^5 - b + 4)$

Solution $= (3b^2)(2b^5) + (3b^2)(-b) + (3b^2)(4)$ Distribute $3b^2$ times each term in parentheses

$= 6b^7 - 3b^3 + 12b^2$ Multiply monomials

1. $a(2a^2 - 3a + 4)$ **2.** $2x(5x^2 - 3x + 7)$ **3.** $-2y(3y^2 - 5y + 4)$

4. $-4x(2x^2 - 3x - 9)$ **5.** $3a^2(4a^2 - 2ab + 3b^2)$ **6.** $5x^3(2x^2 - 3x + 4)$

7. $6xy(5x^2y - 4xy^3 + 2xy)$ **8.** $-a^3b(3a^2b^5 - ab^4 - 7a^2b)$ **9.** $-5x^3y^4(2x^2y + 7xy^6 - 3x^2y^5)$

10. $6a^3b^5(-4a^2b + 5a^3b^3 - 8a^5b + 2)$

Perform the indicated multiplication and simplify (a) by multiplying horizontally and (b) by multiplying vertically. See example 3–2 B.

Example $(3x - 1)(2x + 3)$

Solution $= 6x^2 + 9x - 2x - 3$ Distribute multiplication

$= 6x^2 + 7x - 3$ Combine like terms

11. $(a + 5)(a + 3)$ **12.** $(x + 4)(x + 6)$ **13.** $(b + 4)(b - 5)$ **14.** $(x + 7)(x - 8)$

15. $(2x - y)(x + y)$ **16.** $(3a + b)(a + b)$ **17.** $(5x - y)(2x + y)$ **18.** $(3x + y)(4x - y)$

19. $(7x - 5y)(6x + 4y)$ **20.** $(4a - 7b)(3a - 5b)$

Perform the indicated multiplication and simplify. Use any of the special products where possible. See examples 3–2 C, D, and E.

Example $(x + 3y)^2$

Solution $= (x)^2 + 2[(x)(3y)] + (3y)^2$ Special products, square of a binomial
$= x^2 + 6xy + 9y^2$ Standard form

21. $(x + 3)^2$ **22.** $(2a + b)^2$ **23.** $(3x + y)^2$

24. $(5x + 2y)^2$ **25.** $(4x + 3y)^2$ **26.** $(2x - 3)^2$

27. $(2a - 5)^2$ **28.** $(3a - b)^2$ **29.** $(4x - 3y)^2$

30. $(a - 3b)(a + 3b)$ **31.** $(x - 3y)(x + 3y)$ **32.** $(2x - 3yz)(2x + 3yz)$

33. $(5a + 2bc)(5a - 2bc)$ **34.** $(a + 2b)(a^2 - ab + b^2)$ **35.** $(x + y)(x^2 - 3xy + 2y^2)$

36. $(x - 2y)(3x^2 + 4xy - 2y^2)$ **37.** $(3a - 2b)(5a^2 - 7ab + 3b^2)$ **38.** $(x^2 - 2x + 1)(x^2 + 3x + 2)$

39. $(x^2 + 3x + 4)(x^2 - 2x - 3)$ **40.** $(2a^2 - 3a + 5)(a^2 - 4a + 2)$ **41.** $(5b^2 - 3b + 7)(b^2 + b + 3)$

42. $(a - 3b)^3$ **43.** $(x + 2y)^3$ **44.** $(2x - 3y)^3$

45. $(4a - b)^3$ **46.** $(x - 2y)(2x + y)(x - y)$ **47.** $(a - 2b)(2a - b)(a + 2b)$

Perform the indicated operations, remove all grouping symbols, and simplify. See example 3–2 E.

48. $(x + 2)^2 + (x - 4)^2$ **49.** $(a + 5)^2 + (a - 2)^2$

50. $(2x - 1)(x + 3) - (x + 2)^2$ **51.** $(3x - 2)(x + 5) - (x - 4)^2$

52. $(2x + 3)^2 - (3x + 1)(x - 2)$ **53.** $(y + 3)(2y - 4) - (y + 5)(y - 4)$

54. $(a + 3)(a - 4) - (3a - 2)^2$ **55.** $3[b - (3b + 2) + 5]$

56. $-2[5a - (2a + 3) - 3(3a + 7)]$ **57.** $2x[3x - 2(5x + 1)]$

58. $3y[-2y + 3(5y - 4)]$ **59.** $-2\{3x - 2[5x - (7 - 3x)]\}$

60. $-5x\{2x - 4[x + (3x - 1)]\}$

Perform the indicated multiplication and simplify. Assume that all variables used as exponents represent positive integers.

Examples $x^2(x^n + 1)$ $(x^n - 1)(x^n + 3)$

Solutions $= x^2 \cdot x^n + x^2 \cdot 1$ $= x^n \cdot x^n + x^n \cdot 3 + (-1) \cdot x^n + (-1) \cdot 3$
$= x^{n+2} + x^2$ $= x^{n+n} + 3x^n - x^n - 3$
$= x^{2n} + 2x^n - 3$

61. $a^n(a^2 + 3)$ **62.** $b^{2n}(b^3 - 1)$ **63.** $x^n(3x^n + 1)$ **64.** $x^{n+2}(x^{n+1} + x)$

65. $x^{n-2}(x^{n+5} - x^2)$ **66.** $(a^n + 1)(a^n - 2)$ **67.** $(b^n - 3)(b^n - 2)$ **68.** $(2a^n - b^n)^2$

69. $(3x^n + y^n)^2$ **70.** $(x^{2n} + y^n)(x^{2n} - y^n)$ **71.** $(a^n + b^{3n})(a^n - b^{3n})$ **72.** $(3x^{2n} - 2y^{3n})^2$

Solve the following word problems.

73. The area of the shaded region between two circles is $\pi(R + r)(R - r)$. Perform the indicated multiplication. (π is the lowercase Greek letter pi.)

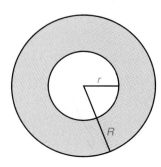

74. When squares of c units on a side are cut from the corners of a square sheet of metal x units on a side and folded up into a tray, its volume is $c(x - 2c)(x - 2c)$ cubic units. Perform the indicated multiplication.

75. The total area of the surface of a cylinder is determined by $A = 2\pi r(h + r)$. Multiply in the right member.

76. The equation for the distance traveled by a rocket fired vertically upward into the air is given by $S = 16t(35 - t)$, where the rocket is S feet from the ground after t seconds. Multiply in the right member.

77. In engineering, the equation for the deflection of a beam is given by

$$Y = \frac{Wx}{48EI}(2x^3 - 3\ell x^2 - \ell^3)$$

Multiply in the right member.

78. In engineering, the equation of transverse shearing stress in a rectangular beam is given in two forms:

(a) $T = \dfrac{V}{8I}(h + 2V_1)(h - 2V_1)$ and

(b) $T = \dfrac{3V}{2A}\left(1 + \dfrac{2V_1}{H}\right)\left(1 - \dfrac{2V_1}{H}\right)$

Multiply in the right member of each equation.

Review exercises

Perform the indicated operations. See section 1–2.

1. $(-4) + (-6)$ **2.** $(-3)(-7)$ **3.** $(-4) - (-8)$ **4.** -4^2

Simplify by using the properties of exponents. See section 3–1.

5. $a^3 \cdot a^5$ **6.** $(x^3)^4$ **7.** $x \cdot x^2 \cdot x^3$ **8.** $(2ab^3)^2$

3–3 ■ *Further properties of exponents*

Quotient property of exponents

Another useful property of exponents can be seen from the following example. Consider the expression

$$\frac{a^5}{a^3}, a \neq 0$$

We use the definition of exponents to write the fraction as

$$\frac{a^5}{a^3} = \frac{a \cdot a \cdot a \cdot a \cdot a}{a \cdot a \cdot a}$$

We reduce the fraction and get

$$\frac{a \cdot a \cdot a \cdot a \cdot a}{a \cdot a \cdot a} = \frac{a \cdot a}{1} = a \cdot a = a^2$$

We reduced by three factors of a, leaving $5 - 3 = 2$ factors of a in the numerator. Therefore

$$\frac{a^5}{a^3} = a^{5-3} = a^2$$

Quotient property of exponents

For all real numbers a, $a \neq 0$, and integers m and n,

$$a^n \div a^m = \frac{a^n}{a^m} = a^{n-m}$$

Concept

To divide quantities having **like bases,** subtract the exponent of the denominator from the exponent of the numerator to get the exponent of the common base in the quotient.

Note If $a = 0$, we have an expression that has no meaning. Therefore $a \neq 0$ indicates that we want our variable to assume no value that would cause the denominator to be zero.

■ *Example 3–3 A*

Perform the indicated operations and simplify. Assume that no variable is equal to zero.

1. $\dfrac{a^9}{a^6} = a^{9-6} = a^3$

2. $b^{12} \div b^4 = b^{12-4} = b^8$

3. $\dfrac{2^8}{2^3} = 2^{8-3} = 2^5 = 32$

   When dividing numbers with like bases raised to a power, the division is carried out by means of subtracting exponents. *The base is unaltered.*

4. $\dfrac{3^3 x^{12} y^9}{3 x^4 y^8} = 3^{3-1} \cdot x^{12-4} \cdot y^{9-8}$ Division of like bases

   $= 3^2 x^8 y^1$ Subtract exponents

   $= 9 x^8 y$ Standard form   ■

Negative exponents

Until now, we have considered only those problems in which the exponent of the numerator is greater than the exponent of the denominator. Consider the example

$$\frac{a^3}{a^5}, a \neq 0$$

By the definition of exponents, this becomes

$$\frac{a^3}{a^5} = \frac{a \cdot a \cdot a}{a \cdot a \cdot a \cdot a \cdot a}$$

and reducing the fraction,

$$\frac{a \cdot a \cdot a}{a \cdot a \cdot a \cdot a \cdot a} = \frac{1}{a \cdot a} = \frac{1}{a^2}$$

Again we reduced by three factors of a, leaving $5 - 3 = 2$ factors of a in the denominator. Hence

$$\frac{a^3}{a^5} = \frac{1}{a^2}$$

However if we use the quotient property of exponents to carry out the division,

$$\frac{a^3}{a^5} = a^{3-5} = a^{-2}$$

Since we should arrive at the same answer regardless of which procedure we use, then a^{-2} must be equivalent to $\frac{1}{a^2}$, that is, $a^{-2} = \frac{1}{a^2}$. This leads us to the definition of a^{-n}.

___ **Definition of negative exponents** _____

For all real numbers a, $a \neq 0$, and positive integers n,

$$a^{-n} = \frac{1}{a^n}$$

Concept
When a symbol is raised to a negative exponent, we can rewrite it as the reciprocal of that symbol to the positive exponent.

■ *Example 3–3 B*

Write the following without negative exponents. Assume that no variable is equal to zero.

1. $3^{-2} = \frac{1}{3^2}$ Rewrite as 1 over 3 to the positive two

    $= \frac{1}{9}$ Standard form

Note In example 1, the value of 3^{-2} cannot be determined until the exponent is made positive.

2. $a^{-4} = \frac{1}{a^4}$

Note From the definition of negative exponents, if a factor is moved either from the numerator to the denominator or from the denominator to the numerator, the sign of its exponent will change. The sign of the base will not be affected by this change.

Alternative procedure:

$a^{-4} = \frac{a^{-4}}{1}$ Rewrite as a fraction

    $= \frac{1}{a^4}$ The sign of the exponent is changed as the factor is moved from the numerator to the denominator

3. $\frac{1}{x^{-4}} = \frac{x^4}{1}$ The sign of the exponent is changed as the factor is moved from the denominator to the numerator

    $= x^4$ Standard form

4. $-b^{-3} = -(b^{-3}) = -\left(\frac{1}{b^3}\right) = -\frac{1}{b^3}$

Note Only the sign of the exponent changes, not the sign of the base. ■

Zero as an exponent

We now examine the situation involving the division of like bases raised to the same power. For example, consider

$$\frac{a^3}{a^3}, \; a \neq 0$$

By the definition of exponents, we have

$$\frac{a^3}{a^3} = \frac{a \cdot a \cdot a}{a \cdot a \cdot a} = \frac{1}{1} = 1$$

and by the quotient property of exponents,

$$\frac{a^3}{a^3} = a^{3-3} = a^0$$

Since

$$\frac{a^3}{a^3} = 1 \text{ and } \frac{a^3}{a^3} = a^0$$

then for consistency, we make the following definition for a^0.

Definition of zero as an exponent

For all real numbers a, $a \neq 0$,

$$a^0 = 1$$

Concept

Any number other than zero raised to the zero power is equal to 1.

Note All the properties of exponents stated so far now apply for any integer used as an exponent.

■ *Example 3–3 C*

Write the following expressions without using zero as an exponent. Assume that no variable is equal to zero.

1. $x^0 = 1$

2. $8^0 = 1$

3. $(-5)^0 = 1$

4. $4x^0 = 4 \cdot x^0$ Only x is raised to the 0 power

        $= 4 \cdot 1$ x^0 is 1

        $= 4$ Standard form

Note The exponent acts only on the symbol immediately to its left. In this example, only x is raised to the zero power. The exponent of 4 is understood to be 1.

5. $-5^0 = -(5^0)$ This is not the same as $(-5)^0$

         $= -(1)$ 5^0 is 1

         $= -1$ Standard form

▶ *Quick check* Write $3b^0$ without using zero as an exponent.

■ *Example 3–3 D*

Perform the indicated operations. Leave the answer with only positive exponents. Assume that no variable is equal to zero.

1. $\dfrac{a^3b^2c^4}{ab^5c^4} = a^{3-1}b^{2-5}c^{4-4} = a^2b^{-3}c^0 = a^2 \cdot \dfrac{1}{b^3} \cdot 1 = \dfrac{a^2}{b^3}$

2. $\dfrac{x^3y^4}{x^5y^9} = x^{3-5} \cdot y^{4-9} = x^{-2}y^{-5} = \dfrac{1}{x^2y^5}$

3. $b^5 \cdot b^{-3} = b^{5+(-3)} = b^2$

4. $\dfrac{x^{-3}y^4}{x^{-7}y^9} = x^{(-3)-(-7)}y^{4-9} = x^4y^{-5} = \dfrac{x^4}{y^5}$

5. $(3a^5)^{-2} = \dfrac{1}{(3a^5)^2} = \dfrac{1}{3^2(a^5)^2} = \dfrac{1}{9a^{10}}$

Alternative solution:

$(3a^5)^{-2} = 3^{-2}(a^5)^{-2} = 3^{-2} \cdot a^{-10} = \dfrac{1}{3^2a^{10}} = \dfrac{1}{9a^{10}}$

In example 5, we see that there can be alternate methods of working problems involving exponents, depending on the order in which the properties are applied.

▶ *Quick check* Perform the indicated operations for $\dfrac{a^{-2}b^3}{a^{-4}b^6}$, leaving the answer with only positive exponents. Assume that no variable is equal to zero. ■

Fraction to a power property of exponents

Our last property of exponents can be derived from the definition of exponents. Consider the expression $\left(\dfrac{a}{b}\right)^3$, $b \ne 0$.

$$\left(\dfrac{a}{b}\right)^3 = \underbrace{\dfrac{a}{b} \cdot \dfrac{a}{b} \cdot \dfrac{a}{b}}_{\text{3 factors of } \frac{a}{b}} = \dfrac{\overbrace{a \cdot a \cdot a}^{\text{3 factors of } a}}{\underbrace{b \cdot b \cdot b}_{\text{3 factors of } b}} = \dfrac{a^3}{b^3}$$

Thus

Fraction raised to a power ⟶ $\left(\dfrac{a}{b}\right)^3 = \dfrac{a^3}{b^3}$ ⟵ Numerator raised to the power ⟵ Denominator raised to the power

---
Fraction to a power property of exponents

For all real numbers a and b, $b \ne 0$, and integers n,

$$\left(\dfrac{a}{b}\right)^n = \dfrac{a^n}{b^n}$$

Concept
Whenever a fraction is raised to a power, we raise the numerator to that power and place it over the denominator raised to that power.

■ Example 3–3 E

Perform the indicated operations and simplify. Leave the answer with only positive exponents. Assume that no variable is equal to zero.

1. $\left(\dfrac{2x}{y}\right)^3 = \dfrac{(2x)^3}{y^3}$

Both numerator and denominator are raised to the third power

$= \dfrac{2^3 x^3}{y^3}$

Each factor in the numerator is raised to the third power

$= \dfrac{8x^3}{y^3}$

2^3 in standard form is 8

2. $\left(\dfrac{x^{-3}y}{z^4}\right)^{-2} = \dfrac{(x^{-3}y)^{-2}}{(z^4)^{-2}}$

Numerator and denominator are raised to the power

$= \dfrac{(x^{-3})^{-2}y^{-2}}{(z^4)^{-2}}$

Numerator has a group of factors to a power

$= \dfrac{x^{(-3)(-2)}y^{-2}}{z^{(4)(-2)}}$

Power of a power

$= \dfrac{x^6 y^{-2}}{z^{-8}}$

Multiply exponents

$= \dfrac{x^6 z^8}{y^2}$

Standard form, factors raised to a negative power are moved to the other side of the fraction bar

3. $\left(\dfrac{3x}{y^3}\right)^2 \left(\dfrac{y^4}{x^3}\right)^3 = \dfrac{(3x)^2}{(y^3)^2} \cdot \dfrac{(y^4)^3}{(x^3)^3}$

Numerators and denominators are raised to the power

$= \dfrac{3^2 x^2}{y^6} \cdot \dfrac{y^{12}}{x^9}$

Group of factors to a power and power of a power

$= \dfrac{9x^2 y^{12}}{x^9 y^6}$

Multiply fractions

$= 9 \cdot x^{2-9} \cdot y^{12-6}$

Division with like bases

$= 9x^{-7} y^6$

Subtract exponents

$= \dfrac{9y^6}{x^7}$

Standard form   ■

Scientific notation

An important use of integer exponents is in scientific areas where we deal with very large or very small numbers. For example, the mass of a hydrogen atom is 0.000 000 000 000 000 000 000 001 67 gram; the mass of an electron is 0.000 000 000 000 000 000 000 000 000 91 gram; the half-life of lead 204 is 14,000,000,000,000,000,000 years.

Working with such numbers becomes quite difficult. Therefore, we convert such numbers into a more manageable form called **scientific notation.** We define the scientific notation form of a positive number Y to be the product

$$Y = a \times 10^n$$

where a is a number greater than or equal to 1 and less than 10, and n is an integer. Use the following steps to achieve this form of the number Y.

Scientific notation

> **Step 1** Move the decimal point to a position immediately following the first nonzero digit in *Y*.
>
> **Step 2** Count the number of places the decimal point has been moved. This is the exponent, *n*, to which 10 is raised.
>
> **Step 3** If
> > a. the decimal point is moved to the *left, n* is *positive*.
> > b. the decimal point is moved to the *right, n* is *negative*.
> > c. the decimal point already follows the first nonzero digit, *n* is *zero*.

■ _Example 3–3 F_

Express the following in scientific notation.

1. $9,000,000 = 9.000000. \times 10^6 = 9 \times 10^6$

2. $0.00000467 = 0.000004.67 \times 10^{-6} = 4.67 \times 10^{-6}$

3. $4.37 = 4.37 \times 10^0$

4. $-0.00341 = -0.003.41 \times 10^{-3} = -3.41 \times 10^{-3}$ ■

Standard form

Sometimes it is necessary to convert a number in scientific notation to its standard form. To do this, we apply the procedure in reverse.

Standard form

> When the exponent in 10^n is
> 1. *positive,* the decimal point is moved to the *right n* places.
> 2. *negative,* the decimal point is moved to the *left n* places.
> 3. *zero,* the decimal point is not moved.

■ _Example 3–3 G_

Express the following in standard form.

1. $6.37 \times 10^4 = 6.3700. = 63,700$

2. $4.81 \times 10^{-5} = 0.00004.81 = 0.0000481$

3. $8.59 \times 10^0 = 8.59$

4. $-3.48 \times 10^{-1} = -0.3.48 = -0.348$ ■

Scientific notation can be used to simplify numerical calculations when the numbers are very large or very small. We first change the numbers to scientific notation and then use the properties of exponents to help perform the indicated operations.

■ *Example 3–3 H*

Perform the indicated operations using scientific notation.

1. $(198,000,000)(0.00347)$

$$= (1.98 \times 10^8)(3.47 \times 10^{-3}) \quad \text{Scientific notation}$$
$$= (1.98 \cdot 3.47) \times (10^8 \cdot 10^{-3}) \quad \text{Commutative and associative properties}$$
$$= 6.8706 \times 10^5 \quad \text{Multiply}$$
$$= 687,060 \quad \text{Standard form}$$

2. $\dfrac{(92,000,000)(0.0036)}{(0.018)(4,000)}$

$$= \frac{(9.2 \times 10^7)(3.6 \times 10^{-3})}{(1.8 \times 10^{-2})(4.0 \times 10^3)} \quad \text{Scientific notation}$$

$$= \frac{(9.2)(3.6)10^7 \cdot 10^{-3}}{(1.8)(4.0)10^{-2} \cdot 10^3} \quad \text{Commutative and associative properties}$$

$$= \frac{(9.2)(3.6)}{(1.8)(4.0)} \cdot \frac{10^4}{10^1} \quad \text{Properties of exponents}$$

$$= 4.6 \times 10^3 \quad \text{Division}$$

$$= 4,600 \quad \text{Standard form}$$

▶ *Quick check* Multiply $(473)(0.0000579)$ using scientific notation. ■

Mastery points

Can you
■ Perform division on factors involving exponents?
■ Perform operations involving negative exponents?
■ Perform operations involving zero as an exponent?
■ Raise a fraction to a power?
■ Use scientific notation?

Exercise 3–3

Assume that all variables or groupings in this exercise represent nonzero real numbers and that all variables used as exponents represent positive integers.

Write each answer with only positive exponents. See examples 3–3 B and C.

Example $3b^0$

Solution $= 3 \cdot b^0$ Only b is raised to the zero power
$ = 3 \cdot 1$ b^0 is 1
$ = 3$ Standard form

1. 5^0 2. $(-3)^0$ 3. $(2a^3 - b)^0$ 4. $(5x^2 + 4y)^0$ 5. $7x^0$

6. $9y^0$ 7. a^{-4} 8. b^{-5} 9. $3a^{-2}$ 10. x^2y^{-5}

11. $ab^{-4}c^{-3}$ 12. $\dfrac{1}{4a^{-3}}$ 13. $\dfrac{1}{2x^{-5}}$ 14. $\dfrac{1}{x^2y^{-4}}$

Perform all indicated operations and leave the answer with only positive exponents. See examples 3–3 A, D, and E.

Example $\dfrac{a^{-2}b^3}{a^{-4}b^6}$

Solution $= a^{(-2)-(-4)}b^{3-6}$ Division of like bases
$= a^2 b^{-3}$ Subtract exponents
$= \dfrac{a^2}{b^3}$ Standard form

15. $\dfrac{a^3}{a^4 a^7}$

16. $\dfrac{x^5 y^3}{xy^5}$

17. $\dfrac{2^3 x^4 y^9}{2xy^3}$

18. $\dfrac{3^3 a^4 b^5}{3^2 a^2 b^3}$

19. $\dfrac{3x^2 y^5}{3^4 x^5 y^5}$

20. $\dfrac{5^2 a^4 b}{5^3 a^9 b^4}$

21. 2^{-2}

22. -2^{-2}

23. -3^{-4}

24. $(-2)^{-3}$

25. $\dfrac{2^{-4}}{2^{-2}}$

26. $\dfrac{3^{-6}}{3^{-4}}$

27. $x^{-5}x^2$

28. $(2a^{-3})(3a^{-5})$

29. $(5x^2)(4x^{-7})$

30. $(3x^{-4}y^3)(2x^2y^{-1})$

31. $(5a^{-3}b^{-4})(2a^2b^5)$

32. $(2a^3)^{-3}$

33. $(3x^4)^{-2}$

34. $(5a^2b^{-3})^{-2}$

35. $(4a^{-3}b^4)^{-3}$

36. $(x^{-3}y^0)^2$

37. $(2a^{-2}b^0)^3$

38. $(5x^{-4}y^0z^2)^{-2}$

39. $(3a^2b^{-3}c^0)^{-3}$

40. $\left(\dfrac{2x}{y^3}\right)^2$

41. $\left(\dfrac{3a^2}{b}\right)^3$

42. $\left(\dfrac{5a^2b}{c^4}\right)^2$

43. $\left(\dfrac{3x^4y^5}{z^9}\right)^3$

44. $\left(\dfrac{2x^5y^2}{4x^3y^7}\right)^4$

45. $\left(\dfrac{9a^5b^4}{18ab^{11}}\right)^3$

46. $\dfrac{x^{-4}y^2}{x^5y^{-3}}$

47. $\dfrac{a^2b^{-2}}{a^{-4}b^5}$

48. $\left(\dfrac{6y^{-3}}{12y^{-5}}\right)^3$

49. $\dfrac{x^{-4}y^2}{x^{-6}y^5}$

50. $\left(\dfrac{y^{-4}z^{-3}}{y^{-5}z^4}\right)^2$

51. $\left(\dfrac{3x^{-2}}{12x^{-4}}\right)\left(\dfrac{16x^{-5}}{x}\right)$

52. $\left(\dfrac{a^{-2}b^3c^0}{a^3b^{-4}c^2}\right)\left(\dfrac{a^{-4}b^{-1}c^2}{a^5b^2}\right)$

53. $\left(\dfrac{3^2a^{-3}b^3}{6a^{-5}b^{-2}}\right)\left(\dfrac{12a^{-7}b^{-3}}{a^2b^5}\right)$

54. $\left(\dfrac{4a^{-3}}{12a^{-5}}\right)^{-3}$

55. $\left(\dfrac{xy^0}{z^{-2}}\right)^{-4}$

56. $\left(\dfrac{2a^3b^{-4}}{6a^5b^2}\right)^{-2}$

57. $\left(\dfrac{4x^2y^3z^{-2}}{12x^5y^{-2}z^0}\right)^{-3}$

58. $(3x^2y^{-2})^2(x^{-4}y^3)^3$

59. $(2a^3b^{-4})^3(a^{-2}b^5)^3$

60. $(x^{-2}y^4)^{-3}(x^{-1}y^{-2})^4$

61. $(a^3b^{-2}c^{-4})^{-2}(a^{-2}b^3c^0)^3$

62. $(ab^{-4}c^2)^{-5}(a^{-2}b^0c^3)^{-3}$

63. $(x^2y^{-4})^{-3}(xy^3z^4)^{-2}$

Perform the indicated operations and leave the answer as a product with no fractions and each variable occurring only once.

Examples $(a^{1-n})^{-3}$ $\qquad\qquad \dfrac{x^{2n-7}}{x^{n-6}}$

Solutions $= a^{(1-n)(-3)}$ $\qquad = x^{(2n-7)-(n-6)}$
$= a^{-3+3n}$ $\qquad\qquad = x^{2n-7-n+6}$
$= a^{3n-3}$ $\qquad\qquad = x^{n-1}$

64. $x^{2n+3}x^{2-n}$

65. $a^{b-4}a^{2b-1}$

66. $(a^{1-2n})^{-2}$

67. $(x^{4-3n})^{-2}$

68. $\dfrac{x^{n-6}}{x^{n-3}}$

69. $\dfrac{a^{2n-1}}{a^{n-5}}$

70. $\dfrac{x^n y^{2n+1}}{x^{n-3}y^{n-4}}$

71. $\dfrac{a^{2n}b^{3n-2}}{a^{n-1}b^{n-5}}$

72. $\left(\dfrac{x^n}{x^{2n-1}}\right)^{-2}$

73. $\left(\dfrac{x^{2n}}{x^{n+1}}\right)^{-3}$

Express the following numbers in scientific notation. See example 3–3 F.

74. 65,000,000   **75.** 155,000   **76.** 0.00012   **77.** 0.0863   **78.** −0.0567

79. The mass of the moon is 8,060,000,000,000,000,000,000 tons.

80. General Motors reported a gross earnings of $7,230,000,000.

81. A cu in. equals 0.0005787 cu ft.

82. 43,560 sq ft equals an acre.

83. A gram equals 0.0022046 pound.

84. 46,656 cu in. equals a cu yd.

85. A kilogram equals 0.001102 ton.

86. A particular gauge of copper wire has a resistance of 0.00006203 ohms/lb.

87. The distance light travels in an hour is 669,600,000 miles.

88. 3,785.3 milliliters equals a gal.

Convert the following numbers in scientific notation to their standard form. See example 3–3 G.

89. -4.37×10^{-2} **90.** 7.61×10^{-7} **91.** 4.99×10^{6} **92.** 7.23×10^{0} **93.** 4.83×10^{4}

Multiply or divide the following using scientific notation. Leave the answer in scientific form, $a \times 10^{n}$, where a is rounded to two decimal places. See example 3–3 H.

Example   (473)(0.0000579)

Solution $= (4.73 \times 10^{2})(5.79 \times 10^{-5})$ Scientific notation
$= (4.73)(5.79) \times (10^{2} \cdot 10^{-5})$ Commutative and associative properties
$= 27.3867 \times 10^{-3}$ Multiplication
$= 2.73867 \times 10^{1} \times 10^{-3}$ Scientific notation
$= 2.74 \times 10^{-2}$ Answer rounded to 2 decimal places

94. $5.23 \times 10^{9} \cdot 1.073 \times 10^{6}$

95. $5.12 \times 10^{6} \cdot 6.2 \times 10^{4}$

96. $8.473 \times 10^{3} \cdot 8.4 \times 10^{-4}$

97. $1.673 \times 10^{-4} \cdot 7.5 \times 10^{-6}$

98. $(6.23 \times 10^{2}) \div (5.73 \times 10^{4})$

99. $(1.47 \times 10^{3}) \div (8.03 \times 10^{1})$

100. $(7.82 \times 10^{-2}) \div (5.6 \times 10^{5})$

101. $(7.23 \times 10^{5}) \div (6.075 \times 10^{-9})$

102. $0.876 \cdot 21.46$

103. $0.000476 \cdot 0.0053$

104. $0.000000089 \cdot 0.145$

Solve the following word problems, leave the answer in scientific notation rounded to two decimal places.

105. The amount of stress on a piece of metal is given by stress $= \dfrac{\text{force producing the stress}}{\text{area of the surface}}$. Find the stress on a piece of metal where the force is 452.6 kg and the area is 0.000763 sq cm.

106. If the volume of a right circular cylinder is given by $V = \pi r^{2}h$, find the volume of a right circular cylinder that has radius $r = 0.0073$ m and height $h = 12$ m.

107. A unit of light intensity is a lumen. If 1 lumen equals 0.001496 watts, how many lumens are there in 4,760,000 watts?

Review exercises

Perform the indicated addition and subtraction. See section 1–6.

1. $5a + 7a$ **2.** $7x^{2} - 4x^{2}$ **3.** $9ab - 3ab$

Perform the indicated multiplication. See section 3–2.

4. $a^{2}(2a + 3)$ **5.** $3a^{2}b(2a^{2} + 3ab - b^{2})$ **6.** $(a + 2b)(x - 3y)$

7. $(3x + 4y)(a - 2b)$ **8.** $(a + 1)(b + 1)$

3–4 ■ *Common factors and factoring by grouping*

Prime numbers

In section 3–2, we studied the procedure for multiplying polynomials. Now we will study the reversal of that process, which is called **factoring.** Factoring polynomials is necessary to deal with algebraic fractions, since, as we have seen with arithmetic fractions, the denominators of the fractions must be in factored form to reduce fractions or to find the **least common denominator** (LCD). Factoring is also useful as a means of finding the solution of certain types of equations.

We saw in chapter 1 that whenever we multiply any two real numbers, a and b, a and b are called **factors** of the **product** ab. For example, 2 and 3 are factors of 6 because $2 \cdot 3 = 6$. The first type of factoring that we will consider is factoring a natural number into its prime factors.

The set of natural numbers greater than 1 can be divided into two types of numbers, prime numbers and composite numbers. First, a **prime number** *is any natural number greater than 1 that is divisible only by itself and by 1.* For example,

$$2, 3, 5, 7, 11, 13$$

are all prime numbers since each is a natural number greater than 1 and divisible only by 1 and the number itself. Second, a **composite number** *is any natural number greater than 1 that is not a prime number.* For example,

$$4, 9, 28, 42$$

are composite numbers since 4 is divisible by 2; 9 is divisible by 3; 28 is divisible by 2, 4, 7, and 14; 42 is divisible by 2, 3, 6, 7, 14, and 21.

When a composite number is stated as a product of only prime numbers, we call this the **prime factor form** or **completely factored form** of the composite number. We can also state negative integers in a completely factored form by first factoring out -1 and then factoring the resulting positive integer into its prime factors.

■ *Example 3–4 A*

Express the following numbers in prime factor form.

1. $28 = 4 \cdot 7 = 2 \cdot 2 \cdot 7 = 2^2 \cdot 7$

Note Whenever a prime number appears as a factor more than once, we write it in exponential form.

2. $42 = 6 \cdot 7 = 2 \cdot 3 \cdot 7$

3. $23 = 23$

Note The number 23 is a prime number, therefore it cannot be expressed as a product of only prime factors. It would be incorrect to write $23 \cdot 1$ since 1 is not a prime number.

4. $120 = 10 \cdot 12 = 2 \cdot 5 \cdot 2 \cdot 2 \cdot 3 = 2^3 \cdot 3 \cdot 5$ ■

The greatest common factor

The first type of factoring that we will do involves finding the **greatest common factor** (GCF) of the polynomial. Recall the statement of the distributive property.

$$a(b + c) = ab + ac$$

$a(b + c)$ is called the *factored form* of $ab + ac$. The greatest common factor consists of the following:

___ *Greatest common factor* ___

1. The greatest integer that is a common factor of all the numerical coefficients, and
2. The variable factor(s) common to every term, each raised to the least power to which they were raised in any of the terms.

Factoring ———————————————————→

Polynomial	**Distributive Property**	**Factored Form**
(terms)	*(determine the GCF)*	*(factors)*
$5x + 15$	$\mathbf{5} \cdot x + \mathbf{5} \cdot 3$	$5(x + 3)$
$12ab - 6ac$	$\mathbf{6a} \cdot 2b - \mathbf{6a} \cdot c$	$6a(2b - c)$
$x^5 + 4x^3 - 2x^2$	$\mathbf{x^2}x^3 + \mathbf{x^2} \cdot 4x - \mathbf{x^2} \cdot 2$	$x^2(x^3 + 4x - 2)$

←——————————————————— *Multiplying*

We were able to determine the greatest common factor of these polynomials by inspection. In some problems, this may not be possible, and the following procedure will be necessary. Factor the polynomial $15x^5y^2 + 30x^3y^4$.

___ *Factoring the greatest common factor* ___

Step 1 Factor each term such that it is the product of primes and variables to powers.

$$15x^5y^2 \qquad 30x^3y^4$$
$$3 \cdot 5 \cdot x^5 \cdot y^2 \qquad 2 \cdot 3 \cdot 5 \cdot x^3 \cdot y^4$$

Step 2 Write down all factors that are common to *every* term.

$$3 \cdot 5 \cdot x \cdot y$$

Note We do not have 2 as part of our greatest common factor since it does not appear in *all* of the terms.

Step 3 Take the factors in step 2 and raise them to the *least* power to which they were raised in any of the terms.

$$3^1 \cdot 5^1 \cdot x^3 \cdot y^2 = 15x^3y^2$$

This is the greatest common factor (GCF).

Step 4 Find the multinomial factor (the polynomial within the parentheses) by dividing each term of the polynomial being factored by the GCF.

$$\frac{15x^5y^2}{15x^3y^2} = x^2 \quad \text{and} \quad \frac{30x^3y^4}{15x^3y^2} = 2y^2$$

$$(x^2 + 2y^2)$$

Step 5 We can now write the polynomial in its factored form.

$$15x^5y^2 + 30x^3y^4$$
$$= 15x^3y^2 \cdot x^2 + 15x^3y^2 \cdot 2y^2$$
$$= 15x^3y^2(x^2 + 2y^2)$$

Completely factored form

Notice that $15x^5y^2 + 30x^3y^4$ could also be factored into

$$15(x^5y^2 + 2x^3y^4) \text{ or } 30\left(\frac{1}{2}x^5y^2 + x^3y^4\right) \text{ or } x^3(15x^2y^2 + 30y^4)$$

This allows room for a given polynomial to be factored in many ways, unless some restrictions are placed on the procedure. We wish to factor each polynomial in a unique manner that will not permit such variations in the results.

— Completely factored form _____

A polynomial with integer coefficients will be considered to be in **completely factored form** when it satisfies the following criteria:

1. The polynomial is written as a product of polynomials with integer coefficients.
2. None of the polynomial factors other than the monomial factor can be factored further.

We see that $15x^3y^2(x^2 + 2y^2)$ is the **completely factored form** of the expression $15x^5y^2 + 30x^3y^4$. All of the coefficients are integers and none of the factors other than the monomial, $15x^3y^2$, can still be written as the product of two polynomials with integer coefficients.

In general, whenever we factor a monomial out of a polynomial, we factor the monomial in such a way that it has a positive coefficient. We should realize that we can also factor out the opposite, or negative, of this common factor. In our previous example, we could have factored out $-15x^3y^2$ and the completely factored form would have been

$$-15x^3y^2(-x^2 - 2y^2)$$

Observe that the only change in our answer when we factor out the opposite of the common factor is that the signs of all the terms inside the parentheses change.

■ *Example 3–4 B*

Write in completely factored form.

1. $12x^5 + 9x^2$

$$= 3x^2(\qquad + \qquad)$$ — Multinomial factor will have as many terms as the original expression

GCF is $3x^2$

$$= 3x^2(4x^3 + 3)$$ Completely factored form

$$\frac{9x^2}{3x^2}$$

$$\frac{12x^5}{3x^2}$$

If we wanted to check the answer, we would apply the distributive property and perform the multiplication as follows:

$$3x^2(4x^3 + 3) = 3x^2 \cdot 4x^3 + 3x^2 \cdot 3 \qquad \text{Distributive property}$$
$$= 12x^5 + 9x^2. \qquad \text{Carry out the multiplication}$$

2. $9r^3s^2 + 15r^2s^4 + 3rs^2$

$\quad = 3^2 \cdot r^3 \cdot s^2 + 3 \cdot 5 \cdot r^2 \cdot s^4 + 3 \cdot r \cdot s^2$ Factor each term

$\quad = 3rs^2 \cdot 3r^2 + 3rs^2 \cdot 5rs^2 + 3rs^2 \cdot 1$ Determine the GCF

$\quad = 3rs^2(3r^2 + 5rs^2 + 1)$ Completely factored form

Note In example 2, the last term in the factored form is 1. This occurs when a term and the GCF are the same, that is, whenever we are able to factor all the numbers and variables out of a given term.

3. $a(x + 2y) + b(x + 2y)$

The quantity $(x + 2y)$ is common to both terms. We then factor the common quantity out of each term and place the remaining factors from each term in a second parentheses.

$a(x + 2y) + b(x + 2y)$

$= (x + 2y)(a + b)$

Common   Remaining
factor   factors

▶ *Quick check* Write $18y^4 + 12y^2$ in completely factored form. ■

Factoring by grouping

Consider the example $ax + ay + bx + by$. We observe that this is a four-term polynomial and we will *try* to factor it by grouping.

$$ax + ay + bx + by = (ax + ay) + (bx + by)$$

There is a common factor of a in the first two terms and a common factor of b in the last two terms.

$$(ax + ay) + (bx + by) = a(x + y) + b(x + y)$$

The quantity $(x + y)$ is a common factor to both terms. Factoring it out, we have

$$a(x + y) + b(x + y) = (x + y)(a + b)$$

Therefore we have factored the polynomial by grouping.

Factoring a four-term polynomial by grouping

1. Arrange the four terms so that the first two terms have a common factor and the last two terms have a common factor.
2. Determine the GCF of each pair of terms and factor it out.
3. If step 2 produces a common binomial factor in each term, factor it out.
4. If step 2 does not produce a common binomial factor in each term, try grouping the terms of the original polynomial in a different way.
5. If step 4 does not produce a common binomial factor in each term, the polynomial will not factor by this procedure.

■ *Example 3–4 C*

Factor completely.

1. $2ac - ad + 4bc - 2bd = (2ac - ad) + (4bc - 2bd)$ Group in pairs
$$= a(2c - d) + 2b(2c - d)$$ Factor out the GCF
$$= (2c - d)(a + 2b)$$ Factor out the common binomial

2. $6ax + by + 2ay + 3bx = 6ax + 2ay + 3bx + by$ Rearrange the terms
$$= (6ax + 2ay) + (3bx + by)$$ Group in pairs
$$= 2a(3x + y) + b(3x + y)$$ Factor out the GCF
$$= (3x + y)(2a + b)$$ Factor out the common binomial

Note   Sometimes the terms must be rearranged such that the pairs will have a common factor, as we did in example 2.

▶ *Quick check*   Factor $3a^3 - a^2 + 9a - 3$ completely.   ■

┌─ **Mastery points** ───

  Can you
  ■ Determine the greatest common factor of a polynomial?
  ■ Factor the greatest common factor from a polynomial?
  ■ Factor a four-term polynomial by grouping?

└───

Exercise 3–4

Express the following numbers in prime factor form. If a number is prime, so state. See example 3–4 A.

1. 12   **2.** 15   **3.** 24   **4.** 28   **5.** 56   **6.** 60   **7.** 41   **8.** 43   **9.** 39

Supply the missing factors or terms.

Examples $-3a + a^3b = -a(? - ?)$ $x^2y - z^3 = -(? + ?)$ $3a^2 - 9b = ?(-a^2 + 3b)$

Solutions $= -a(3 - a^2b)$ $= -(-x^2y + z^3)$ $= -3(-a^2 + 3b)$

10. $5x - 3 = -(? + ?)$ **11.** $2a - 3b = -(? + ?)$

12. $2x^2 - 6y = -2(? + ?)$ **13.** $3a^3 - 9b^2 = -3(? + ?)$

14. $6a - 8b - 12c = 2(? - ? - ?)$ **15.** $-4x^3 - 36xy + 16xy^2 = -4(? + ? - ?)$

16. $5x^2y + 15x^3y^2 + 25x^2y^2 = -5x^2y(? - ? - ?)$ **17.** $3a^3b^2 + 12a^2b^3 + 15a^4b^2 = -3a^2b^2(? - ? - ?)$

18. $-ab^2 - ac^2 = ?(b^2 + c^2)$ **19.** $3x^2 - 9xy = ?(-x + 3y)$

20. $-10a^2b^2 + 15ab - 20a^3b^3 = ?(-2ab + 3 - 4a^2b^2)$

21. $-24RS - 16R + 32R^2 = ?(3S + 2 - 4R)$ **22.** $3x^n - 2x^ny = ?(3 - 2y)$

Write in completely factored form. Assume that all variables used as exponents represent positive integers. See examples 3–4 B and C.

Examples $18y^4 + 12y^2$ $3a^3 - a^2 + 9a - 3$

Solutions $= 2 \cdot 3^2y^4 + 2^2 \cdot 3y^2$ Factor each term $= (3a^3 - a^2) + (9a - 3)$ Group in pairs
$= 6y^2(\ \ + \ \ )$ Determine the GCF $= a^2(3a - 1) + 3(3a - 1)$ Factor out the GCF
$= 6y^2(3y^2 + 2)$ Completely factored form $= (3a - 1)(a^2 + 3)$ Factor out the common binomial

23. $5x^2 + 10xy - 20y$

24. $8a - 12b + 16c$

25. $18ab - 27a + 3ac$

26. $15x^2 - 27y^2 + 12$

27. $15R^2 - 21S^2 + 36T$

28. $8x^3 + 4x^2$

29. $3R^2S - 6RS^2 + 12RS$

30. $16x^3y - 3x^2y^2 + 24x^2y^3$

31. $12x^4y^3 - 8x^4y^2 + 16x^3y$

32. $24a^3b^2 - 3a^2b^2 + 12a^2b^5$

33. $3x^2y - 6xy^4 + 15x^3y^2$

34. $a(x - 3y) + b(x - 3y)$

35. $3a(x - y) + b(x - y)$

36. $5a(3x - 1) + 10(3x - 1)$

37. $14b(a + c) - 7(a + c)$

38. $21x(1 + 2z) - 35y(1 + 2z)$

39. $4ab(2x + y) - 8ac(2x + y)$

40. $2x(y - 16) - (y - 16)$

41. $5a(b + 3) - (b + 3)$

42. $15a^3b^2(x + y) + 30a^2b^5(x + y)$

43. $x^ny^2 + x^nz$

44. $x^{3n} + x^{2n} + x^n$

45. $x^{2n}y^{2n} - x^ny^n$

46. $x^{n+4} + x^4$

47. $y^{n+3} - y^3$

48. $ac + ad - 2bc - 2bd$

49. $2ax + 6bx - ay - 3by$

50. $2ax + bx - 4ay - 2by$

51. $2ax + 3bx + 8ay + 12by$

52. $5ax - 3by + 15bx - ay$

53. $2ax^2 - 3b - bx^2 + 6a$

54. $20x^2 - 3yz + 5xz - 12xy$

55. $3ac - 2bd - 6ad + bc$

56. $6ac + 3bd - 2ad - 9bc$

57. $2x^3 + 15 + 10x^2 + 3x$

58. $3x^3 - 6x^2 + 5x - 10$

59. $8x^3 - 4x^2 + 6x - 3$

Solve the following word problems.

60. The area of the surface of a cylinder is determined by $A = 2\pi rh + 2\pi r^2$, where r is the radius and h is the height. Factor the right member.

61. The total surface area of a right circular cone is given by $A = \pi rs + \pi r^2$, where r is the base radius and s is the slant height. Factor the right member.

62. The equation for the height of a rocket fired vertically upward into the air is given by $S = 560t - 16t^2$, where the rocket is S feet from the ground after t seconds. Factor the right member.

63. If you have P dollars in a savings account where the yearly compounded rate is r percent, then the amount of money in the account at the end of one year can be written as $A = P + Pr$. Factor the right member.

64. In engineering, the equation for deflection of a beam is given by

$$Y = \frac{2wx^4}{48EI} - \frac{3\ell wx^3}{48EI} - \frac{\ell^3 wx}{48EI}$$

Factor the right member.

Review exercises

Perform the indicated multiplication. See section 3–2.

1. $(a + 3)(a + 4)$

2. $(x - 5)(x - 2)$

3. $(x + 4)(x - 6)$

4. $(x - 3)(x + 70)$

5. $(x - 4)(x + 4)$

6. $(a + 5)(a - 5)$

7. $(x + 3)^2$

8. $(x - 4)^2$

3–5 ■ Factoring trinomials of the form $x^2 + bx + c$ and perfect square trinomials

Determining when a trinomial will factor

In section 3–2, we learned how to multiply two binomials as follows:

Factors   Terms

$$(a + 2)(a + 8) = a^2 + 8a + 2a + 16 = a^2 + 10a + 16$$

Multiplying

In this section, we are going to reverse the procedure and factor the trinomial.

Terms   Factors

$$a^2 + 10a + 16 \qquad = \qquad (a + 2)(a + 8)$$

Factoring

The following group of trinomials will enable us to see how a trinomial factors.

1. $a^2 + 10a + 16 \qquad = \qquad (a + 2)(a + 8)$

$16 = 2 \cdot 8$

$10 = 2 + 8$

2. $a^2 - 10a + 16 \qquad = \qquad (a - 2)(a - 8)$

$16 = (-2) \cdot (-8)$

$10 = (-2) + (-8)$

3. $a^2 + 6a - 16 \qquad = \qquad (a - 2)(a + 8)$

$-16 = (-2) \cdot 8$

$6 = (-2) + 8$

4. $a^2 - 6a - 16 \qquad = \qquad (a + 2)(a - 8)$

$-16 = 2 \cdot (-8)$

$-6 = 2 + (-8)$

In general,

$$(x + m)(x + n) = x^2 + (m + n)\, x + m \cdot n$$

The trinomial $x^2 + bx + c$ will factor only if there are two integers, which we will call m and n, such that $m + n = b$ and $m \cdot n = c$.

$$m + n \qquad m \cdot n$$

$$x^2 + bx + c = (x + m)(x + n)$$

Factoring a trinomial of the form $x^2 + bx + c$

1. Factor out the GCF. If there is a common factor, make sure to include it as part of the final factorization.
2. Determine if the trinomial is factorable by finding m and n such that $m + n = b$ and $m \cdot n = c$. If m and n do not exist, we conclude that the trinomial will not factor.
3. Using the m and n values from step 2, write the trinomial in factored form.

The signs (+ or −) for m and n

1. If c is positive, then m and n both have the same sign as b.
2. If c is negative, then m and n have different signs and the one with the greater absolute value has the same sign as b.

■ *Example 3–5 A*

Factor completely each trinomial.

1. $x^2 + 14x + 24$ $m + n = 14$ and $m \cdot n = 24$
Since $b = 14$ and $c = 24$ are both positive, then m and n are both positive.

List of the factorizations of 24	Sum of the factors of 24
$1 \cdot 24$	$1 + 24 = 25$
$2 \cdot 12$	$2 + 12 = 14$ ◄——— Correct sum
$3 \cdot 8$	$3 + 8 = 11$
$4 \cdot 6$	$4 + 6 = 10$

The m and n values are 2 and 12. The factorization is

$$x^2 + 14x + 24 = (x + 2)(x + 12)$$

Note   The commutative property allows us to write the factors in any order. That is $(x + 2)(x + 12) = (x + 12)(x + 2)$.

2. $a^2 - 3a - 18$ $m + n = -3$ and $m \cdot n = -18$
Since $b = -3$ and $c = -18$ are both negative, then m and n have different signs and the one with the greater absolute value is negative.

Factorizations of −18, where the negative sign goes with the factor with the greater absolute value	Sum of the factors of −18
$1 \cdot (-18)$	$1 + (-18) = -17$
$2 \cdot (-9)$	$2 + (-9) = -7$
$3 \cdot (-6)$	$3 + (-6) = -3$ ◄——— Correct sum

The m and n values are 3 and -6. The factorization is

$$a^2 - 3a - 18 = (a + 3)(a - 6)$$

3. $5a - 24 + a^2$
It is easier to identify b and c if we write the trinomial in descending powers of the variable, which is called standard form.

$a^2 + 5a - 24$ $m + n = 5$ and $m \cdot n = -24$

Since $b = 5$ is positive and $c = -24$ is negative, m and n have different signs and the one with the greater absolute value is positive.

Factorizations of −24, where the positive factor is the one with the greater absolute value	Sum of the factors of −24
$(-1) \cdot 24$	$(-1) + 24 = 23$
$(-2) \cdot 12$	$(-2) + 12 = 10$
$(-3) \cdot 8$	$(-3) + 8 = 5$ ◄——— Correct sum
$(-4) \cdot 6$	$(-4) + 6 = 2$

The m and n values are -3 and 8. The factorization is

$$a^2 + 5a - 24 = (a - 3)(a + 8)$$

4. $x^2 + 6x + 12$ $m + n = 6$ and $m \cdot n = 12$
   Since $b = 5$ and $c = 12$ are both positive, m and n are both positive.

Factorizations of 12
   $1 \cdot 12$
   $2 \cdot 6$
   $3 \cdot 4$

Sum of the factors of 12
   $1 + 12 = 13$
   $2 + 6 = 8$
   $3 + 4 = 7$

No sum equals 6

Since none of the factorizations of 12 add to 6, there is no pair of integers (m and n) and the trinomial will not factor using integer coefficients. We call this a **prime polynomial.**

5. $2a^3 - 18a^2 + 40a = 2a(a^2 - 9a + 20)$ common factor of $2a$
   To complete the factorization, we see if the trinomial $a^2 - 9a + 20$ will factor. We need to find m and n that add to -9 and multiply to 20. The values are -4 and -5. The completely factored form is

   $2a^3 - 18a^2 + 40a = 2a(a - 4)(a - 5)$

Note A common error when the polynomial has a common factor is to factor it out but to forget to include it as one of the factors in the completely factored form.

6. $x^2y^2 - 8xy + 15$
   Rewriting the polynomial as $(xy)^2 - 8xy + 15$, we want to find values for m and n that add to -8 and multiply to 15. The values are -3 and -5. The factorization is

   $x^2y^2 - 8xy + 15 = (xy - 3)(xy - 5)$

7. $x^2 + xy - 6y^2$ $m + n = 1$ and $m \cdot n = -6$ m and n are -2 and 3.
   We replace xy with $-2y$ and $3y$. The factorization is

   $x^2 + xy - 6y^2 = (x - 2y)(x + 3y)$

8. $x^2(a - b) + 6x(a - b) + 8(a - b)$ Common factor of $a - b$
       $= (a - b)(x^2 + 6x + 8)$ The trinomial $x^2 + 6x + 8$ has m and n
                                       values of 4 and 2
       $= (a - b)(x + 4)(x + 2)$ Completely factored form

9. $(x + y)^2 + 6(x + y) + 8$
   If we let a represent $x + y$, then using the property of substitution, the factorization would be

$$a^2 \quad + 6 \quad a \quad + 8 = \quad (a \quad + 2) \quad (a \quad + 4)$$

$$(x + y)^2 + 6 (x + y) + 8 = [(x + y) + 2][(x + y) + 4]$$ Substitute $x + y$ for a
$$= (x + y + 2)(x + y + 4)$$ Remove inner parentheses

▶ **Quick check**   Factor $c^2 - 9c + 14$ and
$a^2(x - 2) + 7a(x - 2) + 12(x - 2)$ completely.

Perfect square trinomials

Two of the special products that we studied in section 3–2 were the squares of a binomial. We now restate those special products.

$$a^2 + 2ab + b^2 = (a + b)^2$$

and

$$a^2 - 2ab + b^2 = (a - b)^2$$

The right members of these two equations are called the **squares of binomials,** and the left members are called **perfect square trinomials.** Perfect square trinomials can always be factored by our *m* and *n* procedure from this section if the coefficient of the squared term is 1, or by the procedure in the next section if the coefficient is not 1. However if we observe that the first and last terms of a trinomial are perfect squares, we should see if the trinomial will factor as the square of a binomial. To factor a trinomial as a pefect square trinomial, three conditions need to be met.

Conditions for factoring a perfect square trinomial

1. The first term must have a positive coefficient and be a perfect square, a^2.
2. The last term must have a positive coefficient and be a perfect square, b^2.
3. The middle term must be twice the product of the bases of the first and last terms, $2ab$ or $-2ab$.

We observe that

$$9x^2 + 12x + 4$$
$$= (3x)^2 + 2(3x)(2) + (2)^2.$$

Condition 1   Condition 3   Condition 2

Therefore it is a perfect square trinomial and factors into

$$(3x + 2)^2$$

■ **Example 3–5 B**

The following examples show the factoring of some other perfect square trinomials.

		Condition 1		Condition 3		Condition 2		Square of a binomial
1. $x^2 + 10x + 25$	$=$	$(x)^2$	$+$	$2(x)(5)$	$+$	$(5)^2$	$=$	$(x + 5)^2$
2. $a^2 - 12a + 36$	$=$	$(a)^2$	$-$	$2(a)(6)$	$+$	$(6)^2$	$=$	$(a - 6)^2$
3. $4b^2 + 20b + 25$	$=$	$(2b)^2$	$+$	$2(2b)(5)$	$+$	$(5)^2$	$=$	$(2b + 5)^2$
4. $9x^2 - 24x + 16$	$=$	$(3x)^2$	$-$	$2(3x)(4)$	$+$	$(4)^2$	$=$	$(3x - 4)^2$

▶ **Quick check** Factor $4x^2 + 20x + 25$ and $9x^2 - 6x + 1$ completely. ■

Mastery points

Can you

- Determine two integers whose product is one number and whose sum is another number?
- Recognize when the trinomial $x^2 + bx + c$ will factor and when it will not?
- Factor trinomials of the form $x^2 + bx + c$?
- Always remember to look for the greatest common factor before applying any of the factoring rules?
- Factor perfect square trinomials?

Exercise 3–5

Write in completely factored form. Assume all variables used as exponents represent positive integers. See example 3–5 A.

Example $c^2 - 9c + 14$

Solution $m + n = -9$ and $m \cdot n = 14$
Since $b = -9$ is negative and $c = 14$ is positive, m and n are both negative.

List the factorizations of 14   Sum of the factors of 14

$(-1)(-14)$ $(-1) + (-14) = -15$
$(-2)(-7)$ $(-2) + (-7) = -9$ ◄— Correct sum

The m and n values are -2 and -7. The factorization is

$c^2 - 9c + 14 = (c - 2)(c - 7)$

1. $x^2 + 13x - 30$
2. $a^2 - 14a + 24$
3. $y^2 + 5y - 24$
4. $a^2 + 8a + 12$
5. $x^2 - 2x - 24$
6. $y^2 - y - 12$
7. $3x^2 + 33x - 36$
8. $3y^2 - 18y - 48$
9. $4x^2 - 4x - 24$
10. $5y^2 - 15y - 55$
11. $3y^2 - 27y + 12$
12. $4s^2 - 28s + 12$
13. $x^2y^2 - 4xy - 21$
14. $x^2y^2 - 3xy - 18$
15. $x^2y^2 + 13xy + 12$
16. $a^2b^2 + 13ab - 30$
17. $x^2y^2 - 14xy + 24$
18. $y^2z^2 + 5yz - 24$
19. $4a^2b^2 + 24ab + 32$
20. $3x^2y^2 + 21xy + 30$
21. $-2x^2y^2 + 6xy + 20$
22. $-3x^2y^2 + 6xy + 45$
23. $a^2 + 7ab + 10b^2$
24. $x^2 - 12xy + 32y^2$
25. $x^2 + 11xy + 24y^2$
26. $y^2 - 4yz - 5z^2$
27. $a^2 + 2ab - 35b^2$
28. $y^2 - 4yz - 12z^2$
29. $x^2 + 10xy + 16y^2$
30. $x^{2n} + 5x^n + 6$
31. $x^{2n} + 9x^n + 14$
32. $x^{2n} - 4x^n - 12$
33. $x^{2n} - 8x^n + 15$

Write in completely factored form. See example 3–5 B.

Examples   **Solutions**

	Condition 1		Condition 3		Condition 2	Square of a binomial
$25x^2 + 20x + 4$ =	$(5x)^2$	+	$2(5x)(2)$	+	$(2)^2$	= $(5x + 2)^2$
$9x^2 - 6x + 1$ =	$(3x)^2$	−	$2(3x)(1)$	+	$(1)^2$	= $(3x - 1)^2$

34. $a^2 + 16a + 64$

35. $b^2 - 12b + 36$

36. $x^2 + 14x + 49$

37. $c^2 + 18c + 81$

38. $4a^2 + 20a + 25$

39. $9y^2 + 12y + 4$

40. $x^2 + 10x + 25$

41. $x^2 - 14xy + 49y^2$

42. $9a^2 - 12ab + 4b^2$

43. $4x^2 - 12xy + 9y^2$

44. $25a^2 + 10ab + b^2$

45. $9x^2 + 30xy + 25y^2$

46. $a^2 - 16ab + 64b^2$

Write in completely factored form. See example 3–5 A.

Example $a^2(x - 2) + 7a(x - 2) + 12(x - 2)$

Solution $= (x - 2)(a^2 + 7a + 12)$ Common factor of $x - 2$, m and n values are 3 and 4

$= (x - 2)(a + 3)(a + 4)$ Completely factored form

47. $x^2(a - 2b) + 8x(a - 2b) + 12(a - 2b)$

48. $a^2(x - 3y) + 14a(x - 3y) + 24(x - 3y)$

49. $x^2(y + 2z) - 13x(y + 2z) + 30(y + 2z)$

50. $a^2(2b - c) + a(2b - c) - 12(2b - c)$

51. $x^2(3y - z) + 15x(3y - z) + 36(3y - z)$

52. $a^2(2b + 3c) - 6a(2b + 3c) - 27(2b + 3c)$

53. $a^2(3x - y) - 7a(3x - y) - 60(3x - y)$

54. $x^2(2y + z) + 2x(2y + z) - 63(2y + z)$

55. $(a + b)^2 - 4(a + b) - 12$

56. $(2a - b)^2 + (2a - b) - 30$

57. $(y + 3z)^2 - 5(y + 3z) - 14$

58. $(2a + 3b)^2 + 8(2a + 3b) + 12$

59. $(R + 5S)^2 - 6(R + 5S) + 8$

60. $(x - 4y)^2 - 10(x - 4y) + 21$

Solve the following word problems.

61. If a rectangle has a perimeter of 52 inches and an area of 153 square inches, the dimensions of the rectangle can be found by factoring the expression $W^2 - 26W + 153$. Factor this expression.

62. If the length of a rectangle is 3 meters more than three times its width and its area is 168 square meters, then the dimensions of the rectangle can be found by factoring the expression $3W^2 + 3W - 168$. Factor this expression.

63. When two capacitors are connected in parallel, the total capacitance of the circuit, C_t, is given by $C_t = C_1 + C_2$. If the number of microfarads in C_1 is the square of the number in C_2 and the total capacitance is 42 microfarads, then the expression to find the capacitance C_2 is given by $C_2^2 + C_2 - 42$. Factor this expression.

Review exercises

Factor completely. See section 3–4.

1. $3x(2x - 1) + 5(2x - 1)$

2. $2a(3a + 1) + (3a + 1)$

3. $4x(2x - 3) - (2x - 3)$

4. $5x(x + 4) - 3(x + 4)$

Perform the indicated operations. See section 3–2.

5. $(2x + 1)(3x + 2)$

6. $(5a - 1)(2a + 3)$

7. $(b + 4)(3b - 7)$

8. $(3a + 2)(3a - 2)$

3–6 ■ *Factoring trinomials of the form $ax^2 + bx + c$*

How to factor trinomials

We are going to factor trinomials of the form $ax^2 + bx + c$. This is called the **standard form** of a trinomial, where we have a single variable and the terms of the polynomial are arranged in descending powers of that variable. Letters a, b, and c in our standard form represent integer constants. For example,

$$4x^2 + 20x + 21$$

is a trinomial in standard form, where $a = 4$, $b = 20$, and $c = 21$.

Consider the product

$$(2x + 3)(2x + 7)$$

By multiplying these two quantities together and combining like terms, we get a trinomial.

$$(2x + 3)(2x + 7) = 4x^2 + 14x + 6x + 21$$
$$= 4x^2 + 20x + 21$$

To completely factor the trinomial $4x^2 + 20x + 21$ entails reversing the procedure to get

$$(2x + 3)(2x + 7)$$

The trinomial $ax^2 + bx + c$ will factor with integer coefficients if we can find a pair of integers (m and n) whose sum is equal to b, and whose product is equal to $a \cdot c$. In the trinomial $4x^2 + 20x + 21$, b is equal to 20, and $a \cdot c$ is $4 \cdot 21 = 84$. Therefore we want $m + n = 20$ and $m \cdot n = 84$. The values for m and n are 14 and 6.

If we observe the multiplication process in our example, we see that m and n appear as the coefficients of the middle terms that are to be combined for our final answer.

$$(2x + 3)(2x + 7) = 4x^2 + 14x + 6x + 21$$
$$= 4x^2 + 20x + 21$$

This is precisely what we do with the m and n values. We replace the coefficient of the middle term in the trinomial with these values. In our example, m and n are 14 and 6, and we replace 20 with these numbers.

$$4x^2 + 20x + 21 = 4x^2 + \overbrace{14x + 6x}^{20x} + 21$$

The next step is to group the first two terms and the last two terms.

$$(4x^2 + 14x) + (6x + 21)$$

Now we factor out what is common in each pair. We see that the first two terms contain the common factor $2x$ and the last two terms contain the common factor 3.

$$2x(2x + 7) + 3(2x + 7)$$

When we reach this point, what is inside the parentheses in each term will be the same. Since the quantity $(2x + 7)$ represents just one number and this number is common to both terms, we can factor it out.

$$2x(2x + 7) + 3(2x + 7)$$

Common to both terms

Having factored out what is common, what is left in each term is placed in a second parentheses.

$$2x(2x + 7) + 3(2x + 7)$$
$$(2x + 7)(2x + 3)$$

Common Remaining
factor   factors

The trinomial is factored.

A summary of the steps follows:

Factoring a trinomial

Step 1 Determine if the trinomial $ax^2 + bx + c$ is factorable by finding m and n such that $m \cdot n = a \cdot c$ and $m + n = b$. If m and n do not exist, we conclude that the trinomial will not factor.

Step 2 Replace the middle term, bx, by the sum of mx and nx.

Step 3 Place parentheses around the first and second terms and around the third and fourth terms. Factor out what is common to each pair.

Step 4 Factor out the common quantity of each term and place the remaining factors from each term in a second parentheses.

We determine the signs ($+$ or $-$) for m and n in a fashion similar to that of the previous section.

The signs ($+$ or $-$) for m and n

1. If $a \cdot c$ is positive, then m and n both have the same sign as b.
2. If $a \cdot c$ is negative, then m and n have different signs and the one with the greater absolute value has the same sign as b.

■ **Example 3–6 A**

Express the following trinomials in completely factored form. If the trinomial will not factor, so state.

1. $6x^2 + 13x + 6$

Step 1 $m \cdot n = 6 \cdot 6 = 36$ and $m + n = 13$

We determine by inspection that m and n are 9 and 4.

$$13x$$

Step 2 $= 6x^2 + 9x + 4x + 6$ Replace $13x$ with $9x$ and $4x$

Step 3 $= (6x^2 + 9x) + (4x + 6)$ Group the first two terms and the last two terms

$= 3x(2x + 3) + 2(2x + 3)$ Factor out what is common to each pair

Step 4 $= (2x + 3)(3x + 2)$ Factor out the common quantity

Note   The order in which we place m and n into the problem will not change the answer.

Alternate

$$\text{Step 2} = 6x^2 + \overbrace{4x + 9x}^{13x} + 6$$ Replace $13x$ with $4x$ and $9x$

$$\text{Step 3} = (6x^2 + 4x) + (9x + 6)$$ Group the first two terms and the last two terms

$$= 2x(3x + 2) + 3(3x + 2)$$ Factor out what is common to each pair

$$\text{Step 4} = (3x + 2)(2x + 3)$$ Factor out the common quantity

We see that the outcome in step 4 is the same regardless of the order of m and n in the problem.

2. $3x^2 + 14x + 8$

Step 1 $m \cdot n = 3 \cdot 8 = 24$ and $m + n = 14$
  m and n are 2 and 12.

$$\text{Step 2} = 3x^2 + \overbrace{2x + 12x}^{14x} + 8$$ Replace $14x$ with $2x$ and $12x$

$$\text{Step 3} = (3x^2 + 2x) + (12x + 8)$$ Group the first two terms and the last two terms

$$= x(3x + 2) + 4(3x + 2)$$ Factor out what is common to each pair

$$\text{Step 4} = (3x + 2)(x + 4)$$ Factor out the common quantity

3. $6a^2 - 11a + 4$

Step 1 $m \cdot n = 6 \cdot 4 = 24$ and $m + n = -11$
  m and n are -3 and -8.

$$\text{Step 2} = 6a^2 - \overbrace{3a - 8a}^{-11a} + 4$$ Replace $-11a$ with $-3a$ and $-8a$

$$\text{Step 3} = (6a^2 - 3a) + (-8a + 4)$$ Group the first two terms and the last two terms

$$= 3a(2a - 1) - 4(2a - 1)$$ Factor out what is common to each pair

In the last two terms, we have 4 or -4 as the greatest common factor. We factor out -4 so that we will have the same quantity inside the second parentheses.

$$\text{Step 4} = (2a - 1)(3a - 4)$$ Factor out the common quantity

Note   In the third step, if the third term is preceded by a minus sign, we will usually factor out the negative factor.

4. $4a^2 - 9a - 6$
  $m \cdot n = 4 \cdot (-6) = -24$ and $m + n = -9$
  The m and n values are not obvious by inspection.

Note   If you cannot determine the m and n values by inspection, then you should use the following systematic procedure to list all the possible factorizations of $a \cdot c$. This way you will either find m and n or verify that the trinomial will not factor.

1. Take the natural numbers 1,2,3,4, · · · and divide them into the $a \cdot c$ product. Those that divide in evenly we write as a factorization using the correct m and n signs.

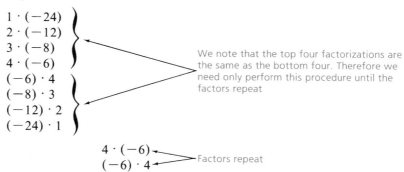

Factorization of -24, where the negative sign goes with the factor with the greater absolute value

$$1 \cdot (-24)$$
$$2 \cdot (-12)$$
$$3 \cdot (-8)$$
$$4 \cdot (-6)$$
$$(-6) \cdot 4$$
$$(-8) \cdot 3$$
$$(-12) \cdot 2$$
$$(-24) \cdot 1$$

We note that the top four factorizations are the same as the bottom four. Therefore we need only perform this procedure until the factors repeat

$$4 \cdot (-6)$$
$$(-6) \cdot 4$$
Factors repeat

2. Find the sum of the factorizations of $a \cdot c$. If there is a sum equal to b, the trinomial will factor. If there is no sum equal to b, then the trinomial will not factor with integer coefficients.

Factorizations of -24 Sum of the factors of -24

$$1 \cdot (-24) \qquad\qquad 1 + (-24) = -23$$
$$2 \cdot (-12) \qquad\qquad 2 + (-12) = -10$$
$$3 \cdot (-8) \qquad\qquad\; 3 + (-8)\; = -5$$
$$4 \cdot (-6) \qquad\qquad\; 4 + (-6)\; = -2$$

Passed -9

No sum equals -9

Since none of the factorizations of -24 also add to -9, there is no pair of integers (m and n) and the trinomial will not factor. Therefore, $4a^2 - 9a - 6$ is a **prime polynomial.**

Note   Regardless of the signs of m and n, the column of values of the sum of the factors will either be increasing or decreasing. Therefore once the desired value has been passed, the process can be stopped and the trinomial will not factor.

▶ *Quick check*   Express $3x^2 + 5x + 2$ and $4x^2 - 11x + 6$ in completely factored form, if possible.   ■

Factoring by inspection—an alternative approach

In the first part of this section, we studied a systematic procedure for determining if a trinomial will factor and how to factor it. In many instances, we are able to determine how the trinomial will factor by inspecting the problem rather than applying this procedure.

Factoring by inspection is accomplished as follows: Factor $8a + 3a^2 + 5$.

Step 1 Write the trinomial in standard form.

$$3a^2 + 8a + 5 \qquad \text{Arrange terms in descending powers of } a$$

Step 2 Determine the possible combinations of first-degree factors of the first term.

$(3a \quad)(a \quad)$ The only factorization of $3a^2$ is $3a \cdot a$

Step 3 Combine with the factors of step 2 all the possible factors of the last term.

$(3a \quad 5)(a \quad 1)$
$(3a \quad 1)(a \quad 5)$ The only factorization of 5 is $5 \cdot 1$

Step 4 Determine the possible symbol ($+$ or $-$) between the terms in each binomial.

$(3a + 5)(a + 1)$
$(3a + 1)(a + 5)$

The rules of signed numbers given in chapter 1 provide the answer to step 4.

1. If the third term is preceded by a $+$ sign and the middle term is preceded by a $+$ sign, then the symbols will be

$(\quad + \quad)(\quad + \quad)$.

2. If the third term is preceded by a $+$ sign and the middle term is preceded by a $-$ sign, then the symbols will be

$(\quad - \quad)(\quad - \quad)$.

3. If the third term is preceded by a $-$ sign, then the symbols will be

$(\quad + \quad)(\quad - \quad)$
or
$(\quad - \quad)(\quad + \quad)$.

Note It is assumed that the first term is preceded by a $+$ sign or no sign. If it is preceded by a $-$ sign, these rules could still be used if a (-1) is first factored out of all the terms.

Step 5 Determine which factors, if any, yield the correct middle term.

$(3a + 5)(a + 1)$

$+5a$ $(+5a) + (+3a) = +8a$ Correct middle term

$+3a$

$(3a + 1)(a + 5)$

$+a$ $(+a) + (15a) = +16a$

$+15a$

The first set of factors gives us the correct middle term. Therefore $(3a + 5)(a + 1)$ is the factorization of $3a^2 + 8a + 5$.

■ *Example 3–6 B*

Factor the following trinomials by inspection.

1. $x^2 + 7x + 12$

Step 1 $x^2 + 7x + 12$ Standard form

Step 2 $(x \quad)(x \quad)$ The only factorization of x^2 is $x \cdot x$

Step 3 $(x \quad 12)(x \quad 1)$ The factorizations of 12 are $1 \cdot 12$, $6 \cdot 2$, and $3 \cdot 4$
$(x \quad 6)(x \quad 2)$
$(x \quad 3)(x \quad 4)$

Step 4 $(x + 12)(x + 1)$ Using the rules of signed numbers, determine the possible signs between the terms
$(x + 6)(x + 2)$
$(x + 3)(x + 4)$

Step 5 $(x + 12)(x + 1)$
$\quad +12x$ $(+12x) + (+x) = +13x$
$\quad +x$

$(x + 6)(x + 2)$
$\quad +6x$ $(+6x) + (+2x) = +8x$
$\quad +2x$

$(x + 3)(x + 4)$
$\quad +3x$ $(+3x) + (+4x) = +7x$
$\quad +4x$ Correct middle term

The factorization of $x^2 + 7x + 12$ is $(x + 3)(x + 4)$.

2. $6x^2 + 17x + 7$

Step 1 $6x^2 + 17x + 7$ Standard form

Step 2 $(6x \quad)(x \quad)$ $6x^2 = 6x \cdot x$ or $3x \cdot 2x$
$(3x \quad)(2x \quad)$

Step 3 $(6x \quad 7)(x \quad 1)$ The only factorization of 7 is $7 \cdot 1$
$(6x \quad 1)(x \quad 7)$
$(3x \quad 7)(2x \quad 1)$
$(3x \quad 1)(2x \quad 7)$

Step 4 $(6x + 7)(x + 1)$ Using the rules of signed numbers, determine the possible signs between the terms
$(6x + 1)(x + 7)$
$(3x + 7)(2x + 1)$
$(3x + 1)(2x + 7)$

Step 5 $(6x + 7)(x + 1)$

$+7x$

$+6x$

$(+7x) + (+6x) = +13x$

$(6x + 1)(x + 7)$

$+x$

$+42x$

$(+x) + (+42x) = +43x$

$(3x + 7)(2x + 1)$

$+14x$

$+3x$

$(+14x) + (+3x) = +17x$ Correct middle term

$(3x + 1)(2x + 7)$

$+2x$

$+21x$

$(+2x) + (+21x) = +23x$

Hence, $(3x + 7)(2x + 1)$ is the factorization of $6x^2 + 17x + 7$.

3. $19x - 7 + 6x^2$

Step 1 $6x^2 + 19x - 7$ Standard form

Step 2 $(6x\ \ )(x\ \ )$ $6x^2 = 6x \cdot x$ or $2x \cdot 3x$
$(2x\ \ )(3x\ \ )$

Step 3 $(6x\ \ 7)(x\ \ 1)$ The only factorization of 7
$(6x\ \ 1)(x\ \ 7)$ is $7 \cdot 1$
$(2x\ \ 7)(3x\ \ 1)$
$(2x\ \ 1)(3x\ \ 7)$

Step 4 $(6x + 7)(x - 1)$ or $(6x - 7)(x + 1)$ Using the rules of signed
$(6x + 1)(x - 7)$ or $(6x - 1)(x + 7)$ numbers, determine the possible
$(2x + 7)(3x - 1)$ or $(2x - 7)(3x + 1)$ signs between the terms
$(2x + 1)(3x - 7)$ or $(2x - 1)(3x + 7)$

Step 5 $(6x + 7)(x - 1)$ or $(6x - 7)(x + 1)$

$+7x$ $-7x$

$-6x$ $+6x$

$(+7x) + (-6x) = +x$ or $(-7x) + (+6x) = -x$

$(6x + 1)(x - 7)$ or $(6x - 1)(x + 7)$

$+x$ $-x$

$-42x$ $+42x$

$(+x) + (-42x) = -41x$ or $(-x) + (+42x) = +41x$

$$(2x + 7)(3x - 1) \text{ or } (2x - 7)(3x + 1)$$

$$+21x \qquad\qquad -21x$$

$$-2x \qquad\qquad +2x$$

$$(+21x) + (-2x) = +19x \text{ or } (-21x) + (+2x) = -19x$$

Correct middle term

$$(2x + 1)(3x - 7) \text{ or } (2x - 1)(3x + 7)$$

$$+3x \qquad\qquad -3x$$

$$-14x \qquad\qquad +14x$$

$$(+3x) + (-14x) = -11x \text{ or } (-3x) + (+14x) = +11x$$

The factorization of $6x^2 + 19x - 7$ is $(2x + 7)(3x - 1)$. ■

___ **Mastery points** ___

Can you

- Determine two integers whose product is one number and whose sum is another number?
- Recognize when a trinomial will factor and when it will not?
- Factor trinomials of the form $ax^2 + bx + c$?
- Always remember to look for the greatest common factor before applying any of the factoring rules?

Exercise 3–6

Factor completely each trinomial. If a trinomial will not factor, state that as the answer. See examples 3–6 A and B.

Example $3x^2 + 5x + 2$

Solution **Step 1** $m \cdot n = 3 \cdot 2 = 6$ and $m + n = 5$
m and n are 2 and 3.

$$\overbrace{}^{5x}$$

Step 2 $= 3x^2 + 2x + 3x + 2$ Replace $5x$ with $2x$ and $3x$
Step 3 $= (3x^2 + 2x) + (3x + 2)$ Group the first two terms and the last two terms
    $= x(3x + 2) + 1(3x + 2)$ Factor out what is common to each pair
Step 4 $= (3x + 2)(x + 1)$ Factor out the common quantity

Example $4x^2 - 11x + 6$

Solution **Step 1** $m \cdot n = 4 \cdot 6 = 24$ and $m + n = -11$
m and n are -3 and -8.

$$\overbrace{}^{-11x}$$

Step 2 $= 4x^2 - 3x - 8x + 6$ Replace $-11x$ with $-3x$ and $-8x$
Step 3 $= (4x^2 - 3x) + (-8x + 6)$ Group the first two terms and the last two terms
    $= x(4x - 3) - 2(4x - 3)$ Factor out what is common to each pair
Step 4 $= (4x - 3)(x - 2)$ Factor out the common quantity

1. $3a^2 + 7a - 6$ 2. $2y^2 + y - 6$ 3. $4x^2 - 5x + 1$ 4. $2z^2 + 3z + 1$

5. $x^2 + 18x + 32$ 6. $2a^2 - 7a + 6$ 7. $2y^2 - y - 1$ 8. $5x^2 - 7x - 6$

9. $8x^2 - 17x + 2$ 10. $9x^2 - 6x + 1$ 11. $2a^2 - 11a + 12$ 12. $5y^2 + 4y + 6$

13. $2x^2 + 13x + 18$ 14. $6x^2 + 13x + 6$ 15. $7x^2 + 20x - 3$ 16. $4a^2 + 20a + 21$

17. $4x^2 - 4x - 3$ 18. $4z^2 - 2z + 5$ 19. $6x^2 - 23x - 4$ 20. $9x^2 - 21x - 8$

21. $10z^2 + 9z + 2$ 22. $10y^2 + 7y - 6$ 23. $7x^2 - 3x + 6$ 24. $2x^2 - 9x + 10$

25. $5x^2 - 9x - 2$ **26.** $6x^2 + 21x + 18$ 27. $6x^2 - 17x + 12$ 28. $4y^2 + 10y + 6$

29. $3x^2 + 12x + 12$ 30. $6x^2 + 7x - 3$ 31. $3x^2 + 8x - 4$ 32. $2x^2 + 6x - 20$

33. $6x^2 - 38x + 40$ 34. $18x^2 + 15x - 18$ **35.** $-4x^2 - 12x - 9$ 36. $-9z^2 + 30z - 25$

37. $-7x^2 + 36x - 5$ 38. $-3x^2 - 2x - 4$ 39. $12x^2 + 13x - 4$ 40. $15x^2 + 2x - 1$

41. $4x^3 + 10x^2 + 4x$ 42. $2x^3 - 6x^2 - 20x$ 43. $18x^2 - 42x - 16$ 44. $2x^3 + 15x^2 + 7x$

45. $8x^2 - 18x + 9$ 46. $8y^2 - 14y - 15$

Review exercises

Perform the indicated multiplication. See section 3-2.

1. $(3a - b)(3a + b)$ 2. $(x - 2y)(x + 2y)$ 3. $(3x - 4y)(3x + 4y)$ 4. $(x^2 + 5)(x^2 - 5)$

5. $(x^2 + 4)(x - 2)(x + 2)$ 6. $(a + 2b)(a^2 - 2ab + 4b^2)$

7. $(x - 3y)(x^2 + 3xy + 9y^2)$ 8. $(3a - 2b)(9a^2 + 6ab + 4b^2)$

3-7 ■ Other methods of factoring

The difference of two squares

We saw in section 3-2 that the product $(a + b)(a - b)$ is $a^2 - b^2$. We refer to the indicated product $(a + b)(a - b)$ as the *product* of the *sum* and *difference* of the same two terms. Notice that in one factor we *add* the terms and in the other we find the *difference* between these same terms. The product will *always* be the *difference of the squares of the two terms.*

To factor the **difference of two squares,** we reverse the equation from section 3-2.

> **The difference of two squares factors as** ⎯⎯⎯⎯⎯⎯
> $$a^2 - b^2 = (a + b)(a - b)$$
>
> **Concept**
> Verbally we have
>
> (1st term)^2 − (2nd term)^2 factors into
> (1st term + 2nd term) · (1st term − 2nd term)

To use this factoring procedure, we must be able to recognize **perfect squares.** For example,

$$16, \quad x^2, \quad 25a^2, \quad 9y^4$$

are all perfect squares since

$$16 = 4 \cdot 4 = (4)^2$$
$$x^2 = x \cdot x = (x)^2$$
$$25a^2 = 5a \cdot 5a = (5a)^2$$
$$9y^4 = 3y^2 \cdot 3y^2 = (3y^2)^2$$

We see from the examples that *a perfect square is any quantity that can be written as an exact square of a rational quantity.*

This is the procedure we use for factoring the difference of two squares.

Factoring the difference of two squares

Step 1 Identify that we have a perfect square minus another perfect square.

Step 2 Rewrite the problem as a first term squared minus a second term squared.

$$(\text{1st term})^2 - (\text{2nd term})^2$$

Step 3 Factor the problem into the first term plus the second term times the first term minus the second term.

$$(\text{1st term} + \text{2nd term})(\text{1st term} - \text{2nd term})$$

■ *Example 3–7 A*

Write in completely factored form.

1. $x^2 - 16$ Identify as the difference of squares
   $= (x)^2 - (4)^2$ Rewrite as squares
   $= (x + 4)(x - 4)$ Factor the binomial

2. $25a^2 - 4b^2$ Identify as the difference of squares
   $= (5a)^2 - (2b)^2$ Rewrite as squares
   $= (5a + 2b)(5a - 2b)$ Factor the binomial

3. $2x^2 - 18y^2$ Always look for the greatest common factor first

   $= 2(x^2 - 9y^2)$ Common factor of 2
   $= 2[(x)^2 - (3y)^2]$ Rewrite as squares
   $= 2(x + 3y)(x - 3y)$ Factor the binomial

4. $x^4 - 81$ Identify as the difference of squares
   $= (x^2)^2 - (9)^2$ Rewrite as squares
   $= (x^2 + 9)(x^2 - 9)$ Factor the binomial and inspect the factors
   $= (x^2 + 9)[(x)^2 - (3)^2]$ Identify the second factor as the difference of squares
   $= (x^2 + 9)(x + 3)(x - 3)$ Factor the second binomial

Our first factorization of $x^4 - 81$ yielded the factors $x^2 + 9$ and $x^2 - 9$. On further inspection, we observe that $x^2 - 9$ is the difference of two squares.

Note In example 4, $x^2 + 9$ is called the sum of two squares and will not factor using real numbers.

5. $x^{2n} - 4$ Identify as the difference of squares
   $= (x^n)^2 - (2)^2$ Rewrite as squares
   $= (x^n + 2)(x^n - 2)$ Factor the binomial

▶ *Quick check* Write $3b^2 - 12c^2$ in completely factored form. ■

The difference of two cubes

To factor problems involving the difference of two squares, we identified the two terms as squares and applied the procedure. Now we will factor the **difference of two cubes** in a similar fashion.

Consider the indicated product $(a - b)(a^2 + ab + b^2)$. If we carry out the multiplication, we have

$$(a - b)(a^2 + ab + b^2) = a^3 + a^2b + ab^2 - a^2b - ab^2 - b^3$$
$$= a^3 - b^3$$

Therefore, $(a - b)(a^2 + ab + b^2) = a^3 - b^3$ and $(a - b)(a^2 + ab + b^2)$ is the factored form of $a^3 - b^3$.

> ### The difference of two cubes factors as
>
> $$a^3 - b^3 = (a - b)(a^2 + ab + b^2)$$
>
> #### Concept
> If we are able to write a two-term polynomial as a first term cubed minus a second term cubed, then it will factor as the difference of two cubes.
>
> $$(\text{1st term})^3 - (\text{2nd term})^3 =$$
> $$(\text{1st term} - \text{2nd term})[(\text{1st term})^2 + (\text{1st term} \cdot \text{2nd term}) + (\text{2nd term})^2]$$

This is the procedure we use for factoring the difference of two cubes.

> ### Factoring the difference of two cubes
>
> **Step 1** Identify that we have a perfect cube minus another perfect cube.
>
> **Step 2** Rewrite the problem as a first term cubed minus a second term cubed.
>
> $$(\text{1st term})^3 - (\text{2nd term})^3$$
>
> **Step 3** Factor the problem into the first term minus the second term times the first term squared plus the first term times the second term plus the second term squared.
>
> $$(\text{1st term} - \text{2nd term})[(\text{1st term})^2 + (\text{1st term} \cdot \text{2nd term}) + (\text{2nd term})^2]$$

■ Example 3–7 B

Factor completely.

1. $a^3 - 8$

We rewrite both the terms as perfect cubes.

$$a^3 - 8 = (a)^3 - (2)^3$$

The first term is a and the second term is 2. Then we write down the procedure for factoring the difference of two cubes.

$$(\quad - \quad)[(\quad)^2 + (\quad)(\quad) + (\quad)^2]$$

1st 2nd 1st 1st 2nd 2nd

Now substitute a where the first term is and 2 where the second term is.

$(a - 2)[(a)^2 + (a)(2) + (2)^2]$

↑ ↑ ↑   ↑ ↑   ↑
1st 2nd 1st 1st 2nd 2nd

Finally, we simplify the second factor.

$(a - 2)(a^2 + 2a + 4)$

Therefore $a^3 - 8 = (a - 2)(a^2 + 2a + 4)$.

2. $40x^3 - 135y^3$ Always look for the greatest common factor first

$= 5(8x^3 - 27y^3)$ 5 is a common factor, identify the binomial

$= 5[(2x)^3 - (3y)^3]$ Rewrite as cubes

$5(\quad - \quad)[(\quad)^2 + (\quad)(\quad) + (\quad)^2]$ Factoring procedure ready for substitution

$5(2x - 3y)[(2x)^2 + (2x)(3y) + (3y)^2]$ The first term is $2x$, the second term is $3y$

$= 5(2x - 3y)(4x^2 + 6xy + 9y^2)$ Simplify within the second group

3. $a^{21} - 64b^{15}$ Identify the difference of two cubes

$= (a^7)^3 - (4b^5)^3$ Rewrite as cubes

$(\quad - \quad)[(\quad)^2 + (\quad)(\quad) + (\quad)^2]$ Factoring procedure ready for substitution

$(a^7 - 4b^5)[(a^7)^2 + (a^7)(4b^5) + (4b^5)^2]$ The first term is a^7, the second term is $4b^5$

$= (a^7 - 4b^5)(a^{14} + 4a^7b^5 + 16b^{10})$ Simplify within the second group

Note In example 3, we observe that a variable raised to a power that is a multiple of 3 can be written as a cube by dividing the exponent by 3. The quotient is the exponent of the number inside the parentheses and the 3 is the exponent outside the parentheses. For example,

$$y^{12} = (y^4)^3 \text{ and } z^{24} = (z^8)^3$$ ∎

The sum of two cubes

If we carry out the indicated multiplication in $(a + b)(a^2 - ab + b^2)$, we have

$$(a + b)(a^2 - ab + b^2) = a^3 - a^2b + ab^2 + a^2b - ab^2 + b^3$$
$$= a^3 + b^3$$

Therefore $(a + b)(a^2 - ab + b^2) = a^3 + b^3$ and $(a + b)(a^2 - ab + b^2)$ is the factored form of $a^3 + b^3$.

The sum of two cubes factors as

$$a^3 + b^3 = (a + b)(a^2 - ab + b^2)$$

Concept

If we are able to write a two-term polynomial as a first term cubed plus a second term cubed, then it will factor as the sum of two cubes.

$$(\text{1st term})^3 + (\text{2nd term})^3 =$$
$$(\text{1st term} + \text{2nd term})[(\text{1st term})^2 - (\text{1st term})(\text{2nd term}) + (\text{2nd term})^2]$$

This is the procedure we use for factoring the sum of two cubes.

___ *Factoring the sum of two cubes* _____

Step 1 Identify that we have a perfect cube plus another perfect cube.
Step 2 Rewrite the problem as a first term cubed plus a second term cubed.

$$(\text{1st term})^3 + (\text{2nd term})^3$$

Step 3 Factor the problem into the first term plus the second term times the first term squared minus the first term times the second term plus the second term squared.

$$(\text{1st term} + \text{2nd term})[(\text{1st term})^2 - (\text{1st term})(\text{2nd term}) + (\text{2nd term})^2]$$

■ *Example 3–7 C*

Factor completely.

1. $x^3 + 27$ Identify the sum of two cubes

     $= (x)^3 + (3)^3$ Rewrite as cubes

     $(\ \ + \ \ )[(\ \ )^2 - (\ \ )(\ \ ) + (\ \ )^2]$ Factoring procedure ready for substitution

       ↑ ↑ ↑   ↑ ↑   ↑
      1st 2nd 1st   1st 2nd   2nd

     $(x + 3)[(x)^2 - (x)(3) + (3)^2]$ The first term is x, the second term is 3

      ↑ ↑ ↑   ↑ ↑   ↑
     1st 2nd 1st 1st 2nd   2nd

     $= (x + 3)(x^2 - 3x + 9)$ Simplify within the second group

2. $8x^3 + y^9$ Identify the sum of two cubes

     $= (2x)^3 + (y^3)^3$ Rewrite as cubes

     $(\ \ + \ \ )[(\ \ )^2 - (\ \ )(\ \ ) + (\ \ )^2]$ Factoring procedure ready for substitution

     $(2x + y^3)[(2x)^2 - (2x)(y^3) + (y^3)^2]$ The first term is $2x$, the second term is y^3

     $= (2x + y^3)(4x^2 - 2xy^3 + y^6)$ Simplify within the second group

3. $a^3b^3 + 27c^3$ Identify the sum of two cubes

     $= (ab)^3 + (3c)^3$ Rewrite as cubes

     $(\ \ + \ \ )[(\ \ )^2 - (\ \ )(\ \ ) + (\ \ )^2]$ Factoring procedure ready for substitution

     $(ab + 3c)[(ab)^2 - (ab)(3c) + (3c)^2]$ The first term is ab, the second term is $3c$

     $= (ab + 3c)(a^2b^2 - 3abc + 9c^2)$ Simplify within the second group

▶ *Quick check* Factor $27b^3 + c^{12}$ completely.   ■

___ *Mastery points* _____

Can you
■ Factor the difference of two squares?
■ Factor the sum and the difference of two cubes?

Exercise 3–7

Write in completely factored form. Assume all variables used as exponents represent positive integers. See examples 3–7 A, B, and C.

Example $3b^2 - 12c^2$

Solution $3(b^2 - 4c^2)$ Common factor of 3 and identify the binomial
$3[(b)^2 - (2c)^2]$ Rewrite as squares
$3(b + 2c)(b - 2c)$ Factor the binomial

Example $27b^3 + c^{12}$

Solution $(3b)^3 + (c^4)^3$ Identify and rewrite as cubes
$(\ + \ )[(\ )^2 - (\ )(\ ) + (\ )^2]$ Factoring procedure ready for substitution
$(3b + c^4)[(3b)^2 - (3b)(c^4) + (c^4)^2]$ The first term is $3b$, the second term is c^4
$(3b + c^4)(9b^2 - 3bc^4 + c^8)$ Simplify within the second group

1. $a^2 - 49$
2. $36 - R^2$
3. $64 - S^2$
4. $4a^2 - 25b^2$
5. $36x^2 - y^4$
6. $x^4 - 4$
7. $16a^2 - 49b^2$
8. $144x^2 - 121y^2$
9. $8a^2 - 32b^2$
10. $3x^2 - 27y^2$
11. $50 - 2x^2$
12. $128a^2 - 2b^2$
13. $98a^2b^2 - 50x^2y^2$
14. $x^4 - y^4$
15. $a^4 - 81$
16. $16R^4 - 1$
17. $x^4 - 16y^4$
18. $x^{2n} - 1$
19. $a^{2n} - 4$
20. $x^{4n} - 25$
21. $x^{2n} - y^{2n}$
22. $x^{4n} - 16$
23. $x^{4n} - 81$
24. $a^2(x + 2y) - b^2(x + 2y)$
25. $4x^2(a + 5b) - y^2(a + 5b)$
26. $2a^2(x + y) - 8b^2(x + y)$
27. $3a^2(3x - y) - 27(3x - y)$
28. $4a(x^2 - y^2) - 8b(x^2 - y^2)$
29. $9x(4a^2 - b^2) - 3y(4a^2 - b^2)$
30. $(a + b)^2 - (x - y)^2$
31. $(2a + b)^2 - (x - 2y)^2$
32. $(2x - y)^2 - (x + y)^2$
33. $(3a - b)^2 - (2a - b)^2$
34. $(4x + 3y)^2 - (4x - 2y)^2$
35. $(x + 2y)^2 - (x - 3y)^2$
36. $R^3 - 8$
37. $27a^3 - b^3$
38. $8y^3 - 27$
39. $x^3 + y^3$
40. $27 + x^3$
41. $a^3 + 8b^3$
42. $64a^3 + b^3$
43. $8y^3 - 27x^3$
44. $64a^3 - 8$
45. $81a^3 - 3b^3$
46. $2a^3 + 16$
47. $3x^3 + 24$
48. $2a^3 - 16$
49. $64z^3 + 125$
50. $a^5 + 27a^2b^3$
51. $2a^3 - 54b^3$
52. $8x^2y^3 - x^5$
53. $R^5 + 64R^2S^3$
54. $54y^3 + 2z^3$
55. $3x^5 - 81x^2y^3$
56. $a^3b^9 - c^{15}$
57. $x^3y^{12} + z^9$
58. $x^{15}y^6 - 8z^9$
59. $a^{18}b^9 - 27c^3$
60. $2a^3b^3 + 16$
61. $3x^3y^6 + 81z^3$
62. $(a + 2b)^3 - c^3$
63. $(x + 3y)^3 + z^3$
64. $(2a - b)^3 + 8c^3$
65. $(4a + b)^3 - 27c^3$
66. $(2a + b)^3 - (x - y)^3$
67. $(a - 2b)^3 - (2x + y)^3$
68. $(3a + b)^3 - (a - 2b)^3$
69. $(3x - y)^3 + (2x + 3y)^3$
70. $(a - 2b)^3 - (a + b)^3$
71. $(2x - y)^3 + (x + y)^3$
72. $(5x - 2y)^3 + (3x + 2y)^3$

73. In engineering, the equation of transverse shearing stress in a rectangular beam is given in two forms:

(a) $T = \dfrac{V}{8I}(h^2 - 4V_1{}^2)$ and

(b) $T = \dfrac{3V}{2A}\left(1 - \dfrac{4V_1{}^2}{H^2}\right)$

Factor the right member of each equation.

Review exercises

Write in completely factored form. See sections 3–4, 3–5, and 3–6.

1. $a^2 - 7a + 10$

2. $5ax + 2bx - 10ay - 4by$

3. $x^2 + 4xy + 4y^2$

4. $6a^2 - ab - b^2$

5. $5a^3 - 40a^2 + 75a$

6. $6x^2 + x - 12$

3–8 ■ Factoring: A general strategy

In this section, we will review the different methods of factoring that we have studied in the previous sections. The following outline gives a general strategy for factoring polynomials.

  I. Factor out the greatest common factor.

    Examples

     1. $5x^3 - 25x^2 = 5x^2(x - 5)$

     2. $x(a - 2b) + 2y(a - 2b) = (a - 2b)(x + 2y)$

  II. Count the number of terms.

    A. Two terms: Check to see if the polynomial is the difference of two squares, the difference of two cubes, or the sum of two cubes.

      Examples

       1. $x^2 - 25y^2 = (x + 5y)(x - 5y)$ Difference of two squares

       2. $8a^3 - b^3 = (2a - b)(4a^2 + 2ab + b^2)$ Difference of two cubes

       3. $m^3 + 64n^3 = (m + 4n)(m^2 - 4mn + 16n^2)$ Sum of two cubes

    B. Three terms: Check to see if the polynomial is a perfect square trinomial. If it is not, use one of the general methods for factoring a trinomial.

      Examples

       1. $m^2 + 6m + 9 = (m + 3)^2$ Perfect square trinomial

       2. $m^2 - 5m - 14 = (m - 7)(m + 2)$ General trinomial, leading coefficient of 1

       3. $6x^2 + 7x - 20 = (2x + 5)(3x - 4)$ General trinomial, leading coefficient other than 1

    C. Four terms: Check to see if we can factor by grouping.

      Examples

       1. $ax + 3a - 2bx - 6b = (a - 2b)(x + 3)$

       2. $x^3 + 2x^2 - 3x - 6 = (x^2 - 3)(x + 2)$

 III. Check to see if any of the factors we have written can be factored further. Any common factors that were missed in part I can still be factored out here.

    Examples

     1. $x^4 - 11x^2 + 28 = (x^2 - 4)(x^2 - 7)$

        $= (x - 2)(x + 2)(x^2 - 7)$ Difference of two squares

     2. $4a^2 - 36b^2 = (2a + 6b)(2a - 6b)$

        $= 2(a + 3b)2(a - 3b) = 4(a + 3b)(a - 3b)$ Overlooked common factor

The following examples illustrate our strategy for factoring polynomials.

■ *Example 3–8 A*

Completely factor the following polynomials.

1. $12a^3 - 3ab^2$

I. First, we look for the greatest common factor.

$$3a(4a^2 - b^2)$$ Common factor of $3a$

II. The factor $4a^2 - b^2$ has two terms and is the difference of two squares.

$$3a(4a^2 - b^2) = 3a(2a + b)(2a - b)$$ Factoring the binomial

III. After checking to see if any of the factors will factor further, we conclude that $3a(2a + b)(2a - b)$ is the completely factored form. Therefore

$$12a^3 - 3ab^2 = 3a(2a + b)(2a - b)$$

2. $3x^2 + 7xy - 6y^2$

I. There is no common factor (other than 1 or -1).

II. Since this trinomial is not a perfect square, we use our general method for factoring trinomials.

$m \cdot n = -18$ and $m + n = 7$, m and n are -2 and 9.

$$
\begin{aligned}
3x^2 + 7xy - 6y^2 \\
= 3x^2 - 2xy + 9xy - 6y^2 &\quad \text{Replace } 7xy \text{ with } -2xy \text{ and } 9xy \\
= x(3x - 2y) + 3y(3x - 2y) &\quad \text{Group the first two terms and the last two terms and factor out what is common to each pair} \\
= (3x - 2y)(x + 3y) &\quad \text{Factor out the common quantity}
\end{aligned}
$$

III. None of the factors will factor further.

$$3x^2 + 7xy - 6y^2 = (3x - 2y)(x + 3y)$$

3. $2am^2 + bm^2 - 2an^2 - bn^2$

I. There is no common factor (other than 1 or -1).

II. The polynomial has four terms and we factor by grouping.

$$
\begin{aligned}
&= (2am^2 + bm^2) + (-2an^2 - bn^2) &\quad \text{Group in pairs} \\
&= m^2(2a + b) - n^2(2a + b) &\quad \text{Factor out what is common to each pair} \\
&= (2a + b)(m^2 - n^2) &\quad \text{Factor out the common quantity}
\end{aligned}
$$

III. The factor $m^2 - n^2$ is the difference of two squares.

$$2am^2 + bm^2 - 2an^2 - bn^2 = (2a + b)(m + n)(m - n)$$ Completely factored form

4. $5x^3 + 5x^2 + 20x$

I. The greatest common factor is $5x$.

$$5x(x^2 + x + 4)$$ Common factor of $5x$

II. The trinomial $x^2 + x + 4$ will not factor.

III. None of the factors will factor further.

$$5x^3 + 5x^2 + 20x = 5x(x^2 + x + 4)$$ Completely factored form

5. $a^8 + 8a^2b^3$

I. The greatest common factor is a^2.

$$a^2(a^6 + 8b^3)$$ Common factor of a^2

II. The factor $a^6 + 8b^3$ is the sum of two cubes.

$a^2(a^6 + 8b^3) = a^2(a^2 + 2b)(a^4 - 2a^2b + 4b^2)$ Factoring the sum of cubes

III. None of the factors will factor further.

$a^8 + 8a^2b^3 = a^2(a^2 + 2b)(a^4 - 2a^2b + 4b^2)$ Completely factored form

6. $a^2(4 - x^2) + 8a(4 - x^2) + 16(4 - x^2)$

I. The greatest common factor is $4 - x^2$.

$(4 - x^2)(a^2 + 8a + 16)$ Common factor of $4 - x^2$

II. The factor $4 - x^2$ is the difference of two squares. The factor $a^2 + 8a + 16$ is a perfect square trinomial.

$(2 + x)(2 - x)(a + 4)^2$ Factoring the binomial and the trinomial

III. None of the factors will factor further.

$a^2(4 - x^2) + 8a(4 - x^2) + 16(4 - x^2)$
$= (2 + x)(2 - x)(a + 4)^2$ Completely factored form

▶ *Quick check*   Factor $4x^2 - 36y^2$ completely.   ■

Mastery points

Can you
- Factor out the greatest common factor?
- Factor the difference of two squares?
- Factor the difference of two cubes?
- Factor the sum of two cubes?
- Factor trinomials?
- Factor a four-term polynomial?
- Use the general strategy for factoring polynomials?

Exercise 3–8

Completely factor the following polynomials. If a polynomial will not factor, so state. See the outline of the general strategy for factoring polynomials and example 3–8 A.

Example $4x^2 - 36y^2$

Solution   I. $4x^2 - 36y^2 = 4(x^2 - 9y^2)$ Common factor of 4
   II. $= 4(x + 3y)(x - 3y)$ Factoring the binomial
   III. $4x^2 - 36y^2 = 4(x + 3y)(x - 3y)$ Completely factored form

1. $m^2 - 49$ **2.** $81 - x^2$ **3.** $x^2 + 6x + 5$

4. $a^2 + 11a + 10$ **5.** $7a^2 + 36a + 5$ **6.** $3x^2 + 13x + 4$

7. $2a^2 + 15a + 18$ **8.** $5b^2 + 16b + 12$ **9.** $a^2b^2 + 2ab - 8$

10. $x^2y^2 - 5xy - 14$ **11.** $27a^3 + b^3$ **12.** $x^3 + 64y^3$

13. $25x^2(3x + y) + 5a(3x + y)$ **14.** $36x^2(2a - b) - 12x(2a - b)$ **15.** $10x^2 - 20xy + 10y^2$

16. $6a^2 - 24ab - 48b^2$ **17.** $4m^2 - 16n^2$ **18.** $9x^2 - 36y^2$

19. $(a - b)^2 - (2x + y)^2$

20. $(3a + b)^2 - (x - y)^2$

21. $3x^6 - 81y^3$

22. $32a^4 - 4ab^9$

23. $12x^3y^2 - 18x^2y^2 + 16xy^4$

24. $9x^5y - 6x^3y^3 + 3x^2y^2$

25. $4x^2 - 9y^2$

26. $36 - a^2b^2$

27. $3ax + 6ay - bx - 2by$

28. $6am + 4bm - 3an - 2bn$

29. $27a^9 - b^3c^3$

30. $x^{12} - y^3z^6$

31. $5a^2 - 32a - 21$

32. $7a^2 + 16a - 15$

33. $a^4 - 5a^2 + 4$

34. $x^4 - 37x^2 + 36$

35. $4a^2 - 4ab - 15b^2$

36. $6x^2 + 7xy - 3y^2$

37. $y^4 - 16$

38. $z^4 - 81$

39. $4a^2 + 10a + 4$

40. $6x^2 + 18x - 60$

41. $(x + y)^2 - 8(x + y) - 9$

42. $(a - 2b)^2 + 7(a - 2b) + 10$

43. $6a^2 + 7a - 5$

44. $4x^2 + 17x - 15$

45. $4ab(x + 3y) - 8a^2b^2(x + 3y)$

46. $3x^2y(m - 4n) + 15xy^2(m - 4n)$

47. $4a^2 - 20ab + 25b^2$

48. $9m^2 - 30mn + 25n^2$

49. $80x^5 - 5x$

50. $3b^5 - 48b$

51. $3a^5b - 18a^3b^3 + 27ab^5$

52. $3x^3y^3 + 6x^2y^4 + 3xy^5$

53. $9a^2 - (x + 5y)^2$

54. $7x(a^2 - 4b^2) + 14(a^2 - 4b^2)$

55. $3x^2 + 8x - 91$

56. $3x^2 - 32x - 91$

57. $3x^5y^9 + 81x^2z^6$

58. $a^8b^4 + 27a^2b^7$

59. $a^2(9 - x^2) - 6a(9 - x^2) + 9(9 - x^2)$

60. $x^2(16 - b^2) + 10x(16 - b^2) + 25(16 - b^2)$

Review exercises

Find the solution set of the following equations. See section 2–1.

1. $x - 7 = 0$

2. $2x + 3 = 0$

3. $5x - 4 = 0$

4. $3x = 0$

5. $6 - 5x = 0$

6. $4 - 2x = 0$

Chapter 3 lead-in problem

The amount A of a radioactive substance remaining after time t can be found using the formula $A = A_0(0.5)^{t/n}$, where A_0 represents the original amount of radioactive material and n is the half-life given in the same units of time as t. If the half-life of radioactive carbon 14 is 5,770 years, how much radioactive carbon 14 will remain after 11,540 years if we start with 100 grams?

Solution

$A = A_0(0.5)^{t/n}$	Original formula
$A = (\ \ )(0.5)^{(\ )/(\ )}$	Formula ready for substitution
$A = (100)(0.5)^{11,540/5770}$	Substitute
$A = (100)(0.5)^2$	Simplify the exponent
$A = (100)(0.25)$	Powers
$A = 25$	Multiply

There will be 25 grams of radioactive carbon 14 remaining after 11,540 years.

Chapter 3 summary

1. In the expression a^n, a is called the **base** and n the **exponent.**
2. The following are properties and definitions involving exponents.

   $$\overbrace{a^n = a \cdot a \cdot a \cdots a}^{n \text{ factors of } a}$$

   a. $a^n = \overbrace{a \cdot a \cdot a \cdots a}$
   b. $a^m \cdot a^n = a^{m+n}$
   c. $(a^m)^n = a^{mn}$
   d. $(ab)^n = a^n b^n$
   e. $a^m \div a^n = \dfrac{a^m}{a^n} = a^{m-n}, a \neq 0$
   f. $a^{-n} = \dfrac{1}{a^n}, a \neq 0$
   g. $a^0 = 1, a \neq 0$
   h. $\left(\dfrac{a}{b}\right)^n = \dfrac{a^n}{b^n}, b \neq 0$

3. When multiplying two multinomials, we multiply each term of the first multinomial by each term of the second multinomial and then combine like terms.
4. Three **special products** are
   $(a + b)^2 = a^2 + 2ab + b^2$
   $(a - b)^2 = a^2 - 2ab + b^2$
   $(a + b)(a - b) = a^2 - b^2$
5. The **scientific notation** form of a number Y is $Y = a \times 10^n$, where a is a number greater than or equal to 1 and less than 10, and n is an integer.

6. **A prime number** is any natural number greater than 1 that is divisible only by itself and by 1.
7. A polynomial with integer coefficients will be considered to be in **completely factored form** when it is expressed as the product of polynomials with integer coefficients and none of the factors except a monomial can still be written as the product of two polynomials with integer coefficients.
8. We try to factor **four-term polynomials** by grouping.
9. The **trinomial** $x^2 + bx + c$ will factor only if we can find a pair of integers, m and n, whose product is c and whose sum is b.
10. The **trinomial** $ax^2 + bx + c$ will factor only if we can find a pair of integers, m and n, whose product is $a \cdot c$ and whose sum is b.
11. Perfect square trinomials factor as
    $$a^2 + 2ab + b^2 = (a + b)^2$$
    and
    $$a^2 - 2ab + b^2 = (a - b)^2$$
12. The **difference of two squares** factors as
    $$a^2 - b^2 = (a + b)(a - b)$$
13. The **sum and difference of two cubes** factors as
    $$a^3 + b^3 = (a + b)(a^2 - ab + b^2)$$
    and
    $$a^3 - b^3 = (a - b)(a^2 + ab + b^2)$$

Chapter 3 error analysis

1. Exponential notation
   Example: $(-2)^4 = -(2 \cdot 2 \cdot 2 \cdot 2) = -16$
   Correct answer: 16
   What error was made? (*see page 16*)
2. Product of like bases
   Example: $a \cdot a^3 \cdot a^5 = a^{3+5} = a^8$
   Correct answer: a^9
   What error was made? (*see page 99*)
3. Power to a power
   Example: $(x^3)^5 = x^{3+5} = x^8$
   Correct answer: x^{15}
   What error was made? (*see page 99*)
4. Power to a power
   Example: $(4ab^3)^2 = 4a^2(b^3)^2 = 4a^2 b^6$
   Correct answer: $16a^2 b^6$
   What error was made? (*see page 100*)
5. Squaring a binomial
   Example: $(x + 9)^2 = (x)^2 + (9)^2 = x^2 + 81$
   Correct answer: $x^2 + 18x + 81$
   What error was made? (*see page 106*)

6. Factoring the sum of two cubes
   Example: $x^3 + 8 = (x + 2)^3$
   Correct answer: $(x + 2)(x^2 - 2x + 4)$
   What error was made? (*see page 144*)
7. Negative exponents
   Example: $5^{-2} = -(5)^2 = -25$
   Correct answer: $\dfrac{1}{25}$
   What error was made? (*see page 113*)
8. Negative exponents
   Example: $-6^{-3} = 6^3 = 216$
   Correct answer: $-\dfrac{1}{216}$
   What error was made? (*see page 113*)
9. Zero exponents
   Example: $5x^0 = 1$
   Correct answer: 5
   What error was made? (*see page 114*)
10. Completely factoring an expression
    Example: $6x^2 - 18x - 24 = (x - 4)(6x + 6)$
    Correct answer: $6(x - 4)(x + 1)$
    What error was made? (*see page 147*)

Chapter 3 critical thinking

Given the problem 48^2, determine a method by which you can square the 48 mentally.

Chapter 3 review

[3–1]

Perform the indicated operations.

1. $(-5x^2)(3x^3)$

2. $(2a^2b)(3ab^4)$

3. $(2a^3b^4c)^3$

4. $(xy^3z)^2(x^2y)^3$

5. $(3a^2)^2a^3 + (2a)^3a^4$

6. $(2b^5)^2 - (3b^2)^3b^2$

[3–2]

Perform the indicated multiplication and simplify.

7. $5a^2b(2a^2 - 3ab + 4b^2)$

8. $(2a - b)(a + b)$

9. $(y - 7)(y + 7)$

10. $(2a + b)^2$

11. $(a + b)(a^2 - 3ab + 2b^2)$

12. $(3x - 2)(x + 4) - (x - 3)^2$

[3–3]

Perform all indicated operations and leave the answers with only positive exponents. Assume that all variables represent nonzero real numbers.

13. $\dfrac{2a^2b^4}{2^3a^5b}$

14. $(3x^{-4})(2x^{-5})$

15. $(3x^{-2}y^3z^0)^2$

16. $\left(\dfrac{16x^4y^3}{4xy^8}\right)^3$

17. $\dfrac{a^5b^{-2}}{a^{-4}b}$

18. $\dfrac{2^{-3}a^0b^{-3}c^2}{4^{-1}a^{-3}b^{-2}c^{-5}}$

19. $(2x^3y^{-4})^2(x^{-2}y)^3$

[3–4]

Write in completely factored form.

20. $12x^3 - 18x^2$

21. $12a^4b - 4a^2b^3 + 24a^3b^2$

22. $x(a + b) - 2y(a + b)$

23. $10x^2(x - 3z) + 5x(x - 3z)$

24. $6ax - 3ay - 2bx + by$

25. $4ax + 6by + 8ay + 3bx$

26. $4ax + 2ay + 6bx + 3by$

27. $ax + 3bx - 2ay - 6by$

28. $6ax + by - 2bx - 3ay$

29. $2ax + 3by + 6bx + ay$

[3–5]

Write in completely factored form. If a polynomial will not factor, so state.

30. $a^2 + 14a + 24$

31. $b^2 - 9b + 14$

32. $2a^3 - 8a^2 - 10a$

33. $x^3 - x^2 - 6x$

34. $x^2y^2 + 10xy + 24$

35. $a^2b^2 - 8ab - 20$

36. $(x + 3y)^2 + (x + 3y) - 2$

37. $(x + 2y)^2 - 14(x + 2y) + 49$

38. $a^2 - ab - 2b^2$

39. $x^2 + 5xy + 6y^2$

[3–6]

Write in completely factored form. If a polynomial will not factor, so state.

40. $2a^2 - 7a + 6$

41. $8x^2 - 14x + 5$

42. $6y^2 - 5y - 4$

43. $3a^2 + 2a - 1$

44. $4x^2 + 11x - 3$

45. $4x^2 + 4x + 1$

46. $24a^2 + 22a + 3$

47. $8x^2 - 18x + 9$

48. $2x^2 + 15x + 18$

49. $6x^2 + 51x + 24$

50. $-2x^2 + 11x + 6$

[3–7]

Write in completely factored form.

51. $x^2 - 81$

52. $4a^2 - 36b^2$

53. $3a^2 - 27b^2$

54. $y^4 - 81$

55. $x^2(a + 2b) - y^2(a + 2b)$

56. $4a^2(x - 3z) - 9b^2(x - 3z)$

57. $a^3 + 8b^3$

58. $27x^3 - y^3$

59. $64a^3 + 8b^3$

60. $27x^5 + x^2y^3$

61. $2x^3 - 16b^3$

62. $x^{12}y^{15} - z^9$

[3–8]

Write in completely factored form.

63. $9x^2 + 24x + 16$

64. $a^6 - b^9$

65. $3a^3 - 75a$

66. $a^2(9 - x^2) + 12a(9 - x^2) + 36(9 - x^2)$

67. $6a^2 + 17a - 3$

68. $4mx - 8my + 3nx - 6ny$

69. $40x^4 + 5xy^3$

70. $a^3b^2 - 4a^2b^3 + 4ab^4$

71. $x^2 + 2xy - 8y^2$

72. $15a^2 + 4a - 4$

Chapter 3 cumulative test

Determine if the following statements are true or false.

[1–1] **1.** $|-10| < 0$

[1–2] **2.** $-3^2 = (-3)^2$

[1–1] **3.** $J \subseteq Q$

Perform the indicated operations, if possible, and simplify.

[1–2] **4.** $\dfrac{(-8)}{(-4)}$

[1–2] **5.** $\dfrac{(-7)}{0}$

[1–2] **6.** $(-2) - (-6)$

[1–2] **7.** $(-2)(4)(0)(-6)$

[1–2] **8.** $(-3)^4$

[1–4] **9.** $48 - 24 \div 6 - 3 - 2^2$

[1–4] **10.** $2[-3(10 - 7) - 12 + 4]$

Solve the following equations and inequalities.

[2–5] **11.** $3x - 4 \geq x + 10$

[2–1] **12.** $2(3x - 4) + 7 = 8x - 11$

[2–4] **13.** $|3x + 4| = 5$

[2–6] **14.** $|8 - 3x| < 9$

[2–6] **15.** $|6x + 5| - 4 > 2$

[2–5] **16.** $-4 < 3 - 2x < 6$

[2–5] **17.** $-3 \leq 2x + 7 \leq 4$

[2–3] **18.** Three times a number is subtracted from 43 and this result is less than 16. Find all numbers that satisfy this condition.

[2–3] **19.** Earl has invested \$8,000 at a 9% rate. How much more must he invest at 12% to make the total income for one year from both investments a 10% rate?

Perform the indicated operations and simplify. Assume that all variables represent nonzero real numbers.

[3–1] **20.** $(2a^2b)(3a^3b^4)$

[3–1] **21.** $(2x^3y^4)^4$

[3–3] **22.** $\dfrac{27a^5b^9c}{18a^2b^4c^7}$

[3–3] **23.** $\dfrac{x^{-4}y^3}{x^2y^{-1}}$

[3–3] **24.** $\left(\dfrac{3x^2y^4}{9x^5y}\right)^3$

[1–6] **25.** $(5x^2 - 3x + 4) - (x^2 - x + 7)$

[3–3] **26.** $(3a^{-3}b^2c^{-1})^{-3}$

[1–6] **27.** $4ab - \{2a + 3b - [a - (2b - 3ab)]\}$

[3–2] **28.** $(3a - b)^2$

[3–1] **29.** $(2a^2b)^3(a^3b^2)^4$

[3–2] **30.** $6x^4y^3(2x^2 - 3x^2y + 4y^2)$

Write in completely factored form. If a polynomial will not factor, so state.

[3–7] **31.** $x^2 - 36$

[3–7] **32.** $x^3 + 27y^3$

[3–6] **33.** $2a^2 - 15a + 18$

[3–6] **34.** $y^2 + 7y + 6$

[3–5] **35.** $x^2y^2 - 2xy - 8$

[3–6] **36.** $7x^2 - 34x - 5$

[3–4] **37.** $2ax - 6ay + 3bx - 9by$

[3–4] **38.** $15x^2(2a - b) + 5x(2a - b)$

[3–5] **39.** $4x^2 - 20xy + 25y^2$

[3–7] **40.** $3a^2 - 12b^2$

4

Rational Expressions

Doug Nance opens the valve to fill his new swimming pool. Without his knowledge, the drain to empty the pool has been left open. If the pool would normally fill in 10 hours and the drain can empty the full pool in 15 hours, how long does it take to fill the pool?

4–1 ■ Fundamental principle of rational expressions

The rational expression

In chapter 1, we defined a rational number to be any number that can be expressed as a quotient of two integers where the denominator is not zero. That is, the set of rational numbers are the elements in the set

$$\left\{ \frac{p}{q} \,\middle|\, p \text{ and } q \text{ are integers, } q \neq 0 \right\}$$

If this definition is extended to polynomials, the result is called a **rational expression.**

┌─ Definition ─────────────────────────────

A **rational expression** is any algebraic expression that can be written as a quotient of two polynomials, where the denominator is not equal to zero.

That is, rational expressions are the elements in the set

$$\left\{ \frac{P}{Q} \,\middle|\, P \text{ and } Q \text{ are polynomials, } Q \neq 0 \right\}$$

The following are rational expressions:

$$\frac{2x}{3}, \quad \frac{2x-3}{x^2+6x-1}, \frac{5x+1}{3}, \quad \text{and} \quad 5x^3 - 3x + 1$$

Note $5x^3 - 3x + 1$ is a rational expression since it can be written

$$\frac{5x^3 - 3x + 1}{1}$$

Recall that division by zero is not defined. The rational expression becomes meaningless for those values of the variable for which the denominator is zero. For example, the rational expression

$$\frac{3x - 2}{x - 6}$$

is meaningless if $x = 6$, because then $x - 6 = 6 - 6 = 0$. Thus we say that

$$\frac{3x - 2}{x - 6}, \; x \neq 6$$

is a rational expression.

Note   The restrictions on the variable are found by setting all factors of the denominator that contain the variable equal to zero and solving the resulting equation(s) for the variable.

Domain of a rational expression

We call the set of all those values that the variable in a rational expression can assume the **domain** of the rational expression.

___ *Definition of the domain* _____

The **domain** of a rational expression is the set of all replacement values of the variable for which the expression is defined.

Thus, given the rational expression

$$\frac{3x - 2}{x - 6}$$

the domain is defined to be the set of all real numbers except 6. In set-builder notation, we symbolize the domain as follows:

$$\text{domain} = \{x \,|\, x \in R, \, x \neq 6\}$$

When the factored denominator consists of more than one factor containing the variable, we use the following property to determine restrictions on the variable.

___ *Zero product property* _____

If P and Q are polynomials, then

$$P \cdot Q = 0$$

if and only if $P = 0$ or $Q = 0$.

Concept
The product of two polynomials is zero provided one or both of the polynomials is equal to zero.

■ *Example 4–1 A*

Determine the domain of the following rational expressions. Express this in set-builder notation.

1. $\dfrac{x - 5}{x + 7}$

   $x + 7 = 0$ Set the denominator equal to 0
   $x = -7$ Solve the equation

   The number -7 cannot be used as a replacement for x.

   Domain $= \{x | x \in R, x \neq -7\}$

2. $\dfrac{4x - 3}{x^2 - 16}$

   $x^2 - 16 = (x + 4)(x - 4)$ Factor the denominator
   $x + 4 = 0$ or $x - 4 = 0$ Set each factor equal to 0
   $x = -4$ $x = 4$

   The numbers -4 and 4 cannot be used as replacements for x.

   Domain $= \{x | x \in R, x \neq -4, x \neq 4\}$

3. $\dfrac{5y^2}{y^2 + 9}$

   Since $y^2 + 9$ is positive for every value of y, then $y^2 + 9$ is *never* equal to zero. Hence, the domain $= \{y | y \in R\}$.

▶ *Quick check* Determine the domain of $\dfrac{3y - 2}{y^2 - 1}$ and express the domain in set-builder notation.

Reducing to lowest terms

The methods used to simplify rational expressions by reducing them to lowest terms are similar to those used to simplify fractions. A rational expression is reduced to lowest terms when there is no polynomial factor (other than 1 or -1) that is common to both the numerator and the denominator. This is accomplished by factoring the numerator and the denominator and removing a factor of 1. Recall

$$\frac{8}{8} = 1; \quad \frac{-9}{-9} = 1; \quad \frac{x - 2}{x - 2} = 1, \text{ if } x \neq 2$$

Given polynomials P, Q, and R,

$$\frac{PR}{QR} = \frac{P \cdot R}{Q \cdot R} = \frac{P}{Q} \cdot \frac{R}{R} = \frac{P}{Q} \cdot 1 = \frac{P}{Q}$$

Thus,

$$\frac{(x + 2)(x - 3)}{(2x - 1)(x - 3)} = \frac{x + 2}{2x - 1} \cdot \frac{x - 3}{x - 3} = \frac{x + 2}{2x - 1} \cdot 1 = \frac{x + 2}{2x - 1} \quad \left(x \neq \frac{1}{2} \text{ or } 3\right)$$

We have applied the fundamental principle of rational expressions which is now stated.

Fundamental principle of rational expressions ————————————

If P, Q, and R are polynomials, $Q \neq 0$ and $R \neq 0$, then

$$\frac{P}{Q} = \frac{P \cdot R}{Q \cdot R} \quad \text{and} \quad \frac{P \cdot R}{Q \cdot R} = \frac{P}{Q}$$

Concept

Both the numerator and the denominator of a rational expression may be multiplied or divided by the same nonzero polynomial without changing the value of the expression.

By this principle, the rational expression

$$\frac{x - 3}{x^2 - 9} = \frac{x - 3}{(x + 3)(x - 3)} \qquad \text{Factor the denominator}$$

$$= \frac{1}{x + 3} \ (x \neq -3, x \neq 3) \ \text{Divide the numerator and the denominator by } x - 3$$

Note   The domain of the expression excludes -3 and 3 even though the factor $x - 3$ is not in the resulting expression.

We use the **fundamental principle of rational expressions** to simplify rational expressions by reducing them to lowest terms. This is accomplished by dividing the numerator and the denominator by all common factors.

To reduce a rational expression to lowest terms ————————————

1. Completely factor the numerator and the denominator.
2. Divide the numerator and the denominator by the greatest common factor.

■ **Example 4-1 B**

Simplify each rational expression by reducing it to lowest terms. State any necessary restrictions on the variables.

Greatest common factor

1. $\dfrac{16x^2}{24x^4} = \dfrac{2(8x^2)}{3x^2(8x^2)}$ Factor the numerator and the denominator

Greatest common factor

$$= \frac{2}{3x^2} \ (x \neq 0)$$ Divide the numerator and the denominator by $8x^2$

2. $\dfrac{6x - 18}{5x - 15} = \dfrac{6(x - 3)}{5(x - 3)}$ Factor the numerator and the denominator

$$= \frac{6}{5} \ (x \neq 3)$$ Divide the numerator and the denominator by the common factor $x - 3$

3. $\dfrac{y - 4}{4 - y}$

We note that the numerator and the denominator are exactly the same except for the signs. That is, $4 - y = -y + 4 = (-1)(y - 4)$. The given expression can be written in lowest terms by multiplying the numerator and the denominator by -1.

$$= \frac{(-1)(y - 4)}{(-1)(4 - y)}$$ Multiply numerator and denominator by -1

$$= \frac{(-1)(y - 4)}{y - 4}$$ Multiply in the denominator, $(-1)(4 - y) = y - 4$

$$= \frac{-1}{1}$$ Reduce by the common factor $y - 4$

$$= -1 \quad (y \neq 4)$$

4. $-\dfrac{x^2 - x - 12}{20 - x - x^2} = -\dfrac{(x - 4)(x + 3)}{(4 - x)(5 + x)}$ Factor the numerator and the denominator

We note opposite factors $x - 4$ and $4 - x$ in the numerator and the denominator.

$$= -\frac{(-1)(x - 4)(x + 3)}{(-1)(4 - x)(5 + x)}$$ Multiply numerator and denominator by -1

$$= -\frac{(-1)(x - 4)(x + 3)}{(x - 4)(5 + x)}$$ $(-1)(4 - x) = x - 4$

$$= -\frac{(-1)(x + 3)}{1(5 + x)}$$ Reduce by $x - 4$

$$= \frac{x + 3}{x + 5} \quad (x \neq 4, x \neq -5)$$ $-\dfrac{-a}{b} = \dfrac{a}{b}$ and $5 + x = x + 5$

▶ **Quick check** Simplify $\dfrac{x^2 + 3x - 28}{x^2 - 16}$ by reducing to lowest terms. State any necessary restrictions on the variables. ■

┌─ *Mastery points* ──────────────────────────────────────┐

Can you
- Determine the domain of a rational expression?
- Apply the fundamental principle of rational expressions to reduce a rational expression to lowest terms?

└──┘

Exercise 4–1

State the domain of the given rational expression in set-builder notation. See example 4–1 A.

Example $\dfrac{3y - 2}{y^2 - 1}$

Solution $= \dfrac{3y - 2}{(y + 1)(y - 1)}$ Factor the denominator

$y + 1 = 0$ or $y - 1 = 0$ Set each factor in the denominator equal to 0
$y = -1$ $y = 1$ Solve each equation

The numbers 1 and -1 cannot be used as replacements for y.

Domain $= \{y | y \in R, y \neq -1, y \neq 1\}$

1. $\dfrac{3}{x - 4}$ **2.** $\dfrac{5}{x - 8}$ **3.** $\dfrac{x}{x + 1}$ **4.** $\dfrac{5y}{y + 9}$ **5.** $\dfrac{3z}{2z - 3}$

6. $\dfrac{5x}{8x - 5}$ **7.** $\dfrac{x - 3}{7x + 4}$ **8.** $\dfrac{p + 9}{6p + 5}$ **9.** $\dfrac{2x - 3}{x^2 - 4x}$ **10.** $\dfrac{m + 5}{3m^2 + 6m}$

11. $\dfrac{4x - 7}{x^2 + 8x + 16}$ **12.** $\dfrac{7 - 5x}{x^2 - 14x + 49}$ **13.** $\dfrac{2y - 5}{4y^2 - 25}$ **14.** $\dfrac{3x + 4}{9x^2 - 16}$

15. $\dfrac{8x^2 + 1}{2x^2 - 5x - 3}$ **16.** $\dfrac{p^2 - 7p + 1}{4p^2 + p - 3}$ **17.** $\dfrac{4x^2 - 2x + 1}{x^2 + 4}$ **18.** $\dfrac{x - 3}{x^2 + 5}$

Simplify the given rational expression by reducing to lowest terms. State any necessary restrictions on the variable. See example 4–1 B.

Example $\dfrac{x^2 + 3x - 28}{x^2 - 16}$

Solution $= \dfrac{(x - 4)(x + 7)}{(x - 4)(x + 4)}$ Greatest common factor   Factor the numerator and the denominator   Greatest common factor

$= \dfrac{x + 7}{x + 4} \; (x \neq -4)$ Reduce by $x - 4$

19. $\dfrac{25x^3}{15x^4}$ **20.** $\dfrac{36y^4}{-42y^2}$ **21.** $\dfrac{p^3q^5}{pq^7}$ **22.** $\dfrac{-m^5n^3}{-m^8n}$ **23.** $\dfrac{a^2b^4c^3}{-abc^7}$

24. $\dfrac{-x^4y^6z}{x^3y^5z}$ **25.** $-\dfrac{-27mn^3p^7}{36m^4np^5}$ **26.** $\dfrac{72p^3qr^4}{-63pqr}$ **27.** $\dfrac{3x - 6}{7x - 14}$ **28.** $\dfrac{8a + 12}{6a + 9}$

29. $\dfrac{15a}{5a^2 - 10a}$ **30.** $\dfrac{-36x}{42x^3 + 24x}$ **31.** $\dfrac{12x^4 - 12x^3}{6x^3}$ **32.** $\dfrac{24x^2 - 36x}{32x^2}$ **33.** $\dfrac{8y^2 - 8}{6y - 6}$

34. $\dfrac{6x + 6}{3x - 3}$ **35.** $\dfrac{5x - 5y}{y - x}$ **36.** $\dfrac{9b - 9a}{a - b}$ **37.** $\dfrac{a^2 - 9}{4a + 12}$ **38.** $\dfrac{6a + 3b}{4a^2 - b^2}$

39. $\dfrac{4y^2 - 1}{1 + 2y}$ **40.** $\dfrac{64 - 49p^2}{7p - 8}$ **41.** $\dfrac{x^3 + 8}{x + 2}$ **42.** $\dfrac{a^3 - 64}{a - 4}$ **43.** $\dfrac{3x - 3y}{2y^3 - 2x^3}$

44. $\dfrac{5a^3 + 5b^3}{7b + 7a}$ **45.** $\dfrac{y^2 - 49}{y^2 + 14y + 49}$ **46.** $\dfrac{a^2 - 10a + 25}{a^2 - 25}$ **47.** $\dfrac{m^2 - 4m - 12}{m^2 - m - 6}$

48. $\dfrac{n^2 - 4n + 3}{n^2 - 5n + 6}$ **49.** $\dfrac{y^2 - 4y - 32}{y^2 - y - 20}$ **50.** $\dfrac{x^2 - x - 42}{x^2 + 12x + 36}$ **51.** $\dfrac{2a^2 - 3a + 1}{2a^2 + a - 1}$

52. $\dfrac{3b^2 - 10b + 3}{3b^2 - 7b + 2}$ **53.** $\dfrac{2x^2 + 3x - 9}{4x^2 + 13x + 3}$ **54.** $\dfrac{8y^2 - 22y + 5}{4y^2 - 15y - 4}$ **55.** $\dfrac{12m^2 + 17m + 6}{8m^2 + 10m + 3}$

56. $\dfrac{6x^2 + 17x + 7}{12x^2 + 13x - 35}$

57. $\dfrac{4a^2 + 2a - 12}{6a^2 - 7a - 3}$

58. $\dfrac{6x^2 - x - 1}{4x^2 + 18x - 10}$

59. $\dfrac{5y^2 - 10y - 15}{4y^2 - 36}$

60. $\dfrac{7x^2 - 7}{3x^2 - 15x + 12}$

61. $\dfrac{3a^2 + 16a - 12}{6 - 7a - 3a^2}$

62. $\dfrac{a^2 + 2ab - 24b^2}{a^2 - 8ab + 16b^2}$

63. $\dfrac{p^2 + 7pq + 12q^2}{p^2 + 5pq + 6q^2}$

Review exercises

Perform the indicated operations. See section 1–4.

1. $\dfrac{3}{4} - \dfrac{5}{6}$

2. $\dfrac{5}{6} \div \dfrac{10}{12}$

3. $\dfrac{7}{8} + \dfrac{3}{4}$

Find the solution set of the following equations. See section 2–1.

4. $4(x - 2) + 3 = 2x - 1$

5. $\dfrac{3}{2}x - 2 = \dfrac{5}{6}x + 1$

6. Solve $4x + 2a = 6x + 5$ for x. See section 2–2.

4–2 ■ Multiplication and division of rational expressions

Multiplication

Recall that the product of two rational numbers is obtained by multiplying the numerators and placing this product over the product of the denominators. This process is indicated by the **multiplication property of rational numbers** stated here.

> ─ **Multiplication property of rational numbers** ─────
>
> If a, b, c, and d are real numbers, $b \neq 0$ and $d \neq 0$, then
>
> $$\frac{a}{b} \cdot \frac{c}{d} = \frac{a \cdot c}{b \cdot d}$$

To illustrate,

$$\frac{3}{4} \cdot \frac{5}{7} = \frac{3 \cdot 5}{4 \cdot 7} = \frac{15}{28}$$

To multiply two or more rational expressions we use the **multiplication property of rational expressions.**

> ─ **Multiplication property of rational expressions** ─────
>
> If P, Q, R, and S are polynomials, $Q \neq 0$ and $S \neq 0$, then
>
> $$\frac{P}{Q} \cdot \frac{R}{S} = \frac{P \cdot R}{Q \cdot S}$$
>
> **Concept**
> When multiplying two rational expressions, multiply the two numerators to determine the numerator of the product and multiply the two denominators to determine the denominator of the product.

For example,

$$\frac{x-3}{x+1} \cdot \frac{x+2}{x-1} = \frac{(x-3)(x+2)}{(x+1)(x-1)} = \frac{x^2-x-6}{x^2-1}$$

Since we want the resulting product in its simplest form, any possible reduction by common factors should be performed **before** the multiplication of the numerators and the denominators takes place.

To multiply rational expressions

1. Completely factor the numerators and the denominators where possible.
2. Write the factors in the numerators and the denominators as a single rational expression.
3. Divide the numerator and the denominator by all common factors.
4. Multiply the remaining factors in the numerator and in the denominator to obtain a rational expression in lowest terms.

■ *Example 4–2 A*

Find the indicated products reduced to lowest terms. Place restrictions on the variables.

1. $\dfrac{6x^2}{15} \cdot \dfrac{10}{3x^3} = \dfrac{6x^2 \cdot 10}{15 \cdot 3x^3}$

$\quad = \dfrac{4 \cdot 15x^2}{3x \cdot 15x^2}$ Greatest common factor / Factor numerator and denominator / Greatest common factor

$\quad = \dfrac{4}{3x} \quad (x \neq 0)$ Divide numerator and denominator by the common factor $15x^2$

2. $\dfrac{3x^2-6x}{x+5} \cdot \dfrac{x^2-25}{x^2-10x+25}$

$\quad = \dfrac{3x(x-2)}{x+5} \cdot \dfrac{(x+5)(x-5)}{(x-5)(x-5)}$ Factor where possible

$\quad = \dfrac{3x(x-2)(x+5)(x-5)}{(x+5)(x-5)(x-5)}$ Multiply numerators and denominators

$\quad = \dfrac{3x(x-2)}{x-5}$ Divide numerator and denominator by the common factors $x+5$ and $x-5$

$\quad = \dfrac{3x^2-6x}{x-5} \quad (x \neq -5,5)$ Multiply in the numerator

Note You may leave your answer in factored form $\dfrac{3x(x-2)}{x-5}$ or multiply as indicated to obtain $\dfrac{3x^2-6x}{x-5}$, we will show both answers.

3. $\dfrac{3x - 1}{x - 4} \cdot \dfrac{16 - x^2}{3x^2 + 14x - 5}$

$= \dfrac{3x - 1}{x - 4} \cdot \dfrac{(4 - x)(4 + x)}{(3x - 1)(x + 5)}$

$= \dfrac{(3x - 1)(4 - x)(4 + x)}{(x - 4)(3x - 1)(x + 5)}$

Since there are opposite factors $4 - x$ and $x - 4$, we multiply the numerator and the denominator by -1.

$= \dfrac{(-1)(4 - x)(3x - 1)(4 + x)}{(-1)(x - 4)(3x - 1)(x + 5)}$

$= \dfrac{(-1)(4 - x)(3x - 1)(4 + x)}{(4 - x)(3x - 1)(x + 5)}$ $(-1)(x - 4) = 4 - x$

$= \dfrac{(-1)(4 + x)}{x + 5}$ Reduce by the common factors $4 - x$ and $3x - 1$

$= \dfrac{-x - 4}{x + 5}$ $\left(x \neq 4, \dfrac{1}{3}, -5\right)$

▶ **Quick check** Find the product of $\dfrac{4x^2 - 12x}{x + 5} \cdot \dfrac{x^2 - 25}{x^2 - 6x + 9}$ and reduce to lowest terms. Place restrictions on the variable. ■

Division of rational expressions

Recall that the reciprocal of any nonzero real number b is the real number $\dfrac{1}{b}$ found by "inverting" the number. In like fashion, we invert rational expressions to obtain their reciprocals. That is,

$$\dfrac{x - 1}{x + 3} \quad \text{and} \quad \dfrac{x + 3}{x - 1} \quad \text{are reciprocals.}$$

We use reciprocals in the division of rational expressions as we did when dividing rational numbers.

Division property of rational numbers

If a, b, c, and d are real numbers and $b \neq 0$, $c \neq 0$, and $d \neq 0$, then

$$\dfrac{a}{b} \div \dfrac{c}{d} = \dfrac{a}{b} \cdot \dfrac{d}{c} = \dfrac{a \cdot d}{b \cdot c}$$

To illustrate,

$$\dfrac{5}{8} \div \dfrac{3}{7} = \dfrac{5}{8} \cdot \dfrac{7}{3} = \dfrac{5 \cdot 7}{8 \cdot 3} = \dfrac{35}{24}$$

We can extend this property to rational expressions.

Division property of rational expressions

If P, Q, R, and S are polynomials, $Q \neq 0$, $R \neq 0$, and $S \neq 0$, then

$$\dfrac{P}{Q} \div \dfrac{R}{S} = \dfrac{P}{Q} \cdot \dfrac{S}{R} = \dfrac{P \cdot S}{Q \cdot R}$$

Concept
When dividing two rational expressions, the quotient is obtained by multiplying the first expression by the reciprocal of the second expression.

■ *Example 4–2 B*

Find each of the indicated quotients reduced to lowest terms. Assume no denominator equals zero.

1.
$$\frac{36p^3}{7q} \div \frac{15p}{14q^2} = \frac{36p^3}{7q} \cdot \frac{14q^2}{15p}$$
Multiply by the reciprocal of $\frac{15p}{14q^2}$

$$= \frac{36 \cdot 14 \cdot p^3q^2}{7 \cdot 15 \cdot pq}$$
Multiply numerators and denominators

$$= \frac{3 \cdot 12 \cdot 7 \cdot 2 \cdot p^3q^2}{7 \cdot 3 \cdot 5 \cdot pq}$$
Factor numerator and denominator

Greatest common factor

$$= \frac{24p^2q \cdot (21pq)}{5 \cdot (21pq)}$$

Greatest common factor

$$= \frac{24p^2q}{5}$$
Reduce by 21pq

2.
$$\frac{2x + 1}{x^2 - 9} \div \frac{2x^2 + 7x + 3}{x - 3}$$

$$= \frac{2x + 1}{x^2 - 9} \cdot \frac{x - 3}{2x^2 + 7x + 3}$$
Multiply by the reciprocal of $\frac{2x^2 + 7x + 3}{x - 3}$

$$= \frac{(2x + 1)(x - 3)}{(x + 3)(x - 3)(2x + 1)(x + 3)}$$
Factor and multiply numerators and denominators

$$= \frac{1}{(x + 3)(x + 3)}$$
Reduce by $(x - 3)(2x + 1)$

$$= \frac{1}{x^2 + 6x + 9}$$
Multiply as indicated

3.
$$\frac{x^2 + 2x}{5x} \div (2x^2 + x - 6)$$

Since $2x^2 + x - 6 = \frac{2x^2 + x - 6}{1}$, then

$$\frac{x^2 + 2x}{5x} \div \frac{2x^2 + x - 6}{1} = \frac{x^2 + 2x}{5x} \cdot \frac{1}{2x^2 + x - 6}$$
Multiply by the reciprocal of $\frac{2x^2 + 6 - 6}{1}$

$$= \frac{x(x + 2)}{5x(2x - 3)(x + 2)}$$
Factor and multiply numerators and denominators
Reduce by $x(x + 2)$

$$= \frac{1}{5(2x - 3)}$$

$$= \frac{1}{10x - 15}$$
Multiply as indicated

▶ *Quick check* Find the quotient of $\dfrac{4x + 3}{x^2 - 49} \div \dfrac{4x^2 - 13x - 12}{x + 7}$ and reduce to lowest terms. Assume no denominator equals zero. ∎ ∎

Mastery points

Can you
■ Multiply rational expressions?
■ Divide rational expressions?

Exercise 4–2

Find each indicated product in lowest terms. Place restrictions on the variable. See example 4–2 A.

Example $\dfrac{4x^2 - 12x}{x + 5} \cdot \dfrac{x^2 - 25}{x^2 - 6x + 9}$

Solution $= \dfrac{4x(x - 3)}{x + 5} \cdot \dfrac{(x + 5)(x - 5)}{(x - 3)(x - 3)}$ Factor the numerator and the denominator

$= \dfrac{4x(x - 3)(x + 5)(x - 5)}{(x + 5)(x - 3)(x - 3)}$ Multiply the numerators and the denominators

Greatest common factors

$= \dfrac{4x(x - 5)(x - 3)(x + 5)}{(x - 3)(x - 3)(x + 5)}$ Commute

Greatest common factors

$= \dfrac{4x(x - 5)}{x - 3}$ Reduce by $(x - 3)(x + 5)$

$= \dfrac{4x^2 - 20x}{x - 3}$ $(x \neq -5, 3)$ Multiply in the numerator

1. $\dfrac{8}{9x^2} \cdot \dfrac{12x}{4}$

2. $\dfrac{12b}{7} \cdot \dfrac{28}{4b^3}$

3. $\dfrac{5x^2}{9y^3} \cdot \dfrac{18y}{20x^4}$

4. $\dfrac{15n}{28m} \cdot \dfrac{14m^3}{30n^2}$

5. $\dfrac{16c}{42ab^2} \cdot \dfrac{3a^3b}{8c^4}$

6. $\dfrac{12xy}{25z^3} \cdot \dfrac{5z}{18x^3y^2}$

7. $16a^2b^3 \cdot \dfrac{3}{24ab}$

8. $8x^3yz^2 \cdot \dfrac{15}{36xyz}$

9. $\dfrac{4a + 8}{3a - 12} \cdot \dfrac{a - 2}{5a + 10}$

10. $\dfrac{y + 3}{8y - 16} \cdot \dfrac{3y - 6}{4y + 12}$

11. $(a + 3) \cdot \dfrac{a - 6}{6a + 18}$

12. $\dfrac{x - 3}{5x - 25} \cdot (5 - x)$

13. $\dfrac{p^2 + 2p + 1}{1 - 4p} \cdot \dfrac{16p^2 - 1}{p^2 - 1}$

14. $\dfrac{m^2 - 9}{3m + 4} \cdot \dfrac{9m^2 - 16}{m^2 + 6m + 9}$

15. $\dfrac{x^2 - 9x + 20}{x^2 - 5x + 6} \cdot \dfrac{x^2 - 3x + 2}{x^2 - 5x + 4}$

16. $\dfrac{y^2 + 3y - 4}{y^2 + 2y - 3} \cdot \dfrac{y^2 - y - 6}{y^2 + y - 12}$

17. $\dfrac{2a^2 - a - 6}{3a^2 - 4a + 1} \cdot \dfrac{3a^2 + 7a + 2}{2a^2 + 7a + 6}$

18. $\dfrac{4b^2 + 8b + 3}{2b^2 - 5b - 12} \cdot \dfrac{b^2 - 16}{2b^2 + 7b + 3}$

19. $\dfrac{x^3 + 8}{x^2 - 4} \cdot \dfrac{x^2 - 4x + 4}{2x^2 + 3x - 14}$

20. $\dfrac{4x^2 - 49}{8x^3 + 27} \cdot \dfrac{4x^2 + 12x + 9}{2x^2 - 13x + 21}$

21. $\dfrac{y - x}{x^2 + 3xy + 2y^2} \cdot \dfrac{x^2 + 2xy + y^2}{x^2 - y^2}$

22. $\dfrac{b^2 - a^2}{3a^2 + ab - 2b^2} \cdot \dfrac{6a^2 - ab - 2b^2}{a^2 - 2ab + b^2}$

Find each of the indicated quotients in lowest terms. Assume all denominators are nonzero. See example 4-2 B.

Example $\dfrac{4x + 3}{x^2 - 49} \div \dfrac{4x^2 - 13x - 12}{x + 7}$

Solution $= \dfrac{4x + 3}{x^2 - 49} \cdot \dfrac{x + 7}{4x^2 - 13x - 12}$ Multiply by the reciprocal

$= \dfrac{(4x + 3)(x + 7)}{(x + 7)(x - 7)(4x + 3)(x - 4)}$ Factor the denominator and multiply

$= \dfrac{(4x + 3)(x + 7)}{(4x + 3)(x + 7)(x - 7)(x - 4)}$ Commute

$= \dfrac{1}{(x - 7)(x - 4)}$ Reduce by GCF $(4x + 3)(x + 7)$

$= \dfrac{1}{x^2 - 11x + 28}$ Multiply in the denominator

23. $\dfrac{9a^2b}{8ab^3} \div \dfrac{18ab^3}{16a^2b^2}$

24. $\dfrac{7ab^3}{10x^2y^3} \div \dfrac{21a^2b^2}{15x^3y^2}$

25. $\dfrac{r + 4}{r^2 - 1} \div \dfrac{r^2 - 16}{r + 1}$

26. $\dfrac{4x + 4y}{x - 3} \div \dfrac{x + y}{x^2 - 9}$

27. $\dfrac{6 - 2x}{2x + 8} \div (9 - 3x)$

28. $\dfrac{x^2 - 25}{2x + 10} \div (x^2 - 10x + 25)$

29. $(x^2 - 2x - 3) \div \dfrac{4x - 12}{x^2 - 1}$

30. $(4x^2 - 9) \div \dfrac{4x + 6}{x + 3}$

31. $\dfrac{x^3 - 125}{2x + 1} \div \dfrac{x^2 - 25}{2x^2 - 9x - 5}$

32. $\dfrac{3x^2 + 10x - 8}{5x - 15} \div \dfrac{x^3 + 64}{x^2 - 9}$

33. $\dfrac{56 - x - x^2}{x^2 - 6x - 7} \div \dfrac{x^2 + 4x - 21}{x^2 - 4x - 5}$

34. $\dfrac{a^2 + 16a + 63}{20 - a - a^2} \div \dfrac{a^2 - 5a - 84}{a^2 - a - 12}$

35. $\dfrac{2x^2 + 5x - 12}{3x^2 - 8x - 16} \div \dfrac{2x^2 + 3x - 9}{3x^2 + 13x + 12}$

36. $\dfrac{6x^2 - 5x - 4}{8x^2 + 10x + 3} \div \dfrac{3x^2 + 14x - 24}{4x^2 - 12x - 7}$

Find each indicated product or quotient in lowest terms. Assume all denominators are nonzero. See examples 4-2 A and B.

37. $\dfrac{a^2 - b^2}{2a + 4b} \cdot \dfrac{a + 2b}{a + b}$

38. $\dfrac{4x + 4y}{x - 3y} \cdot \dfrac{x^2 - 9y^2}{x + y}$

39. $\dfrac{n^2 - m^2}{2m - 3n} \div \dfrac{m - n}{4m^2 - 9n^2}$

40. $\dfrac{p^2 - q^2}{q - 3p} \div \dfrac{p^2 + 2pq + q^2}{9p^2 - q^2}$

41. $\dfrac{a^2 - 8a + 15}{a^2 + 7a + 6} \cdot \dfrac{a^2 - 6a - 7}{a^2 - 9} \div \dfrac{a^2 - 12a + 35}{a^2 - 36}$

42. $\dfrac{x^2 + 3x - 28}{x^2 - 8x + 7} \div \dfrac{x^2 - 7x + 12}{x^2 - x - 42} \cdot \dfrac{2x - 8}{x^2 - 36}$

43. $\dfrac{2x^2 + 5x - 3}{3x^2 - 10x - 8} \div \dfrac{2x^2 - 7x + 3}{3x^2 - x - 2} \cdot \dfrac{x^2 - 6x + 9}{x^2 - 2x + 1}$

44. $\dfrac{2x^2 + x - 6}{5x^2 + 7x - 6} \cdot \dfrac{20x^2 - 7x - 3}{6x^2 - 25x + 4} \div \dfrac{4x^2 - 11x - 3}{6x^2 - 19x + 3}$

45. $\dfrac{27y^3 - 8x^3}{x + y} \div \dfrac{2x - 3y}{x^3 + y^3}$

46. $\dfrac{a^3 + 64b^3}{16b^2 - a^2} \cdot \dfrac{a + 4b}{a^2 + 3ab - 4b^2}$

47. $\dfrac{ab - a + 3b - 3}{ab + a - 4b - 4} \div \dfrac{ab - 2a + 3b - 6}{ab + 2a - 4b - 8}$

48. $\dfrac{mn - n + 4m - 4}{mn - 3n + 4m - 12} \cdot \dfrac{mn - 3n + 2m - 6}{mn + 2m + n + 2}$

49. $\dfrac{xz - xw + yz - yw}{xz + xw - yz - yw} \cdot \dfrac{xz - xw - yz + yw}{xz + xw + yz + yw}$

50. $\dfrac{pr + ps + qr + qs}{pr - ps - qr + qs} \div \dfrac{xr + xs - yr - ys}{xr - xs + yr - ys}$

Review exercises

Add or subtract the following fractions. See section 1–4.

1. $\dfrac{11}{12} + \dfrac{7}{8}$

2. $\dfrac{14}{15} - \dfrac{3}{5}$

Find the solution set of the following inequalities. See section 2–5.

3. $5y - 2 \le y + 6$

4. $-7 \le 2y - 1 < 1$

Find the following products. See section 3–2.

5. $4x^3(2x^2 - x + 6)$

6. $(4y + 1)(y - 4)$

4–3 ■ Addition and subtraction of rational expressions

When we add or subtract a pair of rational numbers that have the same denominator, such as $\dfrac{3}{4} + \dfrac{5}{4}$, we use the property of adding or subtracting fractions.

> **Property of adding or subtracting rational numbers**
>
> If a, b, and c are real numbers, $b \ne 0$, then
>
> $$\frac{a}{b} + \frac{c}{b} = \frac{a + c}{b} \quad \text{and} \quad \frac{a}{b} - \frac{c}{b} = \frac{a - c}{b}$$

For example,

$$\frac{3}{4} + \frac{5}{4} = \frac{3 + 5}{4} = \frac{8}{4} = 2 \quad \text{and} \quad \frac{7}{8} - \frac{1}{8} = \frac{7 - 1}{8} = \frac{6}{8} = \frac{3}{4}$$

We can extend this property to rational expressions.

> **Property of adding or subtracting rational expressions**
>
> If P, Q, and R are polynomials, $Q \ne 0$, then
>
> $$\frac{P}{Q} + \frac{R}{Q} = \frac{P + R}{Q} \quad \text{and} \quad \frac{P}{Q} - \frac{R}{Q} = \frac{P - R}{Q}$$
>
> **Concept**
> To add, or subtract, two rational expressions having the same denominators, add, or subtract, the numerators and place this sum or difference over the common denominator.

■ *Example 4–3 A*

Find each indicated sum or difference in lowest terms. Assume all denominators are nonzero.

1. $\dfrac{3x + 2}{x - 2} + \dfrac{4x - 5}{x - 2}$

$$= \frac{(3x + 2) + (4x - 5)}{x - 2} \qquad \text{Add the numerators}$$

$$= \frac{3x + 2 + 4x - 5}{x - 2} \qquad \text{Remove parentheses}$$

$$= \frac{7x - 3}{x - 2} \qquad \text{Combine like terms in the numerator}$$

2. $\dfrac{3x - 2y}{x^2 - y^2} - \dfrac{2x - 3y}{x^2 - y^2}$

$$= \frac{(3x - 2y) - (2x - 3y)}{x^2 - y^2} \qquad \text{Subtract the numerators}$$

$$= \frac{3x - 2y - 2x + 3y}{x^2 - y^2} \qquad \text{Remove parentheses}$$

$$= \frac{x + y}{x^2 - y^2} \qquad \text{Combine like terms}$$

$$= \frac{x + y}{(x + y)(x - y)} \qquad \text{Factor the denominator}$$

$$= \frac{1}{x - y} \qquad \text{Reduce by } (x + y) \text{ (\textit{always} reduce where possible)}$$

Note When we write the sum or the difference of the numerators, it is important that we write each numerator in parentheses to avoid making a common mistake when we subtract. Failure to do this would have caused the numerator in example 2 to become $3x - 2y - 2x - 3y = x - 5y$.

▶ *Quick check* Find the difference of $\dfrac{4x - 3}{x^2 - 4} - \dfrac{3x - 1}{x^2 - 4}$ in lowest terms. Assume the denominator is nonzero.   ■

When adding, or subtracting, two rational expressions $\dfrac{P}{Q}$ and $\dfrac{R}{S}$ that *do not* have the same denominator, we can add or subtract them in the following way:

┌─ *Property of adding or subtracting rational expressions* ──────
│ *with different denominators*
│   If P, Q, R, and S are polynomials, $Q \neq 0$ and $S \neq 0$, then
│
│ $$\frac{P}{Q} + \frac{R}{S} = \frac{P \cdot S + Q \cdot R}{Q \cdot S} \quad \text{and} \quad \frac{P}{Q} - \frac{R}{S} = \frac{P \cdot S - Q \cdot R}{Q \cdot S}$$

The following examples illustrate the use of this property for rational expressions.

1. $\dfrac{x + 1}{x - 2} + \dfrac{x - 3}{x + 4} = \dfrac{(x + 1)(x + 4) + (x - 2)(x - 3)}{(x - 2)(x + 4)}$

$= \dfrac{(x^2 + 5x + 4) + (x^2 - 5x + 6)}{(x - 2)(x + 4)}$ Multiply as indicated

$= \dfrac{2x^2 + 10}{(x - 2)(x + 4)} \quad (x \neq 2, x \neq -4)$ Combine like terms

2. $\dfrac{x}{x - 7} - \dfrac{2x + 1}{x - 9} = \dfrac{x \cdot (x - 9) - (x - 7)(2x + 1)}{(x - 7)(x - 9)}$

$= \dfrac{(x^2 - 9x) - (2x^2 - 13x - 7)}{(x - 7)(x - 9)}$ Multiply as indicated

$= \dfrac{x^2 - 9x - 2x^2 + 13x + 7}{(x - 7)(x - 9)}$ Remove parentheses and change signs

$= \dfrac{-x^2 + 4x + 7}{(x - 7)(x - 9)} \quad (x \neq 7, x \neq 9)$ Combine like terms

Note We do not multiply in the denominator to aid in recognizing common factors for reducing to lowest terms. Always check the numerator for common factors.

This method is useful when adding or subtracting only two rational expressions and when the denominators have no common factors. When rational numbers or rational expressions do not have the same denominator, we apply the fundamental principle of rational expressions to obtain equivalent expressions (expressions that name the same number) with a common denominator. To do this, we must find the *least common multiple* (LCM) of the denominators.

Least common multiple

The least common multiple (LCM) of a set of expressions is the least expression that is exactly divisible by the expressions.

We use the following procedure to find the LCM of a set of polynomials.

Finding the LCM of a set of polynomials

1. Completely factor the polynomials.
2. The LCM consists of the product of all distinct factors of the polynomials, each raised to the greatest power to which it appears in any one of the factorizations.

■ *Example 4–3 B*

Find the least common multiple.

1. $24x^3, 15x^2, 10x$

$\left. \begin{array}{l} 24x^3 = 2^3 \cdot 3 \cdot x^3 \\ 15x^2 = 3 \cdot 5 \cdot x^2 \\ 10x = 2 \cdot 5 \cdot x \end{array} \right\}$ Completely factor each polynomial

Choose the greatest power of each different factor.

The LCM $= 2^3 \cdot 3 \cdot 5 \cdot x^3 = 120x^3$.

2. $2x - 4,\ x^2 - 4,\ x^2 + 4x + 4$

$$\left.\begin{array}{l}2x - 4 = 2 \cdot (x - 2) \\ x^2 - 4 = (x - 2) \cdot (x + 2) \\ x^2 + 4x + 4 = (x + 2)^2\end{array}\right\} \text{Completely factor each polynomial}$$

Choose the greatest power of each different factor.

The LCM $= 2(x - 2)(x + 2)^2$.

▶ **Quick check** Find the LCM of $3x + 9,\ x^2 - 9,\ x^2 - 6x + 9$. ∎

We now apply the fundamental principle of rational expressions to "build" a given expression to an equivalent rational expression. To do this, we must **multiply** the numerator and the denominator by the same nonzero polynomial.

■ *Example 4–3 C*

Write each rational expression as an equivalent expression with the indicated denominator. Assume all denominators are nonzero.

1. $\dfrac{13}{3a}$, denominator $12a^2$

We obtain the factor with which to build by dividing $\dfrac{12a^2}{3a} = 4a$.

$$\begin{aligned}\frac{13}{3a} &= \frac{13 \cdot 4a}{3a \cdot 4a} \qquad \text{Multiply the numerator and the denominator by } 4a \\ &= \frac{52a}{12a^2} \qquad \text{Multiply in the numerator and the denominator}\end{aligned}$$

2. $\dfrac{3x - 2}{2x - 1}$, denominator $4x^2 - 1$

Since $4x^2 - 1$ factors to $(2x - 1)(2x + 1)$, the building factor is $2x + 1$.

$$\begin{aligned}\frac{3x - 2}{2x - 1} &= \frac{(3x - 2)(2x + 1)}{(2x - 1)(2x + 1)} \qquad \text{Multiply the numerator and the denominator by } (2x + 1) \\ &= \frac{6x^2 - x - 2}{4x^2 - 1} \qquad \text{Multiply in the numerator and the denominator}\end{aligned}$$

3. $\dfrac{3}{1 - x}$, denominator $x - 1$

Since $x - 1 = (-1)(1 - x)$, multiply the numerator and the denominator by -1.

$$\frac{3}{1 - x} = \frac{3(-1)}{(1 - x)(-1)} = \frac{-3}{x - 1}$$

▶ **Quick check** Write $\dfrac{5z - 3}{3z + 4}$ as an equivalent expression with $9z^2 - 16$ as a denominator. Assume the denominator is nonzero. ∎

We now add and subtract rational expressions that have unlike denominators by building the expressions to equivalent rational expressions having a common denominator—the LCM of the denominators (called the least common denominator, LCD).

■ *Example 4–3 D*

Find the indicated sum or difference in lowest terms. Assume no denominator equals zero.

1. $\dfrac{6}{5a} + \dfrac{7}{6b}$

The LCM of $5a$ and $6b$ is $5a \cdot 6b = 30ab$.

$$= \dfrac{6}{5a} \cdot \dfrac{6b}{6b} + \dfrac{7}{6b} \cdot \dfrac{5a}{5a}$$ — Building factors

Multiply by building factors $6b$ and $5a$

$$= \dfrac{36b}{30ab} + \dfrac{35a}{30ab}$$

Equivalent expressions with LCD $30ab$

$$= \dfrac{36b + 35a}{30ab}$$

Add the numerators

2. $\dfrac{5}{16a} - \dfrac{7}{24a^2}$

$\left.\begin{array}{l} 16a = 2^4 \cdot a \\ 24a^2 = 2^3 \cdot 3 \cdot a^2 \end{array}\right\}$ The LCM is $2^4 \cdot 3 \cdot a^2 = 48a^2$.

$$= \dfrac{5}{16a} \cdot \dfrac{3a}{3a} - \dfrac{7}{24a^2} \cdot \dfrac{2}{2}$$ — Building factors

Multiply by building factors $3a$ and 2

$$= \dfrac{15a}{48a^2} - \dfrac{14}{48a^2}$$

Multiply as indicated

$$= \dfrac{15a - 14}{48a^2}$$

Subtract the numerators

3. $\dfrac{10}{x - 3} + \dfrac{9}{x^2 + 3x - 18}$

$\left.\begin{array}{l} x - 3 = (x - 3) \\ x^2 + 3x - 18 = (x - 3)(x + 6) \end{array}\right\}$ The LCM is $(x - 3)(x + 6)$.

$$= \dfrac{10}{x - 3} + \dfrac{9}{(x - 3)(x + 6)}$$

Factor denominators

$$= \dfrac{10}{x - 3} \cdot \dfrac{x + 6}{x + 6} + \dfrac{9}{(x - 3)(x + 6)}$$ — Building factor

Multiply by building factor $x + 6$

$$= \dfrac{10(x + 6)}{(x - 3)(x + 6)} + \dfrac{9}{(x - 3)(x + 6)}$$

$$= \dfrac{10x + 60 + 9}{(x - 3)(x + 6)}$$

Add the numerators

$$= \dfrac{10x + 69}{(x - 3)(x + 6)}$$

Combine like terms

4. $\dfrac{y + 3}{y^2 + 4y - 21} - \dfrac{y + 5}{y^2 + 6y - 27}$

$\left.\begin{array}{l} y^2 + 4y - 21 = (y - 3)(y + 7) \\ y^2 + 6y - 27 = (y + 9)(y - 3) \end{array}\right\}$ The LCM is $(y + 9)(y - 3)(y + 7)$.

$$= \frac{y + 3}{(y - 3)(y + 7)} - \frac{y + 5}{(y + 9)(y - 3)} \qquad \text{Factor denominators}$$

Building factors

$$= \frac{y + 3}{(y - 3)(y + 7)} \cdot \frac{y + 9}{y + 9} - \frac{y + 5}{(y + 9)(y - 3)} \cdot \frac{y + 7}{y + 7}$$

$$= \frac{(y + 3)(y + 9)}{(y - 3)(y + 7)(y + 9)} - \frac{(y + 5)(y + 7)}{(y + 9)(y - 3)(y + 7)} \qquad \begin{array}{l}\text{Multiply in}\\\text{numerator and}\\\text{denominator}\end{array}$$

$$= \frac{(y^2 + 12y + 27)}{(y - 3)(y + 7)(y + 9)} - \frac{(y^2 + 12y + 35)}{(y + 9)(y - 3)(y + 7)} \qquad \begin{array}{l}\text{Multiply in the}\\\text{numerators}\end{array}$$

$$= \frac{(y^2 + 12y + 27) - (y^2 + 12y + 35)}{(y - 3)(y + 9)(y + 7)} \qquad \begin{array}{l}\text{Subtract the}\\\text{numerators}\end{array}$$

$$= \frac{y^2 + 12y + 27 - y^2 - 12y - 35}{(y - 3)(y + 9)(y + 7)} \qquad \begin{array}{l}\text{Remove}\\\text{parentheses and}\\\text{change signs}\end{array}$$

$$= \frac{-8}{(y - 3)(y + 9)(y + 7)} \qquad \text{Combine like terms}$$

5. $\dfrac{x^2}{x - 8} + \dfrac{7x + 8}{8 - x}$

Since the denominators are the opposites of each other, we multiply the numerator and the denominator of $\dfrac{7x + 8}{8 - x}$ by -1.

$$\frac{x^2}{x - 8} + \frac{7x + 8}{8 - x} \cdot \frac{-1}{-1} = \frac{x^2}{x - 8} + \frac{-7x - 8}{x - 8} \qquad \text{Multiply as indicated}$$

$$= \frac{x^2 - 7x - 8}{x - 8} \qquad \text{Add numerators}$$

$$= \frac{(x + 1)(x - 8)}{x - 8} \qquad \text{Factor the numerator}$$

$$= x + 1 \qquad \text{Reduce by } x - 8$$

▶ *Quick check*   Find the difference $\dfrac{2y + 5}{y - 2} - \dfrac{4y + 12}{y^2 + y - 6}$ in lowest terms. Assume no denominator equals zero.   ■

We now generalize the procedure for adding or subtracting rational expressions.

⎡ *To add or subtract rational expressions* ⎤

1. If the denominators are like, add or subtract the numerators and place the result over the common denominator.
2. If the denominators are different, find the least common multiple (LCM) of the denominators.
3. Multiply the numerator and the denominator of each rational expression by all factors present in the LCM but missing in the denominator of the particular rational expression to obtain equivalent rational expressions.
4. Proceed as in step 1. Reduce the results to lowest terms to simplify the answer.

Mastery points

Can you
- Find the least common multiple (LCM) of a set of polynomials?
- Build rational expressions to equivalent rational expressions?
- Add or subtract rational expressions that have the same denominators?
- Add or subtract rational expressions that have different denominators?

Exercise 4–3

Find the least common multiple of the set of polynomials. See example 4–3 B.

Example $3x + 9, x^2 - 9, x^2 - 6x + 9$

Solution $3x + 9 = 3(x + 3)$
$x^2 - 9 = (x + 3)(x - 3)$ } Factor the denominators
$x^2 - 6x + 9 = (x - 3)^2$

The LCM is $3(x + 3)(x - 3)^2$.

1. $14x, 35$

2. $42y, 36$

3. $10k, 16k^2$

4. $18x^3, 21x^2$

5. $32a^3b, 9ab^2$

6. $48x^2y^2, 30x^4y$

7. $xy^3, 6x^2y^2, 15xy^4$

8. $4mn^4, 14m^2n^3, 35mn$

9. $4a, 2a - 4$

10. $9b, 15b + 30$

11. $x - 7, 3x - 21, 6x$

12. $x + 9, 4x + 36, 8x$

13. $p^2 - p - 12, p^2 + 6p + 9, 3p - 12$

14. $n^2 - 9, n^2 - n - 6, 8n^2 + 24n$

15. $a^2 - 25, 10a + 50, 5a - 25$

16. $x^2 + 6x + 9, 2x + 6, x^2 + x - 6$

17. $q^2 - 49, q^2 + 5q - 14, q - 7$

18. $2y + 10, y^2 - 25, y^2 - 10y + 25$

19. $a - b, a^2 - b^2, 5a + 5b$

20. $m - n, 4m + 4n, m^2 - 2mn + n^2$

21. $2a^2 - 13a - 7, 6a^2 + a - 1, a^2 - 49$

22. $4m^2 + 5m - 6, 3m^2 + m - 10, 16m^2 - 9$

Build the given rational expression to equivalent rational expression with the given denominator. Assume all denominators are nonzero. See example 4–3 C.

Example $\dfrac{5z - 3}{3z + 4}$, denominator $9z^2 - 16$

Solution Since $9z^2 - 16 = (3z + 4)(3z - 4)$, the building factor is $3z - 4$.

$= \dfrac{(5z - 3)(3z - 4)}{(3z + 4)(3z - 4)}$ Multiply numerator and denominator by $3z - 4$

$= \dfrac{15z^2 - 29z + 12}{9z^2 - 16}$ Multiply in numerator and denominator

23. $\dfrac{4}{7x}$, denominator $21x^3$

24. $\dfrac{8}{9y^2}$, denominator $72y^5$

25. $\dfrac{5x}{8y}$, denominator $24x^2y^2$

26. $\dfrac{-3a}{7b^2}$, denominator $35a^2b^3$

27. $4p$, denominator $p - 3$

28. $a + 3$, denominator $2a - 1$

29. $\dfrac{p - 3}{p + 2}$, denominator $p^2 - 4$

30. $\dfrac{n + 7}{n - 9}$, denominator $n^2 - 81$

31. $\dfrac{4x}{2x - 3}$, denominator $8x^3 - 27$

32. $\dfrac{9x}{3x + 1}$, denominator $27x^3 + 1$

33. $\dfrac{2n - 1}{n + 7}$, denominator $n^2 + 2n - 35$

34. $\dfrac{4x + 3}{x - 9}$, denominator $x^2 - 4x - 45$

35. $\dfrac{2y - 5}{4y - 1}$, denominator $4y^2 + 7y - 2$

36. $\dfrac{3x + 4}{2x - 9}$, denominator $8x^2 - 30x - 27$

37. $-\dfrac{3m}{4 - 5m}$, denominator $25m^2 - 16$

38. $\dfrac{9 - a}{4 - a}$, denominator $a^2 + a - 20$

39. $\dfrac{-9}{9 - a}$, denominator $a - 9$

40. $-\dfrac{-12}{10 - b}$, denominator $b - 10$

Find each indicated sum or difference in lowest terms. Assume all denominators are nonzero. See example 4-3 A.

Example $\dfrac{4x - 3}{x^2 - 4} - \dfrac{3x - 1}{x^2 - 4}$

Solution $= \dfrac{(4x - 3) - (3x - 1)}{x^2 - 4}$ Subtract numerators

$= \dfrac{4x - 3 - 3x + 1}{x^2 - 4}$ Remove parentheses and subtract (change signs)

$= \dfrac{x - 2}{x^2 - 4}$ Combine like terms

$= \dfrac{x - 2}{(x - 2)(x + 2)}$ Factor the denominator

$= \dfrac{1}{x + 2}$ Reduce by $(x - 2)$

41. $\dfrac{5}{q} + \dfrac{8}{q}$

42. $\dfrac{7}{x} - \dfrac{17}{x}$

43. $\dfrac{4y}{y + 4} - \dfrac{7y}{y + 4}$

44. $\dfrac{8a}{a - 5} + \dfrac{7}{a - 5}$

45. $\dfrac{5y}{y + 2} + \dfrac{10}{y + 2}$

46. $\dfrac{3x}{x - 4} - \dfrac{12}{x - 4}$

47. $\dfrac{7x - 1}{3x + 4} - \dfrac{4x - 5}{3x + 4}$

48. $\dfrac{3y - 2}{4y + 3} + \dfrac{5y + 8}{4y + 3}$

49. $\dfrac{3x - 4}{x^2 + 5x + 6} - \dfrac{2x - 6}{x^2 + 5x + 6}$

50. $\dfrac{2b + 1}{b^2 - 4} + \dfrac{1 - b}{b^2 - 4}$

51. $\dfrac{2x - y}{x + y} - \dfrac{x - 2y}{x + y}$

52. $\dfrac{3x - 2y}{2x + y} - \dfrac{x + y}{2x + y}$

Find the indicated sum or difference. Assume all denominators are nonzero. See example 4–3 D.

Example $\dfrac{2y+5}{y-2} - \dfrac{4y+12}{y^2+y-6}$

Solution $\left.\begin{array}{l} y-2=(y-2) \\ y^2+y-6=(y-2)(y+3) \end{array}\right\}$ Factor denominators

The LCM of the denominators is $(y-2)(y+3)$.

$= \dfrac{2y+5}{y-2} - \dfrac{4y+12}{(y-2)(y+3)}$ Factor denominators

Building factor

$= \dfrac{2y+5}{y-2} \cdot \dfrac{y+3}{y+3} - \dfrac{4y+12}{(y-2)(y+3)}$ Multiply by building factor

$= \dfrac{(2y+5)(y+3)}{(y-2)(y+3)} - \dfrac{4y+12}{(y-2)(y+3)}$ Multiply as indicated

$= \dfrac{2y^2+11y+15}{(y-2)(y+3)} - \dfrac{4y+12}{(y-2)(y+3)}$

$= \dfrac{(2y^2+11y+15)-(4y+12)}{(y-2)(y+3)}$ Subtract numerators

$= \dfrac{2y^2+11y+15-4y-12}{(y-2)(y+3)}$ Remove parentheses and subtract (change signs)

$= \dfrac{2y^2+7y+3}{(y-2)(y+3)}$ Combine like terms

$= \dfrac{(2y+1)(y+3)}{(y-2)(y+3)}$ Factor numerator

$= \dfrac{2y+1}{y-2}$ Reduce by $(y+3)$

53. $\dfrac{8}{3x}+\dfrac{5}{4x}$
54. $\dfrac{7}{5a}-\dfrac{9}{6a}$
55. $\dfrac{5}{z}-\dfrac{2}{z^2}$
56. $\dfrac{7}{y^2}+\dfrac{2}{3y}$

57. $\dfrac{6}{3x}+\dfrac{5}{9x^2}$
58. $\dfrac{8}{4x^2}-\dfrac{3}{2x}$
59. $\dfrac{2}{3x^2y}-\dfrac{4}{9xy^2}$
60. $\dfrac{7}{2ab^2}+\dfrac{6}{6a^2b}$

61. $\dfrac{4a}{3a+5}+\dfrac{2a}{2a-3}$
62. $\dfrac{5m}{m-8}-\dfrac{3m-8}{2m+7}$
63. $\dfrac{17}{5y-10}+\dfrac{19}{2y+4}$
64. $\dfrac{10}{4a-6}-\dfrac{13}{3a+9}$

65. $\dfrac{x+2}{x-9}-\dfrac{x-6}{9-x}$
66. $\dfrac{2a-1}{2a-3}+\dfrac{4-a}{3-2a}$
67. $\dfrac{7}{x^2-5x-6}+\dfrac{9}{x^2-1}$
68. $\dfrac{14}{a^2-7a-18}-\dfrac{8}{a^2-4}$

69. $\dfrac{5x}{x^2-2xy-3y^2}-\dfrac{2y}{x^2+2xy+y^2}$
70. $\dfrac{8q}{4q^2-9p^2}+\dfrac{5q}{4q^2-12pq+9p^2}$

71. $\dfrac{a-7}{2a^2+9a-5}+\dfrac{4-a}{4a^2+23a+15}$
72. $\dfrac{b+9}{12b^2-5b-2}-\dfrac{8-2b}{3b^2-17b+10}$

73. $(4a-3)-\dfrac{2a+5}{5a-2}$
74. $\dfrac{5x+4}{3x+1}+(8x-5)$

75. $\dfrac{5}{8p}+\dfrac{6p-5}{4p^2-8p-60}$
76. $\dfrac{9p-2}{2p^2-2p-84}+\dfrac{7}{6p}$

77. $\dfrac{3y}{y^2 + 5y + 6} - \dfrac{5}{4 - y^2}$

78. $\dfrac{6a}{6 + a - a^2} - \dfrac{7a}{a^2 + 7a + 10}$

79. $\dfrac{b + 3}{b + 2} + \dfrac{2b}{5b^2 - 20}$

80. $\dfrac{3}{6b^2 - 4bc} - \dfrac{4}{9bc - 6c^2}$

81. $\dfrac{2x - 3y}{x^2 - 4xy - 12y^2} - \dfrac{2y - x}{x^2 - 12xy + 36y^2}$

82. $\dfrac{a - b}{a^2 - 3ab - 4b^2} + \dfrac{2b - 5a}{a^2 - 16b^2}$

83. $\dfrac{x - 3}{8x^2 - 26x + 15} + \dfrac{3x + 2}{6x^2 - 13x - 5}$

84. $\dfrac{2p - 3}{8p^2 - 18p - 5} - \dfrac{5 - 7p}{4p^2 - 27p - 7}$

85. $\dfrac{5m - n}{8m^2 + 15mn - 2n^2} + \dfrac{3m + n}{5m^2 + 6mn - 8n^2}$

86. $\dfrac{b - 2a}{3a^2 - 2ab - 8b^2} - \dfrac{3a - 5b}{9a^2 + 6ab - 8b^2}$

Solve the following word problems.

87. Workers A, B, C, and D can complete a given job in p, q, r, and s hours, respectively. Working together they can complete in one hour

$$\frac{1}{p} + \frac{1}{q} + \frac{1}{r} + \frac{1}{s}$$

of the job. Obtain a single expression for what they can do together in one hour.

88. In electricity, the total resistance of any parallel circuit is given by

$$\frac{1}{R_t} = \frac{1}{R_1} + \frac{1}{R_2} + \frac{1}{R_3} + \frac{1}{R_4}$$

Combine the fractions in the right member.

89. Given the focal lengths f_1 and f_2 of two thin lenses that are a distance d apart, the focal length F of the system of lenses is given by

$$\frac{1}{F} = \frac{1}{f_1} + \frac{1}{f_2} - \frac{d}{f_1 f_2}$$

Combine the fractions in the right member.

90. In electricity, the true current reading I_t of a current meter is given by

$$I_t = I_r + \frac{R_m}{R_t} \cdot I_r$$

where I_r = the meter reading, R_m = the resistance of the current meter, and R_t = the total resistance of the circuit without the meter. Perform the indicated operations in the right member and write as a single expression in I_r, R_m, and R_t.

91. A lens maker's equation for making a lens is given by

$$\frac{1}{f} = (n - 1)\left(\frac{1}{R_1} + \frac{1}{R_2}\right)$$

where f is the focal length of the lens, n is the index of refraction, and R_1 and R_2 are the radii of curvature of the surfaces. Simplify the right member by performing the indicated operations and obtaining a single rational expression.

92. The expression

$$V_1\left(1 + \frac{T_2 - T_1}{T_1}\right) - V_2\left(\frac{T_2 - T_1}{T_2} - 1\right)$$

gives the volume change of a gas under constant pressure. Simplify the expression by performing the indicated operations.

Review exercises

Completely factor the following expressions. See section 3–8.

1. $8x^3 - 2x^2 + 12x - 3$

2. $4x^2 - 12x - 7$

3. $9y^2 - 49$

4. Simplify the expression $4[3 - 2(5 - 1) + 8]$. See section 1–4.

5. It costs Hank $24 to rent a mower for one hour. Write a statement for the cost to rent the mower for k hours. See section 2–1.

4–4 ■ *Complex rational expressions*

The previous sections of this chapter have dealt with simple rational numbers and simple rational expressions—rational numbers and rational expressions that have a single integer, or single polynomial, in the numerator and the denominator. We now consider a **complex rational expression**—a rational expression whose numerator or denominator, or both, contain rational expressions. To illustrate,

$$\dfrac{\dfrac{2}{x}}{\dfrac{3}{x^2}}, \quad \dfrac{\dfrac{2}{x}-3}{4+\dfrac{1}{x}}, \quad \dfrac{\dfrac{5}{y}-\dfrac{3}{y}}{y-3}, \quad \text{and} \quad \dfrac{\dfrac{p+1}{p-3}}{\dfrac{4-p}{p}}$$

are all complex rational expressions. To simplify such expressions, we eliminate the rational expression within the numerator and the denominator to obtain a simple rational expression.

Consider the complex rational expression

$$\dfrac{\left.\dfrac{p+1}{p-3}\right\}\text{Primary numerator}}{\left.\dfrac{4-p}{p}\right\}\text{Primary denominator}}$$

For the sake of discussion, we call

$$\dfrac{p+1}{p-3} \longleftarrow \text{Secondary denominator}$$

the **primary numerator** and

$$\dfrac{4-p}{p} \longleftarrow \text{Secondary denominator}$$

the **primary denominator.** The expressions $p-3$ and p are called the **secondary denominators.** Thus, to simplify the complex rational expression, we must eliminate the secondary denominators. This can be accomplished in either one of the two ways.

Simplifying a complex rational expression

Method 1 Form a single rational expression in the primary numerator and the primary denominator and divide the primary numerator by the primary denominator.

Method 2 Multiply the primary numerator and the primary denominator by the LCM of the secondary denominators and reduce the resulting fraction to lowest terms.

■ *Example 4–4 A*

Simplify each complex rational expression using method 1. Assume all denominators are nonzero.

1. $\dfrac{\dfrac{2}{x}}{\dfrac{3}{x^2}} = \dfrac{2}{x} \div \dfrac{3}{x^2}$

$\quad = \dfrac{2}{x} \cdot \dfrac{x^2}{3}$ Multiply by the reciprocal of $\dfrac{3}{x^2}$

$\quad = \dfrac{2x^2}{3x}$ Multiply numerators and denominators

$\quad = \dfrac{2x}{3}$ Reduce by x

2. $\dfrac{\dfrac{2}{x} - 3}{4 + \dfrac{1}{x}}$

$\quad = \dfrac{\dfrac{2}{x} - \dfrac{3x}{x}}{\dfrac{4x}{x} + \dfrac{1}{x}} = \dfrac{\dfrac{2 - 3x}{x}}{\dfrac{4x + 1}{x}}$ Perform the indicated addition (in the denominator) and the indicated subtraction (in the numerator)

$\quad = \dfrac{2 - 3x}{x} \cdot \dfrac{x}{4x + 1}$ Multiply by the reciprocal of $\dfrac{4x + 1}{x}$

$\quad = \dfrac{x(2 - 3x)}{x(4x + 1)}$ Multiply numerators and denominators

$\quad = \dfrac{2 - 3x}{4x + 1}$ Reduce by x

▶ *Quick check*   Simplify $\dfrac{\dfrac{1}{x^2} - \dfrac{1}{y^2}}{\dfrac{1}{x} + \dfrac{1}{y}}$ using method 1. Assume all denominators are nonzero.

A second method to accomplish this same end is to find the least common multiple (LCM) of the *secondary denominators* and to apply the fundamental principle of rational expressions by multiplying the primary numerator and the primary denominator by the LCM, thus eliminating the secondary denominators.

■ *Example 4–4 B*

Simplify each complex rational expression using method 2. Assume all denominators are nonzero.

1. $\dfrac{\dfrac{7}{8}}{\dfrac{5}{6}}$

The LCM of the secondary denominators, 8 and 6, is 24.

$$\frac{\dfrac{7}{8}}{\dfrac{5}{6}} = \frac{\dfrac{7}{8} \cdot 24}{\dfrac{5}{6} \cdot 24}$$

Multiply the primary numerator and the primary denominator by 24

$$= \frac{7 \cdot 3}{5 \cdot 4}$$

Reduce in the numerator and the denominator

$$= \frac{21}{20}$$

Multiply in the numerator and the denominator

2. $\dfrac{\dfrac{p+1}{p-3}}{\dfrac{4-p}{p}}$

The LCM of the secondary denominators of $p - 3$ and p is $p(p-3)$.

$$\frac{\dfrac{p+1}{p-3}}{\dfrac{4-p}{p}} = \frac{\dfrac{p+1}{p-3} \cdot p(p-3)}{\dfrac{4-p}{p} \cdot p(p-3)}$$

Multiply the primary numerator and the primary denominator by the LCM $p(p-3)$

$$= \frac{(p+1)p}{(4-p)(p-3)}$$

Reduce in numerator by $p - 3$ and in denominator by p

$$= \frac{p^2 + p}{-p^2 + 7p - 12}$$

Multiply in numerator and denominator

3. $\dfrac{\dfrac{5}{y} - \dfrac{3}{y}}{y - 3}$

The LCM of the secondary denominators is y.

$$\frac{\dfrac{5}{y} - \dfrac{3}{y}}{y - 3} = \frac{\left(\dfrac{5}{y} - \dfrac{3}{y}\right) \cdot y}{(y-3) \cdot y}$$

Multiply primary numerator and primary denominator by y

$$= \frac{\dfrac{5}{y} \cdot y - \dfrac{3}{y} \cdot y}{y(y-3)}$$

Apply distributive property

$$= \frac{5 - 3}{y(y-3)}$$

Reduce by y in the numerator

$$= \frac{2}{y(y-3)}$$

Subtract in the numerator

4. $\dfrac{(y + 2) + \dfrac{7}{y - 6}}{(y - 1) + \dfrac{4}{y - 6}}$

The LCM of the secondary denominators is $y - 6$.

$$\dfrac{(y + 2) + \dfrac{7}{y - 6}}{(y - 1) + \dfrac{4}{y - 6}} = \dfrac{\left[(y + 2) + \dfrac{7}{y - 6}\right](y - 6)}{\left[(y - 1) + \dfrac{4}{y - 6}\right](y - 6)}$$

Multiply primary numerator and primary denominator by $y - 6$

$$= \dfrac{(y + 2)(y - 6) + \dfrac{7}{y - 6} \cdot (y - 6)}{(y - 1)(y - 6) + \dfrac{4}{y - 6} \cdot (y - 6)}$$

Apply distributive property

$$= \dfrac{y^2 - 4y - 12 + 7}{y^2 - 7y + 6 + 4}$$

Perform indicated operations

$$= \dfrac{y^2 - 4y - 5}{y^2 - 7y + 10}$$

Combine like terms

$$= \dfrac{(y - 5)(y + 1)}{(y - 5)(y - 2)}$$

Factor numerator and denominator

$$= \dfrac{y + 1}{y - 2}$$

Reduce by $y - 5$

5. $\dfrac{y^{-1} - x^{-1}}{y^{-1} + x^{-1}}$

By definition, $y^{-1} = \dfrac{1}{y}$ and $x^{-1} = \dfrac{1}{x}$. Thus,

$$\dfrac{y^{-1} - x^{-1}}{y^{-1} + x^{-1}} = \dfrac{\dfrac{1}{y} - \dfrac{1}{x}}{\dfrac{1}{y} + \dfrac{1}{x}}$$

The LCM of the secondary denominators is xy.

$$= \dfrac{\dfrac{1}{y} \cdot xy - \dfrac{1}{x} \cdot xy}{\dfrac{1}{y} \cdot xy + \dfrac{1}{x} \cdot xy}$$

Multiply the primary numerator and primary denominator by xy using the distributive property

$$= \dfrac{x - y}{x + y}$$

Reduce in each term

▶ **Quick check**   Simplify $\dfrac{\dfrac{x - 3}{x + 2}}{\dfrac{x + 1}{x - 4}}$ using method 2. Assume all denominators are nonzero.   ■

Note   A reminder again—always check the factorability of the numerator and the denominator to *reduce* the simple rational expression to lowest terms.

Exercise 4–4

Simplify each complex rational expression using method 1. Assume all denominators are nonzero. See example 4–4 A.

Example $\dfrac{\dfrac{1}{x^2} - \dfrac{1}{y^2}}{\dfrac{1}{x} + \dfrac{1}{y}}$

Solution $= \dfrac{\dfrac{y^2 - x^2}{x^2 y^2}}{\dfrac{y + x}{xy}}$ Subtract in the numerator and add in the denominator

$= \dfrac{y^2 - x^2}{x^2 y^2} \cdot \dfrac{xy}{y + x}$ Multiply by the reciprocal of $\dfrac{y + x}{xy}$

$= \dfrac{(y - x)(y + x)xy}{x^2 y^2 (y + x)}$ Factor and multiply numerators and denominators

$= \dfrac{y - x}{xy}$ Reduce by common factors xy and $y + x$

1. $\dfrac{\dfrac{3}{5}}{\dfrac{4}{7}}$

2. $\dfrac{\dfrac{5}{6}}{\dfrac{7}{8}}$

3. $\dfrac{\dfrac{2}{3} + \dfrac{1}{4}}{\dfrac{1}{2} - \dfrac{1}{4}}$

4. $\dfrac{\dfrac{1}{3} + \dfrac{1}{2}}{\dfrac{2}{3} - \dfrac{5}{6}}$

5. $\dfrac{4 - \dfrac{2}{5}}{3 + \dfrac{3}{10}}$

6. $\dfrac{\dfrac{4}{x}}{\dfrac{8}{x^2}}$

7. $\dfrac{\dfrac{7}{m^2}}{\dfrac{8}{m}}$

8. $\dfrac{\dfrac{5}{b}}{\dfrac{9}{b - 3}}$

9. $\dfrac{\dfrac{-4}{y + 1}}{\dfrac{3}{y}}$

10. $\dfrac{\dfrac{8}{a - 3}}{\dfrac{-6}{a + 2}}$

11. $\dfrac{\dfrac{x}{x + 7}}{\dfrac{x}{x - 3}}$

12. $\dfrac{\dfrac{m - 3}{4}}{\dfrac{2m + 7}{6}}$

13. $\dfrac{\dfrac{1}{a} + \dfrac{1}{b}}{\dfrac{1}{a^2} - \dfrac{1}{b^2}}$

14. $\dfrac{x - y}{\dfrac{1}{x} - \dfrac{1}{y}}$

15. $\dfrac{\dfrac{x}{y} + 2}{\dfrac{x}{y} - \dfrac{4y}{x}}$

16. $\dfrac{\dfrac{2}{q} - \dfrac{3}{r}}{2 + \dfrac{1}{qr}}$

Simplify each complex rational expression by multiplying the numerator and the denominator by the LCM of the secondary denominators. See example 4–4 B.

Example

$$\frac{\dfrac{x-3}{x+2}}{\dfrac{x+1}{x-4}}$$

Solution

$$= \frac{\dfrac{x-3}{x+2}\cdot(x+2)(x-4)}{\dfrac{x+1}{x-4}\cdot(x+2)(x-4)}$$ Multiply primary numerator and primary denominator by LCM $(x+2)(x-4)$

$$= \frac{(x-3)(x-4)}{(x+1)(x+2)}$$ Reduce by common factors

$$= \frac{x^2-7x+12}{x^2+3x+2}$$ Perform indicated multiplication

17. $\dfrac{\dfrac{5}{9}}{\dfrac{3}{4}}$

18. $\dfrac{\dfrac{1}{6}-\dfrac{7}{8}}{\dfrac{2}{3}+\dfrac{1}{4}}$

19. $\dfrac{\dfrac{5}{y^2}}{\dfrac{7}{y}}$

20. $\dfrac{\dfrac{3}{x-2}}{\dfrac{9}{x}}$

21. $\dfrac{\dfrac{7}{a+9}}{\dfrac{-4}{a-5}}$

22. $\dfrac{\dfrac{n-9}{n+2}}{\dfrac{n}{3n-5}}$

23. $\dfrac{\dfrac{4x^2-y^2}{3x}}{\dfrac{2x+y}{4x}}$

24. $\dfrac{\dfrac{x+3}{y}}{\dfrac{x^2-9y^2}{6y}}$

25. $\dfrac{4-\dfrac{3}{x+3}}{5+\dfrac{6}{x-1}}$

26. $\dfrac{\dfrac{7}{y-3}+8}{9-\dfrac{1}{2y+3}}$

27. $\dfrac{\dfrac{1}{x}+\dfrac{1}{y^2}}{\dfrac{1}{x^2}-\dfrac{1}{x}}$

28. $\dfrac{\dfrac{1}{a}-\dfrac{1}{b}}{\dfrac{1}{a^2}+\dfrac{1}{b^2}}$

29. $\dfrac{m+n}{\dfrac{1}{m}-\dfrac{1}{n}}$

30. $\dfrac{x-y}{\dfrac{1}{x}+\dfrac{1}{y}}$

31. $\dfrac{\dfrac{5}{p^2}+\dfrac{4}{q}}{p-q}$

32. $\dfrac{\dfrac{1}{m^2}-\dfrac{1}{n^2}}{m+n}$

33. $\dfrac{(a+5)+\dfrac{3}{a+4}}{(a+3)-\dfrac{5}{a+4}}$

34. $\dfrac{(x-3)+\dfrac{7}{2x+1}}{(x+9)-\dfrac{3}{2x+1}}$

35. $\dfrac{y-\dfrac{3}{4y-3}}{(y+2)+\dfrac{3}{y+5}}$

36. $\dfrac{(m-3)+\dfrac{6}{2m+3}}{m-\dfrac{9}{m-6}}$

37. $\dfrac{\dfrac{t^2-2t-8}{t^2+7t+6}}{\dfrac{t^2-t-6}{t^2+2t+1}}$

38. $\dfrac{\dfrac{y^2-5y-14}{y^2+3y-10}}{\dfrac{y^2-8y+7}{y^2+6y+5}}$

39. $\dfrac{\dfrac{3}{x^2-x-6}}{\dfrac{2}{x+2}-\dfrac{4}{x-3}}$

40. $\dfrac{\dfrac{9}{a-7}+\dfrac{8}{2a+3}}{\dfrac{10}{2a^2-11a-21}}$

41. $\dfrac{5+\dfrac{4}{b-1}}{\dfrac{7}{b+5}-\dfrac{3}{b-1}}$

42. $\dfrac{\dfrac{6}{x+5}-7}{\dfrac{8}{x+5}-\dfrac{9}{x+3}}$

43. $\dfrac{\dfrac{1}{1-x}-\dfrac{1}{1+x}}{\dfrac{1+x}{1-x}-\dfrac{1-x}{1+x}}$

44. $\dfrac{\dfrac{a-1}{a-b} - \dfrac{a-b}{a-1}}{\dfrac{1}{a-1} - \dfrac{1}{a-b}}$

45. $\dfrac{\dfrac{3}{ab} + \dfrac{4}{bc} - \dfrac{2}{ac}}{\dfrac{5}{abc}}$

46. $\dfrac{\dfrac{x}{yz} - \dfrac{y}{xz} + \dfrac{z}{xy}}{\dfrac{1}{x^2y^2} - \dfrac{1}{x^2z^2} + \dfrac{1}{y^2z^2}}$

47. $\dfrac{x^{-1} + y^{-1}}{x^{-1} - y^{-1}}$

48. $\dfrac{x^{-2} - y^{-2}}{x^{-1} + y^{-1}}$

49. $\dfrac{x^{-1}}{x^{-1} + y^{-1}}$

50. $\dfrac{x^{-2}}{x^{-2} - y^{-2}}$

51. $\dfrac{p^{-2} + q^{-2}}{q^{-2}}$

52. $\dfrac{2x^{-1} - y^{-1}}{x^{-1} + 5y^{-1}}$

53. $(p^{-1} - q^{-1})^{-1}$

54. $(x^{-1} + y^{-1})^{-1}$

Solve the following word problems.

55. In electricity, a relationship for the current, i, in a capacitor of "size" C is given by

$$i = \dfrac{V_s - \dfrac{it}{C}}{R}$$

Simplify the right member.

56. In electricity, the voltage between two adjacent nodes, denoted by $V_{AA'}$, is given by Millman's Theorem,

$$V_{AA'} = \dfrac{\dfrac{V_{S1}}{R_1} + \dfrac{V_{S2}}{R_2} + \dfrac{V_{S3}}{R_3}}{\dfrac{1}{R_1} + \dfrac{1}{R_2} + \dfrac{1}{R_3}}$$

where V_{S1}, V_{S2}, V_{S3} represent equivalent voltages of the branches between nodes A and A', whereas R_1, R_2, and R_3 are equivalent resistances of the branches between nodes A and A'. Simplify the right member.

57. A refrigeration coefficient, CP, of performance formula for the ideal refrigerator is given by

$$CP = \dfrac{1}{\dfrac{T_2}{T_1} - 1}$$

Simplify the right member.

58. The capacitance C of a circuit connecting three capacitances C_1, C_2, and C_3 in series is given by

$$C = \dfrac{1}{\dfrac{1}{C_1} + \dfrac{1}{C_2} + \dfrac{1}{C_3}}$$

Simplify the right member.

59. When making a round trip whose one-way distance is d, the average rate (speed) traveled, r, is given by

$$r = \dfrac{2d}{\dfrac{d}{r_1} + \dfrac{d}{r_2}}$$

where r_1 is the average rate going and r_2 is the average rate coming back. Simplify the right member.

Review exercises

Find the solution set of the following equations. See sections 2–1 and 2–4.

1. $\dfrac{1}{2}x + 3 = \dfrac{2}{3}$

2. $|x - 3| = 5$

3. $5(2x - 4) + 3x = 2x + 2$

Find the solution set of the following inequalities. See section 2–6.

4. $|x - 2| < 4$

5. $|2x + 5| \geq 3$

6. Perform the indicated subtraction. See section 3–2.
$(3a^3 + 2a^2 - 3) - (a^3 - 6a^2 + a + 1)$

4–5 ■ *Quotients of polynomials*

Division of a polynomial by a monomial

The first type of division that we will study is that of a polynomial divided by a monomial. Consider the indicated division

$$\frac{4a^3 - 12a^2 + 8a}{2a}$$

Using the meaning of division, we can rewrite this as

$$(4a^3 - 12a^2 + 8a) \cdot \frac{1}{2a}$$

and applying the distributive property, we have

$$\frac{4a^3 - 12a^2 + 8a}{2a} = \frac{4a^3}{2a} - \frac{12a^2}{2a} + \frac{8a}{2a}$$
$$= 2a^2 - 6a + 4$$

Dividing a polynomial by a monomial

For monomials $a_1, a_2, \cdots, a_n$, and d, where $d \neq 0$,

$$\frac{a_1 + a_2 + \cdots + a_n}{d} = \frac{a_1}{d} + \frac{a_2}{d} + \cdots + \frac{a_n}{d}$$

Concept
To divide a polynomial by a monomial, simply divide each term of the polynomial by the monomial and write the resulting quotients.

■ *Example 4–5 A*

Perform the indicated division. Assume that no denominator equals zero.

1. $\dfrac{4a^6 + 16a^4 - 6a}{4a^2} = \dfrac{4a^6}{4a^2} + \dfrac{16a^4}{4a^2} - \dfrac{6a}{4a^2}$ Divide each term by monomial $4a^2$

     $= a^4 + 4a^2 - \dfrac{3}{2a}$ Reduce to lowest terms

2. $\dfrac{x(3y - 2) + 2(3y - 2)}{3y - 2} = \dfrac{x(3y - 2)}{(3y - 2)} + \dfrac{2(3y - 2)}{(3y - 2)}$ Divide each term by $(3y - 2)$

     $= x + 2$ Reduce to lowest terms

Note A common error in this type of problem is demonstrated in the example

$$\frac{x^3 + x^2}{x^2} = \frac{x^3 + \overset{1}{\cancel{x^2}}}{\underset{1}{\cancel{x^2}}} = \frac{x^3 + 1}{1} = x^3 + 1$$

It is tempting to "cancel" the x^2 in the numerator with the x^2 in the denominator, but the correct procedure is

$$\frac{x^3 + x^2}{x^2} = \frac{x^3}{x^2} + \frac{x^2}{x^2} = x + 1$$

▶ *Quick check* Simplify $\dfrac{5x^3 - 10x^2 + 25x}{5x^2}$. Assume the denominator is nonzero.

Division of a polynomial by a polynomial

The second type of division that we will study is that of a polynomial divided by another polynomial. Consider the example

$$\frac{2x^2 - x - 15}{x - 3}$$

The example is set up so that the divisor, $x - 3$, and the dividend, $2x^2 - x - 15$, are arranged in descending powers of the variable.

$$x - 3 \overline{\smash{\big)}\, 2x^2 - x - 15}$$

Now we divide the first term of the dividend, $2x^2$, by the first term of the divisor, x. The result is $\dfrac{2x^2}{x} = 2x$, and we place the $2x$ above the division line.

$$\begin{array}{r} 2x \\ x - 3 \overline{\smash{\big)}\, 2x^2 - x - 15} \end{array}$$

Next, we multiply $x - 3$ and $2x$, placing the product below the dividend, and subtract.

$$\begin{array}{r} 2x \\ x - 3 \overline{\smash{\big)}\, 2x^2 - x - 15} \\ \underline{2x^2 - 6x} \end{array} \qquad 2x(x - 3) = 2x^2 - 6x$$

Recall that when we subtract, we change the signs and add. Then $(2x^2 - x) - (2x^2 - 6x) = 2x^2 - x - 2x^2 + 6x = 5x$.

$$\begin{array}{r} 2x \\ x - 3 \overline{\smash{\big)}\, 2x^2 - x - 15} \\ \underline{-2x^2 - 6x} \\ 5x \end{array}$$

Now we bring down the -15 and repeat the same process.

$$\begin{array}{r} 2x \\ x - 3 \overline{\smash{\big)}\, 2x^2 - x - 15} \\ \underline{2x^2 - 6x} \\ 5x - 15 \end{array}$$

Divide the $5x$ by x, which results in 5, and multiply 5 and $x - 3$.

$$\begin{array}{r} 2x + 5 \\ x - 3 \overline{\smash{\big)}\, 2x^2 - x - 15} \\ \underline{2x^2 - 6x} \\ 5x - 15 \\ \underline{5x - 15} \end{array}$$

Subtract $(5x - 15) - (5x - 15) = 5x - 15 - 5x + 15 = 0$.

$$\begin{array}{r} 2x + 5 \\ x - 3 \overline{\smash{\big)}\, 2x^2 - x - 15} \\ \underline{2x^2 - 6x} \\ 5x - 15 \\ \underline{5x - 15} \\ 0 \end{array}$$

We subtract and get a remainder of zero, so the quotient is $2x + 5$.

$$\frac{2x^2 - x - 15}{x - 3} = 2x + 5$$

To check the problem, multiply the quotient times the divisor to get the dividend.

$$(2x + 5)(x - 3) = 2x^2 - x - 15$$

Note   *It is important* when dividing a polynomial by a polynomial that both the divisor and the dividend have their terms arranged in descending powers of the variable.

■ *Example 4–5 B*

Find the indicated quotients. Assume no divisor is equal to zero.

1. $(2x^2 - 11x + 15) \div (x - 3)$

$$
\begin{array}{r}
2x - 5 \\
x - 3 \overline{\smash{\big)}\, 2x^2 - 11x + 15} \\
\underline{2x^2 - 6x} \\
-5x + 15 \\
\underline{-5x + 15} \\
0
\end{array}
$$

Subtract (change signs and add)

Subtract (change signs and add)

Therefore $\dfrac{2x^2 - 11x + 15}{x - 3} = 2x - 5$.

2. $(2y^3 - y^2 + 5y + 5) \div (2y + 1)$

$$
\begin{array}{r}
y^2 - y + 3 \\
2y + 1 \overline{\smash{\big)}\, 2y^3 - y^2 + 5y + 5} \\
\underline{2y^3 + y^2} \\
-2y^2 + 5y \\
\underline{-2y^2 - y} \\
6y + 5 \\
\underline{6y + 3} \\
2
\end{array}
$$

Subtract (change signs and add)

Subtract (change signs and add)

Subtract (change signs and add)

Remainder

When there is a remainder, as in this example, we write the remainder over the divisor.

$$\frac{2y^3 - y^2 + 5y + 5}{2y + 1} = y^2 - y + 3 + \frac{2}{2y + 1}$$

3. $(2x^3 - x + 5) \div (x - 1)$

There is no term in the dividend that contains x^2. Therefore we will insert $0x^2$ as a placeholder so that all powers of the variable x are present in descending order. The value of the dividend has not changed since we added $0x^2$, which is another name for 0.

$$
\begin{array}{r}
2x^2 + 2x + 1 \\
x - 1 \overline{\smash{\big)}\, 2x^3 + 0x^2 - x + 5} \\
\underline{2x^3 - 2x^2} \\
2x^2 - x \\
\underline{2x^2 - 2x} \\
x + 5 \\
\underline{x - 1} \\
6
\end{array}
$$

Subtract (change signs and add)

Subtract (change signs and add)

Subtract (change signs and add)

Remainder

Hence $\dfrac{2x^3 - x + 5}{x - 1} = 2x^2 + 2x + 1 + \dfrac{6}{x - 1}$.

▶ **Quick check** Divide $\dfrac{3y^3 + 5y^2 - 11y + 6}{3y - 1}$. Assume the denominator is nonzero. ∎

Mastery points

Can you
- Divide a polynomial by a monomial?
- Divide a polynomial by a polynomial?

Exercise 4–5

Perform the indicated divisions. Assume that no divisor is equal to zero. See example 4–5 A.

Example $\dfrac{5x^3 - 10x^2 + 25x}{5x^2}$

Solution $= \dfrac{5x^3}{5x^2} - \dfrac{10x^2}{5x^2} + \dfrac{25x}{5x^2}$ Divide denominator into each term

$= x - 2 + \dfrac{5}{x}$ Divide in each term by common factors

1. $\dfrac{25x^2 - 15x + 10}{5}$

2. $\dfrac{2a^4 - 3a^2 + a}{a}$

3. $\dfrac{4x^4 - 8x^3 + 12x}{-4x}$

4. $\dfrac{15y^5 + 25y^2 + 10y}{-5y^2}$

5. $\dfrac{ac^2 - ac}{ac}$

6. $\dfrac{bx - b^2x^2}{bx}$

7. $\dfrac{30x^3y^4 + 21x^2y^2 - 18x^2y^4}{3x^2y^2}$

8. $\dfrac{36x^4y^2z^3 - 24x^2y^5z + 18x^2y^2z^2}{6x^2y^3z}$

9. $\dfrac{21a^7b^2c^3 - 35a^5b^5c^3 + 49a^4b^2c^3}{7abc^3}$

10. $\dfrac{x(y - 2) - z(y - 2)}{y - 2}$

11. $\dfrac{2a(b - 4) + 3c(b - 4)}{b - 4}$

12. $\dfrac{x^2y(z + 3) - 3x^4y^3(z + 3)}{xy(z + 3)}$

See example 4–5 B.

Example $\dfrac{3y^3 + 5y^2 - 11y + 6}{3y - 1}$

Solution

$$
\begin{array}{r}
y^2 + 2y - 3 \\
3y - 1 \overline{)\,3y^3 + 5y^2 - 11y + 6} \\
\underline{3y^3 - y^2} \\
6y^2 - 11y \\
\underline{6y^2 - 2y} \\
-9y + 6 \\
\underline{-9y + 3} \\
3
\end{array}
$$

Subtract (change signs and add)

Subtract (change signs and add)

Subtract (change signs and add)

Remainder

$$\dfrac{3y^3 + 5y^2 - 11y + 6}{3y - 1} = y^2 + 2y - 3 + \dfrac{3}{3y - 1}$$

13. $(a^2 - 2a - 8) \div (a + 2)$ **14.** $(y^2 + 2y - 3) \div (y + 3)$ **15.** $(y^2 + 7y + 11) \div (y + 5)$

16. $(2x - 5 + x^2) \div (x + 4)$ **17.** $(10 - 7x + x^2) \div (x - 3)$ **18.** $\dfrac{a^2 - 5a + 1}{a - 1}$

19. $\dfrac{a^3 + a^2 - 2a + 12}{a + 3}$ **20.** $\dfrac{y^3 + 3y^2 - y - 6}{y + 2}$ **21.** $\dfrac{2y^3 - y^2 - 2y + 3}{y + 1}$

22. $\dfrac{3x^3 - 4x^2 - 5x - 4}{x + 2}$ **23.** $\dfrac{5x^2 - 11x + 2x^3 + 4}{2x - 1}$ **24.** $\dfrac{3a^3 - 2 - 5a - a^2}{3a + 2}$

25. $\dfrac{4a^3 + 8a^2 - 5a + 1}{2a + 1}$ **26.** $\dfrac{6y^3 - y^2 - 11y + 10}{3y - 2}$ **27.** $\dfrac{9y^3 + 11y + 6}{3y + 1}$

28. $\dfrac{27x^3 - 1}{3x - 1}$ **29.** $\dfrac{a^3 + 27}{a + 3}$ **30.** $\dfrac{x^4 - 16}{x - 2}$

31. $\dfrac{3x^4 - 2x^3 + x - 1}{x + 1}$ **32.** $\dfrac{2x^3 + 5x^2 + 5x + 3}{x^2 + x + 1}$ **33.** $\dfrac{3a^3 - 4a^2 + 10a - 3}{a^2 - a + 3}$

34. $\dfrac{a^4 - 2a^2 - 3a - 1}{a^2 - 2a - 1}$ **35.** $\dfrac{y^4 + 4y^3 + 3y^2 - 2y - 1}{y^2 + 3y + 1}$ **36.** $\dfrac{y^4 - y^3 - 11y^2 + 10y + 2}{y^2 - 4y + 2}$

37. $\dfrac{x^4 + 4x^3 + x^2 - 10x - 12}{x^2 + x - 4}$ **38.** $\dfrac{2x^4 - x^3 + 5x^2 - x + 3}{x^2 + 1}$ **39.** $\dfrac{3a^4 - 2a^3 + 2a^2 - 4a - 8}{a^2 + 2}$

40. $\dfrac{a^4 + 4a^3 + a^2 - 10a - 9}{a^2 + 3a + 2}$ **41.** $\dfrac{2y^4 - 3y^3 + 8y^2 - 9y + 8}{2y^2 - 3y + 2}$ **42.** $\dfrac{3y^4 + y^3 - 8y^2 - 3y - 5}{3y^2 + y + 1}$

Solve the following word problems.

43. Evaluate $2y^3 - y^2 - 2y + 3$ when $y = -1$ and compare this answer to the remainder found in exercise 21.

44. Evaluate $3x^3 - 4x^2 - 5x - 4$ when $x = -2$ and compare this answer to the remainder found in exercise 22.

45. The area of a rectangle is found by multiplying the length times the width. If the area of a rectangle is $6x^2 - 17x + 12$ and the length is $3x - 4$, find the width.

46. A contractor uses the expression $x^2 + 6x + 8$ to represent the square footage of a room. If she decides that the length of the room will be represented by $x + 4$, what will the width of the room be in terms of x?

47. The volume of a box is found by multiplying the length times the width times the height. If the volume of a box is $6x^3 + 11x^2 - 19x + 6$, the height is $x + 3$, and the width is $2x - 1$, find the length.

48. An electrician uses the expression $x^2 + 5x + 6$ to determine the amount of wire to order when wiring a house. If the formula comes from multiplying the number of rooms times the number of outlets and he knows the number of rooms to be $x + 2$, find the number of outlets in terms of x.

Review exercises

Simplify the following expressions. Assume all denominators are nonzero. Answer with positive exponents only. See sections 3–1 and 3–3.

1. $\dfrac{x^3y}{x^2y}$

2. $\dfrac{x^{-3}}{x^{-6}}$

3. $(2a^{-4}b^0)^3$

4. Simplify the expression $\dfrac{\dfrac{3}{4} - \dfrac{1}{2}}{\dfrac{2}{3} + \dfrac{1}{6}}$ See section 4–4.

Perform the indicated operations. Assume all denominators are nonzero. See sections 4–2 and 4–3.

5. $\dfrac{2x-1}{x^2-9} - \dfrac{x+1}{x+3}$

6. $\dfrac{2x-1}{3x-1} \div \dfrac{4x}{6x-2}$

7. Find the solution set of the equation $\dfrac{1}{5}x - \dfrac{2}{3} = \dfrac{3}{5}$. See section 2–1.

4–6 ■ Synthetic division, the remainder theorem, and the factor theorem

In section 4–5, we studied the procedure for dividing a polynomial by another polynomial. Many times the divisor is a binomial of the form $x - k$, k is a constant. We can shorten the process considerably by using **synthetic division.**

Consider the following example: $(2x^3 + 5x^2 + x - 1) \div (x + 2)$. Performing the indicated division, we obtain

$$
\begin{array}{r}
2x^2 + x - 1 \quad \text{Quotient} \\
x + 2\overline{)2x^3 + 5x^2 + x - 1} \quad \text{Dividend} \\
\underline{2x^3 + 4x^2} \\
x^2 + x \\
\underline{x^2 + 2x} \\
-x - 1 \\
\underline{-x - 2} \\
1 \quad \text{Remainder}
\end{array}
$$

Divisor

We can write the exact same problem using only the coefficients of the terms.

Coefficients from the divisor → 1

$$
\begin{array}{r}
2\ \ 1\ \ -1 \\
2\overline{)2\quad 5\quad 1\quad -1} \\
②\quad 4 \\
1\quad 1 \\
①\quad 2 \\
-1\quad -1 \\
\ominus\quad -2 \\
1
\end{array}
$$

Coefficients from the quotient
Coefficients from the dividend

Remainder

Observe that the circled numbers are repetitions of the numbers directly above them. We can rewrite the problem without them.

$$
\begin{array}{r}
\quad 2\quad\ \ 1\quad -1 \\
1\qquad 2\ \big)\overline{2\quad\ \ 5\quad\ \ 1\quad -1} \\
\underline{4} \\
1\quad\ \textcircled{1} \\
2 \\
\underline{-1\quad \textcircled{-1}} \\
-2 \\
1
\end{array}
$$

The circled numbers are again the same as the numbers directly above them. Therefore we can omit them.

$$
\begin{array}{r}
\quad 2\quad\ \ 1\quad -1 \\
1\qquad 2\ \big)\overline{2\quad\ \ 5\quad\ \ 1\quad\quad -1} \\
\underline{4} \\
1 \\
2 \\
\underline{-1} \\
-2 \\
1
\end{array}
$$

All of the numbers can be condensed. We omit the top row of numbers since it duplicates the bottom set of numbers. We shall also omit the 1 at the upper left.

$$
\begin{array}{r}
2\ \big)\overline{2\quad\ \ 5\quad\ \ 1\quad -1} \\
\underline{4\quad\ \ 2\quad -2} \\
2\quad\ \ 1\quad -1\quad\ \ 1
\end{array}
$$

By changing the 2 at the upper left to its additive inverse, -2, and adding the additive inverse in each step, instead of subtracting, we obtain the same result. The following is the final form to be used when performing synthetic division.

$$
\begin{array}{r}
-2\ \big)\overline{2\quad\ \ 5\quad\ \ 1\quad -1} \\
\underline{-4\quad -2\quad\ \ 2} \\
2\quad\ \ 1\quad -1\quad\ \ 1
\end{array}
$$

$\uparrow\ x^2\qquad \uparrow\ x \qquad \uparrow\ \text{Constant}\qquad \searrow\ \text{Remainder}$

The bottom line of this synthetic division process represents the coefficients of the terms of the quotient along with the remainder. That is, 2   1 -1 1 represents the quotient $2x^2 + x - 1$ with a remainder of 1.

Note The degree 2 of the quotient is one less than the degree 3 of the dividend. This will *always* be the case.

■ *Example 4–6 A*

Divide the following using synthetic division. Assume no denominator equals zero.

1. $(3x^2 + 7x + 6) \div (x + 1)$

   To use synthetic division, the divisor *must* be of the form $x - k$. Therefore we write $x + 1$ as $x - (-1)$ so $k = -1$. Now write the coefficients of the terms in $3x^2 + 7x + 6$ with -1 to the left.

   $$-1 \overline{)\,3 \quad 7 \quad 6}$$

   First, we bring down 3 and multiply that by -1. This product, -3, is added to 7.

   $$\begin{array}{r} -1 \overline{)\,3 \quad\; 7 \quad\; 6} \\ \downarrow \;\; -3 \quad\quad \\ \hline 3 \quad\; 4 \quad\quad \end{array}$$

   The 4 is now multiplied by -1 and the product, -4, is added to 6.

   $$\begin{array}{r} -1 \overline{)\,3 \quad\; 7 \quad\; 6} \\ -3 \quad -4 \\ \hline 3 \quad\; 4 \quad\; 2 \end{array}$$

   We can now read the coefficients of the quotient and the remainder from the last row. The answer is $3x + 4 + \dfrac{2}{x + 1}$. (2 is the remainder.)

2. $(2x^3 - 7x^2 - x + 12) \div (x - 3)$

   The divisor, $x - 3$, is already in the form $x - k$, where $k = 3$. We set up the problem as follows:

   $$3 \overline{)\,2 \quad -7 \quad -1 \quad 12}$$

   We bring down 2 and begin the repetitive process of multiplying the last number in the bottom row times k, that is, 3, and adding this to the value in the next column.

   $$\begin{array}{r} 3 \overline{)\,2 \quad\; -7 \quad\; -1 \quad\;\; 12} \\ \downarrow \quad\; 6 \quad\; -3 \quad -12 \\ \hline 2 \quad -1 \quad -4 \quad\quad 0 \end{array}$$

   Since we have a zero in the last position of the bottom row, there is no remainder and the quotient is $2x^2 - x - 4$.

3. $(x^3 - 11x + 8) \div (x - 3)$

   Since the x^2 term is missing in the dividend, we think of the expression as $x^3 + 0x^2 - 11x + 8$ when writing down the coefficients.

   $$\begin{array}{r} 3 \overline{)\,1 \quad\;\; 0 \quad\; -11 \quad\;\; 8} \\ \downarrow \quad\; 3 \quad\quad 9 \quad -6 \\ \hline 1 \quad\; 3 \quad\; -2 \quad\quad 2 \end{array}$$

   The answer is then $x^2 + 3x - 2 + \dfrac{2}{x - 3}$.

4. $(2x^4 - 10x^3 + 11x^2 - 11x + 8) \div (x - 2)$

$$
\begin{array}{r|rrrrr}
2) & 2 & -10 & 11 & -11 & 8 \\
  & \downarrow & 4 & -12 & -2 & -26 \\
\hline
  & 2 & -6 & -1 & -13 & -18
\end{array}
$$

The answer is $2x^3 - 6x^2 - x - 13 - \dfrac{18}{x - 2}$.

▶ *Quick check* Use synthetic division to divide $(x^3 + 10x - 2) \div (x + 2)$. ■

Remainder and factor theorems

From our work with synthetic division, we can see that division by a polynomial in x, $P(x)$, by a polynomial $x - r$ results in a quotient $Q(x)$ and a constant remainder R. That is,

$$\frac{P(x)}{x - r} = Q(x) + R$$

and so

$$P(x) = (x - r)Q(x) + R$$

Using this equation, we evaluate $P(r)$ when $x = r$.

$$
\begin{aligned}
P(x) &= (x - r)Q(x) + R \\
P(r) &= (r - r)Q(r) + R \\
     &= 0 \cdot Q(r) + R \\
     &= 0 + R \\
     &= R
\end{aligned}
$$

Thus, $P(r) = R$ and we find that the value of the polynomial when $x = r$ is the remainder R. We have just proved the **Remainder Theorem.**

__ *Remainder theorem* _____

If a polynomial $P(x)$ is divided by $x - r$, where r is a real number, the remainder R is $P(r)$.

■ *Example 4–6 B*

1. Determine the remainder when $P(x) = 2x^3 - 5x^2 + 4x - 2$ is divided by $x - 3$. Evaluate $P(3)$ using substitution.

 a. Using synthetic division,

$$
\begin{array}{r|rrrr}
3) & 2 & -5 & 4 & -2 \\
  & \downarrow & 6 & 3 & 21 \\
\hline
  & 2 & 1 & 7 & 19 \longleftarrow \text{Remainder}
\end{array}
$$

Thus, $P(3) = 19$

b. Using substitution, we must show that $P(3) = 19$.

$$P(x) = 2x^3 - 5x^2 + 4x - 2$$
$$P(3) = 2(3)^3 - 5(3)^2 + 4(3) - 2$$
$$= 54 - 45 + 12 - 2$$
$$= 19$$

Thus, we have shown that $R = 19$ is the result in either case.

▶ *Quick check* Using synthetic division and the remainder theorem, find the remainder when $P(x) = 3x^3 - 2x^2 + 4x - 5$ is divided by $x + 4$. Evaluate $P(-4)$ using substitution. ∎

The **factor theorem,** which is a direct result of the remainder theorem (a corollary), applies when the remainder $R = 0$. This theorem shows there is a close relationship between the *factors* of a polynomial $P(x)$ and the values of x for which $P(x) = 0$.

⌐ **Factor theorem** ────────────────────────────

Given polynomial $P(x)$ and real number r, then

$$x - r \text{ is a factor of } P(x) \text{ if and only if } P(r) = 0$$

■ *Example 4–6 C*

1. Show that $x + 2$ is a factor of $P(x) = x^3 - 3x^2 - 6x + 8$.
   By the factor theorem, $x + 2$ is a factor of $P(x)$ if $P(-2) = 0$. Using synthetic division, we show that the remainder $R = 0$.

$$
\begin{array}{r|rrrr}
-2 & 1 & -3 & -6 & 8 \\
   & \downarrow & -2 & 10 & -8 \\
\hline
   & 1 & -5 & 4 & \boxed{0}
\end{array}
$$

   Thus, by the remainder theorem, $P(-2) = 0$ and by the factor theorem $x + 2$ is a factor of $P(x)$.

Note Alternatively, we can evaluate

$$P(-2) = (-2)^3 - 3(-2)^2 - 6(-2) + 8$$
$$= -8 - 12 + 12 + 8$$
$$= 0$$

2. Find the solution set of the equation $x^3 - 3x^2 - 6x + 8 = 0$.
   In example 1, we determined that $x + 2$ is a factor of $x^3 - 3x^2 - 6x + 8$ and the quotient $\dfrac{P(x)}{x + 2} = x^2 - 5x + 4$.

   Thus, we have

   $$(x + 2)(x^2 - 5x + 4) = 0$$
   $$(x + 2)(x - 4)(x - 1) = 0 \quad \text{Factor the left member}$$

   Since $x = 4$ when $x - 4 = 0$ and $x = 1$ when $x - 1 = 0$, the solution set is $\{-2, 4, 1\}$.

Note We have used the zero product property stated in section 4–1 to find the remaining solutions.

3. Find a polynomial $P(x)$ of degree 3 whose zeros are 2, -1, and -3.
By the factor theorem, $(x - 2)$, $(x + 1)$, and $(x + 3)$ are the factors of $P(x)$. Thus,

$$P(x) = (x - 2)(x + 1)(x + 3) = x^3 + 2x^2 - 5x - 6$$

is a polynomial of degree 3 having the given zeros.

Note This polynomial is *not unique,* since multiplying $P(x)$ by any nonzero real number will yield another polynomial having the same zeros. To illustrate,

$$P(x) = 4x^3 + 8x^2 - 20x - 24$$

also has zeros 2, -1, and -3.

▶ *Quick check* Show that $x + 5$ is a factor of $P(x) = x^3 + 4x^2 - 7x - 10$. Find the solution set of the equation $x^3 + 4x^2 - 7x - 10 = 0$. ∎

Given

$$P(x) = 4x^2 - 12x + 9$$

factoring the polynomial, $P(x) = (2x - 3)(2x - 3)$ and the zeros are $\dfrac{3}{2}$ and $\dfrac{3}{2}$. Thus, we may have repeated zeros. In this case, $\dfrac{3}{2}$ is repeated twice and we say the $P(x)$ has zero $\dfrac{3}{2}$ of *multiplicity 2.* In general, if a factor $x - r$ is repeated k times in a polynomial, we say r is a *zero of multiplicity k.*

■ *Example 4–6 D*

1. Given $P(x) = (x + 2)^2(x - 5)^3(x - 1)^4$, the distinct zeros are -2, 5, and 1. Since each is repeated more than once, we say -2 is a zero of multiplicity 2, 5 is a zero of multiplicity 3, and 1 is a zero of multiplicity 4.

2. Given 3 is a zero of the polynomial $P(x) = 2x^3 - 7x^2 - 7x + 30$, find the other zeros.
Since 3 is a zero, then $x - 3$ is a factor of $P(x)$ by the factor theorem. Using synthetic division,

$$
\begin{array}{r|rrrr}
3) & 2 & -7 & -7 & 30 \\
 \downarrow & & 6 & -3 & -30 \\
\hline
 & 2 & -1 & -10 & 0
\end{array}
$$

Thus, $P(x) = (x - 3)(2x^2 - x - 10)$ and factoring $2x^2 - x - 10$,

$$P(x) = (x - 3)(2x - 5)(x + 2)$$

Since $x = 5/2$ when $2x - 5 = 0$ and $x = -2$ when $x + 2 = 0$, the zeros of $P(x)$ are 3, 5/2, and -2.

3. Given -1 is a solution (root) of the equation

$$3x^4 - 7x^3 - 33x^2 - 33x - 10 = 0$$

of multiplicity 2, find the solution set.

Given -1 is a solution of multiplicity 2, then $(x + 1)^2 = x^2 + 2x + 1$ is a factor of the left member of the equation. We divide

$$
\begin{array}{r}
3x^2 - 13x - 10 \\
x^2 + 2x + 1\overline{)3x^4 - 7x^3 - 33x^2 - 33x - 10} \\
\underline{3x^4 + 6x^3 + 3x^2} \\
-13x^3 - 36x^2 - 33x \\
\underline{-13x^3 - 26x^2 - 13x} \\
-10x^2 - 20x - 10 \\
\underline{-10x^2 - 20x - 10} \\
0
\end{array}
$$

Thus, $(x + 1)^2(3x^2 - 13x - 10) = 0$ and if we factor $3x^2 - 13x - 10$, we have

$$(x + 1)^2(3x + 2)(x - 5) = 0$$

The solution set is $\left\{-1, -\dfrac{2}{3}, 5\right\}$.

▶ *Quick check* Given the equation $2x^4 + 5x^3 - 51x^2 + 80x - 28 = 0$ has a solution 2 of multiplicity 2, find the solution set of the equation. Find a polynominal $P(x)$ of lowest degree whose zeros are 3, -2, and 1 of multiplicity 2. ◼

Mastery points

Can you

- Divide $P(x)$ by $x - r$ using synthetic division?
- Use the remainder theorem to find the remainder R when polynomial $P(x)$ is divided by polynomial divisor $x - r$?
- Use the factor theorem to determine if polynomial $x - r$ is a factor of $P(x)$?
- Use the factor theorem to solve an equation $P(x) = 0$ where $P(x)$ is of degree greater than or equal to 3?
- Use the factor theorem to find an equation when given the real roots of the equation?
- Determine the multiplicity of a factor $x - r$?

Exercise 4-6

Perform the indicated divisions using synthetic division. Assume that no divisor is equal to zero.
See example 4-6 A.

Example $(x^3 + 10x - 2) \div (x + 2)$

Solution   Since there is no x^2 term, we insert $0x^2$ in the dividend to obtain $(x^3 + 0x^2 + 10x - 2) \div (x + 2)$.

$$
\begin{array}{r|rrrr}
-2 & 1 & 0 & 10 & -2 \leftarrow \text{Coefficients} \\
   & \downarrow & -2 & 4 & -28 \leftarrow \text{Add} \\
\hline
   & 1 & -2 & 14 & -30
\end{array}
$$

The answer is $x^2 - 2x + 14 - \dfrac{30}{x + 2}$.

1. $\dfrac{a^2 + 7a + 10}{a + 5}$

2. $\dfrac{x^2 + 7x + 6}{x + 1}$

3. $\dfrac{3a - 10 + a^2}{a - 2}$

4. $\dfrac{y^2 - 6 + y}{y - 2}$

5. $\dfrac{x^2 + 5x + 9}{x - 2}$

6. $\dfrac{y^2 - 6y - 4}{y - 3}$

7. $\dfrac{2a + 3a^2 - 1}{a + 1}$

8. $\dfrac{5 - 3x + 2x^2}{x + 2}$

9. $\dfrac{a^3 + 2a^2 - a - 4}{a - 1}$

10. $\dfrac{y^3 - 3y^2 + y + 2}{y + 1}$

11. $\dfrac{x^3 - 1}{x - 1}$

12. $\dfrac{a^3 + 1}{a + 1}$

13. $\dfrac{y^3 + 3y^2 - 4}{y - 2}$

14. $\dfrac{x^3 + 5x^2 - 1}{x - 1}$

15. $\dfrac{2a^3 - 9a^2 - 8a + 17}{a - 5}$

16. $\dfrac{3x^3 + 11x^2 + 12}{x + 4}$

17. $\dfrac{x^4 - 5x^3 + 5x^2 + 6x - 10}{x - 3}$

18. $\dfrac{2a^4 + 6a^3 + a^2 + 2a - 5}{a + 3}$

19. $\dfrac{3y^4 - 7y^3 - 10y^2 + 11y + 6}{y - 3}$

20. $\dfrac{a^5 + 2a^4 + 3a^3 + 6a^2 + a + 2}{a + 2}$

Given $P(x) = 3x^4 - 2x^3 + x^2 - 4x + 1$, in problems 21-26, find the following by (a) using the remainder theorem and (b) by substituting the indicated value of x. See example 4-6 B.

Example   Using synthetic division and the remainder theorem, find the remainder when $P(x) = 3x^3 - 2x^2 + 4x - 5$ is divided by $(x + 4)$. Evaluate $P(-4)$ using substitution.

Solution   a. Using synthetic division,

$$
\begin{array}{r|rrrr}
-4 & 3 & -2 & 4 & -5 \\
   &   & -12 & 56 & -240 \\
\hline
   & 3 & -14 & 60 & -245 \leftarrow \text{Remainder}
\end{array}
$$

b. $P(x) = 3x^3 - 2x^2 + 4x - 5$

$P(-4) = 3(-4)^3 - 2(-4)^2 + 4(-4) + 5$ Replace x with -4
$ = -192 - 32 - 16 - 5$
$ = -245$
$P(-4) = -245$

21. $P(1)$ 22. $P(-1)$ 23. $P(-3)$ 24. $P(3)$ 25. $P(0)$ 26. $P(4)$

27. Given $P(x) = 5x^5 - 2x^3 + 1$, use the remainder theorem to find $P(2)$.

28. Given $P(x) = 8x^6 + 3x^2 - 2x - 5$, use the remainder theorem to find $P(-2)$.

29. Given $P(x) = 24x^4 - 16x^2 + 14x - 6$, use the remainder theorem to find $P\left(\dfrac{1}{2}\right)$.

Use the factor theorem to determine if the given binomial is a factor of the polynomial $P(x)$. See example 4–6 C–1.

Example Show that $x + 5$ is a factor of $P(x) = x^3 + 4x^2 - 7x - 10$.

Solution Using synthetic division,

$$
\begin{array}{r|rrrr}
-5 & 1 & 4 & -7 & -10 \\
   &   & -5 & 5 & 10 \\
\hline
   & 1 & -1 & -2 & 0 \leftarrow x + 5 \text{ is a factor}
\end{array}
$$

30. $P(x) = 3x^2 - 4x - 4;\ x - 2$

31. $P(x) = 2x^2 + 3x - 5;\ x + 3$

32. $P(x) = x^3 + x^2 - 7x - 10;\ x + 2$

33. $P(x) = x^3 - x^2 + 2x - 8;\ x - 2$

34. $P(x) = 4x^4 - 13x^3 - 13x^2 - 4x + 12;\ x + 2$

35. $P(x) = x^4 - 9x^3 + 18x^2 - 3;\ x + 1$

36. $P(x) = x^4 - 81;\ x - 3$

37. $P(x) = x^3 + 64;\ x + 4$

38. $P(x) = 3x^5 - 3x^4 + 5x^2 - 13x - 6;\ x - 3$

39. $P(x) = 3x^5 + 4x^2 - 7;\ x + 1$

Find the solution set of the following equations using the factor theorem and the given root. See example 4–6 C–2 and D–3.

Example Given the equation $2x^4 + 5x^3 - 51x^2 + 80x - 28 = 0$ has a solution 2 of multiplicity 2, find the solution set of the equation.

Solution Given a solution 2 of multiplicity 2, then $(x - 2)^2 = x^2 - 4x + 4$ is a factor of the equation. We divide $2x^4 + 5x^3 - 51x^2 + 80x - 28$ by $x^2 - 4x + 4$.

$$
\require{enclose}
\begin{array}{r}
2x^2 + 13x - 7 \\
x^2 - 4x + 4 \enclose{longdiv}{2x^4 + 5x^3 - 51x^2 + 80x - 28} \\
\underline{2x^4 - 8x^3 + 8x^2} \\
13x^3 - 59x^2 + 80x \\
\underline{13x^3 - 52x^2 + 52x} \\
-7x^2 + 28x - 28 \\
\underline{-7x^2 + 28x - 28} \\
0
\end{array}
$$

The quotient $2x^2 + 13x - 7 = (2x - 1)(x + 7)$, and if we set each factor equal to zero, we obtain $x = \dfrac{1}{2}$ and $x = -7$. The solution set is $\left\{-7, 2, \dfrac{1}{2}\right\}$.

40. $x^3 + 7x^2 - 2x - 12 = 0; -2$

41. $3x^3 + 19x^2 - 38x + 16 = 0; 1$

42. $x^3 - 5x^2 - 2x + 24 = 0; 3$

43. $2x^3 + x^2 - 61x + 30 = 0; 5$

44. $12x^3 + 29x^2 + 8x - 4 = 0; -2$

45. $x^4 - 2x^2 + 1 = 0; 1$ has multiplicity 2

46. $x^4 + 6x^3 - 3x^2 - 52x - 60 = 0;$
-2 has multiplicity 2

Find a polynomial $P(x)$ of lowest degree having the given zeros. See example 4-6 C-3.

Example Find a polynomial $P(x)$ of lowest degree whose zeros are 3, -2, and 1 of multiplicity 2.

Solution Since $P(x)$ has a zero
1. 3, then $x - 3$ is a factor
2. -2, then $x + 2$ is a factor
3. 1 of multiplicity 2, then $(x - 1)^2 = x^2 - 2x + 1$ is a factor
Thus, $P(x) = (x - 3)(x + 2)(x^2 - 2x + 1) = x^4 - 3x^3 - 3x^2 + 11x - 6.$

47. $2, -1, 4$

48. $5, -3, 3$

49. $4, 0, -1$

50. $-2, 2, -3, 3$

51. $-8, 8, -1, 1$

52. $4, -3, -7, 7$

53. $5, -5, 6, 2$

54. 3 of multiplicity 2; -3 of multiplicity 2

55. -1 of multiplicity 2; 4 of multiplicity 2

56. 0 of multiplicity 3; $-2, 4$

57. $-5, -2, 0$ of multiplicity 2

In problems 58–61, find the zeros of the given polynomial, indicating multiplicity where necessary.
See example 4-6 D-1.

58. $P(x) = (x - 2)^2(x + 3)(x - 4)^2$

59. $P(x) = (x - 2)^2(x + 2)^2(x - 3)^3(x + 4)^2$

60. $P(x) = x^3(x - 3)^2(x + 5)$

61. $P(x) = x^2(x + 4)^4(x - 3)^2(x - 6)$

62. Show that $x - 2$ is a factor of $P(x) = x^{10} - 1024.$

63. Find the remainder when
$4x^{98} - 8x^{47} + 9x^{28} - 3x^{17} + 1$ is divided by
$x + 1.$

64. Find all values of k such that
$P(x) = k^2x^3 - 7kx + 6$ is divisible by $x - 1.$

65. Use synthetic division to decide whether or not 3/2 is
a solution of the equation
$2z^4 - 5z^3 + 11z^2 - 14z + 3 = 0.$

66. Use synthetic division to decide whether or not 1/3 is
a solution of the equation
$6x^4 + x^3 - 4x^2 + 13x - 4 = 0.$

Review exercises

Find the solution set of the following equations. See section 2–1.

1. $5(x - 4) + 12 = 3x + 2x - 6$

2. $\frac{1}{2}x - 1 = \frac{1}{3}x + 3$

3. $\frac{7}{12}y - 4 = \frac{5}{6}y - 5$

4. One number is 27 more than another number. The
lesser number is one-fourth of the greater number.
Find the two numbers. See section 2–3.

5. Solve $V = \frac{1}{3}\pi h^2(3R - h)$ for R. See section 2–2.

4–7 ■ *Equations containing rational expressions*

Rational equations

In chapter 2, we studied how to find the solution set of a linear equation. We now consider equations that contain at least one term that is a rational expression. We call these equations **rational equations.** The same properties of real numbers that we used in chapter 2 will apply to solve rational equations once the polynomial denominators are eliminated.

Recall that an equivalent equation is obtained when each term of an equation is multiplied by the same nonzero number. The following examples demonstrate the use of this property to solve rational equations.

■ *Example 4–7 A*

Find the solution set of the following rational equation.

1. $\dfrac{x}{3} + \dfrac{x}{4} - \dfrac{x}{6} = 35$

The LCM of 3, 4, and 6 is 12.

$$12 \cdot \dfrac{x}{3} + 12 \cdot \dfrac{x}{4} - 12 \cdot \dfrac{x}{6} = 12 \cdot 35 \qquad \text{Multiply each term by the LCM 12}$$
$$4x + 3x - 2x = 420 \qquad \text{Simplify each term}$$
$$5x = 420 \qquad \text{Combine like terms}$$
$$x = 84 \qquad \text{Divide by 5}$$

The solution set is $\{84\}$.

When the rational equation contains rational expressions with the variable in the denominator, multiplying all terms by the LCM of the denominators does *not always* produce an equation that is equivalent to the original equation.

2. $1 - \dfrac{2a}{a + 1} = \dfrac{2}{a + 1}$

$$(a + 1) \cdot 1 - (a + 1) \cdot \dfrac{2a}{a + 1} = (a + 1) \cdot \dfrac{2}{a + 1} \qquad \text{Multiply each term by the LCM } (a + 1)$$
$$a + 1 - 2a = 2 \qquad \text{Reduce in each term}$$
$$1 - a = 2 \qquad \text{Combine like terms}$$
$$-a = 1 \qquad \text{Subtract 1 from each member}$$
$$a = -1 \qquad \text{Multiply each member by } -1$$

However, in the original equation, -1 *is not in the domain of the rational expressions involved.* Therefore -1 cannot be a solution and we conclude that the equation has no solution. Hence the solution set is $\emptyset$.

▶ *Quick check* Find the solution set of $3 - \dfrac{4x}{x - 2} = \dfrac{-8}{x - 2}$.   ■

Extraneous solutions

The possible solution -1 in example 2 is called an **extraneous solution**—a solution of the equation obtained by multiplying each term by the LCM of the denominators but *not a solution of the original equation.*

Note   It is important to observe the domain of the rational expressions in the rational equation to know where extraneous solutions may occur.

To further illustrate, consider the rational equation

$$\frac{3x}{x-3} = 2 + \frac{9}{x-3}$$

Note that the domain of the rational expressions is all real numbers *except 3.* That is, $x \neq 3$. If we proceed to solve the equation, we *do* obtain $x = 3$. Thus 3 is an extraneous solution, the equation has no solution, and the solution set is $\emptyset$.

   The procedure for solving rational equations is stated here.

—— *To solve rational equations* ————————————————————

   1. Find the LCM of the denominators.
   2. Multiply each term by the LCM of the denominators.
   3. Solve the resulting equation.
   4. Check for extraneous solutions if the LCM contains a variable.

   Equations and formulas containing rational expressions and more than one variable are common in scientific fields. It is often desirable to solve such equations for one variable in terms of the other variables in the equation. We use the same procedure that we used in solving preceding equations of this section.

■ *Example 4–7 B*

Solve each of the following equations for the indicated variable. Assume all denominators are nonzero.

1. $\dfrac{3x}{5} - \dfrac{4y}{3} = 6$ for x

$$15 \cdot \frac{3x}{5} - 15 \cdot \frac{4y}{3} = 15 \cdot 6 \qquad \text{Multiply by the LCM of 5 and 3, namely 15}$$
$$3 \cdot 3x - 5 \cdot 4y = 90 \qquad \text{Simplify each term}$$
$$9x - 20y = 90$$
$$9x = 90 + 20y \qquad \text{Add } 20y \text{ to each member}$$
$$x = \frac{90 + 20y}{9} \qquad \text{Divide each member by 9}$$

2. $\dfrac{3}{x} - 7 = 9y + \dfrac{2y}{3x}$ for y

$$3x \cdot \frac{3}{x} - 3x \cdot 7 = 3x \cdot 9y + 3x \cdot \frac{2y}{3x} \qquad \text{Multiply each term by LCM of } x \text{ and } 3x, 3x$$
$$3 \cdot 3 - 21x = 27xy + 2y \qquad \text{Simplify each term}$$
$$9 - 21x = 27xy + 2y$$

Since we cannot combine the terms containing y in the right member, we must *factor* the common factor, y, from each term.

$$9 - 21x = (27x + 2)y \qquad \text{Factor } y \text{ in the right member}$$

$$\frac{9 - 21x}{27x + 2} = y \qquad \text{Divide each member by the coefficient of } y, (27x + 2)$$

3. In physics, the rule governing the speeds of two gears, one the driver gear A and the other the driven gear B, is given by

$$\frac{T_A}{T_B} = \frac{R_B}{R_A}$$

where $T =$ the number of teeth in the gear and $R =$ the revolutions per minute of the gear. Solve for the revolutions per minute of the driven gear, R_B.

$$T_B R_A \cdot \frac{T_A}{T_B} = T_B R_A \cdot \frac{R_B}{R_A} \qquad \text{Multiply each term by the LCM of } T_B \text{ and } R_A, T_B R_A$$

$$R_A \cdot T_A = T_B \cdot R_B \qquad \text{Reduce in each term}$$

$$R_B = \frac{R_A T_A}{T_B} \qquad \text{Divide each member by } T_B \text{ and interchange the two members}$$

▶ **Quick check** Given $\dfrac{6}{p - 3} + \dfrac{2}{pq} = \dfrac{5}{4p}$, solve for p. ■

Mastery points

Can you
- Solve a rational equation in one variable?
- Solve a rational equation in two or more variables for one of the variables?

Exercise 4–7

Find the solution set of each of the following equations. Where extraneous solutions exist, so state. See example 4–7 A.

Example Find the solution set of $3 - \dfrac{4x}{x - 2} = \dfrac{-8}{x - 2}$.

Solution The LCM is $(x - 2)$.

$$(x - 2)3 - (x - 2)\frac{4x}{x - 2} = (x - 2)\frac{-8}{x - 2} \qquad \text{Multiply each term by LCM } x - 2$$

$$3x - 6 - 4x = -8 \qquad \text{Reduce and multiply}$$

$$-x - 6 = -8 \qquad \text{Combine like terms}$$

$$-x = -2 \qquad \text{Add 6 to each member}$$

$$x = 2 \qquad \text{Multiply by } -1$$

Since the domain of the rational expressions is all real numbers except 2, the solution set is $\emptyset$ and 2 is an extraneous solution.

1. $\dfrac{x + 7}{4} = \dfrac{2x}{12}$

2. $\dfrac{4b - 3}{10} = \dfrac{2b - 1}{6}$

3. $\dfrac{3x}{5} - \dfrac{4x}{3} = 1$

4. $\dfrac{m}{3} + 4 = \dfrac{7m}{4}$

5. $\dfrac{4}{a} - \dfrac{6}{3a} = \dfrac{3}{5}$

6. $4 - \dfrac{5}{9b} = \dfrac{7}{6b}$

7. $\dfrac{b-3}{10} + \dfrac{2b+1}{15} = 2$

8. $\dfrac{3y-4}{16} - \dfrac{2-3y}{12} = 1$

9. $\dfrac{x-3}{3x} = \dfrac{2x+3}{9x}$

10. $\dfrac{5-b}{8b} = \dfrac{3b+7}{6b}$

11. $\dfrac{5}{x-2} = \dfrac{4}{2x+1}$

12. $\dfrac{-3}{x+7} = \dfrac{2}{3x-1}$

13. $\dfrac{5}{4a+2} = \dfrac{7}{2a+1}$

14. $\dfrac{11}{3y-2} = \dfrac{8}{12y-8}$

15. $\dfrac{6}{6x+3} = \dfrac{3}{2x+1} + 5$

16. $1 + \dfrac{5}{3m-9} = \dfrac{10}{m-3}$

17. $\dfrac{x-1}{x^2-4} = \dfrac{6}{x-2}$

18. $\dfrac{5}{y+3} = \dfrac{2y+1}{y^2-9}$

19. $\dfrac{x}{x-2} + \dfrac{2}{3} = \dfrac{2}{x-2}$

20. $\dfrac{3}{2} - \dfrac{1}{x-4} = \dfrac{-2}{2x-8}$

21. $4 - \dfrac{2x}{5-x} = \dfrac{6}{x-5}$

22. $\dfrac{5y}{3-y} + \dfrac{8}{y-3} = 4$

23. $\dfrac{8b}{b^2-16} = \dfrac{3}{b+4} + \dfrac{5}{4-b}$

24. $\dfrac{10}{5-a} = \dfrac{7}{a+5} - \dfrac{6}{a^2-25}$

25. $\dfrac{5}{a-3} - \dfrac{1}{a+2} = \dfrac{6a}{a^2-a-6}$

26. $\dfrac{7}{x+6} = \dfrac{9}{x-9} - \dfrac{2x-1}{x^2-3x-54}$

27. $\dfrac{4}{2x-6} - \dfrac{12}{4x+12} = \dfrac{12}{x^2-9}$

28. $\dfrac{11x-3}{10x^2-3x-4} - \dfrac{4}{5x-4} = \dfrac{9}{2x+1}$

29. $\dfrac{6}{q^2+q-6} = \dfrac{5}{q^2+3q-10}$

30. $\dfrac{9}{y^2-6y+8} = \dfrac{14}{y^2-16}$

31. $\dfrac{13}{n^2+2n-15} - \dfrac{1}{n^2+10n+25} = 0$

32. $\dfrac{2}{2n^2-7n-4} + \dfrac{6}{6n^2-5n-4} = 0$

33. $\dfrac{4}{x^2-9} = \dfrac{7}{x^2-7x+12} - \dfrac{5}{x^2-x-12}$

34. $\dfrac{-2}{a^2+3a-4} - \dfrac{6}{a^2-1} = \dfrac{1}{a^2+5a+4}$

35. $\dfrac{6}{2n^2+n-3} - \dfrac{5}{4n^2-9} = \dfrac{6}{2n^2-5n+3}$

36. $\dfrac{6}{8x^2-14x-15} + \dfrac{1}{8x^2-26x+15} = \dfrac{9}{16x^2-9}$

Solve the given equations and formulas for the indicated variable. Assume all denominators are nonzero. See example 4–7 B.

Example Given $\dfrac{6}{p-3} + \dfrac{2}{pq} = \dfrac{5}{4p}$, solve for p.

Solution The LCM of $4p$, pq, and $p-3$ is $4pq(p-3)$.

$$4pq(p-3)\dfrac{6}{p-3} + 4pq(p-3)\dfrac{2}{pq} = 4pq(p-3)\dfrac{5}{4p}$$ Multiply each term by LCM $4pq(p-3)$

$$4pq \cdot 6 + 4(p-3)2 = q(p-3)5$$ Reduce in each term

$$24pq + 8(p-3) = 5q(p-3)$$ Commute and multiply

$$24pq + 8p - 24 = 5pq - 15q$$ Distributive property

$$19pq + 8p = 24 - 15q$$ Add 24 to and subtract $5pq$ from each member

$$(19q+8)p = 24 - 15q$$ Factor p in right member

$$p = \dfrac{24-15q}{19q+8}$$ Divide each member by $19q+8$

$$p = \dfrac{3(8-5q)}{19q+8}$$ Factor numerator (not reducing)

37. $\dfrac{4}{a} - \dfrac{3}{b} = 7$ for b

38. $\dfrac{6}{x} - 4 = \dfrac{7}{y}$ for y

39. $\dfrac{3}{p} + 4 = \dfrac{6q}{2p} - 3a$ for p

40. $\dfrac{a+6}{3} - \dfrac{b-2}{4} = \dfrac{c}{12}$ for a

41. $\dfrac{y-3}{x+2} = \dfrac{5}{3}$ for y

42. $\dfrac{-7}{3} = \dfrac{y+1}{x-5}$ for x

43. $S = \dfrac{a}{1-r}$ for r (progression formula)

44. $S = \dfrac{n(a+b)}{2}$ for a (progression formula)

45. $A = \dfrac{1}{2} h(b_1 + b_2)$ for b_1 (area of a trapezoid)

46. $C = \dfrac{5}{9}(F - 32)$ for F (temperature conversion)

See example 4–7 B–3.

47. The coefficient of linear expansion, k, of a solid when heated is given by

$$k = \dfrac{L_t - L_0}{L_0 t}$$

where L_t is the length at $t\,°C$, L_0 is the length at $0\,°C$, and t is any given temperature in Celsius. Solve for t. Solve for L_0.

48. A formula for unknown resistance R_x in a battery is given by

$$R_x = R_m \left(\dfrac{E_1}{E_2} - 1 \right)$$

Solve for E_1.

49. The interest rate, r, on a given amount of money, P, over a given period of time, t, which pays interest, I, is given by

$$r = \dfrac{I}{Pt}$$

Solve the equation for t.

50. The kinetic energy of a body, KE, is computed by

$$KE = \dfrac{Wv^2}{2g}$$

where W is the weight in pounds, v is the velocity expressed in feet per second, and g is the acceleration due to gravity. Solve for g.

51. The total resistance R of a parallel circuit is given by

$$\dfrac{1}{R} = \dfrac{1}{R_1} + \dfrac{1}{R_2} + \dfrac{1}{R_3}$$

where R_1, R_2, and R_3 are the resistances of the respective circuits. Solve for R_3.

52. The capacitance of capacitors C_1, C_2, and C_3 in a series circuit is given by

$$\dfrac{1}{C} = \dfrac{1}{C_1} + \dfrac{1}{C_2} + \dfrac{1}{C_3}$$

Solve for C.

53. Given a large piston and a small piston on which forces F and f, respectively, are applied, then

$$\dfrac{F}{f} = \dfrac{A}{a}$$

where A is the area of the large piston and a is the area of the small piston. Solve for A.

54. As gas expands when heated, the relationship between pressure P, volume V, and absolute temperature T is given by

$$\dfrac{P_1 V_1}{T_1} = \dfrac{P_2 V_2}{T_2}$$

called Charles' Law. Solve for V_2.

55. The formula

$$\dfrac{R_2}{R_1} = \dfrac{M + T_2}{M + T_1}$$

gives a relationship for the increase in the resistance of a circuit caused by a rise in temperature. Solve for T_2. Solve for R_1.

56. The lens maker's formula for the focal length, f, of a lens is given by

$$\dfrac{1}{f} = (\mu_m - 1)\left(\dfrac{1}{R_1} + \dfrac{1}{R_2} \right)$$

where μ_m is the index of refraction and R_1 and R_2 are radii of curvature. Solve for f. Solve for μ_m.

57. If Ann can paint the house in 10 hours, write a rational expression for what part of the house she can paint in (a) 1 hour, (b) 5 hours, (c) t hours.

58. If John can bake a batch of cookies in 3 hours, write a rational expression for what part of the batch he can bake in (a) 1 hour, (b) t hours, (c) 20 minutes.

59. An automobile travels 200 miles. Write a rational expression for the speed (rate of travel) the automobile is traveling if the distance is covered in (a) 5 hours, (b) 4 hours, (c) t hours. (*Hint:* distance = rate · time)

60. Write a rational expression for the time t a boat travels 20 miles downstream if the boat is traveling downstream at (a) 4 mph, (b) 12 mph, (c) r mph.

61. If one number is three times as large as a second number, and if n represents the second number, (a) write a rational expression for the reciprocal of the two numbers, (b) write a rational expression for five times the reciprocal of the second number.

62. If a number n is added to the numerator and the denominator of the rational number $\frac{3}{5}$, write a rational expression for the new number.

63. If the same number n is added to the numerator and taken away from the denominator of $\frac{5}{6}$, write a rational expression for the new number.

Review exercises

Completely factor the following expressions. See sections 3–5, 3–6, and 3–7.

1. $4p^2 - 25q^2$
      2. $x^2 - 24x + 144$
      3. $2y^2 - 15y - 8$

4. Dene invests $20,000, part at 6% and the rest at 8% interest. If the total income from the investments in one year is $1,360, how much did she invest at each rate? See section 2–3.

5. Five times a number increased by 13 gives 53. Find the number. See section 2–3.

6. Given polynomial $P(x) = 2x^2 + x - 3$, find $P(-3)$. See section 1–5.

4-8 ■ Problem solving with rational equations

Now that we can solve rational equations, let us use that skill to solve some types of word problems that result in rational equations when we use problem-solving techniques.

■ Example 4-8 A

Work problem

1. Jan can produce a part in 3 hours and Joe can produce the same part in 4 hours. How long would it take them to produce the part working together?

Since Jan can produce the part in 3 hours and Joe can produce the part in 4 hours, then Jan can produce $\frac{1}{3}$ of the part in 1 hour and Joe can produce $\frac{1}{4}$ of the part in 1 hour.

Let x = the time in hours necessary to produce the part working together.

Then they can produce $\dfrac{1}{x}$ of the part in 1 hour.

Jan in 1 hour	+	Joe in 1 hour	=	together in 1 hour
$\dfrac{1}{3}$	$+$	$\dfrac{1}{4}$	$=$	$\dfrac{1}{x}$

The LCM of the denominators is $12x$.

$$12x \cdot \frac{1}{3} + 12x \cdot \frac{1}{4} = 12x \cdot \frac{1}{x} \qquad \text{Multiply each term by the LCM } 12x$$
$$4x + 3x = 12 \qquad \text{Reduce in each term}$$
$$7x = 12 \qquad \text{Add in left member}$$
$$x = \frac{12}{7} \text{ or } 1\frac{5}{7} \text{ hours} \qquad \text{Divide by 7}$$

Working together, Jan and Joe can produce the part in $1\dfrac{5}{7}$ hours.

Uniform motion problems

2. An automobile can travel 200 miles in the same time that a truck can travel 150 miles. If the automobile travels at an average rate of 15 miles per hour faster than the truck, find the average rate of each. We use the relationship between distance traveled, rate of travel, and time traveled— distance (d) = rate (r) · time (t).

The following table will be helpful in finding the equation necessary to solve this type of problem.

	d	r	t
Automobile			
Truck			

Let r = the rate of the truck. Then $r + 15$ = the rate of the automobile.

	d	r	t	
Automobile	200	$r + 15$	$\dfrac{200}{r + 15}$	
Truck	150	r	$\dfrac{150}{r}$	Times are equal

Note To find the time, we used the relationship

$$\text{time} = \frac{\text{distance}}{\text{rate}}.$$

Since both vehicles traveled the same length of time, we obtain the equation

$$\frac{200}{r + 15} = \frac{150}{r}$$

$$r(r + 15)\frac{200}{r + 15} = r(r + 15)\frac{150}{r}$$ Multiply each term by the LCM of r and $(r + 15)$, $r(r + 15)$

$$200r = (r + 15)150$$
$$200r = 150r + 2{,}250$$
$$50r = 2{,}250$$
$$r = 45$$
$$r + 15 = 60$$

Thus the average rate of the truck is 45 miles per hour and of the automobile is 60 miles per hour.

3. John's boat travels at 20 miles per hour in still water. If the speed of the current is 5 miles per hour, how far can John travel downstream if it takes him a total time of 3 hours to go downstream and back upstream?

Let x = the distance traveled downstream (and back upstream since they are the same distance). Using $t = \dfrac{d}{r}$, we summarize the information in the table.

		=		×
	d	**r**		**t**
Downstream	x	$20 + 5 = 25$		$\dfrac{x}{25}$
Upstream	x	$20 - 5 = 15$		$\dfrac{x}{15}$

Total time is 3 hours

Note When traveling downstream, we *add* the speed in still water to the speed of the current. When traveling upstream, we *subtract* the speed of the current from the speed in still water.

Since total time downstream and back upstream is 3 hours, the equation is

$$\frac{x}{25} + \frac{x}{15} = 3 \longleftarrow \text{Total time}$$

Downstream ⎯⎯⎯⎯⎯⎯⎯⎯⎯⎯⎯⏋   ⎿⎯⎯⎯ Upstream

$$\frac{x}{25} \cdot 75 + \frac{x}{15} \cdot 75 = 3 \cdot 75$$ Multiply each term by the LCM of 25 and 15, 75
$$3x + 5x = 225$$ Multiply as indicated
$$8x = 225$$ Combine like terms
$$x = \frac{225}{8}$$ Divide each member by 8
$$= 28\frac{1}{8}$$

John can travel $28\dfrac{1}{8}$ miles downstream.

Number and reciprocal problem

4. One number is twice another number. The sum of their reciprocals is $\dfrac{9}{2}$. Find the numbers.

Let x = one number, then $2x$ = the other number. Their reciprocals are then $\dfrac{1}{x}$ and $\dfrac{1}{2x}$. The sum of the reciprocals is $\dfrac{9}{2}$, so we have the equation

$$\frac{1}{x} + \frac{1}{2x} = \frac{9}{2}$$

$$2x \cdot \frac{1}{x} + 2x \cdot \frac{1}{2x} = 2x \cdot \frac{9}{2} \qquad \text{Multiply each term by the LCM of } x, 2, \text{ and } 2x, 2x$$

$$2 + 1 = 9x$$
$$3 = 9x$$
$$x = \frac{3}{9} = \frac{1}{3}$$
$$2x = \frac{2}{3}$$

The two numbers are $\dfrac{1}{3}$ and $\dfrac{2}{3}$.

Check: Since the sum of the reciprocals $3 + \dfrac{3}{2} = \dfrac{6}{2} + \dfrac{3}{2} = \dfrac{9}{2}$, the answer is correct.

▶ *Quick check* Pam can clean her house in 2 hours. Her daughters, Linnea and Jean, can do the job in 2 hours and 3 hours, respectively. At these rates, how long would it take the three of them to clean the house working together? ■

```
┌─ Mastery points ──────────────────────────────────────────────┐
│                                                                │
│  Can you                                                       │
│  ■ Set up and solve work problems?                             │
│  ■ Set up and solve uniform motion problems?                   │
│  ■ Set up and solve number and reciprocal problems?            │
│                                                                │
└────────────────────────────────────────────────────────────────┘
```

Exercise 4–8

Solve the following work problems. See example 4–8 A–1.

Work problems

Example Pam can clean her house in 2 hours. Her daughters, Linnea and Jean, can do the job in 2 hours and 3 hours, respectively. At these rates, how long would it take the three of them to clean the house working together?

Solution Let x = time in hours to clean the house when the three of them are working together.

Then 1. Pam can clean $\dfrac{1}{2}$ of the house in 1 hour.

2. Linnea can clean $\dfrac{1}{2}$ of the house in 1 hour.

3. Jean can clean $\frac{1}{3}$ of the house in 1 hour.

4. Working together, they can clean $\frac{1}{x}$ of the house in 1 hour.

part done by Pam		part done by Linnea		part done by Jean		part done by all three working together
$\frac{1}{2}$	$+$	$\frac{1}{2}$	$+$	$\frac{1}{3}$	$=$	$\frac{1}{x}$

The equation is $\frac{1}{2} + \frac{1}{2} + \frac{1}{3} = \frac{1}{x}$.

$$6x \cdot \frac{1}{2} + 6x \cdot \frac{1}{2} + 6x \cdot \frac{1}{3} = 6x \cdot \frac{1}{x}$$ Multiply each term by the LCM of 2, 3, and x, $6x$

$$3x + 3x + 2x = 6$$

$$8x = 6$$ Combine like terms

$$x = \frac{6}{8}$$ Divide each member by 8

$$x = \frac{3}{4}$$ Simplify

It would take Pam and her daughters $\frac{3}{4}$ of an hour to clean the house working together.

1. Pete Hansen milks his herd of cattle in 2 hours, whereas his son John milks them in 3 hours. How many hours and minutes would it take them to do the same job working together?

2. Jake has two hay balers, A and B. If Jake can completely bale a given field in 3 hours using baler A and in $3\frac{3}{4}$ hours using baler B, how long would it take him to bale the field working both balers together?

3. During a "clean-up, paint-up, fix-up week" in Sault Sainte Marie, Jim, Toni, and Ken clean up a given vacant lot. If it takes each person $1\frac{1}{2}$ hours, 2 hours, and 3 hours, respectively, to do the job alone, how long would it take them working together to clean up the lot?

4. Tanya, Bill, and Neysa work in a bakery and can mix 12 loaves of bread individually in 36 minutes, 40 minutes, and 30 minutes, respectively. If they worked together, in how many minutes could they mix the 12 loaves?

5. If two water pipes can fill a swimming pool in 18 hours when both pipes are open and one of the pipes can fill the pool alone in 30 hours, how long would it take the other pipe to fill the pool alone?

6. In a machine shop, machine A can produce certain parts in 1 hour and 20 minutes. If machines A and B working together can produce the parts in 50 minutes, how long would it take machine B to do the job alone?

7. Two inlet pipes can fill a water basin in 10 hours and 12 hours, respectively, when open individually. If an outlet pipe can empty the basin alone in 9 hours, how long would it take to fill the basin if all three pipes are open simultaneously?

8. An open sink drain can empty a sink full of water in 2 minutes. If the cold water and hot water faucets can, when open fully, fill the sink in $3\frac{1}{2}$ and 3 minutes, respectively, how long would it take to fill the sink if all three are open simultaneously?

9. It takes pump A twice as long to unload an oil tanker as it takes pump B to do the job. If the two pumps working together can unload the tanker in 12 hours, how long will it take each to do the job?

10. If one microprocessor can process a set of inputs in three-fifths of the time that a second microprocessor can do the job, and together the machines can do the job in 2 milliseconds, how long will it take each microprocessor to process the set of inputs individually?

11. It takes tug *A* one-half as long to push a series of barges up a river as it takes tug *B* to do the same job. If it takes tug *C* two times as long to do the job as it does tug *B* and all three tugs working together can do the job in 4 hours, how long would it take tug *A* to do the job alone?

12. A portable gasoline-powered generator, a solar cell, and a wind generator fully charge a dead storage battery in 9 hours when charging simultaneously. If the wind generator takes twice as long to fully charge the battery as the solar cell takes and the gasoline-powered generator takes three-fourths as long as the solar cell, how long would it take the solar cell to fully charge the battery alone?

Solve the given uniform motion problems. See example 4–8 A–2 and 3.

Uniform motion problems

13. An excursion boat moves at 16 miles per hour in still water. If the boat travels 20 miles downstream in the same time it takes to travel 14 miles upstream, what is the speed of the current? (*Hint:* Let x = the speed of the current. Then $16 + x$ = the speed of the boat downstream and $16 - x$ = the speed of the boat upstream.)

14. An airplane can cruise at 300 miles per hour in still air. If the airplane takes the same time to fly 950 miles with the wind as it does to fly 650 miles against the wind, what is the speed of the wind?

15. A boat travels 40 kilometers upstream in the same time that it takes the same boat to travel 60 kilometers downstream. If the stream is flowing at 6 kilometers per hour, what is the speed of the boat in still water?

16. The speed of a wind is 15 miles per hour. Find the speed of an airplane in still air if it flies 200 miles against the wind in the same time that it flies 300 miles with the wind.

17. On a trip from Detroit to Los Angeles, a TWA plane takes one-third of an hour longer to make the trip than an American Airlines plane does. If the TWA plane flies at 300 miles per hour and the American Airlines plane flies at 320 miles per hour, how far is it from Detroit to Los Angeles?

18. It takes Bonnie 2 minutes longer to jog a certain distance than it does Maryann. What distance did they run if Bonnie can jog at 5 miles per hour and Maryann can jog at 7 miles per hour?

19. Mary can row her boat 3 miles per hour in still water. If the current of the river is 2 miles per hour, how far downstream can she row if it takes her 2 hours to go down and back?

20. Peter can average 10 miles per hour riding his bike to deliver his papers. By car he can average 30 miles per hour. If it takes him $\frac{1}{2}$ hour less time by car, how long is Peter's paper route?

Solve the following number and reciprocal problems. See example 4–8 A–4.

Number and reciprocal problems

21. One number is three times another number. The sum of their reciprocals is 2. Find the numbers.

22. One number is twice another number. The sum of their reciprocals is $\frac{15}{8}$. Find the numbers.

23. If the same number is added to the numerator and the denominator of the fraction $\frac{3}{7}$, the result is $\frac{4}{5}$. Find the number.

24. If the same number is added to the numerator and the denominator of the fraction $\frac{1}{4}$, the result is $\frac{2}{3}$. Find the number.

25. What number must be added to the denominator of $\frac{6}{7}$ to obtain $\frac{3}{5}$?

26. What number must be added to the numerator and subtracted from the denominator of $\frac{5}{8}$ to obtain its reciprocal?

The following word problems are a mixture of the previous types.

27. A boat can go 6 miles upstream in the same time as it can go 10 miles downstream. If the boat goes 12 miles per hour in still water, what is the speed of the current?

28. What number must be added to the numerator of the fraction $\frac{11}{15}$ to obtain the fraction $\frac{7}{5}$?

29. Jane can type an English term paper in 3 hours. If Pamela helps her, together they can type the paper in 2 hours. How long would it take Pamela to type the paper alone?

30. A water tank can be filled by its inlet pipe in 45 minutes. If its outlet pipe can empty the tank in 30 minutes, how long would it take to empty a full tank if both pipes are open?

31. Fran traveled 120 miles to visit her aunt. If she drove twice as fast returning home and the return trip took 2 hours less time, what was her speed going to her aunt's?

32. If one number is three times a second number and the sum of their reciprocals is $\frac{2}{9}$, find the numbers.

33. Erin can jog 2 miles per hour faster than her cousin Sarah. Erin can jog 9 miles in the same time Sarah can jog 6 miles. How fast does each girl jog?

Review exercises

Perform the indicated operations. See section 3–2.

1. $4x(2x^2 - 3x + 1)$
2. $(3x - 1)(x - 5)$
3. $(5z - 4)^2$
4. $(2y + 3)(2y - 3)$

Simplify the following expressions. Express the answer with positive exponents. See section 3–3.

5. $y^{-3} \cdot y^2 \cdot y^0$
6. $\frac{x^3y^2}{x^{-1}y^{-1}}$

7. What number(s) when squared yield (yields) 49? See section 3–1.
8. What number cubed yields -125? See section 3–1.

Chapter 4 lead-in problem

Doug Nance opens the valve to fill his new swimming pool. Without his knowlege, the drain to empty the pool has been left open. If the pool would normally fill in 10 hours and the drain can empty the full pool in 15 hours, how long does it take to fill the pool?

Solution

Let t represent the time to fill the pool. In 1 hour,

a. the valve would fill $\frac{1}{10}$ of the pool.

b. the drain would empty $\frac{1}{15}$ of the pool.

c. with both open, $\frac{1}{t}$ of the pool would be filled.

We *subtract* the drain time from the fill time to get the equation

$$\frac{1}{10} - \frac{1}{15} = \frac{1}{t}$$

$$30t\left(\frac{1}{10} - \frac{1}{15}\right) = 30t \cdot \frac{1}{t} \quad \text{Multiply each member by the LCD } 30t$$

$$30t\left(\frac{1}{10}\right) - 30t\left(\frac{1}{15}\right) = 30 \quad \text{Distribute in left member and multiply}$$

$$3t - 2t = 30 \quad \text{Simplify in left member}$$

$$t = 30 \quad \text{Combine like terms}$$

It takes 30 hours to fill the pool.

Chapter 4 summary

1. A **rational expression** is any algebraic expression that can be written as a quotient of two polynomials where the denominator does not equal zero.

2. The **domain** of a rational expression is the set of all replacement values of the variable for which the expression is defined.

3. The **fundamental principle of rational expressions** states that we obtain an equivalent expression when we multiply or divide the numerator and the denominator of a rational expression by the same nonzero polynomial.

4. A rational expression is **reduced to its lowest terms** when the numerator and the denominator have no common polynomial factors other than 1 or -1.

5. To **multiply** two or more rational expressions, multiply the numerators to determine the numerator of the product and multiply the denominators to determine the denominator of the product.

6. Given polynomials P, Q, R, and S,

$$\frac{P}{Q} \div \frac{R}{S} = \frac{P}{Q} \cdot \frac{S}{R} = \frac{P \cdot S}{Q \cdot R}$$

where $Q \neq 0$, $R \neq 0$, and $S \neq 0$.

7. Given polynomials P, Q, and R, $Q \neq 0$,

$$\frac{P}{Q} + \frac{R}{Q} = \frac{P + R}{Q}$$

and

$$\frac{P}{Q} - \frac{R}{Q} = \frac{P - R}{Q}$$

8. Given polynomials P, Q, R, and S, $Q \neq 0$ and $S \neq 0$,

$$\frac{P}{Q} + \frac{R}{S} = \frac{P \cdot S + Q \cdot R}{Q \cdot S}$$

and

$$\frac{P}{Q} - \frac{R}{S} = \frac{P \cdot S - Q \cdot R}{Q \cdot S}$$

9. To find the least common multiple (LCM) of a set of polynomials,
   a. completely factor the polynomials,
   b. the LCM consists of the product of all distinct factors of the polynomials, each raised to the greatest power to which it appears in any of the factorizations.

10. Rational expressions can be added or subtracted by building each expression to an equivalent rational expression such that they have a common denominator—the least common denominator (LCD).

11. A **complex rational expression** is a rational expression whose numerator or denominator, or both, contain rational expressions.

12. In the complex rational expression

$$\frac{\dfrac{4}{5}}{\dfrac{7}{8}}$$

we call $\dfrac{4}{5}$ the **primary numerator,** $\dfrac{7}{8}$ the **primary denominator,** and 5 and 8 the **secondary denominators.**

13. To simplify a complex rational expression, multiply the primary numerator and the primary denominator by the least common multiple (LCM) of the secondary denominators.

14. To *divide* a polynomial by a monomial, simply divide each term of the polynomial by the monomial and add or subtract the resulting quotients.

15. When the divisor is a binomial of the form $x - k$, we can use synthetic division to divide a polynomial by a polynomial.

16. The **remainder theorem** and the **factor theorem** uses synthetic division to find (a) the remainder when a polynomial is divided by $x - k$ and (b) the solution set of a polynomial equation.

17. A **rational equation** is an equation that contains at least one expression of the form $\dfrac{P}{Q}$, $Q \neq 0$, where P and Q are polynomials.

18. To obtain a simple equation, an equation containing no rational expression, multiply both members of the rational equation by the LCM of the denominators.

Chapter 4 error analysis

1. Placing restrictions on the variable in a rational expression
Example: What restrictions must be placed on the variable in $\dfrac{3y + 2}{4y^2 + 3y}$?
Since $4y^2 + 3y = 0$ when $y = 0$, then $y \neq 0$.
Correct answer: $y \neq 0$ or $y \neq -\dfrac{3}{4}$
What error was made? (*see page 155*)

2. Finding the domain of a rational expression
Example: $\dfrac{2x - 1}{2x^2 + 5x - 3} = \dfrac{2x - 1}{(2x - 1)(x + 3)} = \dfrac{1}{x + 3}$
Domain $= \{x \mid x \in R, x \neq -3\}$
Correct answer: Domain $= \left\{ x \mid x \in R, x \neq -3, \dfrac{1}{2} \right\}$
What error was made? (*see page 155*)

3. Subtracting rational expressions
Example: $\dfrac{4y + 1}{y^2 + 3} - \dfrac{2y - 4}{y^2 + 3}$
$= \dfrac{4y + 1 - 2y - 4}{y^2 + 3} = \dfrac{2y - 3}{y^2 + 3}$
Correct answer: $\dfrac{2y + 5}{y^2 + 3}$
What error was made? (*see page 167*)

4. Adding rational expressions
Example: $\dfrac{2x}{x - 2} + \dfrac{x - 7}{2 - x} = \dfrac{2x + x - 7}{x - 2} = \dfrac{3x - 7}{x - 2}$
Correct answer: $\dfrac{x + 7}{x - 2}$
What error was made? (*see page 171*)

5. Simplifying complex rational expression to lowest terms
Example: Simplify $\dfrac{1 + \dfrac{1}{x}}{x - \dfrac{1}{x}}$.
$\dfrac{\left(1 + \dfrac{1}{x}\right) \cdot x}{\left(x - \dfrac{1}{x}\right) \cdot x} = \dfrac{x + 1}{x^2 - 1}$
Correct answer: $\dfrac{1}{x - 1}$
What error was made? (*see page 176*)

6. Reducing a rational expression to lowest terms
Example: $\dfrac{2y^3 + y}{y} = \dfrac{2y^3 + \cancel{y}}{\cancel{y}} = \dfrac{2y^3 + 1}{1} = 2y^3 + 1$
Correct answer: $2y^2 + 1$
What error was made? (*see page 157*)

7. Dividing a polynomial by a polynomial
Example: $(3x^3 + 2x^2 - x + 1) \div (x - 1)$

$$x - 1 \overline{\smash{)}\begin{array}{l} x^2 + x - 2 \\ x^3 + 2x^2 - x + 1 \end{array}} = x^2 + x - 2 + \dfrac{3}{x - 2}$$
$$\begin{array}{r} \underline{x^3 - x^2} \\ x^2 - x \\ \underline{x^2 - x} \\ -2x + 1 \\ \underline{-2x + 2} \\ 3 \end{array}$$

Correct answer: $x^2 + 3x + 2 + \dfrac{3}{x - 1}$
What error was made? (*see page 185*)

8. Solving rational equations
Example: Find the solution set of $\dfrac{3x}{x - 4} = \dfrac{12}{x - 4} + 1$.
$(x - 4) \cdot \dfrac{3x}{x - 4} = (x - 4) \cdot \dfrac{12}{x - 4} + (x - 4) \cdot 1$
$3x = 12 + x - 4$
$2x = 8$
$x = 4 \quad \{4\}$
Correct answer: The solution set is $\emptyset$.
What error was made? (*see page 199*)

9. Using synthetic division
Example: Divide $3x^3 - 2x^2 + 4x - 1$ by $x + 3$.

$$\begin{array}{r} 3 \,\underline{\rvert\, \begin{array}{rrrr} 3 & -2 & 4 & -1 \\ & 9 & 21 & 75 \end{array}} \\ \begin{array}{rrrr} 3 & 7 & 25 & 74 \end{array} \end{array} = 3x^2 + 7x + 25 + \dfrac{74}{x + 3}$$

Correct answer: $3x^2 - 11x + 37 - \dfrac{112}{x + 3}$
What error was made? (*see page 190*)

10. Finding the remainder using synthetic division
Example: Given $P(x) = 4x^2 - 2x + 1$, find $P(3)$.

$$\begin{array}{r} 3 \,\underline{\rvert\, \begin{array}{rrr} 4 & -2 & 1 \\ & 12 & -42 \end{array}} \\ \begin{array}{rrr} 4 & -14 & 43 \end{array} \end{array} \quad P(3) = 43$$

Correct answer: $P(3) = 31$
What error was made? (*see page 191*)

Chapter 4 critical thinking

If n is an integer, for what values of n will $4n^2 + 4n - 3$ represent an odd integer?

Chapter 4 review

[4–1]

State the domain of the given rational expression in set-builder notation.

1. $\dfrac{5}{x + 7}$

2. $\dfrac{2x + 1}{3x - 4}$

3. $\dfrac{4 - 2x}{x^2 - 10x + 25}$

4. $\dfrac{4a^2 + 1}{9a^2 - 16}$

5. $\dfrac{3z - 1}{6z^2 + 13z - 5}$

6. $\dfrac{4y - 11}{9y^2 - 12y + 4}$

Reduce each rational expression to lowest terms. Assume all denominators are nonzero.

7. $\dfrac{a^3b^4c}{a^2bc^3}$

8. $\dfrac{-10m^3n^3p}{35m^4np^5}$

9. $\dfrac{5x - 10}{6x - 12}$

10. $\dfrac{24a}{8a^2 - 16a}$

11. $\dfrac{5x - 10y}{4y^2 - x^2}$

12. $\dfrac{y^3 - 64}{y^2 - 16}$

13. $\dfrac{a^2 - 24a + 144}{a^2 - 11a - 12}$

14. $\dfrac{4x^2 - 5x - 6}{5x^2 - 11x + 2}$

15. $\dfrac{10y^2 + 9y - 9}{12 - 2y - 30y^2}$

[4–2]

Perform the indicated multiplication or division. Place necessary restrictions on the variables.

16. $\dfrac{14x}{3y} \cdot \dfrac{9y^2}{7x^2}$

17. $\dfrac{16a^2}{7y} \div \dfrac{4a}{21y^2}$

18. $\dfrac{p^2 - 16}{4p - 3} \cdot \dfrac{16p^2 - 9}{3p + 12}$

19. $\dfrac{z^2 + 1}{2z - 6} \div \dfrac{z^4 - 1}{z^2 - 6z + 9}$

20. $\dfrac{m^3 - 8}{m^2 - 3m} \cdot \dfrac{m^2 + 3m - 18}{m^2 + 3m - 10}$

21. $\dfrac{5a^2 + 17a + 6}{10a^2 - a - 2} \cdot \dfrac{2a^2 + 13a - 7}{a^2 + 10a + 21}$

22. $\dfrac{x^2 - 49}{8x^3 + 1} \div \dfrac{x^2 - 5x - 14}{4x^2 - 1}$

23. $\dfrac{4x^3 - 5x^2}{4x + 5} \div (16x^2 - 25)$

24. $\dfrac{7y^2 + 10y - 8}{y^2 + 4y + 4} \cdot \dfrac{y^2 + 6y + 9}{7y^2 + 17y - 12}$

25. $\dfrac{mq - mp - nq + np}{2mq - 2mp - nq + np} \div \dfrac{qm + qn + pm + pn}{2mp - np + 2qm - nq}$

[4–3]

Find the least common multiple (LCM) of the set of polynomials.

26. $15xy^2, 20x^3y, 9x^2y^3$

27. $2x + 4, 6x^2, x^2 - 16$

28. $a^2 + 3a - 10, a^2 - 2a, 3a + 15$

29. $p^2 - 25, p^2 + 10p + 25, p^2 - 5p$

Perform the indicated additions and/or subtractions. Assume all denominators are nonzero.

30. $\dfrac{3x}{4y} + \dfrac{8x}{3y}$

31. $\dfrac{n - 7}{n + 4} - \dfrac{3n + 8}{n - 1}$

32. $\dfrac{10}{p^2 + 7p - 18} + \dfrac{9}{p^3 - 4p}$

33. $(2b + 7) - \dfrac{b + 9}{3b - 2}$

34. $\dfrac{3y - 4}{y^2 - 49} + \dfrac{2y + 1}{y - 7}$

35. $\dfrac{2x}{2x^2 - 4x - 48} - \dfrac{5x + 1}{x^2 - 36}$

36. $\dfrac{7}{2a - 4} + \dfrac{9}{a - 2} - \dfrac{6}{2 - a}$

37. $\dfrac{x + 2}{x^2 - 3x - 28} - \dfrac{3x - 7}{4x + 16} - \dfrac{5 - x}{8x - 56}$

38. $\dfrac{a + b}{a^2 - ab - 2b^2} + \dfrac{3a - b}{a^2 - 4b^2}$

39. In electricity, the total resistance R_t of any parallel circuit is given by

$$\frac{1}{R_t} = \frac{I_1}{E_1} + \frac{I_2}{E_2} + \frac{I_3}{E_3}$$

Combine the expression in the right member.

[4–4]

Simplify each complex rational expression. Assume all denominators are nonzero.

40. $\dfrac{\dfrac{4}{a^2}}{\dfrac{2}{a}}$

41. $\dfrac{\dfrac{5}{x-3}}{\dfrac{4}{x}}$

42. $\dfrac{\dfrac{3x}{x-5}}{\dfrac{x}{x+2}}$

43. $\dfrac{7-\dfrac{3}{b-5}}{8+\dfrac{6}{b-5}}$

44. $\dfrac{\dfrac{x-y}{2}}{\dfrac{2}{x}+\dfrac{3}{y}}$

45. $\dfrac{(p-3)+\dfrac{2}{p-1}}{(p+1)-\dfrac{3}{p-1}}$

[4–5]

Perform the indicated divisions. Assume that no divisor is equal to zero.

46. $\dfrac{25a^7+15a^3+10a}{5a}$

47. $\dfrac{36a^5b^5c^3-18a^2b^3c^4+6a^2b^3c}{6a^2b^3c}$

48. $(3x^3+x+4)\div(x+1)$

49. $\dfrac{2x^4+x^3-3x^2+3x-1}{2x-1}$

[4–6]

Use synthetic division and the remainder theorem to evaluate each polynomial for the given value of the variable.

50. $P(x)=x^3-2x^2+8x-3;\ x=-2$

51. $P(y)=4y^3+2y^2-y+6;\ y=1$

52. $P(z)=-3z^4-z^2+5z-3;\ z=-1$

Use synthetic division to determine if the given number is a solution of the equation.

53. $x^3+2x^2+3x+18=0;\ -3$

54. $2y^3+4y^2+y-3=0;\ -1$

55. $4z^3-10z^2-z-6=0;\ 2$

[4–7]

Find the solution set of each of the following rational equations.

56. $6-\dfrac{10}{9a}=\dfrac{5}{12a}$

57. $\dfrac{4m-3}{15}+\dfrac{2-5m}{20}=1$

58. $\dfrac{4}{a+6}-\dfrac{9}{2a+12}=3$

59. $\dfrac{9}{4a^2-1}-\dfrac{6}{2a-1}=\dfrac{4}{2a+1}$

60. $\dfrac{8}{6n^2-13n-5}+\dfrac{6}{10n^2-25n}=0$

Solve the given equations for the indicated variable.

61. $\dfrac{2+m}{6}-\dfrac{n-3}{4}=\dfrac{p}{8}$ for p

62. $F=\dfrac{9}{5}C+32$ for C (Temperature conversion)

63. $\dfrac{P_1V_1}{T_1}=\dfrac{P_2V_2}{T_2}$ for V_1 (Physics formula)

64. In electricity, the total resistance R_t for two resistors connected in parallel is given by

$$R_t=\dfrac{R_1R_2}{R_1+R_2}$$

Solve for R_2.

[4–8]

Set up an equation to solve the following problems.

65. If Erin can do a job in 3 days and Sarah takes 6 days to do the same job, how long would it take the girls to do the job working together?

66. A train travels 120 miles in the same time that an automobile travels 80 miles. If the train travels 30 miles per hour faster than the automobile, what is the speed of each?

67. A boat travels 14 miles downstream in the same amount of time that it takes to travel 10 miles upstream. If the boat travels at 18 miles per hour in still water, what is the speed of the current?

68. The reciprocal of four times a number is equal to the reciprocal of twice the number added to 7. Find the number.

Chapter 4 cumulative test

Evaluate the following expressions when $x = 2$ and $y = -3$.

[1–5] **1.** $x^2 - 2xy + y^2$ **[1–5]** **2.** $3(2x - 1) - 2(3x + 2)$ **[1–5]** **3.** $\dfrac{3x^2}{5y^2}$

[4–4] **4.** $\dfrac{\dfrac{1}{x} - \dfrac{1}{y}}{\dfrac{1}{x} + \dfrac{1}{y}}$ **[1–5]** **5.** $\dfrac{2}{3x} - \dfrac{4}{5y} + \dfrac{1}{xy}$

Perform the indicated operations and simplify. Assume all denominators are nonzero.

[1–4] **6.** $3(2x - 1) + 4(x + 5) - (3x + 6)$ **[3–3]** **7.** $(4ab)(-2a^2b)(3a^3b^3)$

[3–2] **8.** $(x + 4)(x^2 - 2x + 5)$ **[4–2]** **9.** $\dfrac{10xy}{9x^2y} \cdot \dfrac{27x^3y^2}{15x^2y^2}$

[4–2] **10.** $\dfrac{x^2 + 5x - 24}{x^2 - 2x} \div \dfrac{x^2 - 9}{x^2 - 4x + 4}$ **[4–3]** **11.** $\dfrac{2y - 3}{8} - \dfrac{4y + 1}{6}$

[3–2] **12.** $(5x - 3)(2x + 9)$ **[3–2]** **13.** $(4x - 5)^2$

[3–2] **14.** $(3y - 5)(3y + 5)$ **[4–5]** **15.** $(2x^3 + 3x^2 - 1) \div (x - 3)$

[4–2] **16.** $\dfrac{6a^2 - a - 2}{a^2 + 4a - 21} \cdot \dfrac{2a^2 - a - 15}{8a^2 - 6a - 5}$ **[4–3]** **17.** $\dfrac{4}{2x^2 - x - 3} - \dfrac{5}{2x^2 + 3x + 1}$

State the domain of the given rational expression.

[4–1] **18.** $\dfrac{4y + 1}{2y - 3}$ **[4–1]** **19.** $\dfrac{3x}{4x^2 - 1}$ **[4–1]** **20.** $\dfrac{x - 5}{x^2 - 25}$

Find the solution set of each equation.

[2–1] **21.** $4(y + 3) - 3(3y + 1) + (4y + 5) = 0$ **[4–7]** **22.** $\dfrac{3}{4x} - 2 = \dfrac{5}{3x}$

[4–7] **23.** $\dfrac{3}{y + 2} + \dfrac{2y + 1}{y^2 - 4} = \dfrac{-3}{y - 2}$

Simplify each complex rational expression. Assume all denominators are nonzero.

[4–4] **24.** $\dfrac{\dfrac{1}{4} + \dfrac{2}{5}}{\dfrac{3}{10} - 1}$ **[4–4]** **25.** $\dfrac{4 - \dfrac{3}{x + 2}}{5 + \dfrac{2}{x - 2}}$

Find the solution set of the following inequalities.

[2–5] **26.** $5 - 3x \geq 8$ **[2–5]** **27.** $3(4y - 1) < 2(5y + 7)$

[2–5] **28.** $\dfrac{z + 1}{4} - \dfrac{z - 3}{6} > \dfrac{1}{3}$ **[2–5]** **29.** $\dfrac{3}{4}(x + 1) - 2(x - 3) \leq \dfrac{2}{3}(x - 2)$

Reduce the following rational expressions to lowest terms. Assume all denominators are nonzero.

[4–1] **30.** $\dfrac{36a^2b^3}{-24ab^5}$ **[4–1]** **31.** $\dfrac{p^2 + p - 12}{p^2 - 9}$ **[4–1]** **32.** $\dfrac{12 - 6y}{4y - 8}$

[4–7] **33.** Find the solution set of the rational equation
$$\dfrac{4}{x} - 3 = 6 + \dfrac{5}{x}.$$

[4–8] **34.** The sum of twice a number and six times its reciprocal is 13. Find the number.

[4–6] **35.** Using synthetic division, find $P(-2)$ when $P(x) = x^4 - x^3 + 2x^2 - 5$.

Exponents, Roots, and Radicals

The formula for approximating the velocity *V* in miles per hour of a car based on the length of its skid marks *S* (in feet) on wet pavement is given by

$$V = 2\sqrt{3S}$$

If the skid marks are 147 feet long, what was the velocity of the car?

5–1 ■ Roots and rational exponents

The nth root

In chapter 3, we were concerned with raising some real number to a power. For example,

if $a = -4$, then $a^3 = (-4)^3 = (-4)(-4)(-4) = -64$ (read "-4 raised to the third power equals -64");
if $a = 2$, then $a^4 = (2)^4 = (2)(2)(2)(2) = 16$ (read "2 raised to the fourth power equals 16").

In this chapter, we will reverse that process. That is, we will start with a power of a real number *a* and find *a*.

Definition of nth root

For every pair of real numbers *a* and *b* and every positive integer *n* greater than 1, if

$$a^n = b$$

then *a* is called an *n*th root of *b*.

Concept
An *n*th root of a number is one of *n* equal factors that, when multiplied, equal the number.

■ *Example 5–1 A*

1. 2 is a fourth root of 16 since $2^4 = 16$.

2. -4 is a third root of -64 since $(-4)^3 = -64$.

3. 3 is a square root of 9 since $3^2 = 9$.

4. -3 is also a square root of 9 since $(-3)^2 = 9$.   ■

We observe from these examples that both 3 and -3 are second roots (square roots) of 9 since $3^2 = 9$ and $(-3)^2 = 9$. To avoid the ambiguity of two different values for the same symbol, we now define the **principal *n*th root** of a number.

Definition of principal nth root ─────────────────

If *n* is a positive integer greater than 1 and the *n*th root is a real number, then the principal *n*th root of a nonzero real number *b*, denoted by $\sqrt[n]{b}$, has the same sign as the number itself. Also, the principal *n*th root of 0 is 0.

Since we most often talk about the principal *n*th root, we usually will eliminate the word "principal" and just say the "*n*th root," with the understanding that we are referring to the principal *n*th root.

Note   In the expression $\sqrt[n]{b}$, the $\sqrt{}$ is called the radical symbol,* *b* is called the radicand, *n* is called the index, and $\sqrt[n]{b}$ is called the radical.

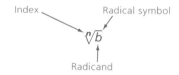

Table 5–1 lists the most common principal roots that we use in this book.

■ Table 5–1

Square roots[a]		Cube roots		Fourth roots
$\sqrt{1} = 1$	$\sqrt{64} = 8$	$\sqrt[3]{1} = 1$	$\sqrt[3]{-1} = -1$	$\sqrt[4]{1} = 1$
$\sqrt{4} = 2$	$\sqrt{81} = 9$	$\sqrt[3]{8} = 2$	$\sqrt[3]{-8} = -2$	$\sqrt[4]{16} = 2$
$\sqrt{9} = 3$	$\sqrt{100} = 10$	$\sqrt[3]{27} = 3$	$\sqrt[3]{-27} = -3$	$\sqrt[4]{81} = 3$
$\sqrt{16} = 4$	$\sqrt{121} = 11$	$\sqrt[3]{64} = 4$	$\sqrt[3]{-64} = -4$	$\sqrt[4]{256} = 4$
$\sqrt{25} = 5$	$\sqrt{144} = 12$	$\sqrt[3]{125} = 5$	$\sqrt[3]{-125} = -5$	$\sqrt[4]{625} = 5$
$\sqrt{36} = 6$	$\sqrt{169} = 13$	$\sqrt[3]{216} = 6$	$\sqrt[3]{-216} = -6$	
$\sqrt{49} = 7$	$\sqrt{196} = 14$			

[a]When we write a square root, $\sqrt{}$, the index is understood to be 2.

*The $\sqrt{}$ is the radical symbol and the $\overline{}$ means that what is under the radical symbol is a grouping.

Note Whenever we want the negative square root of a number, we indicate this by placing a minus sign in front of the square root symbol. That is, $-\sqrt{4} = -2$ and $-\sqrt{49} = -7$. Likewise, if we wanted the negative fourth root of 81, we would indicate it as $-\sqrt[4]{81} = -3$, or indicate the negative eighth root of 256 as $-\sqrt[8]{256} = -2$, and so forth.

In our definition of the principal *n*th root, we made the requirement that the *n*th root is a real number. This is because not all real numbers have a real *n*th root. Consider the example

$$\sqrt{-4} = \text{what?}$$

We know that all real numbers are positive, negative, or zero, and if we raise a real number to an even power (use it as a factor an even number of times), our answer will never be negative. Therefore the square root of a negative number does not exist in the set of real numbers, and, in general, **an even root of a negative number does not exist in the set of real numbers.**

Summary of nth roots

The symbol $\sqrt[n]{b}$ always represents the principal *n*th root of *b*.

	***n* is even**	***n* is odd**
If *b* is positive $b > 0$	Two real *n*th roots Principal *n*th root is positive	One real *n*th root Principal *n*th root is positive
If *b* is negative $b < 0$	NO REAL *n*th ROOTS	One real *n*th root Principal *n*th root is negative
If *b* is zero $b = 0$	One real *n*th root Principal *n*th root is zero	One real *n*th root Principal *n*th root is zero

---

In chapter 1, we defined the set of irrational numbers to be those numbers that cannot be represented by a terminating decimal or a nonterminating repeating decimal. Thus the *n*th root of *b*, $\sqrt[n]{b}$, is irrational if it cannot be expressed by a terminating decimal or a nonterminating repeating decimal. Some examples of irrational numbers are

$$\sqrt{8}, \quad \sqrt[5]{12}, \quad \sqrt[3]{-4}, \quad -\sqrt{10}, \quad \sqrt[4]{17}$$

Whenever we are working with irrational numbers in a problem, we may have to approximate the number to as many decimal places as are required in the problem by using a calculator or an appropriate table.

■ ***Example 5–1 B***   Find the decimal approximation to three decimal places by using a calculator.

1. $\sqrt{2} \approx 1.414$ **2.** $\sqrt{15} \approx 3.873$ **3.** $-\sqrt{11} \approx -3.317$

Note "$\approx$" is read "is approximately equal to" and is used when our answer is not exact.

▶ ***Quick check*** Find the decimal approximation to three decimal places by using a calculator. $\sqrt{17}$ ■

Rational exponents

In chapter 3, we developed a set of properties that guided our use of integers as exponents. We will now define rational exponents so that those same properties that apply to integer exponents will also apply to rational exponents as well.

Consider the equation

$$b^{1/3} = a$$

If we raise both members of the equation to the third power, we have

$$(b^{1/3})^3 = a^3$$

Since the left member is a power to a power, we multiply the exponents to get

$$b^{1/3 \cdot 3} = a^3$$
$$b^1 = a^3$$
$$b = a^3$$

Since $a^3 = b$, then by the definition of the principal nth root, $a = \sqrt[3]{b}$. Since we originally stated that $a = b^{1/3}$, then $b^{1/3}$ must be the same as $\sqrt[3]{b}$. That is,

$$b^{1/3} = \sqrt[3]{b}$$

Definition of $a^{1/n}$

For every real number a and positive integer n greater than 1,
$$a^{1/n} = \sqrt[n]{a}$$
whenever the principal nth root of a is a real number.

Concept
The expression $a^{1/n}$ is equivalent to $\sqrt[n]{a}$ and represents the principal nth root of a.

■ **Example 5–1 C**

Rewrite the following in radical notation. Use a calculator or table 5–1 to simplify where possible. Round the answer to three decimal places when the value is irrational.

1. $6^{1/3} = \sqrt[3]{6} \approx 1.817$
2. $81^{1/4} = \sqrt[4]{81} = 3$
3. $19^{1/2} = \sqrt{19} \approx 4.359$
4. $b^{1/7} = \sqrt[7]{b}$ ■

Next, we need to decide how to define the expression $a^{m/n}$, where m and n are positive integers and a is a real number such that the nth root of a is also a real number. Before we determine the meaning for the expression $a^{m/n}$, we shall place a further restriction on the fraction $\dfrac{m}{n}$ such that $\dfrac{m}{n}$ is reduced to lowest terms. When the fraction $\dfrac{m}{n}$ is reduced to lowest terms, we say that m and n are **relatively prime.** That is, m and n contain no common positive integer factors other than 1. We can write the fraction $\dfrac{m}{n}$ as

$$\frac{m}{n} = \frac{1}{n} \cdot m$$

Hence we have the following definition of $a^{m/n}$.

Definition of $a^{m/n}$

For every real number a and relatively prime positive integers m and n, if the principal nth root of a is a real number, then

$$a^{m/n} = (a^{1/n})^m$$

or, equivalently,

$$a^{m/n} = (\sqrt[n]{a})^m$$

In our previous definition, the fractional exponent $\dfrac{m}{n}$ was rewritten as

$$\frac{1}{n} \cdot m$$

If we apply the commutative property of multiplication, we can write $\dfrac{1}{n} \cdot m$ as

$$m \cdot \frac{1}{n}$$

This fact leads us to the property of $(a^{1/n})^m$.

Property of $(a^{1/n})^m$

For every real number a and relatively prime positive integers m and n, if the principal nth root of a is a real number, then

$$(a^{1/n})^m = (a^m)^{1/n}$$

or, equivalently,

$$(\sqrt[n]{a})^m = \sqrt[n]{a^m}$$

Concept

If we are dealing with a rational exponent that is reduced to lowest terms and the nth root of a is a real number, then raising the nth root of a to the mth power is equivalent to finding the nth root of a^m.

Note   When we have a rational exponent such as

$$\frac{m}{n}$$

the numerator (m) indicates the power to which the base is to be raised and the denominator (n) indicates the root to be taken.

Numerator ⟷ Exponent
Denominator ⟷ Index

$$a^{m/n} = \sqrt[n]{a^m}$$

■ **Example 5–1 D**

Rewrite the following in radical form and use table 5–1 to simplify where possible. Assume that all variables represent positive real numbers.

1. $x^{3/4} = (\sqrt[4]{x})^3$ or $\sqrt[4]{x^3}$ Denominator becomes index, numerator becomes exponent

2. $27^{2/3} = (\sqrt[3]{27})^2 = (3)^2 = 9$
  Alternate: $27^{2/3} = \sqrt[3]{27^2} = \sqrt[3]{729} = 9$

3. $(-8)^{2/3} = (\sqrt[3]{-8})^2 = (-2)^2 = 4$

4. $-8^{2/3} = -(\sqrt[3]{8})^2 = -(2)^2 = -4$

▶ *Quick check* Simplify $(-27)^{2/3}$.   ■

■ *Example 5–1 E*

Rewrite the following with rational exponents. Assume that all variables represent positive real numbers.

1. $\sqrt[3]{a^2} = a^{2/3}$ The exponent becomes the numerator, the index becomes the denominator

2. $\sqrt[5]{x} = x^{1/5}$ The power of x is understood to be 1

3. $\sqrt{5} = 5^{1/2}$ The index of the radical is understood to be 2

▶ *Quick check* Rewrite with rational exponents $\sqrt[7]{a}$ ■

In the previous definition of rational exponents, we required that the values of m and n be relatively prime. The following example illustrates what happens when m and n have a common factor of 2 and the radicand is negative. Consider the expression

$$[(-4)^2]^{1/4}$$

Raising -4 to the second power, we have

$$[(-4)^2]^{1/4} = (16)^{1/4}$$

and using table 5–1 to determine the fourth root, we have

$$(16)^{1/4} = \sqrt[4]{16} = 2$$

If we take the original expression and write it as

$$[(-4)^2]^{1/4} = (-4)^{2/4} = (-4)^{1/2} = \sqrt{-4}$$

we see that the result, $\sqrt{-4}$, has no real answer. Therefore, depending on the method that we use, two different results are possible. For this reason, we have the following definition of $(a^m)^{1/n}$, when $a < 0$.

┌─ **Definition of $(a^m)^{1/n}$, $a < 0$** ─────────────────────────────

If $a < 0$, and m and n are positive even integers,

$$(a^m)^{1/n} = |a|^{m/n}$$

Note In our example,

$$[(-4)^2]^{1/4} = |-4|^{2/4} = 4^{1/2} = \sqrt{4} = 2$$

If m and n are equal ($m = n$) and are even, the definition becomes

$$(a^n)^{1/n} = |a|^{n/n} = |a|^1 = |a|$$

or, equivalently,

$$(a^n)^{1/n} = \sqrt[n]{a^n} = |a|$$

when n is even. If $n = 2$, then

$$(a^2)^{1/2} = \sqrt{a^2} = |a|$$

In general, we make the following definition for $\sqrt[n]{a^n}$.

___ **Definition of** $\sqrt[n]{a^n}$ _____

For every real number a and positive integer n, where $n > 1$,

$$(a^n)^{1/n} = \sqrt[n]{a^n} = \begin{cases} |a| & \text{if } n \text{ is even} \\ a & \text{if } n \text{ is odd} \end{cases}$$

■ Example 5–1 F

Simplify the following. Variables represent all real numbers.

1. $(x^2)^{1/2} = \sqrt{x^2} = |x|$ Even index, absolute value is necessary

2. $b^{5/5} = \sqrt[5]{b^5} = b$ Odd index, do not need absolute value

3. $\sqrt[3]{-8} = \sqrt[3]{(-2)^3} = -2$ Odd index, do not need absolute value

4. $[(-2)^2]^{1/2} = \sqrt{(-2)^2} = |-2| = 2$ Even index, absolute value of -2 is 2

5. $\sqrt{a^2 + 2ab + b^2} = \sqrt{(a + b)^2}$ Factor the perfect square trinomial
$= |a + b|$ Even index, absolute value is necessary

▶ **Quick check** Simplify $(a^2)^{1/2}$ and $\sqrt{x^2 + 2xy + y^2}$. Variables represent all real numbers.

We now make the following definition of $a^{-m/n}$.

___ **Definition of** $a^{-m/n}$ _____

For every real number a, $a \neq 0$, and positive integers m and n, if the principal nth root of a is a real number then,

$$a^{-m/n} = \frac{1}{a^{m/n}}$$

■ Example 5–1 G

Rewrite the following using positive exponents and use table 5–1 to simplify where possible.

1. $16^{-3/4} = \dfrac{1}{16^{3/4}} = \dfrac{1}{(\sqrt[4]{16})^3} = \dfrac{1}{(2)^3} = \dfrac{1}{8}$ Rewrite the expression with a positive exponent, rewrite the expression in radical form, use table 5–1 to simplify

2. $(-8)^{-2/3} = \dfrac{1}{(-8)^{2/3}} = \dfrac{1}{(\sqrt[3]{-8})^2} = \dfrac{1}{(-2)^2} = \dfrac{1}{4}$ Rewrite the expression with a positive exponent, rewrite the expression in radical form, use table 5–1 to simplify

Note It is only the sign of the exponent that changes, not the sign of the base.

Mastery points

Can you
- Find the decimal approximation for a root that is an irrational number?
- Find the principal nth root of a number?
- Express rational exponents in radical form?
- Express radicals in rational exponent form?

Exercise 5–1

Find the decimal approximation to three decimal places by using a calculator. See example 5–1 B.

Example $\sqrt{17}$

Solution ≈ 4.123 Fourth decimal place is 4 or less, round down

1. $\sqrt{18}$ 2. $\sqrt{19}$ 3. $-\sqrt{33}$ 4. $-\sqrt{14}$

Rewrite the following in radical notation and use table 5–1 to simplify wherever possible. Assume that all variables represent positive real numbers. See examples 5–1 C, D, and G.

Example $(-27)^{2/3}$

Solution $= (\sqrt[3]{-27})^2$ Rewrite in radical form
$= (-3)^2$ Simplify the radical, table 5–1
$= 9$ Standard form

5. $9^{1/3}$ 6. $2^{1/2}$ 7. $x^{1/2}$ 8. $a^{1/3}$ 9. $b^{4/5}$ 10. $a^{3/4}$

11. $64^{1/3}$ 12. $(-8)^{1/3}$ 13. $(-64)^{2/3}$ 14. $16^{3/4}$ 15. $81^{3/4}$ 16. $64^{2/3}$

17. $16^{3/2}$ 18. $(-27)^{-1/3}$ 19. $8^{-1/3}$ 20. $49^{-1/2}$ 21. $16^{-1/4}$ 22. $16^{-3/4}$

23. $27^{-2/3}$ 24. $(-8)^{-2/3}$ 25. $(-32)^{-3/5}$ 26. $(-27)^{-2/3}$ 27. $x^{-3/4}$ 28. $a^{-2/3}$

Rewrite the following with rational exponents. See example 5–1 E.

Example $\sqrt[7]{a}$

Solution $= a^{1/7}$ The power of a is understood to be 1; this is the numerator of the rational exponent; the index, 7, is the denominator

29. $\sqrt[7]{a^4}$ 30. $\sqrt[9]{b}$ 31. $\sqrt[5]{x}$

Simplify the following. Variables represent *all* real numbers. See example 5–1 F.

Examples $(a^2)^{1/2}$ $\sqrt{x^2 + 2xy + y^2}$

Solutions $= \sqrt{a^2}$ Index is even, absolute value is necessary $= \sqrt{(x + y)^2}$ Factor the perfect square trinomial
$= |a|$ $= |x + y|$ Index is even, absolute value is necessary

32. $(x^7)^{1/7}$ 33. $(-8)^{3/3}$ 34. $[(-3)^2]^{1/2}$ 35. $[(-4)^2]^{1/2}$

36. $\sqrt{a^2 - 2ab + b^2}$ 37. $\sqrt{4x^2 - 4xy + y^2}$ 38. $\sqrt{a^2 + 2a + 1}$

Solve the following word problems.

39. Find the number whose principal fourth root is 4.

40. Find the number whose principal cube root is -3.

41. The electric-field intensity on the axis of a uniform charged ring is given by

$$E = \frac{T}{(x^2 + r^2)^{3/2}}$$

where T is the total charge on the ring and r is the radius of the ring. Express the rational exponent in radical form.

42. To find the velocity of the center of mass of a rolling cylinder, we use the equation

$$v = \left(\frac{4}{3}gh\right)^{1/2}$$

Write the expression in radical form.

43. The formula for approximating the velocity V in miles per hour of a car based on the length of its skid marks S (in feet) on dry pavement is given by

$$V = 2\sqrt{6S}$$

If the skid marks are 24 feet long, what was the velocity of the car?

44. When a gas is compressed with no gain or loss of heat, the pressure and the volume of the gas are related by the formula

$$p = kv^{-7/5}$$

where p represents the pressure, v represents the volume, and k is a constant. Express the formula in radical form.

Review exercises

Perform the indicated operations. Assume that all variables represent nonzero real numbers. Write each answer with only positive exponents. See sections 3–1 and 3–3.

1. $a^2 \cdot a^4 \cdot a$

2. $(x^2)^5$

3. $(2a^3b^4)^2$

4. $a^3 \div a^7$

5. $3x^0$

6. -3^2

7. $a^{-3} \cdot a^{-4}$

8. $x^{-8} \div x^{-6}$

5–2 ■ Operations with rational exponents

We have now developed the concept of rational exponents so that the same properties that applied to integer exponents can now be extended to rational exponents. The following is a restatement of those properties and definitions.

$$a^n = \overbrace{a \cdot a \cdot a \cdots a}^{n \text{ factors of } a}$$

$$a^m \cdot a^n = a^{m+n}$$

$$(a^m)^n = a^{mn}$$

$$(ab)^n = a^n b^n$$

$$a^m \div a^n = \frac{a^m}{a^n} = a^{m-n}, a \neq 0$$

$$a^{-n} = \frac{1}{a^n}, a \neq 0$$

$$a^0 = 1, a \neq 0$$

$$\left(\frac{a}{b}\right)^n = \frac{a^n}{b^n}, b \neq 0$$

■ *Example 5–2 A*

Perform the indicated operations and simplify. Assume that all variables represent positive real numbers. Leave the answer with only positive exponents.

1. $5^{1/3} \cdot 5^{1/3} = 5^{\frac{1}{3} + \frac{1}{3}}$ Multiplication of like bases

   $= 5^{\frac{2}{3}}$ Add exponents

2. $3^{1/2} \cdot 3^{1/3} = 3^{\frac{1}{2} + \frac{1}{3}}$ Multiplication of like bases

   $= 3^{\frac{3}{6} + \frac{2}{6}}$ Least common denominator is 6

   $= 3^{\frac{5}{6}}$ Add exponents

3. $\dfrac{6^{1/2}}{6^{1/4}} = 6^{\frac{1}{2} - \frac{1}{4}}$ Division of like bases

   $= 6^{\frac{2}{4} - \frac{1}{4}}$ Least common denominator is 4

   $= 6^{\frac{1}{4}}$ Subtract exponents

4. $a^{2/3} \cdot a^{3/4} = a^{\frac{2}{3} + \frac{3}{4}}$ Multiplication of like bases

   $= a^{\frac{8}{12} + \frac{9}{12}}$ Least common denominator is 12

   $= a^{\frac{17}{12}}$ Add exponents

5. $(x^{4/3})^{1/2} = x^{\frac{4}{3} \cdot \frac{1}{2}}$ Power of a power

   $= x^{\frac{2}{3}}$ Multiply exponents

6. $\dfrac{y^{1/3}}{y^{1/4}} = y^{\frac{1}{3} - \frac{1}{4}}$ Division of like bases

   $= y^{\frac{4}{12} - \frac{3}{12}}$ Least common denominator is 12

   $= y^{\frac{1}{12}}$ Subtract exponents

7. $(2^3 a^{15} b^{21})^{1/3} = (2^3)^{\frac{1}{3}} (a^{15})^{\frac{1}{3}} (b^{21})^{\frac{1}{3}}$ Group of factors to a power

   $= 2^{3 \cdot \frac{1}{3}} a^{15 \cdot \frac{1}{3}} b^{21 \cdot \frac{1}{3}}$ Power of a power

   $= 2^1 a^5 b^7$ Multiply exponents

   $= 2a^5 b^7$ Standard form

8. $\dfrac{x^{-1/4}y^{2/5}}{x^{3/4}y^{-4/5}} = x^{\frac{-1}{4} - \frac{3}{4}}y^{\frac{2}{5} - \frac{-4}{5}}$ Division of like bases

$= x^{\frac{-4}{4}}y^{\frac{6}{5}}$ Subtract exponents

$= x^{-1}y^{\frac{6}{5}}$ Simplify the rational exponent

$= \dfrac{y^{\frac{6}{5}}}{x^{1}}$ Rewrite with positive exponents

$= \dfrac{y^{\frac{6}{5}}}{x}$ Standard form

▶ *Quick check* Simplify $7^{\frac{1}{5}} \cdot 7^{\frac{1}{5}}$ and $(a^{\frac{5}{8}})^{\frac{2}{3}}$.   ■

┌─ *Mastery points* ───┐

Can you
■ Apply the properties of exponents to rational exponents?

└──┘

Exercise 5–2

Perform the indicated operations and simplify. Assume that all variables represent positive real numbers. Leave the answer with all exponents positive. See example 5–2 A.

Examples $7^{1/5} \cdot 7^{1/5}$ $(a^{5/8})^{2/3}$

Solutions $= 7^{1/5 + 1/5}$ Multiplication of like bases $= a^{5/8 \cdot 2/3}$ Power of a power
$= 7^{2/5}$ Add exponents $= a^{5/12}$ Multiply exponents

1. $2^{1/2} \cdot 2^{1/2}$ 2. $a^{1/3} \cdot a^{2/3}$ 3. $b^{3/4} \cdot b^{2/3}$ 4. $x^{1/2} \cdot x^{5/4}$ 5. $5^{3/2} \cdot 5^{-1/2}$ 6. $x^{3/4} \cdot x^{-1/4}$

7. $a^{1/3} \cdot a^{-1/4}$ 8. $y^{1/2} \cdot y^{-3/4}$ 9. $(a^{2/3})^{4/5}$ 10. $(b^{2/3})^{1/2}$ 11. $(x^{3/4})^4$ 12. $(a^{1/2})^{1/2}$

13. $(x^{-1/4})^4$ 14. $(a^{-3/4})^{-1/3}$ 15. $(b^{-2/3})^{-1/2}$ 16. $(a^{1/2})^{-2/3}$ 17. $(x^{-1/3})^{-3/4}$ 18. $(x^{-2/3})^{-3/2}$

19. $(16y^4)^{3/4}$ 20. $(a^3b^6)^{1/3}$ 21. $(a^3b)^{2/3}$ 22. $(8a^6b^{12})^{2/3}$ 23. $(16a^{10}b^2)^{3/4}$ 24. $(4a^2b^4)^{-1/2}$

25. $(27a^{12}b^3)^{-1/3}$ 26. $(9x^{-2}y^4)^{-3/2}$ 27. $\dfrac{y^{1/4}}{y^{1/3}}$ 28. $\dfrac{a^{1/3}}{a^{5/6}}$ 29. $\dfrac{b^{3/4}}{b}$ 30. $\dfrac{x^{1/3}}{x}$

31. $\dfrac{x^{3/2}}{x^{-1/2}}$ 32. $\dfrac{y^{-2/3}}{y^{1/3}}$ 33. $\dfrac{a^{-2/3}}{a^{-4/3}}$ 34. $\dfrac{x^{-1/4}}{x^{-1/3}}$ 35. $\dfrac{a^{3/4}b^{1/2}}{a^{1/4}b^{1/4}}$ 36. $\dfrac{xy^{3/4}}{x^{1/2}y^{1/4}}$

37. $\dfrac{ab}{a^{1/2}b^{1/2}}$ 38. $\dfrac{x^{-1/2}x}{x^{1/3}}$ 39. $\dfrac{b^2b^{1/3}}{b^{1/2}}$ 40. $\dfrac{c^{2/3}c^{3/4}}{c^{-1/3}}$ 41. $\dfrac{a^{-2/3}b^{1/2}}{a^{-1/3}b^{3/4}}$ 42. $\dfrac{x^{-1/2}y^{5/4}}{x^{-2/3}y^{3/4}}$

Solve the following word problems.

43. A square-shaped television picture tube has an area of 169 square inches. What is the length of the side of the tube? (*Hint:* Area of a square is found by squaring the length of a side, $A = s^2$.)

44. A garden in the shape of a square is 196 square feet. What is the length of a side? (See exercise 43.)

45. The formula for approximating the velocity V in miles per hour of a car based on the length of its skid marks S (in feet) on wet pavement is given by

$$V = 2\sqrt{3S}$$

If the skid marks are 75 feet long, what was the velocity of the car?

46. A tank whose shape is a cube holds 216 cubic meters of water. What is the length of an edge of the cube? (*Hint:* The volume of a cube is found by raising the length of an edge to the third power, $V = e^3$.)

47. At an altitude of h feet above the sea or level ground, the distance d in miles that a person can see an object is found by using the equation

$$d = \sqrt{\frac{3h}{2}}$$

How far can someone see who is in a tower 216 feet above the ground?

48. How can you find the fourth root of a number on a calculator using only the square root key?

49. How can you find the eighth root of a number on a calculator using only the square root key?

Review exercises

Perform the indicated multiplication. See section 3–1.

1. $(8a^3)(4a^4)$

2. $(4x^2y^2)(4x^2y)$

3. $(5x^2y^2)(75xy^2)$

4. $(3ab^2)(18a^2b^3)$

5. $(25a^5b^4)(15ab^4)$

6. $(8ab)(4a^5b^5)$

5–3 ■ Simplifying radicals—I

Product property for radicals

In this section, we will develop some properties for simplifying radicals. We will consider several forms of simplification that involve radicals. The first type of simplified radical is as follows:

The radicand (the quantity under the radical symbol) contains no factors that can be written to a power greater than or equal to the index.

The following property, called the **product property for radicals,** is useful for this type of simplification.

> ### Product property for radicals
>
> For all nonnegative real numbers a and b and positive integer n greater than 1,
>
> $$\sqrt[n]{ab} = \sqrt[n]{a}\,\sqrt[n]{b}$$
>
> #### Concept
> The nth root of a product is equal to the product of the nth roots of the factors.

To utilize this property in simplifying radicals, we look for factors that are perfect nth roots, that is, factors that are raised to the nth power. We have a perfect nth root when the value of a radical expression is a rational number. The following are examples of perfect nth roots:

$$\sqrt{25} = \sqrt{5^2} = 5 \qquad \sqrt[3]{64} = \sqrt[3]{4^3} = 4$$

$$\sqrt[4]{81} = \sqrt[4]{3^4} = 3 \qquad \sqrt[5]{32} = \sqrt[5]{2^5} = 2$$

Simplifying the principal nth root ——————————————

1. If the radicand is an nth power, write the corresponding nth root.
2. If possible, factor the radicand so that at least one factor is an nth power. Write the corresponding nth root as a coefficient of the radical.
3. The nth root is in simplest form when the radicand has no nth power factors other than 1.

■ *Example 5–3 A*

Simplify the following. Assume that all variables represent nonnegative real numbers.

1. $\sqrt{18}$

Since this is a square root, we are looking for factors that are raised to the second power. Since 9 is a factor of 18 and 9 is 3^2, we have

$$\sqrt{18} = \sqrt{9 \cdot 2} = \sqrt{9}\sqrt{2} = 3\sqrt{2}$$

Hence, $3\sqrt{2}$ is the simplified form of $\sqrt{18}$.

2. $\sqrt[3]{32}$

The index is 3, therefore we are looking for factors raised to the third power. Since 8 is a factor of 32 and $8 = 2^3$, our problem becomes

$$\sqrt[3]{32} = \sqrt[3]{8 \cdot 4} = \sqrt[3]{8}\sqrt[3]{4} = 2\sqrt[3]{4}$$

Hence, $2\sqrt[3]{4}$ is the simplified form of $\sqrt[3]{32}$.

3. $\sqrt[3]{a^5}$

We are looking for factors raised to the third power, and a^5 can be written as $a^3 \cdot a^2$. Hence

$$\sqrt[3]{a^5} = \sqrt[3]{a^3 a^2} = \sqrt[3]{a^3}\sqrt[3]{a^2} = a\sqrt[3]{a^2}$$

4. $\sqrt[5]{a^{10}} = \sqrt[5]{a^5 \cdot a^5} = \sqrt[5]{a^5}\sqrt[5]{a^5} = a \cdot a = a^2$

Note In example 4, the exponent, 10, was a multiple of the index, 5, and the radical was eliminated. When the exponent of a factor is a multiple of the index, that factor will no longer remain under the radical symbol.

▶ *Quick check* Simplify $\sqrt[3]{16}$.   ■

The symmetric property from chapter 1 allows us to restate the product property for radicals as follows:

$$\sqrt[n]{a}\sqrt[n]{b} = \sqrt[n]{ab}$$

This means that the product of nth roots can be written as the nth root of the product.

■ *Example 5–3 B*

Multiply the following radicals and simplify where possible. Assume that all variables represent nonnegative real numbers.

1. $\sqrt{2}\sqrt{10}$

Since this is a square root times a square root, we can multiply the radicals together.

$$\sqrt{2}\sqrt{10} = \sqrt{2 \cdot 10} = \sqrt{20}$$

But $\sqrt{20}$ can be simplified.

$$\sqrt{20} = \sqrt{4 \cdot 5} = \sqrt{4}\sqrt{5} = 2\sqrt{5}$$

Therefore the simplified form of $\sqrt{2}\sqrt{10}$ is $2\sqrt{5}$.

2. $\sqrt[3]{9a^2b}\sqrt[3]{9ab}$ Indices are the same

$$= \sqrt[3]{9a^2b \cdot 9ab} \quad\quad \text{Multiply radicands}$$

$$= \sqrt[3]{81a^3b^2} \quad\quad\quad \text{Simplify}$$

$$= \sqrt[3]{27 \cdot 3a^3b^2} = \sqrt[3]{27}\sqrt[3]{3}\sqrt[3]{a^3}\sqrt[3]{b^2}$$

$$= 3\sqrt[3]{3} \cdot a\sqrt[3]{b^2} = 3a\sqrt[3]{3b^2}$$

▶ *Quick check* Multiply and simplify $\sqrt{3}\sqrt{15}$ ■

A second type of radical simplification can be seen in the following example. When we try to simplify the radical

$$\sqrt[6]{27}$$

we find that no factors of 27 can be written to a power greater than or equal to the index. Therefore it appears that no simplification is possible. But if we express the radical in rational exponent form, we observe

$$\sqrt[6]{27} = \sqrt[6]{3^3} = 3^{3/6} = 3^{1/2} = \sqrt{3}$$

We found that we could start with a radical whose index is 6 and reduce the index to 2 (square root). In reducing the index of the radical, it has been simplified. Therefore a second way that a radical is simplified is when **the exponent of the radicand and the index of the radical have no common factor other than 1.** That is, the exponent and the index are relatively prime.

Property $\sqrt[kn]{a^{km}}$ _____

If a is a positive real number, m is an integer, and n and k are positive integers greater than 1, then

$$\sqrt[kn]{a^{km}} = \sqrt[n]{a^m}$$

Concept

We can divide out a common factor between the index and the exponent of the radicand.

■ *Example 5–3 C*

Simplify the following. Assume that all variables represent positive real numbers.

1. $\sqrt[6]{9} = \sqrt[6]{3^2}$

Since the exponent of the radicand and the index of the radical both have a common factor of 2, we can divide out (cancel) the common factor.

$$\sqrt[6]{3^2} = 3^{2/6} = 3^{1/3} = \sqrt[3]{3^1} = \sqrt[3]{3}$$

2. $\sqrt[9]{x^3y^6}$

The radicand can be written as $(xy^2)^3$, and we see from this that there is a common factor of 3 in the index and exponent.

$$\sqrt[9]{x^3y^6} = \sqrt[9]{(xy^2)^3} = (xy^2)^{3/9} = (xy^2)^{1/3} = \sqrt[3]{(xy^2)^1} = \sqrt[3]{xy^2}$$

Note When the radicand contains two or more different factors, we can reduce the index *only* if there is a common factor in the index and each of the exponents of the *different* factors. For example, $\sqrt[12]{a^4b^3c^6}$ is in simplest form because 1 is the only common factor between the index and the three exponents.　■

┌─ **Mastery points** ──

Can you
- Simplify radicals by using the product property for radicals?
- Multiply radicals with the same indices?
- Simplify radicals by reducing the index of the radical?

└──

Exercise 5–3

Simplify the following. Assume that all variables represent positive real numbers. See examples 5–3 A, B, and C.

Examples　$\sqrt[3]{16}$

Solutions
$= \sqrt[3]{8 \cdot 2}$　　$8 = 2^3$ and is a factor of 16
$= \sqrt[3]{8}\sqrt[3]{2}$　　Product property
$= 2\sqrt[3]{2}$　　Simplified form

$\sqrt{3}\sqrt{15}$

$= \sqrt{3 \cdot 15}$　　Indices are the same, multiply the radicands
$= \sqrt{45}$　　Simplify
$= \sqrt{9 \cdot 5}$　　$9 = 3^2$ and 45 can be written $9 \cdot 5$
$= \sqrt{9}\sqrt{5}$　　Product property
$= 3\sqrt{5}$　　Simplified form

1. $\sqrt{20}$　　　**2.** $\sqrt{63}$　　　**3.** $\sqrt[3]{24}$　　　**4.** $\sqrt[3]{32}$　　　**5.** $\sqrt[4]{a^5}$

6. $\sqrt[5]{a^7}$　　　**7.** $\sqrt[9]{a^{18}}$　　　**8.** $\sqrt[5]{b^{10}}$　　　**9.** $\sqrt[5]{c^5}$　　　**10.** $\sqrt[9]{9x^2y^5}$

11. $\sqrt{25x^3y^9}$　　**12.** $\sqrt{32a^4b^7}$　　**13.** $\sqrt{50a^6bc^9}$　　**14.** $\sqrt[3]{8x^5y^4}$　　**15.** $\sqrt[3]{27a^3b^2c^{12}}$

16. $\sqrt[3]{16a^4b^5}$　　**17.** $\sqrt[3]{81a^5b^{11}}$　　**18.** $\sqrt[5]{64x^{10}y^{14}}$　　**19.** $\sqrt[5]{32a^{10}b^4c^{12}}$　　**20.** $\sqrt[5]{8a^7b^{15}c^3}$

21. $\sqrt{x^2 + 6x + 9}$　　**22.** $\sqrt{a^2 + 10a + 25}$　　**23.** $\sqrt{9a^2 + 6a + 1}$　　**24.** $\sqrt{x^2 + y^2}$

25. $\sqrt{6}\sqrt{27}$　　**26.** $\sqrt{10}\sqrt{20}$　　**27.** $\sqrt{7}\sqrt{7}$　　**28.** $\sqrt{12}\sqrt{12}$

29. $\sqrt{6}\sqrt{24}$　　**30.** $\sqrt[3]{4}\sqrt[3]{4}$　　**31.** $\sqrt[3]{6}\sqrt[3]{12}$　　**32.** $\sqrt[3]{10}\sqrt[3]{4}$

33. $\sqrt{3a}\sqrt{15a}$　　**34.** $\sqrt{7b}\sqrt{14b}$　　**35.** $\sqrt[3]{a^2}\sqrt[3]{a^2}$　　**36.** $\sqrt[5]{b^4}\sqrt[5]{b^3}$

37. $\sqrt[4]{x}\sqrt[4]{x^3}$　　**38.** $\sqrt[3]{2x^2}\sqrt[3]{2x}$　　**39.** $\sqrt[5]{8x^4}\sqrt[5]{4x^3}$　　**40.** $\sqrt[3]{4a^2b}\sqrt[3]{4a^2b^2}$

41. $\sqrt[3]{5a^2b}\sqrt[3]{75a^2b^2}$ **42.** $\sqrt[3]{3ab^2}\sqrt[3]{18a^2b^2}$ **43.** $\sqrt[3]{25x^5y^7}\sqrt[3]{15xy^3}$ **44.** $\sqrt[3]{16a^{11}b^4}\sqrt[3]{12a^4b^6}$

45. $\sqrt[4]{8xy}\sqrt[4]{4x^3y^3}$ **46.** $\sqrt[6]{a^3}$ **47.** $\sqrt[10]{y^5}$ **48.** $\sqrt[6]{b^{10}}$

49. $\sqrt[8]{y^{14}}$ **50.** $\sqrt[4]{4x^2}$ **51.** $\sqrt[6]{8y^3}$ **52.** $\sqrt[9]{27x^6y^6}$

53. $\sqrt[9]{8a^3b^6}$

Simplify the following. Variables represent *all* real numbers. See example 5–1 F.

Example $\sqrt{25a^2}$

Solution $= \sqrt{25}\sqrt{a^2}$ Product property
$= 5|a|$ Index is even, absolute value is necessary

Example $\sqrt{x^2 + 2xy + y^2}$

Solution $= \sqrt{(x + y)^2}$ Factor the trinomial
$= |x + y|$ Index is even, absolute value is necessary

54. $\sqrt{9a^2}$ **55.** $\sqrt{16x^2}$ **56.** $\sqrt{36a^2b^4}$ **57.** $\sqrt{49b^2c^2}$

58. $\sqrt{x^2 + 6x + 9}$ **59.** $\sqrt{a^2 - 8a + 16}$ **60.** $\sqrt{9x^2 + 6xy + y^2}$ **61.** $\sqrt{a^2b}$

62. $\sqrt[3]{8x^3y}$ **63.** $\sqrt[3]{27ab^3}$ **64.** $\sqrt[4]{16x^3y^4}$ **65.** $\sqrt[4]{81a^4b^3}$

66. For what values of x is $\sqrt[4]{4x^2} = \sqrt{2x}$ a false statement?

Solve the following word problems.

67. The moment of inertia for a rectangle is given by the formula

$$I = \frac{bh^3}{12}$$

If we know the values of I and b, we can solve for h as follows:

$$h = \sqrt[3]{\frac{12I}{b}}$$

Find h if $I = 27$ in.⁴ and $b = 4$ in.

68. Use exercise 67 to find h if $I = 2$ in.⁴ and $b = 3$ in.

69. The moment of inertia for a circle is given by the formula

$$I = \frac{\pi r^4}{4}$$

If we know the value of I, we can solve for r as follows:

$$r = \sqrt[4]{\frac{4I}{\pi}}$$

Find r if $I = 63.585$ in.⁴ and we use 3.14 for π.

70. Use exercise 69 to find r if $I = 12.56$ in.⁴.

71. The formula for finding the length of an edge e of a cube when the volume v is known is $e = \sqrt[3]{v}$. What is the length of the edge of a cube whose volume is 216 cubic units?

72. What is the length of the edge of a cube whose volume is 729 cubic units? (Refer to exercise 71.)

73. The current I (amperes) in a circuit is found by the formula

$$I = \sqrt{\frac{watts}{ohms}}$$

What is the current of a circuit that has 3-ohms resistance and uses 450 watts?

74. What is the current of a circuit that has 2-ohms resistance and uses 1,728 watts? (Refer to exercise 73.)

The following problems will make use of an important property of right triangles called the **Pythagorean Theorem.**

In a right triangle, the square of the hypotenuse (the side opposite the right angle) is equal to the sum of the squares of the lengths of the legs (the sides that form the right angle). If c is the hypotenuse and a and b are the lengths of the legs, this property can be stated as:

$$c^2 = a^2 + b^2 \text{ or } c = \sqrt{a^2 + b^2}$$
Also as $a^2 = c^2 - b^2$ or $a = \sqrt{c^2 - b^2}$
Also as $b^2 = c^2 - a^2$ or $b = \sqrt{c^2 - a^2}$

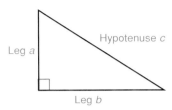

Example   Find the length of the hypotenuse of a right triangle whose legs are 6 cm and 8 cm.

Solution   We want to find c when $a = 6$ cm and $b = 8$ cm.

By the Pythagorean Theorem,

$$c = \sqrt{a^2 + b^2}$$
$$= \sqrt{6^2 + 8^2}$$
$$= \sqrt{36 + 64}$$
$$= \sqrt{100} = 10. \text{ Hence } c = 10 \text{ cm.}$$

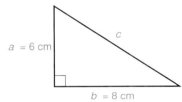

Example   Find the second leg of a right triangle whose hypotenuse has length 13 in. and the first leg is 5 in. long.

Solution   We want to find b given $c = 13$ in. and $a = 5$ in. Using one of the forms of the Pythagorean Theorem,

$$b = \sqrt{c^2 - a^2}$$
$$= \sqrt{13^2 - 5^2}$$
$$= \sqrt{169 - 25}$$
$$= \sqrt{144} = 12. \text{ Hence } b = 12 \text{ in.}$$

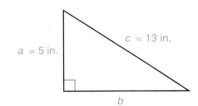

Find the length of the unknown side in the following right triangles.

75. $a = 3$ m, $b = 4$ m

76. $a = 8$ ft, $c = 10$ ft

77. $a = 12$ in., $b = 5$ in.

78. $a = 15$ cm, $b = 8$ cm

79. $a = 5$ in., $b = 4$ in.

80. $a = 6$ yd, $c = 10$ yd

81. $a = 6$ ft, $b = 9$ ft

82. $b = 16$ m, $c = 20$ m

83. $a = 12$ mm, $b = 16$ mm

84. $a = 10$ in., $b = 24$ in.

85. $a = 4$ cm, $c = 10$ cm

Solve the following word problems.

86. A 17-foot ladder is placed against the wall of a house. If the bottom of the ladder is 8 feet from the house, how far from the ground is the top of the ladder?

87. Find the width of a rectangle whose diagonal is 13 feet and length is 12 feet.

88. Find the diagonal of a rectangle whose length is 8 meters and whose width is 6 meters.

89. Under ideal conditions, the velocity v in meters per second of an object falling freely from a height h is given by $v = \sqrt{2gh}$, where g is the acceleration due to gravity. Use a calculator to find the velocity when h is 100 m. Round to two decimal places. Use $g = 9.8$ m/s².

Review exercises

Simplify the following expressions. See section 5–1.

1. $\sqrt{81}$ **2.** $\sqrt[3]{64}$ **3.** $\sqrt[5]{32}$ **4.** $\sqrt[3]{a^3}$ **5.** $\sqrt[5]{x^5}$ **6.** $\sqrt[5]{a^2 \cdot a^3}$ **7.** $\sqrt[3]{2 \cdot 4}$ **8.** $\sqrt[7]{x^5 \cdot x^2}$

5–4 ■ Simplifying radicals—II

The quotient property for radicals

In this section, we continue to develop some of the properties for simplifying radicals. A third way that a radical is simplified is when **the radicand contains no fractions.** The following property, called the **quotient property for radicals,** is useful for this type of simplification.

— **Quotient property for radicals** —————————

For all nonnegative real numbers a and b, where $b \neq 0$, and positive integer n greater than 1,

$$\sqrt[n]{\frac{a}{b}} = \frac{\sqrt[n]{a}}{\sqrt[n]{b}}$$

Concept

The nth root of a quotient can be written as the nth root of the numerator divided by the nth root of the denominator.

■ **Example 5–4 A**

Simplify the following. Assume that all variables represent positive real numbers.

1. $\sqrt{\dfrac{5}{16}} = \dfrac{\sqrt{5}}{\sqrt{16}}$ No simplification can be done inside the radical, apply quotient property

$= \dfrac{\sqrt{5}}{4}$ Simplify radical in denominator

2. $\sqrt{\dfrac{a^3}{b^2}} = \dfrac{\sqrt{a^3}}{\sqrt{b^2}}$ No simplification can be done inside the radical, apply quotient property

$= \dfrac{a\sqrt{a}}{b}$ Simplify radicals

3. $\sqrt[3]{\dfrac{x^5 y^2}{x^2}} = \sqrt[3]{x^3 y^2}$ Reduce fraction by the common factor x^2

$= \sqrt[3]{x^3}\sqrt[3]{y^2}$ Product property

$= x\sqrt[3]{y^2}$ Simplify first radical   ■

Rationalizing a denominator that has a single term

When the problem has been simplified and a radical still remains in the denominator, the evaluation of the problem usually is an involved process. For this reason, the fourth way that a radical is simplified is when **no radicals appear in the denominator.** This procedure is called **rationalizing the denominator,** because it changes the denominator from a radical (irrational number) to a rational number.

Consider the example

$$\sqrt{\frac{4}{5}} = \frac{\sqrt{4}}{\sqrt{5}} = \frac{2}{\sqrt{5}}$$

We would like to eliminate the radical symbol in the denominator. To remove this radical, we multiply the numerator and the denominator by a radical that yields a perfect square radicand and thereby allows us to eliminate the radical in the denominator.

$$\frac{2}{\sqrt{5}} \cdot \frac{\sqrt{5}}{\sqrt{5}} = \frac{2\sqrt{5}}{\sqrt{25}} = \frac{2\sqrt{5}}{5}$$

Since we multiplied the numerator and the denominator by the same number, using the fundamental principle of fractions, our answer is equivalent to the original fraction.

■ *Example 5–4 B*

Simplify the following. Assume that all variables represent positive real numbers.

1. $\sqrt{\dfrac{2}{3}} = \dfrac{\sqrt{2}}{\sqrt{3}}$ Quotient property

$= \dfrac{\sqrt{2}}{\sqrt{3}} \cdot \dfrac{\sqrt{3}}{\sqrt{3}} = \dfrac{\sqrt{2 \cdot 3}}{\sqrt{3 \cdot 3}}$ Multiply by $\dfrac{\sqrt{3}}{\sqrt{3}}$ to rationalize denominator

$= \dfrac{\sqrt{6}}{\sqrt{9}}$ We now have a perfect square root and can eliminate the radical in the denominator.

$= \dfrac{\sqrt{6}}{3}$ Expression is simplified

2. $\sqrt{\dfrac{a^3}{b}} = \dfrac{\sqrt{a^3}}{\sqrt{b}}$ Quotient property

$= \dfrac{a\sqrt{a}}{\sqrt{b}}$ Simplify numerator: product property

$= \dfrac{a\sqrt{a}}{\sqrt{b}} \cdot \dfrac{\sqrt{b}}{\sqrt{b}}$ Multiply by $\dfrac{\sqrt{b}}{\sqrt{b}}$

$= \dfrac{a\sqrt{ab}}{\sqrt{b^2}}$ Multiply radicals

$= \dfrac{a\sqrt{ab}}{b}$ Eliminate radical in denominator   ■

The following example will help us develop a general procedure for rationalizing a denominator that has a single term.

$$\sqrt[3]{\frac{1}{a}} = \frac{\sqrt[3]{1}}{\sqrt[3]{a}} = \frac{1}{\sqrt[3]{a}}$$

At this point, a radical remains in the denominator. We must now determine what we can do to the fraction to remove the radical from the denominator.

Observations:

1. We can multiply the numerator and the denominator by the same number and form equivalent fractions.
2. If we multiply by a radical, the indices must be the same to carry out the multiplication.

3. To bring a factor out from under the radical symbol and not leave any of the factor behind, the exponent must be a multiple of the index.

With these observations in mind, we rationalize the fraction.

$$= \frac{1}{\sqrt[3]{a}} \cdot \frac{\sqrt[3]{a^2}}{\sqrt[3]{a^2}}$$

The indices are the same and we multiply the numerator and the denominator by the same number

$$= \frac{\sqrt[3]{a^2}}{\sqrt[3]{a \cdot a^2}} = \frac{\sqrt[3]{a^2}}{\sqrt[3]{a^3}}$$

The sum of the exponents of a in the denominator adds to the index, forming a perfect nth root (cube root)

$$= \frac{\sqrt[3]{a^2}}{a}$$

The radical is eliminated from the denominator

Procedure for rationalizing a denominator of one term

1. We multiply the numerator and the denominator by a radical with the same index as the radical that we wish to eliminate from the denominator.
2. The exponent of the factor under the radical must be such that when we add it to the original exponent of the factor under the radical in the denominator, the sum will be equal to, or a multiple of, the index of the radical.
3. We carry out the multiplication and reduce the fraction if possible.

■ *Example 5–4 C*

Simplify the following. Assume that all variables represent positive real numbers.

1. $\dfrac{1}{\sqrt[5]{x^2}}$

To eliminate the radical, we multiply by another 5th root where the exponents of x will add up to 5 ($x^2 \cdot x^3 = x^{2+3} = x^5$)

$$= \frac{1}{\sqrt[5]{x^2}} \cdot \frac{\sqrt[5]{x^3}}{\sqrt[5]{x^3}} = \frac{\sqrt[5]{x^3}}{\sqrt[5]{x^2 \cdot x^3}}$$

The resulting denominator is a perfect nth root (5th root) and the radical symbol is eliminated

$$= \frac{\sqrt[5]{x^3}}{\sqrt[5]{x^5}}$$

$$= \frac{\sqrt[5]{x^3}}{x}$$

2. $\dfrac{x}{\sqrt[4]{x}} = \dfrac{x}{\sqrt[4]{x}} \cdot \dfrac{\sqrt[4]{x^3}}{\sqrt[4]{x^3}} = \dfrac{x\sqrt[4]{x^3}}{\sqrt[4]{x^4}} = \dfrac{x\sqrt[4]{x^3}}{x} = \sqrt[4]{x^3}$

3. $\dfrac{\sqrt[5]{a^3}}{\sqrt[5]{b^2}} = \dfrac{\sqrt[5]{a^3}}{\sqrt[5]{b^2}} \cdot \dfrac{\sqrt[5]{b^3}}{\sqrt[5]{b^3}} = \dfrac{\sqrt[5]{a^3b^3}}{\sqrt[5]{b^5}} = \dfrac{\sqrt[5]{a^3b^3}}{b}$

4. $\dfrac{1}{\sqrt[5]{a^2b}}$

Since there are two different factors under the radical, the radicand of the radical that we multiply by must contain a and b with exponents such that the resulting radicand in the denominator is a^5b^5. ($a^2b \cdot a^3b^4 = a^{2+3}b^{1+4} = a^5b^5$) Therefore, we multiply the numerator and the denominator by $\sqrt[5]{a^3b^4}$.

$$= \frac{1}{\sqrt[5]{a^2b}} \cdot \frac{\sqrt[5]{a^3b^4}}{\sqrt[5]{a^3b^4}} = \frac{\sqrt[5]{a^3b^4}}{\sqrt[5]{a^2ba^3b^4}} = \frac{\sqrt[5]{a^3b^4}}{\sqrt[5]{a^5b^5}}$$

Eliminate the radical in the denominator

$$= \frac{\sqrt[5]{a^3b^4}}{ab}$$

▶ *Quick check* Assume that all variables represent positive real numbers.
Simplify $\dfrac{a}{\sqrt[3]{a}}$ and $\sqrt{\dfrac{x^5}{y}}$. ■

The following is a summary of the conditions necessary for a radical expression to be in **simplest form,** also called **standard form.**

___ *Standard form for a radical expression* _____

1. The radicand contains no factors that can be written to an exponent greater than or equal to the index. ($\sqrt[3]{a^4}$ violates this.)
2. The exponent of the radicand and the index of the radical have no common factor other than 1. ($\sqrt[9]{a^6}$ violates this.)
3. The radicand contains no fractions. $\left(\sqrt{\dfrac{a}{b}} \text{ violates this.}\right)$
4. No radicals appear in the denominator. $\left(\dfrac{1}{\sqrt{a}} \text{ violates this.}\right)$

___ *Mastery points* _____

Can you
- Simplify radicals containing fractions by using the quotient property for radicals?
- Rationalize fractions whose denominators are a single term?

Exercise 5–4

Simplify the following expressions leaving no radicals in the denominator. Assume that all variables represent positive real numbers. See examples 5–4 A, B, and C.

Examples $\sqrt{\dfrac{x^5}{y}}$ $\dfrac{a}{\sqrt[3]{a}}$

Solutions

$= \dfrac{\sqrt{x^5}}{\sqrt{y}}$ Quotient property

$= \dfrac{x^2\sqrt{x}}{\sqrt{y}}$ Product property

$= \dfrac{x^2\sqrt{x}}{\sqrt{y}} \cdot \dfrac{\sqrt{y}}{\sqrt{y}}$ Multiply by $\dfrac{\sqrt{y}}{\sqrt{y}}$

$= \dfrac{x^2\sqrt{xy}}{\sqrt{y^2}}$ Product property

$= \dfrac{x^2\sqrt{xy}}{y}$ Radical symbol is eliminated from the denominator

$= \dfrac{a}{\sqrt[3]{a}} \cdot \dfrac{\sqrt[3]{a^2}}{\sqrt[3]{a^2}}$ Multiply by $\dfrac{\sqrt[3]{a^2}}{\sqrt[3]{a^2}}$ to rationalize

$= \dfrac{a\sqrt[3]{a^2}}{\sqrt[3]{a^3}}$

$= \dfrac{a\sqrt[3]{a^2}}{a}$ Eliminate radical symbol from denominator

$= \sqrt[3]{a^2}$ Reduce fraction

1. $\sqrt{\dfrac{16}{25}}$ **2.** $\sqrt{\dfrac{25}{49}}$ **3.** $\sqrt{\dfrac{7}{9}}$ **4.** $\sqrt{\dfrac{3}{4}}$ **5.** $\sqrt[3]{\dfrac{8}{27}}$

6. $\sqrt[3]{\dfrac{27}{64}}$ **7.** $\sqrt{\dfrac{a^6}{9}}$ **8.** $\sqrt[3]{\dfrac{4x^9}{y^6}}$ **9.** $\sqrt[3]{\dfrac{2x}{y^{15}}}$ **10.** $\sqrt[5]{\dfrac{32a^3}{b^{15}}}$

11. $\sqrt[7]{\dfrac{x^{21}}{y^7 z^{14}}}$ **12.** $\sqrt[3]{\dfrac{a^7 b^2}{ab^5}}$ **13.** $\sqrt[3]{\dfrac{16x^4}{2xy^6}}$ **14.** $\sqrt[5]{\dfrac{2a^{12}b^4}{64b^9}}$ **15.** $\sqrt[5]{\dfrac{64x^{14}y^6}{x^4 y}}$

16. $\sqrt[4]{\dfrac{b^4 c^9}{a^{11}}}$ **17.** $\sqrt{\dfrac{1}{2}}$ **18.** $\sqrt{\dfrac{1}{3}}$ **19.** $\sqrt{\dfrac{9}{10}}$ **20.** $\sqrt{\dfrac{4}{11}}$

21. $\sqrt{\dfrac{1}{6}}$ **22.** $\sqrt{\dfrac{1}{10}}$ **23.** $\sqrt{\dfrac{9}{50}}$ **24.** $\sqrt{\dfrac{7}{12}}$ **25.** $\dfrac{2}{\sqrt{2}}$

26. $\dfrac{6}{\sqrt{3}}$ **27.** $\dfrac{12}{\sqrt{8}}$ **28.** $\dfrac{18}{\sqrt{27}}$ **29.** $\sqrt[3]{\dfrac{27}{4}}$ **30.** $\sqrt[3]{\dfrac{9}{25}}$

31. $\sqrt[5]{\dfrac{32}{81}}$ **32.** $\sqrt[3]{\dfrac{27}{16}}$ **33.** $\sqrt[4]{\dfrac{81}{64}}$ **34.** $\sqrt[4]{\dfrac{2}{9}}$ **35.** $\sqrt{\dfrac{x^2}{y}}$

36. $\sqrt{\dfrac{1}{b}}$ **37.** $\sqrt{\dfrac{1}{c}}$ **38.** $\sqrt{\dfrac{x^4}{y^3}}$ **39.** $\sqrt[3]{\dfrac{a^3}{b^2}}$ **40.** $\sqrt[3]{\dfrac{b^9}{c}}$

41. $\sqrt[3]{\dfrac{a}{b^2}}$ **42.** $\sqrt[3]{\dfrac{a}{b}}$ **43.** $\sqrt[5]{\dfrac{32x^5}{y^2}}$ **44.** $\dfrac{x^2}{\sqrt[3]{x^2}}$ **45.** $\dfrac{ab}{\sqrt[5]{b^4}}$

46. $\sqrt{\dfrac{a}{bc}}$ **47.** $\sqrt{\dfrac{2x}{yz}}$ **48.** $\sqrt[3]{\dfrac{x^3}{y^2 z}}$ **49.** $\sqrt[3]{\dfrac{8x}{y^2 z}}$ **50.** $\sqrt[3]{\dfrac{1}{2ab^2}}$

51. $\sqrt[7]{\dfrac{1}{16x^2 y^3}}$ **52.** $\sqrt[4]{\dfrac{16}{a^3 b^2}}$ **53.** $\dfrac{x}{\sqrt[5]{x^2 y^4}}$ **54.** $\dfrac{a}{\sqrt[4]{ab^3}}$ **55.** $\dfrac{xy}{\sqrt[3]{xy^2}}$

56. $\dfrac{ab^2}{\sqrt[5]{a^4 b^{12}}}$ **57.** $\dfrac{b^2 c}{\sqrt[7]{b^4 c^3}}$ **58.** $\dfrac{y^2 z^3}{\sqrt[5]{y^7 z^2}}$ **59.** $\sqrt{\dfrac{2y}{x}}\sqrt{\dfrac{x^2}{8}}$ **60.** $\sqrt{\dfrac{3a}{b^3}}\sqrt{\dfrac{ab}{27}}$

61. $\sqrt[3]{\dfrac{16a^7}{b^4}}\sqrt[3]{\dfrac{b}{2a}}$ **62.** $\sqrt[3]{\dfrac{3y}{x^7}}\sqrt[3]{\dfrac{x}{81y^4}}$ **63.** $\sqrt[5]{\dfrac{x^2 y}{z^7}}\sqrt[5]{\dfrac{y^9 z^3}{x^7}}$ **64.** $\sqrt[5]{\dfrac{a^3 b}{c^8}}\sqrt[5]{\dfrac{b^4 c^4}{a^8}}$

Solve the following word problems. Assume that all variables represent positive real numbers.

65. If we wish to construct a sphere of specific volume, V, we can find the length of the radius, r, necessary by the formula

$$r = \sqrt[3]{\dfrac{3V}{4\pi}}$$

Find the radius necessary for a sphere to have a volume of 904.32 cubic units. (Use 3.14 for π.)

66. Use exercise 65 to find r if $V = 113.04$ cubic units. (Use 3.14 for π.)

67. To find the velocity of the center of mass of a rolling cylinder, we use the equation

$$v = \left(\dfrac{4}{3}gh\right)^{1/2}$$

Write the expression in radical form and leave the answer in standard form.

68. When a gas is compressed with no gain or loss of heat, the pressure and volume of the gas are related by the formula

$$p = kv^{-7/5}$$

where p represents pressure, v represents volume, and k is a constant. Express the formula in radical form and leave the answer in standard form.

69. The formula below gives the length s of the side of an isosceles right triangle with hypotenuse c. Express the radical in standard form.

$$s = \sqrt{\dfrac{c^2}{2}}$$

70. If we know the length of the diagonal d of a square, we can find the length of the side s of the square using the formula

$$s = \sqrt{\dfrac{d^2}{2}}$$

Express the radical in standard form.

71. The formula below gives the diagonal length d of a regular hexagon, where f is the distance across the flats. Express the radical in standard form.

$$d = \sqrt{\frac{4f^2}{3}}$$

72. The resonant frequency f of an AC series circuit is given by

$$f = \frac{1}{2\pi\sqrt{LC}}$$

Express the radical in standard form.

73. The average speed v of a molecule of an ideal gas is given by

$$v = \sqrt{\frac{8kT}{\pi m}}$$

where m is the mass, T is the absolute temperature, and k is the Boltzmann constant. Express the radical in standard form.

74. The formula below is used to find the potential energy in a truss. Express the radical in standard form.

$$C = \frac{3KA}{\sqrt[3]{1{,}024L}}$$

Review exercises

Perform the indicated addition or subtraction. See sections 1–6 and 4–3.

1. $5x^2 - 2x + 3x + 4x^2$

2. $5a^2b - ab^2 - 3a^2b$

3. $a^3 + 3a^2 + 4a^3 - a^2$

4. $x^2y - 3xy^2 + 2x^2y - xy^2$

5. $\dfrac{3}{4} + \dfrac{5}{2}$

6. $\dfrac{1}{2x} + \dfrac{3}{x}$

7. $\dfrac{15}{9a} - \dfrac{1}{a}$

8. $\dfrac{5}{3x} + \dfrac{2}{4x}$

5–5 ■ Sums and differences of radicals

We have learned that we can only combine like terms in addition and subtraction. This same idea applies when we are dealing with radicals. *We can only add or subtract like radicals.*

To have like radicals, the following must be true:

1. The radicals must have the same index.
2. The radicands must be the same.

For example, the expressions $-3\sqrt[5]{11}$ and $4\sqrt[5]{11}$ are like radicals since the indices, 5, are the same and the radicands, 11, are the same. The expressions $3\sqrt[5]{19}$ and $3\sqrt[4]{19}$ are not like radicals because the indices are different ($4 \neq 5$), and the expressions $7\sqrt[3]{13}$ and $7\sqrt[3]{14}$ are not like radicals because the radicands are different ($13 \neq 14$).

Addition and subtraction of radicals follow the same procedure as addition and subtraction of algebraic expressions.

To combine like radicals

1. Perform any simplification within the terms.
2. Use the distributive property to combine terms that have like radicals.

■ *Example 5–5 A*

Perform the indicated addition and subtraction. Assume that all variables represent nonnegative real numbers.

1. $3\sqrt{5} + 2\sqrt{5}$

   Since we have like radicals, we can perform the addition by using the distributive property.

   $$3\sqrt{5} + 2\sqrt{5} = (3 + 2)\sqrt{5} = 5\sqrt{5}$$

2. $8\sqrt[5]{x^2} - 2\sqrt[5]{x^2} + 3\sqrt[5]{x^2} = (8 - 2 + 3)\sqrt[5]{x^2} = 9\sqrt[5]{x^2}$

3. $4\sqrt{a} + \sqrt{b} - 2\sqrt{a} + 3\sqrt{b}$

   Using the commutative and associative properties, we group the like radicals

   $$= (4\sqrt{a} - 2\sqrt{a}) + (\sqrt{b} + 3\sqrt{b})$$

   and applying the distributive property, we perform the addition and subtraction.

   $$= (4 - 2)\sqrt{a} + (1 + 3)\sqrt{b}$$
   $$= 2\sqrt{a} + 4\sqrt{b}$$

   Since $\sqrt{a}$ and $\sqrt{b}$ are not like radicals, no further simplification can be performed. ■

   Consider the problem

   $$\sqrt{27} + 4\sqrt{3}$$

It appears that the indicated addition cannot be performed since we do not have like radicals. However we should have observed that the $\sqrt{27}$ can be simplified.

$$\sqrt{27} = \sqrt{9 \cdot 3} = 3\sqrt{3}$$

Our problem then becomes

$$\sqrt{27} + 4\sqrt{3} = 3\sqrt{3} + 4\sqrt{3} = (3 + 4)\sqrt{3} = 7\sqrt{3}$$

Therefore *whenever we are working with radicals, we must be certain that all radicals are in simplest form.*

■ *Example 5–5 B*

Perform the indicated addition and subtraction.

1. $5\sqrt{2} + \sqrt{18}$

   Since $\sqrt{18}$ can be simplified, we have

   $$= 5\sqrt{2} + \sqrt{9 \cdot 2} = 5\sqrt{2} + \sqrt{9}\sqrt{2} = 5\sqrt{2} + 3\sqrt{2}$$

   and applying the distributive property,

   $$= (5 + 3)\sqrt{2} = 8\sqrt{2}$$

2. $\sqrt{27} + \sqrt{12}$ Simplify radicals

   $$= \sqrt{9 \cdot 3} + \sqrt{4 \cdot 3}$$ Look for factors that are squares
   $$= 3\sqrt{3} + 2\sqrt{3}$$ Add like radicals
   $$= 5\sqrt{3}$$

3. $4\sqrt[3]{81} - \sqrt[3]{24}$ Simplify radicals

$$= 4\sqrt[3]{27 \cdot 3} - \sqrt[3]{8 \cdot 3}$$ Look for factors that are cubes
$$= 4 \cdot 3\sqrt[3]{3} - 2\sqrt[3]{3}$$
$$= 12\sqrt[3]{3} - 2\sqrt[3]{3}$$ Subtract like radicals
$$= 10\sqrt[3]{3}$$

▶ **Quick check**   Add $3\sqrt{3} + \sqrt{12}$.   ■

When we perform addition and subtraction of fractions that contain radicals, our first step is to simplify all radicals involved. We can then find the least common denominator and perform the indicated addition and subtraction.

■ *Example 5–5 C*

Perform the indicated addition and subtraction.

1. $\dfrac{2}{3} + \dfrac{1}{\sqrt{3}}$ Simplify the second fraction by rationalizing the denominator.

$$= \frac{2}{3} + \frac{1}{\sqrt{3}} \cdot \frac{\sqrt{3}}{\sqrt{3}} = \frac{2}{3} + \frac{\sqrt{3}}{\sqrt{9}} = \frac{2}{3} + \frac{\sqrt{3}}{3}$$

Since the denominators are the same, we can add the numerators and write the sum over the common denominator.

$$\frac{2}{3} + \frac{\sqrt{3}}{3} = \frac{2 + \sqrt{3}}{3}$$

2. $\dfrac{3}{\sqrt{2}} - \dfrac{2}{\sqrt{5}} = \dfrac{3}{\sqrt{2}} \cdot \dfrac{\sqrt{2}}{\sqrt{2}} - \dfrac{2}{\sqrt{5}} \cdot \dfrac{\sqrt{5}}{\sqrt{5}}$ Rationalize the denominators

$$= \frac{3\sqrt{2}}{\sqrt{4}} - \frac{2\sqrt{5}}{\sqrt{25}} = \frac{3\sqrt{2}}{2} - \frac{2\sqrt{5}}{5}$$

The least common denominator is 10. Therefore we multiply the first fraction by $\dfrac{5}{5}$ and the second fraction by $\dfrac{2}{2}$.

$$\frac{3\sqrt{2}}{2} \cdot \frac{5}{5} - \frac{2\sqrt{5}}{5} \cdot \frac{2}{2} = \frac{15\sqrt{2}}{10} - \frac{4\sqrt{5}}{10}$$

We now have a common denominator and can finish the problem.

$$\frac{15\sqrt{2}}{10} - \frac{4\sqrt{5}}{10} = \frac{15\sqrt{2} - 4\sqrt{5}}{10}$$

▶ **Quick check**   Add $\dfrac{1}{\sqrt{5}} + \dfrac{1}{5}$.   ■

Mastery points

Can you
- Identify like radicals?
- Add and subtract like radicals?
- Add and subtract fractions containing radicals?

Exercise 5–5

Perform the indicated operations and simplify. Assume that all variables represent positive real numbers. See examples 5–5 A, B, and C.

Example $3\sqrt{3} + \sqrt{12}$

Solution
$\begin{aligned}
&= 3\sqrt{3} + \sqrt{4 \cdot 3} && \text{Look for factors that are squares} \\
&= 3\sqrt{3} + 2\sqrt{3} && \text{Product property} \\
&= (3 + 2)\sqrt{3} = 5\sqrt{3} && \text{Combine like radicals}
\end{aligned}$

Example $\dfrac{1}{\sqrt{5}} + \dfrac{1}{5}$

Solution
$\begin{aligned}
&= \dfrac{1}{\sqrt{5}} \cdot \dfrac{\sqrt{5}}{\sqrt{5}} + \dfrac{1}{5} = \dfrac{\sqrt{5}}{\sqrt{25}} + \dfrac{1}{5} = \dfrac{\sqrt{5}}{5} + \dfrac{1}{5} && \text{Rationalize the denominator} \\
&= \dfrac{\sqrt{5} + 1}{5} && \text{Common denominator, add fractions}
\end{aligned}$

1. $7\sqrt{5} + 4\sqrt{5}$
2. $9\sqrt{11} - 2\sqrt{11}$
3. $3\sqrt{3} + 4\sqrt{3}$
4. $10\sqrt{6} - 6\sqrt{6}$
5. $5\sqrt{5} + 7\sqrt{5} - 4\sqrt{5}$
6. $6\sqrt{2} - 4\sqrt{2} + 8\sqrt{2}$
7. $\sqrt{10} + 4\sqrt{10} - 6\sqrt{10}$
8. $8\sqrt{13} - 11\sqrt{13} - 4\sqrt{13}$
9. $5\sqrt[3]{4} + 2\sqrt[3]{4}$
10. $8\sqrt[5]{3} - 4\sqrt[5]{3} + 7\sqrt[5]{3}$
11. $10\sqrt[4]{3} + \sqrt[4]{3} - 5\sqrt[4]{3}$
12. $7\sqrt[3]{11} - 3\sqrt[3]{7} + 2\sqrt[3]{11}$
13. $\sqrt[5]{12} - \sqrt[5]{16} + 4\sqrt[5]{12}$
14. $\sqrt{2a} - 3\sqrt{a} + 4\sqrt{a}$
15. $2\sqrt{3x} - 4\sqrt{2x} + 2\sqrt{3x}$
16. $\sqrt{12} + 4\sqrt{3}$
17. $\sqrt{20} - 3\sqrt{5}$
18. $\sqrt{8} - 4\sqrt{2}$
19. $\sqrt{12} - \sqrt{75}$
20. $2\sqrt{48} - 3\sqrt{27}$
21. $5\sqrt{3} + 4\sqrt{12}$
22. $4\sqrt{7} - 5\sqrt{63}$
23. $2\sqrt{3} + \sqrt{27} - 2\sqrt{12}$
24. $2\sqrt{8} - \sqrt{50} + 5\sqrt{2}$
25. $\sqrt[3]{16} + \sqrt[3]{54}$
26. $\sqrt[3]{81} - \sqrt[3]{24}$
27. $3\sqrt[3]{16} + 5\sqrt[3]{24}$
28. $4\sqrt[3]{54} - 7\sqrt[3]{16}$
29. $\sqrt[3]{81} + 2\sqrt[3]{250}$
30. $\sqrt{50x} + \sqrt{8x}$
31. $\sqrt{32x} - \sqrt{18x}$
32. $4\sqrt{9x} - 5\sqrt{4x}$
33. $6\sqrt{4a^2b} + 5\sqrt{25a^2b}$
34. $-3\sqrt{8a} - 4\sqrt{50a} + 10\sqrt{2a}$
35. $7\sqrt{36a^2b} + 4\sqrt{49a^2b} - 11\sqrt{2b}$
36. $\sqrt[3]{8x^2} - \sqrt[3]{27x^2}$
37. $\sqrt[4]{16a} + \sqrt[4]{81a}$
38. $\sqrt[4]{256b^3} - \sqrt[4]{81b^3}$
39. $-5\sqrt[3]{27a^2} - 4\sqrt[3]{8a^2}$
40. $\sqrt[3]{64x^2y} - 2\sqrt[3]{27x^2y}$
41. $\sqrt[3]{a^6b} + 3a^2\sqrt[3]{b}$
42. $2\sqrt{x^3y} + 5x\sqrt{xy}$
43. $3a^2\sqrt{ab} - a\sqrt{a^3b}$
44. $4xy^2\sqrt{x^3y} + 2x^2y\sqrt{xy^3}$
45. $3a^2b\sqrt{ab^3} - ab^2\sqrt{a^3b}$
46. $\dfrac{1}{2} + \dfrac{1}{\sqrt{2}}$
47. $\dfrac{1}{5} + \dfrac{2}{\sqrt{5}}$
48. $\dfrac{3}{4} + \dfrac{5}{\sqrt{2}}$
49. $\dfrac{4}{9} - \dfrac{2}{\sqrt{3}}$
50. $\dfrac{1}{\sqrt{3}} + \dfrac{1}{\sqrt{5}}$
51. $\dfrac{2}{\sqrt{5}} + \dfrac{3}{\sqrt{6}}$
52. $\dfrac{4}{\sqrt{7}} - \dfrac{2}{\sqrt{14}}$
53. $\dfrac{6}{\sqrt{5}} + \dfrac{2}{\sqrt{10}}$
54. $\dfrac{5}{\sqrt{3}} + \dfrac{2}{\sqrt{6}}$
55. $\dfrac{4}{\sqrt{12}} - \dfrac{1}{\sqrt{48}}$
56. $\dfrac{2}{\sqrt{18}} - \dfrac{4}{\sqrt{50}}$
57. $\dfrac{7}{\sqrt{4x}} - \dfrac{3}{\sqrt{x}}$
58. $\dfrac{5}{\sqrt{9a}} + \dfrac{2}{\sqrt{a}}$
59. $\dfrac{5}{\sqrt{xy}} - \dfrac{4}{\sqrt{x}}$
60. $\dfrac{2}{\sqrt{a}} + \dfrac{3}{\sqrt{ab}}$

Solve the following word problems.

61. We can find the height h of the given figure by finding b from the formula $b = \sqrt{c^2 - s^2}$. If $c = 13$ units and $s = 5$ units, find h.

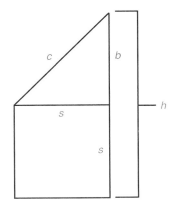

62. Use exercise 61 to find the height of the figure if $c = 10$ feet and $s = 6$ feet.

63. If two hallways of widths h_1 and h_2 meet at right angles, the longest board L that can be carried horizontally around the corner is given by the formula

$$L = \sqrt{(\sqrt[3]{h_1^2} + \sqrt[3]{h_2^2})^3}$$

If h_1 is 8 ft and h_2 is 27 ft, find L. Leave the answer in radical form and also rounded to two decimal places.

64. Use exercise 63 to find L if $h_1 = 6.859$ ft and $h_2 = 21.952$ ft. Leave the answer in radical form and also rounded to two decimal places.

65. The ideal keel length L (in feet) for a hang glider weighing 60 pounds with a pilot weighing P pounds is given by

$$L = \sqrt{\sqrt{2}(60 + P)}$$

If the pilot weighs 175 pounds, find the ideal keel length. Round to two decimal places.

Review exercises

Perform the indicated operations. See section 3-2.

1. $3a(2a - 4)$

2. $(3x - y)(2x + y)$

3. $(a - b)^2$

4. $(a + b)^2$

5. $(2a + b)^2$

6. $(x - y)(x + y)$

7. $(3a - b)(3a + b)$

8. $(4x + 3y)(4x - 3y)$

5–6 ■ *Further operations with radicals*

Multiplying radicals

In section 5–3, we learned the procedure for multiplying two radicals. We now combine those ideas along with the **distributive property,** $a(b + c) = ab + ac$, to perform multiplication of radical expressions that contain more than one term.

■ *Example 5–6 A*

Perform the indicated operations and simplify.

1. $\sqrt{2}(5 + \sqrt{2})$

$= \sqrt{2} \cdot 5 + \sqrt{2} \cdot \sqrt{2}$	Apply distributive property
$= 5\sqrt{2} + \sqrt{4}$	Look for factors that are squares
$= 5\sqrt{2} + 2$	Simplify

2. $\sqrt{3}(\sqrt{15} - \sqrt{21})$

$= \sqrt{3}\sqrt{15} - \sqrt{3}\sqrt{21}$	Apply distributive property
$= \sqrt{45} - \sqrt{63}$	Multiply
$= \sqrt{9 \cdot 5} - \sqrt{9 \cdot 7}$	Look for factors that are squares
$= 3\sqrt{5} - 3\sqrt{7}$	Simplify

3. $(\sqrt{2} + \sqrt{3})(\sqrt{2} + 5\sqrt{3})$

Note In this example, we are multiplying groups together. Therefore, as we did in chapter 3, we will *multiply each term in the first set of parentheses with each term in the second set of parentheses.*

$= \sqrt{2}\sqrt{2} + \sqrt{2} \cdot 5\sqrt{3} + \sqrt{3}\sqrt{2} + \sqrt{3} \cdot 5\sqrt{3}$	Distributive property
$= \sqrt{4} + 5\sqrt{6} + \sqrt{6} + 5\sqrt{9}$	Multiply
$= 2 + 5\sqrt{6} + \sqrt{6} + 5 \cdot 3$	Look for factors that are squares
$= 2 + 5\sqrt{6} + \sqrt{6} + 15$	Multiply
$= (2 + 15) + (5\sqrt{6} + \sqrt{6})$	Combine like terms
$= 17 + 6\sqrt{6}$	

4. $(3 - \sqrt{2})(3 + \sqrt{2})$

$= 3 \cdot 3 + 3\sqrt{2} - 3\sqrt{2} - \sqrt{2}\sqrt{2}$	Apply distributive property
$= 9 + 3\sqrt{2} - 3\sqrt{2} - \sqrt{4}$	Look for factors that are squares
$= 9 + 3\sqrt{2} - 3\sqrt{2} - 2$	
$= 7 + 0$	Combine like terms
$= 7$	

We observe that when we added and subtracted the like terms, there were no longer any radicals in the problem.

▶ *Quick check* Perform the indicated operations and simplify.
$\sqrt{2}(\sqrt{6} - \sqrt{10})$ ■

Conjugate factors and rationalizing denominators

In example 4, we see that there are no radicals in the final answer. The type of factors that we are multiplying, called **conjugate factors,** are derived from the factorization of the special product, called the difference of two squares $[(a + b)(a - b) = a^2 - b^2]$. Conjugate factors are used to rationalize the denominator of a fraction when the denominator has two terms where one or both terms contain a square root. When multiplying conjugate factors, we can simply write our answer as the square of the second term subtracted from the square of the first term.

If we recognize that we are multiplying the factors of the difference of two squares in example 4, the multiplication can be performed as follows:

$$(3 - \sqrt{2})(3 + \sqrt{2}) = (3)^2 - (\sqrt{2})^2 = 9 - 2 = 7$$

We observe that when we combined the like terms, there were no longer any radicals in the answer. Because of this fact, we use the following procedure to rationalize a denominator with two terms where at least one of the terms contains a square root.

> **Rationalizing a denominator that contains square roots and has two terms**
>
> Multiply the numerator and the denominator by the conjugate of the denominator.

To determine the conjugate of a given factor, we write the two terms of the factor and change the sign between them, that is, addition to subtraction, or subtraction to addition.

■ Example 5–6 B

Form the conjugate of the given expressions.

1. $5 - 2\sqrt{3}$ The conjugate is $5 + 2\sqrt{3}$

2. $\sqrt{a} + \sqrt{b}$ The conjugate is $\sqrt{a} - \sqrt{b}$

3. $-3 - \sqrt{7}$ The conjugate is $-3 + \sqrt{7}$ ■

If we wish to rationalize the denominator of the fraction

$$\frac{1}{3 - \sqrt{2}}$$

we recall from example 5–6 A, example 4 that when we multiplied $3 - \sqrt{2}$ by $3 + \sqrt{2}$, there were no radicals left in our product, and this is precisely what we want to occur in our denominator. Therefore to rationalize this fraction, we apply the fundamental principle of fractions and multiply the numerator and the denominator by $3 + \sqrt{2}$, which is the conjugate of the denominator.

$$
\begin{aligned}
\frac{1}{3 - \sqrt{2}} &= \frac{1}{3 - \sqrt{2}} \cdot \frac{3 + \sqrt{2}}{3 + \sqrt{2}} \\
&= \frac{1(3 + \sqrt{2})}{(3)^2 - (\sqrt{2})^2} \\
&= \frac{3 + \sqrt{2}}{9 - 2} \\
&= \frac{3 + \sqrt{2}}{7}
\end{aligned}
$$

■ *Example 5–6 C*

Simplify the following by rationalizing the denominator. Assume that all variables represent positive real numbers, and that no denominator is equal to zero.

1. $\dfrac{5}{\sqrt{13} - \sqrt{3}}$

The conjugate is $\sqrt{13} + \sqrt{3}$

$$= \dfrac{5}{\sqrt{13} - \sqrt{3}} \cdot \dfrac{\sqrt{13} + \sqrt{3}}{\sqrt{13} + \sqrt{3}}$$

Use the special product

$$= \dfrac{5(\sqrt{13} + \sqrt{3})}{(\sqrt{13})^2 - (\sqrt{3})^2} = \dfrac{5(\sqrt{13} + \sqrt{3})}{13 - 3} = \dfrac{5(\sqrt{13} + \sqrt{3})}{10}$$

There is a common factor of 5 and we can reduce the fraction.

$$= \dfrac{5(\sqrt{13} + \sqrt{3})}{5 \cdot 2} = \dfrac{\sqrt{13} + \sqrt{3}}{2}$$

2. $\dfrac{\sqrt{3}}{5 - 2\sqrt{3}}$

The conjugate is $5 + 2\sqrt{3}$

$$= \dfrac{\sqrt{3}}{5 - 2\sqrt{3}} \cdot \dfrac{5 + 2\sqrt{3}}{5 + 2\sqrt{3}}$$

Use the special product

$$= \dfrac{\sqrt{3}(5 + 2\sqrt{3})}{(5)^2 - (2\sqrt{3})^2} = \dfrac{\sqrt{3}(5 + 2\sqrt{3})}{25 - 2^2(\sqrt{3})^2} = \dfrac{\sqrt{3}(5 + 2\sqrt{3})}{25 - 4 \cdot 3}$$

$$= \dfrac{\sqrt{3}(5 + 2\sqrt{3})}{25 - 12} = \dfrac{\sqrt{3}(5 + 2\sqrt{3})}{13}$$

Multiply in the numerator

$$= \dfrac{5\sqrt{3} + 2\sqrt{9}}{13} = \dfrac{5\sqrt{3} + 2 \cdot 3}{13} = \dfrac{5\sqrt{3} + 6}{13}$$

3. $\dfrac{\sqrt{ab}}{\sqrt{a} + \sqrt{b}}$

The conjugate is $\sqrt{a} - \sqrt{b}$

$$= \dfrac{\sqrt{ab}}{\sqrt{a} + \sqrt{b}} \cdot \dfrac{\sqrt{a} - \sqrt{b}}{\sqrt{a} - \sqrt{b}}$$

Use the special product

$$= \dfrac{\sqrt{ab}(\sqrt{a} - \sqrt{b})}{(\sqrt{a})^2 - (\sqrt{b})^2} = \dfrac{\sqrt{ab}(\sqrt{a} - \sqrt{b})}{a - b}$$

Multiply in numerator and simplify

$$= \dfrac{\sqrt{a^2 b} - \sqrt{ab^2}}{a - b} = \dfrac{a\sqrt{b} - b\sqrt{a}}{a - b}$$

▶ *Quick check* Simplify $\dfrac{3}{\sqrt{15} - 3}$ by rationalizing the denominator. ■

── *Mastery points* ──────────

Can you
■ Multiply radical expressions containing more than one term?
■ Form conjugate factors?
■ Multiply conjugate factors?
■ Rationalize a denominator that has two terms in which one or both terms contain a square root?

Exercise 5-6

Perform the indicated operations and simplify. Assume that all variables represent positive real numbers and no denominator is equal to zero. See examples 5-6 A, B, and C.

Examples $\sqrt{2}(\sqrt{6} - \sqrt{10})$ $\dfrac{3}{\sqrt{15} - 3}$

Solutions $= \sqrt{2}\sqrt{6} - \sqrt{2}\sqrt{10}$ Distributive property

$= \sqrt{12} - \sqrt{20}$ Simplify

$= \sqrt{4 \cdot 3} - \sqrt{4 \cdot 5}$ Look for factors that are squares

$= 2\sqrt{3} - 2\sqrt{5}$ Simplify radicals

$= \dfrac{3}{\sqrt{15} - 3} \cdot \dfrac{\sqrt{15} + 3}{\sqrt{15} + 3}$ The conjugate is $\sqrt{15} + 3$

$= \dfrac{3(\sqrt{15} + 3)}{(\sqrt{15})^2 - (3)^2} = \dfrac{3(\sqrt{15} + 3)}{15 - 9}$ Use the special products property

$= \dfrac{3(\sqrt{15} + 3)}{6} = \dfrac{\sqrt{15} + 3}{2}$ Common factor of 3, reduce the fraction

1. $3(\sqrt{5} + \sqrt{3})$

2. $5(\sqrt{2} - \sqrt{3})$

3. $4(3\sqrt{7} + \sqrt{2})$

4. $\sqrt{5}(2\sqrt{11} - 3\sqrt{7})$

5. $\sqrt{3}(\sqrt{2} + \sqrt{5})$

6. $\sqrt{6}(\sqrt{7} - \sqrt{5})$

7. $\sqrt{2}(3\sqrt{3} - \sqrt{11})$

8. $2\sqrt{3}(3\sqrt{2} + 4\sqrt{5})$

9. $5\sqrt{5}(7\sqrt{2} - 4\sqrt{3})$

10. $\sqrt{6}(\sqrt{2} + \sqrt{3})$

11. $\sqrt{5}(\sqrt{15} - \sqrt{20})$

12. $\sqrt{14}(\sqrt{35} + \sqrt{10})$

13. $2\sqrt{7}(\sqrt{35} - 4\sqrt{14})$

14. $5\sqrt{3}(2\sqrt{3} - \sqrt{15})$

15. $\sqrt{x}(\sqrt{x} + \sqrt{y})$

16. $\sqrt{a}(\sqrt{ab} - \sqrt{b})$

17. $3\sqrt{x}(2\sqrt{xy} - 5\sqrt{x})$

18. $6\sqrt{ab}(2\sqrt{a} - 3\sqrt{b})$

19. $5\sqrt{xy}(\sqrt{x} + 4\sqrt{y})$

20. $(5 + \sqrt{2})(3 - \sqrt{2})$

21. $(3 + \sqrt{3})(2 + \sqrt{3})$

22. $(6 - \sqrt{6})(6 - \sqrt{6})$

23. $(4 + \sqrt{x})(5 + \sqrt{x})$

24. $(3 - 4\sqrt{x})(4 - 2\sqrt{x})$

25. $(1 + 2\sqrt{y})(3 - 4\sqrt{y})$

26. $(-3 + 5\sqrt{a})(-2 - \sqrt{a})$

27. $(4 - \sqrt{2})(4 + \sqrt{2})$

28. $(4 - \sqrt{5})(4 + \sqrt{5})$

29. $(3 - 2\sqrt{3})(3 + 2\sqrt{3})$

30. $(5 + 4\sqrt{5})(5 - 4\sqrt{5})$

31. $(\sqrt{5} - \sqrt{7})(\sqrt{5} + \sqrt{7})$

32. $(\sqrt{10} - \sqrt{5})(\sqrt{10} + \sqrt{5})$

33. $(\sqrt{a} - b)(\sqrt{a} + b)$

34. $(2\sqrt{a} - \sqrt{b})(2\sqrt{a} + \sqrt{b})$

35. $(3\sqrt{x} - 4\sqrt{y})(3\sqrt{x} + 4\sqrt{y})$

36. $(3 + \sqrt{5})^2$

37. $(2 - \sqrt{7})^2$

38. $(4\sqrt{3} + 2)^2$

39. $(5\sqrt{3} - 4)^2$

40. $(\sqrt{x} - 2y)^2$

41. $(2\sqrt{a} + b)^2$

42. $(3\sqrt{a} + 2\sqrt{b})(\sqrt{a} - 3\sqrt{b})$

43. $(4\sqrt{x} - \sqrt{y})(5\sqrt{x} + \sqrt{y})$

44. $(2\sqrt{a} + \sqrt{b})(3\sqrt{a} - \sqrt{b})$

45. $\dfrac{1}{\sqrt{3} + 2}$

46. $\dfrac{1}{\sqrt{5} - 2}$

47. $\dfrac{6}{4 + \sqrt{6}}$

48. $\dfrac{6}{3 - \sqrt{6}}$

49. $\dfrac{4}{\sqrt{10} - \sqrt{6}}$

50. $\dfrac{5}{\sqrt{11} - \sqrt{6}}$

51. $\dfrac{3}{4 - 2\sqrt{3}}$

52. $\dfrac{5}{6 - 2\sqrt{5}}$

53. $\dfrac{\sqrt{6}}{\sqrt{2} - 2\sqrt{3}}$

54. $\dfrac{\sqrt{10}}{\sqrt{5} + 2\sqrt{2}}$

55. $\dfrac{\sqrt{14}}{3\sqrt{7} - 2\sqrt{2}}$

56. $\dfrac{\sqrt{2}}{3\sqrt{6} - \sqrt{3}}$

57. $\dfrac{\sqrt{x}}{\sqrt{x} + \sqrt{y}}$

58. $\dfrac{\sqrt{a}}{\sqrt{a} + \sqrt{b}}$

59. $\dfrac{\sqrt{ab}}{\sqrt{ab} + \sqrt{b}}$

60. $\dfrac{\sqrt{a} + b}{\sqrt{a} - b}$

61. $\dfrac{\sqrt{x} - y}{\sqrt{x} + y}$

62. $\dfrac{a + \sqrt{b}}{a - \sqrt{b}}$

63. $\dfrac{a}{\sqrt{ab} - \sqrt{a}}$

64. $\dfrac{x^2 y}{x\sqrt{x} - \sqrt{xy}}$

65. $\dfrac{2a}{\sqrt{a} - \sqrt{ab}}$

66. $\dfrac{3x}{\sqrt{x} - \sqrt{xy}}$

67. $\dfrac{\sqrt{a}}{\sqrt{ab} - \sqrt{a}}$

68. $\dfrac{2\sqrt{x}}{3\sqrt{x} - 2\sqrt{y}}$

69. The electric-field intensity on the axis of a uniform charged ring is given by

$$E = \frac{T}{(x^2 + r^2)^{3/2}}$$

where T is the total charge on the ring and r is the radius of the ring. Express the rational exponent in radical form and leave the answer in standard form.

Review exercises

Perform the indicated operations. See section 3–2.

1. $(2a + b)(a - b)$ **2.** $(a - 2b)(a - b)$ **3.** $(2a + 3b)(2a - 3b)$ **4.** $(a + 3)^2$

Simplify the following. See section 5–3.

5. $\sqrt{16}$ **6.** $\sqrt{20}$ **7.** $\sqrt{6}\sqrt{6}$ **8.** $\sqrt{24}\sqrt{3}$

5–7 ■ Complex numbers

Imaginary numbers

In this section, we will examine what happens when we try to take the square root of a negative number. The expression $\sqrt{-4}$ has no meaning in the system of real numbers because there is no real number that when multiplied by itself equals a negative number, in this case, -4. However there are situations in which an answer for such a problem is required. For example, we need to find the square root of a negative number if we want to find the solution to the equation $x^2 + 4 = 0$. In electronics, the impedance of a circuit, which is the total effective resistance to the flow of current caused by a combination of elements in the circuit, requires that we find the square root of a negative number. We define a new number to provide the required result.

— **Definition of i** —

The number i is a number such that

$$i = \sqrt{-1}$$

and

$$i^2 = -1$$

We can use this definition to define the square root of any negative number.

Definition of $\sqrt{-b}$ _____

For any positive real number b, we define
$$\sqrt{-b} = i\sqrt{b}$$

We now define the system of imaginary numbers as the set of all numbers that can be expressed in the form bi, where b is an element of the set of real numbers.

■ *Example 5–7 A*

Simplify the following.

1. $\sqrt{-4} = i\sqrt{4}$ The first step is always to rewrite $\sqrt{-b}$ as $i\sqrt{b}$
    $= 2i$ Simplify the radical

2. $\sqrt{-2} = i\sqrt{2}$ Rewrite $\sqrt{-b}$ as $i\sqrt{b}$, the $\sqrt{2}$ will not simplify

Note   Whenever we are dealing with the square root of a negative number, we must express our problem in terms of i before proceeding.   ■

If we wish to check our results, we can square the answer to get the original radicand back.

■ *Example 5–7 B*

Simplify the following.

1. $(2i)^2 = 2^2 \cdot i^2$ Square each factor
    $= 4 \cdot (-1)$ i^2 is replaced with -1
    $= -4$ Simplify

2. $(\sqrt{2}\,i)^2 = (\sqrt{2})^2 \cdot i^2$ Square each factor
    $= 2(-1) = -2$ Replace i^2 with -1 and simplify

3. $\sqrt{-2}\sqrt{-8} = i\sqrt{2} \cdot i\sqrt{8}$ Rewrite $\sqrt{-b}$ as $i\sqrt{b}$
    $= i^2\sqrt{16}$ Multiply $i \cdot i$ and multiply radicals
    $= (-1)(4) = -4$ Replace i^2 with -1 and simplify

4. $\sqrt{-6}\sqrt{-3} = i\sqrt{6} \cdot i\sqrt{3}$ Rewrite $\sqrt{-b}$ as $i\sqrt{b}$
    $= i^2\sqrt{18}$ Multiply $i \cdot i$ and multiply radicals
    $= (-1) \cdot 3\sqrt{2} = -3\sqrt{2}$ Replace i^2 with -1 and simplify radical

▶ *Quick check*   Simplify. $\sqrt{-5}\sqrt{-15}$ ■

If we apply the properties of exponents to different exponents of i, we have

$$
\text{A cycle} \begin{cases} i = i \\ i^2 = -1 \\ i^3 = i^2 i = (-1)i = -i \\ i^4 = i^2 i^2 = (-1)(-1) = 1 \end{cases}
\qquad
\text{A cycle} \begin{cases} i^5 = i^4 i = 1 \cdot i = i \\ i^6 = i^4 i^2 = 1(-1) = -1 \\ i^7 = i^4 i^3 = 1(-i) = -i \\ i^8 = i^4 i^4 = 1 \cdot 1 = 1 \end{cases}
$$

It can be seen that the powers of i go through the cycle of i, -1, $-i$, and 1. Using this fact, it is possible to simplify i raised to any positive integer power.

■ *Example 5–7 C*

Simplify.

1. $i^{10} = i^4 \cdot i^4 \cdot i^2 = 1 \cdot 1 \cdot (-1) = -1$

2. $i^{15} = i^4 \cdot i^4 \cdot i^4 \cdot i^3 = 1 \cdot 1 \cdot 1 \cdot (-i) = -i$

3. $i^{20} = i^4 i^4 i^4 i^4 i^4 = 1 \cdot 1 \cdot 1 \cdot 1 \cdot 1 = 1$

Replace i^4 with 1
Replace i^3 with $-i$
Replace i^2 with -1 ■

From these examples, we can see that **when we simplify i to a positive integer power, the resulting power of i is the remainder when we divide the original power by 4.**

■ *Example 5–7 D*

Simplify.

1. $i^{50} = i^2 = -1$ because $50 \div 4 = 12$ Remainder 2

2. $i^{79} = i^3 = -i$ because $79 \div 4 = 19$ Remainder 3

3. $i^{21} = i$ because $21 \div 4 = 5$ Remainder 1   ■

Complex numbers

Now let us define a new type of number that combines the system of real numbers and the system of imaginary numbers. These new numbers are called **complex numbers** and are composed of a real part denoted by a and an imaginary part denoted by b.

Definition of a complex number

A complex number is any number that can be written in the form $a + bi$, where a and b are real numbers and i represents $\sqrt{-1}$.

$$a + bi \longleftarrow \sqrt{-1}$$

Real part   Imaginary part

$a + bi$ is called the **standard form** of a complex number.

If $a = 0$, $0 + bi = bi$ Imaginary number

If $b = 0$, $a + 0i = a$ Real number

Figure 5–1 shows the relationship among the various sets of numbers that we have studied.

Operations with complex numbers

The commutative, associative, and distributive properties for real numbers are also valid for complex numbers. If we wish to perform addition and subtraction of complex numbers, we do so by combining the real parts and combining the imaginary parts.

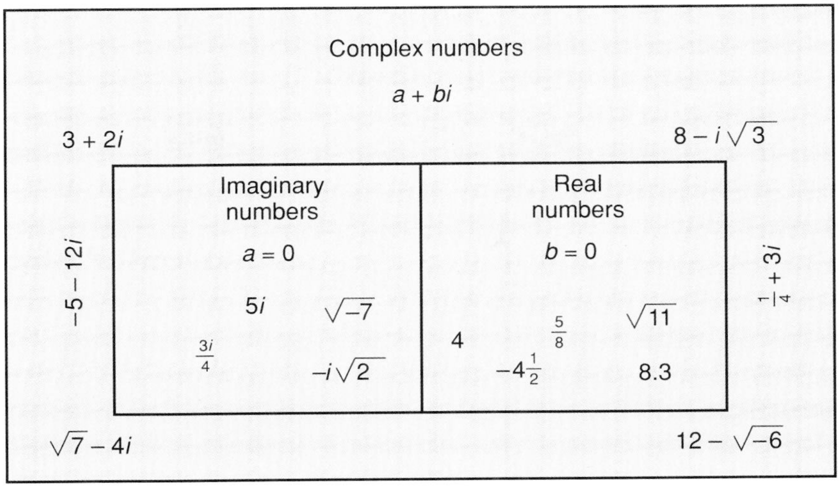

Complex numbers

$a + bi$

$3 + 2i$ $8 - i\sqrt{3}$

Imaginary numbers

$a = 0$

$5i$ $\sqrt{-7}$

$\frac{3i}{4}$ $-i\sqrt{2}$

Real numbers

$b = 0$

4 $\frac{5}{8}$ $\sqrt{11}$

$-4\frac{1}{2}$ 8.3

$-5 - 12i$ $\frac{1}{4} + 3i$

$\sqrt{7} - 4i$ $12 - \sqrt{-6}$

Figure 5–1

Addition or subtraction of complex numbers

1. Combine the real parts.
2. Combine the imaginary parts.
3. Leave the result in the form $a + bi$.

$$(a + bi) + (c + di) = (a + c) + (b + d)i$$
$$(a + bi) - (c + di) = (a - c) + (b - d)i$$

■ *Example 5–7 E*

Perform the indicated addition and subtraction.

1. $(4 + 5i) + (2 + 3i)$
To perform the addition, we add the real parts $(4 + 2)$ and the imaginary parts $(5 + 3)$.

$$= (4 + 2) + (5 + 3)i$$
$$= 6 + 8i$$

2. $(6 + 11i) - (5 + 2i)$
To perform the subtraction, we subtract the real parts $(6 - 5)$ and the imaginary parts $(11 - 2)$.

$$= (6 - 5) + (11 - 2)i$$
$$= 1 + 9i$$

▶ *Quick check*   Add $(7 + 6i) + (1 + 2i)$ ■

When we are multiplying two complex numbers such as

$$(a + bi)(c + di)$$

we multiply each term in the first parentheses with each term in the second parentheses.

> ── *Multiplication of two complex numbers* ──────
> 1. Multiply the numbers as if they are two binomials.
> 2. Substitute −1 for i^2.
> 3. Combine the like terms and leave the result in the form $a + bi$.

■ *Example 5–7 F*

Perform the indicated multiplication.

1. $(2 + 3i)(5 + 2i)$

$= 2 \cdot 5 + 2 \cdot 2i + 3i \cdot 5 + 3i \cdot 2i$ — Distribute the multiplication

$= 10 + 4i + 15i + 6i^2$ — Simplify

$= 10 + 4i + 15i - 6$ — Replace i^2 with −1, then $6i^2 = 6(-1) = -6$

$= (10 - 6) + (4 + 15)i$ — Combine the real parts, combine the imaginary parts

$= 4 + 19i$ — Standard form

2. $(3 - 4i)(3 + 4i)$

$= 3 \cdot 3 + 3 \cdot 4i - 4i \cdot 3 - 4i \cdot 4i$ — Distribute the multiplication

$= 9 + 12i - 12i - 16i^2$ — Simplify

$= 9 + 12i - 12i + 16$ — Replace i^2 with −1, then $-16i^2 = (-16)(-1) = 16$

$= (9 + 16) + (12 - 12)i$ — Combine the real parts, combine the imaginary parts

$= 25 + 0i$ — The imaginary part is 0

$= 25$ — The product is a real number

▶ *Quick check* Multiply $(1 + 2i)(6 + 3i)$ ■

The factors $(3 - 4i)$ and $(3 + 4i)$ are called **complex conjugates** and their product will be a real number. We use complex conjugates to find the quotient of two complex numbers. Consider the quotient

$$(2 + 3i) \div (3 - 4i) = \frac{2 + 3i}{3 - 4i}$$

We would like to perform the division of the two complex numbers and leave the answer in the standard form $a + bi$. First of all, we will eliminate the i in the denominator, since this is just another form of the radical $\sqrt{-1}$. To rationalize the denominator, we multiply by the conjugate of $3 - 4i$, which is $3 + 4i$.

$$\frac{2 + 3i}{3 - 4i} \cdot \frac{3 + 4i}{3 + 4i} = \frac{2 \cdot 3 + 2 \cdot 4i + 3i \cdot 3 + 3i \cdot 4i}{(3)^2 - (4i)^2}$$

$$= \frac{6 + 8i + 9i + 12i^2}{9 - 16i^2}$$

Replacing i^2 with −1, $12i^2$ becomes −12 and $-16i^2$ becomes 16.

$$= \frac{6 + 8i + 9i - 12}{9 + 16}$$

Adding the like terms, we have

$$= \frac{-6 + 17i}{25}$$

Since the answer is to be stated in the form $a + bi$, we divide each term in the numerator by 25 to obtain

$$= \frac{-6}{25} + \frac{17}{25}i$$

This is the quotient of the two complex numbers stated in standard form.

> ___ *Division of one complex number by another complex* ___
> *number*
> 1. Write the division as a fraction.
> 2. Multiply the numerator and the denominator by the conjugate of the denominator.
> 3. Multiply and simplify in the numerator. Use the special product property to simplify the denominator to a real number.
> 4. Write the result in the form $a + bi$.

■ *Example 5–7 G*

Perform the indicated division.

1. $\dfrac{1 + \sqrt{-4}}{3 - \sqrt{-9}}$

We must first simplify the radicals before carrying out the division.

$$= \frac{1 + i\sqrt{4}}{3 - i\sqrt{9}} = \frac{1 + 2i}{3 - 3i}$$ The conjugate of the denominator is $3 + 3i$

$$= \frac{1 + 2i}{3 - 3i} \cdot \frac{3 + 3i}{3 + 3i} = \frac{3 + 3i + 6i + 6i^2}{(3)^2 - (3i)^2}$$

$$= \frac{3 + 3i + 6i - 6}{9 - 9i^2} = \frac{-3 + 9i}{9 + 9} = \frac{-3 + 9i}{18}$$

$$= \frac{-3}{18} + \frac{9}{18}i = \frac{-1}{6} + \frac{1}{2}i$$

2. $\dfrac{-3 - 2i}{i} = \dfrac{-3 - 2i}{i} \cdot \dfrac{-i}{-i}$ The denominator is an imaginary number and can be written as $0 + i$, its conjugate is $0 - i$ or just $-i$

$$= \frac{3i + 2i^2}{-i^2} = \frac{3i - 2}{1}$$ Multiply and simplify

$$= \frac{3i}{1} - \frac{2}{1} = 3i - 2 = -2 + 3i$$ Standard form   ■

___ *Mastery points* ___

Can you
- Simplify the square root of a negative number?
- Simplify i raised to a positive integer power?
- Add, subtract, multiply, and divide complex numbers?

Exercise 5–7

Simplify the following. See examples 5–7 A, B, C, and D.

Example $\sqrt{-5}\sqrt{-15}$

Solution $= i\sqrt{5} \cdot i\sqrt{15}$
$= i^2\sqrt{75}$
$= (-1)5\sqrt{3}$
$= -5\sqrt{3}$

1. $\sqrt{-9}$ **2.** $\sqrt{-16}$ **3.** $\sqrt{-12}$ **4.** $\sqrt{-18}$ **5.** $(3i)^2$

6. $(4i)^2$ **7.** $(\sqrt{3}\,i)^2$ **8.** $(\sqrt{3}\,i)^2$ **9.** $\sqrt{-3}\sqrt{-5}$ **10.** $\sqrt{-7}\sqrt{-11}$

11. $\sqrt{-2}\sqrt{-2}$ **12.** $\sqrt{-6}\sqrt{-6}$ **13.** $(\sqrt{-5})^2$ **14.** $(\sqrt{-4})^2$ **15.** $\sqrt{-3}\sqrt{-12}$

16. $\sqrt{-2}\sqrt{-18}$ **17.** $\sqrt{-3}\sqrt{-15}$ **18.** $\sqrt{-7}\sqrt{-14}$ **19.** i^{10} **20.** i^{15}

21. i^{44} **22.** i^{27} **23.** i^{19} **24.** i^{60}

Perform the indicated operations and leave the answer in standard form. See examples 5–7 E, F, and G.

Example $(7 + 6i) + (1 + 2i)$

Solution $= (7 + 1) + (6 + 2)i$ Add the real parts, add the imaginary parts
$= 8 + 8i$ Standard form

Example $(1 + 2i)(6 + 3i)$

Solution $= 1 \cdot 6 + 1 \cdot 3i + 2i \cdot 6 + 2i \cdot 3i$ Distributive property
$= 6 + 3i + 12i + 6i^2$ Simplify
$= 6 + 3i + 12i - 6$ Replace i^2 with -1, then $6i^2 = 6(-1) = -6$
$= (6 - 6) + (3 + 12)i$ Combine the real parts, combine the imaginary parts
$= 0 + 15i$ The real part is 0
$= 15i$ The product is an imaginary number

25. $(6 + 3i) + (2 + 4i)$ **26.** $(4 + 3i) + (1 + i)$

27. $(6 - 2i) - (8 - 4i)$ **28.** $(4 - 5i) - (3 - 7i)$

29. $(2 + \sqrt{-49}) - (1 - \sqrt{-1})$ **30.** $(9 + \sqrt{-64}) - (9 - \sqrt{-9})$

31. $(4 - \sqrt{-25}) - (4 - \sqrt{-36})$ **32.** $[(2 + 5i) + (3 - 2i)] + (3 - i)$

33. $[(-2 - i) + (3 + 2i)] + (4 - 3i)$ **34.** $[(8 - 5i) - (5 + 4i)] + (6 - 7i)$

35. $[(9 - i) - (6 - 4i)] + (5 + 5i)$

36. $(4 + 3i)(1 + i)$ **37.** $(3 - 2i)(3 + 2i)$ **38.** $(4 + 5i)(4 - 5i)$

39. $(3 + i)(5 - 4i)$ **40.** $(2 + \sqrt{-16})(3 - \sqrt{-25})$ **41.** $(7 + \sqrt{-1})(3 + \sqrt{-4})$

42. $(5 - \sqrt{-25})(4 + \sqrt{-16})$ **43.** $(5 - \sqrt{-9})(5 + \sqrt{-9})$ **44.** $(2 + i)^2$

45. $(4 - 3i)^2$ **46.** $(3 - \sqrt{-9})^2$ **47.** $(2 + \sqrt{-4})^2$ **48.** $\dfrac{3 - 2i}{i}$ **49.** $\dfrac{4 + 5i}{i}$

50. $\dfrac{6 - 2i}{3i}$ **51.** $\dfrac{2 + 4i}{2i}$ **52.** $\dfrac{4 - 9i}{\sqrt{-1}}$ **53.** $\dfrac{5 + 7i}{\sqrt{-9}}$ **54.** $\dfrac{4 - 3i}{1 + i}$

55. $\dfrac{5 - 2i}{5 - i}$ **56.** $\dfrac{4 - 5i}{2 + i}$ **57.** $\dfrac{3 - i}{3 + i}$ **58.** $\dfrac{5 - i}{5 + i}$ **59.** $\dfrac{2 + 5i}{3 - 2i}$

60. $\dfrac{4 + 3i}{3 - i}$ **61.** $\dfrac{6 + 3i}{3 + 4i}$ **62.** $\dfrac{5 - \sqrt{-4}}{3 + \sqrt{-9}}$ **63.** $\dfrac{2 + \sqrt{-16}}{4 - \sqrt{-1}}$ **64.** $\dfrac{7 - \sqrt{-25}}{3 + \sqrt{-36}}$

Solve the following word problems.

65. The impedance of an electrical circuit is the measure of the total opposition to the flow of an electric current. The impedance Z in a series RCL circuit is given by

$$Z = R + i(X_L - X_C)$$

Determine the impedance if $R = 30$ ohms, $X_L = 16$ ohms, and $X_C = 40$ ohms.

66. Use the formula in exercise 65 to find Z if $R = 28$ ohms, $X_L = 16$ ohms, and $X_C = 38$ ohms.

67. The impedance Z in an AC circuit is given by the formula

$$Z = \dfrac{V}{I}$$

where V is the voltage and I is the current. Find Z when $V = 0.3 + 1.2i$ and $I = 2.1i$. Round all values to three decimal places.

68. Use the formula in exercise 67 to find Z if $V = 2.2 - 0.3i$ and $I = -1.1i$. Round all values to three decimal places.

69. The total impedance Z_T of an AC circuit containing impedances Z_1 and Z_2 in parallel is given by the formula

$$Z_T = \dfrac{Z_1 Z_2}{Z_1 + Z_2}$$

Find Z_T when $Z_1 = 3 - i$ and $Z_2 = 2 + i$.

70. Use the formula in exercise 69 to find Z_T if $Z_1 = 4 - i$ and $Z_2 = 3 + i$.

71. If three resistors in an AC circuit are connected in parallel, the total impedance Z_T is given by the formula

$$Z_T = \dfrac{Z_1 Z_2 Z_3}{Z_1 Z_2 + Z_1 Z_3 + Z_2 Z_3}$$

Find Z_T when $Z_1 = 3 - i$, $Z_2 = 3 + i$, and $Z_3 = 2i$.

72. Use the formula in exercise 71 to find Z_T if $Z_1 = 2 - i$, $Z_2 = 2 + i$, and $Z_3 = 2i$.

73. For what values of x does the expression $\sqrt{5 - x}$ represent a real number?

74. For what values of x does the expression $\sqrt{x + 4}$ represent a real number?

75. For what values of x does the expression $\sqrt{x + 11}$ represent an imaginary number?

76. For what values of x does the expression $\sqrt{8 - x}$ represent an imaginary number?

Review exercises

Factor completely. See section 3-8.

1. $x^2 - 12x + 36$ **2.** $3x^2 + 9x$ **3.** $x^2 - 16$ **4.** $9x^2 - 36$

5. $x^2 - 7x + 10$ **6.** $2x^2 - x - 3$ **7.** $5x^2 - 14x - 3$ **8.** $6x^2 - 23x - 4$

Chapter 5 lead-in problem

The formula for approximating the velocity V in miles per hour of a car based on the length of its skid marks S (in feet) on wet pavement is given by

$$V = 2\sqrt{3S}$$

If the skid marks are 147 feet long, what was the velocity of the car?

Solution

$V = 2\sqrt{3S}$	Original formula
$V = 2\sqrt{3(\quad)}$	Formula ready for substitution
$V = 2\sqrt{3(147)}$	Substitute
$V = 2\sqrt{441}$	Simplify under radical
$V = 2 \cdot 21$	Simplify radical
$V = 42$	Multiply

Hence, the velocity of the car was 42 miles per hour.

Chapter 5 summary

1. $a^{1/n} = \sqrt[n]{a}$, whenever the principal n^{th} root of a is a real number.
2. $a^{m/n} = (\sqrt[n]{a})^m = \sqrt[n]{a^m}$, if the principal n^{th} root of a is a real number.
3. For all nonnegative real numbers a and b and positive integer n greater than 1,

$$\sqrt[n]{ab} = \sqrt[n]{a}\sqrt[n]{b}$$

$$\sqrt[n]{\frac{a}{b}} = \frac{\sqrt[n]{a}}{\sqrt[n]{b}}, \qquad b \neq 0$$

and

$$\sqrt[kn]{a^{km}} = \sqrt[n]{a^m}$$

4. We eliminate radicals from the denominator of a fraction by **rationalizing the denominator.**
5. We can only add or subtract *like* radicals.
6. **Conjugate factors** are used to rationalize the denominator of a fraction when the denominator has two terms where one or both terms contain a square root.
7. We define $i = \sqrt{-1}$, so that $i^2 = -1$.
8. A complex number is any number that can be written in the form $a + bi$, where a and b are real numbers and i represents $\sqrt{-1}$.

Chapter 5 error analysis

1. Principal nth root
   Example: $\sqrt{81} = 9$ or -9
   Correct answer: 9
   What error was made? (*see page 216*)
2. Principal nth roots
   Example: $\sqrt[3]{-8}$ does not exist in the real numbers.
   Correct answer: -2
   What error was made? (*see page 216*)
3. Rational number exponents
   Example: $(32)^{-1/5} = -(32)^{1/5} = -2$
   Correct answer: $\dfrac{1}{2}$
   What error was made? (*see page 221*)
4. Operations with rational exponents
   Example: $2^{1/3} \cdot 2^{1/3} = 2^{1/9}$
   Correct answer: $2^{2/3}$
   What error was made? (*see page 224*)
5. Product of radicals
   Example: $\sqrt[4]{3} \cdot \sqrt[4]{2} = \sqrt[4]{5}$
   Correct answer: $\sqrt[4]{6}$
   What error was made? (*see page 228*)
6. Reducing the index
   Example: $\sqrt[6]{x^3y} = \sqrt{xy}$
   Correct answer: $\sqrt[6]{x^3y}$
   What error was made? (*see page 229*)
7. Rationalizing the denominator
   Example: $\dfrac{1}{\sqrt[4]{x}} = \dfrac{\sqrt[4]{x}}{x}$
   Correct answer: $\dfrac{\sqrt[4]{x^3}}{x}$
   What error was made? (*see page 234*)
8. Sum of radicals
   Example: $\sqrt{5} + \sqrt{5} = \sqrt{10}$
   Correct answer: $2\sqrt{5}$
   What error was made? (*see page 238*)
9. Radical of a sum
   Example: $\sqrt{16 + 9} = \sqrt{16} + \sqrt{9} = 4 + 3 = 7$
   Correct answer: 5
   What error was made? (*see page 238*)
10. Multiplying radicals
    Example: $\sqrt{6}(\sqrt{6} - 3) = 3$
    Correct answer: $6 - 3\sqrt{6}$
    What error was made? (*see page 242*)

Chapter 5 critical thinking

Choose a two-digit number whose one's digit is 5 (such as 35).
You can find the square of this number if you multiply the
ten's digit by one more than the ten's digits and place 25 after
this product. For example:

35^2

Ten's digit   One more than the ten's digit   Product   Place 25 after the product

$3 \cdot 4 = 12 \ 25$

Therefore, $35^2 = 1225$
Why is this true?

Chapter 5 review

Assume that all variables represent positive real numbers and no denominator is equal to zero.

[5–1]
Rewrite the following in radical notation and use table 5–1 to simplify.

1. $36^{1/2}$ **2.** $16^{-3/4}$ **3.** $(-27)^{2/3}$

[5–2]
Perform the indicated operations and simplify. Leave the answer with all exponents positive.

4. $a^{2/3} \cdot a^{1/4}$ **5.** $(c^{3/4})^{1/3}$ **6.** $(27x^3)^{2/3}$ **7.** $(b^{-2/3})^{-3}$

8. $\dfrac{a^{1/2}}{a^{1/3}}$ **9.** $(8x^{12}y^6)^{2/3}$ **10.** $\dfrac{x^3 x^{2/3}}{x^{1/2}}$ **11.** $\dfrac{a^{2/3}}{a^{-1/2}}$

[5–3]
Simplify the following.

12. $\sqrt{12}$ **13.** $\sqrt{10}\ \sqrt{15}$ **14.** $\sqrt[5]{x^4}\ \sqrt[5]{x^3}$

15. $\sqrt[3]{6ab^2}\ \sqrt[3]{4a^2b^2}$ **16.** $\sqrt[4]{a^6}$ **17.** $\sqrt[9]{8a^3b^6}$

18. If the hypotenuse of a right triangle is 10 inches long and one
of the legs is 6 inches long, find the length of the other leg.

[5–4]
Simplify the following expressions leaving no radicals in the denominator.

19. $\sqrt{\dfrac{49}{64}}$ **20.** $\sqrt{\dfrac{16}{27}}$ **21.** $\sqrt[3]{\dfrac{16x^3y^2}{z^6}}$ **22.** $\sqrt{\dfrac{1}{8}}$ **23.** $\dfrac{6}{\sqrt{2}}$

24. $\sqrt[3]{\dfrac{8}{25}}$ **25.** $\sqrt{\dfrac{x^2}{y}}$ **26.** $\sqrt[3]{\dfrac{a^2}{b}}$ **27.** $\dfrac{x}{\sqrt[5]{x^2}}$ **28.** $\sqrt[3]{\dfrac{a}{bc^2}}$

29. $\dfrac{a}{\sqrt[4]{ab^2}}$ **30.** $\sqrt{\dfrac{3x}{y^2}}\ \sqrt{\dfrac{xy}{3}}$

[5–5]
Perform the indicated operations and simplify.

31. $4\sqrt{3} - 3\sqrt{3} + 7\sqrt{3}$ **32.** $\sqrt{18} + \sqrt{50}$ **33.** $\sqrt{8a} + 9\sqrt{18a}$

34. $5x^2\sqrt{xy} - 2x\sqrt{x^3y}$ **35.** $\dfrac{5}{\sqrt{6}} - \dfrac{1}{\sqrt{3}}$ **36.** $\dfrac{2}{\sqrt{ab}} - \dfrac{1}{\sqrt{a}}$

[5–6]

Simplify the following expressions leaving no radicals in the denominator.

37. $\sqrt{2}(\sqrt{6} - \sqrt{10})$

38. $2\sqrt{a}(\sqrt{ab} + 2\sqrt{a})$

39. $(5 - \sqrt{5})^2$

40. $(\sqrt{10} - \sqrt{7})(\sqrt{10} + \sqrt{7})$

41. $(2\sqrt{a} + 3\sqrt{b})(2\sqrt{a} - 3\sqrt{b})$

42. $(3\sqrt{x} + y)^2$

43. $\dfrac{1}{\sqrt{6} + 2}$

44. $\dfrac{10}{4 + \sqrt{6}}$

45. $\dfrac{\sqrt{3}}{\sqrt{6} - \sqrt{3}}$

46. $\dfrac{a^2 b}{a\sqrt{a} - \sqrt{ab}}$

[5–7]

Simplify the following.

47. $\sqrt{-49}$

48. $\sqrt{-28}$

49. $(2i)^2$

50. $(\sqrt{7}i)^2$

51. $\sqrt{-3}\,\sqrt{-12}$

52. $(\sqrt{-3})^2$

53. $\sqrt{-2}\,\sqrt{-3}$

54. i^{37}

Simplify the following and leave the answer in standard form.

55. $(4 + 2i) + (3 + 5i)$

56. $(2 - \sqrt{-36}) - (3 + \sqrt{-25})$

57. $(2 - \sqrt{-9})(3 + \sqrt{-16})$

58. $(2 + 5i)^2$

59. $\dfrac{3 + 4i}{i}$

60. $\dfrac{7 - 6i}{\sqrt{-9}}$

61. $\dfrac{4 - i}{2 + i}$

62. $\dfrac{9 - \sqrt{-4}}{7 - \sqrt{-9}}$

Chapter 5 cumulative test

Factor completely.

[3–5]   **1.** $a^2 - 7a - 8$

[3–4]   **2.** $4x^2 - 3x$

[3–7]   **3.** $9x^2 - 36$

[3–6]   **4.** $2x^2 + 11x + 12$

[3–6]   **5.** $3a^2 - 11a - 20$

[3–6]   **6.** $6x^2 + 17x + 12$

[1–5]   **7.** Evaluate the expression $b^2 - 4ac$ at
(a) $a = 1$, $b = 4$, and $c = -3$;
(b) $a = 2$, $b = -4$, and $c = 3$.

Find the solution set.

[2–1]   **8.** $3(2x - 1) + 4 = x + 3$

[2–5]   **9.** $5x + 7 > 2x - 4$

[2–2]   **10.** $3x - 2y = 4(x + y)$

[2–4]   **11.** $|3x - 1| = 4$

[2–6]   **12.** $|2x + 3| > 8$

[2–1]   **13.** $x(x + 1) - (x + 3)^2 = 4$

[2–6]   **14.** $|1 - 4x| \le 7$

Simplify the following and leave in standard form. Assume that all variables represent positive real numbers.

[5–3]   **15.** $\sqrt[5]{64a^{10}b^7}$

[5–7]   **16.** $(3 - 4i)(2 + i)$

[5–3]   **17.** $\sqrt{48}$

[5–2]   **18.** $a^{1/3} \cdot a^{1/4}$

[5–3]   **19.** $\sqrt[3]{8a^4b^6}$

[3–1]   **20.** $(2a^3b^4c)^3$

[5–7]   **21.** $\sqrt{-18}$

[5–6]   **22.** $\dfrac{4}{\sqrt{10} + \sqrt{6}}$

[5–7]   **23.** $\dfrac{1 - 3i}{2 + 3i}$

[5–2]   **24.** $(4a^6)^{1/2}$

[3–3]   **25.** $\dfrac{3x^{-2}y^4}{9x^{-5}y^2}$

[5–4]   **26.** $\sqrt[3]{\dfrac{a^2}{4bc^2}}$

Solve the following word problems.

[2–3] **27.** When the length of a side of a square is increased by 4 inches, the area is increased by 72 square inches. Find the original length of a side.

[2–3] **28.** A metallurgist wishes to form 1,000 kg of an alloy that is 62% copper. This alloy is to be obtained by fusing some alloy that is 80% copper and some that is 50% copper. How many kilograms of each alloy must be used?

[5–1] **29.** In the theory of ballistics, the ballistic limit v of a material is approximated by the formula

$$v = kT^{6/5}$$

where T is the thickness of a sheet of material and k is a constant that is dependent on the material being used. Compute the ballistic limit if $k = 24,000$ and $T = 0.03125$.

[5–1] **30.** Use the formula in exercise 29 to find v if $k = 25,000$ and $T = 0.07776$.

6

Quadratic Equations and Inequalities

The University of Minnesota wishes to set up a rectangular botanical garden. They have 300 meters of fence to enclose 5,000 square meters for the garden. What are the dimensions of the garden?

6–1 ■ *Solution by factoring and extracting roots*

Solution by factoring

So far we have discussed linear (first-degree) equations in one variable having at most one solution. We now consider the solution set of a *quadratic* (second-degree) *equation* in one variable that will have at most *two possible solutions.*

Example

In electrical circuits, the flow of current varies with time. In a certain circuit, this relationship is expressed as

$$I = 12 - 12t^2$$

where I is the current in amperes and t is the time in seconds. When will the current flow be 0? To answer this question, it is necessary to set $I = 0$ in the equation. This yields

$$0 = 12 - 12t^2$$

which is a quadratic equation in the variable t.

┌─── Definition of a quadratic equation in one variable ───────

A **quadratic equation in one variable** is any second-degree equation that can be written in the form

$$ax^2 + bx + c = 0$$

where a, b, and c are real numbers, $a > 0$. We call this the **standard form** of a quadratic equation in one variable, x.

Note It is necessary to restrict a so that it is not equal to zero. If $a = 0$, then we have

$$0 \cdot x^2 + bx + c = 0 \quad \text{or} \quad bx + c = 0$$

and we have a first-degree (linear) equation.

If $b = 0$ or $c = 0$, then the equation is of the form

$$ax^2 + c = 0 \quad \text{or} \quad ax^2 + bx = 0$$

and we still have a second-degree (quadratic) equation. The following are examples of quadratic equations:

$$2x^2 - x + 5 = 0; \quad 5x^2 + 2x = 0; \quad 4x^2 - 12 = 0$$

When the quadratic expression $ax^2 + bx + c$ can be factored, we use the **zero product property,** stated in section 4–1, to solve the quadratic equation. The following procedure outlines the steps for solving quadratic equations using the zero product property.

To solve a quadratic equation by factoring

1. Write the equation in standard form, if it is not given in this form.
2. Completely factor the quadratic expression.
3. Set each factor containing the variable equal to 0 and solve the resulting linear equations.

■ **Example 6–1 A**

Find the solution set of the following quadratic equations.

1. $x^2 + 6 = 7x$

   Since the equation is not in standard form, we subtract $7x$ from each member to obtain the standard form.

   $$x^2 - 7x + 6 = 0$$
   $$(x - 6)(x - 1) = 0 \qquad \text{Factor the left member}$$
   $$x - 6 = 0 \quad \text{or} \quad x - 1 = 0 \qquad \text{Set each factor equal to 0}$$
   $$x = 6 \qquad\qquad x = 1 \qquad \text{Solve resulting linear equations}$$

   The solution set is $\{1,6\}$.

   To check our solutions, as we should always do, substitute 6 for x and then 1 for x in the original equation.

When $x = 6$		When $x = 1$	
$x^2 + 6 = 7x$		$x^2 + 6 = 7x$	
$(6)^2 + 6 = 7(6)$	Replace x with 6	$(1)^2 + 6 = 7(1)$	Replace x with 1
$36 + 6 = 42$		$1 + 6 = 7$	
$42 = 42$	(True)	$7 = 7$	(True)

   We will not show a check in the remaining examples, but we should *always* check the solutions.

2. $6p^2 - 5p = 0$

The equation is already in standard form, so we factor the common factor p.

$$p(6p - 5) = 0 \qquad \text{Factor the left member}$$
$$p = 0 \quad \text{or} \quad 6p - 5 = 0 \qquad \text{Set each factor equal to 0}$$
$$p = 0 \qquad\qquad 6p = 5 \qquad \text{Add 5 to each member}$$
$$p = 0 \qquad\qquad p = \frac{5}{6} \qquad \text{Divide each member by 6}$$

The solution set is $\left\{ 0, \dfrac{5}{6} \right\}$.

Note A common error is to forget the factor p. That is, the solution $p = 0$ is sometimes omitted by students. Be careful!

3. $3q^2 + 4q + \dfrac{4}{3} = 0$

Since we have a rational equation (an equation containing at least one rational term), we clear the denominator by multiplying all terms in each member of the equation by 3.

$$9q^2 + 12q + 4 = 0 \qquad \text{Multiply each member by 3}$$
$$(3q + 2)^2 = 0 \qquad \text{Factor } 9q^2 + 12q + 4$$
$$3q + 2 = 0 \qquad \text{Set the factor equal to 0}$$
$$3q = -2 \qquad \text{Subtract 2 from each member}$$
$$q = -\frac{2}{3} \qquad \text{Divide each member by 3}$$

The solution set is $\left\{ -\dfrac{2}{3} \right\}$.

The linear factor $3q + 2$ appears twice. When this occurs, we say that $-\dfrac{2}{3}$ is a solution of *multiplicity two*.

4. $(2x - 1)(x + 2) = -3$
$$2x^2 + 3x - 2 = -3 \qquad \text{Multiply as indicated}$$
$$2x^2 + 3x + 1 = 0 \qquad \text{Add 3 to each member}$$
$$(2x + 1)(x + 1) = 0 \qquad \text{Factor the left member}$$
$$2x + 1 = 0 \quad \text{or} \quad x + 1 = 0 \qquad \text{Set each factor equal to 0}$$
$$x = -\frac{1}{2} \qquad\qquad x = -1 \qquad \text{Solve for } x$$

The solution set is $\left\{ -1, -\dfrac{1}{2} \right\}$.

5. The power output P of an 80-volt electric generator is defined by $P = 80I - 5I^2$, where I is the current in amperes. What current I is necessary for the power output of 140 watts?

Given $P = 140$, we substitute to obtain the quadratic equation

$$140 = 80I - 5I^2$$

Add $5I^2 - 80I$ to both members.

$5I^2 - 80I + 140 = 0$ Write in standard form
$5(I^2 - 16I + 28) = 0$ Factor the left member
$5(I - 14)(I - 2) = 0$
$I - 14 = 0$ or $I - 2 = 0$ Set each factor equal to 0
$\quad\quad I = 14 \quad\quad\quad\quad I = 2$

The generator will produce 140 watts when $I = 14$ amperes or when $I = 2$ amperes.

Note We do not set 5 equal to 0 since the factor 5 does not contain a variable.

▶ *Quick check* Find the solution set of $3x^2 - 14x = 0$.   ■

Solution by extracting the roots

Consider now the quadratic equation $x^2 - 16 = 0$. Factoring, we obtain

$$(x - 4)(x + 4) = 0$$

Since $x = 4$ when $x - 4 = 0$ and $x = -4$ when $x + 4 = 0$, the solution set is $\{4, -4\}$. We can also solve this equation by writing it in the form $x^2 = 16$. Since 16 is greater than or equal to zero, x is a number that, when squared, yields 16. This can be accomplished when

$$x = \sqrt{16} \quad \text{or} \quad x = -\sqrt{16}$$

since $(\sqrt{16})^2 = 16$ and $(-\sqrt{16})^2 = 16$. Thus the solutions of the quadratic equation $x^2 = 16$ are

$$x = \sqrt{16} = 4 \quad \text{or} \quad x = -\sqrt{16} = -4$$

which are the same values that we determined by factoring.

This example illustrates the square root property.

Square root property

Given real number p and $x^2 = p$, then
$$x = \sqrt{p} \quad \text{or} \quad x = -\sqrt{p}$$

We use this property to solve certain types of quadratic equations by **extracting the roots**.

Note We can write $x = \sqrt{p}$ or $x = -\sqrt{p}$ as $x = \pm\sqrt{p}$, which we read "x equals plus or minus the square root of p."

■ *Example 6–1 B*

Find the solution set by extracting the roots.

1. $x^2 = 13$

$\qquad x = \sqrt{13} \quad$ or $\quad x = -\sqrt{13}$ Extract the roots

The solution set is $\{\sqrt{13}, -\sqrt{13}\}$.

Note We can write the solution set as $\{\pm \sqrt{13}\}$, where "$\pm$" is read "plus or minus."

2. $(3x + 2)^2 = 7$

$\qquad 3x + 2 = \sqrt{7} \quad$ or $\quad 3x + 2 = -\sqrt{7}$ Extract the roots
$\qquad\qquad 3x = -2 + \sqrt{7} \quad$ or $\quad 3x = -2 - \sqrt{7}$ Add -2 to each member
$\qquad\qquad x = \dfrac{-2 + \sqrt{7}}{3} \quad$ or $\quad x = \dfrac{-2 - \sqrt{7}}{3}$ Divide each member by 3

The solution set is $\left\{ \dfrac{-2 - \sqrt{7}}{3}, \dfrac{-2 + \sqrt{7}}{3} \right\}$.

3. $(x - 3)^2 = -4$

$\qquad x - 3 = \sqrt{-4} \quad$ or $\quad x - 3 = -\sqrt{-4}$ Extract the roots

However $\sqrt{-4}$ is not a real number and the equation has *no real number solution*. We have learned that $\sqrt{-4}$ is equal to $2i$, where $i = \sqrt{-1}$. Then

$\qquad x - 3 = 2i \qquad$ or $x - 3 = -2i$ $\sqrt{-4} = 2i$
$\qquad\qquad x = 3 + 2i \qquad\qquad x = 3 - 2i$ Add 3 to each member

The solution set is $\{3 + 2i, 3 - 2i\}$.

Note When the equation is in the form $(kx + \ell)^2 = p$, *do not* perform the indicated multiplication in the left member.

▶ *Quick check* Find the solution set of $(4z + 5)^2 = 11$. ■

Many times when we translate a word problem into mathematical language, we obtain a quadratic equation. Following are some examples.

■ *Example 6–1 C*

Solve the following problems by setting up an equation and solving it.

1. Find the length of each side of a square if the diagonal is 12 centimeters long.

Let $s =$ the length of the side of the square. Using the Pythagorean Theorem for right triangles, we obtain the equation

$\qquad s^2 + s^2 = (12)^2$
$\qquad\quad 2s^2 = 144$
$\qquad\quad\; s^2 = 72$

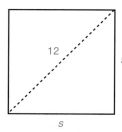

$\qquad s = \sqrt{72} = 6\sqrt{2} \quad$ or $\quad s = -\sqrt{72} = -6\sqrt{2}$ Extract the roots

We reject the negative solution since we are finding the length of a side. Thus $s = 6\sqrt{2}$ centimeters.

2. The lengths of the three sides of a right triangle are three consecutive integers. Find the lengths of the three sides.

Let x = the length of the shortest side, then $x + 1$ = the length of the next longer side, and $x + 2$ = the length of the longest side.

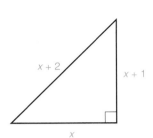

(side)²	plus	(side)²	equals	(hypotenuse)²
x^2	$+$	$(x + 1)^2$	$=$	$(x + 2)^2$

$$x^2 + (x + 1)^2 = (x + 2)^2 \qquad \text{Pythagorean Theorem}$$
$$x^2 + x^2 + 2x + 1 = x^2 + 4x + 4 \qquad \text{Square each term}$$
$$2x^2 + 2x + 1 = x^2 + 4x + 4 \qquad \text{Combine like terms}$$
$$x^2 - 2x - 3 = 0 \qquad \text{Write in standard form}$$
$$(x - 3)(x + 1) = 0 \qquad \text{Factor the left member}$$
$$x - 3 = 0 \quad \text{or} \quad x + 1 = 0 \qquad \text{Set each factor equal to 0}$$
$$x = 3 \qquad\qquad x = -1$$

Reject -1 as an answer since we are finding the length of a side. Then

$$x = 3, x + 1 = 4, \text{ and } x + 2 = 5$$

and the lengths of the sides of the triangle are 3, 4, and 5 units.

3. The sum of the squares of two consecutive even integers is 52. Find the integers.

Let x = the first even integer, then $x + 2$ = the next consecutive even integer.

(integer)²	sum	(integer)²	is	52
x^2	$+$	$(x + 2)^2$	$=$	52

$$x^2 + (x + 2)^2 = 52 \qquad \text{Sum of the squares equals 52}$$
$$x^2 + x^2 + 4x + 4 = 52$$
$$2x^2 + 4x + 4 = 52$$
$$2x^2 + 4x - 48 = 0 \qquad \text{Write in standard form}$$
$$2(x^2 + 2x - 24) = 0 \qquad \text{Factor the left member}$$
$$2(x + 6)(x - 4) = 0$$
$$x + 6 = 0 \quad \text{or} \quad x - 4 = 0 \qquad \text{Set each factor equal to zero (\textit{Note: } 2 \neq 0)}$$
$$x = -6 \qquad\qquad x = 4$$

When $x = -6$, then $x + 2 = -4$. When $x = 4$, then $x + 2 = 6$. The integers are -6 and -4 or 4 and 6.

▶ **Quick check** The sum of the squares of two consecutive *odd* integers is 130. Find the integers.

Mastery points

Can you
- Solve a quadratic equation by factoring?
- Solve a quadratic equation by extracting the roots?
- Solve word problems whose mathematical language yields quadratic equations?

Exercise 6–1

Find the solution set of the following quadratic equations by factoring. See example 6–1 A.

Example $3x^2 - 14x = 0$

Solution

$$x(3x - 14) = 0 \qquad \text{Factor the left member}$$
$$x = 0 \quad \text{or} \quad 3x - 14 = 0 \qquad \text{Set each factor equal to 0}$$
$$x = 0 \qquad\qquad\qquad 3x = 14 \qquad \text{Solve for } x$$
$$x = 0 \qquad\qquad\qquad x = \frac{14}{3}$$

The solution set is $\left\{0, \frac{14}{3}\right\}$.

1. $(x - 3)(x + 4) = 0$
2. $(x - 7)(x + 1) = 0$
3. $(3y - 1)(2y + 5) = 0$
4. $(2z - 3)(3z - 4) = 0$
5. $x^2 - 5x + 6 = 0$
6. $x^2 - 7x - 8 = 0$
7. $y^2 - 10y + 25 = 0$
8. $y^2 - 18y + 81 = 0$
9. $p^2 - 5p = 24$
10. $z^2 - 12 = 4z$
11. $m^2 - m = 0$
12. $q^2 + 3q = 0$
13. $-3y^2 + 27 = 0$
14. $-9z^2 + 144 = 0$
15. $2x^2 - 3x - 2 = 0$
16. $2n^2 + 11n - 6 = 0$
17. $4y^2 + 5y = 6$
18. $3x^2 - 8 = 10x$
19. $x - 1 - \frac{x^2}{4} = 0$
20. $\frac{x^2}{6} - \frac{x}{3} - \frac{1}{2} = 0$
21. $\frac{x}{2} + \frac{7}{2} = \frac{4}{x}$
22. $x + \frac{1}{3} = \frac{4}{3x}$
23. $(y + 6)(y - 2) = -7$
24. $(p + 4)(p - 6) = -16$
25. $(3m + 2)(m - 1) = 4m$
26. $(3x + 2)(x - 1) = -(7x - 7)$
27. $3x(x - 3) = (x - 5)(x - 3)$
28. $(y - 1)(y + 4) = 2y(y + 4)$
29. $(x - 1)^2 = (2x + 5)(x - 1)$

Find the solution set of the following equations by extracting the roots. Express radicals in simplest form. See example 6–1 B.

Example $(4z + 5)^2 = 11$

Solution $4z + 5 = \sqrt{11} \qquad\qquad \text{or} \qquad 4z + 5 = -\sqrt{11} \qquad\qquad \text{Extract the roots}$
$$4z = -5 + \sqrt{11} \qquad \text{or} \qquad 4z = -5 - \sqrt{11} \qquad \text{Add } -5 \text{ to each member}$$
$$z = \frac{-5 + \sqrt{11}}{4} \qquad \text{or} \qquad z = \frac{-5 - \sqrt{11}}{4} \qquad \text{Divide by 4}$$

The solution set is $\left\{\frac{-5 + \sqrt{11}}{4}, \frac{-5 - \sqrt{11}}{4}\right\}$.

30. $x^2 = 81$
31. $x^2 = 121$
32. $3y^2 = 27$
33. $5z^2 = 245$
34. $p^2 = 20$
35. $q^2 = 32$
36. $m^2 - 40 = 0$
37. $y^2 - 72 = 0$
38. $16p^2 - 400 = 0$
39. $7y^2 - 56 = 0$
40. $9x^2 - 162 = 0$
41. $2n^2 - 100 = 0$
42. $(x + 3)^2 = 16$
43. $(y + 7)^2 = 36$
44. $(x - 9)^2 = -144$
45. $(x - 12)^2 = -121$
46. $(x - 6)^2 = 12$
47. $(y + 10)^2 = 48$
48. $(3x - 2)^2 = 25$
49. $(4x + 1)^2 = 81$
50. $(9y + 1)^2 = -24$
51. $(10p - 3)^2 = -84$
52. $(x - 7)^2 = a^2, a > 0$
53. $(y + 8)^2 = b^2, b > 0$

Solve the following equations for x.

Example $x^2 - 3ax - 4a^2 = 0$

Solution Factoring the left member, we have $(x - 4a)(x + a) = 0$. This is true when

$$x - 4a = 0 \text{ or } x + a = 0$$

Solving each equation for x, we obtain

$$x = 4a \text{ or } x = -a$$

54. $x^2 + 4ax + 3a^2 = 0$ **55.** $x^2 - 10bx - 24b^2 = 0$ **56.** $3x^2 - 13xy + 4y^2 = 0$

57. $4x^2 - ax - 14a^2 = 0$ **58.** $12x^2 = 8ax + 15a^2$ **59.** $12xy - 6y^2 = 6x^2$

60. $5x^2 - 6y^2 = 7xy$ **61.** $6x^2 + 5xy = 4y^2$

Solve the following word problems. See example 6–1 A–5 and 6–1 C.

Example The sum of the squares of two consecutive odd integers is 130. Find the integers.

Solution Let $x =$ the first odd integer, then $x + 2 =$ the next consecutive odd integer.

(integer)²	sum	(integer)²	is	130
x^2	$+$	$(x + 2)^2$	$=$	130

$x^2 + (x + 2)^2 = 130$	Sum of the squares equals 130
$x^2 + x^2 + 4x + 4 = 130$	Multiply in left member
$2x^2 + 4x + 4 = 130$	Combine like terms
$2x^2 + 4x - 126 = 0$	Subtract 130 from each member
$2(x^2 + 2x - 63) = 0$	Factor left member
$2(x + 9)(x - 7) = 0$	Factor the trinomial
$x + 9 = 0 \quad$ or $\quad x - 7 = 0$	Set each containing the variable factor equal to 0
$x = -9 \qquad\qquad x = 7$	Solve each equation

When $x = -9$, $x + 2 = -7$ and when $x = 7$, $x + 2 = 9$.
The consecutive odd integers are -9 and -7 or 7 and 9.

62. Given $P = 100I - 5I^2$, find I in amperes, $I > 0$, when (a) $P = 420$, (b) $P = 0$.

63. A ball rolls down a slope and travels a distance d defined by the equation $d = 6t + \dfrac{t^2}{2}$ feet in t seconds. How long does it take the ball to roll (a) $d = 32$ feet, (b) $d = 14$ feet?

64. The height h that an object will reach in t seconds if it is propelled vertically upward with an initial velocity V_0 feet per second is defined by the equation

$$h = -16t^2 + V_0 t$$

When will the object hit the ground?

65. An object with an initial velocity V_0 accelerates at rate a in time t. The displacement of the object for this time is given by the equation

$$s = V_0 t + \frac{1}{2}at^2$$

If $V_0 = 3$ and $a = 6$, when will the displacement s be 6 feet?

66. The sum S of the first n even positive integers is given by $S = n(n + 1)$. Find n when $S = 30$.

67. The sum S of the first n of the numbers 5, 8, 11, . . . , $3n + 2$ is given by $S = \dfrac{1}{2}n(3n + 7)$.

Find n when $S = 98$.

68. If the diagonal of a square is 32 feet, find the length of each side of the square.

69. If the diagonal of a square is $7\sqrt{2}$ meters long, find the length of each side of the square.

70. The three sides of a right triangle are three consecutive even integers. Find the lengths of the three sides.

71. One leg of a right triangle is 7 inches longer than the other leg. If the hypotenuse is 9 inches longer than the shortest leg, find the lengths of the three sides of the triangle.

72. The longest side of a right triangle is 1 yard longer than twice the length of the shortest side. If the third side measures 15 yards, find the lengths of the other two sides.

73. The square of the sum of two consecutive even integers is 100. Find the integers.

74. The square of the sum of two consecutive odd integers is 144. Find the integers.

75. The sum of the squares of two consecutive odd integers is 130. Find the integers.

76. One integer is 1 more than twice the other integer. Their product is 105. Find the integers.

Review exercises

Simplify the following expressions. Assume all variables are nonzero. Answer with all exponents positive. See section 3–3.

1. $\dfrac{x^{-2}y^3}{x^2y^{-1}}$

2. $(4x^{-3}y^2)^{-2}$

Find the following products. See section 3–2.

3. $(2x - 3)^2$

4. $(x + 7)^2$

Find the solution set of the following equations and inequalities. See sections 2–5 and 6–1.

5. $-3 \leq 2x - 1 < 5$

6. $4x^2 - 13x = -3$

6–2 ■ Solution by completing the square

Completing the square

Finding the solution set of quadratic equations by factoring and by extracting the roots required special types of the quadratic equation. Now we develop a method that can be applied to any quadratic equation. The method, called **completing the square,** involves transforming the standard quadratic equation

$$ax^2 + bx + c = 0, a > 0$$

into the form

$$(x + k)^2 = d$$

where k and d are constants. This latter equation can then be solved by extracting the roots as we did in section 6–1.

Consider the identities

$$(x + 8)^2 = x^2 + 16x + 64$$
$$(x - 7)^2 = x^2 - 14x + 49$$

Observe first that the coefficient of x^2 in each case is 1. This is necessary for what we do next. We then consider the relationship between the second term (the linear term) and the third term (the constant term) of the trinomial. Notice that the constant term in each case is the *square of one-half of the coefficient of* the middle (linear) term, x.

1. In $x^2 + 16x + 64$, the constant term, 64, is the square of one-half of the coefficient of the middle (linear) term, 16.

$$\left[\frac{1}{2}(16)\right]^2 = 8^2 = 64$$

2. In $x^2 - 14x + 49$, the constant term, 49, is the square of one-half of the coefficient of the middle (linear) term, -14.

$$\left[\frac{1}{2}(-14)\right]^2 = (-7)^2 = 49$$

Further, we see that the constant term in the binomial square is the number that is one-half of the coefficient of the middle (linear) term of the perfect square trinomial.

1. Given $x^2 + 16x + 64 = (x + 8)^2$

$$\left[\frac{1}{2}(16)\right] = 8$$

2. Given $x^2 - 14x + 49 = [x + (-7)]^2 = (x - 7)^2$

$$\left[\frac{1}{2}(-14)\right] = -7$$

We now use these observations to "build" perfect square trinomials by *completing the square* and thereby obtain their equivalent perfect squares.

■ *Example 6–2 A*

Determine what number must be added to each expression to make it a perfect square trinomial. State the equivalent square of a binomial.

1. $x^2 + 10x$
The coefficient of the linear term, $10x$, is 10. Now we square one-half of 10 to obtain

$$\left[\frac{1}{2}(10)\right]^2 = 5^2 = 25$$

Adding 25 to the given expression, we have

$x^2 + 10x + 25 = (x + 5)^2$

2. $x^2 - 7x$
The coefficient of the linear term is -7. The square of one-half of -7 is

$$\left[\frac{1}{2}(-7)\right]^2 = \left(-\frac{7}{2}\right)^2 = \frac{49}{4}$$

Then

$$x^2 - 7x + \frac{49}{4} = \left(x - \frac{7}{2}\right)^2$$

▶ *Quick check* What must be added to $z^2 - 9z$ to make a perfect square trinomial?

Solving by completing the square

We now use this procedure to obtain the solution set of a quadratic equation by completing the square.

■ *Example 6–2 B*

Find the solution set by completing the square.

1. $x^2 - 6x + 5 = 0$

Isolate $x^2 - 6x$ by subtracting 5 from each member.

$$x^2 - 6x = -5$$

$$x^2 - 6x + \left[\frac{1}{2}(-6)\right]^2 = -5 + \left[\frac{1}{2}(-6)\right]^2 \qquad \text{Complete the square}$$

$$x^2 - 6x + 9 = -5 + 9 \qquad \left[\frac{1}{2}(-6)\right]^2 = 9$$

$$(x - 3)^2 = 4 \qquad \begin{array}{l}\text{Factor the left member;}\\ \text{combine in the right member}\end{array}$$

$$x - 3 = 2 \quad \text{or} \quad x - 3 = -2 \qquad \text{Extract the roots}$$
$$x = 5 \qquad\qquad x = 1$$

The solution set is $\{1,5\}$.

Note A common error when solving by this method is *to **fail** to add the same number to each member* when completing the square. Failure to do so changes the equation.

2. $x^2 + 8x - 2 = 0$

$$x^2 + 8x = 2 \qquad \text{Isolate variable terms}$$

$$x^2 + 8x + \left[\frac{1}{2}(8)\right]^2 = 2 + \left[\frac{1}{2}(8)\right]^2 \qquad \text{Complete the square}$$

$$x^2 + 8x + 16 = 2 + 16 \qquad \left[\frac{1}{2}(8)\right]^2 = 16$$

$$(x + 4)^2 = 18 \qquad \begin{array}{l}\text{Factor the left member and}\\ \text{add in the right member}\end{array}$$

$$x + 4 = \sqrt{18} \quad \text{or} \quad x + 4 = -\sqrt{18} \qquad \text{Extract the roots}$$
$$x + 4 = 3\sqrt{2} \qquad x + 4 = -3\sqrt{2} \qquad \sqrt{18} = 3\sqrt{2}$$
$$x = -4 + 3\sqrt{2} \qquad x = -4 - 3\sqrt{2} \qquad \text{Solve for } x$$

The solution set is $\{-4 - 3\sqrt{2}, -4 + 3\sqrt{2}\}$.

Completing the square can be used *only* when the coefficient of x^2 is 1. When the coefficient is other than 1, we often divide each member of the equation by that coefficient.

3.

$$3x^2 - 12x - 9 = 0$$
$$x^2 - 4x - 3 = 0 \qquad \begin{array}{l}\text{Divide each term by 3 to}\\ \text{obtain leading coefficient of 1}\end{array}$$
$$x^2 - 4x = 3 \qquad \text{Isolate variable terms}$$

$$x^2 - 4x + \left[\frac{1}{2} \cdot (-4)\right]^2 = 3 + \left[\frac{1}{2} \cdot (-4)\right]^2 \qquad \text{Complete the square}$$

$$x^2 - 4x + 4 = 3 + 4$$
$$(x - 2)^2 = 7 \qquad \begin{array}{l}\text{Factor the left member;}\\ \text{combine in the right member}\end{array}$$
$$x - 2 = \sqrt{7} \quad \text{or} \quad x - 2 = -\sqrt{7} \qquad \text{Extract the roots}$$
$$x = 2 + \sqrt{7} \qquad x = 2 - \sqrt{7}$$

The solution set is $\{2 - \sqrt{7}, 2 + \sqrt{7}\}$.

▶ *Quick check* Find the solution set of $3y^2 - 6y - 3 = 0$ by completing the square. ■

In summary, to find the solution set of the quadratic equation $ax^2 + bx + c = 0$, $a > 0$, by completing the square, we follow this procedure.

___ *To solve quadratic equations by completing the square* ___

1. If $a = 1$, proceed to step 2. If $a \neq 1$, divide each term of the equation by a, if necessary to complete the square.
2. Write the equation with the variable terms in the left member and the constant in the right member.
3. Add to each member of the equation the square of one-half of the coefficient of the linear term.
4. Write the left member as a perfect square and combine in the right member.
5. Extract the roots and solve the resulting linear equations.

___ *Mastery points* ___

Can you
■ Complete the square of a binomial of the form $x^2 + kx$?
■ Find the solution set of a quadratic equation by completing the square?

Exercise 6–2

Determine what number must be added to each expression to make a perfect square trinomial. State the equivalent binomial square. See example 6–2 A.

Example $z^2 - 9z$

Solution $\left[\frac{1}{2}(-9)\right]^2 = \left(-\frac{9}{2}\right)^2 = \frac{81}{4}$ Square one-half of the coefficient of z

$z^2 - 9z + \frac{81}{4} = \left(z - \frac{9}{2}\right)^2$ Add $\frac{81}{4}$ to the given expression and factor

1. $x^2 + 4x + ?$
2. $x^2 + 8x + ?$
3. $y^2 - 18y + ?$
4. $z^2 - 24z + ?$
5. $p^2 + 2p + ?$
6. $m^2 - 30m + ?$
7. $x^2 + 3x + ?$
8. $y^2 + y + ?$
9. $w^2 - 11w + ?$
10. $q^2 - 5q + ?$
11. $x^2 + 13x + ?$
12. $y^2 - 15y + ?$

Find the solution set by completing the square. See example 6–2 B.

Example $3y^2 - 6y - 3 = 0$

Solution

$$y^2 - 2y - 1 = 0 \qquad \text{Divide each term by 3}$$
$$y^2 - 2y = 1 \qquad \text{Isolate variable terms}$$
$$y^2 - 2y + \left[\frac{1}{2} \cdot (-2)\right]^2 = 1 + \left[\frac{1}{2} \cdot (-2)\right]^2 \qquad \text{Complete the square}$$
$$y^2 - 2y + 1 = 1 + 1 \qquad \left[\frac{1}{2} \cdot (-2)\right]^2 = 1$$
$$(y - 1)^2 = 2 \qquad \text{Factor left member and add in right member}$$
$$y - 1 = \sqrt{2} \quad \text{or} \quad y - 1 = -\sqrt{2} \qquad \text{Extract the roots}$$
$$y = 1 + \sqrt{2} \qquad y = 1 - \sqrt{2} \qquad \text{Solve for } y$$

The solution set is $\{1 - \sqrt{2}, 1 + \sqrt{2}\}$.

13. $x^2 + 12x + 11 = 0$
14. $x^2 + 5x + 4 = 0$
15. $y^2 - 11y + 10 = 0$
16. $p^2 - 4p = 8$
17. $n^2 + 8n = -25$
18. $x^2 + 6x = -10$
19. $x^2 - 8x = 0$
20. $x^2 + 4x = 0$
21. $y^2 = 3 - y$
22. $x^2 + 2 = -4x$
23. $-2x^2 + 4 = -6x$
24. $2n = 4 - n^2$
25. $2x^2 + 3x - 2 = 0$
26. $2y^2 - 4y - 3 = 0$
27. $3x^2 - 12x + 3 = 0$
28. $1 - z^2 = 3z$
29. $4u^2 - 4u = 3$
30. $4x^2 + 12x + 4 = 0$
31. $5m^2 - 5m + 1 = 0$
32. $5q^2 + 4q + 1 = 0$
33. $(x + 2)(x - 3) = 1$
34. $(2y - 1)(y + 5) = -3$
35. $(3x + 1)^2 = (x - 2)^2$
36. $(2y - 3)^2 = (y + 4)^2$
37. $p(2p - 1) = p + 1$
38. $8p(p + 3) = 2p - 5$
39. $\frac{1}{2}x^2 - \frac{3}{4}x = 1$
40. $y^2 - \frac{1}{3}y = \frac{2}{3}$
41. $\frac{1}{2}x^2 - \frac{2}{3}x - 1 = 0$
42. $z^2 + \frac{1}{2}z - \frac{3}{4} = 0$
43. $x^2 - \frac{3}{2} = 2x$
44. $\frac{1}{5}n^2 = 1 - 2n$
45. $\frac{5}{x} - 2x + 3 = 0$
46. $\frac{3}{x} - 2 = 2x$
47. $5 - \frac{2}{t} = \frac{3}{t^2}$
48. $3 + \frac{5}{p} = \frac{1}{p^2}$

Solve the following word problems by completing the square.

49. When an object is thrown downward with an initial velocity of 9 feet per second, the relationship between the distance s it travels in time t is given by
$$s = 9t + 16t^2$$
How long does it take for the object to fall 100 feet?

50. When an object is dropped from rest, the distance s that the object falls is given by the equation
$$s = 16t^2$$
How long does it take the object to hit the bottom of a gorge when it is dropped a distance of 205 feet from a bridge across the gorge?

51. The supply equation for a specific product is given by
$$S = 32p + p^2$$
where p cents is the price per unit of the product. What should the price to the nearest cent be when the supply S is 500 units?

52. The demand equation for a specific product is given by
$$D = 36p + p^2$$
where p is the price per 1,000 units. What should the price to the nearest cent per 1,000 units be when the demand D is 100,000 units?

53. The radius r of a circular arch having height h and span b is given by
$$r = \frac{(b^2 + 4h^2)}{8h}$$
Find h when $b = 10$ and $r = 13$.

54. The square of a number added to three times the number is 6. Find the number.

55. The square of twice a number less the number is 4. Find the number.

56. The square of the difference between four times a number and 5 is 24. Find the number.

57. The formula for the volume of a cylinder with height h and radius r is

$$V = \pi r^2 h$$

Find r when V is 235 cubic meters and $h = 10$ meters.

Review exercises

Perform the indicated multiplications. See section 5–6.

1. $\sqrt{3}(2 - \sqrt{5})$

2. $(2 - \sqrt{3})(2 + \sqrt{3})$

3. $(\sqrt{2} + \sqrt{3})(2\sqrt{3} - 3\sqrt{2})$

4. $(3\sqrt{2} - 3)^2$

Divide the following expressions. See section 4–5.

5. $\dfrac{16x^3 - 8x^2 + 4x}{4x^2}$

6. $(4x^2 - 3x - 2) \div (x - 3)$

Evaluate $b^2 - 4ac$ given the following values of a, b, and c. See section 1–5.

7. $a = 2, b = -3, c = 1$

8. $a = 3, b = 5, c = -3$

9. $a = 1, b = 2, c = 4$

6–3 ■ Solution by quadratic formula

The quadratic formula

In the previous sections, we have found the solution set of quadratic equations by factoring, extracting the roots, and completing the square. Even though the solution set of any quadratic equation can be found by completing the square, this can be a time-consuming chore. In this section, we will use the method of completing the square to develop a general formula that will always find the solution set. We call this formula the **quadratic formula.**

Given the quadratic equation in standard form,

$$ax^2 + bx + c = 0 \text{ (assume } a > 0)$$

where a, b, and c are real numbers, we can solve for x by completing the square.

$$x^2 + \frac{b}{a}x + \frac{c}{a} = 0 \qquad \text{Divide each term by the coefficient } a \text{ of } x^2$$

$$x^2 + \frac{b}{a}x = -\frac{c}{a} \qquad \text{Subtract } \tfrac{c}{a} \text{ from each member}$$

$$\left[\frac{1}{2}\left(\frac{b}{a}\right)\right]^2 = \left(\frac{b}{2a}\right)^2 = \frac{b^2}{4a^2} \qquad \text{Square one-half of the coefficient of } x$$

$$x^2 + \frac{b}{a}x + \frac{b^2}{4a^2} = \left(-\frac{c}{a}\right) + \left(\frac{b^2}{4a^2}\right) \qquad \text{Add } \left(\frac{1}{2}\cdot\frac{b}{a}\right)^2 = \frac{b^2}{4a^2} \text{ to each member}$$

$$\left(x + \frac{b}{2a}\right)^2 = \frac{b^2 - 4ac}{4a^2} \qquad \text{Factor and subtract}$$

$$x + \frac{b}{2a} = \pm\sqrt{\frac{b^2 - 4ac}{4a^2}} \qquad \text{Extract the roots}$$

$$x = -\frac{b}{2a} \pm \frac{\sqrt{b^2 - 4ac}}{2a} \qquad \text{Add } -\frac{b}{2a} \text{ to each member}$$

$$x = \frac{-b \pm \sqrt{b^2 - 4ac}}{2a} \qquad \text{Add and subtract over the common denominator}$$

The solution set is

$$\left\{ \frac{-b + \sqrt{b^2 - 4ac}}{2a}, \frac{-b - \sqrt{b^2 - 4ac}}{2a} \right\} \text{ or } \left\{ \frac{-b \pm \sqrt{b^2 - 4ac}}{2a} \right\}.$$

___ *The quadratic formula* _____

If $ax^2 + bx + c = 0$, then

$$x = \frac{-b \pm \sqrt{b^2 - 4ac}}{2a}, a > 0$$

where a is the coefficient of the second-degree term, b of the first-degree term, and c is the constant.

Solving by quadratic formula

To find the solution set of any quadratic equation by using the quadratic formula, we only need to substitute the numerical values for a, b, and c into the formula and to simplify the result. Consider the equation

$$x^2 + 4x - 12 = 0$$

Since $a = 1$, $b = 4$, and $c = -12$, we substitute in the quadratic formula to obtain

$$x = \frac{-4 \pm \sqrt{(4)^2 - 4(1)(-12)}}{2(1)}$$
$$= \frac{-4 \pm \sqrt{16 + 48}}{2}$$
$$= \frac{-4 \pm \sqrt{64}}{2}$$
$$= \frac{-4 \pm 8}{2}$$

$$x = \frac{-4 + 8}{2} = \frac{4}{2} = 2 \quad \text{or} \quad x = \frac{-4 - 8}{2} = \frac{-12}{2} = -6$$

The solution set is $\{-6, 2\}$.

We now summarize the procedure for solving a quadratic equation by using the quadratic formula.

___ *To solve quadratic equations using the quadratic formula* ___

1. Write the equation in standard form (if necessary).
2. Identify the numerical values of a, b, and c.
3. Substitute these values into the quadratic formula.
4. Simplify the resulting expression.

■ *Example 6–3 A*

Find the solution set using the quadratic formula.

1. $x^2 = 5 - 3x$

Write the equation in standard form.

$$x^2 + 3x - 5 = 0 \qquad \text{Add } 3x - 5 \text{ to each member}$$

$$a = 1, b = 3, c = -5. \qquad \text{Identify } a, b, \text{ and } c$$

$$x = \frac{-(3) \pm \sqrt{(3)^2 - 4(1)(-5)}}{2(1)}$$ Replace a with 1, b with 3, and c with -5

$$x = \frac{-3 \pm \sqrt{9 + 20}}{2}$$ Perform indicated operations

$$x = \frac{-3 \pm \sqrt{29}}{2}$$ Simplify

The solution set is $\left\{ \dfrac{-3 + \sqrt{29}}{2}, \dfrac{-3 - \sqrt{29}}{2} \right\}$.

2. $2y^2 - 4y + 1 = 0$

$a = 2, b = -4, c = 1.$ Identify a, b, and c

$$y = \frac{-(-4) \pm \sqrt{(-4)^2 - 4(2)(1)}}{2(2)}$$ Replace a with 2, b with -4, and c with 1

$$y = \frac{4 \pm \sqrt{16 - 8}}{4}$$ Perform indicated operations

$$y = \frac{4 \pm \sqrt{8}}{4}$$ Simplify

$$y = \frac{4 \pm 2\sqrt{2}}{4}$$ $\sqrt{8} = \sqrt{4 \cdot 2} = 2\sqrt{2}$

$$y = \frac{2(2 \pm \sqrt{2})}{4}$$ Factor 2 in the numerator

$$y = \frac{2 \pm \sqrt{2}}{2}$$ Reduce by 2

The solution set is $\left\{ \dfrac{2 - \sqrt{2}}{2}, \dfrac{2 + \sqrt{2}}{2} \right\}$.

3. $4 - \dfrac{3}{x} + \dfrac{4}{x^2} = 0 \quad (x \neq 0)$

Multiply each term by the LCM of x and x^2, which is x^2.

$$x^2 \cdot 4 - x^2 \cdot \frac{3}{x} + x^2 \cdot \frac{4}{x^2} = x^2 \cdot 0$$ Multiply each term by x^2

$$4x^2 - 3x + 4 = 0$$ Clear the denominators

$a = 4, b = -3,$ and $c = 4.$ Identify a, b, and c

$$x = \frac{-(-3) \pm \sqrt{(-3)^2 - 4(4)(4)}}{2(4)}$$ Replace a with 4, b with -3, and c with 4

$$= \frac{3 \pm \sqrt{9 - 64}}{8}$$

$$= \frac{3 \pm \sqrt{-55}}{8}$$

$$= \frac{3 \pm i\sqrt{55}}{8}$$ $\sqrt{-55} = i\sqrt{55}$

The solution set is $\left\{ \dfrac{3 + i\sqrt{55}}{8}, \dfrac{3 - i\sqrt{55}}{8} \right\}$.

▶ ***Quick check*** Find the solution set of $3y^2 - 6y + 2 = 0$ using the quadratic formula.

To solve any quadratic equation, we use the following steps.

> ### Solving a quadratic equation
> 1. Write the equation in standard form. Clear fractions if necessary.
> 2. Check to see if the polynomial expression factors. If so, solve by factoring.
> 3. If $b = 0$, solve by extracting the roots.
> 4. Use the quadratic formula.

The discriminant

The general quadratic equation $ax^2 + bx + c = 0$ has two solutions x_1 and x_2 such that, using the quadratic formula,

$$x_1 = \frac{-b + \sqrt{b^2 - 4ac}}{2a} \quad \text{and} \quad x_2 = \frac{-b - \sqrt{b^2 - 4ac}}{2a}$$

We can determine the nature of the solutions x_1 and x_2 (that is, rational, irrational, or complex) by using the radicand, $b^2 - 4ac$, where a, b, and c are **rational** numbers. For this reason, we call $b^2 - 4ac$ the **discriminant** of the quadratic equation $ax^2 + bx + c = 0$. The nature of the solutions can be determined as follows:

> ### Nature of the solutions of a quadratic equation
> When the discriminant $b^2 - 4ac$ is
>
> | Zero | Positive | Negative |
> |------|----------|----------|
> | $b^2 - 4ac = 0$ | $b^2 - 4ac > 0$ | $b^2 - 4ac < 0$ |
> | One *rational* solution of multiplicity two, namely $-\dfrac{b}{2a}$ | If $b^2 - 4ac$ is
 a. a perfect square, *two* distinct *rational* solutions
 b. not a perfect square, *two* distinct *irrational* solutions | *Two* distinct *complex* (nonreal) solutions |

■ *Example 6–3 B*

Use the discriminant to decide the number and the nature of the solutions of the given quadratic equation.

1. $4x^2 + 12x + 9 = 0$

Since $a = 4$, $b = 12$, and $c = 9$, then

$$b^2 - 4ac = (12)^2 - 4(4)(9) = 144 - 144 = 0$$

There is only *one rational* solution of multiplicity two and that solution is

$$\frac{-b}{2a} = \frac{-12}{2(4)} = \frac{-3}{2} = -\frac{3}{2}$$

2. $3x^2 - 4 = 4x$

Write the equation in standard form, $3x^2 - 4x - 4 = 0$. Thus $a = 3$, $b = -4$, and $c = -4$. Then

$$b^2 - 4ac = (-4)^2 - 4(3)(-4) = 16 + 48 = 64$$

Since $64 = 8^2$, the discriminant is positive, a perfect square, and there are *two distinct rational* solutions.

Note   The discriminant can be used to determine if the trinomial is factorable. In examples 1 and 2, where there is either one rational solution or two rational solutions, the equation factors,

$$4x^2 + 12x + 9 = (2x + 3)^2 \quad \text{and}$$
$$3x^2 - 4x - 4 = (3x + 2)(x - 2)$$

and the discriminant is a perfect square.

3. $x^2 = 4x + 6$

We must first write the equation in standard form, $x^2 - 4x - 6 = 0$. Then $a = 1$, $b = -4$, and $c = -6$, and

$$b^2 - 4ac = (-4)^2 - 4(1)(-6) = 16 + 24 = 40$$

Since 40 is not a perfect square, but is positive, there are *two distinct irrational* solutions, and the polynomial $x^2 - 4x - 6$ does not factor.

4. $2y^2 - 3y + 5 = 0$

Since $a = 2$, $b = -3$, and $c = 5$, then

$$b^2 - 4ac = (-3)^2 - 4(2)(5) = 9 - 40 = -31$$

The discriminant is negative, the equation has *two distinct complex* (nonreal) solutions, and the polynomial does not factor.

▶ *Quick check*   Decide the number and the nature of the solutions of $3x^2 + 2x - 3 = 0$ using the discriminant.   ■

Mastery points

Can you
- Identify the numerical values of *a, b,* and *c* in a standard quadratic equation?
- Solve a quadratic equation by using the quadratic formula?
- Use the discriminant to determine the nature of the solutions?

Exercise 6–3

Find the solution set of each quadratic equation using the quadratic formula in exercises 1–10. See example 6–3 A. In exercises 11–37, use any convenient method.

Example $3y^2 - 6y + 2 = 0$

Solution $y = \dfrac{-(-6) \pm \sqrt{(-6)^2 - 4(3)(2)}}{2(3)}$ Replace a with 3, b with -6, and c with 2

$= \dfrac{6 \pm \sqrt{36 - 24}}{6}$ Multiply in radicand

$= \dfrac{6 \pm \sqrt{12}}{6}$ Subtract in radicand

$= \dfrac{6 \pm 2\sqrt{3}}{6}$ $\sqrt{12} = 2\sqrt{3}$

$= \dfrac{2(3 \pm \sqrt{3})}{2 \cdot 3}$ Factor the common factor of 2 from the numerator and the denominator

$= \dfrac{3 \pm \sqrt{3}}{3}$ Reduce to lowest terms

The solution set is $\left\{ \dfrac{3 + \sqrt{3}}{3}, \dfrac{3 - \sqrt{3}}{3} \right\}$.

1. $p^2 = -5p - 7$

2. $y^2 + 6 = 2y$

3. $3x - 5 = x^2$

4. $18 = 10x - x^2$

5. $2x^2 - 7x + 6 = 0$

6. $3y^2 - 5y - 6 = 0$

7. $4z^2 - 8z + 1 = 0$

8. $9p^2 - 8p + 7 = 0$

9. $3q^2 - 2q + 7 = 0$

10. $3z^2 - 4z = -3$

11. $y^2 + 6y - 16 = 0$

12. $z^2 + 6z + 9 = 0$

13. $p^2 - 14p + 49 = 0$

14. $x^2 - 28 = 0$

15. $3y^2 = 20$

16. $4z^2 - 3z = 0$

17. $2x = 5x^2$

18. $m^2 - 2m = 4$

19. $9x^2 - 12x + 4 = 0$

20. $4y^2 + 20y + 25 = 0$

21. $9y^2 - 12y = -5$

22. $2t^2 = 6t - 5$

23. $11 - 6m = 9m^2$

24. $2v = 5v^2 - 2$

25. $x = 2x^2 + 7$

26. $3x - \dfrac{2}{x} + 5 = 0$

27. $2y^2 - \dfrac{7}{2} = \dfrac{y}{2}$

28. $\dfrac{2}{3}x^2 - \dfrac{1}{3} = x$

29. $\dfrac{2}{3}y - \dfrac{1}{3} = \dfrac{4}{9}y^2$

30. $\dfrac{2p}{3} - \dfrac{p^2}{4} = 1$

31. $\dfrac{3}{4}q^2 = \dfrac{1}{2}q + 4$

32. $\dfrac{1}{x + 2} + \dfrac{1}{x - 3} - 2 = 0$

33. $\dfrac{3}{y - 5} - \dfrac{2}{y + 1} + 3 = 0$

34. $\dfrac{1}{2} - \dfrac{3}{2x + 3} = \dfrac{3}{x - 4}$

35. $(z - 3)(z + 2) = 2z - 3$

36. $(x + 6)(x - 5) = 10 - x$

37. $(2x + 1)^2 = (x - 3)^2$

Solve the following equations for x in terms of the other variables or constants. Assume that all other variables are positive real numbers.

Example $2x^2 - 3xy - 5y^2 = 0$

Solution   Here $a = 2$, $b = -3y$ (coefficient of x), and $c = -5y^2$ (term not containing x). Then

$x = \dfrac{-(-3y) \pm \sqrt{(-3y)^2 - 4(2)(-5y^2)}}{2(2)}$ Replace a with 2, b with $-3y$, and c with $-5y^2$

$= \dfrac{3y \pm \sqrt{9y^2 + 40y^2}}{4}$ Simplify

$$= \frac{3y \pm \sqrt{49y^2}}{4}$$

$$= \frac{3y \pm 7y}{4}$$

Thus

$$x = \frac{3y + 7y}{4} = \frac{10y}{4} = \frac{5}{2}y \text{ or } x = \frac{3y - 7y}{4} = \frac{-4y}{4} = -y$$

Therefore $x = \dfrac{5}{2}y$ or $x = -y$.

38. $x^2 - xy - 2y^2 = 0$ **39.** $x^2 - 3xy - 18y^2 = 0$ **40.** $2x^2 - 3ax + 5a^2 = 0$

41. $4x^2 + 2x - 3y = 0$ **42.** $2x^2 - 3x + 4a = 0$ **43.** $x^2 - 4ax + 3a = 0$

Using the discriminant $b^2 - 4ac$, determine the number and the nature of the solutions of each quadratic equation. See example 6–3 B.

Example $3x^2 + 2x - 3 = 0$

Solution $b^2 - 4ac = (2)^2 - 4(3)(-3)$ Replace *a* with 3, *b* with 2, and *c* with -3
$\qquad\qquad\qquad = 4 + 36$ Multiply as indicated
$\qquad\qquad\qquad = 40$

Since 40 is positive and not a perfect square, then the equation has *two distinct irrational* solutions.

44. $x^2 + 4x + 1 = 0$ **45.** $x^2 - 3x - 5 = 0$ **46.** $2y^2 - y + 1 = 0$

47. $3x^2 + 3x + 4 = 0$ **48.** $4y^2 - 4y + 1 = 0$ **49.** $9x^2 - 30x + 25 = 0$

50. $4y^2 - y = 10$ **51.** $3m^2 - 5m = 2$ **52.** $y^2 + 6 = -2y$

53. $x^2 = -4 + 6x$ **54.** $y^2 = 20$ **55.** $m^2 - 18 = 4m$

56. $5t^2 - 3t = 0$ **57.** $7p^2 = 8p + 2$ **58.** $x^2 - \dfrac{1}{2}x + \dfrac{3}{5} = 0$

59. $x^2 - 4x + \dfrac{9}{4} = 0$ **60.** $\dfrac{y^2}{2} + 5y = 1$ **61.** $\dfrac{m^2}{4} + \dfrac{3m}{2} = 5$

Solve the following using the quadratic formula.

62. The distance s through which an object will fall in t seconds is given by

$$s = \frac{1}{2}gt^2$$

where $g = 32$ feet per second per second (32 ft/sec²). Find t to the nearest tenth of a second when (a) $s = 96$ feet, (b) $s = 60$ feet.

63. A metal strip is shaped into a right triangle as shown in the diagram. If $a = x$, $b = x + 3$, and $c = x + 6$, find x.
(*Hint:* Use the Pythagorean Theorem previously discussed.)

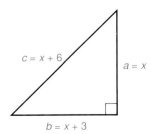

64. In a given electric circuit, the relationship between I (in amperes), E (in volts), and R (in ohms) is given by

$$I^2R + IE = 8{,}000$$

Find I ($I > 0$) when $R = 4$ and $E = 100$.

65. Using the formula

$$s = v_0t + \frac{1}{2}at^2$$

find t when (a) $s = 8$, $v_0 = 3$, $a = 4$;
(b) $s = 80$, $v_0 = 36$, $a = 32$. ($t > 0$)

66. In chemistry, an equation in connection with equilibrium in liquid flow is given by

$$k = \frac{x^2}{(a - x)(b - x)}$$

Solve for x if $a = 1$, $b = 2$, and $k = 3$.

67. If P dollars are invested at $r\%$ interest compounded annually, after two years the worth A in dollars is given by

$$A = P(1 + 0.01r)^2$$

If the amount $A = \$1{,}200$ after two years when $P = \$1{,}000$ is invested, find the rate of interest r, to the nearest tenth of a percent.

Review exercises

Simplify the following radical expressions. See sections 5–4, 5–5, and 5–6.

1. $4\sqrt{7} - 5\sqrt{63}$

2. $\sqrt{\dfrac{9}{25}}$

3. $(3 - \sqrt{5})^2$

4. Multiply $(3 - 2i)(4 + 5i)$. See section 5–7.

5. Subtract $(2 - \sqrt{-3}) - (4 + \sqrt{-3})$. See section 5–7.

6. Subtract $(2x^5)^2 - (3x^2)^3x^4$. See section 3–3.

6–4 ■ Applications of quadratic equations

A number of physical situations generate quadratic equations. Because of this, there may be two answers to the problem. Sometimes, because of the nature of the situation, only one of the answers is logical. For example, it would not be feasible to accept -25 as the measurement of a dimension of a room or $\frac{7}{6}$ as the number of books on a shelf. These answers are not physically logical and would, therefore, be rejected as solutions to an applied problem.

■ **Example 6–4 A**

1. When a ball is thrown straight upward into the air, the equation

$$s = -16t^2 + 80t + 44$$

gives the distance s in feet that the ball is above the ground t seconds after it is thrown. How long does it take for the ball to hit the ground?

The ball hits the ground when $s = 0$. Thus we have

$0 = -16t^2 + 80t + 44$	Replace s with 0
$16t^2 - 80t - 44 = 0$	Multiply each member by -1
$4(4t^2 - 20t - 11) = 0$	Factor in the left member
$4(2t - 11)(2t + 1) = 0$	
$2t - 11 = 0$ or $2t + 1 = 0$	Set each variable factor equal to 0
$t = \dfrac{11}{2}$ $\qquad$ $t = -\dfrac{1}{2}$	

Since t represents time, we reject $t = -\dfrac{1}{2}$, so $t = \dfrac{11}{2}$ or $5\dfrac{1}{2}$. The ball hits the ground in $5\dfrac{1}{2}$ seconds.

Right triangle problem

2. Find the length of side b of the given right triangle if c (the hypotenuse) $= 13$ units and side b is 7 units longer than side a.

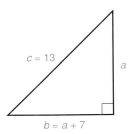

Use the Pythagorean Theorem, $a^2 + b^2 = c^2$. Since b is 7 units longer than a, $b = a + 7$.

$$a^2 + (a + 7)^2 = (13)^2 \qquad \text{Replace } b \text{ with } a + 7 \text{ and } c \text{ with } 13$$
$$a^2 + a^2 + 14a + 49 = 169$$
$$2a^2 + 14a - 120 = 0$$
$$2(a^2 + 7a - 60) = 0$$
$$2(a + 12)(a - 5) = 0$$
$$a + 12 = 0 \quad \text{or} \quad a - 5 = 0$$
$$\text{Then } a = -12 \text{ or } a = 5.$$

We reject $a = -12$, since the dimensions of a triangle cannot be negative.

$a = 5$ and $b = a + 7 = 5 + 7 = 12$. Side b is 12 units long.

Geometric problem

3. A rectangle has an area, A, of 65 square centimeters. If the length ℓ of the rectangle is 3 centimeters more than twice the width w, find the dimensions of the rectangle ($A = \ell w$).

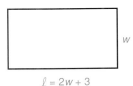

Let $w =$ the width of the rectangle, then $\ell = 2w + 3$.

$$65 = (2w + 3)w \qquad \text{In } A = \ell w, \text{ replace } A \text{ with } 65 \text{ and } \ell \text{ with } 2w + 3$$

$$65 = 2w^2 + 3w$$
$$2w^2 + 3w - 65 = 0$$
$$(2w + 13)(w - 5) = 0$$
$$\text{Then } 2w + 13 = 0 \quad \text{or} \quad w - 5 = 0.$$

Since $w = -\dfrac{13}{2}$ when $2w + 13 = 0$ and $w = 5$ when $w - 5 = 0$, then

$$w = -\dfrac{13}{2} \text{ or } w = 5.$$

Reject $w = -\dfrac{13}{2}$ since we cannot have a negative width. Then $w = 5$ and $\ell = 2w + 3 = 2(5) + 3 = 13$. The rectangle is 5 centimeters wide and 13 centimeters long.

Work problem

4. Two pipes when opened can fill a tank in 5 hours. If one pipe takes 2 hours less to fill the tank than the other one does, how long will it take each pipe to fill the tank alone? (Round off to the nearest tenth.)

Let $x =$ time in hours that the faster pipe takes to fill the tank. Then $x + 2 =$ time in hours that the slower pipe takes to fill the tank.

Faster pipe fills $\dfrac{1}{x}$ of the tank per hour. Slower pipe fills $\dfrac{1}{x + 2}$ of the tank per hour. Together they fill $\dfrac{1}{5}$ of the tank per hour.

$$
\begin{array}{ccccc}
\text{faster pipe} & + & \text{slower pipe} & = & \text{together} \\[4pt]
\dfrac{1}{x} & + & \dfrac{1}{x + 2} & = & \dfrac{1}{5}
\end{array}
$$

$$5x(x + 2) \cdot \frac{1}{x} + 5x(x + 2) \cdot \frac{1}{x + 2} = 5x(x + 2) \cdot \frac{1}{5} \qquad \text{Multiply each term by the LCM } 5x(x + 2)$$

$$
\begin{aligned}
5(x + 2) + 5x &= x(x + 2) \\
5x + 10 + 5x &= x^2 + 2x \\
10x + 10 &= x^2 + 2x \\
x^2 - 8x - 10 &= 0 \qquad \text{Write in standard form}
\end{aligned}
$$

Since $x^2 - 8x - 10$ is not factorable, we use the quadratic formula.

$$x = \frac{-(-8) \pm \sqrt{(-8)^2 - 4(1)(-10)}}{2(1)} \qquad \text{Replace } a \text{ with 1, } b \text{ with } -8, \text{ and } c \text{ with } -10$$

$$= \frac{8 \pm \sqrt{64 + 40}}{2}$$

$$= \frac{8 \pm \sqrt{104}}{2}$$

$$= \frac{8 \pm 2\sqrt{26}}{2} \qquad \begin{aligned}\sqrt{104} &= \sqrt{4 \cdot 26} \\ &= 2\sqrt{26}\end{aligned}$$

$$= \frac{2(4 \pm \sqrt{26})}{2}$$

$$= 4 \pm \sqrt{26}$$

Since $4 - \sqrt{26}$ is a negative time, we reject that solution. So

$$
\begin{aligned}
x &= 4 + \sqrt{26} \approx 4 + 5.1 \approx 9.1 \\
x + 2 &= (4 + \sqrt{26}) + 2 \approx 6 + 5.1 \approx 11.1
\end{aligned}
$$

It would take the faster pipe approximately 9.1 hours and the slower pipe approximately 11.1 hours to fill the tank alone.

▶ *Quick check*   a. When a ball is thrown straight up into the air, the equation $s = -16t^2 + 80t + 96$ gives the distance s that the ball is above the ground t seconds after it is thrown. How long does it take for the ball to hit the ground?
b. A rectangular plot of ground has an area of 161 square meters. If the length of the rectangle is two more than three times the width, what are the dimensions of the rectangle?   ■

Mastery points

Can you

- Substitute and solve physical formulas that are quadratic?
- Solve word problems involving the use of a right triangle and the relationship $a^2 + b^2 = c^2$?
- Solve word problems involving the areas of geometric figures?
- Solve word problems involving economic relationships?
- Solve work problems?

Exercise 6–4

Solve the following using quadratic equations. Compute all answers to the nearest tenth where necessary. See example 6–4 A–1.

Example   When a ball is thrown straight up into the air, the equation $s = -16t^2 + 80t + 96$ gives the distance s that the ball is above the ground t seconds after it is thrown. How long does it take for the ball to hit the ground?

Solution

$$0 = -16t^2 + 80t + 96 \qquad \text{Replace } s \text{ with } 0$$
$$16t^2 - 80t - 96 = 0 \qquad \text{Multiply equation by } -1 \text{ and interchange members}$$
$$16(t^2 - 5t - 6) = 0 \qquad \text{Factor the left member}$$
$$16(t - 6)(t + 1) = 0$$
$$t - 6 = 0 \quad \text{or} \quad t + 1 = 0 \qquad \text{Set each factor equal to } 0$$
$$t = 6 \qquad\qquad t = -1$$

Reject $t = -1$ because time cannot be negative. The ball will strike the ground 6 seconds after being thrown.

1. The distance s a body falls when air resistance is neglected is given by

$$s = v_0 t + 16t^2$$

where s is in feet, t is in seconds, and v_0 in feet per second is the initial velocity of the body. How long will it take a body to fall 80 feet if it has an initial velocity v_0 of 64 feet per second?

2. Using the information in exercise 1, how long will it take the body in a vacuum to fall 80 feet if the body starts from rest? (*Hint:* $v_0 = 0$ feet per second.)

3. Using the formula in exercise 1, how long will it take a body in a vacuum to fall 240 feet if the initial velocity is $v_0 = 32$ feet per second?

4. An object fired vertically into the air with an initial velocity of v_0 feet per second will be at a distance h in feet, t seconds after launching, determined by the equation

$$h = v_0 t - 16t^2$$

If the initial velocity is 96 feet per second, how long will it take the object to reach a height of 80 feet?

5. Using the formula in exercise 4, how long will it take for the object to return to the ground? (*Hint:* $h = 0$ when the object is on the ground.)

6. Using exercise 4, find the time when the object will be 12 feet off the ground.

7. An object is dropped from the top of the Washington Monument (555 feet tall). How long will it take the object to strike the ground? (*Hint:* Use exercise 1.)

8. The current in an electric circuit flows according to the equation

$$I = 18t - 12t^2$$

where I is the current in amperes and t is the time in seconds. In how many seconds will there be 6 amperes of current?

9. Using the formula in exercise 8, in how many seconds t will there be no current?

10. In a given circuit, the relationship between I (in amperes), E (in volts), and R (in ohms) is given by

$$I^2R + IE = 6,000$$

Find I when $E = 60$ volts and $R = 3$ ohms.

11. Using exercise 10, find I when $E = 50$ volts and $R = 5$ ohms.

Business problems

14. The demand equation for a specific commodity is given by

$$D = \frac{2,000}{p}$$

where D is the demand for the specific commodity at price p dollars per unit. If the supply equation is given by

$$S = 300p - 400$$

where S is the quantity of the commodity that the supplier is willing to supply at p dollars per unit, find the equilibrium price.
(*Note:* Equilibrium price occurs when $D = S$.)

15. Suppose that a manufacturer of ballpoint pens finds the demand equation to be

$$D = 24 - p^2$$

and the supply equation to be

$$S = p^2 + 2p$$

where p is the price of each pen in dollars. What is the equilibrium price? (See note in exercise 14.)

16. In exercise 15, at what price, to the nearest cent where necessary, (a) is there no demand, (b) will the quantity that the supplier is willing to sell be zero?

12. The formula for rating engine horsepower, hp, based on an average effective pressure on the piston of 67 pounds per square inch at a piston speed of 1,000 feet per minute is given by

$$hp = 0.4D^2 \times N$$

where D is the diameter of the piston bore and N is the number of pistons. Find D when an 8-cylinder engine has 36 horsepower.

13. A polygon is a closed geometric plane figure that has n sides, where $n \geq 3$. A diagonal of a polygon is a line segment connecting any two nonadjacent vertices of the polygon. The number of diagonals D of a polygon of n sides is given by

$$D = \frac{n(n - 3)}{2}$$

How many sides does the polygon have that has 405 diagonals?

17. In exercise 15, (a) how many pens will be supplied at the equilibrium price, (b) what is the demand at that price?

18. A small manufacturer finds the total cost C for a solar energy device to be

$$C = 50x^2 - 24,000$$

and the total revenue R at a price $100 per unit to be

$$R = 100x$$

where x is the number of units manufactured and sold. What is the break-even point (that is, where total cost = total revenue) to the nearest unit?

19. As in exercise 18, a steel producer's annual cost and revenue are given by

$$C = 20 - 0.4x^2 \text{ and } R = 0.8x$$

where x is the number of units produced and sold. To the nearest unit, what is the break-even point? (See exercise 18.)

20. The profit P in dollars in the manufacture and sale of a product is given by

$$P = \frac{1}{100}n^2 - 20n$$

where n is the number of units manufactured. How many units of the product must be manufactured to make a profit of $20,000?

21. A baker makes a profit P in cents according to the equation

$$P = -n^2 + 240n$$

where n is the number of cakes baked and sold. How many cakes should be made to realize a profit of $144?

22. What is the profit if the baker makes (a) 100 cakes, (b) 120 cakes, (c) 200 cakes? Can we draw any conclusion from these results? (Use the formula given in exercise 21.)

See example 6–4 A–3.

Geometric problems

Example A rectangular plot of ground has an area of 161 square meters. If the length of the rectangle is two more than three times the width, what are the dimensions of the rectangle?

Solution Let w = the width of the rectangle then $\ell = 3w + 2$. Using $A = \ell w$,

$$
\begin{array}{ll}
161 = (3w + 2)w & \text{Replace } A \text{ with 161 and } \ell \text{ with } 3w + 2 \\
161 = 3w^2 + 2w & \text{Multiply in right member} \\
3w^2 + 2w - 161 = 0 & \text{Write in standard form} \\
(3w + 23)(w - 7) = 0 & \text{Factor left member} \\
3w + 23 = 0 \quad \text{or} \quad w - 7 = 0 & \text{Set each factor equal to 0} \\
w = -\dfrac{23}{3} \qquad\qquad w = 7 &
\end{array}
$$

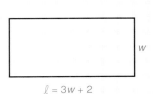
$\ell = 3w + 2$

Reject $-\dfrac{23}{3}$ since the side of a rectangle cannot be negative. Then $w = 7$ and

$\ell = 3w + 2 = 3(7) + 2 = 23$. The rectangle is 7 meters wide and 23 meters long.

24. If a rectangular-shaped playground has a length 5 feet less than three times the width and the distance from one corner to the opposite corner is 100 feet, find the dimensions of the playground.

25. Given a rectangular plot of land whose length is 30 feet longer than its width, what are the dimensions of the plot if a diagonal path across the plot is 240 feet long? (*Hint:* The diagonal is the hypotenuse of a right triangle.)

26. The diagonal of a square piece of metal is 50 centimeters long. Find the length of each side of the piece of metal.

27. The diagonal of a square flower bed is $5\sqrt{2}$ feet. Find the length of each side.

28. The area of a rectangular floor is 24 square meters. Find the dimensions of the floor if the width is 2 meters less than the length. ($A = \ell w$.)

29. A triangular-shaped plate has an altitude that is 5 inches longer than its base. If the area of the plate is 52 square inches, what is the altitude of the triangle? $\left(A = \dfrac{1}{2}bh. \right)$

30. A rectangle that is 6 centimeters long and 3 centimeters wide has its dimensions increased by the same amount. The area of this new rectangle is three times that of the old rectangle. What are the dimensions of the new rectangle? (*Hint:* Let x be the increase in the length and width.)

31. A rectangular field is twice as long as it is wide. Find the dimensions of the field if it contains 5,000 square yards.

32. Find the length of the radius r of a circular disk whose area A is 154 square feet.

(*Note:* $A = \pi r^2$ and use $\pi = \dfrac{22}{7}$ as an approximation of π.)

23. A rug company determines that its marginal profit MP is given by

$$MP = -n^2 + 40n - 80$$

where n is the number of rugs produced in thousands. The maximum profit is where MP is zero. Find n, to the nearest tenth.

33. Find the length of the diameter D of a circular gear whose area A is 12.56 square feet. (*Note:* $A = \pi r^2$ and $D = 2r$. Use $\pi = 3.14$ as an approximation of π.)

34. The base and altitude of a triangle are 3 inches and 5 inches, respectively. If the base is increased by twice as much as the altitude, the area of the new triangle is twice that of the old triangle. What are the lengths of the base and the altitude of the new triangle?

Use the Pythagorean Theorem. See example 6–4 A–2.

Right triangle problems

35. Find the length of the shortest side of a right triangle whose sides are y inches, $y + 1$ inches, and $y + 8$ inches.

36. The hypotenuse (longest side) of a right triangle is 10 millimeters long. One leg is 2 millimeters longer than the other. What are the lengths of the two legs?

37. The hypotenuse of a right triangle is 1 inch longer than the longer of the two legs. The shorter leg is 7 inches. Find the length of the hypotenuse.

38. A plot of ground has the shape of a right triangle. If the longer of the two legs is 9 dekameters longer than the shorter leg, and the hypotenuse is 8 dekameters longer than the longer leg, find the lengths of the three sides.

See example 6–4 A–4.

39. It takes Debbie 39 minutes longer to do a job than it takes Lisa to do the same job. Working together, they can do the job in 40 minutes. How long would it take each girl to do the job working alone?

40. Working together, Tom Roggenbeck and Amy Miyazaki can mow a lawn in 1 hour. Working alone, it would take Amy 90 minutes longer than it would take Tom. How long would it take Amy to mow the lawn alone?

41. A water tank has an inlet pipe (to fill the tank) and an outlet pipe (to empty the tank). The tank will fill in 8 hours when both pipes are open. If it takes 2 hours longer to empty the tank than it does to fill the tank, how long does it take the outlet pipe to empty the tank?
[*Hint:* (rate of inlet) − (rate of outlet) = (rate to fill).]

42. Tim can paint a room in 6 hours less time than Tom. If they can paint the room in 4 hours working together, how long would it take each boy to paint the room working alone?

Review exercises

1. Square $(2\sqrt{3} - 4)^2$. See section 5–6.

Find the solution set of the following equations. See sections 2–1 and 6–1.

2. $3y - 2 = 4(6y + 1)$

3. $y^2 - 5y = 24$

Simplify the following expressions. See section 5–1.

4. $(16)^{3/4}$

5. $x^{1/3} \cdot x^{1/2} \cdot x$

6. $\dfrac{x^{2/3}}{x^{1/2}}$

7. Rationalize the denominator of $\dfrac{\sqrt{2}}{\sqrt{3}}$. See section 5–4.

6–5 ■ *Equations involving radicals*

Equations in which at least one member contains a radical expression that has a variable in the radicand are called **radical equations.** When the radicals are of the second order (involving square root), we use the methods we have learned for solving equations together with the following property to find the solution set.

___ *Property of n*th **power** _____

Given a natural number $n > 1$ and real algebraic expressions P and Q, all of the solutions of the equation

$$P = Q$$

are contained in the solution set of the equation

$$P^n = Q^n$$

Concept
If each member of an equation is raised to some natural number power greater than one, the solution set of the original equation is a subset of the solution set of the resulting equation.

From this property, we can see that the equation $P^n = Q^n$ *may* contain solutions that are not solutions of the equation $P = Q$. Such solutions are called **extraneous solutions.** To illustrate, consider the equation $x - 3 = 5$, whose solution is 8. If we square each member of the equation, we obtain

$$(x - 3)^2 = 5^2$$
$$x^2 - 6x + 9 = 25$$
$$x^2 - 6x - 16 = 0 \quad \text{Write in standard form}$$
$$(x - 8)(x + 2) = 0$$
$$x - 8 = 0 \quad \text{or} \quad x + 2 = 0$$
$$x = 8 \quad \text{or} \qquad x = -2$$
$$S = \{8, -2\}$$

Since -2 is not a solution of the original equation, $x - 3 = 5$, we call -2 an extraneous solution.

Since the application of the *n*th power property may produce extraneous solutions, *each solution* obtained through the use of the property *must be checked* into the original equation. Now consider some radical equations where we apply this property.

■ *Example 6–5 A*

Find the solution set of each equation. Identify extraneous solutions, if they exist.

1. $\sqrt{x} - 4 = 5$
   $\sqrt{x} = 9$ Add 4 to each member
   $(\sqrt{x})^2 = (9)^2$ Square each member
   $x = 81$

Check our solution in the *original* equation.

$\sqrt{x} - 4 = 5$
$\sqrt{81} - 4 = 5$ Replace x with 81
$9 - 4 = 5$
$5 = 5$ (True)

No extraneous solutions exist and the solution set is $\{81\}$.

2.
$$\sqrt{2x - 5} = 3$$
$$(\sqrt{2x - 5})^2 = 3^2 \qquad \text{Square each member}$$
$$2x - 5 = 9$$
$$2x = 14$$
$$x = 7$$

We must check our possible solution.

$$\sqrt{2x - 5} = 3$$
$$\sqrt{2(7) - 5} = 3$$
$$\sqrt{9} = 3$$
$$3 = 3 \quad \text{(True)}$$

There is no extraneous solution and the solution set is $\{7\}$.

3.
$$\sqrt{x} = x - 12$$
$$(\sqrt{x})^2 = (x - 12)^2 \qquad \text{Square each member}$$
$$x = x^2 - 24x + 144$$

— Don't forget the middle term

Note Remember, $(x - 12)^2 = x^2 - 24x + 144$. A common error we can make is to say $(x - 12)^2 = x^2 - 12^2 = x^2 - 144$. $(x - 12)^2 \neq x^2 - 144$.

$$0 = x^2 - 25x + 144 \qquad \text{Subtract } x \text{ from each member}$$
$$0 = (x - 16)(x - 9)$$
$$x - 16 = 0 \quad \text{or} \quad x - 9 = 0$$
$$x = 16 \qquad\qquad x = 9$$

Check:

Let $x = 16$, then
$$\sqrt{16} = 16 - 12$$
$$4 = 16 - 12$$
$$4 = 4 \quad \text{(True)}$$

Let $x = 9$, then
$$\sqrt{9} = 9 - 12 \qquad \text{Replace } x \text{ with 16, 9}$$
$$3 = 9 - 12$$
$$3 = -3 \quad \text{(False)}$$

Therefore 9 is an extraneous solution and the solution set of the equation $\sqrt{x} = x - 12$ is $\{16\}$.

4. $\sqrt{x + 4} - \sqrt{x - 3} = 1$
When two terms contain radical expressions, we must rewrite the equation with a radical expression in each member. Add $\sqrt{x - 3}$ to each member.

$$\sqrt{x + 4} = \sqrt{x - 3} + 1$$
$$(\sqrt{x + 4})^2 = (\sqrt{x - 3} + 1)^2 \qquad \text{Square each member}$$
$$x + 4 = x - 3 + 2\sqrt{x - 3} + 1 \qquad \text{Watch this step carefully}$$
Then $x + 4 = x - 2 + 2\sqrt{x - 3}$

Isolate the remaining radical expression in one member by adding $-x + 2$ to both members. Then

$$x + 4 - x + 2 = 2\sqrt{x - 3}$$
$$6 = 2\sqrt{x - 3} \qquad \text{Divide each member by 2}$$
$$3 = \sqrt{x - 3}$$
$$(3)^2 = (\sqrt{x - 3})^2 \qquad \text{Square each member}$$
$$9 = x - 3$$
$$x = 12$$

Check: Let $x = 12$, then

$$\sqrt{12 + 4} - \sqrt{12 - 3} = 1$$
$$\sqrt{16} - \sqrt{9} = 1$$
$$4 - 3 = 1$$
$$1 = 1 \quad \text{(True)}$$

There is no extraneous solution and the solution set is $\{12\}$.

5.
$$\sqrt[3]{2x - 3} = 2$$
$$(\sqrt[3]{2x - 3})^3 = 2^3 \qquad \qquad \text{Cube each member}$$
$$2x - 3 = 8$$
$$2x = 11$$
$$x = \frac{11}{2}$$

Note Extraneous roots occur only when the radical involves even roots, so there is no need to check the answer.

The solution set is $\left\{\dfrac{11}{2}\right\}$.

6.
$$\sqrt{x} + 1 = 0$$
$$\sqrt{x} = -1 \qquad \qquad \text{Subtract 1 from each member}$$
$$x = 1 \qquad \qquad \text{Square each member}$$

Check: Let $x = 1$, then

$$\sqrt{1} + 1 = 0$$
$$1 + 1 = 0$$
$$2 = 0 \quad \text{(False)}$$

Since 1 does not check, the solution set is $\emptyset$, and 1 is an extraneous solution.

▶ *Quick check* Find the solution set of $\sqrt{z} = z - 6$. Identify any extraneous solutions, if they exist. ◼

In general, we use this procedure to solve a radical equation.

___ **To solve radical equations** _____

1. Isolate one radical term alone in one member of the equation.
2. Raise each member of the equation to the power that is the same as the index of the radical.
3. Solve the resulting equation. If a radical term remains, repeat steps 1 and 2.
4. Check all possible solutions in the original equation if the equation involves an even root.

___ **Mastery points** _____

Can you
■ Identify extraneous solutions of a radical equation?
■ Find the solution set of a radical equation?

Exercise 6–5

Find the solution set of each equation. Identify extraneous solutions, if they exist. See example 6–5 A.

Example $\sqrt{z} = z - 6$

Solution

$$(\sqrt{z})^2 = (z - 6)^2 \qquad \text{Square each member}$$
$$z = z^2 - 12z + 36 \qquad \text{Multiply in right member}$$
$$z^2 - 13z + 36 = 0 \qquad \text{Write equation in standard form}$$
$$(z - 9)(z - 4) = 0 \qquad \text{Factor left member}$$
$$z - 9 = 0 \quad \text{or} \quad z - 4 = 0 \qquad \text{Set each factor equal to 0}$$
$$z = 9 \qquad\qquad z = 4$$

Check:

$$z = 9 \qquad\qquad z = 4$$
$$\sqrt{9} = 9 - 6 \qquad \sqrt{4} = 4 - 6$$
$$3 = 3 \quad \text{(True)} \qquad 2 = -2 \quad \text{(False)}$$

4 is an extraneous solution and the solution set is $\{9\}$.

1. $\sqrt{x} - 6 = 3$
2. $\sqrt{x} + 9 = 13$
3. $\sqrt{x} + 5 = 2$
4. $\sqrt{x + 2} = 5$
5. $\sqrt{y - 4} = 7$
6. $\sqrt{3k - 1} - 5 = 0$
7. $\sqrt{5z + 1} - 11 = 0$
8. $\sqrt{9a + 5} = \sqrt{3a - 1}$
9. $\sqrt{2p + 5} = \sqrt{3p + 4}$
10. $\sqrt{r + 6} = \sqrt{3r + 2}$
11. $2\sqrt{3z} = \sqrt{5z + 7}$
12. $3\sqrt{x - 3} = \sqrt{2x - 5}$
13. $2\sqrt{2z - 1} = \sqrt{4z}$
14. $\sqrt{p}\,\sqrt{p - 8} = 3$
15. $\sqrt{y}\,\sqrt{x - 5} = 6$
16. $\sqrt{m}\,\sqrt{2m + 4} = 4$
17. $\sqrt{2z}\,\sqrt{z - 3} = 6$
18. $\sqrt{x^2 + 3} - 2 = 0$
19. $\sqrt{y^2 + 7} - 3 = 0$
20. $\sqrt{w^2 - 6w} = 4$
21. $\sqrt{t^2 + 8t} = 3$
22. $\sqrt{z^2 + 12} + 3 = 0$
23. $\sqrt{y^2 + 7} - 4 = 0$
24. $x\sqrt{2} = \sqrt{6 - 4x}$
25. $y\sqrt{3} = \sqrt{9y + 30}$
26. $z\sqrt{3} = \sqrt{z + 1}$
27. $t\sqrt{2} = \sqrt{5t - 2}$
28. $\sqrt{z} + 12 = z$
29. $\sqrt{2x} = x - 4$
30. $p = \sqrt{6 - p}$
31. $q = \sqrt{5q - 4}$
32. $\sqrt{x - 2} = x - 2$
33. $\sqrt{u - 1} = u - 3$
34. $\sqrt{3x + 10} - 3x = 4$
35. $2t - \sqrt{5 - 2t} = 5$
36. $1 + \sqrt{5x + 9} = x$
37. $\sqrt{5x + 1} - 1 = \sqrt{3x}$
38. $\sqrt{v + 5} = 5 - \sqrt{v}$
39. $\sqrt{2n + 3} - \sqrt{n - 2} = 2$
40. $3 - \sqrt{y + 4} = \sqrt{y + 7}$
41. $\sqrt{p + 1} = \sqrt{2p + 9} - 2$
42. $(2y + 3)^{1/2} - (4y - 1)^{1/2} = 0$
   [*Hint:* $(2y + 3)^{1/2} = \sqrt{2y + 3}$.]
43. $(2x^2 + 3x - 5)^{1/2} = (2x^2 - x - 2)^{1/2}$
44. $(4x + 2)^{1/2} - (2x)^{1/2} = 0$
45. $(1 - 2y)^{1/2} + (y + 5)^{1/2} = 4$
46. $(x - 2)^{1/2} = (5x + 1)^{1/2} - 3$
47. $\sqrt[3]{x - 7} = 3$
48. $\sqrt[4]{2y - 3} = 1$
49. $\sqrt[3]{x^2 - 6x - 8} = 2$
50. $\sqrt[4]{2x^2 - 3x - 8} = -1$
51. $(4p - 3)^{1/3} = -2$
52. $(2q - 5)^{1/5} = -1$

Solve the following equations and formulas for the indicated variable. Assume all denominators are nonzero.

Example $3x\sqrt{x + y} = 2$ for y

Solution

$$(3x\sqrt{x + y})^2 = 2^2 \qquad \text{Square each member}$$
$$9x^2(x + y) = 4$$
$$9x^3 + 9x^2y = 4 \qquad \text{Perform the indicated multiplication}$$
$$9x^2y = 4 - 9x^3 \qquad \text{Subtract } 9x^3 \text{ from both members}$$
$$y = \frac{4 - 9x^3}{9x^2} \qquad \text{Divide each member by } 9x^2$$

53. $r = \sqrt{\dfrac{A}{4\pi h}}$ for A

54. $t = \sqrt{\dfrac{2s}{g}}$ for s

55. $r = \sqrt{\dfrac{A}{\pi} - R^2}$ for A

56. $r = \sqrt{\dfrac{V}{\pi h}}$ for h

57. $D = \sqrt[3]{\dfrac{6A}{\pi}}$ for A

58. $v = \sqrt{\dfrac{2gKE}{W}}$ for W

59. At an altitude of h ft above the sea or level ground, the distance d in miles that a person can see an object is given by

$$d = \sqrt{\dfrac{3h}{2}}$$

How tall must a person be to see an object 3 miles away?

60. The formula for approximating the velocity V in miles per hour of a car based on the length of its skid marks S (in feet) on dry pavement is given by

$$V = 2\sqrt{6S}$$

If the velocity is 48 mph, how long will the skid marks be?

61. On wet pavement, the formula in exercise 60 is given by

$$V = 2\sqrt{3S}$$

How long will the skid marks be if the car is traveling at 30 mph on wet pavement?

62. Find the number whose principal square root is $3i$. (*Hint:* Let $\sqrt{x} = 3i$.)

63. Find the number whose principal third root is -3.

64. Find the number whose principal fourth root is 4.

Review exercises

Find the solution set of the following equations. See sections 6-1 and 6-3.

1. $y^2 + 6y = 16$

2. $3x^2 + 2x - 2 = 0$

Find the solution set of the following inequalities. See section 2-5.

3. $4y - 1 \geq 2y + 7$

4. $-4 < 3x + 2 \leq 5$

5. A number plus one-half the number plus one-third the number is 33. Find the number. See section 2-3.

6. Write an inequality to state that the temperature on a given day had a low of $21°$ and a high of $62°$. See section 2-5.

6-6 ■ Equations that are quadratic in form

There are a number of equations that are not quadratic equations but they can, nevertheless, be written in **quadratic form**

$$au^2 + bu + c = 0, \qquad a > 0$$

and solved as we solved quadratic equations. The variable u in the equation represents some expression in another variable. Such equations are said to be reducible to quadratic equations by making a substitution. We then solve the resulting equation by the methods of this chapter.

■ *Example 6–6 A*

Find the solution set of each equation.

1. $x^4 + 4x^2 - 12 = 0$

$$x^4 \quad + 4x^2 \quad - 12 = 0$$
$$\downarrow \qquad \downarrow \qquad \downarrow \quad \downarrow$$
$$(x^2)^2 + 4(x^2) - 12 = 0 \qquad \text{Replace } x^4 \text{ with } (x^2)^2$$
$$\downarrow \qquad \downarrow \qquad \downarrow \quad \downarrow$$
$$u^2 \quad + 4u \quad - 12 = 0 \qquad \text{Replace } x^2 \text{ with } u \text{ to make factoring easier}$$

Solve the equation $u^2 + 4u - 12 = 0$.

$$(u + 6)(u - 2) = 0 \qquad \text{Factor the left member}$$
$$u + 6 = 0 \quad \text{or} \quad u - 2 = 0 \qquad \text{Set each factor equal to 0}$$
$$u = -6 \qquad\qquad u = 2$$

THESE ARE *NOT* THE SOLUTIONS FOR *x*. We *must* now replace *u* with x^2 to get back to the original equation.

$$x^2 = -6 \quad \text{or} \quad x^2 = 2 \qquad \text{Replace } u \text{ with } x^2$$
$$x = \pm\sqrt{-6} \qquad\quad x = \pm\sqrt{2} \qquad \text{Extract the roots}$$
$$x = \pm i\sqrt{6} \qquad\qquad\qquad\qquad \sqrt{-6} = i\sqrt{6}$$

The solution set is $\{i\sqrt{6}, -i\sqrt{6}, \sqrt{2}, -\sqrt{2}\}$.

Note   The equation is of fourth degree, and we obtained four solutions. The degree of an equation indicates the *maximum* number of solutions we can expect to find.

2. $y + 3\sqrt{y} - 4 = 0$

$$y \qquad + 3\sqrt{y} - 4 = 0$$
$$\downarrow \qquad\quad \downarrow \qquad \downarrow$$
$$(\sqrt{y})^2 + 3\sqrt{y} - 4 = 0 \qquad \text{Replace } y \text{ with } (\sqrt{y})^2$$
$$\downarrow \qquad\quad \downarrow \qquad \downarrow$$
$$u^2 \quad + 3u \quad - 4 = 0 \qquad \text{Replace } \sqrt{y} \text{ with } u$$

Note   We now have a quadratic equation and will have, at most, two solutions.

$$(u + 4)(u - 1) = 0 \qquad \text{Factor the left member}$$
$$u = -4 \text{ or } u = 1$$

THESE ARE *NOT* THE SOLUTIONS FOR *y*. We *must* now replace *u* with $\sqrt{y}$.

$$\sqrt{y} = -4 \quad \text{or} \quad \sqrt{y} = 1$$
$$(\sqrt{y})^2 = (-4)^2 \qquad (\sqrt{y})^2 = (1)^2 \qquad \text{Square each member}$$
$$y = 16 \qquad\qquad y = 1$$

Since we squared each member of each equation to get these *possible* solutions, we *must* check the original equation $y + 3\sqrt{y} - 4 = 0$ for extraneous solutions.

Check:

Let $y = 16$, then

$$16 + 3\sqrt{16} - 4 = 0$$
$$16 + 3(4) - 4 = 0$$
$$16 + 12 - 4 = 0$$
$$28 - 4 = 0$$
$$24 = 0 \quad \text{(False)}$$

Let $y = 1$, then

$$1 + 3\sqrt{1} - 4 = 0$$
$$1 + 3 - 4 = 0$$
$$4 - 4 = 0$$
$$0 = 0 \quad \text{(True)}$$

Thus 16 is an extraneous solution and the solution set is $\{1\}$.

Note We could have predicted that 16 would be extraneous since it came from the equation $\sqrt{y} = -4$ and the principal square root of a number is *never* negative.

3. $3x^{-2} + 7x^{-1} - 6 = 0$

$$3x^{-2} \quad + 7x^{-1} - 6 = 0$$
$$\downarrow \qquad \downarrow \qquad \downarrow$$
$$3(x^{-1})^2 + 7x^{-1} - 6 = 0 \qquad \text{Replace } x^{-2} \text{ with } (x^{-1})^2$$
$$\downarrow \qquad \downarrow \qquad \downarrow$$
$$3u^2 \quad + 7u \quad - 6 = 0 \qquad \text{Replace } x^{-1} \text{ with } u$$

$$(3u - 2)(u + 3) = 0$$
$$3u - 2 = 0 \quad \text{or} \quad u + 3 = 0$$
$$u = \frac{2}{3} \qquad\qquad u = -3$$

THESE ARE *NOT* THE SOLUTIONS FOR x. We *must* replace u with x^{-1} or $\frac{1}{x}$.

$$\frac{1}{x} = \frac{2}{3} \quad \text{or} \quad \frac{1}{x} = -3 \qquad \text{Replace } u \text{ with } \frac{1}{x}$$
$$2x = 3 \qquad\quad -3x = 1 \qquad \text{Multiply by the LCD}$$
$$x = \frac{3}{2} \qquad\qquad x = -\frac{1}{3}$$

The solution set is $\left\{ \frac{3}{2}, -\frac{1}{3} \right\}$.

Note This equation could be written $\frac{3}{x^2} + \frac{7}{x} - 6 = 0$ and solved as a rational equation.

Note To recognize each of these as being the quadratic type, you must observe the characteristic that each of these equations has in common. Notice that the square of the variable factor of the middle term yields the variable factor of the first term. That is,

1. $(x^2)^2 = x^4$
2. $(\sqrt{y})^2 = y$
3. $(x^{-1})^2 = x^{-2}$

▶ *Quick check* Find the solution set of $5z^{-2} + 6z^{-1} - 8 = 0$.

```
┌─ Mastery points ──────────────────────────────┐
│                                                 │
│  Can you                                        │
│   ■ Identify a quadratic-type equation?         │
│   ■ Solve any quadratic-type equation?          │
│                                                 │
└─────────────────────────────────────────────────┘
```

Exercise 6–6

Find the solution set of each equation. Identify extraneous solutions, if they exist. See example 6–6 A.

Example $5z^{-2} + 6z^{-1} - 8 = 0$

Solution $5z^{-2} + 6z^{-1} - 8 = 0$

$$5(z^{-1})^2 + 6z^{-1} - 8 = 0 \qquad \text{Replace } z^{-2} \text{ with } (z^{-1})^2$$

$$5u^2 + 6u - 8 = 0 \qquad \text{Replace } z^{-1} \text{ with } u$$

We now solve the equation $5u^2 + 6u - 8 = 0$.

$$(5u - 4)(u + 2) = 0 \qquad \text{Factor the left member}$$
$$5u - 4 = 0 \quad \text{or} \quad u + 2 = 0 \qquad \text{Set each factor equal to 0}$$
$$u = \frac{4}{5} \qquad\qquad u = -2$$

Replace u with z^{-1} or $\frac{1}{z}$.

$$\frac{1}{z} = \frac{4}{5} \quad \text{or} \quad \frac{1}{z} = -2 \qquad \text{Replace } u \text{ with } \frac{1}{z}$$
$$4z = 5 \qquad\quad -2z = 1 \qquad \text{Multiply by the LCD}$$
$$z = \frac{5}{4} \qquad\quad z = -\frac{1}{2}$$

The solution set is $\left\{ -\frac{1}{2}, \frac{5}{4} \right\}$.

1. $x^4 - 6x^2 + 5 = 0$
2. $y^4 + 3y^2 - 28 = 0$
3. $3z^4 + z^2 = 2$
4. $4p^4 = 25p^2 - 6$
5. $(x - 2)^4 + 9(x - 2)^2 + 8 = 0$
6. $(m + 5)^4 - 4(m + 5)^2 + 4 = 0$
7. $(p^2 - 2p)^2 - 7(p^2 - 2p) - 8 = 0$
8. $(x + 3)^2 - 6(x + 3) + 8 = 0$
9. $q - 6\sqrt{q} - 27 = 0$
10. $x + \sqrt{x} = 12$
11. $2y + 3\sqrt{y} + 1 = 0$
12. $3t - 4\sqrt{t} = 4$
13. $p - 5p^{1/2} = -4$
14. $2x = 9 - 3x^{1/2}$
15. $(x^2 - 3) - 3\sqrt{x^2 - 3} - 28 = 0$
16. $(m^2 + 1) + \sqrt{m^2 + 1} = 20$
17. $(y - 6) - \sqrt{y - 6} - 2 = 0$
18. $(z + 1) - 8\sqrt{z + 1} + 7 = 0$
19. $y^{2/3} + 3y^{1/3} - 10 = 0$
20. $2y^{2/3} - 3y^{1/3} = 2$
21. $5p^{2/3} + 11p^{1/3} + 2 = 0$
22. $u^{3/2} - 7u^{3/4} - 8 = 0$
23. $x^{3/2} - 2x^{3/4} + 1 = 0$
24. $p^{3/4} - 3p^{3/8} + 2 = 0$
25. $z^{-2} - z^{-1} - 12 = 0$
26. $r^{-2} - 7r^{-1} + 6 = 0$
27. $2y^{-2} = 9y^{-1} - 4$
28. $x^{-4} = 5x^{-2} - 4$
29. $4y^{-4} + 4 = 17y^{-2}$
30. $(t^2 - t)^2 - 4(t^2 - t) - 12 = 0$
31. $(x^2 + 3x)^2 - 8(x^2 + 3x) = 20$

32. Given the equation $x - 2\sqrt{x} - 15 = 0$, find the solution set by the method in (a) section 6–5 and (b) section 6–6.

33. Find the solution set of the equation $y + 4\sqrt{y} - 12 = 0$ using the method of (a) section 6–5 and (b) section 6–6.

Review exercises

Perform the indicated operations. See sections 4–2 and 4–3.

1. $\dfrac{3x}{x^2 - 4} - \dfrac{4x}{x - 2}$

2. $\dfrac{3x + 1}{x - 4} \div \dfrac{3x^2 - 2x - 1}{x^2 - 16}$

3. Given $y = 5x - 2$, find y when (a) $x = 2$, (b) $x = -5$, (c) $x = 0$. See section 1–5.

4. Find the solution set of $4x - 3 \geq 2(x + 1)$. See section 2–5.

Find the solution set of the following. See sections 2–4 and 2–6.

5. $|2x - 3| = 5$

6. $|x - 3| \leq 4$

7. $|3x + 1| > 2$

6-7 ■ Quadratic and rational inequalities

We have solved linear inequalities in chapter 2 and quadratic equations in this chapter. The methods learned there are now used to solve **quadratic inequalities.**

> A quadratic inequality is any inequality that can be written in the form
>
> $$ax^2 + bx + c < 0, \qquad ax^2 + bx + c > 0,$$
> $$ax^2 + bx + c \leq 0, \quad \text{or} \quad ax^2 + bx + c \geq 0,$$
>
> where a, b, and c are real numbers, $a > 0$.

The method used to solve quadratic inequalities is shown in the following example. Given

$$x^2 + 2x - 3 > 0$$

factor the left member to get

$$(x + 3)(x - 1) > 0$$

We now set each factor in the left member equal to 0 and solve each equation.

$$x + 3 = 0 \quad \text{or} \quad x - 1 = 0$$
$$x = -3 \qquad \qquad x = 1$$

The roots -3 and 1, called *critical numbers,* divide the real number line into three regions as shown in figure 6–1.

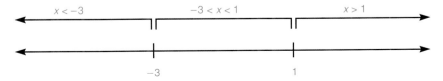

Figure 6–1

It can be shown that if one number in a given region makes the product $(x + 3)(x - 1)$ positive (as we want it to be), then all numbers in that region will make the product positive.

We now choose a number within each region as a *test number* to see if it satisfies the inequality. Suppose we choose the numbers

$$-4 \text{ in } x < -3; \quad 0 \text{ in } -3 < x < 1; \quad 2 \text{ in } x > 1$$

Note *Never test a region with the test number as a critical number.*

Substitute these numbers into the inequality.

When $x = -4$, then $(x + 3)(x - 1) = (-4 + 3)(-4 - 1)$
$= (-1)(-5) = 5$ Positive
When $x = 0$, then $(x + 3)(x - 1) = (0 + 3)(0 - 1)$
$= (3)(-1) = -3$ Negative
When $x = 2$, then $(x + 3)(x - 1) = (2 + 3)(2 - 1)$
$= (5)(1) = 5$ Positive

Thus the solution set includes all points in the regions $x < -3$ or $x > 1$. The solution set of the inequality $x^2 + 2x - 3 > 0$ is

$$\{x \mid x < -3 \text{ or } x > 1\}$$

The solution set is graphed on the number line in figure 6–2.

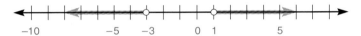

Figure 6–2

To solve a quadratic inequality

1. Write the inequality in the form

$$ax^2 + bx + c < 0, \qquad ax^2 + bx + c > 0,$$
$$ax^2 + bx + c \leq 0, \quad \text{or} \quad ax^2 + bx + c \geq 0$$

2. Factor the quadratic trinomial $ax^2 + bx + c$. If not factorable, then set $ax^2 + bx + c = 0$ and solve for x.

3. Find the *critical numbers* by setting each factor equal to 0 and solving for x.

4. Divide the number line into regions using the critical numbers. Write the regions using the same inequality symbol as the original problem.

5. Choose a *test number* (other than a critical number) within each region obtained and check the *sign* of the product at the test number.

6. Choose the region(s) that satisfy the conditions of the original inequality.

■ **Example 6–7 A**

Find the solution set of the following quadratic inequalities. Graph the solution set on the number line.

1. $x^2 - x - 6 \leq 0$

$$(x - 3)(x + 2) \leq 0 \qquad \text{Factor the left member}$$
$$x - 3 = 0 \qquad x + 2 = 0 \qquad \text{Set each factor equal to 0}$$
$$x = 3 \qquad x = -2 \qquad \text{Solve each equation}$$

The critical numbers are -2 and 3. We write the regions shown here using the same inequality symbols.

Choose the test numbers -3 in $x < -2$; 0 in $-2 < x < 3$; 4 in $x > 3$. Substitute the test numbers into the factors of the inequality $(x - 3)(x + 2) \leq 0$.

If $x = -3$, then $(x - 3)(x + 2) = (-3 - 3)(-3 + 2)$
$$= (-6)(-1) = 6 \qquad \text{Positive}$$
If $x = 0$, then $(x - 3)(x + 2) = (0 - 3)(0 + 2)$
$$= (-3)(2) = -6 \qquad \text{Negative}$$
If $x = 4$, then $(x - 3)(x + 2) = (4 - 3)(4 + 2)$
$$= (1)(6) = 6 \qquad \text{Positive}$$

Since we want those values of x that make the quadratic less than (negative) or equal to zero, the numbers in the region $-2 \leq x \leq 3$ satisfy the inequality. The solution set is $\{x \mid -2 \leq x \leq 3\}$.

2. $3x^2 - 4x - 4 > 0$

$$(3x + 2)(x - 2) > 0 \qquad \text{Factor the left member}$$
$$3x + 2 = 0 \qquad x - 2 = 0 \qquad \text{Set each factor equal to 0}$$
$$x = -\frac{2}{3} \qquad x = 2 \qquad \text{Solve each equation}$$

The critical numbers are $-\dfrac{2}{3}$ and 2. We write the regions as shown here.

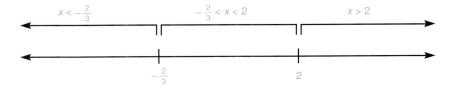

Choose the test numbers -1 in $x < -\dfrac{2}{3}$; 0 in $-\dfrac{2}{3} < x < 2$; 3 in $x > 2$.

Substitute these test numbers in the inequality $(3x + 2)(x - 2) > 0$.

If $x = -1$, then $(3x + 2)(x - 2) = [3(-1) + 2](-1 - 2)$
$\qquad\qquad\qquad\qquad\qquad = (-1)(-3) = 3$ <small>Positive</small>
If $x = 0$, then $(3x + 2)(x - 2) \ = [3(0) + 2](0 - 2)$
$\qquad\qquad\qquad\qquad\qquad = (2)(-2) = -4$ <small>Negative</small>
If $x = 3$, then $(3x + 2)(x - 2) \ = [3(3) + 2](3 - 2)$
$\qquad\qquad\qquad\qquad\qquad = (11)(1) = 11$ <small>Positive</small>

The solution set is $\left\{ x \mid x < -\dfrac{2}{3} \text{ or } x > 2 \right\}$.

▶ *Quick check* Find and graph the solution set of $y^2 + 2y - 8 \le 0$.

We can apply this same method in solving a *rational* inequality such as

$$\frac{z}{z - 5} \ge 3$$

Our first inclination is to multiply each member by the denominator to clear the denominator. However we do not know whether the denominator represents a *positive* or a *negative* number. Recall that this affects the order symbol involved. An easier approach is the method used in the following example.

3. $\dfrac{z}{z - 5} \ge 3$

Write the corresponding equation and solve it.

$$\begin{aligned}
\frac{z}{z - 5} &= 3 &&\text{Change} \ge \text{to} = \\
z &= 3(z - 5) &&\text{Multiply each member by } z - 5 \\
z &= 3z - 15 &&\text{Solve the equation} \\
-2z &= -15 \\
z &= \frac{15}{2}
\end{aligned}$$

Set the denominator equal to zero and solve the equation.

$$\begin{aligned}
z - 5 &= 0 \\
z &= 5
\end{aligned}$$

The critical numbers are 5 and $\dfrac{15}{2}$. They define these regions.

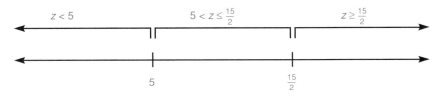

Choose the test numbers 0 in $z < 5$; 6 in $5 < z < \dfrac{15}{2}$; 8 in $z > \dfrac{15}{2}$.

Note   Since $z - 5$ is in the denominator, then $z \neq 5$ and we use $z < 5$ rather than $z \leq 5$.

Test the numbers in the given inequality.

$$\text{If } z = 0, \text{ then } \frac{(0)}{(0) - 5} = 0 \geq 3 \text{ is false.}$$

$$\text{If } z = 6, \text{ then } \frac{(6)}{(6) - 5} = 6 \geq 3 \text{ is true.}$$

$$\text{If } z = 8, \text{ then } \frac{(8)}{(8) - 5} = \frac{8}{3} \geq 3 \text{ is false.}$$

The solution set is $\left\{ z \,\middle|\, 5 < z \leq \dfrac{15}{2} \right\}$.

To solve a rational inequality

1. Write the inequality as an equation and solve it.
2. Set the denominator equal to zero and solve it.
3. The solutions from steps 1 and 2 are the critical numbers. Use these numbers to divide the number line into regions.
4. Test a number in each region by substituting it into the original inequality to determine the region(s) that satisfy it.
5. Exclude any numbers that make the denominator equal to zero.

▶ **Quick check**   Find and graph the solution set. $\dfrac{x}{x + 3} \leq 2$ ■

Mastery points

Can you
- Find the solution set of a quadratic inequality?
- Graph the solution set of a quadratic inequality?
- Find and graph the solution set of a rational inequality?

Exercise 6–7

Find the solution set and graph the solution set for each of the following inequalities. Write the answer in set-builder notation. See example 6–7 A–1 and 2.

Example $y^2 + 2y - 8 \le 0$

Solution $(y + 4)(y - 2) \le 0$ Factor the left member
$\qquad y + 4 = 0 \qquad y - 2 = 0$ Set each factor equal to 0
$\qquad\qquad y = -4 \qquad\quad y = 2$ Solve each equation

The critical numbers are -4 and 2. They define the regions shown here.

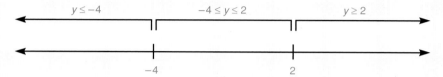

Choose the test numbers -5 in $y < -4$; 0 in $-4 < y < 2$; 3 in $y > 2$.

If $y = -5$, then $(y + 4)(y - 2) = (-5 + 4)(-5 - 2)$
$\qquad\qquad\qquad\qquad\qquad\qquad = (-1)(-7) = 7$ Positive
If $y = 0$, then $(y + 4)(y - 2) \quad = (0 + 4)(0 - 2)$
$\qquad\qquad\qquad\qquad\qquad\qquad = (4)(-2) = -8$ Negative
If $y = 3$, then $(y + 4)(y - 2) \quad = (3 + 4)(3 - 2)$
$\qquad\qquad\qquad\qquad\qquad\qquad = (7)(1) = 7$ Positive

Since we want ≤ 0 (negative or zero), the solution set is $\{y | -4 \le y \le 2\}$.

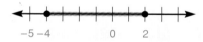

1. $(x + 3)(x - 1) > 0$ 2. $(y - 4)(y - 5) \ge 0$ 3. $(p + 7)(p + 2) \le 0$ 4. $(z - 6)(z + 8) < 0$

5. $r(r - 1) \ge 0$ 6. $q(3q + 4) < 0$ 7. $x^2 - 5x + 4 < 0$ 8. $m^2 + 6m + 5 \ge 0$

9. $q^2 - 3q \le 18$ 10. $t^2 + t > 30$ 11. $x^2 + 2 < 3x$ 12. $w^2 - 8 \le 2w$

13. $u^2 \ge 12 - 4u$ 14. $x^2 - 1 < 0$ 15. $y^2 - 4 > 0$ 16. $p^2 \ge 5$

17. $5x^2 - 15 < 0$ 18. $6y^2 \ge 42$ 19. $x^2 - 5x < 0$ 20. $2m^2 - 3m < 0$

21. $2w^2 \le -6w$ 22. $2x^2 - 7x - 4 > 0$ 23. $2y^2 + y - 6 < 0$ 24. $4z^2 + 7z + 3 \le 0$

25. $3w^2 + 16w + 5 \ge 0$ 26. $2p^2 - 3p < 9$ 27. $9v^2 - 8 < 6v$ 28. $6w^2 - 7 \le -11w$

29. $5y^2 - y \ge 4$ 30. $(x - 3)(x + 1)(x - 2) > 0$ 31. $(y + 4)(y - 5)(y + 6) < 0$

32. $r(3r + 5)(r - 6) \le 0$ 33. $t(t + 5)(3t - 8) \ge 0$

Find the solution set of the following rational inequalities. Graph the solution set. Write the answer in set-builder notation. See example 6–7 A–3.

Example $\dfrac{x}{x + 3} \leq 2$

Solution Write the inequality as an equation and solve the equation.

$$\dfrac{x}{x + 3} = 2 \qquad \text{Replace} \leq \text{with} =$$

$$x = 2(x + 3) \qquad \text{Multiply each member by } x + 3$$

$$x = 2x + 6$$

$$x = -6$$

Set the denominator equal to zero and solve the equation.

$$x + 3 = 0$$

$$x = -3$$

The critical numbers are -6 and -3. They define the following regions.

Choose the test numbers -7 in $x < -6$; -5 in $-6 < x < -3$; and 0 in $x > -3$.

If $x = -7$, then $\dfrac{(-7)}{(-7) + 3} = \dfrac{-7}{-4} = \dfrac{7}{4} \leq 2$ (True)

If $x = -5$, then $\dfrac{(-5)}{(-5) + 3} = \dfrac{-5}{-2} = \dfrac{5}{2} \leq 2$ (False)

If $x = 0$, then $\dfrac{(0)}{(0) + 3} = 0 \leq 2$ (True)

The solution set is $\{x \mid x \leq -6 \text{ or } x > -3\}$.

Note Since $x \neq -3$, we must use $x > -3$ instead of $x \geq -3$.

34. $\dfrac{x - 4}{x - 2} \leq 0$

35. $\dfrac{p + 3}{p + 1} \geq 0$

36. $\dfrac{3q - 2}{2q + 5} > 0$

37. $\dfrac{5t + 4}{7t - 3} < 0$

38. $\dfrac{1}{x} > 2$

39. $\dfrac{5}{y} < 3$

40. $\dfrac{3}{u + 4} \geq 2$

41. $\dfrac{-5}{v - 6} \leq 1$

42. $\dfrac{1}{2r - 3} < -3$

43. $\dfrac{-4}{3t + 8} > 1$

44. $\dfrac{-1}{2x - 7} < -2$

45. $\dfrac{x}{x + 5} > 3$

46. $\dfrac{y}{y - 3} > 4$

47. $\dfrac{x + 6}{x - 7} \leq 4$

48. $\dfrac{y - 4}{2y + 5} < -1$

49. $\dfrac{2z - 5}{z + 2} > 1$

50. $\dfrac{2p}{p-2} \le p$ **51.** $\dfrac{q}{q+4} \ge 2q$ **52.** $\dfrac{3r}{2r+1} > -2r$ **53.** $\dfrac{5t}{3t-2} < -3t$

Review exercises

1. Solve the formula $A = 2\pi r^2$ for $r > 0$. See section 6–1.

2. Simplify the complex rational expression $\dfrac{x - \dfrac{4}{x}}{1 + \dfrac{2}{x}}$.

See section 4–4.

3. Find the solution set of the equation $x + \dfrac{2}{x} = 3$. See section 4–6.

4. Given $y = 4x - 3$, find y when (a) $x = 5$, (b) $x = -2$, (c) $x = 0$. See section 1–5.

5. Evaluate the rational expression $\dfrac{a-b}{c-d}$ when $a = 2$, $b = -3$, $c = -1$, and $d = 2$. See section 1–5.

6. The width of a rectangle is 3 meters less than the length. If the rectangle has perimeter 110 meters, find the dimensions of the rectangle. See section 2–3.

Chapter 6 lead-in problem

The University of Minnesota wishes to set up a rectangular botanical garden. They have 300 meters of fence to enclose 5,000 square meters for the garden. What are the dimensions of the garden?

Solution

Since the perimeter of a rectangle is given by

$$\text{perimeter} = 2(\text{length}) + 2(\text{width})$$

and the perimeter in this case is 300 meters, then

$$300 = 2(\text{length}) + 2(\text{width})$$
$$150 = \text{length} + \text{width} \quad \text{Divide each term by 2}$$

Let x represent the length of the rectangle, then $150 - x$ represents the width of the rectangle. Using Area of a rectangle = length · width, where the area is 5,000 square meters, we substitute to obtain the equation

$$5{,}000 = x(150 - x)$$
$$5{,}000 = 150x - x^2 \quad \text{Distribute in right member}$$
$$x^2 - 150x + 5{,}000 = 0 \quad \text{Write equation in standard form}$$
$$(x - 100)(x - 50) = 0 \quad \text{Factor the left member}$$

$$x - 100 = 0 \quad \text{or} \quad x - 50 = 0$$
$$x = 100 \qquad\qquad x = 50$$

When the length is 100, the width is $150 - x = 150 - 100 = 50$. When the length is 50, the width is $150 - x = 150 - 50 = 100$. The dimensions of the rectangular garden are 50 meters by 100 meters.

Chapter 6 summary

1. The **standard form** of a **quadratic equation** in one variable is given by
$ax^2 + bx + c = 0, a > 0$
2. To solve a quadratic equation by factoring, we use the property $P \cdot Q = 0$, if and only if $P = 0$ or $Q = 0$, where P and Q are polynomials.
3. If $x^2 = p$, then $x = \pm\sqrt{p}$.
4. The quadratic equation $ax^2 + bx + c = 0$ can be solved by **completing the square** by writing the equation in the form $(x + k)^2 = d$.
5. The quadratic equation $ax^2 + bx + c = 0$ can be solved using the **quadratic formula**
$$x = \frac{-b \pm \sqrt{b^2 - 4ac}}{2a}$$

6. Given the equation $P = Q$, where P and Q are polynomials, the solution set of the equation $P = Q$ is a subset of the solution set of the equation $P^n = Q^n$, where n is a positive integer.
7. Solutions of the equation $P^n = Q^n$, where n is a positive integer, that *are not* solutions of the equation $P = Q$ are called **extraneous solutions.**
8. A **quadratic inequality** is any inequality written in the form
$ax^2 + bx + c < 0$, $ax^2 + bx + c > 0$,
$ax^2 + bx + c \le 0$, or $ax^2 + bx + c \ge 0$, where
$a \ne 0$.
9. A **rational inequality** is an inequality containing at least one term that is a rational expression.

Chapter 6 error analysis

1. Solving quadratic equations
Example: $5x^2 + 25x = 0$
$5x(x + 5) = 0$
The solution set is $\{-5,0,5\}$.
Correct answer: The solution set is $\{-5,0\}$.
What error was made? (*see page 260*)
2. Solving a quadratic equation by extracting the roots
Example: $(2x - 1)^2 = 7$
$2x - 1 = \sqrt{7}$
$2x = 1 + \sqrt{7}$
$x = \frac{1 + \sqrt{7}}{2}$ $\left\{\frac{1 + \sqrt{7}}{2}\right\}$
Correct answer: The solution set is $\left\{\frac{1 - \sqrt{7}}{2}, \frac{1 + \sqrt{7}}{2}\right\}$.
What error was made? (*see page 262*)
3. Solve a quadratic equation by completing the square
Example:
$x^2 + 2x - 5 = 0$
$x^2 + 2x = 5$
$x^2 + 2x + 1 = 5$
$(x + 1)^2 = 5$
$x + 1 = \pm\sqrt{5}$
$x = -1 \pm \sqrt{5}$
$\{-1 - \sqrt{5}, -1 + \sqrt{5}\}$
Correct answer: $\{-1 - \sqrt{6}, -1 + \sqrt{6}\}$
What error was made? (*see page 268*)
4. Identifying a, b, and c of a quadratic equation
Example: $3x^2 - 4 = 2x$; $a = 3, b = 2, c = -4$
Correct answer: $a = 3, b = -2, c = -4$
What error was made? (*see page 272*)

5. Solving radical equations
Example: $\sqrt{x + 1} = 5 - \sqrt{x - 4}$
$(\sqrt{x + 1})^2 = (5 - \sqrt{x - 4})^2$
$x + 1 = 25 - x + 4$
$2x = 28$
$x = 14$ $\{14\}$
Correct answer: $\{8\}$
What error was made? (*see page 286*)
6. Solving quadratic-type equations
Example: $x^4 - 3x^2 + 2 = 0$, let $u = x^2$
$u^2 - 3u + 2 = 0$
$(u - 1)(u - 2) = 0$
$u - 1 = 0$ or $u - 2 = 0$
$u = 1$ $u = 2$ $\{1,2\}$
Correct answer: $\{-1, 1, -\sqrt{2}, \sqrt{2}\}$
What error was made? (*see page 290*)
7. Solving rational inequalities
Example: $\frac{x - 2}{x + 3} \le 2$

a. $\frac{x - 2}{x + 3} = 2$ b. $x + 3 = 0$
$x - 2 = 2(x + 3)$ $x = -3$
$x - 2 = 2x + 6$
$-8 = x$
Critical numbers are -8 and -3. Test numbers are -9 in $x \le -8$, -7 in $-8 \le x \le -3$, and 0 in $x \ge -3$.
If $x = -9$, $\frac{-9 - 2}{-9 + 3} = \frac{11}{6} \le 2$ (true)
If $x = -7$, $\frac{-7 - 2}{-7 + 3} = \frac{9}{4} \le 2$ (false)
If $x = 0$, $\frac{0 - 2}{0 + 3} = -\frac{2}{3} \le 2$ (true)
The solution set is $\{x | x \le -8 \text{ or } x \ge -3\}$.
Correct answer: $\{x | x \le -8 \text{ or } x > -3\}$
What error was made? (*see page 296*)

8. Factoring the sum of two squares
   Example: $x^2 + 49 = (x)^2 + (7)^2 = (x + 7)^2$
   Correct answer: $x^2 + 49$ is not factorable in the set of real numbers.
   What error was made? (*see page 142*)

9. Product of two complex numbers
   Example: $\sqrt{-2} \cdot \sqrt{-8} = \sqrt{-2 \cdot -8} = \sqrt{16} = 4$
   Correct answer: -4
   What error was made? (*see page 250*)

10. Solving quadratic equations using the quadratic formula
    Example: Find the solution set of $x^2 - 2x - 7 = 0$ using $a = 1, b = -2, c = -7$.

    $$x = -(-2) + \frac{\sqrt{(-2)^2 - 4(1)(-7)}}{2(1)}$$

    $$= 2 + \frac{\sqrt{30}}{2}$$

    The solution set is $\left\{ \dfrac{4 - \sqrt{30}}{2}, \dfrac{4 + \sqrt{30}}{2} \right\}$

    Correct answer: $\left\{ \dfrac{2 - \sqrt{30}}{2}, \dfrac{2 + \sqrt{30}}{2} \right\}$

    What error was made? (*see page 272*)

Chapter 6 critical thinking

Which integers can be written as the sum of 4 consecutive integers?

Example $18 = 3 + 4 + 5 + 6$

Chapter 6 review

[6–1]

Find the solution set of the following quadratic equations by factoring or extracting the roots.

1. $x^2 - 9x - 10 = 0$
2. $y^2 - 4y = 32$
3. $4p^2 = 7p$
4. $9x^2 - 36 = 0$
5. $4z^2 - 12z + 5 = 0$
6. $2n^2 = 3n + 5$
7. $\dfrac{y}{2} + \dfrac{1}{3y} = \dfrac{7}{6}$
8. $(5x - 1)(x + 3) - 2x(2x + 6) = 0$
9. $(m + 7)^2 = 64$
10. $(3z - 4)^2 = 16$

11. A projectile is fired vertically upward with an initial velocity of 640 ft/sec. The distance S (in feet) above the ground after t seconds is given by
    $$S = -16t^2 + 640t$$
    When will the projectile reach a height of 1,600 feet? When will the projectile hit the ground?

[6–2]

Find the solution set of the given quadratic equations by completing the square.

12. $y^2 + 3y - 8 = 0$
13. $2p^2 - 3p + 1 = 0$
14. $5p^2 + 3p = -1$
15. $(4p + 3)(2p - 1) = p(p + 3)$
16. $\dfrac{2}{3}x^2 - \dfrac{1}{4}x + 1 = 0$

[6–3]

Find the solution set of the given quadratic equation by using the quadratic formula.

17. $x^2 - 11x + 10 = 0$
18. $3y^2 + 3y - 2 = 0$
19. $2z^2 - 3 = 1$
20. $2z^2 = -z - 4$
21. $3x - 2 = \dfrac{5}{x}$
22. $(2y - 3)^2 = 3y - 2$

Solve the following word problems.

23. Solve the equation $4x^2 + xy - 2y^2 = 0$ for x in terms of y using the quadratic formula, where y is positive.

24. A beam of length L has a maximum displacement at a distance x from the end. This is expressed by the equation

$$2x^2 - 3xL + L^2 = 0$$

Solve the equation for x by using the quadratic formula.

[6–4]

Solve the following by using quadratic equations.

25. An object fired vertically into the air with an initial velocity v_0 feet per second will be at a distance h feet in the air at t seconds after launching, according to the equation

$$h = v_0 t - 16t^2$$

How long will it take the object to reach a height of 75 feet if the initial velocity is 120 feet per second (to the nearest tenth)?

26. A rectangular solar panel has an area of 1.0 square meter. If the length is 150 centimeters more than the width, find the dimensions of the panel. (*Hint:* Use $A = \ell w$ and 1 meter = 100 centimeters.)

27. The shortest leg of a right triangular brace is 9.0 feet shorter than the hypotenuse and 7.0 feet shorter than the other leg. What are the lengths of the sides of the brace?

28. Mary can paint the exterior of a house in 2 hours less time than Dick can. Working together, they can paint the house in 12 hours. How long would it take each person to paint the house working alone?

[6–5]

Find the solution set of each radical equation. Indicate any extraneous solutions.

29. $y = \sqrt{y + 20}$

30. $\sqrt{3x + 5} + 1 = 3x$

31. $\sqrt{y + 2} + 1 = \sqrt{y + 6}$

32. $\sqrt{x^2 - x - 5} = 1$

33. $\sqrt{5x + 2x} = \sqrt{20x + 5}$

34. The radius r of a sphere is determined by the equation

$$r = \sqrt[3]{\frac{3V}{4\pi}}$$

Solve the equation for V.

[6–6]

Find the solution set of each equation. Identify extraneous solutions, if they exist.

35. $x^4 - 5x^2 - 14 = 0$

36. $(y + 1)^2 - 6(y + 1) + 5 = 0$

37. $q + 9\sqrt{q} + 8 = 0$

38. $3p - 5p^{1/2} - 2 = 0$

39. $2x^{2/3} + x^{1/3} - 6 = 0$

40. $z^{-2} = 10z^{-1} + 11$

41. $5y^{-4} - 8y^{-2} = 4$

[6–7]

Find the solution set and graph each inequality. Write the answer in set-builder and interval notation.

42. $(x - 3)(x + 7) > 0$

43. $(2x - 1)(3x + 4) \leq 0$

44. $2y^2 - 3y \geq 0$

45. $9z^2 - 4 < 0$

46. $m^2 - 7 > 6m$

47. $4p^2 - 5p < 6$

48. $5x^2 \geq 45$

49. $y^2 - 14y \leq 32$

50. $(y - 2)(y + 4)(y - 5) < 0$

51. $\dfrac{m + 3}{m - 1} \geq 0$

52. $\dfrac{x - 3}{x + 7} \leq 2$

Chapter 6 cumulative test

Perform the indicated operations and simplify. Assume all variables are nonzero real numbers. Leave all answers with positive exponents.

[1–2] 1. $\dfrac{(-18)(-4)}{-6}$

[1–4] 2. $16 - 8[3 - 4(12 - 8) + 14] - 6$

[1–4] 3. $56 - 28 \div 7 - 5 + 3^2$

[1–6] 4. $(4xy^2 - 5xy + 2x^2y) - (-3xy^2 + xy - 4x^2y)$

[3–2] 5. $(3x + 2y)^2$

[3–2] 6. $(4y + 1)(4y - 1)$

[3–2] 7. $(x - 2)(9x^2 + 2x + 4)$

[3–1] 8. $(5xy^2)^3(-3xy)^3$

[3–3] 9. $\left(\dfrac{a^2b^0}{a^{-3}}\right)^4$

[3–3] 10. $\dfrac{-24a^{-3}b^2}{12a^2b^{-3}}$

[1–6] 11. Given $P(x) = 3x^2 - 2x + 1$, find (a) $P(-1)$, (b) $P(0)$, (c) $P(4)$.

Reduce the following rational expressions to lowest terms. Assume all denominators are nonzero.

[4–1] 12. $\dfrac{12a - 12b}{b - a}$

[4–1] 13. $\dfrac{6a^2 - 6}{3a^2 - 15a + 12}$

[4–1] 14. $\dfrac{3b^2 - 12}{4b^2 - 16}$

Perform the indicated operations and simplify. All denominators are nonzero.

[4–2] 15. $\dfrac{x + 2}{x - 1} \cdot \dfrac{x^2 - 1}{x^2 - x - 6}$

[4–2] 16. $\dfrac{x^2 - 2x + 1}{x^2 - 49} \div \dfrac{x^2 - 7x + 6}{2x^2 - 15x + 7}$

[4–3] 17. $\dfrac{15}{p^2 - 7p - 18} - \dfrac{3}{p^2 - 4}$

[4–3] 18. $\dfrac{4a + 1}{a - 6} + \dfrac{3a - 4}{6 - a}$

Find the solution set of the following equations and inequalities.

[2–4] 19. $|4x + 7| = 5$

[2–6] 20. $|2x - 5| \le 5$

[2–6] 21. $|7 - 2x| > 4$

[2–1] 22. $4(2x + 1) - 3(x - 1) = 4x$

[2–5] 23. $4x - 3 \le 5(x + 6)$

[4–7] 24. $\dfrac{3}{x} - \dfrac{4}{x} = \dfrac{6}{10}$

[2–2] 25. Solve $P = 2\ell + 2w$ for w.

[4–4] 26. Simplify the complex fraction $\dfrac{5 - \dfrac{6}{3y}}{2 + \dfrac{5}{2y}}$.

Perform the indicated operations and simplify.

[5–6] 27. $\sqrt{3}(6 - \sqrt{3})$

[5–5] 28. $3\sqrt{75} + \sqrt{27} - \sqrt{12}$

[5–5] 29. $\sqrt[3]{81} - 3\sqrt[3]{24}$

[5–7] 30. $(3 - 2i)(3 + 2i)$

[5–6] 31. $(2\sqrt{5} - 3\sqrt{3})^2$

Rationalize the denominator.

[5–6] 32. $\sqrt{\dfrac{16}{5}}$

[5–6] 33. $\dfrac{\sqrt{2} - \sqrt{3}}{\sqrt{2} + \sqrt{3}}$

[5–7] 34. $\dfrac{i}{2 - 5i}$

Find the solution sets of the following equations and inequalities.

[6–1] 35. $x^2 - 15x + 14 = 0$

[6–3] 36. $5y^2 - y + 3 = 0$

[6–3] 37. $3y^2 + 1 = y - 2$

[6–7] 38. $z^2 - 2z \le 3$

[6–6] 39. $p^4 - 5p^2 - 50 = 0$

[6–7] 40. $\dfrac{2y - 3}{y + 7} < 1$

[4–6] 41. Divide $4x^5 - 3x^2 + x^2 - x + 1$ by $x + 1$ using synthetic division.

7

Linear Equations and Inequalities in Two Variables

A company that manufactures a heating unit can produce 20 units for $13,900 while it would cost $7,500 to manufacture 10 units. Assume the cost and number of units produced are related by the linear equation of a straight line. Let y be the total cost to manufacture x units. Find the linear equation of the straight line.

7–1 ■ The rectangular coordinate system

In chapter 2, we considered the solution set of linear equations (first-degree equations) in one variable. That is, equations of the form

$$ax + b = 0$$

where a and b are real numbers, $a \neq 0$. The solution sets of these equations are sets of real numbers.

In this chapter, we expand our work with equations to consider **linear equations in two variables,** x and y. Such equations are of the form

$$ax + by = c$$

where a, b, and c are real numbers, not both a and b equal to zero. The equations

$$3x + y = 4, \quad 4y - x = 0, \quad y = 2x - 1, \quad \text{and} \quad x = y - 4$$

are examples of linear equations in two variables.

In an equation in two variables, x and y, the x and y are replaced by a pair of numbers. If that pair of numbers makes the equation true, we say the pair of numbers satisfies the equation. Any pair of numbers that satisfies the equation is a solution of that equation. Consider the linear equation $3x - y = 5$. Let $x = 2$ and $y = 1$. If we substitute 2 for x and 1 for y in the equation, we have

$$3(2) - (1) = 5$$
$$6 - 1 = 5$$
$$5 = 5 \quad \text{(True)}$$

Therefore the values 2 for x and 1 for y form a solution of the equation $3x - y = 5$. The pair of numbers $x = 2$ and $y = 1$ that form this solution are usually written in the form (2,1). This pair of numbers is called an **ordered pair of real numbers** because the numbers are written in a specific order, x first and then y, (x,y). We call x the **first component** and y the **second component** of the ordered pair (x,y).

To graph an ordered pair, we use two real number lines that intersect at right angles with each other at their zero points. The point of intersection is called the **origin.** The two lines, one *horizontal* and the other *vertical,* are called **axes.** The horizontal line, called the **x-axis,** is associated with the first number of the ordered pair, and the vertical line, called the **y-axis,** is associated with the second number of the ordered pair. The x-axis and the y-axis form the **rectangular coordinate system** and partition the plane into four equal regions called **quadrants.** The quadrants are numbered I, II, III, and IV in a counterclockwise direction. See figure 7–1.

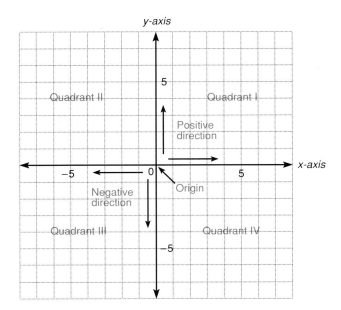

Figure 7–1

Plotting Ordered Pairs

To locate the point in the plane that corresponds to the ordered pair (3,5), we start at the origin and move 3 units to the *right* (the positive direction) along the (horizontal) x-axis, and then we move 5 units *up* (the positive direction) along the y-axis (vertical). See figure 7–2(a). To locate the point in the plane that corresponds to the ordered pair $(-5,-4)$, we start at the origin and move 5 units to the *left* (the negative direction) along the x-axis (horizontal) and then we move 4 units *down* (the negative direction) along the y-axis (vertical). See figure 7–2(b).

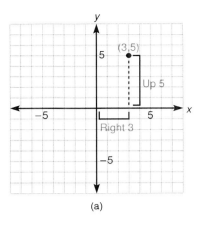

(a)

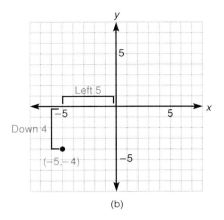

(b)

Figure 7-2

In the ordered pair $(-5,-4)$, we call the numbers -5 and -4 the **coordinates of the point,** where the first number, -5, is called the **abscissa,** or x-coordinate, of the point and the second number, -4, is called the **ordinate,** or y-coordinate, of the point.

Note   The coordinates of the origin are (0,0).

Points are usually named by capital letters and/or their coordinates. When we use the notation $P(x,y)$, we mean the point P whose coordinates are x and y. For example, in figure 7-3 the points $A(4,4)$, $B(0,3)$, $C(-4,2)$, $D(-3,0)$, $E(-6,-2)$, $F(0,-5)$, $G(2,-3)$, $H(7,0)$, and $I(0,0)$ have been located in the plane. Point A lies in quadrant I, point C lies in quadrant II, point E lies in quadrant III, and point G lies in quadrant IV.

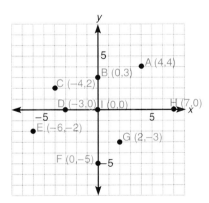

Figure 7-3

To find ordered pairs that are solutions of a linear equation in two variables, we use the following procedure.

┌─ *Finding ordered pair solutions* ─────────────────

  1. Choose any real number value of one of the variables (usually x).
  2. Replace that variable with the chosen value and solve the resulting equation in one variable.
  3. Write the solution as an ordered pair (x,y).

■ *Example 7–1 A*

Find the missing component of the ordered pair for the equation $3x + 2y = 6$.

1. $(0, \quad)$ $x = 0$

$$3x + 2y = 6 \quad \text{Original equation}$$
$$3(0) + 2y = 6 \quad \text{Replace } x \text{ with } 0$$
$$2y = 6 \quad \text{Solve for } y$$
$$y = 3$$

The ordered pair is $(0,3)$.

2. $(\quad ,0)$ $y = 0$

$$3x + 2y = 6 \quad \text{Original equation}$$
$$3x + 2(0) = 6 \quad \text{Replace } y \text{ with } 0$$
$$3x = 6 \quad \text{Solve for } x$$
$$x = 2$$

The ordered pair is $(2,0)$.

3. $(4, \quad)$ $x = 4$

$$3x + 2y = 6 \quad \text{Original equation}$$
$$3(4) + 2y = 6 \quad \text{Replace } x \text{ with } 4$$
$$12 + 2y = 6 \quad \text{Solve for } y$$
$$2y = -6$$
$$y = -3$$

The ordered pair is $(4,-3)$.

▶ *Quick check* Find the missing components for the ordered pairs for the equation $4x - 3y = 12$; $(0, \quad), (\quad ,0), (6, \quad)$. ■

Graphing an equation

To graph the equation $3x + 2y = 6$, we could first graph the ordered pairs $(0,3)$, $(2,0)$, and $(4,-3)$ found in example 7–1 A. The points appear to lie on a straight line. In fact, if *all* ordered pair solutions of $3x + 2y = 6$ were plotted, the points would lie on this straight line. See figure 7–4 for the graph of $3x + 2y = 6$.

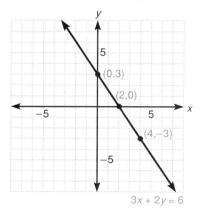

Figure 7–4

> The graph of a linear equation of the form
> $$ax + by = c \text{ (a and b not both 0)}$$
> is a straight line.

Since a straight line is determined by any two distinct points on the line, finding two points on the line is sufficient to graph the equation. Two points most easily found are the

1.   *x*-intercept—the point (if any exists) where the graph crosses the *x*-axis. This occurs when $y = 0$.
2.   *y*-intercept—the point (if any exists) where the graph crosses the *y*-axis. This occurs when $x = 0$.

To guard against arithmetic error, it is wise to find a third point to act as a checkpoint.

___ *Graphing a linear equation in two variables* _____

1.   Let $y = 0$ and solve for *x* to find the *x*-intercept, the point (*x*,0).
2.   Let $x = 0$ and solve for *y* to find the *y*-intercept, the point (0,*y*).
3.   Find a third point as a checkpoint.

■ *Example 7–1 B*

Find the *x*- and *y*-intercepts and sketch the graph of each equation.

1. $3x + 5y = 15$

   a.   Let $y = 0$, then $3x + 5(0) = 15$ Replace *y* with 0
   $$3x + 0 = 15 \qquad \text{Solve for } x$$
   $$3x = 15$$
   $$x = 5$$

   The *x*-intercept is the point (5,0).

   b.   Let $x = 0$, then $3(0) + 5y = 15$ Replace *x* with 0
   $$0 + 5y = 15 \qquad \text{Solve for } y$$
   $$5y = 15$$
   $$y = 3$$

   The *y*-intercept is the point (0,3).

   c.   Checkpoint: Let $x = 1$.

   $$3(1) + 5y = 15 \qquad \text{Replace } x \text{ with 1}$$
   $$3 + 5y = 15 \qquad \text{Solve for } y$$
   $$5y = 12$$
   $$y = \frac{12}{5}$$

   The checkpoint is $\left(1, \dfrac{12}{5}\right)$.

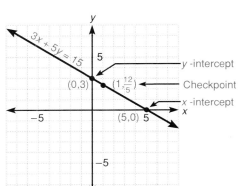

Note   In future examples, we will not always show the checkpoint; however we should *always* find a third point as a check.

2. $4x = y$

Let $x = 0$, then $4(0) = y$ Replace x with 0
$$y = 0$$

The y-intercept (and the x-intercept) is the point $(0,0)$, which is the origin. Since both the intercepts are the same point, we must find two more distinct points.

Let $x = 1$ Let $x = -1$
$4(1) = y$ $4(-1) = y$ Replace x with 1 and -1
$4 = y$ $-4 = y$

Two additional points on the line are $(1,4)$, and $(-1,-4)$.

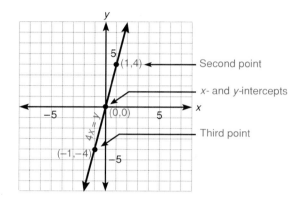

3. $y = -3$

We can write this equation as

$$0 \cdot x + y = -3$$

and for any value of x that we might choose, y is *always* -3. That is,

$$(-5,-3), (-2,-3), (0,-3), (1,-3), \text{ and } (6,-3)$$

are all solutions of the equation. Plotting these points and drawing a straight line through them, the graph is a horizontal line (parallel to the x-axis) having a y-intercept of $(0,-3)$ and no x-intercept.

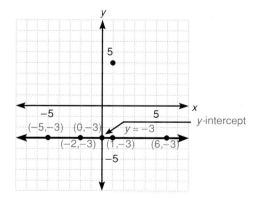

4. $x = 2$

We can write the equation as

$$x + 0 \cdot y = 2$$

and for any value of y that we choose, x is always 2. If we choose two solutions, say $(2,-3)$ and $(2,0)$, and draw a straight line through these points, we have the graph of $x = 2$. The graph is a vertical line (parallel to the y-axis) having an x-intercept of $(2,0)$ and no y-intercept.

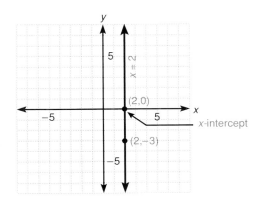

▶ *Quick check*   Find the x- and y-intercepts and sketch the graph for $2x + 7y = 14$. ■

In general, from examples 3 and 4, we see that the following is true:

─── *If k is some real number, k ≠ 0, the graph of* ───────

1. $x = k$ is a vertical line with an x-intercept $(k,0)$ and no y-intercept.
2. $y = k$ is a horizontal line with a y-intercept $(0,k)$ and no x-intercept.

─── *Mastery points* ───

Can you

■ Plot the graph of an ordered pair of real numbers?
■ Determine in what quadrant a point lies?
■ Find the x- and y-intercepts of a linear equation in two variables?
■ Sketch the graph of a linear equation in two variables?
■ Graph equations $x = k$ and $y = k$, k is a constant?

Exercise 7–1

Plot the graph of the following ordered pairs of real numbers. State the quadrant in which the point lies.

1. $(2,4)$

2. $(5,2)$

3. $(-4,3)$

4. $(-6,5)$

5. $(-1,-3)$

6. $(-4,-1)$

7. $(4,0)$

8. $(-6,0)$

9. $(0,-1)$

10. $(0,5)$

11. $\left(\dfrac{1}{2},3\right)$

12. $\left(-2,\dfrac{3}{2}\right)$

13. $\left(-\dfrac{7}{2},-\dfrac{5}{2}\right)$

For each equation, find the missing value in the ordered pairs. Sketch the graph of the equation using these ordered pairs. See example 7–1 A.

Example $4x - 3y = 12$; $(0, \ )$, $(\ , 0)$, $(6, \ )$

Solution   a. $(0, \ )$

$$4(0) - 3y = 12$$ Replace x with 0
$$-3y = 12$$ Solve for y
$$y = -4$$ Ordered pair $(0, -4)$

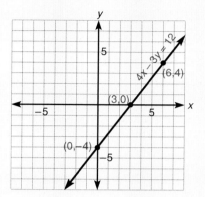

   b. $(\ , 0)$

$$4x - 3(0) = 12$$ Replace y with 0
$$4x - 0 = 12$$ Solve for x
$$4x = 12$$ Ordered pair $(3, 0)$
$$x = 3$$

   c. $(6, \ )$

$$4(6) - 3y = 12$$ Replace x with 6
$$24 - 3y = 12$$ Solve for y
$$-3y = -12$$
$$y = 4$$ Ordered pair $(6, 4)$

14. $3x + y = -1$; $(-3, \ )$, $(-1, \ )$, $(0, \ )$, $(\ , 0)$

15. $2x + y = 3$; $(-2, \ )$, $(0, \ )$, $(2, \ )$, $(\ , 0)$

16. $x - y = 2$; $(-2, \ )$, $(0, \ )$, $(2, \ )$, $(\ , 0)$

17. $x + y = 4$; $(-5, \ )$, $(-3, \ )$, $(0, \ )$, $(\ , 0)$

18. $x + 2y = 4$; $(-2, \ )$, $(0, \ )$, $(2, \ )$, $(\ , 0)$

19. $y - x = -2$; $(-3, \ )$, $(0, \ )$, $(2, \ )$, $(\ , 0)$

20. $2x + 5y = 20$; $(-5, \ )$, $(0, \ )$, $(5, \ )$, $(\ , 0)$

21. $x - 3y = 1$; $(-1, \ )$, $(0, \ )$, $(1, \ )$, $(\ , 0)$

22. $4x - 3y = 6$; $(-6, \ )$, $(-3, \ )$, $(0, \ )$, $(\ , 0)$

23. $3x + 2y = 8$; $(-2, \ )$, $(0, \ )$, $(2, \ )$, $(\ , 0)$

Plot the x- and y-intercepts for the graph of each equation, if they exist. Sketch the lines. See example 7–1 B.

Example $2x + 7y = 14$

Solution   a. Let $x = 0$, then

$$2(0) + 7y = 14$$ Replace x with 0
$$7y = 14$$ Solve for y
$$y = 2$$

y-intercept is $(0, 2)$.

   b. Let $y = 0$, then

$$2x + 7(0) = 14$$ Replace y with 0
$$2x = 14$$ Solve for x
$$x = 7$$

x-intercept is $(7, 0)$.

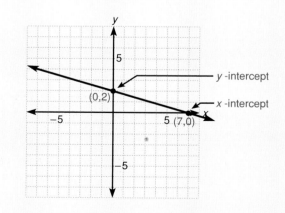

24. $x + 2y = 4$

25. $x - 3y = -6$

26. $4x - 5y = 20$

27. $5x + 2y = 20$

28. $4x + y = 8$

29. $5x - y = -10$

30. $x = 3y$

31. $x = -2y$

32. $y - 3x = 0$ **33.** $y + 2x = 0$ **34.** $4y - x = 0$ **35.** $5y - 3x = 6$

36. $3x + 2y = 8$ **37.** $y = -2$ **38.** $y = 6$ **39.** $x = -1$

40. $x = 8$ **41.** $x = 0$ **42.** $y = 0$

Translate each of the following statements into an equation and graph the equation.

43. If 4 is added to x, the result is y.

44. Two times the value of x less 3 is equal to y.

45. Three times x less two times y is 12.

46. If 3 is added to the y-value, the result is four times the x-value.

47. Temperature measured in Fahrenheit degrees can be converted to Celsius degrees using the equation

$$C = \frac{5}{9}(F - 32)$$

Let the horizontal axis represent F and the vertical axis represent C. Graph the equation.

48. Graph the lines $y = x + 2$ and $y = x - 1$ on the same rectangular coordinate system. What appears to be true of the lines? From what you observe, the graph of $y = x + 6$ will be where in the plane?

49. Graph the lines $y = x + 2$ and $y = 2x - 3$ on the same rectangular coordinate system. At what point do the lines appear to cross?

Review exercises

Perform the indicated operations. Write your answer in the standard form $a + bi$. See section 5–7.

1. $(2 + 3i)(4 - i)$

2. $(4 + i)(4 - i)$

3. $(2 + \sqrt{-9})(2 - 2\sqrt{-9})$

4. $\dfrac{3 + i}{3 - i}$

5. Solve the formula $P = 2\ell + 2w$ for w. See section 2–2.

6. Divide $(x^4 - 1)$ by $(x - 1)$. See section 4–6.

7. Evaluate $\dfrac{p - q}{r - s}$ when $p = 2, q = 4, r = -3$, and $s = -5$. See section 1–5.

7–2 ■ The distance formula and the slope of a line

Distance formula

In section 7–1, we studied the graph of a linear equation that is a straight line. If we choose any two points on that line, the portion of the line between the two points is called a **line segment.** A line has no length, while a line segment has a specific length. We cannot determine the length of a line since it continues indefinitely in both directions, but we can determine the length of a line segment. The length of the line segment is defined as the **distance** between the two points. See figure 7–5.

Figure 7–5

Given line L containing points P_1 and P_2, the length of the line segment from P_1 to P_2 is then called the *distance* from P_1 to P_2. Let us consider three specific examples and then develop the distance formula.

Consider three points in the plane, $(3,2)$, $(3,-3)$, and $(-2,-3)$, shown in figure 7–6.

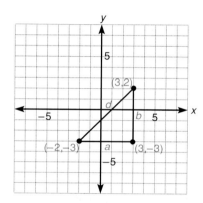

Figure 7–6

The points form a right triangle having sides a, b, and d. The length of

1. side a, which is parallel to the x-axis is

$$|-2 - 3| = |3 - (-2)| = 5 \text{ units long}$$

2. side b, which is parallel to the y-axis, is

$$|2 - (-3)| = |-3 - 2| = 5 \text{ units long}$$

To find the length of side d, we use the Pythagorean Theorem mentioned in chapters 5 and 6 which states

$$d^2 = a^2 + b^2$$

Thus, $$d^2 = 5^2 + 5^2$$
$$= 25 + 25$$
$$= 50$$

Extracting the roots, $$d = \pm\sqrt{25} = \pm 5\sqrt{2}$$

Since distance is nonnegative, the distance $d = 5\sqrt{2}$ units.

Now consider the distance between arbitrary points $P_1(x_1,y_1)$ and $P_2(x_2,y_2)$, denoted by $d(P_1P_2)$. See figure 7–7.

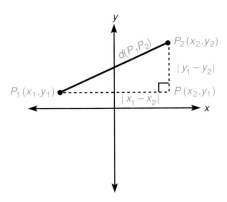

Figure 7-7

We have drawn a horizontal dashed line segment from P_1 and a vertical dashed line segment from P_2 so that these segments meet at point $P(x_2,y_1)$, thus forming a right triangle. The lengths of the dashed line segments by definition are $d(P_1P) = |x_2 - x_1|$ and $d(P_2P) = |y_2 - y_1|$. To find the distance from P_1 to P_2, denoted by d, we use the **Pythagorean Theorem.**

Since the square of any number is never negative, then

$$|x_2 - x_1|^2 = (x_2 - x_1)^2 \quad \text{and} \quad |y_2 - y_1|^2 = (y_2 - y_1)^2$$

and so

$$d^2 = (x_2 - x_1)^2 + (y_2 - y_1)^2$$

Distance is never negative, so we use the principal, or positive, square root to find d.

___ **Distance formula** _____

The distance between any two points (x_1,y_1) and (x_2,y_2) is given by

$$d = \sqrt{(x_2 - x_1)^2 + (y_2 - y_1)^2}$$

Note   When working with two points on a line, it doesn't make any difference which point is labeled (x_1,y_1) and which point is labeled (x_2,y_2).

■ *Example 7-2 A*

Find the distance d from $(3,2)$ to $(6,6)$.

Let $(x_1,y_1) = (3,2)$ and $(x_2,y_2) = (6,6)$.

$d = \sqrt{(6 - 3)^2 + (6 - 2)^2}$ Replace x_1 with 6, x_2 with 3, y_1 with 6, and y_2 with 2

$\quad = \sqrt{(3)^2 + (4)^2}$

$\quad = \sqrt{9 + 16}$

$\quad = \sqrt{25} = 5$

Thus $d = 5$ units.

▶ *Quick check*   Find the distance d from $(-4,3)$ to $(5,-6)$.

The midpoint of a line segment

Sometimes we must find the midpoint of a line segment. We now give the formula for finding the coordinates of this point. Using similar triangles, it can be shown that the coordinates of the point midway between two given points are found by averaging the *x*-coordinates and the *y*-coordinates of the points.

Midpoint of a line segment

The **midpoint** of the line segment joining points (x_1,y_1) and (x_2,y_2) has coordinates

$$\left(\frac{x_1 + x_2}{2}, \frac{y_1 + y_2}{2} \right)$$

■ *Example 7–2 B*

Find the midpoint of the line segment whose endpoints are $(6,-4)$ and $(4,2)$. Let $(x_1,y_1) = (6,-4)$ and $(x_2,y_2) = (4,2)$.

$$\left(\frac{x_1 + x_2}{2}, \frac{y_1 + y_2}{2} \right) = \left(\frac{(6) + (4)}{2}, \frac{(-4) + (2)}{2} \right)$$

Replace x_1 with 6, x_2 with 4, y_1 with -4, and y_2 with 2

$$= \left(\frac{10}{2}, \frac{-2}{2} \right)$$

$$= (5,-1)$$

The midpoint of the line segment is the point $(5,-1)$.

▶ *Quick check*   Find the midpoint of the line segment whose endpoints are $(-4,3)$ and $(5,-6)$.   ■

The slope of a line

Now consider the portions of two lines as one moves from point P_1 to point P_2 on each incline. See figure 7–8.

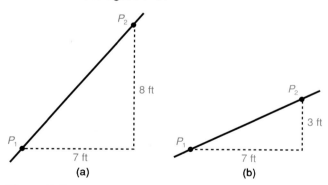

(a)   (b)

Figure 7–8

The incline in figure 7–8(a) is "steeper" than the incline in figure 7–8(b). That is, the inclination in figure 7–8(a) is greater than the inclination in figure 7–8(b). In moving from point P_1 to P_2, the horizontal change is 7 feet in both cases, but the vertical change in (a) is 8 feet and the vertical change in (b) is 3 feet. If we measure this "steepness," or inclination, by the quotient

$$\text{steepness} = \frac{\text{vertical change}}{\text{horizontal change}}$$

the "steepness" of the line in

$$\text{(a) is } \frac{8 \text{ feet}}{7 \text{ feet}} = \frac{8}{7}$$

and of the line in

$$\text{(b) is } \frac{3 \text{ feet}}{7 \text{ feet}} = \frac{3}{7}$$

Note $\frac{8}{7}$ is greater than $\frac{3}{7}$, so the line in (a) is "steeper" than the line in (b).

When applying this concept to any straight line, this "steepness" is called the **slope** of the line. Therefore the slope of a line is given by

$$\text{slope} = \frac{\text{vertical change}}{\text{horizontal change}}$$

See figure 7-9.

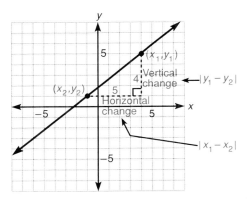

Figure 7-9

The vertical change is 4 units and the horizontal change is 5 units. Then the slope of the line is given by

$$\text{slope} = \frac{\text{vertical change}}{\text{horizontal change}} = \frac{4}{5}$$

Note   This means that for every 5 units moved to the right from a point on the line, there must be a rise of 4 units to get back to the line.

We denote the slope of a line by m.

___ *Definition of the slope of a straight line* _____

If $(x_1 \neq x_2)$, the slope (m) of the line containing points (x_1, y_1) and (x_2, y_2) is defined by

$$m = \frac{y_2 - y_1}{x_2 - x_1}$$

Concept
The slope of a line is obtained by dividing the change in y-values by the corresponding change in x-values.

Note It is a common mistake to write $m = \dfrac{y_1 - y_2}{x_2 - x_1}$ or $m = \dfrac{y_2 - y_1}{x_1 - x_2}$. The order in which the x-values are subtracted *must be the same* as the order in which the y-values are subtracted.

The slope of a line can alternately be defined by

$$m = \frac{\text{rise}}{\text{run}}$$

where the vertical change is the **rise** and the horizontal change is the **run.** The slope m is the amount of rise or fall for each unit of run. See figure 7.10.

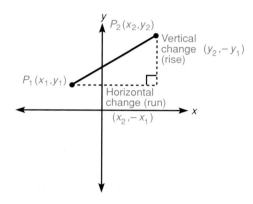

Figure 7–10

■ *Example 7–2 C*

Find the slope of the line passing through the given points. Sketch the line.

1. (4,1) and (8,3)

Let $(x_1, y_1) = (4,1)$ and $(x_2, y_2) = (8,3)$.

$$m = \frac{y_2 - y_1}{x_2 - x_1} = \frac{(3) - (1)}{(8) - (4)}$$ Replace y_2 with 3, y_1 with 1, x_2 with 8, and x_1 with 4

$$= \frac{2}{4}$$

$$= \frac{1}{2}$$

The slope of the line is $\dfrac{1}{2}$.

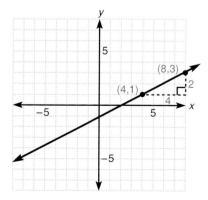

Note The x- and y-values may be subtracted in any order so long as the coordinates of each point are in the same position in the numerator and the denominator. That is, we could use

$$m = \frac{y_1 - y_2}{x_1 - x_2}$$
$$= \frac{(1) - (3)}{(4) - (8)} = \frac{-2}{-4} = \frac{1}{2}$$

2. $(-5,4)$ and $(2,1)$

Let $(x_1,y_1) = (-5,4)$ and $(x_2,y_2) = (2,1)$.

$$m = \frac{y_2 - y_1}{x_2 - x_1} = \frac{(1) - (4)}{(2) - (-5)}$$

Replace y_2 with 1, y_1 with 4, x_2 with 2, and x_1 with -5

$$= \frac{-3}{7} = -\frac{3}{7}$$

Thus the slope of the line is $-\dfrac{3}{7}$.

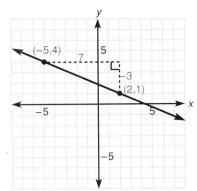

Note For every 7 units moved to the right from a point on the line, there must be a "fall" of 3 units to get back to the line.

We see that the graph of a line having a positive slope (example 1) "slants" *up* from left to right, and the graph of a line having negative slope (example 2) "slants" *down* from left to right. This will always be the case.

3. $(-6,2)$ and $(1,2)$

Let $(x_1,y_1) = (-6,2)$ and $(x_2,y_2) = (1,2)$.

$$m = \frac{y_2 - y_1}{x_2 - x_1} = \frac{(2) - (2)}{(1) - (-6)}$$

Replace y_2 with 2, y_1 with 2, x_2 with 1, and x_1 with -6

$$= \frac{0}{7}$$
$$= 0$$

Thus the slope of the line is 0.

The points lie on a horizontal line, and the slope $m = 0$.

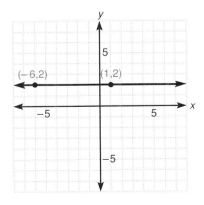

Slope of a horizontal line

The slope m of any horizontal line is 0.

4. $(-3,-1)$ and $(-3,4)$

Let $(x_1,y_1) = (-3,-1)$ and $(x_2,y_2) = (-3,4)$.

$$m = \frac{y_2 - y_1}{x_2 - x_1} = \frac{4 - (-1)}{-3 - (-3)}$$

Replace y_2 with 4, y_1 with -1, x_2 with -3, and x_1 with -3

$$= \frac{5}{-3 + 3}$$

$$= \frac{5}{0} = \text{undefined}$$

The slope of the line is undefined.

Note In our definition of slope, we placed the restriction $x_1 \neq x_2$. In this example, $x_1 = x_2 = -3$.

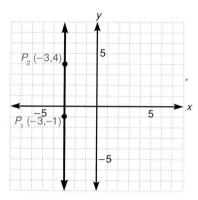

The points lie on a vertical line, and the slope is undefined.

Slope of a vertical line

The slope m of a vertical line is undefined.

▶ **Quick check** Find the slope of the line passing through $(-4,3)$ and $(5,-6)$. Sketch the line. ■

On the basis of the discussion and examples, we can summarize the slope of a line as follows.

The slope of a line is

1. positive if the line slants up from left to right.
2. negative if the line slants down from left to right.
3. zero if the line is horizontal (parallel to the x-axis).
4. undefined if the line is vertical (parallel to the y-axis).

Parallel lines

Parallel lines are defined to be straight lines in the same plane that never meet. For this to be the case, the lines must have the same slope. See figure 7–11.

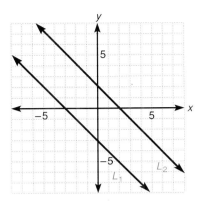

Figure 7–11

___ **Definition of parallel lines** _____

Given line L_1 has slope m_1 and line L_2 has slope m_2,

1. L_1 is parallel to L_2 if $m_1 = m_2$.
2. $m_1 = m_2$ if L_1 is parallel to L_2.

Concept
Two nonvertical lines are parallel if and only if their slopes are the same.

Note *All* vertical lines are parallel even though the slope of a vertical line is undefined.

■ *Example 7–2 D*

Determine if the line containing the points $(-3,2)$ and $(5,1)$ is parallel to the line containing the points $(1,5)$ and $(-7,6)$.

Using $m = \dfrac{y_2 - y_1}{x_2 - x_1}$,

a. Let $(x_1,y_1) = (-3,2)$ and $(x_2,y_2) = (5,1)$.

$$m_1 = \frac{1-2}{5-(-3)} = \frac{-1}{8} = -\frac{1}{8}$$

b. Let $(x_1,y_1) = (1,5)$ and $(x_2,y_2) = (-7,6)$.

$$m_2 = \frac{6-5}{-7-1} = \frac{1}{-8} = -\frac{1}{8}$$

Both the slopes are $-\dfrac{1}{8}$ so the lines are parallel.

▶ *Quick check*   Determine if the line containing the points $(1,2)$ and $(-3,6)$ is parallel to the line containing the points $(5,-6)$ and $(-2,1)$.

Perpendicular lines

It can be shown that the *product* of the slopes of two nonvertical perpendicular lines (lines that form right angles) is -1. For example, two lines whose slopes are $\dfrac{2}{3}$ and $-\dfrac{3}{2}$, respectively, are perpendicular since

$$\left(\frac{2}{3}\right)\left(-\frac{3}{2}\right) = -1$$

Notice that $\dfrac{2}{3}$ and $-\dfrac{3}{2}$ are *negative reciprocals* of one another. This gives rise to the following definition.

Definition of perpendicular lines

Given line L_1 has slope m_1 and line L_2 has slope m_2,

1. L_1 is perpendicular to L_2 if $m_1 \cdot m_2 = -1$.
2. $m_1 \cdot m_2 = -1$ if L_1 is perpendicular to L_2.

Concept

Two nonvertical lines are perpendicular if and only if the product of their slopes is -1; that is, provided their slopes are negative reciprocals.

See figure 7–12.

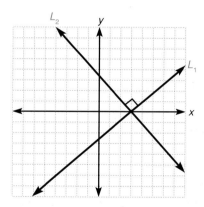

Figure 7–12

Note Any horizontal line is perpendicular to any vertical line.

■ *Example 7–2 E*

1. Determine if the line containing points $(1,5)$ and $(-2,3)$ is perpendicular to the line containing the points $(5,-3)$ and $(3,0)$.

Using $m = \dfrac{y_2 - y_1}{x_2 - x_1}$,

a. Let $(x_1,y_1) = (1,5)$ and $(x_2,y_2) = (-2,3)$.

$$m_1 = \frac{3-5}{-2-1} = \frac{-2}{-3} = \frac{2}{3}$$

b. Let $(x_1,y_1) = (5,-3)$ and $(x_2,y_2) = (3,0)$.

$$m_2 = \frac{0 - (-3)}{3 - 5} = \frac{3}{-2} = -\frac{3}{2}$$

Since $\dfrac{2}{3}\left(-\dfrac{3}{2}\right) = -1$, the slopes are negative reciprocals and the lines are perpendicular.

2. Determine if the graphs of the equations $2x - 4y = 4$ and $4x + 2y = -5$ are perpendicular.

   If we choose two ordered pairs that are in the solution set of each equation, we can determine the slopes of each line. Find the intercepts of each graph.
   a. For $2x - 4y = 4$, the x-intercept is 2 and the y-intercept is -1. Then $(2,0)$ and $(0,-1)$ are solutions and

   $$m_1 = \frac{0 - (-1)}{2 - 0} = \frac{1}{2}$$

   b. For $4x + 2y = -5$, the x-intercept is $-\dfrac{5}{4}$ and the y-intercept is $-\dfrac{5}{2}$. Therefore $\left(-\dfrac{5}{4},0\right)$ and $\left(0,-\dfrac{5}{2}\right)$ are solutions and

   $$m_2 = \frac{0 - \left(-\dfrac{5}{2}\right)}{-\dfrac{5}{4} - 0} = \frac{\dfrac{5}{2}}{-\dfrac{5}{4}} = \frac{5}{2}\cdot\left(-\frac{4}{5}\right) = -2$$

   Thus the graphs of the equations $2x - 4y = 4$ and $4x + 2y = -5$ are perpendicular lines since $m_1 m_2 = \dfrac{1}{2}(-2) = -1$ and the slopes are negative reciprocals.

▶ ***Quick check***   Determine if the line containing the points $(1,-2)$ and $(3,5)$ is perpendicular to the line containing the points $(-1,-1)$ and $(6,-3)$.   ■

--- *Mastery points* ---

Can you
   ■ Find the distance between two points in the rectangular coordinate plane?
   ■ Find the coordinates of the midpoint of a line segment?
   ■ Find the slope of a straight line given two points on the line?
   ■ Determine if two lines are parallel?
   ■ Determine if two lines are perpendicular?

Exercise 7–2

In exercises 1–15, find the distance between the given pairs of points and the slope of the line containing the points. Find the midpoint in exercises 1–10. See examples 7–2 A, B, and C.

Example Let $(x_1, y_1) = (-4, 3)$ and $(x_2, y_2) = (5, -6)$.

Solution a. Using $d = \sqrt{(x_2 - x_1)^2 + (y_2 - y_1)^2}$,

$$
\begin{aligned}
d &= \sqrt{[(5) - (-4)]^2 + [(-6) - (3)]^2} \\
&= \sqrt{(9)^2 + (-9)^2} \\
&= \sqrt{81 + 81} = \sqrt{162} = \sqrt{81 \cdot 2} = 9\sqrt{2}
\end{aligned}
$$

Replace x_1 with -4, x_2 with 5, y_1 with 3, and y_2 with -6

The distance between the points is $9\sqrt{2}$ units.

b. Using $m = \dfrac{y_1 - y_2}{x_1 - x_2}$,

$$
\begin{aligned}
m &= \frac{(3) - (-6)}{(-4) - (5)} \\
&= \frac{9}{-9} = -1
\end{aligned}
$$

Replace x_1 with -4, x_2 with 5, y_1 with 3, and y_2 with -6

The slope is -1.

c. Using $\left(\dfrac{x_1 + x_2}{2}, \dfrac{y_1 + y_2}{2} \right)$,

$$
\begin{aligned}
\text{the midpoint is } &\left(\frac{(-4) + (5)}{2}, \frac{(3) + (-6)}{2} \right) \\
&= \left(\frac{1}{2}, \frac{-3}{2} \right) \\
&= \left(\frac{1}{2}, -\frac{3}{2} \right)
\end{aligned}
$$

Replace x_1 with -4, x_2 with 5, y_1 with 3, and y_2 with -6

The midpoint of the line segment is $\left(\dfrac{1}{2}, -\dfrac{3}{2} \right)$.

1. $(2,2)$ and $(6,7)$
2. $(4,1)$ and $(1,4)$
3. $(-1,5)$ and $(-3,0)$
4. $(-4,6)$ and $(-1,2)$
5. $(-1,4)$ and $(-1,9)$
6. $(2,-6)$ and $(2,1)$
7. $(3,6)$ and $(-4,6)$
8. $(-1,-3)$ and $(5,-3)$
9. $(-3,-1)$ and $(-4,0)$
10. $(3,5)$ and $(-2,6)$
11. $(7,-3)$ and $(-4,4)$
12. $(-2,-2)$ and $(6,-3)$
13. $(4,-5)$ and $(2,4)$
14. $(0,8)$ and $(0,-1)$
15. $(0,-6)$ and $(0,4)$

Determine if lines L_1 and L_2 are parallel. See example 7-2 D.

Example   Determine if the line containing the points $(1,2)$ and $(-3,6)$ is parallel to the line containing the points $(5,-6)$ and $(-2,1)$.

Solution   Using $m = \dfrac{y_2 - y_1}{x_2 - x_1}$,

a. Let $(x_1,y_1) = (1,2)$ and $(x_2,y_2) = (-3,6)$.

$$m_1 = \frac{6 - 2}{-3 - 1} = \frac{4}{-4} = -1$$

b. Let $(x_1,y_1) = (5,-6)$ and $(x_2,y_2) = (-2,1)$.

$$m_2 = \frac{1 - (-6)}{-2 - 5} = \frac{7}{-7} = -1$$

Since the slope of each line is the same, -1, the lines are parallel.

16. L_1 contains $(4,2)$ and $(1,-1)$
L_2 contains $(1,1)$ and $(0,0)$

17. L_1 contains $(5,-2)$ and $(4,1)$
L_2 contains $(0,6)$ and $(2,0)$

18. L_1 contains $(5,1)$ and $(-4,2)$
L_2 contains $(4,-3)$ and $(2,1)$

19. L_1 contains $(-6,3)$ and $(-2,-5)$
L_2 contains $(-4,1)$ and $(7,-6)$

Determine whether lines L_1 and L_2 are perpendicular. See example 7-2 E-1.

Example   Determine if the lines containing the points $(1,-2)$ and $(3,5)$ and containing points $(-1,-1)$ and $(6,-3)$ are perpendicular.

Solution   Using $m = \dfrac{y_2 - y_1}{x_2 - x_1}$,

a. Let $(x_1,y_1) = (1,-2)$ and $(x_2,y_2) = (3,5)$.

$$m_1 = \frac{5 - (-2)}{3 - 1} = \frac{7}{2}$$

b. Let $(x_1,y_1) = (-1,-1)$ and $(x_2,y_2) = (6,-3)$.

$$m_2 = \frac{-3 - (-1)}{6 - (-1)} = \frac{-2}{7} = -\frac{2}{7}$$

Since $\left(\dfrac{7}{2}\right)\left(-\dfrac{2}{7}\right) = -1$, the slopes are negative reciprocals and the lines are perpendicular.

20. L_1 contains $(2,-3)$ and $(4,3)$
L_2 contains $(1,0)$ and $(-2,1)$

21. L_1 contains $(5,-2)$ and $(-3,-3)$
L_2 contains $(4,6)$ and $(5,-2)$

22. L_1 contains $(1,1)$ and $(4,1)$
L_2 contains $(-2,2)$ and $(3,-3)$

23. L_1 contains $(-6,-6)$ and $(-1,-1)$
L_2 contains $(4,-4)$ and $(-4,4)$

Determine if the graphs of the given pairs of equations are parallel, perpendicular, or neither. See example 7-2 E-2.

24. $x + y = 1$ and $3x + 3y = -6$

25. $2x + y = -4$ and $6x + 3y = 12$

26. $y - x = -5$ and $4x + 4y = 8$

27. $4y - 3x = 12$ and $4x + 3y = 24$

28. $2y - 3x = 1$ and $3y + 2x = 5$

29. $x + 4y = 5$ and $2x - 5y = 2$

30. $3y - 4x = 1$ and $8y + 3x = 6$

31. $2y - x = 1$ and $6x + 3y = 0$

32. $x + 5y = 0$ and $3x + 15y = 2$

33. $3x - 7y = 0$ and $3y + 5x = 0$

Determine if the lines through the given sets of points are parallel, perpendicular, or neither.

34. (1,3) and (2,4); (7,2) and (8,3)

35. (−2,0) and (1,4); (5,2) and (9,5)

36. (1,5) and (−2,−3); (0,1) and (3,2)

37. (5,6) and (1,3); (−1,6) and (2,1)

38. (7,2) and (−3,4); (−1,−2) and (5,2)

Solve the following word problems.

39. The *pitch* of a roof is the slope of a roof. If the roof of a house rises vertically a distance of 9 feet through a horizontal distance of 15 feet, what is the pitch of the roof?

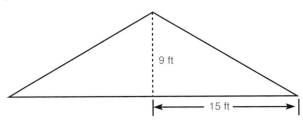

9 ft

15 ft

40. The roof of a school building rises 10 feet through a horizontal run of 35 feet. What is the pitch of the roof? (Refer to exercise 39.)

41. The guy wire of a telephone pole is attached to the pole 5 meters above the ground and attached to the ground 7.5 meters from the base of the pole. What is the slope of the guy wire?

42. A ladder leaning against a house touches the building at a point 14 feet above the ground. If the foot of the ladder is 9 feet from the base of the house, what is the slope of the ladder?

43. If a company's profits (P) are related to the number of items produced (x) by a linear equation, what is the slope of the graph of the equation if the profits rise by \$25,000 for every 175 items produced?

44. A company's profits (P) are related to increases in the workers' average pay (x) by a linear equation. If the company's profits drop by \$25,000 per year for every increase of \$550 per year in the workers' average pay, what is the slope of the graph of the equation?

45. The increase in a jogger's heartbeat in beats per minute is related to her increase in speed in feet per second by a linear equation. What is the slope of the graph of the equation if an increase in speed of 2 feet per second causes an increase of 10 heartbeats per minute and an increase in speed of 4 feet per second causes an increase of 25 heartbeats per minute? [*Hint:* Use ordered pairs (2,10) and (4,25).]

46. The bacteria count in a culture is related to the hours it exists by a linear equation. What is the slope of the graph if after $1\frac{1}{2}$ hours the bacteria count is 1,000,000 and after 2 hours the bacteria count is 4,500,000?

47. The decay of a substance is related by a linear equation to the time in years that the substance is left in the open air. If after 10 years there are 75 grams remaining and after 25 years there are 40 grams remaining, what is the slope of the equation?

48. The vertices of a triangle in the plane are at the points (−2,4), (5,2), and (0,−4). Find the perimeter (distance around) of the triangle.

49. Show that the points (4,2), (3,0), (−1,0), and (0,2) are the vertices of a parallelogram. Find the perimeter of the parallelogram. (*Hint:* A parallelogram is a four-sided figure whose opposite sides are parallel.)

50. Show that the points (−2,1), (5,3), (3,4), and (0,0) are the vertices of a parallelogram. Find the perimeter of the parallelogram. (Refer to exercise 49.)

51. Show that the points (4,2), (−2,−3), and (4,−3) are the vertices of a right triangle. (*Hint:* Show that two sides are perpendicular.)

52. Show that the points (2,−3), (5,1), and (−2,0) are the vertices of a right triangle by (a) using the Pythagorean Theorem and (b) showing two sides are perpendicular.

53. A trapezoid is a four-sided figure with one pair of opposite sides that are parallel. Show that the points (−3,2), (−1,−4), (5,2), and (9,−4) are the vertices of a trapezoid.

54. Three points that lie on the same straight line are said to be **collinear.** Three points are collinear if the sum of the distances between two pairs of points is equal to the distance between the third pair of points and if the slopes between all pairs of points are the same. Show that the points (6,7), (5,2), and (3,−8) are collinear using (a) slopes and (b) the distance formula.

55. Show that the points $(1,2)$, $(-3,-4)$, and $(-5,-7)$ are collinear using (a) slopes and (b) the distance formula. (Refer to exercise 54.)

56. Find the abscissa of the points whose ordinate is -6 if the points are at a distance of $5\sqrt{5}$ from the point $(4,5)$.

57. Find the ordinate of the points whose abscissa is 4 if the points are at a distance of $2\sqrt{13}$ from the point $(-2,-7)$.

58. Given one endpoint of a line segment is $(-2,3)$ and the midpoint of the line segment is $(2,-3)$, what are the coordinates of the other endpoint?

59. Given the midpoint of a line segment is the point $(1,5)$, find the other endpoint of the line segment if one endpoint is $(4,-2)$.

Review exercises

Solve the following equations for y. See section 2–2.

1. $3x + 2y = 4$

2. $4y - 3x = 8$

Solve the following inequalities for y. See section 2–2 and 2–5.

3. $4y + x < 8$

4. $x - 2y \geq 4$

Find the solution set of the following linear equations. See section 2–1.

5. $\dfrac{1}{2}x - 5 = \dfrac{2}{3}x + 1$

6. $3x = \dfrac{1}{2}(x - 2)$

7–3 ■ Finding the equation of a line

In section 7–1, we discussed the graph of a linear equation in two variables of the form $ax + by = c$ ($a \neq 0$ or $b \neq 0$). The graph of the equation is a straight line. In this section, we will determine the **equation of the graph** of a straight line. That is, we want an equation that is satisfied only by the coordinates of the points on the line. Thus the equation must be such that, for any arbitrary point P,

1. if P is on the graph, then its coordinates satisfy the equation, and

2. if P is not on the graph, then its coordinates do not satisfy the equation.

Consider a line in the plane having slope m that passes through a given point $P_1(x_1,y_1)$. Let $P(x,y)$ be any other arbitrary point on the line. See figure 7–13.

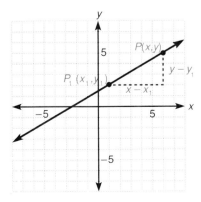

Figure 7–13

By the definition of the slope of a line, provided that $x \neq x_1$, we have

$$m = \frac{y - y_1}{x - x_1}$$

When we multiply each member by $x - x_1$, we obtain

$$y - y_1 = m(x - x_1)$$

We call this the **point-slope** form of the equation of a line.

Point-slope form

The point-slope form of the equation of a nonvertical line having slope m and passing through the known point (x_1, y_1) is given by

$$y - y_1 = m(x - x_1)$$

■ **Example 7–3 A**

1. Find the equation of the line having a slope of $-\dfrac{1}{2}$ and passing through the point $(-1, 4)$.

Use the point-slope form $y - y_1 = m(x - x_1)$.

$$y - (4) = \left(-\frac{1}{2}\right)[x - (-1)] \quad \text{Replace } m \text{ with } -\frac{1}{2}, y_1 \text{ with 4, and } x_1 \text{ with } -1$$

$$y - 4 = -\frac{1}{2}(x + 1)$$
$$2y - 8 = -1(x + 1) \quad \text{Multiply each member by 2}$$
$$2y - 8 = -x - 1$$
$$x + 2y = 7 \quad \text{Add } x + 8 \text{ to each member}$$

We say that the equation $x + 2y = 7$ is written in **standard form.**

Standard form

The standard form of the equation of a line is written as

$$ax + by = c \ (a > 0)$$

where a, b, and c are real numbers and not both a and b equal to 0.

---

2. Find the equation of the line passing through the points $(2, 4)$ and $(-5, -2)$. Write the equation in standard form.

We must first find the slope m of the line. Using $m = \dfrac{y_2 - y_1}{x_2 - x_1}$, let $(x_1, y_1) = (2, 4)$ and $(x_2, y_2) = (-5, -2)$.

$$m = \frac{-2 - 4}{-5 - 2} = \frac{-6}{-7} = \frac{6}{7}$$

Using the point-slope form $y - y_1 = m(x - x_1)$, the slope, and *one* of the given points on the line, $(2, 4)$, we obtain

$$y - (4) = \left(\frac{6}{7}\right)[x - (2)]$$ Replace m with $\frac{6}{7}$, y_1 with 4, and x_1 with 2

$$7y - 28 = 6(x - 2)$$ Multiply each member by 7

$$7y - 28 = 6x - 12$$

$$-6x + 7y = 16$$ Add $-6x + 28$ to each member

$$6x - 7y = -16$$ Multiply each member by -1

The equation in standard form is $6x - 7y = -16$.

▶ *Quick check* Find the equation, in standard form, of the line passing through the points $(0, -7)$ and $(6,4)$. ■

Consider again the point-slope form of the equation of a nonvertical line. Given

$$y - y_1 = m(x - x_1)$$

suppose we let the known point be $(0,b)$, the y-intercept. Substituting 0 for x_1 and b for y_1, we have

$$y - b = m(x - 0)$$
$$y - b = mx$$
$$y = mx + b$$

We call this the **slope-intercept** form of the equation of a line.

Slope-intercept form ─────────────────────────

The slope-intercept form of the equation of a nonvertical line is written as

$$y = mx + b$$

where m is the slope and the point $(0,b)$ is the y-intercept.

■ *Example 7–3 B*

1. Find the slope and y-intercept of the line whose equation is $2y + 3x = 4$.

   To find the slope and y-intercept, we will write the equation in slope-intercept form, which is accomplished by solving the equation for y. Thus

   $$2y + 3x = 4$$
   $$2y = -3x + 4$$ Add $-3x$ to each member
   $$y = -\frac{3}{2}x + 2$$ Divide each term by 2

   Then the slope is $-\frac{3}{2}$ (the coefficient of x) and the y-intercept is the point $(0,2)$.

2. Find the equation of the line whose slope is $-\dfrac{5}{3}$ and whose y-intercept is $(0,3)$. Write the equation in standard form.

Using the slope-intercept form of the equation, $y = mx + b$, where $m = -\dfrac{5}{3}$ and $b = 3$,

$$y = mx + b$$
$$y = -\frac{5}{3}x + 3 \qquad \text{Replace } m \text{ with } -\frac{5}{3} \text{ and } b \text{ with } 3$$
$$3y = -5x + 9 \qquad \text{Multiply each member by 3}$$
$$5x + 3y = 9 \qquad \text{Add } 5x \text{ to each member}$$

▶ *Quick check*   Find the slope and y-intercept of the line whose equation is $8x + 5y = 10$.

We saw in the previous section that the slopes of two nonvertical *parallel* lines are the same and the slopes of two nonvertical *perpendicular* lines are negative reciprocals. We use these facts in the next two examples.

3. Find the equation of the line through the point $(3,1)$ that is parallel to the line whose equation is $3x - 2y = 4$.

For the lines to be parallel, the line we want must have the same slope as the line $3x - 2y = 4$. We solve this equation for y to obtain the slope-intercept form.

$$3x - 2y = 4$$
$$-2y = -3x + 4 \qquad \text{Add } -3x \text{ to each member}$$
$$y = \frac{3}{2}x - 2 \qquad \text{Divide each member by } -2$$

Slope $m = \dfrac{3}{2}$

Using the point-slope form,

$$y - y_1 = m(x - x_1)$$
$$y - (1) = \left(\frac{3}{2}\right)[x - (3)] \qquad \text{Replace } m \text{ with } \frac{3}{2}, y_1 \text{ with 1, and } x_1 \text{ with 3}$$
$$2y - 2 = 3(x - 3) \qquad \text{Multiply each member by 2}$$
$$2y - 2 = 3x - 9$$
$$2y - 3x = -7 \qquad \text{Multiply each member by } -1$$
$$3x - 2y = 7 \qquad \text{Standard form}$$

The equation of the line is $3x - 2y = 7$.

4. Find the equation of the line that is perpendicular to the line $4x - 2y = 5$ and passes through the point $(-3,4)$.

We write the equation $4x - 2y = 5$ in slope-intercept form to find its slope.

$$4x - 2y = 5$$
$$-2y = -4x + 5 \qquad \text{Add } -4x \text{ to each member}$$
$$y = 2x - \frac{5}{2} \qquad \text{Divide each member by } -2$$

Slope $m = 2$

The slope of the line we desire is the negative reciprocal of 2, which is $-\dfrac{1}{2}$.

We want the equation of the line with slope $m = -\dfrac{1}{2}$ and passing through $(-3,4)$. Use the point-slope form.

$$y - y_1 = m(x - x_1)$$

$$y - (4) = \left(-\dfrac{1}{2}\right)[x - (-3)] \qquad \text{Replace } m \text{ with } -\dfrac{1}{2}, y_1 \text{ with 4, and } x_1 \text{ with } -3$$

$$y - 4 = -\dfrac{1}{2}(x + 3)$$

$$2y - 8 = -1(x + 3) \qquad \text{Multiply each member by 2}$$

$$2y - 8 = -x - 3$$

$$x + 2y = 5$$

The equation of the line is $x + 2y = 5$.

5. Determine if the graphs of the equations $2x - 4y = 5$ and $4x + 2y = -5$ are perpendicular, using the slope-intercept form.

We must write each equation in slope-intercept form and compare the slopes. Solve for y.

a. $2x - 4y = 5$

$$-4y = -2x + 5 \qquad \text{Add } -2x \text{ to each member}$$

$$y = \dfrac{-2x + 5}{-4} \qquad \text{Divide each term by } -4$$

$$y = \dfrac{1}{2}x - \dfrac{5}{4} \qquad \text{Write in slope-intercept form}$$

$$m_1 = \dfrac{1}{2}$$

b. $4x + 2y = -5$

$$2y = -4x - 5 \qquad \text{Add } -4x \text{ to each member}$$

$$y = \dfrac{-4x - 5}{2} \qquad \text{Divide each term by 2}$$

$$y = -2x - \dfrac{5}{2} \qquad \text{Write in slope-intercept form}$$

$$m_2 = -2$$

Since $\left(\dfrac{1}{2}\right)(-2) = -1$ (the slopes are negative reciprocals), the lines are perpendicular.

▶ *Quick check* Find the equation of the line through the point $(6, -7)$ that is parallel to the line whose equation is $4x - 5y = 3$. ■

The slope-intercept form of the equation of a line can be used to graph a linear equation in two variables.

■ *Example 7–3 C*

Graph the following equations using the slope and the *y*-intercept.

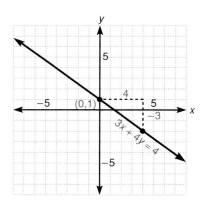

1. $3x + 4y = 4$
   Write the equation in slope-intercept form.

   $$3x + 4y = 4$$
   $$4y = -3x + 4 \qquad \text{Add } -3x \text{ to each member}$$
   $$y = -\frac{3}{4}x + 1 \qquad \text{Divide each term by 4}$$

   The slope is $m = -\dfrac{3}{4}$ and the *y*-intercept is $(0,1)$.

   Recall that the definition of the slope of a line is

   $$m = \frac{\text{vertical change}}{\text{horizontal change}} = \frac{\text{rise}}{\text{run}} = -\frac{3}{4} = \frac{-3}{4}$$

   a. Plot the *y*-intercept $(0,1)$.
   b. From this point, move 4 units to the *right* (horizontal change) and then 3 units *down* (vertical change that is negative) to obtain another point on the graph.
   c. Draw the line through the two points.

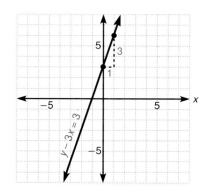

2. $y - 3x = 3$
   Write the equation in slope-intercept form.

   $$y = 3x + 3$$
   $$m = \frac{\text{vertical change}}{\text{horizontal change}} = \frac{\text{rise}}{\text{run}} = 3 = \frac{3}{1}$$

   a. Plot the *y*-intercept $(0,3)$.
   b. From this point, move 1 unit to the *right* (horizontal change) and then 3 units *up* (vertical change that is positive) to obtain a second point.
   c. Draw the line through the two points.

▶ *Quick check* Graph $5y + 8x = 10$ using the slope and *y*-intercept. ■

We now summarize the different forms of a linear equation.

$$ax + by = c(a > 0) \qquad \text{Standard form}$$
$$y - y_1 = m(x - x_1) \qquad \text{Point-slope form}$$
$$y = mx + b \qquad \text{Slope-intercept form}$$
$$x = k \qquad \text{Equation of a vertical line}$$
$$y = k \qquad \text{Equation of a horizontal line}$$

Mastery points

Can you

- Find the equation of a line using the point-slope form
  $y - y_1 = m(x - x_1)$?
- Write the equation of a line in standard form $ax + by = c, a > 0$?
- Write the equation of a line in slope-intercept form $y = mx + b$?
- Find the slope, m, and the y-intercept, b, of a line given its equation?
- Sketch the graph of a linear equation in two variables using the slope and y-intercept?

Exercise 7–3

Find the equation of the line that satisfies the given conditions. Write the equation in standard form. See example 7–3 A.

Example Through points $(0,-7)$ and $(6,4)$

Solution We first find the slope using the two points.

$$m = \frac{y_2 - y_1}{x_2 - x_1}$$

$$= \frac{(4) - (-7)}{(6) - (0)}$$ Replace y_2 with 4, y_1 with -7, x_2 with 6, and x_1 with 0

$$= \frac{4 + 7}{6}$$ Definition of subtraction

$$= \frac{11}{6}$$

Using the point-slope form,

$$y - y_1 = m(x - x_1)$$
$$y - (4) = \left(\frac{11}{6}\right)[x - (6)]$$ Replace m with $\frac{11}{6}$, y_1 with 4, and x_1 with 6 [using point (6,4)]
$$6y - 24 = 11(x - 6)$$ Multiply each member by 6
$$6y - 24 = 11x - 66$$ Distribute in the right member
$$-11x + 6y = -42$$ Add $-11x$ and 24 to each member
$$11x - 6y = 42$$ Multiply each term by -1

1. Slope $m = \dfrac{1}{2}$ and passing through $(3,4)$

2. Slope $m = -\dfrac{5}{4}$ and passing through $(-1,-6)$

3. Slope $m = 5$ and passing through $(0,7)$

4. Slope $m = -\dfrac{3}{2}$ and having x-intercept -5

5. Slope $m = -\dfrac{5}{6}$ and passing through $(0,0)$

6. Slope $m = -6$ and having y-intercept 2

7. Slope $m = 1$ and having y-intercept -3

8. Horizontal line through $(5,-3)$

9. Horizontal line through $(1,4)$

10. Horizontal line through $(0,0)$

11. Slope is undefined and passing through $(-5,6)$

12. Slope is undefined and passing through $(1,2)$

13. Vertical line passing through $(5,-4)$

14. Vertical line passing through $(-6,2)$

15. Vertical line passing through $(0,0)$

16. Slope $m = 0$ and passing through $(3,5)$

17. Slope $m = 0$ and passing through $(0,-7)$

Find the equation of the line through the given points. Write the equation in standard form.

18. $(1,2)$ and $(5,4)$ **19.** $(3,6)$ and $(1,1)$ **20.** $(-3,4)$ and $(5,-1)$ **21.** $(2,6)$ and $(-1,-4)$

22. $(5,0)$ and $(-2,-3)$ **23.** $(4,0)$ and $(5,-2)$ **24.** $(3,-6)$ and $(2,-6)$ **25.** $(5,4)$ and $(-2,4)$

26. $(-2,-2)$ and $(4,4)$ **27.** $(-4,4)$ and $(3,-3)$ **28.** $(-5,5)$ and $(1,-1)$ **29.** $(4,-3)$ and $(4,-7)$

30. $(-2,6)$ and $(-2,-5)$

Write each equation in slope-intercept form $y = mx + b$ and identify the slope m and the y-intercept b. Sketch the graph using the slope and y-intercept. See examples 7–3 B and C.

Example $5y + 8x = 10$

Solution   We first solve the equation for y.

$$5y + 8x = 10 \qquad \text{Original equation}$$
$$5y = -8x + 10 \qquad \text{Add } -8x \text{ to each member}$$
$$y = -\frac{8}{5}x + 2 \qquad \text{Divide each term by 5}$$

The slope is $m = -\dfrac{8}{5}$ and the y-intercept is $(0,2)$.

1. Plot the y-intercept $(0,2)$.

2. From this point, since $m = \dfrac{-8}{5}$ move 5 units to the right (horizontal change) and 8 units *down*
   (vertical change is *negative*) to obtain a second point.

3. Draw the straight line through these two points.

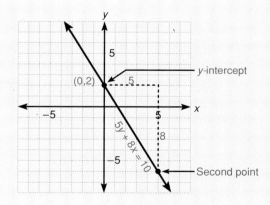

31. $2x + y = 5$ **32.** $3x + y = -3$ **33.** $4x - y = 6$ **34.** $3x - 7y = 0$

35. $5y - 2x = 0$ **36.** $2y + 9x = 0$ **37.** $8x + 3y = 0$ **38.** $x - y = 0$

39. $3y + 5 = 0$ **40.** $2y - 7 = 0$

Find the equation of the line satisfying the given conditions. Write each equation in standard form. See example 7–3 B–3 and 4.

Example Through the point $(6,-7)$ and parallel to the line $4x - 5y = 3$

Solution Find the slope of the given line. Solve for y.

$$4x - 5y = 3$$ Original equation
$$-5y = -4x + 3$$ Subtract $4x$ from each member
$$y = \frac{4}{5}x - \frac{3}{5}$$ Divide each term by -5

The slope $m = \frac{4}{5}$. Since we want a line parallel to the given line, the slope will be the same.

Using the point-slope form,

$$y - y_1 = m(x - x_1)$$

$$y - (-7) = \left(\frac{4}{5}\right)[x - (6)]$$ Replace m with $\frac{4}{5}$, y_1 with -7, and x_1 with 6

$$y + 7 = \frac{4}{5}(x - 6)$$ Double negative property in the left member

$$5y + 35 = 4(x - 6)$$ Multiply each member by 5
$$5y + 35 = 4x - 24$$ Distribute in right member
$$-4x + 5y = -59$$ Subtract $4x$ and 35 from each member
$$4x - 5y = 59$$ Multiply each term by -1

41. Parallel to $3x + y = 6$ and passing through $(1,2)$

42. Perpendicular to $2x + 5y = 3$ and passing through $(1,7)$

43. Passing through $(-3,-1)$ and parallel to $3x - 2y = 9$

44. Passing through $(-7,4)$ and perpendicular to $7y - 2x = 0$

45. Passing through $(0,0)$ and perpendicular to $9x - 2y = 1$

46. Passing through $(5,-3)$ and parallel to $y - 2 = 0$

47. Passing through $(4,0)$ and perpendicular to $y = 5$

48. Find the equation of the perpendicular bisector of the line segment whose endpoints are $(-2,3)$ and $(4,7)$. (*Hint:* We want the line perpendicular at the midpoint.)

49. Find the equation of the perpendicular bisector of the line segment whose endpoints are $(3,-4)$ and $(3,6)$.

Using the slope-intercept form of the equation of a line, determine if the given pairs of equations represent parallel lines, perpendicular lines, or neither. See example 7–3 B–5.

50. $x + 2y = 5$ and $6x - 3y = 4$

51. $3y + 2x = 5$ and $2y + 3x = -1$

52. $2y - 5x = -3$ and $y + 3x = 4$

53. $4y - 5x = -1$ and $8x + 10y = 4$

54. $3x + 8y = 2$ and $6x - 16y = 5$

55. $8x - 5y = 1$ and $16x - 10y = 7$

Solve the following word problems.

56. Given the point-slope form of the equation of a line,

$$y - y_1 = m(x - x_1)$$

if the points $P_1(x_1,y_1)$ and $P_2(x_2,y_2)$ lie on the graph, then

$$m = \frac{y_2 - y_1}{x_2 - x_1}$$

and if we substitute, we obtain the **two-point** form of the equation of a line given by

$$y - y_1 = \frac{y_2 - y_1}{x_2 - x_1}(x - x_1)$$

Using this form, find the equation in standard form for the line passing through the given points (a) (3,2) and (4,1); (b) (5,−1) and (−3,0).

57. Given the two-point form of the equation of a line found in exercise 56, use the points $(a,0)$ and $(0,b)$ to show that

$$\frac{x}{a} + \frac{y}{b} = 1$$

is an equation of the line with x-intercept $(a,0)$ and y-intercept $(0,b)$. We call this the **intercept** form of the equation of a line.

58. Write each equation in **intercept** form and determine the x- and y-intercepts. (Refer to exercise 57.)

  a. $4x + 3y = 12$ b. $2x + 5y = 10$
  c. $3x - y = 3$ d. $3x - 5y = 6$
  e. $5y - 2x = 8$ f. $6x + 5y = 12$

59. Find the slope and y-intercept of the equation $ax + by = c$ in terms of a, b, and c.

60. What is the slope of a line that is parallel to the line with the equation $ax + by = c$?

61. What is the slope of a line that is perpendicular to the line with the equation $ax + by = c$?

There are a number of real-life situations that can be described using linear equations in two variables. If two pairs of values are known, we can use the *two-point* form demonstrated in exercise 56. Express each equation in standard form.

62. A company produces 300 boxes of cereal for $150 and 600 boxes of the same cereal for $250. Let x be the number of boxes of cereal and y be the total cost.

63. John Doe makes $150 profit on 4 waterfront paintings that he does and $400 profit on 7 paintings. Let x be the number of paintings and y be the profit on the paintings.

64. A company found its total sales were $35,000 in the third year of operation and $105,000 in the fifth year of operation. Let x be the year of operation and y the total sales that year.

65. Use the resulting equation of exercise 64 to predict the total sales in the sixth year.

66. Use the resulting equation of exercise 62 to determine the approximate cost of producing 1,000 boxes.

Review exercises

Solve the following inequalities. See sections 2–2 and 2–5.

1. $3x - 2 \le x - 1$

2. $4y - 3x > 6$ (for y)

Graph the following equations. See section 7–1.

3. $x - 3y = 6$

4. $2y + x = -4$

5. Find the solution set of the quadratic equation $2x^2 - 3x = 4$. See section 6–3.

6. Find the solution set of the radical equation $\sqrt{x + 3} = x - 3$. See section 6–5.

7–4 ■ *Graphs of linear inequalities*

Linear inequalities in two variables

In chapter 2, we discussed the solution set of linear inequalities in one variable such as

$$x + 5 < 3, \quad 2x - 3 \le 1, \quad 4x - 3 > 0, \quad \text{and} \quad -4 \le 2x + 1 < 3$$

In this chapter, we have discussed the graph of the solution set of linear equations in two variables such as

$$2x - 3y = 5, \quad x - 2y = 6, \quad y = -4, \quad \text{and} \quad x = 5$$

Now we extend these ideas to consider the solution set and the graph of **linear inequalities in two variables** such as

$$y \le x, \quad 3x - 2y > 4, \quad \text{and} \quad y < 2x + 3$$

> **Linear inequalities in two variables** _____
>
> Any inequality of the form
>
> $$ax + by < c, \qquad ax + by > c,$$
> $$ax + by \le c, \quad \text{or} \quad ax + by \ge c,$$
>
> where a, b, and c are real numbers, a and b not both zero, is called a linear inequality in two variables.

The graph of any linear inequality in two variables will be a *half-plane*, as illustrated in the following examples.

■ *Example 7–4 A*

Graph the following linear inequalities.

1. $2x + y \le 4$
   To graph the inequality $2x + y \le 4$, we first graph the straight line $2x + y = 4$.

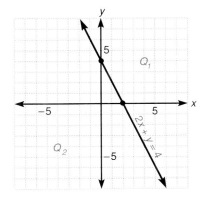

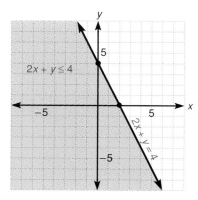

The graph of the inequality includes the points of this line together with all points of the plane either *above* or *below* this line. To decide which half-plane is the solution set, choose *any point not on the line* $2x + y = 4$ and substitute these values into the inequality $2x + y \leq 4$. *The origin, (0,0), is most often a good choice.* Replacing both x and y with 0 in the inequality $2x + y \leq 4$,

$$2(0) + (0) \leq 4 \qquad \text{Replace } x \text{ with 0 and } y \text{ with 0}$$
$$0 \leq 4 \qquad \text{(True)}$$

Since the result is true, (0,0) does satisfy the inequality and the solution set includes all points on the side of the line where (0,0) lies. We then shade that half-plane.

Note If our inequality had been the *strict* inequality $2x + y < 4$, instead of the *weak* inequality $2x + y \leq 4$, the line $2x + y = 4$ would be *dashed* since the points of the line are not in the solution set anymore.

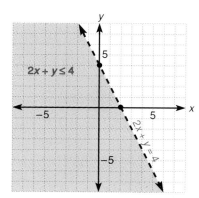

2. $3x + y > 6$

Graph the equation $3x + y = 6$ and make the line dashed since the order symbol, $>$, does not include equality. Since (0,0) does not lie on the line, choose this as a test point and substitute 0 for x and 0 for y in the inequality.

$$3x + y > 6 \qquad \text{Original inequality}$$
$$3(0) + (0) > 6 \qquad \text{Replace } x \text{ with 0 and } y \text{ with 0}$$
$$0 + 0 > 6$$
$$0 > 6 \qquad \text{(False)}$$

Since (0,0) does not satisfy the inequality, shade the half-plane that *does not* contain the origin.

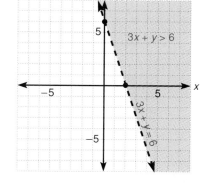

▶ *Quick check* Graph $4x + y > 8$. ■

To graph a linear inequality in two variables

1. Replace the inequality symbol by the equality symbol.
2. Graph this line making it (1) a solid line if the inequality symbol is $\leq$ or $\geq$ (which includes the line in the solution set) or (2) a dashed line if the inequality symbol is $<$ or $>$ (which does _not_ include the line in the solution set).
3. Choose some test point that is _not_ on the line [if possible, the origin (0,0) since the arithmetic is easiest for this point].
4. Substitute the coordinates of the test point in the inequality.
5. If the test point's coordinates satisfy the inequality, shade the half-plane containing that point for the solution set; if the test point's coordinates _do not_ satisfy the inequality, shade the other half-plane for the solution set.

■ _Example 7–4 B_

Graph the following inequalities.

1. $x > -3$

   Graph the vertical line $x = -3$ with a dashed line since the order symbol is $>$ and does not include equality. Choose test point (0,0) and substitute 0 for x in the inequality.

   $x > -3$
   $0 > -3$ Replace x with 0 (True)

   Shade the half-plane containing the origin.

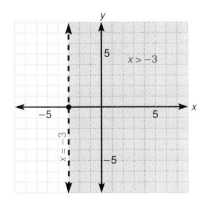

2. $5x - 3y \leq 0$

   Graph the line $5x - 3y = 0$ and make it a solid line. The point (0,0) is on the line since the graph passes through the origin. Therefore we cannot use (0,0) as a test point and we arbitrarily choose some point that does not lie on the line. Suppose we choose (3,0). We substitute 3 for x and 0 for y in the inequality.

   $$5x - 3y \leq 0$$
   $$5(3) - 3(0) \leq 0$$
   $$15 - 0 \leq 0$$
   $$15 \leq 0 \quad \text{(False)}$$

   Since the statement is false, shade the half-plane that _does not_ contain (3,0).

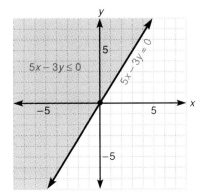

3. $|x| < 1$

In section 2–5, we learned that if

a. $|x| < 1$, then $-1 < x < 1$

b. $|x| = 1$, then $x = 1$ or $x = -1$.

Thus, $x = 1$ and $x = -1$ are boundary lines (dashed), and we want all points whose first coordinate (x-value) lies between 1 and -1. Determine this by checking points in all three regions $x < -1$, $-1 < x < 1$, and $x > 1$.

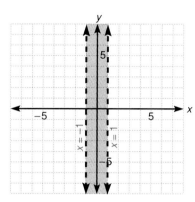

4. $|y - 2| \geq 3$

In section 2–5, we learned that if

a. $|y - 2| \geq 3$, then $y - 2 \geq 3$ or $y - 2 \leq -3$
$$y \geq 5 \qquad\qquad y \leq -1$$

b. $y - 2 = 3$, then $y - 2 = 3$ or $y - 2 = -3$
$$y = 5 \qquad\qquad y = -1$$

Thus, $y = 5$ and $y = -1$ are boundary lines (solid), and we want all points whose second coordinate (y-value) is greater than 5 or less than -1. *Check all three regions $y < -1$, $-1 < y < 5$, and $y > 5$.*

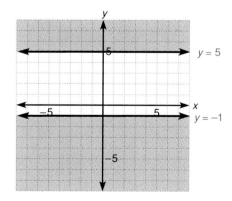

▶ *Quick check* Graph $|x + 3| \geq 1$.

> **Mastery points**
>
> *Can you*
> - Determine when the boundary line is solid and when it is dashed?
> - Graph the solution set of a linear inequality in two variables?
> - Graph an absolute value inequality?

Exercise 7–4

Graph the solution set of each linear inequality. See examples 7–4 A and B.

Example $4x + y > 8$

Solution   1. Graph the line $4x + y = 8$ as a dashed line since we have $>$.
2. Choose test point $(0,0)$ since the boundary line does not go through the origin.

$$4(0) + 0 > 8 \quad \text{Replace } x \text{ with 0 and } y \text{ with 0}$$
$$0 + 0 > 8 \quad \text{Multiply in left member}$$
$$0 > 8 \quad \text{(False)}$$

3. Shade the half-plane that *does not* contain $(0,0)$.

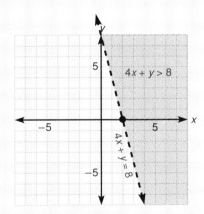

1. $y < 3$	**2.** $y \leq -4$	**3.** $x \geq 1$	**4.** $x > -5$	**5.** $y > 2$
6. $y \geq -1$	**7.** $x - 7 \geq -3$	**8.** $x + y > 5$	**9.** $x + y < -1$	**10.** $y < 3x - 1$
11. $y > 4 - 3x$	**12.** $x \leq 2y$	**13.** $3y > 4x$	**14.** $2y \geq -x$	**15.** $2x - 3y \leq 0$
16. $y - x > 0$	**17.** $5x + 2y \geq -10$	**18.** $3y - 5x > -15$		

Graph the following absolute value inequalities in two variables. See example 7–4 B–3 and 4.

Example $|x + 3| \geq 1$

Solution By property in section 2–5, if $|x + 3| \geq 1$, then $x + 3 \geq 1$ or $x + 3 \leq -1$
$$x \geq -2 \qquad\qquad x \leq -4$$
If $|x + 3| = 1$, then $x + 3 = 1$ or $x + 3 = -1$
$$x = -2 \quad \text{and} \quad x = -4$$
The boundary lines are the solid lines $x = -2$ and $x = -4$. The solutions are all points whose first component (x-value) is less than -4 or greater than -2.

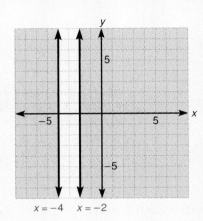

$x = -4$ $x = -2$

19. $|x| < 2$ **20.** $|x| > 3$ **21.** $|x| \geq 1$ **22.** $|x| \leq 4$ **23.** $|y| < 1$

24. $|y| > 2$ **25.** $|y| \geq 3$ **26.** $|y| \leq 5$ **27.** $|x - 1| \leq 3$ **28.** $|x + 2| < 1$

29. $|x - 5| > 2$ **30.** $|3 - x| \geq 4$ **31.** $|y - 4| \leq 3$ **32.** $|y + 6| < 2$ **33.** $|y + 7| > 1$

34. $|5 - y| \geq 3$ **35.** $|2x - 3| \leq 3$ **36.** $|3y + 1| < 5$

Review exercises

Graph the following set of equations on the same coordinate axes. Determine the point of intersection. See section 7–1.

1. $2x - y = 3$
$x + y = 3$

2. $x - 3y = 1$
$ x + 2y = 6$

3. Given $P(x) = 4x^2 - 2x + 1$, find $P(-1)$ and $P(2)$.
See section 1–5.

4. Simplify the expression
$2^2 - [3(5 - 1) + 4(6 + 2)]$. See section 1–4.

Perform the indicated operations. Reduce to lowest terms. See sections 4–2 and 4–3.

5. $\dfrac{3x - 1}{x + 3} - \dfrac{x - 1}{x^2 - 9}$

6. $\dfrac{4x^3}{9y^2} \div \dfrac{16x}{3y}$

Chapter 7 lead-in problem

A company that manufactures a heating unit can produce 20 units for $13,900 while it would cost $7,500 to manufacture 10 units. Assume the cost and number of units produced are related by the linear equation of a straight line. Let y be the total cost to manufacture x units. Find the linear equation of the straight line.

Solution

Using ordered pairs (x,y), we are given the known ordered pairs (20, 13,900) and (10, 7,500). We first find the slope of the line using the ordered pairs.

$$m = \frac{y_1 - y_2}{x_1 - x_2}$$

$$= \frac{13,900 - 7,500}{20 - 10} \quad \text{Replace } y_1 \text{ with } 13,900, y_2 \text{ with } 7,500, x_1 \text{ with } 20, \text{ and } x_2 \text{ with } 10$$

$$= \frac{6,400}{10}$$

$$= 640$$

Using the point-slope form of a line and one of the ordered pairs, (10, 7,500), we find the equation of the line.

$$y - y_1 = m(x - x_1)$$

$$y - 7,500 = 640(x - 10) \quad \text{Replace } m \text{ with } 640, y_1 \text{ with } 7,500, \text{ and } x_1 \text{ with } 10$$

$$y - 7,500 = 640x - 6,400 \quad \text{Distribute in the right member}$$

$$640x - y = -1,100 \quad \text{Write in standard form}$$

Chapter 7 summary

1. An **ordered pair** of real numbers is written as (x,y), where x and y are real numbers.
2. The **rectangular coordinate system** is formed by two real number lines, called **axes,** drawn in the plane—one horizontal (x-axis) and the other vertical (y-axis)—that intersect at the **origin** 0 of each line.
3. The axes divide the plane into four regions called **quadrants.**
4. Each point in the rectangular coordinate system is associated with only one ordered pair of real numbers, called the **coordinates of the point.** The point is called the **graph** of the ordered pair.
5. In the ordered pair (x,y), x is called the **first component** and y is called the **second component** of the ordered pair.
6. The **abscissa** of a point in the plane is the first component of the ordered pair and the **ordinate** of a point is the second component of the ordered pair.
7. The **graph of an equation** is the set of points in the plane associated with the solutions of the equation.
8. The graph of the linear equation in two variables $ax + by = c$ is a **straight line.**
9. The abscissa of the point at which a line intersects the x-axis is called the **x-intercept** of the line.
10. The **y-intercept** of a line is the **ordinate** of the point at which the line intersects the y-axis.
11. The distance between two points in the plane, $P_1(x_1,y_1)$ and $P_2(x_2,y_2)$, denoted by d, is given by

$$d = \sqrt{(x_2 - x_1)^2 + (y_2 - y_1)^2}$$

We call this the **distance formula.**
12. The **slope** m of the line containing $P_1(x_1,y_1)$ and $P_2(x_2,y_2)$ is given by

$$m = \frac{y_2 - y_1}{x_2 - x_1} \text{ or } m = \frac{y_1 - y_2}{x_1 - x_2} \quad (x_1 \neq x_2)$$

13. Two nonvertical lines are **parallel** if and only if they have the same slopes.
14. Two nonvertical lines are **perpendicular** if and only if their slopes are negative reciprocals.
15. The **point-slope** form of the equation of a nonvertical line having slope m and passing through (x_1,y_1) is given by $y - y_1 = m(x - x_1)$.
16. The **slope-intercept** form of the equation of a nonvertical line is written as $y = mx + b$, where m is the slope and b is the y-intercept.
17. The graph of a linear inequality in two variables, $ax + by > c$, $ax + by \geq c$, $ax + by < c$, or $ax + by \leq c$, is the **half-plane** on one side of the line $ax + by = c$. The graph includes the line if we have a weak inequality ($\leq$ or $\geq$), otherwise, it does not.

Chapter 7 error analysis

1. Finding the midpoint of a line segment
    Example: Endpoints are $(2, -3)$ and $(-1, 5)$
    $$\text{Midpoint} = \left(\frac{2 - (-1)}{2}, \frac{-3 - 5}{2}\right) = \left(\frac{3}{2}, -4\right)$$
    Correct answer: $\left(\frac{1}{2}, 1\right)$
    What error was made? (*see page 316*)

2. Finding the slope of a line
    Example: Containing points $(2, -5)$ and $(6, -3)$
    $$m = \frac{-5 - (-3)}{6 - 2} = \frac{-5 + 3}{4} = \frac{-2}{4} = -\frac{1}{2}$$
    Correct answer: $m = \frac{1}{2}$
    What error was made? (*see page 318*)

3. The slope of a line
    Example: The slope of the line through $(5, -2)$ and $(5, 3)$ is $m = 0$ since $x_1 = x_2 = 5$.
    Correct answer: m is undefined.
    What error was made? (*see page 320*)

4. Graphing linear inequalities
    Example: The graph of $2y - 3x \leq -6$ is

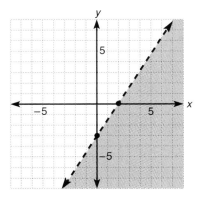

    Correct answer:

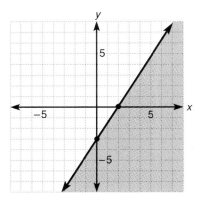

    What error was made? (*see page 338*)

5. Finding the distance between two points
    Example: Find the distance between the points $(-1, 3)$ and $(2, 5)$.
    $$d = \sqrt{(-1 + 2)^2 + (3 + 5)^2} = \sqrt{1^2 + 8^2} = \sqrt{65}$$
    Correct answer: $\sqrt{13}$
    What error was made? (*see page 315*)

6. Squaring a radical binomial
    Example: $(3\sqrt{2} - 5\sqrt{3})^2 = (3\sqrt{2})^2 - (5\sqrt{3})^2$
    $$= 18 - 75 = -57$$
    Correct answer: $93 - 30\sqrt{6}$
    What error was made? (*see page 242*)

7. Combining like terms
    Example: $a^2b - 2ab^2 - 6a^2b + ab^2$
    $= (1 + 6)a^2b - (2 + 1)ab^2 = 7a^2b - 3ab^2$
    Correct answer: $-5a^2b - ab^2$
    What error was made? (*see page 41*)

8. Solving absolute value inequalities
    Example: $|2x - 3| > 7$
    $$-7 < 2x - 3 < 7$$
    $$-4 < 2x < 10$$
    $$\{x | -2 < x < 5\}$$
    Correct answer: $\{x | x < -2 \text{ or } x > 5\}$
    What error was made? (*see page 88*)

9. Perpendicular lines
    Example: The lines $3x - y = 7$ and $6x + 2y = 5$ are perpendicular.
    Correct answer: The lines are not perpendicular.
    What error was made? (*see page 322*)

10. Reducing a rational expression
    Example: $\dfrac{3x^2 - x^3}{x^3} = \dfrac{3x^2 - \cancel{x^3}^{1}}{\cancel{x^3}_{1}} = 3x^2 - 1$
    Correct answer: $\dfrac{3 - x}{x}$
    What error was made? (*see page 157*)

Chapter 7 critical thinking

If there are n points in a plane where no three points lie on the same line, how many lines can be drawn through the n points?

Chapter 7 review

[7–1]
Plot the graph of each ordered pair of real numbers and state the quadrant in which the point lies, if it lies in a quadrant.

1. $(-3,5)$ **2.** $(-2,-7)$ **3.** $(0,-1)$ **4.** $(5,0)$

5. $\left(1,-\dfrac{3}{2}\right)$ **6.** $\left(\dfrac{1}{2},\dfrac{11}{2}\right)$ **7.** $(0,4)$ **8.** $\left(-\dfrac{5}{2},-4\right)$

Find the x- and y-intercepts of the graph for each equation (if they exist).

9. $x - 3y = 6$ **10.** $2x - y = 8$ **11.** $5x - 2y = 10$ **12.** $x = -6$

13. $y = 5$ **14.** $4y + 2x = 7$

Sketch the graph of each equation using the x- and y-intercepts (if they exist).

15. $y = 2x - 3$ **16.** $y = -5x + 1$ **17.** $y = 3x$ **18.** $x - 2y = 4$

19. $3y - 2x = 9$ **20.** $y = -6$ **21.** $x = 1$

[7–2]
Find the distance, midpoint, and slope between each pair of points.

22. $(3,2)$ and $(0,4)$ **23.** $(-2,1)$ and $(-2,4)$ **24.** $(1,5)$ and $(-3,5)$

Determine if the lines L_1 and L_2 containing the given points are parallel, perpendicular, or neither.

25. L_1 contains $(-3,2)$ and $(1,3)$
L_2 contains $(5,-1)$ and $(3,2)$

26. L_1 contains $(0,-1)$ and $(2,3)$
L_2 contains $(3,2)$ and $(1,3)$

27. L_1 contains $(4,0)$ and $(-3,-4)$
L_2 contains $(5,6)$ and $(-2,2)$

28. L_1 contains $(0,0)$ and $(9,-3)$
L_2 contains $(-1,-3)$ and $(8,2)$

Determine if the graphs of each pair of equations are parallel lines, perpendicular lines, or neither.

29. $2x + y = 1$ and $4x + 2y = -6$

30. $x - 3y = -2$ and $2x + 3y = 0$

31. $2x - 5y = 3$ and $5x + 2y = 1$

32. $3x - 2y = 1$ and $6x + 9y = 3$

33. A brace for a wall shelf is attached to the wall 9 inches below where the shelf meets the wall and to the bottom of the shelf $6\dfrac{1}{2}$ inches from the wall. What is the slope of the brace?

34. Show that the points $(-3,1)$, $(4,1)$, and $(-3,4)$ are the vertices of a right triangle.

[7–3]
Find the equation of the line satisfying the given conditions. Write the equation in standard form $ax + by = c$, $a > 0$.

35. Slope $m = \dfrac{2}{3}$ and passing through $(-1,5)$

36. Horizontal line through $(-2,4)$

37. Undefined slope and passing through $(1,-9)$

38. Passing through points $(2,-5)$ and $(3,3)$

Write each equation in slope-intercept form $y = mx + b$, identify the slope m and y-intercept b, and sketch the graph of the equation using the slope and y-intercept.

39. $3x - y = 4$ **40.** $2x + 3y = 9$ **41.** $3x - 2y = 0$ **42.** $2y + 3 = 0$

43. Find the equation of the line (in standard form) that passes through the point $(-7,2)$ and is parallel to the line $2y - x = 3$.

44. Find the equation of the line (in standard form) that is perpendicular to the line $4x + 5y = -2$ and passing through the point $(0,3)$.

[7–4]

Sketch the graph of each linear inequality in two variables.

45. $x \geq 0$ **46.** $y < -1$ **47.** $x + 3 < 0$ **48.** $y - 4 \geq 0$

49. $x + y \geq 4$ **50.** $3x - y < 6$ **51.** $x + 4y \geq 8$ **52.** $2x + 7y \leq 14$

Sketch the graph of each absolute value inequality.

53. $|x| \leq 5$ **54.** $|y| > 1$ **55.** $|x + 3| < 2$ **56.** $|x - 4| \geq 1$

57. $|y + 3| \leq 1$ **58.** $|2x - 3| \geq 1$

Chapter 7 cumulative test

Perform the indicated operations and simplify. Assume all variables are nonzero real numbers. Answer with positive exponents only.

[3–1] **1.** $(3a^2b^3)(-6ab^2)$

[3–3] **2.** $\dfrac{x^{-3}y^2}{x^2y^{-4}}$

[3–3] **3.** $(4a^{-3}b^2c^{-1})^{-2}$

[1–6] **4.** $5y^2 - \{5y - [4y^2 + 2y - 3]\}$

[3–2] **5.** $(x - 2y)^2 - (2x + y)^2$

[3–2] **6.** $6xy(2x^2 - 3y^2 + x^2y - x^3y^3)$

Completely factor the following expressions.

[3–7] **7.** $3a^2 - 12b^2$

[3–5] **8.** $8x^2 - 55x - 7$

[3–7] **9.** $8x^3 + 27y^3$

[3–5] **10.** $16x^2 - 40xy + 25y^2$

Find the solution set of the following equations and inequalities.

[2–1] **11.** $3(2a - 5) - 5a = 5(a + 6)$

[2–4] **12.** $|2x + 3| = 5$

[2–6] **13.** $|2y - 1| < 4$

[2–6] **14.** $|y + 3| \geq 1$

[6–1] **15.** $4x^2 - x = 0$

[6–1] **16.** $x^2 - 7x - 18 = 0$

[6–3] **17.** $4y^2 = 2y + 6$

[2–5] **18.** $-5 < 4 - 3x \leq 5$

[6–3] **19.** $2x^2 - 3x = 7$

Perform the indicated operations and simplify. Assume all denominators are nonzero.

[4–2] **20.** $\dfrac{y + 2}{y^2 - y - 20} \cdot \dfrac{y^2 - 16}{y^2 - 2y - 8}$

[4–3] **21.** $\dfrac{7y - 3}{2y^2 - 14y} - \dfrac{5y}{y^2 - 49}$

[4–4] **22.** Simplify the complex fraction $\dfrac{\dfrac{4}{y} - \dfrac{3}{x}}{\dfrac{4x - 3y}{xy}}$.

[4–7] **23.** Find the solution set of the equation $\dfrac{3}{8x - 1} = \dfrac{2}{2x + 3}$.

[4–6] **24.** Divide $3x^3 + 8x^2 - 7x - 12$ by $(x + 3)$. Use the results to (a) find $P(-3)$ when $P(x) = 3x^3 + 8x^2 - 7x - 12$ and (b) determine if $x + 3$ is a factor of $3x^3 + 8x^2 - 7x - 12$.

Perform the indicated operations, simplify, and rationalize all denominators.

[5–3] **25.** $\sqrt{18} + \sqrt{50} - \sqrt{8}$ **[5–6]** **26.** $(4 + \sqrt{5})^2$ **[5–4]** **27.** $\sqrt{\dfrac{9}{5}}$

[5–4] **28.** $\dfrac{3}{2 - \sqrt{3}}$ **[5–7]** **29.** $(3 + 2i)(3 - 2i)$

[5–7] **30.** $(4 - 3i) - (2 + 4i)$ **[6–5]** **31.** Solve the equation $\sqrt{x - 3} - 1 = \sqrt{x + 2}$. Name any extraneous solutions.

In problems 32–34, find the equation of the line (in standard form) satisfying the following conditions.

[7–3] **32.** Passing through points $(1, -3)$ and $(2, 4)$ **[7–3]** **33.** Passing through $(-1, 2)$ and parallel to $2x - 3y = 1$

[7–3] **34.** A vertical line passing through $(5, -3)$ **[7–3]** **35.** Find the slope and y-intercept of the line $3x - 5y = 10$. Sketch the graph using these facts.

[7–3] **36.** Are the lines $2x - y = 6$ and $4x + 8y = 1$ parallel, perpendicular, or neither?   **[7–2]** **37.** Find the distance between the points $(1, -3)$ and $(2, 5)$. Find the coordinates of the midpoint of the line segment.

Functions

8

The manufacturer of television sets has fixed costs of $2,500,000 and an additional cost of $350 for each set manufactured. The total cost function of manufacturing x sets is given by

$$C(x) = 2,500,000 + 350x$$

Find the cost of manufacturing 125 sets.

8–1 ■ Relations and functions

Relations

In our study of mathematics, and in the physical world, it is often useful to describe one quantity in terms of another quantity. The following examples illustrate this:

1. the demand for a product is related to the price of the product and
2. the cost to send mail is related to the weight of the piece of mail.

The relationships between these quantities are often stated as ordered pairs. Such relationships are called **relations.**

Definition of a relation

A **relation** is any set of ordered pairs.

■ **Example 8–1 A**

1. The set of ordered pairs $\{(1,2), (-3,5), (0,6), (-7,9)\}$ is a relation consisting of four ordered pairs.

2. If it costs 15¢ per mile to operate an automobile, some ordered pairs that form this relation are

   $(10,1.50), (100,15), (200,30),$ and $(500,75)$

   where the first component is the miles traveled and the second component is the cost in dollars.

3. $\{(x,y)|y = 2x + 1\}$ forms a relation that contains all ordered pairs such that x is a real number and y is found by adding 1 to twice the value of x. This relation contains the following ordered pairs.

$(-4,-7), (-3,-5), (-1,-1), (0,1), (2,5)$

There are infinitely many more ordered pairs in this relation. ▪

We see in examples 2 and 3 that the value of the second component of each ordered pair is dependent on the value given in the first component. For this reason, in example 8-1 A3, for instance, we call x the **independent variable** and y the **dependent variable,** since its value is dependent on what value of x we choose. Whenever two variables are involved in the definition of a relation, one of them will be independent and the other will be dependent.

▪ **Example 8-1 B**

Given that it costs 15¢ per mile to operate an automobile, a set that defines this relation is

$\{(d,C)|C = 0.15d\}$

where C is the cost to travel d miles. Then C is the dependent variable and d is the independent variable. ▪

We give special names to the set of first and the set of second components of the ordered pairs forming a relation.

┌─ **Definition of domain and range** ────────────────────

1. The **domain** of a relation is the set of all possible first components of the relation.
2. The **range** of a relation is the set of all possible second components of the relation.

▪ **Example 8-1 C**

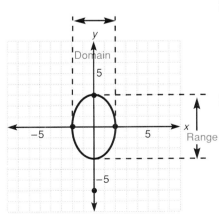

Find the domain and range of the following relations.

1. $\{(1,2), (-3,5), (0,6), (-7,9)\}$
The domain is $\{1,-3,0,-7\}$.
The range is $\{2,5,6,9\}$.

2. $\{(3,-1), (5,-1), (-7,-1), (0,-1)\}$
The domain is $\{3,5,-7,0\}$.
The range is $\{-1\}$.

The graph of a relation is the graph of its ordered pairs. By graphing the ordered pairs satisfying the relation, we can identify its domain and range.

3. $9x^2 + 4y^2 = 36$
The graph of this relation is shown on the left. From the graph, we can see

a. the domain being the set of all possible values of x, the domain is
$\{x|-2 \leq x \leq 2\} = [-2,2]$.

b. the range being the set of all possible values of y, the range is
$\{y|-3 \leq y \leq 3\} = [-3,3]$.

4. $y = x^2 - 3$

The graph of this relation is shown below.

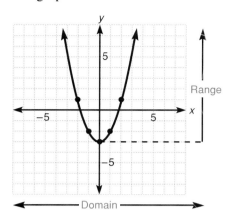

From the graph, we can see that

a. x can take on any real number value, so the domain is
$\{x \mid x \in R\} = (-\infty, \infty)$.

b. y takes on any value that is greater than or equal to -3, so the range is
$\{y \mid y \geq -3\} = [-3, \infty)$.

▶ *Quick check* 1. Find the domain and range of the relation
$\{(-6,0), (-3,-2), (0,4), (7,5)\}$.
2. Find the domain and range of the relation whose graph is given.

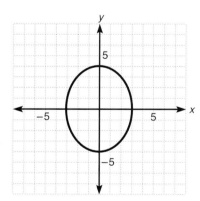

Functions

In all phases of mathematics, from the most elementary to the most
sophisticated, the idea of a function is a cornerstone for each mathematical
development.

Consider the distance d in miles that an automobile travels in time t hours.
Suppose that the automobile is traveling at an average rate of 50 miles per hour.
The variables d and t are related by the equation

$$d = 50t$$

We say that "*d* is a function of *t*" since a change in the value of *t* will cause a change in the value of *d*. For example,

$$\text{when } t = 2 \text{ hours, then } d = 50(2) = 100 \text{ miles}$$
$$\text{when } t = 5 \text{ hours, then } d = 50(5) = 250 \text{ miles}$$
$$\text{when } t = 10 \text{ hours, then } d = 50(10) = 500 \text{ miles}$$

Note *t* is the independent variable and *d* is the dependent variable, since the value of *d* is dependent on the chosen value of *t*.

Notice that for any chosen value of the independent variable *t*, we get a *unique* (one and only one) value of the dependent variable *d*. For this reason, the equation (or correspondence) $d = 50t$ defines *d* as a function of *t*.

A function

___ *Definition of a function* _____

A **function** is a relation that associates with each first component of the ordered pairs *exactly one* value of the second component.

Concept
A function is a set of ordered pairs (a relation) in which no two distinct ordered pairs have the same first component.

The variables *x* and *y* are generally used when defining a mathematical function. By the above definition, for *y to be a function of x*, the two variables must be related so that for each value of *x*, there is assigned a unique (one and only one) corresponding value of *y*. Then **x is an element of the domain and y is an element of the range of the function.** This is a very important concept because when we use an equation to determine the outcome in a given situation, we want the equation to be of the type that produces only one answer.

■ *Example 8-1 D*

Determine if the following relations are functions.

1. $A = \{(1,2), (3,4), (-4,8), (0,-5)\}$ is a function since no two ordered pairs have the same first component. $B = \{(1,3), (-4,3), (9,3), (0,2)\}$ is a function since no two ordered pairs have the same first component. We see that three of the ordered pairs in the set *B* have the same second component, but this does not violate the definition of a function.

2. The relation $C = \{(3,0), (2,9), (3,-1), (-1,7)\}$ is *not* a function since the two ordered pairs (3,0) and (3,-1) have the same first component but different second components.

3. $\{(x,y) | y = 2x + 1\}$
This equation yields infinitely many ordered pairs. However, for each value of *x* we choose, we *will* get only one value of *y*. No two ordered pairs will have the same first component. The equation does define a function.

4. $\{(x,y)|y^2 = x + 1\}$

If we extract the roots, we obtain

$$y = \pm\sqrt{x + 1}$$

Let $x = 3$, then

$$y = \pm\sqrt{3 + 1} = \pm\sqrt{4} = \pm 2$$

We obtain two distinct ordered pairs, $(3,-2)$ and $(3,2)$, having the *same first component*. The equation *does not* define a function.

▶ *Quick check* Determine if the given relation is or is not a function.
1. $\{(3,2), (3,1), (5,2)\}$ 2. $y = -2x + 6$ ∎

Since a function defines a set of ordered pairs and is a special type of a relation, it has a *domain* and a *range*. The domain of a function must always be stated or implied by the nature of the equation defining the function. **If the domain is not stated, we will assume the domain to be the set of all real numbers for which the function is defined.** A function is defined for all real numbers *except* when a number will result in a zero in the denominator or when a number will result in a negative value under the even root symbol.

■ *Example 8–1 E*

Determine the domain of the function defined by the following.

1. $\{(1,-5), (-2,3), (4,1), (-6,-7)\}$
The domain (set of all first components) is $\{1,-2,4,-6\}$.

2. $\{(x,y)|3x + y = 4\}$, $x \in \{-3,-1,0,1,3\}$
Since replacement values of x make up the domain of a function, the domain in this case is restricted to $\{-3,-1,0,1,3\}$.

3. $\left\{(x,y)|y = \dfrac{3}{2x - 1}\right\}$
We want the domain to be the set of all real numbers, unless for some reason we must restrict the values chosen for x. Thus, since division by 0 is undefined and the denominator of the equation $2x - 1 = 0$ when $x = \dfrac{1}{2}$, we determine the domain to be all real numbers except $\dfrac{1}{2}$. In set-builder and interval notation, the

$$\text{domain} = \left\{x|x \in R, x \neq \frac{1}{2}\right\} = \left(-\infty, \frac{1}{2}\right) \cup \left(\frac{1}{2},\infty\right)$$

4. $\{(x,y)|y = \sqrt{x - 1}\}$
Since $\sqrt{x - 1}$ is a real number only when the radicand $x - 1$ is nonnegative, we must restrict the domain of the function to values of x for which $x - 1 \geq 0$; that is, $x \geq 1$. Then the

$$\text{domain} = \{x|x \geq 1\} = [1,\infty)$$

▶ *Quick check* Determine the domain of the function defined by $y = \dfrac{5}{3x + 2}$.∎

___ *Finding the domain of a function* _____

1. The domain is the set of all real numbers except where restrictions are necessary.
2. If the function is defined by $y = \dfrac{P(x)}{Q(x)}$, factor $Q(x)$, set each factor containing the variable equal to zero, and solve for the variable. This yields the values of the variable *not* in the domain of the function.
3. If the function is defined by $y = \sqrt{P(x)}$, solve the inequality $P(x) \geq 0$. This yields the values of the variable that *are* in the domain of the function.

Recall that two or more points having the same first component lie on the same vertical line. Remember that a function cannot have two or more ordered pairs having the same first component (the abscissa of the point). These facts lead us to a visual test for determining if a particular graph does or does not represent a function.

___ *Vertical line test for a function* _____

If every vertical line drawn in the plane intersects the graph of a relation in *at most one point,* the relation is a function.

■ *Example 8–1 F*

Determine by the vertical line test if the following graphs represent functions.

1.

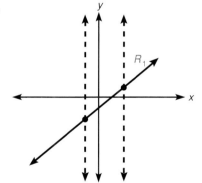

Since any vertical line drawn in the plane will intersect the graph of relation R_1 in *only one point,* R_1 is a function.

2.

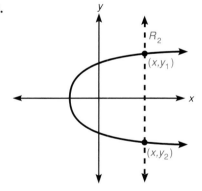

Many lines can be drawn in the plane that will intersect the graph of relation R_2 *in two points,* so R_2 *is not* a function, assuming x is the independent variable.

▶ *Quick check* Determine by the vertical line test if this graph represents a function.

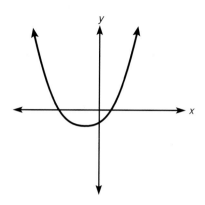

┌─ **Mastery points** ──┐

Can you
- Determine if a relation defines a function?
- Find the domain and range of a relation?
- Determine if a given graph of a relation represents a function by using the vertical line test?

└───┘

Exercise 8–1

Determine the domain and range of the following relations. In problems 9–16, state the answer in set-builder and interval notation. See example 8–1 C.

Examples 1. $\{(-6,0), (-3,-2), (0,4), (7,5)\}$

2.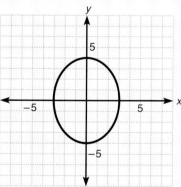

Solutions Domain (set of first components) $= \{-6,-3,0,7\}$ Domain $= \{x \mid -3 \le x \le 3\} = [-3,3]$
Range (set of second components) $= \{0,-2,4,5\}$ Range $= \{y \mid -4 \le y \le 4\} = [-4,4]$

1. $\{(8,0), (5,4), (9,3), (6,4)\}$ 2. $\{(-5,4), (0,7), (-3,-2), (-2,4)\}$
3. $\{(-4,1), (1,2), (-4,3), (1,9)\}$ 4. $\{(-4,3), (5,4), (-4,4), (1,1)\}$
5. $\{(6,-1), (1,1), (2,-1), (3,1)\}$ 6. $\{(5,4), (7,6), (3,4), (-2,6)\}$
7. $\{(5,3), (6,3), (7,3), (6,-4)\}$ 8. $\{(1,4), (2,4), (-6,4), (9,4)\}$

9.

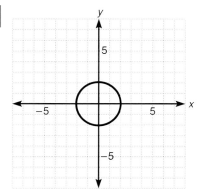

10.

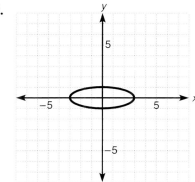

11.

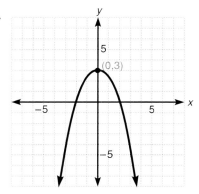

12.

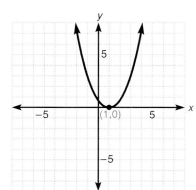

13.

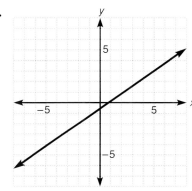

14.

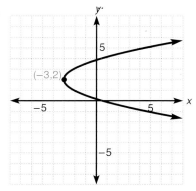

15.

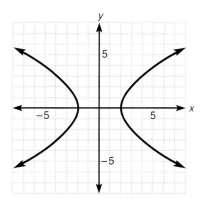

16.

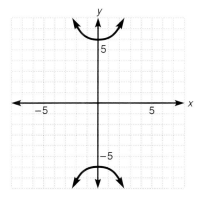

Determine whether or not the given relation defines a function. If not, show why not by an example.
See example 8–1 D.

Examples   1. $\{(3,2), (3,1), (5,2)\}$

2. $y = -2x + 6$

Solutions   Since the ordered pairs $(3,2)$ and $(3,1)$ are distinct and have the same first component, the relation is *not* a function.

Since for any given value of x we would obtain a unique value of y, the relation is a function.

17. $\{(1,4), (3,6), (-1,5), (8,3)\}$

18. $\{(-4,0), (0,0), (6,7), (8,-6)\}$

19. $\{(1,2), (2,3), (-1,6), (2,-7)\}$

20. $\{(-6,2), (4,7), (7,4), (-6,0)\}$

21. $\{(-1,2), (-1,6), (-1,8), (-1,0)\}$

22. $\{(1,4), (2,4), (-3,4), (-6,4)\}$

23. $\{(-3,1),\ (1,1),\ (-7,1),\ (9,1)\}$

24. $\{(1,1),\ (2,2),\ (3,3),\ (4,4)\}$

25. $\{(x,y)\,|\,y = x + 7\}$ 26. $\{(x,y)\,|\,y = 3 - 4x\}$ 27. $\{(x,y)\,|\,y = x^2\}$ 28. $\{(x,y)\,|\,y = x^2 - x + 1\}$

29. $\{(x,y)\,|\,x = y^2\}$ 30. $\{(x,y)\,|\,x = y^2 + 2\}$ **31.** $\{(x,y)\,|\,y = -3\}$ 32. $\{(x,y)\,|\,y = 4\}$

33. $\{(x,y)\,|\,x = -10\}$ 34. $\{(x,y)\,|\,x = 0\}$ 35. $\{(x,y)\,|\,x - 3 = 0\}$ 36. $\{(x,y)\,|\,x + 4 = 0\}$

Determine the domain of each of the given functions. Write the answers in set-builder and interval notation where possible. See example 8–1 E.

Example $y = \dfrac{5}{3x + 2}$

Solution Since division by 0 is undefined, and since $3x + 2 = 0$ when $x = -\dfrac{2}{3}$, the domain is the set of all

real numbers *except* $-\dfrac{2}{3}$. In set-builder notation,

$$\text{domain} = \left\{ x\,|\,x \in R,\ x \neq -\frac{2}{3} \right\} = \left(-\infty, -\frac{2}{3}\right) \cup \left(-\frac{2}{3}, \infty\right)$$

37. $\{(x,y)\,|\,y = 2x - 3\};\ x \in \{-3, -1, 0, 1, 3\}$

38. $\{(x,y)\,|\,y = 4 - 3x\};\ x \in \{-4, -2, 0, 2, 4\}$

39. $\{(x,y)\,|\,y = x\};\ x \in \{-5, -3, 0, 3, 5\}$

40. $\{(x,y)\,|\,y = -x\};\ x \in \{-2, -1, 0, 7, 8\}$

41. $\left\{(x,y)\,|\,y = \dfrac{1}{x}\right\};\ x \in \{-5, -1, 1, 2, 4\}$

42. $\{(x,y)\,|\,y = 4x - 3\}$ 43. $\{(x,y)\,|\,x + y = 8\}$ 44. $\{(x,y)\,|\,y = 3x^2\}$

45. $\{(x,y)\,|\,y = x^2 + 2x + 1\}$ **46.** $\{(x,y)\,|\,xy = 2\}$ 47. $\left\{(x,y)\,|\,y = \dfrac{1}{x}\right\}$

48. $\left\{(x,y)\,|\,y = \dfrac{3}{x + 7}\right\}$ 49. $\left\{(x,y)\,|\,y = \dfrac{5}{2x + 3}\right\}$ 50. $\{(x,y)\,|\,y = \sqrt{x - 9}\}$

51. $\{(x,y)\,|\,y = \sqrt{3x + 4}\}$

Use the vertical line test to identify which of the following graphs represent functions where y is a function of x. Explain the answers. See example 8–1 F.

Example

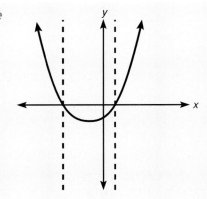

Solution Since any vertical line drawn in the plane will intersect the graph in *only one point,* the graph is a function.

52.

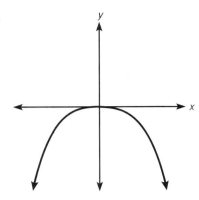

53.

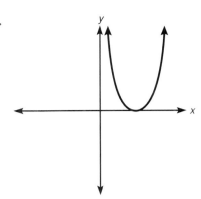

54.

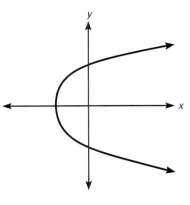

55.

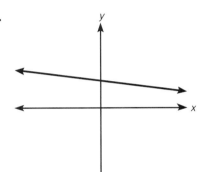

56.

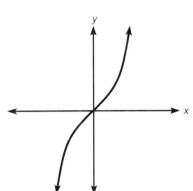

57.

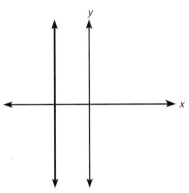

58.

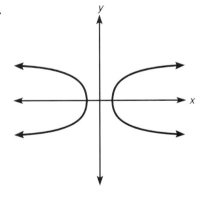

59.

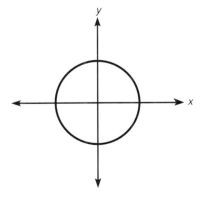

Review exercises

Simplify the following expressions. Assume all variables are nonzero. Express answers with positive exponents only. See sections 3–3.

1. $(2x^4y^{-3})^2$

2. $x^3 \cdot x^2 \cdot x \cdot x^0$

3. $\dfrac{x^{-2}y^3}{x^{-2}y^{-1}}$

4. Does $x^{y+1} = x^y \cdot y$? Explain. See section 3–1.

5. Factor $2x^2 - 32y^2$. See section 3–7.

6. The length of the diagonal of a square is $6\sqrt{3}$ units long. Find the length of each side. See section 6–4.

7. A rectangular field is twice as long as it is wide. Find the dimensions if the area is 7,200 square meters. See section 6–4.

8–2 ■ *Functional notation and algebra of functions*

The lowercase letters f, g, and h are commonly used to denote a function. For example, the function defined by the equation $y = 2x - 5$, where x is the independent variable, is often written $f(x) = 2x - 5$. We read the symbol "$f(x)$" as "f at x" or "f of x," which means "the value of the function at x."

Note We have replaced the dependent variable y with the symbol "$f(x)$." This new symbol represents a value of the function. Remember, $f(x)$ *does not* mean f times x.

The function has been given the name f and since $f(x)$ replaced y in the equation, $f(x)$ represents the element in the range of the function f that is associated with the chosen domain element, x.

To illustrate, given the function $y = 2x - 5$ and the domain element $x = 3$,

$$
\begin{aligned}
y &= 2x - 5 & f(x) &= 2x - 5 \\
&= 2(3) - 5 & f(3) &= 2(3) - 5 & \text{Replace } x \text{ with 3} \\
&= 6 - 5 & &= 6 - 5 \\
&= 1 & &= 1
\end{aligned}
$$

Thus, for the domain element $x = 3$, the corresponding range element is $y = f(3) = 1$ and the ordered pair $(3,1)$ is an element of the function f.

Note We read "$f(3) = 1$" as "the value of the function f when $x = 3$ is 1."

■ *Example 8–2 A*

Given $g(x) = 3x^2 - 4x + 1$, find

1. $g(3)$

$$
\begin{aligned}
g(3) &= 3(3)^2 - 4(3) + 1 & \text{Replace } x \text{ with 3} \\
&= 27 - 12 + 1 \\
&= 16
\end{aligned}
$$

Therefore $g(3) = 16$ and $(3,16)$ is an element of function g.

2. $g(1)$

$$
\begin{aligned}
g(1) &= 3(1)^2 - 4(1) + 1 & \text{Replace } x \text{ with 1} \\
&= 3 - 4 + 1 \\
&= 0
\end{aligned}
$$

$g(1) = 0$ and $(1,0)$ is an element of function g.

3. $g(a)$

$$
\begin{aligned}
g(a) &= 3(a)^2 - 4(a) + 1 & \text{Replace } x \text{ with } a \\
&= 3a^2 - 4a + 1
\end{aligned}
$$

4. $g(x + h)$

$$
\begin{aligned}
g(x + h) &= 3(x + h)^2 - 4(x + h) + 1 & \text{Replace } x \text{ with } (x + h) \\
&= 3(x^2 + 2xh + h^2) - 4x - 4h + 1 \\
&= 3x^2 + 6xh + 3h^2 - 4x - 4h + 1
\end{aligned}
$$

5. $g(x + h) - g(x)$

We have already determined that

$$g(x + h) = 3x^2 + 6xh + 3h^2 - 4x - 4h + 1.$$

Then

$$
\begin{aligned}
g(x + h) - g(x) &= (3x^2 + 6xh + 3h^2 - 4x - 4h + 1) - (3x^2 - 4x + 1) \\
&= 3x^2 + 6xh + 3h^2 - 4x - 4h + 1 - 3x^2 + 4x - 1 \\
&= 6xh + 3h^2 - 4h \qquad \text{Combine like terms}
\end{aligned}
$$

6. $\dfrac{g(x + h) - g(x)}{h}$ $(h \neq 0)$

We have already determined that

$$g(x + h) - g(x) = 6xh + 3h^2 - 4h$$

Then

$$
\begin{aligned}
\frac{g(x + h) - g(x)}{h} &= \frac{6xh + 3h^2 - 4h}{h} \qquad &&\text{Divide each member by } h \\
&= \frac{h(6x + 3h - 4)}{h} \qquad &&\text{Factor out } h \\
&= 6x + 3h - 4 \qquad &&\text{Reduce by } h
\end{aligned}
$$

▶ *Quick check* Given $f(x) = 3x^2 - 4$, find (a) $f(-3)$; (b) $f(x + h)$; (c) $f(x + h) - f(x)$; (d) $\dfrac{f(x + h) - f(x)}{h}$, $h \neq 0$. ■

Algebra of functions

In business, the revenue R that is received from the sale of a particular item is related to the cost C of producing the item and the profit P that is realized from the sale of the item by the equation

$$R = C + P$$

or the profit is related to the revenue and the cost by

$$P = R - C$$

Since these quantities are often dependent on the number x of the items produced, then P, R, and C are functions of x and we can state these by

$$R(x) = C(x) + P(x) \qquad \text{or} \qquad P(x) = R(x) - C(x)$$

Therefore, we can see that some business applications of the mathematics of functions involve the *addition* and *subtraction* of functions. In like fashion, there are business applications of *multiplication* and *division* of functions. We call this the *algebra of functions* where we combine functions by the following definitions.

Definition of the sum, difference, product, ───────
and quotient of functions

Given two functions f and g, then for all values of x for which $f(x)$ and $g(x)$ exist, the functions $f + g$, $f - g$, fg, and $\dfrac{f}{g}$ are defined by

$$(f + g)(x) = f(x) + g(x)$$
$$(f - g)(x) = f(x) - g(x)$$
$$(fg)(x) = f(x) \cdot g(x)$$
$$\left(\frac{f}{g}\right)(x) = \frac{f(x)}{g(x)} \quad [g(x) \neq 0]$$

■ *Example 8–2 B*

1. Given functions f and g defined by $f(x) = 2x - 5$ and $g(x) = x^2 - 2$, then

   a. $(f + g)(2) = f(2) + g(2) = [2(2) - 5] + [(2)^2 - 2]$ Replace x with 2 in functions f and g

   $$= (4 - 5) + (4 - 2)$$
   $$= -1 + 2$$
   $$= 1$$

   Thus, $(f + g)(2) = 1$ and the ordered pair $(2,1)$ is an element in the function $f + g$.

   b. $(f - g)(-1) = f(-1) - g(-1)$
   $$= [2(-1) - 5] - [(-1)^2 - 2] \quad \text{Replace } x \text{ with } -1 \text{ in functions } f \text{ and } g$$
   $$= (-2 - 5) - (1 - 2)$$
   $$= -7 - (-1)$$
   $$= -6$$

   Thus, $(f - g)(-1) = -6$ and the ordered pair $(-1,-6)$ is an element in the function $f - g$.

   c. $(fg)(0) = f(0) \cdot g(0) = [2(0) - 5] \cdot [(0)^2 - 2]$ Replace x with 0 in functions f and g

   $$= (0 - 5) \cdot (-2)$$
   $$= (-5)(-2)$$
   $$= 10$$

   Thus, $(fg)(0) = 10$ and an ordered pair in the function fg is $(0,10)$.

   d. $\left(\dfrac{f}{g}\right)(3) = \dfrac{f(3)}{g(3)}$

   $$= \frac{2(3) - 5}{(3)^2 - 2} \quad \text{Replace } x \text{ with 3 in functions } f \text{ and } g$$

   $$= \frac{6 - 5}{9 - 2}$$

   $$= \frac{1}{7}$$

   Thus, $\left(\dfrac{f}{g}\right)(3) = \dfrac{1}{7}$ and the ordered pair $\left(3, \dfrac{1}{7}\right)$ is an element in the function $\dfrac{f}{g}$.

2. Given functions f and g defined by $f(x) = 4x + 5$ and $g(x) = 2x^2 - 3$, find

   a. $(f + g)(x) = f(x) + g(x) = (4x + 5) + (2x^2 - 3)$
                                    $= 2x^2 + 4x + 2$

     Thus, the function $f + g$ is defined by $(f + g)(x) = 2x^2 + 4x + 2$.

   b. $(f - g)(x) = f(x) - g(x) = (4x + 5) - (2x^2 - 3)$
                                      $= 4x + 5 - 2x^2 + 3$
                                      $= -2x^2 + 4x + 8$

     The function $f - g$ is defined by $(f - g)(x) = -2x^2 + 4x + 8$.

   c. $(fg)(x) = f(x) \cdot g(x) = (4x + 5)(2x^2 - 3)$
                             $= 8x^3 + 10x^2 - 12x - 15$

     The function fg is defined by $(fg)(x) = 8x^3 + 10x^2 - 12x - 15$.

   d. $\left(\dfrac{f}{g}\right)(x) = \dfrac{f(x)}{g(x)} = \dfrac{4x + 5}{2x^2 - 3}$

     The function $\dfrac{f}{g}$ is defined by $\left(\dfrac{f}{g}\right)(x) = \dfrac{4x + 5}{2x^2 - 3}$ $\left(x \neq \pm\dfrac{\sqrt{6}}{2}\right)$.

Note   The domain of the resulting function is the *intersection* of the domains of the two functions that are involved with necessary restrictions.

▶ **Quick check**   Given functions f and g defined by $f(x) = 5x + 3$ and $g(x) = 2x^2 - 1$, find (a) $(f - g)(3)$; (b) $(fg)(-1)$; (c) $(f + g)(x)$.   ■

    A final important operation on functions f and g that produces a third function is called the *composition* of the two functions. This produces what we call a *composite function*.

    — Definition of the composition of two functions ——————————

    Given functions f and g, the composite function of
    a. f and g, denoted by $f \circ g$, is defined by $(f \circ g)(x) = f[g(x)]$ for all x in the domain of g such that $g(x)$ is in the domain of f
    b. g and f, denoted by $g \circ f$, is defined by $(g \circ f)(x) = g[f(x)]$ for all x in the domain of f such that $f(x)$ is in the domain of g.

Note   We read the symbols "$(f \circ g)(x)$" and "$f[g(x)]$" "f composition g at x."

■ *Example 8-2 C*

Given $f(x) = 3x - 1$ and $g(x) = x^2 + 5$, find the following.

1. $(f \circ g)(2) = f[g(2)]$
   $f[g(2)]$ means to evaluate $f(x)$ when $x = g(2)$. We first evaluate $g(2)$.

    $g(2) = (2)^2 + 5$ Replace x with 2 in $x^2 + 5$
          $= 4 + 5$
          $= 9$

Thus we want

$$f(9) = 3(9) - 1 \qquad \text{Replace } x \text{ with 9 in } 3x - 1$$
$$= 27 - 1$$
$$= 26$$

Therefore,

$$(f \circ g)(2) = f[g(2)] = f(9) = 26$$

2. $(g \circ f)(-3) = g[f(-3)]$

$g[f(-3)]$ means to evaluate $g(x)$ when $x = f(-3)$. We first find

$$f(-3) = 3(-3) - 1 \qquad \text{Replace } x \text{ with } -3 \text{ in } 3x - 1$$
$$= -9 - 1$$
$$= -10$$

Then we evaluate

$$g(-10) = (-10)^2 + 5 \qquad \text{Replace } x \text{ with } -10 \text{ in } x^2 + 5$$
$$= 100 + 5$$
$$= 105$$

Therefore,

$$(g \circ f)(-3) = g[f(-3)] = g(-10) = 105$$

3. $(f \circ g)(x)$

$$(f \circ g)(x) = f[g(x)] = f(x^2 + 5)$$
$$= 3(x^2 + 5) - 1 \qquad \text{Replace } x \text{ with } x^2 + 5 \text{ in } 3x - 1$$
$$= 3x^2 + 15 - 1$$
$$= 3x^2 + 14$$

Thus, the function $f \circ g$ is defined by $(f \circ g)(x) = 3x^2 + 14$.

Note $(f \circ g)(2) = 3(2)^2 + 14 = 3(4) + 14 = 26$, which we determined in example 1.

4. $(g \circ f)(x)$

$$(g \circ f)(x) = g[f(x)] = g(3x - 1)$$
$$= (3x - 1)^2 + 5 \qquad \text{Replace } x \text{ with } 3x - 1 \text{ in } x^2 + 5$$
$$= 9x^2 - 6x + 1 + 5$$
$$= 9x^2 - 6x + 6$$

Thus, the function $g \circ f$ is defined by $(g \circ f)(x) = 9x^2 - 6x + 6$.

Note $(g \circ f)(-3) = 9(-3)^2 - 6(-3) + 6 = 81 + 18 + 6 = 105$, which we determined in example 2.

▶ **Quick check** Given $f(x) = 5x + 3$ and $g(x) = 2x^2 - 1$, find (a) $(f \circ g)(3)$, (b) $(g \circ f)(-1)$, (c) $(f \circ g)(x)$. ■

— *Mastery points* —

Can you
- Evaluate $f(x)$ for any value of x given the function f?
- Find the sum, difference, product, and quotient of two functions?
- Find the composition $(f \circ g)(x)$ and the composition $(g \circ f)(x)$?

Exercise 8-2

Find the value of each of the following if $f(x) = 3x - 2$ and $g(x) = x^2 + 2x - 5$. See example 8-2 A.

Example Given $f(x) = 3x^2 - 4$, find (a) $f(-3)$; (b) $f(x + h)$; (c) $f(x + h) - f(x)$;
(d) $\dfrac{f(x + h) - f(x)}{h}$, $h \neq 0$.

Solution Given $f(x) = 3x^2 - 4$,

   a. $f(-3) = 3(-3)^2 - 4$ Replace x with -3
   $\qquad = 3(9) - 4$
   $\qquad = 23$

   b. $f(x + h) = 3(x + h)^2 - 4$ Replace x with $x + h$
   $\qquad = 3(x^2 + 2xh + h^2) - 4$
   $\qquad = 3x^2 + 6xh + 3h^2 - 4$

   c. $f(x + h) - f(x) = (3x^2 + 6xh + 3h^2 - 4) - (3x^2 - 4)$
   $\qquad = 3x^2 + 6xh + 3h^2 - 4 - 3x^2 + 4$
   $\qquad = 6xh + 3h^2$

   d. $\dfrac{f(x + h) - f(x)}{h} = \dfrac{6xh + 3h^2}{h}$

   $\qquad = \dfrac{h(6x + 3h)}{h}$ Factor h from each term

   $\qquad = 6x + 3h$ Reduce by h

1. $f(0)$ **2.** $f(-6)$ **3.** $f\left(\dfrac{2}{3}\right)$ **4.** $g(0)$ **5.** $g(7)$

6. $g(-3)$ **7.** $g\left(\dfrac{1}{2}\right)$ **8.** $f(a)$ **9.** $f(a + 1)$ **10.** $f\left(\dfrac{1}{a}\right)$

11. $f(a^2)$ **12.** $f(5) - f(2)$ **13.** $f(6) - f(-3)$ **14.** $g(3) - g(1)$ **15.** $g(4) - g(-4)$

16. $f(x + h)$ **17.** $f(x + h) - f(x)$ **18.** $\dfrac{f(x + h) - f(x)}{h}$, $h \neq 0$

19. $g(x + h)$ **20.** $g(x + h) - g(x)$ **21.** $\dfrac{g(x + h) - g(x)}{h}$, $h \neq 0$

22. Given $f(x) = 4x^2$, find $\dfrac{f(x + h) - f(x)}{h}$, $h \neq 0$.   **23.** Given $g(x) = 2x^2 + 3x + 2$, find
$\dfrac{g(x + h) - g(x)}{h}$, $h \neq 0$.

24. Given $f(x) = 5$, find $\dfrac{f(x + h) - f(x)}{h}$, $h \neq 0$.

Find the indicated values of each given function. Write the answers as ordered pairs. See example 8-2 A-1 and 2.

25. $f(x) = 3x - 2$; find $f(-5), f(0), f\left(\dfrac{2}{3}\right)$ **26.** $f(x) = 5x - 6$; find $f(-2), f(0), f(2)$

27. $h(x) = 3x^2 - 2x + 1$; find $h\left(-\dfrac{1}{2}\right), h(0), h(3)$ **28.** $h(x) = x^2 - 5$; find $h(-5), h(0), h(\sqrt{5})$

29. $g(x) = 10$; find $g(-15)$, $g(0)$, $g\left(\dfrac{6}{5}\right)$

30. $g(x) = -7$; find $g\left(-\dfrac{7}{8}\right)$, $g(0)$, $g(25)$

31. $h(x) = x^3 + 4$; find $h(-5)$, $h(0)$, $h\left(\dfrac{1}{2}\right)$

Given $f(x) = -x^2 + 2x - 1$ and $g(x) = 5x - 2$, find the following. See example 8–2 B.

> **Example** Given $f(x) = 5x + 3$ and $g(x) = 2x^2 - 1$, find (a) $(f - g)(3)$; (b) $(fg)(-1)$; (c) $(f + g)(x)$.
>
> **Solution** a. $(f - g)(3) = f(3) - g(3)$
> $\qquad\qquad\quad = [5(3) + 3] - [2(3)^2 - 1]$ Replace x with 3
> $\qquad\qquad\quad = 18 - 17$
> $\qquad\qquad\quad = 1$
>
> $\qquad (f - g)(3) = 1$ and the ordered pair $(3,1)$ is in the function $f - g$.
>
>   b. $(fg)(-1) = [f(-1)][g(-1)]$
> $\qquad\qquad\quad = [5(-1) + 3][2(-1)^2 - 1]$ Replace x with -1
> $\qquad\qquad\quad = (-2)(1)$
> $\qquad\qquad\quad = -2$
>
> $\qquad (fg)(-1) = -2$ and the ordered pair $(-1,-2)$ is in the function fg.
>
>   c. $(f + g)(x) = f(x) + g(x)$
> $\qquad\qquad\quad = (5x + 3) + (2x^2 - 1)$
> $\qquad\qquad\quad = 2x^2 + 5x + 2$
>
> The function $f + g$ is defined by $(f + g)(x) = 2x^2 + 5x + 2$.

32. $(f + g)(-3)$ **33.** $(f + g)(5)$ **34.** $(f - g)(0)$ **35.** $(f - g)(-2)$

36. $(fg)(2)$ **37.** $(fg)(-5)$ **38.** $\left(\dfrac{f}{g}\right)(4)$ **39.** $\left(\dfrac{f}{g}\right)(0)$

40. $(f + g)(a)$ **41.** $(f - g)(2z)$ **42.** $\left(\dfrac{f}{g}\right)(a - 1)$ **43.** $\left(\dfrac{f}{g}\right)(2b)$

44. $(f + g)(x)$ **45.** $(f - g)(x)$ **46.** $(fg)(x)$ **47.** $\left(\dfrac{f}{g}\right)(x)$

Given $f(x) = 3x^2 - 5$ and $g(x) = 4x + 2$, find the following compositions and simplify the results. See example 8–2 C.

> **Example** Given $f(x) = 5x + 3$ and $g(x) = 2x^2 - 1$, find (a) $(f \circ g)(3)$; (b) $(g \circ f)(-1)$; (c) $(f \circ g)(x)$.
>
> **Solution** a. $(f \circ g)(3) = f[g(3)]$
> $\qquad\qquad\quad = f[2(3)^2 - 1]$ Replace x with 3 in $2x^2 - 1$
> $\qquad\qquad\quad = f(17)$
> $\qquad\qquad\quad = 5(17) + 3$ Replace x with 17 in $5x + 3$
> $\qquad\qquad\quad = 88$
>
>   b. $(g \circ f)(-1) = g[f(-1)]$
> $\qquad\qquad\quad = g[5(-1) + 3]$ Replace x with -1 in $5x + 3$
> $\qquad\qquad\quad = g(-2)$
> $\qquad\qquad\quad = 2(-2)^2 - 1$ Replace x with -2 in $2x^2 - 1$
> $\qquad\qquad\quad = 7$

c. $(f \circ g)(x) = f[g(x)]$
$= f(2x^2 - 1)$
$= 5(2x^2 - 1) + 3$ Replace x with $2x^2 - 1$ in $5x + 3$
$= 10x^2 - 5 + 3$
$= 10x^2 - 2$

48. $f[g(-1)]$ **49.** $f[g(4)]$ **50.** $g[f(1)]$ **51.** $g[f(-2)]$

52. $f[g(x)]$ **53.** $g[f(x)]$ **54.** $f[f(x)]$ **55.** $g[g(x)]$

Solve the following word problems.

56. The area A of a circle can be expressed as a function of its radius r by $A = g(r) = \pi r^2$, where π is a constant. Find $g\left(\dfrac{1}{2}\right)$, $g(2.1)$, and $g(8)$. (Leave the answers in terms of π.) Write the answers as ordered pairs $(r, g(r))$. What is the domain of g?

57. The temperature in degrees Celsius C can be expressed as a function of degrees Fahrenheit F by $C = g(F)$. If $g(F) = \dfrac{5}{9}(F - 32)$, find $g(14)$, $g(32)$, and $g(212)$. Express the answers as ordered pairs $(F, g(F))$.

58. The temperature in degrees Fahrenheit F can be expressed as a function of degrees Celsius C by $F = f(C)$. If $f(C) = \dfrac{9}{5}C + 32$, find $f(0)$, $f(100)$, and $f(-10)$. Express the answers as ordered pairs $(C, f(C))$.

59. The cost C in cents of sending a first-class letter can be expressed as a function of its weight w in ounces by $C = h(w) = 29 + 20(w - 1)$. Find $h(2)$, $h(3)$, and $h(5)$. Express the answers as ordered pairs $(w, h(w))$.

60. The cost C in dollars of gasoline can be expressed as a function of the number of gallons n by $C = f(n) = 0.96n$. Find $f(10)$, $f(12)$, and $f(25)$. Express the answers as ordered pairs $(n, f(n))$.

61. The volume V of a cube can be expressed as a function of its side s by $V = f(s) = s^3$. Find $f(1)$, $f(3)$, and $f(5)$. Write the answers as ordered pairs $(s, f(s))$. What is the domain of f?

62. A company manufactures pens that sell for \$2.50 each. They find that the cost function, C, in dollars, to produce the pens is given by the function

$$C(x) = 50{,}000 + 5\sqrt{x}$$

where x is the number of pens produced.
   a. Write an equation defining the revenue function, R, when x pens are produced—in terms of $R(x)$.
   b. Write an equation defining the profit function, P, when x pens are produced—in terms of $P(x)$. *Hint:* Recall, $P(x) = R(x) - C(x)$.
   c. Find the profit when 40,000 pens are produced; that is, find $P(40{,}000)$.

63. The Mag Pie Company bakes pies that sell for \$3 per pie. The company finds that the cost function, C in dollars, to produce the pies is given by

$$C(x) = 1{,}500 + 3\sqrt[3]{x}$$

where x is the number of pies baked.
   a. Write an equation defining the revenue function, R, when x pies are baked—in terms of $R(x)$.
   b. Write an equation defining the profit function, P, when x pies are baked—in terms of $P(x)$.
   c. Find the profit when 1,000 pies are baked—that is, find $P(1{,}000)$.

For problems 64–67, show that $(f \circ g)(x) = x$ and $(g \circ f)(x) = x$.

64. $f(x) = 3x + 1$ and $g(x) = \dfrac{x - 1}{3}$

65. $f(x) = \sqrt{x - 2}$ and $g(x) = x^2 + 2, x \geq 2$

66. $f(x) = \dfrac{5x - 2}{4}$ and $g(x) = \dfrac{4x + 2}{5}$

67. $f(x) = x^3 - 7$ and $g(x) = \sqrt[3]{x + 7}$

68. Given $f(x) = 3x - k$ and $g(x) = 2x + 5$, where k is a constant, find a value of k such that $f \circ g$ and $g \circ f$ are the same function.

69. We know that $a(b + c) = ab + ac$ by the distributive property. In like fashion, for functions f, g, and h will
   a. $f \cdot (g + h) = f \cdot g + f \cdot h$?
   b. $f \circ (g + h) = f \circ g + f \circ h$?

70. We know that $a \cdot (b \cdot c) = (a \cdot b) \cdot c$ by the associative property. For functions f, g, and h, will $f \circ (g \circ h) = (f \circ g) \circ h$?

Review exercises

1. Subtract $(4x^2 + 6x - 9) - (-2x^2 - x + 5)$. See section 1–6.

2. Given $P(x) = 2x + 1$, $Q(x) = x^2 - 1$, and $R(x) = 5x - 4$, find $P(x) - Q(x) - R(x)$. See section 1–5.

3. Find the slope and the y-intercept of the linear equation $4x - 5y = -20$. See section 7–3.

4. What is the degree of the polynomial $5x^4 - 3x^3 + 1$? See section 1–5.

5. Divide $(3x^3 - 4x^2 + 2x + 3) \div (x - 1)$. See section 4–5.

8–3 ■ Special functions and their graphs

In previous chapters, we discussed and graphed such equations as

$$y = 3x + 2 \quad \text{and} \quad y = 4 - 5x$$

We found that the graph was a straight line. These equations were called linear equations and define linear functions. In this section, we will study the linear functions as well as other types of functions.

The linear function

A linear equation yields a straight line graph. All straight lines, except vertical lines, represent **linear functions.**

> **Linear function** ───────────────
>
> A function that can be written in the form
>
> $$f(x) = mx + b$$
>
> where m and b are real numbers, is called a **linear function.**

Recall in chapter 7 we found that in the linear equation $y = mx + b$, m was the slope of the line and b determined the y-intercept to be the point $(0,b)$. In like fashion, in the linear function $f(x) = mx + b$, m and b define the slope and y-intercept, respectively, of the graph of function f and they can be used to graph a linear function.

■ **Example 8–3 A**

Write the equation $y = 3x - 2$ as a function and graph the function. Replacing y with $f(x)$, we have

$$f(x) = 3x - 2$$

Then $m = 3 = \dfrac{3}{1}$ and b (y-intercept) $= -2$. From the y-intercept, the point $(0,-2)$, move 1 unit to the right and 3 units up to obtain a second point. The following graph represents the graph of the linear function $f(x) = 3x - 2$.

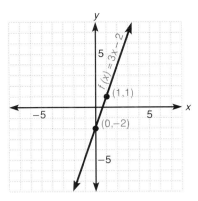

▶ *Quick check*   Write the equation $y = -4x + 1$ as a function and graph the function.   ■

The constant function

Recall that the graph of any equation of the form $y = k$, where k is a real number constant, is a horizontal straight line. Because the value of y is constant (always the same) for any value of x that is chosen, this equation defines a special linear function called a **constant function.**

___ **Constant function** _____

Any function f that can be written in the form
$$f(x) = k$$
where k is a real number, is called a **constant function.**

■ *Example 8–3 B*

Write the equation $y - 4 = 0$ as a function and graph the function.

If we solve this equation for y, we obtain $y = 4$. Then $f(x) = 4$. The graph is a horizontal straight line through the point (0,4).

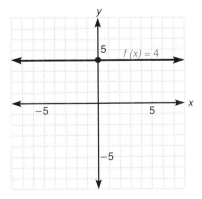

■

The linear and constant functions are special cases of a more general class of functions called the polynomial function that we will discuss in a later chapter.

The square root function

We now define a special function called the **square root function**. It is defined as the principal square root of a polynomial expression.

> ### Square root function
>
> Any function that can be stated in the form
> $$f(x) = \sqrt{P}$$
> where P is a polynomial in x, $P \geq 0$, is called a **square root function**.

■ *Example 8–3 C*

Graph the square root function $f(x) = \sqrt{x - 3}$.

Since the radicand $x - 3$ must be greater than or equal to zero, we solve the inequality $x - 3 \geq 0$ to determine the domain of f.

$$x - 3 \geq 0 \qquad \text{Set the radicand} \geq 0$$
$$x \geq 3 \qquad \text{Solve the inequality}$$

The domain consists of all real numbers $x \geq 3$ so we choose values of x accordingly. See the following table.

x	$f(x) = \sqrt{x - 3}$
3	0
4	1
7	2
12	3

Note Since we can choose any value for $x \geq 3$, it is convenient to choose values of x such that $x - 3$ is a perfect square. This we have done.

If we plot the points $[x, f(x)]$ and draw a smooth curve through, we obtain the following graph.

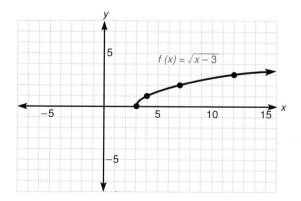

The quadratic function

The graph of an equation of the form $y = ax^2 + bx + c$, $a \neq 0$, is a curve called a *parabola*. This equation is called a quadratic equation in two variables and defines the quadratic function that we will discuss in detail in a later chapter.

Quadratic function

Any function *f* that can be written in the form

$$f(x) = ax^2 + bx + c$$

where *a*, *b*, and *c* are real numbers, $a \neq 0$, is called a **quadratic function.**

■ *Example 8–3 D*

Write each equation as a quadratic function and graph the function.

1. $y = x^2 + 3x$

To graph the function $f(x) = x^2 + 3x$, we form a table of related values found by choosing values of *x* and finding the corresponding values of $f(x)$, or *y*. We then plot the corresponding points and connect them by a smooth curve.

x	$f(x) = x^2 + 3x$
-4	4
-3	0
-2	-2
-1	-2
0	0
1	4
2	10

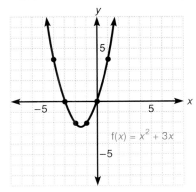

2. $y = -x^2 + 2x + 3$

We graph the function $f(x) = -x^2 + 2x + 3$ in the same manner as we did example 1.

x	$f(x) = -x^2 + 2x + 3$
-2	-5
-1	0
0	3
1	4
2	3
3	0

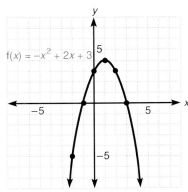

Note In general, when graphing the quadratic function, we choose values for *x* starting with $x = 0$ and then 2 or 3 negative values and 2 or 3 positive values.

Piecewise functions

Some functions consist of two or more different functions defined on differing intervals. This often causes the graph to be broken into two or more pieces. Such functions are called **piecewise functions.**

■ *Example 8–3 E*

Graph the function f defined by

$$f(x) = \begin{cases} -4 & \text{for } x \leq -2 \\ x + 1 & \text{for } -2 < x \leq 1 \\ 2 & \text{for } x > 1 \end{cases}$$

For $x \leq -2$, $f(x) = -4$ and for $x > 1$, $f(x) = 2$, and function f is a constant function (whose graph is a horizontal line), while for $-2 < x \leq 1$, function f is a linear function defined by $f(x) = x + 1$. This means:

1. For every x less than or equal to -2, $f(x) = -4$ and the graph is part of the horizontal line $y = -4$, starting with a closed dot at $(-2,-4)$ and running in the negative direction.

2. For every x greater than -2 and less than or equal to 1, $f(x) = x + 1$ and the graph is a line running from $f(-2) = (-2) + 1 = -1$, the point $(-2,-1)$ to $f(1) = 1 + 1 = 2$, the point $(1,2)$.

Note The point $(-2,-1)$ is an *open* dot since $f(-2) = -4$ (not -1) while the point $(1,2)$ is a closed dot because of ≤ 1.

3. For every x greater than 1, $f(x) = 2$ and the graph is part of the horizontal line $y = 2$, starting with the point $(1,2)$ and running in the positive direction.

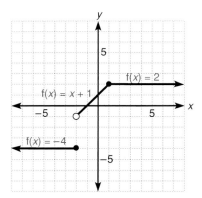

Note Whenever there is a "break" in the graph, the function is said to be *discontinuous*. Thus, the function f is discontinuous at $x = -2$.

▶ *Quick check* Graph the piecewise function defined by

$$f(x) = \begin{cases} x - 2 & \text{for } x < -1 \\ 0 & \text{for } x = -1 \\ x + 3 & \text{for } x > -1 \end{cases}$$

■

Increasing and decreasing functions

A function can *increase* or *decrease* for every value of x in its domain or the function can increase or decrease over certain sets of values of x (over certain intervals). In figure 8–1, the function f is said to be increasing for all $x \leq a$ and for all $x \geq b$ since the values of the function increase (the graph rises) as the value of x increases. The function is said to be decreasing (the graph falls) for all x between a and b since the values of the function decrease as the value of x increases. That is, f increases on the intervals $(-\infty, a]$ and $[b, \infty)$, and decreases on the interval $[a, b]$.

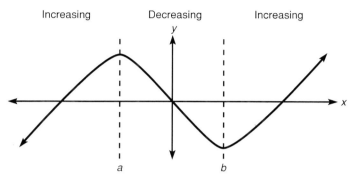

Figure 8–1

More formally, we define increasing and decreasing functions as follows.

Definition of increasing and decreasing functions

A function f is
a. *increasing* on interval I if $f(a) < f(b)$ whenever $a < b$ for every a and b in I.
b. *decreasing* on interval I if $f(a) > f(b)$ whenever $a < b$ for every a and b in I.

■ **Example 8–3 F**

Find the intervals over which the given function is increasing or decreasing.

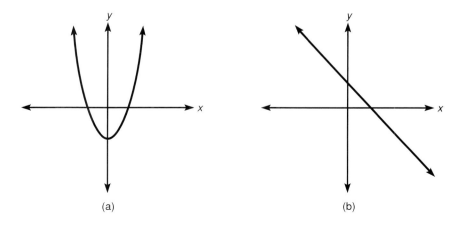

(a)   (b)

a. Since the values of y decrease (the graph falls) as the values of x increase for all $x \leq 0$ and the values of y increase (the graph rises) as the values of x increase for all $x \geq 0$, the function is decreasing on the interval $(-\infty,0]$ and increasing on the interval $[0,+\infty)$.

b. Since the values of y decrease (the graph falls) as the values of x increase *for all* x, the function is decreasing on the interval $(-\infty,\infty)$. That is, the function decreases everywhere.

▶ *Quick check* Find the intervals over which the graphed function is increasing or decreasing.

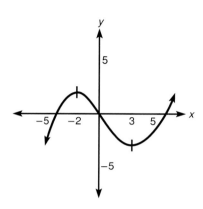

Even and odd functions

Functions can be classified as even, odd, or neither even nor odd.

─── **Definition of even or odd functions** ───

A function f is said to be

$$\begin{array}{ll} even & \text{if } f(x) = f(-x) \\ odd & \text{if } f(-x) = -f(x) \end{array}$$

for all x and $-x$ in the domain of f.

■ *Example 8–3 G*

Determine if the given function is even, odd, or neither even nor odd.

1. $f(x) = x^4 - 3x^2$
We first find $f(-x)$.

$$\begin{aligned} f(-x) &= (-x)^4 - 3(-x)^2 \quad \text{Replace } x \text{ with } -x \\ &= x^4 - 3x^2 \end{aligned}$$

Since $f(x) = x^4 - 3x^2 = f(-x)$ for each x in the domain of the function, f is an even function.

2. $g(x) = 3x^3 + 2x$
We first find $g(-x)$.

$$\begin{aligned} g(-x) &= 3(-x)^3 + 2(-x) \quad \text{Replace } x \text{ with } -x \\ &= -3x^3 - 2x \end{aligned}$$

Since $3x^3 + 2x \neq -3x^2 - 2x$ then $g(x) \neq g(-x)$ and g is not even.

Now $-g(x) = -(3x^3 + 2x)$
$$= -3x^3 - 2x$$

Since $g(-x) = -3x^3 - 2x = -g(x)$, g is an odd function.

3. $h(x) = 2x^2 - x + 1$
We first find $h(-x)$.

$h(-x) = 2(-x)^2 - (-x) + 1$ Replace x with $-x$
$$= 2x^2 + x + 1$$

Since $2x^2 - x + 1 \neq 2x^2 + x + 1$, then $h(x) \neq h(-x)$ and h is not even.

Now $-h(x) = -(2x^2 - x + 1)$
$$= -2x^2 + x - 1$$

Since $2x^2 + x + 1 \neq -2x^2 + x - 1$, then $h(-x) \neq -h(x)$ and h is not odd. Function h is neither even nor odd.

▶ **Quick check**   Determine if the function defined by $f(x) = 5x^2 - 3$ is even, odd, or neither even nor odd.   ■

Mastery points

Can you
■ Identify a linear function, a constant function, a quadratic function, a square root function, and a piecewise function?
■ Graph each of the special functions stated above?
■ Determine intervals over which a function increases or decreases?
■ Determine if a function is even, odd, or neither even nor odd?

Exercise 8-3

Identify each function and sketch its graph. See examples 8-3 A, B, C, and D.

Example $f(x) = -4x + 1$

Solution   Since $f(x) = -4x + 1$ is of the form $f(x) = mx + b$, the function is linear and the graph is a straight line.

1. $b = 1$, so the y-intercept is the point $(0,1)$.

2. From $(0,1)$, since $m = -4 = \dfrac{-4}{1}$, we move 1 unit to the *right* (horizontal distance) and then

4 units *down* (vertical distance) to obtain a second point, $(1,-3)$. Draw a straight line through the points $(0,1)$ and $(-1,-3)$ to graph function f.

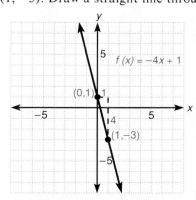

1. $f(x) = x + 5$

2. $f(x) = x - 2$

3. $f(x) = 4 - 3x$

4. $f(x) = 5 - x$

5. $f(x) = 6x - 1$

6. $f(x) = 2x + 3$

7. $f(x) = \frac{1}{2}x + 2$

8. $f(x) = \frac{3}{4}x - 1$

9. $f(x) = -\frac{2}{3}x - 3$

10. $f(x) = -\frac{5}{2}x + 4$

11. $f(x) = 2$

12. $f(x) = -3$

13. $f(x) = 0$

14. $f(x) = (x - 1)^2$

15. $f(x) = (x + 3)^2$

16. $f(x) = x^2 - 4x$

17. $f(x) = 5x^2 + 2x$

18. $f(x) = x^2 - x - 6$

19. $f(x) = x^2 + 8x + 12$

20. $f(x) = -x^2 - 6x - 7$

21. $f(x) = -x^2 + 2x + 15$

22. $f(x) = 2x^2 - x - 3$

23. $f(x) = -3x^2 + 2x + 1$

24. $f(x) = \sqrt{x - 5}$

25. $f(x) = \sqrt{4 - x}$

26. $f(x) = \sqrt{2 - x}$

27. $f(x) = \sqrt{4x - 1}$

28. $f(x) = \sqrt{2x + 3}$

29. $f(x) = \sqrt{x} - 1$

30. $f(x) = \sqrt{x} + 5$

31. $f(x) = -\sqrt{x + 7}$

32. $f(x) = -\sqrt{x - 6}$

33. $f(x) = 2\sqrt{x + 1}$

34. $f(x) = -3\sqrt{x - 2}$

Graph the following piecewise functions. State the values of x where the function is discontinuous.
See example 8–3 E.

Example   Graph the piecewise function f defined by $f(x) = \begin{cases} x - 2 & \text{for } x < -1 \\ 0 & \text{for } x = -1 \\ x + 3 & \text{for } x > -1 \end{cases}$

Solution   For $x < -1$ $f(x) = x - 2$ and for $x > -1$ $f(x) = x + 3$ so we have linear functions. The graphs on those intervals will be rays (half-lines) starting at the point $(-1, -3)$, an open dot, for $x < -1$ and at the point $(-1, 2)$, an open dot, for $x > -1$. Graph $f(x) = x - 2$ from the point $(-1, -3)$ choosing values of $x < -1$. Graph $f(x) = x + 3$ from the point $(-1, 2)$ choosing values of $x > -1$. The graph will have a point at $(-1, 0)$ since $f(-1) = 0$ and will be discontinuous at $x = -1$.

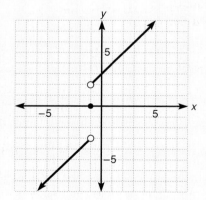

35. $f(x) = \begin{cases} -2 & \text{for } x < 0 \\ 3 & \text{for } x \geq 0 \end{cases}$

36. $f(x) = \begin{cases} -5 & \text{for } x \leq -1 \\ 2 & \text{for } x > -1 \end{cases}$

37. $f(x) = \begin{cases} -1 & \text{for } x \leq 2 \\ x - 2 & \text{for } x > 2 \end{cases}$

38. $f(x) = \begin{cases} 2x + 1 & \text{for } x < 3 \\ 4 & \text{for } x \geq 3 \end{cases}$

39. $f(x) = \begin{cases} 4 & \text{for } x < -3 \\ x + 4 & \text{for } -3 \le x < 0 \\ -5 & \text{for } x \ge 0 \end{cases}$

40. $f(x) = \begin{cases} 2 & \text{for } x \le 0 \\ 4 - x & \text{for } 0 < x \le 3 \\ 1 & \text{for } x > 3 \end{cases}$

41. $f(x) = \begin{cases} x - 1 & \text{for } x \le 2 \\ x^2 & \text{for } x > 2 \end{cases}$

42. $f(x) = \begin{cases} -3 & \text{for } x < 4 \\ \sqrt{x} & \text{for } x \ge 4 \end{cases}$

43. $f(x) = \begin{cases} 2 & \text{if } x < -1 \\ x^2 + 1 & \text{if } -1 \le x < 2 \\ x & \text{if } x \ge 2 \end{cases}$

44. $f(x) = \begin{cases} -\sqrt{1 - x} & \text{for } x < 1 \\ -5 & \text{for } 1 \le x \le 2 \\ 2 - x^2 & \text{for } x > 2 \end{cases}$

For the following functions, find $f(-2), f(-1), f(0), f(1),$ and $f(2)$. Plot the points and connect them to graph each function. These are examples of the *absolute value function.*

Example $f(x) = |x| + 1$

Solution $f(-2) = |-2| + 1 = 2 + 1 = 3; (-2,3)$
$f(-1) = |-1| + 1 = 1 + 1 = 2; (-1,2)$
$f(0) = |0| + 1 = 0 + 1 = 1; (0,1)$
$f(1) = |1| + 1 = 1 + 1 = 2; (1,2)$
$f(2) = |2| + 1 = 2 + 1 = 3; (2,3)$

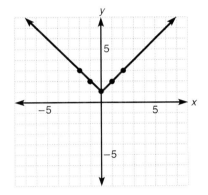

45. $f(x) = |x| - 1$

46. $f(x) = |x| + 3$

47. $f(x) = |x|$

48. $f(x) = -|x|$

49. $f(x) = 5 - |x|$

50. $f(x) = 3 - |x|$

51. Given $f(x) = |x - 2|$, find values of $f(x)$ when x takes on values $-1, 0, 1, 2, 3,$ and 4. Plot the points and connect them to get the graph of $f(x) = |x - 2|$.

52. Graph $f(x) = -|x - 2|$ using the same values of x as in exercise 51.

For each of the following graphed functions, indicate on which intervals it is increasing and on which intervals it is decreasing. See example 8-3 F.

Example Find the intervals over which the graphed function is increasing or decreasing.

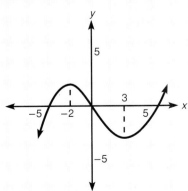

Solution Since the graph rises for all values of $x \le -2$ and $x \ge 3$, the function increases on these intervals, and since the graph *falls* for all values of x such that $-2 \le x \le 3$, the function decreases on this interval. Thus, f increases on $(-\infty, -2]$ and on $[3, \infty)$ while f decreases on $[-2, 3]$.

53.

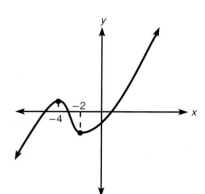

54.

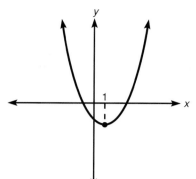

55.

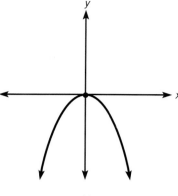

56.

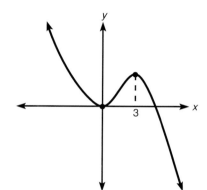

57.

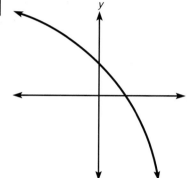

58.

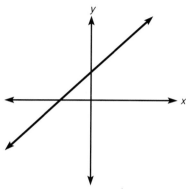

For each of the following functions, determine if the function is even, odd, or neither even nor odd. See example 8–3 G.

Example Determine if the function f defined by $f(x) = 5x^2 - 3$ is even, odd, or neither even nor odd.

Solution We first find $f(-x)$.

$$f(-x) = 5(-x)^2 - 3 = 5x^2 - 3$$

Thus, $f(x) = 5x^2 - 3 = f(-x)$. Since $f(x) = f(-x)$ for all x in the domain of f, the function is an even function.

59. $f(x) = 2x$ **60.** $f(x) = x^3 - x$ **61.** $f(x) = x^2 - x - 3$ **62.** $f(x) = x^2 + 2$

63. $f(x) = 5x + 1$ **64.** $f(x) = x^4 - x^2 + 4$ **65.** $f(x) = 3$

Review exercises

1. Simplify the complex rational expression. See section 4–4.

$$\dfrac{\dfrac{25a^2 - b^2}{4a}}{\dfrac{5a + b}{7a}}$$

2. Solve the inequality $z^2 - 2z \geq 3$. Write the answer in interval notation. See section 6–7.

3. Reduce $\dfrac{3x^2 - 12}{4x^2 - 16}$ to lowest terms. See section 4–1.

4. Simplify the expression $\sqrt{\dfrac{16}{5}}$. See section 5–4.

Perform the indicated operations. See sections 5–6 and 5–7.

5. $(3\sqrt{2} - 2\sqrt{3})^2$

6. $(4 - 3i)(4 + 3i)$

8–4 ■ *Inverse functions*

Consider the function f defined by $f(x) = 3x - 5$. If we let x take on the values $-4, -2, 0, 2$, and 4, we can determine that the ordered pairs

$$(-4,-17), (-2,-11), (0,-5), (2,1), \text{ and } (4,7)$$

belong to the function f. Now consider the function g defined by

$$g(x) = \frac{x + 5}{3}$$

If we let x take on the values $-17, -11, -5, 1$, and 7 (range elements of f), we substitute to find that

1. $g(-17) = \dfrac{(-17) + 5}{3} = \dfrac{-12}{3} = -4$

2. $g(-11) = \dfrac{(-11) + 5}{3} = \dfrac{-6}{3} = -2$

3. $g(-5) = \dfrac{(-5) + 5}{3} = \dfrac{0}{3} = 0$

4. $g(1) = \dfrac{(1) + 5}{3} = \dfrac{6}{3} = 2$

5. $g(7) = \dfrac{(7) + 5}{3} = \dfrac{12}{3} = 4$

Thus the ordered pairs

$$(-17,-4), (-11,-2), (-5,0), (1,2), \text{ and } (7,4)$$

are elements of the function g.

Observing the functions f and g, we see that when the components of the ordered pairs in f are interchanged, we have the ordered pairs in function g and vice versa. If we continued to find other ordered pairs in the two functions, we would find that the same relationship would hold. When this relationship exists between two sets of ordered pairs that define functions, we say the functions are **inverses** of one another. Thus function g is the inverse of function f, and function f is the inverse of function g.

Given a function f, we denote the inverse of f by the symbol "f^{-1}" (read "the inverse of f" or "f inverse"). Thus $g = f^{-1}$ and $f = g^{-1}$ in the previous examples.

Note The -1 in the symbol "f^{-1}" *should not* be interpreted as an exponent. Rather, it is a necessary part of the symbol as a whole and denotes the inverse of the function.

It is important to note that the inverse of a function is *not necessarily a function*. Given a function defined by a set of ordered pairs, we can determine if the function has an inverse function by interchanging components.

■ *Example 8–4 A*

Determine if each function has an inverse function.

1. Given function f defined by

$$f = \{(1,2), (4,5), (6,7), (-1,4)\}$$

For every second component, there corresponds only one first component. Thus the inverse of f is a function and f^{-1} is defined by

$$f^{-1} = \{(2,1), (5,4), (7,6), (4,-1)\}$$

2. Given function g defined by

$$g = \{(1,2), (2,2), (3,-6), (4,-6)\}$$

If we interchange range and domain, we obtain the set of ordered pairs

$$\{(2,1), (2,2), (-6,3), (-6,4)\}$$

which *does not* define a function since there are two distinct ordered pairs that have the same first component, 2 (and also -6).

▶ *Quick check* Determine if the inverse of the function defined by $g = \{(-1,2), (-3,-2), (4,1), (2,2)\}$ is a function. ■

We then conclude that the inverse of a function is also a function if for each value of y there is *exactly one* value of x. We call this a **one-to-one function.**

Definition of a one-to-one function _____

A **one-to-one function** is any function that associates a unique value of x with each distinct value of y.

Concept
A function is a one-to-one function if no two ordered pairs in the function have the *same second components.*

Thus the inverse of a function will be a function provided the function itself is one-to-one. We use this to determine when a function has an inverse function.

Note We have y as a function of x if for each x there is a unique y, and the function is one-to-one if for each y there is a unique x.

Recall that we have the vertical line test to determine if a graph represents a function. Since any two ordered pairs having the same second component will lie on a horizontal line, it follows that any horizontal line drawn in the plane must intersect the graph in only one point if the function is one-to-one. We call this the **horizontal line test** for a one-to-one function, $y = f(x)$.

■ *Example 8–4 B*

Use the horizontal line test to determine if the given graphed function is one-to-one.

1.

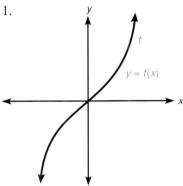

2.

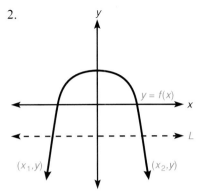

Since any horizontal line drawn in the plane will intersect the graph in only one point, the graph represents a one-to-one function.

Since there exists at least one horizontal line L that can be drawn in the plane which will intersect the graph in two points, the graph does not represent a one-to-one function.

▶ *Quick check*   Use the horizontal line test to determine if the graphed function is one-to-one.

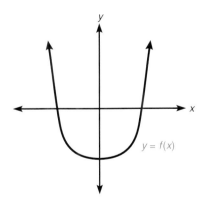

We have seen that to obtain the inverse of a function defined by a set of ordered pairs, we interchange the components (x- and y-values) of the ordered pairs.

$$f = \{(1,2),\ (-2,6),\ (3,0),\ (-8,7)\}$$
$$f^{-1} = \{(2,1),\ (6,-2),\ (0,3),\ (7,-8)\}$$

When a function is defined by an equation, the equation of the inverse of the function is found by using the following procedure.

> ─── *Finding the inverse of function y = f(x)* ───────────
>   1. Replace $f(x)$ with y.
>   2. Interchange x and y in the equation.
>   3. Solve the resulting equation for y.
>   4. Replace y with $f^{-1}(x)$.

■ *Example 8–4 C*

Find the inverse of the given function.

1. $f(x) = 4x - 3$

   a. $(y) = 4x - 3$ Replace $f(x)$ with y

   b. $x = 4y - 3$ Interchange x and y

   c. Solve the equation for y.

   $$x + 3 = 4y$$ Add 3 to each member

   $$\frac{x + 3}{4} = y$$ Divide each member by 4

   d. $f^{-1}(x) = \dfrac{x + 3}{4}$ Replace y with $f^{-1}(x)$

2. $g(x) = x^3 - 3$

   a. $(y) = x^3 - 3$ Replace $g(x)$ with y

   b. $x = y^3 - 3$ Interchange x and y

   c. Solve the equation for y.

   $$x + 3 = y^3$$ Add 3 to each member
   $$y = \sqrt[3]{x + 3}$$ Take the principal cube root of each member

   d. $g^{-1}(x) = \sqrt[3]{x + 3}$ Replace y with $g^{-1}(x)$

▶ *Quick check* Find the inverse of the function $f(x) = 2x + 9$.   ■

   There is another important fact that we should know about inverse functions f and f^{-1}. Suppose the ordered pair (a,b) belongs to the function f. Then the ordered pair (b,a) belongs to the function f^{-1}. Consider their positions in the plane relative to the graph of the equation $y = x$. To illustrate, suppose we let $(2,4)$ be in f and then $(4,2)$ is in f^{-1}. See figure 8–2.
   The line segment connecting points $(2,4)$ and $(4,2)$ is perpendicular to, and cut in half by, the line $y = x$. The points $(2,4)$ and $(4,2)$ are "mirror images" of each other with respect to the line $y = x$. Thus, to graph the function f^{-1}, we locate mirror images of each point in the graph of function f with respect to the line $y = x$.
   In figure 8–3 we show the graphs of two one-to-one functions and the graphs of their inverse functions. The line segments (dashed because they are not part of either graph) join "mirror image" points in the graphs of f and f^{-1}, g and g^{-1}.

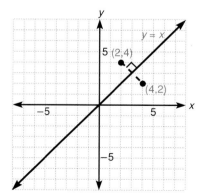

Figure 8–2

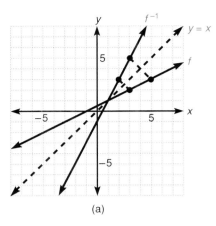

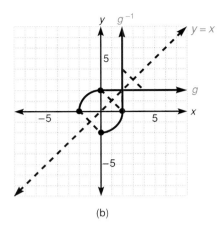

(a)   (b)

Figure 8-3

■ *Example 8-4 D*

Sketch the graph of f^{-1} on the same coordinate plane as $f(x) = x^3 - 1$. First we graph $f(x) = x^3 - 1$ by finding a number of ordered pairs and plotting them on a graph.

x	$f(x) = x^3 - 1$
-2	-9
-1	-2
0	-1
1	0
2	7

To determine points on the graph of f^{-1} that correspond to the points on f that we used, draw a line perpendicular to $y = x$ from each point on f. The corresponding point on f^{-1} must be at the same distance from $y = x$ on this perpendicular line as is the distance from the point on f to $y = x$.

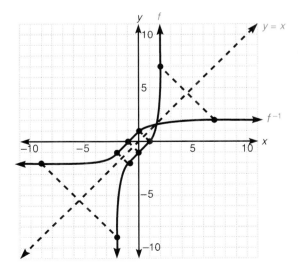

> ── *Mastery points* ──
>
> *Can you*
> - Determine if a given function is one-to-one and has an inverse function?
> - Use the horizontal line test to determine if a graph represents a one-to-one function?
> - Find the inverse function of a one-to-one function?
> - Sketch the graph of f^{-1}, using the graph of f?

Exercise 8–4

Determine if the given functions are one-to-one functions and thus have an inverse function. Explain the answer. See example 8–4 A.

Example $g = \{(-1,2), (-3-2), (4,1), (2,2)\}$

Solution The function contains two ordered pairs, $(-1,2)$ and $(2,2)$, having the same second component so the function is not one-to-one and does not have an inverse function.

1. $f = \{(-4,3), (2,1), (-6,-4), (0,0)\}$ 2. $g = \{(-8,0), (1,2), (7,8), (1,11)\}$
3. $h = \{(1,-6), (2,3), (4,-6), (5,-7)\}$ 4. $F = \{(0,-7), (3,-4), (7,6), (1,-4)\}$

Example $f(x) = 5x - 2$

Solution Since for any value of x we choose we will obtain *only one* value of $y = f(x)$, and any horizontal line drawn in the plane will intersect its graph in only one point, the function is one-to-one and has an inverse function.

5. $f(x) = 2x - 7$ 6. $g(x) = 8 - 3x$ 7. $G(x) = x^2 - 3x + 1$
8. $H(x) = -x^2 + 2x + 4$ 9. $f(x) = |x - 3|$ 10. $g(x) = |2x + 4|$
11. $h(x) = \sqrt{x - 3}$ 12. $F(x) = \sqrt{4 - 2x}$

See example 8–4 B.

Example

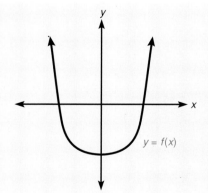

$y = f(x)$

Solution Since there are many horizontal lines that would intersect the graph in two points, the function is not one-to-one and the function has no inverse function.

13.

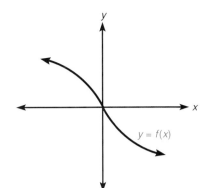

14.

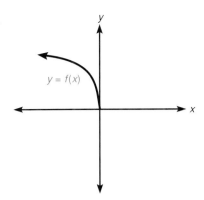

15.

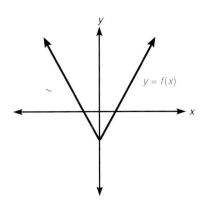

16.

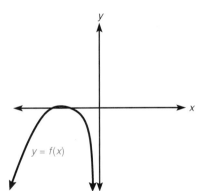

17.

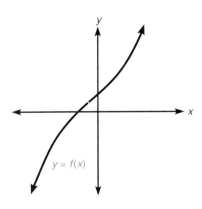

18.

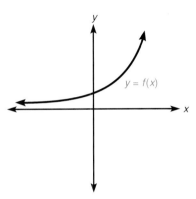

Find the inverse function of each given one-to-one function. See example 8–4 C.

Example $f(x) = 2x + 9$

Solution   1. $(y) = 2x + 9$ .... Replace $f(x)$ with y
   2. $x = 2y + 9$ .... Interchange x and y
   3. Solve the equation for y.
   $x - 9 = 2y$ .... Subtract 9 from each member
   $y = \dfrac{x - 9}{2}$ .... Divide each member by 2
   4. $f^{-1}(x) = \dfrac{x - 9}{2}$ .... Replace y with $f^{-1}(x)$

19. $f = \{(-3,4), (2,-3), (0,7)\}$
20. $g = \{(10,-4), (-6,3), (8,2)\}$
21. $h(x) = 5x$

22. $F(x) = -4x$
23. $G(x) = 3x + 7$
24. $H(x) = 5x - 1$

25. $f(x) = x^3 + 2$
26. $g(x) = 2x^3 - 3$
27. $h(x) = \sqrt{x + 5}, x \geq -5$

28. $F(x) = \sqrt{3x - 4}, x \geq \dfrac{4}{3}$
29. $G(x) = \sqrt{x + 2}, x \geq -2$
30. $H(x) = \sqrt{x} - 5, x \geq 0$

31. $f(x) = \sqrt[3]{x - 2}$
32. $g(x) = \sqrt[3]{5x + 2}$
33. $h(x) = \sqrt[3]{x} + 7$

34. $F(x) = \sqrt[3]{2x} - 1$

Sketch the graphs of f and f^{-1} on the same coordinate plane. See example 8–4 D.

35. $f(x) = -4x$
36. $f(x) = 2x$
37. $f(x) = x + 4$

38. $f(x) = x - 2$
39. $f(x) = 5x - 6$
40. $f(x) = 2x - 3$

41. $f(x) = -\sqrt{3 - 2x}, x \leq \dfrac{3}{2}$ **42.** $f(x) = \sqrt{x + 2}, x \geq -2$ **43.** $f(x) = x^3 + 1$

44. $f(x) = x^3$ **45.** $f(x) = x^2 - 2, x \geq 0$ **46.** $f(x) = x^3 - 4$

47. $f(x) = -x^2 + 4, x \leq 0$

Prove that f and g are inverse functions by showing that $f[g(x)] = x$ and $g[f(x)] = x$.

Example $f(x) = 2x - 5$ and $g(x) = \dfrac{x + 5}{2}$

Solution   1. $f[g(x)] = f\left[\dfrac{x + 5}{2}\right]$ Replace $g(x)$ with $\dfrac{x + 5}{2}$

$\qquad\qquad = 2\left(\dfrac{x + 5}{2}\right) - 5$ Replace x in $f(x)$ with $\dfrac{x + 5}{2}$

$\qquad\qquad = x + 5 - 5$

$\qquad\qquad = x$

2. $g[f(x)] = g(2x - 5)$ Replace $f(x)$ with $2x - 5$

$\qquad\qquad = \dfrac{(2x - 5) + 5}{2}$ Replace x in $g(x)$ with $2x - 5$

$\qquad\qquad = \dfrac{2x}{2}$

$\qquad\qquad = x$

Thus, $f[g(x)] = x = g[f(x)]$ and the functions are inverses.

48. $f(x) = 5x$ and $g(x) = \dfrac{x}{5}$ **49.** $f(x) = 4x + 7$ and $g(x) = \dfrac{x - 7}{4}$

50. $f(x) = x^3 - 3$ and $g(x) = \sqrt[3]{x + 3}$ **51.** $f(x) = \sqrt{x - 2}, x \geq 2$ and $g(x) = x^2 + 2$

Review exercises

Find the solution set of the following equations. See sections 2–1 and 4–7.

1. $3y - 2 = 4$

2. $\dfrac{y}{y - 3} + \dfrac{4}{5} = \dfrac{3}{y - 3}$

3. Find the equation of the line passing through $(-2,3)$ and $(4,1)$. Write the equation in standard form $ax + by = c$. See section 7–3.

4. Simplify the radical expression $\sqrt{12} - \sqrt{75} + \sqrt{48}$. See section 5–5.

5. Solve the formula $S = \dfrac{n}{2}(a + \ell)$ for a.

     See section 2–2.

8–5 ■ Variation

If two variables are so related that as one variable changes, a change will also take place with the other variable (as with a function), then we have what is called a **variation.** This relationship is the basis for many formulas that are used in the scientific and physical world. Such formulas often determine functions. For example, consider the formula for the circumference C of a circle, $C = 2\pi r$, where 2π is a constant and r is the radius of the circle. As the radius r varies, the circumference C will also vary. That is,

1. as the radius r becomes longer, the circumference C becomes longer; and
2. as the radius r becomes shorter, the circumference C becomes shorter.

This variation is an example of a **direct variation** and we say "C varies directly as r."

Definition of direct variation

A variable y is said to *vary directly* as a variable x if

$$y = kx$$

where k is a positive constant.

Concept
Two variables vary directly if one variable, y, is some positive multiple of the other variable, x.

In our previous example, $C = 2\pi r$, $k = 2\pi$. Another, more general, example of direct variation is where y varies directly as the nth power of x if

$$y = kx^n, \quad n > 0, \quad k > 0$$

The constant k is called the **constant of variation** (or the **constant of proportionality**).

■ **Example 8–5 A**

The formula for the area A of the surface of a sphere having radius r is given by $A = 4\pi r^2$. In this example, the constant of variation $k = 4\pi$ and "A varies directly as the square of r." An alternate terminology is "A is *directly proportional* to the second power of r."

Given the direct variation $y = kx$ and a pair of corresponding values of x and y, we can determine the value of the constant of variation k.

■ **Example 8–5 B**

Find the constant of variation k if the variable y is directly proportional to the variable x, and $y = 6$ when $x = 12$.

Since y is directly proportional to x,

$$y = kx$$
$$(6) = k \cdot (12) \qquad \text{Replace } y \text{ with 6 and } x \text{ with 12}$$
$$k = \frac{6}{12}$$
$$= \frac{1}{2}$$

The constant of variation is $\frac{1}{2}$.

▶ *Quick check*   The variable y is directly proportional to the variable x where $y = 12$ when $x = 16$. Find the constant of variation.

If we substitute the value for the constant of variation in the equation $y = kx$, we can then find the value of y given a specific value of x, and vice versa.

■ *Example 8–5 C*

1. Let y vary directly as x and suppose $y = 20$ when $x = 5$. Then find y when $x = 16$.

Since y varies directly as x,

$$y = kx$$

To find the constant of proportionality k,

$$\begin{aligned}(20) &= k \cdot (5) \\ 4 &= k\end{aligned}$$ Replace y with 20 and x with 5

Then we have the equation

$$\begin{aligned}y &= (4)x \\ y &= 4(16) \\ &= 64\end{aligned}$$ Replace k with 4 in $y = kx$
                    Replace x with 16

Therefore $y = 64$ when $x = 16$.

2. The distance d in feet that a body falls from rest is directly proportional to the square of the time t in seconds (disregarding air resistance). If an object falls 144 feet in 3 seconds, how far will the object fall in 5 seconds?

Since d is directly proportional to the square of t,

$$\begin{aligned}d &= kt^2 \\ (144) &= k(3)^2 \\ 144 &= k \cdot 9 \\ k &= 16\end{aligned}$$ Replace d with 144 and t with 3

Then
$$\begin{aligned}d &= (16)t^2 \\ d &= 16(5)^2 \\ &= 16(25) = 400\end{aligned}$$ Replace k with 16
                    Replace t with 5

The object falls 400 feet in 5 seconds.   ■

A second kind of variation between two variables occurs when substitution of increasing positive values of one variable results in decreasing positive values of the other variable and vice versa. Such a variation is called **inverse variation.**

── *Definition of inverse variation* ──────────

A variable y **varies inversely** as the variable x if

$$y = \frac{k}{x}, \, k > 0$$

and inversely as the nth power of x if

$$y = \frac{k}{x^n}, \, k > 0, n > 0$$

■ *Example 8–5 D*

Boyle's law for gases states that the volume V of a gas varies inversely as the pressure P, provided the mass and temperature are constant. Then

$$V = \frac{k}{p}$$

Note Alternately, if we solve the formula for P and write the formula as $P = \frac{k}{V}$, then "P varies inversely with V." ■

Again we can determine the value of the constant of variation k given a pair of corresponding values of the variables. Using this, we can find the value of one of the variables, given a value of the other.

■ *Example 8–5 E*

1. Given $V = \frac{k}{P}$, (a) find the constant of variation k if $V = 12P$ cubic feet when $P = 100$ pounds per square foot; (b) find V when $P = 75$ pounds per square foot.

   a. $V = \dfrac{k}{P}$

   $(12) = \dfrac{k}{(100)}$ Replace V with 12 and P with 100

   $k = 1{,}200$

   b. $V = \dfrac{(1{,}200)}{P}$ Replace k with 1,200

   $V = \dfrac{1{,}200}{(75)}$ Replace P with 75

   $= 16$

   The volume $V = 16$ cu ft when pressure $P = 75$ lb per ft².

   Note *Decreasing* the pressure P from 100 pounds per square foot to 75 pounds per square foot caused the volume V to *increase* from 12 cubic feet to 16 cubic feet. This is what we expect with inverse variation.

2. The resistance R, measured in ohms, of an electrical circuit in a given length of wire is inversely proportional to the square of the diameter of the wire measured in centimeters. If the wire has resistance $R = 0.50$ ohms when the diameter $d = 0.02$ cm, what is the resistance of the same length of wire when $d = 0.01$ cm.
   Since R is inversely proportional to the square of d,

   $R = \dfrac{k}{d^2}$

   $(0.50) = \dfrac{k}{(0.02)^2}$ Replace R with 0.50 and d with 0.02

   $0.50 = \dfrac{k}{0.0004}$

   $k = 0.0002$ Multiply each member by 0.0004

Then $R = \dfrac{(0.0002)}{d^2}$ Replace k with 0.0002 in $R = \dfrac{k}{d^2}$

$ = \dfrac{0.0002}{(0.01)^2}$ Replace d with 0.01

$ = \dfrac{0.0002}{0.0001} = 2$

The circuit has resistance $R = 2$ ohms when $d = 0.01$ cm.

▶ *Quick check*   If y is inversely proportional to the cube of x, find y when $x = 4$ if $y = 6$ when $x = 2$. ■

A third variation relates one variable to two or more other variables. We call this a **joint variation.**

Definition of joint variation _____

A variable z is said to **vary jointly** as variables x and y if

$$z = kxy, \; k > 0$$

■ *Example 8–5 F*

1. The formula for the area A of a triangle with base b and altitude h is given by

$$A = \frac{1}{2}bh$$

The constant of variation is $\dfrac{1}{2}$ and "A varies jointly as b and h."

Alternately, we could say that "A varies directly as b and h" or "A is directly proportional to the product of b and h."

2. Coulomb's law says that the magnitude F of a force between two charges of electricity is given by

$$F = k \cdot \frac{q_a \cdot q_b}{r^2}$$

where q_a and q_b are two electrical charges that are at a distance r apart. Then the force "F varies directly as the product of q_a and q_b and inversely as the square of r." ■

In these examples, we can determine the constant k if we are given values of the variables. We can then evaluate any one of the variables by knowing the values of the others.

■ *Example 8–5 G*

1. The diametrical pitch P of a gear is directly proportional to the number of teeth N and is inversely proportional to the pitch diameter D. Find k if $P = 10$ when $N = 12$ and $D = 4$.

If P is directly proportional to N and is inversely proportional to D, then

$$P = \frac{kN}{D}$$

$$(10) = \frac{k \cdot (12)}{(4)} \qquad \text{Replace } P \text{ with 10, } N \text{ with 12, and } D \text{ with 4}$$

$$40 = k \cdot 12 \qquad \text{Multiply each member by 4}$$

$$k = \frac{40}{12} = \frac{10}{3} \qquad \text{Divide each member by 12}$$

2. We can now use the previous information to find N when $P = 9$ and $D = 20$.

$$(9) = \frac{\frac{10}{3}N}{(20)} \qquad \text{Replace } P \text{ with 9, } D \text{ with 20, and } k \text{ with } \frac{10}{3}$$

$$180 = \frac{10}{3}N \qquad \text{Multiply each member by 20}$$

$$N = 180 \cdot \frac{3}{10} \qquad \text{Multiply each member by } \frac{3}{10}$$

$$= 18 \cdot 3$$

$$= 54$$

Therefore $N = 54$ when $P = 9$ and $D = 20$.

Mastery points

Can you
- Write an equation expressing a direct, inverse, or joint variation between variables, using the constant of variation k?
- Find the constant of variation under stated conditions?
- Find the value of one of the variables, knowing the constant k and the value(s) of the other variable(s)?

Exercise 8–5

Express the given statements as equations using the constant of variation k. See examples 8–5 A, D, and F.

1. The speed S of a falling object varies directly as the time t.

2. The perimeter P of an equilateral triangle is directly proportional to the side s.

3. The time t required for an automobile to travel a fixed distance is inversely proportional to the rate r.

4. The resistance R to the flow of electricity in a conductor is inversely proportional to the diameter d of the wire.

5. The momentum M of a body is directly proportional to the product of its mass m and its velocity v.

6. The current I varies directly as the product of the voltage E and the resistance R.

7. The maximum force F exerted on the vane of a wind generator varies jointly as the area A of the vane and the square of the wind velocity v.

8. The resistance R of an electrical conductor varies directly as the length ℓ and inversely as the cross-section area A.

9. The maximum torsinal stress S_{max} on a circular shaft is directly proportional to the torque T and is inversely proportional to the cube of the radius r.

10. The ideal gas law states that the pressure P of a gas varies directly with the absolute temperature T of the gas and inversely as the volume V of the gas.

11. The force of impact F varies directly as the product of the mass m and the velocity v and inversely as the product of the acceleration of gravity g and time t.

Find the constant of variation for the stated conditions. See examples 8–5 B and C.

Example The variable y is directly proportional to the variable x where $y = 12$ when $x = 16$. Find the constant of variation.

Solution Since y is directly proportional to x, then

$$y = kx$$
$$(12) = k \cdot (16) \qquad \text{Replace } y \text{ with 12 and } x \text{ with 16}$$
$$k = \frac{12}{16} \qquad \text{Divide each member by 16}$$
$$= \frac{3}{4} \qquad \text{Reduce to lowest terms}$$

The constant of variation is $\frac{3}{4}$.

12. y is directly proportional to x, and $x = 20$ when $y = 5$.

13. p is directly proportional to h, and $p = 36$ when $h = 9$.

14. A varies directly as the square of r, and $A = 154$ when $r = 7$.

15. V is directly proportional to the cube of s, and $V = 81$ when $s = 3$.

16. P is inversely proportional to V, and $P = 120$ when $V = 5$.

17. s is inversely proportional to T, and $s = 48$ when $T = 2.5$.

18. n varies inversely as the square of p, and $n = 14$ when $p = 9$.

19. p varies directly as T and inversely as V, and $p = 24$ when $T = 6$ and $V = 5$.

20. A is directly proportional to the square of b and inversely proportional to c, and $A = 42$ when $b = 3$ and $c = 3$.

21. A varies jointly as h and $(a + b)$, and $A = 36$ when $h = 9$, $a = 2$, and $b = 6$.

Find the value of the indicated variable. See examples 8–5 C, E, and G.

Example If y is inversely proportional to the cube of x, find y when $x = 4$ if $y = 6$ when $x = 2$.

Solution Since y is inversely proportional to the cube of x, then

$$y = \frac{k}{x^3}$$
$$(6) = \frac{k}{(2)^3} \qquad \text{Replace } y \text{ with 6 and } x \text{ with 2}$$
$$k = 6 \cdot 8 = 48$$
$$y = \frac{48}{x^3} \qquad \text{Replace } k \text{ with 48}$$
$$= \frac{48}{(4)^3} \qquad \text{Replace } x \text{ with 4}$$
$$= \frac{48}{64} = \frac{3}{4}$$

Thus, $y = \frac{3}{4}$ when $x = 4$.

22. x varies directly as y. If $x = 36$ when $y = 4$, find x when $y = 13$.

23. w is directly proportional to ℓ. If $w = 9$ when $\ell = 6$, find w when $\ell = 15$.

24. R_1 is inversely proportional to R_2. If $R_1 = 48$ when $R_2 = 4$, find R_1 when $R_2 = 12$.

25. If y varies inversely as the square of z, find y when $z = 3$ if $y = 12$ when $z = 2$.

26. Find A when $r = 3$ if A is directly proportional to the square of r, and $A = 48$ when $r = 4$.

27. If v varies directly as s and inversely as t, find v when $s = 8$ and $t = 6$ if $v = 20$ when $s = 4$ and $t = 2$.

28. If T is directly proportional to the square of s and is inversely proportional to the cube of m, find T when $s = 2$ and $m = 3$, if $T = 14$ when $s = 3$ and $m = 6$.

29. If z varies jointly as x and the cube of y, find z when $x = 4$ and $y = 3$ if $z = 72$ when $x = 3$ and $y = 2$.

30. If P varies jointly as the square of q and x, find P when $q = 7$ and $x = 2$ if $P = 150$ when $q = 5$ and $x = 3$.

Solve the following word problems.

31. The volume of sales V varies inversely as the unit price P of a commodity. If $V = 4{,}000$ units when $P = \$1.50$, find P when $V = 3{,}000$ units.

32. The velocity v of a soundwave varies directly as the product of frequency n and the wavelength ℓ. If $v = 12$ feet per second when $n = 2$ and $\ell = 4$, find v when $n = 6$ and $\ell = 8$.

33. The intensity of illumination on a surface E in foot-candles varies inversely as the square of the distance d of the light source from the surface. If $E = 6.4$ foot-candles when $d = 8$ feet, find E when $d = 6$ feet (round to nearest tenth).

34. The field intensity H of a magnetic field varies directly as the force F acting on it and inversely as the strength of the pole m. If $H = 3$ oersteds when $F = 750$ dynes and $m = 200$, find H when $F = 500$ dynes and $m = 175$.

35. The electrical resistance R of a wire is directly proportional to the length ℓ of the wire and inversely proportional to the square of the diameter d of the wire. If $R = 8$ ohms when $\ell = 8$ feet and $d = 1$ inch, find R when $\ell = 6$ feet and $d = 2$ inches.

36. The simple interest I earned by a savings deposit in a given time t varies jointly as the principal P and the interest rate r. If $I = \$180$ when $P = \$1{,}000$ and $r = 0.06$, find I when $P = \$750$ and $r = 0.06$ in the same amount of time.

37. The maximum force F that a rectangular cantilever beam (a beam supported at one end) can withstand at its free end varies directly as the beam's width w and the square of its height h and inversely as its length ℓ. If $F = 5{,}000$ pounds when $w = 0.5$ feet, $h = 0.7$ feet, and $\ell = 12$ feet, find F when $w = 0.45$ feet, $h = 0.6$ feet, and $\ell = 18$ feet (round to nearest whole number).

Review exercises

1. Find the solution set of the equation $y + 3\sqrt{y} - 10 = 0$. See section 6–6.

2. Find the standard equation of a line through $(2,5)$ and perpendicular to the line $4x - y = 3$. See section 7–3.

3. Find the solution set of the inequality $|3 - x| < 5$. See section 2–6.

4. Find the value of C when $C = \dfrac{C_1 \, C_2}{C_1 + C_2}$, $C_1 = 2$ and $C_2 = 10$. See section 1–5.

5. Simplify $\dfrac{-36a^{-2}b^4}{24a^3b^{-1}}$. Assume $a \neq 0$, $b \neq 0$. Answer with positive exponents only. See section 3–3.

Chapter 8 lead-in problem

The manufacturer of television sets has fixed costs of $2,500,000 and an additional cost of $350 for each set manufactured. The total cost function of manufacturing x sets is given by $C(x) = 2,500,000 + 350x$. Find the cost of manufacturing 125 sets.

Solution

Given the total cost function $C(x) = 2,500,000 + 350x$, we want $C(125)$. Now

$$C(125) = 2,500,000 + 350(125) \qquad \text{Replace } x \text{ with 125}$$
$$= 2,500,000 + 43,750$$
$$= 2,543,750$$

It costs $2,543,750 to manufacture 125 sets.

Chapter 8 summary

1. A **relation** is any set of ordered pairs.
2. A **function** is a relation in which no two distinct ordered pairs have the same first component. For each x there is a unique y.
3. The **domain** of a function is the set of the first components of the ordered pairs in the function.
4. The **range** of a function is the set of the second components of the ordered pairs in the function.
5. A graph represents a function if any vertical line drawn in the plane intersects the graph in at most one point.
6. A **one-to-one function** is a function in which no two ordered pairs have the same second component. For each y there is a unique x.
7. A graph represents a one-to-one function if any *horizontal line* drawn in the plane intersects the graph in at most one point.
8. The inverse of a function f, denoted by f^{-1}, is obtained by interchanging the components in each ordered pair of the function f.
9. Given function f, f^{-1} is a function if and only if f is a one-to-one function.
10. The graphs of f and f^{-1} are mirror images (reflections) of each other with respect to the line $y = x$.
11. The composition of f and g is given by $(f \circ g)(x) = f[g(x)]$ and the composition of g and f is given by $(g \circ f)(x) = g[f(x)]$.
12. Given functions f and g
    a. $(f + g)(x) = f(x) + g(x)$
    b. $(f - g)(x) = f(x) - g(x)$
    c. $(fg)(x) = f(x)\, g(x)$
    d. $\left(\dfrac{f}{g}\right)(x) = \dfrac{f(x)}{g(x)}[g(x) \neq 0]$

13. The following are special functions.
    a. A **linear function** is any function of the form $f(x) = mx + b$.
    b. A **quadratic function** is any function of the form $f(x) = ax^2 + bx + c, a \neq 0$.
    c. A **constant function** is any function of the form $f(x) = k$, where k is a real number.
    d. A **square root function** is any function of the form $f(x) = \sqrt{P}(P \geq 0)$, where P is a polynomial in x.
    e. A **piecewise function** is any function whose graph is defined by two or more other functions.
14. A function f is said to be (a) *increasing* on an interval I if $f(a) < f(b)$ when $a < b$ and (b) *decreasing* on interval I if $f(a) > f(b)$ when $a < b$ for every a and b in I.
15. Function f is said to be (a) *even* if $f(x) = f(-x)$ and (b) *odd* if $f(-x) = -f(x)$ for all x and $-x$ in the domain of f.
16. Two variables x and y **vary directly** (are directly proportional) if $y = kx$ or $x = ky$ for some positive constant k.
17. Two variables x and y **vary inversely** (are inversely proportional) if $y = \dfrac{k}{x}$ or $x = \dfrac{k}{y}$ for some positive constant k.
18. The variable z **varies jointly** as variables x and y if $z = kxy$ for some $k > 0$.

Chapter 8 error analysis

1. Defining functions
   Example: The relation $A = \{(2,-1), (-1,-1), (3,-1), (9,-1)\}$ does not define a function.
   Correct answer: Relation A *does* define a function.
   What error was made? (*see page 351*)

2. Domain and range of a relation
   Example: Given the relation whose graph is

   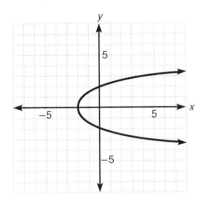

   The domain $= \{x|x \geq -2\}$
   range $= \{y|-2 \leq y \leq 2\}$
   Correct answer: domain $= \{x|x \geq -2\}$
   range $= \{y|y \in R\}$
   What error was made? (*see page 349*)

3. Operations on functions
   Example: Given $f(x) = 4x - 1$ and $g(x) = 3 - 2x$, then $(f - g)(-2) = -8$.
   Correct answer: $(f - g)(-2) = -16$
   What error was made? (*see page 360*)

4. Composition of functions
   Example: Given $f(x) = x^2 + 2$ and $g(x) = x - 7$, then $(f \circ g)(x) = x^2 + x - 5$.
   Correct answer: $(f \circ g)(x) = x^2 - 14x + 51$
   What error was made? (*see page 361*)

5. Evaluating a function
   Example: Given $f(x) = 3x - 4$, $f(x + h) = 3h - 4$
   Correct answer: $f(x + h) = 3x + 3h - 4$
   What error was made? (*see page 358*)

6. Even and odd functions
   Example: Function f defined by $f(x) = 3x^3 - x + 1$ is an odd function.
   Correct answer: f is neither even nor odd.
   What error was made? (*see page 372*)

7. Inverse functions
   Example: Given $f(x) = 5x - 2$, then $f^{-1}(x) = \dfrac{x - 2}{5}$.
   Correct answer: $f^{-1}(x) = \dfrac{x + 2}{5}$.
   What error was made? (*see page 380*)

8. One-to-one functions
   Example: The graph

   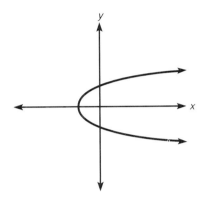

   represents a one-to-one function by the horizontal line test.
   Correct answer: Not a one-to-one function
   What error was made? (*see page 378*)

9. Squaring a complex number
   Example: $(3 - 2i)^2 = 5$
   Correct answer: $5 - 12i$
   What error was made? (*see page 272*)

10. Subtracting rational expressions
    Example: $\dfrac{3x - 2}{x + 4} - \dfrac{5x - 4}{x + 4} = \dfrac{3x - 2 - 5x - 4}{x + 4}$
    $= \dfrac{-2x - 6}{x + 4}$
    Correct answer: $\dfrac{-2x + 2}{x + 4}$
    What error was made? (*see page 167*)

Chapter 8 critical thinking

Pick any integer from 1 to 99 and double it. Then add 8 and divide that sum by 2. If you subtract the original number from this result, your answer is always 4. Why is this true?

Chapter 8 review

[8–1]

Determine whether or not each relation defines a function. If not, explain why not.

1. $\{(1,3), (2,-4), (1,0)\}$
2. $\{(4,3), (-2,3), (1,3)\}$
3. $\{(0,7), (-7,1), (6,-3)\}$
4. $\{(x,y)|2x + y = 3\}$
5. $\{(x,y)|y = 5\}$
6. $\{(x,y)|x = -7\}$

Determine the domain and range of each relation.

7. $\{(-3,4), (4,2), (0,0), (-2,7)\}$
8. $\{(x,y)|y = -4x + 1\}; x \in \{-3,-2,-1,0,1,2,3\}$
9. $\{(x,y)|y = x^2 + 2x - 1\}; x \in \{-3,0,1,4\}$
10. $\{(x,y)|y = -7\}; x \in \{-8,-2,0,3,7\}$

Determine the domain of each relation. Write the answer in interval notation.

11. $\{(x,y)|y = 2x + 7\}$
12. $\{(x,y)|y = 2x^2 + 9\}$
13. $\left\{(x,y)|y = \dfrac{-3}{x}\right\}$
14. $\left\{(x,y)|y = \dfrac{3}{3x - 4}\right\}$
15. $\{(x,y)|y = \sqrt{x + 2}\}$

[8–2]

Given $f(x) = x + 2$ and $g(x) = x^2 - 3$, find each of the following.

16. $f(-2)$
17. $f(-4)$
18. $g(0)$
19. $f(3) - g(2)$
20. $f[g(-3)]$
21. $g[f(0)]$
22. $\dfrac{f(x + h) - f(x)}{h}, h \neq 0$

For each of the pairs of functions, define $f + g, f - g, fg,$ and $\dfrac{f}{g}$. State the domain of each new function.

23. $f(x) = 7 - 2x$ and $g(x) = 5x + 9$
24. $f(x) = 3x^2 - 7$ and $g(x) = x + 2$

Given functions f and g defined by $f(x) = 7x + 3$ and $g(x) = 3x^2 - 1$, find the following.

25. $(f + g)(-5)$
26. $(f - g)(1)$
27. $(fg)(-3)$
28. $\left(\dfrac{f}{g}\right)(0)$
29. $(g - f)(-4)$
30. $\left(\dfrac{g}{f}\right)(2)$

[8–3]

Identify each function by name and sketch the graph.

31. $f(x) = 3x + 5$
32. $f(x) = -7$
33. $f(x) = x^2 - 6x + 5$
34. $f(x) = \sqrt{x + 5}$
35. $f(x) = |x| + 7$
36. $f(x) = \begin{cases} x - 1 & \text{if } x \leq 0 \\ 3 & \text{if } x > 0 \end{cases}$

Determine the intervals on which the given graphed function is increasing or decreasing.

37.

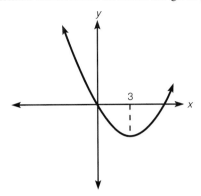

38.

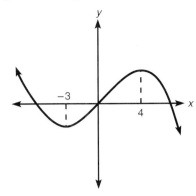

Determine if the given function is even, odd, or neither even nor odd.

39. $f(x) = 2x^2 - x + 1$ **40.** $g(x) = x^2 - 2$ **41.** $h(x) = 4x^3$

[8–4]

In each of the following, determine if the given function is one-to-one. Use the horizontal line test if necessary.

42. $f = \{(-3,1), (4,3), (2,1)\}$ **43.** $g = \{(0,2), (4,-3), (6,3)\}$ **44.** $f(x) = 5 - 4x$

45. $g(x) = x^2 + 1$ **46.** $h(x) = |2x - 3|$ **47.** $F(x) = \sqrt{x + 4}$

Find the inverse function, $f^{-1}(x)$, of each given one-to-one function.

48. $f(x) = 4x + 3$ **49.** $f(x) = x^2 - 2, x \geq 0$ **50.** $f(x) = \sqrt{2x - 3}, x \geq \dfrac{3}{2}$

Show that functions f and g are inverse functions of each other by showing that $f[g(x)] = x$ and $g[f(x)] = x$.

51. $f(x) = 5 - 2x$ and $g(x) = \dfrac{5 - x}{2}$ **52.** $f(x) = 3x^3 - 4$ and $g(x) = \sqrt[3]{\dfrac{x + 4}{3}}$

Sketch the graph of each function f and its inverse f^{-1} on the same set of coordinate axes.

53. $f(x) = 3x - 2$ **54.** $f(x) = \sqrt{4 - x}$ **55.** $f(x) = x^2 - 7, x \geq 0$ **56.** $f(x) = |x|, x \geq 0$

[8–5]

Express each statement as an equation using a constant of variation k.

57. The current I is inversely proportional to the resistance R.

58. x is directly proportional to the square of y.

59. The resistance R of a conductor is directly proportional to the length ℓ and is inversely proportional to the area A.

60. The angular momentum H of a particle varies jointly as the linear velocity v and the radius of rotation r.

Find the value of the indicated variable.

61. If y varies inversely as the square of x, find k if $y = 24$ when $x = 3$.

62. If s varies directly as w^3, find the constant of variation k if $s = 16$ when $w = 2$.

63. P varies jointly as R and I^2. If $P = 1,500$ when $R = 20$ and $I = 5$, find P when $R = 20$ and $I = 4$.

64. If R varies directly as ℓ and inversely as A, find k if $R = 0.45$ when $\ell = 5.0$ and $A = 0.055$.

65. d varies jointly as a and b and inversely as c. If $d = 15$ when $a = 3$, $b = 7$, and $c = 14$, find d when $a = 5$, $b = 8$, and $c = 20$.

Chapter 8 cumulative test

[1–4] **1.** Perform the indicated operations and simplify.
$$\frac{4(2 + 5) - 4^2 + 9}{(-2)(-7)}$$

[1–5] **2.** Given $P(x) = x^3 - 4x + 1$, find (a) $P(1)$, (b) $P(-3)$, (c) $P(0)$.

[1–5] **3.** Given $S = V_0t + \frac{1}{2}at^2$, find S when $V_0 = 20$, $t = 3, a = 32$.

Perform the indicated operations. Assume all variable exponents are positive integers. Assume all variables are nonzero.

[3–1] **4.** $(x^4y^5)^2(-x^5y^8)$

[3–3] **5.** $\frac{x^{n-1}}{x^{2-3n}}$

[3–3] **6.** $(2x - 1)^0$

[3–1] **7.** $\frac{x^{-4}y^{-5}}{x^2y^2}$

[3–3] **8.** 4^{-2}

[3–3] **9.** $\left(\frac{1}{2}\right)^{-3}$

[3–3] **10.** $\frac{4^{-6}}{4^{-4}}$

Completely factor the following expressions.

[3–7] **11.** $3a^2 - 27b^2$

[3–7] **12.** $y^4 - x^4$

[3–4] **13.** $5x(2x + 3) - 10(2x + 3)$

[3–6] **14.** $6a^2 + 7a - 3$

[3–6] **15.** $2a^2b^2 - ab - 1$

[3–7] **16.** $2x^3 - 54y^3$

[2–2] **17.** Given $ax - by = 2x + 1$, solve for x.

Find the solution set of the following equations and inequalities.

[2–1] **18.** $-3(a + 5) + 4(2a - 1) = 4$

[2–5] **19.** $-3 \leq 2x + 1 < 5$

[6–7] **20.** $x^2 - 6x - 16 \geq 0$

[4–7] **21.** $\frac{3}{y} + \frac{2}{y} = \frac{7}{3}$

[6–1] **22.** $y - \frac{3}{y} = 2$

Perform the indicated operations. Assume all denominators are nonzero. Answer with positive exponents only.

[5–2] **23.** $(a^{-3/4})^{1/2}$

[5–2] **24.** $\frac{b^{5/4}}{b}$

[5–2] **25.** $(x^4y^6)^{1/2}$

Given $f(x) = 5 - 6x$ and $g(x) = x^2 + x - 1$, find

[8–2] **26.** $f(-3)$

[8–2] **27.** $f[g(2)]$

[8–2] **28.** $g[f(-1)]$

[8–4] **29.** Given $f(x) = 4x + 3$, find $f^{-1}(x)$.

[8–4] **30.** Given $f(x) = x^3 + 1$ and $g(x) = \sqrt[3]{x - 1}$, show that f and g are inverse functions by verifying that (a) $f[g(x)] = x$, (b) $g[f(x)] = x$.

[8–2] **31.** Given $f(x) = x + 1$ and $g(x) = x^2 + 2$, define (a) $(f + g)(x)$, (b) $(f - g)(x)$, (c) $(fg)(x)$, (d) $\left(\frac{f}{g}\right)(x)$.

Identify each function by name. Sketch its graph.

[8–3] **32.** $f(x) = 7$

[8–3] **33.** $f(x) = 4x - 6$

[8–3] **34.** $f(x) = 3x^2 - 11x - 4$

[8–3] **35.** $f(x) = \sqrt{x + 9}$

[8–3] **36.** $f(x) = |x| - 3$

[8–3] **37.** $f(x) = \begin{cases} 2x - 1 & \text{if } x \leq 1 \\ x + 3 & \text{if } x > 1 \end{cases}$

[8–5] **38.** If x varies directly as y and is inversely proportional to z, find the constant of variation if $x = 10$ when $y = 4$ and $z = 10$. Find x when $y = 5$ and $z = 2$.

[8–4] **39.** Determine if the following functions f have an inverse function. If so, state the inverse f^{-1}.
(a) $f = \{(-2,1), (3,4), (0,8), (-9,2)\}$,
(b) $f(x) = x^3 - 1$, (c) $f(x) = 2x^2 + 3$

9

Conic Sections

If a gun is fired upward with an initial velocity of 288 ft/sec, the bullet's height h after t seconds is given by the equation $h = 288t - 16t^2$. Find the maximum height attained by the bullet.

In chapter 7, we graphed first-degree equations in two variables of the form $ax + by = c$, where a and b are not both 0. We called these linear equations since the graph was a straight line. In this chapter, we will consider the graphs of second-degree equations in which one or more terms are second-degree. These graphs result from slicing a cone with a plane as shown in figure 9–1. For this reason, the graphs are called the **conic sections.**

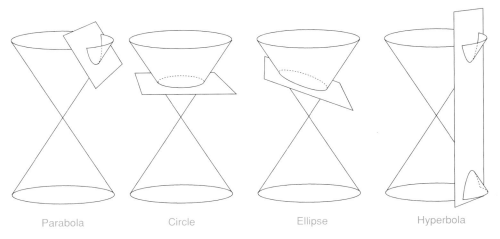

Parabola   Circle   Ellipse   Hyperbola

Figure 9–1

9–1 ■ *Quadratic functions—the parabola*

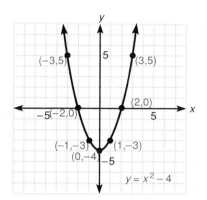

Figure 9–2

If coordinates are assigned to the cutting plane, each section can be described by a second-degree (quadratic) equation. The first of these equations we will consider is $y = ax^2 + bx + c$ $(a \neq 0)$, which graphs into a figure called a *parabola*. The equation defines the quadratic function

$$y = f(x) = ax^2 + bx + c \quad (a \neq 0)$$

which we discussed in section 8–3.

Consider the quadratic equation in two variables

$$y = x^2 - 4$$

If we choose the values for x in the set $\{-3, -2, -1, 0, 1, 2, 3\}$ and compute the corresponding values of y, the ordered pairs we obtain are

$$(-3, 5), (-2, 0), (-1, -3), (0, -4), (1, -3), (2, 0), \text{ and } (3, 5)$$

Plotting these ordered pairs and connecting the points with a smooth curve, we obtain the parabola shown in figure 9–2. This parabola is the graph of the equation $y = x^2 - 4$.

Similarly, consider the graph of the equation $y = -x^2 + 1$. Choosing values of x and then finding the corresponding values of y, we get the results shown in this table of related values.

x	-2	-1	0	1	2
y	-3	0	1	0	-3

Plot the points and connect them with a smooth curve to obtain the parabola shown in figure 9–3.

As we examine the parabolas in figures 9–2 and 9–3, we see some of the basic properties of the parabola.

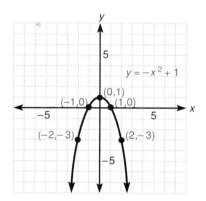

Figure 9–3

1. For each x in the domain, there is a unique value of y in the range.
2. There is a *lowest* point [$(0, -4)$ in figure 9–2] or a *highest* point [$(0, 1)$ in figure 9–3] of each graph. We call this the **vertex** of the parabola.
3. A vertical line through the vertex, in this case the y-axis, divides the parabola into two identical (mirror-image) parts. This line is called the **axis of symmetry.** For each point on the parabola to the right of the axis of symmetry, there is a corresponding point to the left of the axis of symmetry equidistant from the axis of symmetry and having the same second component.
4. When the coefficient, a, of the squared term is
   a. positive, $(a > 0)$ as in $y = x^2 - 4$ where $a = 1$, the parabola opens upward and the vertex is the lowest point of the graph;
   b. negative, $(a < 0)$ as in $y = -x^2 + 1$ where $a = -1$, the parabola opens downward and the vertex is the highest point of the graph.

── **Equation for a parabola with vertical axis of symmetry** ──
$$y = ax^2 + bx + c$$
where a, b, and c are real numbers, $a \neq 0$.

Now let us consider the graph of the general quadratic equation $y = ax^2 + bx + c$. We could choose a number of values for x and obtain the corresponding values for y as we did before. However the parabola has certain special points (when they exist) that we can use to sketch the graph when we know the general shape of the curve. These points are (1) the vertex and (2) the x- and y-intercepts. These points, together with the previously observed characteristics of the parabola, enable us to obtain a reasonably accurate sketch of the curve.

The vertex

To obtain the coordinates of the vertex, we use the procedure of completing the square that we learned in chapter 6. Completing the square in the right member, we can write the equation $y = ax^2 + bx + c$ in the form

$$y = a(x - h)^2 + k \qquad (a \neq 0)$$

where the point (h,k) is the vertex of the parabola and the line $x = h$ is the axis of symmetry. To show this is true, when $a > 0$, then $a(x - h)^2 \geq 0$ and the least value of y occurs when $a(x - h)^2 = 0$, or when $x = h$. If $x = h$, then $y = a(h - h)^2 + k = a \cdot 0 + k = k$. The lowest point (the vertex) is at (h,k). A similar argument can be made when $a < 0$, where the vertex (h,k) is the highest point.

Forms of the equation of a parabola _____

Given a parabola that opens upward when $a > 0$ and downward when $a < 0$,

1. The *general form* of the equation of the parabola is
   $y = ax^2 + bx + c$.
2. The *standard form* of the equation of the parabola is
   $y = a(x - h)^2 + k$ where the point (h,k) is the vertex of the parabola.

■ *Example 9–1 A*

Determine the coordinates of the vertex of the parabola. State whether the parabola opens upward or downward.

1. $y = (x - 2)^2 + 3$
   Since $h = 2$ and $k = 3$, the vertex is the point $(2,3)$ and since $a = 1$, the parabola opens upward.

2. $y = -3(x + 4)^2 - 6$
   The equation is not exactly in the form $y = a(x - h)^2 + k$, so we rewrite the equation as

   $$y = -3[x - (-4)]^2 + (-6)$$

   Now $h = -4$ and $k = -6$, so the vertex is the point $(-4,-6)$. Since $a = -3$, the parabola opens downward.

3. $y = x^2 + 4x - 12$
   Rewrite the equation in the form

   $$y = (x^2 + 4x + ?) - 12 - (?)$$

   Complete the square inside parentheses as we learned in section 6–2.

   $$y = (x^2 + 4x + 4) - 12 - (4)$$

Note To complete the square, we added 4. Thus, we must *subtract* 4 so that the equation is not changed.

$$y = (x + 2)^2 - 16$$

Write the trinomial as a binomial square and combine

$$y = [x - (-2)]^2 + (-16)$$

Write in the form $a(x - h)^2 + k$

Thus, $h = -2$ and $k = -16$ so the vertex is the point $(-2, -16)$. Since $a = 1$, the parabola opens upward.

▶ *Quick check* Determine the coordinates of the vertex of $y = x^2 + 2x - 8$. ∎

The x-intercept(s)

There can be two, one, or no x-intercepts in the graph of the quadratic equation $y = ax^2 + bx + c$. The x-intercept(s) (if there are any) are found by replacing y with 0 and solving for x. This is the same as we did with linear equations.

The y-intercept

There will always be *one* y-intercept in the graph of $y = ax^2 + bx + c$. As with linear equations, let $x = 0$ and solve for y. Given

$$\begin{aligned} y &= ax^2 + bx + c \\ &= a(0)^2 + b(0) + c \\ &= c \end{aligned}$$

Replace x with 0

The y-intercept of the quadratic equation in general form $y = ax^2 + bx + c$ is always the point $(0, c)$.

∎ *Example 9–1 B*

Find the x- and y-intercepts of the parabola $y = x^2 + 2x - 8$.

1. Let $y = 0$.

$$\begin{aligned} (0) &= x^2 + 2x - 8 \\ 0 &= (x + 4)(x - 2) \end{aligned}$$

Replace y with 0

Factor the right member

$$x + 4 = 0 \quad \text{or} \quad x - 2 = 0$$

Set each factor equal to 0

$$x = -4 \qquad\qquad x = 2$$

Solve for x

The x-intercepts are the points $(-4, 0)$ and $(2, 0)$.

2. Since $c = -8$, the y-intercept is the point $(0, -8)$.

▶ *Quick check* Determine the x- and y-intercepts of the parabola $y = x^2 + 7x - 8$. ∎

We now outline the procedure for graphing a quadratic equation of the form $y = ax^2 + bx + c, a \neq 0$.

To graph the quadratic equation $y = ax^2 + bx + c, a \neq 0$

1. Determine whether the parabola opens upward $(a > 0)$ or downward $(a < 0)$.
2. Find the x-intercept(s).
3. Find the y-intercept.
4. Determine the coordinates of the vertex by completing the square to get the equation in the standard form $y = a(x - h)^2 + k$. The vertex is the point (h,k).
5. Find and plot other points to the right and left of the axis of symmetry, $x = h$, as needed.

■ **Example 9–1 C**

Sketch the graphs of the following equations using the vertex and the x- and y-intercepts.

1. $y = x^2 - 4x - 5$
   a. Since $a = 1$, which is positive, the parabola opens upward.
   b. Let $y = 0$, then

   $$(0) = x^2 - 4x - 5 \qquad \text{Replace } y \text{ with } 0$$
   $$0 = (x - 5)(x + 1) \qquad \text{Factor the right member}$$
   $$x - 5 = 0 \quad \text{or} \quad x + 1 = 0 \qquad \text{Set each factor equal to } 0$$
   $$x = 5 \qquad\qquad x = -1$$

   The x-intercepts are $(5,0)$ and $(-1,0)$.
   c. Since $c = -5$, the y-intercept is $(0,-5)$.
   d. Given $y = x^2 - 4x - 5$
   $$= (x^2 - 4x \quad) - 5 - (\quad)$$
   $$= (x^2 - 4x + 4) - 5 - (4) \qquad \text{Complete the square}$$
   $$= (x - 2)^2 - 9 \qquad \text{Write binomial square}$$
   $$= (x - 2)^2 + (-9) \qquad \text{Write in standard form } y = a(x - h)^2 + k$$

   The vertex is the point $(2,-9)$.
   e. Using the y-intercept $(0,-5)$ and the axis of symmetry $x = 2$, choose point $(4,-5)$ that is the same distance from the axis of symmetry (two units) and has the same second component, -5.

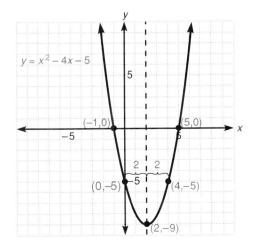

2. $y = -x^2 + 2x + 3$

a. Since $a = -1$ (negative), the parabola opens downward.

b. Let $y = 0$.

$$
\begin{aligned}
(0) &= -x^2 + 2x + 3 && \text{Replace } y \text{ with } 0 \\
&= x^2 - 2x - 3 && \text{Multiply each member by } -1 \\
&= (x - 3)(x + 1) && \text{Factor the right member}
\end{aligned}
$$

$$x - 3 = 0 \quad \text{or} \quad x + 1 = 0 \qquad \text{Set each factor equal to 0}$$
$$x = 3 \qquad\qquad x = -1$$

The x-intercepts are $(3,0)$ and $(-1,0)$.

c. Since $c = 3$, the y-intercept is $(0,3)$.

d. Given $y = -x^2 + 2x + 3$

$$
\begin{aligned}
&= -1(x^2 - 2x \quad) + 3 && \text{Factor } -1 \text{ to obtain } x^2 \\
&= -1(x^2 - 2x + 1) + 3 + 1 && \text{Complete the square} \\
&= -1(x - 1)^2 + 4
\end{aligned}
$$

The vertex is the point $(1,4)$.

e. The point $(2,3)$ is symmetrical to the y-intercept $(0,3)$ about the axis of symmetry $x = 1$, since it is the same distance from $x = 1$ as the y-intercept and has the same second component.

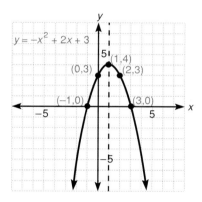

Let us now use the method of completing the square to write the equation $y = ax^2 + bx + c$ in the form $y = a(x - h)^2 + k$ to determine the coordinates of the vertex.

$$
\begin{aligned}
y &= ax^2 + bx + c \\
&= a\left(x^2 + \frac{b}{a}x \quad\right) + c && \text{Factor } a \text{ from the first two terms} \\
&= a\left(x^2 + \frac{b}{a}x + \frac{b^2}{4a^2}\right) + c - \frac{b^2}{4a} && \text{Complete the square} \\
&= a\left(x + \frac{b}{2a}\right)^2 + c - \frac{b^2}{4a} && \text{Factor the trinomial} \\
&= a\left[x - \left(-\frac{b}{2a}\right)\right]^2 + \left(c - \frac{b^2}{4a}\right)
\end{aligned}
$$

The vertex is the point with coordinates $\left(-\dfrac{b}{2a}, c - \dfrac{b^2}{4a}\right)$.

An alternate way of finding the vertex of a parabola is to

1. let $x = -\dfrac{b}{2a}$ to determine the x-value of the vertex and the equation of the axis of symmetry,

2. replace x with this value in the original equation and solve for y to find the y-value of the vertex.

This method is demonstrated in example 3.

3. $y = -2x^2 - 4x - 3$

a. Since $a = -2$ (negative), the parabola opens downward.

b. Let $y = 0$

$$(0) = -2x^2 - 4x - 3 \qquad \text{Replace } y \text{ with } 0$$
$$0 = 2x^2 + 4x + 3 \qquad \text{Multiply each member by } -1$$

We cannot factor $2x^2 + 4x + 3$, we use the quadratic formula to solve for x.

$$x = \frac{-b \pm \sqrt{b^2 - 4ac}}{2a} = \frac{-4 \pm \sqrt{(4)^2 - 4(2)(3)}}{2(2)} \qquad \begin{array}{l}\text{Replace } a \text{ with } 2,\\ b \text{ with } 4, \text{ and}\\ c \text{ with } 3\end{array}$$
$$= \frac{-4 \pm \sqrt{-8}}{4} \qquad \text{Simplify}$$

Since $\sqrt{-8}$ is not a real number and the rectangular coordinate system graphs ordered pairs of *real numbers,* there are no x-intercepts and the graph does not cross the x-axis.

c. Since $c = -3$, the y-intercept is $(0, -3)$.

d. Since $a = -2$ and $b = -4$,

$$(1) \quad x = -\frac{b}{2a} = -\frac{(-4)}{2(-2)} = -1 \qquad \text{Replace } a \text{ with } -2 \text{ and } b \text{ with } -4$$
$$(2) \quad y = -2(-1)^2 - 4(-1) - 3 \qquad \text{Replace } x \text{ with } -1$$
$$= -2 + 4 - 3 \qquad \text{Simplify}$$
$$= -1$$

The vertex is the point $(-1, -1)$.

e. The point $(-2, -3)$ is symmetrical to the y-intercept $(0, -3)$ about the axis of symmetry $x = -1$.

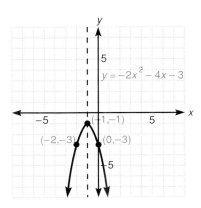

▶ **Quick check** Sketch the graph of $y = x^2 + 4x + 3$.

When considering the quadratic equation $y = ax^2 + bx + c$, we see that the graph has a *low point* (or *minimum value of y*) when $a > 0$ and a *high point* (or *maximum value of y*) when $a < 0$. We can use these values of y to answer questions in a physical situation.

■ *Example 9–1 D*

1. A projectile is fired upward from the ground so that its distance s in feet above the ground t seconds after firing is given by $s = -16t^2 + 96t$. Find the maximum height it will reach and the number of seconds it takes to reach that height.

   Since $a = -16$, the graph will have a high point, or maximum value of s. We must find the vertex. Using $t = -\dfrac{b}{2a}$, where $a = -16$, and $b = 96$,

   $$t = -\frac{(96)}{2(-16)} = \frac{96}{32} = 3 \qquad \text{Replace } a \text{ with } -16 \text{ and } b \text{ with } 96$$

   $$\text{then} \quad s = -16(3)^2 + 96(3) \qquad \text{Replace } t \text{ with } 3$$
   $$= -144 + 288 = 144$$

   The vertex of the parabola is (3,144) and the projectile will reach the maximum height s of 144 feet in $t = 3$ seconds after firing.

2. Harry has 48 feet of fencing to use to fence off a rectangular area behind his house for his dog Nappy. What are the dimensions of the largest area that he can fence off with his 48 feet of fencing? (*Note:* One side of the fenced area is the house.)

   Let x = the length of one of the equal sides. Then $48 - 2x$ = the length of the side opposite the house. (Harry has 48 feet of fencing and the other two sides are x in length.) Using area (A) = length (ℓ) · width (w), where width (w) = x and length (ℓ) = $48 - 2x$, we have the equation

   $$A = x(48 - 2x)$$
   $$= -2x^2 + 48x$$

   Completing the square in the right member,

   $$A = -2x^2 + 48x$$
   $$= -2(x^2 - 24x + \quad) + (\quad) \qquad \text{Factor } -2 \text{ from each term}$$
   $$= -2(x^2 - 24x + 144) + (288) \qquad \text{Complete the square}$$
   $$= -2(x - 12)^2 + 288$$

   The vertex is the point (12,288) and the maximum area is 288 square feet when $x = 12$. The dimensions of the rectangle are 12 feet by $48 - 2(12) = 24$ feet. ■

Mastery points

Can you

- Find the *x*- and *y*-intercepts of a quadratic equation in two variables?
- Find the coordinates of the vertex of the parabola?
- Sketch the graph of a quadratic equation in two variables?
- Find the maximum and minimum values of a quadratic equation of the form $y = ax^2 + bx + c$?

Exercise 9–1

Find the coordinates of the vertex of the parabola for the graph of each quadratic equation. See example 9–1 A.

Example $y = x^2 + 2x - 8$

Solution Given $y = x^2 + 2x - 8$

$$y = (x^2 + 2x + \quad) - 8 - (\quad)$$
$$y = (x^2 + 2x + 1) - 8 - (1) \qquad \text{Complete the square}$$
$$y = (x + 1)^2 - 9 \qquad \text{Write the square of a binomial and combine}$$
$$y = [x - (-1)]^2 + (-9) \qquad \text{Write in the form } y = a(x - h)^2 + k$$

The vertex is the point $(-1, -9)$.

1. $y = (x - 3)^2 + 4$
2. $y = 3(x + 1)^2 - 7$
3. $y = x^2 - 16$
4. $y = x^2 + 4$
5. $y = (x - 5)^2$
6. $y = (x + 6)^2$
7. $y = x^2 + 4x - 5$
8. $y = x^2 - 6x + 5$
9. $y = -x^2 + 2x + 3$
10. $y = -x^2 + 3x - 2$
11. $y = 2x^2 - 7x + 3$
12. $y = -5x^2 + 6x - 1$

Find the x- and y-intercepts of the following parabolas. See example 9–1 B.

Example $y = x^2 + 7x - 8$

Solution 1. Let $y = 0$.
$$(0) = x^2 + 7x - 8 \qquad \text{Replace } y \text{ with } 0$$
$$= (x + 7)(x - 1) \qquad \text{Factor in right member}$$
$$x + 7 = 0 \quad \text{or} \quad x - 1 = 0 \qquad \text{Set each factor equal to 0}$$
$$x = -7 \qquad\qquad x = 1$$

The x-intercepts are the points $(-7,0)$ and $(1,0)$.
2. Since $c = -8$, the y-intercept is the point $(0,-8)$.

13. $y = (x - 3)^2 + 4$
14. $y = 3(x + 1)^2 - 7$
15. $y = x^2 - 16$
16. $y = x^2 + 4$
17. $y = (x - 5)^2$
18. $y = (x + 6)^2$
19. $y = x^2 + 4x - 5$
20. $y = x^2 - 6x + 5$
21. $y = -x^2 + 2x + 3$
22. $y = -x^2 + 3x - 2$
23. $y = 2x^2 - 7x + 3$
24. $y = -5x^2 - 6x - 1$

Sketch the graph of each given quadratic equation using the intercepts and the coordinates of the vertex where possible. See example 9–1 C.

Example $y = x^2 + 4x + 3$

Solution 1. Since $a = 1$ (positive), the parabola opens upward.
2. Let $y = 0$.
$$(0) = x^2 + 4x + 3 \qquad \text{Replace } y \text{ with } 0$$
$$= (x + 3)(x + 1) \qquad \text{Factor the right member}$$
$$x + 3 = 0 \quad \text{or} \quad x + 1 = 0 \qquad \text{Set each factor equal to 0}$$
$$x = -3 \qquad\qquad x = -1$$

The x-intercepts are $(-3,0)$ and $(-1,0)$.
3. Since $c = 3$, the y-intercept is $(0,3)$.

4. Given $y = x^2 + 4x + 3$

$$y = (x^2 + 4x \quad) + 3 - (\quad)$$
$$y = (x^2 + 4x + 4) + 3 - (4) \qquad \text{Complete the square and balance}$$
$$y = (x + 2)^2 - 1 \qquad \text{Write as the square of a binomial and combine}$$
$$y = [x - (-2)]^2 + (-1) \qquad \text{Write in the form } y = a(x - h)^2 + k$$

The vertex is the point $(-2, -1)$.

5. The point $(-4, 3)$ is symmetrical to the y-intercept $(0, 3)$ about the axis of symmetry, the vertical line $x = -2$.

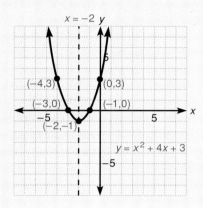

25. $y = x^2 - 4$

26. $y = -x^2 + 9$

27. $y = x^2 - 2x$

28. $y = 3x^2 + 2x$

29. $y = x^2 - 4x - 5$

30. $y = x^2 + 6x + 5$

31. $y = x^2 - x - 6$

32. $y = x^2 + 3x + 4$

33. $y = -x^2 - 2x + 3$

34. $y = -x^2 + 7x - 6$

35. $y = x^2 - 4x + 4$

36. $y = x^2 + 2x + 1$

37. $y = 2x^2 - x - 3$

38. $y = 3x^2 + 4x + 1$

39. $y = -2x^2 + 3x - 1$

40. $y = -4x^2 + 4x + 3$

41. $y = x^2 + 2x + 5$

42. $y = x^2 - 3x + 6$

43. $y = 2x^2 - 4x + 5$

44. $y = 3x^2 + 6x + 4$

45. $y = -2x^2 + 4x + 5$

46. $y = -4x^2 + 12x - 9$

47. $y = -x^2 - 2x - 1$

48. $y = x^2 - 6x + 4$

See example 9–1 D–1.

49. The equation $s = 32t - 16t^2$ determines the distance s in feet that an object thrown vertically upward is above the ground in time t seconds. Find the time at which the object reaches its greatest height. Sketch the graph of the equation for $0 \leq t \leq 2$. (*Hint:* The greatest height is at the vertex.) When will the object hit the ground again? (*Hint:* When $s = 0$)

50. Given the equation $s = 64t - 8t^2$, do as instructed in exercise 49.

51. An arrow is shot vertically into the air with an initial velocity of 96 feet per second. If the height h in feet of the arrow at any time t seconds is given by

$$h = 96t - 16t^2$$

find the maximum height that the arrow will attain. When will the arrow come back to the ground?

52. A company's profit P when producing x units of a commodity in a given week is given by $P = -x^2 + 100x - 1,000$. How many units must be produced to attain maximum profit? (*Hint:* The vertex point will give the maximum profit.)

53. Helen Nance owns a dress shop. She finds the profits from the shop are approximately given by

$$P = -x^2 + 16x + 42$$

where P is the profit when x dresses are sold daily. How many dresses must Helen sell daily to produce the maximum profit?

54. Russell Sanderson owns a pizza shop. From past results, he determines that the cost C of running the shop is given by the equation

$$C = 3n^2 - 36n + 140$$

where n is the number of pizzas sold daily. Find the number of pizzas Russell must sell to produce the lowest cost.

55. Tim Wesner sets up a Kool-Aid stand. He finds that the cost C of operating the stand is given by the equation

$$C = 2x^2 - 20x + 100$$

where x denotes the number of glasses of Kool-Aid sold. How many glasses of Kool-Aid should he sell for his cost to be the lowest?

56. The velocity distribution of natural gas flowing smoothly in a pipeline is given by

$$V = 6x - x^2$$

where V is the velocity in meters per second and x is the distance in meters from the inside wall of the pipe. What is the maximum velocity of the gas? (*Hint:* We want the second component V of the vertex of the graph of the equation.)

57. The power output P of an automobile alternator that generates 14 volts and has an internal resistance of 0.20 ohms is given by

$$P = 14I - 0.20I^2$$

At what current I does the generator generate maximum power and what is the maximum power?

See example 9-1 D-2.

58. A farmer wishes to fence in a rectangular piece of ground along a river bank for grazing his cattle. If he has 400 feet of fencing, what should be the dimensions of the rectangle to fence in the greatest area, using the river as one side of the rectangle?

59. Find the two numbers whose sum is 56 and whose product is the greatest possible value. (*Hint:* Let x be one number and $56 - x$ be the other number. Maximize their product.)

60. The sum of the length and the width of a rectangle is 48 inches. Find the length of the rectangle that would yield the greatest area. (*Hint:* Let $x =$ the length and $48 - x =$ the width. Use the formula $A = \ell w$.)

61. What are the dimensions of the largest rectangular plot of ground that can be enclosed by 42 meters of fence? (*Hint:* $2\ell + 2w = 42$. Solve for either ℓ or w and substitute in the formula $A = \ell w$. Maximize A.)

62. Find the maximum area of a rectangular plot of ground that can be enclosed with 28 rods of fencing.

63. Given that the difference between two numbers is 8, what is the minimum product of the two numbers? (*Hint:* Let x be one number and $x - 8$ be the other number.)

64. Find the minimum product of two numbers whose difference is 10.

Review exercises

Factor the following expressions. See section 3-8.

1. $9y^2 - 6y + 1$ **2.** $4x^2 + 4x - 3$ **3.** $6y^2 - 24x^2$

Perform the indicated operations on rational expressions. See sections 4-2 and 4-3.

4. $\dfrac{4y}{y^2 - 25} - \dfrac{y}{y + 5}$ **5.** $\dfrac{16x^2}{7y} \div \dfrac{4x}{21y^2}$

6. Simplify the complex rational expression $\dfrac{\dfrac{x}{x + 4}}{\dfrac{x}{x - 3}}$. See section 4-4.

9–2 ■ *Quadratic relations—more about parabolas*

If we interchange the variables of the quadratic equation $y = ax^2 + bx + c$, we obtain another quadratic equation of the form

$$x = ay^2 + by + c \quad (a \neq 0)$$

The graph of this equation is also a parabola. However, these parabolas

1. open right when $a > 0$ (positive),
2. open left when $a < 0$ (negative).

The vertex will be the point with the least, or greatest, value of x.

We graph this equation in the same way, using the vertex and the intercepts. In this case, to find the coordinates of the vertex, we can write the equation in the form

$$x = a(y - k)^2 + h$$

and the point (h,k) is the vertex of the parabola. The equation of the axis of symmetry is $y = k$. Alternately, we can find the coordinates of the vertex by using

$$y = -\frac{b}{2a}$$

and substituting for y in $x = ay^2 + by + c$ to find the corresponding value of x in the vertex.

■ *Example 9–2 A*

1. Determine the vertex, the x- and y-intercepts, and graph the equation $x = y^2 + 4y - 5$.
   a. Since $a = 1$ (positive), the parabola opens right.
   b. The x-intercept is $(-5,0)$ since $c = -5$.
   c. Let $x = 0$.

   $$(0) = y^2 + 4y - 5 \qquad \text{Replace } x \text{ with } 0$$
   $$0 = (y + 5)(y - 1) \qquad \text{Factor the right member}$$
   $$y + 5 = 0 \quad \text{or} \quad y - 1 = 0 \qquad \text{Set each factor equal to } 0$$
   $$y = -5 \qquad\qquad y = 1$$

   The y-intercepts are the points $(0,-5)$ and $(0,1)$.

   d. Using $y = -\dfrac{b}{2a}$ where $a = 1$ and $b = 4$,

   $$y = -\frac{(4)}{2(1)} = -2 \qquad \text{Replace } a \text{ with 1 and } b \text{ with 4}$$

   Then

   $$x = (-2)^2 + 4(-2) - 5 \qquad \text{Replace } y \text{ with } -2$$
   $$= 4 - 8 - 5$$
   $$= -9$$

   The vertex is the point $(-9,-2)$. The axis of symmetry is $y = -2$.
   e. The point $(-5,-4)$ is symmetric to the x-intercept about the axis of symmetry, $y = -2$, since it has the same first component, -5, and is the same distance, 2 units, below $y = -2$ as the x-intercept.

2. Graph the quadratic equation $x = -y^2 + 3y + 4$.

   a. The parabola opens left since $a < 0$.

   b. Since $c = 4$, the x-intercept is the point $(4,0)$.

   c. Let $x = 0$.

   $$(0) = -y^2 + 3y + 4 \qquad \text{Replace } x \text{ with } 0$$
   $$0 = y^2 - 3y - 4 \qquad \text{Multiply by } -1$$
   $$0 = (y - 4)(y + 1) \qquad \text{Factor the right member}$$
   $$y - 4 = 0 \quad \text{or} \quad y + 1 = 0 \qquad \text{Set each factor equal to 0}$$
   $$y = 4 \qquad\qquad y = -1$$

   The y-intercepts are the points $(0,4)$ and $(0,-1)$.

   d. Using $y = -\dfrac{b}{2a}$ where $a = -1$ and $b = 3$,

   $$y = -\frac{(3)}{2(-1)} = \frac{3}{2} \qquad \text{Replace } a \text{ with } -1 \text{ and } b \text{ with } 3$$

   Then

   $$x = -\left(\frac{3}{2}\right)^2 + 3\left(\frac{3}{2}\right) + 4 \qquad \text{Replace } y \text{ with } \frac{3}{2}$$
   $$= -\frac{9}{4} + \frac{9}{2} + 4$$
   $$= \frac{25}{4}$$

   The vertex is the point $\left(\dfrac{25}{4}, \dfrac{3}{2}\right)$. The axis of symmetry is $y = \dfrac{3}{2}$.

   e. The point $(4,3)$ is symmetric to the x-intercept $(4,0)$ about the axis of symmetry since it has the same first component, 4, and is the same distance above $y = \dfrac{3}{2}$, $1\dfrac{1}{2}$ units, as the x-intercept is below.

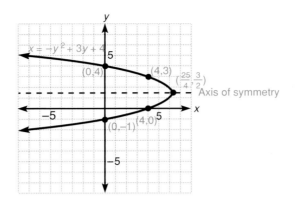

Note The quadratic equation $x = ay^2 + by + c$ *does not define a function. Why not?*

--- *Forms of the equation of a parabola* ---

Given the parabola opens right when $a > 0$ and opens left when $a < 0$,

1. The *general form* of the equation of the parabola is
$$x = ay^2 + by + c$$
2. The *standard form* of the equation of the parabola is
$$x = a(y - k)^2 + h$$

▶ **Quick check**   Graph the quadratic equation $x = -y^2 - 2y + 3$.   ■

--- **Mastery points** ---

Can you
- Find the coordinates of the vertex of the graph of a parabola of the form $x = ay^2 + by + c$?
- Find the x- and y-intercepts of the graph of a parabola of the form $x = ay^2 + by + c$?
- Graph a general quadratic equation of the form $x = ay^2 + by + c$?

Exercise 9–2

Find the coordinates of the vertex and the x- and y-intercepts of the following parabolas. See example 9–2 A.

Example $x = y^2 + y - 2$

Solution 1. Using $y = -\dfrac{b}{2a}$ where $a = 1$ and $b = 1$,

$$y = -\frac{(1)}{2(1)} = -\frac{1}{2}$$ Replace *a* with 1 and *b* with 1

Then

$$x = \left(-\frac{1}{2}\right)^2 + \left(-\frac{1}{2}\right) - 2 \qquad \text{Replace } y \text{ with } -\frac{1}{2}$$

$$= \frac{1}{4} - \frac{1}{2} - 2$$

$$= -\frac{1}{4} - 2$$

$$= -\frac{9}{4}$$

The vertex is $\left(-\frac{9}{4}, -\frac{1}{2}\right)$.

2. Since $c = -2$, the x-intercept is the point $(-2, 0)$.
3. Set $x = 0$ to find the y-intercept.

$$(0) = y^2 + y - 2 \qquad \text{Replace } x \text{ with } 0$$
$$0 = (y + 2)(y - 1) \qquad \text{Factor the right member}$$
$$y + 2 = 0 \quad \text{or} \quad y - 1 = 0 \qquad \text{Set each factor equal to 0}$$
$$y = -2 \qquad\qquad y = 1$$

The y-intercepts are $(0, -2)$ and $(0, 1)$.

1. $x = (y - 3)^2 + 2$
2. $x = (y + 4)^2 - 3$
3. $x = 3(y + 5)^2$
4. $x = -1\left(y - \frac{1}{2}\right)^2$
5. $x = y^2 - 2y - 3$
6. $x = y^2 + 4y - 5$
7. $x = -y^2 - y + 2$
8. $x = -y^2 + 2y - 1$
9. $x = 4y^2 + 2y - 3$
10. $x = 2y^2 - 5y + 3$
11. $x = -5y^2 - 4y + 1$
12. $x = -3y^2 - 5y + 2$

Graph the following parabolas. State the equation of the axis of symmetry. See example 9–2 A.

Example $x = -y^2 - 2y + 3$

Solution   1. Since $a = -1$ (negative), the graph opens to the left.
2. Since $c = 3$, the x-intercept is the point $(3, 0)$.
3. Let $x = 0$.

$$(0) = -y^2 - 2y + 3 \qquad \text{Replace } x \text{ with } 0$$
$$0 = y^2 + 2y - 3 \qquad \text{Multiply each member by } -1$$
$$0 = (y + 3)(y - 1) \qquad \text{Factor the right member}$$
$$y + 3 = 0 \quad \text{or} \quad y - 1 = 0 \qquad \text{Set each factor equal to 0}$$
$$y = -3 \qquad\qquad y = 1 \qquad \text{Solve for } y$$

The y-intercepts are the points $(0, -3)$ and $(0, 1)$.
4. Given $x = -y^2 - 2y + 3$

$$x = -1(y^2 + 2y \quad) + 3 - (\ ) \qquad \text{Factor } -1 \text{ from } -y^2 - 2y$$
$$x = -1(y^2 + 2y + 1) + 3 - (-1) \qquad \text{Complete the square}$$
$$x = -1(y + 1)^2 + 4 \qquad \text{Write as the square of a binomial and combine}$$
$$x = -1[y - (-1)]^2 + 4 \qquad \text{Write in form } x = a(y - k)^2 + h$$

The vertex is the point $(4, -1)$ and the axis of symmetry is $y = -1$.
5. The point $(3, -2)$ is symmetrical to the x-intercept, $(3, 0)$, with respect to the axis of symmetry, $y = -1$, since it has the same first component, 3, and is the same distance below $y = -1$, 1 unit, as the x-intercept is above.

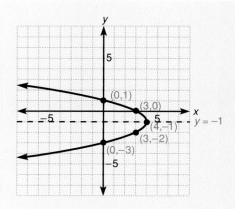

13. $x = 3(y - 1)^2 + 2$

14. $x = 2(y + 3)^2 - 1$

15. $x = -2(y - 3)^2 - 2$

16. $x = -1(y + 2)^2 + 1$

17. $x = y^2 - 3y + 2$

18. $x = y^2 - 2y - 8$

19. $x = -y^2 + 2y + 3$

20. $x = 3y^2 - y - 6$

21. $x = 5y^2 + 2y - 3$

22. $x = -2y^2 - 5y + 3$

23. $x = y^2 - 2$

24. $x = -y^2 + 4$

25. $x = -9y^2$

26. $x = 2y^2 - 3y$

27. $x = 4y^2 - y$

28. $x = y^2$

29. $x = -y^2 + 6y$

30. $x = 5y^2$

31. $x = -y^2$

Review exercises

Simplify the following expressions. Assume all variables are nonzero. Write the answers with positive exponents only. See section 3–3.

1. $(2a^{-2}b^3c^0)^{-3}$

2. $\left[\dfrac{8x^{-5}}{16x^3}\right]^3$

Find the x- and y-intercepts of the following equations. See section 7–1.

3. $x + 2y = 4$

4. $4x + 4y = 12$

Find the distance between the following sets of points. See section 7–2.

5. $(-3,2)$ and $(1,7)$

6. $(6,-5)$ and (x,y)

7. (x,y) and (h,k)

9–3 ■ The circle

In sections 9–1 and 9–2, we discussed second-degree equations having just one second-degree term. Now we will consider second-degree equations that contain two second-degree terms, x^2 and y^2. The circle is one of the conic sections that contains these terms in the equation.

┌─ **Definition of a circle** ──────────────────────────

A **circle** is defined to be the set of all points in a plane that are at a fixed distance from a fixed point.

The fixed point is called the *center* of the circle, denoted by C, and the fixed distance from the center to each point on the circle is called the *radius* of the circle, denoted by r. See figure 9-4.

Recall the distance formula we discussed in chapter 7.

$$\sqrt{(x_2 - x_1)^2 + (y_2 - y_1)^2} = d$$

If we square each member of this equation, we have

$$(x_2 - x_1)^2 + (y_2 - y_1)^2 = d^2$$

Replace the point (x_2, y_2) with the point on the circle $P(x, y)$, the point (x_1, y_1) with the center $C(h, k)$, and the distance d with the radius r to obtain the *standard form* of the equation of the circle in figure 9-4.

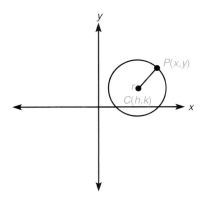

Figure 9–4

Equation of a circle

The standard form of the equation of a circle is

$$(x - h)^2 + (y - k)^2 = r^2$$

where (h, k) is the center of the circle and r is the length of the radius.

■ *Example 9–3 A*

Find the equation for the circle with center at $C(3,2)$ and radius $r = 4$ units.

We are given $h = 3$, $k = 2$, and $r = 4$.

$$(x - h)^2 + (y - k)^2 = r^2 \qquad \text{Standard form}$$
$$[x - (3)]^2 + [y - (2)]^2 = (4)^2 \qquad \text{Replace } h \text{ with 3, } k \text{ with 2, and } r \text{ with 4}$$
$$(x - 3)^2 + (y - 2)^2 = 16$$

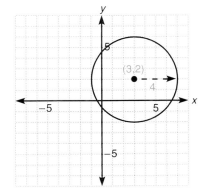

▶ *Quick check* Find the standard equation of the circle with center at $C(4, -3)$ and radius $r = 5$ units. ∎

A special case exists when the center is at the origin. Then

$$(h,k) = (0,0)$$

and the equation of the circle is given by

$$(x - 0)^2 + (y - 0)^2 = r^2$$
$$x^2 + y^2 = r^2$$

Equation of a circle with center at origin ──────────

The equation of the circle with the center at the origin, $C(0,0)$, and radius r is

$$x^2 + y^2 = r^2$$

■ *Example 9–3 B*

Find the equation of the circle with center $C(0,0)$ and radius $r = 5$. Graph the circle.

Using $x^2 + y^2 = r^2$, we substitute to obtain

$x^2 + y^2 = (5)^2$ Replace r with 5
$x^2 + y^2 = 25$ $5^2 = 25$

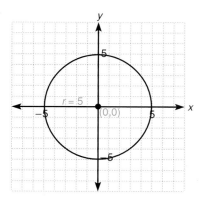

∎

The equation of a circle can also be stated in the form

$$ax^2 + ay^2 + bx + cy + d = 0$$

We call this the *general form* of the equation of a circle. When the equation of a circle is of this form, the coordinates of the center and the length of the radius are found as illustrated in the following examples.

Note In the equation of a circle, *both* squared terms must have the same coefficient.

■ *Example 9–3 C*

Determine the coordinates of the center and the length of the radius of the given circle. Sketch the graph.

1. $x^2 + y^2 - 10x - 8y - 23 = 0$

We first rewrite this into the form $(x - h)^2 + (y - k)^2 = r^2$ by completing the square in both x and y.

$$(x^2 - 10x) + (y^2 - 8y) - 23 = 0$$ Group terms having same variables

$$(x^2 - 10x + \quad) + (y^2 - 8y + \quad) = 23 + (\quad) + (\quad)$$ Add 23 to each member

Completing the square in both x and y (adding same quantities to both members), we have

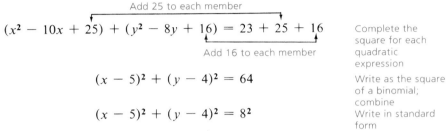

$$(x^2 - 10x + 25) + (y^2 - 8y + 16) = 23 + 25 + 16$$ Complete the square for each quadratic expression

Add 16 to each member

$$(x - 5)^2 + (y - 4)^2 = 64$$ Write as the square of a binomial; combine

$$(x - 5)^2 + (y - 4)^2 = 8^2$$ Write in standard form

The circle has center $C(5,4)$ and radius $r = 8$.
Plot the center $(5,4)$ and draw a circle that has a radius of 8 units.

2. $3x^2 + 3y^2 - 18x + 12y + 27 = 0$

To complete the square, we must make the coefficients of the squared terms 1. To accomplish this, divide *all* terms by 3 to obtain

$$x^2 + y^2 - 6x + 4y + 9 = 0$$ Divide each term by 3
$$(x^2 - 6x) + (y^2 + 4y) + 9 = 0$$ Group terms having same variable
$$(x^2 - 6x) + (y^2 + 4y) = -9$$ Subtract 9 from each member

Completing the square and adding the same quantities to each member of the equation,

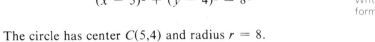

$$(x^2 - 6x + 9) + (y^2 + 4y + 4) = -9 + 9 + 4$$ Complete the square for each quadratic expression

Add 4 to each member

$$(x - 3)^2 + (y + 2)^2 = 4$$ Write as the square of a binomial; combine

$$(x - 3)^2 + (y + 2)^2 = 2^2$$

$$(x - 3)^2 + [y - (-2)]^2 = 2^2$$ Write in standard form

Therefore, the circle has center $C(3,-2)$ and radius $r = 2$.
Plot the center $(3,-2)$ and draw a circle that has a radius of 2 units.

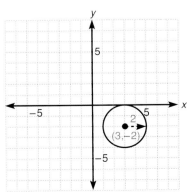

3. Given the circle with center $C(4,2)$ and radius $r = 6$ units, write the equation of the circle in *general form* $x^2 + y^2 + bx + cy + d = 0$.

Given $C(4,2)$ and $r = 6$, using the standard form $(x - h)^2 + (y - k)^2 = r^2$,

$$[x - (4)]^2 + [y - (2)]^2 = (6)^2 \qquad \text{Replace } h \text{ with 4, } k \text{ with 2, and } r \text{ with 6}$$
$$x^2 - 8x + 16 + y^2 - 4y + 4 = 36 \qquad \text{Perform indicated operations}$$
$$x^2 + y^2 - 8x - 4y + 20 = 36 \qquad \text{Reorder and combine}$$
$$x^2 + y^2 - 8x - 4y - 16 = 0 \qquad \text{Subtract 36 from each member}$$

▶ *Quick check*   Determine the center and radius of the circle $x^2 + y^2 + 2x - 4y - 4 = 0$. ■

Mastery points

Can you
- Write the equation of the circle given the coordinates of its center and radius?
- Find the coordinates of the center and the length of the radius of a circle when given the equation?
- Given the general form of the equation of a circle, find the standard form of the equation of the circle?
- Graph the equation of a circle?

Exercise 9–3

Write the standard equation for each of the following circles. See example 9–3 A.

Example   Center at $C(4,-3)$ and radius $r = 5$ units

Solution   We are given $h = 4$, $k = -3$, and $r = 5$.
Using $(x - h)^2 + (y - k)^2 = r^2$,
$$[x - (4)]^2 + [y - (-3)]^2 = (5)^2 \qquad \text{Replace } h \text{ with 4, } k \text{ with } -3, \text{ and } r \text{ with 5}$$
$$(x - 4)^2 + (y + 3)^2 = 25$$

1. Center $C(1,2)$ and radius 2
2. Center $C(-5,7)$ and radius 4
3. Center $C(4,-3)$ and radius $\sqrt{6}$
4. Center $C(-4,-5)$ and radius $2\sqrt{2}$
5. Center at the origin and radius 3
6. Center at the origin and radius $\sqrt{5}$

Write the equation in the form $x^2 + y^2 + bx + cy + d = 0$, where b, c, and d are integers.

Example   Center $C(3,-2)$ and radius $r = 4$

Solution   Using the standard form $(x - h)^2 + (y - k)^2 = r^2$,

$$[x - (3)]^2 + [y - (-2)]^2 = (4)^2 \qquad \text{Replace } h \text{ with 3, } k \text{ with } -2, \text{ and } r \text{ with 4}$$
$$(x - 3)^2 + (y + 2)^2 = 16$$
$$x^2 - 6x + 9 + y^2 + 4y + 4 = 16 \qquad \text{Perform indicated multiplications}$$
$$x^2 + y^2 - 6x + 4y + 13 = 16$$
$$x^2 + y^2 - 6x + 4y - 3 = 0 \qquad \text{Subtract 16 from each member}$$

7. Center $C(-5,2)$ and radius 1
8. Center $C(1,-3)$ and radius $\sqrt{10}$
9. Center at origin and radius 6
10. Center $C(0,7)$ and radius $\sqrt{7}$

Determine the coordinates of the center and the radius of each of the following circles. See example 9-3 C.

Example $x^2 + y^2 + 2x - 4y - 4 = 0$

Solution   We must write the equation in standard form $(x - h)^2 + (y - k)^2 = r^2$.

$$(x^2 + 2x) + (y^2 - 4y) - 4 = 0 \qquad \text{Group terms with same variables}$$
$$(x^2 + 2x \quad) + (y^2 - 4y \quad) = 4 + (\quad) + (\quad) \qquad \text{Add 4 to each member}$$
$$(x^2 + 2x + 1) + (y^2 - 4y + 4) = 4 + 1 + 4 \qquad \text{Add 1 and 4 to each member to complete the square}$$
$$(x + 1)^2 + (y - 2)^2 = 9 \qquad \text{Write as binomial squares}$$
$$[x - (-1)]^2 + (y - 2)^2 = 3^2 \qquad \text{Write in standard form}$$

The center is at $C(-1,2)$ and the radius is 3 units.

11. $(x - 3)^2 + (y - 2)^2 = 49$

12. $(x - 1)^2 + (y - 4)^2 = 16$

13. $(x - 5)^2 + (y + 3)^2 = 8$

14. $(x + 1)^2 + (y - 7)^2 = 11$

15. $(x + 1)^2 + (y + 9)^2 = 25$

16. $(x + 6)^2 + (y + 2)^2 = 24$

17. $x^2 + y^2 = 36$

18. $x^2 + y^2 = 1$

19. $x^2 + y^2 = 2$

20. $x^2 + y^2 = 18$

21. $x^2 + y^2 + 4x - 6y - 23 = 0$

22. $x^2 + y^2 + 4x - 4y - 8 = 0$

23. $x^2 + y^2 + 6x - 8y = 0$

24. $x^2 + y^2 - 16x + 30y = 0$

25. $2x^2 + 2y^2 + 8x - 12y = 74$

26. $3x^2 + 3y^2 - 24x + 12y - 87 = 0$

27. $3x^2 + 3y^2 + 18x - 6y = 45$

28. $2x^2 + 2y^2 - 4x + 4y = 22$

Graph each of the following circles. See examples 9-3 A, B, and C.

Example   Graph the equation $x^2 + y^2 + 2x - 4y - 4 = 0$.

Solution   In the previous example, we determined the center to be at $(-1,2)$ and the radius 3 units long.

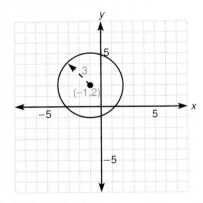

29. $x^2 + y^2 = 1$

30. $x^2 + y^2 = 36$

31. $3x^2 + 3y^2 = 12$

32. $4x^2 + 4y^2 = 24$

33. $(x - 4)^2 + (y + 3)^2 = 4$

34. $(x + 2)^2 + (y - 5)^2 = 5$

35. $x^2 + y^2 - 2x = 15$

36. $x^2 + y^2 - 4x = 0$

37. $x^2 + y^2 - 2x + 4y - 20 = 0$

38. $x^2 + y^2 + 6x - 2y - 39 = 0$

39. $2x^2 + 2y^2 - 12x + 4y = -12$

40. $4x^2 + 4y^2 - 16x + 24y = 92$

41. Find the equation of the circle passing through the origin with center $C(0,5)$.

42. Find the equation of the circle passing through the origin with center $C(-2,0)$.

43. Find the equation of the circle passing through the origin with center $C(3,-2)$.

Identify the following equations as parabolas, lines, or circles and graph each equation.

44. $y = x^2 - x - 2$

45. $3x - 2y = -6$

46. $3x^2 = 12 - 3y^2$

47. $x - y^2 = 4y + 3$

Review exercises

1. Graph the equation $3x - y = 6$ using the slope and the y-intercept. See section 7–3.

2. Combine $\sqrt{8xy^3} - \sqrt{32xy^3}$ and simplify. See section 5–5.

3. Given $P(x) = 4x^2 - 2x + 8$, find $P(-1)$. See section 1–5.

4. Write the set $\{x \mid -2 \le x < 4\}$ in interval notation. See section 2–5.

5. Given the line $2x - 3y = -6$, find the equation of the line through $(2,-1)$ that is perpendicular to the given line. See section 7–3.

9–4 ■ The ellipse and the hyperbola

The ellipse

Two other conics containing x^2 and y^2 terms are the ellipse and the hyperbola.

— Definition of an ellipse —

An **ellipse** is the set of all points in the plane such that the sum of the distances from each point on the ellipse to two fixed points in the plane is a positive constant, k.

In figure 9–5, for every point on the ellipse, $d_1 + d_2 = k$ (positive constant). We call the fixed points the *foci* of the ellipse. Each fixed point is a **focus** of the ellipse, which we denote by F_1 and F_2. See figure 9–5.

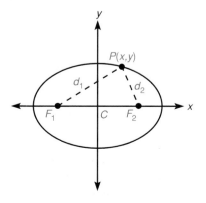

Figure 9–5

Given the x-intercepts of the ellipse at $(a,0)$ and $(-a,0)$ and the y-intercepts at $(0,b)$ and $(0,-b)$ (a and b are both positive), it can be shown that the equation of the ellipse with its center at the origin $(0,0)$ is given as follows:

___ *Equation of an ellipse with center at the origin* _____

The **standard form** of the equation of the ellipse with center at the origin is

$$\frac{x^2}{a^2} + \frac{y^2}{b^2} = 1 \qquad (a \neq b)$$

where the intercepts are $(a,0)$, $(-a,0)$, $(0,b)$, and $(0,-b)$.

To graph an ellipse with its center at $(0,0)$, plot the four intercepts $(a,0)$, $(-a,0)$, $(0,b)$, $(0,-b)$ and sketch the ellipse through the intercepts. See figure 9–6.

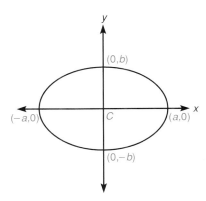

Figure 9–6

In figure 9–6, the longer line segment with endpoints $(-a,0)$ and $(a,0)$ is called the *major axis* of the ellipse. The shorter line segment with endpoints $(0,-b)$ and $(0,b)$ is called the *minor axis*. The endpoints of the major and minor axes provide points used to sketch the ellipse.

■ *Example 9–4 A*

Find the coordinates of the intercepts of the following ellipses and sketch the graph.

1. $\dfrac{x^2}{25} + \dfrac{y^2}{9} = 1$

   a. $a^2 = 25$, so $a = 5$.
   b. $b^2 = 9$, so $b = 3$.

   The x-intercepts are $(-5,0)$ and $(5,0)$ while the y-intercepts are $(0,-3)$ and $(0,3)$.
   The major axis is the line segment with endpoints $(-5,0)$ and $(5,0)$ and the minor axis is the line segment with endpoints $(0,-3)$ and $(0,3)$. The major axis is 10 units long and the minor axis is 6 units long.

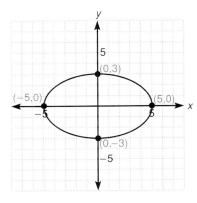

2. $16x^2 + 9y^2 = 144$

To obtain the standard form of the equation, we must divide each term by 144, the constant in the right member, since that constant in the standard form must be 1.

$$\frac{16x^2}{144} + \frac{9y^2}{144} = \frac{144}{144}$$ Divide each term by 144

$$\frac{x^2}{9} + \frac{y^2}{16} = 1$$ Reduce in each term

a. $a^2 = 9$, so $a = 3$.
b. $b^2 = 16$, so $b = 4$.

The x-intercepts are $(3,0)$ and $(-3,0)$ while the y-intercepts are $(0,4)$ and $(0,-4)$.

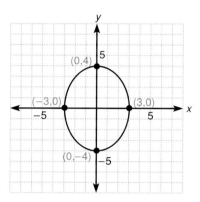

The major axis is 8 units long, $(0,-4)$ to $(0,4)$, and the minor axis is 6 units long, $(-3,0)$ to $(3,0)$.

▶ *Quick check*   Find the intercepts and graph the equation $25x^2 + 4y^2 = 100$. ■

Just as a circle may have its center at a point other than the origin, so may an ellipse have its center at the point (h,k).

┌─ *Equation of an ellipse with center at (h,k)* ─────────

The standard form of the equation of an ellipse with center at (h,k) and with major axis parallel to the x-axis is

$$\frac{(x-h)^2}{a^2} + \frac{(y-k)^2}{b^2} = 1$$

and is
$$\frac{(y-k)^2}{b^2} + \frac{(x-h)^3}{a^2} = 1 \qquad (a \neq b)$$

when the major axis is parallel to the y-axis.

■ *Example 9–4 B*

Find the coordinates of the center and the *x*- and *y*-intercepts. Sketch the graph of the equation.

1. $\dfrac{(x + 2)^2}{16} + \dfrac{(y - 1)^2}{9} = 1$

The equation is in standard form so
a. the center is at $(-2,1)$
b. $a^2 = 16$, so $a = 4$.
c. $b^2 = 9$, so $b = 3$.

From the center $(-2,1)$,
a. locate two points 4 units each way on a horizontal line, the points $(-6,1)$ and $(2,1)$,
b. locate two points 3 units each way on a vertical line, the points $(-2,4)$ and $(-2,-2)$.
c. draw a smooth closed curve through these four points.

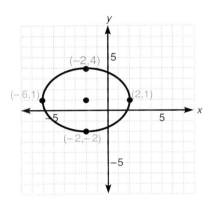

Note   The major axis is $2(4) = 8$ units long and the minor axis is $2(3) = 6$ units long.

2. $9x^2 + 4y^2 - 18x - 16y = 11$

We must write the equation in standard form. Do this by completing the square in each variable.

$$9x^2 + 4y^2 - 18x - 16y = 11$$
$$(9x^2 - 18x) + (4y^2 - 16y) = 11 \qquad \text{Group like variables}$$
$$9(x^2 - 2x) + 4(y^2 - 4y) = 11 \qquad \text{Factor the common factor from each expression}$$

$$9(x^2 - 2x + 1) + 4(y^2 - 4y + 4) = 11 + 9 + 16 \qquad \text{Complete the square and balance}$$

$$9(x - 1)^2 + 4(y - 2)^2 = 36 \qquad \text{Factor trinomials}$$
$$\frac{(x - 1)^2}{4} + \frac{(y - 2)^2}{9} = 1 \qquad \text{Divide each term by 36}$$

The center is at $(1,2)$ and since
a. $a^2 = 4$, then $a = 2$.
b. $b^2 = 9$, then $b = 3$.

From $(1,2)$, locate two points 3 units each way on a vertical line, the points $(1,5)$ and $(1,-1)$, and 2 units each way on a horizontal line, the points $(-1,2)$ and $(3,2)$. Draw a smooth closed curve through these four points.

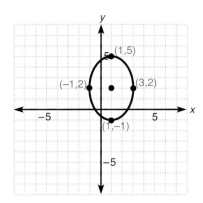

Note The major axis is 2(3) = 6 units long and the minor axis is 2(2) = 4 units long.

▶ *Quick check*   Graph the ellipse $4x^2 + y^2 - 16x - 4y = -16$.   ∎

The hyperbola

The equation of the last conic section, the hyperbola, is similar to the equation of the ellipse.

— **Definition of a hyperbola** ——————————————

A **hyperbola** is the set of all points in the plane such that the absolute value of the *difference* between the distances from each point on the hyperbola to two fixed points is a positive constant, *k*.

The fixed points are again called the **foci** of the hyperbola, which we denote by F_1 and F_2. See figure 9–7. The points of intersection on the *x*-axis, the points $(-a,0)$ and $(a,0)$, are called the *vertices* of the parabola.

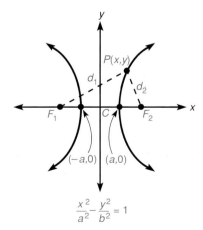

$$\frac{x^2}{a^2} - \frac{y^2}{b^2} = 1$$

In figure 9–7, for every point on the hyperbola, $|d_1 - d_2| = k$ (positive constant). Thus $|d_1 - d_2| =$ a positive constant for any point (x,y) on the hyperbola. Using the above relationship, it can be shown (but will not do so here) that the standard form of the equation of a hyperbola is as follows:

Equations of hyperbolas

The **standard form** of the equation of a hyperbola with the center at the origin is given by

$$\frac{x^2}{a^2} - \frac{y^2}{b^2} = 1$$

with x-intercepts of $(a,0)$ and $(-a,0)$ and no y-intercepts (see figure 9–7) or

$$\frac{y^2}{b^2} - \frac{x^2}{a^2} = 1$$

with y-intercepts of $(0,b)$ and $(0,-b)$ and no x-intercepts (see figure 9–8).

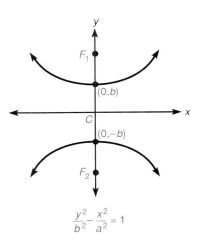

$$\frac{y^2}{b^2} - \frac{x^2}{a^2} = 1$$

Figure 9–8

Note   The points $(0,-b)$ and $(0,b)$ are the vertices of the parabola when there are y-intercepts.

Solving the equations $\dfrac{x^2}{a^2} - \dfrac{y^2}{b^2} = 1$ and $\dfrac{y^2}{b^2} - \dfrac{x^2}{a^2} = 1$ for y in terms of a, b, and x, it can be shown that the graph of this hyperbola will approach (get closer and closer to) but not cross the lines whose equations are

$$y = \frac{b}{a}x \quad \text{and} \quad y = -\frac{b}{a}x$$

We call such lines **asymptotes.** They are indicated by dashed lines. These lines are useful for sketching the graph of the hyperbola and are found in the following way:

Asymptotes of a hyperbola

The asymptotes of the hyperbolas $\dfrac{x^2}{a^2} - \dfrac{y^2}{b^2} = 1$ and $\dfrac{y^2}{b^2} - \dfrac{x^2}{a^2} = 1$ are the extended diagonals of the rectangle whose corners are (a,b), $(a,-b)$, $(-a,b)$, and $(-a,-b)$.

To illustrate, given the equation

$$\frac{x^2}{4} - \frac{y^2}{25} = 1$$

then, (1) $a^2 = 4$, so $a = 2$
(2) $b^2 = 25$, so $b = 5$.

We can now construct a rectangle with corners $(2,5)$, $(2,-5)$, $(-2,5)$, and $(-2,-5)$. To obtain the asymptotes for this hyperbola, draw the diagonals. See figure 9–9.

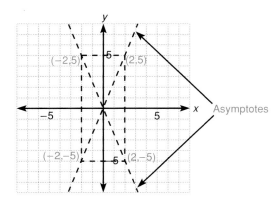

Figure 9–9

Using the x-intercepts $(-2,0)$ and $(2,0)$ and the asymptotes, we can sketch the graph of a hyperbola with its center at the origin $(0,0)$. The graph of the equation is shown in figure 9–10. We call the line segment whose endpoints are the vertices, V' and V, the *transverse axis*. The line segment whose endpoints are W' and W is called the *conjugate axis*.

Since $a = 2$, $b = 5$, and the equations of the asymptotes are $y = \pm\dfrac{b}{a}$, then

$$y = -\frac{b}{a}x = -\frac{5}{2}x \quad \text{and} \quad y = \frac{b}{a}x = \frac{5}{2}x$$

when we replace a with 2 and b with 5.

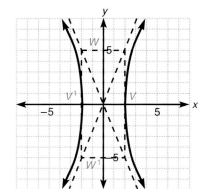

Figure 9–10

In summary, to graph a hyperbola of either of the two standard forms, we follow this procedure:

To graph a hyperbola

Step 1 Write the equation in standard form.
Step 2 Locate the x-intercepts at $(-a,0)$ and $(a,0)$ if the coefficient of x^2 is positive and the y-intercepts at $(0,-b)$ and $(0,b)$ if the coefficient of y^2 is positive.
Step 3 Draw the rectangle (in dashed lines) with corners at (a,b), $(a,-b)$, $(-a,b)$, and $(-a,-b)$.
Step 4 Sketch the asymptotes (in dashed lines) that are extensions of the diagonals of the rectangle.
Step 5 Sketch the hyperbola through the intercepts and approaching the lines that are the asymptotes.

■ *Example 9–4 C*

Determine the intercepts and equations of the asymptotes. Sketch the graph.

1. $\dfrac{x^2}{16} - \dfrac{y^2}{4} = 1$

   a. $a^2 = 16$, so $a = 4$.
   b. $b^2 = 4$, so $b = 2$.

   The x-intercepts (the vertices) are $(-4,0)$ and $(4,0)$ and there are no y-intercepts. The transverse axis is the line segment from $(-4,0)$ to $(4,0)$ while the conjugate axis is the line segment from $(0,-2)$ to $(0,2)$. The rectangle is formed by the points $(4,2)$, $(4,-2)$, $(-4,2)$, and $(-4,-2)$. Draw the diagonals of this rectangle to obtain the asymptotes. The equations of the asymptotes are $y = \dfrac{1}{2}x$ and

   $y = -\dfrac{1}{2}x$.

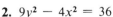

2. $9y^2 - 4x^2 = 36$
   To get the equation into one of the standard forms, divide each term by 36. Then

   $\dfrac{y^2}{4} - \dfrac{x^2}{9} = 1$

   a. $b^2 = 4$, so $b = 2$.
   b. $a^2 = 9$, so $a = 3$.

The y-intercepts (the vertices) are $(0,-2)$ and $(0,2)$. There are no x-intercepts. The transverse axis is the line segment from $(0,-2)$ to $(0,2)$ while the conjugate axis is the line segment from $(-3,0)$ to $(3,0)$. The rectangle is formed by the points $(3,2)$, $(3,-2)$, $(-3,2)$, and $(-3,-2)$. Draw the diagonals of the rectangle to obtain the asymptotes. The equations of the asymptotes are $y = \dfrac{2}{3}x$ and $y = -\dfrac{2}{3}x$.

▶ **Quick check**   Determine the intercepts, equations of the asymptotes, and graph the equation $9y^2 - 9x^2 = 81$.   ■

As with the ellipse, the standard equation of a hyperbola with center at the point (h,k) can be developed.

--- **Equation of a hyperbola with center at (h,k)** ---

The standard form of the equation of a hyperbola with center at (h,k) and foci on a line parallel to the x-axis is

$$\frac{(x - h)^2}{a^2} - \frac{(y - k)^2}{b^2} = 1$$

and with the foci on a line parallel to the y-axis is

$$\frac{(y - k)^2}{b^2} - \frac{(x - h)^2}{a^2} = 1$$

The following examples illustrate the graphing of hyperbolas with center at (h, k). We use the endpoints of the transverse axis and the conjugate axis to construct the asymptote rectangle.

■ *Example 9–4 D*

Find the coordinates of the center and graph the following hyperbolas.

1. $\dfrac{(x - 3)^2}{4} - \dfrac{(y + 2)^2}{4} = 1$

The equation is in standard form so the center is at $(3,-2)$. Since (a) $a^2 = 4$, then $a = 2$ and since (b) $b^2 = 4$, then $b = 2$. As illustrated in the accompanying graph, the transverse axis is parallel to the x-axis and the vertices are $(3 \pm 2, -2)$; that is, the points $(5,-2)$ and $(1,-2)$. The endpoints of the conjugate axis are $(3,-2 \pm 2)$; that is, the points $(3,0)$ and $(3,-4)$. Use these points to construct the rectangle whose extended diagonals serve as the asymptotes. Sketch the hyperbola through the vertices $(5,-2)$ and $(1,-2)$ and approaching the asymptotes.

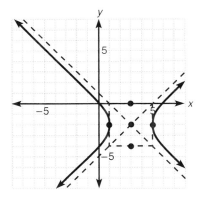

2. $y^2 - 4x^2 + 6y + 32x = 59$

We must first write the equation in standard form by completing the square in each variable.

$$y^2 - 4x^2 + 6y + 32x = 59$$
$$(y^2 + 6y) - (4x^2 - 32x) = 59 \qquad \text{Group like variables}$$
$$(y^2 + 6y) - 4(x^2 - 8x) = 59 \qquad \text{Factor the common factor}$$
$$(y^2 + 6y + 9) - 4(x^2 - 8x + 16) = 59 + 9 - 64 \qquad \text{Complete the square and balance}$$
$$(y + 3)^2 - 4(x - 4)^2 = 4 \qquad \text{Factor the trinomials}$$
$$\frac{(y + 3)^2}{4} \frac{(x - 4)^2}{1} = 1 \qquad \text{Divide each term by 4}$$

The center is at $(4, -3)$. Since $b^2 = 4$, then $b = 2$ and since $a^2 = 1$, then $a = 1$. The transverse axis is parallel to the y-axis and the vertices are at $(4, -3 \pm 2)$; that is, the points $(4, -5)$ and $(4, -1)$. The endpoints of the conjugate axis are at $(4 \pm 1, -3)$; that is, the points $(5, -3)$ and $(3, -3)$. Construct the asymptote rectangle through these four points and, using the extended diagonals as asymptotes, sketch the hyperbola through the vertices $(4, -5)$ and $(4, -1)$, approaching the asymptotes.

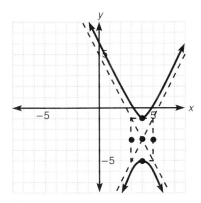

▶ *Quick check*   Graph the hyperbola $x^2 + 2x - y^2 + 2y = -4$.   ∎

We now summarize the characteristics of the conic sections.

Equation	Conic section	Characteristics
$y = ax^2 + bx + c$	Parabola	If $a > 0$, opens up If $a < 0$, opens down
$x = ay^2 + by + c$	Parabola	If $a > 0$, opens right If $a < 0$, opens left
$(x - h)^2 + (y - k)^2 = r^2$ $(x^2 + y^2 + ax + by + c = 0)$	Circle	Center at (h,k) with radius r
$\dfrac{x^2}{a^2} + \dfrac{y^2}{b^2} = 1 \ (a \neq b)$	Ellipse	x-intercepts, $(-a,0)$ and $(a,0)$ y-intercepts, $(0,-b)$ and $(0,b)$
$\dfrac{(x - h)^2}{a^2} + \dfrac{(y - k)^2}{b^2} = 1$ $(a \neq b)$	Ellipse	Center at (h,k)
$\dfrac{x^2}{a^2} - \dfrac{y^2}{b^2} = 1$	Hyperbola	x-intercepts, $(-a,0)$ and $(a,0)$
$\dfrac{y^2}{b^2} - \dfrac{x^2}{a^2} = 1$	Hyperbola	y-intercepts, $(0,-b)$ and $(0,b)$
$\dfrac{(x - h)^2}{a^2} - \dfrac{(y - k)^2}{b^2} = 1$	Hyperbola	Center at (h,k); vertices on a line parallel to x-axis
$\dfrac{(y - k)^2}{b^2} - \dfrac{(x - h)^2}{a^2} = 1$	Hyperbola	Center at (h,k); vertices on a line parallel to the y-axis

■ *Example 9–4 E*

Identify each equation as a parabola, a circle, an ellipse, or a hyperbola.

1. $x^2 - 2y = 0$
   Since the equation contains the second power of x and the first power of y, then y is linear and x is quadratic so the graph is a **parabola.**

2. $4x^2 = 12 - 3y^2$
   Adding $3y^2$ to each member, we obtain the equation

   $$4x^2 + 3y^2 = 12 \quad \text{or} \quad \frac{x^2}{3} + \frac{y^2}{4} = 1$$

   The equation of an **ellipse.**

3. $3y^2 = 9 + 3x^2$
   Adding $-3x^2$ to each member, we obtain

   $$-3x^2 + 3y^2 = 9 \quad \text{or} \quad \frac{y^2}{3} - \frac{x^2}{3} = 1$$

   The equation of a **hyperbola.**

┌───┐
│ ─ *Mastery points* ─   │
│   │
│ *Can you*   │
│ ■ Find the coordinates of the *x*- and *y*-intercepts for an ellipse and a │
│   hyperbola?   │
│ ■ Sketch the graphs of ellipses and hyperbolas?   │
│ ■ Find the equations of the asymptotes of a hyperbola and graph │
│   them?   │
│ ■ Identify the equation of a parabola, a circle, an ellipse, and a │
│   hyperbola and sketch their graphs?   │
└───┘

Exercise 9–4

Find the *x*- and *y*-intercepts of each given ellipse. Sketch the graph of each equation. See example 9–4 A.

Example $25x^2 + 4y^2 = 100$

Solution   To get the equation in standard form, divide each member by 100.

$$\frac{x^2}{4} + \frac{y^2}{25} = 1$$

1. $a^2 = 4$, so $a = 2$.
2. $b^2 = 25$, so $b = 5$.
The *x*-intercepts are $(-2,0)$
and $(2,0)$. The *y*-intercepts
are $(0,-5)$ and $(0,5)$.

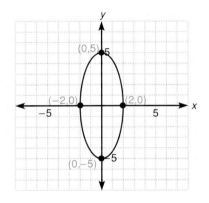

1. $\dfrac{x^2}{9} + \dfrac{y^2}{25} = 1$

2. $\dfrac{x^2}{4} + \dfrac{y^2}{9} = 1$

3. $\dfrac{x^2}{49} + \dfrac{y^2}{25} = 1$

4. $\dfrac{x^2}{121} + \dfrac{y^2}{36} = 1$

5. $x^2 + \dfrac{y^2}{9} = 1$

6. $x^2 + \dfrac{y^2}{4} = 1$

7. $\dfrac{x^2}{16} + y^2 = 1$

8. $36x^2 + 9y^2 = 324$

9. $x^2 + 25y^2 = 100$

10. $16x^2 + y^2 = 64$

11. $3x^2 + 4y^2 = 12$

12. $9x^2 + 2y^2 = 18$

13. $8x^2 + y^2 = 16$

14. $x^2 + 3y^2 = 27$

Graph the following ellipses. See example 9–4 B.

Example $4x^2 + y^2 - 16x - 4y = -16$

Solution Write the equation in standard form using completing the square.

$$4x^2 + y^2 - 16x - 4y = -16$$

$$\begin{array}{ll}
(4x^2 - 16x) + (y^2 - 4y) = -16 & \text{Group like variables} \\
4(x^2 - 4x) + (y^2 - 4y) = -16 & \text{Factor the common factor from } 4x^2 - 16x \\
4(x^2 - 4x + 4) + (y^2 - 4y + 4) = -16 + 16 + 4 & \text{Complete the square} \\
4(x - 2)^2 + (y - 2)^2 = 4 & \text{Factor trinomials and combine} \\
\dfrac{(x - 2)^2}{1} + \dfrac{(y - 2)^2}{4} = 1 & \text{Divide each term by 4}
\end{array}$$

The center is at (2,2). Since $a^2 = 1$, then $a = 1$ and since $b^2 = 4$, then $b = 2$. The ellipse passes through the points (1,2), (3,2), (2,4), and (2,0).

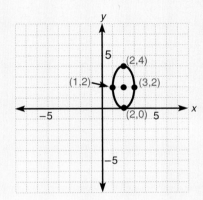

15. $\dfrac{(x - 1)^2}{4} + \dfrac{(y - 2)^2}{9} = 1$

16. $\dfrac{(x + 2)^2}{9} + \dfrac{(y - 3)^2}{16} = 1$

17. $4x^2 + y^2 + 8x - 10y - 7 = 0$

18. $x^2 + 2y^2 - 8x - 4y + 14 = 0$

19. $2x^2 + y^2 + 4x + 2y - 1 = 0$

20. $4x^2 + 9y^2 + 24x - 18y + 9 = 0$

Find the x-intercepts, or y-intercepts, and the equations of the asymptotes of the hyperbola for the given equation. Sketch the graph of the equation. See example 9–4 C.

Example $9y^2 - 9x^2 = 81$

Solution Divide each member by 81 to get the equation in standard form.

$$\frac{9y^2}{81} - \frac{9x^2}{81} = \frac{81}{81}$$

$$\frac{y^2}{9} - \frac{x^2}{9} = 1$$

1. $b^2 = 9$, so $b = 3$.
2. $a^2 = 9$, so $a = 3$.

y-intercepts $(0,-3)$, $(0,3)$; no x-intercepts

The rectangle has vertices (3,3), (3,−3), (−3,3), and (−3,−3).

The equations of the asymptotes are $y = -x$ and $y = x$.

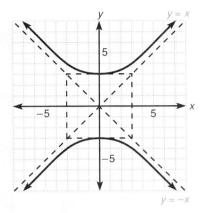

21. $\dfrac{x^2}{16} - \dfrac{y^2}{9} = 1$

22. $\dfrac{x^2}{4} - \dfrac{y^2}{25} = 1$

23. $\dfrac{y^2}{9} - \dfrac{x^2}{4} = 1$

24. $\dfrac{y^2}{16} - \dfrac{x^2}{25} = 1$

25. $y^2 - \dfrac{x^2}{9} = 1$

26. $y^2 - \dfrac{x^2}{4} = 1$

27. $x^2 - \dfrac{y^2}{16} = 1$

28. $9x^2 - y^2 = 36$

29. $y^2 - 25x^2 = -25$

30. $y^2 - 16x^2 = 64$

31. $25y^2 - 16x^2 = 400$

32. $16y^2 - 9x^2 = 144$

33. $25x^2 - 4y^2 = -100$

34. $2y^2 - 3x^2 = -18$

35. $2x^2 - 9y^2 = -36$

36. $6x^2 - 6y^2 = -1$

Find the coordinates of the center and graph the following hyperbolas. See example 9–4 D.

Example $x^2 + 2x - y^2 + 2y = -4$

Solution   Write the equation in standard form by completing the square.

$$x^2 + 2x - y^2 + 2y = -4$$
$$(x^2 + 2x) - (y^2 - 2y) = -4 \qquad \text{Group like variables}$$
$$(x^2 + 2x + 1) - (y^2 - 2y + 1) = -4 + 1 - 1 \qquad \text{Complete the square}$$
$$(x + 1)^2 - (y - 1)^2 = -4 \qquad \text{Factor the trinomials and combine}$$
$$\dfrac{(y - 1)^2}{4} - \dfrac{(x + 1)^2}{4} = 1 \qquad \text{Divide each term by } -4$$

The center is at $(-1,1)$. Since $b^2 = 4$, then $b = 2$ and since $a^2 = 4$, then $a = 2$. The vertices of the parabola are the points $(-1,3)$ and $(-1,-1)$. The endpoints of the conjugate axis that help form the rectangle to obtain the asymptotes are $(-3,1)$ and $(1,1)$.

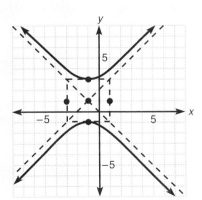

37. $\dfrac{(x-1)^2}{9} - \dfrac{(y+3)^2}{25} = 1$ 38. $\dfrac{(y-5)^2}{4} - \dfrac{(x+1)^2}{9} = 1$

39. $x^2 - y^2 - 6x - 4y + 4 = 0$ 40. $y^2 - x^2 - 8x + 2y - 16 = 0$

41. $y^2 - 4x^2 - 8x - 6y = 11$ 42. $4x^2 - 9y^2 + 8x - 36y = -4$

Write each of the following equations in standard form. Identify as a parabola, a circle, an ellipse, or a hyperbola. See example 9–4 E.

43. $x^2 = 9 - y^2$ 44. $9x^2 = 36 + 4y^2$ **45.** $x^2 + 3 = y$

46. $2x^2 + y^2 = 8$ 47. $4y^2 = 25 - 4x^2$ 48. $y^2 - x^2 = 25$

49. $y^2 = 121 - x^2$ 50. $8y^2 = 24 + 8x^2$ 51. $x^2 + y = 8$

52. $18 - 2y^2 = 3x^2$ 53. $x^2 + y^2 - 2x + 4y = 0$ 54. $2x^2 - 8x + 5y^2 + 20y = 12$

55. $4x^2 - 8x - y^2 - 6y = 21$

Review exercises

Perform the indicated operations on the following rational expressions. See sections 4–2 and 4–3.

1. $\dfrac{2y+1}{y-3} \cdot \dfrac{y^2 - 2y - 3}{2y^2 - 5y - 3}$ 2. $\dfrac{4z}{3z+2} - \dfrac{6z}{2z-1}$

3. Simplify the complex rational expression $\dfrac{\dfrac{1}{x} - \dfrac{1}{y}}{\dfrac{1}{x^2} - \dfrac{1}{y^2}}$. See section 4–4.

Find the solution set of the following quadratic equations. See section 6–1.

4. $4x^2 = 20$ 5. $5x^2 - 4x = 0$

Perform the indicated operations. See sections 5–5 and 5–6.

6. $(\sqrt{3} - 2\sqrt{2}) - (3\sqrt{3} + \sqrt{2})$ 7. $(2\sqrt{5} + \sqrt{3})(3\sqrt{5} - \sqrt{3})$

Chapter 9 lead-in problem

If a gun is fired upward with an initial velocity of 288 ft/sec, the bullet's height h after t seconds is given by the equation $h = 288t - 16t^2$. Find the maximum height attained by the bullet.

Solution

The path the bullet will follow is a parabola. We want the coordinates (t,h) of the vertex. The first coordinate of the vertex, t, is given by

$$t = -\frac{b}{2a}$$

where $b = 288$ and $a = -16$. Then,

$$t = -\frac{(288)}{2(-16)} \qquad \text{Replace } b \text{ with 288 and } a \text{ with } -16$$

$$= \frac{288}{32} = 9$$

Maximum height is attained in 9 seconds. Now we want h.

$$\begin{aligned} h &= 288t - 16t^2 \\ &= 288(9) - 16(9)^2 \qquad \text{Replace } t \text{ with 9} \\ &= 2{,}592 - 1{,}296 \\ &= 1{,}296 \end{aligned}$$

The maximum height attained by the bullet is 1,296 feet.

Chapter 9 summary

1. **Conic sections**—the parabola, the circle, the ellipse, and the hyperbola—are obtained by slicing a cone with a plane in four different ways.
2. The general form of the equations of conic sections is

   $$ax^2 + by^2 + cx + dy + e = 0$$

3. The graph of a **quadratic equation** of the form $y = ax^2 + bx + c$ is a parabola opening upward if $a > 0$ and the vertex is a minimum and opening downward if $a < 0$ and the vertex is a maximum.
4. The graph of a *quadratic* equation of the form

   $$x = ay^2 + by + c$$

   is a parabola that opens right when $a > 0$ and opens left when $a < 0$.
5. An equation of the **circle** with center $C(h,k)$ and having radius r is given by

   $$(x - h)^2 + (y - k)^2 = r^2$$

   When C is at $(0,0)$, the equation is given by

   $$x^2 + y^2 = r^2$$

6. The general form of the equation of a circle is given by

   $$ax^2 + ay^2 + bx + cy + d = 0$$

7. The equation of the ellipse with center $C(0,0)$ is given by

   $$\frac{x^2}{a^2} + \frac{y^2}{b^2} = 1$$

   where the x-intercepts are $(a,0)$ and $(-a,0)$ and the y-intercepts are $(0,b)$ and $(0,-b)$.
8. The standard form of the equation of an ellipse with center at (h,k) is

   $$\frac{(x - h)^2}{a^2} + \frac{(y - k)^2}{b^2} = 1$$

9. The equation of the hyperbola with center $C(0,0)$ and having x-intercepts $(a,0)$ and $(-a,0)$ is given by

   $$\frac{x^2}{a^2} - \frac{y^2}{b^2} = 1$$

   and having y-intercepts $(0,b)$ and $(0,-b)$ is given by

   $$\frac{y^2}{b^2} - \frac{x^2}{a^2} = 1$$

10. The equations of the **asymptotes** for the graph of either hyperbola are given by

$$y = \frac{b}{a}x \text{ and } y = -\frac{b}{a}x$$

They are the diagonals of a rectangle whose vertices are (a,b), $(a,-b)$, $(-a,b)$, and $(-a,-b)$.

11. The standard form of the equation of a hyperbola with center at (h,k) is given by

$$\frac{(x-h)^2}{a^2} - \frac{(y-k)^2}{b^2} = 1$$

when the vertices are on a line parallel to the x-axis and

$$\frac{(y-k)^2}{b^2} - \frac{(x-h)^2}{a^2} = 1$$

when the vertices are on a line parallel to the y-axis.

Chapter 9 error analysis

1. The graph of a parabola
   Example: The graph of $y = -2x^2 + 2x - 3$ opens upward since $a = -2$.
   Correct answer: The graph opens downward.
   What error was made? (*see page 399*)

2. The vertex of a parabola
   Example: The vertex of the parabola $y = (x + 3)^2 - 4$ is $(3,4)$.
   Correct answer: $(-3,-4)$
   What error was made? (*see page 399*)

3. Finding the vertex of a parabola
   Example: Given $y = -x^2 + 4x - 3$, then
   $$y = -1(x^2 - 4x \quad) - 3$$
   $$= -1(x^2 - 4x + 4) - 3 - 4$$
   $$= -1(x - 2)^2 - 7$$
   The vertex is the point $(2,-7)$.
   Correct answer: $(2,1)$
   What error was made? (*see page 399*)

4. Finding the vertex of a parabola
   Example: Given $x = y^2 - 4y - 12$, then
   $$x = (y^2 - 4y \quad) - 12$$
   $$= (y^2 - 4y + 4) - 12 - 4$$
   $$= (y - 2)^2 - 16$$
   The vertex is the point $(2,-16)$.
   Correct answer: $(-16,2)$
   What error was made? (*see page 408*)

5. Finding the intercepts of a parabola
   Example: Given $y = 3x^2 - 7x + 2$,
   1. Let $y = 0$, then
   $$0 = 3x^2 - 7x + 2$$
   $$= (3x - 2)(x - 1)$$
   $$3x - 2 = 0 \quad \text{or} \quad x - 1 = 0$$
   $$x = \frac{2}{3} \qquad x = 1$$
   The x-intercepts are $\left(\frac{2}{3},0\right)$ and $(1,0)$.

   2. Let $x = 0$, then $y = 2$. The y-intercept is $(0,2)$.

   Correct answer: x-intercepts, $\left(\frac{1}{3},0\right)$ and $(2,0)$;

   y-intercept, $(0,2)$
   What error was made? (*see page 400*)

6. The center and radius of a circle
   Example: Given $x^2 + y^2 - 6x + 8y - 2 = 0$,
   $$(x^2 - 6x \quad) + (y^2 + 8y \quad) = 2$$
   $$(x^2 - 6x + 9) + (y^2 + 8y + 16) = 2$$
   $$(x - 3)^2 + (y + 4)^2 = 2$$
   Center is at $(3,4)$ and radius $r = \sqrt{2}$.
   Correct answer: Center at $(3,-4)$ and radius $r = 3\sqrt{3}$.
   What error was made? (*see page 415*)

7. Intercepts of an ellipse
   Example: Given $4x^2 + 8y^2 = 4$, then $x^2 + 2y^2 = 1$; the x-intercepts are $(-1,0)$ and $(1,0)$ and the y-intercepts are $(0,-2)$ and $(0,2)$.
   Correct answer: x-intercepts are $(-1,0)$ and $(1,0)$;
   y-intercepts are $\left(0,-\frac{\sqrt{2}}{2}\right), \left(0,\frac{\sqrt{2}}{2}\right)$.
   What error was made? (*see page 420*)

8. Equations of the asymptotes of a hyperbola
   Example: Given $\frac{x^2}{4} - \frac{y^2}{9} = 1$, the equations of the asymptotes are $y = -\frac{4}{9}x$ and $y = \frac{4}{9}x$.
   Correct answer: $y = -\frac{3}{2}x$ and $y = \frac{3}{2}x$
   What error was made? (*see page 423*)

9. The intercepts of a hyperbola
   Example: Given $\frac{x^2}{4} - \frac{y^2}{9} = 1$,
   The x-intercepts are $(-2,0)$ and $(2,0)$ and the y-intercepts are $(0,-3)$ and $(0,3)$.
   Correct answer: x-intercepts, $(-2,0)$, $(2,0)$; y-intercepts, none
   What error was made? (*see page 423*)

10. Identifying conic sections
    Example: The equation $3x + 3y^2 = 12$ is the equation of a circle.
    Correct answer: parabola
    What error was made? (*see page 428*)

Chapter 9 critical thinking

Write an algebraic expression for the following relationship.
Will this relationship always be true?

$$\left(\frac{1}{3}\right)^2 + \left(1 - \frac{1}{3}\right) = \left(\frac{1}{3}\right) + \left(1 - \frac{1}{3}\right)^2$$
$$(0.7)^2 + (1 - 0.7) = (0.7) + (1 - 0.7)^2$$

Chapter 9 review

[9–1]
Determine the vertex, the x-intercept(s), the y-intercept(s), and graph the following parabolas.

1. $y = -3(x + 1)^2$ **2.** $y = -x^2 + 4x + 5$ **3.** $y = x^2 - 2$

4. $y = -x^2 - 3x + 4$ **5.** $y = x^2 + 3x + 1$

[9–2]
Graph the following quadratic equations.

6. $x = y^2 - 1$ **7.** $x = y^2 - 4y - 5$ **8.** $x = -y^2 + y + 2$

[9–3]
Write the equation for each of the following circles in the (1) standard form $(x - h)^2 + (y - k)^2 = r^2$ and (2) general form $x^2 + y^2 + bx + cy + d = 0$, where $b, c,$ and d are integers.

9. Center $C(-2,5)$ and radius 5 **10.** Center $C(4,0)$ and radius $\sqrt{11}$

11. Center at the origin and radius 9

Determine the coordinates of the center and the length of the radius of the following circles.

12. $(x - 5)^2 = 36 - (y + 1)^2$ **13.** $y^2 = 7 - x^2$

14. $x^2 + y^2 + 6x - 8y + 1 = 0$ **15.** $x^2 + y^2 - 2x + 4y - 3 = 0$

16. $4x^2 + 4y^2 - 16y + 8x - 20 = 0$ **17.** $x^2 + y^2 - 4y - 15 = 0$

Graph each of the following circles.

18. $x^2 + y^2 = 2$ **19.** $5x^2 = 20 - 5y^2$ **20.** $(x - 3)^2 + (y + 4)^2 = 1$

21. $x^2 + y^2 + 6x = 0$ **22.** $x^2 - 8y = 6x - y^2$ **23.** $x^2 + y^2 + 6x - 4y - 3 = 0$

[9–4]
Determine the x- and y-intercepts for each given ellipse. Graph the equation.

24. $\dfrac{x^2}{16} + \dfrac{y^2}{100} = 1$ **25.** $x^2 + \dfrac{y^2}{25} = 1$ **26.** $16x^2 + 9y^2 = 144$

27. $4x^2 + 8y^2 = 16$ **28.** $6x^2 + y^2 = 24$ **29.** $x^2 + 8y^2 = 8$

Graph the following ellipses.

30. $\dfrac{(x + 3)^2}{9} + \dfrac{(y + 1)^2}{4} = 1$ **31.** $\dfrac{(x - 1)^2}{16} + \dfrac{(y - 2)^2}{25} = 1$

32. $9x^2 + 4y^2 - 54x + 16y + 61 = 0$ **33.** $4x^2 + y^2 + 8x - 6y + 9 = 0$

Determine the x-intercepts, or y-intercepts, and the equations of the asymptotes for the given hyperbola. Graph the equation.

34. $\dfrac{x^2}{36} - \dfrac{y^2}{49} = 1$ **35.** $\dfrac{y^2}{25} - \dfrac{x^2}{9} = 1$ **36.** $x^2 - \dfrac{y^2}{36} = 1$

37. $y^2 = 1 + \dfrac{x^2}{9}$ **38.** $16x^2 - 25y^2 = 400$ **39.** $x^2 - 9y^2 = -9$

Graph the following hyperbolas.

40. $\dfrac{(x-2)^2}{4} - \dfrac{(y+3)^2}{16} = 1$ **41.** $\dfrac{(y-1)^2}{25} - \dfrac{(x+2)^2}{9} = 1$

42. $x^2 - y^2 - 6x - 4y + 4 = 0$ **43.** $4x^2 - y^2 - 8x + 4y = 4$

Identify each equation as a parabola, a circle, an ellipse, or a hyperbola.

44. $x^2 - 16 = y^2$ **45.** $y + 4x^2 = 8x - 3$ **46.** $4x^2 = 100 - 25y^2$

47. $9y^2 = 225 - 9x^2$ **48.** $2y^2 = 8 - 2x^2$ **49.** $x^2 = y - 2x + 3$

50. $2y^2 + y = x - 2$ **51.** $\dfrac{(x+1)^2}{1} - \dfrac{(y-1)^2}{4} = 1$ **52.** $4x^2 + 9y^2 - 8x + 54y + 84 = 0$

Chapter 9 cumulative test

Given $x = 3$, $y = -5$, and $z = -2$, evaluate the following expressions.

[1–5]   **1.** $2x - 3y + 4z$ **[1–5]**   **2.** $4x^2 + 2y - 5$ **[1–5]**   **3.** $4x^2 + y^2 - z^2$

[1–5]   **4.** Given $C = \dfrac{5}{9}(F - 32)$, find C when $F = 212$.   **[1–5]**   **5.** What is the degree of the polynomial $4x^4 - 3x^2 + 7$?

[1–6]   **6.** From $4x^2 + 6x - 9$ subtract $-2x^2 - x + 5$.

Given $P(x) = x^2 + 3x - 1$, $Q(x) = 4x - 1$, and $R(x) = x^2 + 3$ find

[1–6]   **7.** $P(x) - Q(x) + R(x)$ **[1–6]**   **8.** $P(x) + Q(x) - R(x)$

[1–5]   **9.** $P(3)$ **[1–5]** **10.** $Q(-4)$ **[1–5]** **11.** $R\left(\dfrac{1}{4}\right)$

[3–2] **12.** Simplify the expression $x^{2+n}x^{3n-1}$ by performing the indicated operation.

Perform the indicated operations.

[3–2] **13.** $(2a - 1)(3a^2 + 4a - 1)$ **[4–5]** **14.** $\dfrac{28a^2b^3c^5 - 35ab^3c^3}{7abc}$

[4–5] **15.** $(3x^3 - 4x^2 - 5x - 4) \div (x + 2)$ **[4–3]** **16.** $(2x + 9) - \dfrac{x+3}{x-5}$

[4–2] **17.** $\dfrac{16 - b^2}{2b+1} \div (b + 4)$ **[4–2]** **18.** $\dfrac{x^2 - x - 6}{x^2 + x - 12} \cdot \dfrac{x^2 + 3x - 4}{x^2 + 2x - 3}$

[4–4] **19.** Simplify the complex rational expression $\dfrac{\dfrac{1}{y} - \dfrac{4}{x}}{\dfrac{2x - 8y}{xy}}$.   **[4–6]** **20.** Using synthetic division, find $P(-1)$ given $P(x) = x^4 - 2x^3 + x - 1$.

Find the solution set of the following equations.

[4–7] **21.** $\dfrac{y}{y-3} + \dfrac{4}{5} = \dfrac{3}{y-3}$

[6–3] **22.** $3x^2 - 2x + 2 = 0$

[2–4] **23.** $|4x - 3| = 7$

[6–6] **24.** $x^4 - 3x^2 - 10 = 0$

[6–5] **25.** $\sqrt{2x+1} - 3 = 0$

[7–4] **26.** Graph the inequality $3y - 2x \le 12$.

In problems 27 and 28, find the slope of the given line.

[7–2] **27.** Passing through $(-5,6)$ and $(1,2)$

[7–2] **28.** Whose equation is $3x - 5y = 10$

[7–2] **29.** Determine if the lines $2x + 4y = 3$ and $4x - 8y = 6$ are parallel, perpendicular, or neither.

[7–3] **30.** Find the equation of the horizontal line passing through the point $(-1,7)$.

[8–2] **31.** Given $f(x) = 2x - 5$ and $g(x) = x^2 + 5$, find
a. $f(-4)$ b. $(f + g)(x)$ c. $(f \circ g)(x)$
d. $\dfrac{f(a + h) - f(a)}{h}$ $(h \ne 0)$

[8–4] **32.** Given $f(x) = 8x - 1$, find $f^{-1}(x)$.

[8–5] **33.** If x varies directly as the square of y, find constant of variation k if $x = 20$ when $y = 3$. Find x when $y = 7$.

[9–3] **34.** Given the circle $x^2 + y^2 - 6x + 4y - 3 = 0$, find the coordinates of the center and the length of the radius.

[9–1] **35.** Find the vertex and intercepts of the parabola $y = 2x^2 - 4x + 3$.

[9–3] **36.** Find the x- and y-intercepts of the ellipse $3x^2 = 12 - 2y^2$.

[9–4] **37.** Find the intercepts and equations of the asymptotes of the hyperbola $\dfrac{x^2}{9} - \dfrac{y^2}{4} = 1$.

[9–4] **38.** Find the center of the ellipse $x^2 + 2y^2 - 2x + 4y = 1$.

Identify each equation as a circle, a parabola, an ellipse, or a hyperbola.

[9–1] **39.** $4x^2 - y = x + 3$

[9–4] **40.** $4y^2 = 2x^2 + 8$

[9–3] **41.** $3x^2 + 3y^2 = 12$

[9–4] **42.** $9y^2 = 36 - x^2$

[9–4] **43.** $16x^2 + 9y^2 + 64x + 54y + 1 = 0$

Polynomial and Rational Functions

A particle moving along a straight line is displaced at time t in seconds according to the function

$$s(t) = t^3 - 9t^2 + 20t - 12$$

where s is in centimeters. At $t = 2$ seconds, the displacement is 0. At what other times will the displacement be 0?

In chapter 8, we discussed the linear function defined by

$$f(x) = ax + b$$

whose graph was a straight line, and the quadratic function defined by

$$f(x) = ax^2 + bx + c, a \neq 0$$

whose graph was a curve called a parabola. At that time, we said that these are special examples of a larger class of functions called **polynomial functions.**

Definition of a polynomial function

A polynomial function of degree n is any function that can be written in the form

$$f(x) = a_n x^n + a_{n-1} x^{n-1} + a_{n-2} x^{n-2} + \cdots + a_1 x + a_0, a_n \neq 0$$

where the coefficients $a_n, a_{n-1}, a_{n-2}, \ldots, a_1, a_0$ are complex or real numbers.

If the polynomial function in the definition is set equal to zero, we obtain the **polynomial equation of degree n**

$$a_n x^n + a_{n-1} x^{n-1} + a_{n-2} x^{n-2} + \cdots + a_1 x + a_0 = 0, a_n \neq 0$$

Our attention in this chapter will be directed to finding the solutions of the polynomial equation. The solutions (or roots) of the equation are known as the **zeros of the polynomial**

$$a_n x^n + a_{n-1} x^{n-1} + a_{n-2} x^{n-2} + \cdots + a_1 x + a_0$$

10–1 ■ *Descartes' rule of signs and bound of zeros*

In section 4–6, we used the factor theorem and synthetic division to find solutions of polynomial equations. These solutions were the *zeros of the polynomial* forming the equation. Before working with more complicated polynomials, we need to know whether or not a polynomial has a zero. Carl Friedrich Gauss (1777–1855) answered that question when he proved the **fundamental theorem of algebra.**

The fundamental theorem of algebra

Given polynomial $P(x)$ with positive degree, then $P(x)$ has *at least one* complex zero.

To illustrate, by this theorem,

$$5x - 2, \quad z^5 - 5z + 1, \quad \text{and} \quad y^{25} + 2iy - 6$$

all have at least one complex zero. In this section and the following sections, we will study several theorems that provide guidelines and procedures for finding those zeros. The first theorem tells us how many zeros we can expect for a given polynomial.

Theorem

A polynomial $P(x)$ with degree $n > 0$ has *at most n* distinct complex zeros.

To illustrate, by this theorem, the polynomial

$$P(x) = x^4 + 4x^3 + 6x^2 + 4x + 1$$

has at most four distinct complex zeros.

Note Factoring $P(x)$, we obtain $P(x) = (x + 1)^4$ and $P(x)$ has one zero, -1. However, there are four factors of $x + 1$, so -1 is a zero four times. We call -1 a **zero of multiplicity 4.** In general, any zero that occurs k times is called a **zero of multiplicity k.**

Consider the solutions of the quadratic equation $x^2 - 2x + 2 = 0$. Using the quadratic formula (since $x^2 - 2x + 2$ does not factor),

$$x = \frac{-(-2) \pm \sqrt{(-2)^2 - 4(1)(2)}}{2(1)}$$
$$= \frac{2 \pm \sqrt{4 - 8}}{2} = \frac{2 \pm \sqrt{-4}}{2}$$
$$= \frac{2 \pm 2i}{2} = 1 \pm i$$

The solutions are $1 + i$ and $1 - i$. The complex solutions are *conjugates* of each other. Such conjugate pairs always occur. This is stated in the **conjugate zeros theorem.**

> **Conjugate zeros theorem**
>
> Given polynomial $P(x)$ having only real number coefficients, if $a + bi$ is a zero of $P(x)$, then $a - bi$ is also a zero of $P(x)$.
>
> **Concept**
>
> Complex zeros of polynomials with real coefficients occur in conjugate pairs.

Note Let $P(x)$ be a polynomial with rational coefficients. If $a + b\sqrt{c}$ is a zero of $P(x)$ where a, b, and c are rational and $\sqrt{c}$ is irrational, then $a - b\sqrt{c}$ is also a zero of $P(x)$.

■ **Example 10–1 A**

1. Find a polynomial $P(x)$ of lowest degree having real number coefficients and zeros 1 and $2 - i$.

By the conjugate zeros theorem, since $2 - i$ is a zero, then $2 + i$ is also a zero of $P(x)$. Thus, the only zeros of $P(x)$ are 1, $2 - i$, and $2 + i$. By the factor theorem, there must then be three factors $x - 1$, $x - (2 - i)$, and $x - (2 + i)$. Thus,

$$
\begin{aligned}
P(x) &= (x - 1)[x - (2 - i)][x - (2 + i)] \\
&= (x - 1)[(x - 2) + i][(x - 2) - i] \\
&= (x - 1)[(x - 2)^2 - i^2] \\
&= (x - 1)[x^2 - 4x + 4 - (-1)] \\
&= (x - 1)(x^2 - 4x + 5) \\
&= x^3 - 5x^2 + 9x - 5
\end{aligned}
$$

Note Other polynomials such as $3(x^3 - 5x^2 + 9x - 5)$ and $-6(x^3 - 5x^2 + 9x - 5)$ also satisfy the conditions.

2. Find all zeros of $P(x) = x^4 - 6x^3 + 15x^2 - 18x + 10$ if $2 - i$ is a zero of $P(x)$.

Since $2 - i$ is a zero of $P(x)$ and $P(x)$ has real coefficients, then $2 + i$ must also be a zero. We found out in example 1 that $[x - (2 - i)][x - (2 + i)] = x^2 - 4x + 5$. Using long division,

$$
\begin{array}{r}
x^2 - 2x + 2 \\
x^2 - 4x + 5 \overline{\smash{)}\; x^4 - 6x^3 + 15x^2 - 18x + 10} \\
\underline{x^4 - 4x^3 + 5x^2} \\
-2x^3 + 10x^2 - 18x \\
\underline{-2x^3 + 8x^2 - 10x} \\
2x^2 - 8x + 10 \\
\underline{2x^2 - 8x + 10} \\
0
\end{array}
$$

Now find the zeros of the quadratic polynomial quotient $x^2 - 2x + 2$. Since the polynomial is not factorable, we use the quadratic formula.

$$
\begin{aligned}
x &= \frac{-(-2) \pm \sqrt{(-2)^2 - 4(1)(2)}}{2(1)} \\
&= \frac{2 \pm \sqrt{4 - 8}}{2} = \frac{2 \pm \sqrt{-4}}{2} = \frac{2 \pm 2i}{2} \\
&= 1 \pm i
\end{aligned}
$$

The zeros of $P(x)$ are $2 - i$, $2 + i$, $1 + i$, and $1 - i$.

▶ *Quick check* Find all zeros of $P(x) = x^4 - 7x^3 + 18x^2 - 22x + 12$ if $1 - i$ is a zero.

Our objective here is to determine the number of *positive* real zeros and the number of *negative* real zeros of a given polynomial having *only* real number coefficients.

Given a polynomial with real coefficients whose terms are written in descending order, a *variation in sign* occurs when two consecutive terms have different signs. To illustrate, the polynomial

$$P(x) = 3x^4 - 2x^3 - x^2 + 3x - 1$$

has three variations in sign.

Note We ignore terms having zero coefficients (missing terms) when determining the number of variations in sign.

The French mathematician René Descartes, who lived from 1596 to 1650, gave us a rule that is useful in determining the nature of the real zeros of a polynomial. The rule relates the variations in sign to the number of positive and the number of negative zeros of a polynomial.

___ *Descartes' rule of signs* _____

Given polynomial $P(x)$ with real coefficients, leading coefficient positive, constant term nonzero, and terms in descending order, the number of

a. *positive real zeros* of $P(x)$ is either equal to the number of variations in sign of $P(x)$ or is less than the number of variations in sign by a positive even number,

b. *negative real zeros* of $P(x)$ is either equal to the number of variations in sign of $P(-x)$ or is less than the number of variations in sign by a positive even number.

We can now acquire some information about the zeros of a polynomial $P(x)$ without actually finding the zeros.

■ *Example 10-1 B*

Use Descartes' rule of signs to determine the possible number of positive and negative real zeros of each polynomial.

1. $P(x) = x^3 + 2x^2 + 3x + 1$
   a. Since there are no variations of sign in $P(x)$, there are *no positive real zeros*.
   b. Replacing x by $-x$,

$$P(-x) = (-x)^3 + 2(-x)^2 + 3(-x) + 1$$
$$= -x^3 + 2x^2 - 3x + 1$$

Since there are three variations of sign in $P(-x)$, the polynomial $P(x)$ has three or one negative real zeros.

2. $P(x) = 4x^4 + 2x^2 - x - 3$

$$P(x) = 4x^4 + \underbrace{2x^2 - x}_{1} - 3$$

a. Since there is one variation of sign in $P(x)$, the polynomial has one positive real zero.

b. Replacing x by $-x$,

$$P(-x) = 4(-x)^4 + 2(-x)^2 - (-x) - 3$$
$$= 4x^4 + 2x^2 + \underbrace{x - 3}_{1}$$

Since there is one variation of sign in $P(-x)$, the polynomial $P(x)$ has one negative real zero.

Note Since $P(x)$ has degree 4, there are four zeros and two of them are real, then two of them must be nonreal zeros.

3. $P(x) = \underbrace{x^5 - }_{1}\underbrace{2x^3 + }_{2}\underbrace{x - }_{3} 1$

a. There are three variations of sign so the number of positive real zeros is three or one.

b. Replacing x with $-x$,

$$P(-x) = (-x)^5 - 2(-x)^3 + (-x) - 1$$
$$= \underbrace{-x^5 + }_{1}\underbrace{2x^3 - }_{2} x - 1$$

There are two variations in sign so the number of negative real zeros is two or none.

Note The number of possible combinations of positive, negative, and nonreal complex zeros is given in the following table.

Number of positive zeros	Number of negative zeros	Number of nonreal complex zeros
3	2	0
3	0	2
1	2	2
1	0	4

4. $P(x) = x^4 + 4x^2 + 1$

a. There are no variations in sign, so there are no positive real zeros.

b. Replacing x with $-x$,

$$P(-x) = (-x)^4 + 4(-x)^2 + 1$$
$$= x^4 + 4x^2 + 1$$

There are no variations in sign, so there are no negative real zeros.

Note The polynomial then has four nonreal complex zeros and they will occur in conjugate pairs.

▶ *Quick check*   Determine the number of possible positive and negative real zeros of the polynomial $P(x) = 4x^4 - 3x^3 - 2x^2 + x - 4$.   ∎

We can see that sometimes the nature of the zeros can be determined, while in other polynomials, we can merely restrict the nature to a few possibilities. We should also note that in each case the number of nonreal zeros is *always even*.

A second rule that can be used in conjunction with Descartes' rule is used to determine the **upper** and **lower bounds** of the finite interval that contains all real zeros of a polynomial. Given polynomial $P(x)$ with real number coefficients,

1. a *lower bound* for the real zeros of $P(x)$ is any real number that is less than or equal to all the real zeros of $P(x)$,

2. an *upper bound* for the real zeros of $P(x)$ is any real number that is greater than or equal to all the real zeros of $P(x)$.

Note   There is nothing unique about an upper or lower bound since we said *"any* number."

Thus, if a real number a is a lower bound, a real number b is an upper bound, and r is a real zero of $P(x)$, then

$$a \le r \le b$$

___ **Rule for bounds for real zeros of polynomials** _____

Let $P(x)$ be a polynomial with real number coefficients and a positive leading coefficient. Divide $P(x)$ synthetically by $x - c$.

1. If $c > 0$ and all numbers in the *third row* of the division process are either positive or zero, then c is an *upper bound* of the real zeros of $P(x)$.
2. If $c < 0$ and if all numbers in the *third row* of the division process alternate in sign (with 0 considered positive or negative as needed), then c is a *lower bound* of the real zeros of $P(x)$.

Note   If 0 appears in the quotient, 0 can be regarded as being either positive or negative to help the signs alternate in part 2 of the rule.

■ *Example 10–1 C*

Use the upper- and lower-bound rule to find integers that are upper and lower bounds for the real zeros of

$$P(x) = x^4 - x^3 - 12x^2 - 2x + 3$$

If we divide synthetically by $x + 2$,

```
-2 | 1   -1   -12   -2     3
   |      -2     6   12   -20
   ┗━━━━━━━━━━━━━━━━━━━━━━━━━━━
     1   -3    -6   10   -17
```

Since the signs do not alternate, -2 is not a lower bound. We now try -3.

```
-3 | 1   -1   -12   -2   3
   |      -3    12    0   6
   ┗━━━━━━━━━━━━━━━━━━━━━━━
     1   -4     0   -2   9
```

If we consider 0 in the quotient to be positive $(+)$, the signs alternate so -3 is a lower bound of the real zeros of $P(x)$.

To find the upper bound, we try 3, so we synthetically divide by $x - 3$.

$$
\begin{array}{r|rrrrr}
3 & 1 & -1 & -12 & -2 & 3 \\
  &   & 3 & 6   & -18 & -60 \\
\hline
  & 1 & 2 & -6 & -20 & -57
\end{array}
$$

Since not every sign in the third row is positive, 3 is not an upper bound. We try 4 and 5.

$$
\begin{array}{r|rrrrr}
4 & 1 & -1 & -12 & -2 & 3 \\
  &   & 4 & 12 & 0 & -8 \\
\hline
  & 1 & 3 & 0   & -2 & -5
\end{array}
\qquad
\begin{array}{r|rrrrr}
5 & 1 & -1 & -12 & -2 & 3 \\
  &   & 5 & 20 & 40 & 190 \\
\hline
  & 1 & 4 & 8   & 38 & 193
\end{array}
$$

Since division by $x - 5$ yields a quotient with positive coefficients, then 5 is an upper bound of the real zeros of $P(x)$. Thus, the real zeros of

$$P(x) = x^4 - x^3 - 12x^2 - 2x + 3$$

are in the open interval $(-3,5)$.

Note The upper- and lower-bound rule will give us an interval that is guaranteed to contain all real zeros of $P(x)$, but it may not be the *smallest* interval that contains all of the zeros.

As you can see, determining upper and lower bounds is a trial-and-error process of choosing values of c such that $c = 1, 2, 3, \ldots$ and $c = -1, -2, -3, \ldots$.

▶ *Quick check* Use the upper- and lower-bound rule to find integers that are upper and lower bounds of the real zeros of $P(x) = 18x^3 - 3x^2 - 37x + 12$. ■

Mastery points

Can you

- Find a polynomial $P(x)$ of lowest degree having real number coefficients with given complex zeros?
- Find all zeros of a polynomial $P(x)$, given one zero of the polynomial?
- Use Descartes' rule of signs to determine the number of positive and negative real zeros of $P(x)$?
- Determine a lower and an upper bound of the real zeros of polynomial $P(x)$?

Exercise 10–1

Find the polynomial of lowest degree with real number coefficients having the given zeros. See example 10–1 A–1.

Example $1 - i$ and 2

Solution Since $1 - i$ is a zero, then $1 + i$ is a zero by the conjugate zeros theorem. Thus,

$$\begin{aligned} P(x) &= (x - 2)[x - (1 - i)][x - (1 + i)] \\ &= (x - 2)[(x - 1) + i][(x - 1) - i] \\ &= (x - 2)[(x - 1)^2 - i^2] \\ &= (x - 2)[x^2 - 2x + 1 - (-1)] \\ &= (x - 2)(x^2 - 2x + 2) \\ &= x^3 - 4x^2 + 6x - 4 \end{aligned}$$

1. i and 2
2. $-i$ and 5
3. $2 + i$ and -3
4. $3 - i$ and 4

5. $3 - 2i$ and 1
6. $4 + 3i$ and -4
7. $4 + i$ and $1 - i$
8. $5 - i$ and $6 + i$

9. $2 + 3i$ and $1 - 2i$
10. $5 + 4i$ and $-2 - 3i$
11. $2 + i, 1 - i,$ and -1
12. $4 + 3i, 2 - i,$ and 3

13. $4 - 3i,$ 2 and 1
14. $5 + 2i, -1,$ and 5
15. $1 - \sqrt{3}$ and 4
16. $3 + \sqrt{2}$ and -3

17. $2 + \sqrt{3}, 1 - \sqrt{2},$ and 1
18. $5 + \sqrt{5}, 3 - \sqrt{3},$ and -2

19. $4 + 3i$ and -3 (multiplicity 2)
20. $-3 + 4i$ and 2 (multiplicity 2)

Find all zeros of each polynomial when one zero is given. See example 10–1 A–2.

Example $P(x) = x^4 - 7x^3 + 18x^2 - 22x + 12; 1 - i$

Solution Since $1 - i$ is a zero and all coefficients are real, then $1 + i$ is a zero by the conjugate zeros theorem.
Now $[x - (1 - i)][x - (1 + i)] = [(x - 1) + i][(x - 1) - i]$
$$\begin{aligned} &= (x - 1)^2 - i^2 \\ &= x^2 - 2x + 1 - (-1) = x^2 - 2x + 2. \end{aligned}$$

Using long division, we divide

$$\begin{array}{r} x^2 - 5x + 6 \\ x^2 - 2x + 2\overline{)x^4 - 7x^3 + 18x^2 - 22x + 12} \\ \underline{x^4 - 2x^3 + 2x^2} \\ -5x^3 + 16x^2 - 22x \\ \underline{-5x^3 + 10x^2 - 10x} \\ 6x^2 - 12x + 12 \\ \underline{6x^2 - 12x + 12} \\ 0 \end{array}$$

We now find the zeros of the quadratic polynomial quotient $x^2 - 5x + 6$.

$x^2 - 5x + 6 = (x - 3)(x - 2)$ Factor
$x - 3 = 0$ or $x - 2 = 0$ Set each factor equal to 0
$x = 3$ $x = 2$ Solve each equation

The zeros of $P(x)$ are $1 - i, 1 + i,$ 2, and 3.

21. $P(x) = x^3 - 7x^2 + 14x - 8;$ 4
22. $P(x) = x^3 + 8x^2 + 11x - 18;$ -3

23. $P(x) = x^3 + x^2 - 20x - 50;$ $-3 - i$
24. $P(x) = x^3 - 13x^2 + 59x - 87;$ $5 - 2i$

25. $P(x) = 2x^3 + 2x^2 - 3x - 1;$ 1
26. $P(x) = 3x^3 + 8x^2 - 8x - 15;$ -3

27. $P(x) = x^4 + 10x^3 + 38x^2 + 66x + 45; -2 + i$ 28. $P(x) = x^4 - 3x^3 + 6x^2 + 2x - 60; 1 + 3i$

29. $P(x) = x^4 + 8x^3 + 26x^2 + 72x + 135; 3i$ 30. $P(x) = x^4 - 6x^3 - 3x^2 - 24x - 28; -2i$

31. $P(x) = 2x^4 - 17x^3 + 137x^2 - 57x - 65; 4 + 7i$ 32. $P(x) = 4x^4 - 25x^3 + 19x^2 + 14x - 12; 3 - 2i$

Using Descartes' rule of signs, determine the number of possible positive and negative real zeros of each polynomial. See example 10–1 B.

Example $P(x) = 4x^4 - 3x^3 - 2x^2 + x - 4$

Solution   a. $P(x) = 4x^4 - 3x^3 - 2x^2 + x - 4$

Since there are three variations of sign, the number of positive real zeros is three or one.

b. Replace x with $-x$.

$$P(-x) = 4(-x)^4 - 3(-x)^3 - 2(-x)^2 + (-x) - 4$$
$$= 4x^4 + 3x^3 - 2x^2 - x - 4$$

Since there is one variation of sign, the polynomial has one negative real zero.

33. $P(x) = x^3 + 2x^2 - 3x - 1$ 34. $P(x) = x^3 + 5x^2 - 3x - 4$

35. $P(x) = x^3 - 5x^2 - 6x + 10$ 36. $P(x) = x^3 - x^2 - 2x + 1$

37. $P(x) = x^3 + 4x^2 + x + 5$ 38. $P(x) = -x^3 - 4x^2 - 3x - 2$

39. $P(x) = x^4 - 3x^3 + 4x^2 - x - 1$ 40. $P(x) = 5x^4 + 3x^3 - 2x^2 + 2x + 5$

41. $P(x) = -x^4 + 3x^3 + 5x - 6$ 42. $P(x) = -3x^4 - 3x^2 + 5x - 7$

In exercises 43–46, use Descartes' rule of signs to find the number of real positive, real negative, and nonreal complex solutions of each equation.

43. $4x^3 + 2x^2 - x + 1 = 0$ 44. $-2x^3 - 7x^2 - 5x - 4 = 0$

45. $4x^4 = -6$ 46. $x^4 + 5x^2 - 3x = 4$

Find an upper and a lower bound of the real zeros of each polynomial. See example 10–1 C.

Example $P(x) = 18x^3 - 3x^2 - 37x + 12$

Solution   We try several synthetic divisions. Using $x - 1$ and $x - 2$,

```
1 ) 18    -3    -37     12
          18     15    -22
    ─────────────────────────
    18    15    -22    -10  ← Contains some negative numbers
```

```
2 ) 18    -3    -37     12
          36     66     58
    ─────────────────────────
    18    33     29     70  ← All positive numbers
```

Since division by $x - 2$ yields all positive numbers in the third row, 2 is an upper bound of the real zeros of $P(x)$.

To find a lower bound, we try $x + 1$ and $x + 2$.

$$
\begin{array}{r|rrrr}
-1 & 18 & -3 & -37 & 12 \\
   &   & -18 & 21 & 16 \\
\hline
   & 18 & -21 & -16 & 28 \leftarrow \text{Signs do not alternate}
\end{array}
$$

$$
\begin{array}{r|rrrr}
-2 & 18 & -3 & -37 & 12 \\
   &   & -36 & 78 & -82 \\
\hline
   & 18 & -39 & 41 & -70 \leftarrow \text{Signs alternate}
\end{array}
$$

Since the signs alternate in the third row when dividing by $x + 2$, -2 is a lower bound of the real zeros of $P(x)$. The real zeros of $P(x)$ are in the open interval $\{x \mid -2 < x < 2\} = (-2,2)$.

47. $P(x) = x^2 - 6x - 7$

48. $P(x) = 3x^2 + x - 3$

49. $P(x) = x^3 - 4x^2 - 5x + 7$

50. $P(x) = 2x^3 - 5x^2 + 4x - 8$

51. $P(x) = x^4 - x^3 - 2x^2 + 3x + 6$

52. $P(x) = 3x^4 - 4x^3 + 5x^2 - 2x - 4$

53. $P(x) = x^5 + x^4 - 8x^3 - 8x^2 + 15x + 15$

54. $P(x) = x^5 + 2x^4 + x^3 - 2x^2 + x - 1$

55. $P(x) = 2x^5 - 13x^3 + 2x - 5$

56. $P(x) = 3x^5 + 2x^4 - x^3 - 8x - 7$

57. How many possible complex solutions does the equation $x^9 = 4$ have?

58. How many solutions does the equation $x^{25} = 27$ have?

59. If one solution of the equation $x(4x^5 - 2) = 4x$ is 0, how many other possible solutions does the equation have?

60. Consider the equation $x^4 + ax^2 + cx - d = 0$, where a, c, and d are positive. Show that the equation has one positive, one negative, and two nonreal complex solutions.

Review exercises

Find the solution set of the following quadratic equations. See sections 6–1 and 6–3.

1. $4x^2 - 8x - 5 = 0$

2. $x^2 + 5x + 5 = 0$

3. $-2x^2 + x = 4$

Find the value of the following using synthetic division. See section 4–6.

4. $P(-2)$ when $P(x) = 4x^3 + 2x^2 - x - 5$

5. $P(4)$ when $P(x) = x^4 - x^2 + x + 3$

6. Find the inverse of the function f given $f(x) = x^3 - 4$. See section 8–4.

10-2 ■ Rational zeros of a polynomial function

In section 10–1, Descartes' rule of signs and the upper- and lower-bound rules provide us with useful information about zeros of polynomials. However, they *do not* provide us with a method of determining the *actual numerical values* of the zeros. In this section, we study methods for finding the zeros of a polynomial with real coefficients.

Should the coefficients of polynomial $P(x)$ be rational numbers, it will be necessary that we multiply each term by the least common multiple (LCM) of the denominators to obtain integral coefficients. We can then use the following theorem to find all rational zeros of the polynomial.

> ### Theorem on rational zeros
>
> Let $P(x) = a_n x^n + a_{n-1} x^{n-1} + \cdots + a_1 x + a_0 \ (a_n \neq 0)$ be a polynomial with integral coefficients. If $\dfrac{p}{q}$ is a rational zero of $P(x)$ reduced to lowest terms, then p is a factor of the constant term a_0 and q is a factor of the leading coefficient a_n.

Proof   Given polynomial $P(x) = a_n x^n + a^{n-1} x^{n-1} + \cdots + a_1 x + a_0$, since $\dfrac{p}{q}$ is a zero of $P(x)$, then $P\!\left(\dfrac{p}{q}\right) = 0$ and we have the equation

$$a_n\left(\frac{p}{q}\right)^n + a_{n-1}\left(\frac{p}{q}\right)^{n-1} + \cdots + a_1\left(\frac{p}{q}\right) + a_0 = 0$$

To clear the denominators, we multiply each term by the least common denominator q^n to obtain the equation

$$a_n p^n + a_{n-1} p^{n-1} q + a_{n-2} p^{n-2} q^2 + \cdots + a_1 p q^{n-1} + a_0 q^n = 0$$

Subtracting $a_0 q^n$ from each member of the equation,

$$a_n p^n + a_{n-1} p^{n-1} q + a_{n-2} p^{n-2} q^2 + \cdots + a_1 p q^{n-1} = -a_0 q^n$$

We now factor p from each term in the left member. Then

$$p(a_n p^{n-1} + a_{n-1} p^{n-2} q + \cdots + a_1 q^{n-1}) = -a_0 q^n$$

Since p, q, and all coefficients $a_1, a_2, \ldots, a_n$ are integers, then the quantity in parentheses must be an integer since it is the product and sum of integers. Then p is a factor of the right member $-a_0 q^n$. But p and q have no common factors since $\dfrac{p}{q}$ is reduced to lowest terms. Therefore p must be a factor of a_0. We can also write the original equation in the form

$$q(a_{n-1} p^{n-1} + a_{n-2} p^{n-2} q + \cdots + a_0 q^{n-1}) = -a_n p^n$$

In similar fashion, q is a factor of the left member and so must be a factor of the right member, $-a_n p^n$. But p and q have no common factors, so q must be a factor of a_n. The theorem is proved.

The technique using the rational zero theorem to find all rational zeros of a polynomial with integral coefficients is illustrated in the following examples.

■ Example 10–2 A

1. List all possible rational zeros of the polynomial

$$P(x) = \frac{1}{2}x^4 - \frac{2}{3}x^3 + 2x^2 - \frac{1}{3}x + 2.$$

We first note that there are rational coefficients. To obtain integral coefficients, multiply each term by the LCD, 6. Then

$$6P(x) = 3x^4 - 4x^3 + 12x^2 - 2x + 12$$

Since the zeros of $P(x)$ make $P(x) = 0$, we can ignore the coefficient 6 in the left member. We now list all the possibilities for a_0 and a_4.

a. $a_0 = 12$; the possible numerators (p) are

$\pm 1, \pm 2, \pm 3, \pm 4, \pm 6, \pm 12$

b. $a_4 = 3$; the possible denominators (q) are

$\pm 1, \pm 3$

c. Using factors of 12 divided by ± 1, we get $\pm 1, \pm 2, \pm 3, \pm 4, \pm 6, \pm 12$ and using factors of 12 divided by ± 3, we get $\pm \dfrac{1}{3}, \pm \dfrac{2}{3}, \pm \dfrac{4}{3}$. The possible rational zeros are $\pm 1, \pm 2, \pm 3, \pm 4, \pm 6, \pm 12, \pm \dfrac{1}{3}, \pm \dfrac{2}{3}, \pm \dfrac{4}{3}$.

▶ *Quick check* List all of the possible rational zeros of $P(x) = 5x^4 - 3x^3 + 2x^2 + x - 4$.

2. Find the rational zeros of the polynomial

$P(x) = x^4 + x^3 + x^2 + 3x - 6$.

Since $a_4 = 1$, we need only consider the factors of $a_0 = -6$ as possible rational zeros. The possibilities are $\pm 1, \pm 2, \pm 3, \pm 6$. Because $P(x)$ is of fourth degree, there are four zeros. Using Descartes' rule of signs, there is one positive real zero and three or one negative real zeros.

We now begin checking each of the possibilities

$-6, -3, -2, -1, 1, 2, 3, 6$.

Our strategy is to first check positive rational zeros starting with 1 and working up.

To start, consider $x = 1$.

$$
\begin{array}{r|rrrr}
1 & 1 & 1 & 1 & 3 & -6 \\
  &   & 1 & 2 & 3 & 6 \\
\hline
  & 1 & 2 & 3 & 6 & 0 \;\leftarrow\; x - 1 \text{ is a factor}
\end{array}
$$

Since, by Descartes' rule of signs, there is only one positive real zero and we have found that, 1, we can eliminate 2, 3, and 6 as possibilities. Now, we consider $-6, -3, -2,$ and -1.

We now have $P(x) = (x - 1)(x^3 + 2x^2 + 3x + 6)$ and we consider a possible negative zero, starting with -1 and going down, in $x^3 + 2x^2 + 3x + 6$.

$$
\begin{array}{r|rrrr}
-1 & 1 & 2 & 3 & 6 \\
   &   & -1 & -1 & -2 \\
\hline
   & 1 & 1 & 2 & 4 \;\leftarrow\; \text{Remainder 4, } -1 \text{ not a zero}
\end{array}
$$

$$
\begin{array}{r|rrrr}
-2 & 1 & 2 & 3 & 6 \\
   &   & -2 & 0 & -6 \\
\hline
   & 1 & 0 & 3 & 0 \;\leftarrow\; \text{Remainder 0, } -2 \text{ is a zero}
\end{array}
$$

Thus, $x + 2$ is a factor and we now have

$$P(x) = (x - 1)(x + 2)(x^2 + 3)$$

Set $x^2 + 3 = 0$ and solve for x.

$$x^2 + 3 = 0$$
$$x^2 = -3 \qquad \text{Subtract 3 from each member}$$
$$x = \pm\sqrt{-3} \qquad \text{Extract the roots}$$
$$x = \pm i\sqrt{3} \qquad \sqrt{-3} = i\sqrt{3}$$

The rational zeros of $P(x)$ are 1 and -2, since $-i\sqrt{3}$ and $i\sqrt{3}$ are imaginary numbers.

3. Find the solution set of the equation $x^4 - 2x^3 - 17x^2 + 18x + 72 = 0$.

   To find the solution set of the equation, we want the zeros of the polynomial $P(x) = x^4 - 2x^3 - 17x^2 + 18x + 72$. We first list the possible rational solutions.
   a. p (the numerator) can be ± 1, ± 2, ± 3, ± 4, ± 6, ± 8, ± 9, ± 12, ± 18, ± 24, ± 36, ± 72.
   b. q (the denominator) can be ± 1.
   c. The possible rational solutions are ± 1, ± 2, ± 3, ± 4, ± 6, ± 8, ± 9, ± 12, ± 18, ± 24, ± 36, ± 72.

   We first check for positive rational solutions, of which there are two or none. Let $x = 1$, $x = 2$, $x = 3$, $x = 4$, and so on.

$1)$	1	-2	-17	18	72
		1	-1	-18	0
	1	-1	-18	0	72

$2)$	1	-2	-17	18	72
		2	0	-34	-32
	1	0	-17	-16	40

$3)$	1	-2	-17	18	72
		3	3	-42	-72
	1	1	-14	-24	0 $\leftarrow$ $x - 3$ is a factor

   Thus, 3 is a solution of the equation which can be written $(x - 3)(x^3 + x^2 - 14x - 24) = 0$. Since there are two or none, positive real solutions, we continue with $x = 4$. Divide $x - 4$ into $x^3 + x^2 - 14x - 24$.

$4)$	1	1	-14	-24
		4	20	24
	1	5	6	0 $\leftarrow$ $x - 4$ is a factor

   Now 3 and 4 are positive rational solutions and the equation can be written $(x - 3)(x - 4)(x^2 + 5x + 6) = 0$. Factoring $x^2 + 5x + 6 = (x + 2)(x + 3)$ and so $(x - 3)(x - 4)(x + 2)(x - 3) = 0$. Since $x = -2$ when $x + 2 = 0$ and $x = -3$ when $x + 3 = 0$, the remaining solutions are -2 and -3. The solution set is $\{-3, -2, 3, 4\}$.

4. Using the theorem on rational zeros, argue that $\sqrt{7}$ is not a rational number.

   Given the equation $x^2 - 7 = 0$, the solution set is $\{-\sqrt{7}, \sqrt{7}\}$. However, by the theorem on rational zeros, the possible rational solutions of this equation would be ± 1 and ± 7. Since $\sqrt{7}$ is a solution of this equation but is none of the possible rational solutions of the equation $x^2 - 7 = 0$, $\sqrt{7}$ is not a rational number.

▶ *Quick check*   Find the solution set of the equation

$$x^4 - \frac{1}{6}x^3 + \frac{2}{3}x^2 - \frac{1}{6}x - \frac{1}{3} = 0$$

___ *Mastery points* ___

Can you

- Determine all possible rational zeros of a polynomial with integral coefficients?
- Find the rational zeros of a polynomial $P(x)$ having rational coefficients?
- Find all solutions of a polynomial equation with integral coefficients?

Exercise 10-2

Using the theorem on rational zeros, list all the possible rational zeros of the following polynomials. See example 10-2 A-1.

Example   List the possible rational zeros of the polynomial $P(x) = 5x^4 - 3x^3 + 2x^2 + x - 4$.

Solution   1. Since $a_0 = 4$, the possible numerators (p) are $\pm 1, \pm 2, \pm 4$.
2. Since $a_4 = 5$, the possible denominators (q) are $\pm 1, \pm 5$.

3. The possible rational zeros are $\pm 1, \pm 2, \pm 4, \pm \dfrac{1}{5}, \pm \dfrac{2}{5}, \pm \dfrac{4}{5}$.

1. $P(x) = x^3 + 2x^2 - x - 2$
2. $P(x) = x^3 - x^2 - 10x - 8$
3. $P(x) = x^4 - 10x^3 + 35x^2 - 50x + 24$
4. $P(x) = 2x^4 - 3x^3 + 2x^2 - 3x - 3$
5. $P(x) = 4x^4 - 8x^3 - x^2 + 8x - 8$
6. $P(x) = 3x^4 - 14x^3 + 11x^2 + 16x - 12$

Find all zeros of the following polynomials. See example 10-2 A-2.

Example   Find the rational zeros of $P(x) = x^4 - x^3 + 2x^2 - 4x - 8$.

Solution   Since $a_0 = -8$ (the numerator) and $a_4 = 1$ (the denominator), the possible real zeros are $\pm 1, \pm 2, \pm 4,$ ± 8. Using Descartes' rule of signs, there are three or one possible positive real zeros and one negative real zero. Through trial and error, we find

1. $-1 \overline{)\;1 \quad -1 \quad 2 \quad -4 \quad -8\;}$
   $\phantom{-1 \overline{)\;1}}\;\;-1 \quad 2 \quad -4 \quad 8$
   $\overline{\;\;1 \quad -2 \quad 4 \quad -8 \quad \;\;0}$ $\quad$ $x + 1$ is a factor

2. Using the quotient $x^3 - 2x^2 + 4x - 8$, we try 2.

   $2 \overline{)\;1 \quad -2 \quad 4 \quad -8\;}$
   $\phantom{2 \overline{)\;1}}\;\;\;2 \quad 0 \quad 8$
   $\overline{\;\;1 \quad \;\;0 \quad 4 \quad \;\;0}$ $\quad$ $x - 2$ is a factor

Thus, $P(x) = (x + 1)(x - 2)(x^2 + 4)$ and if we set

$x^2 + 4 = 0$
$\quad x^2 = -4$ $\qquad$ Subtract 4 from each member
$\quad x^2 = \pm\sqrt{-4}$ $\qquad$ Extract the roots
$\quad\; x = \pm 2i$

The rational zeros of $P(x)$ are $-1, 2, -2i,$ and $2i$.

7. $P(x) = x^3 + x^2 - 14x - 24$ 8. $P(x) = x^3 + 2x^2 - x - 2$

9. $P(x) = x^4 - 10x^3 + 35x^2 - 50x + 24$ 10. $P(x) = x^4 + 3x^3 - 13x^2 - 9x + 30$

11. $P(x) = 4x^4 + x^3 + x^2 + x - 3$ 12. $P(x) = 6x^4 - x^3 - 5x^2 + 2$

13. $P(x) = 2x^5 - 13x^4 + 26x^3 - 22x^2 + 24x - 9$ 14. $P(x) = 4x^5 - 12x^4 + 15x^3 - 45x^2 - 4x + 12$

Find the solution set of the following equations. See example 10–2 A–3.

Example $x^4 - \dfrac{1}{6}x^3 + \dfrac{2}{3}x^2 - \dfrac{1}{6}x - \dfrac{1}{3} = 0$

Solution   Multiply each member of the equation by 6 to eliminate fractions. Then

$$6x^4 - x^3 + 4x^2 - x - 2 = 0$$

The possible values of

1. p are ± 1, ± 2,
2. q are ± 1, ± 2, ± 3, ± 6.

3. The possible rational solutions are ± 1, ± 2, $\pm \dfrac{1}{2}$, $\pm \dfrac{1}{3}$, $\pm \dfrac{1}{6}$, and $\pm \dfrac{2}{3}$.

Descartes' rule of signs indicates there are three or one positive rational solutions. Let $x = 1$.

```
1 ⟌ 6   -1   4   -1   -2
          6   5    9    8
    ─────────────────────────
      6   5   9    8    6   ← Remainder 6, 1 is not a solution, 1 is an upper bound of real solutions
```

We try a number less than 1, say $\dfrac{2}{3}$.

```
2/3 ⟌ 6   -1   4   -1   -2
            4    2   4    2
     ─────────────────────────
       6   3    6   3    0   ← Remainder 0, 2/3 is a solution
```

By Descartes' rule of signs, there is one negative rational solution. We now have
$(2x - 3)(6x^3 + 3x^2 + 6x + 3) = 0$ so we divide $6x^3 + 3x^2 + 6x + 3$ by $x + 1$. Let $x = -1$.

```
-1 ⟌ 6    3   6    3
          -6   3   -9
    ─────────────────────────
      6   -3   9   -6   ← Remainder -6, -1 is not a solution, alternating signs so -1 is a lower bound
```

We try a negative number greater than -1, say $-\dfrac{1}{2}$.

```
-1/2 ⟌ 6    3   6    3
            -3   0   -3
     ─────────────────────────
       6    0   6    0   ← Remainder 0, -1/2 is a solution
```

Set the quotient equal to 0 and solve.

$$6x^2 + 6 = 0$$
$$x^2 + 1 = 0 \qquad \text{Divide each term by 6}$$
$$x^2 = -1 \qquad \text{Subtract 1 from each member}$$
$$x = \pm\sqrt{-1} \qquad \text{Extract the roots}$$
$$x = \pm i$$

The solution set is $\left\{ -\dfrac{1}{2}, \dfrac{2}{3}, -i, i \right\}$.

15. $x^3 - 2x^2 - 5x + 6 = 0$

16. $x^3 - 2x^2 - x + 2 = 0$

17. $x^4 + 4x^3 + 6x^2 + 4x + 1 = 0$

18. $x^4 - x^3 - 13x^2 + x + 12 = 0$

19. $x^5 - 2x^4 - 2x^3 + 4x^2 + x - 2 = 0$

20. $x^5 - 6x^4 - x^3 - 2x - 3 = 0$

21. $2x^3 - 3x^2 - 17x + 30 = 0$

22. $2x^3 - 5x^2 + 8x - 3 = 0$

23. $4x^3 - 8x^2 - 15x + 9 = 0$

24. $6x^3 + 19x^2 + x - 6 = 0$

25. $3x^4 - 11x^3 - x^2 + 19x + 6 = 0$

26. $8x^4 + 16x^3 - 26x^2 - 12x + 15 = 0$

27. $4x^4 - x^3 + 5x^2 - 2x - 6 = 0$

28. $16x^4 - 16x^3 - 29x^2 - 12x + 15 = 0$

29. $4x^5 + 12x^4 - 41x^3 - 99x^2 + 10x + 24 = 0$

30. $8x^5 - 4x^4 + 6x^3 - 3x^2 - 2x + 1 = 0$

31. $\frac{1}{4}x^4 + x^3 + \frac{3}{2}x^2 + x + \frac{1}{4} = 0$

32. $\frac{1}{3}x^3 + \frac{1}{6}x^2 - \frac{13}{6}x + 1 = 0$

Using the theorem of rational zeros, argue that the following are not rational numbers. See example 10–2 A–4.

Example Using the theorem of rational zeros, argue that $\sqrt[3]{4}$ is not a rational number.

Solution Given $x^3 - 4 = 0$, the solution set is $\{\sqrt[3]{4}\}$. However, by the theorem, the possible rational solutions are $\pm 1, \pm 2, \pm 4$. Since $\sqrt[3]{4}$ is a solution of this equation but is not one of the possible rational solutions of $x^3 - 4 = 0$. Thus, $\sqrt[3]{4}$ is not a rational number.

33. $\sqrt{2}$ 34. $\sqrt[3]{7}$ 35. $1 + \sqrt{2}$ 36. $2 - \sqrt{3}$

37. Argue that $x^n + c = 0$ has no real solutions if n is an even positive integer and c is a positive constant.

38. Argue that $x^n - c = 0$ has exactly two real number solutions if n is an even positive integer and c is a positive constant.

39. Show that the equation $x^4 + 3x^2 + 2 = 0$ has no rational solutions.

40. Show that the equation $x^5 - x^3 + 3 = 0$ has no rational solutions.

Review exercises

Express the following intervals in set-builder notation. See section 2–5.

1. $[-2,3]$ **2.** $(-\infty,4]$ **3.** $(-6,+\infty)$

Evaluate the following polynomials for the given value. See section 1–5.

4. $3x^2 - x + 2; -2$ **5.** $-4x^2 - 2x + 4; \frac{1}{2}$ **6.** $2y^2 - y + 1; 1.5$

Graph the following equations. See sections 7–1 and 9–1.

7. $3x - 4y = -12$ **8.** $y = x^2 - 2x - 3$

10–3 ■ Irrational zeros of polynomials

In section 10–2, we discussed rational zeros of a polynomial. There are methods useful for finding the zeros of *any* polynomial through the fourth degree. This section will provide techniques for approximating real zeros of polynomials of higher degree, even though the exact values of the zeros cannot be determined. The method for finding these irrational zeros involves starting at a rough approximation and then through successive approximations getting closer and closer to the exact value. This method works well when using a computer or calculator.

To do this, we use a property of polynomials that we call the **change-of-sign property.**

Change-of-sign property for polynomials
Let $P(x)$ be a polynomial with real number coefficients. If a and b are real numbers and $P(a)$ and $P(b)$ have opposite signs then there exists *at least one* real number r between a and b such that $P(r) = 0$.

The proof of this property requires the use of calculus. However, we can intuitively see its validity if we consider the graph of continuous polynomial function $y = f(x)$ in figure 10–1.

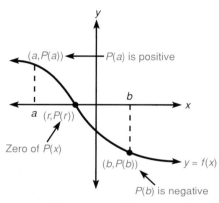

Figure 10–1

If $P(a)$ and $P(b)$ have opposite signs, then points $(a,P(a))$ and $(b,P(b))$ are on opposite sides of the x-axis and the graph of f *must* cross the x-axis at least once at some point $(r,P(r))$ between a and b. Then r is a zero of $P(x)$.

■ *Example 10–3 A*

Show that the polynomial $P(x) = 2x^2 + x - 5$ has a zero in the interval $(1,2)$. We evaluate $P(x)$ when $x = 1$ and when $x = 2$.

1. Let $x = 1$, then

   $$P(1) = 2(1)^2 + (1) - 5 \qquad \text{Replace } x \text{ with 1}$$
   $$= -2 \leftarrow \text{Negative}$$

2. Let $x = 2$, then

   $$P(2) = 2(2)^2 + (2) - 5 \qquad \text{Replace } x \text{ with 2}$$
   $$= 5 \leftarrow \text{Positive}$$

   Since $P(1) = -2$ and $P(2) = 5$, by the change-of-sign property, there exists some r such that $1 < r < 2$ and $P(r) = 0$.

▶ *Quick check* Show that $P(x) = x^3 - 2x^2 - x + 1$ has a zero in the interval $(2,3)$.

We use the change-of-sign property and a procedure we call *bisecting intervals* to find approximations for a zero of the polynomial. To do this, we take the following steps.

1. By trial and error, using synthetic division, find a closed interval $[a,b]$ such that $P(a)$ and $P(b)$ have opposite signs.

2. Using $r = \dfrac{a + b}{2}$, divide the interval (a,b) into sub-intervals $[a,r]$ and (r,b). Then r is the midpoint of $[a,b]$ and so bisects the interval.

3. Evaluate $P(r)$. If $P(r) = 0$, then r is a zero of the polynomial. If $P(a)$ and $P(r)$ have opposite signs, then P has a zero in the interval (a,r). If $P(r)$ and $P(b)$ have opposite signs, then P has a zero in the interval (r,b).

4. If there is a zero in (a,r), then find the midpoint of the interval (a,r), which is $\dfrac{a + r}{2}$ and repeat step 3. If there is a zero in (r,b), do the same.

5. Continue this bisecting of intervals until you get as close an approximation of the zero as you desire.

■ *Example 10-3 B*

Find an approximation of the zero of $P(x) = 2x^2 + x - 5$ on the closed interval $[1,2]$ to the nearest hundredth by bisecting three times.

1. The midpoint of $[1,2]$ is $\dfrac{1 + 2}{2} = \dfrac{3}{2} = 1.5$. Evaluating P when $x = 1, 1.5$, and 2,

$$P(1) = -2 \qquad P(1.5) = 1 \qquad P(2) = 5$$

        Opposite signs   Same signs

2. Consider the midpoint of $[1,1.5]$, which is $\dfrac{1 + 1.5}{2} = 1.25$. We now evaluate P when $x = 1, 1.25$, and 1.5.

$$P(1) = -2 \qquad P(1.25) = -0.625 \qquad P(1.5) = 1$$

        Same signs   Opposite signs

3. Consider the midpoint of $[1.25,1.5]$, which is $\dfrac{1.25 + 1.5}{2} = 1.375$. We now evaluate P when $x = 1.25, 1.375$, and 1.5.

$$P(1.25) = -0.625 \qquad P(1.375) = 0.15625 \qquad P(1.5) = 1$$

        Opposite signs   Same signs

Since 0.15625 is closer to zero than -0.625, we approximate $r \approx 1.375 \approx 1.38$, correct to the nearest hundredth.

▶ *Quick check* Find an approximation, to the nearest hundredth, of the zero of $P(x) = 4x^2 + 2x - 3$ on the closed interval $[0,1]$ by bisecting three times. ■

■ *Example 10–3 C*

Find an approximation of $\sqrt{3}$ correct to the nearest hundredth using bisecting intervals by bisecting three times.

We know that $\sqrt{3}$ is a zero of $P(x) = x^2 - 3$. Note that $P(1) = (1)^2 - 3 = -2$ and $P(2) = (2)^2 - 3 = 1$. Thus, $P(1)$ and $P(2)$ have opposite signs and we know $\sqrt{3}$ lies in the interval $[1,2]$.

1. The midpoint of $[1,2]$ is 1.5. Consider

$$P(1) = -2 \qquad P(1.5) = -0.75 \qquad P(2) = 1$$

Same signs   Opposite signs

2. The midpoint of $[1.5,2]$ is 1.75. Consider

$$P(1.5) = -0.75 \qquad P(1.75) = 0.0625 \qquad P(2) = 1$$

Opposite signs   Same signs

3. The midpoint of $[1.5,1.75]$ is 1.625. Consider

$$P(1.5) = -0.75 \qquad P(1.625) = -0.1523438 \qquad P(1.75) = 0.0625$$

Same signs   Opposite signs

Since 0.0625 is closer to zero than -0.1523438, we approximate $\sqrt{3} \approx 1.75$.

Note Correct to three decimal places $\sqrt{3} \approx 1.732$. By continuing to bisect the interval, we could obtain a closer approximation.

▶ *Quick check* Find an approximation of $\sqrt{5}$, correct to the nearest hundredth, using bisecting three times. ■

┌─ *Mastery points* ──────────────────────

Can you

■ Show that a polynomial has a zero in a given interval using the change-of-sign property?
■ Approximate the value of a zero in an interval using the bisecting intervals method?
■ Approximate the root of a number using the bisecting intervals method?

Exercise 10–3

Show that each given polynomial $P(x)$ has at least one real zero in the given closed interval. Use your calculator where necessary. See example 10–3 A.

Example $P(x) = x^3 - 2x^2 - x + 1$ in $[2,3]$

Solution We evaluate $P(2)$ and $P(3)$ using synthetic division.

```
2 )1   -2   -1    1              3 )1   -2   -1   1
        2    0   -2                    3    3   6
   ─────────────────                ──────────────────
   1    0   -1   -1  ← P(2)         1    1    2   7  ← P(3)
```

Since $P(2) = -1$ and $P(3) = 7$, they have opposite signs and there exists at least one r between 2 and 3 such that $P(r) = 0$.

1. $P(x) = 2x^2 + x - 3$; $[-2,-1]$

2. $P(x) = 2x^3 + 17x^2 + 31x - 20$; $[-1,2]$

3. $P(x) = x^4 - 2x^2 - 3$; $[1,2]$

4. $P(x) = x^4 - 9x^2 + 18$; $[2,3]$

5. $P(x) = 3x^3 + 2x^2 + x - 5$; $[0,1]$

6. $P(x) = 2x^3 - 3x^2 + 2x - 3$; $[1,2]$

7. $P(x) = x^3 - 3x + 1$; $[-2,-1]$

8. $P(x) = x^4 - 3x - 23$; $[-3,-2]$

9. $P(x) = x^5 - 6x^3 + 14$; $[2.1,2.2]$

10. $P(x) = x^5 - 2x^3 - 1$; $[1.5,1.6]$

11. $P(x) = x^4 - 4x^3 - 20x^2 + 32x + 16$; $[-1,0]$

12. $P(x) = x^4 - 4x^3 - 44x^2 + 160x - 80$; $[2,3]$

Approximate the value, correct to the nearest hundredth, of the zero of the given polynomial $P(x)$ in the given interval by bisecting three times. See example 10–3 B.

Example $P(x) = 4x^2 + 2x - 3$; $[0,1]$

Solution   1. The midpoint of $[0,1]$ is $\dfrac{0 + 1}{2} = 0.5$.

$$P(0) = -3 \quad P(0.5) = -1 \quad P(1) = 3$$

Same signs   Opposite signs

2. Consider the interval $[0.5,1]$ whose midpoint is 0.75.

$$P(0.5) = -1 \quad P(0.75) = 0.75 \quad P(1) = 3$$

Opposite signs   Same signs

3. Consider the interval $[0.5,0.75]$ whose midpoint is 0.625.

$$P(0.5) = -1 \quad P(0.625) = -0.1875 \quad P(0.75) = 0.75$$

Same signs   Opposite signs

Since -0.1875 is closer to zero than 0.75, we approximate $r \approx 0.625 \approx 0.63$.

13. $P(x) = 2x^2 + x - 2$; $[-2,-1]$

14. $P(x) = x^4 - 2x^2 - 3$; $[1,2]$

15. $P(x) = 3x^3 + 2x^2 + x - 5$; $[0,1]$

16. $P(x) = x^3 - 3x + 1$; $[-2,-1]$

17. $P(x) = x^4 - 3x - 23$; $[-3,-2]$

18. $P(x) = x^4 - 4x^3 - 20x^2 + 32x + 16$; $[1,2]$

19. $P(x) = 30x^3 - 61x^2 - 39x + 10$; $[-1,0]$

Approximate the following by bisecting three times. Round off to the nearest hundredth. See example 10–3 C.

Example $\sqrt{5}$

Solution $\sqrt{5}$ is a zero of $P(x) = x^2 - 5$. We note that the values $P(2) = (2)^2 - 5 = -1$ and $P(3) = (3)^2 - 5 = 4$. Thus, $P(2)$ and $P(3)$ have opposite signs and there is a zero between 2 and 3.

1. The midpoint of $[2,3]$ is 2.5.

$$P(2) = -1 \quad P(2.5) = 1.25 \quad P(3) = 4$$

Opposite signs   Same signs

2. The midpoint of $[2,2.5]$ is 2.25.

$$P(2) = -1 \quad P(2.25) = 0.0625 \quad P(2.5) = 1.25$$

Opposite signs   Same signs

3. The midpoint of [2,2.25] is 2.125.

$$P(2) = -1 \qquad P(2.125) = -0.484375 \qquad P(2.25) = 0.0625$$

Same signs   Opposite signs

Since 0.0625 is closer to zero than -0.484375, we approximate $\sqrt{5} \approx 2.25$.

20. $\sqrt{2}$ **21.** $\sqrt{7}$ **22.** $\sqrt[3]{10}$ **23.** $\sqrt[3]{25}$

24. Given $P(x) = x^3 + 3x^2 - 2x - 6$, (a) find the possible number of positive and negative real zeros and (b) approximate each zero by bisecting intervals three times.

25. Given $P(x) = 4x^3 - 3x^2 + 4x - 4$, (a) find the possible number of positive and negative real zeros and (b) approximate each zero by bisecting intervals three times.

Review exercises

Graph the following equations. See sections 7–1 and 9–1.

1. $y = \dfrac{1}{2}x - 3$

2. $y = x^2 + 3x + 2$

Find the x- and y-intercepts of the following equations. See sections 7–1, 9–1, and 9–4.

3. $5x - 2y = -20$

4. $y = 3x^2 - x - 3$

5. $x^2 + 3y^2 = 12$

6. Find the equation of the line passing through the points $(-2,3)$ and $(4,1)$. See section 7–3.

7. Find the solution set of the radical equation $\sqrt{x + 2} = x$. See section 6–5.

10–4 ■ *Graphing polynomial functions*

We have learned that the graph of a linear function $y = f(x) = mx + b$ is a straight, nonvertical line and the graph of a quadratic function $y = f(x) = ax^2 + bx + c$ is a parabolic curve that opens up when $a > 0$ and opens down when $a < 0$. Both of these are special cases of a general class of functions called *polynomial functions* of degree n. They take the form

$$f(x) = a_n x^n + a_{n-1} x^{n-1} + \cdots + a_1 x + a_0$$

where $a_n, a_{n-1}, \ldots, a_0$ are real number coefficients and n is nonnegative. In this section, we study the graphs of third- and fourth-degree polynomial functions.

One method used to graph any polynomial function $y = f(x)$ is to make a table of related values x and y, plot the corresponding points in the plane, and connect the points with a smooth curve. In this section, we will use a second method to graph functions which uses:

1. the x- and y-intercepts,
2. a sign chart that shows over what intervals the function is positive or negative,
3. a table of selected ordered pairs that satisfy the equation, and
4. the fact that a polynomial function with real coefficients is a *continuous* curve with no breaks.

A further aid in graphing third- and fourth-degree functions is knowing the type of curve one might expect. Third-degree functions are shown in figure 10–2.

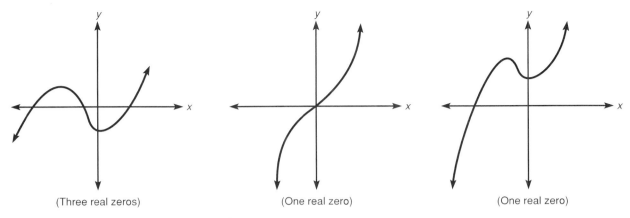

(Three real zeros)   (One real zero)   (One real zero)

Figure 10–2

We illustrate the methods used to graph third-degree functions in the following examples.

■ *Example 10–4 A*

Sketch the graphs of the following third-degree functions.

1. $f(x) = x^3 - x^2 - 4x + 4$
   a. To find the intercepts,
      (1) let $x = 0$, then $y = (0)^3 - (0)^2 - 4(0) + 4 = 4$ [y-intercept is the point $(0,4)$.]
      (2) Let $y = 0$, then $0 = x^3 - x^2 - 4x + 4$
      $$= x^2(x - 1) - 4(x - 1) \quad \text{Factor}$$
      $$= (x^2 - 4)(x - 1)$$
      $$= (x + 2)(x - 2)(x - 1)$$
      Set each factor equal to 0 and solve. We find $x = 2$, $x = -2$, or $x = 1$. [The x-intercepts are the points $(-2,0)$, $(2,0)$, and $(1,0)$.]
   b. We set up a sign chart based on the intervals determined by the x-intercepts. Choose a number in each interval and determine the sign value of each factor and their products using that number.

	Test Number	$x + 2$	$x - 2$	$x - 1$	Product
$x < -2$	-3	$-$	$-$	$-$	$-$ $(y < 0)$
$-2 < x < 1$	0	$+$	$-$	$-$	$+$ $(y > 0)$
$1 < x < 2$	$\dfrac{3}{2}$	$+$	$-$	$+$	$-$ $(y < 0)$
$x > 2$	3	$+$	$+$	$+$	$+$ $(y > 0)$

Note When $y > 0$, the graph is *above* the x-axis and when $y < 0$, the graph is *below* the x-axis.

c. We now choose values of x in the determined intervals and find ordered pairs to aid in graphing the function.

(1) On $x < -2$, let $x = -3$.

$$f(-3) = (-3)^3 - (-3)^2 - 4(-3) + 4 = -20; \ (-3, -20)$$

(2) On $-2 < x < 1$, let $x = 0$.

$$f(0) = (0)^3 - (0)^2 - 4(0) + 4 = 4; \ (0, 4)$$

(3) On $1 < x < 2$, let $x = \dfrac{3}{2}$.

$$f\left(\frac{3}{2}\right) = \left(\frac{3}{2}\right)^3 - \left(\frac{3}{2}\right)^2 - 4\left(\frac{3}{2}\right) + 4 = -\frac{7}{8}; \ \left(\frac{3}{2}, -\frac{7}{8}\right)$$

(4) On $x > 2$, let $x = 3$.

$$f(3) = (3)^3 - (3)^2 - 4(3) + 4 = 10; \ (3, 10)$$

Plot these points and connect them with a smooth curve.

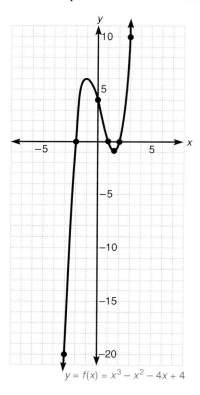

$$y = f(x) = x^3 - x^2 - 4x + 4$$

Note We have chosen a number in each of the intervals determined by the x-intercepts (zeros).

2. $f(x) = x^3 - 1$

   a. Let $x = 0$, then $y = f(0) = (0)^3 - 1 = -1$. [$y$-intercept is the point $(0,-1)$.]

     Let $y = 0$, then $0 = x^3 - 1$
$$= (x - 1)(x^2 + x + 1)$$

     Since $x = 1$ when $x - 1 = 0$, the x-intercept is 1 [the point $(1,0)$].

Note $x^2 + x + 1$ yields two *complex* roots that cannot be graphed on a rectangular coordinate plane.

   b. Set up a sign chart using intervals determined by the x-intercept.

	Test Number	$x - 1$	$x^2 + x + 1$	Product
$x < 1$	0	−	+	− ($y < 0$)
$x > 1$	2	+	+	+ ($y > 0$)

   c. Determine a selected point in each interval.

     (1) Let $x = -1; f(-1) = (-1)^3 - 1 = -2; (-1,-2)$

     (2) Let $x = 2; f(2) = (2)^3 - 1 = 7; (2,7)$

     Plot these points and connect them with a smooth curve.

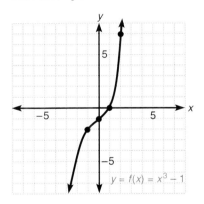

$y = f(x) = x^3 - 1$

     Fourth-degree polynomial functions can take on the curves shown in figure 10–3.

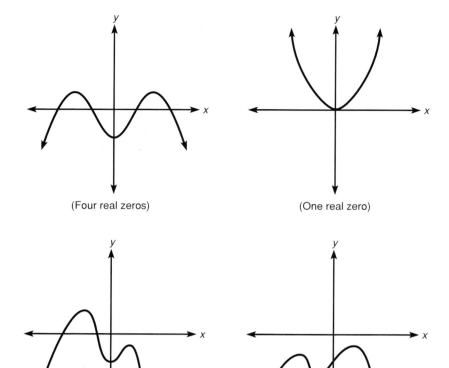

(Four real zeros)

(One real zero)

(Two real zeros)

(No real zeros)

Figure 10–3

We illustrate the graphing of fourth-degree functions by the following examples.

■ *Example 10–4 B*

Sketch the graph of the following fourth-degree functions.

1. $f(x) = x^4 - 4x^2$

a. Let $x = 0$, then $y = f(0) = 0$ [y-intercept is the point $(0,0)$.]

Let $y = 0$, then $0 = x^4 - 4x^2$
$$= x^2(x^2 - 4)$$
$$= x^2(x - 2)(x + 2)$$

Now $x = 0$ when $x^2 = 0$, $x = 2$ when $x - 2 = 0$, and $x = -2$ when $x + 2 = 0$. [The x-intercepts are the points $(0,0)$, $(-2,0)$, and $(2,0)$.]

b. Using the x-intercepts, we set up the following table of signs.

	Test Number	x^2	$x + 2$	$x - 2$	Product	
$x < -2$	-3	$+$	$-$	$-$	$+$	$(y > 0)$
$-2 < x < 0$	-1	$+$	$+$	$-$	$-$	$(y < 0)$
$0 < x < 2$	1	$+$	$+$	$-$	$-$	$(y < 0)$
$x > 2$	3	$+$	$+$	$+$	$+$	$(y > 0)$

c. Determine a selected point in each interval.
   (1) Let $x = -3$; $f(-3) = (-3)^4 - 4(-3)^2 = 45$; $(-3,45)$
   (2) Let $x = -1$; $f(-1) = (-1)^4 - 4(-1)^2 = -3$; $(-1,-3)$
   (3) Let $x = 1$; $f(1) = (1)^4 - 4(1)^2 = -3$; $(1,-3)$
   (4) Let $x = 3$; $f(3) = (3)^4 - 4(3)^2 = 45$; $(3,45)$
   Plot these points and connect them with a smooth curve.

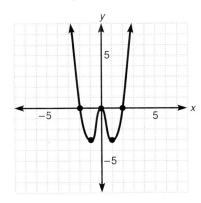

2. $y = x^4 - 9$
   a. Let $x = 0$, then $y = f(0) = -9$ [y-intercept is the point $(0,-9)$.]
      Let $y = 0$, then $0 = x^4 - 9$
      $$= (x^2 - 3)(x^2 + 3)$$
      Set each factor equal to zero and solve for x.

$x^2 - 3 = 0$ $\qquad\qquad$ $x^2 + 3 = 0$
$\quad x^2 = 3$ $\qquad\qquad\quad$ $x^2 = -3$
$\quad x = \pm\sqrt{3}$ $\qquad\qquad$ $x = \pm\sqrt{-3} = \pm i\sqrt{3}$

The x-intercepts are $-\sqrt{3}$ and $\sqrt{3}$. [The points are $(-\sqrt{3},0)$ and $(\sqrt{3},0)$.]

   b. Set up a table of signs using the x-intercepts.

	Test Number	$x + \sqrt{3}$	$x - \sqrt{3}$	$x^2 + 3$	Product
$x < -\sqrt{3}$	-2	$-$	$-$	$+$	$+$ $(y > 0)$
$-\sqrt{3} < x < \sqrt{3}$	0	$+$	$-$	$+$	$-$ $(y < 0)$
$x > \sqrt{3}$	2	$+$	$+$	$+$	$+$ $(y > 0)$

   c. Determine a selected point in each interval.
   (1) Let $x = -2$; $f(-2) = (-2)^4 - 9 = 7$; $(-2,7)$
   (2) Let $x = -1$; $f(-1) = (-1)^4 - 9 = -8$; $(-1,-8)$
   (3) Let $x = 1$; $f(1) = (1)^4 - 9 = -8$; $(1,-8)$
   (4) Let $x = 2$; $f(2) = (2)^4 - 9 = 7$; $(2,7)$
   Plot these points and connect them with a smooth curve.

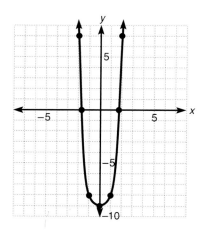

▶ *Quick check*   Sketch the graph of the function $f(x) = x^3 + 2x^2 - 5x - 6$. ■

It is not possible in this book to find the high and the low points in the graphs of many of these functions. This requires the study of calculus.

Mastery points

Can you

- Determine the x- and y-intercepts of a special case third- and fourth-degree polynomial function?
- Set up a table of signs for third- and fourth-degree polynomial functions?
- Sketch the graph of third- and fourth-degree polynomial functions?

Exercise 10–4

Graph the following polynomial functions. See examples 10–4 A and B.

Example $y = f(x) = x^3 + 2x^2 - 5x - 6$

Solution   a. Let $x = 0$, then $y = f(0) = -6$ [The y-intercept is the point $(0,-6)$.] Let $y = 0$, then
$0 = x^3 + 2x^2 - 5x - 6$. Using the theorem on rational zeros, we can determine the factors of
$x^3 + 2x^2 - 5x - 6$ are $(x - 2)(x + 3)(x + 1)$. Thus, $0 = (x - 2)(x + 3)(x + 1)$ and since
$x = 2$ when $x - 2 = 0$, $x = -3$ when $x + 3 = 0$, and $x = -1$ when $x + 1 = 0$, the x-intercepts
are -3, -1, and 2 [the points $(-3,0)$, $(-1,0)$, and $(2,0)$].

b. Set up a table of signs using the x-intercepts.

	Test Number	$x - 2$	$x + 3$	$x + 1$	Product
$x < -3$	-4	$-$	$-$	$-$	$-$ ($y < 0$)
$-3 < x < -1$	-2	$-$	$+$	$-$	$+$ ($y > 0$)
$-1 < x < 2$	0	$-$	$+$	$+$	$-$ ($y < 0$)
$x > 2$	3	$+$	$+$	$+$	$+$ ($y > 0$)

c. Determine a selected point in each interval.
1. Let $x = -4$; $f(-4) = (-4)^3 + 2(-4)^2 - 5(-4) - 6 = -18$; $(-4,-18)$
2. Let $x = -2$; $f(-2) = (-2)^3 + 2(-2)^2 - 5(-2) - 6 = 4$; $(-2,4)$
3. Let $x = 1$; $f(1) = (1)^3 + 2(1)^2 - 5(1) - 6 = -8$; $(1,-8)$
4. Let $x = 3$; $f(3) = (3)^3 + 2(3)^2 - 5(3) - 6 = 24$; $(3,24)$

Plot these points and connect them with a smooth curve.

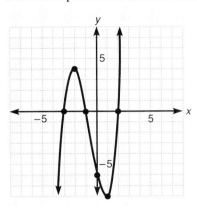

1. $f(x) = x^3 - x$
2. $f(x) = x^3 - 4x$
3. $f(x) = x^3 + x^2 - 2x$
4. $f(x) = -x^3 - x^2 + 6x$
5. $f(x) = x^4 - 16x^2$
6. $f(x) = x^4 - 9x^2$
7. $f(x) = x^4 - 16$
8. $f(x) = x^4 - 2$
9. $f(x) = x^4 - 5x^2 + 4$
10. $f(x) = x^4 - 7x^2 + 6$
11. $f(x) = x^4 - 6x^3 + 8x^2$
12. $f(x) = x^4 - 12x^3 + 27x^2$
13. $f(x) = x^3 - 2x^2 - x + 2$
14. $f(x) = x^3 + 3x^2 - x - 3$
15. $f(x) = x^3 - 3$
16. $f(x) = x^3 + 2$

For each of the following polynomial functions, find (a) the y-intercept, (b) the x-intercepts, and (c) the intervals over which $y = f(x) > 0$ and over which $y = f(x) < 0$. Do not graph the function.

Example $y = f(x) = (x + 4)(x - 1)(x + 5)$

Solution 1. To find the y-intercept, replace x with 0. Then $y = (0 + 4)(0 - 1)(0 + 5) = 4(-1)(5) = -20$
The y-intercept is the point $(0,-20)$.
2. Set each factor $x + 4$, $x - 1$, and $x + 5$ equal to zero and we get $x = -4$, $x = 1$, and $x = -5$. The x-intercepts are the points $(-5,0)$, $(-4,0)$, and $(1,0)$.
3. Set up a table of signs using the x-intercepts.

	Test Number	$x + 5$	$x + 4$	$x - 1$	Product
$x < -5$	-6	$-$	$-$	$-$	$-$ $(y < 0)$
$-5 < x < -4$	$-\dfrac{9}{2}$	$+$	$-$	$-$	$+$ $(y > 0)$
$-4 < x < 1$	0	$+$	$+$	$-$	$-$ $(y < 0)$
$x > 1$	2	$+$	$+$	$+$	$+$ $(y > 0)$

The function is positive on intervals $(-5,-4)$ and $(1,\infty)$ and negative on intervals $(-\infty,-5)$ and $(-4,1)$.

17. $y = f(x) = (x + 2)(x - 3)(x - 4)$

18. $y = f(x) = (x - 7)(x - 1)(x + 3)$

19. $y = f(x) = x^2(x - 2)(x + 6)$

20. $y = f(x) = x^2(x + 4)(x - 8)$

21. $y = f(x) = (x + 2)^2(x - 3)^2$

22. $y = f(x) = (x - 5)^2(x + 3)^2$

23. $y = f(x) = x(x + 5)^2(x - 2)$

24. $y = f(x) = x(x + 1)(x - 4)^2$

Review exercises

State the domain of the following rational expressions. See section 4–1.

1. $\dfrac{x - 1}{x + 3}$

2. $\dfrac{5y - 4}{y^2 - 2y - 35}$

3. $\dfrac{6}{x^2 - 4}$

Evaluate the following rational expressions for the given value. See section 4–1.

4. $\dfrac{3y - 2}{2y + 7}; y = 4$

5. $\dfrac{x^2 - 2x + 1}{x^2 + x - 2}; x = -2$

6. Given the points $(-1,4)$ and $(5,6)$, find (a) the distance between the points and (b) the slope of the line containing the two points. See section 7–2.

7. Find the solution set of the equation $y - 3\sqrt{y} - 10 = 0$. See section 6–6.

10–5 ■ *Graphing rational functions*

A function that can be written in the form

$$f(x) = \frac{P(x)}{Q(x)}, Q(x) \neq 0$$

where $P(x)$ and $Q(x)$ are polynomials, is called a *rational function*. To illustrate,

$$f(x) = \frac{3}{x + 1}, f(x) = \frac{3x}{x^2 - x - 2}, \text{ and } f(x) = \frac{x + 2}{x^3 + 27}$$

are examples of rational functions. It is important to note the domain of a rational function—the set of all real numbers *except* for those real number replacements of the variable that make the denominator equal to zero. For example, given

$$f(x) = \frac{3}{x + 1}$$

since the denominator $x + 1 = 0$ when $x = -1$, the domain of the function is the set of all real numbers *except* -1. That is,

$$\text{domain of } f = \{x | x \in R, x \neq -1\}$$

This excluded number is very important in the graph of rational functions. It is at these excluded numbers that the graph of the function will have breaks in a curve that would otherwise be continuous.

To illustrate this situation, consider the rational function

$$f(x) = \frac{1}{x}$$

whose domain $= \{x | x \in R, x \neq 0\}$. There will be a break in the curve of the graph at $x = 0$ (the y-axis).

Since the number 0 cannot be used for x, let us consider the values of the function, $f(x)$, as we choose values of x that get close to 0. The following table shows what happens.

x	-1	$-.5$	-0.1	-0.01	-0.001	-0.0001
$f(x) = \dfrac{1}{x}$	-1	-2	-10	-100	-1000	$-10,000$
x	1	0.5	0.1	0.01	0.001	0.0001
$f(x) = \dfrac{1}{x}$	1	2	10	100	1000	10,000

We can see that, in either case, as the value of x gets closer and closer to 0, denoted by $x \to 0$, the absolute value of $f(x)$ becomes greater and greater, denoted by $|f(x)| \to \infty$. That is, as

$$x \to 0, \text{ then } |f(x)| \to \infty$$

Since x cannot equal 0, the graph of $f(x) = \dfrac{1}{x}$ will never intersect the line $x = 0$ (the y-axis) but the graph will get closer and closer to this line. The line $x = 0$ is called a *vertical asymptote*.

Note To say that $|f(x)| \to \infty$ means that the absolute value of $f(x)$ *increases without bound.*

In like fashion, as $|x|$ gets greater and greater, denoted by $|x| \to \infty$, $\left| \dfrac{1}{x} \right|$ becomes less and less. In fact, $\left| \dfrac{1}{x} \right|$ approaches, but does not reach, a value of 0, denoted by $\left| \dfrac{1}{x} \right| \to 0$. The graph approaches, but does not intersect, the line $y = 0$ (the x-axis). We call this line $y = 0$ a *horizontal asymptote*. See figure 10-4 for the graph of $f(x) = \dfrac{1}{x}$. This example illustrates the following rule for finding vertical asymptotes.

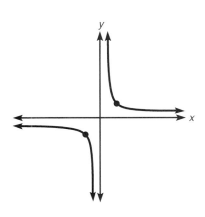

Figure 10-4

___ **Vertical asymptote rule** _____

Given rational function $f(x) = \dfrac{P(x)}{Q(x)}$, if r is a zero of $Q(x)$, then the line $x = r$ is a vertical asymptote for the graph of function f.

To determine the equations of horizontal asymptotes for the graph of a rational function, consider function f defined by

$$f(x) = \frac{3x - 1}{x + 2}$$

If we divide the numerator and the denominator by x (the greatest power of the variable in the function), we obtain

$$f(x) = \frac{3 - \dfrac{1}{x}}{1 + \dfrac{2}{x}}$$

As the absolute value of x takes on greater and greater values, the terms $\dfrac{1}{x}$ and $\dfrac{2}{x}$ become less and less—in fact, they both approach the value of 0. That is,

$$\text{as } |x| \to \infty, \text{ then } \begin{cases} \dfrac{1}{x} \to 0 \\ \dfrac{2}{x} \to 0 \end{cases}$$

Thus, as $|x| \to \infty$, then $f(x)$ approaches $\dfrac{3 - 0}{1 - 0} = \dfrac{3}{1} = 3$. The line $y = 3$ is then a horizontal asymptote for the graph of function f. We now state the rule for finding the equation of a horizontal asymptote for the graph of a rational function.

Horizontal asymptote rule

Given the rational function $f(x) = \dfrac{P(x)}{Q(x)}$, divide the numerator and the denominator by the greatest power of the variable. If $y = f(x)$ approaches some constant value k as $|x|$ becomes arbitrarily greater and greater, then the line $y = k$ is a horizontal asymptote for the graph of f.

■ Example 10–5 A

Find the equations of the vertical and horizontal asymptotes for the graph of the following rational functions.

1. $f(x) = \dfrac{4}{x - 5}$

   a. Since $x = 5$ when the denominator $x - 5 = 0$, the line $x = 5$ is a vertical asymptote.
   b. Since the greatest power of x is 1, divide the numerator and the denominator by x to obtain

   $$f(x) = \frac{\dfrac{4}{x}}{1 - \dfrac{5}{x}}$$

   As $|x|$ gets greater and greater, $x \to \infty$ or $x \to -\infty$, we can see that $\dfrac{4}{x} \to 0$ and $\dfrac{5}{x} \to 0$, so $f(x) \to \dfrac{0}{1 - 0} = \dfrac{0}{1} = 0$. The line $y = 0$ (the x-axis) is a horizontal asymptote.

2. $f(x) = \dfrac{2x^2 + 3}{x^2 - 4}$

   a. Since $x^2 - 4 = (x + 2)(x - 2)$ and $x = -2$ when $x + 2 = 0$ and $x = 2$ when $x - 2 = 0$, the vertical asymptotes are the lines $x = -2$ and $x = 2$.

   b. The greatest power of x is 2, so we divide the numerator and the denominator by x^2. Then

$$f(x) = \dfrac{2 + \dfrac{3}{x^2}}{1 - \dfrac{4}{x^2}}$$

   and as $|x| \to \infty$, $\dfrac{3}{x^2}$ and $\dfrac{4}{x^2} \to 0$. Then $f(x) \to \dfrac{2 + 0}{1 - 0} = \dfrac{2}{1} = 2$.

   The line $y = 2$ is a horizontal asymptote.

Note The coefficients of x^2 form the equation of the asymptote. $y = \dfrac{2}{1} = 2$

3. $f(x) = \dfrac{x^3 - 1}{2x^2 + x - 1}$

   a. Since $2x^2 + x - 1 = (2x - 1)(x + 1)$, $x = \dfrac{1}{2}$ when $2x - 1 = 0$ and $x = -1$ when $x + 1 = 0$, then the vertical asymptotes are the lines $x = \dfrac{1}{2}$ and $x = -1$.

   b. Divide the numerator and the denominator by x^3, the greatest power of x, to obtain

$$f(x) = \dfrac{1 - \dfrac{1}{x^3}}{\dfrac{2}{x} + \dfrac{1}{x^2} - \dfrac{1}{x^3}}$$

   Now $\dfrac{2}{x}$, $\dfrac{1}{x^2}$, and $\dfrac{1}{x^3} \to 0$ as $|x| \to \infty$, so $f(x) \to \dfrac{1 - 0}{0 + 0 - 0} = \dfrac{1}{0}$, which is undefined. Thus, there is no horizontal asymptote.

▶ ***Quick check*** Find the equations of the vertical and horizontal or oblique asymptotes of the rational function $f(x) = \dfrac{5x - 3}{2x + 1}$. ■

Now that we have determined the equations of the vertical and horizontal asymptotes, we need only determine some arbitrary points in the graph and connect them with smooth curves to obtain the graph of the rational function.

■ *Example 10–5 B*

Graph the following rational functions.

1. $f(x) = \dfrac{4}{x - 5}$

In example 10–5 A–1, we found the line $x = 5$ to be a vertical asymptote and the line $y = 0$ (the x-axis) to be a horizontal asymptote. Now we choose some values of x that lie in the intervals bounded by the vertical asymptotes and find their corresponding values of $y = f(x)$.

x	$y = f(x)$
0	$-\dfrac{4}{5}$
3	-2
4	-4
6	4
7	2
8	$\dfrac{4}{3}$

Plot the points, draw the vertical asymptote in a dashed line, and connect the points with a smooth curve.

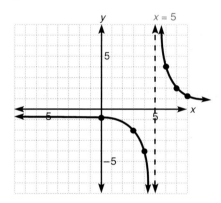

Note The y-intercept is $-\dfrac{4}{5}$ and there is no x-intercept since that is the line $y = 0$, which serves as an asymptote. We shall include the intercepts, if they exist, when choosing points.

2. $f(x) = \dfrac{2x^2 + 3}{x^2 - 4}$

In example 10–5, A–2, we found the lines $x = -2$ and $x = 2$ to be vertical asymptotes while the line $y = 2$ is a horizontal asymptote for the graph of f. We now find some additional points

x	$y = f(x)$
-4	$\dfrac{35}{12}$
-3	$\dfrac{21}{5}$
-1	$-\dfrac{5}{3}$
0	$-\dfrac{3}{4}$
1	$-\dfrac{5}{3}$
3	$\dfrac{21}{5}$
4	$\dfrac{35}{12}$

Plot the points, draw the vertical and horizontal asymptotes in dashed lines, and connect the points with a smooth curve.

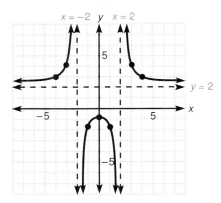

When the degree of the numerator is greater than the degree of the denominator, we perform the indicated polynomial division as in the next example.

■ *Example 10–5 C*

Graph the function $f(x) = \dfrac{x^2 + x - 3}{x - 2}$.

Since $x = 2$ when $x - 2 = 0$, $x = 2$ is a vertical asymptote. Dividing the numerator and the denominator by x^2, we can show that there is no horizontal asymptote. We now divide the numerator by the denominator using synthetic division.

$$
\begin{array}{r|rrr}
2 & 1 & 1 & -3 \\
  &   & 2 & 6 \\
\hline
  & 1 & 3 & 3
\end{array}
$$

Thus, $(x^2 + x - 3) \div (x - 2) = x + 3 + \dfrac{3}{x - 2}$ and $f(x) = x + 3$

$+ \dfrac{3}{x - 2}$. As $|x| \to \infty$, $\dfrac{3}{x - 2} \to 0$ and $y = f(x) \to x + 3$. The graph of f will approach the oblique asymptote $y = x + 3$ as $|x|$ increases without bound. We need a few additional points to graph the function.

x	$y = f(x)$
-5	$-\dfrac{17}{7}$
-4	$-\dfrac{3}{2}$
-3	$-\dfrac{3}{5}$
-2	$\dfrac{1}{4}$
-1	1
0	$\dfrac{3}{2}$
1	1
3	9
4	$\dfrac{17}{2}$
5	9

Plot the points, draw the vertical and oblique asymptotes in dashed lines, and connect the points with a smooth curve.

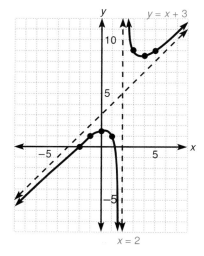

▶ *Quick check* Graph the function $f(x) = \dfrac{3x^2 + 1}{x^2 - 1}$.

We now summarize the steps to be taken in graphing the rational function $f(x) = \dfrac{P(x)}{Q(x)}$ where $P(x)$ and $Q(x)$ have no common factors.

1. *Find the y-intercept* by setting $x = 0$ and solving for y.
2. *Find the x-intercept(s)* by setting $P(x)$ (the numerator) equal to 0. The real solutions of the equation $P(x) = 0$, if there are any, are the x-intercepts.
3. *Find the equation(s) of the vertical asymptote(s)* by setting $Q(x)$ (the denominator) equal to 0. The solutions of the equation $Q(x) = 0$, if there are any, determine the equations of the vertical asymptotes.
4. *Find the equation of the horizontal asymptote* using the following guidelines.
    a. If the degree of $P(x)$ is less than the degree of $Q(x)$, the line $y = 0$ (the x-axis) is a horizontal asymptote.
    b. If the degree of $P(x)$ equals the degree of $Q(x)$, the line $y = \dfrac{p}{q}$, where p is the leading coefficient of $P(x)$ and q is the leading coefficient of $Q(x)$, is a horizontal asymptote.
    c. If the degree of $P(x)$ is greater than the degree of $Q(x)$, there is no horizontal asymptote.
5. When the degree of $P(x)$ is one more than the degree of $Q(x)$, you find an oblique (slanting) asymptote by dividing $P(x) \div Q(x)$. The line $y = mx + b$ (ignoring the remainder if there is any) is an oblique asymptote in the graph.

Mastery points

Can you
■ Find the horizontal, vertical, and oblique asymptotes of a rational function?
■ Graph a rational function?

Exercise 10–5

Find the equations of any vertical, horizontal, or oblique asymptotes for the following rational functions.
See example 10–5 A.

Example $f(x) = \dfrac{5x - 3}{2x + 1}$

Solution a. Since $x = -\dfrac{1}{2}$ when $2x + 1 = 0$, the line $x = -\dfrac{1}{2}$ is a vertical asymptote.

b. Since the degree of the numerator equals the degree of the denominator, the line $y = \dfrac{5}{2}$ [the leading coefficient of $P(x)$ over the leading coefficient of $Q(x)$] is a horizontal asymptote.

c. There is no oblique asymptote since the degree of the numerator is *not* greater than the degree of the denominator.

1. $f(x) = \dfrac{3}{x - 2}$

2. $f(x) = \dfrac{-5}{x + 5}$

3. $f(x) = \dfrac{4}{3x - 2}$

4. $f(x) = \dfrac{8}{2x + 5}$

5. $f(x) = \dfrac{x - 2}{x + 3}$

6. $f(x) = \dfrac{4 - x}{x - 5}$

7. $f(x) = \dfrac{3x - 1}{2x + 9}$

8. $f(x) = \dfrac{2 - 5x}{3x + 2}$

9. $f(x) = \dfrac{5}{x^2 - 3x - 10}$

10. $f(x) = \dfrac{-9}{x^2 + x - 12}$

11. $f(x) = \dfrac{x^2 + 2}{x - 3}$

12. $f(x) = \dfrac{x^2 - 4}{x + 7}$

13. $f(x) = \dfrac{x - 2}{2x^2 - 5x - 3}$

14. $f(x) = \dfrac{3x - 4}{3x^2 - 7x + 2}$

15. $f(x) = \dfrac{x^2 - 5}{x^2 + 2}$

16. $f(x) = \dfrac{2x^2 - 1}{3x^2 + 1}$

Graph the following rational functions. See examples 10–5 A, B, and C.

Example $f(x) = \dfrac{3x^2 - 1}{x^2 - 1}$

Solution a. Since $x^2 - 1 = (x + 1)(x - 1)$, $x = 1$ when $x - 1 = 0$ and $x = -1$ when $x + 1 = 0$, the lines $x = 1$ and $x = -1$ are vertical asymptotes.

b. Since the degree of the numerator and the denominator is the same, the line $y = 3$ [leading coefficient of $P(x)$ over the leading coefficient of $Q(x)$] is a horizontal asymptote.

c. Let $x = 0$, then $y = f(0) = 1$ so the point $(0,1)$ is the y-intercept.

d. To find the x-intercepts, set the numerator $3x^2 - 1 = 0$ and solve.

$$3x^2 - 1 = 0$$
$$3x^2 = 1$$
$$x^2 = \frac{1}{3}$$
$$x = \pm \sqrt{\frac{1}{3}} = \pm\frac{\sqrt{3}}{3} \approx \pm 0.58$$

The x-intercepts are the points $\left(-\dfrac{\sqrt{3}}{3}, 0\right)$ and $\left(\dfrac{\sqrt{3}}{3}, 0\right)$.

e. Find some additional points.

x	$y = f(x)$
-3	$\dfrac{13}{4}$
-2	$\dfrac{11}{3}$
$-\dfrac{1}{2}$	$\dfrac{1}{3}$
$\dfrac{1}{2}$	$\dfrac{1}{3}$
2	$\dfrac{11}{3}$
3	$\dfrac{13}{4}$

Plot the points, draw the vertical and horizontal asymptotes in dashed lines, and connect the points with a smooth curve.

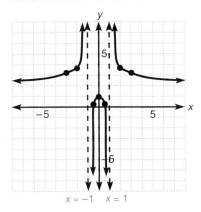

$x = -1$ $x = 1$

17. $f(x) = \dfrac{3}{x - 4}$

18. $f(x) = \dfrac{-2}{x + 1}$

19. $f(x) = \dfrac{-1}{(x + 1)(x - 3)}$

20. $f(x) = \dfrac{5}{(x + 2)(x - 5)}$

21. $f(x) = \dfrac{3x}{x^2 - 1}$

22. $f(x) = \dfrac{-4x}{x^2 - 9}$

23. $f(x) = \dfrac{x + 5}{x - 3}$

24. $f(x) = \dfrac{2x - 1}{x - 1}$

25. $f(x) = \dfrac{4 - x}{x + 2}$

26. $f(x) = \dfrac{2 - 3x}{x - 2}$

27. $f(x) = \dfrac{x^2 - 1}{x^2 - 2x + 1}$

28. $f(x) = \dfrac{x^2 - 4}{x^2 + 4x + 4}$

29. $f(x) = \dfrac{x - 4}{(x + 2)(x - 3)}$

30. $f(x) = \dfrac{x + 1}{(x + 4)(x - 2)}$

31. $f(x) = \dfrac{x^2 - x}{x + 1}$

32. $f(x) = \dfrac{x^2 + 3x}{2x - 3}$

33. $f(x) = \dfrac{3x^2 + 2}{3x - 1}$

34. $f(x) = \dfrac{4x^2 - 5}{2x + 3}$

35. $f(x) = \dfrac{3}{x^2 + 2}$

36. $f(x) = \dfrac{-8}{x^2 + 1}$

37. $f(x) = \dfrac{x^2 - 4x - 12}{x + 2}$

38. $f(x) = \dfrac{x^2 - 5x + 4}{x - 1}$

39. Given a function f defined by $f(x) = \dfrac{ax + b}{cx + d}$, $a \neq 0$ and $c \neq 0$, where a, b, c, and d are integral constants, b and d not both 0, (a) what is the x-intercept, (b) what is the y-intercept, (c) what is the equation of the vertical asymptote, and (d) what is the equation of the horizontal asymptote?

40. Boyle's law of gases relates the pressure P to the volume V of a gas maintained at a constant temperature by the equation

$$P = \dfrac{k}{V}$$

where k is a constant. What is the domain of this rational equation? Labeling the horizontal axis V and the vertical axis P, graph the equation over its domain when $k = 4.0$. As V decreases, what happens to P?

Review exercises

Perform the indicated operations. Assume all variables are nonzero. Answer with positive exponents only. See section 3–3.

1. $(x^3)(x^{-2})(x)$

2. $\dfrac{x^{-3}y^2}{xy^{-1}}$

3. $(-4x^3y)^{-2}$

4. Solve the equation $3x - xy = 4$ for x. See section 2–2.

5. Find the solution set of $\dfrac{3x - 2}{x + 1} = 4$. See Section 4–7.

6. Find the solution set of the inequality $x^2 - 2x - 15 \leq 0$. See section 6–7.

Chapter 10 lead-in problem

A particle moving along a straight line is displaced at time t in seconds according to the function

$$s(t) = t^3 - 9t^2 + 20t - 12$$

where s is in centimeters. At $t = 2$ seconds, the displacement is 0. At what other times will the displacement be 0?

Solution

The displacement $s(t) = 0$ when $t = 2$. Thus, 2 is a zero of the function. Using synthetic division, divide by 2.

$$
\begin{array}{r|rrrr}
2 & 1 & -9 & 20 & -12 \\
  &   & 2 & -14 & 12 \\
\hline
  & 1 & -7 & 6 & 0
\end{array}
$$
← Remainder 0, 2 is a zero

We now find zeros of the quotient $t^2 - 7t + 6$.

$$(t - 6)(t - 1) = 0$$
$$t - 6 = 0 \text{ or } t - 1 = 0$$
$$t = 6 \qquad t = 1$$

The displacement of the particle is 0 when $t = 1$ second and when $t = 6$ seconds in addition to when $t = 2$ seconds.

Chapter 10 summary

1. By the **fundamental theorem of algebra,** any polynomial $P(x)$ with positive degree has at least one complex zero.
2. By theorem, a polynomial $P(x)$ of degree $n > 0$ has at most n distinct complex or real zeros.
3. By the **conjugate zeros theorem,** if $a + bi$ is a zero of $P(x)$ having real coefficients, then $a - bi$ is also a zero of $P(x)$.
4. **Descartes' rule of signs** determines the number of real number zeros of a polynomial $P(x)$ with real number coefficients and a nonzero constant term.
5. By the **rule for bounds for real zeros of polynomials,** if $P(x)$ is divided by $x - c$, then (a) if $c > 0$ and the coefficients of the quotient are all positive numbers or zero, c is an upper bound of the real zeros of $P(x)$ and (b) if $c < 0$ and the coefficients of the quotient are alternating positive and negative numbers (0 is either positive or negative), c is a lower bound of the real zeros of $P(x)$.
6. By the theorem on **rational zeros,** if p/q is a rational zero of $P(x)$ in lowest terms, then p is a factor of the constant term and q is a factor of the leading coefficient.

7. By the **change-of-sign property** for polynomials, if a and b are real numbers and $P(a)$ and $P(b)$ have opposite signs, then there is at least one real zero between a and b.

8. When *graphing a rational function* $f(x) = \dfrac{P(x)}{Q(x)}$, if r is a zero of $Q(x)$, then $x = r$ is a vertical asymptote for the graph of f.

9. Given a rational function $f(x) = \dfrac{P(x)}{Q(x)}$, if $y = f(x)$ approaches some constant k as $|x|$ becomes greater and greater, then the line $y = k$ is a horizontal asymptote for the graph of f.

10. Given $f(x) = \dfrac{P(x)}{Q(x)}$, an oblique asymptote occurs when the degree of $P(x)$ is one more than the degree of $Q(x)$. The equation of the asymptote is found by dividing $P(x) \div Q(x)$.

Chapter 10 error analysis

1. Zeros of polynomials
   Example: If i and 2 are zeros of $P(x)$ with real coefficients, then $P(x) = x^2 - 2x - xi + 2i$.
   Correct answer: $P(x) = x^3 - 2x^2 + x - 2$
   What error was made? (*see page 440*)

2. Polynomials at $-x$
   Example: Given $P(x) = 4x^3 - 2x^2 + x - 3$, then $P(-x) = 4x^3 + 2x^2 - x - 3$.
   Correct answer: $P(-x) = -4x^3 - 2x^2 - x - 3$
   What error was made? (*see page 441*)

3. Rational zeros of a polynomial
   Example: Given $P(x) = 3x^4 - 2x^3 + x^2 - x + 4$, the possible rational zeros of $P(x)$ are ± 1, ± 3,
   $\pm \dfrac{1}{4}, \pm \dfrac{3}{4}, \pm \dfrac{3}{2}$.
   Correct answer: $\pm 1, \pm 2, \pm 4, \pm \dfrac{1}{3}, \pm \dfrac{2}{3}, \pm \dfrac{4}{3}$
   What error was made? (*see page 448*)

4. Solution set of equations
   Example: The solution set of $x^2 + 4 = 0$ is $\{-2, 2\}$.
   Correct answer: $\{-2i, 2i\}$
   What error was made? (*see page 450*)

5. Values of polynomials
   Example: Given $P(x) = 3x^3 - 2x^2 + 4$, using synthetic division, $P(-2) = 20$.
   Correct answer: $P(-2) = -28$
   What error was made? (*see page 191*)

6. Factoring difference of two cubes
   Example: $x^3 - 8 = (x - 2)(x^2 - 4x + 4)$
   Correct answer: $x^3 - 8 = (x - 2)(x^2 + 2x + 4)$
   What error was made? (*see page 145*)

7. Vertical asymptotes of rational functions
   Example: A vertical asymptote for the graph of $f(x) = \dfrac{3x - 2}{x + 2}$ is $x = 3$.
   Correct answer: $x = -2$
   What error was made? (*see page 467*)

8. The x-intercept of a rational function
   Example: Given $f(x) = \dfrac{x - 3}{x + 2}$, the x-intercept is the point $\left(-\dfrac{3}{2}, 0\right)$.
   Correct answer: $(3, 0)$
   What error was made? (*see page 473*)

9. Horizontal asymptotes
   Example: Given $f(x) = \dfrac{3x^2 - 2x + 1}{x + 3}$, a horizontal asymptote is the line $y = 3$.
   Correct answer: $f(x) = \dfrac{3x^2 - 2x + 1}{x + 3}$ has no horizontal asymptote.
   What error was made? (*see page 468*)

Chapter 10 critical thinking

Choose a three-digit number greater than 100 where all three digits are the same (such as 444 or 777). Add the three digits and divide the original number by this sum. No matter what three-digit number you choose, the answer is 37. Why is this true?

Chapter 10 review

[10–1]

Find the polynomial with real coefficients of lowest degree having the given zeros.

1. $-i$ and 6

2. $3 - i$ and $2 + i$

3. $4 - 3i$, i, and -5

Find all zeros of the given polynomial when one zero is given.

4. $P(x) = x^3 - 5x^2 + 17x - 13$; 1

5. $P(x) = 2x^3 - 5x^2 + 6x - 2$; $1 - i$

6. $P(x) = x^4 + 5x^2 + 4$; i

Using Descartes' rule of signs, determine the number of positive and negative real zeros of $P(x)$.

7. $P(x) = x^3 - 3x^2 - x + 2$

8. $P(x) = 3x^3 + 2x^2 + x - 1$

9. $P(x) = 4x^4 + 3x^3 + 2x^2 + x + 3$

Find an upper and a lower bound of the real zeros of $P(x)$.

10. $P(x) = x^3 + 3x^2 - 5x + 3$

11. $P(x) = 3x^3 + x^2 + 2x - 4$

12. $P(x) = x^4 - x^3 + x^2 - 3x + 12$

[10–2]

Find all the zeros of the given polynomial.

13. $P(x) = 3x^3 + 2x^2 + 2x - 1$

14. $P(x) = 6x^3 + 25x^2 + 3x - 4$

15. $P(x) = 2x^3 - 3x^2 - 5x + 6$

Find the solution set of each equation.

16. $x^3 - x^2 - 8x + 12 = 0$

17. $x^3 - 10x - 12 = 0$

18. $3x^4 - x^3 - 8x^2 + 2x + 4 = 0$

[10–3]

Show that each given polynomial $P(x)$ has at least one real zero in the given interval.

19. $P(x) = 3x^2 - 2x - 6$; (1,2)

20. $P(x) = 3x^3 + 7x^2 - 4$; $\left(\dfrac{1}{2}, 1\right)$

21. $P(x) = 5x^3 - 9x^2 - 4x + 9$; $(-1, 2)$

22. Evaluate $\sqrt[3]{53}$ to two decimal places by bisecting three times.

[10–4]

Graph the following polynomial functions.

23. $f(x) = x^3 + x^2 - 6x$

24. $f(x) = x^3 + 2x^2 - x - 2$

25. $f(x) = x^3 - 2x^2 - 11x + 12$

[10–5]

Graph the following rational functions.

26. $f(x) = \dfrac{5}{x - 3}$

27. $f(x) = \dfrac{3x - 1}{3x + 2}$

28. $f(x) = \dfrac{x^2 - x + 2}{x - 3}$

Chapter 10 cumulative test

[1–1]   **1.** Given $A = \{-7,-3,0,8,9\}$ and
$B = \{-3,-2,0,1,7,8\}$, define (a) $A \cup B$ and
(b) $A \cap B$.

[1–3]   **2.** $4(x + 6) = 4 \cdot x + 4 \cdot 6$ is an application of
what property?

[1–4]   **3.** Perform the indicated operations of
$$\frac{3 \cdot 4 - 2^2}{10} - \frac{(-24)}{3}.$$

[1–5]   **4.** Given $S = \frac{1}{2}gt^2$, find S when $g = 32$ and $t = 4$.

[2–1]   **5.** Find the solution set of the equation
$$\frac{7}{12}x - 3 = \frac{5}{6}x + 4.$$

[2–2]   **6.** Solve the formula $S = \frac{n}{2}[a + (n - 1)d]$ for a.

[2–3]   **7.** Jack invested $4,000 at 6% interest. How much
must he invest at 9% interest to realize a net
return of 8% on his investment?

Find the solution set of the following equations and inequalities.

[2–5]   **8.** $-6 \leq 2 - 3x < -1$

[2–4]   **9.** $|4x - 3| = 5$

[2–6]   **10.** $|1 - 2x| < 5$

[6–2]   **11.** $3x^2 - 2x - 8 = 0$

[6–5]   **12.** $\sqrt[4]{2x - 3} - 2 = 0$

[4–6]   **13.** $\dfrac{x}{x + 2} + \dfrac{2}{3} = \dfrac{-1}{x + 2}$

Perform the indicated operations and simplify. Assume $x > 0$.

[5–2]   **14.** $\dfrac{x^{1/2}}{x^{1/4}}$

[5–2]   **15.** $(x^{3/4})^{1/2}$

[5–3]   **16.** $\sqrt{8}\sqrt{3}$

[5–4]   **17.** $\sqrt{\dfrac{4}{27}}$

[5–6]   **18.** $3\sqrt{2}(2\sqrt{3} - \sqrt{2})$

[5–7]   **19.** Multiply $(3 + \sqrt{-4})(5 - \sqrt{-16})$ and write the
answer in the form $a + bi$ or $a - bi$.

[6–4]   **20.** The distance s a body falls when air resistance is
neglected is given by $s = v_0t - 16t^2$, where s is
in feet, t is in seconds, and v_0 is the initial
velocity. When will the body strike the ground if
$v_0 = 32$ feet per second?

[6–7]   **21.** Find the solution set of the quadratic inequality
$x^2 - 2x - 3 \geq 0$. Graph the solution set on the
number line and write it in set-builder notation.

[7–2]   **22.** Determine whether the lines $2x - y = 6$ and
$2y - 4x = -3$ are parallel, perpendicular, or
neither.

[7–2]   **23.** Find the distance between the points $(-3,4)$ and
$(2,7)$. What are the coordinates of the midpoint
of the line segment with these endpoints?

[7–3]   **24.** Find the equation of the line in standard form.
a. containing points $(-3,0)$ and $(1,4)$
b. having slope $-2/3$ and passing through $(0,4)$

[7–4]   **25.** Graph the inequality $|x - 3| \geq 2$ in the
xy-plane.

[8–1]   **26.** Determine if the following relations define a
function.
a. $\{(-3,2), (2,2), (1,6), (0,7)\}$
b. $\{(x,y)|y = 2x - 3\}$
c. $\{(x,y)|x = -4\}$

[8–3]   **27.** Graph the square root function defined by
$f(x) = \sqrt{x - 3}$.

[8–2] 28. Given $f(x) = 2x - 5$ and $g(x) = 1 - 2x^2$, find
 a. $(f - g)(-3)$
 b. $g[f(4)]$
 c. $f(5) + g(-1)$

[8–5] 29. If x varies directly as y and inversely as the square of z, find k if $x = 6$ when $y = 3$ and $z = 4$. What is x when $y = 4$ and $z = 2$?

[9–1] 30. Graph the quadratic function defined by $f(x) = x^2 - x - 2$.

[9–3] 31. Graph the equation $x^2 + y^2 - 2x - 6y - 6 = 0$.

[9–4] 32. Identify each equation as an ellipse, a parabola, a circle, or a hyperbola.
 a. $3x^2 = 2 - 3y^2$
 b. $y = x^2 + 2x - 1$
 c. $\dfrac{x^2}{4} + \dfrac{y^2}{2} = 1$

[4–6] 33. Find $P(-2)$ when $P(x) = 3x^3 - 2x^2 + x - 1$ using synthetic division.

[4–6] 34. Use the remainder and factor theorems to determine if -1 is a solution of the equation $x^3 + x^2 - 3x - 3 = 0$.

[10–2] 35. Find all the rational zeros of the polynomial $P(x) = 3x^3 - 5x^2 - x + 2$.

[10–4] 36. Graph the polynomial function f defined by $f(x) = x^4 - 2x^3 - 9x^2 + 2x + 8$.

[10–5] 37. Graph the rational function defined by $g(x) = \dfrac{5}{4 - x}$.

11

Exponential and Logarithmic Functions

The Richter Scale rating of earthquakes with a given intensity I is given by

$$\log_{10}\left(\frac{I}{I_0}\right)$$

where I_0 is the intensity of a certain (small) size earthquake. Find the Richter Scale rating of an earthquake having intensity $10{,}000{,}000 I_0$.

11–1 ■ The exponential function

The exponential function

Expressions such as 2^x and $(1.05)^x$, where the exponent is a variable and the base is a constant, are important mathematical tools. These expressions can be used

1. to solve problems that involve the growth of bacteria in a culture, or the decay of a substance, and
2. to compute the amount to which a savings account increases when interest is compounded periodically.

We now wish to consider a function that is defined by an expression in which the exponent is a variable and the base is a constant. Such a function is called an **exponential function.**

Definition of exponential function

Given $b > 0$ and $b \neq 1$, **the exponential function with base b** is the function f defined by

$$f(x) = b^x$$

In our work with exponents thus far, we have defined 2^x when x is a rational number. That is, we know

$$2^0 = 1; \quad 2^2 = 4; \quad 2^{-2} = \frac{1}{2^2} = \frac{1}{4}; \quad 2^{-1/2} = \frac{1}{\sqrt{2}}; \quad 2^{1/3} = \sqrt[3]{2}$$

■ 481

What we have not previously defined is

$$2^x \ (x \text{ is irrational})$$

For example, consider the value of

$$2^{\sqrt{2}}$$

The definition of $2^{\sqrt{2}}$ is beyond the scope of this text. For now, we assume that 2^x exists for all real numbers x, rational *and* irrational.

We know the irrational number $\sqrt{2}$ is approximately equal to 1.41421, correct to five decimal places. Then

$$2^1 < 2^{\sqrt{2}} < 2^2$$

and since

$$1.414 < \sqrt{2} < 1.415$$

then

$$2^{1.414} < 2^{\sqrt{2}} < 2^{1.415}$$
$$2.665 < 2^{\sqrt{2}} < 2.667 \qquad 2^{\sqrt{2}} \approx 2.67$$

This same type of argument can be made for any irrational exponent x. We then assume that for $b > 0$, b^x is defined for all real numbers. We also assume that all properties of exponents previously learned extend to the real numbers.

Graph of the exponential function

Exponential functions can be graphed by finding several ordered pairs that belong to the function, plotting the points, and connecting the points with a smooth curve.

■ *Example 11–1 A*

Sketch the graphs of the following exponential functions.

1. $f(x) = 3^x$

We make a table of related values.

x	-3	-2	-1	0	1	2
$y = f(x)$ $= 3^x$	$3^{-3} = \dfrac{1}{27}$	$3^{-2} = \dfrac{1}{9}$	$3^{-1} = \dfrac{1}{3}$	$3^0 = 1$	$3^1 = 3$	$3^2 = 9$

Plot the points and connect them with a smooth curve.

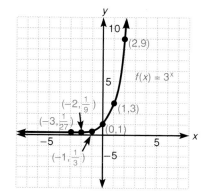

Note As *x* increases in value, *y* = *f*(*x*) also increases in value. Since *y* is always a positive number, the graph never crosses the *x*-axis. We say the *x*-axis is a horizontal asymptote.

2. $g(x) = \left(\dfrac{1}{3}\right)^x$

We make a table of related values.

x	-2	-1	0	1	2	3
$y = g(x)$ $= \left(\dfrac{1}{3}\right)^x$	$\left(\dfrac{1}{3}\right)^{-2} = 9$	$\left(\dfrac{1}{3}\right)^{-1} = 3$	$\left(\dfrac{1}{3}\right)^{0}$ $= 1$	$\left(\dfrac{1}{3}\right)^{1}$ $= \dfrac{1}{3}$	$\left(\dfrac{1}{3}\right)^{2}$ $= \dfrac{1}{9}$	$\left(\dfrac{1}{3}\right)^{3}$ $= \dfrac{1}{27}$

Plot the graphs of the ordered pairs and draw a smooth curve.

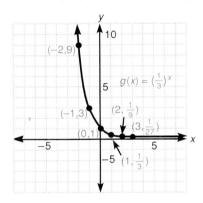

Note As *x* increases in value, *y* = *g*(*x*) decreases in value. Since *y* is always a positive number, the graph never crosses the *x*-axis. We say the *x*-axis is a horizontal asymptote.

3. $f(x) = 2^{x+1}$

We make a table of related values.

x	-3	-2	-1	0	1	2
$y = f(x)$ $= 2^{x+1}$	$2^{-2} = \dfrac{1}{4}$	$2^{-1} = \dfrac{1}{2}$	$2^{0} = 1$	$2^{1} = 2$	$2^{2} = 4$	$2^{3} = 8$

Plot the points and draw a smooth curve through them.

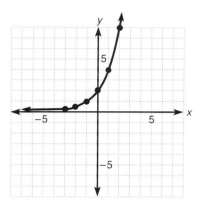

▶ *Quick check* Sketch the graph of $f(x) = 5^x$. ■

From our two graphs, we observe the following characteristics about the exponential function.

> **Characteristics of the exponential function**
> 1. When $b > 1$, $f(x) = b^x$ *increases* in value as x increases in value. This is called **exponential growth.**
> 2. When $0 < b < 1$, $f(x) = b^x$ *decreases* in value as x increases in value. This is called **exponential decay.**
> 3. The graph of $f(x) = b^x$ passes through the point $(0,1)$.
> 4. The graph of $f(x) = b^x$ approaches, but does not cross, the x-axis. The x-axis is an **asymptote.**

Note When $b = 1$, then $f(x) = 1^x = 1$ for *every* real number x. Thus, we exclude $b = 1$. The graph is the horizontal line $y = f(x) = 1$.

In each case, we see that the domain of the function is the set of all real numbers, and since the graph of each function is entirely above the x-axis, the range is the set of positive real numbers.

Note The vertical line test for functions and the horizontal line test for one-to-one functions are both met. Thus the exponential function is a one-to-one function and so it has an inverse that is a function. We will study this inverse function in section 11–2.

The following property of exponents is used to solve some equations involving exponential expressions, called **exponential equations.**

> **Property of exponential equations**
> For any real numbers x, y, and b, $b > 0$, $b \neq 1$,
> $$\text{if } b^x = b^y, \text{ then } x = y$$

The following examples illustrate how this property is used to solve exponential equations.

■ Example 11-1 B

Find the solution set for the following exponential equations.

1. $3^x = 81$

Since $81 = 3^4$, then
$$3^x = (3^4)$$ Replace 81 with 3^4
$$x = 4$$ Property of exponential equations

The solution set is $\{4\}$.

2. $5^{4x + 1} = (25)^{3x}$
Since $25 = 5^2$
$$5^{4x + 1} = (5^2)^{3x}$$ Replace $(25)^{3x}$ with $(5^2)^{3x}$
$$= 5^{2 \cdot 3x}$$ Property of exponents $(x^a)^b = x^{a \cdot b}$
$$5^{4x + 1} = 5^{6x}$$
and $4x + 1 = 6x$ Property of exponential equations
$$1 = 2x$$
$$x = \frac{1}{2}$$

The solution set is $\left\{ \dfrac{1}{2} \right\}$.

▶ **Quick check**   Find the solution set for $5^x = 125$.   ■

We mentioned earlier that exponential equations can be used to solve problems that occur in bacteria growth. The following example illustrates that process.

■ Example 11-1 C

The number of bacteria in a culture that initially contains 3,000 bacteria triples every 15 hours. How many bacteria are there in the culture after (a) 15 hours, (b) 30 hours, (c) 45 hours? Write an equation for the number of bacteria in the culture, N, after t hours.

Since the bacteria *triples* (3 times) every 15 hours, after

a. 15 hours, there are $3,000(3) = 9,000$ bacteria;
b. 30 hours, the number of bacteria triples *twice,* so there are

$$[3,000(3)](3) = 3,000(3)^2 = 3,000(9)$$
$$= 27,000 \text{ bacteria}$$

c. 45 hours, the number of bacteria triples *three* times, so there are

$$[3,000(3)](3)(3) = 3,000(3)^3$$
$$= 3,000(27) = 81,000 \text{ bacteria}$$

In t hours, the number of times the bacteria triples is found by $\dfrac{t}{15}$, since it triples every 15 hours. Thus the number of bacteria, N, is given by

$$N = 3,000(3)^{t/15}$$ ■

┌─ **Mastery points** ─────────────────────────────────────

Can you
■ Sketch the graph of an exponential function?
■ Find the solution set of exponential equations?

Exercise 11–1

Sketch the graph of each given exponential function. See example 11–1 A.

Example Sketch the graph of $f(x) = 5^x$.

Solution Choosing values of $x = -3, -2, -1, 0, 1, 2,$ and 3, we find corresponding values of $y = f(x)$.

x	−3	−2	−1	0	1	2	3
$y = f(x)$	$\dfrac{1}{125}$	$\dfrac{1}{25}$	$\dfrac{1}{5}$	1	5	25	125

Now graph the ordered pairs $\left(-3,\dfrac{1}{125}\right)$, $\left(-2,\dfrac{1}{25}\right)$, $\left(-1,\dfrac{1}{5}\right)$, (0,1), (1,5), (2,25), (3,125) and draw a smooth curve through the points.

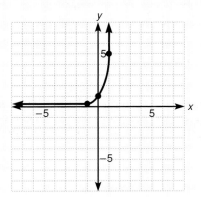

1. $f(x) = 2^x$

2. $g(x) = 4^x$

3. $h(x) = \left(\dfrac{1}{2}\right)^x$

4. $f(x) = \left(\dfrac{1}{4}\right)^x$

5. $g(x) = 3^{-x}$

6. $h(x) = 2^{-x}$

7. $f(x) = 2^{2x}$

8. $g(x) = 3^{2x}$

9. $h(x) = \left(\dfrac{1}{3}\right)^{-x}$

10. $f(x) = 2^{x-1}$

11. $f(x) = 3^{x^2}$

12. $h(x) = 2^{x^2}$

Find the solution set of each exponential equation. See example 11–1 B.

Example $5^x = 125$

Solution Since $125 = 5^3$, then

$$5^x = 5^3 \qquad \text{Replace 125 with } 5^3$$
$$x = 3 \qquad \text{Property of exponential equations}$$

The solution set is {3}.

13. $2^x = 32$

14. $3^x = 27$

15. $4^x = 32$

16. $9^x = 243$

17. $16^x = 64$

18. $2^x = \dfrac{1}{16}$

19. $3^x = \dfrac{1}{9}$

20. $4^x = \dfrac{1}{64}$

21. $5^{-x} = 25$

22. $3^{-x} = 81$

23. $16^{-x} = 32$

24. $9^{2x} = 27$

25. $8^{3x} = 4$

26. $\left(\dfrac{1}{2}\right)^x = 16$

27. $\left(\dfrac{1}{3}\right)^x = 27$

28. $\left(\dfrac{2}{3}\right)^x = \dfrac{4}{9}$

29. $\left(\dfrac{4}{3}\right)^x = \dfrac{64}{27}$

30. $\left(\dfrac{1}{2}\right)^{-x} = 8$

31. $\left(\dfrac{3}{4}\right)^{-x} = \dfrac{64}{27}$

32. $\left(\dfrac{2}{3}\right)^x = \dfrac{9}{4}$

33. $\left(\dfrac{5}{6}\right)^x = \dfrac{216}{125}$

34. $(32)^{x-1} = 8$

35. $(27)^{2x+1} = 9$

36. $(25)^{4x+1} = 125$

See example 11-1 C.

37. The number of bacteria in a culture is initially 1,500 bacteria. If the number of bacteria triples every 12 hours, how many bacteria, N, are there in the culture after (a) 12 hours, (b) 36 hours, (c) 30 hours (use a calculator), (d) t hours?

38. Due to a growing number of industries, the population of University City doubles every 10 years. If the population was 40,000 in 1980, what is the population going to be in the year (a) 2000, (b) 2005, (c) t years from 1980?

39. Money deposited in a savings account will double every $\dfrac{72}{r}$ years, where r is the rate of interest per year. If $5,000 is deposited at 8% interest, how much money is in the account after (a) 18 years, (b) 27 years, (c) t years?

40. A radioactive substance decays according to the equation

$A = 200(3)^{-0.5t}$

where A is the amount in grams and t is the time in months. How many grams of the substance are present after (a) 2 months, (b) 1 year?

41. Due to the introduction of an antibacterial substance into a culture that contains 8,000 bacteria, the bacteria are being destroyed according to the equation

$N = 8,000(2)^{-t}$

where N is the number of bacteria present and t is the time in hours. How many bacteria are present after (a) 3 hours, (b) 5 hours?

42. The production of an oil well is decreasing at a rate according to the equation

$A = 1,000,000(2)^{-0.3t}$

where A is the amount in barrels and t is the time in years. What is the production when (a) $t = 1$ year, (b) $t = 4$ years, (c) $t = 5$ years?

Review exercises

1. Find the inverse of the function $f(x) = 4x - 3$. See section 8-4.

2. Find the domain of the function $f(x) = \dfrac{4x}{3x - 5}$. See section 8-1.

3. Given $f(x) = 2x - 1$ and $g(x) = x^2 + 2$, find $f[g(x)]$. See section 8-2.

Perform the indicated operations. See section 5-1.

4. $x^{2/3} \cdot x^{-1/2}$

5. $(x^3y^9)^{1/3}$

6. $\dfrac{x^{5/3}}{x}$

7. Find the solution set of the equation $z - \dfrac{5}{z} = 4$. See section 6-1.

11–2 ■ *The logarithm*

In section 11–1, we discussed the exponential function f as defined by $f(x) = b^x$, where $b > 0$ and $b \neq 1$. It is a one-to-one function and thus has an inverse that is a function. This inverse function is called the **logarithmic function.** Recall, to find the inverse function of a one-to-one function, we interchange the variables x and y and solve for y. Then, given the exponential function with base b defined by $f(x) = b^x$, we replace $f(x)$ with y to get $y = b^x$ and interchange x and y to obtain the equation $x = b^y$. We introduce a new notation to define the logarithmic function.

Definition of logarithm _____

For every $b > 0$, $b \neq 1$,

$$x = b^y \text{ is equivalent to } y = \log_b x$$

In the definition,

1. to say that $x = b^y$ is equivalent to $y = \log_b x$ is to say that a solution for one equation is a solution for the other equation;
2. we restrict values of b to greater than 0 since, if $b < 0$, then b^x could not be defined for values of x such as $\dfrac{1}{2}$. Also, if $b = 1$, then $b^x = b^y$ for every real number value of x and y, hence we would not have a one-to-one function.

Note The word "**log**" is the abbreviation of the word **logarithm.** The notation $\log_b x$ is read "the logarithm of x to the base b" or, in shortened form, "log base b of x."

Graph of the logarithmic function

We saw in chapter 10 that the graph of an inverse function f^{-1} is the reflection of the graph of function f with respect to the line $y = x$. We can then sketch the graph of the general logarithmic function $f^{-1}(x) = \log_b x$, using the graph of the general exponential function, $f(x) = b^x$ $(b > 0, b \neq 1)$. (See figure 11–1.)

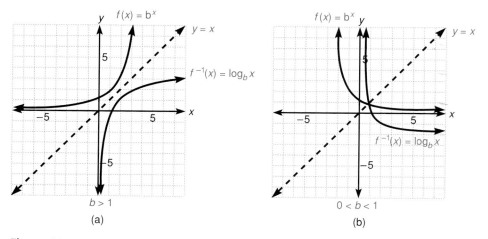

(a)   (b)

Figure 11–1

From the graphs in figure 11-1, we can determine that the logarithmic function has

1. domain $= \{x | x > 0\}$
2. range $\quad = \{y | y \in R\}$

Since, by definition, $x = b^y$ is equivalent to $y = \log_b x$, we use this relationship to write the statements in example 11-2 A.

■ *Example 11-2 A*

1. $8 = 2^3$ is equivalent to $3 = \log_2 8$.

2. $\dfrac{1}{27} = \left(\dfrac{1}{3}\right)^3$ is equivalent to $3 = \log_{1/3}\left(\dfrac{1}{27}\right)$.

3. $\dfrac{1}{16} = 4^{-2}$ is equivalent to $-2 = \log_4\left(\dfrac{1}{16}\right)$.

4. $v = b^u$ is equivalent to $u = \log_b v$.

▶ *Quick check*   Fill in the blanks.
$9 = 3^2$ is equivalent to _____ .

_____ is equivalent to $-2 = \log_7\left(\dfrac{1}{49}\right)$.

$\dfrac{1}{27} = 3^{-3}$ is equivalent to _____ .

_____ is equivalent to $-3 = \log_{1/5}(125)$.

Note   It is important to remember that the **logarithm of a number to base b is the *exponent* to which the base b must be raised to get that number.** Thus **the logarithm of a number is an *exponent*.**

Using the previous equivalencies, we can state the following properties of the logarithmic function.

— *Properties of the logarithmic function* —————————

1. Since $b^1 = b$, then $\log_b b = 1$.
2. Since $b^0 = 1$, then $\log_b 1 = 0$.
3. Since $x = b^y$ is equivalent to $y = \log_b x$, then $b^{\log_b x} = x$.
4. Since $\log_b y = x$ means $y = b^x$, then $\log_b b^x = x$.

■ *Example 11-2 B*

Determine what property is used in each of the following.

1. $\log_3 3 = 1$ (property 1)
    2. $\log_{1/4}\left(\dfrac{1}{4}\right) = 1$ (property 1)

3. $\log_6 1 = 0$ (property 2)
    4. $\log_{1/2} 1 = 0$ (property 2)

5. $5^{\log_5 9} = 9$ (property 3)
    6. $4^{\log_4 7} = 7$ (property 3)

7. $\log_2 2^5 = 5$ (property 4)
    8. $\log_6 6^7 = 7$ (property 4)

Logarithmic equations

We can use the definition $x = b^y$ is equivalent to $y = \log_b x$ to find the solution set of equations involving logarithms, called **logarithmic equations.** The following examples illustrate how this is done.

■ *Example 11–2 C*

Find the solution set for each logarithmic equation.

1. $\log_4 64 = x$
By definition, $\log_4 64 = x$ is equivalent to $4^x = 64$.

Since $64 = 4^3$
$$4^x = (4^3) \qquad \text{Replace 64 with } 4^3$$
$$x = 3 \qquad \text{Property of exponential equations}$$

Thus the solution set is $\{3\}$.

2. $\log_{10} x = 5$
By definition, $\log_{10} x = 5$ is equivalent to $10^5 = x$.

Since $10^5 = 100,000$
$$x = (100,000) \qquad \text{Replace } 10^5 \text{ with } 100,000$$

The solution set is $\{100,000\}$.

3. $\log_x 512 = 3$
By definition, $\log_x 512 = 3$ is equivalent to $x^3 = 512$.

Since $512 = 8^3$
$$x^3 = (8^3) \qquad \text{Replace 512 with } 8^3$$
$$x = 8 \qquad \text{Take principal cube root of each member}$$

Thus the solution set is $\{8\}$.

▶ *Quick check* Find the solution set of the logarithmic equation $\log_x(256) = 4$, $x > 0$. ■

___ **Mastery points** ___

Can you
- Sketch the graph of a logarithmic function?
- Given an exponential equation, write the equivalent logarithmic equation?
- Given a logarithmic equation, write the equivalent exponential equation?
- Evaluate a logarithmic expression?
- Find the solution set of a logarithmic equation?
- Sketch the graph of a logarithmic function using the related exponential function?

Exercise 11–2

Sketch the graph of $f(x) = \log_b x$ using the graph of $y = b^x$.

Example $f(x) = \log_3 x$

Solution   We sketch the graph of $y = 3^x$. Because $g(x) = 3^x$ and $f(x) = \log_3 x$ are inverse functions, we reflect this graph with respect to the line $y = x$.

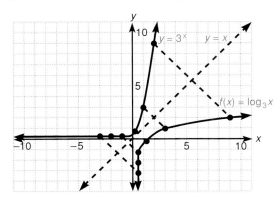

1. $f(x) = \log_2 x$ **2.** $g(x) = \log_4 x$ **3.** $h(x) = \log_5 x$

4. $F(x) = \log_{10} x$ **5.** $G(x) = \log_{1/2} x$ **6.** $H(x) = \log_{1/3} x$

Write the following exponential equations in logarithmic form. See example 11–2 A.

Example $9 = 3^2$

Solution $9 = 3^2$ is equivalent to $2 = \log_3 9$.

7. $81 = 3^4$ **8.** $16 = 4^2$ **9.** $64 = 2^6$ **10.** $512 = 8^3$

11. $\dfrac{1}{8} = \left(\dfrac{1}{2}\right)^3$ **12.** $\dfrac{1}{81} = \left(\dfrac{1}{3}\right)^4$ **13.** $\dfrac{8}{27} = \left(\dfrac{3}{2}\right)^{-3}$ **14.** $\dfrac{36}{25} = \left(\dfrac{5}{6}\right)^{-2}$

Write the following logarithmic equations in exponential form. See example 11–2 B.

Example $-2 = \log_7\left(\dfrac{1}{49}\right)$

Solution $-2 = \log_7\left(\dfrac{1}{49}\right)$ is equivalent to $\dfrac{1}{49} = 7^{-2}$

15. $\log_2 16 = 4$ **16.** $\log_3 81 = 4$ **17.** $\log_{10} 1,000 = 3$ **18.** $\log_5 625 = 4$

19. $\log_4\left(\dfrac{1}{16}\right) = -2$ **20.** $\log_2\left(\dfrac{1}{32}\right) = -5$ **21.** $\log_{10} 0.0001 = -4$ **22.** $\log_{10} 0.00001 = -5$

23. $\log_{1/2} 8 = -3$ **24.** $\log_{1/3} 81 = -4$

Evaluate each logarithmic expression.

Example $\log_{10}1{,}000$

Solution Since $\log_{10}1{,}000 = y$ means $(10)^y = 1{,}000$, we write both members with the same base. Now $1{,}000 = (10)^3$, so we have the equation

$$(10)^y = (10)^3$$
and $\qquad y = 3$
Thus $\log_{10}1{,}000 = 3$

25. $\log_2 32$ **26.** $\log_3 81$ **27.** $\log_5 25$ **28.** $\log_7 1$ **29.** $\log_2\left(\dfrac{1}{4}\right)$

30. $\log_4\left(\dfrac{1}{64}\right)$ **31.** $\log_6\left(\dfrac{1}{216}\right)$ **32.** $\log_5\left(\dfrac{1}{25}\right)$ **33.** $\log_8 8$ **34.** $\log_{1/2}\left(\dfrac{1}{8}\right)$

35. $\log_{1/3}\left(\dfrac{1}{81}\right)$ **36.** $\log_{3/2}\left(\dfrac{27}{8}\right)$ **37.** $\log_{5/4}\left(\dfrac{125}{64}\right)$ **38.** $\log_5\sqrt{5}$ **39.** $\log_7\sqrt[3]{7}$

Find the solution set of each logarithmic equation. See example 11–2 C.

Example $\log_x(256) = 4,\ x > 0$

Solution By definition, $\log_x(256) = 4$ is equivalent to $x^4 = 256$.

Since $256 = 4^4$
$\qquad x^4 = (4^4)$ Replace 256 with 4^4
$\qquad\ x = 4$ Take the principal 4th root of each member

The solution set is $\{4\}$.

40. $\log_x 25 = 2$ **41.** $\log_x 32 = 5$ **42.** $\log_x 243 = 3$ **43.** $\log_x\left(\dfrac{1}{27}\right) = -3$

44. $\log_x\left(\dfrac{1}{64}\right) = -6$ **45.** $\log_x 7 = 1$ **46.** $\log_x 5 = 1$ **47.** $\log_b 1 = 0$

48. $\log_{10} x = -2$ **49.** $\log_3 x = 4$ **50.** $\log_6 x = 3$ **51.** $\log_6 x = -2$

52. $\log_7 x = -3$ **53.** $\log_{1/4} x = 4$ **54.** $\log_2 x = 0$

55. For what values of x is $\log_6(x - 2)$ defined? (*Hint:* The domain of the logarithmic function is the set of positive real numbers.)

56. For what values of p is $\log_7(p + 1)$ defined?

57. For what values of x is $\log_8(x^2 - 1)$ defined?

58. For what values of y is $\log_9(y^2 - y - 12)$ defined?

Review exercises

1. Given $C = \dfrac{5}{9}(F - 32)$, find C when $F = 77$.

See section 1–5.

Perform the indicated operations. Assume all denominators are nonzero. See sections 4–2 and 4–3.

2. $\dfrac{25 - y^2}{2y + 3} \div (y + 5)$ **3.** $(3z + 2) - \dfrac{z + 4}{z - 2}$

4. Simplify $\left(\dfrac{ab^0}{c^{-3}}\right)^{-4}$ and eliminate negative exponents. See section 3-3.

5. Simplify $\dfrac{3(2-5)-3^2+8}{(-3)(-5)}$. See section 1-4.

6. Find the solution set of the equation $\sqrt[3]{3x-1}+1=0$. See section 6-5.

11-3 ■ Properties of logarithms

Logarithms are used to simplify complex numerical computations. In this section, we shall see how logarithms can be used to perform these computations.

Product property of logarithms

If $b>0$, $b\neq 1$, and x and y are positive real numbers, then

$$\log_b(xy)=\log_b x+\log_b y$$

Concept

The logarithm of the product of two positive real numbers is equal to the sum of the logarithms of the numbers. (Base remains same.)

To prove this property, we use the fact that a *logarithm of a number is an exponent*. Let $\log_b x=m$ and $\log_b y=n$, then,

$$b^m=x \text{ and } b^n=y$$

Using substitution,

$$x\cdot y=b^m\cdot b^n=b^{m+n}\qquad\text{Product property of exponents}$$

By definition of logarithms,

$$xy=b^{m+n}\text{ is equivalent to }\log_b(xy)=m+n$$

Substituting $\log_b x$ for m and $\log_b y$ for n,

$$\log_b(xy)=\log_b x+\log_b y$$

■ Example 11-3 A

Use the product property of logarithms to write each logarithmic statement as the sum of logarithms or as the logarithm of a single number. Simplify where possible.

1. $\log_3(7\cdot 4)=\log_3 7+\log_3 4$

2. $\log_5 6+\log_5 9=\log_5(6\cdot 9)=\log_5 54$

3. $\log_2 20=\log_2(4\cdot 5)=\log_2(2\cdot 2\cdot 5)$
$=\log_2 2+\log_2 2+\log_2 5$ Product property
$=1+1+\log_2 5$ $\log_b b=1$
$=2+\log_2 5$ Combine

Note We factored $20=4\cdot 5=2\cdot 2\cdot 5$ since the base is 2 and $\log_2 2=1$. We used prime factors.

▶ **Quick check** Write $\log_6 18$ as a sum of logarithms. ■

Note A common error is to assume that $\log_b(x \pm y)$ is the same as $\log_b x \pm \log_b y$. **This is not true.** To illustrate,

$$\log_{10}(1 + 1) = \log_{10}2 = 0.30103$$

while $\log_{10}1 + \log_{10}1 = 0 + 0 = 0$. Thus, $\log_{10}(1 + 1) \neq \log_{10}1 + \log_{10}1$.

The following three properties state the remaining properties of the logarithms. Their proofs are similar to the one just completed and they are left as exercises.

_____ **Quotient property of logarithms** _____

If $b > 0$, $b \neq 1$, and x and y are positive real numbers, then

$$\log_b\left(\frac{x}{y}\right) = \log_b x - \log_b y$$

Concept
The logarithm of the quotient of two positive real numbers is equal to the logarithm of the numerator minus the logarithm of the denominator. (Base remains same.)

■ *Example 11–3 B*

Write each logarithmic expression as the difference of two logarithms or as the logarithm of a single number. Simplify where possible.

1. $\log_6\left(\dfrac{8}{9}\right) = \log_6 8 - \log_6 9$ Quotient property

2. $\log_3 5 - \log_3 7 = \log_3\left(\dfrac{5}{7}\right)$ Quotient property

3. $\log_5\left(\dfrac{5}{8}\right) = \log_5 5 - \log_5 8$ Quotient property

$\qquad\qquad = 1 - \log_5 8$ Replace $\log_5 5$ with 1   ($\log_b b = 1$)

▶ *Quick check*   Write $\log_2 5 - \log_2 3$ as a single number.   ■

_____ **Power property of logarithms** _____

If $b > 0$, $b \neq 1$, r is any real number and x is a positive real number, then

$$\log_b(x^r) = r \log_b x$$

Concept
The logarithm of any real power of a positive real number is equal to the exponent times the logarithm of the positive real number. (Base remains same.)

■ *Example 11–3 C*

Write each logarithmic expression as a constant times a logarithm or as the logarithm of a single number. Simplify where possible.

1. $\log_3 2^3 = 3 \log_3 2$ Power property

2. $6 \log_5 2 = \log_5(2^6)$ Power property

$\qquad\qquad = \log_5 64$ Replace 2^6 with 64 $(2^6 = 64)$

3. $\log_7(7^2) = 2 \log_7 7$ Power property

$\qquad\qquad = 2 \cdot 1$ Replace $\log_7 7$ with 1 $(\log_b b = 1)$

$\qquad\qquad = 2$ ■

Using these properties and the following property of logarithms, we can find the solution set of other logarithmic equations.

Property of logarithms

If $x > 0$, $y > 0$, and $\log_b x = \log_b y$, then $x = y$ $(b > 0, b \neq 1)$

Note   Since $\log_b^2 x = (\log_b x)^2$, then $\log_b^2 x \neq 2 \log_b x$.

■ *Example 11–3 D*

Find the solution set of the following logarithmic equations.

1. $\log_5(x - 2) = 3$

Writing this equation in its equivalent exponential form,

$x - 2 = 5^3$ $\log_5(x - 2) = 3$ is equivalent to $x - 2 = 5^3$

$x - 2 = 125$

$x = 127$

The solution set is $\{127\}$.

2. $\log_2 x + \log_2(x + 2) = 3$

Since the bases are the same, we can use the property $\log_b(xy) = \log_b x + \log_b y$.

$\log_2 x + \log_2(x + 2) = \log_2 x(x + 2)$

$\qquad\qquad\qquad\qquad = \log_2(x^2 + 2x)$

Now $\log_2(x^2 + 2x) = 3$ is equivalent to $x^2 + 2x = 2^3$.

Then $\qquad\quad x^2 + 2x = 8$

$\qquad x^2 + 2x - 8 = 0$

$\qquad (x + 4)(x - 2) = 0$

$\qquad$ Thus $x = -4$ or $x = 2$.

For $x = -4$, $\log_2(-4 + 2) = \log_2(-2)$, which is undefined since the domain contains only positive real numbers. The solution set is $\{2\}$.

Note   When solving logarithmic equations, we must *always* check our solutions for values that may not be in the domain of the function defined by the expressions in the original equation.

3. $\log_3(6x - 7) + \log_3 x = \log_3 5$

$\log_3(6x - 7) + \log_3 x = \log_3(6x - 7)x$ $\log_b x + \log_b y = \log_b(xy)$

$\log_3(6x - 7)x = \log_3 5$ Replace $\log_3(6x - 7) + \log_3 x$ with $\log_3(6x - 7)x$

$(6x - 7)x = 5$ Property of logarithms

$6x^2 - 7x = 5$

$6x^2 - 7x - 5 = 0$

$(3x - 5)(2x + 1) = 0$ Factor the left member

$3x - 5 = 0$ or $2x + 1 = 0$

$x = \dfrac{5}{3}$ $x = -\dfrac{1}{2}$

Since the domain of the logarithm function allows only positive numbers, $x = -\dfrac{1}{2}$ is not a solution. The solution set is $\left\{\dfrac{5}{3}\right\}$.

▶ *Quick check*   Find the solution set of $\log_3(x + 6) - \log_3(x - 2) = 2$.   ■

Mastery points

Can you
- Write a logarithmic expression in alternate forms using the properties?
- Solve logarithmic equations using the properties
  1. $\log_b(xy) = \log_b x + \log_b y$?
  2. $\log_b\left(\dfrac{x}{y}\right) = \log_b x - \log_b y$?
  3. $\log_b x^r = r \log_b x$?
  4. If $\log_b x = \log_b y$, then $x = y$?

Exercise 11–3

Use the properties of logarithms to write the following logarithms as a sum or difference of two, or more, logarithms, or as a product of a real number times a logarithm or as the sum or difference of a real number and a logarithm. All variables represent positive real numbers. See examples 11–3 A and B.

Example $\log_6 18$

Solution $\log_6 18 = \log_6(6 \cdot 3)$ Factor 18 with base 6 as one factor

$= \log_6 6 + \log_6 3$ Product property of logarithms

$= 1 + \log_6 3$ $\log_6 6 = 1$

1. $\log_b(3 \cdot 5)$ 2. $\log_2(7 \cdot 5)$ 3. $\log_3\left(\dfrac{7}{13}\right)$ 4. $\log_7\left(\dfrac{23}{17}\right)$

5. $\log_b(5^3)$ 6. $\log_6(11^2)$ 7. $\log_{10}(5^4)$ **8.** $\log_b(24)$

9. $\log_{1/2}(36)$

10. $\log_{10}\left(\dfrac{15}{16}\right)$

11. $\log_5\left(\dfrac{24}{25}\right)$

12. $\log_b\left(\dfrac{x^3}{y^2}\right)$

13. $\log_b(xy)^3$

14. $\log_b\sqrt[3]{x^2}$

15. $\log_b(\sqrt{x}\cdot\sqrt[3]{y})$

16. $\log_b\left(\dfrac{\sqrt[3]{x}\cdot\sqrt[4]{y}}{z}\right)$

17. $\log_b(2x + 3y)$

Write the following logarithms as the logarithm of a single number or expression. Assume that all variables are positive real numbers. See examples 11–3 A, B, and C.

Example $\log_2 5 - \log_2 3$

Solution $\log_2 5 - \log_2 3 = \log_2\left(\dfrac{5}{3}\right)$ Quotient property of logarithms

18. $\log_4 5 + \log_4(11)$

19. $\log_3(17) + \log_3 5$

20. $\log_{10}(25) - \log_{10}(12)$

21. $\log_7(11) - \log_7(16)$

22. $2\log_5(11)$

23. $3\log_6 4$

24. $2\log_2 7 + \log_2 5$

25. $3\log_{10}3 + 2\log_{10}4$

26. $4\log_9 2 - \log_9(10)$

27. $\dfrac{1}{5}\log_4(32) + \dfrac{1}{4}\log_4(81)$

28. $\log_5 4 + 2\log_5 3 - 3\log_5 2$

29. $2\log_{10}5 - \dfrac{1}{2}\log_{10}9 + 3\log_{10}2$

30. $4\log_2 3 - \dfrac{1}{3}\log_2 8 + \log_2 7 - 2\log_2 5$

31. $\dfrac{1}{3}\log_b(x^2)$

32. $\dfrac{1}{2}\log_b x + \dfrac{1}{2}\log_b y$

33. $\dfrac{1}{4}\log_b(x^3) - \dfrac{1}{4}\log_b y$

34. $\log_{10}(x + 3) + \log_{10}(x - 4)$

35. $\log_5(x + 6) - 2\log_5(3x) + \log_5(x - 4)$

36. $2\log_b x - \log_b(3x - 1) + \log_b(x + 1)$

Given $\log_b 2 = 0.3010$, $\log_b 3 = 0.4771$, $\log_b 7 = 0.8451$, and $\log_b 10 = 1$, compute the following logarithms and then write the statement in equivalent exponential form.

Example Given $\log_b 2 = 0.3010$ and $\log_b 3 = 0.4771$, find $\log_b 6$.

Solution Since $\log_b 6 = \log_b(2 \cdot 3) = \log_b 2 + \log_b 3$, then we replace $\log_b 2$ by 0.3010 and $\log_b 3$ by 0.4771 to obtain

$$\log_b 6 = \log_b(2 \cdot 3)$$
$$= \log_b 2 + \log_b 3 = 0.3010 + 0.4771 = 0.7781$$

Thus $\log_b 6 = 0.7781$.

37. $\log_b(14)$

38. $\log_b(20)$

39. $\log_b(49)$

40. $\log_b 8$

41. $\log_b(42)$

42. $\log_b\left(\dfrac{27}{4}\right)$

43. $\log_b\left(\dfrac{7}{10}\right)$

44. $\log_b\sqrt[3]{3}$

45. $\log_b\sqrt{7}$

46. $\log_b\sqrt{\dfrac{3}{4}}$

47. $\log_b\sqrt[4]{9}$

48. In exercises 37–47, what is b? (*Hint:* Look at the values given in the directions.)

Find the solution set of the following logarithmic equations. See example 11–3 D.

Example $\log_3(x + 6) - \log_3(x - 2) = 2$

Solution Using the property $\log_b x - \log_b y = \log_b\left(\dfrac{x}{y}\right)$

$$\log_3(x + 6) - \log_3(x - 2) = \log_3\left(\dfrac{x + 6}{x - 2}\right)$$

Thus,

$$\log_3\left(\dfrac{x + 6}{x - 2}\right) = 2 \qquad \text{Replace } \log_3(x + 6) - \log_3(x - 2) \text{ with } \log_3\dfrac{x + 6}{x - 2}$$

$$\dfrac{x + 6}{x - 2} = 3^2 \qquad \log_3\left(\dfrac{x + 6}{x - 2}\right) = 2 \text{ is equivalent to } \dfrac{x + 6}{x - 2} = 3^2$$

$$\dfrac{x + 6}{x - 2} = 9$$

$$x + 6 = 9x - 18 \qquad \text{Multiply each member by } x - 2$$

$$24 = 8x$$

$$x = 3$$

The solution set is $\{3\}$.

49. $\log_4 x + \log_4 5 = 3$

50. $\log_3 x + \log_3 8 = 2$

51. $\log_5 x - 2\log_5 2 = 1$

52. $\log_2 x - 3\log_2 3 = 5$

53. $\log_3(2x + 1) - \log_3 5 = 3$

54. $\log_4(x - 5) - \log_4 3 = 2$

55. $\log_3(x + 6) - \log_3(x - 2) = 2$

56. $\log_{10}(x) - \log_{10}(x + 3) = 1$

57. $\log_{10}(x + 21) + \log_{10} x = 2$

58. $\log_{10}(3x + 2) + \log_{10} 2 = 2$

59. $\log_3 x + \log_3(x + 6) = 3$

60. $\log_4(x - 15) + \log_4 x = 2$

61. $\log_2(x + 1) = \log_2 3 - \log_2(2x - 1)$

62. $\log_5(x - 2) = \log_5 5 - \log_5(x + 3)$

63. $\log_b(x + 2) + \log_b(2x - 1) = \log_b x$

64. $\log_2(2x - 1) = 2\log_2 x$

65. $\log_b(6x - 5) = 2\log_b x$

66. $\dfrac{1}{2}\log_3 x = \log_3(x - 6)$

67. Prove that for $b > 0$, $b \neq 1$, and positive real numbers u and v,

$$\log_b\left(\dfrac{u}{v}\right) = \log_b u - \log_b v$$

68. Prove that for $b > 0$, $b \neq 1$, positive real number u and any real number r,

$$\log_b(u^r) = r\log_b u$$

69. Show by an example that for positive real numbers u and v,

$$\log_b(u + v) \neq \log_b u + \log_b v$$

Review exercises

What type of conic section figure do you get in the graph of the following equations? Find the x- and y-intercepts if they exist. See sections 9–1 to 9–3.

1. $x^2 + 3y^2 = 9$

2. $y = x^2 + x - 20$

3. $\dfrac{y^2}{9} - \dfrac{x^2}{16} = 1$

4. Find the solution set of the equation $\dfrac{4}{z} - \dfrac{3}{z} = \dfrac{3}{4}$.

   See section 4–7.

5. If y varies inversely as the square of x, find x when $y = 3$ if $y = 8$ when $x = 2$. See section 8–5.

6. Evaluate $(-27)^{2/3}$. See section 5–1.

7. Find the solution set of the inequality $|4 - 5x| \geq 2$. See section 2–6.

11–4 ■ *The common logarithms*

We have worked thus far with logarithms to different bases. There are two logarithms that are most commonly used and are of great importance in mathematics. They are logarithms to base 10 and to base e, defined, respectively, by (1) $y = \log_{10}x$ and (2) $y = \log_e x$. Because 10 is the base of our number system, we will begin our discussion with base 10 logarithms. In section 11–5, we will discuss base e logarithms. The two bases have similar properties.

The function values of the logarithmic functions with base 10 are called the **common logarithms.** When we use 10 as a base, it is customary to omit the base when writing a logarithm. That is, we will use

$$\log x \quad \text{instead of} \quad \log_{10}x, \, x > 0$$

and the base will be understood to be 10.

From our previous work with logarithms, we realize that a common logarithm of a number is the power to which 10 must be raised to obtain the number. For example,

$$\log 100 = 2 \quad \text{since} \quad 10^2 = 100$$

In like fashion,

$\log 1{,}000 = 3$	since $10^3 = 1{,}000$
$\log 100 = 2$	since $10^2 = 100$
$\log 10 = 1$	since $10^1 = 10$
$\log 1 = 0$	since $10^0 = 1$
$\log 0.1 = -1$	since $10^{-1} = \dfrac{1}{10} = 0.1$
$\log 0.01 = -2$	since $10^{-2} = \dfrac{1}{100} = 0.01$
$\log 0.001 = -3$	since $10^{-3} = \dfrac{1}{1{,}000} = 0.001$

Use of a calculator

To find the common logarithm of a number, we can use a calculator with a **log** key. The log key means $\log_{10}$, with base 10 being understood. To illustrate, to find log 724, press

$$\boxed{7} \; \boxed{2} \; \boxed{4} \; \boxed{\log}$$

and read 2.8597386 on the display. Thus,

$$\log 724 \approx 2.8597 \text{ (correct to four decimal points)}$$

This means that $10^{2.8597} \approx 724$.

Note In the example, the logarithm of the number, 2.8597386, has two parts. We call

1. the integer number, 2, the **characteristic,** and
2. the decimal number, .8597386, the **mantissa** of the number.

■ *Example 11–4 A*

Find the following common logarithms using a calculator. Round to four decimal places.

1. log 715,000

Using the calculator, press ⑦ ① ⑤ ⓪ ⓪ ⓪ ☐log☐ and read 5.854306 on the display. Rounding off to four decimal places, log 715,000 ≈ 5.8543, and we have found $10^{5.8543} \approx 715{,}000$.

2. log 0.0749

Press ☐.☐ ⓪ ⑦ ④ ⑨ ☐log☐ and read −1.125518 on the display. Thus log 0.0749 = −1.1255 and $10^{-1.1255} \approx 0.0749$.

▶ *Quick check*   Find the log 0.000315 on the calculator.   ■

The antilogarithm

Now suppose we know the common logarithm of a number N and wish to determine N. That is, we know log N and wish to find N. The number N is called the **antilogarithm** (or abbreviated **antilog**) of log N. Thus

$$\text{antilog } x = N \text{ is equivalent to } \log N = x$$

In example 11–4 A,

715,000 is the antilog of 5.8543

while   0.0749 is the antilog of −1.1255

We can use the calculator to find the antilogarithm of x. To find the antilogarithm on a calculator, we use either the ☐10^x☐ key or the ☐INV☐ key followed by the ☐log☐ key. To illustrate, to find the antilogarithm of 2.8597, we press

② . ⑧ ⑤ ⑨ ⑦ ☐10^x☐

or   ② . ⑧ ⑤ ⑨ ⑦ ☐INV☐ ☐log☐

and read 723.93571 on the display. We have found that

log 723.93571 ≈ 2.8597

or $10^{2.8597} \approx 723.93571$

In our examples, we will use only the ☐10^x☐ key.

■ *Example 11–4 B*

Find the antilogarithm of the following.

1. 3.8603

Press

③ . ⑧ ⑥ ⓪ ③ ☐10^x☐

and read 7249.376556 on the display. Thus the antilogarithm ≈ 7249.376556 and we have found that

log 7249.376556 ≈ 3.8603 and $10^{3.8603} \approx 7249.376556$

2. -1.1255
Press

$\boxed{1}\;\boxed{.}\;\boxed{1}\;\boxed{2}\;\boxed{5}\;\boxed{5}\;\boxed{+/-}\;\boxed{10^x}$

and read 0.074903135 on the display. Thus, the antilogarithm ≈ 0.074903135 and we have found that

$\log 0.074903135 \approx -1.1255$ and $10^{-1.1255} \approx 0.074903135$

▶ *Quick check*   Find the antilogarithm of -3.5734 on the calculator.

Note   An alternate way (one that is often used) of writing the negative logarithm -1.1255 in example 2 is

$$8.8745 - 10$$

since $-10 + 8.8745 = -1.1255$. This is done to eliminate negative numbers in the logarithm of a number.   ■

┌─ *Mastery points* ─────────────────────────────────

Can you
■ Find the common logarithm of a number using a calculator?
■ Find the antilogarithm of a number using a calculator?

└──

Exercise 11–4

Find each of the following logarithms using a calculator. Then write the equivalent exponential statement. Round off to four decimal places. See example 11–4 A.

Example   log 0.000315

Solution   Press $\boxed{.}\;\boxed{0}\;\boxed{0}\;\boxed{0}\;\boxed{3}\;\boxed{1}\;\boxed{5}\;\boxed{\log}$ and read -3.501689 on the display. Thus

$\log 0.0003 \approx -3.5017$

and we have found that $10^{-3.5017} \approx 0.0003$

1. log 8 | **2.** log 5.7 | **3.** log 53 | **4.** log 547
5. log 80,200 | **6.** log 523,000 | **7.** log 794,000,000 | **8.** log 0.157
9. log 0.00863 | **10.** log 0.0000941 | **11.** log 0.000000107 | **12.** log 0.0431

Round off to four decimal places unless otherwise stated.

13. An artificial earth satellite is orbiting above the earth's surface at a height of 504,000 miles. Find log 504,000.

14. The resistive force F exerted by water on a ship traveling at $1\frac{1}{2}$ statute miles per hour is 244,000 pounds. Find log 244,000.

15. The angular velocity of the earth is 0.000073 radians per second and the earth's moment of inertia is 9.8×10^{37} kilograms per cubic meter. Find (a) log 0.000073 and (b) log (9.8×10^{37}).

16. The current rate of interest on money market certificates is 6.75%. Find log (6.75%). (*Hint:* Write the percent in decimal form first.)

17. Given the formula $H = w \log T$, find (a) H when $w = 110$ and $T = 360$ and (b) w when $H = 16.4$ and $T = 31.1$. (Round to two decimal places.)

18. Given $\log R = \dfrac{D}{10}$, find D when $R = 1.47$.

19. Given $P_H = -\log A_H$, find P_H when $A_H = 0.00167$.

20. The number of decibels of sound, N_{db}, in comparing the "loudness" of two sounds, is given by

$$N_{db} = 10 \log\left(\frac{P_1}{P_2}\right)$$

where P_1 and P_2 are the signal power levels to be compared. Find N_{db} when $P_1 = 950$ and $P_2 = 25$.

21. Using exercise 20, find N_{db} when $P_1 = 1{,}070$ and $P_2 = 35$.

22. A decibel power gain, G_{db}, of an electronic amplifier is given by

$$G_{db} = 10 \log\left(\frac{P_0}{P_i}\right)$$

where $P_0 =$ the power output and $P_i =$ the power input. Find G_{db} when $P_0 = 20$ and $P_i = 0.005$.

23. Using exercise 22, find G_{db} when $P_0 = 18$ and $P_i = 0.006$.

Find the antilogarithm of each number, using a calculator. See example 11–4 B.

Example -3.5734

Solution Press ③ ⎡.⎤ ⑤ ⑦ ③ ④ ⎡+/−⎤ ⎡10ˣ⎤ and read 0.0002671 on the display. Thus, the antilogarithm ≈ 0.0002671 and we have found that

$$\log 0.0002671 \approx -3.5734 \text{ and } 10^{-3.5734} \approx 0.000267$$

24. 0.5416
25. 3.8445
26. 4.4997
27. 7.9138
28. -1.2472
29. -4.7825
30. 8.8340
31. 2.5283
32. -6.0104
33. -2.2301
34. -3.2391
35. -6.4727

Solve the following statements as indicated. Round off to four decimal places.

36. Given $P_H = -\log A_H$, find A_H when $P_H = 7.78$.

37. Given $H = w \log T$, find T when $w = 1.87$ and $H = 3.15$.

38. Given $\log R = \dfrac{D}{10}$, find R when $D = 44.5$.

39. Given $G_{db} = 10 \log\left(\dfrac{P_0}{P_i}\right)$, find P_0 when $G_{db} = 15$ and $P_i = 0.60$. (*Hint:* Use the property $\log\left(\dfrac{P_0}{P_i}\right) = \log P_0 - \log P_i$.)

40. Given $z_0 = 276 \log\left(\dfrac{b}{a}\right)$, find (a) b when $z_0 = 487$ and $a = 0.112$ and (b) a when $z_0 = 552.1$ and $b = 5.7$.

41. A chemist defines the pH (hydrogen potential) of a solution by

$$pH = -\log[H^+]$$

where $[H^+]$ denotes the numerical value for the concentration of hydrogen ions in aqueous solution in moles per liter. Find the pH of the solution with hydrogen concentration, $[H^+]$, as 5.0×10^{-4}.

42. Using exercise 41, find pH with the hydrogen concentration 6.1×10^{-7}.

43. Using exercise 41, find the hydrogen ion concentration with a pH 5.8.

44. Using exercise 41, find the hydrogen ion concentration with a pH 6.8.

Review exercises

Simplify the following expressions.

1. $4 - 2[12 - (7 - 5) - (13 + 2)]$ See section 1–4.

2. $5ab - (4a + 2b - a) - (3b + ab)$
   See section 1–6.

3. $\sqrt{18} + \sqrt{50} - \sqrt{98}$ See section 5–5.

4. $(2 - \sqrt{-36}) - (3 + \sqrt{-49})$ See section 5–7.

5. Jan takes 24 minutes to do a particular job, and her sister Jody takes 30 minutes to do the same job. How long would it take the sisters to do the job working together? See section 4–8.

6. One bottle of battery acid is a 10% solution, and another bottle is a 4% solution. How many cubic centimeters of each solution are necessary to make 100 cm³ of 8% solution? Use systems of linear equations. See section 8–2.

11–5 ■ Logarithms to the base e

Natural logarithms

In section 11–4, we mentioned two important logarithmic bases: base 10 and base *e*. Also in section 11–4, we studied the base 10 logarithms, called the **common logarithms.** In applications, logarithms to base *e* are more frequently used than logarithms to base 10. We call the base *e* logarithms the **natural logarithms.**

The number *e* is an irrational number, as is the irrational number π. Computed $e \approx 2.7182818284 \ldots$, but we generally round off this value and use the value $e \approx 2.718$.

Just as a special symbol was used to represent the logarithm to base 10, that is, "log," we also have a special symbol to represent the natural logarithm of positive number *x*. We denote

$$\log_e x \quad \text{by} \quad \ln x$$

Definition of the natural logarithm

$$\ln x = y \quad \text{is equivalent to} \quad e^y = x$$

Changing the base

To develop methods for finding decimal approximations of $\ln x$, we first develop a special formula for converting logarithms from one base to another. Since we can find the logarithm to the base 10 for any positive number, we will then be able to find the logarithm for any positive number to any other base, such as base *e*.

Change of base property

Let $a > 0$, $b > 0$, $x > 0$, $a \neq 1$, and $b \neq 1$. Then

$$\log_a x = \frac{\log_b x}{\log_b a}$$

and using common logarithms,

$$\log_a x = \frac{\log x}{\log a}$$

■ **Example 11–5 A**

Find the value of each logarithm, using common logarithms. Round off to four decimal places.

1. $\log_5 13$

$$\log_5 13 = \frac{\log (13)}{\log (5)}$$ Replace x with 13 and a with 5 in $\frac{\log x}{\log a}$

On the calculator, press

| 1 | | 3 | | log | | ÷ | | 5 | | log | | = |

and read 1.5936926 on the display. Thus, $\log_5 13 \approx 1.5937$.

2. $\log_{14} 33.5$

$$\log_{14} 33.5 = \frac{\log (33.5)}{\log (14)}$$ Replace x with 33.5 and a with 14 in $\frac{\log x}{\log a}$

On the calculator, press

| 3 | | 3 | | . | | 5 | | log | | ÷ | | 1 | | 4 | | log | | = |

and read 1.330606 on the display. Thus, $\log_{14} 33.5 \approx 1.3306$.

3. $\log_e 0.234$

$$\log_e 0.234 = \frac{\log (0.234)}{\log (e)}$$ Replace x with 0.234 and a with e in $\frac{\log x}{\log a}$

$$= \frac{\log 0.234}{\log (2.718)}$$ Replace e with 2.718

On the calculator, press

| . | | 2 | | 3 | | 4 | | log | | ÷ | | 2 | | . | | 7 | | 1 | | 8 | | log | | = |

and read -1.4525848 on the display.

$$\log_e 0.234 = \ln 0.234 \approx -1.4526$$

▶ **Quick check**   Find the value of $\log_{16} 25.7$ using common logarithms. Round off to four decimal places.   ■

Natural logarithms can be found directly on a calculator that has a key labeled | ln |.

■ *Example 11–5 B*

Find the value of each natural logarithm correct to four decimal places using the calculator.

1. ℓn 57.6
Enter ⑤ ⑦ · ⑥ ℓn and read 4.0535225 on the display. Thus,

ℓn 57.6 ≈ 4.0535

2. ℓn 0.5731
For numbers between 0 and 1, the natural logarithm will be negative as with common logarithms.
Enter · ⑤ ⑦ ③ ① ℓn and read −0.5566951 on the display. Thus,

ℓn 0.5731 ≈ −0.5567

▶ *Quick check*   Find the value of ℓn 0.932, correct to four decimal places using the calculator.   ■

In example 1, we found that ℓn 57.6 ≈ 4.0535. Suppose we want antiln 4.0535 (which is 57.6). We use the e^x key and enter

④ · ⓪ ⑤ ③ ⑤ e^x

into the calculator. Read 57.5987 on the display. Thus, antiln 4.0535 ≈ 57.5987 and

$$e^{4.0535} \approx 57.5987$$

The properties of logarithms that we used with common logarithms also apply to the natural logarithms.

Growth and decay

Problems involving natural growth and natural decay require the use of the exponential and logarithmic functions. The general equations for natural growth and natural decay are given by

growth, $q = q_0 e^{rt}$ $(r > 0)$
decay, $q = q_0 e^{-rt}$ $(r > 0)$

where q is the quantity (or number) present at time $t > 0$, q_0 is the quantity (or number) present when time $t = 0$, and r is the rate of growth or decay.

Note   1. e is the base of the natural logarithm.
2. We assume $r > 0$ at all times. Then $-r < 0$ and this causes the decrease in quantity.

We will use these equations to find the time it takes for a quantity q_0 at time $t = 0$ to *grow* or *decay* to a quantity q. Solving for t, we obtain the following equations:

Growth and decay formulas

$$t = \frac{1}{r} \ell n \left(\frac{q}{q_0}\right), \text{ for growth;} \quad t = -\frac{1}{r} \ell n \left(\frac{q}{q_0}\right), \text{ for decay.}$$

■ *Example 11–5 C*

1. How long will it take the earth's population to double if it continues to grow at the rate of 2 percent per year compounded continuously? Find the answer to the nearest tenth.

Since the earth's population is to double, we are given that $q = 2q_0$. Also we are given $r = 0.02$, and we want time t. Using the equation

$$t = \frac{1}{r} \ln\left(\frac{q}{q_0}\right)$$

$$t = \frac{1}{(0.02)} \ln\left(\frac{2q_0}{q_0}\right) \qquad \text{Replace } r \text{ with 0.02 and } q \text{ with } 2q_0$$

$$= 50 \ln 2 \qquad \text{Reduce by } q_0$$
$$= 50(0.6931) \qquad \text{Perform indicated operations}$$
$$\approx 34.7$$

The population will double in approximately 34.7 years.

2. Radioactive strontium 90 is used in nuclear reactors and decays according to the equation

$$q = q_0 e^{-0.0248t}$$

Find the half-life of strontium 90.

Note **Half-life** is the time necessary for a substance to decay to one-half of its original quantity.

We are given $q = \frac{1}{2}q_0$, $r = 0.0248$, and we want t in years. Using the equation

$$t = -\frac{1}{r} \ln\left(\frac{q}{q_0}\right)$$

$$t = -\frac{1}{(0.0248)} \ln\left(\frac{\frac{1}{2}q_0}{q_0}\right) \qquad \text{Replace } r \text{ with 0.0248 and } q \text{ with } \frac{1}{2}q_0$$

$$= -40.3 \ln\left(\frac{1}{2}\right) \qquad \text{Reduce by } q_0$$

$$= -40.3(-0.6931) \qquad \text{Perform indicated operations}$$
$$\approx 27.9$$

Therefore, the half-life of strontium 90 is approximately 27.9 years. ■

─ *Mastery points* ─

Can you
- Find the logarithm of any positive number to any given base?
- Find the natural logarithm of a positive number using the common logarithms?
- Use natural logarithms in working applications of natural growth and natural decay?

Exercise 11-5

Find each of the following logarithms. Round off to four decimal places. See example 11-5 A.

Example $\log_{16} 25.7$

Solution $\log_{16} 25.7 = \dfrac{\log (25.7)}{\log (16)}$ Replace x with 25.7 and a with 16

On the calculator enter, $\boxed{2}\ \boxed{5}\ \boxed{.}\ \boxed{7}\ \boxed{\log}\ \boxed{\div}\ \boxed{1}\ \boxed{6}\ \boxed{\log}\ \boxed{=}$ and read 1.1709241 on the display. Therefore, $\log_{16} 25.7 \approx 1.1709$.

1. $\log_5 8$ **2.** $\log_7 9$ **3.** $\log_4 25$ **4.** $\log_7 47$

5. $\log_2 0.156$ **6.** $\log_3 0.0324$ **7.** $\log_{23} 45.6$ **8.** $\log_{18} 157$

See example 11-5 B.

Example $\ln 0.932$

Solution Using the calculator, press $\boxed{.}\ \boxed{9}\ \boxed{3}\ \boxed{2}\ \boxed{\ln}$ and read -0.0704225 on the display. Thus $\ln 0.932 \approx -0.0704$.

9. $\log_e 5$ **10.** $\log_e 3$ **11.** $\log_e 2.73$ **12.** $\log_e 107$ **13.** $\ln 6.73$

14. $\ln 5.13$ **15.** $\ln 0.0504$ **16.** $\ln 0.00619$ **17.** $\ln 347$ **18.** $\ln 197$

19. $\log 5e$ **20.** $\log 3e$ **21.** $\ln 7e$ **22.** $\log 6e^3$ **23.** $\ln 3e^3$

Find the antiln of the following logarithms.

24. 1.3415 **25.** 2.6137 **26.** 4.0076 **27.** -0.1234

28. -3.1743 **29.** -2.9145 **30.** -0.0421

Solve the following problems. Round off to the nearest tenth. See example 11-5 C.

31. When money is invested at interest rate r percent per year compounded continuously, after t years it is worth

$q_0 e^{rt/100}$ dollars

where q_0 is the amount invested. How long will it take for an investment of $200,000 to triple at 6% per year compounded continuously?
(*Hint:* Use $q = q_0 e^{rt/100}$, where $r = 6$.)

32. Using exercise 31, how long will it take for $15,000 to become one and one-half times as great?

33. Using the equation for the decay of strontium 90,

$q = q_0 e^{-0.0248t}$

how many years will it take until only one-fourth of the original amount of the substance is left?

34. Using the formula of exercise 33, how long will it take for 90% of the strontium 90 to decay?

35. Radioactive carbon 14 diminishes by radioactive decay according to the equation

$q = q_0 e^{-0.000124t}$

where t is in years. Carbon 14 enters all living tissue through carbon dioxide and is maintained to be constant as long as the plant or animal is alive. The decay takes place when the tissues are dead. Estimate the age of the skull uncovered if 20% of the original amount of carbon 14 is still present.
(*Hint:* $q = 0.2q_0$.)

36. Using the information of exercise 35, estimate the age of the skull if only 5% of the original amount of carbon 14 is still present.

37. For relatively clear bodies of water, light intensity is reduced according to the equation

$$I = I_0 e^{-kd}$$

where I is the intensity at d feet below the surface and I_0 is the intensity at the surface. In a particular body of water, $k = 0.00853$. Find the depth to the nearest tenth of a foot at which the light is reduced to 10% of that at the surface.

38. Using exercise 37, when $k = 0.0585$, find the depth at which the light is reduced to 2% of that at the surface.

39. Suppose a certain species of bees increases in number according to the exponential equation

$$q = 25e^{0.2t}$$

where t is measured in days. In how many days, correct to the nearest tenth, will there be 375 bees?

Review exercises

State the domain of the following functions. See section 8–1.

1. $f = \{(-4,3), (2,3), (0,-8), (1,10)\}$

2. $g(x) = \sqrt{2x - 3}$

3. Is f in exercise 1 a one-to-one function? If not, why not? See section 8–4.

Graph the following functions. See section 8–3.

4. $f(x) = 4x + 3$

5. $g(x) = x^2 - x + 6$

11–6 ■ Exponential equations

In section 11–1, we found the solution set of some exponential equations, that is, equations in which the exponent is a variable. Now we consider some exponential equations whose solution set can be found by using logarithms. To do this, we use a property of logarithms.

> **Property of logarithms**
> Given positive real numbers x, y, and b, $b \neq 1$, $b > 0$,
> $$\text{if } x = y, \text{ then } \log_b x = \log_b y \tag{1}$$

■ **Example 11–6 A**

Find the solution set of each exponential equation, correct to the nearest hundredth.

1. $3^{x+7} = 11$

$$\log(3^{x+7}) = \log(11) \qquad \text{Take the common logarithm of each member}$$
$$(x + 7)\log 3 = \log(11) \qquad \text{Power property of logarithms}$$
$$x + 7 = \frac{\log(11)}{\log 3} \qquad \text{Divide each member by log 3}$$
$$x = \frac{\log(11)}{\log 3} - 7 \qquad \text{Subtract 7 from each member}$$

Using the calculator, enter

| 1 | 1 | log | ÷ | 3 | log | = | − | 7 | = |

and read -4.8173417 on the display. Thus, $x \approx -4.82$ and the solution set is $\{-4.82\}$.

2. An amount of money P (called the principal) is invested at r percent compounded annually. The amount of money A after t years is given by the equation $A = P(1 + r)^t$. How long will it take $P = \$1,000$ to be worth $A = \$2,500$ at 6% compounded annually?

We want t in years when $r = 0.06$, $A = \$2,500$, and $P = \$1,000$.

$$(2,500) = (1,000)[1 + (0.06)]^t \quad \text{Replace } r \text{ with 0.06, } A \text{ with 2,500, and } P \text{ with 1,000}$$
$$2.5 = (1.06)^t \quad \text{Divide each member by 1,000; combine}$$
$$\log(2.5) = \log(1.06)^t \quad \text{Take the log of each member}$$
$$\log(2.5) = t \log(1.06) \quad \text{Power property of logs}$$
$$t = \frac{\log(2.5)}{\log(1.06)} \quad \text{Divide each member by log(1.06)}$$

Using the calculator, enter

$\boxed{2}\,\boxed{.}\,\boxed{5}\,\boxed{\log}\,\boxed{\div}\,\boxed{1}\,\boxed{.}\,\boxed{0}\,\boxed{6}\,\boxed{\log}\,\boxed{=}$ and read 15.725209 on the display. Thus, $t \approx 15.73$

Therefore an investment of \$1,000 would take approximately 15.73 years to grow to \$2,500 when compounded annually at 6%.

Note $t = \dfrac{\ln (2.5)}{\ln (1.06)}$ will work just as well.

▶ ***Quick check*** Find the solution set of $4^{x-2} = 21$, correct to the nearest hundredth. ∎

--- **Mastery points** ---

Can you
■ Find the solution set of an exponential equation by using common logarithms (or natural logarithms)?
■ Solve applications of exponential equations?

Exercise 11–6

Find the solution set of the given exponential equation. Find solutions correct to the nearest hundredth.

Example $4^{x-2} = 21$

Solution
$$\log(4^{x-2}) = \log(21) \quad \text{Take the log of each member}$$
$$(x - 2)\log 4 = \log(21) \quad \text{Power property of logs}$$
$$x - 2 = \frac{\log(21)}{\log 4} \quad \text{Divide each member by log 4}$$
$$x = \frac{\log(21)}{\log 4} + 2 \quad \text{Add 2 to each member}$$

Using the calculator, enter

$\boxed{2}\,\boxed{1}\,\boxed{\log}\,\boxed{\div}\,\boxed{4}\,\boxed{\log}\,\boxed{=}\,\boxed{+}\,\boxed{2}\,\boxed{=}$ and read 4.1961587 on the display.

Thus, $x \approx 4.20$ and the solution set is $\{4.20\}$.

1. $2^x = 9$

2. $7^x = 8$

3. $4^{x+1} = 6$

4. $5^{x-2} = 11$

5. $3^{-x} = 10$

6. $9^{-x} = 19$

7. $6^{2x-1} = 12$

8. $8^{3x+1} = 9$

9. $3^{x+1} = 4$

10. $5^{x+2} = 3$

11. $16^{1-x} = 13$

12. $19^{2-x} = 17$

13. $4^{5-2x} = 9$

14. $2^{x^2} = 7$

15. $\left(\dfrac{1}{3}\right)^{1-x} = 5$

16. $\left(\dfrac{3}{5}\right)^{2x} = 2$

17. Solve the equation $y = x^{3n}$ for n using the common logarithms, log.

18. Solve the equation $y = e^{kt}$ for t using the natural logarithms, $\ln$.

19. What rate of interest, to the nearest tenth of a percent, is required for $50 to yield $61.50 after 5 years, compounded annually? [*Hint:* Use $A = P(1 + r)^t$.]

20. If $2, compounded annually, yields $2.98 after 8 years, what is the rate of interest, to the nearest hundredth of a percent?

21. If $65 yields $82.50 when the rate of interest is 6.25% compounded annually, how many years was the money invested? (to the nearest tenth)

22. For how many years, to the nearest hundredth, must $1,050 be invested at 8.5% interest compounded annually to have a return of $1,760?

23. A certain bacteria divides every 15 minutes to produce two new bacteria. If the number increases according to the equation

$$A = A_0 2^{4t}$$

where t is in hours, how long would it take, to the nearest tenth of an hour, for 2,000 bacteria to increase to 750,000 bacteria?

24. Work exercise 23 for 5,000 bacteria to increase to 1,000,000 bacteria.

25. Using the formula of exercise 23, how long would it take for the bacteria to triple?

26. Radium disintegrates in such a way that of q_0 milligrams present at a given time,

$$q = q_0(0.96)^t$$

milligrams will remain after t centuries. What is the half-life of radium, to the nearest tenth of a century?

27. Using exercise 26, in how many centuries (to the nearest tenth) will 200 milligrams of radium disintegrate to 125 milligrams?

28. A machine depreciates according to the formula

$$V = V_0(1 - r)^t$$

where V is the value after t years, V_0 is the original value, and r percent is the constant rate of depreciation. In how many years (to the nearest tenth) will the machine depreciate to one-half its original value at a depreciation rate of 15%?

29. Using exercise 28, in how many years (to the nearest hundredth) will the value depreciate from a value of $6,500 to a value of $5,000 at a depreciation rate of 12%?

30. Using exercise 28, what is the rate of depreciation (to the nearest tenth) if a machine depreciates to one-fourth its original value in 9 years?

31. Using exercise 28, what is the rate of depreciation (to the nearest tenth) if the machine reduces in value from $1,600 to $1,280 in 6 years?

Review exercises

Factor the following expressions. See section 3–8.

1. $5x^2 - x - 4$

2. $25z^2 - 36$

3. $5x^3 - 2x^2 + x$

4. $9z^2 - 12z + 4$

5. Find the solution set of the quadratic-type equation $w^4 - 5w^2 = 14$. See section 6–6.

6. Find the solution set of the rational inequality $\dfrac{5}{y} \le 3$. See section 6–7.

7. Find the distance in the plane from $(-4,2)$ to $(6,1)$. What are the coordinates of the midpoint of the line segment with these endpoints? See section 7–2.

Chapter 11 lead-in problem

The Richter Scale rating of earthquakes with a given intensity I is given by $\log_{10}\dfrac{I}{I_0}$, where I_0 is the intensity of a certain (small) size earthquake. Find the Richter Scale rating of an earthquake having intensity $10,000,000 I_0$.

Solution

We want $\log_{10}\dfrac{I}{I_0}$ when $I = 10,000,000 I_0$. Thus,

$$\text{Richter rating} = \log_{10}\frac{10,000,000 I_0}{I_0} \quad \text{Replace } I \text{ with } 10,000,000 I_0$$

$$= \log_{10} 10,000,000 \quad \text{Reduce by } I_0$$

$$= \log_{10} 10^7 \quad \text{Write } 10,000,000 \text{ as exponential form } 10^7$$

$$= 7$$

The Richter Scale rating of the earthquake was 7.

Chapter 11 summary

1. An **exponential function** with base b is a function f defined by $f(x) = b^x$, $b > 0$, $b \neq 1$.
2. The **logarithmic function** with base b, $b > 0$, $b \neq 1$, is the function defined by $f(x) = \log_b x$ such that $f(x) = y = \log_b x$ if and only if $x = b^y$, $x > 0$.
3. The exponential and logarithmic functions are inverse functions of each other.
4. The logarithm of a number to base b is the exponent to which the base b must be raised to get that number.
5. If x and y are positive real numbers, $b > 0$, $b \neq 1$, then
   a. $\log_b(xy) = \log_b x + \log_b y$
   b. $\log_b\left(\dfrac{x}{y}\right) = \log_b x - \log_b y$
   c. $\log_b(x^r) = r \log_b x$ (r is a real number)
   d. $\log_b 1 = 0$
   e. $\log_b b = 1$
   f. $b^{\log_b x} = x$
   g. $\log_b b^x = x$
6. The **common logarithms** are logarithms with base 10, denoted by log.
7. The **natural logarithms** are logarithms with base e, denoted by ℓn.
8. $\log_a x = \dfrac{\log_b x}{\log_b a}$, $a > 0$, $b > 0$, $x > 0$, $a \neq 1$, $b \neq 1$.

Chapter 11 error analysis

1. Direct and inverse variation
   Example: Given $P = \dfrac{kV}{T}$, > 0, as the value of V increases, the value of P will decrease, and as the value of T increases, the value of P will increase.
   Correct answer: As V increases P will increase and as T increases P will decrease.
   What error was made? (*see page 388*)
2. Exponential equations
   Example: Find the solution set of the equation
   $$4^{x-2} = 16^{5x+1}$$
   $$4^{x-2} = (4^2)^{5x+1}$$
   $$4^{x-2} = 4^{10x+2}$$
   $$x - 2 = 10x + 2$$
   $$-4 = 9x$$
   $$-\frac{4}{9} = x \qquad \left\{-\frac{4}{9}\right\}$$
   Correct answer: $\left\{-\dfrac{2}{3}\right\}$
   What error was made? (*see page 485*)

3. Inverses of functions
   Example: The function defined by $f(x) = x^2$ is one-to-one and has an inverse function.
   Correct answer: $f(x) = x^2$ is not one-to-one so it does not have an inverse function.
   What error was made? (*see page 378*)
4. Logarithms
   Example: The symbol "$\log_b y$" is read "logarithm to the base y of b."
   Correct answer: "logarithm to the base b of y"
   What error was made? (*see page 488*)
5. Equivalent logarithm statements
   Example: The equation $81 = 3^4$ is equivalent to $\log_4 81 = 3$.
   Correct answer: $\log_3 81 = 4$
   What error was made? (*see page 489*)
6. Logarithms
   Example: $\log_4 2 = \log_4(1 + 1) = \log_4 1 + \log_4 1$
   Correct answer: $\log_4 2 = \dfrac{1}{2}$
   What error was made? (*see page 493*)

7. Equivalent logarithm statements

   Example: $\log_2 3 - \log_2 5 + 3\log_2 4 = \log_2\left(\dfrac{3 \cdot 3^4}{5}\right)$

   $= \log_2\left(\dfrac{3^5}{5}\right) = \log_2\left(\dfrac{243}{5}\right)$

   Correct answer: $\log_2\left(\dfrac{192}{5}\right)$

   What error was made? (*see page 493*)

8. Logarithmic equation solutions

   Example: Find the solution set of

   $\log_4 x + \log_4(x - 15) = 2$

   $\log_4 x(x - 15) = 2$

   $x(x - 15) = 4^2$

   $x^2 - 15x = 16$

   $x^2 - 15x - 16 = 0$

   $(x - 16)(x + 1) = 0$

   $x - 16 = 0$ or $x + 1 = 0$

   $x = 16 \qquad x = -1$

   $\{16, -1\}$

Correct answer: $\{16\}$

What error was made? (*see page 495*)

9. Exponential equations

   Example: Find the solution set of $3^x = 7$ correct to four decimal places.

   $\log 3^x = \log 7$

   $x \log 3 = \log 7$

   $x = \dfrac{\log 7}{\log 3} = \log 7 - \log 3$

   $\qquad = 0.8451 - 0.4771$

   $\qquad = 0.3680$

   $\{0.3680\}$

   Correct answer: $\{1.7713\}$

   What error was made? (*see page 508*)

10. Squaring a radical binomial

    Example:

    $(x - \sqrt{2x + 1})^2 = (x)^2 - (\sqrt{2x + 1})^2$

    $\qquad = x^2 - 2x + 1$

    Correct answer: $x^2 - 2x\sqrt{2x + 1} + 2x + 1$

    What error was made? (*see page 286*)

Chapter 11 critical thinking

Given three consecutive integers, the product of the first and third is always 1 less than the square of the middle one. Why is this true?

Chapter 11 review

[11–1]

1. Sketch the graph of the exponential function (a) $f(x) = 3^x$,
   (b) $g(x) = \left(\dfrac{1}{3}\right)^x$, (c) $h(x) = 4^{-x}$.

2. Find the solution set of each exponential equation
   (a) $5^x = 625$, (b) $16^{2x-1} = 8$.

[11–2]

3. Sketch the graph of $f(x) = \log_6 x$ by using the graph of $y = 6^x$.

4. Write the exponential equation $4^{-3} = \dfrac{1}{64}$ as a logarithmic equation.

5. Write the logarithmic equation $\log_{1/3} 27 = -3$ as an exponential equation.

Evaluate each logarithmic expression.

6. $\log_6 216$

7. $\log_{1/4}(256)$

8. $\log_9 \sqrt[3]{9}$

Find the solution set of each logarithmic equation.

9. $\log_x 125 = 3$

10. $\log_3 x = -5$

11. $\log_2 x = \dfrac{5}{2}$

12. For what values of x is $\log_{10}(x^2 + 2x - 15)$ defined? (*Hint:* $x^2 + 2x - 15 > 0$)

[11–3]

Use the properties of logarithms to write each logarithmic expression as a sum or difference of logarithms, or as a product of a real number times a logarithm. Use the prime factor form of each number. All variables are positive real numbers.

13. $\log_b(56)$

14. $\log_4\left(\dfrac{5}{6}\right)$

15. $\log_5\left(\dfrac{9}{4}\right)$

Write each of the following logarithms as the logarithm of a single number. All variables are positive real numbers.

16. $\log_5 7 + \log_5 4$

17. $\log_6 15 - \log_6 3$

18. $4 \log_b x + 2 \log_b y - \log_b z$

19. $\log_4(x + 3) - \log_4(x - 4)$

20. $3 \log_b x + \log_b(2x - 1) - 3 \log_b(x + 1)$

Given $\log_b 2 = 0.3010$, $\log_b 3 = 0.4771$, and $\log_b 11 = 1.0413$, compute the following logarithms and then write the statement in equivalent exponential form. Round off to four decimal places.

21. $\log_b 22$

22. $\log_b 121$

23. $\log_b 96$

24. $\log_b \sqrt[3]{11}$

25. $\log_b \left(\dfrac{3}{22} \right)$

26. $\log_b \left(\dfrac{27}{\sqrt[4]{11}} \right)$

Find the solution set of each logarithmic equation.

27. $\log_4 x - \log_4 3 = 2$

28. $2 \log_2 3 + \log_2 x = 1$

29. $\log_5(2x + 1) + \log_5 2 = 2$

30. $\log_3(x - 3) - \log_3(x + 2) = -3$

[11–4]

Using a calculator, find the following common logarithms.

31. $\log 342$

32. $\log 507,000$

33. $\log 0.00736$

34. Given the formula pH $= -\log[\text{H}^+]$, find pH when $[\text{H}^+] = 6.00312$.

35. The formula for compound interest is given by $A = P(1 + i)^n$, where A is the amount after n years that the principal P will grow to at interest rate i percent (per year). Use common logarithms to find how many years it will take for the principal to triple ($A = 3P$) at 7.5% interest compounded annually. (Round off to the nearest tenth.)

[11–5]

Use a calculator and $\log_a x = \dfrac{\log x}{\log a}$ to evaluate the following logarithms (correct to two decimal places).

36. $\log_6 7$

37. $\log_{32} 4.73$

38. $\ln 47.3$

39. The number of a radioactive substance, present at time t, is given by $q = q_0 e^{-0.6t}$, where t is measured in days. What is the half-life of the substance? (Round to the nearest tenth.)

40. Using the formula in exercise 39, how long will it take for 2.3 grams of a substance to decay to 0.7 grams? (Round to the nearest tenth.)

[11–6]

Find the solution set of each exponential equation (correct to two decimal places).

41. $5^x = 17$

42. $4^{-x} = 3$

43. $(15)^{x-2} = 9$

44. $(4)^{2x} = 3$

45. $\left(\dfrac{4}{5} \right)^{3x} = 2$

46. A certain bacteria increases in number according to the equation $A = A_0(3)^{2t}$, where t is in hours, A_0 is the initial number, and A is the number of bacteria at time t. How long will it take for 1,000 bacteria to increase to 125,000 bacteria?

Chapter 11 cumulative test

[1–1] **1.** Given $\{x \,|\, -4 < x \leq 5\}$, list the integers in the set.

[1–4] **2.** Simplify the expression $4 - 12 \div 3 + 2^3 - 3^2 \cdot 4$.

Perform the indicated operations and simplify.

[1–4] **3.** $3[-6(9 - 5) - 3 \cdot 4 + 9]$

[3–2] **4.** $(2a - b)^2$

[3–2] **5.** $(5x - 3y)(5x + 3y)$

[3–2] **6.** $(4y + 2)(3y - 5)$

Perform the indicated operations. Give the answer with positive exponents.

[3–1]   **7.** $(3a^2b)(-6a^3b^2)$

[3–3]   **8.** $(-3a^{-1}b^2c^{-3})^3$

[3–3]   **9.** $\left(\dfrac{4a^2b^{-3}}{12a^{-4}b^2}\right)^2$

Find the solution set of the following equations or inequalities.

[2–1]   **10.** $3(x+2) - 4x = 5(2x-1)$

[2–5]   **11.** $4y - 3 \le 2y + 1$

[2–5]   **12.** $-5 \le 2x - 1 < 4$

[2–4]   **13.** $|4y - 3| = 2$

[2–5]   **14.** $|2 - 3x| \le 4$

[2–5]   **15.** $|x + 5| > 6$

[6–1]   **16.** $2y^2 - y = 3$

[4–7]   **17.** $\dfrac{3}{x} - \dfrac{2}{x} = \dfrac{4}{5}$

[4–3]   **18.** Add $\dfrac{5}{4ab^2} + \dfrac{6}{8a^2b}$.

[4–3]   **19.** Subtract $\dfrac{6y}{y-7} - \dfrac{9y}{7-y}$.

[4–2]   **20.** Divide $\dfrac{x^2 - 5x + 4}{x^2 - 25} \div \dfrac{x^2 - 4x + 3}{x^2 - 10x + 25}$.

[4–4]   **21.** Simplify the complex fraction $\dfrac{\dfrac{4}{x} - \dfrac{3}{y}}{\dfrac{4y - 3x}{xy}}$.

Perform the indicated operations and simplify.

[5–6]   **22.** $(2\sqrt{3} - 1)^2$

[5–7]   **23.** $(4 + 3i)(4 - 3i)$

[5–7]   **24.** i^{23}

[5–5]   **25.** $\sqrt{50} - 3\sqrt{8} + 2\sqrt{32}$

[5–6]   **26.** Rationalize the denominator of the expression $\dfrac{4}{\sqrt{10} - 2}$.

Find the solution set of the following equations.

[6–3]   **27.** $y^2 - 3y = 7$

[6–5]   **28.** $\sqrt{5x + 9} = x - 1$

[6–7]   **29.** Find the solution set of the inequality $z^2 - 3z \le 40$. Graph the solution set on the number line.

[7–3]   **30.** Find the equation of the line that satisfies the given conditions. Write the answer in standard form.
(a) Passes through the points $(-1,4)$ and $(0,6)$
(b) Passes through the point $(-9,8)$ and parallel to $2x - 3y = 1$
(c) Passes through $(5,-3)$ and having slope $\dfrac{4}{3}$

[8–2]   **31.** Given $f(x) = 5x + 1$ and $g(x) = x^3 + 2$, find
(a) $f(5)$, (b) $g(-3)$, (c) $f[f(-2)]$,
(d) $\dfrac{f(x + h) - f(x)}{h}$, $h \ne 0$.

[7–3]   **32.** Sketch the graph of the equation $2x - 3y = -12$ using the slope and y-intercept.

[9–2]   **33.** Sketch the graph of the equation $x = y^2 - 4y + 3$.

[9–4]   **34.** Determine if the graph of the given equation is a parabola, a circle, an ellipse, or a hyperbola:
(a) $3y^2 + 3x^2 = 14$, (b) $y^2 + 2y = x$,
(c) $4y^2 = 3x^2 + 9$, (d) $x^2 + 4y^2 = 4$.

[10–2] **35.** Find all rational zeros of the polynomial $P(x) = 2x^4 + 7x^3 - 12x^2 - 28x + 16$.

[10–4] **36.** Sketch the graph of the polynomial function $f(x) = x^3 - 2x^2 - 11x + 12$.

[11–6] **37.** Find the solution set of the exponential equation $4^x = 16$.

[11–3] **38.** Find x, given (a) $\log_x 32 = 5$,
(b) $\log_6 x = -2$.

[11–5] **39.** Evaluate $\log_7 14$ using natural logarithms and a calculator. (Round off to four decimal places.)

[11–3] **40.** Solve the equation $\log_3(x - 1) - \log_3(x - 2) = 2$.

[11–3] **41.** Write the expression $\log_b 4 - 3\log_b 5 + \log_b 2$ as the logarithm of a single expression.

12

Systems of Linear Equations

A total of 12,000 persons paid $240,375 to attend a rock concert. If only two types of tickets were sold, one selling for $17.50 and the other selling for $25.00, how many of each type of ticket were sold?

12–1 ■ Systems of linear equations in two variables

In chapter 7, we learned that the graph of a linear equation in two variables is a straight line. In this chapter, we will consider the relationship that exists between two or more linear equations involving the same variables. These equations form what we call a **system of linear equations.** To illustrate, the following is a system of linear equations.

$$x + y = 12$$
$$y = x + 6$$

A solution of a system of linear equations is any ordered pair that satisfies *both* equations. Consider the ordered pair (3,9) and the system $x + y = 12$
$$y = x + 6.$$

$x + y = 12$	$y = x + 6$
$(3) + (9) = 12$ Replace x with 3 and y with 9	$(9) = (3) + 6$
$12 = 12$ (True)	$9 = 9$ (True)

The ordered pair (3,9) is called a *solution* of the system of linear equations. The solution set of this system of linear equations is
$\{(x,y) | x + y = 12\} \cap \{(x,y) | y = x + 6\} = \{(3,9)\}.$

■ *Example 12–1 A*

Determine if the given ordered pair is a solution of the system of linear equations.

1. $(x,y) = (2,5)$ and the system $x + 2y = 12$
$\qquad\qquad\qquad\qquad\qquad\qquad\quad 3x - y = 1.$

$$x + 2y = 12 \qquad\qquad\qquad 3x - y = 1$$
$$(2) + 2(5) = 12 \qquad\qquad 3(2) - (5) = 1 \qquad \text{Replace } x \text{ with 2 and } y \text{ with 5}$$
$$2 + 10 = 12 \qquad\qquad\qquad 6 - 5 = 1$$
$$12 = 12 \quad \text{(True)} \qquad\qquad\qquad 1 = 1 \quad \text{(True)}$$

Therefore $(2,5)$ is a solution of the system of linear equations since it satisfies both of the original equations.

2. $(x,y) = (-5,-2)$ and the system $3x - 2y = -11$
$\qquad\qquad\qquad\qquad\qquad\qquad\qquad\qquad 2x + y = 10.$

$$3x - 2y = -11 \qquad\qquad\qquad 2x + y = 10$$
$$3(-5) - 2(-2) = -11 \qquad\qquad 2(-5) + (-2) = 10 \qquad \text{Replace } x \text{ with } -5$$
$$\qquad\qquad\qquad\qquad\qquad\qquad\qquad\qquad\qquad\qquad\qquad\text{and } y \text{ with } -2$$
$$-15 + 4 = -11 \qquad\qquad\qquad -10 + (-2) = 10$$
$$-11 = -11 \quad \text{(True)} \qquad\qquad\qquad -12 = 10 \quad \text{(False)}$$

The ordered pair $(-5,-2)$ is *not* a solution of the system of linear equations since it does not satisfy both equations.

▶ *Quick check* Determine if the ordered pair $(3,-1)$ is a solution of the system of linear equations $2x - y = 7$
$\qquad\qquad\qquad\qquad\qquad\qquad\qquad\qquad x - 5y = 2.$ ■

If we wish to represent a system of linear equations graphically, the graph will be a pair of straight lines. Two straight lines, L_1 and L_2, can be related in one of three ways. Solutions of a system of linear equations are determined by these relationships. See figure 12–1.

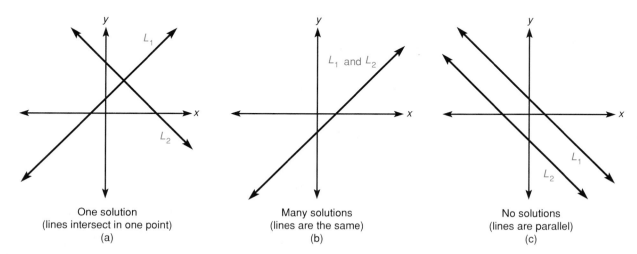

One solution
(lines intersect in one point)
(a)

Many solutions
(lines are the same)
(b)

No solutions
(lines are parallel)
(c)

Figure 12–1

Graphs of systems of linear equations

1. The graphs intersect in a single point. [12–1(a)]. The solution is given by the coordinates of the point of intersection. This system is called **consistent and independent.**
2. The graph is the same line [12–1(b)]. Any solution of one equation is a solution of the other equation. The system is called **consistent and dependent.**
3. The graphs are parallel lines [12–1(c)]. There is no solution and the system is called **inconsistent.**

Graphing systems of equations can be used only as a good estimate of the solution since the coordinates of the point of intersection are not always obvious.

■ *Example 12–1 B*

Find the solution set of the system of linear equations $4x - y = 1$ by graphing.
$$x + y = 4$$

Using the x- and y-intercepts, we graph each equation.

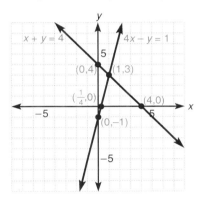

The lines appear to intersect at the point $(1,3)$, so the system is consistent and independent. The solution set of the system is $\{(1,3)\}$, which we can write in set-builder notation by

$$\{(x,y)|4x - y = 1\} \cap \{(x,y)|x + y = 4\} = \{(1,3)\}$$

▶ *Quick check*   Find the solution set of the system of linear equations
$4x - y = 0$ by graphing.
$x + y = 5$ ■

Solution by elimination

Since it is time-consuming to graph and it can be difficult to read exact solutions from the graph, we usually use other, algebraic, methods to solve a system. One such method is called the **elimination** (or **addition**) method. The following examples demonstrate this method in which we eliminate one of the variables through addition.

■ *Example 12–1 C*

Find the solution set of the following systems of linear equations by elimination.

1. $4x - 3y = 5$ (1)
   $2x + 3y = 7$ (2)

The elimination method involves adding the respective members of the two equations so that one of the variables is eliminated.

$4x - 3y = 5$
$\underline{2x + 3y = 7}$
$6x \quad\quad = 12$ Add left and right members
$\quad\quad x = 2$ Divide each member by 6

We now substitute 2 for x in either equation and solve for y. Using the equation $4x - 3y = 5$,

$4(2) - 3y = 5$ Replace x with 2
$\quad 8 - 3y = 5$
$\quad\quad -3y = -3$ Subtract 8 from each member
$\quad\quad\quad y = 1$ Divide each member by -3

The solution of the system is $x = 2$ and $y = 1$, which we write as the ordered pair (2,1). The solution set is $\{(2,1)\}$. This can be checked by substituting 2 for x and 1 for y into each equation.

(1) $4(2) - 3(1) = 5$ (2) $2(2) + 3(1) = 7$ Replace x with 2 and y with 1

$\quad\quad 8 - 3 = 5$ $4 + 3 = 7$

$\quad\quad\quad 5 = 5$ (True) $7 = 7$ (True)

Note We will not check our solution in future examples, but this is something we should get into the habit of doing automatically.

We *eliminated* one variable (in this case y) through *addition*. This was accomplished because the *coefficients* of y were *additive inverses,* or opposites (in this case -3 and 3). We then obtain a linear equation in one variable that we can easily solve. Sometimes it is necessary to multiply the terms of one equation, or both equations, by some nonzero number to get additive inverse, or opposite, coefficients so that one of the variables is eliminated when we add respective members.

2. $3x = 5y - 1$
   $2x - 3y = 8$

We must first write the first equation in standard form.

$3x - 5y = -1$ Subtract 5y from each member

The system of linear equations is now

$3x - 5y = -1$ (1)
$2x - 3y = 8$ (2)

Adding the respective members of these equations yields $5x - 8y = 7$ and no variable has been eliminated. We must multiply by the appropriate number(s) to eliminate either x or y when the equations are added.

One way to eliminate one variable (and there are other ways) is to multiply equation (1) by 2 and equation (2) by -3.

$$\left. \begin{array}{ll} 3x - 5y = -1 & \text{Multiply by 2} \quad\rightarrow \quad 6x - 10y = -2 \\ 2x - 3y = 8 & \text{Multiply by } -3 \rightarrow \underline{-6x + 9y = -24} \end{array} \right\} \begin{array}{l} \text{Coefficients of } x \text{ are} \\ \text{additive inverses} \end{array}$$

$$\begin{array}{lr} \text{Add} & -y = -26 \\ & y = 26 \end{array}$$

Substitute 26 for y in equation (2) and solve for x.

$$\begin{array}{ll} 2x - 3(26) = 8 & \text{Replace } y \text{ with } 26 \\ 2x - 78 = 8 & \\ 2x = 86 & \\ x = 43 & \end{array}$$

The solution set is $\{(43,26)\}$.

Note   We could have multiplied equation (1) by 3 and equation (2) by -5, or equation (1) by -3 and equation (2) by 5. In either case, we would eliminate y through adding.

Occasionally, when we solve a system of linear equations, we find the system to be *inconsistent* (no solutions) or *dependent* (many solutions).

3. $3x - 4y = 1$
$6x - 8y = -5$

$$\begin{array}{ll} 3x - 4y = 1 & \text{Multiply by } -2 \rightarrow -6x + 8y = -2 \\ 6x - 8y = -5 & \phantom{\text{Multiply by } -2 \rightarrow} \underline{6x - 8y = -5} \end{array}$$

$$\begin{array}{lr} & 0 = -7 \qquad \text{Add members} \end{array}$$

We obtain a *false* statement. This indicates the system has *no solution*. The system is *inconsistent* and the solution set is $\emptyset$. The lines are parallel.

4. $3x - 2y = 1$
$9x - 6y = 3$

$$\begin{array}{ll} 3x - 2y = 1 & \text{Multiply by } -3 \rightarrow -9x + 6y = -3 \\ 9x - 6y = 3 & \phantom{\text{Multiply by } -3 \rightarrow} \underline{9x - 6y = 3} \end{array}$$

$$\begin{array}{lr} & 0 = 0 \qquad \text{Add members} \end{array}$$

The resulting statement is *true*, which indicates that every solution of one equation is also a solution of the other equation. The system is *consistent and dependent* and the solution set is

$$\{(x,y)|3x - 2y = 1\} \text{ or } \{(x,y)|9x - 6y = 3\}$$

The lines are the same.

▶ *Quick check*   Find the solution set of the system $2x - 3y = 2$
$3x + 2y = 3$
by elimination.   ■

To summarize the elimination method of solving a system of two linear equations in two variables, we use the following procedure.

> **To solve a system of two linear equations by elimination**
>
> 1. Write the system so that each equation is in standard form $ax + by = c$.
> 2. Multiply one equation (or both equations), if necessary, by a number to obtain additive inverse coefficients of one of the variables.
> 3. Add the respective members of the two equations and solve the resulting equation in *one variable*.
> 4. In step 3, if we get
>   a. a false statement, the system is inconsistent, there are no solutions, and the lines are parallel.
>   b. the equation $0 = 0$, the system is consistent and dependent, there are infinitely many solutions, the solution set is the solution set of either equation, and the lines are the same.
> 5. Substitute the resulting value of the variable found in step 3 into one of the *original* equations and solve for the other variable.

Solution by substitution

A second algebraic method that is used to solve a system of linear equations involves solving one of the equations for one of the variables and substituting for that variable into the other equation to obtain an equation in one variable. We call this the **substitution** method for solving a system of linear equations.

Note Substitution can be used when nonlinear equations appear in the system.

■ *Example 12–1 D*

Find the solution set of the following systems of linear equations by substitution.

1. $4x + 3y = 14$ (1)
        $y = 3x - 4$ (2)

Notice that equation (2) is solved for y. Substitute $3x - 4$ for y in equation (1).

$$4x + 3(3x - 4) = 14 \qquad \text{Replace } y \text{ with } 3x - 4$$
$$4x + 9x - 12 = 14 \qquad \text{Multiply in the left member}$$
$$13x - 12 = 14$$
$$13x = 26$$
$$x = 2$$

Since $y = 3x - 4$, substitute 2 for x *in* this equation.

$$y = 3(2) - 4 \qquad \text{Replace } x \text{ with } 2$$
$$= 6 - 4$$
$$y = 2$$

The solution set is $\{(2,2)\}$.
Check the solution by substituting 2 for x and 2 for y in both equations.

2. $x - 3y = 4$ (1)
$3x + 4y = 1$ (2)

Solving the equation $x - 3y = 4$ for x, we have the system

$x = 3y + 4$ Add $3y$ to each member of (1)
$3x + 4y = 1$

Substituting $3y + 4$ for x into the equation $3x + 4y = 1$, we obtain

$3(3y + 4) + 4y = 1$ Replace x with $3y + 4$
$9y + 12 + 4y = 1$
$13y + 12 = 1$
$13y = -11$
$y = -\dfrac{11}{13}$

To find x, substitute $-\dfrac{11}{13}$ for y into one of the original equations, say $x - 3y = 4$.

$x - 3\left(-\dfrac{11}{13}\right) = 4$ Replace y with $-\dfrac{11}{13}$

$x + \dfrac{33}{13} = 4$

$x = 4 - \dfrac{33}{13} = \dfrac{52}{13} - \dfrac{33}{13}$

$x = \dfrac{19}{13}$

The solution set of the system is $\left\{\left(\dfrac{19}{13}, -\dfrac{11}{13}\right)\right\}$.

▶ *Quick check*   Find the solution set of the system of linear equations
$y - 3x = 2$ by substitution.
$2x + 5y = -7$ ▪

To summarize the substitution method of solving a system of two linear equations in two variables, we use the following procedure.

— *To solve a system of two linear equations by substitution* —

1. Solve one of the equations for one of the variables. (If one of the variables has a coefficient of 1 or -1, solve for that variable.)

2. Substitute the expression obtained in step 1 for that variable in the *other equation,* and solve the resulting equation in one variable.

3. In step 2,
   a. if we get a false statement, the solution set is $\emptyset$, the system is inconsistent, and the lines are parallel.
   b. if we get a statement that is always true, the solution set is the solution set of either equation, the system is consistent and dependent, and the lines are the same.

4. Substitute the value for the variable into the equation obtained in step 1 and solve for the other variable.

┌───┐
Mastery points

Can you
- Determine whether an ordered pair is a solution of a system?
- Solve a system of linear equations in two variables using elimination?
- Solve a system of linear equations in two variables using substitution?
└───┘

Exercise 12–1

Determine whether the given ordered pair is a solution of the system of linear equations. See example 12–1 A.

Example $2x - y = 7$
$x - 5y = 2; (3,-1)$

Solution

$2x - y = 7$	$x - 5y = 2$	Original equation
$2(3) - (-1) = 7$	$(3) - 5(-1) = 2$	Replace x with 3 and y with -1
$6 + 1 = 7$	$3 + 5 = 2$	
$7 = 7$ (True)	$8 = 2$ (False)	

The ordered pair $(3,-1)$ is not a solution of the system since it does not satisfy the equation $x - 5y = 2$.

1. $x + y = 6$
   $x - y = -2; (2,4)$

2. $3x - y = 1$
   $x + y = -5; (-1,-4)$

3. $2x + 3y = 6$
   $x - 2y = 3; (3,0)$

4. $x - 6y = 0$
   $2x + 3y = 8; (4,1)$

5. $3x - 5y = -12$
   $x - y = -7; (-1,2)$

6. $5x + 8y = 32$
   $9x - 2y = -8; (0,4)$

Find the solution set of the following systems of linear equations by graphing. If the system is inconsistent or dependent, so state. See example 12–1 B.

Example $4x - y = 0$ (1)
$x + y = 5$ (2)

Solution We graph each equation on the same rectangular coordinate plane using the slope and intercept for equation (1) and the x- and y-intercepts for equation (2).

(1) $y = 4x + 0$

$m = 4 = \dfrac{4}{1}$ and $b = 0$

(2) $x + y = 5$
x-intercept, $(5,0)$
y-intercept, $(0,5)$

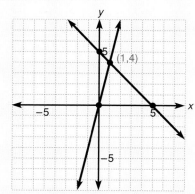

The graphs intersect at the point $(1,4)$ so the solution set is $\{(1,4)\}$.

7. $2x - y = 4$
$x + y = 5$

8. $2x + 3y = 12$
$x - y = 1$

9. $2x - y = -5$
$x + 2y = 0$

10. $6x - 2y = -8$
$3x - y = 4$

11. $x - 2y = 6$
$-2x + 4y = -12$

12. $2x - y = -1$
$3x - y = -1$

Find the solution set of each system of linear equations by elimination. If the system is inconsistent or dependent, so state. See example 12–1 C.

Example $2x - 3y = 2$
$3x + 2y = 3$

Solution $2x - 3y = 2$ Multiply by 2 → $\left.\begin{array}{r} 4x - 6y = 4 \\ 9x + 6y = 9 \end{array}\right\}$ Coefficients of y are opposites
$3x + 2y = 3$ Multiply by 3 →

$$13x \quad\quad = 13 \quad\quad \text{Add members}$$
$$x = 1 \quad\quad \text{Divide by 13}$$

Using $2x - 3y = 2$,

$2(1) - 3y = 2$ Replace x with 1
$2 - 3y = 2$
$-3y = 0$
$y = 0$

The solution set is $\{(1,0)\}$.

13. $x - y = 3$
$x + y = -7$

14. $3x + y = 1$
$2x - y = 9$

15. $x + 4y = 10$
$x + 2y = 4$

16. $x - y = 3$
$3x - y = -7$

17. $5x - 3y = 1$
$2x - 3y = -5$

18. $-5x - y = 4$
$-5x + 2y = 7$

19. $3x + 2y = 11$
$x - y = 2$

20. $3x + 4y = 18$
$2x - y = 1$

21. $4x + y = 7$
$2x + 3y = 6$

22. $4x - 3y = 7$
$3x - 2y = 6$

23. $8x - y = -4$
$4x + 7y = -32$

24. $3x - y = 10$
$6x - 2y = 5$

25. $3x + y = 2$
$9x + 3y = 6$

26. $-x + 2y = -7$
$-2x + 6y = 1$

27. $6x - 5y = 0$
$12x - 10y = -3$

28. $x - y = 9$
$-4x + 4y = -36$

29. $4x - 2y = 0$
$3x + 3y = 5$

30. $-2x + y = 6$
$4x + y = 1$

31. $\dfrac{1}{2}x + \dfrac{1}{3}y = 1$
$\dfrac{1}{4}x - \dfrac{2}{3}y = \dfrac{1}{12}$

32. $\dfrac{2}{3}x - \dfrac{1}{4}y = 4$
$x + \dfrac{1}{3}y = 2$

33. $\dfrac{3}{2}x + \dfrac{2}{5}y = \dfrac{9}{10}$
$\dfrac{1}{2}x + \dfrac{6}{5}y = \dfrac{3}{10}$

34. $\dfrac{5}{7}x - \dfrac{4}{5}y = \dfrac{9}{10}$
$\dfrac{2}{7}x - \dfrac{2}{5}y = \dfrac{3}{10}$

35. $\dfrac{5}{2}x - y = \dfrac{-17}{2}$
$\dfrac{2}{3}x + \dfrac{2}{3}y = 1$

36. $(0.3)x - (0.8)y = 1.6$
$(0.1)x + (0.4)y = 1.2$

37. $(0.1)x + (0.3)y = -0.3$
$(0.5)x + (0.4)y = 0.7$

38. $x - (0.5)y = 3$
$(0.9)x + y = -0.2$

Find the solution set of the following systems of linear equations by substitution. If the system is dependent or inconsistent, so state. See example 12–1 D.

Example $y - 3x = 2$
$2x + 5y = -7$

Solution   Solve the equation $y - 3x = 2$ for y.

$y - 3x = 2$
$\quad y = 3x + 2$ $\qquad$ Add $3x$ to each member

Substitute $3x + 2$ for y in $2x + 5y = -7$ and solve for x.

$2x + 5(3x + 2) = -7$ $\qquad$ Replace y with $3x + 2$
$2x + 15x + 10 = -7$ $\qquad$ Solve the resulting equation
$17x + 10 = -7$
$17x = -17$
$x = -1$

Substitute -1 for x in one of the original equations.

$y - 3x = 2$
$y - 3(-1) = 2$ $\qquad$ Replace x with -1
$y + 3 = 2$
$y = -1$

The solution set is $\{(-1,-1)\}$.

39. $2x + y = 10$
$\quad y = -x + 3$

40. $3x + 2y = 9$
$\quad y = 2x - 3$

41. $x = y + 5$
$\quad x + 5y = -4$

42. $5x + y = 10$
$\quad x = -2 + 3y$

43. $y = 3 - 6x$
$\quad 2x + 5y = -13$

44. $3x - 5y = 4$
$\quad x + 2y = -2$

45. $x - y = 3$
$\quad 2x + 2y = -10$

46. $3x + 3y = 0$
$\quad x + y = 5$

47. $5x - y = 8$
$\quad 2x - y = -4$

48. $-x + 3y = 4$
$\quad -x - 8y = 1$

49. $4x - 3y = 7$
$\quad y = -5$

50. $2x + 7y = 8$
$\quad y = 1$

51. $5x - 6y = -6$
$\quad x = 6$

52. $7x + 6y = -9$
$\quad x = -8$

53. $-4x - 3y = 4$
$\quad y = 0$

Find the solution set of each system of linear equations by either elimination or substitution. Try to choose the most suitable method. See examples 12–1 C and D.

54. $-6x + 3y = 4$
$\quad -12x - 6y = 1$

55. $3x + 2y = -6$
$\quad -6x - 4y = 1$

56. $5x - 10y = 5$
$\quad x - 2y = 1$

57. $2x - y = 7$
$\quad 6x - 3y = 21$

58. $2x - 3y = 5$
$\quad 3x + 4y = 1$

59. $4x - 3y = 4$
$\quad -2x + 4y = 3$

60. $7x - 8y = 14$
$\quad 5x + 3y = -4$

61. $-\dfrac{1}{3}x + y = 4$

$\quad x = -\dfrac{1}{2}y - 1$

62. $\dfrac{2}{3}x - \dfrac{1}{3}y = 4$

$\quad x = \dfrac{1}{3}y - 1$

63. $\dfrac{2}{5}x - \dfrac{3}{5}y = 3$

$\quad y = \dfrac{5}{2}x - 3$

64. $\dfrac{1}{6}x + y = -7$

$\quad y = \dfrac{-2}{3}x + 2$

Find the solution set of the following systems. Let $p = \dfrac{1}{x}$ and $q = \dfrac{1}{y}$. Substitute, solve for p and q, then solve for x and y.

65. $\dfrac{1}{x} + \dfrac{1}{y} = 3$

$\dfrac{1}{x} - \dfrac{1}{y} = -5$

66. $\dfrac{2}{x} - \dfrac{3}{y} = 1$

$\dfrac{1}{x} + \dfrac{2}{y} = 2$

67. $\dfrac{4}{x} + \dfrac{5}{y} = 0$

$\dfrac{3}{x} + \dfrac{2}{y} = 1$

68. $\dfrac{-3}{x} + \dfrac{1}{y} = 5$

$\dfrac{2}{x} - \dfrac{4}{y} = -1$

Using the variables x and y, write an algebraic equation for each verbal statement.

Example   The sum of two numbers is 36.

Solution   Let $x =$ one of the numbers and $y =$ the other number.
The algebraic equation is $x + y = 36$.

69. The sum of two numbers is 502.

70. Jane invested a total of $4,000 in two accounts.

71. One number is 6 more than another number.

72. One number is 3 less than twice a second number.

73. The length of a rectangle is 4 more than three times the width.

74. The length of a rectangle is 5 less than twice the width.

75. The difference in two electric currents is 33 amperes.

76. A boat takes $3\dfrac{1}{2}$ hours to go from point A to point B and back.

Review exercises

Perform the indicated operations. See sections 1–6 and 3–2.

1. $(4x^3 - 2x^2 + 1) - (5x^3 + 3x^2 + x - 4)$

2. $(x + 2)(x^2 - 2x + 4)$

3. $(5y - 2z)^2$

4. Find the solution set of the radical equation $\sqrt{x}\sqrt{x + 8} = 3$. See section 6–5.

5. One number is 27 more than another number. The smaller number is one-fourth of the larger number. Find the two numbers. See section 2–3.

6. If three times a number is increased by 15, the result is 51. What is the number? See section 2–3.

12–2 ■ *Applied problems using systems of linear equations*

Being able to solve a system of linear equations lets us solve many word problems using two variables as opposed to the one-variable approach we used in chapter 2.

> **How to solve a word problem using a system of equations**
> 1. Read the problem carefully and completely. Write down what information is given and what information we wish to find.
> 2. Whenever possible, draw a diagram showing the relationships in the problem.
> 3. Choose different variables for each unknown quantity.
> 4. Use word statements in the problem to write a system of linear equations. Generally there are *as many equations as there are different variables.*
> 5. Solve the system of linear equations using one of the methods we have learned.
> 6. Check the results in the original statement of the problem.

■ *Example 12–2 A*

$\ell = 2w + 11$

1. The perimeter of a rectangular plot of ground is 238 meters. If the length of the rectangle is 11 meters longer than twice the width, what are the dimensions of the rectangle? [Use perimeter = 2 (length) + 2 (width).]

Let w = the width of the rectangle and ℓ = the length of the rectangle.

length of the rectangle	is	11 meters	longer than	twice the width
ℓ	$=$	11	$+$	$2 \cdot w$

Thus $\ell = 2w + 11$.

We obtain the second equation by using the formula $P = 2\ell + 2w$ and the given information that $P = 238$.

$(238) = 2\ell + 2w$ Replace P with 238

We form the system of linear equations $\ell = 2w + 11$
                                           $2\ell + 2w = 238$

Since one equation is already solved for ℓ, we use the substitution method.

$$2(2w + 11) + 2w = 238$$ Replace ℓ with $2w + 11$ in $2\ell + 2w = 238$
$$4w + 22 + 2w = 238$$
$$6w + 22 = 238$$
$$6w = 216$$
$$w = 36$$

Since $\ell = 2w + 11$
        $\ell = 2(36) + 11$ Replace w with 36
        $\ell = 72 + 11$
        $\ell = 83$

the rectangle is 36 meters wide and 83 meters long.

Check: 1. $2(36) + 2(83) = 72 + 166 = 238$ (True)
           2. 83 is 11 more than $2(36)$ (True)

Note We will not show checks in the future. However, you should always check your answers.

2. Arlene wishes to invest $5,000. If she invests part at 7% simple interest, part at 6% simple interest, and receives a total interest of $332 after one year, how much does she invest at each rate?

Note   We use the formula $i = prt$ where i is the simple interest received when principal p is invested at rate r for t years. Time will always be one year in these problems so we may use the simplified form $i = pr$.

Let $x =$ the amount invested at 7% and $y =$ the amount invested at 6%. We use the following data table to set up equations.

	Amount invested	Rate of interest	Amount of interest
First investment	x	7%	0.07x
Second investment	y	6%	0.06y
Total investment	5,000		332

The equations are found by reading *down* the data table (see arrows). The system of equations is then

$$x + y = 5,000$$
$$0.07x + 0.06y = 332$$

To eliminate decimal numbers in the second equation, multiply each term by 100. Then, we have

$$\begin{array}{l} x + y = 5,000 \\ 7x + 6y = 33,200 \end{array} \quad \text{Multiply by } -7 \rightarrow \quad \begin{array}{r} -7x - 7y = -35,000 \\ \underline{7x + 6y = 33,200} \\ -y = -1,800 \\ y = 1,800 \end{array}$$

Add members

Since $x + y = 5,000$, then

$$x + (1,800) = 5,000 \quad \text{Replace } y \text{ with 1,800}$$
$$x = 3,200$$

Arlene invests $1,800 at 6% simple interest and $3,200 at 7% simple interest.

▶ *Quick check*   a. The perimeter of a rectangular flower garden is 82 feet. If the length of the rectangle is 5 feet longer than twice the width, what are the dimensions of the rectangle?
b. Jim wishes to invest $7,500. If he invests part at 6% simple interest, part at 8% simple interest, and receives $536 total interest after one year, how much did he invest at each rate?

Mastery points

Can you
- Solve a word problem using systems of linear equations in two variables?

Exercise 12–2

Set up a system of two linear equations and solve. See example 12–2 A–1.

Example The perimeter of a rectangular flower garden is 82 feet. If the length of the rectangle is 5 feet longer than twice the width, what are the dimensions of the rectangle?

Solution Let w = the width of the rectangle and ℓ = the length of the rectangle.

length	is	5 feet	longer than	twice the width
ℓ	$=$	5	$+$	$2 \cdot w$

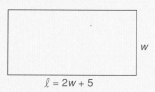

$\ell = 2w + 5$

Using $P = 2\ell + 2w$, and we are given that the perimeter $P = 82$, then we have the system of linear equations

$$2\ell + 2w = (82) \qquad \text{Replace } P \text{ with } 82$$
$$\ell = 2w + 5$$

We solve the system using substitution.

$$2(2w + 5) + 2w = 82 \qquad \text{Replace } \ell \text{ with } 2w + 5 \text{ in } 2\ell + 2w = 82$$
$$4w + 10 + 2w = 82 \qquad \text{Multiply in left member}$$
$$6w + 10 = 82 \qquad \text{Combine like terms}$$
$$6w = 72 \qquad \text{Subtract 10 from each member}$$
$$w = 12 \qquad \text{Divide each member by 6}$$

Using the equation

$$\ell = 2w + 5$$
$$\ell = 2(12) + 5 \qquad \text{Replace } w \text{ with } 12$$
$$\ell = 29$$

The dimensions of the rectangle are 12 feet wide and 29 feet long.

---

1. If twice the length of a rectangular floor is increased by three times the width, the sum is 48 feet. The perimeter of the room is 40 feet. What are the dimensions of the floor?

2. The distance around a rectangular flower garden is 64 feet. If the length is three times the width, what are the dimensions?

3. The perimeter of a rectangular plot of ground is 30 meters. Three times the length minus four times the width is 3 meters. Find the length and width of the plot.

4. A 20-foot board must be cut into two pieces so that one piece is 4 feet longer than the other piece. How long is each piece?

5. A 21-foot piece of pipe must be cut into two pieces so that one piece is 9 feet longer than the other piece. How long is each piece of pipe?

6. The sum of two electric currents is 96 amperes. If one current is 22 amperes less than the other, how many amperes are there in each current?

7. The sum of two voltages in an electric circuit is 47 volts and their difference is 25 volts. Find the two voltages.

8. The difference in the number of teeth in two gears is 14 and their sum is 72. How many teeth are there in each gear?

See example 12–2 A–2.

Example   Jim wishes to invest $7,500. If he invests part at 6% simple interest and part at 8% simple interest, and receives $536 total interest after one year, how much did he invest at each rate?

Solution   Let x = the amount invested at 6% interest and y = the amount invested at 8% interest.

	Amount invested	Rate of interest	Amount of interest
First investment	x	6%	0.06x
Second investment	y	8%	0.08y
Total investment	7,500		536

The system of linear equations we obtain is

$$x + y = 7,500$$
$$0.06x + 0.08y = 536$$

Multiply the terms of the second equation by 100 to clear the decimal points.

$$
\begin{array}{ll}
x + y = 7,500 & \text{Multiply by } -8 \rightarrow \quad -8x - 8y = -60,000 \\
6x + 8y = 53,600 & \qquad\qquad\qquad\qquad \underline{6x + 8y = 53,600} \\
& \text{Add members} \qquad -2x \quad\quad = -6,400 \\
& \qquad\qquad\qquad\qquad\quad x = 3,200
\end{array}
$$

Using the equation $x + y = 7,500$

$$
\begin{array}{ll}
(3,200) + y = 7,500 & \text{Replace } x \text{ with 3,200} \\
y = 4,300 &
\end{array}
$$

Jim invested $3,200 at 6% interest and $4,300 at 8% interest.

9. Phoebe wishes to invest $20,000, part at 7% interest and the rest at $6\frac{1}{2}$%. If her total income from the two investments for one year is $1,370, how much should she invest at each rate?

10. The income from two investments for one year is $1,485. If $19,000 is invested, part at 8% and the rest at $7\frac{1}{2}$%, how much is invested at each rate?

11. The income from an 8% investment is $300 more than the income from a 6% investment. How much is invested at each rate if a total of $30,000 is invested?

12. Jamie invests a total of $16,000, part at 7% interest and part at 9% interest. If the income from each investment is the same, how much money does Jamie invest at each rate?

13. The income from two investments is the same. If $36,000 is invested, part at 7% and the rest at 8%, how much is invested at each rate?

14. Juanita invests a total of $22,000. She suffers a net loss of $160 on her two investments. If one investment made her a 12% profit and the other investment caused an 8% loss, how much was each investment?

15. Simone makes a 15% profit on one investment but takes a 12% loss on a second investment. If she invests a total of $25,000 and realizes a net gain of $1,320 on her investments, how much is each investment?

Example   Sam's Pizza Emporium sells two kinds of small pizzas at $6.50 and $8.00 per pizza. If Sam sells 43 s:
pizzas for $306.50, how many of each kind of pizza does he sell?

Solution   Let x = number of $6.50 pizzas sold and y = number of $8.00 pizzas sold.
We now set up a table to help us determine the equations.

	Number of pizzas	Cost per pizza	Amount of income	
$6.50 pizzas	x	6.50	6.50x	
$8.00 pizzas	y	8.00	8.00y	
Total number	43		306.50	

The system of linear equations is then

$$x + y = 43$$
$$6.50x + 8.00y = 306.50$$

Multiply the second equation by 10 to clear the decimal points.

$$
\begin{array}{ll}
x + y = 43 & \text{Multiply by } -65\rightarrow \quad -65x - 65y = -2{,}795 \\
65x + 80y = 3{,}065 & \qquad\qquad\qquad\qquad\quad \underline{65x + 80y = 3{,}065} \\
& \text{Add members} \qquad\qquad\qquad 15y = 270 \\
& \qquad\qquad\qquad\qquad\qquad\quad y = 18
\end{array}
$$

Using the equation $x + y = 43$,

$$
\begin{array}{ll}
x + y = 43 & \\
x + (18) = 43 & \text{Replace } y \text{ with } 18 \\
x = 25 &
\end{array}
$$

Sam sold 25 pizzas at $6.50 per pizza and 18 pizzas at $8.00 per pizza.

16. A keypunch operator at a local firm works for $9 per hour and an entry-level typist works for $6.50 per hour. The total pay for an 8-hour day is $476, and there are two more typists than keypunch operators. How many keypunch operators does the firm employ?

17. A clothing store sells men's suits at $152 and $205 per suit. The store sells 32 suits and takes in $5,659. How many suits at each price does the store sell?

18. A hardware supply company sells two types of doorknobs. The chromium-plated knob sells at $8 per knob and the solid brass knob sells for $11.50 per knob. The company sold 420 doorknobs for $3,622.50. How many of each type were sold?

19. A road construction crew consists of cat operators working at $90 per day and laborers working at $50 per day. The total payroll per day is $1,600. If there are 3 laborers doing odd jobs and 4 laborers are assigned to work with each cat operator, how many laborers are there in the crew?

20. Skilled and unskilled workers are employed by a construction firm. If 5 skilled workers and 8 unskilled workers are employed, the total wages per day are $948. When 3 skilled and 5 unskilled workers are employed, the total wages per day are $580. What is the daily rate of pay of each type of worker?

21. The tickets for a puppet show cost $3.50 for adults and $1.25 for children. If $853.75 in tickets are sold to an audience of 503, how many children's tickets were sold?

22. A movie theater sold 323 tickets for $831.50. If adult tickets cost $3 and children's tickets cost $1.75, how many tickets of each type were sold?

23. Fernando has saved 43 coins in dimes and quarters. If he has saved a total of $7.15, how many dimes and how many quarters does he have?

24. Pam has a collection of fifty-two 13-cent and 20-cent stamps. If the face value of her collection is $9.28, how many 20-cent stamps does she have?

Example An auto mechanic has two bottles of battery acid. One bottle contains a 10% pure acid solution and the other contains a 4% pure acid solution. How many cubic centimeters (cm³) of each solution are needed to make 120 cubic centimeters of a 6% pure acid solution?

Solution Let x = the number of cubic centimeters of 10% pure acid solution and y = the number of cubic centimeters of 4% pure acid solution.

> **Note** 10% pure acid solution means that 10% of the solution is acid and the rest, 90%, is water. Thus, the 10% pure acid solution is

$$(0.10)x = 0.10x \text{ cu. cm of acid}$$
$$(0.90)x = 0.90x \text{ cu. cm of water}$$

	Volume	% acid	Amount of acid
First solution	x	10%	$0.10x$
Second solution	y	4%	$0.04y$
Total mixture	120	6%	$0.06(120) = 7.2$

The system of linear equations is then

$$x + y = 120$$
$$0.10x + 0.04y = 7.2$$

To clear the decimal points in the second equation, multiply each term by 100.

$$
\begin{array}{ll}
x + y = 120 & \text{Multiply by } -4 \rightarrow \quad -4x - 4y = -480 \\
10x + 4y = 720 & \qquad\qquad\qquad\qquad\; \underline{10x + 4y = 720} \\
& \text{Add} \qquad\qquad 6x \qquad\quad = 240 \\
& \qquad\qquad\qquad\qquad\quad x = 40
\end{array}
$$

Using $x + y = 120$,

$$
\begin{array}{ll}
x + y = 120 & \\
(40) + y = 120 & \text{Replace } x \text{ with } 40 \\
y = 80 &
\end{array}
$$

Thus, 40 cm³ of 10% pure acid solution must be mixed with 80 cm³ of 4% pure acid solution to obtain 120 cm³ of 6% acid solution.

---

25. A metallurgist wishes to form 2,000 kilograms of an alloy that is 80% copper. He is to obtain this alloy by fusing together some alloy that is 60% copper with some alloy that is 85% copper. How many kilograms of each alloy must be used?

26. How many grams of silver that is 60% pure must be mixed together with silver that is 35% pure to obtain a mixture of 90 grams of silver that is 45% pure?

27. How many liters of a 3.5% solution and a 6% solution of acid must a chemist mix together to form 800 liters of a 4.5% acid solution?

28. How much of each substance must be mixed together if a jeweler wishes to form 16 ounces of 65% pure gold from sources that are 50% and 70% pure gold?

29. How much of an 18% salt solution must be mixed with 40 ml of a 30% salt solution to obtain a 25% salt solution?

30. How much pure salt must be mixed with 9 cubic centimeters of 20% salt solution to obtain a 40% salt solution?

31. How much pure antifreeze must be added to a 4% antifreeze mixture to obtain a 20% antifreeze mixture to fill an automobile radiator that holds 12 liters?

32. A grocer wishes to mix two candies, one selling for $2 per pound and the other selling for $3 per pound. How much of each candy must he mix to obtain 50 pounds of mix selling for $1.75 per pound?

Example A boat can travel 24 miles downstream in 2 hours and 16 miles upstream in the same amount of time. What is the speed of the boat in still water and what is the speed of the current? [*Hint:* Use distance (d) = rate (r) × time (t).]

Solution Let x = speed of the boat in still water and y = speed of the current.
$x + y$ is the speed of the boat with the current (downstream).
$x - y$ is the speed of the boat against the current (upstream).
We set up a distance–rate–time table.

	Distance	Rate	Time	$d = r \cdot t$
Downstream	24	$x + y$	2	→ $24 = (x + y) \cdot 2$
Upstream	16	$x - y$	2	→ $16 = (x - y) \cdot 2$

$2(x + y) = 24$ $2(x - y) = 16$
$x + y = 12$ Divide each member by 2 $x - y = 8$

We must solve the system of linear equations.

$x + y = 12$
$\underline{x - y = 8}$
$2x = 20$ Add members
$x = 10$

Using $x + y = 12$,
$(10) + y = 12$ Replace x with 10
$y = 2$

The boat speed in still water is 10 mph, and the current is 2 mph.

33. A jogger runs a given distance and then catches a ride back to his home by car. If the round trip of 10 miles takes 1 hour, the car travels at 40 miles per hour, and he jogs at 5 miles per hour, how long did the jogger run?

34. An airplane can fly at 268 miles per hour against the wind and 380 miles per hour with the wind. What is the speed of the airplane in still air and what is the speed of the wind?

35. Terry travels upstream in his motorboat at top speed to town, a distance of 24 miles, in $1\frac{1}{2}$ hours. If the return trip downstream takes 1 hour, what is the top speed of Terry's motorboat and what is the speed of the stream?

36. Two trains leave the same city at 2:00 P.M., traveling in opposite directions. If one train travels at 48 mph and the other at 60 mph, at what time will they be 594 miles apart?

37. A mother and her daughter set out at the same time from their home, jogging in opposite directions. Maintaining their normal rate, the two women are 12 miles apart after 2 hours. What is the rate of each if the daughter jogs twice as fast as her mother?

38. Two canoeists make a 30-mile trip in 7 hours. If they paddle at a rate of 4.5 miles per hour part of the time and 4 miles per hour for the remaining time, how many hours did they travel at each rate?

39. A cyclist and a pedestrian are 40 miles apart. If they travel toward each other, they will meet in $2\frac{1}{4}$ hours. But if they travel in the same direction, the cyclist will overtake the pedestrian in 5 hours. At what rate is each traveling?

40. Two automobiles start from towns 450 miles apart and travel toward each other. They meet after 5 hours. If one automobile travels 12 miles per hour faster than the other, what is the speed of each automobile?

41. Two automobiles are 150 miles apart. If they drive toward each other, they will meet in $1\frac{1}{2}$ hours; if they drive in the same direction, they will meet in 3 hours. What are their speeds?

Find equations for the two lines, write them in standard form, and solve the system.

42. Find the equation of the line passing through points $(0,5)$ and $(-6,2)$ and the equation of the line through points $(1,1)$ and $(5,-7)$. Find their point of intersection by solving the resulting system of linear equations. [*Hint:* Use the point-slope form of an equation of a line $y - y_1 = m(x - x_1)$.]

43. Find the equation of the line passing through points $(-1,-2)$ and $(3,4)$ and the equation of the line through points $(4,1)$ and $(2,-4)$. Find their point of intersection by solving the resulting system of linear equations.

44. Find the point of intersection of line L_1 having slope 0 and passing through the point $(-4,-3)$ and line L_2 having slope -4 and y-intercept 2.

45. Find the point of intersection of line L_1 having slope 2 and y-intercept -6 and line L_2 having slope -3 and passing through the point $(1,2)$.

46. Find the point of intersection of line L_1 having x-intercept 3 and y-intercept -1 and line L_2 having slope 5 and passing through the point $(2,2)$.

Review exercises

Find the equation of the line, written in standard form $ax + by = c$, satisfying the following conditions. See section 7–3.

1. Through points $(1,3)$ and $(-2,1)$

2. Having y-intercept $(0,-3)$ and slope $\frac{1}{2}$

3. Through the point $(2,-1)$ and parallel to the line $2x + y = 4$

Evaluate the following radicals. See sections 5–1 and 5–7.

4. $-\sqrt{64}$

5. $\sqrt[3]{-27}$

6. $\sqrt{-4}$

7. Simplify the radical $\sqrt{8x^2y^3}$. See section 5–3.

12–3 ■ *Systems of linear equations in three variables*

Consider the equation $x + 2y - 4z = 12$, which involves three variables x, y, and z. Such an equation is called a **linear** (first-degree) **equation in three variables.** A solution of this equation is an **ordered triple** of real numbers, (x,y,z), if the resulting statement is true when the variables x, y, and z are replaced by real numbers. Then the set of all ordered triples of real numbers that satisfy the equation is the **solution set** of the equation.

To illustrate, the ordered triple $(4,0,-2)$ is a solution of the equation $x + 2y - 4z = 12$ since when we replace x with 4, y with 0, and z with -2, we obtain

$$(4) + 2(0) - 4(-2) = 12$$
$$4 + 0 + 8 = 12$$
$$12 = 12 \quad \text{(True)}$$

Other ordered triples that satisfy the equation are

$$(2,1,-2), (10,1,0), (-4,0,-4), \text{ and } (-8,4,-3)$$

In this section, we will discuss the solution set of a system of three linear equations in three variables such as

$$3x + 2y - z = 4$$
$$x - 3y + z = -1$$
$$2x + y - z = 0$$

The graph of a linear equation in three variables is a **plane.** Graphing this requires three-dimensional graphing, which is beyond the scope of this course. As with linear equations in two variables, there are a number of possible solutions.

1. The planes can intersect in one point. See figure 12–2(a).

2. The planes can have no common solutions. Two or more of the planes are parallel. See figure 12–2(b).

3. The planes can intersect in a common line. See figure 12–2(c).

4. The three planes can all be the same and the solutions are all points of the plane. See figure 12–2(d).

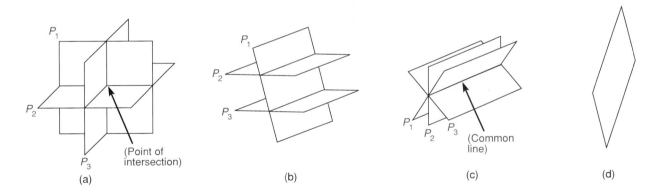

Figure 12–2

Since solving a system of three linear equations in three variables by graphing is difficult and impractical, we find the simultaneous solution by *eliminating* variables as we did with two linear equations in two variables.

■ *Example 12–3 A*

Find the solution set of each system of linear equations.

1.
$$2x - 5y - z = -8 \quad (1)$$
$$-x + 2y + 3z = 13 \quad (2)$$
$$x + 3y - z = 5 \quad (3)$$

The first objective is to obtain an equivalent system of two linear equations in two variables by eliminating one of the variables. Let us eliminate x from two of the equations.

To begin, use equations (2) and (3). We can eliminate x by adding the respective members of the two equations.

$$\begin{array}{ll} -x + 2y + 3z = 13 & (2) \\ \underline{x + 3y - z = 5} & (3) \\ 5y + 2z = 18 & (4) \end{array} \quad \text{Add members}$$

We obtain equation (4) involving the variables y and z. We must form another equation involving the variables y and z. Using equation (1) (the equation that has not been involved thus far) and one of the other equations, say equation (2), we must again eliminate x. Multiply the terms of equation (2) by 2 and add the respective members of the two equations.

$$\begin{array}{ll} 2x - 5y - z = -8 & (1) \\ \underline{-2x + 4y + 6z = 26} & \text{Multiply each member of (2) by 2} \\ -y + 5z = 18 & (5) \quad \text{Add members} \end{array}$$

Now we solve the system of two linear equations in two variables formed by equations (4) and (5).

$$\begin{array}{ll} 5y + 2z = 18 & (4) \\ -y + 5z = 18 & (5) \end{array}$$

To do this, use one of the methods that we learned in section 12–1. Multiply the terms of equation (5) by 5 to obtain the system

$$\begin{array}{ll} 5y + 2z = 18 & \\ \underline{-5y + 25z = 90} & \text{Multiply (5) by 5} \\ 27z = 108 & \text{Add members} \\ z = 4 & \end{array}$$

Substitute 4 for z in equation (5) [or in equation (4)].

$$\begin{array}{ll} -y + 5(4) = 18 & \text{Replace } z \text{ with 4} \\ -y + 20 = 18 & \\ -y = -2 & \\ y = 2 & \text{Multiply each member by } -1 \end{array}$$

To find the third variable, x, we substitute 2 for y and 4 for z into equation (1), (2), or (3). Using equation (3),

$$\begin{array}{ll} x + 3y - z = 5 & \\ x + 3(2) - (4) = 5 & \text{Replace } y \text{ with 2 and } z \text{ with 4} \\ x + 6 - 4 = 5 & \\ x + 2 = 5 & \\ x = 3 & \end{array}$$

We have found the ordered triple $(3,2,4)$ to be the only solution of the system. The solution set is $\{(3,2,4)\}$.

2. $2x + z = 7$ (1)
$x + y = 2$ (2)
$y - z = -2$ (3)

Note that one variable is missing in each equation and it is not the same variable. Suppose we use equations (1) and (2). Since equation (3) does not contain x, we must eliminate x when working with equations (1) and (2) to obtain a system of two linear equations in two variables. Multiply the members of equation (2) by -2 and add.

$$\begin{array}{ll} 2x + z = 7 & (1) \\ \underline{-2x - 2y = -4} & \text{Multiply members of equation (2) by } -2 \\ z - 2y = 3 & (4) \qquad \text{Add members} \end{array}$$

We now solve the system of linear equations involving equations (3) and (4) for y and z.

$$\begin{array}{ll} y - z = -2 & (3) \\ \underline{-2y + z = 3} & (4) \\ -y = 1 & \text{Add members} \\ y = -1 & \text{Multiply each member by } -1 \end{array}$$

Replace y with -1 in equation (3).

$$\begin{array}{ll} y - z = -2 & (3) \\ (-1) - z = -2 & \text{Replace } y \text{ with } -1 \text{ in (3)} \\ -z = -1 & \\ z = 1 & \end{array}$$

Finally, replace z with 1 in equation (1).

$$\begin{array}{ll} 2x + z = 7 & (1) \\ 2x + (1) = 7 & \text{Replace } z \text{ with 1 in (1)} \\ 2x = 6 & \\ x = 3 & \end{array}$$

Thus the ordered triple $(3, -1, 1)$ is the only solution and the solution set is $\{(3, -1, 1)\}$.

Note Alternately, we *could* have added equations (1) and (3) to eliminate z. Study each system for the easiest variable to eliminate before beginning your solution. ∎

A system of three linear equations in three variables may be an inconsistent or a dependent system. The following examples illustrate these cases.

■ *Example 12–3 B*

Find the solution set of the following systems of linear equations.

1. $x - 2y + 3z = 1$ (1)
$2x - 4y + 6z = 5$ (2)
$-x + 3y - 2z = -1$ (3)

Using equations (1) and (3), we add to eliminate x.

$$\begin{array}{ll} x - 2y + 3z = 1 & (1) \\ \underline{-x + 3y - 2z = -1} & (3) \\ y + z = 0 & \text{Add members} \end{array}$$

Using equations (1) and (2), multiply equation (1) by -2 and add the result to equation (2).

$$-2x + 4y - 6z = -2 \qquad \text{Multiply members of equation (1) by } -2$$
$$\underline{2x - 4y + 6z = 5} \qquad (2)$$
$$0 = 3 \qquad\qquad\quad \text{Add members}$$

The resulting false statement indicates there is no common solution. The system is inconsistent and at *least two* of the planes would be parallel. The solution set is ∅.

2. $\quad 3x + 12y - 3z = 9 \qquad (1)$
$\qquad x + 4y - z = 3 \qquad (2)$
$\quad -2x - 8y + 2z = -6 \qquad (3)$

If we multiply each member of equation (2) by -3 and add this to equation (1), we obtain

$$3x + 12y - 3z = 9$$
$$\underline{-3x - 12y + 3z = -9}$$
$$0 = 0 \qquad \text{Add members}$$

Thus, the system of linear equations is dependent and the solution set is $\{(x,y,z)|x + 4y - z = 3\}$

Note Any one of the three equations could be used to state the solution set. ∎

In general, to find the solution set of a system of three linear equations in three variables, we use the following procedure.

To solve linear systems in three variables by elimination

Step 1 Eliminate any one variable from any two of the given three equations to obtain a linear equation in two variables.

Step 2 Eliminate the *same* variable using the equation not yet involved and one of the other two equations. The result is another equation in the same two variables as in step 1.

Step 3 Solve the resulting system of linear equations in the two variables found in steps 1 and 2. (If this system is dependent or inconsistent, our given system is also dependent or inconsistent.)

Step 4 Substitute the values of the variables found in step 3 into any one of the three original equations to find the value of the third variable.

▶ *Quick check* Find the solution set of the system of linear equations
$x - y + z = 2$
$2x - y + z = 3$
$x + 2y - 3z = -4$. ∎

Now we will consider word problems with three unknown quantities and set up a system of three linear equations in three variables.

■ *Example 12–3 C*

The sum of the measures of the three angles of a triangle is 180 degrees. The middle-sized angle has measure 8° less than twice the measure of the smallest angle, and the largest angle has measure 20° less than the sum of the measures of the other two angles. Find the measures of the three angles of the triangle.

Let x = the measure of the smallest angle, y = the measure of the middle-sized angle, and z = the measure of the largest angle.

From "the sum of the measures of the three angles of a triangle is 180 degrees,"

$$x + y + z = 180 \qquad (1)$$

From "the middle-sized angle (y) has measure 8° less than twice the measure of the smallest angle (x),"

$$y = 2x - 8 \qquad (2)$$

From "the largest angle has measure 20° less than the sum of the measures of the other two angles,"

$$z = x + y - 20 \qquad (3)$$

Thus the answer to our problem is the solution of the system of linear equations

$$
\begin{aligned}
x + y + z &= 180 \qquad &(1)\\
y &= 2x - 8 \qquad &(2)\\
z &= x + y - 20 \qquad &(3)
\end{aligned}
$$

Rewrite the system in the standard form.

$$
\begin{aligned}
x + y + z &= 180 \qquad &(1)\\
-2x + y &= -8 \qquad &(2)\\
-x - y + z &= -20 \qquad &(3)
\end{aligned}
$$

We solve the system by elimination. Since the variable z is missing from equation (2), we use equations (1) and (3) and eliminate z. Multiplying equation (3) by -1, we have the system

$$
\begin{array}{ll}
x + y + z = 180 & (1)\\
\underline{x + y - z = 20} & \text{\small Multiply each member of (3) by } -1\\
2x + 2y = 200 \quad (4) & \text{\small Add members}
\end{array}
$$

We now solve the system of linear equations in two variables involving equations (2) and (4).

$$
\begin{array}{ll}
-2x + y = -8 & (2)\\
\underline{2x + 2y = 200} & (4)\\
3y = 192 & \text{\small Add members}\\
y = 64 &
\end{array}
$$

Using $-2x + y = -8$, replace y with 64 and solve for x.

$$
\begin{array}{ll}
-2x + (64) = -8 & \text{\small Replace } y \text{ with } 64\\
-2x = -72 &\\
x = 36 &
\end{array}
$$

Replacing x with 36 and y with 64 in equation (1),

$$
\begin{array}{ll}
(36) + (64) + z = 180 & \text{\small Replace } x \text{ with } 36 \text{ and } y \text{ with } 64\\
z = 80 &
\end{array}
$$

The three angles of the triangle have measure 36°, 64°, and 80°.

▶ *Quick check* The sum of the measures of the three angles of a triangle is 180°. The middle-sized angle has measure 2° less than twice the measure of the smallest angle, and the largest angle has measure 32° less than the sum of the measures of the other two angles. Find the measures of the three angles. ■

--- Mastery points ---

Can you

■ Find the solution set of a system of three linear equations in three variables?

■ Determine when a system of three linear equations in three variables is dependent or inconsistent?

■ Solve a word problem by setting up a system of three linear equations in three variables and solving the system?

Exercise 12–3

Find the solution set of the given system of linear equations. If the system is dependent or inconsistent, state. See examples 12–3 A and B.

Example
$$x - y + z = 2 \quad (1)$$
$$2x - y + z = 3 \quad (2)$$
$$x + 2y - 3z = -4 \quad (3)$$

Solution Using equations (1) and (2), we eliminate z.

$$-x + y - z = -2$$ Multiply each member of (1) by -1
$$\underline{2x - y + z = 3} \quad (2)$$
$$x \qquad = 1$$ Add members

Both y and z were eliminated and $x = 1$ was obtained. Replace x with 1 in equations (1) and (3) and solve the resulting system.

$$(1) - y + z = 2 \quad (1)$$ Replace x with 1
$$\underline{(1) + 2y - 3z = -4} \quad (3)$$ Replace x with 1

$$-y + z = 1 \quad (4)$$ Subtract 1 from each member
$$2y - 3z = -5 \quad (5)$$ Subtract 1 from each member

$$-2y + 2z = 2$$ Multiply each member of equation (4) by 2
$$\underline{2y - 3z = -5} \quad (5)$$
$$-z = -3$$ Add members
$$z = 3$$ Multiply each member by -1

Using equation (1),

$$(1) - y + (3) = 2$$ Replace x with 1 and z with 3
$$4 - y = 2$$
$$-y = -2$$
$$y = 2$$

The solution set is $\{(1,2,3)\}$.

1. $x + y + z = 6$
$x - 2y - z = -1$
$x + y - z = 2$

2. $x + y - z = 9$
$x + y + z = 5$
$x - y - z = 1$

3. $2x + 3y - z = 7$
$x + y + z = 6$
$3x - y - z = 6$

4. $x + y + z = 1$
$2x - y + 3z = 2$
$2x - y - z = 2$

5. $-2x + y + 4z = 3$
$x + y - 3z = 2$
$x + y - 2z = 3$

6. $3x - y + z = -8$
$4x - 2y - 3z = 3$
$2x + 3y - 2z = -1$

7. $-x + y - z = -6$
$2x + 3y - z = 1$
$x + 2y + 2z = 5$

8. $x + 2y + 3z = 5$
$-x + y - z = -6$
$2x + y + 4z = 4$

9. $-x + y + z = -3$
$3x + 9y + 5z = 5$
$x + 3y + 2z = 4$

10. $3x - 2y + 3z = 11$
$2x + 3y - 2z = -5$
$x + 4y - z = -5$

11. $6x + 4y + 2z = -1$
$3x + 2y + z = 3$
$x - 3y + z = 4$

12. $x - 2y + 3z = 4$
$3x - 3y + 4z = 5$
$2x - y + z = 1$

13. $x - 4y + z = -5$
$3x - 12y + 3z = -15$
$-2x + 8y - 2z = 10$

14. $5x - 2y + z = 6$
$-2x - 3y + 4z = -2$
$4x + 6y - 8z = 4$

15. $-5x + 3y - 2z = -13$
$4x - 2y + 5z = 13$
$2x + 4y - 3z = -9$

16. $4x + 6y + 5z = 2$
$-2x + 3y - 2z = -10$
$5x + 2y + 3z = -2$

17. $7x + 8y - 2z = -5$
$-2x + 5y + z = -3$
$5x + 14y - z = -11$

18. $x - y = -1$
$x + z = -2$
$y - z = 2$

19. $x + 2y = 10$
$-x + 3z = -23$
$4y - z = 9$

20. $x + z = 1$
$5x + 3y = 4$
$3y - 4z = 4$

21. $2x + z = 0$
$-4x + y = 1$
$3y + z = -7$

22. $2y + z = -4$
$y = -3$
$x - 3y + 2z = 9$

23. $4x + y - 2z = 1$
$x - y = 5$
$z = -2$

24. $3x + 2y = -3$
$-2x + 5z = 38$
$4y + 3z = 12$

25. $2x - 5y = 2$
$-4x + 3z = 5$
$3y + 4z = 12$

26. $2x - y + 3z = -5$
$x + y = -4$
$2x - y + 2z = 6$

27. $x + y - z = 8$
$-x + y + z = -3$
$y + z = 5$

28. $3x + y + 5z = 3$
$5x + y - z = -3$
$8x + 2y - z = -5$

29. $2x + 8y - 2z = 6$
$3x + 12y - 3z = 9$
$-x - 4y + z = -3$

30. $x + y - 3z = 4$
$-3x + y - z = 2$
$2x + 2y - 6z = 5$

In each of the following exercises, set up a system of three linear equations in three variables and solve. See example 12–3 C.

Example The sum of the measures of the three angles of a triangle is 180°. The middle-sized angle has measure 2° less than twice the measure of the smallest angle, and the largest angle has measure 32° less than the sum of the measures of the other two angles. Find the measures of the three angles.

Solution Let x = the measure of the smallest angle, y = the measure of the middle-sized angle, and z = the measure of the largest angle.

From "the sum of the measures of the three angles of a triangle is 180°,"

$$x + y + z = 180$$

From "the middle-sized angle measures 2° less than twice the measure of the smallest angle,"

$$y = 2x - 2$$

From "the largest angle has measure 32° less than the sum of the measures of the other two angles,"

$$z = x + y - 32$$

The system of linear equations written in standard form is

$$x + y + z = 180 \qquad (1)$$
$$-2x + y = -2 \qquad (2)$$
$$x + y - z = 32 \qquad (3)$$

Using equations (1) and (3), we eliminate z.

$$x + y + z = 180 \qquad (1)$$
$$\underline{x + y - z = 32} \qquad (3)$$
$$2x + 2y = 212 \qquad\qquad\text{Add members}$$
$$x + y = 106 \qquad (4) \qquad \text{Divide each member by 2}$$

We now solve the linear system involving equations (2) and (4).

$$-2x + y = -2 \qquad (2)$$
$$\underline{-x - y = -106} \qquad\qquad \text{Multiply each member of (4) by } -1$$
$$-3x = -108 \qquad\qquad \text{Add members}$$
$$x = 36$$

Using $y = 2x - 2$, replace x with 36 and solve for y.

$$y = 2(36) - 2 \qquad\qquad \text{Replace } x \text{ with 36}$$
$$y = 72 - 2$$
$$y = 70$$

Using equation (1), we solve for z.

$$x + y + z = 180$$
$$(36) + (70) + z = 180 \qquad\qquad \text{Replace } x \text{ with 36 and } y \text{ with 70}$$
$$106 + z = 180$$
$$z = 74$$

The three angles have measure $36°$, $70°$, and $74°$.

31. The sum of the measures of the three angles of a triangle is 180°. If the largest angle is 20° more than the sum of the other two angles and the smallest angle is 67° less than the larger angle, find the measure of the three angles of the triangle.

32. The sum of the measures of the three angles of a triangle is 180°. The sum of the smallest angle and the largest angle is 120°. If the middle-sized angle has measure 30° more than the smallest angle, what is the measure of the three angles of the triangle?

33. The perimeter of a triangular-shaped garden is 122 meters. If the length of the longest side is equal to the sum of the lengths of the other two sides, and twice the length of the shortest side is 11 meters less than the length of the longest side, find the lengths of the three sides. (*Note:* Perimeter is the distance around the triangle.)

34. The longest side of a triangle is twice the length of the shortest side, and the middle-length side is 9 inches longer than the shortest side. If the perimeter is 65 inches, what are the lengths of the three sides?

35. Tickets for a Harry Belafonte concert have three prices, "expensive," "middle-priced," and "cheapest." The "middle-priced" tickets cost $4 more than the "cheapest," and the "expensive" tickets cost $6 more than the "middle-priced" tickets. If the "expensive" tickets cost $1 less than twice the "cheapest" tickets, find the price of each kind of ticket.

36. A used-car salesman must sell a quota of cars before receiving a bonus. The cars are placed in three different price categories—A, B, and C. He must sell two more at price B than at price A and three times as many at price C as at price A. If the number at price C is one more than twice the number at price B, find the number of each car that he must sell.

37. Bill has a special stamp collection that is worth approximately $151,000 on the market. The stamps are separated into three different approximate price categories—$750, $1,500, and $25,000 per stamp. If the number of $750 stamps is four times the number of $25,000 stamps and the number of $1,500 stamps is ten more than the number of $750 stamps, how many of each kind does Bill's collection contain?

38. Find the values of *a*, *b*, and *c* so that the points $(0,5)$, $(-1,2)$, and $(2,17)$ lie on the graph of $y = ax^2 + bx + c$. (*Hint:* Substitute the coordinates of each point into the equation to obtain three linear equations in variables *a*, *b*, and *c*.)

39. Find the values of *a*, *b*, and *c* so that the points $(1,-2)$, $(0,2)$, and $(-2,13)$ lie on the graph of $y = ax^2 + bx + c$.

40. Find the values of *a*, *b*, and *c* so that the points $(2,-8)$, $(-1,-2)$, and $(3,-22)$ lie on the graph of $y = ax^2 + bx + c$.

41. Jay has 33 bills in denominations of fives, tens, and twenties. If he has $360 total and the number of five-dollar bills is two more than the number of twenty-dollar bills, how many of each denomination does he have?

42. Erin has a collection of pennies, nickels, and dimes in her piggy bank. She has twice as many pennies as dimes and eight more nickels than dimes. If she has $2.44 altogether, how many of each coin does she have?

Review exercises

1. Write 0.000247 in scientific notation. See section 3–3.

2. Subtract $\dfrac{4y - 3}{y^2 - 2y - 3} - \dfrac{2y + 5}{y^2 - 1}$. See section 4–3.

3. Evaluate $3a - 2b + c$ when $a = 4$, $b = -5$, and $c = -6$. See section 1–5.

4. Graph the linear inequality $2x - y \le 4$. See section 7–4.

5. Find the distance from the point $(1,-3)$ to the point $(-6,1)$. What are the coordinates of the midpoint of the line segment with these endpoints? See section 7–2.

6. Find the solution set of the quadratic-type equation $3x^4 - x^2 - 4 = 0$. See section 6–6.

12–4 ■ Systems of nonlinear equations

In sections 12–1 and 12–2, we discussed the solution(s) of systems of linear equations. Now we will consider systems of two equations in which at least one of the equations is quadratic, that is, has degree 2. These are called *systems of nonlinear equations*. The methods used to solve systems of linear equations can be used to solve systems involving quadratic equations. The following examples illustrate this kind of system.

■ **Example 12–4 A**

Find the solution set of the given system of equations.

$$y = x^2 - 2x + 1 \quad (1)$$
$$x + y = 3 \quad (2)$$

We first solve equation (2) for *y* to obtain the equivalent system.

$$y = x^2 - 2x + 1 \quad (1)$$
$$y = 3 - x \quad (3)$$

Substitute $3 - x$ for y in equation (1) and solve for x.

$$y = x^2 - 2x + 1 \qquad (1)$$
$$(3 - x) = x^2 - 2x + 1 \qquad \text{Replace } y \text{ with } 3 - x$$
$$0 = x^2 - x - 2 \qquad \text{Add } x \text{ and subtract 3 from each member}$$
$$x^2 - x - 2 = 0 \qquad \text{Write the equation in standard form}$$
$$(x - 2)(x + 1) = 0 \qquad \text{Factor left member}$$
$$x - 2 = 0 \quad \text{or} \quad x + 1 = 0 \qquad \text{Set each factor equal to 0}$$
$$x = 2 \qquad\qquad x = -1$$

Using equation (2), the simpler of the two equations, substitute for x and solve for y.

(1)   Let $x = 2$.   (2)   Let $x = -1$.
$$x + y = 3 \qquad\qquad\qquad x + y = 3$$
$$(2) + y = 3 \quad \text{Replace } x \text{ with } 2 \qquad (-1) + y = 3 \quad \text{Replace } x \text{ with } -1$$
$$y = 1 \qquad\qquad\qquad\qquad y = 4$$

The simultaneous solutions are the ordered pairs $(2,1)$ and $(-1,4)$. The solution set is $\{(2,1), (-1,4)\}$.

Equation (1) is a parabola and equation (2) is a straight line. Graphing (1) using the vertex and the intercepts and equation (2) using the intercepts, we see that the two graphs intersect at the points $(-1,4)$ and $(2,1)$.

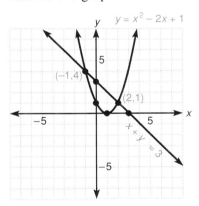

▶ **Quick check**   Find the solution set of the system $x^2 + y^2 = 9$
$$2x - y = 3.$$

When both equations making up the system are quadratic equations, we use the method of solution by **elimination.** The result is then a quadratic equation in one variable, and we proceed as in our preceding example.

■ *Example 12–4 B*

Find the solution set of the system.

1. $x^2 + y^2 = 25 \qquad (1)$
   $x^2 + 4y^2 = 52 \qquad (2)$

To eliminate the variable x^2, we multiply equation (1) by -1 and add to equation (2).

$$x^2 + y^2 = 25 \qquad \text{Multiply by } -1 \rightarrow \quad -x^2 - y^2 = -25$$
$$x^2 + 4y^2 = 52 \qquad\qquad\qquad\qquad\qquad \underline{x^2 + 4y^2 = 52}$$
$$\text{Add} \qquad\qquad\qquad\qquad 3y^2 = 27$$
$$y^2 = 9$$
$$\text{Extract the roots} \qquad\qquad y = \pm 3$$

Substituting 3 and -3 for y in either original equation, say $x^2 + y^2 = 25$, we obtain corresponding values of x.

When $y = 3$,
$x^2 + (3)^2 = 25$ Replace y with 3
$x^2 + 9 = 25$
$x^2 = 16$
$x = \pm 4$

When $y = -3$,
$x^2 + (-3)^2 = 25$ Replace y with -3
$x^2 + 9 = 25$
$x^2 = 16$
$x = \pm 4$

Thus, (a) $x = 4$ when $y = 3$, (b) $x = -4$ when $y = 3$, (c) $x = 4$ when $y = -3$, and (d) $x = -4$ when $y = -3$. The solution set is $\{(4,3), (-4,3), (4,-3), (-4,-3)\}$.

Equation (1) is a circle and equation (2) is an ellipse with centers at the origin, $(0,0)$. If we graph both equations using their x- and y-intercepts, we see the graphs intersect at the points $(-4,3)$, $(-4,-3)$, $(4,3)$, and $(4,-3)$.

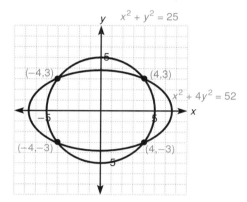

▶ *Quick check*   Find the solution set of the system of nonlinear equations
$x^2 + y^2 = 18$
$2x^2 - y^2 = -6$.

┌─ **Mastery points** ─────────────────────────────

Can you

■ Solve a system of equations that contain one linear and one quadratic equation?

■ Solve a system of equations that contain two quadratic equations?

Exercise 12–4

Find the solution set of each system of equations by substitution. Sketch the graphs of the systems in exercises 1–6. See example 12–4 A.

Example $x^2 + y^2 = 9$ (1)
            $2x - y = 3$ (2)

Solution   We first solve equation (2) for y.

$$2x - y = 3$$
$$-y = -2x + 3 \qquad \text{Subtract } 2x \text{ from each member}$$
$$y = 2x - 3 \qquad \text{Multiply each member by } -1$$

We now substitute $2x - 3$ for y in equation (1).

$$x^2 + y^2 = 9$$
$$x^2 + (2x - 3)^2 = 9 \qquad \text{Replace } y \text{ with } 2x - 3$$
$$x^2 + 4x^2 - 12x + 9 = 9 \qquad \text{Square in left member}$$
$$5x^2 - 12x = 0 \qquad \text{Combine like terms and subtract 9 from each member}$$
$$x(5x - 12) = 0 \qquad \text{Factor in left member}$$
$$x = 0 \quad \text{or} \quad 5x - 12 = 0 \qquad \text{Set each factor equal to 0}$$
$$x = 0 \quad \text{or} \qquad x = \frac{12}{5}$$

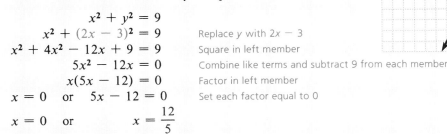

Substitute 0 and $\frac{12}{5}$ for x in equation (2) and solve for y.

Let $x = 0$.
$$2x - y = 3$$
$$2(0) - y = 3 \qquad \text{Replace } x \text{ with } 0$$
$$-y = 3$$
$$y = -3$$

Let $x = \frac{12}{5}$.
$$2x - y = 3$$
$$2\left(\frac{12}{5}\right) - y = 3 \qquad \text{Replace } x \text{ with } \frac{12}{5}$$
$$\frac{24}{5} - y = 3$$
$$-y = 3 - \frac{24}{5}$$
$$-y = -\frac{9}{5}$$
$$y = \frac{9}{5}$$

The solutions are $(0, -3)$ and $\left(\frac{12}{5}, \frac{9}{5}\right)$ and the solution set is $\left\{ (0, -3), \left(\frac{12}{5}, \frac{9}{5}\right) \right\}$.

1. $x^2 + y^2 = 5$
   $x + y = 1$

2. $x^2 + y^2 = 1$
   $x + 2y = 2$

3. $x^2 - y^2 = 24$
   $x + 2y = -3$

4. $x^2 - y^2 = 9$
   $x + y = 5$

5. $y = x^2 - 2x + 1$
   $y = 3 - x$

6. $y = x^2 - 4x + 4$
   $x + y = 2$

7. $x^2 + 2y^2 = 12$
   $2x = y + 2$

8. $x^2 + 3y^2 = 3$
   $x - 3y = 0$

9. $x^2 - y^2 = 15$
   $xy = 4$

10. $x^2 - y^2 = 35$
    $xy = 6$

11. $x^2 + y^2 = 8$
    $xy = 4$

12. $3x^2 - 4y^2 = 12$
    $x = 4$

13. $2x^2 + y^2 = 4$
    $y = -1$

14. $y = x^2 - 6x - 8$
    $y = 10$

Find the solution set of each system of equations by elimination or by a combination of elimination and substitution. See example 12–4 B.

Example $x^2 + y^2 = 18$
$2x^2 - y^2 = -6$

Solution We can eliminate y by adding members.

$$x^2 + y^2 = 18 \quad (1)$$
$$\underline{2x^2 - y^2 = -6} \quad (2)$$
$$3x^2 \qquad = 12 \qquad \text{Add members}$$
$$x^2 = 4 \qquad \text{Divide each member by 3}$$
$$x = \pm 2 \qquad \text{Extract the roots}$$

Substitute 2 and -2 in equation (1) and solve for y.

Let $x = 2$.
$x^2 + y^2 = 18$
$(2)^2 + y^2 = 18$ Replace x with 2
$4 + y^2 = 18$
$y^2 = 14$
$y = \pm\sqrt{14}$

Let $x = -2$.
$x^2 + y^2 = 18$
$(-2)^2 + y^2 = 18$ Replace x with -2
$4 + y^2 = 18$
$y^2 = 14$
$y = \pm\sqrt{14}$

The solutions are $(2,\sqrt{14})$, $(2,-\sqrt{14})$, $(-2,\sqrt{14})$, and $(-2,-\sqrt{14})$ and the solution set is $\{(2,\sqrt{14}), (2,-\sqrt{14}), (-2,\sqrt{14}), (-2,-\sqrt{14})\}$.

15. $x^2 + y^2 = 9$
    $x^2 - y^2 = 9$

16. $x^2 - y^2 = 24$
    $x^2 + y^2 = 8$

17. $x^2 + 4y^2 = 64$
    $x^2 + y^2 = 25$

18. $x^2 + 2y^2 = 22$
    $2x^2 + y^2 = 17$

19. $x^2 - 2y^2 = -9$
    $x^2 + y^2 = 18$

20. $3x^2 + 4y^2 = 16$
    $x^2 - y^2 = 8$

21. $2x^2 - 9y^2 = 18$
    $4x^2 + 9y^2 = 36$

22. $3x^2 + 4y^2 = 39$
    $5x^2 - 2y^2 = -13$

23. $x^2 + y^2 = 4$
    $2x^2 + 3y^2 = 5$

24. $x^2 + y^2 = 10$
    $x^2 - 3xy + y^2 = 1$

25. $x^2 + y^2 = 6$
    $2x^2 + 3xy + 2y^2 = 21$

26. $x^2 + xy - y^2 = 1$
    $x^2 - y^2 = 3$

For each of the following exercises, translate the verbal statements into a system of equations and solve.

Example Find the dimensions of a rectangular plot of ground if the perimeter is 22 meters and the area is 24 square meters.

Solution We use the formulas perimeter $P = 2\ell + 2w$ and area $A = \ell w$, where ℓ is the length and w is the width of the rectangle. Replacing P with 22 and A with 24, we obtain the system of equations

$2\ell + 2w = (22)$
$\ell w = (24)$

Solving the second equation for ℓ and substituting into the first equation, we obtain

$$\ell = \frac{24}{w}$$

$$2\left(\frac{24}{w}\right) + 2w = 22 \qquad \text{Replace } \ell \text{ with } \frac{24}{w}$$

$$\frac{48}{w} + 2w = 22$$

$$48 + 2w^2 = 22w \qquad \text{Multiply each term by } w$$

$$2w^2 - 22w + 48 = 0 \qquad \text{Write in standard form}$$

$$w^2 - 11w + 24 = 0 \qquad \text{Divide each member by 2}$$

$$(w - 3)(w - 8) = 0 \qquad \text{Factor the left member}$$

Then $w = 3$ or $w = 8$. $\qquad$ Set each factor equal to zero and solve

Substituting for w in the equation $\ell = \dfrac{24}{w}$, when

$$\begin{array}{ccc}
w = 3 & \text{or} & w = 8 \\
\ell = \dfrac{24}{3} & & \ell = \dfrac{24}{8} \\
\ell = 8 & & \ell = 3
\end{array}$$

In either case, the dimensions of the rectangle are 8 meters by 3 meters.

27. Find the dimensions of a rectangle whose area is 120 square inches and whose perimeter is 46 inches.

28. The area of a rectangle is 80 square centimeters and the perimeter is 42 centimeters. What are the dimensions of the rectangle?

29. The sum of two numbers is 16 and the difference between their squares is 32. Find the two numbers.

30. The sum of two numbers is 12 and the sum of their squares is 80. Find the two numbers.

31. A manufacturer determines that the relationship between the demand (x) and the price (p) for their commodity is $xp = 6$ and the relationship between the supply (x) and price (p) of the same commodity is $4x = 23 - 5p$. The **equilibrium point** is where the supply equals the demand. Solve the system of equations to find the equilibrium point.

32. A piece of cardboard is in the form of a rectangle and has an area of 260 square inches. A square 2 inches on each side is cut out of each corner and an open box is formed by folding up the ends and sides. If the volume of the resulting box is 288 cubic inches, find the length and width of the cardboard.

Review exercises

Evaluate the following expressions. See section 1–5.

1. $2x - 3y$ when $x = 3$ and $y = 4$

2. $4x^2 + 2y^2$ when $x = -1$ and $y = 2$

3. Find the solution set of the inequality $3y - 2 \geq 0$. See section 2–5.

Given $P(x) = 2x - 5$ and $Q(x) = x^2 + x - 3$, find

4. $P(-6)$

5. $Q(2)$

6. $P(2) - Q(1)$

See section 1–5.

7. Find the solution set of the inequality $|2x - 5| \leq 3$. See section 2–6.

12–5 ■ *Systems of linear inequalities and linear programming*

As shown in chapter 7, the solution set of an inequality in two variables contains infinitely many solutions and the graph is a region of the rectangular coordinate plane. The solution set of a *system of linear inequalities* such as

$$x + y > 2$$
$$x - y \leq -2$$

is the intersection of the solution sets of the individual inequalities. This solution set is best illustrated in a graph. Figure 12–3(a) shows the solution set of $x + y > 2$, figure 12–3(b) shows the solution set of $x - y \leq -2$, and figure 12–3(c) shows the intersection of the two graphs. Therefore, figure 12–3(c) represents the graph of the system of linear inequalities $x + y > 2$
$$x - y \leq -2.$$

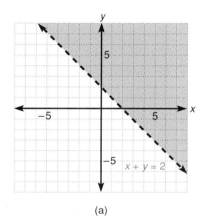
(a)

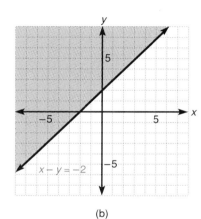
(b)

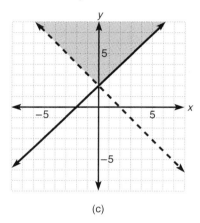
(c)

Figure 12–3

Remember that dashed lines indicate the points on the line *are not* included in the solution set and solid lines indicate the points on the line *are* included in the solution set.

Note The point of intersection of the two boundary lines, (0,2), is *not* in the solution set of the system since it is *not* in the solution of $x + y > 2$.

In the next examples, only the final solution set of the system will be shown.

■ *Example 12–5 A*

Graph the solution set of the given system of linear inequalities.

1. $\begin{cases} x \geq 1 \\ y < -2 \end{cases}$

Recall that we are working in the rectangular coordinate system even though one variable is missing in each of the inequalities. The system could be written

$$\begin{cases} x + 0y \geq 1 \\ 0x + y < -2 \end{cases}$$

The graph contains all points on and to the right of the vertical line $x = 1$ *and* below the horizontal line $y = -2$.

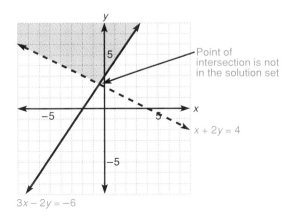

Note   The point of intersection of the boundary lines is *not* in the solution set.

2. $\begin{cases} 3x - 2y \le -6 \\ x + 2y > 4 \end{cases}$

    a. The graph of $3x - 2y \le -6$ contains all points *on or above* the line $3x - 2y = -6$.

    b. The graph of $x + 2y > 4$ contains all points *above* the line $x + 2y = 4$.

The final graph contains all points on or above the line $3x - 2y = -6$ and above the line $x + 2y = 4$.

3. $\begin{cases} 5x + 2y \le 10 \\ x + 3y \le 6 \\ x \ge 0 \\ y \ge 0 \end{cases}$

The solution set is the intersection of the solution sets of the four inequalities. Given $x \ge 0$ and $y \ge 0$, all points in this solution set must lie on or to the *right* of the y-axis and on or *above* the x-axis (the first quadrant).

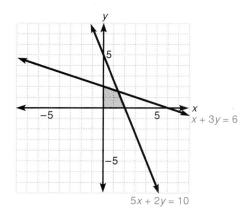

$5x + 2y = 10$

Notice all points lie in quadrant I, on or below the line $5x + 2y = 10$, and on or below the line $x + 3y = 6$ together with the points on that portion of the x- and y-axes that bound the region.

Note All four corner points of the region *are* in the solution set.

▶ *Quick check* Graph the solution set of the system of linear inequalities
$$\begin{cases} 3x + 2y \le 12 \\ x + 2y \ge 2 \\ x \ge 0 \\ y \ge 0. \end{cases}$$

Linear programming

Systems of linear inequalities provide the basis for an important application of mathematics to business and social sciences, called **linear programming.** Linear programming is used to find such information as maximizing the profit on a certain manufactured item and/or minimizing the cost of an item.

Typically, the quantity to be maximized or minimized is a linear function in two or more variables, called the *objective function.* The solution to a linear programming problem will be dependent on certain limitations, called *constraints.* Any solution to the objective function that satisfies these constraints is called a *feasible solution.* In linear programming, the constraints are expressed as a system of linear inequalities and the objective function has the form $f = ax + by + c$.

■ *Example 12–5 B*

Find the maximum value of the objective function $f = 2x + 3y$, subject to the constraints provided by the system of linear inequalities

$$\begin{cases} x + y \le 6 \\ 2x + 4y \le 16 \\ x \ge 0 \\ y \ge 0 \end{cases}$$

Under the constraints $x \ge 0$ and $y \ge 0$, the region containing the solution set of the system of inequalities will lie in quadrant I and that portion of the axes that bound the region. We now graph the system to show the region that contains the *set* of feasible solutions.

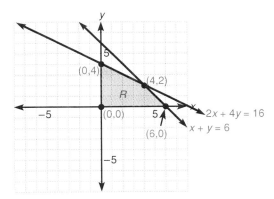

We have shown the coordinates of vertices of the polygon (points of intersection of the lines and axes) that borders the shaded region containing all feasible solutions. These vertices are important in maximizing the objective function. The following theorem refers to this importance.

Theorem _____

The maximum (or minimum) value of the objective function that is subject to the constraints of a linear programming problem are attained at a *vertex* of the region R containing the set of feasible solutions.

Using this theorem, to find the maximum value of the objective function f, we test the value at each of the vertices $(0,4)$, $(4,2)$, $(6,0)$, and the origin $(0,0)$.

Vertex	$(0,4)$	$(4,2)$	$(6,0)$	$(0,0)$
$f = 2x + 3y$	12	14	12	0

The maximum value 14 is attained at the vertex $(4,2)$, when $x = 4$ and $y = 2$.

▶ *Quick check*   Find the maximum and minimum values of the objective function $f = x + 3y$ using the given region R of feasible solutions.

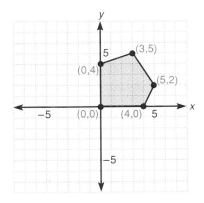

The following examples illustrate linear programming in the business field.

■ *Example 12–5 C*

1. Erin and Sarah produce products A and B in a business. Erin has at most 24 hours and Sarah has at most 16 hours per week to work on the products. They realize $120 profit per unit of A and $100 profit per unit of B. Erin must work 3 hours and Sarah 1 hour on product A, while Erin works 2 hours and Sarah works 2 hours on product B. How many of each item should they make to maximize their profits each week?

Let x = the number of units of product A the girls make and y = the number of units of product B.
The objective function P is given by

$$P = \frac{\text{profit}}{\text{on A}} \cdot \frac{\text{number of}}{\text{units of A}} + \frac{\text{profit}}{\text{on B}} \cdot \frac{\text{number of}}{\text{units of B}}$$
$$= 120 \cdot x + 100 \cdot y$$

Thus, $P = 120x + 100y$.

Since Erin works 3 hours on product A and 2 hours on product B per unit and can work at most 24 hours, then

$$3x + 2y \leq 24$$

is the constraint on Erin's time. In like fashion, Sarah works 1 hour on product A and 2 hours on product B per unit and has at most 16 hours to give, then

$$x + 2y \leq 16$$

is the constraint on Sarah's time. Therefore, the system of inequalities

$$\begin{cases} 3x + 2y \leq 24 \\ x + 2y \leq 16 \\ x \geq 0 \\ y \geq 0 \end{cases}$$

form the constraints on the objective function $P = 120x + 100y$. We now graph region R of feasible solutions.

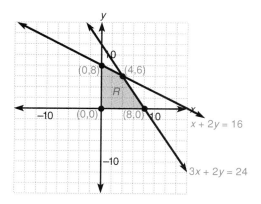

The vertices of R are (0,8), (4,6), (8,0), and (0,0). The values of these points when substituted into the objective function are shown in the following table.

Vertex	(0,8)	(4,6)	(8,0)	(0,0)
$P = 120x + 100y$	800	1,080	960	0

Thus, the maximum profit of $1,080 occurs for a weekly production of 4 units of product A and 6 units of product B.

2. A manufacturer of a certain product stores the product in two warehouses, W_1 and W_2. There are 90 units of the product stored in W_1 and 80 units stored in W_2. Customer C_1 orders 70 units and customer C_2 orders 45 units. If the shipping costs are $10 per unit from W_1 to C_1, $12 per unit from W_1 to C_2, $12 per unit from W_2 to C_1, and $16 per unit from W_2 to C_2, how should the order be filled to minimize the cost of shipping the products from warehouse to customer?

Let x = the number of units shipped from W_1 to C_1 and y = the number of units shipped from W_1 to C_2. Then $70 - x$ = the number of units shipped from W_2 to C_1 and $45 - y$ = the number of units shipped from W_2 to C_2.

Since warehouse W_1 holds 90 units and warehouse W_2 holds 80 units, then two constraints are

$$x + y \leq 90$$
$$(70 - x) + (45 - y) \leq 80$$

Simplifying the second inequality, we obtain

$$x + y \leq 90$$
$$x + y \geq 35$$

Additionally, since C_1 wishes 70 units and C_2 wishes 45 units, then

$$0 \leq x \leq 70$$
$$0 \leq y \leq 45$$

We now graph the region R of feasible solutions using the constraints

$$\begin{cases} x + y \leq 90 \\ x + y \geq 35 \\ 0 \leq x \leq 70 \\ 0 \leq y \leq 45 \end{cases}$$

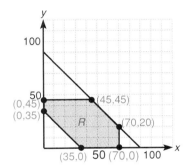

The objective function, the cost function C, is

$$C = 10x + 12y + 12(70 - x) + 16(45 - y)$$
$$= 10x + 12y + 840 - 12x + 720 - 16y$$
$$= 1,560 - 2x - 4y$$

We now test the vertices $(0,35)$, $(0,45)$, $(45,45)$, $(70,20)$, $(70,0)$, and $(35,0)$ for the minimum cost.

Vertex	(0,35)	(0,45)	(45,45)	(70,20)	(70,0)	(35,0)
$C = 1,560 - 2x - 4y$	1,420	1,380	1,290	1,340	1,420	1,490

We can see the minimum cost for shipping is \$1,290. This occurs when $x = 45$ units are shipped from W_1 to C_1 and $y = 45$ units are shipped from W_1 to C_2. The other 25 units for C_1 are shipped from W_2, but C_2 has received full shipment from W_1.

▶ *Quick check* A company manufactures products P_1 and P_2. It takes 3 hours of working time on machine A and 1 hour of working time on machine B to produce P_1; 2 hours on machine A and 1 hour on machine B to produce P_2. Machine A is available for at most 120 hours per week and machine B is available for at most 50 hours per week. How many of product P_1 and how many of product P_2 should be produced each week to maximize profit if the profit on P_1 is \$3.50 each and the profit on P_2 is \$2 each? ■

Mastery points

Can you
- Graph a region R containing the set of feasible solutions?
- Maximize (or minimize) an objective function P subject to the constraints of a system of linear inequalities?
- Set up the set of constraints, determine the objective function, graph the region R of feasible solutions, and maximize (or minimize) the objective function of a linear programming problem?

Exercise 12–5

Graph the solution set of the given systems of linear inequalities. See example 12–5 A.

Example
$$\begin{cases} 3x + 2y \le 12 \\ x + 2y \ge 2 \\ x \ge 0 \\ y \ge 0 \end{cases}$$

Solution The solution set will lie in quadrant I and that portion of the axes that bound the region since $x \ge 0$ and $y \ge 0$. All points will lie on or below the line $3x + 2y = 12$ and on or above the line $x + 2y = 2$.

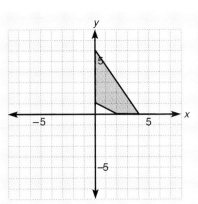

1. $\begin{cases} x \le 0 \\ y \ge 0 \end{cases}$

2. $\begin{cases} 2x + y \le 12 \\ x + 2y \le 6 \end{cases}$

3. $\begin{cases} x + y \quad \le 5 \\ 3x - 2y \le 6 \end{cases}$

4. $\begin{cases} x - y \ge 4 \\ \quad y \le 1 \end{cases}$

5. $\begin{cases} x - 2y \ge 2 \\ \quad x \ge 1 \end{cases}$

6. $\begin{cases} 4x + y \le 8 \\ x + y \ge 1 \end{cases}$

7. $\begin{cases} x + y \le 4 \\ x + y \ge 2 \\ \quad x \ge 0 \\ \quad y \ge 0 \end{cases}$

8. $\begin{cases} 2x + y \le 6 \\ x + 3y \le 6 \\ \quad x \ge 0 \\ \quad y \ge 0 \end{cases}$

9. $\begin{cases} 3x + 2y \le 12 \\ x + 2y \ge 4 \\ \quad x \ge 0 \\ \quad y \ge 0 \end{cases}$

10. $\begin{cases} x + y \le 8 \\ 2x + 3y \ge 6 \\ \quad x \ge 0 \\ \quad y \ge 0 \end{cases}$

11. $\begin{cases} 2x + y \ge 4 \\ x + y \le 6 \\ \quad x \ge 0 \\ \quad y \ge 0 \end{cases}$

12. $\begin{cases} y \ge 1 - x \\ y \le 2 - x \\ x \ge 0 \\ y \ge 0 \end{cases}$

Find the maximum and minimum values of the given objective function using the indicated region R of feasible solutions. See example 12–5 B.

Example $f = x + 3y$

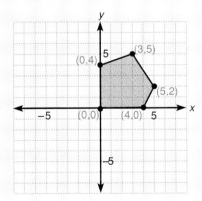

Solution

Vertex	(0,4)	(3,5)	(5,2)	(4,0)	(0,0)
$f = x + 3y$	12	18	11	4	0

The maximum value is 18 at vertex (3,5) and the minimum value is 0 at vertex (0,0).

13. $f = 3x + 2y$

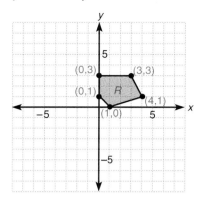

14. $f = 6x + y$

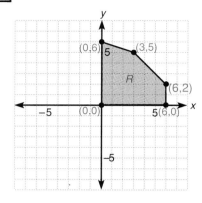

15. $f = 30x + 25y$

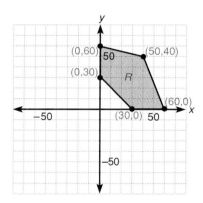

16. $f = 45x + 100y$

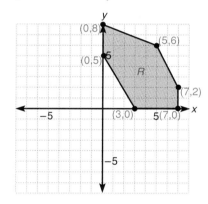

17. $f = x + 4y$

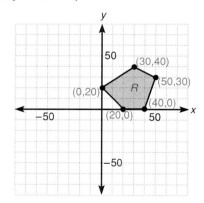

18. $f = 70x + 30y$

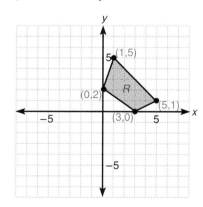

Using graphical methods, maximize (or minimize) as indicated the objective function f under the given set of constraints.

Example   Maximize the objective function $f = 3x + 2y$
under the constraints
$$x + 2y \leq 5 \text{ using graphs.}$$
$$2x + y \leq 4$$
$$x \geq 0$$
$$y \geq 0$$

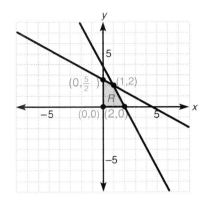

Solution

Vertex	$\left(0, \dfrac{5}{2}\right)$	$(1,2)$	$(2,0)$	$(0,0)$
$f = 3x + 2y$	5	7	6	0

The objective function $f = 3x + 2y$ maximizes at the vertex $(1,2)$ when $x = 1$ and $y = 2$.

19. $f = 3x + 5y$ (maximize)
$$\begin{cases} x + 2y \leq 4 \\ 2x + y \leq 6 \\ x \geq 0 \\ y \geq 0 \end{cases}$$

20. $f = 5x - y$ (maximize)
$$\begin{cases} 3x - y \leq 4 \\ 2x + y \leq 2 \\ x \geq 0 \\ y \geq 0 \end{cases}$$

21. $f = 4x + 3y$ (minimize)
$$\begin{cases} 3x + 4y \leq 12 \\ x + y \geq 2 \\ x \geq 0 \\ y \geq 0 \end{cases}$$

22. $f = 4y - x$ (minimize)
$$\begin{cases} x + y \leq 80 \\ x + y \geq 60 \\ 0 \leq x \leq 25 \\ 0 \leq y \leq 3 \end{cases}$$

See example 12–5 C.

Example   A company manufactures products P_1 and P_2. It takes 3 hours of working time on machine A and 1 hour of working time on machine B to produce P_1; 2 hours on machine A and 1 hour on machine B to produce P_2. Machine A is available for at most 120 hours per week and machine B is available for at most 50 hours per week. How many of product P_1 and how many of product P_2 should be produced each week to maximize profit if the profit on P_1 is $3.50 each and the profit on P_2 is $2 each.

Solution   Let $x =$ the number of P_1 and $y =$ the number of P_2.
The profit function is then $P = 3.50x + 2y$. The constraints on the problem are given by the following inequalities:

$$\begin{cases} 3x + 2y \leq 120 \\ x + y \leq 50 \\ x \geq 0 \\ y \geq 0 \end{cases}$$
Machine A available at most 120 hours
Machine B available at most 50 hours

We graph these inequalities to obtain the set of feasible solutions as indicated by the shaded region R.

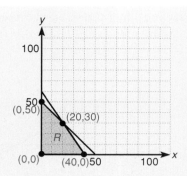

The vertices are at (0,50), (20,30), (40,0), and (0,0).

Vertex	(0,50)	(20,30)	(40,0)	(0,0)
$P = 3.50x + 2y$	100	130	140	0

The maximum profit of $140 is realized by producing 40 units of P_1 and no units of P_2.

23. Two machines A and B work 24 hours a day manufacturing chocolate chip and peanut butter cookies. The profit on a case of chocolate chip cookies is $100 per case and on peanut butter cookies is $125 per case. To make a case of chocolate chip cookies, machine A must run 4 hours and machine B must run 8 hours. To make a case of peanut butter cookies, machine A must run 6 hours and machine B must run 4 hours. How many cases of each cookie must be produced each day to maximize the profits?

24. A bakery makes cakes and cookies. It requires 2 hours of baking and 3 hours of decorating and packaging to make a batch of cakes. It requires 30 minutes of baking and 40 minutes of decorating and packaging to make a batch of cookies. The ovens are available 16 hours per day for baking and the decorating and packaging rooms are available 12 hours per day. If the profit per batch of cakes is $40 and the profit per batch of cookies is $20, find the number of batches of each that are needed to maximize the profit. $\left(\textit{Hint: } 30 \text{ min} = \frac{1}{2} \text{ hour}; \right.$

$\left. 40 \text{ min} = \frac{2}{3} \text{ hour} \right)$

25. A manufacturer of golf clubs makes two models—A and B. The daily production of model A is no more than 50 sets while no more than 20 sets of model B can be produced per day. The manufacturer makes a profit of $40 per set of model A and $60 per set

of model B. If the daily production of golf club sets cannot be more than 40 sets, how many sets of each model should be manufactured per day to maximize profit? What is the profit?

26. Farmer Smith has 400 acres of farmland for planting wheat and corn. He is able to plant, care for, and harvest at most 200 acres of wheat and 300 acres of corn. If his profit is $100 per acre of corn and $80 per acre of wheat, how many acres of each crop should he plant to maximize his profit? What is his profit?

27. In problem 26, if it costs $10 per acre to fertilize the corn and $12 per acre to fertilize the wheat, how many acres of each crop should he plant to minimize his cost of planting?

28. A manufacturer of chain saws can manufacture 200 14-inch saws and 140 16-inch saws in a given month. It takes 6 hours to manufacture each 14-inch saw and 10 hours to manufacture each 16-inch saw. In one month, there are 3,200 hours available. If the profit is $45 per 14-inch saw and $60 per 16-inch saw, how many saws of each kind should he manufacture to maximize the profit? What is the maximum profit?

29. In problem 28, if each 14-inch saw costs $35 per unit to produce and each 16-inch saw costs $50 per unit to produce, how many saws of each kind should be manufactured to minimize the cost?

30. Ben and John each can spend 14 hours per week to make furniture. If each table yields $50 profit and each chair yields $10 profit, how many of each item should they produce each week to maximize the profit given that Ben must work 3 hours and John 2 hours to make a chair while Ben must work 2 hours and John 6 hours to make a table.

Review exercises

1. Find all real solutions of the equation $x^3 + x^2 + x + 1 = 0$. See section 10–2.

2. Given $p = a + (n - 1)d$, find p when $a = 3$, $n = 16$, and $d = -3$. See section 1–5.

3. The diagonal of a square piece of metal is $\sqrt{98}$ inches long. Find the length of each side of the piece of metal. See section 6–4.

4. Solve the logarithmic equation $\log_2(x - 1) - \log_2(2x + 1) = 3$. See section 11–3.

5. Given functions f and g defined by $f(x) = 3x + 1$ and $g(x) = x^2 - 4$, find
   a. $f[g(-2)]$ b. $g[f(x)]$ c. $f(-3) - f(2)$
   See section 8–2.

Chapter 12 lead-in problem

A total of 12,000 persons paid $240,375 to attend a rock concert. If only two types of tickets were sold, one selling for $17.50 and the other selling for $25.00, how many of each type of ticket were sold?

Solution

Let x represent the number of $17.50 tickets sold and y represent the number of $25.00 tickets sold. We set up a table to help determine the equations.

	Number sold	Cost/ticket	Total income
First type	x	17.50	17.50x
Second type	y	25.00	25.00y
Totals	12,000		240,375

The system of equations is

$$x + y = 12{,}000 \qquad (1)$$
$$17.50x + 25.00y = 240{,}375 \qquad (2)$$

Multiply each member of (2) by 10 to clear decimal points. We then get

$$x + y = 12{,}000 \qquad \text{Multiply by } -175 \rightarrow \qquad -175x - 175y = -2{,}100{,}000$$
$$175x + 250y = 2{,}403{,}750 \qquad\qquad\qquad \underline{175x + 250y = 2{,}403{,}750}$$
$$75y = 303{,}750 \qquad \text{Add}$$
$$y = 4{,}050$$

Using the equation (1),

$$x + y = 12{,}000$$
$$x + (4{,}050) = 12{,}000 \qquad \text{Replace } y \text{ with 4,050}$$
$$x = 7{,}950$$

Thus, 4,050 tickets were sold at $25.00 and 7,950 were sold at $17.50.

Chapter 12 summary

1. Two or more linear equations that involve the same variables are called a **system of linear equations.**
2. A system of linear equations is **consistent and independent** if the system has only one solution.
3. A system of linear equations is **consistent and dependent** if all solutions of one equation are also solutions of the other equation(s).

4. A system of linear equations is **inconsistent** if the system has no solution.
5. A system of equations is **nonlinear** when one or both equations have degree greater than one.
6. **Linear programming** is the method of using a system of linear inequalities to maximize profit or minimize cost in business.

Chapter 12 error analysis

1. Solving a system of linear equations
   Example:

   $x - 2y = 3$ Multiply by $-2 \rightarrow$ $-2x + 4y = 3$
   $2x - 3y = 5$ $\underline{2x - 3y = 5}$
   $y = 8$

   $x - 2(8) = 3$
   $x = 19$
   The solution set is $\{(19,8)\}$.
   Correct answer: $\{(1,-1)\}$
   What error was made? (*see page 518*)

2. Solving a system of linear equations
   Example: $x - 2y = 3$
   $y = x - 1$

   Using substitution,
   $x - 2(x - 1) = 3$ $y = (-5) - 1$
   $x - 2x - 2 = 3$ $y = -6$
   $-x = 5$
   $x = -5$
   The solution set is $\{(-5,-6)\}$.
   Correct answer: $\{(-1,-2)\}$
   What error was made? (*see page 520*)

3. Solutions of systems of linear equations
   Example: The ordered triplet $(1,2,3)$ is a solution of the equation $x - y + z = 6$.
   Correct answer: $(1,-2,3)$
   What error was made? (*see page 534*)

4. Solving a system of nonlinear equations
   Example: $x^2 + y^2 = 34$
   $x = y + 2$

   Using substitution,
   $(y + 2)^2 + y^2 = 34$
   $y^2 + 4 + y^2 = 34$
   $2y^2 = 30$
   $y^2 = 15$
   $y = \pm\sqrt{15}$
   $x = 2 + \sqrt{15}$ or $2 - \sqrt{15}$
   The solution set is $\{(2 + \sqrt{15},\sqrt{15})$, $(2 - \sqrt{15},-\sqrt{15})\}$.
   Correct answer: $\{(-3,-5), (5,3)\}$
   What error was made? (*see page 542*)

5. Graphing inequalities in two variables
   Example: The graph of $3x - 2y < 6$ is

   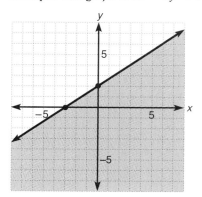

   Correct answer:

   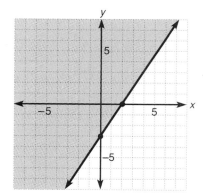

   What errors were made? (*see page 337*)

6. Region of feasible solutions
   Example: The region of feasible solutions of the system of constraints

$$\begin{cases} x + y \le 5 \\ x + y \ge 3 \\ x \ge 0 \\ y \ge 0 \end{cases}$$

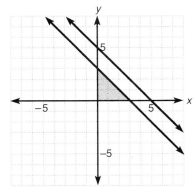

Correct answer:

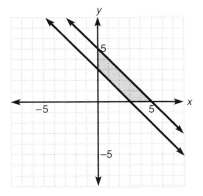

What error was made? (*see page 548*)

7. Quotients of polynomial expressions
   Example: $\dfrac{15a^3 - 10a^2 + 5a}{5a} = 3a^2 - 2a$
   Correct answer: $3a^2 - 2a + 1$
   What error was made? (*see page 183*)

8. Synthetic division
   Example: Using synthetic division, divide
   $$(4x^2 - 3x + 1) \div (x - 2) = 4x - 11 + \frac{23}{x - 2}$$

   $$-2\overline{)\begin{array}{rrr} 4 & -3 & 1 \\ & -8 & 22 \\ \hline 4 & -11 & 23 \end{array}}$$

   Correct answer: $4x + 5 + \dfrac{11}{x - 2}$
   What error was made? (*see page 190*)

9. Solving absolute value inequalities
   Example: Find the solution set of
   $|2x - 1| \ge 3$.
   $-3 \le 2x - 1 \le 3$
   $-2 \le 2x \le 4$
   $-1 \le x \le 2$
   $\{x \mid -1 \le x \le 2\}$
   Correct answer: $\{x \mid x \le -1 \text{ or } x \ge 2\}$
   What error was made? (*see page 88*)

10. Perpendicular lines
   Example: The equation of the line through the point $(1, -2)$ and perpendicular to $3x - 2y = 1$ is
   $3x - 2y = 7$.
   Correct answer: $2x + 3y = -4$
   What error was made? (*see page 322*)

Chapter 12 critical thinking

Write an algebraic expression for the following relationship.
Will this relationship always be true?

$(1)^2 + (2)^2 = (3)^2 - (2)^2$
$(2)^2 + (3)^2 = (7)^2 - (6)^2$
$(3)^2 + (4)^2 = (13)^2 - (12)^2$
$(4)^2 + (5)^2 = (21)^2 - (20)^2$

Chapter 12 review

[12–1]

Find the solution set of each system of linear equations by elimination. If the system is inconsistent or dependent, so state. Verify the solution set by graphing in problems 1–3.

1. $x + y = 4$
   $x - y = 2$

2. $2x + 3y = 2$
   $2x - 3y = 0$

3. $2x + 3y = 4$
   $4x + 6y = 8$

4. $x - 4y = 5$
   $3x - 12y = 0$

5. $x + y = 1$
   $4x - 4y = 6$

6. $\dfrac{1}{2}x + y = 4$

   $x - \dfrac{1}{2}y = -5$

7. Forces F_1 and F_2 on a structure yield the system of linear equations

$$\frac{1}{3}F_1 + \frac{2}{3}F_2 = 3$$

$$\frac{2}{3}F_1 - \frac{1}{3}F_2 = 5$$

Find the forces F_1 and F_2.

Find the solution set of each system of linear equations by substitution. If the system is inconsisent or dependent, so state.

8. $3x + 2y = 1$
   $y = -4$

9. $4x - y = 6$
   $x = -1$

10. $5y - x = 1$
    $y = 3x + 1$

11. $6x + 2y = -3$
    $x = 4 - y$

12. $x - 4y = 1$
    $2y - 2x = 3$

[12–2]

Set up a system of linear equations and solve each problem.

13. Find the point of intersection of the two lines L_1 and L_2 if L_1 contains the points $(-3,2)$ and $(1,0)$ and L_2 contains the points $(5,1)$ and $(-2,-3)$.

14. Noel Doe wishes to enclose her rectangular yard with 180 feet of fencing. If she wishes to make the enclosed yard twice as long as it is wide, what are the dimensions of the yard? (*Hint:* The perimeter $P = 2\ell + 2w$. Set up two equations in ℓ and w and solve.)

15. The total number of fire alarms in Detroit on a given day is 30. If the number of real alarms is two more than six times the number of false alarms, how many of each alarm are sounded?

16. A woman has $3,000 to invest. If she invests part at 7% simple interest and the rest at $6\frac{1}{2}$% simple interest, and the total income for the year is $201, how much does she invest at each rate?

17. Solder made of 20% tin is to be melted with solder made of 5% tin to produce 50 grams of solder containing 10% tin. How many grams of each should be used?

18. George bought a suit and a topcoat for $177. If one-fifth of the cost of the suit is $9 more than one-sixth of the cost of the topcoat, what is the price of each article of clothing?

19. Two airplanes leave from Detroit, one flying East and the other flying West. After 5 hours and 12 minutes the two planes are 2,886 miles apart. How fast was each plane flying if one plane flies 55 miles faster than the other plane?

[12–3]

Find the solution set of each system of three linear equations. If the system is inconsistent or dependent, so state.

20. $x - y + 2z = 5$
    $x + y + z = 6$
    $2x - y - z = -3$

21. $7x - 2y + 9z = -3$
    $4x - 6y + 8z = -5$
    $8x + y + z = 1$

22. $2u + v + 3w = -2$
    $5u + 2v = 5$
    $2v - 3w = -7$

23. $p - r = -1$
    $-q + r = 2$
    $-2p + q = -4$

24. Find the values of a, b, and c such that the points $(0,2)$, $(1,-3)$, and $(2,-2)$ lie on the graph of $y = ax^2 + bx + c$.

25. Three forces on a beam are related by the system of linear equations

$$0.2F_1 + 0.3F_2 = 2$$
$$0.2F_1 - 0.1F_3 = 1$$
$$0.4F_2 + 0.2F_3 = 3$$

Find forces F_1, F_2, and F_3.

26. If the middle-sized angle of a triangle measures 16° more than the smallest angle and the largest angle is twice the size of the smallest angle, find the measures of the three angles of the triangle.

27. Ron has a total of 285 coins—nickels, dimes, and quarters—in his piggy bank. If there is a total of $26.25 in the bank and there are four times as many dimes as quarters, how many of each coin does he have in the bank?

[12–4]

Find the solution set of the following systems of nonlinear equations.

28. $x^2 + y^2 = 2$
$x + y = 2$

29. $x^2 + y^2 = 25$
$x = 7 - y$

30. $x^2 + y^2 = 5$
$x = 3$

31. $2x^2 - 3y^2 = 4$
$y = -2$

32. $x^2 - y^2 = -1$
$x + y = 3$

Find the solution set of each system of equations by elimination or a combination of elimination and substitution.

33. $2x^2 + y^2 = 4$
$x^2 - y^2 = 8$

34. $x^2 - y^2 = 16$
$x^2 + 2y^2 = 25$

35. Find the length and the width of a rectangle whose perimeter is 52 inches and whose area is 153 square inches.

[12–5]

36. Graph the region R of feasible solutions defined by the constraints
$$\begin{cases} 2x + 4y \leq 8 \\ 4x + 3y \leq 12 \\ x \geq 0 \\ y \geq 0 \end{cases}$$

37. Find the maximum and minimum values of the objective function $f = 3x - y$ using the given region R of feasible solutions.

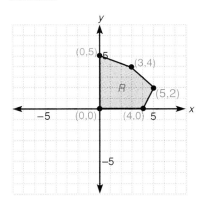

38. Suzie Q raises pigs and geese on her farm. She wants to raise at most 12 geese and at most 16 animals all together. It costs $2 to raise a goose and $5 to raise a pig. If she has $50 to spend on the animals, find the maximum profit she can make if she makes a profit of $4 for each pig and $8 for each goose.

Chapter 12 cumulative test

[1–4]   **1.** Simplify $3[-4(12 - 3) - 14 + 6(3 - 9)]$.

Evaluate the following formulas.

[1–5]   **2.** $V = k + gt$ when $k = 14$, $g = 32$, and $t = 3$

Perform the indicated operations and simplify.

[4–5]   **3.** $\dfrac{3a^3 - 11a^2 + 12a - 3}{a - 3}$

[4–5]   **4.** $\dfrac{30a^7 - 25a^5 + 15a^3}{5a^2}$

[3–2]   **5.** $(5x + 1)(x - 9) - (x + 6)^2$

[3–3]   **6.** $\dfrac{3^{-1}x^0 y^{-2}}{6^{-1}x^{-4}y^3}$

Find the solution set of the following equations and inequalities.

[2–1] 7. $5(6y - 1) - 3(4y + 3) = 5y$

[6–3] 8. $3x^2 - 2x - 4 = 0$ (Use the quadratic formula.)

[6–2] 9. $x^2 + 6x - 5 = 0$ (Use completing the square.)

[2–5] 10. $3(z + 1) - 1 \leq 2(3z + 3)$

[6–7] 11. $x^2 + 2x - 3 \leq 0$

[2–4] 12. $|2x - 1| = 4$

[2–6] 13. $|5 - 6x| > 2$

[4–1] 14. Reduce the expression $\dfrac{8y + 12}{4y^2 - 9}$ to lowest terms.

Perform the indicated operations and simplify. Assume all denominators are nonzero.

[4–2] 15. $\dfrac{x^2 - 16}{4x - 3} \cdot \dfrac{16x^2 - 9}{3x + 12}$

[4–3] 16. $\dfrac{11}{x^2 - x - 20} - \dfrac{7}{x^2 + x - 12}$

[4–3] 17. $\dfrac{3y}{y - 7} + \dfrac{y}{7 - y}$

[4–2] 18. $\dfrac{p^3 q}{16ab^2} \div \dfrac{7pq^2}{24a^2b^3}$

Simplify the following. Leave all answers with positive exponents. Assume all variables are positive.

[5–1] 19. $(-27)^{2/3}$

[5–1] 20. $(y^{3/2})^{-2}$

[5–1] 21. $\dfrac{a^{3/4}}{a^{-1/2}}$

[5–6] 22. $\sqrt{8} \cdot \sqrt{10}$

[5–6] 23. $(\sqrt{6} + 3)(\sqrt{2} - 4)$

[5–7] 24. $(4 + 5i)^2$

[5–3] 25. $\sqrt[3]{8xy^3} - \sqrt[3]{64xy^3}$

[6–6] 26. Find the solution set of the equation $p + 7\sqrt{p} + 6 = 0$. Identify any extraneous solutions that exist.

[7–1] 27. Graph the equation $4x + 5y = -20$ using the intercepts.

[7–3] 28. Graph the equation $3x - y = 6$ using the slope and y-intercept.

[7–3] 29. Find the equation in standard form of the line through the points $(0,4)$ and $(-7,9)$.

[7–3] 30. Find the equation in standard form of the line through $(1,3)$ and perpendicular to the line $4x - y = 5$.

[7–4] 31. Sketch the graph of the inequality $2y + x < 2$.

[8–2] 32. Given $f(x) = 2x - 3$ and $g(x) = x^2 + 5$, find
a. $f(-3)$
b. $(f - g)(4)$
c. $(f \circ g)(-1)$

[8–3] 33. The function defined by $f(x) = x^2 - 2x + 1$ is even, odd, or neither even nor odd.

[8–4] 34. Find the inverse of function f defined by $f(x) = 4x - 3$.

[9–4] 35. Identify each equation as a parabola, a circle, an ellipse, or a hyperbola.
a. $x^2 + y^2 - 4x = 0$
b. $3y^2 + 14 - 4x^2 = 0$
c. $x + y^2 - 3y + 1 = 0$

[10–2] 36. Find the solution set of the equation $x^4 - 4x^3 - 2x^2 - 12x - 15 = 0$.

[10–5] 37. Find the equations of the horizontal and vertical asymptotes in the graph of $f(x) = \dfrac{3x - 2}{4x + 3}$.

[11–1] 38. Find the solution set of the exponential equation $4^{2x + 1} = 32$.

[11–3] 39. Find the solution set of the logarithmic equation $\log_3(x + 1) - \log_3 x = 2$.

Find the solution set of the following systems of linear equations.

[12–1] 40. $4x + 3y = 0$
$3x - 2y = 1$

[12–3] 41. $x + 4y - z = -3$
$-2x + y + 2z = 0$
$3x - 2y + z = 1$

[12–4] 42. Find the solution set of the system of nonlinear equations
$x^2 + 2y^2 = 8$
$x^2 - y^2 = -19$

[12–5] 43. Maximize the objective function $f = 4y + 5x$ under the constraints
$6x + 5y \leq 45$
$5x + 12y \leq 6$
$x \geq 0$
$y \geq 0$

Matrices and Determinants

As a tenant, the manager of an apartment complex pays two-thirds of the rent that each of the remaining 9 tenants pay. Each month the landlord collects a total of $6,090. How much does each tenant pay?

13–1 ■ *Matrices and determinants*

In sections 12–1 and 12–3, we solved systems of linear equations in two and three variables by using algebraic methods involving elimination of a variable and by substituting an expression for one of the variables. We now consider another method for solving these linear systems by using **determinants.** A determinant is the number associated with an array of numbers called a **matrix.**

⌐ *Definition of a matrix* _____

A matrix is any rectangular array of numbers.

The following rectangular array of numbers is called a "three by two" matrix (denoted 3 × 2) because there are three rows and two columns in the matrix.

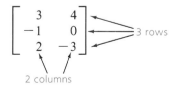

$$\begin{bmatrix} 3 & 4 \\ -1 & 0 \\ 2 & -3 \end{bmatrix} \quad \text{3 rows}$$

2 columns

Note The array of numbers making up a matrix is enclosed within a set of brackets. The numbers are called the *elements* of the matrix.

When the matrix has the same number of rows as columns, the array is called a *square matrix*. An example of a 2 × 2 square matrix and a 3 × 3 square matrix is shown.

$$\begin{bmatrix} 1 & 2 \\ 3 & 4 \end{bmatrix} \qquad \begin{bmatrix} 4 & -7 & 5 \\ 2 & 7 & 4 \\ 0 & -1 & 6 \end{bmatrix}$$

$$2 \times 2 \qquad\qquad 3 \times 3$$

Associated with every square matrix having real number entries is a real number called its **determinant.** To denote the determinant, we enclose the array between two vertical line segments.

Definition

If a, b, c, and d are real numbers, then the **determinant** of the square matrix

$$\begin{bmatrix} a & b \\ c & d \end{bmatrix} \text{ is denoted by } \begin{vmatrix} a & b \\ c & d \end{vmatrix}$$

The determinant of the

$$2 \times 2 \text{ matrix } \begin{bmatrix} 1 & 2 \\ 3 & 4 \end{bmatrix} \text{ is written } \begin{vmatrix} 1 & 2 \\ 3 & 4 \end{vmatrix}$$

$$3 \times 3 \text{ matrix } \begin{bmatrix} 4 & -7 & 5 \\ 2 & 7 & 4 \\ 0 & -1 & 6 \end{bmatrix} \text{ is written } \begin{vmatrix} 4 & -7 & 5 \\ 2 & 7 & 4 \\ 0 & -1 & 6 \end{vmatrix}$$

The value of a 2 × 2 determinant is now given.

Value of a 2 × 2 determinant

$$\begin{vmatrix} a_1 & b_1 \\ a_2 & b_2 \end{vmatrix} = a_1 b_2 - a_2 b_1$$

■ *Example 13–1 A*

Find the value of each determinant.

1. $\begin{vmatrix} 1 & -2 \\ 3 & 4 \end{vmatrix} = (1)(4) - (3)(-2) = 4 + 6 = 10$

2. $\begin{vmatrix} 0 & 3 \\ -4 & -1 \end{vmatrix} = (0)(-1) - (-4)(3) = 0 + 12 = 12$

▶ *Quick check* Find the value of the determinant. $\begin{vmatrix} 3 & 4 \\ -1 & 2 \end{vmatrix}$

The value of the determinant of a 3×3 matrix is found by rewriting the determinant in terms of 2×2 determinants, which we call *minors*. We now define the minor of an element, or entry, of a 3×3 determinant.

Definition of a minor

The **minor** of an element of a 3×3 determinant is defined to be the 2×2 determinant that remains after the row and column in which the element appears have been deleted.

Given the determinant $\begin{vmatrix} 4 & -7 & 5 \\ 2 & 7 & -4 \\ 0 & -1 & 6 \end{vmatrix}$, to find the minor of

1. 4 in the first row, eliminate the row and column that contain 4.

$$\begin{vmatrix} 4 & -7 & 5 \\ 2 & 7 & -4 \\ 0 & -1 & 6 \end{vmatrix}$$

The minor of 4 is $\begin{vmatrix} 7 & -4 \\ -1 & 6 \end{vmatrix}$.

2. 2 in the second row, eliminate the row and column that contain 2.

$$\begin{vmatrix} 4 & -7 & 5 \\ 2 & 7 & -4 \\ 0 & -1 & 6 \end{vmatrix}$$

The minor of 2 is $\begin{vmatrix} -7 & 5 \\ -1 & 6 \end{vmatrix}$.

3. 0 in the third row, eliminate the row and column that contain 0.

$$\begin{vmatrix} 4 & -7 & 5 \\ 2 & 7 & -4 \\ 0 & -1 & 6 \end{vmatrix}$$

The minor of 0 is $\begin{vmatrix} -7 & 5 \\ 7 & -4 \end{vmatrix}$.

The value of this 3×3 determinant is found by *expanding by minors* about the elements of one row or one column. To do this, we multiply each element of that row (or column) by its minor and then combine using the following pattern:

$$\begin{vmatrix} + & - & + \\ - & + & - \\ + & - & + \end{vmatrix}$$

Thus, using the elements of the first column in the expansion,

$$\begin{vmatrix} 4 & -7 & 5 \\ 2 & 7 & -4 \\ 0 & -1 & 6 \end{vmatrix} = +4\begin{vmatrix} 7 & -4 \\ -1 & 6 \end{vmatrix} - 2\begin{vmatrix} -7 & 5 \\ -1 & 6 \end{vmatrix} + 0\begin{vmatrix} -7 & 5 \\ 7 & -4 \end{vmatrix}$$

$$\begin{vmatrix} + & - & + \\ - & + & - \\ + & - & + \end{vmatrix}$$

$$= 4[(7)(6) - (-1)(-4)] - 2[(-7)(6) - (-1)(5)]$$
$$+ 0[(-7)(-4) - (7)(5)]$$
$$= 4(42 - 4) - 2(-42 + 5) + 0(28 - 35)$$
$$= 4(38) - 2(-37) + 0(-7)$$
$$= 152 + 74 + 0$$
$$= 226$$

The value of the 3×3 determinant is 226.

Note Had we chosen to evaluate the determinant about the elements of the *second row,*

$$\begin{vmatrix} 4 & -7 & 5 \\ 2 & 7 & -4 \\ 0 & -1 & 6 \end{vmatrix} = -2 \begin{vmatrix} -7 & 5 \\ -1 & 6 \end{vmatrix} + 7 \begin{vmatrix} 4 & 5 \\ 0 & 6 \end{vmatrix} - (-4) \begin{vmatrix} 4 & -7 \\ 0 & -1 \end{vmatrix}$$

$$\begin{vmatrix} + & - & + \\ - & + & - \\ + & - & + \end{vmatrix}$$

$$= -2(-42 + 5) + 7(24 - 0) + 4(-4 - 0)$$
$$= -2(-37) + 7(24) + 4(-4)$$
$$= 74 + 168 - 16$$
$$= 226 \qquad \text{The same result}$$

In general, to expand a 3×3 determinant by minors about the elements of the first column, we use the following procedure.

$$\begin{vmatrix} a_1 & b_1 & c_1 \\ a_2 & b_2 & c_2 \\ a_3 & b_3 & c_3 \end{vmatrix} = +a_1 \begin{vmatrix} b_2 & c_2 \\ b_3 & c_3 \end{vmatrix} - a_2 \begin{vmatrix} b_1 & c_1 \\ b_3 & c_3 \end{vmatrix} + a_3 \begin{vmatrix} b_1 & c_1 \\ b_2 & c_2 \end{vmatrix}$$

■ *Example 13–1 B*

Expand the determinant by minors about the first column.

$$\begin{vmatrix} 2 & -3 & -1 \\ 1 & 2 & 3 \\ -4 & -2 & 1 \end{vmatrix} = +2 \begin{vmatrix} 2 & 3 \\ -2 & 1 \end{vmatrix} - 1 \begin{vmatrix} -3 & -1 \\ -2 & 1 \end{vmatrix} + (-4) \begin{vmatrix} -3 & -1 \\ 2 & 3 \end{vmatrix}$$
$$= 2(2 + 6) - 1(-3 - 2) + (-4)(-9 + 2)$$
$$= 2(8) - 1(-5) + (-4)(-7)$$
$$= 16 + 5 + 28$$
$$= 49$$

The value of the 3×3 determinant is 49.

An alternate way of evaluating a 3×3 determinant is by *diagonal expansion.* We augment the given determinant on the right with the first two columns of the given determinant and compute the diagonal products p_1, p_2, p_3, p_4, p_5, and p_6 as indicated by arrows. We demonstrate this method on the determinant in example 13–1 B.

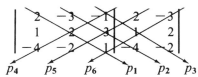

The value of the determinant is

$$p_1 + p_2 + p_3 - p_4 - p_5 - p_6$$

Upper left to lower right diagonals are *added*   Upper right to lower left diagonals are *subtracted*

Thus,

$$
\begin{vmatrix}
2 & -3 & -1 \\
1 & 2 & 3 \\
-4 & -2 & 1
\end{vmatrix}
\begin{matrix}
2 & -3 \\
1 & 2 \\
-4 & -2
\end{matrix}
$$

$$= (2)(2)(1) + (-3)(3)(-4) + (-1)(1)(-2)$$
$$- [(-4)(2)(-1)] - [(-2)(3)(2)] - [(1)(1)(-3)]$$
$$= 4 + 36 + 2 - 8 - (-12) - (-3)$$
$$= 4 + 36 + 2 - 8 + 12 + 3$$
$$= 49 \qquad \text{As we determined previously}$$

▶ **Quick check**   Find the value of the determinant. $\begin{vmatrix} 3 & 2 & 1 \\ 0 & 4 & -2 \\ -1 & 2 & -3 \end{vmatrix}$ ∎

The expansion of higher-order determinants by minors can be accomplished in the same way as with third-order determinants. The pattern of alternating signs in the sign array extends to higher-order determinants. The determinants that are the minors in each expansion will be of order one less than the order of the original determinant. The following is the sign array of a 4 × 4 determinant.

$$
\begin{matrix}
+ & - & + & - \\
- & + & - & + \\
+ & - & + & - \\
- & + & - & +
\end{matrix}
$$

Note   The sign array *always* begins with a + in the upper-left corner and alternates signs from there on.

■ *Example 13–1 C*

Expand the determinant by minors about the first row.

$$
\begin{vmatrix}
-1 & -2 & 3 & 2 \\
0 & 1 & 4 & -2 \\
3 & -1 & 4 & 0 \\
2 & 1 & 0 & 3
\end{vmatrix}
$$

$$
= +(-1)\begin{vmatrix} 1 & 4 & -2 \\ -1 & 4 & 0 \\ 1 & 0 & 3 \end{vmatrix} - (-2)\begin{vmatrix} 0 & 4 & -2 \\ 3 & 4 & 0 \\ 2 & 0 & 3 \end{vmatrix} + 3\begin{vmatrix} 0 & 1 & -2 \\ 3 & -1 & 0 \\ 2 & 1 & 3 \end{vmatrix}
$$

$$
- 2\begin{vmatrix} 0 & 1 & 4 \\ 3 & -1 & 4 \\ 2 & 1 & 0 \end{vmatrix}
$$

Evaluating each 3 × 3 by expanding by minors about the first column, we obtain the following.

$$= -1\left[1\begin{vmatrix}4 & 0\\0 & 3\end{vmatrix} - (-1)\begin{vmatrix}4 & -2\\0 & 3\end{vmatrix} - 1\begin{vmatrix}4 & -2\\4 & 0\end{vmatrix}\right]$$

$$+ 2\left[0\begin{vmatrix}4 & 0\\0 & 3\end{vmatrix} - 3\begin{vmatrix}4 & -2\\0 & 3\end{vmatrix} + 2\begin{vmatrix}4 & -2\\4 & 0\end{vmatrix}\right]$$

$$+ 3\left[0\begin{vmatrix}-1 & 0\\1 & 3\end{vmatrix} - 3\begin{vmatrix}1 & -2\\1 & 3\end{vmatrix} + 2\begin{vmatrix}1 & -2\\-1 & 0\end{vmatrix}\right]$$

$$- 2\left[0\begin{vmatrix}-1 & 4\\1 & 0\end{vmatrix} - 3\begin{vmatrix}1 & 4\\1 & 0\end{vmatrix} + 2\begin{vmatrix}1 & 4\\-1 & 4\end{vmatrix}\right]$$

$$= -1[12 + 12 + 8] + 2[0 - 36 + 8] + 3[0 - 15 - 4]$$
$$- 2[0 + 12 + 16]$$
$$= -1(32) + 2(-28) + 3(-19) - 2(28)$$
$$= -201$$

Mastery points

Can you
■ Evaluate 2 × 2 and 3 × 3 determinants?

Exercise 13–1

Evaluate the following determinants. See example 13–1 A.

Example $\begin{vmatrix}3 & 4\\-1 & 2\end{vmatrix}$

Solution $\begin{vmatrix}3 & 4\\-1 & 2\end{vmatrix} = (3)(2) - (-1)(4) = 6 + 4 = 10$

1. $\begin{vmatrix}1 & -3\\2 & 5\end{vmatrix}$
2. $\begin{vmatrix}4 & 1\\2 & 7\end{vmatrix}$
3. $\begin{vmatrix}4 & -2\\-6 & -3\end{vmatrix}$
4. $\begin{vmatrix}2 & 5\\4 & 10\end{vmatrix}$

5. $\begin{vmatrix}0 & -5\\1 & 2\end{vmatrix}$
6. $\begin{vmatrix}4 & 0\\2 & -7\end{vmatrix}$
7. $\begin{vmatrix}5 & 4\\0 & -7\end{vmatrix}$
8. $\begin{vmatrix}1 & 2\\3 & 0\end{vmatrix}$

Evaluate the following determinants using expansion by minors about any column or row. See example 13–1 B.

Example
$$\begin{vmatrix} 3 & 2 & 1 \\ 0 & 4 & -2 \\ -1 & 2 & -3 \end{vmatrix}$$

Solution   Expand about the first column.

$$\begin{vmatrix} 3 & 2 & 1 \\ 0 & 4 & -2 \\ -1 & 2 & -3 \end{vmatrix} = +3\begin{vmatrix} 4 & -2 \\ 2 & -3 \end{vmatrix} - 0\begin{vmatrix} 2 & 1 \\ 2 & -3 \end{vmatrix} + (-1)\begin{vmatrix} 2 & 1 \\ 4 & -2 \end{vmatrix}$$

$$= 3(-12 + 4) - 0(-6 - 2) - 1(-4 - 4)$$
$$= 3(-8) - 0(-8) - 1(-8) = -24 - 0 + 8$$
$$= -16$$

9. $\begin{vmatrix} 1 & 2 & 3 \\ 3 & 2 & 1 \\ 2 & 1 & 3 \end{vmatrix}$

10. $\begin{vmatrix} -2 & 1 & 3 \\ 1 & 0 & 4 \\ -1 & 4 & 3 \end{vmatrix}$

11. $\begin{vmatrix} -3 & 2 & 1 \\ -1 & 3 & 2 \\ 1 & 4 & 5 \end{vmatrix}$

12. $\begin{vmatrix} 3 & 0 & 1 \\ 2 & 0 & 4 \\ 3 & 0 & -1 \end{vmatrix}$

13. $\begin{vmatrix} 3 & -1 & 2 \\ 0 & 0 & 0 \\ 4 & 3 & 1 \end{vmatrix}$

14. $\begin{vmatrix} 2 & 0 & 0 \\ 0 & 2 & 0 \\ 0 & 0 & 2 \end{vmatrix}$

15. $\begin{vmatrix} 2 & 3 & 1 \\ 1 & 2 & 3 \\ 0 & 1 & 1 \end{vmatrix}$

16. $\begin{vmatrix} -1 & -1 & -1 \\ 2 & 2 & 2 \\ 3 & -3 & 3 \end{vmatrix}$

17. $\begin{vmatrix} -1 & 4 & 0 \\ -2 & 1 & -3 \\ -3 & 7 & 2 \end{vmatrix}$

18. $\begin{vmatrix} 0 & 1 & 7 \\ 3 & 0 & 2 \\ 4 & 0 & -1 \end{vmatrix}$

19. $\begin{vmatrix} -4 & 0 & 1 \\ 2 & 0 & 3 \\ -5 & -1 & -2 \end{vmatrix}$

20. $\begin{vmatrix} 5 & 10 & 15 \\ 1 & -1 & 0 \\ 2 & 1 & 0 \end{vmatrix}$

21. $\begin{vmatrix} 2 & 2 & -2 \\ 3 & -3 & 3 \\ -1 & 1 & 1 \end{vmatrix}$

22. $\begin{vmatrix} 0 & -1 & 2 \\ 1 & 0 & 2 \\ -5 & 6 & 0 \end{vmatrix}$

23. $\begin{vmatrix} a & 0 & a \\ 0 & a & 0 \\ 0 & 0 & a \end{vmatrix}$

24. $\begin{vmatrix} 0 & x & 0 \\ 0 & 0 & x \\ x & 0 & 0 \end{vmatrix}$

25. $\begin{vmatrix} x & y & 0 \\ 0 & x & y \\ y & x & 0 \end{vmatrix}$

26. $\begin{vmatrix} a & b & 0 \\ b & 0 & a \\ 0 & a & b \end{vmatrix}$

Expand and evaluate each determinant about row one using the sign array $+ - + -$. See example 13–1 C.

27. $\begin{vmatrix} 1 & 2 & 3 & -1 \\ 2 & 0 & 1 & 3 \\ -2 & 1 & 0 & -1 \\ 0 & 3 & 2 & 0 \end{vmatrix}$

28. $\begin{vmatrix} 0 & 5 & 3 & 1 \\ 2 & 0 & 4 & -2 \\ -1 & -1 & 0 & 1 \\ 2 & 1 & 3 & 0 \end{vmatrix}$

29. $\begin{vmatrix} 4 & 0 & 1 & 0 \\ -5 & -1 & 0 & -1 \\ 0 & 1 & 0 & -2 \\ 2 & 0 & -3 & 1 \end{vmatrix}$

30. $\begin{vmatrix} 0 & 0 & 1 & 4 \\ -5 & 6 & 0 & -3 \\ 2 & 0 & 1 & 0 \\ 4 & 2 & 0 & 7 \end{vmatrix}$

Review exercises

1. Find the solution set of the system of equations by elimination. $3x - y = -8$
$$2x + 3y = 2$$
See section 12–1.

Graph the following equations. See section 7–1.

2. $y = 3x - 2$

3. $x = -2$

4. $2x + 3y = -6$

5. Given $P(x) = x^2 - 6x + 9$, find $P(-1)$.
See section 1–5.

6. Find the solution set of the equation
$y - 3\sqrt{y} + 2 = 0$. See section 6–6.

13–2 ■ Solutions of systems of linear equations by determinants

Determinants can be used to solve systems of linear equations. Consider the system of two linear equations in two variables

$$a_1x + b_1y = c_1$$
$$a_2x + b_2y = c_2 \qquad \text{Written in standard form}$$

Suppose we solve the system of equations by elimination. To eliminate x, multiply each term of the first equation by $-a_2$ and each term of the second equation by a_1 to obtain the equivalent system

$$-a_1a_2x - a_2b_1y = -a_2c_1$$
$$a_1a_2x + a_1b_2y = a_1c_2$$

Adding the terms, the resulting equation in y is given by

$$a_1b_2y - a_2b_1y = a_1c_2 - a_2c_1$$

Factoring y from the terms in the left member, we obtain

$$(a_1b_2 - a_2b_1)y = a_1c_2 - a_2c_1$$

and dividing each member by $a_1b_2 - a_2b_1$, we have

$$y = \frac{a_1c_2 - a_2c_1}{a_1b_2 - a_2b_1} \qquad (a_1b_2 - a_2b_1 \neq 0) \qquad (1)$$

In like fashion, if we solve the system for x, we obtain

$$x = \frac{c_1b_2 - c_2b_1}{a_1b_2 - a_2b_1} \qquad (a_1b_2 - a_2b_1 \neq 0) \qquad (2)$$

Now by definition,

$$D = \begin{vmatrix} a_1 & b_1 \\ a_2 & b_2 \end{vmatrix} = a_1b_2 - a_2b_1 \qquad \text{Denominator of equations (1) and (2)}$$

$$D_x = \begin{vmatrix} c_1 & b_1 \\ c_2 & b_2 \end{vmatrix} = c_1b_2 - c_2b_1 \qquad \text{Numerator of equation (2)}$$

$$D_y = \begin{vmatrix} a_1 & c_1 \\ a_2 & c_2 \end{vmatrix} = a_1c_2 - a_2c_1 \qquad \text{Numerator of equation (1)}$$

Note   1. D contains the coefficients of the variables in order as they appear in the equations, when written in standard form.
2. D_x contains the coefficients of the variables replacing the x-coefficients with the column of constants. (Note the color type.)
3. D_y contains the coefficients of the variables replacing the y-coefficients with the column of constants. (Note the color type.)

To solve this system of linear equations for x and for y, we obtain the following determinant ratios used to solve a system of linear equations by determinants (called **Cramer's Rule**), as seen in finding x and y in equations (1) and (2).

Cramer's Rule for 2 × 2 linear systems _____

Given the system of linear equations $a_1x + b_1y = c_1$
$\qquad\qquad\qquad\qquad\qquad\qquad\qquad a_2x + b_2y = c_2$

then

$$x = \frac{D_x}{D} = \frac{\begin{vmatrix} c_1 & b_1 \\ c_2 & b_2 \end{vmatrix}}{\begin{vmatrix} a_1 & b_1 \\ a_2 & b_2 \end{vmatrix}} \quad \text{and} \quad y = \frac{D_y}{D} = \frac{\begin{vmatrix} a_1 & c_1 \\ a_2 & c_2 \end{vmatrix}}{\begin{vmatrix} a_1 & b_1 \\ a_2 & b_2 \end{vmatrix}}$$

where $D = a_1b_2 - a_2b_1 \neq 0$.

■ *Example 13-2 A*

1. Use Cramer's Rule to find the solution set of the system of linear equations.
$2x - 3y = 4$
$x + 4y = -1$

By Cramer's Rule, $x = \dfrac{D_x}{D}$ and $y = \dfrac{D_y}{D}$. To find D, we use the coefficients of the variables, so

$$D = \begin{vmatrix} 2 & -3 \\ 1 & 4 \end{vmatrix} = 2(4) - 1(-3) = 8 + 3 = 11$$

To find D_x, replace the column $\begin{matrix} 2 \\ 1 \end{matrix}$ of D by the constants $\begin{matrix} 4 \\ -1 \end{matrix}$.

$$D_x = \begin{vmatrix} 4 & -3 \\ -1 & 4 \end{vmatrix} = 4(4) - (-1)(-3) = 16 - 3 = 13$$

To find D_y, replace the column $\begin{matrix} -3 \\ 4 \end{matrix}$ of D by the constants $\begin{matrix} 4 \\ -1 \end{matrix}$.

$$D_y = \begin{vmatrix} 2 & 4 \\ 1 & -1 \end{vmatrix} = 2(-1) - 1(4) = -2 - 4 = -6$$

By Cramer's Rule,

$$x = \frac{D_x}{D} = \frac{13}{11}$$

$$y = \frac{D_y}{D} = \frac{-6}{11} = -\frac{6}{11}$$

The solution set of the system is $\left\{ \left(\frac{13}{11}, -\frac{6}{11} \right) \right\}$.

Note If $D = 0$, the system of equations is then either inconsistent or dependent. When

$$D = 0 \text{ and } D_y \neq 0 \text{ and/or } D_x \neq 0$$

the system of equations is *inconsistent*. If

$$D = 0 \text{ and } D_y = D_x = 0$$

the system of equations is *dependent*.

2. Use Cramer's Rule to find the solution set.

$$4x - y = 3 \qquad (1)$$

$$2x - \frac{y}{2} = 0 \qquad (2)$$

We first clear the denominator of equation (2).

$$2x - \frac{y}{2} = 0$$

$$2 \cdot 2x - 2 \cdot \frac{y}{2} = 2 \cdot 0 \qquad \text{Multiply each member by 2}$$

$$4x - y = 0$$

We now solve the system of linear equations.

$$4x - y = 3$$
$$4x - y = 0$$

Using determinants,

$$D = \begin{vmatrix} 4 & -1 \\ 4 & -1 \end{vmatrix} = (-4) - (-4) = 0$$

$$D_x = \begin{vmatrix} 3 & -1 \\ 0 & -1 \end{vmatrix} = (-3) - (0) = -3$$

Since $D = 0$ and at least one of D_x or D_y (in this case D_x) is not zero, we have no solution. The system is inconsistent and the solution set is $\emptyset$.

▶ **Quick check**   Use Cramer's Rule to find the solution set of the system of linear equations. $3x - 4y = 1$
$$x + 2y = -4$$

Now we consider a system of three linear equations in three variables. The procedure is similar to that used to solve a system of two linear equations in two variables.

—— **Cramer's Rule for 3 × 3 linear systems** ——————————

Given the system of linear equations

$$a_1x + b_1y + c_1z = d_1$$
$$a_2x + b_2y + c_2z = d_2 \qquad \text{Written in standard form}$$
$$a_3x + b_3y + c_3z = d_3$$

where

$$D = \begin{vmatrix} a_1 & b_1 & c_1 \\ a_2 & b_2 & c_2 \\ a_3 & b_3 & c_3 \end{vmatrix} \qquad \text{Determinant of coefficients}$$

$$D_x = \begin{vmatrix} d_1 & b_1 & c_1 \\ d_2 & b_2 & c_2 \\ d_3 & b_3 & c_3 \end{vmatrix} \qquad \text{Constants replacing } x\text{-coefficients}$$

$$D_y = \begin{vmatrix} a_1 & d_1 & c_1 \\ a_2 & d_2 & c_2 \\ a_3 & d_3 & c_3 \end{vmatrix} \qquad \text{Constants replacing } y\text{-coefficients}$$

$$D_z = \begin{vmatrix} a_1 & b_1 & d_1 \\ a_2 & b_2 & d_2 \\ a_3 & b_3 & d_3 \end{vmatrix} \qquad \text{Constants replacing } z\text{-coefficients}$$

Then $x = \dfrac{D_x}{D}, \qquad y = \dfrac{D_y}{D}, \qquad z = \dfrac{D_z}{D}, \qquad D \neq 0.$
If $D = 0$, the system is either inconsistent or dependent.

■ **Example 13–2 B**

Use Cramer's Rule to find the solution set of the system of linear equations.

1. $2x - 2y + z = 0$
$x + 5y - 7z = 3$
$x - y - 3z = -7$
We first evaluate the denominator D by minors about the first column.

$$D = \begin{vmatrix} 2 & -2 & 1 \\ 1 & 5 & -7 \\ 1 & -1 & -3 \end{vmatrix} = 2\begin{vmatrix} 5 & -7 \\ -1 & -3 \end{vmatrix} - 1\begin{vmatrix} -2 & 1 \\ -1 & -3 \end{vmatrix} + 1\begin{vmatrix} -2 & 1 \\ 5 & -7 \end{vmatrix}$$
$$= 2(-15 - 7) - 1(6 + 1) + 1(14 - 5)$$
$$= -42$$

$$D_x = \begin{vmatrix} 0 & -2 & 1 \\ 3 & 5 & -7 \\ -7 & -1 & -3 \end{vmatrix} = 0 \begin{vmatrix} 5 & -7 \\ -1 & -3 \end{vmatrix} - 3 \begin{vmatrix} -2 & 1 \\ -1 & -3 \end{vmatrix}$$

$$+ (-7) \begin{vmatrix} -2 & 1 \\ 5 & -7 \end{vmatrix}$$

$$= 0(-15 - 7) - 3(6 + 1) - 7(14 - 5)$$
$$= -84$$

$$D_y = \begin{vmatrix} 2 & 0 & 1 \\ 1 & 3 & -7 \\ 1 & -7 & -3 \end{vmatrix} = 2 \begin{vmatrix} 3 & -7 \\ -7 & -3 \end{vmatrix} - 1 \begin{vmatrix} 0 & 1 \\ -7 & -3 \end{vmatrix} + 1 \begin{vmatrix} 0 & 1 \\ 3 & -7 \end{vmatrix}$$

$$= 2(-9 - 49) - 1(0 + 7) + 1(0 - 3)$$
$$= -126$$

$$D_z = \begin{vmatrix} 2 & -2 & 0 \\ 1 & 5 & 3 \\ 1 & -1 & -7 \end{vmatrix} = 2 \begin{vmatrix} 5 & 3 \\ -1 & -7 \end{vmatrix} - 1 \begin{vmatrix} -2 & 0 \\ -1 & -7 \end{vmatrix} + 1 \begin{vmatrix} -2 & 0 \\ 5 & 3 \end{vmatrix}$$

$$= 2(-35 + 3) - 1(14 - 0) + 1(-6 - 0)$$
$$= -84$$

$$x = \frac{D_x}{D} \qquad\qquad y = \frac{D_y}{D} \qquad\qquad z = \frac{D_z}{D}$$

$$= \frac{-84}{-42} = 2, \qquad = \frac{-126}{-42} = 3, \qquad = \frac{-84}{-42} = 2$$

The solution set is $\{(2,3,2)\}$. *Be sure to check the solution.*

2. $x - y + z = -9$
$\quad 3x + 4y = 6$
$\quad\quad 2y - z = 10$

When setting up D, D_x, D_y, and D_z, we insert zeros wherever a variable is missing from the equation. We expand about the first column.

$$D = \begin{vmatrix} 1 & -1 & 1 \\ 3 & 4 & 0 \\ 0 & 2 & -1 \end{vmatrix} = 1 \begin{vmatrix} 4 & 0 \\ 2 & -1 \end{vmatrix} - 3 \begin{vmatrix} -1 & 1 \\ 2 & -1 \end{vmatrix} + 0 \begin{vmatrix} -1 & 1 \\ 4 & 0 \end{vmatrix}$$

$$= 1(-4 - 0) - 3(1 - 2) + 0(0 - 4)$$
$$= -4 + 3 + 0 = -1$$

$$D_x = \begin{vmatrix} -9 & -1 & 1 \\ 6 & 4 & 0 \\ 10 & 2 & -1 \end{vmatrix} = -9 \begin{vmatrix} 4 & 0 \\ 2 & -1 \end{vmatrix} - 6 \begin{vmatrix} -1 & 1 \\ 2 & -1 \end{vmatrix} + 10 \begin{vmatrix} -1 & 1 \\ 4 & 0 \end{vmatrix}$$

$$= -9(-4 - 0) - 6(1 - 2) + 10(0 - 4)$$
$$= 36 + 6 - 40 = 2$$

$$D_y = \begin{vmatrix} 1 & -9 & 1 \\ 3 & 6 & 0 \\ 0 & 10 & -1 \end{vmatrix} = 1 \begin{vmatrix} 6 & 0 \\ 10 & -1 \end{vmatrix} - 3 \begin{vmatrix} -9 & 1 \\ 10 & -1 \end{vmatrix} + 0 \begin{vmatrix} -9 & 1 \\ 6 & 0 \end{vmatrix}$$

$$= 1(-6 - 0) - 3(9 - 10) + 0(0 - 6)$$
$$= -6 + 3 + 0 = -3$$

$$D_z = \begin{vmatrix} 1 & -1 & -9 \\ 3 & 4 & 6 \\ 0 & 2 & 10 \end{vmatrix} = 1\begin{vmatrix} 4 & 6 \\ 2 & 10 \end{vmatrix} - 3\begin{vmatrix} -1 & -9 \\ 2 & 10 \end{vmatrix} + 0\begin{vmatrix} -1 & -9 \\ 4 & 6 \end{vmatrix}$$

$$= 1(40 - 12) - 3(-10 + 18) + 0(-6 + 36)$$

$$= 28 - 24 + 0 = 4$$

$$x = \frac{D_x}{D} = \frac{2}{-1} = -2, \qquad y = \frac{D_y}{D} = \frac{-3}{-1} = 3, \qquad z = \frac{D_z}{D} = \frac{4}{-1} = -4$$

The solution set is $\{(-2, 3, -4)\}$.

▶ *Quick check*   Use Cramer's Rule to find the solution set of the system of linear equations. $x - 2y + z = 0$
$2x - y + 3z = 4$
$x - y - 2z = -2$ ■

```
┌─ Mastery points ─────────────────────────────────────────────────┐

  Can you
    ■ Use Cramer's Rule to find the solution set of systems of linear
      equations?

└──────────────────────────────────────────────────────────────────┘
```

Exercise 13-2

Using Cramer's Rule, find the solution set of the following systems of linear equations. If the system is inconsistent or dependent, so state. See example 13–2 A.

Example $3x - 4y = 1$
$x + 2y = -4$

Solution   Using determinants,

$$D = \begin{vmatrix} 3 & -4 \\ 1 & 2 \end{vmatrix} = 3(2) - (1)(-4) = 6 + 4 = 10$$

Replace $\begin{matrix} 3 \\ 1 \end{matrix}$ with the constants $\begin{matrix} 1 \\ -4 \end{matrix}$ to find D_x.

$$D_x = \begin{vmatrix} 1 & -4 \\ -4 & 2 \end{vmatrix} = 1(2) - (-4)(-4) = 2 - 16 = -14$$

Replace $\begin{matrix} -4 \\ 2 \end{matrix}$ with constants $\begin{matrix} 1 \\ -4 \end{matrix}$ to find D_y.

$$D_y = \begin{vmatrix} 3 & 1 \\ 1 & -4 \end{vmatrix} = 3(-4) - 1(1) = -12 - 1 = -13$$

By Cramer's Rule,

$$x = \frac{D_x}{D} = \frac{-14}{10} = -\frac{7}{5}$$

$$y = \frac{D_y}{D} = \frac{-13}{10} = -\frac{13}{10}$$

The solution set is $\left\{ \left(-\frac{7}{5}, -\frac{13}{10} \right) \right\}$.

1. $x - y = 2$
    $2x + y = 3$

2. $3x + y = 8$
    $x - 2y = -1$

3. $3x + 5y = 6$
    $2x - 3y = -4$

4. $4x - y = 3$
    $8x - 2y = 1$

5. $10x - 2y = -3$
    $5x - y = 0$

6. $x + 2y = 7$
    $3x + 6y = 21$

7. $-4x - 6y = 2$
    $2x + 3y = -1$

8. $6x - 2y = 7$
    $x - 6y = 11$

9. $-3x - y = 1$
    $4x + 5y = -5$

10. $2x - 3y = 1$
    $-3x + 5y = 2$

11. $6x - 2y = 7$
    $2y = 8$

12. $x - 7y = -3$
    $-y = 8$

13. $4x - 5y = -33$
    $-x + 3y = 17$

14. $3x + 15y = -1$
    $x + 5y = 6$

15. $\dfrac{1}{3}x - \dfrac{3}{2}y = 6$

   $-\dfrac{2}{3}x + \dfrac{1}{2}y = -7$

16. $-\dfrac{1}{4}x + 3y = -8$

   $x + \dfrac{2}{3}y = -6$

See example 13–2 B.

Example $\quad x - 2y + z = 0$
      $2x - y + 3z = 4$
      $x - y - 2z = -2$

Solution Using minors about column 1,

$$D = \begin{vmatrix} 1 & -2 & 1 \\ 2 & -1 & 3 \\ 1 & -1 & -2 \end{vmatrix} = 1\begin{vmatrix} -1 & 3 \\ -1 & -2 \end{vmatrix} - 2\begin{vmatrix} -2 & 1 \\ -1 & -2 \end{vmatrix} + 1\begin{vmatrix} -2 & 1 \\ -1 & 3 \end{vmatrix}$$

$$= 1(2 + 3) - 2(4 + 1) + 1(-6 + 1)$$
$$= -10$$

$$D_x = \begin{vmatrix} 0 & -2 & 1 \\ 4 & -1 & 3 \\ -2 & -1 & -2 \end{vmatrix} = 0\begin{vmatrix} -1 & 3 \\ -1 & -2 \end{vmatrix} - 4\begin{vmatrix} -2 & 1 \\ -1 & -2 \end{vmatrix} + (-2)\begin{vmatrix} -2 & 1 \\ -1 & 3 \end{vmatrix}$$

$$= 0 - 4(4 + 1) - 2(-6 + 1)$$
$$= -10$$

$$D_y = \begin{vmatrix} 1 & 0 & 1 \\ 2 & 4 & 3 \\ 1 & -2 & -2 \end{vmatrix} = 1\begin{vmatrix} 4 & 3 \\ -2 & -2 \end{vmatrix} - 2\begin{vmatrix} 0 & 1 \\ -2 & -2 \end{vmatrix} + 1\begin{vmatrix} 0 & 1 \\ 4 & 3 \end{vmatrix}$$

$$= 1(-8 + 6) - 2(0 + 2) + 1(0 - 4)$$
$$= -10$$

$$D_z = \begin{vmatrix} 1 & -2 & 0 \\ 2 & -1 & 4 \\ 1 & -1 & -2 \end{vmatrix} = 1\begin{vmatrix} -1 & 4 \\ -1 & -2 \end{vmatrix} - 2\begin{vmatrix} -2 & 0 \\ -1 & -2 \end{vmatrix} + 1\begin{vmatrix} -2 & 0 \\ -1 & 4 \end{vmatrix}$$

$$= 1(2 + 4) - 2(4 - 0) + 1(-8 - 0)$$
$$= -10$$

$$x = \frac{D_x}{D} = \frac{-10}{-10} = 1, \qquad y = \frac{D_y}{D} = \frac{-10}{-10} = 1, \qquad z = \frac{D_z}{D} = \frac{-10}{-10} = 1$$

The solution set is $\{(1,1,1)\}$.

17. $4x - y + 2z = 0$
$2x + y + z = 3$
$3x - y + z = -2$

18. $x - y - z = 6$
$2x + 3y - z = 1$
$x + 2y + 2z = 5$

19. $x - 2y - z = 0$
$2x - y + z = 0$
$4x + 2y - 2z = 0$

20. $2x + y - z = 0$
$x - 3y + 3z = 0$
$-3x + 2y - 2z = 0$

21. $3x - y - 6z = 5$
$-6x + 2y - 2z = -4$
$3x - y + z = 2$

22. $6x - 9y + 12z = 24$
$-4x + 6y - 8z = -16$
$2x - 3y + 4z = 8$

23. $2x - y + 3z = 5$
$3x - y + 2z = 10$
$3x - 2y + 7z = 3$

24. $4x + 2y - 3z = 2$
$6x + y - 2z = 0$
$2x + 3y - z = 0$

25. $3x + 2y + 4z = 3$
$3x - y - 2z = 0$
$3x + y + 2z = 2$

26. $2x - 7y + 3z = 1$
$4x - y - 6z = -6$
$-2x + 5y - 3z = -1$

27. $3x - 2y + 4z = -6$
$2x - 2y + 4z = -1$
$x - y + 2z = 4$

28. $2x + 3y + 4z = 13$
$-x - y + 5z = 4$
$x + y - 7z = -6$

29. $-3x - y + 4z = 1$
$5x - y + 2z = 10$
$7x + 2y - 2z = -4$

30. $x + y = 1$
$y + 2z = -2$
$2x - z = 0$

31. $2y + z = 6$
$3x + 4z = 14$
$3x - y = 4$

32. $x - 4y + z = -4$
$4x + 2y - 3z = 6$
$-x + 2z = 2$

33. $x - 2y + z = -2$
$3x + y = 7$
$2x - z = 0$

34. $y + 4z = 6$
$4x + z = 0$
$5y - z = 9$

35. $3x + 4y - 2z = -25$
$2x - y + 3z = -5$
$-x + z = 6$

36. $x - y + z = -9$
$3x + 4y = 6$
$2y - z = 10$

In exercises 37–45, select two variables to represent two unknowns, set up a system of two linear equations in two variables and solve the resulting system using Cramer's Rule.

Example   The perimeter of a rectangle is 72 inches. If twice the length is added to three times the width, the sum is 88 inches. What are the dimensions of the rectangle?

Solution   Let w = the width and ℓ = the length of the rectangle. Using $2\ell + 2w = p$ (perimeter), we obtain the equation

$2\ell + 2w = 72$ Replace p with 72

From the statement of the problem,

twice the length added to three times the width is 88 inches
  2ℓ $+$ $3w$ $= 88$

We must solve the system of linear equations $2\ell + 2w = 72$
$2\ell + 3w = 88.$

Using Cramer's Rule,

$$D = \begin{vmatrix} 2 & 2 \\ 2 & 3 \end{vmatrix} = 6 - 4 = 2$$

$$D_\ell = \begin{vmatrix} 72 & 2 \\ 88 & 3 \end{vmatrix} = 216 - 176 = 40$$

$$D_w = \begin{vmatrix} 2 & 72 \\ 2 & 88 \end{vmatrix} = 176 - 144 = 32$$

$$\ell = \frac{D_\ell}{D} = \frac{40}{2} = 20 \text{ and } w = \frac{D_w}{D} = \frac{32}{2} = 16$$

The rectangle is 20 inches long and 16 inches wide.

37. The length of a rectangular garden is two feet longer than twice the width. Find the dimensions of the garden if the perimeter is 46 feet.

38. The sum of two numbers is 102. If their difference is 48, find the two numbers.

39. A donut shop sells 5 cream-filled and 7 jelly-filled donuts for $3.64 and 3 cream-filled and 9 jelly-filled donuts for $3.72. Find the cost of a single cream-filled and a single jelly-filled donut.

40. A bank teller receives 145 bills, consisting of $5 and $10 bills. If the bills total $1,190 in value, how many of each bill did she receive?

41. The sum of two numbers is 34 and the difference of the two numbers is 8. Find the numbers.

42. The sum of two numbers is 19. The greater number is 1 more than twice the lesser number. Find the numbers.

43. How many ounces of an 8% solution must be mixed with a 15% solution to obtain 100 ounces of a 12.2% solution?

44. How many pounds of nuts costing $2 per pound must be mixed with nuts costing $4 per pound to obtain 60 pounds of nuts worth $3 per pound?

45. A party store sold 3 cartons of pop and 4 pounds of candy for $12. The following week 3 cartons of pop and 2 pounds of candy cost $9. What was the price per carton of pop and per pound of candy?

In exercises 46–53, select three variables to represent three unknowns, set up a system of three linear equations in three variables and solve the resulting system using Cramer's Rule.

Example   The sum of three numbers is 25. The third number is 1 more than twice the first number and the second number is 4 more than the first number. Find the numbers.

Solution   Let x = the first number, y = the second number, and z = the third number. Then

1. from

the sum of three numbers   is   25

$$x + y + z = 25$$

2. from

the third number   is   one more than   twice the first

$$z = 1 + 2x$$

3. from

the second number   is   four more than   the first

$$y = 4 + x$$

We then have the system of linear equations

$$x + y + z = 25 \qquad\qquad x + y + z = 25$$
$$z = 1 + 2x \quad \text{which becomes} \quad -2x + z = 1$$
$$y = x + 4 \qquad\qquad -x + y = 4$$

Using Cramer's Rule and expanding about the first column,

$$D = \begin{vmatrix} 1 & 1 & 1 \\ -2 & 0 & 1 \\ -1 & 1 & 0 \end{vmatrix} = 1\begin{vmatrix} 0 & 1 \\ 1 & 0 \end{vmatrix} - (-2)\begin{vmatrix} 1 & 1 \\ 1 & 0 \end{vmatrix} + (-1)\begin{vmatrix} 1 & 1 \\ 0 & 1 \end{vmatrix}$$
$$= 1(0 - 1) + 2(0 - 1) - 1(1 - 0) = -1 - 2 - 1 = -4$$

$$D_x = \begin{vmatrix} 25 & 1 & 1 \\ 1 & 0 & 1 \\ 4 & 1 & 0 \end{vmatrix} = 25\begin{vmatrix} 0 & 1 \\ 1 & 0 \end{vmatrix} - 1\begin{vmatrix} 1 & 1 \\ 1 & 0 \end{vmatrix} + 4\begin{vmatrix} 1 & 1 \\ 0 & 1 \end{vmatrix}$$
$$= 25(-1) - 1(-1) + 4(1) = -25 + 1 + 4 = -20$$

$$D_y = \begin{vmatrix} 1 & 25 & 1 \\ -2 & 1 & 1 \\ -1 & 4 & 0 \end{vmatrix} = 1 \begin{vmatrix} 1 & 1 \\ 4 & 0 \end{vmatrix} - (-2) \begin{vmatrix} 25 & 1 \\ 4 & 0 \end{vmatrix} + (-1) \begin{vmatrix} 25 & 1 \\ 1 & 1 \end{vmatrix}$$

$$= 1(-4) + 2(-4) - 1(24) = -4 - 8 - 24 = -36$$

$$D_z = \begin{vmatrix} 1 & 1 & 25 \\ -2 & 0 & 1 \\ -1 & 1 & 4 \end{vmatrix} = 1 \begin{vmatrix} 0 & 1 \\ 1 & 4 \end{vmatrix} - (-2) \begin{vmatrix} 1 & 25 \\ 1 & 4 \end{vmatrix} + (-1) \begin{vmatrix} 1 & 25 \\ 0 & 1 \end{vmatrix}$$

$$= 1(-1) + 2(-21) - 1(1) = -1 - 42 - 1 = -44$$

$$x = \frac{D_x}{D} = \frac{-20}{-4} = 5, \qquad y = \frac{D_y}{D} = \frac{-36}{-4} = 9, \qquad z = \frac{D_z}{D} = \frac{-44}{-4} = 11$$

The three numbers are 5, 9, and 11.

46. A collection of 18 nickels, dimes, and quarters is worth $2.65. There are as many dimes as there are quarters. How many of each coin are there?

47. The sum of three angles of a triangle is 180°. If angle A is equal to angle B and twice angle C is 5° less than angle A, find the measures of the three angles.

48. The sum of three numbers is 47. If the first number is doubled, the sum is 56. If the second number is tripled, the sum is 81. Find the three numbers.

49. Tickets for a city band concert cost $3, $4, and $5. The total receipts from sales of 325 tickets is $1,302. If there were two more $5 tickets sold than there were $3 tickets sold, how many of each ticket were sold?

50. A bank teller has $565 in a stack of $5, $10, and $20 bills. The number of $10 bills is twice the number of $5 and $20 bills put together. There are six more $20 bills than $5 bills. How many of each bill are there if there are 48 bills in all? How much money is in the stack?

51. The sum of three numbers is 36. If the second number is 1 more than the first number and the last number is $1\frac{1}{2}$ times the first, find the three numbers.

52. Sarah has a collection of 23 stuffed animals—elephants, bears, and dogs. She has four times as many bears as she has elephants and the number of dogs is two more than twice the number of elephants. How many of each animal does she have?

53. The perimeter of a triangular plot is 74 feet. The longest side is 2 feet shorter than the sum of the other two sides and the shortest side is 2 feet longer than one-third the length of the middle-length side. Find the length of the three sides.

Review exercises

1. Given the linear equation $3x + y = 6$, complete the given ordered pairs and graph the equation. See section 7-1.
$(-2, \ )$; $(-1, \ )$; $(0, \ )$; $(1, \ )$; $(2, \ )$

2. Find the solution set of the quadratic inequality $2x^2 + 5x - 3 > 0$. See section 6-7.

3. Find the solution set of the quadratic equation $3x^2 - 16x = -5$. See section 6-3.

4. Find the solution set of the equation $\frac{3}{x} - x = 2$. See section 6-1.

Perform the indicated operations. See sections 5-6 and 5-7.

5. $(2\sqrt{3} - 5)^2$

6. $(4\sqrt{2} - 3)(4\sqrt{2} + 3)$

7. i^{13}

13–3 ■ *Solving linear systems of equations by the augmented matrix method*

We have learned several methods for solving linear systems of equations. Now we will consider a method for solving these systems using matrices which were mentioned in section 13–1. This method is easily adapted to a computer.

With every system of linear equations, we can associate a matrix that consists of the coefficients and constant terms. To illustrate, consider the system

$$6x + 5y = -3$$
$$2x + y = -7$$

with which we can associate the matrix

$$\begin{bmatrix} 6 & 5 & | & -3 \\ 2 & 1 & | & -7 \end{bmatrix}$$

We call this the **augmented matrix** of the system. The coefficients of the system are placed to the left of the vertical bar and the constants to the right. We now operate on the rows of the augmented matrix just as we did with the equations of the system. Using the following **elementary row operations,** we produce new matrices that lead to systems containing solutions of the original system.

Elementary row operations

1. Any two rows of the augmented matrix can be interchanged.
2. Any row can be multiplied by a nonzero constant.
3. Any row of the augmented matrix can be changed by adding a nonzero multiple of another row to that row.

Note We used these same operations when solving a system of linear equations by the elimination (addition) method.

Use row operations to rewrite the matrix until it is a system whose solution is easy to find. The object is to obtain the first column $\begin{matrix} 1 \\ 0 \end{matrix}$ and the second column constant $\begin{matrix} \\ 1 \end{matrix}$ that now represent the coefficients of the variables x and y. We can then easily solve the resulting system.

■ *Example 13–3 A*

Find the solution set of the following system using the augmented matrix and elementary row operations.
$$6x + 5y = -3$$
$$2x + y = -7$$
The augmented matrix of the system is
$$\begin{bmatrix} 6 & 5 & | & -3 \\ 2 & 1 & | & -7 \end{bmatrix}$$

To start with, we want to make sure that there is a 1 in the first row, first column position. We multiply the first row by $\dfrac{1}{6}$ to obtain the matrix

$$\begin{bmatrix} 1 & \dfrac{5}{6} & \Big| & -\dfrac{1}{2} \\ 2 & 1 & \Big| & -7 \end{bmatrix}$$

Next, we must get 0s in every position below the first position. To get 0 in row two, column one, use the third elementary row operation: add to the numbers in row two the results of multiplying the numbers in row one by -2. We obtain

$$\begin{bmatrix} 1 & \dfrac{5}{6} & \Big| & -\dfrac{1}{2} \\ 2 + 1(-2) & 1 + \dfrac{5}{6}(-2) & \Big| & -7 + \left(-\dfrac{1}{2}\right)(-2) \end{bmatrix}$$

Original $+\ (-2)$ times
number   number from row one

$$\begin{bmatrix} 1 & \dfrac{5}{6} & \Big| & -\dfrac{1}{2} \\ 0 & -\dfrac{2}{3} & \Big| & -6 \end{bmatrix}$$

We now have 1 in the first row, first column position and 0s in every position below that. Now we go to the second column and obtain 1 in the second row, second column position. To get this, use the second row operation and multiply row two by $-\dfrac{3}{2}$.

$$\begin{bmatrix} 1 & \dfrac{5}{6} & \Big| & -\dfrac{1}{2} \\ 0 & 1 & \Big| & 9 \end{bmatrix}$$

This augmented matrix yields the system

$$x + \dfrac{5}{6}y = -\dfrac{1}{2}$$
$$0x + 1y = 9 \text{ (or } y = 9)$$

Thus $y = 9$ and substituting 9 for y in the equation $x + \dfrac{5}{6}y = -\dfrac{1}{2}$, we obtain

$$x + \dfrac{5}{6}(9) = -\dfrac{1}{2} \qquad \text{Replace } y \text{ with } 9$$
$$x + \dfrac{15}{2} = -\dfrac{1}{2}$$
$$x = -\dfrac{1}{2} - \dfrac{15}{2}$$
$$= -\dfrac{16}{2}$$
$$= -8$$

The solution set of the system is $\{(-8,9)\}$.

To solve systems with three linear equations, use elementary row operations to get 1s down the diagonal from left to right and 0s below each 1. The following example demonstrates this method.

■ *Example 13–3 B*

Find the solution set of the following system using the augmented matrix and elementary row operations.

$$2x - y + 2z = -8$$
$$x + 2y - 3z = 9$$
$$3x - y - 4z = 3$$

Write the augmented matrix of the system.

$$\begin{bmatrix} 2 & -1 & 2 & | & -8 \\ 1 & 2 & -3 & | & 9 \\ 3 & -1 & -4 & | & 3 \end{bmatrix}$$

To obtain 1 in row one, column one, we interchange rows one and two.

$$\begin{bmatrix} 1 & 2 & -3 & | & 9 \\ 2 & -1 & 2 & | & -8 \\ 3 & -1 & -4 & | & 3 \end{bmatrix}$$

We must now get 0s in column one below the first row. Add to row two the results of multiplying row one by -2.

$$\begin{bmatrix} 1 & 2 & -3 & | & 9 \\ 0 & -5 & 8 & | & -26 \\ 3 & -1 & -4 & | & 3 \end{bmatrix}$$

Add to row three the results of multiplying row one by -3.

$$\begin{bmatrix} 1 & 2 & -3 & | & 9 \\ 0 & -5 & 8 & | & -26 \\ 0 & -7 & 5 & | & -24 \end{bmatrix}$$

We must now obtain a 0 in row three, column two. Add to row three the results of multiplying row two by $-\dfrac{7}{5}$.

$$\begin{bmatrix} 1 & 2 & -3 & | & 9 \\ 0 & -5 & 8 & | & -26 \\ 0 & 0 & -\dfrac{31}{5} & | & \dfrac{62}{5} \end{bmatrix}$$

Multiply row three by $-\dfrac{5}{31}$ (the reciprocal of $-\dfrac{31}{5}$) to get 1 in the row three, column three position.

$$\begin{bmatrix} 1 & 2 & -3 & | & 9 \\ 0 & -5 & 8 & | & -26 \\ 0 & 0 & 1 & | & -2 \end{bmatrix}$$

Multiply each member of row two by $-\dfrac{1}{5}$.

$$\begin{bmatrix} 1 & 2 & -3 & \bigm| & 9 \\ 0 & 1 & -\dfrac{8}{5} & \bigm| & \dfrac{26}{5} \\ 0 & 0 & 1 & \bigm| & -2 \end{bmatrix}$$

We now have the system

$$x + 2y - 3z = 9$$
$$y - \frac{8}{5}z = \frac{26}{5}$$
$$z = -2$$

Replace z by -2 in $y - \dfrac{8}{5}z = \dfrac{26}{5}$ and solve for y.

$$y - \frac{8}{5}(-2) = \frac{26}{5}$$
$$y + \frac{16}{5} = \frac{26}{5}$$
$$y = \frac{26}{5} - \frac{16}{5} = \frac{10}{5} = 2$$

Replace y by 2 and z by -2 in the equation $x + 2y - 3z = 9$.

$$x + 2(2) - 3(-2) = 9$$
$$x + 4 + 6 = 9$$
$$x + 10 = 9$$
$$x = -1$$

The solution set of the system is $\{(-1, 2, -2)\}$.

When systems of equations are inconsistent (no solutions) or dependent (infinitely many solutions), the augmented matrix yields (1) a false statement or (2) a statement that is always true.

■ Example 13–3 C

Find the solution set of the following systems of linear equations using an augmented matrix and elementary row operations.

1. $3x - 2y = 6$
$6x - 4y = 1$

$$\begin{bmatrix} 3 & -2 & \bigm| & 6 \\ 6 & -4 & \bigm| & 1 \end{bmatrix} = \begin{bmatrix} 1 & -\dfrac{2}{3} & \bigm| & 2 \\ 6 & -4 & \bigm| & 1 \end{bmatrix} \qquad \text{Multiply row 1 by } \frac{1}{3}$$

$$= \begin{bmatrix} 1 & -\dfrac{2}{3} & \bigm| & 2 \\ 0 & 0 & \bigm| & -11 \end{bmatrix} \qquad \text{Multiply row 1 by } -6 \text{ and add to row 2}$$

The second row yields $0 = -11$ (false) and the system is inconsistent. The solution set is $\varnothing$.

2. $4x + 6y = -2$
$2x + 3y = -1$

$$\begin{bmatrix} 4 & 6 & | & -2 \\ 2 & 3 & | & -1 \end{bmatrix} = \begin{bmatrix} 1 & \dfrac{3}{2} & | & -\dfrac{1}{2} \\ 2 & 3 & | & -1 \end{bmatrix}$$ Multiply row 1 by $\dfrac{1}{4}$

$$= \begin{bmatrix} 1 & \dfrac{3}{2} & | & -\dfrac{1}{2} \\ 0 & 0 & | & 0 \end{bmatrix}$$ Multiply row 1 by -2 and add to row 2

The second row yields $0 = 0$ (true) and the system is dependent. The solution set is $\{(x,y) | 2x + 3y = -1\}$.

▶ *Quick check*   Find the solution set of the system of equations using an augmented matrix and elementary row operations.
$4x - 5y = -33$
$-x + 3y = 17$ ■

┌─ **Mastery points** ───┐

Can you

■ Solve a system of two linear equations in two variables using the augmented matrix?

■ Solve a system of three linear equations in three variables using the augmented matrix?

└───┘

Exercise 13–3

Solve the following systems of equations using the augmented matrix. See examples 13–3 A and B.

Example $4x - 5y = -33$
$-x + 3y = 17$

Solution   The augmented matrix is

$$\begin{bmatrix} 4 & -5 & | & -33 \\ -1 & 3 & | & 17 \end{bmatrix}$$

Multiply row 2 by 3.

$$\begin{bmatrix} 4 & -5 & | & -33 \\ -3 & 9 & | & 51 \end{bmatrix}$$

Add numbers in row 2 to the numbers in row 1.

$$\begin{bmatrix} 1 & 4 & | & 78 \\ -3 & 9 & | & 51 \end{bmatrix}$$

Multiply row 1 by 3 and add to row 2.

$$\begin{bmatrix} 1 & 4 & | & 18 \\ 0 & 21 & | & 105 \end{bmatrix}$$

Multiply row 2 by $\frac{1}{21}$.

$$\begin{bmatrix} 1 & 4 & | & 18 \\ 0 & 1 & | & 5 \end{bmatrix}$$

The augmented matrix yields the system

$$x + 4y = 18$$
$$y = 5$$

Replace y with 5 in

$$x + 4y = 18$$
$$x + 4(5) = 18$$
$$x + 20 = 18$$
$$x = -2$$

The solution set is $\{(-2,5)\}$.

1. $x + 3y = 11$
   $2x - y = 1$

2. $x - 5y = 11$
   $2x + 3y = -4$

3. $x - 4y = -6$
   $3x + y = -5$

4. $x + 6y = -14$
   $5x - 3y = -4$

5. $2x + y = 5$
   $3x - 5y = 14$

6. $3x - 2y = 16$
   $4x + 2y = 12$

7. $5x - y = 0$
   $2x + 3y = -1$

8. $-3x + 2y = 1$
   $x - y = 4$

9. $-x - 2y = 4$
   $x + 6y = -2$

10. $-x - y = 4$
   $2x + 2y = -1$

11. $4x - 2y = 1$
   $-8x + 4y = -2$

12. $x + 3y = 1$
   $3x + 9y = 3$

13. $-3x - y = 1$
   $4x + 5y = -5$

14. $x + 3y - z = 5$
   $3x - y + 2z = 5$
   $x + y + 2z = 7$

15. $x - 2y + 3z = -11$
   $2x + 3y - z = 6$
   $3x - y - z = 3$

16. $2x - y + z = 8$
   $x - 2y - 3z = 4$
   $3x + 3y - z = -4$

17. $x - 2y - 2z = 4$
   $2x + y - 3z = 7$
   $x - y - z = 3$

18. $2x - y - z = -4$
   $x + 3y - 4z = 12$
   $x + y + z = -5$

19. $2x + 3y + z = 11$
   $3x - y - z = 11$
   $x - 2y - 5z = 2$

20. $2x - y = -1$
   $2y - z = 6$
   $x + z = 1$

21. $x - y = 1$
   $2x - z = 0$
   $2y - z = -2$

Translate each problem into a system of linear equations and solve the system using the augmented matrix.

22. A plane travels 525 miles with the wind in $1\frac{1}{2}$ hours. The return trip against the wind takes $2\frac{1}{10}$ hours. Find the speed of the plane and the speed of the wind.

23. Find the values of a and b so that the points $(-2,1)$ and $(1,3)$ lie on the graph of $ax + by = 4$.

24. A total of $5,000 is invested by Erin, part at 8% and part at 12%. How much did she invest at each rate if the return from both investments is the same?

25. Salesman Jim Connor must sell 24 new cars to meet his sales quota. He must sell an equal number of intermediate-sized and large-sized cars. If he must sell 3 more small-sized cars than intermediate-sized cars, how many of each size must he sell?

Review exercises

Find the x- and y-intercepts of the following equations. See section 7–3.

1. $y = 2x + 8$ **2.** $3x - 2y = -9$ **3.** $y = 6$

Complete the square of the following to obtain a binomial square. See section 6–2.

4. $x^2 + 8x$ **5.** $y^2 - 5y$ **6.** $z^2 - \dfrac{1}{2}z$

Factor the following expressions. See section 3–7.

7. $y^2 - 14y + 49$ **8.** $x^2 + 10x + 25$

Chapter 13 lead-in problem

As a tenant, the manager of an apartment complex pays two-thirds of the rent that each of the remaining 9 tenants pay. Each month the landlord collects a total of $6,090. How much does each tenant pay?

Solution

Let $x =$ the amount each of the 9 tenants pay and $y =$ the amount the manager pays. Since the manager pays two-thirds of what each of the tenants pay,

$$y = \frac{2}{3}x$$

9 tenants pay	+	manager pays	=	total income
$9x$	+	y	=	$6,090$

We solve the system of linear equations $y = \dfrac{2}{3}x$

$$9x + y = 6,090.$$

Writing the first equation in standard form, we get the system

$$2x - 3y = 0$$
$$9x + y = 6,090$$

Using determinants,

$$D = \begin{vmatrix} 2 & -3 \\ 9 & 1 \end{vmatrix}$$
$$= 2 - (-27) = 29$$

$$D_x = \begin{vmatrix} 0 & -3 \\ 6,090 & 1 \end{vmatrix}$$
$$= 0 - (-18,270) = 18,270$$

$$D_y = \begin{vmatrix} 2 & 0 \\ 9 & 6,090 \end{vmatrix}$$
$$= 12,180 - 0 = 12,180$$

$$x = \frac{D_x}{D} = \frac{18,270}{29} = 630,$$

$$y = \frac{D_y}{D} = \frac{12,180}{29} = 420$$

Each tenant pays $630 per month and the manager pays $420 per month.

Chapter 13 summary

1. A **matrix** is an ordered array of rows and columns of numbers.
2. A **determinant** is a real number value of its square matrix.
3. The 2×2 determinant $\begin{vmatrix} a_1 & b_1 \\ a_2 & b_2 \end{vmatrix} = a_1b_2 - a_2b_1$
4. By definition, the determinant
$$\begin{vmatrix} a_1 & b_1 & c_1 \\ a_2 & b_2 & c_2 \\ a_3 & b_3 & c_3 \end{vmatrix} = a_1 \begin{vmatrix} b_2 & c_2 \\ b_3 & c_3 \end{vmatrix} - a_2 \begin{vmatrix} b_1 & c_1 \\ b_3 & c_3 \end{vmatrix} + a_3 \begin{vmatrix} b_1 & c_1 \\ b_2 & c_2 \end{vmatrix}$$
5. By Cramer's Rule, given the system of linear equations
$$a_1x + b_1y = c_1$$
$$a_2x + b_2y = c_2$$
$$x = \frac{D_x}{D} \text{ and } y = \frac{D_y}{D}$$
where $D = \begin{vmatrix} a_1 & b_1 \\ a_2 & b_2 \end{vmatrix} (D \neq 0)$, $D_x = \begin{vmatrix} c_1 & b_1 \\ c_2 & b_2 \end{vmatrix}$, and
$$D_y = \begin{vmatrix} a_1 & c_1 \\ a_2 & c_2 \end{vmatrix}$$
The system is inconsistent or dependent when $D = 0$.

6. The augmented matrix of the system of linear equations $a_1x + b_1y = c_1$ is given by $a_2x + b_2y = c_2$
$$\begin{bmatrix} a_1 & b_1 & c_1 \\ a_2 & b_2 & c_2 \end{bmatrix}$$
7. The **elementary row operations** are:
   a. Any two rows of the augmented matrix can be interchanged.
   b. Any row can be multiplied by a nonzero constant.
   c. Any row of the augmented matrix can be changed by adding a nonzero multiple of another row to that row.

Chapter 13 error analysis

1. Evaluating a determinant
Example: $\begin{vmatrix} 4 & -3 \\ 5 & 2 \end{vmatrix} = 8 - 15 = -7$
Correct answer: 23
What error was made? (*see page 566*)
2. Evaluating a 3×3 determinant
Example:
$$\begin{vmatrix} 1 & 0 & -2 \\ -2 & 1 & -1 \\ 3 & 4 & 5 \end{vmatrix}$$
$$= 1 \begin{vmatrix} 1 & -1 \\ 4 & 5 \end{vmatrix} - 2 \begin{vmatrix} 0 & -2 \\ 4 & 5 \end{vmatrix} + 3 \begin{vmatrix} 0 & -2 \\ 1 & -1 \end{vmatrix}$$
$$= 1(5 + 4) - 2(0 + 8) + 3(0 + 2)$$
$$= -1$$
Correct answer: 31
What error was made? (*see page 567*)

3. Solving a system of linear equations using Cramer's Rule
Example: Given the system of linear equations
$$x - 2y = 6$$
$$3x + y = 2$$
Then $D = \begin{vmatrix} 1 & -2 \\ 3 & 1 \end{vmatrix} = 7$
$$D_x = \begin{vmatrix} 1 & 6 \\ 3 & 2 \end{vmatrix} = -16$$
$$D_y = \begin{vmatrix} 6 & -2 \\ 2 & 1 \end{vmatrix} = 10$$
Correct answer: $D = 7, D_x = 10, D_y = -16$
What error was made? (*see page 573*)

4. Solving a system of linear equations using the augmented matrix

Example: Given the system of linear equations

$$6x - y = 3$$
$$2x + 3y = -4$$

The augmented matrix is

$$\begin{bmatrix} 6 & -1 & 3 \\ 2 & 3 & -4 \end{bmatrix}$$

and using row operations, we obtain

$$\begin{bmatrix} 1 & -\dfrac{1}{6} & \dfrac{1}{2} \\ 0 & 1 & -\dfrac{50}{3} \end{bmatrix}$$

Correct answer:

$$\begin{bmatrix} 1 & -\dfrac{1}{6} & \dfrac{1}{2} \\ 0 & 1 & -\dfrac{3}{2} \end{bmatrix}$$

What error was made? (*see page 582*)

5. Simplifying multiple operations

Example: $2^2 - 12 \div 4 + 6 \cdot 3 = 12$

Correct answer: 19

What error was made? (*see page 27*)

6. Squaring a binomial

Example: $(4 + 2\sqrt{3})^2 = 4^2 + (2\sqrt{3})^2 = 28$

Correct answer: $28 + 16\sqrt{3}$

What error was made? (*see page 106*)

7. Domain of a rational expression

Example: The domain of the rational expression $\dfrac{x}{x^2 - x}$

is $\{x | x \, \epsilon \, R, \, x \neq 1\}$.

Correct answer: $\{x | x \, \epsilon \, R, \, x \neq 0,1\}$

What error was made? (*see page 155*)

8. Complex roots of a polynomial

Example: The roots of a polynomial $p(x)$ can be i and 4.

Correct answer: $i, -i,$ and 4

What error was made? (*see page 440*)

9. Simplifying a complex fraction

Example: $\dfrac{\dfrac{3}{x} - \dfrac{2}{y}}{\dfrac{1}{y} + \dfrac{4}{x}} = \dfrac{2x - 3y}{x + 4y}$

Correct answer: $\dfrac{3y - 2x}{x + 4y}$

What error was made? (*see page 177*)

10. Solution set of an absolute value inequality

Example: The solution set of $|x - 3| \leq 4$ is $x \leq -1$ or $x \geq 7$.

Correct answer: $-1 \leq x \leq 7$

What error was made? (*see page 87*)

Chapter 13 critical thinking

How many diagonals does a convex polygon have?

Chapter 13 review

[13–1]

Evaluate each determinant.

1. $\begin{vmatrix} 1 & 4 \\ -1 & 3 \end{vmatrix}$

2. $\begin{vmatrix} 0 & -6 \\ 4 & 3 \end{vmatrix}$

3. $\begin{vmatrix} 5 & 7 \\ -8 & -3 \end{vmatrix}$

Evaluate each determinant using expansion about any row or column.

4. $\begin{vmatrix} 3 & 4 & 3 \\ -2 & 2 & 0 \\ 1 & -5 & 6 \end{vmatrix}$

5. $\begin{vmatrix} -3 & 3 & 2 \\ -1 & 0 & -4 \\ -2 & 0 & 5 \end{vmatrix}$

6. $\begin{vmatrix} 0 & 1 & -3 \\ -2 & 0 & 5 \\ 6 & 7 & 8 \end{vmatrix}$

[13–2]

Use Cramer's Rule to find the solution set of each system of linear equations.

7. $2x - y = 3$
$\quad x - 3y = 4$

8. $4x + 5y = 0$
$\quad 2x - 4y = -1$

9. $-4x + 2y = 3$
$\quad y = 9$

10. $6x - 3y = -2$
$\quad 3x = 8$

11. $x - 3y + 2z = 0$
$\quad 2x - y + z = 0$
$\quad x + 4y - 3z = 0$

12. $-4x + y - 3z = -2$
$\quad 3x - 2y + z = 4$
$\quad x + 3y - 2z = 1$

[13–3]
Use the augmented matrix and elementary row operations to solve the following systems.

13. $3x - y = 2$
$x + 2y = 0$

14. $x - y + 2z = 3$
$x + 3y - z = 1$
$2x - y + 2z = 0$

Solve the following problems using a system of linear equations. Solve by determinants or by augmented matrix and elementary row operations.

15. Erica's piggy bank contains 44 coins in pennies, nickels, and dimes worth $1.99. There are three times as many dimes as there are nickels. How many of each coin are there in the piggy bank?

16. Mrs. Jones invested a sum of money, part at 8% and part at 7%. Her total yearly return from both investments is $258. If she interchanged her investments, the total yearly income would be $252. How much did Mrs. Jones invest at each rate?

Chapter 13 cumulative test

[1–1] **1.** Given the set $\left\{-9, -\frac{3}{2}, -\sqrt{2}, -\frac{1}{9}, 0, \sqrt{3}, 4, \frac{9}{2}\right\}$
    a. List the integers in the set.
    b. List the rational numbers in the set.
    c. List the irrational numbers in the set.

[1–4] **2.** Perform the indicated operations and simplify.
$3 - [2^3 + 8 \div 2 - 6 \cdot 3]$

[2–2] **3.** Solve the equation $3x - 2 = xy + 3$ for x.

[3–2] **4.** Perform the indicated multiplications.
    a. $(3y - 1)(2y + 7)$
    b. $(6x - 5)^2$
    c. $(3x + 2y)(3x - 2y)$
    d. $(y - 1)(y^2 + y + 1)$

Completely factor each expression.

[3–4] **5.** $6x^2y + 3xy^2 - 12x^3y^3$

[3–6] **6.** $7x^2 - 12x - 4$

[3–7] **7.** $9y^2 - 30xy + 25x^2$

[3–8] **8.** $3x^2 - 75y^2$

[3–8] **9.** $8a^3 + b^3$

[3–8] **10.** $4ab - 2ac - 2b^2 + bc$

Find the solution set of the following equations and inequalities.

[2–1] **11.** $2(4x - 1) - 2x = 3(x + 7)$

[2–5] **12.** $5(y - 1) \leq 2y + 6$

[2–4] **13.** $|3 - 2x| = 5$

[2–6] **14.** $|5 - 4x| \leq 7$

[2–6] **15.** $|3y + 2| > 4$

[6–1] **16.** $5x^2 - 8x = 4$

[4–7] **17.** $\frac{2}{y + 3} - 3 = \frac{6}{y + 3}$

[6–6] **18.** $x^{1/2} - 2x^{1/4} = 3$

Perform the indicated operations.

[4–3] **19.** $\frac{5}{3x - 1} - \frac{7}{2x + 5}$

[4–3] **20.** $\frac{x - 3}{x^2 - x - 30} + \frac{x + 1}{x^2 - 25}$

[4–2] **21.** $\frac{a^3 - b^3}{a^2 - 2a + 1} \div \frac{a^2 + ab + b^2}{a^2 + 4a - 5}$

[4–4] **22.** Simplify the complex rational expression
$\dfrac{\frac{1}{x} - \frac{2}{y}}{\frac{3}{y} - \frac{2}{x}}$.

Perform the indicated operations.

[5-5] 23. $(\sqrt{3} - 2) - (4\sqrt{3} + 1)$

[5-5] 24. $3\sqrt{50} + \sqrt{32} - 5\sqrt{8}$

[5-6] 25. $(3\sqrt{2} + 3)(3\sqrt{2} - 3)$

[5-7] 26. $(3 - 2i)(2 + i)$

[5-7] 27. $\dfrac{3 - i}{4 + 3i}$ (Rationalize the denominator.)

[6-5] 28. Find the solution set of the radical equation $\sqrt{x + 34} = x + 4$. Indicate any extraneous solutions.

[6-7] 29. Find the solution set of the inequality $\dfrac{x - 2}{x + 1} \le 3$. Write the answer in interval notation.

Find an equation of the line having the following conditions. Write the equation in standard form $ax + by = c$ where $a > 0$.

[7-3] 30. Through points $(-2,1)$ and $(4,3)$

[7-3] 31. Passing through $(5,-3)$ and parallel to the line $3x - y = 5$

[7-3] 32. Through $(1,-7)$ and perpendicular to the line $4x + 3y = 1$

[7-3] 33. Graph the line $y - 3x = -1$ using the slope and the y-intercept.

[7-4] 34. Graph the solution set of the inequality $4y - 3x > 12$.

[8-2] 35. Given function f defined by $f(x) = x^2 - 2x + 3$, find
a. $f(2)$
b. $f(-3)$
c. $\dfrac{f(x + h) - f(x)}{h}$, $h \ne 0$

[8-4] 36. Find the inverse of the function g defined by $g(x) = x^3 - 3$.

[8-2] 37. Given functions f and g defined by $f(x) = 2x - 1$ and $g(x) = x^2 + 1$, find
a. $(f + g)(-4)$
b. $(f/g)(x)$
c. $f[g(3)]$

[9-4] 38. Determine if each of the following is a parabola, a circle, an ellipse, or a hyperbola.
a. $x - y^2 + 2y = 1$
b. $3y^2 = 4 - x^2$
c. $x^2 = y^2 + 7$
d. $x^2 + y^2 - 4x - 6y + 1 = 0$

[9-4] 39. Graph the hyperbola $9x^2 - 4y^2 = 36$.

[4-6] 40. Using synthetic division, find all zeros of the polynomial $P(x) = x^3 - 3x^2 - 10x + 24$.

[10-5] 41. Graph the rational function $f(x) = \dfrac{3x - 2}{x + 4}$.

[11-3] 42. Find the solution set of the equation $\log_3(x + 3) - \log_3(2x - 1) = 2$.

[11-6] 43. Find the solution set of the exponential equations.
a. $3^{2x-1} = 9$
b. $2^{x+1} = 5$ (to the nearest tenth)

Find the solution set of the following systems of linear equations.

[12-1] 44. $2x - y = 6$
$x + 3y = -1$

[13-2] 45. $2x - y + z = 4$
$3x - 2y + 2z = 3$ (Using Cramer's Rule)
$x - y + 3z = 2$

14

Sequences, Series and Mathematical Induction

A sky diver falls 10 meters during the first second, 20 meters during the second second, 30 meters during the third second, and so on. How many meters will the diver fall during the eleventh second?

14–1 ■ Sequences

A sequence

In chapter 8, we discussed the concept of a function. Now we wish to consider a very special function—a function whose domain is the set of positive integers. Such a function is called a **sequence.** There are two kinds of sequences, finite and infinite.

> ### Definition of a sequence
>
> An **infinite sequence** is a function whose domain is the set of positive integers $\{1,2,3,\cdots\}$.
> A **finite sequence** is a sequence whose domain is the first n positive integers.

Examples of sequences are

$$2,5,8,11,14,\cdots$$

⎰ The three dots indicate an *infinite*
⎱ sequence (never ends)

where each term of the sequence is found by adding 3 to the preceding term, and

$$10,20,30,40,50$$

⎰ Indicates a *finite*
⎱ sequence (ends at 50)

where each term of the sequence is found by adding 10 to the preceding term.

Sequences play an important role in the fields of science, finance, and mathematics. For example, the periodic interest earned on a savings account when the interest is compounded is a special kind of sequence.

The numbers that make up the sequence are called the *terms* of the sequence. A general sequence is written,

$$a_1, a_2, a_3, a_4, \cdots, a_n, \cdots$$

where

$a_1 =$ first term
$a_2 =$ second term
$a_3 =$ third term
$\vdots$
$a_n = n$th term (often called the *general term* of the sequence).
$\vdots$

Note The *subscript* in each term represents the number of the term.

When we think of each term of a sequence as a *function of n,* where n is the term number and *n is a positive integer,* we write

$$a_n = f(n)$$

■ *Example 14–1 A*

1. The following function defines a sequence.

$$a_n = f(n) = 3n + 1$$

Then $a_1 = f(1) = 3(1) + 1 = 4$ Replace n with 1
$a_2 = f(2) = 3(2) + 1 = 7$ Replace n with 2
$a_3 = f(3) = 3(3) + 1 = 10$ Replace n with 3
$\vdots$
$a_9 = f(9) = 3(9) + 1 = 28$ Replace n with 9
$\vdots$

The infinite sequence generated is $4, 7, 10, \cdots, 28, \cdots$

2. Given $a_n = f(n) = \dfrac{n-2}{3}$, find the first five terms of the sequence

$$a_1 = f(1) = \frac{1-2}{3} = -\frac{1}{3}$$

$$a_2 = f(2) = \frac{2-2}{3} = 0$$

$$a_3 = f(3) = \frac{3-2}{3} = \frac{1}{3}$$

$$a_4 = f(4) = \frac{4-2}{3} = \frac{2}{3}$$

$$a_5 = f(5) = \frac{5-2}{3} = 1$$

The first five terms of the sequence are $-\dfrac{1}{3}, 0, \dfrac{1}{3}, \dfrac{2}{3}, 1$, which represents a finite sequence.

3. Given $a_n = 5n - 7$, find a_7.
   Since we want a_7, then $n = 7$ and
   $$a_7 = 5(7) - 7 \qquad \text{Replace } n \text{ with } 7$$
   $$= 35 - 7$$
   $$= 28$$

   $$a_7 = 28$$

▶ **Quick check** 1. Given $a_n = 2n + 5$, find the first five terms.
2. Given $a_n = 4n - 3$, find a_9. ■

Finding the general term

On occasion, we may be given the first few terms of a sequence and asked to find an expression for the general term, a_n. There are no rules for finding this from the first few terms of the sequence. We usually do this by inspection or trial and error. However, we should be aware of a clue to the coefficient of n as demonstrated in the following examples. Consider the sequences

1. $a_n = 5n + 2$, whose terms are $7, 12, 17, 22, 27, \cdots$; each term of the sequence differs by 5, and the coefficient of n is 5;

2. $a_n = (-1)^n(n + 7)$, whose terms are $-8, 9, -10, 11, -12, \cdots$; each term alternates in sign, caused by the factor -1, to some positive integer power; the numerical value of each term differs by 1 and the coefficient of n is 1.

Thus, the coefficient of n for some sequences is the difference that is common between the terms of the sequences.

■ *Example 14–1 B*

Find an expression for the general term of the given sequence.

1. $5, 7, 9, 11, \cdots$
   The difference between each term is 2, so we conclude that $2n$ must be part of the general term. Now if we consider the first term a_1, let $n = 1$ and ask ourselves, $2(1) + $ (what number) $= 5$? Since $2(1) + 3 = 5$, we think $a_n = 2n + 3$. Check this with the succeeding terms.

   $$a_2 = 2(2) + 3 = 4 + 3 = 7 \qquad \text{(True)}$$
   $$a_3 = 2(3) + 3 = 6 + 3 = 9 \qquad \text{(True)}$$

   Therefore $a_n = 2n + 3$.

2. $\dfrac{3}{4}, -\dfrac{9}{8}, \dfrac{27}{16}, -\dfrac{81}{32}, \cdots$
   First, the signs alternate and the first term a_1 is positive. We must then have a factor of -1 to an even power when $n = 1$. Therefore one factor of the nth term could be $(-1)^{n+1}$ since $(-1)^{1+1} = (-1)^2 = 1$. Inspection shows us that the numerator of each term is a power of 3, starting with 3^1. Inspecting the denominator, we see that each denominator is a power of 2, starting with 2^2. When $n = 1$, the denominator of the first term could be

   $$2^{n+1} = 2^{1+1} = 2^2 = 4$$

The denominator of the second term, when $n = 2$, is then

$$2^{n+1} = 2^{2+1} = 2^3 = 8$$

and so on. Therefore we conclude that the numerator is 3^n and the denominator is 2^{n+1}. Thus

$$a_n = (-1)^{n+1} \frac{3^n}{2^{n+1}}$$

▶ **Quick check**   Find an expression for the general term of the sequence $4, 9, 14, 19, \ldots$ ■

--- **Mastery points** ---

Can you

■ Find the terms of a sequence, given the general term?
■ Find any given term of the sequence?
■ Find an expression for the general term, given a sequence?

Exercise 14–1

Write the first five terms of the sequence whose general term a_n is given. See example 14–1 A.

Example $a_n = 2n + 5$

Solution $a_1 = 2(1) + 5 = 7$ Replace n with 1
$a_2 = 2(2) + 5 = 9$ Replace n with 2
$a_3 = 2(3) + 5 = 11$ Replace n with 3
$a_4 = 2(4) + 5 = 13$ Replace n with 4
$a_5 = 2(5) + 5 = 15$ Replace n with 5

The first five terms of the sequence are $7, 9, 11, 13,$ and 15.

1. $a_n = 4n + 3$

2. $a_n = 5n - 4$

3. $a_n = \dfrac{2}{3n}$

4. $a_n = \dfrac{5}{2n + 1}$

5. $a_n = \dfrac{n + 5}{3n - 4}$

6. $a_n = \dfrac{4n}{5n + 2}$

7. $a_n = \dfrac{2^n}{5n}$

8. $a_n = \dfrac{3^{n+1}}{n + 2}$

9. $a_n = (-1)^n(6n - 5)$

10. $a_n = (-1)^{n+1}(n + 1)^2$

11. $a_n = (-1)^{n-1}\dfrac{3^n}{2^n + 1}$

12. $a_n = (-1)^n\dfrac{3}{4^n + 2}$

13. $a_n = (-1)^{2n}$

14. $a_n = (-1)^{3n-1}$

Find the indicated term of the given sequence. See example 14–1 A–3.

Example   Given $a_n = 4n - 3$, find a_9.

Solution   Given $a_n = 4n - 3$,

then $a_9 = 4(9) - 3$ Replace n with 9
$a_9 = 36 - 3$
$a_9 = 33$

Thus, $a_9 = 33$

15. $a_n = 3n + 5$, find a_9. **16.** $a_n = 7 - 2n$, find a_{11}. **17.** $a_n = \dfrac{1}{3n}$, find a_{37}.

18. $a_n = \dfrac{3}{5n}$, find a_{51}. **19.** $a_n = \dfrac{4n - 3}{2n + 7}$, find a_{17}. **20.** $a_n = \dfrac{9 - 5n}{-6 - 3n}$, find a_{12}.

21. $a_n = (-1)^n(6n + 5)$, find a_{14}. **22.** $a_n = (-1)^{n-1}(5n - 6)$, find a_{15}.

23. $a_n = (-1)^{n+2}n(n + 6)$, find a_{13}. **24.** $a_n = 2n^2(3n - 1)$, find a_8.

Given the terms of the following sequences, find an expression for the general term a_n. See example 14–1 B.

Example Find an expression for the general term a_n of the sequence $4, 9, 14, 19, \cdots$.

Solution The common difference between each term is 5, so $5n$ is part of the general term. Let $n = 1$ and consider

$$a_1 = 5(1) + \text{(what number)}?$$
$$= 5(1) + (-1)$$
$$= 5 - 1$$
$$= 4$$

Thus, we conclude $a_n = 5n - 1$. To check this,

$$a_2 = 5(2) - 1 = 10 - 1 = 9$$
$$\text{and } a_3 = 5(3) - 1 = 15 - 1 = 14$$

25. $6, 8, 10, 12, 14, \cdots$ **26.** $1, 4, 7, 10, 13, \cdots$ **27.** $2, 7, 12, 17, 22, \cdots$

28. $1, 4, 9, 16, \cdots$ **29.** $4, 9, 16, 25, 36, \cdots$ **30.** $\dfrac{1}{3}, \dfrac{1}{9}, \dfrac{1}{27}, \dfrac{1}{81}, \dfrac{1}{243}, \cdots$

31. $1, \dfrac{1}{2}, \dfrac{1}{4}, \dfrac{1}{8}, \dfrac{1}{16}, \cdots$ **32.** $\dfrac{2}{3}, \dfrac{3}{4}, \dfrac{4}{5}, \dfrac{5}{6}, \dfrac{6}{7}, \cdots$ **33.** $\dfrac{3}{5}, \dfrac{4}{7}, \dfrac{5}{9}, \dfrac{6}{11}, \dfrac{7}{13}, \cdots$

34. $3, \dfrac{8}{3}, \dfrac{7}{3}, 2, \dfrac{5}{3}, \cdots$ **35.** $-6, 10, -14, 18, -22, \cdots$ **36.** $-\dfrac{3}{2}, 3, -\dfrac{9}{2}, 6, -\dfrac{15}{2}, \cdots$

Solve the following word problems.

37. A culture of bacteria triples every hour. If the original culture has 1,000 bacteria, how many bacteria were there after (a) 3 hours, (b) 5 hours, (c) n hours?

38. A pendulum swings a distance of 20 inches on its first swing. If each subsequent swing back is three-fourths of the previous swing, how far does it swing on the (a) third swing, (b) seventh swing?

39. A ball is dropped from a height of 10 feet. If the ball rebounds one-half the height of its previous fall, how high does it rebound on the (a) second bounce, (b) sixth bounce, (c) nth bounce?

40. Jim Jarrett gives his son an allowance each month of 10¢ on the first day, 15¢ on the second day, 20¢ on the third day, and so on. Write a sequence for the first ten days of the month. Write an expression for the amount received on the nth day of the month. How much does he receive on the thirtieth day?

41. Steve Navarro begins a new job at a starting yearly salary of $16,000, with a promise of a salary increase of $1,500 per year for the first 6 years. (a) Write a sequence showing his salary during the first 6 years. (b) Write the general term for the sequence. (c) If this yearly pay increase were to continue beyond 6 years, what would his salary be after 20 years?

Review exercises

1. Given $f(x) = 5x + 4$, find (a) $f(-2)$, (b) $f(0)$, (c) $f(2)$. See section 8–2.

2. Evaluate $\begin{vmatrix} 1 & 2 & 3 \\ 0 & -1 & 4 \\ -2 & -3 & 0 \end{vmatrix}$. See section 13–1.

3. Find the solution set of the system of equations
$2x - 3y = 4$
$x + 2y = 0$
using Cramer's Rule. See section 13–2.

Find the equation of the following lines. See section 7–3.

4. Vertical line passing through $(-3,4)$

5. Parallel to $x + 3y = 2$ and passing through $(-1,-1)$

6. Find the solution set of the exponential equation $3^{2x+3} = 9$. See section 11–6.

14–2 ■ Series

A series

Associated with any sequence is the sum of the terms of the sequence, called a **series.**

Definition of a series

A **series** is the indicated sum of the terms in a sequence.

If the sequence is

1. finite, we have a finite series, such as

$a_1 + a_2 + a_3 + a_4 + \cdots + a_n$

2. infinite, we have an infinite series

$a_1 + a_2 + a_3 + \cdots + a_n + \cdots$

To illustrate, given the sequence whose general term is $a_n = 2n + 1$, the *sequence* is

$3,5,7,9, \cdots$

and the *series* of this sequence is

$3 + 5 + 7 + 9 + \cdots + (2n + 1) + \cdots$

Definition of a partial sum

A **partial sum** of a series, denoted by S_n, is the sum of a finite number of consecutive terms of the series starting with a_1.

$$\text{Thus, } S_1 = a_1 \qquad \text{First partial sum}$$
$$S_2 = a_1 + a_2 \qquad \text{Second partial sum}$$
$$S_3 = a_1 + a_2 + a_3 \qquad \text{Third partial sum}$$
$$\vdots$$
$$S_n = a_1 + a_2 + a_3 + \cdots + a_n \qquad n\text{th partial sum}$$

■ **Example 14–2 A**

Find S_1, S_2, S_3, and S_4 for the following sequences.

1. $a_n = 5n + 4$

Now $a_1 = 5(1) + 4 = 9$
$\quad a_2 = 5(2) + 4 = 14$
$\quad a_3 = 5(3) + 4 = 19$
$\quad a_4 = 5(4) + 4 = 24$

Then, $S_1 = a_1 = 9$
$\quad S_2 = a_1 + a_2 = 9 + 14 = 23$
$\quad S_3 = a_1 + a_2 + a_3 = 9 + 14 + 19 = 42$
$\quad S_4 = a_1 + a_2 + a_3 + a_4 = 9 + 14 + 19 + 24 = 66$

2. $a_n = (-1)^n(n - 2)$

Now $a_1 = (-1)^1(1 - 2) = 1$
$\quad a_2 = (-1)^2(2 - 2) = 0$
$\quad a_3 = (-1)^3(3 - 2) = -1$
$\quad a_4 = (-1)^4(4 - 2) = 2$

Then, $S_1 = 1$
$\quad S_2 = 1 + 0 = 1$
$\quad S_3 = 1 + 0 + (-1) = 0$
$\quad S_4 = 1 + 0 + (-1) + (2) = 2$

▶ **Quick check**   Given $a_n = 2n - 7$, find S_3.   ■

Sigma notation

A compact way of representing a sum when the general term is known is by
sigma (or **summation**) **notation.** To do this, we use the Greek letter sigma, Σ, in
conjunction with the general term of the related sequence. To illustrate, consider
the sequence

$$3,7,11,15,19,23,27, \cdots$$

We can determine that the general term of this sequence is $4n - 1$. Suppose we
want the sum of the first seven terms of the sequence (the seventh **partial sum**).
We want

$$S_7 = a_1 + a_2 + a_3 + a_4 + a_5 + a_6 + a_7$$
$$= (4 \cdot 1 - 1) + (4 \cdot 2 - 1) + (4 \cdot 3 - 1) + (4 \cdot 4 - 1)$$
$$+ (4 \cdot 5 - 1) + (4 \cdot 6 - 1) + (4 \cdot 7 - 1)$$
$$= 3 + 7 + 11 + 15 + 19 + 23 + 27 = 105$$

This partial sum can be written in sigma notation by

$$S_7 = \sum_{i=1}^{7} (4i - 1)$$
$$= 105$$

where an expression for the general term, $4n - 1$, becomes $4i - 1$ when n is replaced by i. We call the letter i, as used in this situation, the **index of summation.** Other letters often used for this purpose are j and k.

Note   This use of i has no connection with its use in our work with complex numbers.

We read the expression $\sum\limits_{i=1}^{7} (4i - 1)$ "the summation as i goes from 1 to 7 of $4i - 1$." The first and last integers used to replace the index of summation, in this case 1 and 7, are called the **lower and upper limits of summation,** respectively.

Note   If $a_i = c$ (constant) for all i, the $\sum\limits_{i=1}^{n} c = \underbrace{c + c + \cdots + c}_{n \text{ terms}} = nc.$

■ *Example 14–2 B*

Expand the following indicated partial sums. Find the indicated sum.

1. $\sum\limits_{j=1}^{4} (5j + 2)$

   Now the index of summation is j, and we successively replace j by the integers 1, 2, 3, and 4. Thus

   $$\sum\limits_{j=1}^{4} (5j + 2) = [5(1) + 2] + [5(2) + 2] + [5(3) + 2] + [5(4) + 2]$$
   $$= 7 + 12 + 17 + 22$$
   $$= 58$$

   Therefore, $\sum\limits_{j=1}^{4} (5j + 2) = 58$

2. $\sum\limits_{k=1}^{3} (-1)^k(2k + 5)$

   We successively replace the index of summation k by the integers 1, 2, and 3. Therefore

   $$\sum\limits_{k=1}^{3} (-1)^k(2k + 5) = (-1)^1[2(1) + 5] + (-1)^2[2(2) + 5]$$
   $$+ (-1)^3[2(3) + 5]$$
   $$= (-1)(7) + 1(9) + (-1)(11)$$
   $$= -7 + 9 - 11$$
   $$= -9$$

   Thus, $\sum\limits_{k=1}^{3} (-1)^k(2k + 5) = -9$

3. $\sum\limits_{i=1}^{5} 6 = 5(6) = 6 + 6 + 6 + 6 + 6 = 5(6) = 30$

   Therefore, $\sum\limits_{i=1}^{5} 6 = 30$

▶ *Quick check*   Expand the partial sum $\sum\limits_{i=1}^{5} (2i - 7)$. Find the indicated sum.   ■

Reversing the procedure, it is sometimes desirable to express a sum in the compact sigma notation form.

■ **Example 14–2 C**

Write the following sums in sigma notation.

1. $-5 + 25 - 125 + 625$

There are four terms, so we can use the limits of summation 1 and 4. Since the operations alternate with a negative first term, the general term will contain -1 to a power of j so that $j = 1$ gives -1. Use $(-1)^j$ for the first factor of a general term. By inspection, the numerical value of the terms of the series are powers of 5. Thus

$$-5 + 25 - 125 + 625 = \sum_{j=1}^{4} (-1)^j 5^j$$

2. $6 - 10 + 14 - 18 + 22 - 26$

There are six terms so we use limits of summation 1 and 6. Since the signs alternate and the first term is positive, the general term will contain (-1) to a power of $i + 1$ so that $i = 1$ gives 1, $(-1)^{i+1}$. The common difference between terms is 4 so the general term will contain $4n$. Since the first term, 6, can be obtained by $4(1) + 2$, then $a_n = 4n + 2$ and

$$6 - 10 + 14 - 18 + 22 - 26 = \sum_{i=1}^{6} (-1)^{i+1}(4i + 2)$$

3. $\dfrac{2}{3} + \dfrac{4}{7} + \dfrac{6}{11} + \dfrac{8}{15} + \dfrac{10}{19}$

The limits of summation can be 1 and 5, since there are five terms. Inspection tells us the general term of the numerator is $2n$ and of the denominator is $4n - 1$. Then

$$a_n = \frac{2n}{4n - 1} \text{ and}$$

$$\frac{2}{3} + \frac{4}{7} + \frac{6}{11} + \frac{8}{15} + \frac{10}{19} = \sum_{k=1}^{5} \frac{2k}{4k - 1}$$

We should note that there is nothing unique about the way the above-indicated sums have been stated using sigma notation. To illustrate,

$-5 + 25 - 125 + 625$ could be written $\displaystyle\sum_{j=2}^{5} (-1)^{j-1}(5)^{j-1}$ where the limits of

summation have been changed to 2 and 5.

▶ **Quick check** Write the partial sum $1 + 3 + 5 + 7$ in sigma notation. ■

┌─ **Mastery points** ─────────────────────────────────

Can you
- Find the *n*th partial sum of an infinite series?
- Expand a partial sum that is written using sigma notation?
- Write a partial sum of a series in sigma notation?

Exercise 14–2

Expand the following indicated partial sums. Find the indicated sum. See example 14–2 A.

Example Given $a_n = 2n - 7$, find S_3.

Solution Now $a_1 = 2(1) - 7 = -5$ Replace n with 1
$\qquad\qquad a_2 = 2(2) - 7 = -3$ Replace n with 2
$\qquad\qquad a_3 = 2(3) - 7 = -1$ Replace n with 3

$\qquad$ since $S_3 = a_1 + a_2 + a_3$
$\qquad\qquad S_3 = (-5) + (-3) + (-1)$ Replace a_1 with -5, a_2 with -3, a_3 with -1
$\qquad\qquad S_3 = -9$

$\qquad$ Thus, $S_3 = -9$.

1. $a_n = 4n + 3; S_5$

2. $a_n = 5n - 1; S_4$

3. $a_n = \dfrac{3}{n + 2}; S_3$

4. $a_n = \dfrac{4}{(2n + 3)}; S_2$

5. $a_n = (-1)(6n - 1); S_3$

6. $a_n = (-1)(n - 9); S_4$

7. $a_n = \dfrac{2n - 1}{4 - n}; S_3$

8. $a_n = 2^n + 3; S_3$

See example 14–2 B.

Example Expand $\displaystyle\sum_{i=1}^{5} (2i - 7)$.

Solution $\displaystyle\sum_{i=1}^{5} (2i - 7) = a_1 + a_2 + a_3 + a_4 + a_5$

$\qquad\qquad = [2(1) - 7] + [2(2) - 7] + [2(3) - 7] + [2(4) - 7] + [2(5) - 7]$
$\qquad\qquad = (-5) + (-3) + (-1) + 1 + 3$
$\qquad\qquad = -5$

$\qquad$ Therefore, $\displaystyle\sum_{i=1}^{5} (2i - 7) = -5$.

9. $\displaystyle\sum_{j=1}^{5} j^2$

10. $\displaystyle\sum_{k=1}^{4} k^3$

11. $\displaystyle\sum_{i=1}^{6} (2i + 3)$

12. $\displaystyle\sum_{i=1}^{7} (i - 2)$

13. $\displaystyle\sum_{k=1}^{5} k(k - 3)$

14. $\displaystyle\sum_{j=1}^{6} (j^2 + 2)$

15. $\displaystyle\sum_{i=1}^{4} i(2i - 1)$

16. $\displaystyle\sum_{i=1}^{4} (i + 1)(i - 2)$

17. $\displaystyle\sum_{j=1}^{5} (j - 3)(j + 2)$

18. $\displaystyle\sum_{k=1}^{4} k^2(k + 2)$

19. $\displaystyle\sum_{k=1}^{5} \dfrac{1}{k + 3}$

20. $\displaystyle\sum_{i=1}^{4} \dfrac{3}{3i - 1}$

21. $\displaystyle\sum_{k=1}^{5} \dfrac{2k + 1}{k + 3}$

22. $\displaystyle\sum_{j=1}^{4} \dfrac{j^2}{3j - 2}$

23. $\displaystyle\sum_{i=1}^{5} \dfrac{4}{i^2}$

24. $\displaystyle\sum_{j=1}^{5} (-1)^j \cdot \dfrac{3}{2j}$

25. $\displaystyle\sum_{k=1}^{4} (-1)^{k + 1} \cdot \dfrac{1}{3k}$

26. $\displaystyle\sum_{i=1}^{5} (-1)^i i$

27. $\displaystyle\sum_{j=1}^{3} (-1)^{j - 1}(j)^j$

28. $\displaystyle\sum_{k=1}^{5} (-1)^{k + 1}(k + 1)^k$

29. $\displaystyle\sum_{i=1}^{9} 8$

30. $\displaystyle\sum_{j=1}^{10} (-3)$

Expand and find the following indicated partial sums.

Example $\displaystyle\sum_{i=2}^{5} (3i + 1)$

Solution $\displaystyle\sum_{i=2}^{5} (3i + 1) = [3(2) + 1] + [3(3) + 1] + [3(4) + 1] + [3(5) + 1]$
$= 7 + 10 + 13 + 16$
$= 46$

Thus, $\displaystyle\sum_{i=2}^{5} (3i + 1) = 46$.

31. $\displaystyle\sum_{i=3}^{8} (i + 2)$ **32.** $\displaystyle\sum_{j=0}^{4} (2j + 1)$ **33.** $\displaystyle\sum_{k=2}^{5} \frac{1}{k}$ **34.** $\displaystyle\sum_{i=4}^{9} (i^2 + 3)$

35. $\displaystyle\sum_{j=0}^{5} \frac{2j + 1}{j + 1}$ **36.** $\displaystyle\sum_{k=2}^{6} (-1)^k$ **37.** $\displaystyle\sum_{i=3}^{7} (-1)^i(3i - 2)$ **38.** $\displaystyle\sum_{i=4}^{7} (-1)^i\frac{2}{2i + 3}$

Write the following partial sums in sigma notation. See example 14–2 C.

Example $1 + 3 + 5 + 7$

Solution   There are four terms so the limits of summation can be 1 to 4. Inspection tells us there is a common difference of 2 so $2n$ is part of the general term. Now

$a_1 = 2(1) + ?$
$= 2(1) + (-1)$
$= 2(1) - 1$
$= 1$

The general term is $a_n = 2n - 1$, so

$1 + 3 + 5 + 7 = \displaystyle\sum_{i=1}^{4} (2i - 1)$ Replace n with i

39. $1 + 2 + 3 + 4 + 5$ **40.** $3 + 6 + 9 + 12$

41. $1 + 8 + 27 + 64 + 125 + 216$ **42.** $5 + 9 + 13 + 17 + 21$

43. $\dfrac{2}{3} + \dfrac{3}{4} + \dfrac{4}{5} + \dfrac{5}{6}$ **44.** $\dfrac{1}{3} + \dfrac{3}{5} + \dfrac{5}{7} + \dfrac{7}{9} + \dfrac{9}{11}$

45. $2 + \dfrac{4}{3} + \dfrac{6}{9} + \dfrac{8}{27}$ **46.** $\dfrac{1}{1} + \dfrac{2}{2} + \dfrac{3}{4} + \dfrac{4}{8} + \dfrac{5}{16}$

47. $\dfrac{5}{4} + \dfrac{7}{7} + \dfrac{9}{10} + \dfrac{11}{13}$ **48.** $2 - 5 + 8 - 11 + 14$

49. $-2 + 4 - 8 + 16 - 32 + 64$ **50.** $\dfrac{1}{2} - \dfrac{2}{3} + \dfrac{3}{4} - \dfrac{4}{5} + \dfrac{5}{6}$

Review exercises

Evaluate the following expressions when $a = 2$, $n = 4$, and $r = 3$. See section 1–5.

1. $a + (n - 1)r$ **2.** $\dfrac{n}{2}(a + r)$ **3.** $\dfrac{n}{2}[a + (n - 1)r]$ **4.** ar^{n-1}

5. Simplify the complex rational fraction $\dfrac{\dfrac{1}{3} - \dfrac{2}{5}}{1 + \dfrac{1}{5}}$.

See section 4–4.

6. Find the solution set of the equation $3^{x-1} = 4$ using common logarithms, correct to three decimal places. See section 11–6.

14–3 ■ Arithmetic sequences

An arithmetic sequence

Consider the sequence defined by

$$3, 7, 11, 15, 19, \cdots, 4n - 1, \cdots$$

whose major characteristic is that the difference between any two successive terms is 4. Recall that this determines the coefficient of n in the general term. Such a sequence is called an **arithmetic sequence** (or **arithmetic progression**).

> ### Definition of an arithmetic sequence
> An **arithmetic sequence** is a sequence in which each term after the first differs from the preceding term by the same constant number.

We call this constant number the **common difference,** to be denoted by d. Thus in the above sequence, the common difference $d = 4$. That is, for any arithmetic sequence,

$$d = a_{n+1} - a_n$$

where d is the common difference and a_n and a_{n+1} are successive terms of the sequence.

■ Example 14–3 A

1. The sequence $7, 11, 15, 19$ is arithmetic since

$$11 - 7 = 4, \quad 15 - 11 = 4, \quad 19 - 15 = 4$$

and the common difference $d = 4$.

2. $-10, -4, 2, 8, 14$ is an arithmetic sequence since

$$
\begin{aligned}
-4 - (-10) &= -4 + 10 = 6 \\
2 - (-4) &= 2 + 4 = 6 \\
8 - 2 &= 6 \\
\text{and } 14 - 8 &= 6
\end{aligned}
$$

The common difference $d = 6$.

3. $1, 3, 9, 12, 36$ is not an arithmetic sequence since $3 - 1 = 2$ whereas $9 - 3 = 6$

▶ **Quick check** Is $8, 5, 2, -1, \cdots$ an arithmetic sequence? If so, find the common difference.

■

The general term of an arithmetic sequence

To find an expression for the nth (general) term of an arithmetic sequence, a_n, look at the terms $a_1, a_2, a_3, \cdots$.

$$a_1, \quad a_2 = a_1 + d, \quad a_3 = a_1 + 2d, \quad a_4 = a_1 + 3d, a_5 = a_1 + 4d$$

This suggests the following:

General term of an arithmetic sequence
The general term a_n of an arithmetic sequence with first term a_1 and common difference d is given by
$$a_n = a_1 + (n - 1)d$$

■ *Example 14–3 B*

1. Find the twenty-first term of the arithmetic sequence whose first term $a_1 = 3$ and common difference $d = 4$.

   Using $a_n = a_1 + (n - 1)d$, we want a_{21} when $a_1 = 3$, $d = 4$, and $n = 21$.

   $\begin{aligned} a_{21} &= (3) + [(21) - 1](4) \qquad \text{Replace } a_1 \text{ with 3, } n \text{ with 21, and } d \text{ with 4} \\ &= 3 + (20)4 \\ &= 3 + 80 \\ &= 83 \end{aligned}$

   The twenty-first term is $a_{21} = 83$.

2. Given the arithmetic sequence $5, -1, -7, -13, \cdots$, find a_{20}.

   Since $-1 - 5 = -6$, then $d = -6$. Now $a_1 = 5$ and $n = 20$.

   Using $a_n = a_1 + (n - 1)d$, we want a_{20}.

   $\begin{aligned} a_{20} &= (5) + [(20) - 1](-6) \qquad \text{Replace } a_1 \text{ with 5, } n \text{ with 20, and } d \text{ with } -6 \\ &= 5 + (19)(-6) \\ &= 5 + (-114) \\ &= -109 \end{aligned}$

   The twentieth term of the sequence is $a_{20} = -109$.

▶ *Quick check* Find the nineteenth term of the arithmetic sequence whose first term $a_1 = -2$ and the common difference $d = 3$. ■

   Given a finite arithmetic sequence, it is possible to determine the number of terms in the sequence, n, if we can determine d and a_1.

■ *Example 14–3 C*

Find the number of terms in the finite arithmetic sequence $-9, -4, 1, 6, \cdots, 111$.

From the given terms of the sequence, we can determine $a_1 = -9$, $d = -4 - (-9) = -4 + 9 = 5$, and $a_n = 111$. We want the value of n. Using the formula $a_n = a_1 + (n - 1)d$, we get

$\begin{aligned} (111) &= (-9) + (n - 1)(5) \qquad \text{Replace } a_n \text{ with 111, } a_1 \text{ with } -9, \text{ and } d \text{ with 5} \\ 111 &= -9 + 5n - 5 \\ 111 &= 5n - 14 \\ 125 &= 5n \\ 25 &= n \end{aligned}$

Thus the finite sequence has $n = 25$ terms.

▶ *Quick check* Find the number of terms in the finite arithmetic sequence $-5, -1, 3, 7, \cdots, 115$. ■

Arithmetic series

Now consider the *sum* of the terms in an arithmetic sequence having n terms. We denote this by S_n (called the *nth* partial sum). This sum can be written

$$S_n = a_1 + (a_1 + d) + (a_1 + 2d) + (a_1 + 3d) + \cdots + [a_1 + (n-1)d]$$

Another way to write this sum would be to add in reverse order, starting with the *nth* term, a_n, and subtracting multiples of the common difference d. Then we obtain

$$S_n = a_n + (a_n - d) + (a_n - 2d) + (a_n - 3d) + \cdots + [a_n - (n-1)d]$$

Now if we add the corresponding terms of both members of the two equations, we obtain

$$S_n = a_1 + (a_1 + d) + (a_1 + 2d) + \cdots + [a_1 + (n-1)d]$$
$$+ \; S_n = a_n + (a_n - d) + (a_n - 2d) + \cdots + [a_n - (n-1)d]$$
$$2S_n = \underbrace{(a_1 + a_n) + (a_1 + a_n) + (a_1 + a_n) + \cdots + (a_1 + a_n)}_{n \text{ terms of } (a_1 + a_n)}$$

We can write the right member as the product $n(a_1 + a_n)$ and the sum of these two equations is given by

$$2S_n = n(a_1 + a_n)$$

Dividing each member of the equation by 2, we obtain the sum of the first n terms.

> ### Sum of the first n terms of an arithmetic sequence
>
> The sum of the first n terms of an arithmetic sequence is given by
>
> $$S_n = \frac{n}{2}(a_1 + a_n)$$

Note If we replace a_n in the above formula with $a_1 + (n-1)d$, this formula becomes $S_n = \frac{n}{2}[2a + (n-1)d]$.

■ *Example 14–3 D*

1. Find the sum of the first twenty-five terms of the arithmetic sequence whose general term is $a_n = 3n + 5$.

    Given $a_n = 3n + 5$, we can determine $a_1 = 3(1) + 5 = 8$, $d = 3$ (the coefficient of *n*) and, using $a_n = a_1 + (n-1)d$,

    $$a_{25} = (8) + [(25) - 1](3) \qquad \text{Replace } a_1 \text{ with 8, } n \text{ with 25, and } d \text{ with 3}$$
    $$= 8 + (24)(3)$$
    $$= 80$$

Since $n = 25$, using the formula

$$S_n = \frac{n}{2}(a_1 + a_n)$$

$$S_{25} = \frac{(25)}{2}[(8) + (80)]$$ Replace n with 25, a_{25} with 80, and a_1 with 8

$$= \frac{25}{2}(88)$$

$$= 1{,}100$$

The sum of the first twenty-five terms is $S_{25} = 1{,}100$.

2. Find $\displaystyle\sum_{i=1}^{13}(3i - 1)$.

Since $\displaystyle\sum_{i=1}^{13}(3i - 1) = [3(1) - 1] + [3(2) - 1] + [3(3) - 1] + \cdots$

$$= 2 + 5 + 8 + \cdots$$

we can see the common difference $d = 3$. We want S_{13} (the thirteenth partial sum). Now $n = 13$, $a_1 = 2$, and $d = 3$ (the coefficient of i). We substitute into the formula $a_n = a_1 + (n - 1)d$ to determine a_{13}.

$$a_{13} = (2) + [(13) - 1](3)$$ Replace a_1 with 2, n with 13, and d with 3
$$= 2 + 12(3)$$
$$= 38$$

Using $S_n = \dfrac{n}{2}(a_1 + a_n)$,

$$S_{13} = \frac{(13)}{2}[(2) + (38)]$$ Replace n with 13, a_1 with 2, and a_{13} with 38

$$= \frac{13}{2}(40)$$

$$= 260$$

Therefore $\displaystyle\sum_{i=1}^{13}(3i - 1) = 260$.

3. Find the sum of the first twelve terms of the arithmetic sequence whose first term $a_1 = -11$ and common difference $d = 4$.

Using the formula $S_n = \dfrac{n}{2}[2a_1 + (n - 1)d]$, we want S_{12}.

$$S_{12} = \frac{(12)}{2}[2(-11) + (12 - 1)(4)]$$ Replace n with 12, a_1 with -11, and d with 4

$$= 6[-22 + (11)4]$$
$$= 6[-22 + 44]$$
$$= 6(22)$$
$$= 132$$

Thus the sum of the first twelve terms is $S_{12} = 132$.

▶ *Quick check*   Find S_{29} of the arithmetic sequence if $a_1 = 3$ and $a_{29} = -53$. ∎

Mastery points

Can you
- Determine if a sequence is an arithmetic sequence?
- Find the common difference of an arithmetic sequence?
- Find a specific term of an arithmetic sequence?
- Find the number of terms of a finite arithmetic sequence?
- Find the sum of a given number of terms in an arithmetic sequence?

Exercise 14–3

Determine whether or not the given sequence is arithmetic. If it is arithmetic, find the common difference d.
See example 14–3 A.

Example $8, 5, 2, -1, \cdots$

Solution Since $5 - 8 = -3$, $2 - 5 = -3$, $-1 - 2 = -3$, the sequence is arithmetic and the common difference $d = -3$.

1. $2, 3, 4, 5, 6, \cdots$

2. $1, 6, 11, 16, 21, \cdots$

3. $-10, -8, -6, -4, \cdots$

4. $4, 6, 8, 10, 12, \cdots$

5. $3, 5, 8, 10, 12, \cdots$

6. $-16, -19, -22, -25, \cdots$

7. $\dfrac{3}{2}, 2, \dfrac{5}{2}, 3, \dfrac{7}{2}, \cdots$

8. $-\dfrac{1}{2}, \dfrac{1}{2}, \dfrac{3}{2}, \dfrac{5}{2}, \cdots$

9. $-\dfrac{7}{3}, -\dfrac{2}{3}, 1, \dfrac{8}{3}, \cdots$

10. $1, \dfrac{1}{2}, \dfrac{1}{3}, \dfrac{1}{4}, \dfrac{1}{5}, \cdots$

Find the indicated term of the arithmetic sequence having the following characteristics. See example 14–3 B.

Example Find the nineteenth term of the arithmetic sequence whose first term $a_1 = -2$ and the common difference $d = 3$.

Solution Using $a_n = a_1 + (n - 1)d$, we want a_{19} where $n = 19$.

$$a_{19} = (-2) + [(19) - 1](3) \quad \text{Replace } a_1 \text{ with } -2, n \text{ with } 19, \text{ and } d \text{ with } 3$$
$$= (-2) + (18)3 \quad \text{Multiply and simplify}$$
$$= (-2) + 54$$
$$= 52$$

The nineteenth term of the arithmetic sequence is $a_{19} = 52$.

11. $a_1 = 4$, $d = 5$; find a_{16}.

12. $a_1 = -4$, $d = 2$; find a_{21}.

13. $a_1 = -10$, $d = -3$; find a_{17}.

14. $a_1 = 6$, $d = -7$; find a_{23}.

15. $a_1 = 2$, $d = \dfrac{1}{3}$; find a_{12}.

16. $a_1 = 4$, $d = \dfrac{1}{2}$; find a_{10}.

17. $a_1 = \dfrac{1}{3}$, $d = \dfrac{2}{3}$; find a_{14}.

18. $a_1 = \dfrac{3}{4}$, $d = -\dfrac{1}{2}$; find a_{13}.

19. $5, 9, 13, 17, \cdots$; find a_{16}.

20. $4, 13, 22, 31, \cdots$; find a_{19}.

21. $-15, -10, -5, 0, \cdots$; find a_{25}.

22. $-27, -25, -23, -21, \cdots$; find a_{20}.

Find the number of terms in the given finite arithmetic sequence. See example 14–3 C.

Example Find the number of terms of the finite arithmetic sequence $-5, -1, 3, 7, \cdots, 115$.

Solution We are given $a_1 = -5$ and $a_n = 115$, and we want n. Now $-1 - (-5) = 4$, so the common difference $d = 4$. Using $a_n = a_1 + (n - 1)d$,

$$(115) = (-5) + (n - 1)(4)$$ Replace a_n with 115, a_1 with -5, and d with 4

$$120 = 4n - 4$$ Add 5 to each member and multiply by 4 in the right member

$$124 = 4n$$

$$31 = n$$

There are thirty-one terms in the arithmetic sequence.

23. $14, 29, 44, 59, \cdots, 89$ **24.** $2, 5, 8, \cdots, 83$

25. $-3, -11, -19, -27, \cdots, -115$ **26.** $7, 1, -5, -11, \cdots, -101$

27. $\dfrac{1}{2}, 0, -\dfrac{1}{2}, -1, \cdots, -\dfrac{27}{2}$ **28.** $\dfrac{5}{3}, \dfrac{4}{3}, 1, \dfrac{2}{3}, \cdots, -2$

Find the indicated partial sum for each of the given arithmetic sequences. The last given term is the nth term, a_n. (That is, $a_{16} = 33$ in exercise 29.) See example 14–3 D–1.

Example Find S_{29} of the arithmetic sequence if $a_1 = 3$ and $a_{29} = -53$.

Solution a. Using $a_n = a_1 + (n - 1)d$,

$$(-53) = (3) + [(29) - 1]d$$ Replace a_1 with 3, n with 29, and a_n with -53

$$-56 = 28d$$ Subtract 3 from each member; $29 - 1 = 28$

$$-2 = d$$ Divide each member by 28

b. Using $S_n = \dfrac{n}{2}(a_1 + a_n)$,

$$S_{29} = \dfrac{(29)}{2}[(3) + (-53)]$$ Replace n with 29, a_1 with 3, and a_n with -53

$$= \dfrac{29}{2}(-50)$$ Combine in the right member

$$= 29(-25)$$ Reduce in the right member

$$= -725$$

The sum of the first twenty-nine terms is $S_{29} = -725$.

29. $3, 5, 7, \cdots, 33$; find S_{16}.   **30.** $0, 3, 6, \cdots, 63$; find S_{22}.

31. $7, 4, 1, \cdots, -32$; find S_{14}.   **32.** $1, -7, -15, \cdots, -111$; find S_{15}.

33. $\dfrac{1}{2}, 1, \dfrac{3}{2}, \cdots, 9$; find S_{18}.   **34.** $\dfrac{1}{4}, 1, \dfrac{7}{4}, \cdots, \dfrac{37}{4}$; find S_{13}.

35. $a_n = 2n + 5$; find S_{14}.   **36.** $a_n = 6n - 3$; find S_{15}.

37. $a_n = 5 - n$; find S_{19}.   **38.** $a_n = 7 - 3n$; find S_{17}.

39. $a_1 = 3$ and $d = 2$; find S_{23}.   **40.** $a_1 = -5$ and $d = -3$; find S_{19}.

Find the indicated sum. See example 14–3 D–2.

Example $S_{13} = \sum\limits_{i=1}^{13} (2 - 5i)$

Solution Given $\sum\limits_{i=1}^{13} (2 - 5i)$, we find $a_1 = 2 - 5(1) = -3$, the common difference $d = -5$, and $n = 13$.

Using $S_n = \dfrac{n}{2}[2a_1 + (n - 1)d]$,

$S_{13} = \dfrac{(13)}{2}[2(-3) + (13 - 1)(-5)]$ Replace a_1 with -3, n with 13, and d with -5

$\quad\;\; = \dfrac{13}{2}[-6 + 12(-5)]$

$\quad\;\; = -429$

Therefore, $S_{13} = \sum\limits_{i=1}^{13} (2 - 5i) = -429$.

41. $\sum\limits_{k=1}^{15} (2k - 5)$ **42.** $\sum\limits_{j=1}^{13} (3j + 9)$ **43.** $\sum\limits_{i=1}^{22} (3 - 2i)$ **44.** $\sum\limits_{i=1}^{19} (4 - i)$

45. $\sum\limits_{j=1}^{17} \dfrac{1}{3}j$ **46.** $\sum\limits_{k=1}^{14} \dfrac{1}{2}k$ **47.** $\sum\limits_{j=1}^{10} \left(\dfrac{3}{5}j - 2\right)$ **48.** $\sum\limits_{i=1}^{11} \left(\dfrac{2}{3}i + 4\right)$

Find the indicated partial sum of the terms in the given arithmetic sequence. See example 14–3 D–3.

Example $-9, -5, -1, 3, \cdots$; find S_{27}.

Solution We are given $a_1 = -9$, $n = 27$, and find $d = -5 - (-9) = 4$. Using $S_n = \dfrac{n}{2}[2a_1 + (n - 1)d]$,

$S_{27} = \dfrac{(27)}{2}[2(-9) + (27 - 1)(4)]$ Replace n with 27, a_1 with -9, and d with 4

$\quad\;\; = \dfrac{27}{2}[-18 + 26(4)]$

$\quad\;\; = \dfrac{27}{2}(86)$

$\quad\;\; = 1{,}161$

Therefore, $S_{27} = 1{,}161$.

49. $2, 8, 14, 22, \cdots$; find S_{13}.   **50.** $1, 8, 15, 22, \cdots$; find S_{12}.   **51.** $24, 19, 14, \cdots$; find S_{15}.

52. $-6, -4, -2, \cdots$; find S_{14}.   **53.** $\dfrac{1}{6}, -\dfrac{5}{6}, -\dfrac{11}{6}, \cdots$; find S_{11}.   **54.** $-1, -4, -7, \cdots$; find S_{16}.

Solve the following word problems.

55. A display of cans has 21 cans in the bottom row, 19 cans in the row above, 17 cans in the next row, and so on. How many cans are there if the top row contains 1 can? (*Hint:* $a_1 = 21$ and $d = -2$.)

56. In exercise 55, if there are 8 rows of cans, how many cans are in the display?

57. A stock boy in a grocery store stacks a number of boxes of cereal so that there are 30 boxes in the first row, 27 boxes in the second row, 24 boxes in the third row, and so on. How many boxes of cereal does he have if there are 3 boxes in the top row?

58. In exercise 57, if there are 9 rows of cereal, how many boxes of cereal are there?

59. Find the sum of the even integers from 2 to 116.

60. How many times will a clock strike in 12 hours if it strikes only on the hour?

61. A parachutist in free fall falls vertically 16 feet during the first second, 48 feet during the second second, 80 feet during the third second, and so on. How far will she fall during the eighth second? How far will she fall during the first 10 seconds?

62. Neglecting air resistance, how long would it take before the parachutist pulls the rip cord to break her fall after falling 3,600 feet in exercise 61? (*Hint: $S_n = 3,600$ and find n.*)

63. Ron Line is offered a job as a mechanic starting at $700 per month. If he is guaranteed a pay increase of $10 per month every 3 months, what will his salary be after 8 years?

64. In exercise 63, how many years would it take for his salary to reach $1,000 per month?

65. In exercise 63, what total salary would Ron have earned in 8 years?

66. Find the sum of the odd integers from 1 to 101.

67. Kenny Kranz opened a savings account for his daughter by depositing $50 on the day that she was born. On each subsequent birthday, he deposited $30 more than the previous year. How much money was deposited on his daughter's eighteenth birthday?

68. In exercise 67, how much money (disregarding interest) had Kenny deposited for his daughter after her eighteenth birthday?

Review exercises

1. Subtract $\dfrac{3}{y-6} - \dfrac{2}{6-y}$. See section 4-3.

2. Given $f(x) = 5x - 3$, find $\dfrac{f(x+h) - f(x)}{h}$, $h \neq 0$. See section 8-2.

3. Find the solution set of the logarithmic equation $\log_x 64 = 3$. See section 11-3.

4. Identify each equation as a parabola, circle, ellipse, or hyperbola. See section 9-4.
  a. $x^2 - y = 3x + 1$
  b. $3y^2 = 3x^2 + 1$
  c. $2x^2 + y^2 = 10$

5. Rationalize the denominator of $\dfrac{\sqrt{2}}{\sqrt{2} - \sqrt{3}}$. See section 5-6.

6. Reduce $\dfrac{3a^2 - 3}{3a^2 + 2a - 5}$ to lowest terms. See section 4-1.

14-4 ■ Geometric sequences and series

A geometric sequence

Suppose a man offers to rent you his house under the conditions that the rent will be figured daily as follows:

first day	1¢
second day	2¢
third day	4¢
fourth day	8¢
fifth day	16¢
sixth day	32¢
.	.
.	.
.	.

The daily rent on the house forms the sequence

$$1,2,4,8,16,32, \cdots$$

in which case each term, after the first, is obtained by multiplying the preceding term by the constant 2. Such a sequence is called a **geometric sequence.**

___ *Definition of a geometric sequence* _____

A **geometric sequence** is a sequence having the property that each term after the first term can be obtained by multiplying the preceding term by the same nonzero constant multiplier.

A geometric sequence is also called a **geometric progression.** We call the constant multiplier the **common ratio** since successive terms of the sequence form a

"common ratio." We denote the common ratio by r where $r = \dfrac{a_{n+1}}{a_n}$. That is, r

can be obtained by dividing any term after the first by the preceding term. This common ratio is the characteristic that distinguishes a geometric sequence from any other sequence.

■ *Example 14–4 A*

Determine whether the given sequence is geometric. If it is geometric, find the common ratio r.

1. $4,12,36,108, \cdots$
   Since $12 \div 4 = 3$, $36 \div 12 = 3$, and $108 \div 36 = 3$, the sequence is geometric and the common ratio $r = 3$.

2. $6,12,36,72, \cdots$
   Since $12 \div 6 = 2$ and $36 \div 12 = 3$, the sequence is not geometric.

▶ *Quick check* Determine if the sequence $-1,2,-4,8, \cdots$ is geometric. If so, find the common ratio and the next three terms of the sequence. ■

General term of a geometric sequence

The definition of a geometric sequence shows that the first few terms of a general geometric sequence with first term a_1 and common ratio r take the form

$$a_1, \ a_2 = a_1 r, \ a_3 = a_1 r^2, \ a_4 = a_1 r^3, \cdots$$

from which we can determine the following about the general term of a geometric sequence.

___ *General term of a geometric sequence* _____

The general term of a geometric sequence with first term a_1 and common ratio r is given by

$$a_n = a_1 r^{n-1}$$

■ *Example 14-4 B*

1. Given the sequence $2, 10, 50, 250, \cdots$

   we can determine that this is a geometric sequence with the common ratio
   $r = 5$ because $\dfrac{10}{2} = 5$, $\dfrac{50}{10} = 5$, and $\dfrac{250}{50} = 5$. Since $a_1 = 2$, the general
   term of the geometric sequence is given by

   $a_n = 2(5)^{n-1}$

2. State the terms of the geometric sequence whose nth general term is

   $a_n = 3\left(\dfrac{1}{2}\right)^{n-1}$

   The terms of the sequence are

   $a_1 = 3\left(\dfrac{1}{2}\right)^0 = 3(1) = 3$

   $a_2 = 3\left(\dfrac{1}{2}\right)^1 = \dfrac{3}{2}$

   $a_3 = 3\left(\dfrac{1}{2}\right)^2 = 3\left(\dfrac{1}{4}\right) = \dfrac{3}{4}$

   $a_4 = 3\left(\dfrac{1}{2}\right)^3 = 3\left(\dfrac{1}{8}\right) = \dfrac{3}{8}$

   $\vdots$

   The sequence then is given by

   $3, \dfrac{3}{2}, \dfrac{3}{4}, \dfrac{3}{8}, \cdots, 3\left(\dfrac{1}{2}\right)^{n-1}, \cdots$

   where $a_1 = 3$ and $r = \dfrac{1}{2}$.

3. Find the sixth term, a_6, of the geometric sequence with $a_1 = 24$ and
   $r = -\dfrac{1}{2}$.

   Using $a_n = a_1 r^{n-1}$,

   $a_6 = (24)\left(-\dfrac{1}{2}\right)^{(6)-1}$ Replace n with 6, a_1 with 24, and r with $-\dfrac{1}{2}$

   $= 24\left(-\dfrac{1}{2}\right)^5$

   $= 24\left(-\dfrac{1}{32}\right)$

   $= -\dfrac{3}{4}$

   The sixth term of the geometric sequence is $-\dfrac{3}{4}$.

Note   We could use the calculator and y^x key in example 3. Press
$\boxed{\cdot}\ \boxed{5}\ \boxed{+/-}\ \boxed{y^x}\ \boxed{5}\ \boxed{=}\ \boxed{\times}\ \boxed{2}\ \boxed{4}\ \boxed{=}$ and read "-0.75" on the
display.

4. A man pays rent on his house at the rate of 1¢ the first day, 2¢ the second day, 4¢ the third day, 8¢ the fourth day, and so on. How much rent did he pay on the twentieth day?

We are given $a_1 = 1$, $n = 20$, and find $r = 2$. Using $a_n = a_1 r^{n-1}$,

$$a_{20} = (1)(2)^{(20)-1}$$

Replace n with 20, r with 2, and a_1 with 1

$$= (2)^{19}$$
$$= 524,288 \text{ cents}$$
$$= \$5,242.88$$

The man would pay $5,242.88 rent on the twentieth day.

▶ *Quick check*   Find the sixth term of the geometric sequence with $a_1 = 27$ and $r = -1/3$.   ■

Geometric series

Now consider the sum of the first n terms of a geometric sequence, denoted by S_n, and called the nth partial sum of the geometric sequence. With the geometric sequence

$$a_1, a_1 r, a_1 r^2, a_1 r^3, \cdots, a_1 r^{n-1}$$

is associated the geometric series

$$S_n = a_1 + a_1 r + a_1 r^2 + a_1 r^3 + \cdots + a_1 r^{n-1} \tag{1}$$

If we multiply each member of this equation by the common ratio r, we obtain

$$r S_n = a_1 r + a_1 r^2 + a_1 r^3 + a_1 r^4 + \cdots + a_1 r^n \tag{2}$$

Subtracting term by term equation (2) from equation (1), we obtain

$$S_n = a_1 + a_1 r + a_1 r^2 + a_1 r^3 + \cdots + a_1 r^{n-1}$$
$$r S_n = \phantom{a_1 + {}} a_1 r + a_1 r^2 + a_1 r^3 + \cdots + a_1 r^{n-1} + a_1 r^n$$
$$\overline{S_n - r S_n = a_1 \phantom{+ a_1 r + a_1 r^2 + a_1 r^3 + \cdots + a_1 r^{n-1}} - a_1 r^n}$$

Thus

$$S_n - r S_n = a_1 - a_1 r^n$$
$$(1 - r)S_n = a_1 - a_1 r^n \qquad \text{Factor } S_n$$
$$S_n = \frac{a_1 - a_1 r^n}{1 - r} \qquad r \neq 1$$
$$= \frac{a_1(1 - r^n)}{1 - r} \qquad \text{Factor } a_1 \text{ in the numerator}$$

Note   When $r = 1$, then

$$S_n = \underbrace{a_1 + a_1 + a_1 + \cdots + a_1}_{n \text{ terms}} = n a_1$$

┌─ *Sum of the first* n *terms of a geometric sequence* ────────
│ The sum of the first n terms of a geometric sequence is given by
│
$$S_n = \frac{a_1(1 - r^n)}{1 - r} \quad (r \neq 1)$$
│
│ where a_1 is the first term and r is the common ratio. When $r = 1$, then
$$S_n = na_1$$

■ *Example 14–4 C*

1. Find the sum of the first seven terms of the geometric sequence whose first term $a_1 = 3$ and whose common ratio $r = 2$.

Now we want S_7 where $n = 7$, $a_1 = 3$, and $r = 2$. Using the formula

$$S_n = \frac{a_1(1 - r^n)}{1 - r}$$

$$S_7 = \frac{(3)(1 - 2^7)}{1 - (2)}$$ Replace n with 7, a_1 with 3, and r with 2

$$= \frac{3(1 - 128)}{-1}$$

$$= \frac{3(-127)}{-1}$$

$$= 381$$

The sum of the first seven terms is 381.

2. Find $\sum\limits_{k=1}^{4} 2(3)^k$.

The series is the fourth partial sum of a geometric sequence, where $a_1 = 2 \cdot 3 = 6$ and $r = 3$. We want S_4. Using the formula

$$S_n = \frac{a_1(1 - r^n)}{1 - r}$$

$$S_4 = \frac{(6)(1 - 3^4)}{1 - (3)}$$ Replace a_1 with 6, r with 3, and n with 4

$$= \frac{6(1 - 81)}{-2}$$

$$= \frac{6(-80)}{-2}$$

$$= 240$$

Therefore $S_4 = \sum\limits_{k=1}^{4} 2(3)^k = 240$.

3. Find the sum of the first twenty-five terms of the geometric sequence $3,3,3,3, \cdots$.

The common ratio is $\dfrac{3}{3} = 1$ and $a_1 = 3$. Thus,

$$S_{25} = na_1 = 25(3) = 75$$

▶ *Quick check*   Find S_8 of the geometric sequence where $a_1 = 2$ and $r = 3$.   ■

Mastery points

Can you
- Identify a geometric sequence?
- Find the common ratio of a geometric sequence?
- Write the terms of a geometric sequence?
- Find the general term of a geometric sequence?
- Find the indicated term of a given geometric sequence?
- Find the nth partial sum of a geometric sequence?

Exercise 14–4

Determine if the given terms form a geometric sequence. If they do, find the common ratio and write the next three terms of the sequence. See example 14–4 A.

Example Determine if the sequence $-1, 2, -4, 8, \cdots$ is geometric. If so, find the common ratio and the next three terms of the sequence.

Solution Since $2 \div -1 = -2$, $-4 \div 2 = -2$ and $8 \div -4 = -2$, the sequence is geometric and the common ratio is $r = -2$. Successively multiplying by -2, the next three terms are $-16, 32, -64$.

1. $1, 3, 9, \cdots$ 2. $6, 12, 24, \cdots$ 3. $\dfrac{1}{2}, \dfrac{1}{6}, \dfrac{1}{18}, \cdots$ 4. $\dfrac{1}{3}, \dfrac{1}{2}, \dfrac{3}{5}, \cdots$

5. $4, -2, 1, \cdots$ 6. $-1, \dfrac{1}{2}, -\dfrac{1}{4}, \cdots$ 7. $6, -2, \dfrac{2}{3}, \cdots$ 8. $12, 4, \dfrac{4}{3}, \cdots$

9. $\dfrac{1}{2}, \dfrac{2}{3}, \dfrac{3}{4}, \cdots$ 10. $16, 48, 80, \cdots$

Find the general term, a_n, of the given geometric sequence. See example 14–4 B–1.

Example $-12, 6, -3, \cdots$

Solution The first term $a_1 = -12$ and the common ratio $r = \dfrac{6}{-12} = -\dfrac{1}{2}$. Thus the general term

$$a_n = -12(-\dfrac{1}{2})^{n-1}.$$

11. $3, 6, 12, \cdots$ 12. $8, 12, 18, \cdots$ 13. $27, -18, 12, \cdots$ 14. $-81, 27, -9, \cdots$

15. $1, \sqrt{3}, 3, \cdots$ 16. $32, 16\sqrt{2}, 16, \cdots$ 17. $-\dfrac{1}{15}, \dfrac{1}{5}, -\dfrac{3}{5}, \cdots$

Find the indicated term of the geometric sequence having the following characteristics. See example 14-4 B-3.

Example Find the sixth term of the geometric sequence with $a_1 = 27$ and $r = -\dfrac{1}{3}$.

Solution Using $a_n = a_1 r^{n-1}$

$$a_6 = (27)\left(-\frac{1}{3}\right)^{(6)-1} \qquad \text{Replace } n \text{ with 6, } a_1 \text{ with 27, and } r \text{ with } -\frac{1}{3}$$

$$= 27\left(-\frac{1}{3}\right)^5$$

$$= 3^3\left(-\frac{1}{3^5}\right) \qquad \text{Write 27 as } 3^3$$

$$= -\frac{1}{3^2} \qquad \text{Reduce by } 3^3$$

$$= -\frac{1}{9}$$

The sixth term of the geometric sequence is $-\dfrac{1}{9}$.

18. $a_1 = 3$, $r = 2$; find a_6.

19. $a_1 = 2$, $r = 3$; find a_5.

20. $a_1 = 16$, $r = \dfrac{1}{2}$; find a_7.

21. $a_1 = 81$, $r = \dfrac{1}{3}$; find a_4.

22. $a_1 = 1$, $r = -4$; find a_5.

23. $a_1 = 5$, $r = -2$; find a_6.

24. $a_1 = 25$, $r = -\dfrac{1}{5}$; find a_4.

25. $a_1 = -32$, $r = -\dfrac{1}{4}$; find a_5.

26. $3,18,108,\cdots$; find a_6.

27. $9,18,36,\cdots$; find a_7.

28. $10,-20,40,\cdots$; find a_8.

29. $7,-14,28,\cdots$; find a_9.

Find the indicated partial sum of the geometric sequence having the following characteristics. See example 14-4 C.

Example Find S_8 of the geometric sequence where $a_1 = 2$ and $r = 3$.

Solution Using $S_n = \dfrac{a_1(1 - r^n)}{1 - r}$

$$S_8 = \frac{(2)[1 - (3^8)]}{1 - 3} \qquad \text{Replace } n \text{ with 8, } a_1 \text{ with 2, and } r \text{ with 3}$$

$$= \frac{2(1 - 6{,}561)}{-2}$$

$$= -(-6{,}560)$$

$$= 6{,}560$$

The sum of the first eight terms is 6,560.

30. $a_1 = 8$, $r = 2$; sum of the first six terms

31. $a_1 = 14$, $r = 3$; sum of the first five terms

32. $a_1 = -64$, $r = \dfrac{1}{4}$; sum of the first four terms

33. $a_1 = -10$, $r = \dfrac{1}{5}$; sum of the first four terms

34. $9,18,36,\cdots$; find S_7.

35. $3,18,108,\cdots$; find S_6.

36. $\dfrac{1}{3},\dfrac{1}{9},\dfrac{1}{27},\cdots$; find S_5.

37. $\dfrac{1}{2},\dfrac{1}{4},\dfrac{1}{8},\cdots$; find S_9.

39. $-5,15,-45,\cdots$; find S_5.

38. $\dfrac{4}{3},\dfrac{8}{3},\dfrac{16}{3},\cdots$; find S_7.

See example 14–4 C–2.

40. $\sum\limits_{i=1}^{9} 3^i$

41. $\sum\limits_{j=1}^{8} 4^j$

42. $\sum\limits_{k=1}^{7} (-2)^k$

43. $\sum\limits_{i=1}^{5} (-3)^i$

44. $\sum\limits_{j=1}^{6} \left(\frac{1}{4}\right)^j$

45. $\sum\limits_{k=1}^{7} \left(\frac{2}{3}\right)^k$

46. $\sum\limits_{i=1}^{8} \left(-\frac{1}{3}\right)^i$

47. $\sum\limits_{k=1}^{6} 3\left(\frac{2}{5}\right)^k$

48. $\sum\limits_{j=1}^{8} 4\left(\frac{2}{3}\right)^j$

49. $\sum\limits_{i=1}^{6} -5\left(\frac{3}{5}\right)^i$

50. $\sum\limits_{j=1}^{6} -3\left(-\frac{1}{3}\right)^j$

Solve the following word problems.

51. A ball is dropped from a height of 9 feet. If on each rebound it rises two-thirds of the height from which it fell, what distance has it traveled when it strikes the ground for the sixth time?

52. A basketball rebounds to a height that is three-fourths of the height from which it fell. If the basketball is dropped initially from a height of 4 meters, what distance has it traveled when it strikes the floor for the fifth time?

53. On a visit to Las Vegas, Emmett Broughton doubled his bet each time that he lost. If his first bet was $2 and he lost consecutive bets, how much did he bet on the ninth bet?

54. In exercise 53, if Emmett loses his ninth bet also, how much will he have lost after his ninth loss?

55. In the first paragraph of this section, a man is paid monthly rent for his house at the rate of 1¢ the first day, 2¢ the second day, 4¢ the third day, and so on. At this rate, what would the rent for the house be for a month of 30 days?

56. A certain bacteria culture under a given condition triples in number each hour. If there were originally 1,000 bacteria, how many hours would it take the number of bacteria present in the culture to surpass 1 million?

57. A pump used to expel air from a tank removes one-fifth of what remains in the tank with each stroke. What part of the air in the tank has been removed after the fifth stroke?

58. An automobile depreciates in value each year by one-fifth of its value at the beginning of the year. If an auto is purchased for $8,000, what is its value at the end of the fourth year?

Review exercises

Perform the indicated operations. See section 3–2.

1. $(4x + y)^2$

2. $(3x - 2y)^3$

3. Complete the square of the expression $x^2 - 12x$. See section 6–2.

4. Evaluate the expression $\frac{p}{1 - q}$ when $p = \frac{2}{3}$ and $q = \frac{1}{2}$. See section 1–5.

5. Rationalize the denominator of the expression $\frac{3}{2 - i}$. See section 5–7.

6. Find the solution set of the quadratic equation $2y^2 - y + 4 = 0$. See section 6–3.

14–5 ■ Infinite geometric series

Consider the formula for the sum of the first n terms of a geometric sequence given by

$$S_n = \frac{a_1 - a_1 r^n}{1 - r} = \frac{a_1(1 - r^n)}{1 - r}$$

which can be written in the form

$$S_n = \frac{a_1}{1 - r}(1 - r^n)$$

Now let n get greater and greater. Let $|r| < 1$ and recall that if $|r| < 1$, then $-1 < r < 1$. We can show that as n becomes increasingly large, r^n becomes closer to zero. To illustrate, if $r = \dfrac{1}{3}$, then

$$r^2 = \left(\frac{1}{3}\right)^2 = \frac{1}{9}, \quad r^3 = \left(\frac{1}{3}\right)^3 = \frac{1}{27},$$
$$r^4 = \left(\frac{1}{3}\right)^4 = \frac{1}{81}, \quad \cdots, \quad r^{10} = \frac{1}{59{,}049}, \text{ and so on}$$

As n increases, we can see $r^n = \left(\dfrac{1}{3}\right)^n$ approaches (gets closer and closer to) the value zero.

Now since when $|r| < 1$, r^n approaches the value zero as n increases, then

$$\frac{a_1}{1 - r}(1 - r^n)$$

approaches the value

$$\frac{a_1}{1 - r}(1 - 0) = \frac{a_1}{1 - r}$$

The sum of an infinite geometric series

If $|r| < 1$, the sum of the terms of an infinite geometric series, denoted by S_∞, is given by

$$S_\infty = \frac{a_1}{1 - r}$$

If $|r| \geq 1$, the sum does not exist.

In sigma notation, we write the preceding statement by

$$S_\infty = \sum_{i=1}^{\infty} a_1 r^{i-1} = \frac{a_1}{1 - r}, \quad |r| < 1$$

where the upper limit n is replaced by ∞ in the statement $S_n = \sum_{i=1}^{n} a_1 r^{i-1}$.

■ *Example 14–5 A*

1. Find the sum of the terms of an infinite geometric series such that $a_1 = 2$ and $r = \dfrac{1}{2}$.

We want S_∞ and use the formula

$$S_\infty = \frac{a_1}{1 - r}$$

$$= \frac{(2)}{1 - \left(\dfrac{1}{2}\right)} \qquad \text{Replace } a_1 \text{ with 2 and } r \text{ with } \dfrac{1}{2}$$

$$= \frac{2}{\dfrac{1}{2}}$$

$$= 4$$

The sum of the terms in the infinite geometric series is 4.

2. Find $\displaystyle\sum_{i=1}^{\infty} 2\left(\frac{1}{4}\right)^i$.

We want $S_\infty = \displaystyle\sum_{i=1}^{\infty} 2\left(\frac{1}{4}\right)^i$, in which $a_1 = 2\left(\dfrac{1}{4}\right) = \dfrac{2}{4} = \dfrac{1}{2}$ and $r = \dfrac{1}{4}$.

Using $S_\infty = \dfrac{a_1}{1 - r}$

$$S_\infty = \frac{\left(\dfrac{1}{2}\right)}{1 - \left(\dfrac{1}{4}\right)} \qquad \text{Replace } a_1 \text{ with } \dfrac{1}{2} \text{ and } r \text{ with } \dfrac{1}{4}$$

$$= \frac{\dfrac{1}{2}}{\dfrac{3}{4}}$$

$$= \frac{1}{2} \cdot \frac{4}{3}$$

$$= \frac{2}{3}$$

Thus, $S_\infty = \displaystyle\sum_{i=1}^{\infty} 2\left(\frac{1}{4}\right)^i = \frac{2}{3}$.

▶ *Quick check* Find $\displaystyle\sum_{j=1}^{\infty} 12\left(\frac{1}{4}\right)^j$. ■

The sum of the terms of an infinite geometric series, $|r| < 1$, has some practical uses. We now consider two of the most common applications.

■ *Example 14–5 B*

1. Write the repeating decimal $0.27\overline{27}$ as a rational number in lowest terms.
Now $0.272727 = 0.27 + 0.0027 + 0.000027 + \cdots$. We have an infinite geometric series with $a_1 = 0.27$ and

$$r = \frac{0.0027}{0.27} = 0.01$$

Using the formula $S_\infty = \dfrac{a_1}{1-r}$, we substitute 0.27 for a_1 and 0.01 for r to obtain

$$S_\infty = \dfrac{(0.27)}{1-(0.01)} \qquad \text{Replace } a_1 \text{ with 0.27 and } r \text{ with 0.01}$$

$$= \dfrac{0.27}{0.99}$$

$$= \dfrac{27}{99}$$

$$= \dfrac{3}{11}$$

Thus the repeating decimal $0.2727\overline{27} = \dfrac{3}{11}$.

2. A ball is dropped from a height of 12 meters. If each time it strikes the floor the ball rebounds to a height that is three-fourths of the height from which it fell, find the total distance that the ball travels before it comes to rest on the floor.

Let d be the total distance the ball travels. After the initial 12-meter drop, the ball will travel the same distance up and back down after each striking on the floor. Therefore

$$a_1 = 12\left(\dfrac{3}{4}\right) = 9 \text{ and } r = \dfrac{3}{4} \text{ and using the formula } S_\infty = \dfrac{a_1}{1-r}$$

$$S_\infty = \dfrac{(9)}{1-\left(\dfrac{3}{4}\right)} \qquad \text{Replace } a_1 \text{ with 9 and } r \text{ with } \dfrac{3}{4}$$

$$= \dfrac{9}{\dfrac{1}{4}}$$

$$= 36$$

Then $d = 12 + 2(36) = 12 + 72 = 84$.
  Distance traveled up and down
  Initial drop

The ball would travel a distance of 84 meters before coming to rest.

▶ *Quick check* Write $0.3636\overline{36}$ as a rational number in lowest terms. ■

Mastery points

Can you
- Find the sum of the terms of an infinite geometric series with $|r| < 1$?
- Express a repeating decimal as a rational number using $S_\infty = \dfrac{a_1}{1-r}$, $|r| < 1$?

Exercise 14–5

Find the sum of the terms of the given infinite geometric series. If the series has no sum, indicate that condition.
See example 14–5 A–1.

Example Find $\displaystyle\sum_{j=1}^{\infty} 12\left(\frac{1}{4}\right)^j$.

Solution Using $S_\infty = \dfrac{a_1}{1-r}$

$$S_\infty = \frac{(3)}{1-\left(\dfrac{1}{4}\right)} \qquad \text{Replace } a_1 \text{ with 3 and } r \text{ with } \frac{1}{4}$$

$$= \frac{3}{\dfrac{3}{4}}$$

$$= 3 \cdot \frac{4}{3}$$

$$= 4$$

Thus, $\displaystyle\sum_{j=1}^{\infty} 12\left(\frac{1}{4}\right)^j = 4$.

1. $a_1 = 1, r = \dfrac{2}{3}$

2. $a_1 = 2, r = \dfrac{1}{3}$

3. $a_1 = -3, r = \dfrac{1}{2}$

4. $a_1 = \dfrac{1}{5}, r = \dfrac{1}{10}$

5. $a_1 = \dfrac{3}{5}, r = \dfrac{1}{3}$

6. $a_1 = 4, r = -\dfrac{1}{2}$

7. $a_1 = -\dfrac{5}{6}, r = -\dfrac{2}{3}$

8. $14 + 7 + \dfrac{7}{2} + \cdots$

9. $12 + 4 + \dfrac{4}{3} + \cdots$

10. $3 + \dfrac{3}{4} + \dfrac{3}{16} + \cdots$

11. $4 + \dfrac{4}{5} + \dfrac{4}{25} + \cdots$

12. $1 + \dfrac{2}{3} + \dfrac{4}{9} + \cdots$

13. $6 - 8 + \dfrac{32}{3} - \cdots$

14. $\displaystyle\sum_{i=1}^{\infty} \left(\frac{1}{2}\right)^{i-1}$

15. $\displaystyle\sum_{j=1}^{\infty} \left(\frac{4}{5}\right)^j$

16. $\displaystyle\sum_{k=1}^{\infty} \left(\frac{7}{8}\right)^{k+1}$

17. $\displaystyle\sum_{i=1}^{\infty} \left(-\frac{2}{3}\right)^i$

18. $\displaystyle\sum_{k=1}^{\infty} \left(-\frac{5}{3}\right)^k$

19. $\displaystyle\sum_{i=1}^{\infty} 3\left(\frac{1}{5}\right)^i$

20. $\displaystyle\sum_{j=1}^{\infty} 4\left(\frac{1}{6}\right)^{j-1}$

Express the given repeating decimal as a rational number in lowest terms. See example 14–5 B–1.

Example $0.36\overline{36}$

Solution We can write $0.36\overline{36} = 0.36 + 0.0036 + 0.000036 + \cdots$.

Then $a_1 = 0.36$ and $r = \dfrac{0.0036}{0.36} = 0.01$.

Using $S_\infty = \dfrac{a_1}{1-r}$

$= \dfrac{(0.36)}{1-(0.01)}$ Replace a_1 with 0.36 and r with 0.01

$= \dfrac{0.36}{0.99}$

$= \dfrac{36}{99}$ Multiply the numerator and the denominator by 100

$= \dfrac{4}{11}$ Reduce to lowest terms

The rational number equivalent of $0.36\overline{36}$ is $\dfrac{4}{11}$.

21. $0.33\overline{3}$ **22.** $0.4242\overline{42}$ **23.** $0.28181\overline{81}$ **24.** $0.4727\overline{72}$ **25.** $0.03636\overline{36}$

Solve the following word problems. See example 14–5 B–2.

26. A ball returns to two-thirds of its previous height with each bounce. If the ball is dropped from a height of 6 feet, what is the total distance the ball will travel before coming to rest?

27. When a weight on an attached spring is dropped, it falls a distance of 30 inches before the spring stretches to its limit and the weight springs back up. If the weight rebounds to nine-tenths of the preceding distance it fell, through what total distance does the weight travel before coming to rest?

28. A bob in a pendulum travels an arc length that is seven-eighths of its preceding arc length. If the first arc length is 16 centimeters, how far will the bob move before coming to rest?

29. If the first swing of a pendulum bob is 14 inches and each succeeding swing is five-sixths as long as the preceding one, what is the total distance the bob will travel before coming to rest?

30. A grant from an alumnus of Henry Ford Community College was such that the college was to receive $30,000 the first year and two-thirds of the preceding year's donation each year thereafter. What was the total amount of money the college would receive from the alumnus?

31. Muriel Lakey's cat, Epu, receives 5 milligrams of a medicine at 2 P.M. If Epu is to receive four-fifths of the preceding dose of the medicine every hour thereafter, how many milligrams of medicine does Epu receive altogether?

Review exercises

1. Expand $(x+y)^3$ by multiplying $(x+y)(x+y)(x+y)$. See section 3–2.

Perform the indicated operations. Write the answer in positive exponents only. See section 5–1.

2. $(a^{-4})^{1/2}$ (Assume $a \neq 0$.)

3. $(a^6b^4)^{1/2}$

4. Sketch the graph of $f(x) = \sqrt{x+1}$. See section 8–3.

5. Given $f(x) = x^2 + 2x + 1$ and $g(x) = 2x - 1$, find $f[g(x)]$. See section 8–2.

Completely factor the following. See sections 3–7 and 3–8.

6. $a^4 - b^4$ **7.** $3x^2 - 27y^2$ **8.** $3x^3 + 24y^3$

14–6 ■ *Mathematical induction*

A method of proof used to prove the formulas found in this chapter, called **mathematical induction,** was first used by Italian mathematician Giuseppe Peano in the late nineteenth and early twentieth centuries.

Let us consider an example. Suppose that we want the sum of the first n odd positive integers.

$$
\begin{aligned}
1 &= 1 \\
1 + 3 &= 4 = 2^2 \\
1 + 3 + 5 &= 9 = 3^2 \\
1 + 3 + 5 + 7 &= 16 = 4^2 \\
1 + 3 + 5 + 7 + 9 &= 25 = 5^2
\end{aligned}
$$

We can see a pattern developing that leads us to conclude that

the sum of the first n odd positive integers is n^2

Now the general term for the sequence whose partial sum we are finding would be $2n - 1$. Thus, we might conclude that

$$1 + 3 + 5 + 7 + \cdots + (2n - 1) + \cdots = n^2$$

for any positive integer n.

We have shown this proposition to be true for the first five odd positive integers. How do we know that it will be true for the first 1,000 odd positive integers? To show this, we use proof by *mathematical induction.*

To state the principle we use in proof by mathematical induction, we first introduce some notation. Suppose that for any positive integer n, we denote the proposition to be proved by P_n. Then, in our example above,

P_1 is the proposition that $1 = 1^2$
P_2 is the proposition that $1 + 3 = 2^2$
P_3 is the proposition that $1 + 3 + 5 = 3^2$
$\vdots$

We will write these propositions by

P_1: $1 = 1^2$
P_2: $1 + 3 = 2^2$
P_3: $1 + 3 + 5 = 3^2$
$\vdots$
P_n: $1 + 3 + 5 + 7 + \cdots + (2n - 1) + \cdots = n^2$

Using this notation, we now state the principle of mathematical induction that we use in our proofs.

Principle of mathematical induction

Suppose that for each positive integer n we have a proposition P_n such that the following conditions are true:

1. P_1 is true.
2. For each positive integer k, when P_k is true, then P_{k+1} is true.

Then P_n is true for all positive integers n.

Using the principle of mathematical induction to prove proposition true involves two parts.

1. Show the statement P_n is true when $n = 1$.
2. Show that if P_n is true when $n = k$, then it is also true when $n = k + 1$.

Both of these properties must be established for the proof to be complete. The second part of the proof is the most difficult part. We *assume* that P_k is true and must show that P_{k+1} is true as a result of this assumption. We call the statement P_k the *induction hypothesis.*

■ **Example 14–6 A**

1. Prove that for every positive integer n,

$$P_n: 1 + 2 + 3 + \cdots + n = \frac{n(n + 1)}{2}$$

We are asked to prove that the sum of the first n positive integers is $\dfrac{n(n + 1)}{2}$.

Proof: a. Show that P_1 is true. P_1 is the statement that the first term in the left member, 1, is equal to $\dfrac{n(n + 1)}{2}$ when $n = 1$.

$$P_1: 1 = \frac{(1)(1 + 1)}{2} \qquad \text{Replace } n \text{ with 1}$$

$$1 = \frac{1(2)}{2}$$

$$1 = 1$$

Thus, P_n is true and part 1 is complete.

b. We assume the statement P_k is true when n is replaced by *some* positive integer k. The *induction hypothesis* states

$$P_k: 1 + 2 + 3 + \cdots + k = \frac{k(k + 1)}{2} \text{ for some } k \qquad (1)$$

We must show that

$$P_{k+1}: 1 + 2 + 3 + \cdots + k + (k + 1) = \frac{(k + 1)[(k + 1) + 1]}{2} \qquad (2)$$

is true.

We note that the left member of statement (2) contains the extra term $(k + 1)$. Therefore add $(k + 1)$, to both members of the induction hypothesis, statement (1).

$$1 + 2 + 3 + \cdots + k + (k + 1) = \frac{k(k + 1)}{2} + (k + 1) \qquad (3)$$

We now simplify the right member of statement (3).

$$
\begin{aligned}
1 + 2 + 3 + \cdots + k + (k + 1) &= \frac{k(k + 1)}{2} + \frac{2(k + 1)}{2} \\
&= \frac{k(k + 1) + 2(k + 1)}{2} \\
&= \frac{(k + 1)(k + 2)}{2} \qquad \text{Factor } (k + 1) \\
&= \frac{(k + 1)[(k + 1) + 1]}{2}
\end{aligned}
$$

The final statement is statement (2), which we were trying to prove to be true. Because statement (1), P_k, implies the truth of statement (2), $P_{k + 1}$, the proof is complete and P_n is true for all positive integers n.

Note   By this statement,

$$
\begin{aligned}
1 + 2 + 3 + \cdots + 52 &= \frac{52(52 + 1)}{2} \\
&= \frac{52(53)}{2} \\
&= 1{,}378
\end{aligned}
$$

If we add the first 52 positive integers, we *do* get 1,378.

2. Prove that for every positive integer n,

$$P_n: 1 + 4 + 9 + \cdots + n^2 = \frac{n(n + 1)(2n + 1)}{6}$$

Proof: **a.** Show P_1 is true.

$$
\begin{aligned}
P_1: 1 &= \frac{(1)[(1) + 1][2(1) + 1]}{6} \qquad \text{Replace } n \text{ with 1} \\
&= \frac{1(2)(3)}{6} \\
1 &= 1
\end{aligned}
$$

Thus, P_1 is true.

b. Assume the induction hypothesis P_k is true ($n = k$). Assume

$$P_k: 1 + 4 + 9 + \cdots + k^2 = \frac{k(k + 1)(2k + 1)}{6} \quad (1) \text{ for some } k.$$

We must now show $P_{k + 1}$ is true ($n = k + 1$).

P_{k+1}: $1 + 4 + 9 + \cdots + k^2 + (k + 1)^2$

$$= \frac{(k + 1)[(k + 1) + 1][2(k + 1) + 1]}{6}$$

We now add $(k + 1)^2$ to each member of the induction hypothesis (1).

$1 + 4 + 9 + \cdots + k^2 + (k + 1)^2$

$$= \frac{k(k + 1)(2k + 1)}{6} + (k + 1)^2$$

$$= \frac{k(k + 1)(2k + 1)}{6} + \frac{6(k + 1)^2}{6}$$

$$= \frac{k(k + 1)(2k + 1) + 6(k + 1)^2}{6}$$

Factor the common factor $(k + 1)$ from each term of the numerator.

$$= \frac{(k + 1)[k(2k + 1) + 6(k + 1)]}{6}$$

$$= \frac{(k + 1)(2k^2 + k + 6k + 6)}{6} \qquad \text{Multiply}$$

$$= \frac{(k + 1)(2k^2 + 7k + 6)}{6} \qquad \text{Combine}$$

$$= \frac{(k + 1)[(k + 2)(2k + 3)]}{6} \qquad \text{Factor}$$

$$= \frac{(k + 1)[(k + 1) + 1][2(k + 1) + 1]}{6}$$

The assumed truth of the induction hypothesis P_k thus implies the statement P_{k+1} so the statement P_n is true for all positive integers n.

Note   When $n = 25$, then

$$P_{25}: 1 + 4 + 9 + \cdots + (25)^2 = \frac{25(25 + 1)[2(25) + 1]}{6}$$

$$= \frac{25(26)(51)}{6}$$

$$= 25(13)(17)$$

$$= 5{,}525$$

The sum of the squares of the first 25 positive integers is 5,525.

3. Prove that for every positive integer $n \geq 2$,

     P_n: $3^n > 2n + 1$

Proof: **a.** Since the statement must be true for $n \geq 2$, we first show P_n is true when $n = 2$.

     P_2: $3^{(2)} > 2(2) + 1$

         $9 > 4 + 1$

         $9 > 5$ (True)

Thus, P_n is true when $n = 2$.

b. Assume the induction hypothesis P_k is true. Assume P_k: $3^k > 2k + 1$ for some k. We must show that P_n is true when $n = k + 1$.

P_{k+1}: $3^{k+1} > 2(k + 1) + 1$

By the induction hypothesis,

$3^k > 2k + 1$

Multiply each member of the inequality by 3.

$3^k \cdot 3 > (2k + 1)3$
$3^{k+1} > 6k + 3$ \qquad Property of exponents

But we can write

$6k + 3 = 2(k + 1) + 1 + 4k$

Thus, we have

$3^{k+1} > 2(k + 1) + 1 + 4k$

Since $4k > 1$ for any positive integer k, we add $2(k + 1)$ to each member of this inequality.

Then \qquad\qquad $2(k + 1) + 4k > 2(k + 1) + 1$
and \qquad\qquad $2(k + 1) + 1 + 4k > 2(k + 1) + 4k$
Thus, \qquad\qquad $2(k + 1) + 1 + 4k > 2(k + 1) + 4k > 2(k + 1) + 1$
and \qquad $3^{k+1} > 2(k + 1) + 1 + 4k > 2(k + 1) + 1$

therefore, by the transitive property of inequalities (if $a > b$ and $b > c$, then $a > c$),

$3^{k+1} > 2(k + 1) + 1$

Assuming P_k true implies that P_{k+1} is true and we have proved that

P_n: $3^n > 2n + 1$

for all positive integers $n \geq 2$.

Note Let $n = 7$, then

$$P_7: 3^7 > 2(7) + 1$$
$$2{,}187 > 15$$

which is certainly true.

▶ *Quick check* Prove that $3 + 6 + 9 + \cdots + 3n = \dfrac{3n(n + 1)}{2}$ for all positive integers n. ■

┌─ *Mastery points* ───────────────────────────────

Can you
- Prove a statement of equality using mathematical induction?
- Prove a statement of inequality using mathematical induction?

Exercise 14–6

Verify the following statements are true for $n = 1, 2, 3,$ and 4. See example 14–6 A.

Example P_n: $3 + 6 + 9 + \cdots + 3n = \dfrac{3n(n + 1)}{2}$

Solution   1. When $n = 1$

$$3 = \frac{3(1)(1 + 1)}{3}$$

$$3 = \frac{3(2)}{2}$$

$$3 = 3$$

2. When $n = 2$

$$3 + 6 = \frac{3(2)(2 + 1)}{2}$$

$$9 = \frac{3(2)(3)}{2}$$

$$9 = 9$$

3. When $n = 3$

$$3 + 6 + 9 = \frac{3(3)(3 + 1)}{2}$$

$$18 = \frac{3(3)(4)}{2}$$

$$18 = 18$$

4. When $n = 4$

$$3 + 6 + 9 + 12 = \frac{3(4)(4 + 1)}{2}$$

$$30 = \frac{3(4)(5)}{2}$$

$$30 = 30$$

1. P_n: $1 + 2 + 3 + \cdots + n = \dfrac{n(n + 1)}{2}$

2. P_n: $1^2 + 2^2 + 3^2 + \cdots + n^2 = \dfrac{n(n + 1)(2n + 1)}{6}$

3. P_n: $1^3 + 2^3 + 3^3 + \cdots + n^3 = \dfrac{n^2(n + 1)^2}{4}$

4. P_n: $\dfrac{1}{1 \cdot 2} + \dfrac{1}{2 \cdot 3} + \dfrac{1}{3 \cdot 4} + \cdots + \dfrac{1}{n(n + 1)} = \dfrac{n}{n + 1}$

5. P_n: $n < 2^n$

6. P_n: $\dfrac{1}{2} + \dfrac{1}{4} + \dfrac{1}{8} + \cdots + \dfrac{1}{2}n < 1$

Prove the following statements P_n by mathematical induction. See example 14–6 A.

Example   Prove that for every positive integer n,

$$P_n: 3 + 6 + 9 + \cdots + 3n = \frac{3n(n + 1)}{2}$$

Solution   a. Prove P_n true when $n = 1$.

$$P_1: 3 = \frac{3(1)[(1) + 1]}{2} \qquad \text{Replace } n \text{ with 1}$$

$$= \frac{3 \cdot 2}{2}$$

$$3 = 3$$

Therefore, P_1 is true.

b. Assume P_n true when $n = k$.

Assume P_k: $3 + 6 + 9 + \cdots + 3k = \dfrac{3k(k + 1)}{2}$ for some k.

c. We must prove P_n true when $n = k + 1$.

Prove P_{k+1}: $3 + 6 + 9 + \cdots + 3k + 3(k + 1) = \dfrac{3(k + 1)[(k + 1) + 1]}{2}$

By induction hypothesis,

$$3 + 6 + 9 + \cdots + 3k = \dfrac{3k(k + 1)}{2}$$

We then add the $(k + 1)^{st}$ term, $3(k + 1)$, to each member.

$$
\begin{aligned}
3 + 6 + 9 + \cdots + 3k + 3(k + 1) &= \dfrac{3k(k + 1)}{2} + 3(k + 1) \\
&= \dfrac{3k(k + 1)}{2} + \dfrac{2 \cdot 3(k + 1)}{2} &&\text{Add in right member} \\
&= \dfrac{3(k + 1)(k + 2)}{2} &&\text{Factor } 3(k + 1) \\
&= \dfrac{3(k + 1)[(k + 1) + 1]}{2} &&\text{Write in the form of } P_{k+1}
\end{aligned}
$$

Thus, the assumption that P_k is true implies that P_{k+1} is true and so P_n is true for all positive integers n.

7. P_n: $2 + 4 + 6 + \cdots + 2n = n(n + 1)$

8. P_n: $4 + 8 + 12 + \cdots + 4n = 2n(n + 1)$

9. P_n: $3 + 7 + 11 + \cdots + (4n - 1) = n(2n + 1)$

10. P_n: $1 + 3 + 5 + \cdots + (2n - 1) = n^2$

11. P_n: $3 + 9 + 15 + \cdots + (6n - 3) = 3n^2$

12. P_n: $2 + 5 + 8 + \cdots + (3n - 1) = \dfrac{n(3n + 1)}{2}$

13. P_n: $\dfrac{1}{1 \cdot 2} + \dfrac{1}{2 \cdot 3} + \dfrac{1}{3 \cdot 4}$
$+ \cdots + \dfrac{1}{n(n + 1)} = \dfrac{n}{n + 1}$

14. P_n: $\dfrac{1}{1 \cdot 2 \cdot 3} + \dfrac{1}{2 \cdot 3 \cdot 4} + \dfrac{1}{3 \cdot 4 \cdot 5} + \cdots$
$+ \dfrac{1}{n(n + 1)(n + 2)} = \dfrac{n(n + 3)}{4(n + 1)(n + 2)}$

15. P_n: $1^3 + 2^3 + 3^3 + \cdots + n^3 = \dfrac{n^2(n + 1)^2}{4}$

16. P_n: $1^3 + 3^3 + 5^3 + \cdots + (2n - 1)^3$
$= n^2(2n^2 - 1)$

17. P_n: $n < 2^n$

18. P_n: $\dfrac{1}{2} + \dfrac{1}{4} + \dfrac{1}{8} + \cdots + \dfrac{1}{2}n < 1$

19. P_n: $2^n > 2n$, $n \geq 3$

20. P_n: $1 + 2 + 3 + \cdots + n < \dfrac{1}{8}(2n + 1)^2$

21. Prove that 2 is a factor of $n^2 + n$.

22. Prove that 3 is a factor of $n^3 - n + 3$.

23. Prove that $7^n - 1$ is divisible by 6 for any positive integer n.

24. Prove that $x^n - 1$ has a factor of $x - 1$ for all $n \in N$.

25. Prove that $(ab)^n = a^n b^n$ where a and b are constants.

26. Prove that $a^n > 1$, if $a > 1$.

Review exercises

1. Given sets $A = \{1,2,4,5,7,8\}$ and $B = \{3,5,7,9\}$, find (a) $A \cup B$ and (b) $A \cap B$. See section 1–1.

2. If $n = 4$, evaluate $n(n - 1)(n - 2)(n - 3)$. See section 1–5.

3. Simplify the complex rational expression $\dfrac{\dfrac{1}{x} - \dfrac{1}{y}}{\dfrac{1}{x^2} - \dfrac{1}{y^2}}$.

See section 4–4.

Find the solution set of the following equations.

4. $x^2 - 5x + 6 = 0$. See section 6–3.

5. $\log_2(x + 2) + \log_2 x = 3$. See section 11–3.

Chapter 14 lead-in problem

A sky diver falls 10 meters during the first second, 20 meters during the second second, 30 meters during the third second, and so on. How many meters will he fall during the eleventh second?

Solution

This is an arithmetic sequence where we use the formula $a_n = a_1 + (n - 1)d$, where a_1 is the first term of the sequence, n is the number of the term we wish, a_{11}, and d is the common difference between each second. Thus, $a_1 = 10$, $n = 11$, and $d = 10$.

$a_{11} = (10) + [(11) - 1](10)$ Replace a_1 with 10, n with 11, and d with 10

$= 10 + 10(10)$
$= 10 + 100$
$= 110$

The sky diver falls 110 meters during the eleventh second.

Chapter 14 summary

1. An infinite **sequence** is a function whose domain is the set of the positive integers.
2. A sequence is **finite** when its domain is the set $\{1,2,3, \cdots, n\}$ for some fixed n.
3. A **series** is the sum of the first n terms of a sequence.
4. The sum of the first n terms of a sequence whose general term is a_n, in **sigma notation**, is given by

$$\sum_{i=1}^{n} a_i$$

   where i is the **index of summation**, 1 is the **lower limit,** and n is the **upper limit** of summation.
5. An **arithmetic sequence** is a sequence in which each term after the first differs from the preceding term by the same common difference d.
6. The nth term a_n of an arithmetic sequence is given by

$$a_n = a_1 + (n - 1)d$$

   where a_1 is the first term and d is the common difference.

7. The nth partial sum S_n of the terms of an arithmetic sequence is given by

$$S_n = \frac{n}{2}(a_1 + a_n) = \frac{n}{2}[2a_1 + (n - 1)d]$$

8. A **geometric sequence** is a sequence in which each term after the first term can be obtained by multiplying the preceding term by the same nonzero constant multiplier, called the **common ratio** and denoted by r.
9. The general term of a geometric sequence with first term a_1 and common ratio r is given by

$$a_n = a_1 r^{n-1}$$

10. The nth partial sum S_n of the terms of a geometric sequence is given by

$$S_n = \frac{a_1 - a_1 r^n}{1 - r} = \frac{a_1(1 - r^n)}{1 - r} \quad (r \neq 1)$$

   When $r = 1$,

$$S_n = na$$

11. The sum of the terms in an infinite geometric series is given by

$$S_\infty = \frac{a_1}{1 - r}, \ |r| < 1$$

When $|r| \geq 1$, S_∞ does not exist.

12. To prove a statement using **mathematical induction,** we
   a. Show the statement true when $n = 1$.
   b. Assume the statement is true when $n = k$ for some k (called *induction hypothesis*).
   c. Show the statement is true when $n = k + 1$.

Chapter 14 error analysis

1. Terms of a sequence
   Example: Given $a_n = (-1)^n(2n + 3)$, find $a_1, a_2,$ and a_3
   $a_1 = (-1)[2(1) + 3] = -5$
   $a_2 = (-1)[2(2) + 3] = -7$
   $a_3 = (-1)[2(3) + 3] = -9$
   Correct answer: $-5, 7, -9$
   What error was made? (*see page 594*)

2. Sigma notation
   Example: $\sum\limits_{i=2}^{5} (2i + 5) = 7 + 9 + 11 + 13 + 15$
   $= 55$
   Correct answer: 48
   What error was made? (*see page 600*)

3. Sigma notation
   Example: Write the partial sum $3 + 8 + 13 + 18$ in sigma notation.
   $\sum\limits_{i=1}^{4} (4i - 1)$
   Correct answer: $\sum\limits_{i=1}^{4} (5i - 2)$
   What error was made? (*see page 601*)

4. Arithmetic sequence
   Example: The sequence $3, -1, -6, -12, -19, \cdots$ is arithmetic.
   Correct answer: not arithmetic
   What error was made? (*see page 604*)

5. Geometric sequence
   Example: The sequence $5, \frac{5}{2}, \frac{5}{6}, \frac{5}{24}, \frac{5}{120}, \cdots$ is geometric.
   Correct answer: not geometric
   What error was made? (*see page 613*)

6. Geometric series
   Example: $\sum\limits_{i=1}^{3} 3(2)^i = 3 + 12 + 24 = 39$
   Correct answer: $\sum\limits_{i=1}^{3} 3(2)^i = 42$
   What error was made? (*see page 615*)

7. Geometric series
   Example: Find $\sum\limits_{k=1}^{5} 3(3)^k$, using $S_n = \frac{a_1(1 - r^n)}{1 - r}$, where
   $a_1 = 9, n = 5,$ and $r = 3$.
   $S_5 = \frac{9(1 - 3^5)}{1 - 3} = \frac{9(2^5)}{-2} = -9(2^4) = -144$
   Correct answer: $S_5 = \sum\limits_{k=1}^{5} 3(3)^k = 1,089$
   What error was made? (*see page 615*)

8. Infinite geometric series
   Example: $\sum\limits_{i=1}^{\infty} 3\left(\frac{1}{5}\right)^{i-1} = \frac{\frac{3}{5}}{1 - \frac{1}{5}} = \frac{\frac{3}{5}}{\frac{4}{5}} = \frac{3}{4}$
   Correct answer: $\sum\limits_{i=1}^{\infty} 3\left(\frac{1}{5}\right)^{i-1} = \frac{15}{4}$
   What error was made? (*see page 621*)

9. Sums of radical expressions
   Example: $\sqrt{16 + 36} = \sqrt{16} + \sqrt{36} = 4 + 6 = 10$
   Correct answer: $2\sqrt{13}$
   What error was made? (*see page 237*)

10. Logarithm of a number
   Example: $2^5 = 32$ is equivalent to $\log_5 32 = 2$.
   Correct answer: $\log_2 32 = 5$
   What error was made? (*see page 486*)

Chapter 14 critical thinking

The epsilon fraternity has a ritualistic handshake that the members perform at the beginning and end of each meeting. If there are 20 members at a meeting, how many handshakes will be performed?
Note: This is combinations of 20 hands taken 2 at a time, but the answer could be found by logically thinking about the event.

Chapter 14 review

[14–1]

Write the first five terms of each sequence whose general term a_n is given.

1. $a_n = 4n + 3$

2. $a_n = \dfrac{5n}{2n - 1}$

3. $a_n = (-1)^n \cdot \dfrac{4}{2n + 5}$

4. $a_n = (-1)^{n + 1} \cdot 2^n$

Find the indicated term of the sequence whose general term a_n is given.

5. $a_n = 4 - 3n$, find a_6.

6. $a_n = (-1)^n(3n - 4)$, find a_7.

7. $a_n = \dfrac{3^{n + 1}}{2n}$, find a_9.

8. $a_n = (-1)^{n - 1} \cdot \dfrac{2^n + 1}{3^n}$, find a_{11}.

Given the following sequences, find an expression for the general term a_n.

9. $5,7,9,11, \cdots$

10. $3,8,13,18, \cdots$

11. $\dfrac{2}{3}, \dfrac{3}{7}, \dfrac{4}{11}, \dfrac{5}{15}, \cdots$

12. $-4,9,-14,19, \cdots$

13. Dockage fees for a boat at a marina are $3.00 for the first night, $3.25 for the second night, $3.50 for the third night, and so on. Write an expression for the general term of the sequence. How much did it cost to dock the boat on the eighth night?

[14–2]

Expand each indicated sum and find the sum.

14. $\displaystyle\sum_{i=1}^{4} (4i - 1)$

15. $\displaystyle\sum_{i=1}^{6} i(i + 5)$

16. $\displaystyle\sum_{k=1}^{5} \dfrac{k^2}{k + 1}$

17. $\displaystyle\sum_{j=1}^{6} (-1)^j \cdot \dfrac{4}{5j}$

Write each sum in sigma notation.

18. $5 + 8 + 11 + 14$

19. $\dfrac{4}{5} - \dfrac{5}{6} + \dfrac{6}{7} - \dfrac{7}{8} + \dfrac{8}{9}$

[14–3]

Find the indicated term of each arithmetic sequence having the following characteristics.

20. $a_1 = 5$, $d = 4$; find a_{15}.

21. $a_1 = -3$, $d = 5$; find a_{17}.

22. $3,7,11,15, \cdots$; find a_{21}.

23. $-6,-8,-10,-12$; find a_{19}.

Find the number of terms in each given finite arithmetic sequence.

24. $-2,3,8, \cdots, 68$

25. $4,0,-4,-8, \cdots, -76$

Find the indicated partial sum of each given arithmetic sequence.

26. $-3,3,9, \cdots, 81$; find S_{15}.

27. $10,7,4, \cdots, -50$; find S_{21}.

28. $\displaystyle\sum_{j=1}^{29} \dfrac{2}{3}j$

29. $\displaystyle\sum_{k=1}^{25} \left(\dfrac{1}{2}k + 1 \right)$

30. Company B starts offers of a beginning wage of $12,000 with a raise of $450 each year thereafter. What would the wage be after 11 years?

[14–4]

Find the general term a_n of each given geometric sequence.

31. $3, 6, 12, \cdots$

32. $-\dfrac{3}{4}, \dfrac{9}{16}, -\dfrac{27}{64}, \cdots$

Find the indicated term of each geometric sequence having the following characteristics.

33. $a_1 = 5, r = 3$; find a_5.

34. $a_1 = -24, r = \dfrac{1}{3}$; find a_4.

35. $a_1 = 36, r = -\dfrac{2}{3}$; find a_3.

36. $-9, 18, -36, \cdots$; find a_7.

Find the indicated partial sum of each geometric sequence having the following characteristics.

37. $a_1 = 3, r = 3$; find S_5.

38. $a_1 = -24, r = \dfrac{1}{2}$; find S_6.

39. $\displaystyle\sum_{j=1}^{5} \left(-\dfrac{3}{4}\right)^j$

40. $\displaystyle\sum_{k=1}^{7} 4\left(\dfrac{1}{3}\right)^k$

[14–5]

Find the sum of the terms of each given infinite geometric series.

41. $a_1 = 3, r = \dfrac{3}{4}$

42. $a_1 = -2, r = \dfrac{1}{5}$

43. $\displaystyle\sum_{j=1}^{\infty} 3\left(-\dfrac{1}{4}\right)^j$

44. $\displaystyle\sum_{k=1}^{\infty} \left(-\dfrac{2}{3}\right)\left(-\dfrac{1}{5}\right)^{k+1}$

Use the infinite geometric series to write each repeating decimal as a rational number in lowest terms.

45. $0.353\overline{535}$

46. $0.4323\overline{232}$

47. A boat at anchor experiences a series of waves, each wave having 25% less amplitude (height) than the previous one. If the first wave has amplitude 3 meters, how much vertical distance does the boat travel before coming to rest? (*Hint:* The boat travels *up* and *down* the same distance with each wave.)

[14–6]

48. Verify the statement
$P_n: 2 + 6 + 18 + \cdots + 2 \cdot 3^{n-1} = 3^n - 1$ when $n = 1$, 2, and 3.

Using mathematical induction, prove the following statements true for all positive integers n.

49. $P_n: 2 + 7 + 12 + \cdots + (5n - 3) = \dfrac{n(5n - 1)}{2}$

50. $P_n: 1 + 2n \le 3^n$

Chapter 14 cumulative test

[1–1] **1.** Given $\{y|y$ is an integer between -8 and $4\}$, list the elements in the set.

[1–1] **3.** Given the set $\{x|-4 \leq x < 9\}$, list the integers in the set.

[3–3] **5.** Simplify the following expressions. Leave all answers with positive exponents. Assume $a \neq 0$, $b \neq 0$.
   a. $(-a^3b^{-3})(2a^4b^{-2})$
   b. $(-2a^{-3}b^2)^{-2}$
   c. $\dfrac{a^{-4}b^3}{a^2b^{-1}}$

Completely factor the following expressions.

[3–4] **7.** $12xy - 4x^2y^3 + 8x^3y^2$

[3–5] **9.** $4a^2 - 20ab + 25b^2$

[3–8] **11.** $16a^3 - 2b^3$

Find the solution set of the following equations and inequalities.

[2–1] **13.** $5(3y - 1) + 2y = 8(2 - y)$

[2–5] **15.** $-9 < 4x + 1 \leq 5$

[2–6] **17.** $|5 - 3x| < 4$

[6–1] **19.** $x^2 - 5x = 24$

[4–3] **21.** Add $\dfrac{5}{2a - 1} + \dfrac{6}{4a + 3}$.

[4–2] **23.** Divide $\dfrac{3a^2 - 13a + 4}{a^2 + 2a + 1} \div \dfrac{a^2 - 8a + 16}{a^2 - 1}$.

[4–6] **25.** Divide $(3x^3 + 4x^2 - 2x + 1) \div (x - 3)$ using synthetic division.

[5–7] **27.** Divide $\dfrac{2 + i}{3 - 2i}$ and write the answer in the form $a + bi$.

[6–3] **29.** Find the solution set of the quadratic equation $3p^2 - 2 = 7p$.

[6–7] **31.** Find the solution set of the quadratic inequality $y^2 \leq 4y - 3$.

[1–1] **2.** Given $A = \{-1,2,4,7\}$, $B = \{0,1,2,7,9\}$, and $C = \{-4,-1,0,9\}$, find (a) $A \cap B$, (b) $A \cup C$, (c) $(B \cup C) \cap \emptyset$.

[1–4] **4.** Perform the indicated operations and simplify.
$5 - \{6 - [3 + 18 \div 2 - 3^2] - (4 + 7)\}$

[3–2] **6.** Multiply as indicated.
   a. $(2y + 7)^2$
   b. $(4x + 2y)(4x - 2y)$
   c. $(x - 2)(x^2 + 2x + 4)$

[3–5] **8.** $7y^2 - 34y - 5$

[3–8] **10.** $8x^2 - 50y^2$

[3–8] **12.** $6ax - 3ay - 2bx + by$

[2–5] **14.** $3(2x + 3) < 4(x - 5)$

[2–4] **16.** $|2x - 5| = 3$

[2–6] **18.** $|5y - 4| \geq 6$

[4–7] **20.** $\dfrac{3}{x - 1} - \dfrac{4}{3} = \dfrac{6}{x - 1}$

[4–3] **22.** Subtract $\dfrac{2y + 1}{y^2 - y - 42} - \dfrac{y + 1}{y^2 - 36}$.

[4–4] **24.** Simplify the complex fraction $\dfrac{\dfrac{4}{a} - \dfrac{2}{b}}{\dfrac{1}{b} - \dfrac{2}{a}}$.

[5–5] **26.** Combine the expression $3\sqrt{75} + 2\sqrt{27} - \sqrt{48}$.

[5–6] **28.** Multiply $(3 - 2\sqrt{5})(4 + 3\sqrt{2})$.

[6–5] **30.** Find the solution set of the radical equation $\sqrt{x - 1} = x - 3$. Indicate extraneous solutions.

[7–3] **32.** Find the equation of the line having the following conditions. Write the equation in standard form.
   a. Through points $(1,2)$ and $(-3,-4)$
   b. Through $(4,-3)$ and perpendicular to the line $y - 4x = 3$
   c. Through $(3,-2)$ and having slope $-\dfrac{2}{3}$

[7–2] 33. Find the slope and y-intercept of the line
$5y - 3x = 9$.

[8–2] 34. Given $f(x) = 5x - 3$ and $g(x) = x^2 - x + 1$, find
  a. $f(-3)$
  b. $g(4)$
  c. $\dfrac{f(x + h) - f(x)}{h}$, $h \neq 0$.

Sketch the graph of the following equations and inequalities.

[7–3] 35. $2y - 5x = 10$

[9–1] 36. $y = x^2 - 3x - 10$

[7–4] 37. $4y - 3x < -24$

[9–4] 38. $5x^2 + y^2 = 20$

[9–4] 39. Determine if the given equation represents a circle, a parabola, an ellipse, or a hyperbola.
  a. $x^2 + y^2 - 2x + 4y - 3 = 0$
  b. $2y^2 + 6 = x^2$
  c. $4x - x^2 = y$
  d. $8y^2 = 6 - 5x^2$

[12–1] 40. Find the solution set of the system of linear equations
$3y + 5x = 1$
$2y - 3x = -3$.

[12–4] 41. Find the solution set of the system of nonlinear equations
$x^2 + y^2 = 13$
$3x + 2y = 0$.

[13–1] 42. Evaluate $\begin{vmatrix} 4 & -1 & 3 \\ 5 & 0 & -2 \\ 3 & 6 & 1 \end{vmatrix}$.

[13–2] 43. Find the solution set of the system of equations
$x - 3y = 4$
$2x + 5y = 3$ by determinants.

[11–3] 44. Write the expression $\log_b 5 + \log_b 6 - 3\log_b 2$ as a logarithm of a single number.

[11–4] 45. Find $\log_3 7$ using the common logarithms, correct to four decimal places.

[11–6] 46. Find the solution set of the equation $3^{2-x} = 4$. Round to the nearest tenth.

[11–5] 47. Find $\ln 36$, correct to two decimal places.

[14–2] 48. Write the sum $3 + 9 + 15 + 21 + 27$ in summation notation.

[14–2] 49. Find the indicated sum $\displaystyle\sum_{i=1}^{7} (2i - 1)$.

[14–3] 50. Given $a_1 = -2$ and $d = -3$, find a_{15} of the arithmetic sequence.

[14–3] 51. Find S_{26} of the arithmetic sequence
$\dfrac{1}{2}, 1, \dfrac{3}{2}, 2, \cdots$.

[14–4] 52. Find a_4 of the geometric sequence given
$a_1 = \dfrac{1}{2}$ and $r = -\dfrac{1}{3}$.

[14–5] 53. Find $\displaystyle\sum_{i=1}^{\infty} 2\left(-\dfrac{1}{2}\right)^i$.

[14–5] 54. Find the rational number equivalent of $0.234\overline{234}$.

[14–6] 55. Using mathematical induction, prove P_n: $2^2 + 4^2 + 6^2 + \cdots + (2n)^2 = \dfrac{2n(n + 1)(n + 1)}{3}$.

Counting Techniques, Probability, and the Binomial Theorem

To win the Michigan lottery, a person must correctly select the six integers drawn by the lottery commission from the set of integers from 1 to 47. How many different selections are possible?

15–1 ■ *Fundamental principle of counting*

Multiplication of choices

If a die is rolled and allowed to come to rest, there are six different ways in which this experiment will end (one, two, three, four, five, or six dots on the top face of the die). Chance events such as this are studied in a branch of mathematics called probability. The basic concepts of probability depend on our ability to determine the number of possible ways that an experiment can end. These different possible results of the experiment are called outcomes. If we wish to determine the probability of winning a football pool, getting exactly 7 questions correct on a 10-question test, or winning the lottery, we must know how many possible outcomes there are to the problem. That is, we must know what is possible before we can determine what is probable.

To illustrate the counting process, consider the following example. In the school cafeteria, a quarter-pound hamburger can be ordered cooked rare, medium rare, medium, or well-done. The hamburger can be served on a white, whole wheat, or rye bun. In how many different ways can the hamburger be ordered? The choices can be displayed using a tree diagram as shown in figure 15–1. From the "start" point, we represent by separate lines, or branches of the tree, the four ways that the hamburger can be cooked. The second set of branches represents the type of bun on which the hamburger can be served. We find that (going from left to right) there are altogether twelve different paths along the "branches" of the tree. That is, there are twelve possibilities, called arrangements.

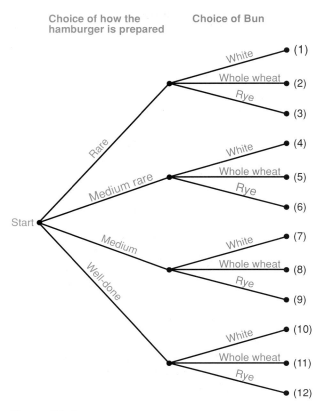

Choice of how the hamburger is prepared   Choice of Bun

Figure 15–1

A tree diagram is a useful means of visualizing the counting process when the number of possible outcomes is small. When the problem has a large number of outcomes, instead of using a tree diagram to list all the possible outcomes, we will be interested only in knowing how many outcomes there will be.

We can see that the answer to the hamburger problem is the product of the number of ways in which the hamburger can be cooked and the number of types of buns on which it can be served ($4 \cdot 3 = 12$). We can now state the following property.

Multiplication of choices property

If a choice consists of two decisions where the first can be made in m ways and for each of these the second can be made in n ways, then the complete choice can be made in $m \cdot n$ ways.

■ *Example 15–1 A*

Determine the number of outcomes.

1. This year the Fine Arts Auditorium is hosting 4 plays and 12 concerts. In how many different ways can a student buy tickets for 1 play and 1 concert?

   Using the multiplication of choices property we have 4 times 12 or 48 different ways.

2. A dance class consists of 7 girls and 5 boys. A pair of 1 girl and 1 boy is to perform at a recital. In how many different ways can the pair be selected?

Since there are 7 girls and 5 boys, we have 7 times 5 or 35 different pairs for the recital.

▶ *Quick check* From an ordinary deck of playing cards, in how many different ways can a person select one red card and one black card? ■

We can generalize the multiplication of choices property so that it will apply to choices involving more than two decisions. We now state the following property.

— *Extended multiplication of choices property* —————————

If a choice consists of k decisions where the first can be made in n_1 ways, for each of these the second can be made in n_2 ways, . . . , and for each of these the kth can be made in n_k ways, then the complete choice can be made in $n_1 \cdot n_2 \cdots \cdot n_k$ ways.

■ *Example 15–1 B*

Determine the number of outcomes.

1. The local newsstand carries 5 different newspapers, 4 different sporting magazines, 6 fashion magazines, and 12 general interest magazines. In how many different ways can a shopper buy one of each of these types of publications?

We multiply as follows:

$5 \cdot 4 \cdot 6 \cdot 12 = 1,440$

There are 1,440 different ways.

2. A computer company has 65 employees of which 7 are in administration, 18 are in research, 25 are in retail sales, and 15 are in commercial sales. If a committee of 4 people is to be formed containing 1 person from each area, how many different committees are possible?

We must choose 1 person from each area, and the total number of possible committees would be

$7 \cdot 18 \cdot 25 \cdot 15 = 47,250$

3. A race has 5 girls in it. In how many different ways can the 5 runners finish the race?

Any 1 of the 5 girls can finish first. Once a girl has finished first, any 1 of the 4 remaining girls can finish second. We then have 3 girls to finish third, 2 left to finish fourth, and 1 left to finish fifth.

$$\underbrace{5}_{1^{st} place} \cdot \underbrace{4}_{2^{nd} place} \cdot \underbrace{3}_{3^{rd} place} \cdot \underbrace{2}_{4^{th} place} \cdot \underbrace{1}_{5^{th} place} = 120$$

Thus, there are 120 different ways to finish the race. ■

Factorial notation

In example 15–1 B number 3, the answer to the problem was obtained by multiplying the consecutive integers from 1 to 5. This type of product occurs often enough in counting problems that we define the following notation.

> **Definition**
>
> Factorial notation for every positive integer n,
> $$n! = n(n - 1)(n - 2) \cdots \cdots 3 \cdot 2 \cdot 1$$
>
> **Concept**
> $n!$ (read n factorial) is the product of all positive integers less than or equal to the positive integer n.

■ **Example 15–1 C**

Evaluate each of the following.

1. 6! (This is read "6 factorial.")

$$6! = 6 \cdot 5 \cdot 4 \cdot 3 \cdot 2 \cdot 1$$
$$= 720$$

2. 10! (read 10 factorial)

$$10! = 10 \cdot 9 \cdot 8 \cdot 7 \cdot 6 \cdot 5 \cdot 4 \cdot 3 \cdot 2 \cdot 1$$
$$= 3{,}628{,}800$$

3. 1!

$$1! = 1 \qquad \text{Since there are no positive integers less than 1} \qquad ■$$

Using example 15–1 C number 1, we see that

$$6! = 6 \cdot 5 \cdot 4 \cdot 3 \cdot 2 \cdot 1$$
$$= 6 \cdot (5 \cdot 4 \cdot 3 \cdot 2 \cdot 1)$$
$$= 6 \cdot 5!$$

and in general,

$$n! = n(n - 1)!$$
$$= n(n - 1)(n - 2)!$$
$$= n(n - 1)(n - 2)(n - 3)!$$

We make use of this fact to simplify some problems involving factorials.

Consider the problem $\dfrac{8!}{6!}$. If we rewrite 8! as $8 \cdot 7 \cdot 6!$, the problem will simplify as follows:

$$\frac{8!}{6!} = \frac{8 \cdot 7 \cdot 6!}{6!} \qquad \text{Divide out the 6!}$$
$$= 8 \cdot 7$$
$$= 56$$

Note When simplifying the fraction $\dfrac{8!}{6!}$, we cannot divide out the common factor of 2 between the 6 and the 8. To do that would be to treat the problem as if 6! and 8! were the same as 6 and 8.

■ *Example 15–1 D*

Reduce the following expressions and leave the answer as a positive integer.

1. $\dfrac{12!}{8!}$

We expand the factorial in the numerator until the last factor is the same as the factorial in the denominator. We then divide out the common factorial.

$$\frac{12!}{8!} = \frac{12 \cdot 11 \cdot 10 \cdot 9 \cdot 8!}{8!}$$
$$= 12 \cdot 11 \cdot 10 \cdot 9$$
$$= 11,880$$

2. $\dfrac{11!}{5!6!}$

When there are two factorials in the denominator, we expand the factorial in the numerator until the last factor is the same as the greater factorial in the denominator. We then divide out the common factorial.

$$\frac{11!}{5!6!} = \frac{11 \cdot 10 \cdot 9 \cdot 8 \cdot 7 \cdot 6!}{5!6!}$$
$$= \frac{11 \cdot 10 \cdot 9 \cdot 8 \cdot 7}{5!}$$

Now we expand the lesser factorial in the denominator and reduce the resulting fraction.

$$= \frac{11 \cdot 10 \cdot 9 \cdot 8 \cdot 7}{5 \cdot 4 \cdot 3 \cdot 2 \cdot 1}$$
$$= 462$$

3. $\dfrac{18!}{16!2!} = \dfrac{18 \cdot 17 \cdot 16!}{16!2!}$

$$= \frac{18 \cdot 17}{2!}$$
$$= \frac{18 \cdot 17}{2 \cdot 1}$$
$$= 153$$

▶ *Quick check* Reduce the following expression and leave the answer as a positive integer. $\dfrac{10!}{8!2!}$

■

┌─ *Mastery points* ─────────────────────────────────

Can you
- Draw a tree diagram?
- Use the multiplication of choices property?
- Use factorial notation?

Exercise 15-1

Show all the possible outcomes. See figure 15–1.

1. Draw a tree diagram of a person selecting a Ford, Chevrolet, or Plymouth that is a two-door or four-door car and has a four- or six-cylinder engine.

2. Draw a tree diagram of a person selecting a VHS, Beta, or eight-millimeter camcorder that is automatic or fixed focus and comes with or without a carrying case.

3. Draw a tree diagram of a person flipping one coin three times.

4. Draw a tree diagram of a person flipping one coin four times.

5. There are five people in a race. Draw a tree diagram of the first and second place finishers. Use A, B, C, D, and E to represent the five people.

6. There are four people in a race. Draw a tree diagram of the finish of the race. Use A, B, C, and D to represent the four people.

7. Draw a tree diagram of a person rolling a pair of dice such that the outcome of the first die is even and the outcome of the second die is odd.

8. Draw a tree diagram of a person rolling a pair of dice such that the dice do not have the same number up.

Determine the number of outcomes. See examples 15–1 A and B.

Example From an ordinary deck of playing cards, in how many different ways can a person select one red card and one black card?

Solution
$$\underset{\substack{\text{Number} \\ \text{of red} \\ \text{cards}}}{26} \cdot \underset{\substack{\text{Number} \\ \text{of black} \\ \text{cards}}}{26} = 676 \qquad \text{Multiplication of choices}$$

Thus, there are 676 different ways.

9. A boy has four pairs of slacks and six shirts. How many different shirt–slack outfits can he wear?

10. When a salesman cannot find the shoe that a customer wants in her size, the woman complains that a good shoe store would stock every style in every color and size. If the store has 50 different styles and each style comes in 12 sizes and 3 colors, how many different pairs of shoes would the store need to stock to have one of each?

11. The Liberal Arts Building has 12 entrances. In how many different ways can a student go in one entrance and out a different one?

12. A class has 12 girls and 15 boys. A girl and a boy are to be selected as student council representatives. In how many different ways can the selection be made?

13. A menu offers a choice of 5 appetizers, 3 salads, 12 entrees, 4 kinds of potatoes, and 5 vegetables. If a meal consists of one of each, in how many different ways can a person select dinner?

14. If a person chooses a glass of wine from a list of 7 and a dessert from a list of 8, how many different complete dinners would there be using the information in exercise 13?

15. In how many different ways can a student answer all the questions on a quiz consisting of 8 true-or-false questions?

16. A quiz contains 5 multiple-choice questions and the number of choices for each question is 4. In how many different ways can a student answer all the questions?

17. From an ordinary deck of playing cards, in how many different ways can a person select one ace, one king, one queen, and one jack?

18. From an ordinary deck of playing cards, in how many different ways can a person select one heart, one diamond, one club, and one spade?

19. In the Mr. Onionhead game, children make up names for Junior Onionhead. If there are 24 first names and 18 middle names, how many different names are possible?

20. In how many different ways can 4 boys and 4 girls sit in a row if the boys and girls must alternate?

21. In how many different ways can 5 boys and 4 girls sit in a row if the boys and girls must alternate?

22. An interior decorator is designing a room. If he has 6 different paint colors, 12 different floor coverings, and 9 different window treatments, with respect to these features, how many different room finishes are possible?

23. GGBB represents a family with 4 children, where the order in which the 4 children were born is 2 girls and then 2 boys. How many different orderings of 4-children families are possible?

24. With respect to exercise 23, how many different orderings of 5-children families are possible?

25. Given the digits $\{1,2,3,4,5\}$:

   a. How many different 4-digit numbers (numerals) can be formed?

   b. How many different 4-digit numbers can be formed without repetition?

   c. How many different 4-digit numbers greater than 3,000 can be formed?

   d. How many different 4-digit numbers greater than 3,000 can be formed without repetition?

   e. How many different 4-digit even numbers can be formed?

   f. How many different 4-digit even numbers can be formed without repetition?

   g. How many different 4-digit odd numbers can be formed if the digits must alternate even, odd, even, odd?

   h. What is the answer to **g** if repetition is not allowed?

   i. How many different 4-digit odd numbers greater than 5,000 can be formed?

   j. What is the answer to **i** if repetition is not allowed?

26. Use the digits $\{1,2,3,4,5,6,7\}$ to answer **a** through **h** of exercise 25.

Evaluate each of the following. Leave your answer as a positive integer. See example 15–1 C.

Example $\dfrac{10!}{8!2!}$

Solution $\dfrac{10!}{8!2!} = \dfrac{10 \cdot 9 \cdot 8!}{8!2!}$

$$= \dfrac{10 \cdot 9}{2!} = \dfrac{10 \cdot 9}{2 \cdot 1}$$

$$= 5 \cdot 9 = 45$$

27. $4!$

28. $5!$

29. $7!$

30. $8!$

31. $\dfrac{10!}{6!}$

32. $\dfrac{8!}{5!}$

33. $\dfrac{9!}{3!}$

34. $\dfrac{10!}{5!}$

35. $\dfrac{9!}{1!}$

36. $\dfrac{6!}{6!}$

37. $\dfrac{7!}{3!4!}$

38. $\dfrac{9!}{5!4!}$

39. $\dfrac{10!}{4!6!}$

40. $\dfrac{14!}{7!7!}$

41. $\dfrac{20!}{10!10!}$

42. $\dfrac{17!}{2!15!}$

43. $\dfrac{40!}{6!34!}$

44. $\dfrac{32!}{6!26!}$

45. $\dfrac{44!}{6!38!}$

46. $\dfrac{46!}{6!40!}$

Review exercises

Completely factor the following expressions. See sections 3–5, 3–6, 3–7, and 3–8.

1. $x^2 + 8x + 15$

2. $x^3 + y^6$

3. $9x^2 - 36$

4. $4x^2 + 11x - 3$

5. $2ax - ay + 6bx - 3by$

6. $8a^3 - 27$

7. $4x^2 - 12x + 9$

8. $6x^2 + 17x - 3$

15–2 ■ *Permutations*

Permutations of distinct elements

Consider the problem where a club with 5 members wants to choose a president and a secretary. If the 5 members are represented by the letters A, B, C, D, and E, we can visualize the problem with a tree diagram as in figure 15–2.

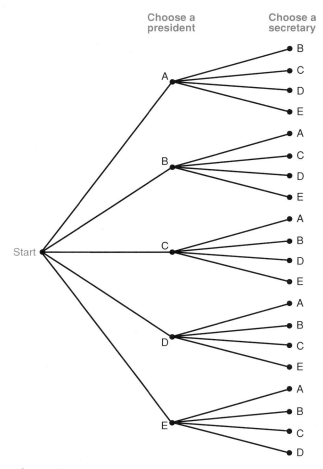

Figure 15–2

Any 1 of the 5 members can fill the position of president and then any 1 of the 4 remaining members can fill the position of secretary. The multiplication of choices property is often used when one or more selections are made from one set and the *order* or *position* of the elements is important. *Ordered arrangements are called permutations.* Each arrangement that can be made by using all or some of the elements of a set *without repetition* is called a permutation of those elements. The phrase "without repetition" means that no element can appear more than once in an arrangement.

In section 15–1, the example of the 5 girls in a race (example 15–1 B number 3) was a permutation of all 5 girls and the answer was 5! or 120. In general, *the number of possible permutations of n distinct objects is n!*.

In the president-secretary example, we formed permutations of the 5 members taken 2 at a time. To generalize the situation where we wish to find the number of permutations of n elements when r of them are taken at a time would be as follows: We will use the symbol $_nP_r$ to represent permutations of n elements taken r at a time.

$$_nP_r = \underset{\text{1}^{\text{st}} \text{ position}}{n} \cdot \underset{\text{2}^{\text{nd}} \text{ position}}{(n-1)} \cdot \underset{\text{3}^{\text{rd}} \text{ position}}{(n-2)} \cdots \underset{r^{\text{th}} \text{ position}}{[n-(r-1)]}$$
$$= n(n-1)(n-2) \cdots (n-r+1)$$

If we multiply the right member of the equation by $\dfrac{(n-r)!}{(n-r)!}$, which is equivalent to 1, we obtain

$$_nP_r = n(n-1)(n-2) \cdots (n-r+1) \cdot \frac{(n-r)!}{(n-r)!}$$
$$= \frac{n(n-1)(n-2) \cdots (n-r+1)(n-r)!}{(n-r)!}$$

Since $n(n-1)(n-2) \cdots (n-r+1)(n-r)! = n!$, we have

$$= \frac{n!}{(n-r)!}$$

We summarize these results as follows.

The number of permutations of a set of n distinct elements selected r at a time, where $r \leq n$, is

$$_nP_r = n(n-1)(n-2) \cdots (n-r+1)$$

or in factorial notation,

$$_nP_r = \frac{n!}{(n-r)!}$$

Note The first formula is useful if we think about the problem as one of filling positions. The second formula is quicker when using a calculator.

Going back to the 5 girls in a race example, the number of ways would be

$$_5P_5 = \frac{5!}{(5-5)!} = \frac{5!}{0!}$$

Since the answer to the problem is 5!, we can see that for the factorial formula to work, 0! must be defined to be equal to 1.

Definition

$$0! = 1$$

■ *Example 15–2 A*

Determine the number of outcomes.

1. During periods of radio silence, messages are sent by means of signal flags. If the signal person has 8 different flags, how many different signals can be sent by placing 3 flags, one above the other, on a flagpole?

Since there is a definite order to the arrangement of the flags (top, middle, and bottom), the problem is a permutation.

By positions: $_8P_3 = \dfrac{8}{\text{top}} \cdot \dfrac{7}{\text{middle}} \cdot \dfrac{6}{\text{bottom}} = 336$

By factorials: $_8P_3 = \dfrac{8!}{(8-3)!} = \dfrac{8!}{5!} = \dfrac{8 \cdot 7 \cdot 6 \cdot 5!}{5!}$
$$= 8 \cdot 7 \cdot 6 = 336$$

2. There are 24 letters in the Greek alphabet. How many fraternities can be specified by choosing three different Greek letters?

By positions: $_{24}P_3 = \dfrac{24}{\text{1}^{st}\text{ letter}} \cdot \dfrac{23}{\text{2}^{nd}\text{ letter}} \cdot \dfrac{22}{\text{3}^{rd}\text{ letter}} = 12{,}144$

By factorials: $_{24}P_3 = \dfrac{24!}{(24-3)!} = \dfrac{24!}{21!} = \dfrac{24 \cdot 23 \cdot 22 \cdot 21!}{21!}$
$$= 24 \cdot 23 \cdot 22 = 12{,}144$$

▶ *Quick check* There are 24 letters in the Greek alphabet. How many fraternities can be specified by choosing 2 different Greek letters? ■

Distinguishable permutations

Consider the possible arrangements of the word NOON. Since there are 4 letters, we might conclude that there are 4! or 24 different arrangements possible. If we switch the two identical *N*'s, the newly formed arrangement, NOON, will not be distinguishable from the original. The same result would occur if we switched the two identical *O*'s. We observe that there are fewer distinguishable permutations of *n* elements when some of those elements are identical than there are when the *n* elements are distinctly different.

For each arrangement, if the *N*'s only are permuted among themselves, there is no actual change. Since the 2 *N*'s can be permuted 2! ways, it follows that we must divide 4! by 2! so as to count only the distinguishable arrangements involving permutations of the *N*'s. Likewise, we must also divide by a second 2! so as to count only the distinguishable arrangements involving permutations of the *O*'s. Therefore, the number of distinguishable permutations of the letters in the word NOON would be

$$\frac{4!}{2!2!} = \frac{4 \cdot 3 \cdot 2!}{2!2!} = \frac{4 \cdot 3}{2!}$$
$$= \frac{4 \cdot 3}{2 \cdot 1} = 6$$

and they are: NOON, NONO, NNOO, ONON, OONN, and ONNO. We can now make the following generalization.

The number of distinct permutations P of the n elements taken n at a time where n_1 are alike of one kind, n_2 are alike of another kind, $\cdots$, and n_k are alike of another kind, is given by

$$P = \frac{n!}{n_1!n_2! \cdots n_k!}$$

where $n_1 + n_2 + \cdots + n_k = n$.

■ *Example 15–2 B*

Determine the number of outcomes.

1. Find the number of distinct permutations of the letters in the word MISSISSIPPI.

   There are eleven letters altogether of which there are one of type M, four of type I, four of type S, and two of type P. The number of distinct permutations would be

   $$P = \frac{11!}{1!4!4!2!} = 34{,}650$$

2. If the signal person has 9 flags of which 3 are red, 3 are white, and 3 are blue, how many different signals can be sent by placing 3 flags, one above the other, on a flagpole?

   $$P = \frac{\overset{\text{9 flags}}{9!}}{\underset{\text{3 red 3 white 3 blue}}{3!\ \ 3!\ \ 3!}} = 1{,}680$$

▶ *Quick check*   Find the number of distinct permutations of the letters in the word COOKBOOK.   ■

┌─ *Mastery points* ───

 Can you
 ■ Determine when an arrangement is a permutation?
 ■ Determine the number of permutations of n distinct elements taken r at a time?
 ■ Determine the number of permutations of n elements taken n at a time where some elements are alike?

└──

Exercise 15–2

Evaluate the following. See example 15–1 D.

1. $_6P_4$ 2. $_7P_2$ 3. $_5P_5$ 4. $_6P_6$ 5. $_{10}P_1$

6. $_{12}P_1$ 7. $_{15}P_4$ 8. $_{20}P_6$ 9. $_{12}P_6$ 10. $_{20}P_{10}$

Determine the number of outcomes. See example 15–2 A.

Example There are 24 letters in the Greek alphabet. How many fraternities can be specified by choosing two different Greek letters?

Solution By positions: $_{24}P_2 = \underset{\text{1st letter}}{24} \cdot \underset{\text{2nd letter}}{23} = 552$

By factorials: $_{24}P_2 = \dfrac{24!}{(24-2)!} = \dfrac{24!}{22!} = \dfrac{24 \cdot 23 \cdot 22!}{22!}$

$= 24 \cdot 23 = 552$

11. In a 9-horse race, how many different first-second-third place finishes are possible?

12. Seven girls are in the final race. How many different first-second-third-fourth place finishes are possible?

13. A president, a vice-president, and a secretary are to be elected from a club with 25 members. In how many different ways can the offices be filled?

14. A basketball team has 15 players. In how many different ways can a captain and a co-captain be chosen from the players?

15. In how many different ways can 8 students be seated in a row of 8 chairs?

16. In how many different ways can 6 different books be arranged on a shelf?

17. In horse racing, a perfecta bet is to pick the first and second place finishers in a race. If a race has 11 horses, how many different perfecta bets are possible?

18. In horse racing, a trifecta bet is to pick the first, second, and third place finishers in a race. If a race has 12 horses, how many different trifecta bets are possible?

19. A contractor wishes to build 8 houses, each different in design. In how many different ways can he build these homes if 5 lots are on one side of the street and 3 lots are on the other side of the street?

20. In horse racing, a superfecta bet is to pick the first, second, third, and fourth place finishes in a race. If a race has 10 horses, how many different superfecta bets are possible?

21. There are 15 players on a baseball team. If the coach selects 9 players for a game, how many different batting orders are possible?

22. A track team has 8 runners. In how many different ways can the coach arrange a relay team of 4 runners?

23. From the set of numbers {1,2,3,4,5,6}, form the following without repetition:
   a. All possible 4-digit numbers
   b. All possible 5-digit numbers
   c. All possible 6-digit numbers
   d. All possible 4-digit numbers less than 2,000
   e. All possible 5-digit numbers greater than 60,000

24. During the first half of the Super Bowl, the producer must arrange 12 different commercials. In how many different ways can this be done?

25. A football coach checks a player for six performance traits and lists the three strongest on a card. If the coach arranges the traits according to the player's level of proficiency, how many different evaluations are possible?

26. Little Lisa is playing with four blocks, each containing one of the letters in her name on all the sides. In how many different ways can she spell her name wrong if she randomly arranges the blocks?

27. A set of reference books has 6 volumes. If the books are randomly arranged on a bookshelf, in how many different ways can the books be arranged out of order?

28. In a race with 11 horses, how many different losing perfecta bets are possible? (See exercise 17.)

29. A witness to a bank robbery said the license number of the get-away car was a 6-digit number of which the first 2 digits were 47. Although she does not recall the remaining digits, she is certain that all the digits in the license number were different. How many different license numbers must the police check?

30. Referring to exercise 29, if the woman was sure that there were no zeros in the license number, how many different license numbers must the police check?

Determine the number of outcomes. See example 15–2 B.

Example Find the number of distinct permutations of the letters in the word COOKBOOK.

Solution There are 8 letters altogether of which there are 1 of type *c*, 1 of type *b*, 2 of type *k*, and 4 of type *o*. The number of distinct permutations would be

$$P = \frac{8!}{1!1!2!4!} = 840$$

31. The secret word of the day is zyzzybalubah. How many different words can be formed using the letters of the secret word?

32. How many different words can be formed using the letters of the word "mammal"?

33. In how many different ways can the monomial $3a^2b^4c^5$ be written without using exponents?

34. Referring to exercise 33, how many different arrangements are possible if the numerical coefficient must be written first?

For exercises 35–40, determine the number of distinguishable permutations of the following words.

35. gamma   36. everywhere   37. TENNESSEE   38. bookkeeper   39. inattention   40. hodgepodge

Review Exercises

Find the solution set of the following linear equations and inequalities. See sections 2–1 and 2–4.

1. $3(2x + 1) = x - 7$
2. $-4x + 6 > -20$
3. $5(x - 4) - 2(2x + 3) = 7$
4. $3(1 - 3x) \leq 2(x + 5)$
5. $5(3x - 2) = -10$
6. $-4 \leq 2x + 5 \leq 8$
7. $4x - 3x + 7 = x + 9$
8. $-7 < 5 - 2x < 3$

15–3 ■ Combinations

Permutations versus combinations

Permutations are ordered arrangements. In the previous section, choosing a president and a secretary was a permutation because we were concerned with assigning a specific position or order to the members. Many times we are not interested in the order of the elements, but rather in what elements make up the selection. For example, if from a club we select two members to serve on a committee, the order in which we choose or list the members is not important.

When a problem requires us to make a selection of elements without regard to the order, then a change in the order of the elements selected does not give a new selection. When this is the situation, we say that we are forming combinations of *n* elements taken *r* at a time. We denote this by $_nC_r$.

If we select A to be president and B to be secretary, this is a different arrangement than if B was president and A was secretary. Whereas if we select a committee made up of A and B, it is the same committee whether we call it the committee of A and B or the committee of B and A.

Combinations

When we are forming combinations, we are forming subsets, since there is no order relationship within a subset. In a subset with r elements, there are $r!$ possible permutations of those elements. Therefore, to determine the number of subsets of r elements (without regard to order) that can be chosen from a set of n objects, we need to divide the number of permutations of n elements taken r at a time by $r!$; that is, $_nP_r \div r!$. We summarize this result as follows.

The number of combinations (subsets) of n distinct elements taken r at a time, where $r \leq n$, is

$$_nC_r = \frac{n(n-1)(n-2) \cdot \cdots \cdot (n-r+1)}{r!},$$

or in factorial notation,

$$_nC_r = \frac{n!}{(n-r)!r!}$$

Note The following are some of the other commonly used notations for permutations and combinations.

$$_nC_r = \binom{n}{r} = C(n,r) = C_{n,r}$$
$$_nP_r = P(n,r) = P_{n,r}$$

The notation $\binom{n}{r}$ is most commonly used when studying the binomial theorem for the expansion of $(a+b)^n$.

■ *Example 15–3 A*

Determine the number of outcomes.

1. A man has twelve shirts and wants to pack three of them for a trip. In how many different ways can this be done?

   Since we are interested in what three shirts are packed and not the order in which they were selected, this is a combination.

   $$_{12}C_3 = \frac{12!}{(12-3)!3!} = \frac{12!}{9!3!} = \frac{12 \cdot 11 \cdot 10 \cdot 9!}{9!3!} = \frac{12 \cdot 11 \cdot 10}{3!} = \frac{12 \cdot 11 \cdot 10}{3 \cdot 2 \cdot 1}$$
   $$= 220$$

2. On a ten-question test, in how many different ways can a student get exactly seven questions correct?

   We wish to work seven questions correctly. We are not interested in which one was worked first or second and so on, therefore, this is a combination.

   $$_{10}C_7 = \frac{10!}{(10-7)!7!} = \frac{10!}{3!7!} = 120$$

   Hence, there are 120 different ways.

▶ *Quick check* From an ordinary deck of playing cards, in how many different ways can a person select a group of three cards? ▪

In the previous example, if instead of asking for the number of different ways that the student could get exactly seven questions correct, we wanted to know in how many different ways the student could get exactly three questions wrong, the answer would be as follows:

$$_{10}C_3 = \frac{10!}{(10-3)!3!} = \frac{10!}{7!3!} = 120$$

From this we can see that the number of combinations of n elements taken r at a time is the same as the number of combinations of n elements taken $n - r$ at a time.

$$_nC_r = {_nC_{n-r}} \text{ or } \binom{n}{r} = \binom{n}{n-r}$$

Further counting problems

Many counting problems cannot be answered by using a single counting property, but rather, several counting properties must be used to solve the problem. The following examples illustrate the strategy necessary for solving such problems.

▪ *Example 15–3 B*

Determine the number of outcomes.

1. A class contains 30 students of which 18 are girls and 12 are boys. How many different committees of 7 students can be formed if there must be 4 girls and 3 boys on the committee?

We use the multiplication of choices property to solve this problem after we have determined the number of ways in which we can select 4 girls from the 18 and 3 boys from the 12. The girls can be selected in $_{18}C_4$ ways and the boys can be selected in $_{12}C_3$ ways. By the multiplication of choices property, each choice of girls can be associated with each choice of boys and we have

$$\underbrace{_{18}C_4}_{\text{choose the girls}} \cdot \underbrace{_{12}C_3}_{\text{choose the boys}}$$
Multiplication of choices

$$= \frac{18!}{(18-4)!4!} \cdot \frac{12!}{(12-3)!3!} = \frac{18!}{14!4!} \cdot \frac{12!}{9!3!}$$
$$= (3,060) \cdot (220) = 673,200$$

Therefore, 673,200 different committees can be formed.

2. How many different 7-card hands from a deck of 52 cards are possible if each hand is to contain 3 hearts, 2 diamonds, and 2 cards that are not a heart or a diamond?

$$\underbrace{_{13}C_3}_{\text{choose the hearts}} \cdot \underbrace{_{13}C_2}_{\text{choose the diamonds}} \cdot \underbrace{_{26}C_2}_{\text{choose the black cards}}$$
Multiplication of choices

$$= \frac{13!}{(13-3)!3!} \cdot \frac{13!}{(13-2)!2!} \cdot \frac{26!}{(26-2)!2!} = \frac{13!}{10!3!} \cdot \frac{13!}{11!2!} \cdot \frac{26!}{24!2!}$$
$$= 286 \cdot 78 \cdot 325 = 7,250,100$$

Hence, there are 7,250,100 different 7-card hands.

3. Two stables are planning a grudge race. The Adams Stable is going to enter 3 of its 7 horses. The Baker Stable is going to enter 3 of its 8 horses. In how many different ways can the 6 horses finish the race?

The Adams horses can be selected in $_7C_3$ ways and the Baker horses can be selected in $_8C_3$ ways. The 6 horses can be selected in $_7C_3 \cdot {}_8C_3$ ways. Each selection of 6 horses can finish the race in $_6P_6$ ways, therefore, the total number of finishes to the race would be

$$_7C_3 \cdot {}_8C_3 \cdot {}_6P_6 = \frac{7!}{(7-3)!3!} \cdot \frac{8!}{(8-3)!3!} \cdot \frac{6!}{(6-6)!}$$
$$= \frac{7!}{4!3!} \cdot \frac{8!}{5!3!} \cdot \frac{6!}{0!} = 35 \cdot 56 \cdot 720 = 1{,}411{,}200$$

Thus, there are 1,411,200 different ways.

▶ *Quick check* A class contains 22 students of which 7 are girls and 15 are boys. How many different committees of 12 students can be formed if there must be 6 girls and 6 boys on the committee? ■

Summary

You should always carefully identify the type of counting problem that you are working. Keep in mind that if there is order in the arrangement, then permutations should be used, but if order does not matter, then combinations should be used. The following table outlines the counting properties.

If a choice consists of k decisions where the first can be made in n_1 ways, for each of these the second can be made in n_2 ways, . . .	Choosing r elements from a set of n distinct elements where repetition is not allowed.	
	Permutations Arrangements of r elements from a set of n where order is important	*Combinations* Subsets of r elements from a set of n where order is not important
Multiplication of choices		
$n_1 \cdot n_2 \cdots \cdots n_k$	$_nP_r = \dfrac{n!}{(n-r)!}$	$_nC_r = \dfrac{n!}{(n-r)!r!}$

┌─ *Mastery points* ─────────────────────

Can you
- Determine the number of combinations of n distinct elements taken r at a time?
- Determine which counting property is to be used to solve a problem?

Exercise 15–3

Evaluate each of the following. Leave your answer as a positive integer. See example 15–1 D.

1. $_{15}C_{10}$
2. $_{18}C_6$
3. $_{12}C_{12}$
4. $_9C_9$
5. $_{18}C_1$
6. $_{20}C_1$
7. $_{42}C_6$
8. $_{44}C_6$
9. $_{20}C_{10}$
10. $_{18}C_9$

Determine the number of outcomes. See example 15–3 A.

Example   From an ordinary deck of playing cards, in how many different ways can a person select a group of three cards?

Solution   Since we are interested in what three cards are selected and not in the order in which they are selected, this is a combination.

$$_{52}C_3 = \frac{52!}{(52-3)!3!} = \frac{52!}{49!3!} = \frac{52 \cdot 51 \cdot 50 \cdot 49!}{49!3!} = \frac{52 \cdot 51 \cdot 50}{3 \cdot 2 \cdot 1} = 22,100$$

Hence, there are 22,100 different ways.

11. A child chooses 4 candies from a box containing 20 different kinds. In how many different ways can she make her selection?

12. Mary Ann's Ice Cream Emporium carries 32 different flavors of ice cream. If each dip must be a different flavor, how many different triple-dip ice cream cones are possible?

13. How many different 5-card hands can be dealt from a normal 52-card deck of playing cards?

14. How many different 13-card hands can be dealt from a normal 52-card deck of playing cards?

15. If on an examination consisting of 12 questions a student may omit 4 questions, in how many different ways can the student select the questions he will attempt to answer?

16. A man has 11 suits and wishes to take 4 with him on a business trip. In how many different ways can this be done?

17. NASA has 19 astronauts it considers suitable for the next mission. If a flight team consists of 6 astronauts, how many different flight teams are possible?

18. How many different 3-member committees can be formed from a club that has 25 members?

19. In how many different ways can a 5-card diamond flush be dealt from an ordinary deck of 52 playing cards?

20. A stockbroker is going to invest in 6 different stocks. If there are 27 different stocks that she is considering, how many different investments are possible?

21. Paul's Pizza Palace has toppings of sausage, green pepper, onion, pepperoni, hamburger, mushroom, black olives, and anchovies. Every Tuesday Paul's special lets you have 4 toppings for the price of two. If Paul does not allow you to have more than one helping of any topping, how many different specials are possible?

22. If you hate anchovies, how many different specials are possible without anchovies? (See exercise 21.)

23. New members at Vince's Video Village are given 4 free rentals from among the 40 newest releases. In how many different ways can a new member make his choice if he cannot rent the same movie more than once?

24. Ten boys wish to form 2 teams of 5 players each. In how many different ways can this be done?

25. Larry's Luncheonettes has restaurants in 14 states. The company decides that it will open restaurants in 8 new states. In how many different ways can the selection be made?

26. A set of 7 distinct points lie on a circle. How many different inscribed triangles can be drawn such that all of their vertices come from this set?

27. How many different lines can be drawn using 15 distinct points, assuming that no 3 points are in a line?

28. An octagon is a polygon with 8 sides. How many different inscribed triangles can be drawn such that all of their vertices come from the vertices of the octagon? (Assume that a side of the octagon can be a side of the triangle.)

29. How many different triangles can be constructed where the vertices are selected from 21 different points in a plane, assuming that no 3 points are in a line?

30. A set of 13 distinct points lie on a circle. How many different chords can be drawn such that their endpoints come from this set?

For exercises 31–40, choose the appropriate counting property and determine the number of possible outcomes. See examples 15–1 A and B, 15–2 A, and 15–3 A.

31. Fourteen children are playing a game of musical chairs. If there is a row of 10 chairs in which the children can sit when the music stops, how many different groups of children could be eliminated from the game? In how many different ways can the seating take place?

32. From an ordinary deck of playing cards, in how many different ways can a person select an ace, a king, a queen, and a jack?

33. There are seventeen players on a baseball team.
   a. In how many different ways can a team of 9 be chosen if every player can play every position.
   b. In how many different ways can a captain and a co-captain be chosen?
   c. How many different batting orders are possible?
   d. In how many different ways can the team membership be reduced to 12 players?

34. In how many different ways can the IRS select 6 tax returns from a group of 30 for a special audit?

35. In how many different ways can 3 boys and 4 girls sit in a row? How many different ways are possible if the boys and girls must alternate?

36. In a race with 6 boys, how many different finishes are possible? How many different first-second-third place finishes are possible?

37. A decagon is a polygon with 10 sides. How many different inscribed triangles can be drawn such that all of their vertices are also vertices of the decagon? (Assume that a side of the decagon can be a side of the triangle.)

38. In how many different ways can a student answer all the questions on a quiz consisting of 10 true-or-false questions?

39. From the set of numbers $\{1,2,3,4,5\}$, form the following:
   a. All possible 3-digit numbers
   b. All possible 3-digit odd numbers
   c. All possible 3-digit numbers where the first and last digit must be even
   d. All possible 3-digit numbers using only odd digits

40. Answer exercise 39 if repetition is not allowed.

Exercises 41–56 may require several counting properties to solve the problem. Determine the number of possible outcomes. See example 15–3 B.

Example A class contains 22 students of which 7 are girls and 15 are boys. How many different committees of 12 students can be formed if there must be 6 girls and 6 boys on the committee?

Solution We use the multiplication of choices property to solve this problem after we have determined the number of ways in which we can select 6 girls from the 7 and 6 boys from the 15. The girls can be selected in $_7C_6$ ways, and the boys can be selected in $_{15}C_6$ ways. By the multiplication of choices property, each choice of girls can be associated with each choice of boys, and we have

$$\underbrace{_7C_6}_{\text{choose the girls}} \cdot \underbrace{_{15}C_6}_{\text{choose the boys}}$$

Multiplication of choices

$$= \frac{7!}{(7-6)!6!} \cdot \frac{15!}{(15-6)!6!} = \frac{7!}{1!6!} \cdot \frac{15!}{9!6!}$$
$$= 7 \cdot 5005 = 35{,}035$$

41. In how many different ways can a group of eight boys and six girls be selected for a group consisting of four boys and three girls?

42. Ten teams are in a league. If each team is required to play every other team twice during the season, what is the total number of league games that will be played?

43. If a group consists of 18 men and 12 women, in how many different ways can a committee of 6 be selected if:

   a. The committee is to have an equal number of men and women?

   b. The committee is to be all women?

   c. There are no restrictions on the membership of the committee?

44. A shopper is choosing 6 different frozen dinners from a selection of 17 and 4 different fruits from a selection of 11. In how many different ways can the selections be made?

For exercises 45–50, determine how many different 7-card hands can be dealt from an ordinary deck of 52 playing cards under the given conditions.

45. There must be 2 red cards and 5 black cards.

46. There are no restrictions.

47. There must be no face cards. (Assume that a face card is a king, a queen, or a jack.)

48. All the cards are of the same suit.

49. No two cards have the same denomination.

50. There are 3 spades, 2 clubs, and 2 red cards.

51. At the beginning and end of every meeting of a certain club, each member must give the ritual handshake to each of the other members. If there are 20 members present at the meeting, how many different handshakes will take place?

52. A test contains 3 groups of questions. Groups A, B, and C contain 5, 4, and 3 questions, respectively. If a student must select 3 questions from group A and 2 from each of the remaining groups, how many different tests are possible?

53. Ten friends are to be seated in a row. In how many different ways can this be done if Don and Brenda must sit next to each other?

54. In horse racing, a double trifecta bet is to pick the first, second, and third place finishers in the first two races. If there are 9 horses in the first race and 8 horses in the second race, how many different bets are possible?

55. A stamp collector has 8 different foreign stamps and 10 different United States stamps. Find the number of ways in which he can select 3 foreign stamps and 3 United States stamps and arrange them in 6 numbered spaces in his album.

56. Two four-member relay teams are being combined into one. If the coach is going to pick two members from each team and then arrange the four runners for a race, in how many different ways can this be done?

Review exercises

Find the solution set for the following absolute value equations and inequalities. See sections 2–4 and 2–6.

1. $|3x - 1| = 4$

2. $|2x + 1| = |x - 2|$

3. $|2x + 5| > 7$

4. $|3x - 4| \leq 5$

5. $|1 - 3x| < 6$

6. $|2x - 5| + 7 = 12$

7. $|3x - 1| \geq 4$

8. $|2 - 5x| \geq 8$

15–4 ■ *Introduction to probability*

Terminology

Probability is a means of measuring uncertainty. The formal study of probability was started in the seventeenth century by the famous French mathematicians Blaise Pascal and Pierre de Fermat. It was originally developed in connection with games of chance, but it is now used for everything from predicting the chance of rain tomorrow to determining the probability that an engine will fail on the space shuttle.

In the study of probability, any happening whose result is uncertain is called an *experiment*. The different possible results of the experiment are called *outcomes,* and the set of all possible outcomes of an experiment is called the *sample space* of the experiment and is denoted by S. (In this textbook, all sample spaces are finite.)

An *event* is a subset of the sample space. If an event is the empty set, it is called the *impossible event*. If an event has only one element, it is called a *simple event*. Any nonempty event that is not simple is called a *compound event*.

■ *Example 15–4 A*

Determine the sample space.

1. Tossing a six-sided die

The sample space consists of six outcomes, which are represented by the integers from 1 to 6.

$S = \{1,2,3,4,5,6\}$

2. Flipping a coin twice

The possible outcomes are
a. a head both flips
b. a head the first flip and a tail the second flip
c. a tail the first flip and a head the second flip
d. a tail both flips

If we use H for a head and T for a tail, the sample space would be

$S = \{HH,HT,TH,TT\}$

In this example, the event of a head and a tail can occur two ways (HT or TH) and is a compound event. The event of getting two heads can occur only one way (HH) and is a simple event. ■

Probability of an event

If we wish to determine the **probability of an event,** we must know how many outcomes make up the event and the sample space. The number of outcomes in event A is represented by $n(A)$ and the number of outcomes in the sample space is represented by $n(S)$. If the outcomes of a sample space are *equally likely,* then every outcome in the sample space has the same chance of occurring.

___ *The probability of an event* _____

If an event A is made up of $n(A)$ equally likely outcomes from a sample space S that has $n(S)$ equally likely outcomes, then the **probability** of the event A, represented by $P(A)$, is

$$P(A) = \frac{n(A)}{n(S)}$$

Since the number of outcomes that make up an event will be less than or equal to the total number of outcomes in the sample space, the probability of an event cannot exceed 1. Furthermore, since the number of outcomes can never be negative, the probability of an event cannot be negative. These observations give rise to some basic principles of probability.

___ *Basic probability principles* _____

If the event A has $n(A)$ equally likely outcomes and the sample space S has $n(S)$ equally likely outcomes, then

a. $0 \le P(A) \le 1$

b. $P(S) = \dfrac{n(S)}{n(S)} = 1$

c. $P(A) = 0$ means event A cannot occur and is called an impossible event.

d. $P(A) = 1$ means event A must occur and is called a certain event.

■ *Example 15–4 B*

When rolling a single six-sided die, name the following events and give their probability.

	Event	*Name*	*Probability*
1.	$\{1\}$	rolling a one	$\dfrac{1}{6}$
2.	$\{2,4,6\}$	rolling an even number	$\dfrac{3}{6} = \dfrac{1}{2}$
3.	$\{7\}$	rolling a 7 (impossible event)	0
4.	$\{1,2,3,4,5,6\}$	rolling a 1,2,3,4,5 or 6 (certain event)	1 ■

In the examples of rolling a single die or flipping a coin twice, it was easy to list all of the possible outcomes in the sample space. When the problem has a large number of possible outcomes, we will not concern ourselves with listing all possible outcomes, instead we will be interested only in knowing how many outcomes there will be. We will use the various counting techniques to determine the number of outcomes that make up an event or the sample space.

■ *Example 15–4 C*

Find the probability of the given event.

1. Five cards are selected from a standard deck of playing cards. What is the probability of a heart flush (all five cards are hearts)?

Since there is no order in the selection of the cards, the number of different five-card hands that is the sample space would be combinations of the 52 cards taken 5 cards at a time.

$$n(S) = {}_{52}C_5 = \frac{52!}{47!5!} = 2{,}598{,}960$$

The number of different five-card hands composed of only hearts that is $n(A)$ would be combinations of the 13 hearts taken 5 hearts at a time.

$$n(A) = {}_{13}C_5 = \frac{13!}{8!5!} = 1{,}287$$

The probability of a heart flush would be

$$P(A) = \frac{n(A)}{n(S)} = \frac{{}_{13}C_5}{{}_{52}C_5} = \frac{1{,}287}{2{,}598{,}960} = \frac{33}{66{,}640} \approx 0.0005$$

2. To win the State of Michigan Lottery, a person must correctly select the six integers drawn by the lottery commission from the set of integers from 1 to 47. If a person buys one ticket, what is the probability of winning?

Since there is no order in the selection of the integers, the number of different sets of 6 integers, which is the sample space, is combinations of the 47 integers taken 6 at a time.

$$n(S) = {}_{47}C_6 = \frac{47!}{41!6!} = 10{,}737{,}573.$$

If a person buys only 1 ticket, then the number of ways of winning, which is $n(A)$, is 1.
The probability of winning the State of Michigan Lottery is

$$P(A) = \frac{n(A)}{n(S)} = \frac{1}{{}_{47}C_6} = \frac{1}{10{,}737{,}573} \approx 0.00000009$$

▶ *Quick check* A class contains 8 men and 6 women. A committee of 4 people is selected. What is the probability that all 4 people will be women? ■

If two events from the same sample space have no outcomes in common, they are called *mutually exclusive* events. For example, if a single die is tossed, the event A of rolling a number less than 3 and the event B of rolling a number greater than 4 are mutually exclusive.

┌─ *Probability of mutually exclusive events* ─────────────

If the events A and B are mutually exclusive, then the probability that A or B will occur is

$$P(A \text{ or } B) = P(A \cup B) = P(A) + P(B)$$

■ *Example 15–4 D*   Find the probability of the given event.

1. A card is drawn from a standard deck of 52 cards. What is the probability of an ace or a jack?

The card that is drawn cannot be an ace and a jack at the same time. Therefore, the two events are mutually exclusive. If $P(A)$ is the probability of an ace and $P(J)$ is the probability of a jack, we have

$$P(A) = \frac{n(A)}{n(S)} = \frac{4}{52} = \frac{1}{13}$$

also

$$P(J) = \frac{n(J)}{n(S)} = \frac{4}{52} = \frac{1}{13}$$

and

$$P(A \cup J) = P(A) + P(J)$$
$$= \frac{1}{13} + \frac{1}{13} = \frac{2}{13}$$

2. A single die is tossed. What is the probability of a 5 or an even number?

Since a 5 is not an even number, these events are mutually exclusive. Let F be the event that 5 appears.

$$P(F) = \frac{n(F)}{n(S)} = \frac{1}{6}$$

Also if we use E to represent the event that an even number appears,

$$P(E) = \frac{n(E)}{n(S)} = \frac{3}{6} = \frac{1}{2}$$

and

$$P(F \cup E) = P(F) + P(E)$$
$$= \frac{1}{6} + \frac{3}{6}$$
$$= \frac{4}{6} = \frac{2}{3}$$

▶ *Quick check*   A coin is flipped 3 times. What is the probability of exactly 2 heads?   ■

Consider the problem where we are asked to find the probability, when rolling a single die, of an even number or a number greater than 4. These two events are not mutually exclusive because the outcome of a 6 is within both the event of an even number and the event of a number greater than 4. So that the probability of getting a 6 is not counted twice, we must subtract its value from the total. This leads us to the following property.

┌─ *General addition rule of probability* ─────────────────

If the events A and B are from the same sample space, then the probability that A or B will occur is

$$P(A \text{ or } B) = P(A \cup B) = P(A) + P(B) - P(A \cap B)$$

Note If the events A and B are mutually exclusive, they cannot occur at the same time, and $P(A \cap B) = 0$.

■ *Example 15–4 E*

Find the probability of the given event.

1. When rolling a single die, what is the probability of getting an even number or a number greater than 4?

   These events have the number 6 in common. Therefore, if E represents the event of an even number, G represents the event of a number greater than 4, and $E \cap G$ represents the outcomes they have in common (the number 6), we use the following:

   $$P(E) = \frac{n(E)}{n(S)} = \frac{3}{6} = \frac{1}{2}$$
   $$P(G) = \frac{n(G)}{n(S)} = \frac{2}{6} = \frac{1}{3} \text{ and}$$
   $$P(E \cap G) = \frac{n(E \cap G)}{n(S)} = \frac{1}{6} \text{ then}$$
   $$P(E \cup G) = P(E) + P(G) - P(E \cap G)$$
   $$= \frac{3}{6} + \frac{2}{6} - \frac{1}{6}$$
   $$= \frac{4}{6} = \frac{2}{3}$$

2. A card is drawn from a standard deck of 52 cards. What is the probability of getting a diamond or a face card?

   We will let D represent the event of a diamond and F represent the event of a face card. The events D and F have the king, queen, and jack of diamonds in common. The probability is found as follows:

   $$P(D) = \frac{n(D)}{n(S)} = \frac{13}{52} = \frac{1}{4}$$
   $$P(F) = \frac{n(F)}{n(S)} = \frac{12}{52} = \frac{3}{13} \text{ and}$$
   $$P(D \cap F) = \frac{n(D \cap F)}{n(S)} = \frac{3}{52} \text{ then,}$$
   $$P(D \cup F) = P(D) + P(F) - P(D \cap F) = \frac{13}{52} + \frac{12}{52} - \frac{3}{52} = \frac{22}{52} = \frac{11}{26}$$

   ▶ *Quick check* A card is drawn from a standard deck of playing cards. What is the probability of a club or a 10?   ■

The *complement of an event A* is the set of all outcomes in the sample space that are not contained in A. We denote the complement of the event A by A'. The following is true.

$$n(A \text{ or } A') = n(A \cup A') = n(S)$$

therefore, $$P(A \text{ or } A') = P(A \cup A') = P(S) = 1$$

Since A and A' are mutually exclusive,

$$P(A \cup A') = P(A) + P(A') = 1$$

From this, we state the following property.

> ### Probability of complementary events
>
> If A is any event of a sample space S, and A' is its complementary event, then
>
> $$P(A) + P(A') = 1$$

Note The following is also true of complementary events A and A'.

$$P(A') = 1 - P(A)$$
$$P(A \cup A') = 1 \text{ and}$$
$$P(A \cap A') = 0$$

■ *Example 15–4 F*

Find the probability of the given event.

1. A card is drawn from a standard deck of 52 cards. What is the probability of not selecting a face card?

   The face cards are the kings, queens, and jacks. If we let F represent the event of picking a face card, then $P(F)$ is found as follows:

   $$P(F) = \frac{n(F)}{n(S)} = \frac{12}{52} = \frac{3}{13}$$

   F' represents the event of not selecting a face card and $P(F')$ is found as follows:

   $$P(F') = 1 - P(F)$$
   $$= 1 - \frac{3}{13} = \frac{10}{13}$$

   Note The event of not selecting a face card is equivalent to selecting a 2, 3, 4, 5, 6, 7, 8, 9, 10, or ace, and the answer could have been found by determining the probability of a 2, 3, 4, 5, 6, 7, 8, 9, 10, or ace.

2. Five cards are selected from a standard deck of 52 playing cards. What is the probability of not getting a heart flush?

   In example 15–4 C–1, we determined that the probability of a heart flush, $P(A)$, was $\frac{33}{66,640}$. Therefore, with this information, we can find the probability of not getting a heart flush as follows:

   $$P(A') = 1 - P(A)$$
   $$= 1 - \frac{33}{66,640}$$
   $$= \frac{66,607}{66,640} \approx 0.9995$$

   ■

Mastery points

Can you
- Determine the number of outcomes in an event?
- Determine the number of outcomes in a sample space?
- Determine the probability of an event?

Exercise 15–4

For exercises 1–20, find the probability of the given event. See examples 15–4 A, B, and D.

Example A coin is flipped 3 times. What is the probability of exactly 2 heads?

Solution From the counting techniques, we know that the sample space will have $2 \cdot 2 \cdot 2 = 8$ outcomes. The sample space is

$$S = \{HHH, HHT, HTH, HTT, THH, THT, TTH, TTT\}$$

Therefore, if A represents the event of exactly 2 heads, then $A = \{HHT, HTH, THH\}$

So, $P(A) = \dfrac{n(A)}{n(S)} = \dfrac{3}{8}$

For exercises 1–4, a coin is tossed 2 times.

1. Exactly one head   **2.** One head and one tail   **3.** At least one head   **4.** No heads

For exercises 5–8, a coin is tossed 3 times.

5. All tails   **6.** No tails   **7.** At least two tails   **8.** At most two tails

For exercises 9–12, a coin is tossed 4 times.

9. Exactly two heads   **10.** Exactly two tails   **11.** Less than two heads   **12.** More than two heads

For exercises 13–20, a card is drawn from a standard deck of playing cards.

13. A 10   **14.** A 7   **15.** A club   **16.** A heart

17. A card from 4 to 9   **18.** A card from 3 to 6   **19.** A card between 3 and 10   **20.** A card between 5 and 9

Find the probability of the given event. See examples 15–4 C, D, E, and F.

Example A card is drawn from a standard deck of playing cards. What is the probability of a club or a 10?

Solution We will let C represent the event of a club and T represent the event of a 10. The events C and T have the 10 of clubs in common. The probability is found as follows:

$$P(C) = \frac{n(C)}{n(S)} = \frac{13}{52} = \frac{1}{4}$$

$$P(T) = \frac{n(T)}{n(S)} = \frac{4}{52} = \frac{1}{13} \text{ and }$$

$$P(C \cap T) = \frac{n(C \cap T)}{n(S)} = \frac{1}{52} \text{ then}$$

$$P(C \cup T) = P(C) + P(T) - P(C \cap T) = \frac{13}{52} + \frac{4}{52} - \frac{1}{52} = \frac{16}{52} = \frac{4}{13}$$

For exercises 21–32, a card is drawn from a standard deck of playing cards.

21. The card is a heart or a 7.

22. The card is a diamond or a queen.

23. The card is from 2 to 6 or a spade.

24. The card is from 5 to 8 or a club.

25. The card is not a 10.

26. The card is not a jack.

27. The card is not from 4 to 10.

28. The card is not from 2 to 5.

29. The card is not a club.

30. The card is not a heart.

31. The card is not red.

32. The card is not black.

For exercises 33–37, use the fact that a roulette wheel contains the numbers from 1 to 36, eighteen numbers are red and eighteen are black. There are two more numbers, 0 and 00, that are green. What is the probability that:

33. The number 10 occurs?

34. The number will be red?

35. The number will not be green?

36. The number will be black or green?

37. The number will be white?

For exercises 38–42, a bowl contains 6 red, 10 blue, and 8 white balls. If one ball is selected, what is the probability of:

38. A red ball?

39. A red or a white ball?

40. A red, white, or blue ball?

41. A black ball?

42. Not a white ball?

Find the probability of the given event. You may need to use some of the counting techniques to determine $n(A)$ and $n(S)$. See examples 15–4 C, D, E, and F.

Example A class contains 8 men and 6 women. A committee of 4 people is selected. What is the probability that all 4 people will be women?

Solution Let W represent the event of selecting 4 women. Since there is no order in the selection of the women, the number of different sets of 4 women would be combinations of the 6 women taken 4 at a time.

$$n(W) = {}_6C_4 = \frac{6!}{2!4!} = 15$$

The number of different 4-people committees would be combinations of the 14 people (6 women plus 8 men) taken 4 people at a time.

$$n(S) = {}_{14}C_4 = \frac{14!}{10!4!} = 1,001$$

Then $P(W) = \dfrac{n(W)}{n(S)} = \dfrac{15}{1,001}$

For exercises 43–48, a class has 10 men and 8 women. A committee of 6 is chosen. What is the probability of:

43. All women?

44. All men?

45. Three men and three women?

46. No men?

47. One man?

48. Four men?

For exercises 49–58, five cards are drawn from a standard deck of playing cards. Find the probability of the following five-card hands.

49. All five cards are spades.

50. All five cards are red.

51. All five cards are black.

52. None of the cards are clubs.

53. None of the cards are face cards.

54. All of the cards are face cards.

55. Three black cards and two red cards.

56. One black card and four red cards.

57. Three clubs and two hearts.

58. Four diamonds and one spade.

For exercises 59–64, six numbers are to be drawn from a group of n numbers. Find the probability of correctly selecting the six numbers from a group of:

59. 44 numbers.　　　　**60.** 45 numbers.　　　　**61.** 46 numbers.

62. 47 numbers.　　　　**63.** 48 numbers.　　　　**64.** 49 numbers.

For exercises 65–68, a businesswoman has meetings with Mr. Allen, Mr. Baker, Ms. Carter, and Mrs. Dunn. If she randomly chooses the order in which to have her meetings, find the probability that:

65. Ms. Carter is seen first.　　　　　　**66.** Mr. Allen is seen last.

67. She sees the men before the women.　　**68.** She sees them in alphabetical order.

Review exercises

Find the solution set for the following quadratic equations and inequalities. See sections 6–1, 6–2, 6–3, and 6–6.

1. $x^2 - 2x - 8 = 0$　　　**2.** $x^2 - 3x - 10 < 0$　　　**3.** $x^2 + 3x + 5 = 0$

4. $4x^2 + 11x - 3 = 0$　　**5.** $x^2 = 3x$　　　　　　　**6.** $2x^2 + x - 4 = 0$

7. $x^2 + 6x + 8 \geq 0$　　　**8.** $x^2 + x < 0$

15–5 ■ The binomial expansion

Consider the indicated product $(x + y)^n$, where n is a positive integer. By performing the indicated multiplication, we can obtain polynomial expressions for the positive integral powers of the binomial expression $x + y$. That is, we can multiply to show that

$$(x + y)^1 = x + y$$
$$(x + y)^2 = x^2 + 2xy + y^2$$
$$(x + y)^3 = x^3 + 3x^2y + 3xy^2 + y^3$$
$$(x + y)^4 = x^4 + 4x^3y + 6x^2y^2 + 4xy^3 + y^4$$
$$(x + y)^5 = x^5 + 5x^4y + 10x^3y^2 + 10x^2y^3 + 5xy^4 + y^5$$

and so on. Each of the polynomials thus obtained is called the **binomial expansion** of the related power of the binomial $x + y$. In this section, we shall develop a formula that will enable us to express any positive integral power of a binomial as a polynomial.

Before doing this, let us investigate the expansions given above to determine the properties that will hold for the expansion of the general binomial $(x + y)^n$.

1. The first term of each expansion is x raised to the power of the binomial itself, x^n.

2. The second term of each expansion is of the form $nx^{n-1}y$.

3. As we proceed term by term from this point, the exponent of x decreases by 1 and the exponent of y increases by 1 with each succeeding term.

4. The next to last term is of the form nxy^{n-1}.

5. The last term of each expansion is y raised to the power of the binomial, y^n.

6. In each term, the sum of the exponents of x and y is always n.

7. There are $n + 1$ terms in each expansion.

Pascal's triangle

If we consider the coefficients of the five expansions stated previously, we can write them in a triangular pattern.

$$(x + y)^1 \qquad\qquad\qquad 1 \qquad 1$$
$$(x + y)^2 \qquad\qquad\quad 1 \quad 2 \quad 1$$
$$(x + y)^3 \qquad\qquad 1 \quad 3 \quad 3 \quad 1$$
$$(x + y)^4 \qquad\quad 1 \quad 4 \quad 6 \quad 4 \quad 1$$
$$(x + y)^5 \quad 1 \quad 5 \quad 10 \quad 10 \quad 5 \quad 1$$

This pattern was used by a seventeenth-century French mathematician named Blaise Pascal (1623–62) and is called **Pascal's Triangle.** When the coefficients are thus arranged, it is possible to determine the coefficients of the next expansion.

Inspection reveals the following characteristics of Pascal's triangle:

1. The coefficients of the first and the last terms are always 1.
2. Each of the other coefficients is obtained by *adding the two numbers above it,* one to the left and one to the right.

To expand the coefficients of $(x + y)^6$, we use the coefficients of $(x + y)^5$ in our triangle to obtain the coefficients shown in figure 15–3.

$$
\begin{array}{ccccccccccccc}
1 & + & 5 & + & 10 & + & 10 & + & 5 & + & 1 \\
\\
1 & & 6 & & 15 & & 20 & & 15 & & 6 & & 1
\end{array}
$$

Figure 15–3

Therefore

$$(x + y)^6 = x^6 + 6x^5y + 15x^4y^2 + 20x^3y^3 + 15x^2y^4 + 6xy^5 + y^6$$

Factorial notation

Although it is possible to use Pascal's Triangle to determine the coefficients in any expansion $(x + y)^n$, where n is a positive integer, we need a more efficient way to do this for greater powers of the binomial. To do this, we must use **factorial notation.**

We now state the general binomial expansion of $(x + y)^n$ for any positive integer n.

$$(x + y)^n = x^n + \frac{n}{1!}x^{n-1}y + \frac{n(n-1)}{2!}x^{n-2}y^2 + \frac{n(n-1)(n-2)}{3!}x^{n-3}y^3$$
$$+ \frac{n(n-1)(n-2)(n-3)}{4!}x^{n-4}y^4 + \cdots + \frac{n}{1!}xy^{n-1} + y^n$$

This statement is called the **binomial expansion** (or **binomial theorem**).

■ *Example 15–5 A*

1. Expand and simplify $(a + 3b)^4$.

   Applying the binomial expansion, we replace n with 4, x with a, and y with $3b$ to obtain the statement

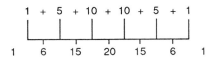

$$(a + 3b)^4 = a^4 + \frac{4}{1!}a^3(3b) + \frac{4 \cdot 3}{2!}a^2(3b)^2 + \frac{4 \cdot 3 \cdot 2}{3!}a(3b)^3 + (3b)^4$$
$$= a^4 + 4a^3(3b) + 6a^2(9b^2) + 4a(27b^3) + 81b^4$$
$$= a^4 + 12a^3b + 54a^2b^2 + 108ab^3 + 81b^4$$

2. Expand and simplify $(3c - 2d^2)^5 = [3c + (-2d^2)]^5$.

From our binomial expansion, $n = 5$, $x = 3c$, and $y = -2d^2$, so we substitute these values to obtain the statement

$$(3c - 2d^2)^5 = (3c)^5 + \frac{5}{1!}(3c)^4(-2d^2) + \frac{5 \cdot 4}{2!}(3c)^3(-2d^2)^2$$
$$+ \frac{5 \cdot 4 \cdot 3}{3!}(3c)^2(-2d^2)^3$$
$$+ \frac{5 \cdot 4 \cdot 3 \cdot 2}{4!}(3c)(-2d^2)^4 + (-2d^2)^5$$
$$= 243c^5 + 5(81c^4)(-2d^2) + 10(27c^3)(4d^4)$$
$$+ 10(9c^2)(-8d^6) + 5(3c)(16d^8) + (-32d^{10})$$
$$= 243c^5 - 810c^4d^2 + 1,080c^3d^4 - 720c^2d^6 + 240cd^8 - 32d^{10}$$

▶ *Quick check* Expand and simplify $(a - 5b)^4$. ∎

Finding the *r*th term of an expansion

Sometimes we wish to find one of the terms in the expansion and we would like to do this without expanding the binomial fully. We determine the *r*th term of the expansion using the following expression:

┌─ *r*th term of the binomial expansion $(x + y)^n$ is ─────────

$$\frac{n!}{[n - (r - 1)]!(r - 1)!}x^{n - (r - 1)}y^{r - 1}$$

In general, in the expansion of $(x + y)^n$, the term containing the variables $x^{n - k}y^k$ has coefficient $\frac{n!}{(n - k)!k!}$, which is sometimes written in the form $\binom{n}{k}$.
That is, for the *r*th term in the expansion

$$\binom{n}{k} = \frac{n!}{(n - k)!k!}$$

where $k = r - 1$. To illustrate, by definition,

$$\binom{9}{5} = \frac{9!}{(9 - 5)!5!} = \frac{9!}{4!5!} = \frac{9 \cdot 8 \cdot 7 \cdot 6 \cdot 5!}{4 \cdot 3 \cdot 2 \cdot 1 \cdot 5!}$$
$$= \frac{9 \cdot 8 \cdot 7 \cdot 6}{4 \cdot 3 \cdot 2 \cdot 1} = 9 \cdot 2 \cdot 7 = 126$$

Note An alternate statement of the binomial expansion is

$$(x + y)^n = \binom{n}{0}x^ny^0 + \binom{n}{1}x^{n - 1}y + \binom{n}{2}x^{n - 2}y^2 + \cdots + \binom{n}{n}x^0y^n$$

■ *Example 15–5 B*

Find the sixth term in the expansion of $(2a + b)^9$.

Here $n = 9$, $r = 6$, $x = 2a$, and $y = b$. Since $r = 6$, then $r - 1 = 5$. We substitute 9 for n, 6 for r, 5 for $r - 1$, $2a$ for x, and b for y to obtain the expression for the sixth term as

$$\frac{9!}{(9 - 5)!5!}(2a)^{9 - 5}(b)^5 = \frac{9!}{4!5!}(2a)^4(b)^5$$
$$= 126(16a^4)(b^5) = 2{,}016a^4b^5$$

▶ *Quick check*   Find the seventh term in the expansion of $(4a - 3b)^{10}$.   ■

We can use the binomial expansion to approximate a power of a decimal number.

■ *Example 15–5 C*

Use the binomial expansion to evaluate $(2.01)^6$ correct to four decimal places. Expand to the first four terms.

Now $(2.01)^6 = (2 + 0.01)^6$ and applying the binomial expansion to this expression and expanding the first four terms,

$$(2 + 0.01)^6 \approx 2^6 + \frac{6}{1!}(2)^5(0.01) + \frac{6 \cdot 5}{2!}(2)^4(0.01)^2 + \frac{6 \cdot 5 \cdot 4}{3!}(2)^3(0.01)^3$$
$$= 64 + 6(32)(0.01) + 15(16)(0.0001) + 20(8)(0.000001)$$
$$= 64 + 192(0.01) + 240(0.0001) + 160(0.000001)$$
$$= 64 + 1.92 + 0.024 + 0.00016 = 65.94416$$

Rounding to four decimal places, $(2.01)^6 \approx 65.9442$.   ■

___ *Mastery points* _____

Can you
- Expand and simplify a binomial expression raised to any positive integer power n?
- Determine a specific term in the expansion of a binomial?
- Use the binomial expansion to approximate the value of a decimal number to some positive integer power n?

Exercise 15–5
_____

Expand and simplify the following binomials using the binomial expansion. See example 15–5 A.

Example $(a - 5b)^4$

Solution In our expansion of $(x + y)^4$, let

$x = a$ and $y = -5b$

and using the coefficients 1, 4, 6, 4, 1,

$$(a - 5b)^4 = 1(a)^4 + 4(a)^3(-5b) + 6(a)^2(-5b)^2 + 4(a)(-5b)^3 + 1(-5b)^4$$
$$= a^4 - 20a^3b + 150a^2b^2 - 500ab^3 + 625b^4$$

1. $(a - 3)^4$ **2.** $(b + 2)^5$ **3.** $(p + q)^6$ **4.** $(a - b)^7$ **5.** $(2a + 3)^4$

6. $(3b + 2)^5$ **7.** $\left(\dfrac{p}{2} - q\right)^6$ **8.** $\left(2r - \dfrac{q}{3}\right)^4$ **9.** $(a^2 + b^2)^5$ **10.** $(x^3 + y)^6$

11. $(x^2 - y^3)^4$ **12.** $(a^3 - b^3)^4$ **13.** $(2a^3 + b)^5$ **14.** $(3a^2 + b^3)^4$

Find the indicated term of each binomial expansion. See example 15–5 B.

Example Find the seventh term in the expansion of $(4a - 3b)^{10}$.

Solution Now $n = 10$, $r = 7$, $x = 4a$, and $y = -3b$. Since $r = 7$, then $r - 1 = 6$. Using

$$\frac{n!}{[n - (r - 1)]!(r - 1)!}x^{n - (r - 1)}y^{r - 1}$$

$$\frac{10!}{(10 - 6)!6!}(4a)^{10 - 6}(-3b)^6 = \frac{10!}{4!6!}(4a)^4(-3b)^6$$

$$= 210(256a^4)(729b^6)$$

$$= 39{,}191{,}040a^4b^6$$

15. $(a + b)^9$, fourth term   **16.** $(a + b)^{13}$, seventh term   **17.** $(a - b)^{14}$, sixth term

18. $(p + 3)^{11}$, eighth term   **19.** $(q - 2)^{12}$, fifth term   **20.** $(r - 2s)^{10}$, fifth term

21. $(6 - k)^9$, seventh term   **22.** $(4 - y)^8$, fifth term   **23.** $(a^2 + b)^8$, fourth term

24. $(x^3 + y^2)^{10}$, fifth term   **25.** $(a^2 - b^2)^{11}$, ninth term

Use the binomial expansion to calculate the following expressions correct to four decimal places. Expand to the first four terms. See example 15–5 C.

26. $(1.01)^8$ **27.** $(1.002)^{13}$ **28.** $(2.02)^7$ **29.** $(0.97)^5$

30. In the expansion of $\left(p^2 - \dfrac{1}{4}\right)^{12}$, find the term involving p^{10}.

31. Find the middle term in the expansion of $(a + \sqrt{a})^{12}$.

Review exercises

Simplify the following. Leave your answer in standard form. Assume that all variables represent positive real numbers. See sections 5–1, 5–2, 5–3, 5–4, and 5–5.

1. $x^{3/4} \cdot x^{1/2}$ **2.** $\sqrt[3]{32x^6y^4}$ **3.** $2\sqrt{8} + 3\sqrt{18}$ **4.** $\dfrac{1}{\sqrt[3]{4ab^2}}$

5. $\dfrac{1}{\sqrt{10}} + \dfrac{1}{\sqrt{6}}$ **6.** $(\sqrt{a} + \sqrt{b})^2$ **7.** $\dfrac{1}{\sqrt{x} - 2}$ **8.** $\dfrac{3}{\sqrt{7} + 2}$

Chapter 15 lead-in problem

To win the Michigan lottery, a person must correctly select the six integers drawn by the lottery commission from the set of integers from 1 to 47. How many different selections are possible?

Solution

Since we are interested in what six integers are picked and not the order in which they were selected, this is a combination of 47 numbers taken 6 at a time.

$$_{47}C_6 = \frac{47!}{(47-6)!6!}$$
$$= \frac{47!}{41!\,6!}$$
$$= 10{,}737{,}573$$

There are 10,737,573 different selections possible.

Chapter 15 summary

1. **Extended multiplication of choices property:** If a choice consists of k decisions where the first can be made in n_1 ways, for each of these the second can be made in n_2 ways, $\cdots$, and for each of these the kth can be made in n_k ways, then the complete choice can be made in $n_1 \cdot n_2 \cdots n_k$ ways.

2. Factorial notation for every positive integer n,
$$n! = n(n-1)(n-2)\cdots\cdots 3\cdot 2\cdot 1$$

3. $0! = 1$

4. The number of permutations of a set of n distinct elements selected r at a time, where $r \le n$, is
$$_nP_r = n(n-1)(n-2)\cdots\cdots(n-r+1)$$
or in factorial notation,
$$_nP_r = \frac{n!}{(n-r)!}$$

5. The number of distinct permutations P of the n elements taken n at a time, where n_1 are alike of one kind, n_2 are alike of another kind, $\cdots$, and n_k are alike of another kind, is given by
$$P = \frac{n!}{n_1!n_2!\cdots n_k!}$$
where $n_1 + n_2 + \cdots + n_k = n$.

6. The number of combinations (subsets) of n distinct elements taken r at a time, where $r \le n$, is
$$_nC_r = \frac{n(n-1)(n-2)\cdots\cdots(n-r+1)}{r!}$$
or in factorial notation,
$$_nC_r = \frac{n!}{(n-r)!r!}$$

7. **The probability of an event:** If an event A is made up of $n(A)$ equally likely outcomes from a sample space S that has $n(S)$ equally likely outcomes, then the **probability** of the event A, represented by $P(A)$, is $P(A) = \dfrac{n(A)}{n(S)}$.

8. **Basic probability principles:** If the event A has $n(A)$ equally likely outcomes and the sample space S has $n(S)$ equally likely outcomes, then
   a. $0 \le P(A) \le 1$
   b. $P(S) = \dfrac{n(S)}{n(S)} = 1$
   c. $P(A) = 0$ means event A cannot occur and is called an impossible event.
   d. $P(A) = 1$ means event A must occur and is called a certain event.

9. **Probability of mutually exclusive events:** If the events A and B are mutually exclusive, then the probability that A or B will occur is
$$P(A \text{ or } B) = P(A \cup B) = P(A) + P(B)$$

10. **General addition rule of probability:** If the events A and B are from the same sample space, then the probability that A or B will occur is
$$P(A \text{ or } B) = P(A \cup B) = P(A) + P(B) - P(A \cap B)$$

11. **Probability of complementary events:** If A is any event of a sample space S, and A' is the complementary event, then
$$P(A) + P(A') = 1$$

12. Expanding a binomial to any nth power

$$(x + y)^n = 1(x)^n(y)^0 + \frac{n}{1}(x)^{n-1}(y)$$
$$+ \frac{n(n-1)}{1 \cdot 2}(x)^{n-2}(y)^2$$
$$+ \frac{n(n-1)(n-2)}{1 \cdot 2 \cdot 3}(x)^{n-3}(y)^3$$
$$+ \cdots + \frac{n}{1}xy^{n-1} + 1x^0y^n$$

13. rth term of the binomial expansion $(x + y)^n$ is

$$\frac{n!}{[n-(r-1)]!(r-1)!}x^{n-(r-1)}y^{r-1}$$

Chapter 15 error analysis

1. Number of possible outcomes
   Example: A woman can buy one of 5 dresses, one of 6 pairs of shoes, one of 4 hats, and one of 2 pairs of gloves in $5 + 6 + 4 + 2 = 17$ different ways.
   Correct answer: 240 different ways
   What error was made? (*see page 639*)

2. Factorial notation
   Example: $\dfrac{10!}{3!} = \dfrac{10 \cdot 9 \cdot 8 \cdot 7 \cdot 6 \cdot 5 \cdot 4 \cdot 3 \cdot 2 \cdot 1}{3}$
   $= 1,209,600$
   Correct answer: 604,800
   What error was made? (*see page 641*)

3. Permutations
   Example: $_{11}P_4 = \dfrac{11!}{4!} = 1,663,200$
   Correct answer: $_{11}P_4 = 7,920$
   What error was made? (*see page 646*)

4. Permutations
   Example: The number of distinct permutations of the letters in the word BELLADONNA is $10! = 3,628,800$
   Correct answer: 453,600
   What error was made? (*see page 647*)

5. Combinations
   Example: $_{10}C_4 = \dfrac{10!}{4!} = 151,200$
   Correct answer: 210
   What error was made? (*see page 650*)

6. Probability of an event
   Example: The probability of drawing a king or a queen from a standard deck of 52 cards is $P(k) \cdot P(q) =$
   $\dfrac{4}{52} \cdot \dfrac{4}{52} = \dfrac{1}{13} \cdot \dfrac{1}{13} = \dfrac{1}{169}$
   Correct answer: $\dfrac{2}{13}$
   What error was made? (*see page 659*)

7. Complementary events
   Example: If the probability of an event E, $P(E)$, is $\dfrac{4}{15}$, then the complementary event, $P(E')$ is $\dfrac{15}{4}$
   Correct answer: $\dfrac{11}{15}$
   What error was made? (*see page 661*)

8. Binomial expansion
   Example: Since the coefficients in the expansion of $(a + b)^5$ are 1   5   10   10   5   1, then the coefficients in the expansion of $(a + b)^6$ are
   1   5   50   100   50   5   1
   Correct answer: 1   6   15   20   15   6   1
   What error was made? (*see page 665*)

9. rth term of the expansion of $(a + b)^n$
   Example: The coefficient of the seventh term in the expansion of $(a + b)^{10}$ is $\dbinom{10}{7} = \dfrac{10!}{7!} = 720$
   Correct answer: $\dbinom{10}{4} = 210$
   What error was made? (*see page 666*)

10. rth term in the expansion of a binomial
    Example: The sixth term in the expansion of $(2a + 3b)^9$ is $126(2a)^3(3b)^6$
    Correct answer: $126(2a)^4(3b)^5$
    What error was made? (*see page 667*)

Chapter 15 critical thinking

Using the digits 2, 3, 4, and 5, how many different four-digit numbers can be formed if each digit can be used only once within each number? What will be the sum of all the possible numbers?

Chapter 15 review

[15–1]

Show all possible outcomes.

1. There are three people in a race. Draw a tree diagram of the finish to the race. Use *A, B,* and *C* to represent the three people.

Determine the number of outcomes.

3. A menu offers a choice of 6 appetizers, 4 salads, 7 entrees, 3 kinds of potatoes, and 3 vegetables. If a meal consists of one of each, in how many different ways can a person select dinner?

4. In how many different ways can a student answer all the questions on a quiz consisting of 10 true or false questions?

[15–2]

Determine the number of outcomes.

7. In a 12-horse race, how many different first-second-third place finishes are possible.

2. Draw a tree diagram of a person rolling a pair of dice such that the first die is odd and the second die is odd.

5. From an ordinary deck of playing cards, in how many different ways can a person select one five, one six, one seven, one eight, and one nine?

6. A class has 10 girls and 12 boys. A girl and a boy are to be selected as student council representatives. In how many different ways can the selection be made?

8. From the set of numbers {1,2,3,4}, form the following without repetition.
   a. All possible 4-digit numbers
   b. All possible 3-digit numbers
   c. All possible 2-digit numbers
   d. All possible 4-digit numbers less than 3,000
   e. All possible 3-digit numbers greater than 400

Determine the number of distinguishable permutations of the following words.

9. COOKBOOK

10. roomette

[15–3]

Determine the number of outcomes.

11. A child chooses 5 toys from a box containing 25 different toys. In how many different ways can she make her selection?

12. How many different 7-card hands can be dealt from a normal 52-card deck of playing cards?

13. If, on an examination consisting of twenty questions, a student may omit six questions, in how many different ways can the student select the problems he will attempt to answer?

14. A man has 12 shirts and wishes to take 3 with him on a business trip. In how many different ways can this be done?

[15–4]

Find the probability of the given event.

15. A coin is tossed 3 times.
   a. Two tails
   b. At least one tail

16. A card is drawn from a standard deck of playing cards.
   a. A 4
   b. A diamond
   c. A card from 2 to 6
   d. A card between 5 and 10
   e. A heart or a ten
   f. Not a spade

[15–5]

Expand and simplify each binomial.

17. $(x + 5)^7$

18. $(2a - 3b)^5$

19. $\left(\dfrac{1}{2}a - 3b\right)^4$

Find the indicated term in the expansion of each given binomial.

20. $(a - 4)^{11}$, fifth term

21. $(3a + b)^{14}$, seventh term

22. $(2x + 3y)^{12}$, the term in which y^2 appears

Final examination

[1–1]   **1.** List the elements in the set $\{x|x$ is an integer between -5 and $6\}$.

[1–1]   **2.** Given $A = \{-4, -2, 6, 9\}$ and $B = \{-6, -4, 0, 1, 6, 9\}$, find
   a. $A \cup B$ b. $A \cap B$ c. $\emptyset \cap B$

[1–4]   **3.** Perform the indicated operations and simplify.
$4 - \{5[4 - 9 \div 3 + 9]\}$

[1–3]   **4.** $4(5 + 6) = 4 \cdot 6 + 4 \cdot 5$ demonstrates the use of what two properties of real numbers?

[3–2]   **5.** Perform the indicated multiplications.
   a. $(3y - 2)(3y + 2)$ b. $(4x - 1)^2$
   c. $(y + 2)(y^2 - 2y + 4)$

[3–3]   **6.** Simplify the following expressions. Assume all denominators are nonzero. Leave all answers with positive exponents.
   a. $(a^{-2}b^3)^{-3}$ b. $(-4ab^2c^{-1})(2a^3bc^4)$
   c. $\dfrac{x^{-3}y^2}{xy^{-4}}$

Find the solution set of the following equations.

[2–1]   **7.** $4(x + 3) - 1 = 2(x - 7) + 3$

[4–7]   **9.** $\dfrac{2}{x + 2} - \dfrac{3}{4} = \dfrac{3}{x + 2}$

[2–5]   **8.** $|5y - 2| = 3$

Completely factor the following expressions.

[3–6]   **10.** $5y^2 + 14y - 3$

[3–4]   **12.** $3xy - xz + 6y^2 - 2yz$

[3–8]   **14.** $2y^2 - 98$

[3–4]   **11.** $3x^2y - 9xy^2 + 15x^3y^3$

[3–5]   **13.** $5x^2 - 10x - 15$

[3–7]   **15.** $9x^2 - 42xz + 49z^2$

Find the solution set of the following equations and inequalities.

[2–5]   **16.** $5(3x - 2) \leq 4(x + 3)$

[2–6]   **18.** $|5 - x| \leq 3$

[6–1]   **20.** $y^2 - 2 = y$

[6–5]   **22.** $\sqrt{x + 15} = x + 3$

[6–7]   **24.** $y^2 - 4y - 32 \geq 0$

[2–5]   **17.** $-2 \leq 4 - 3x \leq 1$

[2–6]   **19.** $|2x - 7| > 3$

[6–3]   **21.** $4x^2 - x - 2 = 0$

[6–6]   **23.** $x^4 - 6x^2 - 16 = 0$

[6–7]   **25.** $\dfrac{x - 1}{x + 2} < 4$

Perform the indicated operations.

[4–3] **26.** $\dfrac{4}{2y-1} + \dfrac{3}{y+5}$

[4–3] **27.** $\dfrac{x}{x^2-4x+3} - \dfrac{2x}{x^2-9}$

[4–2] **28.** $\dfrac{x^2-25}{x^2+6x+5} \div \dfrac{2x^2-9x-5}{x^3+1}$

[4–4] **29.** Simplify the complex fraction $\dfrac{\dfrac{3}{x} - \dfrac{2}{y}}{\dfrac{1}{x} + \dfrac{5}{y}}$.

[5–4] **30.** Combine the expression $4\sqrt{32} - 3\sqrt{18} + \sqrt{8}$.

[5–5] **31.** Multiply
a. $(5 + 3\sqrt{2})(6 - \sqrt{3})$ b. $(4 - 3i)^2$

[5–6] **32.** Rationalize the denominator $\dfrac{1-2i}{2-i}$.

[7–2] **33.** Find the slope and the intercepts of the graph of the equation $5x - 2y = 10$.

[7–3] **34.** Find the equation of the line having the following characteristics. Write each equation in standard form $ax + by = c$, $a > 0$.
a. Passing through the points $(2,3)$ and $(-1,4)$
b. Passing through point $(1,-5)$ and perpendicular to the line $x - 2y = 3$.

[8–2] **35.** Given $f(x) = 3x - 1$ and $g(x) = x^2 - 9$, find
a. $f(-4)$ b. $g(5)$
c. $\dfrac{g(x+h) - g(x)}{h}$, $h \neq 0$

[8–4] **36.** Given $f(x) = 9x - 6$, find the inverse of function f, that is find $f^{-1}(x)$.

[9–1] **37.** Given the quadratic equation $y = x^2 - 6x - 7$, find (a) the x- and y-intercepts, (b) the vertex, (c) the equation of the axis of symmetry in the graph of f.

[9–4] **38.** Given the equation $5x^2 + 2y^2 = 10$, find the x- and y-intercepts. Name the figure that is the graph of the equation.

[9–4] **39.** Determine if the given equation represents a circle, a parabola, an ellipse or a hyperbola.
a. $4x^2 = 9 + y^2$ b. $3 + x = y^2 - 2y$
c. $4y^2 - 9 = -4x^2$

[4–6] **40.** Divide $4x^3 - 3x^2 + 2x - 1$ by $x + 3$ using synthetic division.

[10–2] **41.** Using synthetic division, find the solution set of the equation $x^3 - 6x^2 + 3x + 10 = 0$ if one solution is -1.

[10–4] **42.** Sketch the graph of the polynomial function $f(x) = x^3 - 4x$.

[11–3] **43.** Write the expression $\log_3 5 - \log_3 4 + \log_3 9$ as the logarithm of a single number.

[11–3] **44.** Find the solution set of the logarithmic equation $\log_2(x+1) - \log_2(x-3) = 3$.

[11–6] **45.** Find the solution set of the exponential equation $5^{x-1} = 3$. Round to 2 decimal places.

[11–5] **46.** Find $\ln 52.3$ using the calculator. Round to 2 decimal places.

[12–1] **47.** Find the solution set of the system of equations $\begin{aligned} 2x - y &= 4 \\ 3x + 2y &= -1. \end{aligned}$

[13–1] **48.** Evaluate $\begin{vmatrix} 1 & -2 & 3 \\ 4 & 0 & 1 \\ -6 & 2 & 0 \end{vmatrix}$

[13–2] **49.** Using Cramer's rule, find the solution set of the system of equations
$\begin{aligned} x + 2y &= 1 \\ 3x - y &= -6. \end{aligned}$

[14–1] **50.** Write the sum $5 - 11 + 17 - 23 + 29$ in sigma notation.

[14–2] **51.** Find $\displaystyle\sum_{i=1}^{7} (4i + 3)$.

[14–3] **52.** Find a_{21} of the arithmetic sequence having $a_1 = -4$ and $d = 5$.

[14–3] **53.** Find S_{31} of the sequence $13, 23, 33, 43, \cdots$.

[14–5] **54.** Find $\displaystyle\sum_{i=1}^{\infty} 3\left(-\dfrac{2}{3}\right)^i$.

[14–4] **55.** Find a_5 of the geometric sequence if $a_1 = \dfrac{1}{3}$ and $r = -\dfrac{1}{2}$.

[15–1] **56.** In how many different ways can 5 boys and 6 girls sit in a row if the boys and girls must alternate?

[15–2] **57.** There are 17 players on a baseball team. If the coach selects 9 players for a game, how many different batting orders are possible?

[15–2] **58.** In horse racing, a trifecta bet is to pick the first, second, and third place finishers in a race. If a race has 10 horses, how many different trifecta bets are possible?

[15–3] **59.** How many different 5-member committees can be formed from a club that has 30 members?

[15–3] **60.** If a group consists of 10 men and 15 women, in how many different ways can a committee of 8 be selected if:
   a. The committee is to have an equal number of men and women?
   b. The committee is to be all men?
   c. There are no restrictions on the membership of the committee?

[15–3] **61.** Evaluate $_{11}C_5$.

For problems 62–63, a class has 8 men and 6 women. A committee of 6 is chosen. What is the probability of:

[15–4] **62.** All men?

[15–4] **63.** Three men and three women?

For problems 64–65, five cards are drawn from a standard deck of playing cards. Find the probability of the following five-card hands.

[15–4] **64.** All five cards are hearts.

[15–4] **65.** None of the cards are red.

[15–5] **66.** Find the seventh term of the expansion of $(a - 5)^{12}$.

[15–5] **67.** Expand the binomial $(3x - 2y)^4$.

Answers and Solutions

Chapter 1

Exercise 1–1

Answers to odd-numbered problems

1. {Sunday,Saturday}   **3.** {10,12,14}   **5.** ∅   **7.** $A \subseteq D$
9. ∅ ⊆ B   **11.** {6,7,8}   **13.** true   **15.** true   **17.** false
19. true   **21.** true   **23.** {1,2,3,4}   **25.** {1,2,3,4,7,8,9}   **27.** ∅
29. {2,3,4}   **31.** {1,2,3,4}

33.    **35.**

37.

39. 3   **41.** 0   **43.** −2   **45.** −4   **47.** <   **49.** >   **51.** <
53. <   **55.** $H \cap Q = \emptyset$ **57.** {8} ⊆ N   **59.** {8} ⊄ N   **61.** yes
63. no   **65.** yes   **67.** yes   **69.** no   **71.** yes   **73.** yes
75. yes

Solutions to trial exercise problems

4. ∅. There are no months with less than 28 days.   **11.** {6,7,8}. The elements are inside the braces.

31. {1,2,3,4} ∪ ({2,3,4} ∩ {7,8,9}) = {1,2,3,4} ∪ ∅ = {1,2,3,4}

36.    Using a calculator, we approximate $-\sqrt{3}$ as −1.732 and $\sqrt{2}$ as 1.414.   **43.** $-|-2|$ = −2. The negative sign in front of the absolute value bar remains.
52. 0 > −5. 0 is greater than −5 since 0 is to the right of −5 on the number line.   **58.** 8 ⊄ N. We place a slash mark, /, through ⊆ to negate it.

Exercise 1–2

Answers to odd-numbered problems

1. 2   **3.** −14   **5.** 1   **7.** 18   **9.** −7   **11.** −4   **13.** −20
15. −12   **17.** 13   **19.** −9   **21.** −17   **23.** 10   **25.** −28
27. −27   **29.** −24   **31.** 48   **33.** −240   **35.** 60   **37.** 0
39. −64   **41.** 36   **43.** −64   **45.** −8   **47.** 2   **49.** −4
51. 3   **53.** −9   **55.** indeterminate   **57.** 10   **59.** 0   **61.** 4
63. undefined   **65.** −4   **67.** −$88   **69.** 3   **71.** 5 yd
73. 7(−10) = −70   **75.** −24° F   **77.** $68   **79.** 32
81. 240¢ = $2.40   **83.** 290 ml   **85.** 8,708 feet   **87.** $1.80
89. 34 miles   **91.** 133 over 78   **93.** 27   **95.** 2 mb gain

Solutions to trial exercise problems

9. (−7) − 0 = −7. We can subtract zero from any number and the number will be our answer.   **14.** 6 − 9 + 11 − 8
= 6 + (−9) + 11 + (−8) = (−3) + 11 + (−8) = 8 + (−8)
= 0   **21.** [(−6) − 5] + 4 − (16 − 6)
= [(−6) + (−5)] + 4 − [16 + (−6)] = (−11) + 4 − 10
= (−11) + 4 + (−10) = (−7) + (−10) = −17
36. (−9)(+2)(0)(−4) = 0. When zero is one of the factors, the product will be zero.   **38.** $-5^2 = -(5 \cdot 5) = -(25) = -25$
59. $\dfrac{(-6)(0)}{(-2)} = \dfrac{0}{(-2)} = 0$ **62.** $\dfrac{(-8)(-6)}{(-2)(0)} = \dfrac{48}{0} =$ undefined

Exercise 1–3

Answers to odd-numbers problems

1. $3y + 4x$ **3.** $4 + (2 + 6)$ **5.** $a \cdot 5$ **7.** $2(xy)$ **9.** 0
11. $x + 5$ **13.** $2a - 3$ **15.** 7   **17.** $a \geq 5$ **19.** $a < b$
21. $x > 5$ or $x < 5$ **23.** $a > 6$ **25.** closure property for addition
27. identity property of multiplication   **29.** multiplicative inverse property   **31.** identity property of addition   **33.** associative property of addition   **35.** associative property of multiplication
37. identity property of addition   **39.** commutative property of addition   **41.** associative property of addition   **43.** identity property of addition   **45.** $-6, \dfrac{1}{6}$ **47.** $-5, \dfrac{1}{-5}$ **49.** $-x, \dfrac{1}{x}$

51. $7, -7$ **53.** $3, \dfrac{1}{3}$ **55. a.** given   **b.** closure property of multiplication   **c.** reflexive property of equality   **d.** given
e. substitution property of equality   **57. a.** given   **b.** additive inverse property   **c.** additive inverse property   **d.** uniqueness of additive inverse

Solutions to trial exercise problems

22. $4 \geq y$. From trichotomy, if 4 is not less than y, it must be greater than or equal to y.   **36.** commutative property of addition because the order in which we added the numbers was changed

Exercise 1–4

Answers to odd-numbered problems

1. 2   **3.** 37   **5.** −8   **7.** 41   **9.** 0   **11.** 10   **13.** 1
15. −2   **17.** 25   **19.** 41   **21.** 19   **23.** 9.38   **25.** $-\dfrac{7}{24}$

27. −10.74   **29.** 108.05   **31.** $\dfrac{2}{17}$ **33.** 155   **35.** 45
37. 17   **39.** 5   **41.** 80   **43.** 1   **45.** 100 square meters
47. 5,313.6 pounds per square inch   **49.** 262 square feet   **51.** 77
53. $93   **55.** $176,000   **57.** $2,547   **59.** 4,900 words

Solutions to trial exercise problems

10. $8 + 0(5 - 7) = 8 + 0(-2) = 8 + 0 = 8$ **20.** $12 + 3 \cdot 16$
$\div 4^2 - 5 = 12 + 3 \cdot 16 \div 16 - 5 = 12 + 48 \div 16 - 5$
$= 12 + 3 - 5 = 15 - 5 = 10$ **33.** $5[20 - 3(4 - 6) + 5]$
$= 5[20 - 3(-2) + 5] = 5[20 - (-6) + 5] = 5[26 + 5]$
$= 5[31] = 155$ **41.** $\left[\dfrac{10 + (-2)}{2(-1)}\right]\left[\dfrac{(-10)(-4)}{(-2)}\right]$

$= \left[\dfrac{8}{(-2)}\right]\left[\dfrac{40}{(-2)}\right] = (-4)(-20) = 80$

Exercise 1–5

Answers to odd-numbered problems

1. 3 terms, polynomial **3.** 3 terms, polynomial **5.** 1 term, not a
polynomial since a variable appears under the radical symbol
7. $4, -7, 1$ **9.** $5, 1, -7$ **11.** trinomial, 3rd degree **13.** binomial,
4th degree **15.** monomial, 5th degree **17.** -12 **19.** 60
21. 31 **23.** -7 **25.** -22 **27.** $Q(2) = 2$, $Q(-3) = 22$,
$Q(0) = 4$ **29.** $P(1) = 0$, $P(-2) = -9$, $P(0) = -1$ **31.** 6
33. 31 **35.** -42 **37.** 4 **39.** -9 **41.** 60 **43.** 55
45. 83 **47.** 144 **49.** $\dfrac{79}{19}$ or $4\dfrac{3}{19}$ **51.** 12 **53.** 21
55. 38 m **57.** $\dfrac{1{,}280}{3}$ or $426\dfrac{2}{3}$ rpm **59.** 2.5 amperes **61.** $3x$
63. $x + 8$ **65.** $4(x + 7)$ **67.** $x - 4$
69. let n = the number; $n - 15$ **71.** let n = the number;
$8n + 14$ **73.** let n = the number; $4(n + 3)$ **75.** let n = the
number; $\dfrac{1}{4}n$ **77.** let n = the number; $\dfrac{1}{2}n$

Solutions to trial exercise problems

4. $\dfrac{4a^2 - b^2}{10}$ has one term since the fraction bar is a grouping symbol,
and it is a polynomial because division by a constant is allowed.
8. 3 is the numerical coefficient of x^3, -2 is the numerical coefficient
of x^2, and 1 is understood to be the numerical coefficient of x.
19. $[4(\ \ ) + 3(\ \ )][(\ \ ) - 2(\ \ )] = [4(-3) + 3(2)][(-2) -$
$2(4)] = [(-12) + 6][(-2) - 8] = [(-12) + 6][(-2) + (-8)]$
$= (-6)(-10) = 60$ **24.** $[3(\ \ ) - 2(\ \ )] - [2(\ \ ) + (\ \ )][(\ \ )$
$+ (\ \ )] = [3(-3) - 2(2)] - [2(-3) + (2)][(-2) + (4)] =$
$[(-9) - (4)] - [(-6) + (2)][2] = [(-9) + (-4)] - [-4][2]$
$= [-13] - [-8] = [-13] + [8] = -5$ **36.** $P[Q(2)]$
$= P[2(2) - 1] = P(4 - 1) = P(3) = (3)^2 + 2(3) + 1$
$= 9 + 2(3) + 1 = 9 + 6 + 1 = 15 + 1 = 16$
42. $I = (\ \ )(\ \ )(\ \ ) = (2{,}000)(0.09)(3) = 540$
49. $A = \dfrac{(\ \ )(\ \ ) + (\ \ )(\ \ )}{(\ \ ) + (\ \ )} = \dfrac{(80)(3) + (110)(5)}{(80) + (110)} = \dfrac{240 + 550}{190}$
$= \dfrac{790}{190} = \dfrac{79}{19}$ or $4\dfrac{3}{19}$

Exercise 1–6

Answers to odd-numbered problems

1. $-x^2 + 8x$ **3.** $-5a^3 + a^2 + 5a$ **5.** $4x^2y + 3xy$
7. $-x^2y + 12xy$ **9.** $3b^2$ **11.** $-4x + 22yz - 26z$
13. $-9a^2b + 13ab - 16c$ **15.** $-28x + 37y - 3z$
17. $10x^2 - 7x - 7$ **19.** $2y^2 - xy - 3yz$ **21.** $4x^2 - 3xy - 2y^2$
23. $x^2y - 2x^2y^2 - xy + 8xy^2$ **25.** $-2a^2b^2 + ab + 3$
27. $2x^2 + 5x - 8$ **29.** $8a - 10$ **31.** $9x^2 + 6x + 3$

33. $-9y^2 + 16y + 8$ **35.** $12a^2 - 7$ **37.** $-3a^2 + 13a - 15$
39. $7x^2 - 10xy$ **41.** $-x - 16$ **43.** $3t - 2$ **45.** $2a - 5$
47. $2x - y$ **49.** $-9x + 6y$ **51.** $-3x - 8y$ **53.** $2a + b$
55. $11x$ **57.** $2a - 2b - 3c$ **59.** $8x^2 - 7x$
61. $-8x^2 + 10x + 7$ **63.** $-6x^2y^2 + 3x^2y - 2xy^2$
65. $5x^2 - 2x$ **67.** $-5x^2 + 2x$ **69.** $-5x^2 + 2x$
71. $-5x^2 + 2x$

Solutions to trial exercise problems

13. $-(5a^2b - 6ab + 16c) + (7ab - 4a^2b) = -5a^2b + 6ab - 16c$
$+ 7ab - 4a^2b = (-5a^2b - 4a^2b) + (6ab + 7ab) - 16c = -9a^2b$
$+ 13ab - 16c$
23.

$$\begin{array}{r}(3x^2y - 2x^2y^2 + 3xy^2) \\ -\ (2x^2y + \ \ \ \ xy - 5xy^2)\end{array} = \begin{array}{r}3x^2y - 2x^2y^2 \ \ \ \ \ \ \ \ + 3xy^2 \\ -\ 2x^2y \ \ \ \ \ \ \ \ \ \ - xy + 5xy^2 \\ \hline x^2y - 2x^2y^2 - xy + 8xy^2\end{array}$$

27. $(5x^2 + 3x - 7) - (3x^2 - 2x + 1) = 5x^2 + 3x - 7 - 3x^2 +$
$2x - 1 = 2x^2 + 5x - 8$ **32.** $(6t^2 - 7t + 14) - (8t^2 - 11t + 6)$
$= 6t^2 - 7t + 14 - 8t^2 + 11t - 6 = -2t^2 + 4t + 8$
36. $[(6x^2 + 3x) + (5x^2 - 4x + 2)] - (2x^2 - 9x + 4)$
$= [6x^2 + 3x + 5x^2 - 4x + 2] - (2x^2 - 9x + 4)$
$= (11x^2 - x + 2) - (2x^2 - 9x + 4) = 11x^2 - x + 2 - 2x^2 +$
$9x - 4 = 9x^2 + 8x - 2$ **40.** $[(8a^2 + 11) + (2a^2 - 7a + 6)]$
$- (4a^2 - a - 1) = [8a^2 + 11 + 2a^2 - 7a + 6]$
$- (4a^2 - a - 1) = [10a^2 - 7a + 17] - (4a^2 - a - 1)$
$= 10a^2 - 7a + 17 - 4a^2 + a + 1 = 6a^2 - 6a + 18$
49. $-(3x - 2y) - (x + y) - [2x - y + (3x - 4y)]$
$= -3x + 2y - x - y - [2x - y + 3x - 4y] = -4x + y$
$- [5x - 5y] = -4x + y - 5x + 5y = -9x + 6y$
54. $-\{5x - 3y + [2x - (5x - 7y)] + 4y\} = -\{5x - 3y$
$+ [2x - 5x + 7y] + 4y\} = -\{5x - 3y + [-3x + 7y] + 4y\}$
$= -\{5x - 3y - 3x + 7y + 4y\} = -\{2x + 8y\} = -2x - 8y$
60. $-[-3a^2 + (4a - 3) + 5a] - \{7a^2 + [4a - (a^2 - 3) + 5a^2]\}$
$= -[-3a^2 + 4a - 3 + 5a] - \{7a^2 + [4a - a^2 + 3 + 5a^2]\}$
$= -[-3a^2 + 9a - 3] - \{7a^2 + [4a^2 + 4a + 3]\} = 3a^2 - 9a$
$+ 3 - \{7a^2 + 4a^2 + 4a + 3\} = 3a^2 - 9a + 3 - \{11a^2 + 4a + 3\}$
$= 3a^2 - 9a + 3 - 11a^2 - 4a - 3 = -8a^2 - 13a$
66. $(2x^2 - x + 3) - [(5x - 4) + (3x^2 + 4x - 7)]$
$= 2x^2 - x + 3 - [5x - 4 + 3x^2 + 4x - 7]$
$= 2x^2 - x + 3 - [3x^2 + 9x - 11] = 2x^2 - x + 3 - 3x^2 - 9x$
$+ 11 = -x^2 - 10x + 14$

Chapter 1 review

1. $\{c,o,l,e,g\}$ **2.** $\{January,February,March\}$ **3.** $\emptyset$ **4.** $B \subseteq D$
5. $\emptyset \subseteq A$ **6.** $5 \in C$ **7.** $B \not\subseteq C$ **8.** true **9.** true **10.** false
11. $\{1,2,3,4,5,6\}$ **12.** $\{2,3,4\}$ **13.** $\emptyset$ **14.** $<$ **15.** $>$ **16.** $<$
17. -13 **18.** 8 **19.** -3 **20.** -9 **21.** 6 **22.** -10
23. 5 **24.** -6 **25.** -3 **26.** 0 **27.** -16 **28.** 6
29. -5 **30.** additive inverse property **31.** distributive property
of multiplication over addition **32.** multiplicative inverse property
33. associative property of multiplication **34.** 2 **35.** 4
36. -4 **37.** 29 **38.** -71 **39.** -72 **40.** $-\dfrac{16}{3}$ or $-5\dfrac{1}{3}$
41. 12 **42.** $\dfrac{31}{24}$ or $1\dfrac{7}{24}$ **43.** 13.5 **44.** 9.79 **45.** 4 terms,
polynomial **46.** 1 term, polynomial **47.** -8 **48.** 1 **49.** -60
50. -364 **51.** (a) 5, (b) -1, (c) 5, (d) 11 **52.** 70 **53.** 75
54. $\dfrac{12}{5}$ or $2\dfrac{2}{5}$ **55.** $y + 4$ **56.** $a - 12$ **57.** $7(a + 6)$
58. $5x - 3$ **59.** let n = the number; $5(n + 12)$

60. let n = the number; $\dfrac{n+4}{8}$ **61.** $13y^2 - 9y$

62. $14x^2y + 5x^2y^2 - xy^2$ **63.** $8x + 17$ **64.** $4a - 4$

65. $5x - 2$ **66.** $11c - 6$ **67.** $2x^2 - 5x + 11$ **68.** $8a + 4$

69. $-10x$ **70.** $-2a^2 - 3b$

Chapter 1 test

1. false   **2.** true   **3.** true   **4.** {2,4,6,8}   **5.** {4,6}   **6.** {2,4}

7. -14 **8.** -6 **9.** 9   **10.** $-x - 5$ **11.** -9 **12.** 0

13. $7a$ **14.** 10   **15.** -3 **16.** 16   **17.** 8   **18.** $\dfrac{5}{12}$ **19.** 0

20. $4a + 5b$ **21.** 15   **22.** 5   **23.** 6   **24.** let x = the number;

$x - 3$ **25.** let x = the number; $\dfrac{x+5}{8}$

Chapter 2

Exercise 2–1

Answers to odd-numbered problems

1. {3}   **3.** $\left\{\dfrac{5}{2}\right\}$ **5.** {6}   **7.** {3}   **9.** {12}   **11.** $\left\{\dfrac{32}{3}\right\}$

13. {12}   **15.** $\left\{\dfrac{27}{2}\right\}$ **17.** {3}   **19.** {−1}   **21.** $\left\{\dfrac{5}{3}\right\}$

23. $\left\{\dfrac{1}{3}\right\}$ **25.** $\left\{\dfrac{15}{14}\right\}$ **27.** $\left\{\dfrac{5}{3}\right\}$ **29.** $\emptyset$ **31.** $\left\{-\dfrac{7}{2}\right\}$

33. {8}   **35.** {24}   **37.** $\left\{\dfrac{9}{2}\right\}$ **39.** {2}   **41.** $\left\{\dfrac{31}{3}\right\}$ **43.** $\left\{\dfrac{-6}{7}\right\}$

45. {6}   **47.** {−8.4}   **49.** {8}   **51.** $\emptyset$ **53.** $\ell = 8$

55. $t = 4$ years   **57.** $115c$ **59.** $\dfrac{18}{h}$ **61.** $50h + 25q + 10d + 5n$

63. $3n, 3n - 8$ **65.** $d + 464 - 5m$ **67.** $w + 1$ **69.** $j + 2$

71. $2d + 500$ **73.** $59c + 115b$ **75.** $20(w + 11)$

Solutions to trial exercise problems

18. $4x + 5 = 5$
$4x = 0$
$x = 0$
{0}

44. $5.6z - 22.15 = 24.33$
$5.6z = 46.48$
$z = 8.3$
{8.3}

50. $3(2x - 1) = 6x + 7$
$6x - 3 = 6x + 7$
$-3 = 7$ (false)
$\emptyset$

Review exercises

1. 264   **2.** 600   **3.** 220   **4.** 38.75   **5.** 256   **6.** 38

Exercise 2–2

Answers to odd-numbered problems

1. $R = 9$ **3.** $P = 3{,}000$ **5.** $n = 20$ **7.** $V_1 = 51$ **9.** $t = \dfrac{I}{pr}$

11. $m = \dfrac{E}{c^2}$ **13.** $m = \dfrac{F}{a}$ **15.** $b = \dfrac{A}{h}$ **17.** $R = \dfrac{W}{I^2}$

19. $k = V - gt$ **21.** $g = \dfrac{V - k}{t}$ **23.** $q = \dfrac{D - R}{d}$

25. $\ell = \dfrac{px - m}{p}$ **27.** $W = R + 2bc + b^2$ **29.** $a = \dfrac{V + br^2}{r^2}$

31. $d = \dfrac{2S - 2an}{n^2 - n}$ **33.** $g = \dfrac{2Vt - 2S}{t^2}$ **35.** $d = \dfrac{\ell - a}{n - 1}$

37. $x = \dfrac{12 - 3y}{2}$ **39.** $y = \dfrac{-3x}{7}$ **41.** $x = \dfrac{5y + 6}{10}$

43. $y = \dfrac{4x - 3}{a}$ **45.** $y = \dfrac{ax + 3a + 4b}{b}$ **47.** $v = \dfrac{2s - gt^2}{2t}$

49. $P_1 = \dfrac{nP_2 - P - c}{n}$

Solution to trial exercise problem

26.
$R = W - b(2c + b); \; c$
$R = W - 2bc - b^2$
$2bc + R = W - b^2$
$2bc = W - b^2 - R$
$c = \dfrac{W - b^2 - R}{2b}$

Review exercises

1. $3x$ **2.** $6(a + 7)$ **3.** $\dfrac{y - 2}{4}$ **4.** let n = the number; $5n$

5. let n = the number; $n - 12$ **6.** let n = the number; $\dfrac{n}{8} - 9$

Exercise 2–3

Answers to odd-numbered problems

1. 40,48   **3.** 6,48   **5.** 21,26   **7.** 18   **9.** 84   **11.** -26

13. 17   **15.** 23,58   **17.** 13   **19.** 26,38   **21.** 47,79   **23.** 5,19

25. 22,23,24   **27.** 14,16,18   **29.** $-23, -21, -19$ **31.** 9,36

33. 24   **35.** 29,40   **37.** 21,22   **39.** 8,10,12   **41.** 11,33,5

43. 8,15   **45.** \$10,000 at 8%; \$5,000 at 6%   **47.** \$13,000 at 10%;

\$13,000 at 12%   **49.** \$5,000 at 10%; \$7,000 at 12%

51. \$12,285.71 at 5%; \$5,714.29 at 9%   **53.** \$4,000

55. \$14,000 at 14%; \$12,000 at 10%   **57.** \$10,000 at 14%;

\$8,000 at 9%   **59.** \$19,000 at 12%; \$15,000 at 21%

61. $\ell = 28$ feet, $w = 23$ feet   **63.** $\ell = 48$ feet, $w = 16$ feet

65. 14 cm, 7 cm, 17 cm   **67.** 40 cm³ of 10% solution,

80 cm³ of 4% solution   **69.** 3 ounces of 60% gold, 9 ounces of 80%

gold   **71.** 100 3-grain capsules, 100 2-grain capsules

73. 10 liters of 60% solution, 20 liters of 30% solution

Solutions to trial exercise problems

9. Let n = the number.

a number is divided by 4	increased by	6	is	27
$\dfrac{n}{4}$	$+$	6	$=$	27

$\dfrac{n}{4} + 6 = 27$ Original equation

$4\left(\dfrac{n}{4} + 6\right) = 4 \cdot 27$ Multiply by 4 to clear the fraction

$n + 24 = 108$ Distribute the multiplication in the left member

$n = 84$ Subtract 24

The number is 84.

52. $x =$ the amount invested at 10%.

$5,000 at 8%   more invested at 10%   will be   total at 9%

$5,000(0.08)$ $+ x(0.10)$ $=$ $(5,000 + x)(0.09)$

Solving for x: $5,000(0.08) + 0.10x = (5,000 + x)(0.09)$

$$400 + 0.10x = 450 + 0.09x$$
$$400 + 0.01x = 450$$
$$0.01x = 50$$
$$x = 5,000$$

Therefore $5,000

	Investment at 8%	Investment at 10%	Total investment at 9%
Amount invested	5,000	x	$5,000 + x$
Interest received	$5,000(0.08)$	$x(0.10)$	$(5,000 + x)(0.09)$

Review exercises

1. 12   **2.** 0   **3.** $\dfrac{3}{4}$ **4.** 6   **5.** $\left\{\dfrac{20}{3}\right\}$ **6.** $\{-6\}$ **7.** $\{5\}$

8. $\{-7\}$

Exercise 2–4

Answers to odd-numbered problems

1. true   **3.** false   **5.** false   **7.** true   **9.** $\{9,-9\}$ **11.** $\{4,-4\}$

13. $\{4,-4\}$ **15.** $\{-12,12\}$ **17.** $\{2,-10\}$ **19.** $\{1,5\}$

21. $\left\{\dfrac{-4}{3},4\right\}$ **23.** $\{-8,1\}$ **25.** $\emptyset$ **27.** $\emptyset$ **29.** $\left\{-\dfrac{3}{4},\dfrac{9}{4}\right\}$

31. $\left\{-\dfrac{13}{5},\dfrac{9}{5}\right\}$ **33.** $\{-1,7\}$ **35.** $\{3,9\}$ **37.** $\{-24,4\}$

39. $\left\{-\dfrac{16}{3},\dfrac{32}{3}\right\}$ **41.** $\{3,15\}$ **43.** $\left\{-\dfrac{7}{3},\dfrac{17}{3}\right\}$ **45.** $\left\{0,\dfrac{3}{2}\right\}$

47. $\left\{-5,\dfrac{1}{2}\right\}$ **49.** $\left\{-\dfrac{3}{2},-\dfrac{1}{2}\right\}$ **51.** $\left\{-\dfrac{9}{2},-\dfrac{1}{2}\right\}$

53. $\left\{-\dfrac{2}{3}\right\}$ **55.** $\left\{-\dfrac{1}{2}\right\}$ **57.** $\{-6,6\}$ **59.** $\{-4,4\}$

61. $\{-6,22\}$ **63.** $\{-22,8\}$ **65.** $\left\{-\dfrac{4}{3},4\right\}$ **67.** $\{-21,63\}$

Solutions to trial exercise problems

25. $|5x - 3| = -4$

$\emptyset$ since the absolute value can only be greater than or equal to zero.

43. $|5 - 3x| - 4 = 8$

$$|5 - 3x| = 12$$

$5 - 3x = 12$ or $5 - 3x = -12$

$-3x = 7$ $-3x = -17$

$x = -\dfrac{7}{3}$ $x = \dfrac{17}{3}$

$$\left\{-\dfrac{7}{3},\dfrac{17}{3}\right\}$$

47. $|3x - 7| = |5x + 3|$

$3x - 7 = 5x + 3$ or $3x - 7 = -(5x + 3)$

$-7 = 2x + 3$ $3x - 7 = -5x - 3$

$-10 = 2x$ $8x - 7 = -3$

$-5 = x$ $8x = 4$

$$\left\{-5,\dfrac{1}{2}\right\}$$ $x = \dfrac{4}{8} = \dfrac{1}{2}$

Review exercises

1. $x + 2$ **2.** $y - 6$ **3.** $a - 4$ **4.** let $x =$ the number; $\dfrac{x}{5}$

5. let $x =$ the number; $\dfrac{1}{3}x$ **6.** let $x =$ the number; $\dfrac{x - 2}{8}$

Exercise 2–5

Answers to odd-numbered problems

1. $[-3,1]$

3. $(-5,-1)$

5. $[4,+\infty)$

7. $(-\infty,-1]$

9. $\{x|x > 9\}$ or $(9,+\infty)$ **11.** $\{x|x \geq 12\}$ or $[12,+\infty)$

13. $\{x|x \geq -5\}$ or $[-5,+\infty)$ **15.** $\left\{x|x < \dfrac{3}{2}\right\}$ or $\left(-\infty,\dfrac{3}{2}\right)$

17. $\{x|x > 1\}$ or $(1,+\infty)$ **19.** $\left\{x|x \geq -\dfrac{27}{2}\right\}$ or $\left[\dfrac{-27}{2},+\infty\right)$

21. $\left\{x|x \geq \dfrac{7}{6}\right\}$ or $\left[\dfrac{7}{6},+\infty\right)$ **23.** $\left\{x|x \geq \dfrac{1}{4}\right\}$ or $\left[\dfrac{1}{4},+\infty\right)$

25. $\{x|x < 3\}$ or $(-\infty,3)$ **27.** $\{x|x \geq -1\}$ or $[-1,+\infty)$

29. $\{x|x \geq 6\}$ or $[6,+\infty)$ **31.** $\{x|x \geq -14\}$ or $[-14,+\infty)$

33. $\{x|x \leq -1\}$ or $(-\infty,-1]$ **35.** $\left\{x\left|\dfrac{2}{3} < x < 3\right.\right\}$ or $\left(\dfrac{2}{3},3\right)$

37. $\left\{x\left|\dfrac{2}{7} \leq x \leq \dfrac{8}{7}\right.\right\}$ or $\left[\dfrac{2}{7},\dfrac{8}{7}\right]$

39. $\left\{x\left|\dfrac{-10}{3} \leq x \leq 0\right.\right\}$ or $\left[\dfrac{-10}{3},0\right]$

41. $\{x|-1 < x < 1\}$ or $(-1,1)$ **43.** $\left\{x|4 \leq x \leq \dfrac{13}{2}\right\}$ or $\left[4,\dfrac{13}{2}\right]$

45. $\left\{x|2 \leq x \leq \dfrac{7}{2}\right\}$ or $\left[2,\dfrac{7}{2}\right]$

47. $\left\{x\left|\dfrac{-2}{3}\le x\le\dfrac{2}{3}\right.\right\}$ or $\left[\dfrac{-2}{3},\dfrac{2}{3}\right]$

49. $(x=$ student's score$)$, $x\ge90$

51. $(t=$ temperature$)$, $t\le42$ **53.** $(c=$ number of cars$)$, $c\ge10$

55. $(m=$ number of minutes$)$, $96\le m\le384$ **57.** $\dfrac{3}{2}c\le P\le3c$

59. $5x-6<17$, $\left\{x\left|x<\dfrac{23}{5}\right.\right\}$ or $\left(-\infty,\dfrac{23}{5}\right)$

61. $\dfrac{1}{2}x+16>24$, $\{x|x>16\}$ or $(16,+\infty)$

63. $19-2x\le8$, $\left\{x\left|x\ge\dfrac{11}{2}\right.\right\}$ or $\left[\dfrac{11}{2},+\infty\right)$

65. $\dfrac{66+71+84+x}{4}\ge75$, minimum score is 79, $\{x|x\ge79\}$ or

$[79,+\infty)$ **67.** $12<3x+2<23$, $\left\{x\left|\dfrac{10}{3}<x<7\right.\right\}$ or $\left(\dfrac{10}{3},7\right)$

69. $16<4s<84$, $\{s|4<s<21\}$ or $(4,21)$

Solutions to trial exercise problems

6. $(-\infty,2)$ or

We use a hollow circle or a parenthesis at 2 to denote that we do not contain the endpoint (strict inequality).

13. $-4x\le20$

$\dfrac{-4x}{-4}\ge\dfrac{20}{-4}$

$x\ge-5$

$\{x|x\ge-5\}$ or $[-5,+\infty)$

20. $4-2(3x+1)>8x-12$

$4-6x-2>8x-12$

$2-6x>8x-12$

$2>14x-12$

$14>14x$

$1>x$

$\{x|x<1\}$ or $(-\infty,1)$

40. $2\le1-x\le6$

$1\le-x\le5$

$\dfrac{1}{-1}\ge\dfrac{-x}{-1}\ge\dfrac{5}{-1}$

$-1\ge x\ge-5$

$\{x|-5\le x\le-1\}$ or $[-5,-1]$

54. If $x=$ the temperature, then $18\le x\le41$.

64. If $x=$ the score on the fourth quiz, then

$\dfrac{7+10+8+x}{4}\ge8$

$\dfrac{25+x}{4}\ge8$

$25+x\ge32$

$x\ge7.$

Therefore she must score 7 or more on the fourth quiz to have an average of 8 or more.

Review exercises

1. 21   **2.** 8   **3.** -2 **4.** $2y$ **5.** $a+6$

6. let $x=$ the number; $\dfrac{1}{4}x$ **7.** let $x=$ the number; $x-12$

8. let $x=$ the number; $x+7$

Answers to odd-numbered problems

1. false   **3.** true   **5.** true   **7.** false

9. $\{x|-4<x<4\}$, $(-4,4)$

11. $\{x|x\le-2$ or $x\ge2\}$, $(-\infty,-2]\cup[2,+\infty)$

13. $\{x|-3\le x\le3\}$, $[-3,3]$

15. $\{x|x<-5$ or $x>5\}$, $(-\infty,-5)\cup(5,+\infty)$

17. $\left\{x\left|-\dfrac{2}{3}<x<\dfrac{10}{3}\right.\right\}$, $\left(-\dfrac{2}{3},\dfrac{10}{3}\right)$

19. $\left\{x\left|x\le-\dfrac{4}{5}\text{ or }x\ge2\right.\right\}$, $\left(-\infty,-\dfrac{4}{5}\right]\cup[2,+\infty)$

21. $\left\{x\left|-3\le x\le\dfrac{17}{3}\right.\right\}$, $\left[-3,\dfrac{17}{3}\right]$

23. $\left\{x\left|\dfrac{-19}{5}<x<1\right.\right\}$, $\left(\dfrac{-19}{5},1\right)$

25. $\left\{x\left|-3\le x\le\dfrac{17}{3}\right.\right\}$, $\left[-3,\dfrac{17}{3}\right]$

27. $\{x|-5<x<10\}$, $(-5,10)$ **29.** $\emptyset$ **31.** all real numbers, $\{x|x\in R\}$, $(-\infty,+\infty)$ **33.** $\left\{x\left|\dfrac{2}{7}<x<\dfrac{6}{7}\right.\right\}$, $\left(\dfrac{2}{7},\dfrac{6}{7}\right)$

35. $\left\{x\left|x\le-2\text{ or }x\ge\dfrac{8}{3}\right.\right\}$, $(-\infty,-2]\cup\left[\dfrac{8}{3},+\infty\right)$

37. $\emptyset$ **39.** all real numbers, $\{x|x\in R\}$, $(-\infty,+\infty)$

41. $\left\{x\left|x<\dfrac{1}{5}\text{ or }x>1\right.\right\}$, $\left(-\infty,\dfrac{1}{5}\right)\cup(1,+\infty)$

43. $\{x|x<1.475$ or $x>6.2\}$, $(-\infty,1.475)\cup(6.2,+\infty)$

45. $\{x|-0.5\le x\le2\}$, $[-0.5,2]$ **47.** $\{x|-1<x<7\}$, $(-1,7)$

49. let $x=$ the number; $|x|=12$; $\{-12,12\}$

51. let $x=$ the number; $\left|\dfrac{1}{2}x\right|=7$; $\{-14,14\}$

53. let $x=$ the number; $|4x+3|=19$; $\left\{-\dfrac{11}{2},4\right\}$

55. let $x=$ the number; $|x|\le6$, $\{x|-6\le x\le6\}$, $[-6,6]$

57. let $x=$ the number; $|2x|<14$, $\{x|-7<x<7\}$, $(-7,7)$

59. let $x=$ the number; $|2x+5|>15$, $\{x|x<-10$ or $x>5\}$,

$(-\infty,-10)\cup(5,+\infty)$ **61.** let $x=$ the number; $\left|\dfrac{1}{4}x-8\right|<12$;

$\{x|-16<x<80\}$, $(-16,80)$

Solutions to trial exercise problems

24. $|1 - 2x| \leq 5$

$$-5 \leq 1 - 2x \leq 5$$
$$-6 \leq -2x \leq 4$$
$$\frac{-6}{-2} \geq \frac{-2x}{-2} \geq \frac{4}{-2}$$
$$3 \geq x \geq -2$$
$$\{x| -2 \leq x \leq 3\}, \ [-2,3]$$

28. $|4x - 9| < 0$

$\emptyset$. The absolute value cannot be less than zero.

58. Let x = the number, then $|3x - 4| \geq 11$.

$$3x - 4 \geq 11 \qquad \text{or} \qquad 3x - 4 \leq -11$$
$$3x \geq 15 \qquad\qquad\qquad 3x \leq -7$$
$$x \geq 5 \qquad\qquad\qquad\quad x \leq -\frac{7}{3}$$

$$\left\{x \middle| x \leq -\frac{7}{3} \text{ or } x \geq 5\right\}, \left(-\infty, -\frac{7}{3}\right] \cup [5, +\infty)$$

Review exercises

1. -16 **2.** 16 **3.** -16 **4.** 16 **5.** x^5 **6.** x^3 **7.** x^2
8. xy

Chapter 2 review

1. $\{8\}$ **2.** $\{6\}$ **3.** $\{28\}$ **4.** $\{4\}$ **5.** $\left\{\frac{15}{7}\right\}$ **6.** $\left\{\frac{10}{3}\right\}$ **7.** $\left\{\frac{5}{3}\right\}$

8. $\left\{-\frac{13}{5}\right\}$ **9.** $\{9\}$ **10.** $w = \frac{v}{\ell h}$ **11.** $t = \frac{v - k}{g}$

12. $d = \frac{D - R}{q}$ **13.** $b = \frac{ar^2 - v}{r^2}$ **14.** $v = \frac{2s + gt^2}{2t}$

15. $n = \frac{\ell - a + d}{d}$ **16.** $x = \frac{3y}{2}$ **17.** 12 **18.** 36

19. $\frac{33}{5}, \frac{99}{5}, \frac{3}{5}$ **20.** 11 feet, 30 feet **21.** \$15,000 at 10%;

\$9,000 at 8% **22.** 40 cl of 42% solution, 60 cl of 12% solution

23. $\{-15,15\}$ **24.** $\left\{-\frac{17}{3}, \frac{7}{3}\right\}$ **25.** $\left\{-\frac{3}{2}, \frac{17}{2}\right\}$ **26.** $\{0,3\}$

27. $\left\{-7, -\frac{1}{5}\right\}$ **28.** $\left\{-\frac{11}{2}, -\frac{1}{6}\right\}$ **29.** $\{x| x \leq 6\}, (-\infty, 6]$

30. $\{x| x > 16\}, (16, +\infty)$ **31.** $\left\{x\middle| x > -\frac{9}{2}\right\}, \left(-\frac{9}{2}, +\infty\right)$

32. $\left\{x\middle| x \leq \frac{9}{8}\right\}, \left(-\infty, \frac{9}{8}\right]$ **33.** $\left\{x\middle| x > -\frac{3}{2}\right\}, \left(-\frac{3}{2}, +\infty\right)$

34. $\left\{x\middle| x \leq \frac{25}{3}\right\}, \left(-\infty, \frac{25}{3}\right]$ **35.** $\left\{x\middle| -\frac{7}{2} < x < 1\right\}, \left(-\frac{7}{2}, 1\right)$

36. $\left\{x\middle| -\frac{4}{5} \leq x \leq 0\right\}, \left[-\frac{4}{5}, 0\right]$ **37.** $\{x| -5 < x < -2\},$

$(-5, -2)$ **38.** $\left\{x\middle| \frac{2}{3} < x \leq \frac{10}{3}\right\}, \left(\frac{2}{3}, \frac{10}{3}\right]$

39. $\{x| -1 \leq x < 0\}, [-1,0)$ **40.** let x = the number;
$4x - 5 \geq 19, \{x| x \geq 6\}, [6, +\infty)$ **41.** let x = the number;
$22 < 3x + 7 < 34, \{x| 5 < x < 9\}, (5,9)$
42. $\{x| x \leq -10 \text{ or } x \geq 10\}, (-\infty, -10] \cup [10, +\infty)$

43. $\left\{x\middle| -\frac{11}{2} < x < \frac{1}{2}\right\}, \left(-\frac{11}{2}, \frac{1}{2}\right)$

44. $\left\{x\middle| -\frac{6}{5} \leq x \leq \frac{8}{5}\right\}, \left[-\frac{6}{5}, \frac{8}{5}\right]$

45. $\left\{x\middle| x > \frac{1}{2} \text{ or } x < -4\right\}, (-\infty, -4) \cup \left(\frac{1}{2}, +\infty\right)$

46. $\left\{x\middle| x \leq -\frac{4}{3} \text{ or } x \geq 2\right\}, \left(-\infty, -\frac{4}{3}\right] \cup [2, +\infty)$

47. $\left\{x\middle| -\frac{9}{4} < x < \frac{15}{4}\right\}, \left(-\frac{9}{4}, \frac{15}{4}\right)$

48. all real numbers, $\{x| x \in R\}, (-\infty, +\infty)$

49. $\left\{x\middle| -\frac{3}{2} < x < \frac{5}{2}\right\}, \left(-\frac{3}{2}, \frac{5}{2}\right)$

50. $\left\{x\middle| x \leq -\frac{9}{2} \text{ or } x \geq 2\right\}, \left(-\infty, -\frac{9}{2}\right] \cup [2, +\infty)$ **51.** $\emptyset$

Chapter 2 cumulative test

1. false **2.** false **3.** true **4.** true **5.** true **6.** true
7. true **8.** 24 **9.** -1 **10.** 12 **11.** 0 **12.** 37 **13.** -6
14. -49 **15.** commutative property of multiplication
16. commutative property of multiplication **17.** $\{1,2,3,4,5\}$
18. $\{4,6,8,9,10,11\}$ **19.** $\{10,12\}$ **20.** $\left\{\frac{7}{4}\right\}$

21. $P = -\frac{M}{\ell - x} \text{ or } \frac{M}{x - \ell}$

22. $P_2 = \frac{P + nP_1 + c}{n}$ **23.** $\{x| x \geq -5\}$

24. $\left\{x\middle| x \leq -\frac{11}{4} \text{ or } x \geq -\frac{1}{4}\right\}$ **25.** $\left\{\frac{21}{17}\right\}$ **26.** $\left\{x\middle| x \geq \frac{15}{4}\right\}$

27. $\{-1,1\}$ **28.** $\left\{\frac{16}{7}\right\}$ **29.** $\emptyset$ **30.** $\left\{x\middle| -1 \leq x \leq \frac{11}{3}\right\}$

31. 16,8,48 **32.** \$26,000 at 11%; \$14,000 at 8%

Chapter 3

Exercise 3–1

Answers to odd-numbered problems

1. $(-2)^4$, -2 base, 4 exponent **3.** x^5, x base, 5 exponent
5. $(2x)^4$, $2x$ base, 4 exponent **7.** $(x^2 + 3y)^3$, $x^2 + 3y$ base,
3 exponent **9.** -2^2, 2 base, 2 exponent **11.** a^9 **13.** y^3
15. $(-2)^6 = 64$ **17.** $(-2)^4 = 16$ **19.** -36 **21.** x^8
23. x^7y^5 **25.** $6a^3b^2$ **27.** $6x^5$ **29.** $24x^3y^9$ **31.** $2^6 = 64$
33. -729 **35.** 64 **37.** a^{12} **39.** $x^{10}y^5z^{15}$ **41.** $49s^8t^4$
43. $x^{36}y^{48}z^{32}$ **45.** a^9b^{17} **47.** $x^{13}y^{24}$ **49.** $-x^{10}y^{18}$
51. $-75x^{10}y^{11}$ **53.** $17x^9$ **55.** $-76a^{13}$ **57.** $9x^{10} - 8x^8$
59. $24a^{13} + 18a^{11}$ **61.** $3x^{17} + 96x^{14}$ **63.** a^{9b} **65.** a^{9b}
67. a^{5b+3} **69.** x^{3y+5} **71.** x^{15y^2} **73.** \$5,634.13
75. 2.5 grams

Solutions to trial exercise problems

18. $(-2)(-2^2) = (-2)(-4) = 8$ **32.** $(-2^2)^3 = (-4)^3 = -64$
44. $(3x^2y)^2(2xy^3) = 9x^4y^2 \cdot 2xy^3 = 18x^5y^5$
52. $(2a^2)^2a^3 + (3a)^3a^4 = 4a^4a^3 + 27a^3a^4 = 4a^7 + 27a^7 = 31a^7$
57. $(3x^5)^2 - (2x^2)^3x^2 = 9x^{10} - 8x^6x^2 = 9x^{10} - 8x^8$. The
subtraction cannot be performed because we do not have like terms.
62. $x^{5n} \cdot x^{4n} = x^{5n+4n} = x^{9n}$
66. $x^{2n+1} \cdot x^{n+4} = x^{(2n+1)+(n+4)} = x^{2n+1+n+4} = x^{3n+5}$
70. $(a^{3n})^{4n} = a^{3n \cdot 4n} = a^{12n^2}$

Review exercises

1. -24 **2.** -9 **3.** 0 **4.** 25 **5.** $9ab$ **6.** $a^2 - 2a - 15$
7. $x^2 - 9$ **8.** $x^2 + 2y^2$

Exercise 3–2

Answers to odd-numbered problems

1. $2a^3 - 3a^2 + 4a$ **3.** $-6y^3 + 10y^2 - 8y$ **5.** $12a^4 - 6a^3b$
$+ 9a^2b^2$ **7.** $30x^3y^2 - 24x^2y^4 + 12x^2y^2$ **9.** $-10x^5y^5 - 35x^4y^{10}$
$+ 15x^5y^9$ **11.** $a^2 + 8a + 15$ **13.** $b^2 - b - 20$ **15.** $2x^2 + xy$
$- y^2$ **17.** $10x^2 + 3xy - y^2$ **19.** $42x^2 - 2xy - 20y^2$
21. $x^2 + 6x + 9$ **23.** $9x^2 + 6xy + y^2$ **25.** $16x^2 + 24xy + 9y^2$
27. $4a^2 - 20a + 25$ **29.** $16x^2 - 24xy + 9y^2$ **31.** $x^2 - 9y^2$
33. $25a^2 - 4b^2c^2$ **35.** $x^3 - 2x^2y - xy^2 + 2y^3$
37. $15a^3 - 31a^2b + 23ab^2 - 6b^3$ **39.** $x^4 + x^3 - 5x^2 - 17x - 12$
41. $5b^4 + 2b^3 + 19b^2 - 2b + 21$ **43.** $x^3 + 6x^2y + 12xy^2 + 8y^3$
45. $64a^3 - 48a^2b + 12ab^2 - b^3$ **47.** $2a^3 - a^2b - 8ab^2 + 4b^3$
49. $2a^2 + 6a + 29$ **51.** $2x^2 + 21x - 26$ **53.** $y^2 + y + 8$
55. $-6b + 9$ **57.** $-14x^2 - 4x$ **59.** $26x - 28$
61. $a^{n+2} + 3a^n$ **63.** $3x^{2n} + x^n$ **65.** $x^{2n+3} - x^n$
67. $b^{2n} - 5b^n + 6$ **69.** $9x^{2n} + 6x^ny^n + y^{2n}$ **71.** $a^{2n} - b^{6n}$
73. $\pi R^2 - \pi r^2$ **75.** $2\pi rh + 2\pi r^2$ **77.** $\dfrac{Wx^4}{24EI} - \dfrac{\ell Wx^3}{16EI} - \dfrac{\ell^3 Wx}{48EI}$

Solutions to trial exercise problems

8. $-a^3b(3a^2b^5 - ab^4 - 7a^2b) = -3a^5b^6 + a^4b^5 + 7a^5b^2$
30. $(a - 3b)(a + 3b) = (a)^2 - (3b)^2 = a^2 - 9b^2$
38. $(x^2 - 2x + 1)(x^2 + 3x + 2) = x^4 + 3x^3 + 2x^2 - 2x^3 - 6x^2$
$- 4x + x^2 + 3x + 2 = x^4 + x^3 - 3x^2 - x + 2$ **42.** $(a - 3b)^3$
$= (a - 3b)(a - 3b)(a - 3b) = [(a - 3b)(a - 3b)](a - 3b)$
$= [a^2 - 3ab - 3ab + 9b^2](a - 3b) = [a^2 - 6ab + 9b^2](a - 3b)$
$= a^3 - 3a^2b - 6a^2b + 18ab^2 + 9ab^2 - 27b^3 = a^3 - 9a^2b$
$+ 27ab^2 - 27b^3$ **56.** $-2[5a - (2a + 3) - 3(3a + 7)]$
$= -2[5a - 2a - 3 - 9a - 21] = -2[-6a - 24] = 12a + 48$
64. $x^{n+2}(x^{n+1} + x) = x^{n+2+n+1} + x^{n+2+1} = x^{2n+3} + x^{n+3}$
68. $(2a^n - b^n)^2 = (2a^n - b^n)(2a^n - b^n) = 4a^{n+n} - 2a^nb^n$
$- 2a^nb^n + b^{n+n} = 4a^{2n} - 4a^nb^n + b^{2n}$

Review exercises

1. -10 **2.** 21 **3.** 4 **4.** -16 **5.** a^8 **6.** x^{12} **7.** x^6
8. $4a^2b^6$

Exercise 3–3

Answers to odd-numbered problems

1. 1 **3.** 1 **5.** 7 **7.** $\dfrac{1}{a^4}$ **9.** $\dfrac{3}{a^2}$ **11.** $\dfrac{a}{b^4c^3}$ **13.** $\dfrac{x^5}{2}$ **15.** $\dfrac{1}{a^8}$

17. $4x^3y^6$ **19.** $\dfrac{1}{27x^3}$ **21.** $\dfrac{1}{4}$ **23.** $-\dfrac{1}{81}$ **25.** $\dfrac{1}{4}$ **27.** $\dfrac{1}{x^3}$

29. $\dfrac{20}{x^5}$ **31.** $\dfrac{10b}{a}$ **33.** $\dfrac{1}{9x^8}$ **35.** $\dfrac{a^9}{64b^{12}}$ **37.** $\dfrac{8}{a^6}$ **39.** $\dfrac{b^9}{27a^6}$

41. $\dfrac{27a^6}{b^3}$ **43.** $\dfrac{27x^{12}y^{15}}{z^{27}}$ **45.** $\dfrac{a^{12}}{8b^{21}}$ **47.** $\dfrac{a^6}{b^7}$ **49.** $\dfrac{x^2}{y^3}$ **51.** $\dfrac{4}{x^4}$

53. $\dfrac{18}{a^7b^3}$ **55.** $\dfrac{1}{x^4z^8}$ **57.** $\dfrac{27x^9z^6}{y^{15}}$ **59.** $8a^3b^3$ **61.** $\dfrac{b^{13}c^8}{a^{12}}$

63. $\dfrac{y^6}{x^8z^8}$ **65.** a^{3b-5} **67.** x^{6n-8} **69.** a^{n+4} **71.** $a^{n+1}b^{2n+3}$

73. x^{-3n+3} **75.** 1.55×10^5 **77.** 8.63×10^{-2} **79.** 8.06×10^{21}
81. 5.787×10^{-4} **83.** 2.2046×10^{-3} **85.** 1.102×10^{-3}
87. 6.696×10^8 **89.** -0.0437 **91.** $4,990,000$ **93.** $48,300$
95. 3.17×10^{11} **97.** 1.25×10^{-9} **99.** 1.83×10^1
101. 1.19×10^{14} **103.** 2.52×10^{-6} **105.** 5.93×10^5
107. 3.18×10^9

Solutions to trial exercise problems

9. $3a^{-2} = 3 \cdot \dfrac{1}{a^2} = \dfrac{3}{a^2}$ **12.** $\dfrac{1}{4a^{-3}} = \dfrac{1}{4 \cdot \dfrac{1}{a^3}} = \dfrac{1}{\dfrac{4}{a^3}} = 1 \cdot \dfrac{a^3}{4} = \dfrac{a^3}{4}$

22. $-2^{-2} = -(2^{-2}) = -\left(\dfrac{1}{2^2}\right) = -\dfrac{1}{4}$

28. $(2a^{-3})(3a^{-5}) = 2 \cdot 3 \cdot a^{-3+(-5)} = 6a^{-8} = 6 \cdot \dfrac{1}{a^8} = \dfrac{6}{a^8}$

44. $\left(\dfrac{2x^5y^2}{4x^3y^7}\right)^4 = \left(\dfrac{x^2}{2y^5}\right)^4 = \dfrac{(x^2)^4}{(2y^5)^4} = \dfrac{x^8}{2^4(y^5)^4} = \dfrac{x^8}{16y^{20}}$

51. $\left(\dfrac{3x^{-2}}{12x^{-4}}\right)\left(\dfrac{16x^{-5}}{x}\right) = \left(\dfrac{1}{4} \cdot x^{(-2)-(-4)}\right)(16x^{(-5)-1})$

$= \left(\dfrac{1}{4} x^2\right)(16x^{-6}) = \dfrac{x^2}{4} \cdot \dfrac{16}{x^6} = \dfrac{4}{x^4}$

58. $(3x^2y^{-2})^2 (x^{-4}y^3)^3 = \left(\dfrac{3x^2}{y^2}\right)^2 \left(\dfrac{y^3}{x^4}\right)^3 = \dfrac{3^2(x^2)^2}{(y^2)^2} \cdot \dfrac{(y^3)^3}{(x^4)^3}$

$= \dfrac{9x^4}{y^4} \cdot \dfrac{y^9}{x^{12}} = \dfrac{9y^5}{x^8}$ **89.** $-4.37 \times 10^{-2} = -0.0437$

Review exercises

1. $12a$ **2.** $3x^2$ **3.** $6ab$ **4.** $2a^3 + 3a^2$
5. $6a^4b + 9a^3b^2 - 3a^2b^3$ **6.** $ax + 2bx - 3ay - 6by$
7. $3ax + 4ay - 6bx - 8by$ **8.** $ab + a + b + 1$

Exercise 3–4

Answers to odd-numbered problems

1. $2^2 \cdot 3$ **3.** $2^3 \cdot 3$ **5.** $2^3 \cdot 7$ **7.** prime   **9.** $3 \cdot 13$
11. $-(-2a + 3b)$ **13.** $-3(-a^3 + 3b^2)$
15. $-4(x^3 + 9xy - 4xy^2)$ **17.** $-3a^2b^2(-a - 4b - 5a^2)$
19. $-3x(-x + 3y)$ **21.** $-8R(3S + 2 - 4R)$
23. $5(x^2 + 2xy - 4y)$ **25.** $3a(6b - 9 + c)$ **27.** $3(5R^2 - 7S^2$
$+ 12T)$ **29.** $3RS(R - 2S + 4)$ **31.** $4x^3y(3xy^2 - 2xy + 4)$
33. $3xy(x - 2y^3 + 5x^2y)$ **35.** $(3a + b)(x - y)$
37. $7(2b - 1)(a + c)$ **39.** $4a(b - 2c)(2x + y)$
41. $(5a - 1)(b + 3)$ **43.** $x^n(y^2 + z)$ **45.** $x^ny^n(x^ny^n - 1)$
47. $y^3(y^n - 1)$ **49.** $(a + 3b)(2x - y)$ **51.** $(2a + 3b)(x + 4y)$
53. $(2a - b)(x^2 + 3)$ **55.** $(c - 2d)(3a + b)$
57. $(2x^2 + 3)(x + 5)$ **59.** $(2x - 1)(4x^2 + 3)$
61. $\pi r(s + r)$ **63.** $P(1 + r)$

Solutions to trial exercise problems

4. $28 = 2 \cdot 2 \cdot 7 = 2^2 \cdot 7$ **18.** $-ab^2 - ac^2 = (-a)b^2 + (-a) \cdot c^2$
$= -a(b^2 + c^2)$ **36.** $5a(3x - 1) + 10(3x - 1)$
$= \underline{5(3x - 1)} \cdot a + \underline{5(3x - 1)} \cdot 2 = 5(3x - 1)(a + 2)$ **46.** x^{n+4}
$+ x^4 = x^4 \cdot x^n + x^4 = x^4(x^n + 1)$ **53.** $2ax^2 - 3b - bx^2 + 6a$
$= 2ax^2 - bx^2 + 6a - 3b = (2ax^2 - bx^2) + (6a - 3b)$
$= x^2(2a - b) + 3(2a - b) = (2a - b)(x^2 + 3)$

Review exercises

1. $a^2 + 7a + 12$ **2.** $x^2 - 7x + 10$ **3.** $x^2 - 2x - 24$
4. $x^2 + 67x - 210$ **5.** $x^2 - 16$ **6.** $a^2 - 25$ **7.** $x^2 + 6x + 9$
8. $x^2 - 8x + 16$

Exercise 3–5

Answers to odd-numbered problems

1. $(x + 15)(x - 2)$ **3.** $(y + 8)(y - 3)$ **5.** $(x - 6)(x + 4)$
7. $3(x + 12)(x - 1)$ **9.** $4(x - 3)(x + 2)$
11. $3(y^2 - 9y + 4)$ **13.** $(xy + 3)(xy - 7)$
15. $(xy + 1)(xy + 12)$ **17.** $(xy - 2)(xy - 12)$
19. $4(ab + 4)(ab + 2)$ **21.** $-2(xy - 5)(xy + 2)$
23. $(a + 5b)(a + 2b)$ **25.** $(x + 3y)(x + 8y)$
27. $(a + 7b)(a - 5b)$ **29.** $(x + 8y)(x + 2y)$
31. $(x^n + 7)(x^n + 2)$ **33.** $(x^n - 5)(x^n - 3)$ **35.** $(b - 6)^2$
37. $(c + 9)^2$ **39.** $(3y + 2)^2$ **41.** $(x - 7y)^2$ **43.** $(2x - 3y)^2$
45. $(3x + 5y)^2$ **47.** $(a - 2b)(x + 6)(x + 2)$
49. $(y + 2z)(x - 10)(x - 3)$ **51.** $(3y - z)(x + 12)(x + 3)$
53. $(3x - y)(a - 12)(a + 5)$ **55.** $(a + b - 6)(a + b + 2)$
57. $(y + 3z - 7)(y + 3z + 2)$ **59.** $(R + 5S - 4)(R + 5S - 2)$
61. $(W - 17)(W - 9)$ **63.** $(C_2 + 7)(C_2 - 6)$

Solutions to trial exercise problems

10. $5y^2 - 15y - 55 = 5(y^2 - 3y - 11)$; m and n do not exist,
therefore $5(y^2 - 3y - 11)$ is the completely factored form.
21. $-2x^2y^2 + 6xy + 20 = -2(x^2y^2 - 3xy - 10)$; m and n are -5
and 2. Then $-2(x^2y^2 - 3xy - 10) = -2(xy - 5)(xy + 2)$.
23. $a^2 + 7ab + 10b^2$; m and n are 5 and 2. $(a + 5b)(a + 2b)$
30. $x^{2n} + 5x^n + 6$; m and n are 2 and 3; $(x^n + 2)(x^n + 3)$
47. $x^2(a - 2b) + 8x(a - 2b) + 12(a - 2b) = (a - 2b)(x^2 + 8x + 12)$. In the second parentheses m and n are 6 and 2, and we have
$= (a - 2b)(x + 6)(x + 2)$.   **55.** $(a + b)^2 - 4(a + b) - 12$;
m and n are -6 and 2. $[(a + b) - 6][(a + b) + 2]$
$= (a + b - 6)(a + b + 2)$

Review exercises

1. $(2x - 1)(3x + 5)$ **2.** $(3a + 1)(2a + 1)$ **3.** $(2x - 3)$
$(4x - 1)$ **4.** $(x + 4)(5x - 3)$ **5.** $6x^2 + 7x + 2$
6. $10a^2 + 13a - 3$ **7.** $3b^2 + 5b - 28$ **8.** $9a^2 - 4$

Exercise 3–6

Answers to odd-numbered problems

1. $(3a - 2)(a + 3)$ **3.** $(4x - 1)(x - 1)$ **5.** $(x + 2)(x + 16)$
7. $(2y + 1)(y - 1)$ **9.** $(8x - 1)(x - 2)$ **11.** $(2a - 3)(a - 4)$
13. $(2x + 9)(x + 2)$ **15.** $(7x - 1)(x + 3)$
17. $(2x + 1)(2x - 3)$ **19.** $(6x + 1)(x - 4)$
21. $(2z + 1)(5z + 2)$ **23.** will not factor
25. $(5x + 1)(x - 2)$ **27.** $(3x - 4)(2x - 3)$
29. $3(x + 2)(x + 2) = 3(x + 2)^2$ **31.** will not factor
33. $2(3x - 4)(x - 5)$ **35.** $(2x + 3)(-2x - 3) = -(2x + 3)^2$
37. $(-7x + 1)(x - 5)$ or $(7x - 1)(-x + 5)$ or
$-(7x - 1)(x - 5)$ **39.** $(3x + 4)(4x - 1)$
41. $2x(x + 2)(2x + 1)$ **43.** $2(3x - 8)(3x + 1)$
45. $(4x - 3)(2x - 3)$

Solutions to trial exercise problems

26. $6x^2 + 21x + 18 = 3(2x^2 + 7x + 6)$; m and n are 3 and 4.
$3[2x^2 + 3x + 4x + 6] = 3[(2x^2 + 3x) + (4x + 6)]$
$= 3[x(2x + 3) + 2(2x + 3)] = 3(2x + 3)(x + 2)$
35. $-4x^2 - 12x - 9$; m and n are -6 and -6. $-4x^2 - 6x$
$- 6x - 9 = (-4x^2 - 6x) + (-6x - 9) = -2x(2x + 3) - 3(2x$
$+ 3) = (2x + 3)(-2x - 3) = -(2x + 3)^2$
41. $4x^3 + 10x^2 + 4x = 2x(2x^2 + 5x + 2)$ m and n are 4 and 1.
$2x(2x^2 + 4x + x + 2) = 2x[(2x^2 + 4x) + (x + 2)]$
$= 2x[2x(x + 2) + 1(x + 2)] = 2x(x + 2)(2x + 1)$

Review exercises

1. $9a^2 - b^2$ **2.** $x^2 - 4y^2$ **3.** $9x^2 - 16y^2$ **4.** $x^4 - 25$
5. $x^4 - 16$ **6.** $a^3 + 8b^3$ **7.** $x^3 - 27y^3$ **8.** $27a^3 - 8b^3$

Exercise 3–7

Answers to odd-numbered problems

1. $(a + 7)(a - 7)$ **3.** $(8 + S)(8 - S)$ **5.** $(6x + y^2)(6x - y^2)$
7. $(4a + 7b)(4a - 7b)$ **9.** $8(a + 2b)(a - 2b)$
11. $2(5 + x)(5 - x)$ **13.** $2(7ab + 5xy)(7ab - 5xy)$
15. $(a^2 + 9)(a + 3)(a - 3)$ **17.** $(x^2 + 4y^2)(x + 2y)(x - 2y)$
19. $(a^n + 2)(a^n - 2)$ **21.** $(x^n + y^n)(x^n - y^n)$
23. $(x^{2n} + 9)(x^n + 3)(x^n - 3)$ **25.** $(a + 5b)(2x + y)(2x - y)$
27. $3(3x - y)(a + 3)(a - 3)$ **29.** $3(3x - y)(2a + b)(2a - b)$
31. $(2a + b + x - 2y)(2a + b - x + 2y)$
33. $(3a - b + 2a - b)(3a - b - 2a + b) = (5a - 2b)(a)$
$= a(5a - 2b)$ **35.** $(x + 2y + x - 3y)(x + 2y - x + 3y)$
$= (2x - y)(5y) = 5y(2x - y)$ **37.** $(3a - b)(9a^2 + 3ab + b^2)$
39. $(x + y)(x^2 - xy + y^2)$ **41.** $(a + 2b)(a^2 - 2ab + 4b^2)$
43. $(2y - 3x)(4y^2 + 6xy + 9x^2)$
45. $3(3a - b)(9a^2 + 3ab + b^2)$ **47.** $3(x + 2)(x^2 - 2x + 4)$
49. $(4z + 5)(16z^2 - 20z + 25)$
51. $2(a - 3b)(a^2 + 3ab + 9b^2)$ **53.** $R^2(R + 4S)$
$(R^2 - 4RS + 16S^2)$ **55.** $3x^2(x - 3y)(x^2 + 3xy + 9y^2)$
57. $(xy^4 + z^3)(x^2y^8 - xy^4z^3 + z^6)$ **59.** $(a^6b^3 - 3c)$
$(a^{12}b^6 + 3a^6b^3c + 9c^2)$ **61.** $3(xy^2 + 3z)(x^2y^4 - 3xy^2z + 9z^2)$
63. $(x + 3y + z)(x^2 + 6xy + 9y^2 - xz - 3yz + z^2)$
65. $(4a + b - 3c)(16a^2 + 8ab + b^2 + 12ac + 3bc + 9c^2)$
67. $(a - 2b - 2x - y)(a^2 - 4ab + 4b^2 + 2ax - 4bx + ay$
$- 2by + 4x^2 + 4xy + y^2)$ **69.** $(5x + 2y)(7x^2 - xy + 13y^2)$

71. $9x(x^2 - xy + y^2)$ **73. a.** $\dfrac{V}{8I}(h + 2v_1)(h - 2v_1)$,

b. $\dfrac{3V}{2A}\left(1 + \dfrac{2V_1}{H}\right)\left(1 - \dfrac{2V_1}{H}\right)$

Solutions to trial exercise problems

9. $8a^2 - 32b^2 = 8(a^2 - 4b^2) = 8[(a)^2 - (2b)^2] = 8(a + 2b)$
$(a - 2b)$ **14.** $x^4 - y^4 = (x^2)^2 - (y^2)^2 = (x^2 + y^2)(x^2 - y^2)$
$= (x^2 + y^2)(x + y)(x - y)$ **18.** $x^{2n} - 1 = (x^n)^2 - (1)^2$
$= (x^n + 1)(x^n - 1)$ **24.** $a^2(x + 2y) - b^2(x + 2y)$
$= (x + 2y)(a^2 - b^2) = (x + 2y)(a + b)(a - b)$ **31.** $(2a + b)^2$
$- (x - 2y)^2 = [(2a + b) + (x - 2y)][(2a + b) - (x - 2y)]$
$= (2a + b + x - 2y)(2a + b - x + 2y)$
44. $64a^3 - 8 = 8(8a^3 - 1) = 8[(2a)^3 - (1)^3]$.
Then $8(\ \ - \ ) [(\ )^2 + (\ )(\ ) + (\ )^2]$ and $8(2a - 1)$
$[(2a)^2 + (2a)(1) + (1)^2] = 8(2a - 1)(4a^2 + 2a + 1)$.
55. $3x^5 - 81x^2y^3 = 3x^2(x^3 - 27y^3) = 3x^2[(x)^3 - (3y)^3]$.
Then $3x^2(\ \ - \ ) [(\ )^2 + (\ )(\ ) + (\ )^2]$ and $3x^2(x - 3y)$
$[(x)^2 + (x)(3y) + (3y)^2] = 3x^2(x - 3y)(x^2 + 3xy + 9y^2)$.
62. $(a + 2b)^3 - c^3$. Then $(\ \ - \ ) [(\ )^2 + (\ )(\ ) + (\ )^2]$
and $[(a + 2b) - c] [(a + 2b)^2 + (a + 2b)(c) + (c)^2]$
$= (a + 2b - c)(a^2 + 4ab + 4b^2 + ac + 2bc + c^2)$.

Review exercises

1. $(a - 5)(a - 2)$ **2.** $(5a + 2b)(x - 2y)$
3. $(x + 2y)^2$ **4.** $(3a + b)(2a - b)$ **5.** $5a(a - 3)(a - 5)$
6. $(3x - 4)(2x + 3)$

Exercise 3–8

Answers to odd-numbered problems

1. $(m - 7)(m + 7)$ **3.** $(x + 5)(x + 1)$ **5.** $(7a + 1)(a + 5)$
7. $(2a + 3)(a + 6)$ **9.** $(ab + 4)(ab - 2)$
11. $(3a + b)(9a^2 - 3ab + b^2)$ **13.** $5(3x + y)(5x^2 + a)$
15. $10(x - y)^2$ **17.** $4(m - 2n)(m + 2n)$ **19.** $(a - b - 2x - y)$
$(a - b + 2x + y)$ **21.** $3(x^2 - 3y)(x^4 + 3x^2y + 9y^2)$
23. $2xy^2(6x^2 - 9x + 8y^2)$ **25.** $(2x + 3y)(2x - 3y)$
27. $(x + 2y)(3a - b)$ **29.** $(3a^3 - bc)(9a^6 + 3a^3bc + b^2c^2)$
31. $(5a + 3)(a - 7)$ **33.** $(a - 1)(a + 1)(a - 2)(a + 2)$
35. $(2a + 3b)(2a - 5b)$ **37.** $(y - 2)(y + 2)(y^2 + 4)$
39. $2(a + 2)(2a + 1)$ **41.** $(x + y - 9)(x + y + 1)$
43. $(2a - 1)(3a + 5)$ **45.** $4ab(x + 3y)(1 - 2ab)$
47. $(2a - 5b)^2$ **49.** $5x(4x^2 + 1)(2x + 1)(2x - 1)$
51. $3ab(a^2 - 3b^2)^2$ **53.** $(3a - x - 5y)(3a + x + 5y)$
55. $(3x - 13)(x + 7)$ **57.** $3x^2(xy^3 + 3z^2)(x^2y^6 - 3xy^3z^2 + 9z^4)$
59. $(3 - x)(3 + x)(a - 3)^2$

Review exercises

1. $\{7\}$ **2.** $\left\{-\dfrac{3}{2}\right\}$ **3.** $\left\{\dfrac{4}{5}\right\}$ **4.** $\{0\}$ **5.** $\left\{\dfrac{6}{5}\right\}$ **6.** $\{2\}$

Chapter 3 review

1. $-15x^5$ **2.** $6a^3b^5$ **3.** $8a^9b^{12}c^3$ **4.** $x^8y^9z^2$ **5.** $17a^7$
6. $4b^{10} - 27b^8$ **7.** $10a^4b - 15a^3b^2 + 20a^2b^3$ **8.** $2a^2 + ab - b^2$
9. $y^2 - 49$ **10.** $4a^2 + 4ab + b^2$ **11.** $a^3 - 2a^2b - ab^2 + 2b^3$
12. $2x^2 + 16x - 17$ **13.** $\dfrac{b^3}{4a^3}$ **14.** $\dfrac{6}{x^9}$ **15.** $\dfrac{9y^6}{x^4}$ **16.** $\dfrac{64x^9}{y^{15}}$
17. $\dfrac{a^9}{b^3}$ **18.** $\dfrac{a^3c^7}{2b}$ **19.** $\dfrac{4}{y^5}$ **20.** $6x^2(2x - 3)$
21. $4a^2b(3a^2 - b^2 + 6ab)$ **22.** $(x - 2y)(a + b)$
23. $5x(2x + 1)(x - 3z)$ **24.** $(2x - y)(3a - b)$
25. $(2y + x)(4a + 3b)$ **26.** $(2x + y)(2a + 3b)$
27. $(a + 3b)(x - 2y)$ **28.** $(2x - y)(3a - b)$
29. $(a + 3b)(2x + y)$ **30.** $(a + 2)(a + 12)$
31. $(b - 7)(b - 2)$ **32.** $2a(a - 5)(a + 1)$
33. $x(x - 3)(x + 2)$ **34.** $(xy + 6)(xy + 4)$
35. $(ab - 10)(ab + 2)$ **36.** $(x + 3y - 1)(x + 3y + 2)$
37. $(x + 2y - 7)^2$ **38.** $(a - 2b)(a + b)$
39. $(x + 2y)(x + 3y)$ **40.** $(2a - 3)(a - 2)$
41. $(4x - 5)(2x - 1)$ **42.** $(2y + 1)(3y - 4)$
43. $(3a - 1)(a + 1)$ **44.** $(4x - 1)(x + 3)$
45. $(2x + 1)(2x + 1) = (2x + 1)^2$ **46.** $(6a + 1)(4a + 3)$
47. $(4x - 3)(2x - 3)$ **48.** $(2x + 3)(x + 6)$
49. $3(2x + 1)(x + 8)$ **50.** $(-2x - 1)(x - 6)$ or
$(2x + 1)(-x + 6)$ or $-(2x + 1)(x - 6)$ **51.** $(x + 9)(x - 9)$
52. $4(a + 3b)(a - 3b)$ **53.** $3(a + 3b)(a - 3b)$
54. $(y^2 + 9)(y + 3)(y - 3)$ **55.** $(a + 2b)(x + y)(x - y)$
56. $(x - 3z)(2a + 3b)(2a - 3b)$ **57.** $(a + 2b)(a^2 - 2ab + 4b^2)$
58. $(3x - y)(9x^2 + 3xy + y^2)$ **59.** $8(2a + b)(4a^2 - 2ab + b^2)$
60. $x^2(3x + y)(9x^2 - 3xy + y^2)$ **61.** $2(x - 2b)(x^2 + 2bx + 4b^2)$
62. $(x^4y^5 - z^3)(x^8y^{10} + x^4y^5z^3 + z^6)$ **63.** $(3x + 4)^2$
64. $(a^2 - b^3)(a^4 + a^2b^3 + b^6)$ **65.** $3a(a + 5)(a - 5)$
66. $(a + 6)^2(3 + x)(3 - x)$ **67.** $(6a - 1)(a + 3)$
68. $(x - 2y)(4m + 3n)$ **69.** $5x(2x + y)(4x^2 - 2xy + y^2)$
70. $ab^2(a - 2b)^2$ **71.** $(x + 4y)(x - 2y)$
72. $(3a + 2)(5a - 2)$

Chapter 3 cumulative test

1. false **2.** false **3.** true **4.** 2 **5.** undefined **6.** 4 **7.** 0
8. 81 **9.** 37 **10.** -34 **11.** $\{x|x \geq 7\}$ **12.** $\{5\}$
13. $\left[-3, \dfrac{1}{3}\right]$ **14.** $\left\{x\left|-\dfrac{1}{3} < x < \dfrac{17}{3}\right.\right\}$
15. $\left\{x\left|x < \dfrac{-11}{6} \text{ or } x > \dfrac{1}{6}\right.\right\}$ **16.** $\left\{x\left|-\dfrac{3}{2} < x < \dfrac{7}{2}\right.\right\}$
17. $\left\{x\left|-5 \leq x \leq \dfrac{-3}{2}\right.\right\}$ **18.** let x = the number; $\{x|x > 9\}$
19. \$4,000 **20.** $6a^5b^5$ **21.** $16x^{12}y^{16}$
22. $\dfrac{3a^3b^5}{2c^6}$ **23.** $\dfrac{y^4}{x^6}$ **24.** $\dfrac{y^9}{27x^9}$ **25.** $4x^2 - 2x - 3$ **26.** $\dfrac{a^9c^3}{27b^6}$
27. $7ab - a - 5b$ **28.** $9a^2 - 6ab + b^2$ **29.** $8a^{18}b^{11}$
30. $12x^6y^3 - 18x^6y^4 + 24x^4y^5$ **31.** $(x + 6)(x - 6)$
32. $(x + 3y)(x^2 - 3xy + 9y^2)$ **33.** $(2a - 3)(a - 6)$
34. $(y + 6)(y + 1)$ **35.** $(xy + 2)(xy - 4)$
36. $(7x + 1)(x - 5)$ **37.** $(2a + 3b)(x - 3y)$
38. $5x(2a - b)(3x + 1)$ **39.** $(2x - 5y)^2$
40. $3(a + 2b)(a - 2b)$

Chapter 4

Exercise 4–1

Answers to odd-numbered problems

1. domain $= \{x|x \in R, x \neq 4\}$ **3.** domain $= \{x|x \in R, x \neq -1\}$
5. domain $= \left\{z\left|z \in R, z \neq \dfrac{3}{2}\right.\right\}$ **7.** domain $= \left\{x\left|x \in R, x \neq -\dfrac{4}{7}\right.\right\}$
9. domain $= \{x|x \in R, x \neq 0,4\}$ **11.** domain $= \{x|x \in R, x \neq -4\}$
13. domain $= \left\{y\left|y \in R, y \neq -\dfrac{5}{2}, \dfrac{5}{2}\right.\right\}$
15. domain $= \left\{x\left|x \in R, x \neq -\dfrac{1}{2}, 3\right.\right\}$
17. domain $= \{x|x \in R\}$ **19.** $\dfrac{5}{3x}$ $(x \neq 0)$ **21.** $\dfrac{p^2}{q^2}$ $(p \neq 0, q \neq 0)$
23. $-\dfrac{ab^3}{c^4}$ $(a \neq 0, b \neq 0, c \neq 0)$ **25.** $\dfrac{3n^2p^2}{4m^3}$ $(m \neq 0, n \neq 0, p \neq 0)$
27. $\dfrac{3}{7}$ $(x \neq 2)$ **29.** $\dfrac{3}{a - 2}$ $(a \neq 0,2)$ **31.** $2(x - 1)$ $(x \neq 0)$
33. $\dfrac{4(y + 1)}{3}$ $(y \neq 1)$ **35.** -5 $(x \neq y)$ **37.** $\dfrac{a - 3}{4}$ $(a \neq -3)$
39. $2y - 1$ $\left(y \neq -\dfrac{1}{2}\right)$ **41.** $x^2 - 2x + 4$ $(x \neq -2)$
43. $\dfrac{-3}{2(y^2 + xy + x^2)}$ $(x \neq y)$ **45.** $\dfrac{y - 7}{y + 7}$ $(y \neq -7)$
47. $\dfrac{m - 6}{m - 3}$ $(m \neq -2,3)$ **49.** $\dfrac{y - 8}{y - 5}$ $(y \neq -4,5)$
51. $\dfrac{a - 1}{a + 1}$ $\left(a \neq -1, \dfrac{1}{2}\right)$ **53.** $\dfrac{2x - 3}{4x + 1}$ $\left(x \neq -3, -\dfrac{1}{4}\right)$
55. $\dfrac{3m + 2}{2m + 1}$ $\left(x \neq -\dfrac{3}{4}, -\dfrac{1}{2}\right)$ **57.** $\dfrac{2(a + 2)}{3a + 1}$ $\left(a \neq -\dfrac{1}{3}, \dfrac{3}{2}\right)$
59. $\dfrac{5(y + 1)}{4(y + 3)}$ $(y \neq -3,3)$ **61.** $\dfrac{-(a + 6)}{3 + a}$ $\left(a \neq -3, \dfrac{2}{3}\right)$
63. $\dfrac{p + 4q}{p + 2q}$ $(p \neq -3q, -2q)$

Solutions to trial exercise problems

5. Since $2z - 3 = 0$ when $z = \dfrac{3}{2}$, the domain is $\left\{ z \mid z \in R, z \neq \dfrac{3}{2} \right\}$.

10. Since $3m^2 + 6m = 3m(m + 2)$ and $3m = 0$ when $m = 0$, $m + 2 = 0$ when $m = -2$, the domain is $\{m \mid m \in R, m \neq 0, -2\}$.

13. Since $4y^2 - 25 = (2y + 5)(2y - 5)$ and $2y + 5 = 0$ when $y = -\dfrac{5}{2}$, $2y - 5 = 0$ when $y = \dfrac{5}{2}$, the domain is $\left\{ y \mid y \in R, y \neq -\dfrac{5}{2}, \dfrac{5}{2} \right\}$. **17.** Since $x^2 + 4 \neq 0$ for any value of x, the domain is $\{x \mid x \in R\}$.

30. $\dfrac{-36x}{42x^3 + 24x} = \dfrac{-36x}{6x(7x^2 + 4)} = \dfrac{6x \cdot (-6)}{6x(7x^2 + 4)} = \dfrac{-6}{7x^2 + 4} \ (x \neq 0)$

35. $\dfrac{5x - 5y}{y - x} = \dfrac{5(x - y)}{-(x - y)} = \dfrac{5}{-1} = -5 \ (x \neq y)$

41. $\dfrac{x^3 + 8}{x + 2} = \dfrac{(x + 2)(x^2 - 2x + 4)}{x + 2} = x^2 - 2x + 4 \ (x \neq -2)$

53. $\dfrac{2x^2 + 3x - 9}{4x^2 + 13x + 3} = \dfrac{(2x - 3)(x + 3)}{(x + 3)(4x + 1)} = \dfrac{2x - 3}{4x + 1}$

$\left(x \neq -3, -\dfrac{1}{4} \right)$ **63.** $\dfrac{p^2 + 7pq + 12q^2}{p^2 + 5pq + 6q^2} = \dfrac{(p + 3q)(p + 4q)}{(p + 2q)(p + 3q)}$

$= \dfrac{p + 4q}{p + 2q} \ (p \neq -3q, -2q)$

Review exercises

1. $-\dfrac{1}{12}$ **2.** 1 **3.** $\dfrac{13}{8}$ or $1\dfrac{5}{8}$ **4.** $\{2\}$ **5.** $\left\{ \dfrac{9}{2} \right\}$

6. $x = \dfrac{2a - 5}{2}$

Exercise 4–2

Answers to odd-numbered problems

1. $\dfrac{8}{3x} \ (x \neq 0)$ **3.** $\dfrac{1}{2x^2y^2} \ (x, y \neq 0)$ **5.** $\dfrac{a^2}{7bc^3} \ (a, b, c \neq 0)$

7. $2ab^2 \ (a, b \neq 0)$ **9.** $\dfrac{4a - 8}{15a - 60} \ (a \neq 4, -2)$ **11.** $\dfrac{a - 6}{6} \ (a \neq -3)$

13. $\dfrac{-4p^2 - 5p - 1}{p - 1} \left(p \neq -1, \dfrac{1}{4}, 1 \right)$ **15.** $\dfrac{x - 5}{x - 3} \ (x \neq 1, 2, 3, 4)$

17. $\dfrac{3a^2 - 5a - 2}{3a^2 - 4a + 1} \left(a \neq -2, -\dfrac{3}{2}, \dfrac{1}{3}, 1 \right)$

19. $\dfrac{x^2 - 2x + 4}{2x + 7} \left(x \neq -2, -\dfrac{7}{2}, 2 \right)$ **21.** $\dfrac{-1}{x + 2y} \ (x \neq -2y, -y, y)$

23. $\dfrac{a^2}{b^3}$ **25.** $\dfrac{1}{r^2 - 5r + 4}$ **27.** $\dfrac{1}{3x + 12}$ **29.** $\dfrac{x^3 + x^2 - x - 1}{4}$

31. $\dfrac{x^3 - 125}{x + 5}$ **33.** $\dfrac{-x^2 - 3x + 40}{x^2 + 4x - 21}$ **35.** $\dfrac{x + 4}{x - 4}$ **37.** $\dfrac{a - b}{2}$

39. $-(2m^2 + 5mn + 3n^2)$ **41.** $\dfrac{a - 6}{a + 3}$ **43.** $\dfrac{x^2 - 9}{x^2 - 5x + 4}$

45. $-(9y^2 + 6xy + 4x^2)(x^2 - xy + y^2)$ **47.** $\dfrac{b^2 + b - 2}{b^2 - b - 2}$

49. $\dfrac{z^2 - 2wz + w^2}{z^2 + 2wz + w^2}$

Solutions to trial exercise problems

9. $\dfrac{4a + 8}{3a - 12} \cdot \dfrac{a - 2}{5a + 10} = \dfrac{4(a + 2) \cdot (a - 2)}{3(a - 4) \cdot 5(a + 2)} = \dfrac{4(a - 2)}{15(a - 4)}$

$= \dfrac{4a - 8}{15a - 60} \ (a \neq -2, a \neq 4)$ **17.** $\dfrac{2a^2 - a - 6}{3a^2 - 4a + 1} \cdot \dfrac{3a^2 + 7a + 2}{2a^2 + 7a + 6}$

$= \dfrac{(2a + 3)(a - 2)}{(3a - 1)(a - 1)} \cdot \dfrac{(3a + 1)(a + 2)}{(2a + 3)(a + 2)} = \dfrac{(a - 2)(3a + 1)}{(3a - 1)(a - 1)}$

$= \dfrac{3a^2 - 5a - 2}{3a^2 - 4a + 1} \left(a \neq -2, -\dfrac{3}{2}, \dfrac{1}{3}, 1 \right)$

25. $\dfrac{r + 4}{r^2 - 1} \div \dfrac{r^2 - 16}{r + 1} = \dfrac{(r + 4) \cdot (r + 1)}{(r^2 - 1) \cdot (r^2 - 16)}$

$= \dfrac{(r + 4)(r + 1)}{(r + 1)(r - 1)(r + 4)(r - 4)} = \dfrac{1}{(r - 4)(r - 1)}$

$= \dfrac{1}{r^2 - 5r + 4}$ **33.** $\dfrac{56 - x - x^2}{x^2 - 6x - 7} \div \dfrac{x^2 + 4x - 21}{x^2 - 4x - 5}$

$= \dfrac{-(x^2 + x - 56) \cdot (x^2 - 4x - 5)}{(x^2 - 6x - 7) \cdot (x^2 + 4x - 21)}$

$= \dfrac{-(x - 7)(x + 8) \cdot (x - 5)(x + 1)}{(x - 7)(x + 1) \cdot (x + 7)(x - 3)} = \dfrac{-(x + 8)(x - 5)}{(x + 7)(x - 3)}$

$= \dfrac{-(x^2 + 3x - 40)}{x^2 + 4x - 21} = \dfrac{-x^2 - 3x + 40}{x^2 + 4x - 21}$

39. $\dfrac{n^2 - m^2}{2m - 3n} \div \dfrac{m - n}{4m^2 - 9n^2} = \dfrac{(n^2 - m^2) \cdot (4m^2 - 9n^2)}{(2m - 3n) \cdot (m - n)}$

$= \dfrac{-(m - n)(m + n) \cdot (2m + 3n)(2m - 3n)}{(2m - 3n) \cdot (m - n)}$

$= -(m + n)(2m + 3n) = -(2m^2 + 5mn + 3n^2)$

41. $\dfrac{a^2 - 8a + 15}{a^2 + 7a + 6} \cdot \dfrac{a^2 - 6a - 7}{a^2 - 9} \div \dfrac{a^2 - 12a + 35}{a^2 - 36}$

$= \dfrac{(a^2 - 8a + 15) \cdot (a^2 - 6a - 7) \cdot (a^2 - 36)}{(a^2 + 7a + 6) \cdot (a^2 - 9) \cdot (a^2 - 12a + 35)}$

$= \dfrac{(a - 3)(a - 5) \cdot (a - 7)(a + 1) \cdot (a + 6)(a - 6)}{(a + 6)(a + 1) \cdot (a + 3)(a - 3) \cdot (a - 7)(a - 5)} = \dfrac{a - 6}{a + 3}$

47. $\dfrac{ab - a + 3b - 3}{ab + a - 4b - 4} \div \dfrac{ab - 2a + 3b - 6}{ab + 2a - 4b - 8}$

$= \dfrac{(ab - a + 3b - 3) \cdot (ab + 2a - 4b - 8)}{(ab + a - 4b - 4) \cdot (ab - 2a + 3b - 6)}$

$= \dfrac{(a + 3)(b - 1) \cdot (a - 4)(b + 2)}{(a - 4)(b + 1) \cdot (a + 3)(b - 2)} = \dfrac{(b - 1)(b + 2)}{(b + 1)(b - 2)}$

$= \dfrac{b^2 + b - 2}{b^2 - b - 2}$

Review exercises

1. $\dfrac{43}{24}$ or $1\dfrac{19}{24}$ **2.** $\dfrac{1}{3}$ **3.** $\{y \mid y \leq 2\} = (-\infty, 2]$

4. $\{y \mid -3 \leq y < 1\} = [-3, 1]$ **5.** $8x^5 - 4x^4 + 24x^3$

6. $4y^2 - 15y - 4$

Exercise 4–3

Answers to odd-numbered problems

1. $70x$ **3.** $80k^2$ **5.** $288a^3b^2$ **7.** $30x^2y^4$ **9.** $4a(a - 2)$

11. $6x(x - 7)$ **13.** $3(p + 3)^2(p - 4)$ **15.** $10(a + 5)(a - 5)$

17. $(q + 7)(q - 7)(q - 2)$ **19.** $5(a + b)(a - b)$

21. $(2a + 1)(3a - 1)(a - 7)(a + 7)$ **23.** $\dfrac{12x^2}{21x^3}$ **25.** $\dfrac{15x^3y}{24x^2y^2}$

27. $\dfrac{4p^2 - 12p}{p - 3}$ **29.** $\dfrac{p^2 - 5p + 6}{p^2 - 4}$ **31.** $\dfrac{16x^3 + 24x^2 + 36x}{8x^3 - 27}$

33. $\dfrac{2n^2 - 11n + 5}{n^2 + 2n - 35}$ **35.** $\dfrac{2y^2 - y - 10}{4y^2 + 7y - 2}$ **37.** $\dfrac{15m^2 + 12m}{25m^2 - 16}$

39. $\dfrac{9}{a - 9}$ **41.** $\dfrac{13}{q}$ **43.** $\dfrac{-3y}{y + 4}$ **45.** 5   **47.** 1   **49.** $\dfrac{1}{x + 3}$

51. 1   **53.** $\dfrac{47}{12x}$ **55.** $\dfrac{5z - 2}{z^2}$ **57.** $\dfrac{18x + 5}{9x^2}$ **59.** $\dfrac{6y - 4x}{9x^2y^2}$

61. $\dfrac{14a^2 - 2a}{(3a + 5)(2a - 3)}$ **63.** $\dfrac{129y - 122}{10(y - 2)(y + 2)}$ **65.** $\dfrac{2x - 4}{x - 9}$

67. $\dfrac{16x - 61}{(x - 6)(x + 1)(x - 1)}$ **69.** $\dfrac{5x^2 + 3xy + 6y^2}{(x - 3y)(x + y)^2}$

71. $\dfrac{2a^2 - 16a - 25}{(2a - 1)(4a + 3)(a + 5)}$ **73.** $\dfrac{20a^2 - 25a + 1}{5a - 2}$

75. $\dfrac{17p^2 - 20p - 75}{8p(p - 5)(p + 3)}$ **77.** $\dfrac{3y^2 - y + 15}{(y + 3)(y - 2)(y + 2)}$

79. $\dfrac{5b^2 + 7b - 30}{5(b + 2)\,(b - 2)}$ **81.** $\dfrac{3x^2 - 15xy + 14y^2}{(x - 6y)^2(x + 2y)}$

83. $\dfrac{15x^2 - 9x - 9}{(4x - 3)(2x - 5)(3x + 1)}$

85. $\dfrac{49m^2 - 20mn + 3n^2}{(8m - n)(m + 2n)(5m - 4n)}$ **87.** $\dfrac{qrs + prs + pqs + pqr}{pqrs}$

89. $\dfrac{f_1 + f_2 - d}{f_1 f_2}$ **91.** $\dfrac{(n - 1)(R_2 + R_1)}{R_1 R_2}$

Solutions to trial exercise problems

13. $p^2 - p - 12 = (p - 4)(p + 3)$; $p^2 + 6p + 9 = (p + 3)^2$;
$3p - 12 = 3(p - 4)$. The LCM is $3(p + 3)^2(p - 4)$.

18. $2y + 10 = 2(y + 5)$; $25 - y^2 = -(y + 5)(y - 5)$;
$y^2 - 10y + 25 = (y - 5)^2$. The LCM is $2(y + 5)(y - 5)^2$.

21. $2a^2 - 13a - 7 = (2a + 1)(a - 7)$; $6a^2 + a - 1$
$= (3a - 1)(2a + 1)$; $a^2 - 49 = (a + 7)(a - 7)$.
The LCM is $(2a + 1)(3a - 1)(a - 7)(a + 7)$.

31. Since $8x^3 - 27 = (2x - 3)(4x^2 + 6x + 9)$, then $\dfrac{4x}{2x - 3}$

$= \dfrac{4x \cdot (4x^2 + 6x + 9)}{(2x - 3) \cdot (4x^2 + 6x + 9)} = \dfrac{16x^3 + 24x^2 + 36x}{8x^3 - 27}$.   **37.** Since

$25m^2 - 16 = (5m + 4)(5m - 4)$, then $-\dfrac{3m}{4 - 5m} = -\dfrac{3m}{-(5m - 4)}$

$= \dfrac{3m}{5m - 4} = \dfrac{3m \cdot (5m + 4)}{(5m - 4)(5m + 4)} = \dfrac{15m^2 + 12m}{25m^2 - 16}$.

53. $\dfrac{8}{3x} + \dfrac{5}{4x} = \dfrac{8(4)}{3x(4)} + \dfrac{5(3)}{4x(3)} = \dfrac{32}{12x} + \dfrac{15}{12x} = \dfrac{32 + 15}{12x} = \dfrac{47}{12x}$

65. $\dfrac{x + 2}{x - 9} - \dfrac{x - 6}{9 - x} = \dfrac{x + 2}{x - 9} - \dfrac{x - 6}{-(x - 9)}$

$= \dfrac{x + 2}{x - 9} + \dfrac{x - 6}{x - 9} = \dfrac{(x + 2) + (x - 6)}{x - 9} = \dfrac{x + 2 + x - 6}{x - 9}$

$= \dfrac{2x - 4}{x - 9}$ **70.** $\dfrac{8q}{4q^2 - 9p^2} + \dfrac{5q}{4q^2 - 12pq + 9p^2}$

$= \dfrac{8q}{(2q + 3p)(2q - 3p)} + \dfrac{5q}{(2q - 3p)^2} = \dfrac{8q(2q - 3p)}{(2q + 3p)(2q - 3p)^2}$

$+ \dfrac{5q(2q + 3p)}{(2q - 3p)^2(2q + 3p)} = \dfrac{(16q^2 - 24pq) + (10q^2 + 15pq)}{(2q + 3p)(2q - 3p)^2}$

$= \dfrac{26q^2 - 9pq}{(2q + 3p)(2q - 3p)^2}$

73. $(4a - 3) - \dfrac{2a + 5}{5a - 2} = \dfrac{4a - 3}{1} - \dfrac{2a + 5}{5a - 2}$

$= \dfrac{(4a - 3) \cdot (5a - 2) - 1 \cdot (2a + 5)}{1 \cdot (5a - 2)}$

$= \dfrac{20a^2 - 23a + 6 - 2a - 5}{5a - 2} = \dfrac{20a^2 - 25a + 1}{5a - 2}$

82. $\dfrac{a - b}{a^2 - 3ab - 4b^2} + \dfrac{2b - 5a}{a^2 - 16b^2} = \dfrac{a - b}{(a - 4b)(a + b)}$

$+ \dfrac{2b - 5a}{(a + 4b)(a - 4b)} = \dfrac{(a - b) \cdot (a + 4b)}{(a - 4b)(a + b)(a + 4b)}$

$+ \dfrac{(2b - 5a) \cdot (a + b)}{(a + 4b)(a - 4b)(a + b)}$

$= \dfrac{(a^2 + 3ab - 4b^2) + (2b^2 - 3ab - 5a^2)}{(a + 4b)(a - 4b)(a + b)}$

$= \dfrac{a^2 + 3ab - 4b^2 + 2b^2 - 3ab - 5a^2}{(a + 4b)(a - 4b)(a + b)}$

$= \dfrac{-4a^2 - 2b^2}{(a + 4b)(a - 4b)(a + b)}$

Review exercises

1. $(2x^2 + 3)(4x - 1)$ **2.** $(2x + 1)(2x - 7)$
3. $(3y + 7)(3y - 7)$ **4.** 12   **5.** $24k$

Exercise 4–4

Answers to odd-numbered problems

1. $\dfrac{21}{20}$ **3.** $\dfrac{11}{3}$ **5.** $\dfrac{12}{11}$ **7.** $\dfrac{7}{8m}$ **9.** $\dfrac{-4y}{3y + 3}$ **11.** $\dfrac{x - 3}{x + 7}$

13. $\dfrac{ab}{b - a}$ **15.** $\dfrac{x}{x - 2y}$ **17.** $\dfrac{20}{27}$ **19.** $\dfrac{5}{7y}$ **21.** $\dfrac{35 - 7a}{4a + 36}$

23. $\dfrac{8x - 4y}{3}$ **25.** $\dfrac{4x^2 + 5x - 9}{5x^2 + 16x + 3}$ **27.** $\dfrac{xy^2 + x^2}{y^2 - xy^2}$

29. $\dfrac{mn\,(m + n)}{n - m}$ **31.** $\dfrac{5q + 4p^2}{p^2q\,(p - q)}$ **33.** $\dfrac{a^2 + 9a + 23}{a^2 + 7a + 7}$

35. $\dfrac{4y^3 + 17y^2 - 18y - 15}{4y^3 + 25y^2 + 31y - 39}$ **37.** $\dfrac{t^2 - 3t - 4}{t^2 + 3t - 18}$ **39.** $\dfrac{-3}{2x + 14}$

41. $\dfrac{5b^2 + 24b - 5}{4b - 22}$ **43.** $\dfrac{1}{2}$ **45.** $\dfrac{3c + 4a - 2b}{5}$ **47.** $\dfrac{y + x}{y - x}$

49. $\dfrac{y}{y + x}$ **51.** $\dfrac{q^2 + p^2}{p^2}$ **53.** $\dfrac{pq}{q - p}$ **55.** $\dfrac{V_sC - it}{RC}$

57. $CP = \dfrac{T_1}{T_2 - T_1}$ **59.** $r = \dfrac{2dr_1r_2}{dr_2 + dr_1} = \dfrac{2r_1r_2}{r_2 + r_1}$

Solutions to trial exercise problems

25. $\dfrac{4 - \dfrac{3}{x + 3}}{5 + \dfrac{6}{x - 1}} = \dfrac{4(x + 3)(x - 1) - \dfrac{3}{x + 3} \cdot (x + 3)(x - 1)}{5(x + 3)(x - 1) + \dfrac{6}{x - 1} \cdot (x + 3)(x - 1)}$

$= \dfrac{4(x^2 + 2x - 3) - 3(x - 1)}{5(x^2 + 2x - 3) + 6(x + 3)}$

$= \dfrac{4x^2 + 8x - 12 - 3x + 3}{5x^2 + 10x - 15 + 6x + 18}$

$= \dfrac{4x^2 + 5x - 9}{5x^2 + 16x + 3}$

28. $\dfrac{\dfrac{1}{a} - \dfrac{1}{b}}{\dfrac{1}{a^2} + \dfrac{1}{b^2}} = \dfrac{\dfrac{1}{a} \cdot a^2b^2 - \dfrac{1}{b} \cdot a^2b^2}{\dfrac{1}{a^2} \cdot a^2b^2 + \dfrac{1}{b^2} \cdot a^2b^2} = \dfrac{ab^2 - a^2b}{b^2 + a^2}$

33. $\dfrac{(a+5)+\dfrac{3}{a+4}}{(a+3)-\dfrac{5}{a+4}} = \dfrac{(a+5)(a+4)+\dfrac{3}{a+4}(a+4)}{(a+3)(a+4)-\dfrac{5}{a+4}(a+4)}$

$= \dfrac{a^2+9a+20+3}{a^2+7a+12-5} = \dfrac{a^2+9a+23}{a^2+7a+7}$

39. $\dfrac{\dfrac{3}{x^2-x-6}}{\dfrac{2}{x+2}-\dfrac{4}{x-3}} = \dfrac{\dfrac{3}{(x-3)(x+2)}}{\dfrac{2}{x+2}-\dfrac{4}{x-3}}$

$= \dfrac{\dfrac{3}{(x-3)(x+2)}\cdot(x-3)(x+2)}{\dfrac{2}{x+2}(x-3)(x+2)-\dfrac{4}{x-3}(x-3)(x+2)}$

$= \dfrac{3}{2(x-3)-4(x+2)} = \dfrac{3}{2x-6-4x-8}$

$= \dfrac{3}{-2x-14} = \dfrac{3}{-(2x+14)} = \dfrac{-3}{2x+14}$

Review exercises

1. $\left\{-\dfrac{14}{3}\right\}$ **2.** $\{-2,8\}$ **3.** $\{2\}$ **4.** $\{x\,|\,-2<x<6\}=(-2,6)$
5. $\{x\,|\,x\le -4 \text{ or } x\ge -1\}=(-\infty,-4]\cup[-1,\infty)$
6. $2a^3+8a^2-a-4$

Exercise 4–5

Answers to odd-numbered problems

1. $5x^2-3x+2$ **3.** $-x^3+2x^2-3$ **5.** $c-1$
7. $10xy^2+7-6y^2$ **9.** $3a^6b-5a^4b^4+7a^3b$ **11.** $2a+3c$
13. $a-4$ **15.** $y+2+\dfrac{1}{y+5}$ **17.** $x-4-\dfrac{2}{x-3}$
19. a^2-2a+4 **21.** $2y^2-3y+1+\dfrac{2}{y+1}$ **23.** x^2+3x-4
25. $2a^2+3a-4+\dfrac{5}{2a+1}$ **27.** $3y^2-y+4+\dfrac{2}{3y+1}$
29. a^2-3a+9 **31.** $3x^3-5x^2+5x-4+\dfrac{3}{x+1}$
33. $3a-1$ **35.** y^2+y-1 **37.** $x^2+3x+2-\dfrac{4}{x^2+x-4}$
39. $3a^2-2a-4$ **41.** $y^2+3+\dfrac{2}{2y^2-3y+2}$ **43.** 2, it is the
same. **45.** $2x-3$ **47.** $3x-2$

Solutions to trial exercise problems

10. $\dfrac{x(y-2)-z(y-2)}{y-2} = \dfrac{x(y-2)}{y-2}-\dfrac{z(y-2)}{y-2} = x-z$
16. $(2x-5+x^2)\div(x+4) = (x^2+2x-5)\div(x+4)$

$\begin{array}{r} x-2 \\ x+4{\overline{\smash{\big)}\,x^2+2x-5}} \\ \underline{x^2+4x} \\ -2x-5 \\ \underline{-2x-8} \\ 3 \end{array}$ Answer: $x-2+\dfrac{3}{x+4}$

27. $\dfrac{9y^3+11y+6}{3y+1}$

$\begin{array}{r} 3y^2-y+4 \\ 3y+1{\overline{\smash{\big)}\,9y^3+0y^2+11y+6}} \\ \underline{9y^3+3y^2} \\ -3y^2+11y \\ \underline{-3y^2-y} \\ 12y+6 \\ \underline{12y+4} \\ 2 \end{array}$

Answer: $3y^2-y+4+\dfrac{2}{3y+1}$

32. $\dfrac{2x^3+5x^2+5x+3}{x^2+x+1}$

$\begin{array}{r} 2x+3 \\ x^2+x+1{\overline{\smash{\big)}\,2x^3+5x^2+5x+3}} \\ \underline{2x^3+2x^2+2x} \\ 3x^2+3x+3 \\ \underline{3x^2+3x+3} \\ 0 \end{array}$

Answer: $2x+3$

38. $\dfrac{2x^4-x^3+5x^2-x+3}{x^2+1}$

$\begin{array}{r} 2x^2-x+3 \\ x^2+0x+1{\overline{\smash{\big)}\,2x^4-x^3+5x^2-x+3}} \\ \underline{2x^4+0x^3+2x^2} \\ -x^3+3x^2-x \\ \underline{-x^3-0x^2-x} \\ 3x^2+0x+3 \\ \underline{3x^2+0x+3} \\ 0 \end{array}$

Answer: $2x^2-x+3$

Review exercises

1. x **2.** x^3 **3.** $\dfrac{8}{a^{12}}$ **4.** $\dfrac{3}{10}$ **5.** $\dfrac{-x^2+4x+2}{(x+3)(x-3)}$ **6.** $\dfrac{2x-1}{2x}$
7. $\left\{\dfrac{19}{3}\right\}$

Exercise 4–6

Answers to odd-numbered problems

1. $a+2$ **3.** $a+5$ **5.** $x+7+\dfrac{23}{x-2}$ **7.** $3a-1$
9. $a^2+3a+2-\dfrac{2}{a-1}$ **11.** x^2+x+1
13. $y^2+5y+10+\dfrac{16}{y-2}$ **15.** $2a^2+a-3+\dfrac{2}{a-5}$
17. $x^3-2x^2-x+3-\dfrac{1}{x-3}$ **19.** $3y^3+2y^2-4y-1+\dfrac{3}{y-3}$
21. $P(1)=-1$ **23.** $P(-3)=319$ **25.** $P(0)=1$
27. $P(2)=145$ **29.** $P\left(\dfrac{1}{2}\right)=-\dfrac{3}{2}$ **31.** $P(-3)=4$;
$x+3$ is not a factor **33.** $P(2)=0$; $x-2$ is a factor
35. $P(-1)=25$; $x+1$ is not a factor **37.** $P(-4)=0$; $x+4$ is
a factor **39.** $P(-1)=-6$; $x+1$ is not a factor
41. $\left\{-8,\dfrac{2}{3},1\right\}$ **43.** $\left\{-6,\dfrac{1}{2},5\right\}$ **45.** $\{-1,1\}$
47. $P(x)=x^3-5x^2+2x+8$ **49.** $P(x)=x^3-3x^2-4x$

51. $P(x) = x^4 - 65x^2 + 64$ **53.** $P(x) = x^4 - 8x^3 - 13x^2 + 200x - 300$ **55.** $P(x) = x^4 - 6x^3 + x^2 + 24x + 16$
57. $P(x) = x^4 + 7x^3 + 10x$ **59.** 2 of multiplicity 2; -2 of multiplicity 2; 3 of multiplicity 3; -4 of multiplicity 2   **61.** 0 of multiplicity 2; -4 of multiplicity 4; 3 of multiplicity 2; 6   **63.** 25

65. Since the remainder is 0, $\dfrac{3}{2}$ is a solution.

Solutions to trial exercise problems

8. $\dfrac{5 - 3x + 2x^2}{x + 2} = \dfrac{2x^2 - 3x + 5}{x + 2} = 2x - 7 + \dfrac{19}{x + 2}$

$$-2 \begin{array}{|rrr} 2 & -3 & 5 \\ \downarrow & -4 & 14 \\ \hline 2 & -7 & 19 \end{array}$$
$$\quad\ \ \underset{x}{\uparrow}$$

11. $\dfrac{x^3 - 1}{x - 1} = \dfrac{x^3 + 0x^2 + 0x - 1}{x - 1} = x^2 + x + 1$

$$1 \begin{array}{|rrrr} 1 & 0 & 0 & -1 \\ \downarrow & 1 & 1 & 1 \\ \hline 1 & 1 & 1 & 0 \end{array}$$
$$\quad\ \ \underset{x^2}{\uparrow}\ \underset{x}{\uparrow}$$

18. $\dfrac{2a^4 + 6a^3 + a^2 + 2a - 5}{a + 3} = 2a^3 + a - 1 - \dfrac{2}{a + 3}$

$$-3 \begin{array}{|rrrrr} 2 & 6 & 1 & 2 & -5 \\ \downarrow & -6 & 0 & -3 & 3 \\ \hline 2 & 0 & 1 & -1 & -2 \end{array}$$
$$\quad\ \ \underset{a^3}{\uparrow}\ \ \underset{a^2}{\uparrow}\ \ \underset{a}{\uparrow}$$

23. $-3 \begin{array}{|rrrrr} 3 & -2 & 1 & -4 & 1 \\ \downarrow & -9 & 33 & -102 & 318 \\ \hline 3 & -11 & 34 & -106 & 319 \end{array}$

$P(-3) = 319$

28. $-2 \begin{array}{|rrrrrrr} 8 & 0 & 0 & 0 & 3 & -2 & -5 \\ \downarrow & -16 & 32 & -64 & 128 & -262 & 528 \\ \hline 8 & -16 & 32 & -64 & 131 & -264 & 523 \end{array}$

$P(-2) = 523$

35. $-1 \begin{array}{|rrrrr} 1 & -9 & 18 & 0 & -3 \\ \downarrow & -1 & 10 & -28 & 28 \\ \hline 1 & -10 & 28 & -28 & 25 \end{array}$ $\leftarrow x + 1$ is *not* a factor

40. $-2 \begin{array}{|rrrr} 1 & 7 & 4 & -12 \\ \downarrow & -2 & -10 & 12 \\ \hline 1 & 5 & -6 & 0 \end{array}$

set $x^2 + 5x - 6 = 0$
$(x + 6)(x - 1) = 0$
$x + 6 = 0$ or $x - 1 = 0$
$\quad x = -6$ $\qquad\qquad x = 1$
The solution set is $\{-6, -2, 1\}$.

45. $\begin{array}{r} x^2 + 2x + 1 \\ x^2 - 2x + 1{\overline{\smash{\big)}\,x^4 + 0x^3 - 2x^2 + 0x + 1}} \\ \underline{x^4 - 2x^3 +\ x^2} \\ 2x^3 - 3x^2 + 0x \\ \underline{2x^3 - 4x^2 + 2x} \\ x^2 - 2x + 1 \\ \underline{x^2 - 2x + 1} \\ 0 \end{array}$

$x^2 + 2x + 1 = 0$
$(x + 1)^2 = 0$
$x + 1 = 0$
$\quad x = -1$
The solution set is $\{-1, 1\}$.

47. Since $2, -1$ and 4 are zeros, then
$P(x) = (x - 2)(x + 1)(x - 4) = (x^2 - x - 2)(x - 4)$
$\qquad = x^3 - 5x^2 + 2x + 8$

50. Since $-2, 2, 3$ and -3 are zeros, then
$P(x) = (x + 2)(x - 2)(x - 3)(x + 3)$
$\qquad = (x^2 - 4)(x^2 - 9) = x^4 - 13x^2 + 36$

55. Since -1 of multiplicity 2 and 4 of multiplicity 2 are zeros, then
$P(x) = (x + 1)^2(x - 4)^2 = (x^2 + 2x + 1)(x^2 - 8x + 16)$
$\qquad = x^4 - 6x^3 + x^2 + 24x + 16$

Review exercises

1. $\emptyset$ **2.** $\{24\}$ **3.** $\{4\}$ **4.** 9 and 36   **5.** $R = \dfrac{3V + \pi h^3}{3\pi h^2}$

Exercise 4–7

Answers to odd-numbered problems

1. $\{-21\}$ **3.** $\left\{\dfrac{-15}{11}\right\}$ **5.** $\left\{\dfrac{10}{3}\right\}$ **7.** $\left\{\dfrac{67}{7}\right\}$ **9.** $\{12\}$

11. $\left\{\dfrac{-13}{6}\right\}$ **13.** $\emptyset$ **15.** $\left\{\dfrac{-3}{5}\right\}$ **17.** $\left\{\dfrac{-13}{5}\right\}$

19. $\emptyset$; 2 is extraneous   **21.** $\left\{\dfrac{13}{3}\right\}$ **23.** $\left\{\dfrac{-16}{5}\right\}$ **25.** $\left\{\dfrac{13}{2}\right\}$

27. $\emptyset$; 3 is extraneous   **29.** $\{-15\}$ **31.** $\left\{\dfrac{-17}{3}\right\}$ **33.** $\{26\}$

35. $\left\{\dfrac{-31}{5}\right\}$ **37.** $b = \dfrac{3a}{4 - 7a}$ **39.** $p = \dfrac{3q - 3}{3a + 4}$

41. $y = \dfrac{5x + 19}{3}$ **43.** $r = \dfrac{S - a}{S}$ **45.** $b_1 = \dfrac{2A - b_2 h}{h}$

47. $t = \dfrac{L_t - L_0}{kL_0}$; $L_0 = \dfrac{L_t}{kt + 1}$ **49.** $t = \dfrac{I}{Pr}$

51. $R_3 = \dfrac{RR_1 R_2}{R_1 R_2 - RR_2 - RR_1}$ **53.** $A = \dfrac{aF}{f}$

55. $T_2 = \dfrac{R_2M + R_2T_1 - R_1M}{R_1}$; $R_1 = \dfrac{R_2(M + T_1)}{M + T_2}$ **57. a.** $\dfrac{1}{10}$

b. $\dfrac{1}{2}$ **c.** $\dfrac{t}{10}$ **59. a.** $\dfrac{200}{5} = 40$ mph   **b.** $\dfrac{200}{4} = 50$ mph

c. $\dfrac{200}{t}$ mph   **61. a.** $\dfrac{1}{n}, \dfrac{3}{n}$ **b.** $\dfrac{15}{n}$ **63.** $\dfrac{5 + n}{6 - n}$

Solutions to trial exercise problems

5. $\dfrac{4}{a} - \dfrac{6}{3a} = \dfrac{3}{5}$

Multiply by the LCM, $15a$.

$$15a \cdot \dfrac{4}{a} - 15a \cdot \dfrac{6}{3a} = 15a \cdot \dfrac{3}{5}$$
$$15 \cdot 4 - 5 \cdot 6 = 3a \cdot 3$$
$$60 - 30 = 9a$$
$$30 = 9a$$
$$\dfrac{10}{3} = a$$

The solution set is $\left\{\dfrac{10}{3}\right\}$.

16. $1 + \dfrac{5}{3m - 9} = \dfrac{10}{m - 3}$

Multiply by the LCM, $3(m - 3)$.

$$3(m - 3) \cdot 1 + 3(m - 3) \cdot \dfrac{5}{3(m - 3)} = 3(m - 3) \cdot \dfrac{10}{m - 3}$$
$$3m - 9 + 5 = 3 \cdot 10$$
$$3m - 4 = 30$$
$$3m = 34$$
$$m = \dfrac{34}{3}$$

The solution set is $\left\{\dfrac{34}{3}\right\}$.

21. $4 - \dfrac{2x}{5 - x} = \dfrac{6}{x - 5}$

Since $5 - x = -(x - 5)$, then we have

$$4 - \dfrac{2x}{-(x - 5)} = \dfrac{6}{x - 5}$$

$$4 + \dfrac{2x}{x - 5} = \dfrac{6}{x - 5}. \text{ Multiply by } x - 5.$$

$$4(x - 5) + (x - 5) \cdot \dfrac{2x}{x - 5} = (x - 5) \cdot \dfrac{6}{x - 5}$$
$$4x - 20 + 2x = 6$$
$$6x - 20 = 6$$
$$6x = 26$$
$$x = \dfrac{26}{6} = \dfrac{13}{3}$$

The solution set is $\left\{\dfrac{13}{3}\right\}$.

29. $\dfrac{6}{q^2 + q - 6} = \dfrac{5}{q^2 + 3q - 10}$

Since $q^2 + q - 6 = (q + 3)(q - 2)$ and $q^2 + 3q - 10$ $= (q + 5)(q - 2)$, the LCM is $(q + 3)(q - 2)(q + 5)$.
Multiply each member by the LCM.

$$(q + 3)(q - 2)(q + 5) \cdot \dfrac{6}{(q + 3)(q - 2)} = (q + 3)(q - 2)(q + 5) \cdot \dfrac{5}{(q + 5)(q - 2)}$$
$$(q + 5) \cdot 6 = (q + 3) \cdot 5$$
$$6q + 30 = 5q + 15$$
$$q = -15$$

The solution set is $\{-15\}$.

41. $\dfrac{y - 3}{x + 2} = \dfrac{5}{3}$

Multiply by the LCM, $3(x + 2)$.

$$3(x + 2) \cdot \dfrac{y - 3}{x + 2} = 3(x + 2) \cdot \dfrac{5}{3}$$
$$3(y - 3) = (x + 2) \cdot 5$$
$$3y - 9 = 5x + 10$$
$$3y = 5x + 19$$
$$y = \dfrac{5x + 19}{3}$$

48. $R_x = R_m\left(\dfrac{E_1}{E_2} - 1\right)$

Now $R_x = R_m \cdot \dfrac{E_1}{E_2} - R_m$

$$R_x + R_m = \dfrac{R_m E_1}{E_2}$$

Multiply each member by E_2.
$$E_2(R_x + R_m) = R_m E_1$$
Divide each member by R_m.
$$\dfrac{E_2(R_x + R_m)}{R_m} = E_1$$

57. If Ann can paint the house in 10 hours, she can paint

a. $1 \div 10 = \dfrac{1}{10}$,   **b.** $5 \div 10 = \dfrac{5}{10} = \dfrac{1}{2}$,   **c.** $t \div 10 = \dfrac{t}{10}$ of the house in 1 hour, 5 hours, and t hours.   **61.** Let $n = $ 1st number, then $\dfrac{1}{3}n = \dfrac{n}{3} = $ 2nd number. **a.** The reciprocals are $\dfrac{1}{n}$ and $\dfrac{3}{n}$.

b. Five times the 2nd reciprocal $5 \cdot \dfrac{3}{n} = \dfrac{15}{n}$.

Review exercises

1. $(2p + 5q)(2p - 5q)$ **2.** $(x - 12)^2$ **3.** $(2y + 1)(y - 8)$
4. \$12,000 at 6%; \$8,000 at 8%   **5.** 8   **6.** $P(-3) = 12$

Exercise 4–8

Answers to odd-numbered problems

1. 1 hour 12 minutes   **3.** 40 minutes   **5.** 45 hours

7. $13\dfrac{11}{13}$ hours ≈ 13.8 hours   **9.** 18 hours for pump B;

36 hours for pump A **11.** 7 hours   **13.** $\dfrac{48}{17}$ mph   **15.** 30 km/hr

17. 1,600 miles   **19.** $1\dfrac{2}{3}$ miles   **21.** $\dfrac{2}{3}$ and 2   **23.** 13

25. 3   **27.** 3 mph   **29.** 6 hours   **31.** 30 mph
33. Sarah, 4 mph; Erin, 6 mph

Solutions to trial exercise problems

4. Let x = minutes taken to mix the 12 loaves working together.

$$\frac{1}{36} + \frac{1}{40} + \frac{1}{30} = \frac{1}{x} \qquad \text{Multiply by the LCM, } 360x.$$

$$360x \cdot \frac{1}{36} + 360x \cdot \frac{1}{40} + 360x \cdot \frac{1}{30} = 360x \cdot \frac{1}{x}$$

$$10x + 9x + 12x = 360$$
$$31x = 360$$
$$x = \frac{360}{31} = 11\frac{19}{31}$$

Working together, they could mix the 12 loaves in $11\frac{19}{31}$ minutes
≈ 11.6 minutes.

7. Let x = number of hours necessary to fill the basin with all pipes open.

$$\frac{1}{10} + \frac{1}{12} - \frac{1}{9} = \frac{1}{x} \qquad \text{Multiply by the LCM, } 180x.$$

$$180x \cdot \frac{1}{10} + 180x \cdot \frac{1}{12} - 180x \cdot \frac{1}{9} = 180x \cdot \frac{1}{x}$$

$$18x + 15x - 20x = 180$$
$$13x = 180$$
$$x = \frac{180}{13} = 13\frac{11}{13}$$

It would take $13\frac{11}{13}$ hours ≈ 13.8 hours to fill the basin with all three pipes open.

10. Let x = time for slower microprocessor to do the job. Then $\frac{3}{5}x$ = time for faster microprocessor to do the job.

$$\frac{1}{x} + \frac{1}{\frac{3}{5}x} = \frac{1}{2}$$

$$\frac{1}{x} + \frac{5}{3x} = \frac{1}{2}$$

$$6x \cdot \frac{1}{x} + 6x \cdot \frac{5}{3x} = 6x \cdot \frac{1}{2}$$

$$6 + 2 \cdot 5 = 3x$$
$$16 = 3x$$
$$x = \frac{16}{3}$$
$$\frac{3}{5}x = \frac{16}{5}$$

It would take the microprocessors $\frac{16}{3}$ or $5\frac{1}{3}$ milliseconds and $\frac{16}{5}$ or $3\frac{1}{5}$ milliseconds, respectively, to process the set of inputs individually.

14. Let r = speed of the wind. Then $300 + r$ = speed of the plane with the wind and $300 - r$ = speed of the plane against the wind.

Now t_w (time with the wind) $= \dfrac{950}{300 + r}$

t_a (time against the wind) $= \dfrac{650}{300 - r}$

Then $\dfrac{950}{300 + r} = \dfrac{650}{300 - r}$.

Multiply each member by $(300 + r)(300 - r)$.

$$(300 + r)(300 - r) \cdot \frac{950}{300 + r} = (300 + r)(300 - r) \cdot \frac{650}{300 - r}$$

$$(300 - r) \cdot 950 = (300 + r) \cdot 650$$
$$285{,}000 - 950r = 195{,}000 + 650r$$
$$90{,}000 = 1{,}600r$$
$$r = \frac{90{,}000}{1{,}600}$$
$$r = \frac{900}{16} = 56\frac{1}{4}$$

The speed of the wind is $56\frac{1}{4}$ mph.

21. Let n = one number, then $3n$ = the other number. The reciprocals are $\dfrac{1}{n}$ and $\dfrac{1}{3n}$.

sum of the reciprocals   is   2

$$\frac{1}{n} + \frac{1}{3n} = 2$$

$$3n \cdot \frac{1}{n} + 3n \cdot \frac{1}{3n} = 3n \cdot 2$$

$$3 + 1 = 6n$$
$$4 = 6n$$
$$n = \frac{4}{6} = \frac{2}{3}$$
$$3n = 3 \cdot \frac{2}{3} = 2$$

The numbers are $\frac{2}{3}$ and 2.

27. Let x = speed of current.

	d	r	t
downstream	10	$12 + x$	$\dfrac{10}{12 + x}$
upstream	6	$12 - x$	$\dfrac{6}{12 - x}$

Since times are the same,

$$\frac{10}{12 + x} = \frac{6}{12 - x}$$
$$10(12 - x) = 6(12 + x)$$
$$120 - 10x = 72 + 6x$$
$$16x = 48$$
$$x = 3$$

The current has a speed of 3 mph.

30. Inlet pipe fills $\frac{1}{45}$ of the tank in 1 minute

Outlet pipe empties $\frac{1}{30}$ of the tank in 1 minute.

Let x = number of minutes to empty the tank

$$\frac{1}{30} - \frac{1}{45} = \frac{1}{x}$$

$$90x \cdot \frac{1}{30} - 90x \cdot \frac{1}{45} = 90x \cdot \frac{1}{x}$$

$$3x - 2x = 90$$

$$x = 90$$

It would take 90 minutes to empty the tank.

Review exercises

1. $8x^3 - 12x^2 + 4x$ **2.** $3x^2 - 16x + 5$ **3.** $25z^2 - 40z + 16$

4. $4y^2 - 9$ **5.** $\frac{1}{y}$ **6.** x^4y^3 **7.** -7 or 7 **8.** -5

Chapter 4 review

1. $\{x \mid x \in R, x \neq -7\}$ **2.** $\left\{x \mid x \in R, x \neq \frac{4}{3}\right\}$

3. $\{x \mid x \in R, x \neq 5\}$ **4.** $\left\{a \mid a \in R, a \neq -\frac{4}{3},\frac{4}{3}\right\}$

5. $\left\{z \mid z \in R, z \neq -\frac{5}{2},\frac{1}{3}\right\}$ **6.** $\left\{y \mid y \in R, y \neq \frac{2}{3}\right\}$ **7.** $\frac{ab^3}{c^2}$

8. $\frac{-2n^2}{7mp^4}$ **9.** $\frac{5}{6}$ **10.** $\frac{3}{a-2}$ **11.** $\frac{-5}{2y+x}$

12. $\frac{y^2+4y+16}{y+4}$ **13.** $\frac{a-12}{a+1}$ **14.** $\frac{4x+3}{5x-1}$

15. $\frac{-(2y+3)}{2(3y+2)}$ **16.** $\frac{6y}{x}$ $(x \neq 0, y \neq 0)$

17. $12ay(a \neq 0, y \neq 0)$ **18.** $\frac{(4p+3)(p-4)}{3}$ $\left(p \neq -4,\frac{3}{4}\right)$

19. $\frac{z-3}{2(z+1)(z-1)}$ $(z \neq -1,1,3)$

20. $\frac{(m^2+2m+4)(m+6)}{m(m+5)}$ $(m \neq -5,0,2,3)$

21. $1\left(a \neq -7,-3,-\frac{2}{5},\frac{1}{2}\right)$

22. $\frac{(x+7)(2x-1)}{(4x^2-2x+1)(x+2)}\left(x \neq -2,-\frac{1}{2},\frac{1}{2},7\right)$

23. $\frac{x^2}{(4x+5)^2}\left(x \neq -\frac{5}{4},\frac{5}{4}\right)$ **24.** $\frac{y+3}{y+2}\left(y \neq -3,-2,\frac{4}{7}\right)$

25. $\frac{m-n}{m+n}$ $\left(m \neq -n,\frac{n}{2}; q \neq -p,p\right)$ **26.** $180x^3y^3$

27. $6x^2(x+2)(x+4)(x-4)$ **28.** $3a(a+5)(a-2)$

29. $p(p-5)(p+5)^2$ **30.** $\frac{41x}{12y}$ **31.** $\frac{-2n^2-28n-25}{(n+4)(n-1)}$

32. $\frac{10p^2+29p+81}{p(p+9)(p+2)(p-2)}$ **33.** $\frac{6b^2+16b-23}{3b-2}$

34. $\frac{2y^2+18y+3}{(y+7)(y-7)}$ **35.** $\frac{-(4x^2+15x+4)}{(x-6)(x+6)(x+4)}$

36. $\frac{37}{2(a-2)}$ **37.** $\frac{-5x^2+63x-102}{8(x-7)(x+4)}$ **38.** $\frac{4a+b}{(a-2b)(a+2b)}$

39. $\frac{1}{R_t} = \frac{I_1E_2E_3+I_2E_1E_3+I_3E_1E_2}{E_1E_2E_3}$ **40.** $\frac{2}{a}$ **41.** $\frac{5x}{4x-12}$

42. $\frac{3x+6}{x-5}$ **43.** $\frac{7b-38}{8b-34}$ **44.** $\frac{x^2y-xy^2}{2y+3x}$

45. $\frac{p^2-4p+5}{p^2-4}$ **46.** $5a^6+3a^2+2$ **47.** $6a^3b^2c^2-3c^3+1$

48. $3x^2-3x+4$ **49.** x^3+x^2-x+1 **50.** $P(-2)=-35$

51. $P(1)=11$ **52.** $P(-1)=-12$ **53.** -3 is a solution

54. -1 is not a solution **55.** 2 is not a solution **56.** $\left\{\frac{55}{216}\right\}$

57. $\{66\}$ **58.** $\left\{-\frac{37}{6}\right\}$ **59.** $\left\{\frac{7}{20}\right\}$ **60.** $\left\{-\frac{3}{29}\right\}$

61. $p = \frac{4m-6n+26}{3}$ **62.** $C = \frac{5}{9}(F-32)$

63. $V_1 = \frac{P_2V_2T_1}{P_1T_2}$ **64.** $R_2 = \frac{R_tR_1}{R_1-R_t}$ **65.** 2 days

66. 60 mph, automobile; 90 mph, train **67.** 3 mph **68.** $-\frac{1}{28}$

Chapter 4 cumulative test

1. 25 **2.** -7 **3.** $\frac{4}{15}$ **4.** 5 **5.** $\frac{13}{30}$ **6.** $7x+11$

7. $-24a^6b^5$ **8.** $x^3+2x^2-3x+20$ **9.** 2 **10.** $\frac{x^2+6x-16}{x^2+3x}$

11. $\frac{-10y-13}{24}$ **12.** $10x^2+39x-27$ **13.** $16x^2-40x+25$

14. $9y^2-25$ **15.** $2x^2+9x+27+\frac{80}{x-3}$

16. $\frac{6a^2+11a-10}{4a^2+23a-35}$ **17.** $\frac{-2x+19}{(2x-3)(x+1)(2x+1)}$

18. $\left\{y \mid y \in R, y \neq \frac{3}{2}\right\}$ **19.** $\left\{x \mid x \in R, x \neq \frac{1}{2},-\frac{1}{2}\right\}$

20. $\{x \mid x \in R, x \neq -5,5\}$ **21.** $\{14\}$ **22.** $\left\{-\frac{11}{24}\right\}$ **23.** $\left\{-\frac{1}{8}\right\}$

24. $-\frac{13}{14}$ **25.** $\frac{4x^2-3x-10}{5x^2+2x-16}$ **26.** $\{x \mid x \leq -1\}$

27. $\left\{y \mid y < \frac{17}{2}\right\}$ **28.** $\{z \mid z > -5\}$ **29.** $\left\{x \mid x \geq \frac{97}{23}\right\}$

30. $-\frac{3a}{2b^2}$ **31.** $\frac{p+4}{p+3}$ **32.** $-\frac{3}{2}$ **33.** $\left\{-\frac{1}{9}\right\}$ **34.** $\frac{1}{2}$ or 6

35. $P(-2)=27$

Chapter 5

Exercise 5–1

Answers to odd-numbered problems

1. 4.243 **3.** -5.745 **5.** $\sqrt[3]{9}$ **7.** $\sqrt{x}$ **9.** $\sqrt[5]{b^4}$ **11.** 4

13. 16 **15.** 27 **17.** 64 **19.** $\frac{1}{2}$ **21.** $\frac{1}{2}$ **23.** $\frac{1}{9}$ **25.** $\frac{-1}{8}$

27. $\frac{1}{\sqrt[4]{x^3}}$ **29.** $a^{4/7}$ **31.** $x^{1/5}$ **33.** -8 **35.** $|-4|=4$

37. $|2x-y|$ **39.** 256 **41.** $E = \frac{T}{\sqrt{(x^2+r^2)^3}}$

43. 24 miles per hour

Solutions to trial exercise problems

13. $(-64)^{2/3} = (\sqrt[3]{-64})^2 = (-4)^2 = 16$

18. $(-27)^{1/3} = \dfrac{1}{(-27)^{1/3}} = \dfrac{1}{\sqrt[3]{-27}} = \dfrac{1}{-3} = -\dfrac{1}{3}$

Review exercises

1. a^7 **2.** x^{10} **3.** $4a^6b^8$ **4.** $\dfrac{1}{a^4}$ **5.** 3   **6.** -9 **7.** $\dfrac{1}{a^7}$

8. $\dfrac{1}{x^2}$

Exercise 5–2

Answers to odd-numbered problems

1. 2   **3.** $b^{17/12}$ **5.** 5   **7.** $a^{1/12}$ **9.** $a^{8/15}$ **11.** x^3 **13.** $\dfrac{1}{x}$

15. $b^{1/3}$ **17.** $x^{1/4}$ **19.** $8y^3$ **21.** $a^2b^{2/3}$ **23.** $8a^{15/2}b^{3/2}$

25. $\dfrac{1}{3a^4b}$ **27.** $\dfrac{1}{y^{1/12}}$ **29.** $\dfrac{1}{b^{1/4}}$ **31.** x^2 **33.** $a^{2/3}$ **35.** $a^{1/2}b^{1/4}$

37. $a^{1/2}b^{1/2}$ **39.** $b^{11/6}$ **41.** $\dfrac{1}{a^{1/3}b^{1/4}}$ **43.** 13 in.   **45.** 30 mph

47. 18 miles   **49.** take the square root three times

Solutions to trial exercise problems

9. $(a^{2/3})^{4/5} = a^{2/3 \cdot 4/5} = a^{8/15}$ **13.** $(x^{-1/4})^4 = x^{-1/4 \cdot 4} = x^{-1} = \dfrac{1}{x}$

19. $(16y^4)^{3/4} = (\sqrt[4]{16y^4})^3 = (2y)^3 = 8y^3$ **27.** $\dfrac{y^{1/4}}{y^{1/3}} = y^{1/4 - 1/3}$

$= y^{3/12 - 4/12} = y^{-1/12} = \dfrac{1}{y^{1/12}}$ **41.** $\dfrac{a^{-2/3}b^{1/2}}{a^{-1/3}b^{3/4}} = a^{-2/3 - (-1/3)}b^{1/2 - 3/4}$

$= a^{-1/3}b^{2/4 - 3/4} = a^{-1/3}b^{-1/4} = \dfrac{1}{a^{1/3}b^{1/4}}$

Review exercises

1. $32a^7$ **2.** $16x^4y^3$ **3.** $375x^3y^4$ **4.** $54a^3b^5$ **5.** $375a^6b^8$
6. $32a^6b^6$

Exercise 5–3

Answers to odd-numbered problems

1. $2\sqrt{5}$ **3.** $2\sqrt[3]{3}$ **5.** $a\sqrt[4]{a}$ **7.** a^2 **9.** c **11.** $5xy^4\sqrt{xy}$
13. $5a^3c^4\sqrt{2bc}$ **15.** $3ac^4\sqrt[3]{b^2}$ **17.** $3ab^3\sqrt[3]{3a^2b^2}$ **19.** $2a^2c^2\sqrt[5]{b^4c^2}$
21. $x + 3$ **23.** $3a + 1$ **25.** $9\sqrt{2}$ **27.** 7   **29.** 12   **31.** $2\sqrt[3]{9}$
33. $3a\sqrt{5}$ **35.** $a\sqrt[3]{a}$ **37.** x **39.** $2x\sqrt[5]{x^2}$ **41.** $5ab\sqrt[3]{3a}$
43. $5x^2y^3\sqrt[3]{3y}$ **45.** $2xy\sqrt[4]{2}$ **47.** $\sqrt{y}$ **49.** $y\sqrt[4]{y^3}$ **51.** $\sqrt{2y}$
53. $\sqrt[3]{2ab^2}$ **55.** $4|x|$ **57.** $7|bc|$ **59.** $|a - 4|$ **61.** $|a|\sqrt{b}$
63. $3b\sqrt[3]{a}$ **65.** $3|a|\sqrt[4]{b^3}$ **67.** $3\sqrt[3]{3}$ in.   **69.** 3 in.   **71.** 6 units
73. $5\sqrt{6}$ amperes   **75.** 5 m   **77.** 13 in.   **79.** $\sqrt{41}$ in.
81. $3\sqrt{13}$ ft   **83.** 20 mm   **85.** $2\sqrt{21}$ cm   **87.** 5 feet
89. 44.27 meters per second

Solutions to trial exercise problems

10. $\sqrt{9x^2y^5} = \sqrt{9x^2y^2y^2y} = \sqrt{9}\sqrt{x^2}\sqrt{y^2}\sqrt{y^2}\sqrt{y} = 3xyy\sqrt{y}$
$= 3xy^2\sqrt{y}$ **24.** $\sqrt{x^2 + y^2}$. Will not simplify because x^2 and y^2
are terms, not factors.   **40.** $\sqrt[3]{4a^2b}\sqrt[3]{4a^2b^2} = \sqrt[3]{4a^2b \cdot 4a^2b^2}$
$= \sqrt[3]{16a^4b^3} = \sqrt[3]{8 \cdot 2 \cdot a^3 \cdot a \cdot b^3} = \sqrt[3]{8}\sqrt[3]{2}\sqrt[3]{a^3}\sqrt[3]{a}\sqrt[3]{b^3}$
$= 2 \cdot \sqrt[3]{2} \cdot a \cdot \sqrt[3]{a} \cdot b = 2ab\sqrt[3]{2a}$ **48.** $\sqrt[6]{b^{10}} = b^{10/6} = b^{5/3}$
$= \sqrt[3]{b^5} = \sqrt[3]{b^3}\sqrt[3]{b^2} = b\sqrt[3]{b^2}$ **67.** $h = \sqrt[3]{\dfrac{12I}{b}} = \sqrt[3]{\dfrac{12(27)}{(4)}}$
$= \sqrt[3]{3(27)} = \sqrt[3]{27}\sqrt[3]{3} = 3\sqrt[3]{3}$ in.   **79.** $c = \sqrt{a^2 + b^2}$
$= \sqrt{(5)^2 + (4)^2} = \sqrt{25 + 16} = \sqrt{41}$ in.

Review exercises

1. 9   **2.** 4   **3.** 2   **4.** a **5.** x **6.** a **7.** 2   **8.** x

Exercise 5–4

Answers to odd-numbered problems

1. $\dfrac{4}{5}$ **3.** $\dfrac{\sqrt{7}}{3}$ **5.** $\dfrac{2}{3}$ **7.** $\dfrac{a^3}{3}$ **9.** $\dfrac{\sqrt[3]{2x}}{y^5}$ **11.** $\dfrac{x^3}{yz^2}$

13. $\dfrac{2x}{y^2}$ **15.** $2x^2y\sqrt[5]{2}$ **17.** $\dfrac{\sqrt{2}}{2}$ **19.** $\dfrac{3\sqrt{10}}{10}$ **21.** $\dfrac{\sqrt{6}}{6}$

23. $\dfrac{3\sqrt{2}}{10}$ **25.** $\sqrt{2}$ **27.** $3\sqrt{2}$ **29.** $\dfrac{3\sqrt[3]{2}}{2}$ **31.** $\dfrac{2\sqrt[5]{3}}{3}$

33. $\dfrac{3\sqrt[4]{4}}{4} = \dfrac{3\sqrt{2}}{4}$ **35.** $\dfrac{x\sqrt{y}}{y}$ **37.** $\dfrac{\sqrt{c}}{c}$ **39.** $\dfrac{a\sqrt[3]{b}}{b}$

41. $\dfrac{\sqrt[3]{ab}}{b}$ **43.** $\dfrac{2x\sqrt[5]{y^3}}{y}$ **45.** $a\sqrt[5]{b}$ **47.** $\dfrac{\sqrt{2xyz}}{yz}$

49. $\dfrac{2\sqrt[3]{xyz^2}}{yz}$ **51.** $\dfrac{\sqrt[7]{8x^5y^4}}{2xy}$ **53.** $\dfrac{\sqrt[5]{x^3y}}{y}$ **55.** $\sqrt[3]{x^2y}$

57. $b\sqrt[7]{b^3c^4}$ **59.** $\dfrac{\sqrt{xy}}{2}$ **61.** $\dfrac{2a^2}{b}$ **63.** $\dfrac{y^2\sqrt[5]{z}}{xz}$ **65.** 6 units

67. $\dfrac{2\sqrt{3gh}}{3}$ **69.** $\dfrac{c\sqrt{2}}{2}$ **71.** $\dfrac{2f\sqrt{3}}{3}$ **73.** $\dfrac{2\sqrt{2\pi kmT}}{\pi m}$

Solutions to trial exercise problems

23. $\sqrt{\dfrac{9}{50}} = \dfrac{\sqrt{9}}{\sqrt{50}} = \dfrac{3}{\sqrt{25 \cdot 2}} = \dfrac{3}{5\sqrt{2}} \cdot \dfrac{\sqrt{2}}{\sqrt{2}} = \dfrac{3\sqrt{2}}{5 \cdot 2} = \dfrac{3\sqrt{2}}{10}$

44. $\dfrac{x^2}{\sqrt[3]{x^2}} \cdot \dfrac{\sqrt[3]{x}}{\sqrt[3]{x}} = \dfrac{x^2\sqrt[3]{x}}{x} = x\sqrt[3]{x}$

51. $\sqrt[7]{\dfrac{1}{16x^2y^3}} = \dfrac{\sqrt[7]{1}}{\sqrt[7]{16x^2y^3}} = \dfrac{1}{\sqrt[7]{2^4x^2y^3}} \cdot \dfrac{\sqrt[7]{2^3x^5y^4}}{\sqrt[7]{2^3x^5y^4}} = \dfrac{\sqrt[7]{8x^5y^4}}{2xy}$

59. $\sqrt{\dfrac{2y}{x}}\sqrt{\dfrac{x^2}{8}} = \sqrt{\dfrac{2y}{x} \cdot \dfrac{x^2}{8}} = \sqrt{\dfrac{xy}{4}} = \dfrac{\sqrt{xy}}{\sqrt{4}} = \dfrac{\sqrt{xy}}{2}$

Review exercises

1. $9x^2 + x$ **2.** $2a^2b - ab^2$ **3.** $5a^3 + 2a^2$ **4.** $3x^2y - 4xy^2$
5. $\dfrac{13}{4}$ **6.** $\dfrac{7}{2x}$ **7.** $\dfrac{2}{3a}$ **8.** $\dfrac{13}{6x}$

Exercise 5–5

Answers to odd-numbered problems

1. $11\sqrt{5}$ **3.** $7\sqrt{3}$ **5.** $8\sqrt{5}$ **7.** $-\sqrt{10}$ **9.** $7\sqrt[3]{4}$
11. $6\sqrt[4]{3}$ **13.** $5\sqrt[5]{12} - \sqrt[5]{16}$ **15.** $4\sqrt{3x} - 4\sqrt{2x}$
17. $-\sqrt{5}$ **19.** $-3\sqrt{3}$ **21.** $13\sqrt{3}$ **23.** $\sqrt{3}$ **25.** $5\sqrt[3]{2}$
27. $6\sqrt[3]{2} + 10\sqrt[3]{3}$ **29.** $3\sqrt[3]{3} + 10\sqrt[3]{2}$ **31.** $\sqrt{2x}$
33. $37a\sqrt{b}$ **35.** $70a\sqrt{b} - 11\sqrt{2b}$ **37.** $5\sqrt[4]{a}$
39. $-23\sqrt[3]{a^2}$ **41.** $4a^2\sqrt[3]{b}$ **43.** $2a^2\sqrt{ab}$ **45.** $2a^2b^2\sqrt{ab}$
47. $\dfrac{1 + 2\sqrt{5}}{5}$ **49.** $\dfrac{4 - 6\sqrt{3}}{9}$ **51.** $\dfrac{4\sqrt{5} + 5\sqrt{6}}{10}$
53. $\dfrac{6\sqrt{5} + \sqrt{10}}{5}$ **55.** $\dfrac{7\sqrt{3}}{12}$ **57.** $\dfrac{\sqrt{x}}{2x}$
59. $\dfrac{5\sqrt{xy} - 4y\sqrt{x}}{xy}$ **61.** 17 units   **63.** $13\sqrt{13} \approx 46.87$ feet
65. 18.23 feet

Solutions to trial exercise problems

12. $7\sqrt[3]{11} - 3\sqrt[3]{7} + 2\sqrt[3]{11} = (7\sqrt[3]{11} + 2\sqrt[3]{11}) - 3\sqrt[3]{7}$
$= 9\sqrt[3]{11} - 3\sqrt[3]{7}$ **16.** $\sqrt{12} + 4\sqrt{3} = \sqrt{4 \cdot 3} + 4\sqrt{3}$
$= 2\sqrt{3} + 4\sqrt{3} = 6\sqrt{3}$ **41.** $\sqrt[3]{a^6 b} + 3a^2\sqrt[3]{b}$
$= \sqrt[3]{a^3 a^3 b} + 3a^2\sqrt[3]{b} = a \cdot a\sqrt[3]{b} + 3a^2\sqrt[3]{b} = a^2\sqrt[3]{b} + 3a^2\sqrt[3]{b}$
$= 4a^2\sqrt[3]{b}$ **52.** $\dfrac{4}{\sqrt{7}} - \dfrac{2}{\sqrt{14}} = \dfrac{4}{\sqrt{7}} \dfrac{\sqrt{7}}{\sqrt{7}} - \dfrac{2}{\sqrt{14}} \dfrac{\sqrt{14}}{\sqrt{14}}$
$= \dfrac{4\sqrt{7}}{7} - \dfrac{2\sqrt{14}}{14} = \dfrac{4\sqrt{7}}{7} - \dfrac{\sqrt{14}}{7} = \dfrac{4\sqrt{7} - \sqrt{14}}{7}$

Review exercises

1. $6a^2 - 12a$ **2.** $6x^2 + xy - y^2$ **3.** $a^2 - 2ab + b^2$
4. $a^2 + 2ab + b^2$ **5.** $4a^2 + 4ab + b^2$ **6.** $x^2 - y^2$
7. $9a^2 - b^2$ **8.** $16x^2 - 9y^2$

Exercise 5-6

Answers to odd-numbered problems

1. $3\sqrt{5} + 3\sqrt{3}$ **3.** $12\sqrt{7} + 4\sqrt{2}$ **5.** $\sqrt{6} + \sqrt{15}$
7. $3\sqrt{6} - \sqrt{22}$ **9.** $35\sqrt{10} - 20\sqrt{15}$ **11.** $5\sqrt{3} - 10$
13. $14\sqrt{5} - 56\sqrt{2}$ **15.** $x + \sqrt{xy}$ **17.** $6x\sqrt{y} - 15x$
19. $5x\sqrt{y} + 20y\sqrt{x}$ **21.** $9 + 5\sqrt{3}$ **23.** $20 + 9\sqrt{x} + x$
25. $3 + 2\sqrt{y} - 8y$ **27.** 14 **29.** -3 **31.** -2 **33.** $a - b^2$
35. $9x - 16y$ **37.** $11 - 4\sqrt{7}$ **39.** $91 - 40\sqrt{3}$
41. $-4a + 4b\sqrt{a} + b^2$ **43.** $20x - \sqrt{xy} - y$ **45.** $2 - \sqrt{3}$
47. $\dfrac{12 - 3\sqrt{6}}{5}$ **49.** $\sqrt{10} + \sqrt{6}$ **51.** $\dfrac{6 + 3\sqrt{3}}{2}$
53. $\dfrac{-\sqrt{3} - 3\sqrt{2}}{5}$ **55.** $\dfrac{21\sqrt{2} + 4\sqrt{7}}{55}$ **57.** $\dfrac{x - \sqrt{xy}}{x - y}$
59. $\dfrac{a - \sqrt{a}}{a - 1}$ **61.** $\dfrac{x - 2y\sqrt{x} + y^2}{x - y^2}$ **63.** $\dfrac{\sqrt{ab} + \sqrt{a}}{b - 1}$
65. $\dfrac{2\sqrt{a} + 2\sqrt{ab}}{1 - b}$ **67.** $\dfrac{\sqrt{b} + 1}{b - 1}$ **69.** $\dfrac{T\sqrt{x^2 + r^2}}{(x^2 + r^2)^2}$

Solutions to trial exercise problems

10. $\sqrt{6}(\sqrt{2} + \sqrt{3}) = \sqrt{6 \cdot 2} + \sqrt{6 \cdot 3} = \sqrt{12} + \sqrt{18}$
$= \sqrt{4 \cdot 3} + \sqrt{9 \cdot 2} = 2\sqrt{3} + 3\sqrt{2}$ **27.** $(4 - \sqrt{2})(4 + \sqrt{2})$
$= 16 + 4\sqrt{2} - 4\sqrt{2} - \sqrt{2}\sqrt{2} = 16 - 2 = 14.$
Since these are conjugate factors, we could have
written $(4 - \sqrt{2})(4 + \sqrt{2}) = (4)^2 - (\sqrt{2})^2 = 16 - 2 = 14.$
47. $\dfrac{6}{4 + \sqrt{6}} = \dfrac{6}{4 + \sqrt{6}} \cdot \dfrac{4 - \sqrt{6}}{4 - \sqrt{6}} = \dfrac{6(4 - \sqrt{6})}{(4)^2 - (\sqrt{6})^2}$
$= \dfrac{6(4 - \sqrt{6})}{16 - 6} = \dfrac{6(4 - \sqrt{6})}{10} = \dfrac{3(4 - \sqrt{6})}{5} = \dfrac{12 - 3\sqrt{6}}{5}$
53. $\dfrac{\sqrt{6}}{\sqrt{2} - 2\sqrt{3}} = \dfrac{\sqrt{6}}{\sqrt{2} - 2\sqrt{3}} \cdot \dfrac{\sqrt{2} + 2\sqrt{3}}{\sqrt{2} + 2\sqrt{3}}$
$= \dfrac{\sqrt{6}(\sqrt{2} + 2\sqrt{3})}{(\sqrt{2})^2 - (2\sqrt{3})^2} = \dfrac{\sqrt{6 \cdot 2} + 2\sqrt{3 \cdot 6}}{2 - 4 \cdot 3} = \dfrac{\sqrt{12} + 2\sqrt{18}}{2 - 12}$
$= \dfrac{\sqrt{4 \cdot 3} + 2\sqrt{9 \cdot 2}}{-10} = \dfrac{2\sqrt{3} + 2 \cdot 3\sqrt{2}}{-10} = \dfrac{2(\sqrt{3} + 3\sqrt{2})}{-10}$
$= \dfrac{-(\sqrt{3} + 3\sqrt{2})}{5} = \dfrac{-\sqrt{3} - 3\sqrt{2}}{5}$
60. $\dfrac{\sqrt{a} + b}{\sqrt{a} - b} = \dfrac{\sqrt{a} + b}{\sqrt{a} - b} \cdot \dfrac{\sqrt{a} + b}{\sqrt{a} + b} = \dfrac{(\sqrt{a} + b)(\sqrt{a} + b)}{(\sqrt{a})^2 - (b)^2}$
$\dfrac{\sqrt{a}\sqrt{a} + b\sqrt{a} + b\sqrt{a} + b^2}{a - b^2} = \dfrac{a + 2b\sqrt{a} + b^2}{a - b^2}$

63. $\dfrac{a}{\sqrt{ab} - \sqrt{a}} = \dfrac{a}{\sqrt{ab} - \sqrt{a}} \cdot \dfrac{\sqrt{ab} + \sqrt{a}}{\sqrt{ab} + \sqrt{a}}$
$= \dfrac{a(\sqrt{ab} + \sqrt{a})}{(\sqrt{ab})^2 - (\sqrt{a})^2} = \dfrac{a(\sqrt{ab} + \sqrt{a})}{ab - a} = \dfrac{a(\sqrt{ab} + \sqrt{a})}{a(b - 1)}$
$= \dfrac{\sqrt{ab} + \sqrt{a}}{b - 1}$

Review exercises

1. $2a^2 - ab - b^2$ **2.** $a^2 - 3ab + 2b^2$ **3.** $4a^2 - 9b^2$
4. $a^2 + 6a + 9$ **5.** 4 **6.** $2\sqrt{5}$ **7.** 6 **8.** $6\sqrt{2}$

Exercise 5-7

Answers to odd-numbered problems

1. $3i$ **3.** $2i\sqrt{3}$ **5.** -9 **7.** -3 **9.** $-\sqrt{15}$ **11.** -2
13. -5 **15.** -6 **17.** $-3\sqrt{5}$ **19.** -1 **21.** 1 **23.** $-i$
25. $8 + 7i$ **27.** $-2 + 2i$ **29.** $1 + 8i$ **31.** i **33.** $5 - 2i$
35. $8 + 8i$ **37.** 13 **39.** $19 - 7i$ **41.** $19 + 17i$ **43.** 34
45. $7 - 24i$ **47.** $8i$ **49.** $5 - 4i$ **51.** $2 - i$
53. $\dfrac{7}{3} - \dfrac{5}{3}i$ **55.** $\dfrac{27}{26} - \dfrac{5}{26}i$ **57.** $\dfrac{4}{5} - \dfrac{3}{5}i$
59. $\dfrac{-4}{13} + \dfrac{19}{13}i$ **61.** $\dfrac{6}{5} - \dfrac{3}{5}i$ **63.** $\dfrac{4}{17} + \dfrac{18}{17}i$ **65.** $30 - 24i$
67. $0.571 - 0.143i$ **69.** $\dfrac{7}{5} + \dfrac{1}{5}i$ **71.** $\dfrac{60}{61} + \dfrac{50}{61}i$
73. $x \le 5$ **75.** $x < -11$

Solutions to trial exercise problems

5. $(3i)^2 = 3^2 i^2 = 9(-1) = -9$
28. $(4 - 5i) - (3 - 7i) = 4 - 5i - 3 + 7i = 1 + 2i$
29. $(2 + \sqrt{-49}) - (1 - \sqrt{-1}) = (2 + i\sqrt{49}) - (1 - i)$
$= (2 + 7i) - (1 - i) = 2 + 7i - 1 + i = 1 + 8i$
32. $[(2 + 5i) + (3 - 2i)] + (3 - i) = [2 + 5i + 3 - 2i]$
$+ (3 - i) = [5 + 3i] + (3 - i) = 5 + 3i + 3 - i = 8 + 2i$
44. $(2 + i)^2 = (2 + i)(2 + i) = 4 + 2i + 2i + i^2 = 4 + 4i$
$+ (-1) = 3 + 4i$ **48.** $\dfrac{3 - 2i}{i} = \dfrac{3 - 2i}{i} \dfrac{i}{i} = \dfrac{i(3 - 2i)}{i^2}$
$= \dfrac{3i - 2i^2}{-1} = \dfrac{3i - 2(-1)}{-1} = \dfrac{3i + 2}{-1} = -2 - 3i$
54. $\dfrac{4 - 3i}{1 + i} = \dfrac{4 - 3i}{1 + i} \cdot \dfrac{1 - i}{1 - i} = \dfrac{(4 - 3i)(1 - i)}{(1)^2 - (i)^2}$
$= \dfrac{4 - 4i - 3i + 3i^2}{1 - (-1)} = \dfrac{4 - 7i + 3(-1)}{2} = \dfrac{4 - 7i + (-3)}{2}$
$= \dfrac{1 - 7i}{2} = \dfrac{1}{2} - \dfrac{7}{2}i$

Review exercises

1. $(x - 6)^2$ **2.** $3x(x + 3)$ **3.** $(x + 4)(x - 4)$
4. $9(x + 2)(x - 2)$ **5.** $(x - 5)(x - 2)$ **6.** $(2x - 3)(x + 1)$
7. $(5x + 1)(x - 3)$ **8.** $(6x + 1)(x - 4)$

Chapter 5 review

1. 6   **2.** $\dfrac{1}{8}$ **3.** 9   **4.** $a^{11/12}$ **5.** $c^{1/4}$ **6.** $9x^2$ **7.** b^2

8. $a^{1/6}$ **9.** $4x^8y^4$ **10.** $x^{19/6}$ **11.** $a^{7/6}$ **12.** $2\sqrt{3}$

13. $5\sqrt{6}$ **14.** $x\sqrt[5]{x^2}$ **15.** $2ab\sqrt[3]{3b}$ **16.** $a\sqrt{a}$

17. $\sqrt[3]{2ab^2}$ **18.** 8 in.   **19.** $\dfrac{7}{8}$ **20.** $\dfrac{4\sqrt{3}}{9}$ **21.** $\dfrac{2x\sqrt[5]{2y^2}}{z^2}$

22. $\dfrac{\sqrt{2}}{4}$ **23.** $3\sqrt{2}$ **24.** $\dfrac{2\sqrt[3]{5}}{5}$ **25.** $\dfrac{x\sqrt{y}}{y}$ **26.** $\dfrac{\sqrt[3]{a^2b^2}}{b}$

27. $\sqrt[5]{x^3}$ **28.** $\dfrac{\sqrt[3]{ab^2c}}{bc}$ **29.** $\dfrac{\sqrt[4]{a^3b^2}}{b}$ **30.** $\dfrac{x\sqrt{y}}{y}$ **31.** $8\sqrt{3}$

32. $8\sqrt{2}$ **33.** $29\sqrt{2a}$ **34.** $3x^2\sqrt{xy}$ **35.** $\dfrac{5\sqrt{6}-2\sqrt{3}}{6}$

36. $\dfrac{2\sqrt{ab}-b\sqrt{a}}{ab}$ **37.** $2\sqrt{3}-2\sqrt{5}$ **38.** $2a\sqrt{b}+4a$

39. $30-10\sqrt{5}$ **40.** 3   **41.** $4a-9b$ **42.** $9x+6y\sqrt{x}+y^2$

43. $\dfrac{\sqrt{6}-2}{2}$ **44.** $4-\sqrt{6}$ **45.** $\sqrt{2}+1$ **46.** $\dfrac{a^2b\sqrt{a}+ab\sqrt{ab}}{a^2-b}$

47. $7i$ **48.** $2i\sqrt{7}$ **49.** -4 **50.** -7 **51.** -6 **52.** -3

53. $-\sqrt{6}$ **54.** i **55.** $7+7i$ **56.** $-1-11i$ **57.** $18-i$

58. $-21+20i$ **59.** $4-3i$ **60.** $-2-\dfrac{7}{3}i$ **61.** $\dfrac{7}{5}-\dfrac{6}{5}i$

62. $\dfrac{69}{58}+\dfrac{13}{58}i$

Chapter 5 cumulative test

1. $(a-8)(a+1)$ **2.** $x(4x-3)$ **3.** $9(x-2)(x+2)$
4. $(2x+3)(x+4)$ **5.** $(3a+4)(a-5)$

6. $(3x+4)(2x+3)$ **7.** (a) 28, (b) -8 **8.** $\left\{\dfrac{2}{5}\right\}$

9. $\left\{x\mid x>-\dfrac{11}{3}\right\}$ **10.** $x=-6y$ **11.** $\left\{-1,\dfrac{5}{3}\right\}$

12. $\left\{x\mid x<-\dfrac{11}{2}\text{ or }x>\dfrac{5}{2}\right\}$ **13.** $\left\{-\dfrac{13}{5}\right\}$

14. $\left\{x\mid -\dfrac{3}{2}\le x\le 2\right\}$ **15.** $2a^2b\sqrt[5]{2b^2}$ **16.** $10-5i$

17. $4\sqrt{3}$ **18.** $a^{7/12}$ **19.** $2ab^2\sqrt[3]{a}$ **20.** $8a^9b^{12}c^3$ **21.** $3i\sqrt{2}$

22. $\sqrt{10}-\sqrt{6}$ **23.** $-\dfrac{7}{13}-\dfrac{9}{13}i$ **24.** $2a^3$ **25.** $\dfrac{x^3y^2}{3}$

26. $\dfrac{\sqrt[3]{2a^2b^2c}}{2bc}$ **27.** 7 inches   **28.** 400 kg of 80% copper,
600 kg of 50% copper   **29.** 375 meters per second
30. 1,166.4 meters per second

Chapter 6

Exercise 6–1

Answers to odd-numbered problems

1. $\{3,-4\}$ **3.** $\left\{\dfrac{1}{3},-\dfrac{5}{2}\right\}$ **5.** $\{2,3\}$ **7.** $\{5\}$ **9.** $\{-3,8\}$

11. $\{0,1\}$ **13.** $\{-3,3\}$ **15.** $\left\{-\dfrac{1}{2},2\right\}$ **17.** $\left\{-2,\dfrac{3}{4}\right\}$

19. $\{2\}$ **21.** $\{-8,1\}$ **23.** $\{-5,1\}$ **25.** $\left\{-\dfrac{1}{3},2\right\}$

27. $\left\{-\dfrac{5}{2},3\right\}$ **29.** $\{-6,1\}$ **31.** $\{-11,11\}$ **33.** $\{-7,7\}$

35. $\{-4\sqrt{2},4\sqrt{2}\}$ **37.** $\{-6\sqrt{2},6\sqrt{2}\}$ **39.** $\{-2\sqrt{2},2\sqrt{2}\}$
41. $\{-5\sqrt{2},5\sqrt{2}\}$ **43.** $\{-1,-13\}$ **45.** $\{12+11i,12-11i\}$
47. $\{-10+4\sqrt{3},-10-4\sqrt{3}\}$ **49.** $\left\{2,-\dfrac{5}{2}\right\}$

51. $\left\{\dfrac{3+2i\sqrt{21}}{10},\dfrac{3-2i\sqrt{21}}{10}\right\}$ **53.** $\{-8-b,-8+b\}$

55. $x=-2b,12b$ **57.** $x=-\dfrac{7a}{4},2a$ **59.** $x=y$

61. $x=-\dfrac{4y}{3},\dfrac{y}{2}$ **63.** (a) $t=4$ sec, (b) $t=2$ sec

65. $t=1$ sec   **67.** $n=7$ **69.** 7 meters   **71.** 8,15,17
73. 4,6; $-6,-4$ **75.** 7,9; $-7,-9$

Solutions to trial exercise problems

13. $-3y^2+27=0$
 $-3(y^2-9)=0$
 $-3(y+3)(y-3)=0$
 $y=-3$ when $y+3=0$, $y=3$
 when $y-3=0$
 The solution set is $\{-3,3\}$.

21. $\dfrac{x}{2}+\dfrac{7}{2}=\dfrac{4}{x}$
 Multiply each member by the LCM, $2x$.
 $2x\cdot\dfrac{x}{2}+2x\cdot\dfrac{7}{2}=2x\cdot\dfrac{4}{x}$
 $x^2+7x=8$
 $x^2+7x-8=0$
 $(x+8)(x-1)=0$
 $x=-8$ when $x+8=0$ and
 $x=1$ when $x-1=0$
 The solution set is $\{-8,1\}$.

23. $(y+6)(y-2)=-7$
 $y^2+4y-12=-7$
 $y^2+4y-5=0$
 $(y+5)(y-1)=0$
 $y=-5$ when $y+5=0$
 and $y=1$ when $y-1=0$
 The solution set is $\{-5,1\}$.

44. $(x-9)^2=-144$
 $x-9=\sqrt{-144}=12i$ or $x-9=-\sqrt{-144}=-12i$
 Then $x=9+12i$ or $x=9-12i$
 The solution set is $\{9+12i,\ 9-12i\}$

52. $(x-7)^2=a^2,\ a>0$
 $x-7=\sqrt{a^2}=a$ or $x-7=-\sqrt{a^2}=-a$
 Then $x=7+a$ or $x=7-a$
 $\{7-a,7+a\}$

56. $3x^2-13xy+4y^2=0$
 $(3x-y)(x-4y)=0$
 $x=\dfrac{y}{3}$ when $3x-y=0$ and
 $x=4y$ when $x-4y=0$, so
 $x=\dfrac{y}{3}$ or $x=4y$.

62. a. $P=100I-5I^2$
 $420=100I-5I^2$
 $5I^2-100I+420=0$
 $5(I^2-20I+84)=0$
 $5(I-6)(I-14)=0$
 $I=6$ when $I-6=0$ and $I=14$ when $I-14=0$
 So $P=420$ when $I=6$ amperes or $I=14$ amperes.

Review exercises

1. $\dfrac{y^4}{x^4}$ **2.** $\dfrac{x^6}{16y^4}$ **3.** $4x^2 - 12x + 9$ **4.** $x^2 + 14x + 49$

5. $\{x|-1 \le x < 3\} = [-1,3)$ **6.** $\left\{\dfrac{1}{4},3\right\}$

Exercise 6–2

Answers to odd-numbered problems

1. $x^2 + 4x + 4 = (x + 2)^2$ **3.** $y^2 - 18y + 81 = (y - 9)^2$

5. $p^2 + 2p + 1 = (p + 1)^2$ **7.** $x^2 + 3x + \dfrac{9}{4} = \left(x + \dfrac{3}{2}\right)^2$

9. $w^2 - 11w + \dfrac{121}{4} = \left(w - \dfrac{11}{2}\right)^2$

11. $x^2 + 13x + \dfrac{169}{4} = \left(x + \dfrac{13}{2}\right)^2$ **13.** $\{-11,-1\}$

15. $\{1,10\}$ **17.** $\{-4 + 3i,-4 - 3i\}$ **19.** $\{0,8\}$

21. $\left\{\dfrac{-1 + \sqrt{13}}{2},\dfrac{-1 - \sqrt{13}}{2}\right\}$ **23.** $\left\{\dfrac{3 - \sqrt{17}}{2},\dfrac{3 + \sqrt{17}}{2}\right\}$

25. $\left\{-2,\dfrac{1}{2}\right\}$ **27.** $\{2 - \sqrt{3},2 + \sqrt{3}\}$

29. $\left\{-\dfrac{1}{2},\dfrac{3}{2}\right\}$ **31.** $\left\{\dfrac{5 - \sqrt{5}}{10},\dfrac{5 + \sqrt{5}}{10}\right\}$

33. $\left\{\dfrac{1 + \sqrt{29}}{2},\dfrac{1 - \sqrt{29}}{2}\right\}$ **35.** $\left\{\dfrac{1}{4},-\dfrac{3}{2}\right\}$

37. $\left\{\dfrac{1 - \sqrt{3}}{2},\dfrac{1 + \sqrt{3}}{2}\right\}$ **39.** $\left\{\dfrac{3 + \sqrt{41}}{4},\dfrac{3 - \sqrt{41}}{4}\right\}$

41. $\left\{\dfrac{2 - \sqrt{22}}{3},\dfrac{2 + \sqrt{22}}{3}\right\}$ **43.** $\left\{\dfrac{2 - \sqrt{10}}{2},\dfrac{2 + \sqrt{10}}{2}\right\}$

45. $\left\{\dfrac{5}{2},-1\right\}$ **47.** $\left\{-\dfrac{3}{5},1\right\}$

49. $t = \dfrac{-9 + \sqrt{6,481}}{32}$ sec ≈ 2.23 sec

51. $p = -16 + 6\sqrt{21}$ ¢ ≈ 12¢ **53.** $h = 25$ or $h = 1$

55. $\dfrac{1 + \sqrt{65}}{8};\dfrac{1 - \sqrt{65}}{8}$ **57.** $\sqrt{\dfrac{235}{10\pi}} = \dfrac{\sqrt{94\pi}}{2\pi}$

Solutions to trial exercise problems

7. $\left[\dfrac{1}{2}(3)\right]^2 = \left(\dfrac{3}{2}\right)^2 = \dfrac{9}{4}$

$x^2 + 3x + \dfrac{9}{4} = \left(x + \dfrac{3}{2}\right)^2$

20. $x^2 + 4x = 0$
$x^2 + 4x + 4 = 4$
$(x + 2)^2 = 4$
$x + 2 = \pm 2$
$x = -2 \pm 2$
$x = 0$ or $x = -4$
The solution set is $\{0,-4\}$.

21. $y^2 = 3 - y$
$y^2 + y = 3$
$y^2 + y + \dfrac{1}{4} = 3 + \dfrac{1}{4}$
$\left(y + \dfrac{1}{2}\right)^2 = \dfrac{13}{4}$
$y + \dfrac{1}{2} = \pm \dfrac{\sqrt{13}}{2}$
$y = -\dfrac{1}{2} \pm \dfrac{\sqrt{13}}{2} = \dfrac{-1 \pm \sqrt{13}}{2}$
The solution set is $\left\{\dfrac{-1 + \sqrt{13}}{2},\dfrac{-1 - \sqrt{13}}{2}\right\}$.

33. $(x + 2)(x - 3) = 1$
$x^2 - x - 6 = 1$
$x^2 - x = 7$
$x^2 - x + \dfrac{1}{4} = 7 + \dfrac{1}{4}$
$\left(x - \dfrac{1}{2}\right)^2 = \dfrac{29}{4}$
$x - \dfrac{1}{2} = \pm \dfrac{\sqrt{29}}{2}$
$x = \dfrac{1}{2} \pm \dfrac{\sqrt{29}}{2} = \dfrac{1 \pm \sqrt{29}}{2}$
The solution set is $\left\{\dfrac{1 + \sqrt{29}}{2},\dfrac{1 - \sqrt{29}}{2}\right\}$.

39. $\dfrac{1}{2}x^2 - \dfrac{3}{4}x = 1$ \quad Multiply by the LCM, 4.

$2x^2 - 3x = 4$ \quad Multiply by $\dfrac{1}{2}$.

$x^2 - \dfrac{3}{2}x = 2$

$x^2 - \dfrac{3}{2}x + \dfrac{9}{16} = 2 + \dfrac{9}{16}$

$\left(x - \dfrac{3}{4}\right)^2 = \dfrac{41}{16}$

$x - \dfrac{3}{4} = \pm \dfrac{\sqrt{41}}{4}$

$x = \dfrac{3}{4} \pm \dfrac{\sqrt{41}}{4} = \dfrac{3 \pm \sqrt{41}}{4}$

The solution set is $\left\{\dfrac{3 + \sqrt{41}}{4},\dfrac{3 - \sqrt{41}}{4}\right\}$.

45. $\dfrac{5}{x} - 2x + 3 = 0$

$5 - 2x^2 + 3x = 0$ \quad Multiply by the LCM, x.
$2x^2 - 3x - 5 = 0$ \quad Multiply each member by -1.
$2x^2 - 3x = 5$

$x^2 - \dfrac{3}{2}x = \dfrac{5}{2}$

$x^2 - \dfrac{3}{2}x + \dfrac{9}{16} = \dfrac{5}{2} + \dfrac{9}{16}$

$\left(x - \dfrac{3}{4}\right)^2 = \dfrac{49}{16}$

$x - \dfrac{3}{4} = \pm \dfrac{7}{4}$

$x = \dfrac{3}{4} \pm \dfrac{7}{4} = \dfrac{3 \pm 7}{4}$

The solution set is $\left\{\dfrac{5}{2},-1\right\}$.

49. $s = 100$, so $100 = 9t + 16t^2$.

$$16t^2 + 9t - 100 = 0$$

$$t^2 + \frac{9}{16}t - \frac{100}{16} = 0$$

$$t^2 + \frac{9}{16}t = \frac{25}{4}$$

$$t^2 + \frac{9}{16}t + \frac{81}{1,024} = \frac{25}{4} + \frac{81}{1,024}$$

$$\left(t + \frac{9}{32}\right)^2 = \frac{6,400 + 81}{1,024} = \frac{6,481}{1,024}$$

$$t + \frac{9}{32} = \pm\frac{\sqrt{6,481}}{32}$$

$$t = -\frac{9}{32} \pm \frac{\sqrt{6,481}}{32} = \frac{-9 \pm \sqrt{6,481}}{32}$$

Therefore $t = \dfrac{-9 + \sqrt{6,481}}{32}$ sec ≈ 2.23 sec. (We discard the other value of t since $t > 0$.)

Review exercises

1. $2\sqrt{3} - \sqrt{15}$ **2.** 1 **3.** $-\sqrt{6}$ **4.** $27 - 18\sqrt{2}$

5. $4x - 2 + \dfrac{1}{x}$ **6.** $4x + 9 + \dfrac{25}{x-3}$ **7.** 1 **8.** 61

9. -12

Exercise 6–3

Answers to odd-numbered problems

1. $\left\{\dfrac{-5 + i\sqrt{3}}{2}, \dfrac{-5 - i\sqrt{3}}{2}\right\}$ **3.** $\left\{\dfrac{3 - i\sqrt{11}}{2}, \dfrac{3 + i\sqrt{11}}{2}\right\}$

5. $\left\{2, \dfrac{3}{2}\right\}$ **7.** $\left\{\dfrac{2 + \sqrt{3}}{2}, \dfrac{2 - \sqrt{3}}{2}\right\}$

9. $\left\{\dfrac{1 - 2i\sqrt{5}}{3}, \dfrac{1 + 2i\sqrt{5}}{3}\right\}$ **11.** $\{-8, 2\}$ **13.** $\{7\}$

15. $\left\{\dfrac{-2\sqrt{15}}{3}, \dfrac{2\sqrt{15}}{3}\right\}$ **17.** $\left\{0, \dfrac{2}{5}\right\}$ **19.** $\left\{\dfrac{2}{3}\right\}$

21. $\left\{\dfrac{2 - i}{3}, \dfrac{2 + i}{3}\right\}$ **23.** $\left\{\dfrac{-1 - 2\sqrt{3}}{3}, \dfrac{-1 + 2\sqrt{3}}{3}\right\}$

25. $\left\{\dfrac{1 - i\sqrt{55}}{4}, \dfrac{1 + i\sqrt{55}}{4}\right\}$ **27.** $\left\{\dfrac{1 + \sqrt{113}}{8}, \dfrac{1 - \sqrt{113}}{8}\right\}$

29. $\left\{\dfrac{3 + i\sqrt{3}}{4}, \dfrac{3 - i\sqrt{3}}{4}\right\}$ **31.** $\left\{-2, \dfrac{8}{3}\right\}$

33. $\left\{\dfrac{11 + \sqrt{145}}{6}, \dfrac{11 - \sqrt{145}}{6}\right\}$ **35.** $\left\{\dfrac{3 + \sqrt{21}}{2}, \dfrac{3 - \sqrt{21}}{2}\right\}$

37. $\left\{-4, \dfrac{2}{3}\right\}$ **39.** $x = -3y, x = 6y$

41. $x = \dfrac{-1 \pm \sqrt{1 + 12y}}{4}$ **43.** $x = 2a \pm \sqrt{4a^2 - 3a}$

45. two, irrational **47.** two, complex **49.** one, rational
51. two, rational **53.** two, irrational **55.** two, irrational
57. two, irrational **59.** two, irrational **61.** two, irrational

63. $c = 15, a = 9, b = 12$ **65.** (a) $\dfrac{-3 + \sqrt{73}}{4}$ sec,

(b) $\dfrac{-9 + \sqrt{401}}{8}$ sec **67.** $r = 9.5\%$

Solutions to trial exercise problems

19. $9x^2 - 12x + 4 = 0$

$a = 9, b = -12, c = 4$

$$x = \frac{-(-12) \pm \sqrt{(-12)^2 - 4(9)(4)}}{2(9)}$$

$$= \frac{12 \pm \sqrt{144 - 144}}{18} = \frac{12 \pm \sqrt{0}}{18} = \frac{12}{18} = \frac{2}{3}$$

The solution set is $\left\{\dfrac{2}{3}\right\}$.

26. $3x - \dfrac{2}{x} + 5 = 0$

$3x^2 - 2 + 5x = 0$

$3x^2 + 5x - 2 = 0$

$a = 3, b = 5, c = -2$

$$x = \frac{-5 \pm \sqrt{5^2 - 4(3)(-2)}}{2(3)}$$

$$x = \frac{-5 \pm \sqrt{25 + 24}}{6}$$

$$= \frac{-5 \pm \sqrt{49}}{6} = \frac{-5 \pm 7}{6}$$

$$x = \frac{-5 + 7}{6} = \frac{2}{6} = \frac{1}{3} \text{ or } x = \frac{-5 - 7}{6} = \frac{-12}{6} = -2$$

The solution set is $\left\{-2, \dfrac{1}{3}\right\}$.

27. $2y^2 - \dfrac{7}{2} = \dfrac{y}{2}$

$4y^2 - 7 = y$

$4y^2 - y - 7 = 0$

$a = 4, b = -1, c = -7$

$$y = \frac{-(-1) \pm \sqrt{(-1)^2 - 4(4)(-7)}}{2(4)}$$

$$= \frac{1 \pm \sqrt{1 + 112}}{8} = \frac{1 \pm \sqrt{113}}{8}$$

The solution set is $\left\{\dfrac{1 + \sqrt{113}}{8}, \dfrac{1 - \sqrt{113}}{8}\right\}$.

32. $\dfrac{1}{x + 2} + \dfrac{1}{x - 3} - 2 = 0$

Multiply by the LCM, $(x + 2)(x - 3)$.

$(x - 3) + (x + 2) - 2(x + 2)(x - 3) = 0$

$x - 3 + x + 2 - 2(x^2 - x - 6) = 0$

$2x - 1 - 2x^2 + 2x + 12 = 0$

$-2x^2 + 4x + 11 = 0$

Multiply by -1 to get $2x^2 - 4x - 11 = 0$.

$a = 2, b = -4$, and $c = -11$

$$x = \frac{-(-4) \pm \sqrt{(-4)^2 - 4(2)(-11)}}{2(2)}$$

$$= \frac{4 \pm \sqrt{16 + 88}}{4} = \frac{4 \pm \sqrt{104}}{4}$$

$$= \frac{4 \pm 2\sqrt{26}}{4} = \frac{2(2 \pm \sqrt{26})}{4} = \frac{2 \pm \sqrt{26}}{2}$$

The solution set is $\left\{\dfrac{2 + \sqrt{26}}{2}, \dfrac{2 - \sqrt{26}}{2}\right\}$.

35. $(z - 3)(z + 2) = 2z - 3$

$z^2 - z - 6 = 2z - 3$

$z^2 - 3z - 3 = 0$

$a = 1, b = -3, c = -3$

$z = \dfrac{-(-3) \pm \sqrt{(-3)^2 - 4(1)(-3)}}{2(1)}$

$= \dfrac{3 \pm \sqrt{9 + 12}}{2} = \dfrac{3 \pm \sqrt{21}}{2}$

The solution set is $\left\{\dfrac{3 + \sqrt{21}}{2}, \dfrac{3 - \sqrt{21}}{2}\right\}$.

41. $4x^2 + 2x - 3y = 0$

$a = 4, b = 2, c = -3y$

$x = \dfrac{-2 \pm \sqrt{(2)^2 - 4(4)(-3y)}}{2(4)}$

$= \dfrac{-2 \pm \sqrt{4 + 48y}}{8} = \dfrac{-2 \pm \sqrt{4(1 + 12y)}}{8}$

$= \dfrac{-2 \pm 2\sqrt{1 + 12y}}{8}$

$= \dfrac{-1 \pm \sqrt{1 + 12y}}{4}$

$x = \dfrac{-1 \pm \sqrt{1 + 12y}}{4}$

62. (b) $60 = \dfrac{1}{2}(32)t^2$

$60 = 16t^2$

$16t^2 - 60 = 0$

$a = 16, b = 0, c = -60$

$t = \dfrac{0 \pm \sqrt{0^2 - 4(16)(-60)}}{2(16)} = \dfrac{\pm \sqrt{3,840}}{32}$

$= \dfrac{\pm 16\sqrt{15}}{32} = \dfrac{\pm \sqrt{15}}{2}$

Then $t = \dfrac{\sqrt{15}}{2} \approx 1.9$ sec.

$\left(\text{Discard } t = \dfrac{-\sqrt{15}}{2} \text{ since } t > 0.\right)$

Review exercises

1. $-11\sqrt{7}$　**2.** $\dfrac{3}{5}$　**3.** $14 - 6\sqrt{5}$　**4.** $22 + 7i$

5. $-2 - 2i\sqrt{3}$　**6.** $-23x^{10}$

Exercise 6–4

Answers to odd-numbered problems

1. $t = 1$ sec　**3.** $t = 3$ sec　**5.** $t = 6$ sec　**7.** $t = \dfrac{-8 + \sqrt{619}}{4}$

≈ 4.2 sec　**9.** $t = 0$ sec and $t = 1.5$ sec　**11.** 30 amperes

13. 30 sides　**15.** \$3.00　**17.** (a) 15 pens, (b) 15 pens

19. $-1 + \sqrt{51} \approx 6$ units　**21.** 120 cakes　**23.** $20 \pm 8\sqrt{5}$

≈ 37.9 or 2.1　**25.** $-15 + 15\sqrt{127}$ ft ≈ 154 ft by $15 + 15\sqrt{127}$

ft ≈ 184 ft　**27.** 5 ft　**29.** 13 in.　**31.** 50 yd by 100 yd

33. $D = 4$ ft　**35.** $7 + 4\sqrt{7} \approx 17.6$ in.　**37.** 25 in.

39. Lisa, 1 hr 5 min (65 min); Debbie, 1 hr 44 min (104 min)

41. $-1 + \sqrt{17}$ hr ≈ 3.1 hr

Solutions to trial exercise problems

4. $h = 80$ and $v_0 = 96$, so $80 = 96t - 16t^2$, then

$16t^2 - 96t + 80 = 0$

$16(t^2 - 6t + 5) = 0$

$16(t - 5)(t - 1) = 0$

The object reaches $h = 80$ feet at $t = 1$ sec.

Note: The object is at $h = 80$ feet again at $t = 5$ sec, on its way down to earth.

20. Using $20,000 = \dfrac{1}{100} n^2 - 20\,n$,

$2,000,000 = n^2 - 2,000\,n$

$n^2 - 2,000\,n - 2,000,000 = 0$

$n = \dfrac{-(-2,000) \pm \sqrt{(-2,000)^2 - 4(1)(-2,000,000)}}{2(1)}$

$= \dfrac{2,000 \pm \sqrt{4,000,000 + 8,000,000}}{2}$

$= \dfrac{2,000 \pm \sqrt{12,000,000}}{2} = \dfrac{2,000 \pm 2,000\sqrt{3}}{2}$

Then $n = \dfrac{2,000 + 2,000\sqrt{3}}{2} = 1,000 + 1,000\sqrt{3}$

$\approx 2,732$ $(n > 0)$.

Thus about 2,732 units must be produced to make a \$20,000 profit.

27. Let $s =$ the length of the side of the square.

Using $s^2 + s^2 = (\sqrt{50})^2$

$2s^2 = 50$

so $2s^2 - 50 = 0$

$2(s^2 - 25) = 0$

$2(s - 5)(s + 5) = 0$, so $s = 5$ or $s = -5$.

Then $s = 5$ feet. (*Note:* Reject $s = -5$ since the length of the side cannot be negative.)

30. Let $x =$ the amount the dimensions are increased. Since the area of the original rectangle is 18 cm², and the new rectangle has area $3 \cdot 18 = 54$ cm², then

$(x + 6)(x + 3) = 54$

$x^2 + 9x + 18 = 54$

$x^2 + 9x - 36 = 0$

$(x + 12)(x - 3) = 0$, then $x = -12$ or $x = 3$.

Reject -12, so $x = 3$ cm and the new dimensions are $3 + 6 = 9$ cm long and $3 + 3 = 6$ cm wide.

39. Let x = the time in minutes for Lisa to do the job. Then $x + 39$ = the time in minutes for Debbie to do the job.

Then $\dfrac{1}{x} + \dfrac{1}{x + 39} = \dfrac{1}{40}$

$40(x + 39) + 40x = x(x + 39)$

$40x + 1{,}560 + 40x = x^2 + 39x$

$x^2 - 41x - 1{,}560 = 0$

$x = \dfrac{-(-41) \pm \sqrt{(-41)^2 - 4(1)(-1{,}560)}}{2(1)}$

$= \dfrac{41 \pm \sqrt{1{,}681 + 6{,}240}}{2} = \dfrac{41 \pm \sqrt{7{,}921}}{2} = \dfrac{41 \pm 89}{2}$

Thus $x = \dfrac{41 + 89}{2} = \dfrac{130}{2} = 65$ or $x = \dfrac{41 - 89}{2} = \dfrac{-48}{2}$

$= -24$ (reject this answer).

Therefore Lisa can do the job in 65 minutes and Debbie can do the job in $65 + 39 = 104$ minutes.

Review exercises

1. $28 - 16\sqrt{3}$ **2.** $\left\{-\dfrac{2}{7}\right\}$ **3.** $\{-3,8\}$ **4.** 8   **5.** $x^{11/6}$

6. $x^{1/6}$ **7.** $\dfrac{\sqrt{6}}{3}$

Exercise 6–5

Answers to odd-numbered problems

1. $\{81\}$ **3.** $\emptyset$ **5.** $\{53\}$ **7.** $\{24\}$ **9.** $\{1\}$ **11.** $\{1\}$ **13.** $\{1\}$
15. $\{9\}$; -4 is extraneous   **17.** $\{6\}$; -3 is extraneous
19. $\{-\sqrt{2},\sqrt{2}\}$ **21.** $\{-9,1\}$ **23.** $\{-3,3\}$ **25.** $\{5\}$; -2 is
extraneous   **27.** $\left\{\dfrac{1}{2},2\right\}$ **29.** $\{8\}$; 2 is extraneous   **31.** $\{1,4\}$

33. $\{5\}$; 2 is extraneous   **35.** $\left\{\dfrac{5}{2}\right\}$; 2 is extraneous   **37.** $\{0,3\}$

39. $\{3,11\}$ **41.** $\{0,8\}$ **43.** $\emptyset$; $\dfrac{3}{4}$ is extraneous

45. $\left\{-\dfrac{20}{9},-4\right\}$ **47.** $\{34\}$ **49.** $\{-2,8\}$

51. $\left\{-\dfrac{5}{4}\right\}$ **53.** $A = 4\pi r^2 h$ **55.** $A = \pi r^2 + \pi R^2$

57. $A = \dfrac{\pi D^3}{6}$ **59.** 6 ft   **61.** 75 ft   **63.** -27

Solutions to trial exercise problems

14. $\sqrt{p}\,\sqrt{p - 8} = 3$
Squaring, $p(p - 8) = 9$
$p^2 - 8p = 9$
$p^2 - 8p - 9 = 0$
$(p - 9)(p + 1) = 0$
$p = 9$ or $p = -1$
The solution set is $\{9\}$.
-1 is an extraneous solution.

34. $\sqrt{3x + 10} - 3x = 4$
$\sqrt{3x + 10} = 3x + 4$
Square both sides.
$3x + 10 = 9x^2 + 24x + 16$
$9x^2 + 21x + 6 = 0$
$3(3x^2 + 7x + 2) = 0$
$3(3x + 1)(x + 2) = 0$
$x = -\dfrac{1}{3}$ or $x = -2$
The solution set is $\left\{-\dfrac{1}{3}\right\}$.
-2 is an extraneous solution.

37. $\sqrt{5x + 1} - 1 = \sqrt{3x}$
$\sqrt{5x + 1} = 1 + \sqrt{3x}$
Squaring, $5x + 1 = 1 + 2\sqrt{3x} + 3x$
$2x = 2\sqrt{3x}$
$x = \sqrt{3x}$
$x^2 = 3x$
$x^2 - 3x = 0$
$x(x - 3) = 0$
$x = 0$ or $x = 3$
The solution set is $\{0,3\}$.

42. $(2y + 3)^{1/2} - (4y - 1)^{1/2} = 0$
$\sqrt{2y + 3} = \sqrt{4y - 1}$
Squaring, $2y + 3 = 4y - 1$
$4 = 2y$
$y = 2$
The solution set is $\{2\}$.

47. $\sqrt[3]{x - 7} = 3$
Cubing, $x - 7 = 27$
$x = 34$
The solution set is $\{34\}$.

55. $r = \sqrt{\dfrac{A}{\pi} - R^2}$

$r^2 = \dfrac{A}{\pi} - R^2$

$\pi r^2 = A - \pi R^2$

$A = \pi r^2 + \pi R^2$

62. Let x = the number.
Then $\sqrt{x} = 3i$ (Square each member.)
$x = 9(-1)$
$x = -9$.

Review exercises

1. $\{-8,2\}$ **2.** $\left\{\dfrac{-1 - \sqrt{7}}{3},\dfrac{-1 + \sqrt{7}}{3}\right\}$ **3.** $\{y|y \geq 4\} = [4,\infty)$

4. $\{x|-2 < x \leq 1\} = (-2,1]$ **5.** 18   **6.** $21° \leq t \leq 62°$

Exercise 6–6

Answers to odd-numbered problems

1. $\{-\sqrt{5},\sqrt{5},-1,1\}$ **3.** $\left\{-\dfrac{\sqrt{6}}{3},-i,i,\dfrac{\sqrt{6}}{3}\right\}$ **5.** $\{2 \pm 2i\sqrt{2},2 \pm i\}$

7. $\{-2,1,4\}$ **9.** $\{81\}$; 9 is extraneous   **11.** $\emptyset$ **13.** $\{1,16\}$
15. $\{2\sqrt{13},-2\sqrt{13}\}$; $-\sqrt{19}$ and $\sqrt{19}$ are extraneous

17. $\{10\}$; 7 is extraneous   **19.** $\{-125,8\}$ **21.** $\left\{-\dfrac{1}{125},-8\right\}$

23. $\{1\}$ **25.** $\left\{\dfrac{1}{4},-\dfrac{1}{3}\right\}$ **27.** $\left\{\dfrac{1}{4},2\right\}$ **29.** $\left\{-2,-\dfrac{1}{2},\dfrac{1}{2},2\right\}$

31. $\{-5,-2,-1,2\}$ **33.** $\{4\}$; 36 is extraneous

Solutions to trial exercise problems

5. $(x - 2)^4 + 9(x - 2)^2 + 8 = 0$
   Let $u = (x - 2)^2$
   $u^2 + 9u + 8 = 0$
   $(u + 8)(u + 1) = 0$
   $u = -8$ or $u = -1$
   Substitute $(x - 2)^2$ for u.
   $(x - 2)^2 = -8$ or $(x - 2)^2 = -1$
   $x - 2 = \pm\sqrt{-8}$ or $x - 2 = \pm\sqrt{-1}$
   $x - 2 = \pm 2i\sqrt{2}$ or $x - 2 = \pm i$
   $x = 2 \pm 2i\sqrt{2}$ or $x = 2 \pm i$
   The solution set is $\{2 - 2i\sqrt{2}, 2 + 2i\sqrt{2}, 2 - i, 2 + i\}$.

15. $(x^2 - 3) - 3\sqrt{x^2 - 3} - 28 = 0$
   Let $u = \sqrt{x^2 - 3}$, then $u^2 = (\sqrt{x^2 - 3})^2 = (x^2 - 3)$ and so
   $u^2 - 3u - 28 = 0$
   $(u - 7)(u + 4) = 0$
   So $u = 7$ or $u = -4$. Then substitute $\sqrt{x^2 - 3}$ for u.
   $\sqrt{x^2 - 3} = 7$ or $\sqrt{x^2 - 3} = -4$
   $x^2 - 3 = 49$ $x^2 - 3 = 16$
   $x^2 = 52$ $x^2 = 19$
   $x = \pm\sqrt{52} = \pm 2\sqrt{13}$ $x = \pm\sqrt{19}$
   The solution set is $\{2\sqrt{13},-2\sqrt{13}\}$. *Note:* $\sqrt{19}$ and $-\sqrt{19}$ are
   extraneous solutions.

19. $y^{2/3} + 3y^{1/3} - 10 = 0$
   Let $u = y^{1/3}$, then
   $u^2 + 3u - 10 = 0$
   $(u + 5)(u - 2) = 0$
   $u = -5$ or $u = 2$
   Substitute $y^{1/3}$ for u
   $y^{1/3} = -5$ or $y^{1/3} = 2$
   $y = -125$ or $y = 8$ Cube each member.
   The solution set is $\{-125,8\}$.

30. $(t^2 - t)^2 - 4(t^2 - t) - 12 = 0$
   Let $u = t^2 - t$, then $u^2 = (t^2 - t)^2$ and so
   $u^2 - 4u - 12 = 0$
   $(u - 6)(u + 2) = 0$ so $u = 6$ or $u = -2$.
   Then $t^2 - t = 6$
   $t^2 - t - 6 = 0$
   $(t - 3)(t + 2) = 0$
   So $t = 3$ or $t = -2$.
      or
   $t^2 - t = -2$
   $t^2 - t + 2 = 0$
   $t = \dfrac{-(-1) \pm \sqrt{(-1)^2 - 4(1)(2)}}{2(1)}$
   $= \dfrac{1 \pm \sqrt{1 - 8}}{2} = \dfrac{1 \pm i\sqrt{7}}{2}$
   $\left\{3,-2,\dfrac{1 + i\sqrt{7}}{2},\dfrac{1 - i\sqrt{7}}{2}\right\}$

Review exercises

1. $\dfrac{-4x^2 - 5x}{(x + 2)(x - 2)}$ **2.** $\dfrac{x + 4}{x - 1}$ **3. a.** $y = 8$ **b.** $y = -27$

c. $y = -2$ **4.** $\left\{x\Big|x \geq \dfrac{5}{2}\right\} = \left[\dfrac{5}{2},\infty\right)$ **5.** $\{-1,4\}$

6. $\{x|-1 \leq x \leq 7\} = [-1,7]$

7. $\left\{x\Big|x < -1 \text{ or } x > \dfrac{1}{3}\right\} = (-\infty,-1) \cup \left(\dfrac{1}{3},\infty\right)$

Exercise 6–7

Answers to odd-numbered problems

1. $\{x|x < -3 \text{ or } x > 1\} = (-\infty,-3) \cup (1,\infty)$

3. $\{p|-7 \leq p \leq -2\} = [-7,-2]$

5. $\{r|r \leq 0 \text{ or } r \geq 1\} = (-\infty,0] \cup [1,\infty)$

7. $\{x|1 < x < 4\} = (1,4)$

9. $\{q| -3 \leq q \leq 6\} = [-3,6]$

11. $\{x|1 < x < 2\} = (1,2)$

13. $\{u|u \leq -6 \text{ or } u \geq 2\} = (-\infty,-6] \cup [2,\infty)$

15. $\{y|y < -2 \text{ or } y > 2\} = (-\infty,-2) \cup (2,\infty)$

17. $\{x|-\sqrt{3} < x < \sqrt{3}\} = (-\sqrt{3},\sqrt{3})$

19. $\{x|0 < x < 5\} = (0,5)$

21. $\{w|-3 \leq w \leq 0\} = [-3,0]$

23. $\left\{y\Big|-2 < y < \dfrac{3}{2}\right\} = \left(-2,\dfrac{3}{2}\right)$

25. $\left\{ w \mid w \le -5 \text{ or } w \ge -\dfrac{1}{3} \right\} = (-\infty, -5] \cup \left[-\dfrac{1}{3}, \infty \right)$

27. $\left\{ v \mid -\dfrac{2}{3} < v < \dfrac{4}{3} \right\} = \left(-\dfrac{2}{3}, \dfrac{4}{3} \right)$

29. $\left\{ x \mid x \le -\dfrac{4}{5} \text{ or } x \ge 1 \right\} = \left(-\infty, -\dfrac{4}{5} \right] \cup [1, \infty)$

31. $\{ y \mid y < -6 \text{ or } -4 < y < 5 \} = (-\infty, -6) \cup (-4, 5)$

33. $\left\{ t \mid -5 \le t \le 0 \text{ or } t \ge \dfrac{8}{3} \right\} = [-5, 0] \cup \left[\dfrac{8}{3}, \infty \right)$

35. $\{ p \mid p \le -3 \text{ or } p \ge -1 \} = (-\infty, -3] \cup [-1, \infty)$

37. $\left\{ t \mid -\dfrac{4}{5} < t < \dfrac{3}{7} \right\} = \left(-\dfrac{4}{5}, \dfrac{3}{7} \right)$

39. $\left\{ y \mid y < 0 \text{ or } y > \dfrac{5}{3} \right\} = (-\infty, 0) \cup \left(\dfrac{5}{3}, \infty \right)$

41. $\{ y \mid y \le 1 \text{ or } y > 6 \} = (-\infty, -1] \cup (6, \infty)$

43. $\left\{ t \mid -4 < t < -\dfrac{8}{3} \right\} = \left(-4, -\dfrac{8}{3} \right)$

45. $\left\{ x \mid -\dfrac{15}{2} < x < -5 \right\} = \left(-\dfrac{15}{2}, -5 \right)$

47. $\left\{ x \mid x < 7 \text{ or } x \ge \dfrac{34}{3} \right\} = (-\infty, 7) \cup \left[\dfrac{34}{3}, \infty \right)$

49. $\{ z \mid z < -2 \text{ or } z > 7 \} = (-\infty, -2) \cup (7, \infty)$

51. $\left\{ q \mid q < -4 \text{ or } -\dfrac{7}{2} \le q \le 0 \right\} = (-\infty, -4) \cup \left[-\dfrac{7}{2}, 0 \right]$

53. $\left\{ t \mid t < 0 \text{ or } \dfrac{1}{9} < t < \dfrac{2}{3} \right\} = (-\infty, 0) \cup \left(\dfrac{1}{9}, \dfrac{2}{3} \right)$

Solutions to trial exercise problems

14.
$$x^2 - 1 < 0$$
$$(x + 1)(x - 1) < 0$$
The critical numbers are -1 and 1.
The solution set is $\{ x \mid -1 < x < 1 \} = (-1, 1)$.

	$x + 1$	$x - 1$	product
$x < -1$	$-$	$-$	$+$
$-1 < x < 1$	$+$	$-$	$\ominus$
$x > 1$	$+$	$+$	$+$

16.
$$p^2 \ge 5$$
$$p^2 - 5 \ge 0$$
$$(p + \sqrt{5})(p - \sqrt{5}) \ge 0$$
The critical numbers are $-\sqrt{5}$ and $\sqrt{5}$.
The solution set is $\{ p \mid p \le -\sqrt{5} \text{ or } p \ge \sqrt{5} \} = [-\sqrt{5}, \sqrt{5}]$.

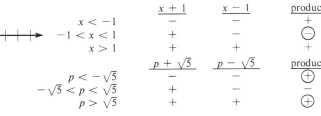

	$p + \sqrt{5}$	$p - \sqrt{5}$	product
$p < -\sqrt{5}$	$-$	$-$	$\oplus$
$-\sqrt{5} < p < \sqrt{5}$	$+$	$-$	$\ominus$
$p > \sqrt{5}$	$+$	$+$	$\oplus$

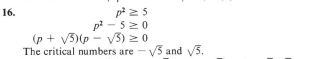

19. $x^2 - 5x < 0$
$x(x - 5) < 0$
The critical numbers are 0 and 5.
The solution set is $\{x | 0 < x < 5\} = (0,5)$.

	x	$x - 5$	product
$x < 0$	$-$	$-$	$+$
$0 < x < 5$	$+$	$-$	$\ominus$
$x > 5$	$+$	$+$	$+$

22. $2x^2 - 7x - 4 > 0$
$(2x + 1)(x - 4) > 0$

The critical numbers are $-\dfrac{1}{2}$ and 4.

	$2x + 1$	$x - 4$	product
$x < -\dfrac{1}{2}$	$-$	$-$	$\oplus$
$-\dfrac{1}{2} < x < 4$	$+$	$-$	$-$
$x > 4$	$+$	$+$	$\oplus$

The solution set is $\left\{x \middle| x < -\dfrac{1}{2} \text{ or } x > 4\right\} = \left(-\infty, -\dfrac{1}{2}\right) \cup (4,\infty)$.

30. $(x - 3)(x + 1)(x - 2) > 0$ The critical numbers are -1, 2, and 3.

	$x - 3$	$x + 1$	$x - 2$	product
$x < -1$	$-$	$-$	$-$	$-$
$-1 < x < 2$	$-$	$+$	$-$	$\oplus$
$2 < x < 3$	$-$	$+$	$+$	$\ominus$
$x > 3$	$+$	$+$	$+$	$\oplus$

The solution set is $\{x | -1 < x < 2 \text{ or } x > 3\} = (-1,2) \cup (3,\infty)$.

38. $\dfrac{1}{x} > 2$

(a) Set the denominator equal to zero. Then $x = 0$.
(b) Replace $>$ with $=$ and solve for x.

$$\frac{1}{x} = 2$$
$$2x = 1$$
$$x = \frac{1}{2}$$

The critical numbers are 0 and $\dfrac{1}{2}$ and the regions are $x < 0$,

$0 < x < \dfrac{1}{2}$, and $x > \dfrac{1}{2}$.

Let $x = -1$, then $\dfrac{1}{-1} = -1 > 2$ (False)

Let $x = \dfrac{1}{4}$, then $\dfrac{1}{\frac{1}{4}} = 4 > 2$ (True)

Let $x = 1$, then $\dfrac{1}{1} = 1 > 2$ (False)

The solution set is $\left\{x \middle| 0 < x < \dfrac{1}{2}\right\} = \left(0, \dfrac{1}{2}\right)$.

45. $\dfrac{x}{x + 5} > 3$

(a) Set $x + 5 = 0$
$x = -5$.
(b) Replace $>$ with $=$.

$$\frac{x}{x + 5} = 3$$
$$3x + 15 = x$$
$$2x = -15$$
$$x = -\frac{15}{2}$$

The critical numbers are $-\dfrac{15}{2}$ and -5 and the regions are

$x < -\dfrac{15}{2}, -\dfrac{15}{2} < x < -5, x > -5$.

Let $x = -8$, then $\dfrac{-8}{-8 + 5} = \dfrac{8}{3} > 3$ (False)

Let $x = -6$, then $\dfrac{-6}{-6 + 5} = 6 > 3$ (True)

Let $x = -4$, then $\dfrac{-4}{-4 + 5} = -4 > 3$ (False)

The solution set is $\left\{x \middle| -\dfrac{15}{2} < x < -5\right\} = \left(-\dfrac{15}{2}, -5\right)$.

50. $\dfrac{2p}{p-2} \le p$ (Note: $p \ne 2$)

  (a) Set $p - 2 = 0$, $p = 2$.

  (b) Replace $\le$ with $=$.

$$\frac{2p}{p-2} = p$$
$$p^2 - 2p = 2p$$
$$p^2 - 4p = 0$$
$$p(p-4) = 0$$
$$p = 0 \text{ or } p = 4$$

The critical numbers are 0, 2, and 4 so the regions are $p < 0$, $0 < p < 2$, $2 < p < 4$, and $p > 4$.

Let $p = -1$, then $\dfrac{2(-1)}{-1-2} = \dfrac{2}{3} \le -1$ (False)

Let $p = 1$, then $\dfrac{2(1)}{1-2} = -2 \le 1$ (True)

Let $p = 3$, then $\dfrac{2(3)}{3-2} = 6 \le 3$ (False)

Let $p = 5$, then $\dfrac{2(5)}{5-2} = \dfrac{10}{3} \le 5$ (True)

The solution set is $\{p \mid 0 \le p < 2 \text{ or } p \ge 4\} = [0,2) \cup [4,\infty)$.

Review exercises

1. $r = \sqrt{\dfrac{A}{2\pi}} = \dfrac{\sqrt{2\pi A}}{2\pi}$ **2.** $x - 2$ **3.** $\{1,2\}$ **4. a.** $y = 17$

b. $y = -11$ **c.** $y = -3$ **5.** $-\dfrac{5}{3}$ **6.** $\ell = 29$ m, $w = 26$ m

Chapter 6 review

1. $\{10,-1\}$ **2.** $\{8,-4\}$ **3.** $\left\{0,\dfrac{7}{4}\right\}$ **4.** $\{-2,2\}$

5. $\left\{\dfrac{1}{2},\dfrac{5}{2}\right\}$ **6.** $\left\{\dfrac{5}{2},-1\right\}$ **7.** $\left\{\dfrac{1}{3},2\right\}$ **8.** $\{-3,1\}$

9. $\{1,-15\}$ **10.** $\left\{0,\dfrac{8}{3}\right\}$ **11.** $20 \pm 10\sqrt{3} \approx 2.7$ sec or

37.3 sec; 40 sec   **12.** $\left\{\dfrac{-3-\sqrt{41}}{2},\dfrac{-3+\sqrt{41}}{2}\right\}$ **13.** $\left\{\dfrac{1}{2},1\right\}$

14. $\left\{\dfrac{-3-i\sqrt{11}}{10},\dfrac{-3+i\sqrt{11}}{10}\right\}$ **15.** $\left\{\dfrac{1-\sqrt{85}}{14},\dfrac{1+\sqrt{85}}{14}\right\}$

16. $\left\{\dfrac{3-5i\sqrt{15}}{16},\dfrac{3+5i\sqrt{15}}{16}\right\}$ **17.** $\{10,1\}$

18. $\left\{\dfrac{-3-\sqrt{33}}{6},\dfrac{-3+\sqrt{33}}{6}\right\}$ **19.** $\{-\sqrt{2},\sqrt{2}\}$

20. $\left\{\dfrac{-1-i\sqrt{31}}{4},\dfrac{-1+i\sqrt{31}}{4}\right\}$ **21.** $\left\{-1,\dfrac{5}{3}\right\}$

22. $\left\{\dfrac{11}{4},1\right\}$ **23.** $x = \dfrac{-y \pm y\sqrt{33}}{8}$ **24.** $x = L$ or $x = \dfrac{1}{2}L$

25. $\dfrac{15 \pm 5\sqrt{6}}{4} \approx 0.7$ sec or 6.8 sec   **26.** 200 cm by 50 cm or 2 m

by $\dfrac{1}{2}$ m   **27.** 15 ft, 8 ft, 17 ft   **28.** Mary, $11 + \sqrt{145} \approx 23$ hr;

Dick, $13 + \sqrt{145} \approx 25$ hr   **29.** $\{5\}$; -4 is extraneous

30. $\left\{\dfrac{4}{3}\right\}$; $-\dfrac{1}{3}$ is extraneous   **31.** $\left\{\dfrac{1}{4}\right\}$ **32.** $\{3,-2\}$ **33.** $\{5\}$;

$\dfrac{5}{4}$ is extraneous   **34.** $V = \dfrac{4\pi r^3}{3}$ **35.** $\{\sqrt{7},-\sqrt{7},i\sqrt{2},-i\sqrt{2}\}$

36. $\{0,4\}$ **37.** $\emptyset$; 1 and 64 are extraneous roots

38. $\{4\}$; $\dfrac{1}{9}$ is extraneous root   **39.** $\left\{-8,\dfrac{27}{8}\right\}$ **40.** $\left\{-1,\dfrac{1}{11}\right\}$

41. $\left\{-\dfrac{\sqrt{2}}{2},\dfrac{\sqrt{2}}{2},\dfrac{-i\sqrt{10}}{2},\dfrac{i\sqrt{10}}{2}\right\}$

42. $\{x \mid x < -7 \text{ or } x > 3\} = (-\infty,-7) \cup (3,\infty)$

43. $\left\{x \mid -\dfrac{4}{3} \le x \le \dfrac{1}{2}\right\} = \left[-\dfrac{4}{3},\dfrac{1}{2}\right]$

44. $\left\{y \mid y \le 0 \text{ or } y \ge \dfrac{3}{2}\right\} = (-\infty,0] \cup \left[\dfrac{3}{2},0\right)$

45. $\left\{z \mid -\dfrac{2}{3} < z < \dfrac{2}{3}\right\} = \left(-\dfrac{2}{3},\dfrac{2}{3}\right)$

46. $\{m \mid m < -1 \text{ or } m > 7\} = (-\infty,-1) \cup (7,\infty)$

47. $\left\{p \mid -\dfrac{3}{4} < p < 2\right\} = \left(-\dfrac{3}{4},2\right)$

48. $\{x \mid x \le -3 \text{ or } x \ge 3\} = (-\infty,-3] \cup [3,\infty)$

49. $\{y \mid -2 \le y \le 16\} = [-2,16]$

50. $\{y \mid y < -4 \text{ or } 2 < y < 5\} = (-\infty,-4) \cup (2,5)$

51. $\{m \mid m \le -3 \text{ or } m > 1\} = (-\infty,-3] \cup (1,\infty)$

52. $\{x \mid x \le -17 \text{ or } x > -7\} = (-\infty,-17] \cup (-7,\infty)$

Chapter 6 cumulative test

1. -12 **2.** 2 **3.** 56 **4.** $7xy^2 - 6xy + 6x^2y$
5. $9x^2 + 12xy + 4y^2$ **6.** $16y^2 - 1$ **7.** $9x^3 - 16x^2 - 8$

8. $-3,375x^6y^9$ **9.** a^{20} **10.** $\dfrac{-2b^5}{a^5}$ **11.** (a) 6, (b) 1, (c) 41

12. -12 **13.** $\dfrac{2(a+1)}{a-4}$ **14.** $\dfrac{3}{4}$ **15.** $\dfrac{x+1}{x-3}$

16. $\dfrac{2x^2 - 3x + 1}{x^2 + x - 42}$ **17.** $\dfrac{12p - 3}{(p-9)(p+2)(p-2)}$ **18.** $\dfrac{a+5}{a-6}$

19. $\left\{-\dfrac{1}{2}, -3\right\}$ **20.** $\{x \mid 0 \le x \le 5\} = [0,5]$

21. $\left\{x \mid x < \dfrac{3}{2} \text{ or } x > \dfrac{11}{2}\right\} = \left(-\infty, \dfrac{3}{2}\right) \cup \left(\dfrac{11}{2}, \infty\right)$ **22.** $\{-7\}$

23. $\{x \mid x \ge -33\} = [-33, \infty)$ **24.** $\left\{-\dfrac{5}{3}\right\}$ **25.** $w = \dfrac{P - 2\ell}{2}$

26. $\dfrac{10y - 4}{4y + 5}$ **27.** $6\sqrt{3} - 3$ **28.** $16\sqrt{3}$ **29.** $-3\sqrt[3]{3}$ **30.** 13

31. $47 - 12\sqrt{15}$ **32.** $\dfrac{4\sqrt{5}}{5}$ **33.** $2\sqrt{6} - 5$ **34.** $\dfrac{-5 + 2i}{29}$ or

$\dfrac{-5}{29} + \dfrac{2}{29}i$ **35.** $\{14, 1\}$ **36.** $\left\{\dfrac{1 + i\sqrt{59}}{10}, \dfrac{1 - i\sqrt{59}}{10}\right\}$

37. $\left\{\dfrac{1 - i\sqrt{35}}{6}, \dfrac{1 + i\sqrt{35}}{6}\right\}$ **38.** $\{z \mid -1 \le z \le 3\} = [-1,3]$
39. $\{i\sqrt{5}, -i\sqrt{5}, \sqrt{10}, -\sqrt{10}\}$ **40.** $\{y \mid -7 < y < 10\} = (-7,10)$

41. $4x^4 - 4x^3 + x^2 - 1 + \dfrac{2}{x+1}$

Chapter 7

Exercise 7–1

Answers to odd-numbered problems

1. $(2,4)$; quadrant I (see graph)
3. $(-4,3)$; quadrant II (see graph)
5. $(-1,-3)$; quadrant III (see graph)
7. $(4,0)$; quadrantal (see graph)

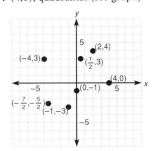

9. $(0,-1)$; quadrantal (see graph)

11. $\left(\dfrac{1}{2}, 3\right)$; quadrant I (see graph)

13. $\left(-\dfrac{7}{2}, -\dfrac{5}{2}\right)$; quadrant III (see graph)

15. $(-2,7), (0,3), (2,-1), \left(\dfrac{3}{2}, 0\right)$;
$y = -2x + 3$

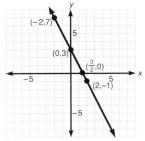

17. $(-5,9), (-3,7), (0,4), (4,0)$;
$y = -x + 4$

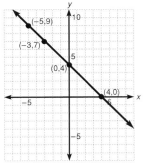

19. $(-3,-5), (0,-2), (2,0), (2,0)$;
$y = x - 2$

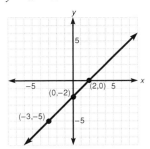

21. $\left(-1, -\dfrac{2}{3}\right), \left(0, -\dfrac{1}{3}\right), (1,0), (1,0)$;

$y = \dfrac{1}{3}x - \dfrac{1}{3}$

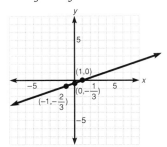

23. $(-2,7),(0,4),(2,1),\left(\dfrac{8}{3},0\right)$;

$$y = -\dfrac{3}{2}x + 4$$

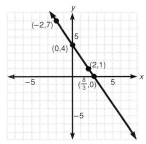

25. x-intercept, -6; y-intercept, 2

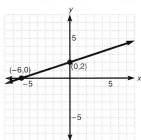

27. x-intercept, 4; y-intercept, 10

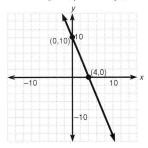

29. x-intercept, -2; y-intercept, 10

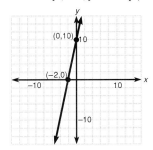

31. x-intercept, 0; y-intercept, 0

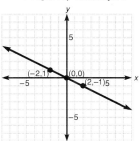

33. x-intercept, 0; y-intercept, 0

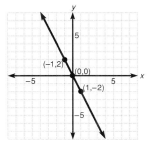

35. x-intercept, -2; y-intercept, $\dfrac{6}{5}$

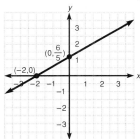

37. no x-intercept; y-intercept, -2

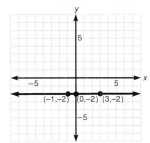

39. x-intercept, -1; no y-intercept

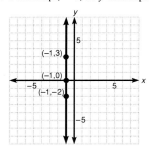

41. x-intercept, 0; all points on y-axis are y-intercepts

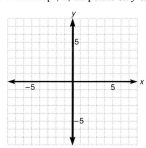

43. $y = x + 4$

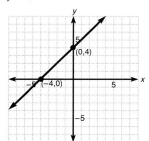

45. $3x - 2y = 12$

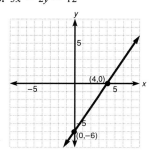

47. $C = \dfrac{5}{9}(F - 32)$

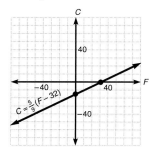

49. (5,7)

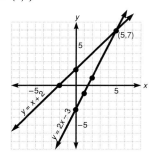

Solutions to trial exercise problems

18. $x + 2y = 4$

Solving for y, $y = \dfrac{4 - x}{2}$.

When $x = -2$, $y = \dfrac{4 - (-2)}{2} = 3$

$x = 0$, $y = \dfrac{4 - 0}{2} = 2$

$x = 2$, $y = \dfrac{4 - 2}{2} = 1$

$y = 0;\; x + 2(0) = 4$

$x = 4$

$(-2,3),(0,2),(2,1),(4,0)$

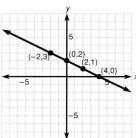

30. $x = 3y$

When $x = 0$, $3y = 0$, $y = 0$,
the x- and y-intercepts are 0.
Choose second point (3,1).

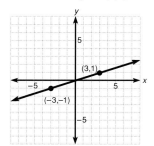

37. $y = -2$
There is no x-intercept
and the y-intercept is -2.

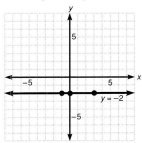

44. Two times x is $2x$, less 3 is $2x - 3$, is equal to y is $y = 2x - 3$.

x	-1	0	1	2	3
y	-5	-3	-1	1	3

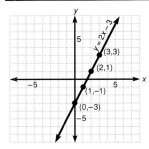

Review exercises

1. $11 + 10i$ **2.** $17 + 0i$ **3.** $22 - 6i$ **4.** $\dfrac{4 + 3i}{5}$

5. $w = \dfrac{P - 2\ell}{2}$ **6.** $x^3 + x^2 + x + 1$ **7.** -1

Exercise 7–2

Answers to odd-numbered problems

1. $d = \sqrt{41}$ units; $m = \dfrac{5}{4}$; midpoint, $\left(4, \dfrac{9}{2}\right)$

3. $d = \sqrt{29}$ units; $m = \dfrac{5}{2}$; midpoint, $\left(-2, \dfrac{5}{2}\right)$

5. $d = 5$ units; undefined slope; midpoint, $\left(-1, \dfrac{13}{2}\right)$

7. $d = 7$ units; $m = 0$; midpoint, $\left(-\dfrac{1}{2}, 6\right)$

9. $d = \sqrt{2}$ units; $m = -1$; midpoint, $\left(-\dfrac{7}{2}, -\dfrac{1}{2}\right)$

11. $d = \sqrt{170}$ units; $m = -\dfrac{7}{11}$

13. $d = \sqrt{85}$ units; $m = -\dfrac{9}{2}$

15. $d = 10$ units; undefined slope

17. $m_1 = -3$, $m_2 = -3$, $m_1 = m_2$; parallel

19. $m_1 = -2$, $m_2 = -\dfrac{7}{11}$, $m_1 \neq m_2$; not parallel

21. $m_1 = \dfrac{1}{8}$, $m_2 = -8$, $m_1 m_2 = -1$; perpendicular

23. $m_1 = 1$, $m_2 = -1$, $m_1 m_2 = -1$; perpendicular

25. $m_1 = -2$, $m_2 = -2$, $m_1 = m_2$; parallel

27. $m_1 = \dfrac{3}{4}$, $m_2 = -\dfrac{4}{3}$, $m_1 m_2 = -1$; perpendicular

29. $m_1 = -\dfrac{1}{4}$, $m_2 = \dfrac{2}{5}$, $m_1 \neq m_2$, $m_1 m_2 \neq -1$; neither

31. $m_1 = \dfrac{1}{2}$, $m_2 = -2$, $m_1 m_2 = -1$; perpendicular

33. $m_1 = \dfrac{3}{7}$, $m_2 = -\dfrac{5}{3}$, $m_1 \neq m_2$, $m_1 m_2 \neq -1$; neither

35. $m_1 = \dfrac{4}{3}$, $m_2 = \dfrac{3}{4}$; neither

37. $m_1 = \dfrac{3}{4}$, $m_2 = -\dfrac{5}{3}$; neither

39. Pitch is $\dfrac{3}{5}$.

41. $m = \dfrac{2}{3}$

43. $m = \dfrac{1,000}{7}$

45. $m = \dfrac{15}{2}$

47. $m = -\dfrac{7}{3}$

49. Slopes of opposite sides are 2 and 0; opposite sides have lengths 4 and $\sqrt{5}$; $p = 8 + 2\sqrt{5}$ units.   **51.** Slopes of two sides are $m_1 = 0$ and m_2 is undefined thus perpendicular; $6^2 + 5^2 = (\sqrt{61})^2$; $36 + 25 = 61$; $61 = 61$.   **53.** Two sides have slope $m_1 = m_2 = 0$, so are parallel, while the other sides have unequal slopes.

55. a. The slope using any pair of points is $\dfrac{3}{2}$.   **b.** Distances between points are $\sqrt{13}$, $\sqrt{52}$, and $\sqrt{117}$; $\sqrt{13} + \sqrt{52} = \sqrt{117}$; $\sqrt{13} + 2\sqrt{13} = 3\sqrt{13}$; $3\sqrt{13} = 3\sqrt{13}$.   **57.** $y = -3$ or -11
59. $(-2, 12)$

Solutions to trial exercise problems

7. $(3,6)$ and $(-4,6)$; midpoint, $\left(\dfrac{3 + (-4)}{2}, \dfrac{6 + 6}{2}\right) = \left(-\dfrac{1}{2}, 6\right)$

distance $= \sqrt{[3 - (-4)]^2 + (6 - 6)^2}$

$\qquad = \sqrt{7^2}$

$\qquad = 7$

$m = \dfrac{6 - 6}{3 - (-4)} = \dfrac{0}{7} = 0$

14. $(0,8)$ and $(0,-1)$

distance $= \sqrt{(0 - 0)^2 + [8 - (-1)]^2}$

$\qquad = \sqrt{0 + 9^2}$

$\qquad = \sqrt{81}$

$\qquad = 9$

$m = \dfrac{8 - (-1)}{0 - 0} = \dfrac{9}{0}$ undefined

18. $m_1 = \dfrac{1 - 2}{5 - (-4)} = \dfrac{-1}{9} = -\dfrac{1}{9}$

$m_2 = \dfrac{-3 - 1}{4 - 2} = \dfrac{-4}{2} = -2$ The lines are *not* parallel.

22. $m_1 = \dfrac{1 - 1}{1 - 4} = \dfrac{0}{-3} = 0$

$m_2 = \dfrac{2 - (-3)}{-2 - 3} = \dfrac{5}{-5} = -1$

Since $m_1 \cdot m_2 = 0 \cdot -1 = 0$, the lines are not perpendicular.

31. $2y - x = 1$ $6x + 3y = 0$

Using $\left(0, \dfrac{1}{2}\right)$ and $(-1,0)$,   Using $(0,0)$ and $(1,-2)$

$m_1 = \dfrac{\dfrac{1}{2} - 0}{0 - (-1)} = \dfrac{\dfrac{1}{2}}{1} = \dfrac{1}{2}$ $m_2 = \dfrac{0 - (-2)}{0 - 1} = \dfrac{2}{-1} = -2.$

Since $m_1 \cdot m_2 = \dfrac{1}{2} \cdot -2 = -1$, the lines are perpendicular.

39. $m = \text{pitch} = \dfrac{9 \text{ ft}}{15 \text{ ft}} = \dfrac{3}{5}$

45. Using $(2,10)$ and $(4,25)$, $m = \dfrac{10 - 25}{2 - 4} = \dfrac{-15}{-2} = \dfrac{15}{2}.$

51. Using $(4,2)$ and $(4,-3)$,

$m = \dfrac{2 - (-3)}{4 - 4} = \dfrac{5}{0}$ (undefined).

(vertical line)

Using $(-2,-3)$ and $(4,-3)$,

$m = \dfrac{-3 - (-3)}{-2 - 4} = \dfrac{-3 + 3}{-6} = \dfrac{0}{-6} = 0.$

(horizontal line)

The two lines are perpendicular so the triangle has one right angle and is a right triangle.

Distance from $(4,2)$ to $(-2,-3)$

$= \sqrt{[4 - (-2)]^2 + [2 - (-3)]^2} = \sqrt{6^2 + 5^2}$

$= \sqrt{36 + 25} = \sqrt{61}$

Distance from $(4,2)$ to $(4,-3)$

$= \sqrt{(4 - 4)^2 + [2 - (-3)]^2} = \sqrt{0^2 + 5^2}$

$= \sqrt{25} = 5$

Distance from $(-2,-3)$ to $(4,-3)$

$= \sqrt{(-2 - 4)^2 + [-3 - (-3)]^2}$

$= \sqrt{(-6)^2 + 0^2} = \sqrt{36} = 6$

Now $5^2 + 6^2 = (\sqrt{61})^2$

$\qquad 25 + 36 = 61$

$\qquad\qquad 61 = 61$

56. Let x be the abscissa. Then using $(x,-6)$ and $(4,5)$,

$5\sqrt{5} = \sqrt{(x - 4)^2 + (-6 - 5)^2}$

$5\sqrt{5} = \sqrt{x^2 - 8x + 16 + 121}$

$(5\sqrt{5})^2 = (\sqrt{x^2 - 8x + 137})^2$

$125 = x^2 - 8x + 137$

$0 = x^2 - 8x + 12$

$0 = (x - 6)(x - 2)$, so $x = 6$ or $x = 2$.

Thus the abscissa is 6 or 2.

58. Let x be the first component of the endpoint.

Let y be the second component of the endpoint.

Then $\dfrac{x + (-2)}{2} = 2$

$\qquad \dfrac{x - 2}{2} = 2$

$\qquad x - 2 = 4$

$\qquad\qquad x = 6$

$\dfrac{y + 3}{2} = -3$

$y + 3 = -6$

$\qquad y = -9$

The other endpoint is $(6,-9)$.

Review exercises

1. $y = \dfrac{-3x + 4}{2}$ **2.** $y = \dfrac{3x + 8}{4}$ **3.** $y < \dfrac{-x + 8}{4}$

4. $y \le \dfrac{x - 4}{2}$ **5.** $\{-36\}$ **6.** $\left\{-\dfrac{2}{5}\right\}$

Exercise 7–3

Answers to odd-numbered problems

1. $x - 2y = -5$ **3.** $5x - y = -7$ **5.** $5x + 6y = 0$

7. $x - y = 3$ **9.** $y = 4$ **11.** $x = -5$ **13.** $x = 5$

15. $x = 0$ **17.** $y = -7$ **19.** $5x - 2y = 3$ **21.** $10x - 3y = 2$

23. $2x + y = 8$ **25.** $y = 4$ **27.** $x + y = 0$ **29.** $x = 4$

31. $y = -2x + 5$; $m = -2$; $b = 5$

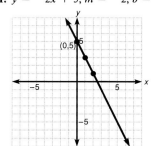

33. $y = 4x - 6$; $m = 4$; $b = -6$

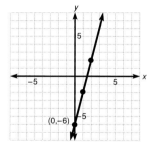

35. $y = \dfrac{2}{5}x; \ m = \dfrac{2}{5}; \ b = 0$

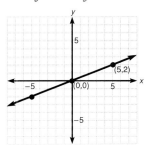

37. $y = -\dfrac{8}{3}x; \ m = -\dfrac{8}{3}; \ b = 0$

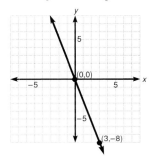

39. $y = -\dfrac{5}{3}; \ m = 0; \ b = -\dfrac{5}{3}$

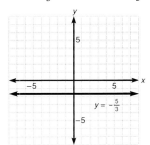

41. $3x + y = 5$ **43.** $3x - 2y = -7$ **45.** $2x + 9y = 0$
47. $x = 4$ **49.** $y = 1$ **51.** neither   **53.** perpendicular

55. parallel   **57.** $y - b = \dfrac{b - 0}{0 - a}(x - 0); \ y - b = -\dfrac{b}{a}x;$

$ay - ab = -bx; \ bx + ay = ab; \ \dfrac{x}{a} + \dfrac{y}{b} = 1$

59. $ax + by = c$
$by = -ax + c$
$y = -\dfrac{a}{b}x + \dfrac{c}{b};$

the slope is $-\dfrac{a}{b}$; the y-intercept is $\dfrac{c}{b}$ **61.** $m = \dfrac{b}{a}$

63. $250x - 3y = 550$ **65.** \$140,000

Solutions to trial exercise problems

3. $m = 5$ and $(x_1, y_1) = (0,7)$
$y - 7 = 5(x - 0)$
$y - 7 = 5x$
$5x - y = -7$

6. Using $y = mx + b$, $m = -6$
and $b = 2$, $y = -6x + 2$
$6x + y = 2.$

8. Horizontal line has slope 0,
then $y - (-3) = 0(x - 5)$
$y + 3 = 0; \ y = -3.$

11. Having undefined slope, the line must be vertical and passing
through $(-5,6)$, the *first component* of every point is -5. So
$x = -5$ is the equation.

22. $(5,0)$ and $(-2,-3)$

Now $m = \dfrac{0 - (-3)}{5 - (-2)} = \dfrac{3}{5 + 2} = \dfrac{3}{7}$

Using point $(5,0)$ and the point-slope form,

$y - 0 = \dfrac{3}{7}(x - 5)$

$y = \dfrac{3}{7}(x - 5)$

$7y = 3x - 15$
$3x - 7y = 15.$

29. $(4,-3)$ and $(4,-7)$
$m = \dfrac{-3 - (-7)}{4 - 4} = \dfrac{4}{0} =$ undefined.

The slope is undefined, so the line is vertical and passes through
$x = 4$. Thus the equation is $x = 4$.

31. $2x + y = 5$
$y = -2x + 5$
$m = -2; \ b = 5$

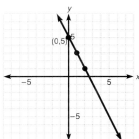

34. $3x - 7y = 0$
$y = \dfrac{3}{7}x + 0$

$m = \dfrac{3}{7}; \ b = 0$

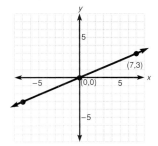

39. $3y + 5 = 0$

$$y = -\frac{5}{3} = 0x - \frac{5}{3}$$

$$m = 0; b = -\frac{5}{3}$$

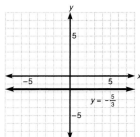

41. $3x + y = 6$

$$y = -3x + 6$$

So $m = -3$. Using $(1,2)$,

$$y - 2 = -3(x - 1)$$
$$y - 2 = -3x + 3$$
$$3x + y = 5.$$

49. We want the line through the midpoint that is perpendicular to the line having midpoint

$$\left(\frac{3+3}{2}, \frac{-4+6}{2}\right) = \left(\frac{6}{2}, \frac{2}{2}\right) = (3,1).$$

Now $m = \dfrac{-4-6}{3-3} = \dfrac{-10}{0}$ (undefined)

So, the line we want has slope $m = 0$ and using

$$y - y_1 = m(x - x)$$
$$y - 1 = 0(x - 3)$$
$$y - 1 = 0$$
$$y = 1.$$

The equation of the line is $y = 1$.

52. $2y - 5x = -3$

$$2y = 5x - 3$$
$$y = \frac{5}{2}x - \frac{3}{2}$$

So $m_1 = \dfrac{5}{2}$

$$y + 3x = 4$$
$$y = -3x + 4$$

So $m_2 = -3$.

The lines are neither parallel nor perpendicular since

$$\frac{5}{2} \neq -3 \text{ and } \frac{5}{2} \cdot -3 \neq -1.$$

56. (a) $(3,2)$ and $(4,1)$

$$y - 2 = \frac{1-2}{4-3}(x - 3)$$
$$y - 2 = \frac{-1}{1}(x - 3)$$
$$y - 2 = -1(x - 3)$$
$$y - 2 = -x + 3$$
$$x + y = 5$$

58. (a) $4x + 3y = 12$

Divide each member by 12.

$$\frac{4x}{12} + \frac{3y}{12} = \frac{12}{12}$$
$$\frac{x}{3} + \frac{y}{4} = 1$$

The x-intercept is 3 and the y-intercept is 4.

(d) $3x - 5y = 6$

Divide each member by 6.

$$\frac{3x}{6} - \frac{5y}{6} = \frac{6}{6}$$
$$\frac{x}{2} + \frac{y}{-\frac{6}{5}} = 1$$

The x-intercept is 2 and the y-intercept is $-\dfrac{6}{5}$

62. $(300,150)$ and $(600,250)$

$$y - 150 = \frac{250 - 150}{600 - 300}(x - 300)$$
$$y - 150 = \frac{100}{300}(x - 300)$$
$$y - 150 = \frac{1}{3}(x - 300)$$
$$3y - 450 = x - 300$$
$$x - 3y = -150$$

Review exercises

1. $\left\{x \,\middle|\, x \leq \dfrac{1}{2}\right\} = \left(-\infty, \dfrac{1}{2}\right]$ **2.** $y > \dfrac{3x + 6}{4}$

3. $x - 3y = 6$

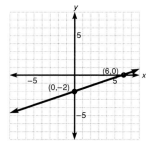

4. $2y + x = -4$

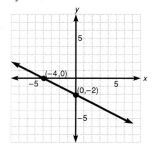

5. $\left\{\dfrac{3 - \sqrt{41}}{4}, \dfrac{3 + \sqrt{41}}{4}\right\}$ **6.** $\{6\}$; 1 is extraneous

Exercise 7–4

Answers to odd-numbered problems

1.
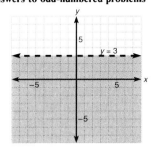

3.
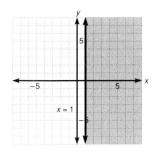

5.
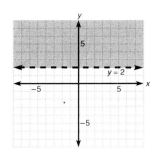

7.
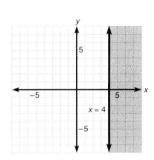

9.
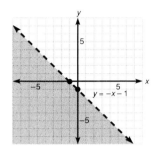

11.
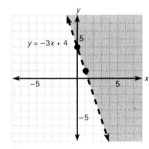

13.
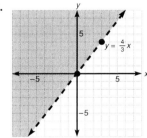

15.
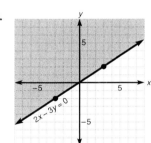

17.
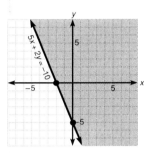

19.
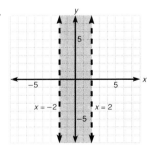

21.
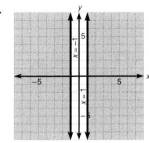

23.
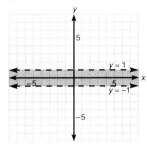

25.
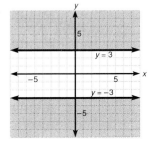

27.
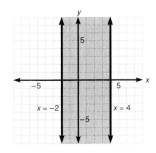

29.
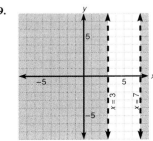

31.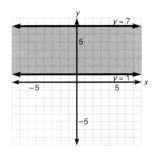

33.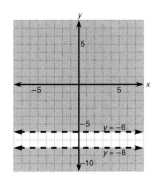

35.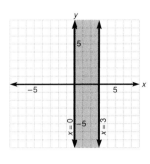

Solutions to trial exercise problems

1. $y < 3$

Since every point having $y < 3$ lies below the line $y = 3$, we dash the horizontal line $y = 3$ and shade the plane below this horizontal line.

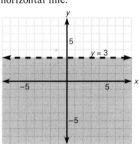

8. $x + y > 5$

Graph the line $x + y = 5$ (dashed) and since for (0,0), $0 + 0 > 5$ is false, shade *above* this line.

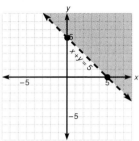

17. $5x + 2y \geq -10$

Graph the line $5x + 2y = -10$ (make this line *solid*). Since for (0,0), $5(0) + 2(0) \geq -10$ is true, shade the plane to the right of this line.

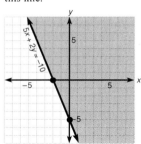

19. $|x| < 2, \ -2 < x < 2$

Shade the plane between the dashed vertical lines, $x = -2$ and $x = 2$.

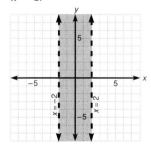

30. $|3 - x| \geq 4$

$3 - x \geq 4$ or $3 - x \leq -4$

$x \leq -1$ or $x \geq 7$

Shade to the left of the solid vertical line $x = -1$ and to the right of the solid vertical line $x = 7$.

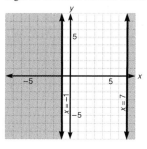

35. $|2x - 3| \leq 3$

$-3 \leq 2x - 3 \leq 3$

$0 \leq 2x \leq 6$

$0 \leq x \leq 3$

Shade between the solid vertical lines $x = 0$ (y-axis) and $x = 3$.

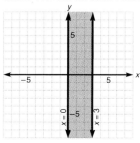

Review exercises

1.

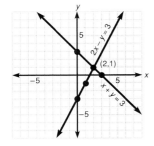

2.

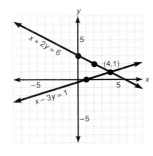

3. $P(-1) = 7$
$P(2) = 13$

4. -40

5. $\dfrac{3x^2 - 11x + 4}{(x - 3)(x + 3)}$

6. $\dfrac{x^2}{12y}$

Chapter 7 review

1. $(-3,5)$; quadrant II (see graph)

2. $(-2,-7)$; quadrant III (see graph)

3. $(0,-1)$; quadrantal (see graph)

4. $(5,0)$; quadrantal (see graph)

5. $\left(1, -\dfrac{3}{2}\right)$; quadrant IV (see graph)

6. $\left(\dfrac{1}{2}, \dfrac{11}{2}\right)$; quadrant I (see graph)

7. $(0,4)$; quadrantal (see graph)

8. $\left(-\dfrac{5}{2}, -4\right)$; quadrant III (see graph)

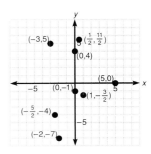

9. x-intercept, 6; y-intercept, -2

10. x-intercept, 4; y-intercept, -8

11. x-intercept, 2; y-intercept, -5

12. x-intercept, -6; no y-intercept

13. no x-intercept; y-intercept, 5

14. x-intercept, $\dfrac{7}{2}$; y-intercept, $\dfrac{7}{4}$

15. x-intercept, $\dfrac{3}{2}$; y-intercept, -3

16. x-intercept, $\dfrac{1}{5}$; y-intercept, 1

17. x-intercept, 0; y-intercept, 0

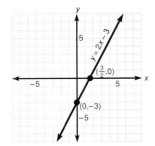

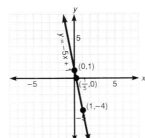

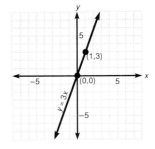

18. x-intercept, 4; y-intercept, -2

19. x-intercept, $-\dfrac{9}{2}$; y-intercept, 3

20. no x-intercept; y-intercept, -6

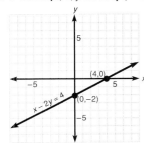

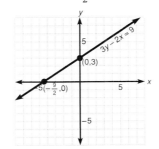

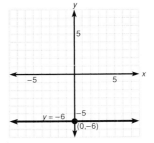

21. x-intercept, 1; no y-intercept

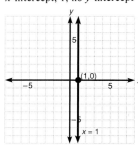

22. $\sqrt{13}$ units midpoint, $\left(\frac{3}{2},3\right)$; $m = -\frac{2}{3}$

23. 3 units; midpoint, $\left(-2,\frac{5}{2}\right)$; m is undefined

24. 4 units midpoint, $(-1,5)$; $m = 0$

25. neither; $m_1 = \frac{1}{4}$, $m_2 = \frac{-3}{2}$

26. perpendicular; $m_1 = 2$, $m_2 = -\frac{1}{2}$

27. parallel; $m_1 = \frac{4}{7} = m_2$

28. neither; $m_1 = -\frac{1}{3}$, $m_2 = \frac{5}{9}$

29. parallel; $m_1 = -2 = m_2$

30. neither; $m_1 = \frac{1}{3}$, $m_2 = -\frac{2}{3}$

31. perpendicular; $m_1 = \frac{2}{5}$, $m_2 = -\frac{5}{2}$

32. perpendicular; $m_1 = \frac{3}{2}$, $m_2 = -\frac{2}{3}$

33. $m = \frac{18}{13}$

34. $7^2 + 3^2 = (\sqrt{58})^2$; $49 + 9 = 58$; $58 = 58$

35. $2x - 3y = -17$

36. $y = 4$

37. $x = 1$

38. $8x - y = 21$

39. $y = 3x - 4$; $m = 3$; y-intercept $= -4$

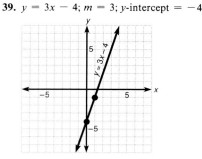

40. $y = \frac{-2}{3}x + 3$; $m = \frac{-2}{3}$; y-intercept $= 3$

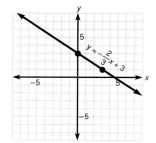

41. $y = \frac{3}{2}x$; $m = \frac{3}{2}$; y-intercept $= 0$

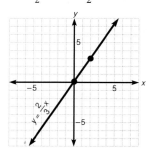

42. $y = \frac{-3}{2}$; $m = 0$; y-intercept $= \frac{-3}{2}$

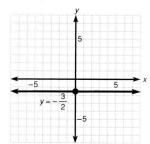

43. $x - 2y = -11$

44. $5x - 4y = -12$

45.

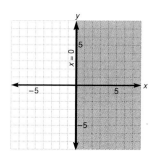

46.

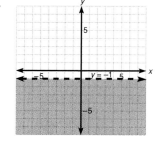

47.

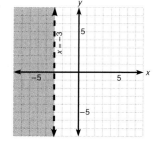

48.

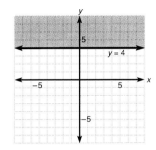

49.

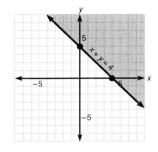

50.

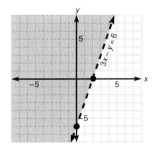

51.

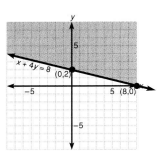

52.

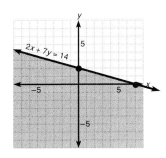

53.

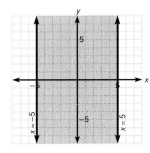

54.

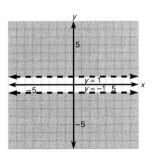

55.

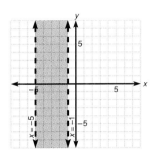

56.

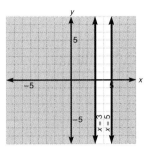

57.

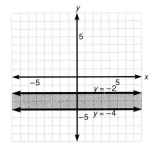

58.

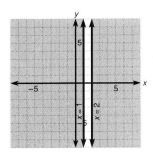

Chapter 7 cumulative test

1. $-18a^3b^5$ **2.** $\dfrac{y^6}{x^5}$ **3.** $\dfrac{a^6c^2}{16b^4}$ **4.** $9y^2 - 3y - 3$

5. $-3x^2 - 8xy + 3y^2$ **6.** $12x^3y - 18xy^3 + 6x^3y^2 - 6x^4y^4$

7. $3(a + 2b)(a - 2b)$ **8.** $(8x + 1)(x - 7)$

9. $(2x + 3y)(4x^2 - 6xy + 9y^2)$ **10.** $(4x - 5y)^2$ **11.** $\left\{-\dfrac{45}{4}\right\}$

12. $\{1, -4\}$ **13.** $\left\{y \mid -\dfrac{3}{2} < y < \dfrac{5}{2}\right\} = \left(-\dfrac{3}{2}, \dfrac{5}{2}\right)$

14. $\{y|y \le -4 \text{ or } y \ge -2\} = (-\infty, -4] \cup [-2, \infty)$

15. $\left\{0, \dfrac{1}{4}\right\}$ **16.** $\{9, -2\}$ **17.** $\left\{\dfrac{3}{2}, -1\right\}$

18. $\left\{x \Big| -\dfrac{1}{3} \le x < 3\right\} = \left[-\dfrac{1}{3}, 3\right)$ **19.** $\left\{\dfrac{3 + \sqrt{65}}{4}, \dfrac{3 - \sqrt{65}}{4}\right\}$

20. $\dfrac{1}{y - 5}$ **21.** $\dfrac{-3y^2 + 46y - 21}{2y(y + 7)(y - 7)}$ **22.** 1   **23.** $\left\{\dfrac{11}{10}\right\}$

24. a. $P(-3) = 0$ **b.** $x + 3$ is a factor of $3x^3 + 8x^2 - 7x = 12$

25. $6\sqrt{2}$ **26.** $21 + 8\sqrt{5}$ **27.** $\dfrac{3\sqrt{5}}{5}$ **28.** $6 + 3\sqrt{3}$ **29.** 13

30. $2 - 7i$ **31.** $\emptyset$; 7 is extraneous   **32.** $7x - y = 10$

33. $2x - 3y = -8$ **34.** $x = 5$ **35.** $m = \dfrac{3}{5}, b = -2$

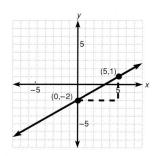

36. perpendicular   **37.** $d = \sqrt{65}$; midpoint, $\left(\dfrac{3}{2}, 1\right)$

Chapter 8

Exercise 8–1

Answers to odd-numbered problems

1. domain = $\{8, 5, 9, 6\}$, range = $\{0, 4, 3\}$ **3.** domain = $\{-4, 1\}$, range = $\{1, 2, 3, 9\}$ **5.** domain = $\{6, 1, 2, 3\}$, range = $\{-1, 1\}$ **7.** domain = $\{5, 6, 7\}$, range = $\{3, -4\}$ **9.** domain = $\{x| -2 \le x \le 2\} = [-2, 2]$, range = $\{y| -2 \le y \le 2\} = [-2, 2]$ **11.** domain = $\{x|x \in R\} = (-\infty, \infty)$, range = $\{y|y \le 3\} = (-\infty, 3]$ **13.** domain = $\{x|x \in R\} = (-\infty, \infty)$, range = $\{y|y \in R\} = (-\infty, \infty)$ **15.** domain = $\{x|x \le -2 \text{ or } x \ge 2\} = (-\infty, -2] \cup [2, \infty)$, range = $\{y|y \in R\} = (-\infty, \infty)$ **17.** function   **19.** not a function because two of the ordered pairs have the same first component $(2, 3)$ and $(2, -7)$ **21.** not a function because two of the ordered pairs have the same first component $(-1, 2)$ and $(-1, 6)$ **23.** function   **25.** function   **27.** function   **29.** not a function because two of the ordered pairs have the same first component $(4, -2)$ and $(4, 2)$ **31.** function   **33.** not a function since two ordered pairs have the same first component, $(-10, 0)$ and $(-10, 1)$ **35.** not a function since two ordered pairs have the same first component, $(3, 1)$ and $(3, 4)$ **37.** domain = $\{-3, -1, 0, 1, 3\}$ **39.** domain = $\{-5, -3, 0, 3, 5\}$ **41.** domain = $\{-5, -1, 1, 2, 4\}$ **43.** domain = $\{x|x \in R\} = (-\infty, \infty)$ **45.** domain = $\{x|x \in R\} = (-\infty, \infty)$ **47.** domain = $\{x|x \in R, = \left(-\infty, -\dfrac{3}{2}\right) \cup \left(-\dfrac{3}{2}, \infty\right)$ **51.** domain = $\left\{x|x \ge -\dfrac{4}{3}\right\} = \left[-\dfrac{4}{3}, \infty\right)$ **53.** function since any vertical line intersects the graph at only one point   **55.** function since any vertical line intersects the graph at only one point   **57.** not a function since at least one vertical line intersects at more than one point on the graph   **59.** not a function since at least one vertical line intersects at more than one point on the graph

Solutions to trial exercise problems

5. The domain (set of first components) = $\{6, 1, 2, 3\}$, the range (set of second components) = $\{-1, 1\}$ **9.** The circle intersects the x-axis at -2 and 2, so the domain = $\{x| -2 \le x \le 2\}$, and the y-axis at -2 and 2, so the range = $\{y| -2 \le y \le 2\}$.   **15.** The hyperbola extends from -2 left and from 2 right, so the domain = $\{x|x \le -2 \text{ or } x \ge 2\}$. The curve extends infinitely far above and below, so the range = $\{y|y \in R\}$.   **22.** does define a function since no two ordered pairs have the same first component   **31.** defines a function—no two ordered pairs have the same first component

46. Since $y = \dfrac{2}{x}$, the function defined is $f(x) = \dfrac{2}{x}$ and the domain = $\{x|x \in R, x \ne 0\} = (-\infty, 0) \cup (0, \infty)$.

Review exercises

1. $\dfrac{4x^8}{y^6}$ **2.** x^6 **3.** y^4 **4.** no, since exponents are *added only* when like bases are multiplied   **5.** $2(x + 4y)(x - 4y)$ **6.** $3\sqrt{6}$ units   **7.** 60 m by 120 m

Exercise 8–2

Answers to odd-numbered problems

1. -2 **3.** 0   **5.** 58   **7.** $-\dfrac{15}{4}$ **9.** $3a + 1$ **11.** $3a^2 - 2$ **13.** 27   **15.** 16   **17.** $3h$ **19.** $x^2 + 2xh + h^2 + 2x + 2h - 5$ **21.** $2x + h + 2$ **23.** $4x + 2h + 3$ **25.** $f(-5) = -17$, $(-5, -17); f(0) = -2, (0, -2); f\left(\dfrac{2}{3}\right) = 0, \left(\dfrac{2}{3}, 0\right)$

27. $h\left(-\dfrac{1}{2}\right) = \dfrac{11}{4}, \left(-\dfrac{1}{2}, \dfrac{11}{4}\right); h(0) = 1, (0, 1); h(3) = 22, (3, 22)$

29. $g(-15) = 10, (-15, 10); g(0) = 10, (0, 10); g\left(\dfrac{6}{5}\right) = 10, \left(\dfrac{6}{5}, 10\right)$ **31.** $h(-5) = -121, (-5, -121); h(0) = 4, (0, 4); h\left(\dfrac{1}{2}\right) = \dfrac{33}{8}, \left(\dfrac{1}{2}, \dfrac{33}{8}\right)$ **33.** 7   **35.** 3   **37.** 972   **39.** $\dfrac{1}{2}$

41. $-4z^2 - 6z + 1$ **43.** $\dfrac{-4b^2 + 4b - 1}{10b - 2}$ **45.** $-x^2 - 3x + 1$

47. $\dfrac{-x^2 + 2x - 1}{5x - 2}$ **49.** 967   **51.** 30   **53.** $48x^2 + 48x + 7$ **55.** $16x + 10$ **57.** $(14, -10), (32, 0), (212, 100)$ **59.** $(2, 49), (3, 69), (5, 109)$ **61.** $(1, 1), (3, 27), (5, 125)$, domain = $\{s|s \ge 0\}$

63. a. $R(x) = 3x$ **b.** $P(x) = R(x) - C(x) = 3x -$
$[1,500 + 3\sqrt[3]{x}]$ **c.** $P(1,000) = 1,470$ **65. a.** $(f \circ g)(x) = f[g(x)]$
$= f(x^2 + 2) = \sqrt{(x^2 + 2) - 2} = \sqrt{x^2} = x$ **b.** $(g \circ f)(x)$
$= g[f(x)] = g(\sqrt{x - 2}) = (\sqrt{x - 2})^2 + 2 = x - 2 + 2 = x$
67. a. $(f \circ g)(x) = f[g(x)] = f(\sqrt[3]{x + 7}) = (\sqrt[3]{x + 7})^3 - 7 = x$
$+ 7 - 7 = x$ **b.** $(g \circ f)(x) = g[f(x)] = g(x^3 - 7) =$
$\sqrt[3]{x^3 - 7 + 7} = \sqrt[3]{x^3} = x$ **69. a.** yes   **b.** no

Solutions to trial exercise problems

9. $f(a + 1) = 3(a + 1) - 2 = 3a + 3 - 2 = 3a + 1$
12. $f(5) = 3(5) - 2 = 15 - 2 = 13$ and $f(2) = 3(2) - 2 = 4$, so
$f(5) - f(2) = 13 - 4 = 9$
22. a. $f(x + h) = 4(x + h)^2 = 4x^2 + 8xh + 4h^2$
  b. $f(x + h) - f(x) = (4x^2 + 8xh + 4h^2) - (4x^2) = 8xh + 4h^2$
  c. $\dfrac{f(x + h) - f(x)}{h} = \dfrac{8xh + 4h^2}{h} = \dfrac{h(8x + 4h)}{h} = 8x + 4h$
30. $g(x) = -7$
  $g\left(-\dfrac{7}{8}\right) = -7; \left(-\dfrac{7}{8}, -7\right)$
  $g(0) = -7; (0, -7)$
  $g(25) = -7; (25, -7)$
33. $(f + g)(5) = f(5) + g(5) = [-(5)^2 + 2(5) - 1] + [5(5) - 2]$
  $= [-25 + 10 - 1] + [25 - 2] = (-16) + 23 = 7$
41. $(f - g)(2z) = f(2z) - g(2z)$
  $= [-(2z)^2 + 2(2z) - 1] - [5(2z) - 2]$
  $= -4z^2 + 4z - 1 - 10z + 2 = -4z^2 - 6z + 1$
49. $f[g(4)] = f[4(4) + 2] = f(18) = 3(18)^2 - 5 = 967$

54. $f[f(x)] = f[3x^2 - 5] = 3(3x^2 - 5)^2 - 5$
  $= 3(9x^4 - 30x^2 + 25) - 5$
  $= 27x^4 - 90x^2 + 75 - 5$
  $= 27x^4 - 90x^2 + 70$

57. $g(F) = \dfrac{5}{9}(F - 32)$
  $g(14) = \dfrac{5}{9}(14 - 32) = \dfrac{5}{9}(-18) = -10; (14, -10)$
  $g(32) = \dfrac{5}{9}(32 - 32) = \dfrac{5}{9}(0) = 0; (32, 0)$
  $g(212) = \dfrac{5}{9}(212 - 32) = \dfrac{5}{9}(180) = 100; (212, 100)$

62. a. $R(x) = 2.50x$
  b. $P(x) = R(x) - C(x) = 2.50x - [50,000 + 5\sqrt{x}]$
    $= 2.50x - 50,000 - 5\sqrt{x}$
  c. $P(40,000) = 2.50(40,000) - 50,000 - 5\sqrt{40,000}$
    $= 100,000 - 50,000 - 5(200) = 50,000 - 1,000$
    $= \$49,000$
65. a. $(f \circ g)(x) = f[g(x)] = f[x^2 + 2] = \sqrt{(x^2 + 2) - 2} = \sqrt{x^2}$
  $= x$ **b.** $(g \circ f)(x) = g[f(x)] = g[\sqrt{x - 2}] = (\sqrt{x - 2})^2 + 2$
  $= x - 2 + 2 = x$

Review exercises

1. $6x^2 + 7x - 14$ **2.** $-x^2 - 3x + 6$ **3.** $m = \dfrac{4}{5}$; y-intercept,
$(0, 4)$ **4.** fourth degree   **5.** $3x^2 - x + 1 + \dfrac{4}{x - 1}$

Exercise 8–3

Answers to odd-numbered problems

1. linear

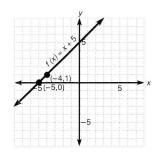

3. linear

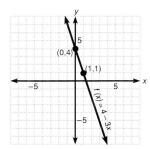

5. linear

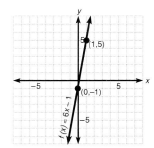

7. linear

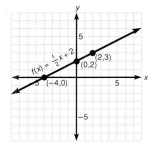

9. linear

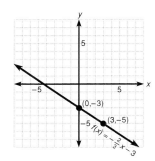

11. constant, linear

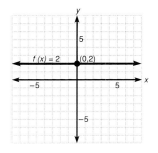

13. constant, linear

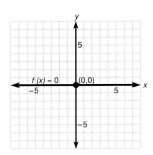

15. quadratic

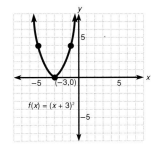

17. quadratic

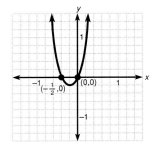

19. quadratic

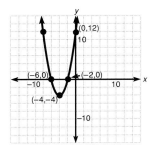

21. quadratic

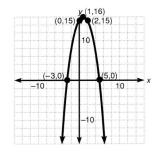

23. quadratic

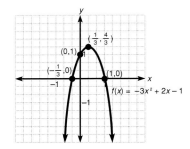

25. square root

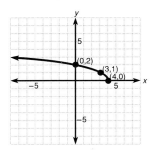

27. square root

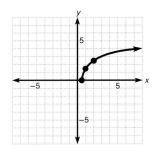

29. square root

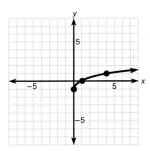

31. square root

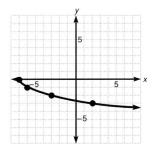

33. square root

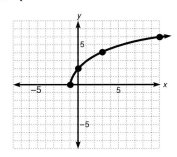

35. discontinuous at $x = 0$

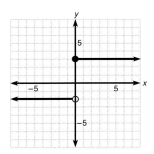

37. discontinuous at $x = 2$

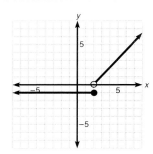

39. discontinuous at $x = -3$ and at $x = 0$

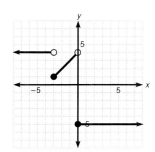

41. discontinuous at $x = 2$

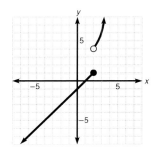

43. discontinuous at $x = 2$

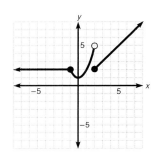

45. $f(-2) = 1, f(-1) = 0, f(0) = -1,$
$f(1) = 0, f(2) = 1$

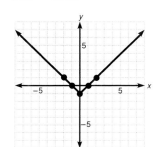

47. $f(-2) = 2, f(-1) = 1, f(0) = 0,$
$f(1) = 1, f(2) = 2$

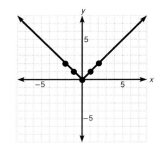

49. $f(-2) = 3, f(-1) = 4, f(0) = 5, f(1) = 4, f(2) = 3$

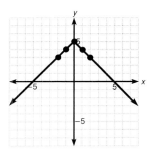

51. $f(-1) = 3, f(0) = 2, f(1) = 1, f(2) = 0, f(3) = 1, f(4) = 2$

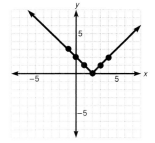

53. increases on $\{x | x \le -4\}$ and $\{x | x \ge -2\}$, decreases on
$\{x | -4 \le x \le -2\}$
55. increases on $\{x | x \le 0\}$, decreases on $\{x | x \ge 0\}$
57. decreases on $\{x | x \in R\}$ **59.** odd   **61.** neither   **63.** neither
65. even

Solutions to trial exercise problems

4. $f(x) = 5 - x$ is a linear function since the largest power of the variable is one. Use y-intercept 5 and slope $m = -1$.

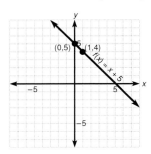

11. $f(x) = 2$ is a constant (linear) function whose graph is a horizontal line through $y = 2$.

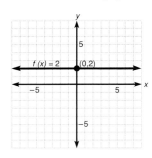

14. $f(x) = (x - 1)^2$ is a quadratic function whose graph is a parabola opening upward since the largest power of the variables is two. Using x-intercept $(1,0)$ and y-intercept $(0,1)$, plot the points and graph the parabola.

x	y
0	1
1	0
2	1
3	4
-1	4

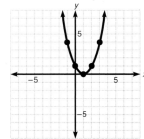

21. $f(x) = -x^2 + 2x + 15$ is a quadratic function whose graph is a parabola opening downward. The y-intercept is 15, x-intercepts are 5 and -3, and the vertex is $(1,16)$.

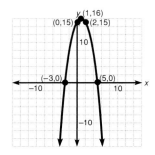

28. $f(x) = \sqrt{2x + 3}$ is a square root function with x-intercept $-\dfrac{3}{2}$, y-intercept $\sqrt{3} \approx 1.7$ and is part of a parabola (above the x-axis) opening to the right.

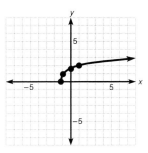

38. For all $x < 3$, we graph the equation $y = 2x + 1$ and for all $x \geq 3$, we graph the equation $y = 4$.
Since there is a break in the graph at $x = 3$, the function is discontinuous at $x = 3$.

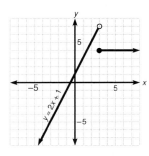

43. For all $x < -1$, we graph $y = 2$; for all x between -1 (inclusive) and 2, we graph $y = x^2 + 1$; for all $x \geq 2$, we graph $y = x$. Since there is a break in the graph at $x = 2$, the function is discontinuous at $x = 2$.

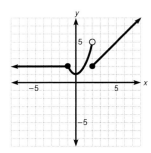

45. $f(x) = |x| - 1$ is an absolute value function with y-intercept -1, x-intercepts 1, -1 and is "V" shaped.

x	y
0	-1
1	0
-1	0
2	1
-2	1

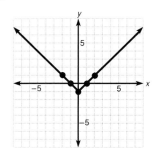

57. Since the graph of f runs from upper left to lower right (as the value of x increases, the value of y decreases), the function is decreasing everywhere, $\{x | x \in R\}$.

63. Since $f(x) = 5x + 1$,
$$f(-x) = 5(-x) + 1 = -5x + 1,$$
and $-f(x) = -(5x + 1) = -5x - 1,$
then $f(x) \neq f(-x)$ and $f(-x) \neq -f(x)$ so f is neither even nor odd.

Review exercises

1. $\dfrac{7(5a - b)}{4}$ **2.** $\{z | z \leq -1 \text{ or } z \geq 3\} = (-\infty, -1] \cup [3, \infty)$

3. $\dfrac{3}{4}$ **4.** $\dfrac{4\sqrt{5}}{5}$ **5.** $30 - 12\sqrt{6}$ **6.** 25

Exercise 8–4

Answers to odd-numbered problems

1. one-to-one function since no two ordered pairs have the same second component   **3.** not a one-to-one function since $(1, -6)$ and $(4, -6)$ have the same second components   **5.** one-to-one function since for every value of x, we obtain only one value of y **7.** not a one-to-one function since $G(1) = -1$ and $G(2) = -1$ **9.** not a one-to-one function since at least two of the ordered pairs have the same second component $(1,2)$ and $(5,2)$ **11.** one-to-one function since $h(x)$ has a different value for every $x \geq 3$ **13.** one-to-one function since any horizontal line drawn in the plane intersects the graph at only one point   **15.** not a one-to-one since a horizontal line intersects the graph in more than one point   **17.** one-to-one function since any horizontal line does not intersect the graph in more than one point   **19.** $f^{-1}(x) = \{(4, -3), (-3, 2), (7, 0)\}$ **21.** $h^{-1}(x) = \dfrac{x}{5}$

23. $G^{-1}(x) = \dfrac{x - 7}{3}$ **25.** $f^{-1}(x) = \sqrt[3]{x - 2}$

27. $h^{-1}(x) = x^2 - 5, x \geq 0$ **29.** $G^{-1}(x) = x^2 - 2, x \geq 0$

31. $f^{-1}(x) = x^3 + 2$ **33.** $h^{-1}(x) = (x - 7)^3$

35.

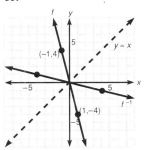

37.

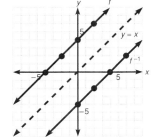

39.

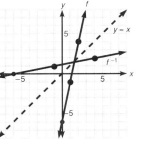

41.

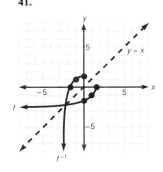

43.

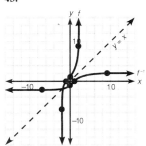

45.

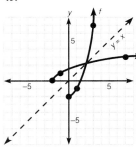

47.

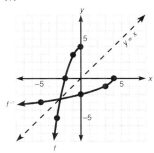

49. a. $f[g(x)] = f\left(\dfrac{x - 7}{4}\right) = 4\left(\dfrac{x - 7}{4}\right) + 7 = x - 7 + 7 = x$

b. $g[f(x)] = g(4x + 7) = \dfrac{(4x + 7) - 7}{4} = \dfrac{4x}{4} = x$

51. a. $f[g(x)] = f(x^2 + 2) = \sqrt{(x^2 + 2) - 2} = \sqrt{x^2} = x$

b. $g[f(x)] = g(\sqrt{x - 2}) = (\sqrt{x - 2})^2 + 2 = x - 2 + 2 = x$

Solutions to trial exercise problems

7. G is not a one-to-one function since $G(1) = 1 - 3(1) + 1 = -1$ and $G(2) = (2)^2 - 3(2) + 1 = 4 - 6 + 1 = -1$. Thus $(1, -1)$ and $(2, -1)$ are points on the graph.   **11.** h is one-to-one since no two ordered pairs have the same second component for $x \geq 3$.

25. Given $f(x) = x^3 + 2$, then $y = x^3 + 2$ and interchanging variables $x = y^3 + 2$, then $y^3 = x - 2$ and $y = \sqrt[3]{x - 2}$. Thus $f^{-1}(x) = \sqrt[3]{x - 2}$.

42. Graph the function $f(x) = \sqrt{x + 2}$. Choose points on this curve for $x \geq -2$ and locate other points on the other side of the line $y = x$, the same distance away from this line. Connect these points to obtain the graph of f^{-1}.

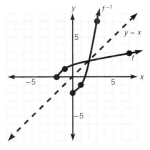

45. Graph the function $f(x) = x^2 - 2$, $x \geq 0$. Choose points on this curve that are on the other side of the line $y = x$, the same distance from this line. Connect these points to obtain the graph of f^{-1}.

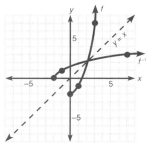

48. $f[g(x)] = f\left(\dfrac{x}{5}\right) = 5\left(\dfrac{x}{5}\right) = x$; $g[f(x)] = g(5x) = \dfrac{5x}{5} = x$

Review exercises

1. $\{2\}$ **2.** $\emptyset$; 3 is extraneous   **3.** $x + 3y = 7$ **4.** $\sqrt{3}$

5. $a = \dfrac{2S - n\ell}{n}$

Exercise 8–5

Answers to odd-numbered problems

1. $S = kt$ **3.** $t = \dfrac{k}{r}$ **5.** $M = kmv$ **7.** $F = kAv^2$

9. $S_{max} = \dfrac{kT}{r^3}$ **11.** $F = \dfrac{kmv}{gt}$ **13.** $p = kh$, $k = 4$

15. $V = ks^3$, $k = 3$ **17.** $s = \dfrac{k}{T}$, $k = 120$

19. $p = \dfrac{kT}{V}$, $k = 20$ **21.** $A = kh(a + b)$, $k = \dfrac{1}{2}$

23. $w = k\ell$, $k = \dfrac{3}{2}$, $w = \dfrac{45}{2}$ **25.** $y = \dfrac{k}{z^2}$, $k = 48$, $y = \dfrac{16}{3}$

27. $v = \dfrac{ks}{t}$, $k = 10$, $v = \dfrac{40}{3}$ **29.** $z = kxy^3$; $z = 324$

31. $V = \dfrac{k}{P}$; $k = 6{,}000$; $P = \$2.00$ **33.** $E = \dfrac{k}{d^2}$, $k = 409.6$,

$E = 11.4$ foot-candles   **35.** $R = \dfrac{k\ell}{d^2}$, $k = 1$, $R = \dfrac{3}{2}$ ohms

37. $F = \dfrac{kwh^2}{\ell}$, $k = 244{,}897.96$, $F = 2{,}204$ lb

Solutions to trial exercise problems

5. Since M varies directly as the product of m and v, $M = kmv$.

8. Since R varies directly as ℓ and inversely as A, $R = \dfrac{k\ell}{A}$.

18. $n = \dfrac{k}{p^2}$, $n = 14$ when $p = 9$, so $14 = \dfrac{k}{(9)^2} = \dfrac{k}{81}$,

$k = 14 \cdot 81 = 1{,}134$.   **27.** $v = \dfrac{ks}{t}$. $v = 20$ when $s = 4$ and

$t = 2$, so $20 = \dfrac{k \cdot 4}{2}$ and $k = 10$. Then $v = \dfrac{10 \cdot s}{t}$ and $s = 8$

when $t = 6$, so $v = \dfrac{10 \cdot 8}{6} = \dfrac{40}{3}$.

Review exercises

1. $\{4\}$; 25 is extraneous   **2.** $x + 4y = 22$ **3.** $\{x \mid -2 < x < 8\}$

4. $C = \dfrac{5}{3}$ **5.** $-\dfrac{3b^5}{2a^5}$

Chapter 8 review

1. not a function since two of the ordered pairs have the same first component $(1,3)$ and $(1,0)$ **2.** function   **3.** function   **4.** function   **5.** function   **6.** not a function since $x = k$; does *not* define a function; (vertical line)   **7.** domain $= \{-3, 4, 0, -2\}$, range $= \{4, 2, 0, 7\}$ **8.** domain $= \{-3, -2, -1, 0, 1, 2, 3\}$, range $= \{13, 9, 5, 1, -3, -7, -11\}$ **9.** domain $= \{-3, 0, 1, 4\}$, range $= \{-1, 2, 23\}$ **10.** domain $= \{-8, -2, 0, 3, 7\}$, range $= \{-7\}$ **11.** domain $= \{x \mid x \in R\} = (-\infty, \infty)$ **12.** domain $= \{x \mid x \in R\} = (-\infty, \infty)$ **13.** domain $= \{x \mid x \in R, x \neq 0\} = (-\infty, 0) \cup (0, \infty)$

14. domain $= \left\{x \mid x \in R, x \neq \dfrac{4}{3}\right\} = \left(-\infty, \dfrac{4}{3}\right) \cup \left(\dfrac{4}{3}, \infty\right)$

15. domain $= \{x \mid x \geq -2\}$ **16.** 0   **17.** -2 **18.** -3
19. $4 = [-2, \infty)$ **20.** 8   **21.** 1   **22.** 1
23. $(f + g)(x) = 3x + 16$, domain $= \{x \mid x \in R\}$;
$(f - g)(x) = -7x - 2$, domain $= \{x \mid x \in R\}$;
$(fg)(x) = -10x^2 + 17x + 63$, domain $= \{x \mid x \in R\}$;
$\left(\dfrac{f}{g}\right)(x) = \dfrac{7 - 2x}{5x + 9}$, domain $= \left\{x \mid x \in R, x \neq -\dfrac{9}{5}\right\}$
24. $(f + g)(x) = 3x^2 + x - 5$, domain $= \{x \mid x \in R\}$;
$(f - g)(x) = 3x^2 - x - 9$, domain $= \{x \mid x \in R\}$;
$(fg)(x) = 3x^3 + 6x^2 - 7x - 14$, domain $= \{x \mid x \in R\}$;
$\left(\dfrac{f}{g}\right)(x) = \dfrac{3x^2 - 7}{x + 2}$, domain $= \{x \mid x \in R, x \neq -2\}$

25. 42   **26.** 8   **27.** -468 **28.** -3 **29.** 72   **30.** $\dfrac{11}{17}$

31. linear

32. constant, linear

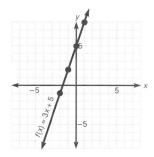

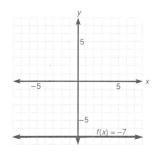

33. quadratic

34. square root

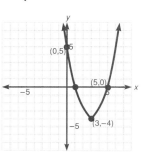

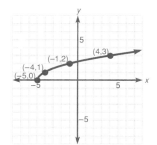

35. absolute value

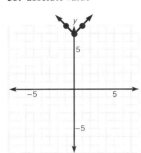

36. piecewise function

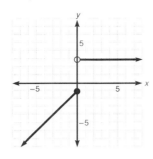

37. decreasing, $\{x|x \leq 3\}$; increasing, $\{x|x \geq 3\}$ **38.** decreasing, $\{x|x \leq -3\}$ and $\{x|x \geq 4\}$; increasing, $\{x|-3 \leq x \leq 4\}$
39. neither   **40.** even   **41.** odd   **42.** not one-to-one function
43. one-to-one function   **44.** one-to-one function   **45.** not one-to-one function   **46.** not one-to-one function   **47.** one-to-one function

48. $f^{-1}(x) = \dfrac{x - 3}{4}$ **49.** $f^{-1}(x) = \sqrt{x + 2}, x \geq -2$

50. $f^{-1}(x) = \dfrac{1}{2}x^2 + \dfrac{3}{2}, x \geq 0$ **51. a.** $f[g(x)]$

$= f\left(\dfrac{5 - x}{2}\right) = 5 - 2\left(\dfrac{5 - x}{2}\right) = 5 - (5 - x) = x$

b. $g[f(x)] = g(5 - 2x) = \dfrac{5 - (5 - 2x)}{2} = \dfrac{2x}{2} = x$

52. a. $f[g(x)] = f\left(\sqrt[3]{\dfrac{x + 4}{3}}\right) = 3\left(\sqrt[3]{\dfrac{x + 4}{3}}\right)^3 - 4$

$= 3\left(\dfrac{x + 4}{3}\right) - 4 = x + 4 - 4 = x$

b. $g[f(x)] = g(3x^3 - 4) = \sqrt[3]{\dfrac{(3x^3 - 4) + 4}{3}} = \sqrt[3]{\dfrac{3x^3}{3}}$

$= \sqrt[3]{x^3} = x$

53.

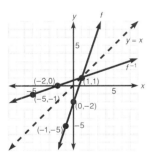

54.

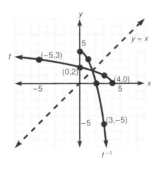

55.

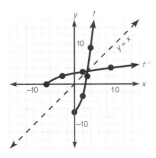

56.

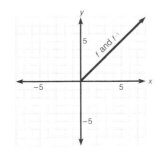

57. $I = \dfrac{k}{R}$ **58.** $x = ky^2$ **59.** $R = \dfrac{k\ell}{A}$ **60.** $H = kvr$

61. $y = \dfrac{k}{x^2}, k = 216$ **62.** $s = kw^3, k = 2$ **63.** $P = kRI^2,$

$k = 3, P = 960$ **64.** $R = \dfrac{k\ell}{A}, k = .00495$ **65.** $d = \dfrac{kab}{c},$

$k = 10, d = 20$

Chapter 8 cumulative test

1. $\dfrac{3}{2}$ **2. a.** -2,   **b.** -14,   **c.** 1 **3.** 204 **4.** $-x^{13}y^{18}$

5. $x^{4n - 3}$ **6.** 1 if $x \neq \dfrac{1}{2}$ **7.** $\dfrac{1}{x^6 y^7}$ **8.** $\dfrac{1}{16}$ **9.** 8 **10.** $\dfrac{1}{16}$

11. $3(a + 3b)(a - 3b)$ **12.** $(x^2 + y^2)(x + y)(y - x)$
13. $5(x - 2)(2x + 3)$ **14.** $(3a - 1)(2a + 3)$
15. $(ab - 1)(2ab + 1)$ **16.** $2(x - 3y)(x^2 + 3xy + 9y^2)$

17. $x = \dfrac{1 + by}{a - 2}$ **18.** $\left\{\dfrac{23}{5}\right\}$ **19.** $\{x|-2 \leq x < 2\}$

20. $\{x|x \leq -2 \text{ or } x \geq 8\}$ **21.** $\left\{\dfrac{15}{7}\right\}$ **22.** $\{3, -1\}$ **23.** $\dfrac{1}{a^{3/8}}$

24. $b^{1/4}$ **25.** $x^2 y^3$ **26.** 23 **27.** -25 **28.** 131

29. $f^{-1}(x) = \dfrac{x - 3}{4}$

30. a. $f[g(x)] = f[\sqrt[3]{x - 1}] = (\sqrt[3]{x - 1})^3 + 1 = x - 1 + 1 = x$
b. $g[f(x)] = g[x^3 + 1] = \sqrt[3]{(x^3 + 1) - 1} = \sqrt[3]{x^3} = x$
31. $(f + g)(x) = x^2 + x + 3; (f - g)(x) = -x^2 + x - 1;$

$(fg)(x) = x^3 + x^2 + 2x + 2; \left(\dfrac{f}{g}\right)(x) = \dfrac{x + 1}{x^2 + 2}$

32. constant, linear   **33.** linear

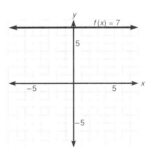

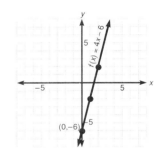

34. quadratic

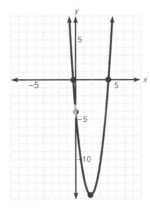

35. square root

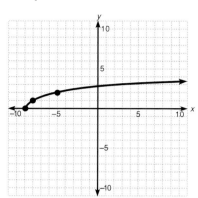

36. absolute value

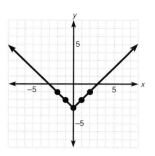

37. piece-wise

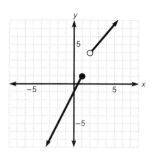

38. $x = \dfrac{ky}{z}$; $k = 25$, $x = 62\dfrac{1}{2}$

39. a. has an inverse; $f^{-1} = \{(1,-2), (4,3), (8,0), (2,-9)\}$
b. has an inverse; $f^{-1} = \sqrt[3]{x+1}$ **c.** has no inverse

Chapter 9

Exercise 9–1

Answers to odd-numbered problems

1. $(3,4)$ **3.** $(0,-16)$ **5.** $(5,0)$ **7.** $(-2,-9)$ **9.** $(1,4)$

11. $\left(\dfrac{7}{4}, -\dfrac{25}{8}\right)$ **13.** no x-intercepts; y-intercept, $(0,13)$

15. x-intercepts, $(-4,0)$, $(4,0)$; y-intercept, $(0,-16)$

17. x-intercept, $(5,0)$; y-intercept, $(0,25)$

19. x-intercepts, $(-5,0)$, $(1,0)$; y-intercept, $(0,-5)$

21. x-intercepts, $(3,0)$, $(-1,0)$; y-intercept, $(0,3)$

23. x-intercepts, $\left(\dfrac{1}{2},0\right)$, $(3,0)$; y-intercept, $(0,3)$

25.

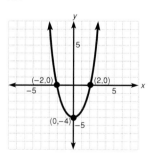

27.

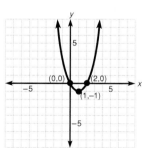

29.

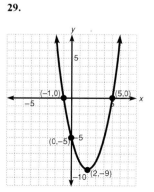

31.

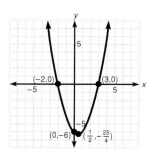

33.

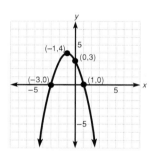

35.

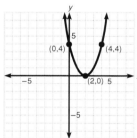

37.

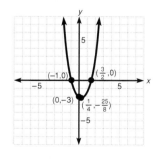

39.

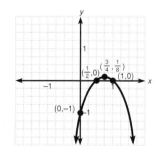

41.

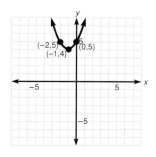

43.

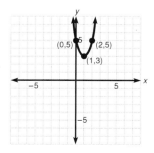

45.

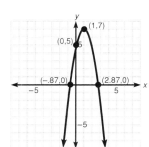

47.

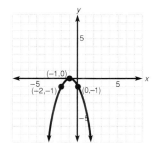

49. 1 second; 2 seconds

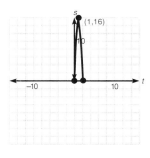

51. The maximum height is 144 feet; the arrow will reach the ground in 6 seconds. **53.** 8 dresses must be sold daily. **55.** 5 glasses of Kool Aid must be sold. **57.** Maximum power $P = 245$ when current $I = 35$. **59.** 28 and 28 **61.** $\ell = w = \dfrac{21}{2}$ or $10\dfrac{1}{2}m$

63. -16

Solutions to trial exercise problems

3. $y = x^2 - 16$; $y = (x - 0)^2 - 16$. The vertex is at $(0,-16)$ and since $a = 1$, $a > 0$, the parabola opens up and the vertex is the lowest point. **9.** $y = -x^2 + 2x + 3$; $x = -\dfrac{b}{2a} = -\dfrac{2}{2(-1)} = 1$; $y = -(1)^2 + 2(1) + 3 = -1 + 2 + 3 = 4$. The vertex is at $(1,4)$ and since $a = -1$, $a < 0$, the parabola opens down. The vertex is the highest point. **11.** $y = 2x^2 - 7x + 3$; $x = -\dfrac{b}{2a} = -\dfrac{-7}{2(2)} = \dfrac{7}{4}$; $y = 2\left(\dfrac{7}{4}\right)^2 - 7\left(\dfrac{7}{4}\right) + 3 = \dfrac{49}{8} - \dfrac{49}{4} + 3 = -\dfrac{49}{8} + \dfrac{24}{8} = -\dfrac{25}{8}$; the vertex is at $\left(\dfrac{7}{4},-\dfrac{25}{8}\right)$. **13.** $y = (x - 3)^2 + 4$ $= x^2 - 6x + 9 + 4 = x^2 - 6x + 13$; let $x = 0$, then $y = 13$ and the y-intercept is $(0,13)$. When $y = 0$, then $0 = x^2 - 6x + 13$ and, using the quadratic formula, $= \dfrac{-(-6) \pm \sqrt{(-6)^2 - 4(1)(13)}}{2(1)}$ $= \dfrac{6 \pm \sqrt{36 - 52}}{2} = \dfrac{6 \pm \sqrt{-16}}{2} = \dfrac{6 \pm 4i}{2} = 3 \pm 2i$. Imaginary x-intercepts mean *no* x-intercept in the real plane. **18.** $y = (x + 6)^2$; when $x = 0$, $y = (0 + 6)^2 = 36$ and the y-intercept is $(0,36)$; when $y = 0$, then $0 = (x + 6)^2$, $x + 6 = 0$, $x = -6$ and the x-intercept is $(-6,0)$. **21.** $y = -x^2 + 2x + 3$; when $x = 0$, then $y = 3$ and the y-intercept is $(0,3)$; when $y = 0$, then $0 = -x^2 + 2x + 3$, $0 = x^2 - 2x - 3 = (x - 3)(x + 1)$. When $x - 3 = 0$, $x = 3$, and when $x + 1 = 0$, $x = -1$. The x-intercepts are $(3,0)$ and $(-1,0)$.

27. a. Let $x = 0$, $y = 0$ and $(0,0)$ is y-intercept
b. Let $y = 0$, $0 = x^2 - 2x$; $0 = x(x - 2)$. x-intercepts are $(0,0)$ and $(2,0)$
c. $x = -\dfrac{b}{2a} = -\dfrac{-2}{2(1)} = 1$
$y = 1^2 - 2(1) = 1 - 2 = -1$
vertex is $(1,-1)$.

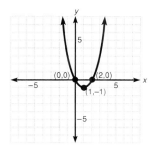

34. a. y-intercept is $(0,-6)$.
b. Let $y = 0$, then
$0 = -x^2 + 7x - 6$
$= x^2 - 7x + 6$
$= (x - 6)(x - 1)$
x-intercepts are $(1,0)$ and $(6,0)$.
c. $x = -\dfrac{b}{2a} = -\dfrac{7}{2(-1)} = \dfrac{7}{2}$
$y = -\left(\dfrac{7}{2}\right)^2 + 7\left(\dfrac{7}{2}\right) - 6$
$= -\dfrac{49}{4} + \dfrac{49}{2} - 6$
$= \dfrac{49}{4} - 6 = \dfrac{25}{4}$
vertex is $\left(\dfrac{7}{2},\dfrac{25}{4}\right)$.

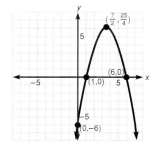

45. a. y-intercept is (0,5).

b. Let $y = 0$.

$$0 = -2x^2 + 4x + 5$$

$$0 = 2x^2 - 4x - 5$$

$$x = \frac{-(-4) \pm \sqrt{(-4)^2 - 4(2)(-5)}}{2(2)}$$

$$= \frac{4 \pm \sqrt{16 + 40}}{4}$$

$$= \frac{4 \pm \sqrt{56}}{4} = \frac{4 \pm 2\sqrt{14}}{4}$$

$$= \frac{2 \pm \sqrt{14}}{2}$$

$$\frac{2 - \sqrt{14}}{2} \approx -0.9; \frac{2 + \sqrt{14}}{2} \approx 2.9$$

The x-intercepts are $(-0.9,0)$ and $(2.9,0)$.

c. $x = -\dfrac{b}{2a} = -\dfrac{4}{2(-2)} = 1$

$y = -2(1)^2 + 4(1) + 5 = 7$

The vertex is (1,7).

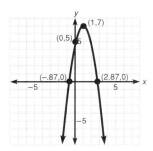

52. $P = -x^2 + 100x - 1,000$

$= -1(x^2 - 100x) - 1,000$

$= -1(x^2 - 100x + 2,500) - 1,000 + 2,500$

$= -1(x - 50)^2 + 1,500$

Since the vertex is at (50;1,500), then 50 units must be produced to attain maximum profit.

59. Let x = one number and $56 - x$ = the other number.

Then $y = x(56 - x) = 56x - x^2$

$= -x^2 + 56x$

$= -1(x^2 - 56x + 784) + 784$

$= -1(x - 28)^2 + 784$

The vertex is at (28,784) and the numbers are 28 and 56 − 28 = 28.

Review exercises

1. $(3y - 1)^2$ **2.** $(2x - 1)(2x + 3)$ **3.** $6(y + 2x)(y - 2x)$

4. $\dfrac{-y^2 + 9y}{(y + 5)(y - 5)}$ **5.** $12xy$ **6.** $\dfrac{x - 3}{x + 4}$

Exercise 9–2

Answers to odd-numbered problems

1. a. vertex, (2,3)   **b.** y-intercepts, none   **c.** x-intercept, (11,0)

3. a. vertex, (0,−5)   **b.** y-intercept, (0,−5)   **c.** x-intercept, (75,0)

5. a. vertex, (2,1)   **b.** y-intercepts, (0,−1) and (0,3)

c. x-intercept, (−3,0)   **7. a.** vertex, $\left(\dfrac{9}{4}, -\dfrac{1}{2}\right)$ **b.** y-intercepts,

(0,−2) and (0,1)   **c.** x-intercept, (2,0)   **9. a.** vertex, $\left(-\dfrac{13}{4}, -\dfrac{1}{4}\right)$

b. y-intercepts, (0,−1.2) and (0,0.7)   **c.** x-intercept, (−3,0)

11. a. vertex, $\left(\dfrac{9}{5}, -\dfrac{2}{5}\right)$ **b.** y-intercepts, (0,−1) and $\left(0,\dfrac{1}{5}\right)$

c. x-intercept, (1,0)

13.

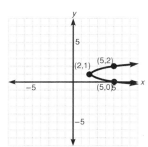

15.

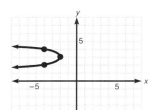

17.

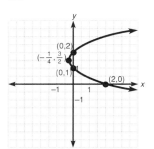

19.

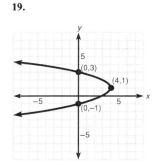

21.

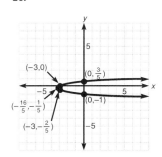

23.

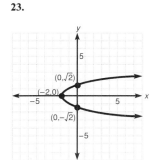

25.

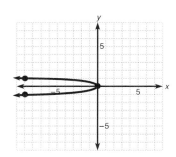

27.

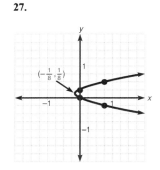

29.

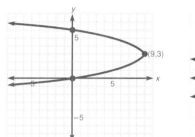

31.

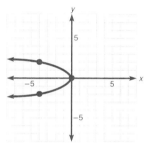

Solutions to trial exercise problems

3. a. Given $x = 3(y + 5)^2$, then $h = 0$ and $k = -5$; the vertex is $(0, -5)$.

  b. Let $x = 0$, then $0 = 3(y + 5)^2$ and $y + 5 = 0$, $y = -5$. The y-intercept is $(0, -5)$.

  c. Let $y = 0$, then $x = 3(0 + 5)^2 = 3(25) = 75$; the x-intercept is $(75, 0)$.

10. a. Let $y = -\dfrac{b}{2a} = -\dfrac{-5}{2(2)} = \dfrac{5}{4}$,

  then $x = 2\left(\dfrac{5}{4}\right)^2 - 5\left(\dfrac{5}{4}\right) + 3$

  $= \dfrac{25}{8} - \dfrac{25}{4} + 3$

  $= -\dfrac{25}{8} + 3 = -\dfrac{1}{8}$

  The vertex is $\left(-\dfrac{1}{8}, \dfrac{5}{4}\right)$.

  b. Let $x = 0$, then $0 = 2y^2 - 5y + 3$; $0 = (2y - 3)(y - 1)$ and the y-intercepts are $\left(0, \dfrac{3}{2}\right)$ and $(0, 1)$.

  c. Let $y = 0$, then $x = 3$; the x-intercept is $(3, 0)$.

17. a. $y = -\dfrac{b}{2a} = -\dfrac{-3}{2(1)} = \dfrac{3}{2}$; $x = \left(\dfrac{3}{2}\right)^2 - 3\left(\dfrac{3}{2}\right) + 2 =$

  $\dfrac{9}{4} - \dfrac{9}{2} + 2 = -\dfrac{1}{4}$; the vertex is $\left(-\dfrac{1}{4}, \dfrac{3}{2}\right)$.

  b. Let $x = 0$, then $0 = y^2 - 3y + 2$; $0 = (y - 2)(y - 1)$. The y-intercepts are $(0, 2)$ and $(0, 1)$.

  c. Let $y = 0$, then $x = 2$; the x-intercept is $(2, 0)$.

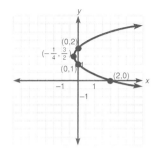

22. a. $x = -2\left(y^2 + \dfrac{5}{2}y + \dfrac{25}{16}\right) + 3 + \dfrac{25}{8}$

  $= -2\left(y + \dfrac{5}{4}\right)^2 + \dfrac{49}{8}$

  The vertex is $\left(\dfrac{49}{8}, -\dfrac{5}{4}\right)$.

  b. Let $x = 0$, then $0 = -2y^2 - 5y + 3$; $0 = 2y^2 + 5y - 3$; $0 = (2y - 1)(y + 3)$. The y-intercepts are $(0, -3)$ and $\left(0, \dfrac{1}{2}\right)$.

  c. Let $y = 0$, then $x = 3$; the x-intercept is $(3, 0)$.

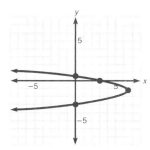

Review exercises

1. $\dfrac{a^6}{8b^9}$ **2.** $\dfrac{1}{8x^{24}}$ **3.** x-intercept, $(4, 0)$; y-intercept, $(0, 2)$

4. x-intercept, $(3, 0)$; y-intercept, $(0, 3)$ **5.** $\sqrt{41}$

6. $\sqrt{(x - 6)^2 + (y + 5)^2}$ **7.** $\sqrt{(x - h)^2 + (y - k)^2}$

Exercise 9–3

Answers to odd-numbered problems

1. $(x - 1)^2 + (y - 2)^2 = 4$ **3.** $(x - 4)^2 + (y + 3)^2 = 6$
5. $x^2 + y^2 = 9$ **7.** $x^2 + y^2 + 10x - 4y + 28 = 0$
9. $x^2 + y^2 = 36$ **11.** $C(3, 2)$, $r = 7$ **13.** $C(5, -3)$, $r = 2\sqrt{2}$
15. $C(-1, -9)$, $r = 5$ **17.** $C(0, 0)$, $r = 6$ **19.** $C(0, 0)$, $r = \sqrt{2}$
21. $C(-2, 3)$, $r = 6$ **23.** $C(-3, 4)$, $r = 5$ **25.** $C(-2, 3)$,
$r = 5\sqrt{2}$ **27.** $C(-3, 1)$, $r = 5$

29.

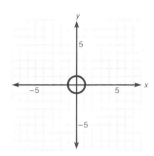

31.

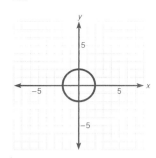

33.

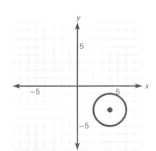

35.

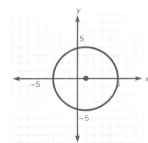

37.

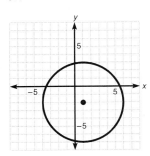

39.

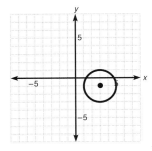

41. $x^2 + y^2 - 10y = 0$ **43.** $x^2 + y^2 + 4y - 6x = 0$
45. straight line   **47.** parabola

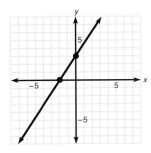

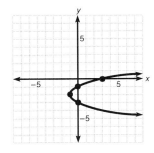

Solutions to trial exercise problems

7. $(x + 5)^2 + (y - 2)^2 = 1^2$; $(x^2 + 10x + 25) + (y^2 - 4y + 4)$
$= 1$; $x^2 + y^2 + 10x - 4y + 28 = 0$ **21.** $x^2 + y^2 + 4x - 6y -$
$23 = 0$; $(x^2 + 4x + 4) + (y^2 - 6y + 9) = 23 + 4 + 9$; $(x + 2)^2$
$+ (y - 3)^2 = 36$; $C(-2,3)$, $r = 6$ **25.** $2x^2 + 2y^2 + 8x - 12y =$
74; $x^2 + y^2 + 4x - 6y = 37$; $(x^2 + 4x + 4) + (y^2 - 6y + 9) =$
$37 + 4 + 9$; $(x + 2)^2 + (y - 3)^2 = 50$; $C(-2,3)$, $r = 5\sqrt{2}$
31. $3x^2 + 3y^2 = 12$; $x^2 + y^2 = 4$; $C(0,0)$, $r = 2$

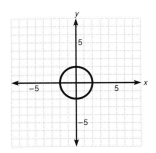

39. $2x^2 + 2y^2 - 12x + 4y = -12$; $x^2 + y^2 - 6x + 2y = -6$;
$(x^2 - 6x + 9) + (y^2 + 2y + 1) = -6 + 9 + 1$; $(x - 3)^2 +$
$(y + 1)^2 = 4$; $C(3,-1)$, $r = 2$

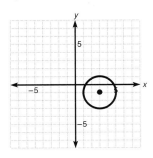

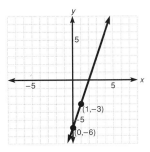

2. $-2y\sqrt{2xy}$ **3.** 14   **4.** $[-2,4)$ **5.** $3x + 2y = 4$

Exercise 9–4

Answers to odd-numbered problems

1. x, $(\pm 3,0)$; y, $(0,\pm 5)$ **3.** x, $(\pm 7,0)$; y, $(0,\pm 5)$

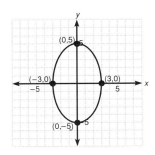

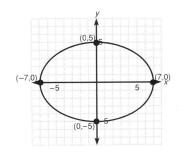

5. x, $(\pm 1,0)$; y, $(0,\pm 3)$ **7.** x, $(\pm 4,0)$; y, $(0,\pm 1)$

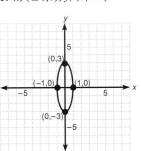

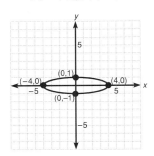

9. x, $(\pm 10,0)$; y, $(0,\pm 2)$ **11.** x, $(\pm 2,0)$; y, $(0,\pm \sqrt{3})$

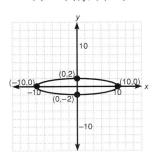

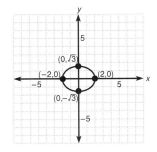

13. x, $(\pm\sqrt{2},0)$; y, $(0,\pm4)$

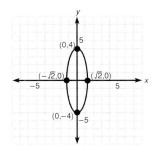

15. $\dfrac{(x-1)^2}{4} + \dfrac{(y-2)^2}{9} = 1$

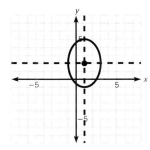

17. $4x^2 + y^2 + 8x - 10y - 7 = 0$

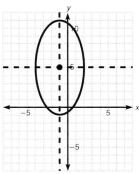

19. $2x^2 + y^2 + 4x + 2y - 1 = 0$

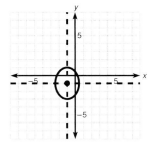

21. x, $(\pm4,0)$; asymptotes, $y = \pm\dfrac{3}{4}x$

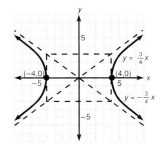

23. y, $(0,\pm3)$; asymptotes, $y = \pm\dfrac{3}{2}x$

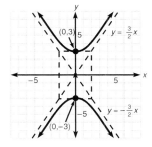

25. y, $(0,\pm1)$; asymptotes, $y = \pm\dfrac{1}{3}x$

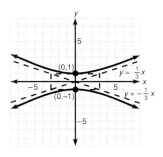

27. x, $(\pm1,0)$; asymptotes, $y = \pm4x$

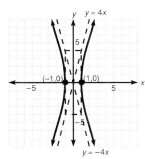

29. x, $(\pm1,0)$; asymptotes, $y = \pm5x$

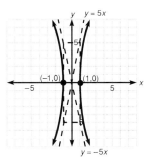

31. y, $(0,\pm4)$; asymptotes, $y = \pm\dfrac{4}{5}x$

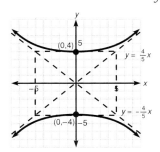

33. y, $(0,\pm5)$; asymptotes, $y = \pm\dfrac{5}{2}x$

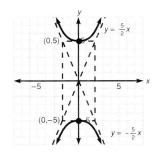

35. y, $(0,\pm2)$; asymptotes, $y = \pm\dfrac{\sqrt{2}}{3}x$

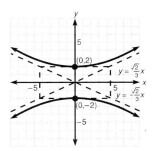

Answers to Exercise 9–4

37. $\dfrac{(x-1)^2}{9} - \dfrac{(y+3)^2}{25} = 1$

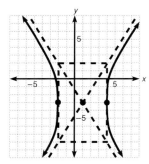

39. $x^2 - y^2 - 6x - 4y + 4 = 0$

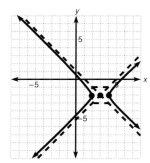

41. $y^2 - 4x^2 - 8x - 6y = 11$

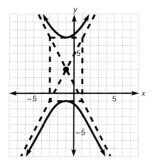

43. $x^2 + y^2 = 9$; circle **45.** $x^2 + 3 = y$; parabola

47. $x^2 + y^2 = \dfrac{25}{4}$; circle **49.** $x^2 + y^2 = 121$; circle

51. $y = -x^2 + 8$; parabola **53.** $\dfrac{(x-1)^2}{5} + \dfrac{(y+2)^2}{5} = 1$; ellipse

55. $\dfrac{(x-1)^2}{4} - \dfrac{(y+3)^2}{16} = 1$; hyperbola

Solutions to trial exercise problems

5. $x^2 + \dfrac{y^2}{9} = 1$; $b^2 = 1$, so $b = 1$; $a^2 = 9$, so $a = 3$; x-intercepts,
$(1,0)$, $(-1,0)$; y-intercepts, $(0,3)$, $(0,-3)$

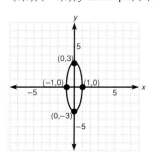

8. $36x^2 + 9y^2 = 324$; $\dfrac{x^2}{9} + \dfrac{y^2}{36} = 1$; $b^2 = 9$, $b = 3$; $a^2 = 36$, $a = 6$;
x-intercepts, $(3,0)$, $(-3,0)$; y-intercepts, $(0,6)$, $(0,-6)$

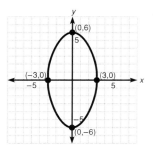

18. $x^2 + 2y^2 - 8x - 4y + 14 = 0$
$(x^2 - 8x + 16) + 2(y^2 - 2y + 1) = -14 + 16 + 2$
$(x - 4)^2 + 2(y - 1)^2 = 4$
$\dfrac{(x-4)^2}{4} + \dfrac{(y-2)^2}{2} = 1$
$C(4,2)$; $a^2 = 4$, $a = 2$, $b^2 = 2$, $b = \sqrt{2} \approx 1.4$

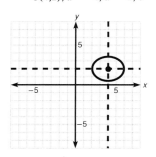

25. $y^2 - \dfrac{x^2}{9} = 1$; $a^2 = 9$, $a = 3$; $b^2 = 1$, $b = 1$; y-intercepts, $(0,1)$,
$(0,-1)$; asymptotes, $y = \dfrac{1}{3}x$ and $y = -\dfrac{1}{3}x$

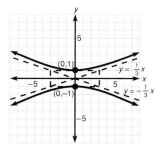

28. $9x^2 - y^2 = 36$; $\dfrac{x^2}{4} - \dfrac{y^2}{36} = 1$; $a^2 = 4$, $a = 2$; $b^2 = 36$, $b = 6$;
x-intercepts, $(2,0)$, $(-2,0)$; asymptotes, $y = 3x$ and $y = -3x$

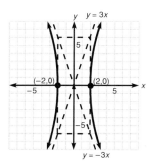

39. $x^2 - y^2 - 6x - 4y + 4 = 0$

$(x^2 - 6x + 9) - (y^2 + 4y + 4) = -4 + 9 - 4$

$(x - 3)^2 - (y + 2)^2 = 1$

$C(3, -2); a^2 = 1, a = 1, b^2 = 1, b = 1$

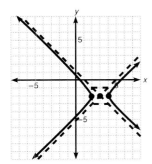

45. Since $x^2 + 3 = y$ has one variable linear and the other squared, it is a parabola.

Review exercises

1. $\dfrac{y + 1}{y - 3}$ **2.** $\dfrac{-2z(5z + 8)}{(3z + 2)(2z - 1)}$ **3.** $\dfrac{xy}{x + y}$

4. $\{-\sqrt{5}, \sqrt{5}\}$ **5.** $\left\{0, \dfrac{4}{5}\right\}$ **6.** $-2\sqrt{3} - 3\sqrt{2}$

7. $27 + \sqrt{15}$

Chapter 9 review

1. vertex, $(-1, 0)$; x-intercept, -1; y-intercept, -3

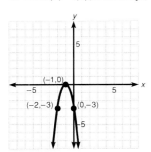

2. vertex, $(2, 9)$; x-intercept, -1 and 5; y-intercepts, 5

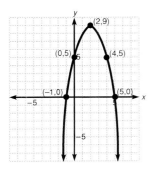

3. vertex, $(0, -2)$; x-intercepts, $-\sqrt{2}$ and $\sqrt{2}$; y-intercept, -2

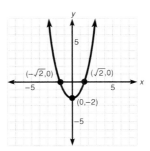

4. vertex, $\left(-\dfrac{3}{2}, \dfrac{25}{4}\right)$; x-intercepts, -4 and 1; y-intercept, 4

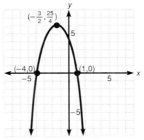

5. vertex, $\left(-\dfrac{3}{2}, -\dfrac{5}{4}\right)$; x-intercepts, -2.6 and -0.5; y-intercept, 1

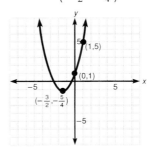

6.

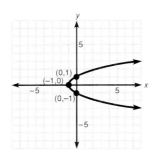

7.

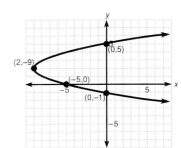

8.

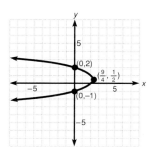

9. a. $(x + 2)^2 + (y - 5)^2 = 25$ **b.** $x^2 + y^2 + 4x - 10y + 4 = 0$ **10. a.** $(x - 4)^2 + y^2 = 11$ **b.** $x^2 + y^2 - 8x + 5 = 0$
11. a. $x^2 + y^2 = 81$ **b.** $x^2 + y^2 - 81 = 0$ **12.** $C(5,-1)$, $r = 6$ **13.** $C(0,0)$, $r = \sqrt{7}$ **14.** $C(-3,4)$, $r = 2\sqrt{6}$
15. $C(1,-2)$, $r = 2\sqrt{2}$ **16.** $C(-1,2)$, $r = \sqrt{10}$
17. $C(0,2)$, $r = \sqrt{19}$

18.

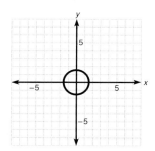

19.

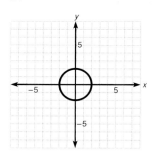

20.

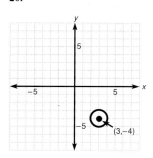

21.

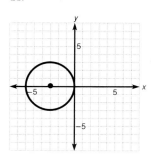

22.

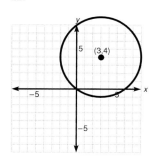

23.

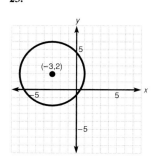

24. x, $(\pm 4,0)$; y, $(0,\pm 10)$

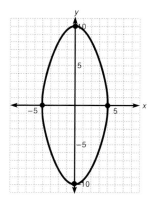

25. x, $(\pm 1,0)$; y, $(0,\pm 5)$

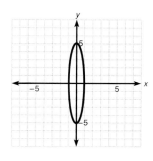

26. x, $(\pm 3,0)$; y, $(0,\pm 4)$

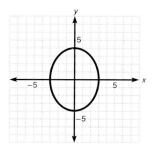

27. x, $(\pm 2,0)$; y, $(0,\pm \sqrt{2})$

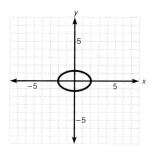

28. x, $(\pm 2,0)$; y, $(0,\pm 2\sqrt{6})$

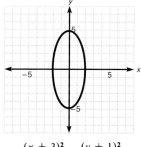

29. x, $(\pm 2\sqrt{2},0)$; y, $(0,\pm 1)$

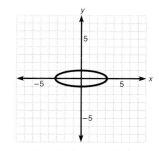

30. $\dfrac{(x + 3)^2}{9} + \dfrac{(y + 1)^2}{4} = 1$

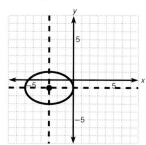

31. $\dfrac{(x - 1)^2}{16} + \dfrac{(y - 2)^2}{25} = 1$

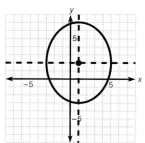

32. $9x^2 + 4y^2 + 16y - 54x + 61 = 0$

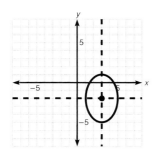

33. $4x^2 + y^2 + 8x - 6y + 9 = 0$

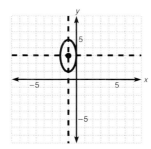

34. x, $(\pm 6,0)$; asymptotes, $y = \pm\dfrac{7}{6}x$

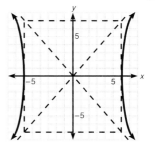

35. y, $(0,\pm 5)$; asymptotes, $y = \pm\dfrac{5}{3}x$

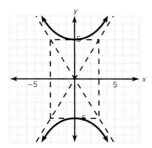

36. x, $(\pm 1,0)$; asymptotes, $y = \pm 6x$

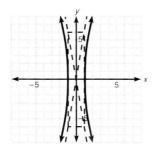

37. y, $(0,\pm 1)$; asymptotes, $y = \pm\dfrac{1}{3}x$

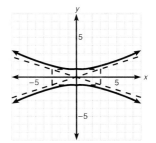

38. x, $(\pm 5,0)$; asymptotes, $y = \pm\dfrac{4}{5}x$

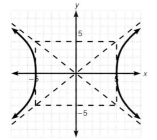

39. y, $(0,\pm 1)$; asymptotes, $y = \pm\dfrac{1}{3}x$

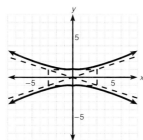

40. $\dfrac{(x-2)^2}{4} - \dfrac{(y+3)^2}{16} = 1$

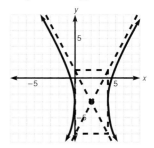

41. $\dfrac{(y-1)^2}{25} - \dfrac{(x+2)^2}{9} = 1$

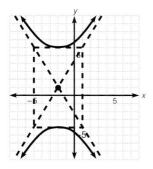

42. $x^2 - y^2 - 6x - 4y + 4 = 0$

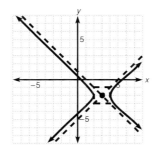

43. $4x^2 - y^2 - 8x + 4y = 4$

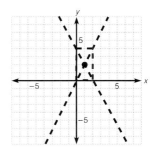

44. hyperbola   **45.** parabola   **46.** ellipse
47. circle   **48.** circle   **49.** parabola
50. parabola   **51.** hyperbola   **52.** ellipse

Chapter 9 cumulative test

1. 13   **2.** 21   **3.** 57   **4.** $C = 100$ **5.** 4th   **6.** $6x^2 + 7x - 14$

7. $2x^2 - x + 3$ **8.** $7x - 5$ **9.** 17   **10.** -17 **11.** $\dfrac{49}{16}$

12. x^{4n+1} **13.** $6a^3 + 5a^2 - 6a + 1$ **14.** $4ab^2c^4 - 5b^2c^2$

15. $3x^2 - 10x + 15 + \dfrac{-34}{x+2}$ **16.** $\dfrac{2x^2 - 2x - 48}{x - 5}$ **17.** $\dfrac{4 - b}{2b + 1}$

18. $\dfrac{x+2}{x+3}$ **19.** $\dfrac{1}{2}$ **20.** 1   **21.** ϕ; 3 is extraneous

22. $\left\{\dfrac{1 + i\sqrt{5}}{3}, \dfrac{1 - i\sqrt{5}}{3}\right\}$ **23.** $\left\{\dfrac{5}{2}, -1\right\}$

24. $\{\sqrt{5}, -\sqrt{5}, i\sqrt{2}, -i\sqrt{2}\}$ **25.** $\{4\}$

26.

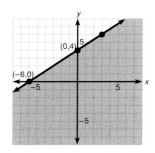

27. $m = -\dfrac{2}{3}$

28. $m = \dfrac{3}{5}$ **29.** neither   **30.** $y = 7$ **31. a.** -13

b. $x^2 + 2x$ **c.** $2x^2 + 5$ **d.** 2   **32.** $f^{-1}(x) = \dfrac{x+1}{8}$

33. $k = \dfrac{20}{9}$; $x = \dfrac{980}{9}$ **34.** $C(3, -2)$, $r = 4$ **35.** vertex, $(1,1)$;

y-intercept, 3; no x-intercepts   **36.** x-intercepts, 2 and -2;
y-intercepts, $\sqrt{6}$ and $-\sqrt{6}$ **37.** x-intercepts, 3 and -3;

asymptotes, $y = \pm\dfrac{2}{3}x$ **38.** $C(1, -1)$ **39.** parabola

40. hyperbola, $\dfrac{y^2}{2} - \dfrac{x^2}{4} = 1$ **41.** circle, $x^2 + y^2 = 4$

42. ellipse   **43.** ellipse

Chapter 10

Exercise 10–1

Answers to odd-numbered problems

1. $P(x) = x^3 - 2x^2 + x - 2$ **3.** $P(x) = x^3 - x^2 - 7x + 15$
5. $P(x) = x^3 - 7x^2 + 19x - 13$
7. $P(x) = x^4 - 10x^3 + 35x^2 - 50x + 34$
9. $P(x) = x^4 - 6x^3 + 26x^2 - 46x + 65$ **11.** $P(x) = x^5 - 5x^4$
$+ 9x^3 - 3x^2 - 8x + 10$ **13.** $P(x) = x^4 - 11x^3 + 51x^2 - 91x$
$+ 50$ **15.** $P(x) = x^3 - 6x^2 + 6x + 8$ **17.** $P(x) = x^5 - 7x^4$
$+ 14x^3 - 6x^2 - 3x + 1$ **19.** $P(x) = x^4 - 2x^3 - 14x^2$
$+ 78x + 225$ **21.** 1,2,4   **23.** $-3 - i, -3 + i, 5$

25. $1, \dfrac{-2 + \sqrt{2}}{2}, \dfrac{-2 - \sqrt{2}}{2}$ **27.** $-2 + i, -2 - i, -3$ of

multiplicity 2   **29.** $3i, -3i, -3, -5$ **31.** $4 + 7i$,

$4 - 7i, -\dfrac{1}{2}, 1$ **33.** 1 positive real zero; 2 or 0 negative real zeros

35. 2 or 0 positive real zeros; 1 negative real zero   **37.** 3 or 1
negative real zeros; no positive real zeros   **39.** 3 or 1 positive real
zeros; 1 negative real zero   **41.** 2 or 0 positive real zeros; no
negative real zeros

43.

Positive	Negative	Nonreal complex
2	1	0
0	1	2

45. 4 nonreal complex solutions   **47.** 7 is an upper bound. -1 is a
lower bound.   **49.** 5 is an upper bound. -1 is a lower bound.
51. 2 is an upper bound. -2 is a lower bound.   **53.** 3 is an upper
bound. -4 is a lower bound.   **55.** 3 is an upper bound. -3 is a
lower bound.   **57.** 9   **59.** 5

Solutions to trial exercise problems

3. The zeros are $2 + i$, $2 - i$, and -3 so we multiply $P(x)$
$= (x + 3)[x - (2 + i)][x - (2 - i)] = (x + 3)(x^2 - 4x + 5)$
$= x^3 - x^2 - 7x + 15$ **12.** The zeros are $4 + 3i$, $4 - 3i$, $2 - i$,
$2 + i$, and 3. $P(x) = (x - 3)[x - (4 + 3i)][x - (4 - 3i)]$
$[x - (2 - i)][x - (2 + i)] = (x - 3)(x^2 - 8x + 25)(x^2 - 4x$
$+ 5) = (x - 3)(x^4 - 12x^3 + 62x^2 - 140x + 125) = x^5 - 15x^4$
$+ 98x^3 - 326x^2 + 545x - 375$ **15.** The zeros are $1 - \sqrt{3}$,
$1 + \sqrt{3}$, and 4. $P(x) = (x - 4)[x - (1 - \sqrt{3})][x - (1 + \sqrt{3})]$
$= (x - 4)(x^2 - 2x - 2) = x^3 - 6x^2 + 6x + 8$ **24.** Since $5 -$
$2i$ is a zero, then $5 + 2i$ is also a zero. Multiply $[x - (5 - 2i)][x -$
$(5 + 2i)] = x^2 - 10x + 29$. Now divide $(x^3 - 13x^2 + 59x - 87)$
$\div (x^2 - 10x + 29) = x - 3$. The zeros are $5 - 2i$, $5 + 2i$, and 3.
27. Since $-2 + i$ is a zero, then $-2 - i$ is a zero. Multiply
$[x - (-2 + i)][x - (-2 - i)] = x^2 + 4x + 5$. Divide
$(x^4 + 10x^3 + 38x^2 + 66x + 45) \div (x^2 + 4x + 5) = x^2 + 6x + 9$.
Now $x^2 + 6x + 9 = (x + 3)^2$ so the zeros are $-2 + i$, $-2 - i$,
-3 with multiplicity 2.
35. $P(x) = x^3 - 5x^2 - 6x + 10 \rightarrow$ There are 2 or 0 positive real
zeros.

   1 change   2 change
$P(-x) = -x^3 - 5x^2 + 6x + 10 \rightarrow$ There is 1 negative real
          zero.

        1 change
38. $P(x) = -x^3 - 4x^2 - 3x - 2 \rightarrow$ No positive real zeros since no
          change of signs
$P(-x) = x^3 - 4x^2 + 3x - 2 \rightarrow$ 3 or 1 negative real zeros

      1   2   3
43. $P(x) = 4x^3 + 2x^2 - x + 1 \rightarrow$ 2 or 0 positive real solutions

        1   2
$P(-x) = -4x^3 + 2x^2 + x + 1 \rightarrow$ 1 negative real solution

         1
Possible combinations:

Positive	Negative	Nonreal complex
2	1	0
0	1	2

49. Using zero 5, 5

$$\begin{array}{r|rrrr} & 1 & -4 & -5 & 7 \\ & & 5 & 5 & 0 \\ \hline & 1 & 1 & 0 & 7 \end{array}$$ ← all nonnegative

5 is an upper bound.

Using $-2, -2$

$$\begin{array}{r|rrrr} & 1 & -4 & -5 & 7 \\ & & -2 & 12 & -14 \\ \hline & 1 & -6 & 7 & -7 \end{array}$$ ← alternating signs

-2 is a lower bound.

52. Using zero 2, 2

$$\begin{array}{r|rrrrr} & 3 & -4 & -5 & -2 & -4 \\ & & 6 & 4 & 18 & 32 \\ \hline & 3 & 2 & 9 & 16 & 28 \end{array}$$ ← all positive

2 is an upper bound.

Using zero $-1, -1$

$$\begin{array}{r|rrrrr} & 3 & -4 & 5 & -2 & -4 \\ & & -3 & 7 & -12 & 14 \\ \hline & 3 & -7 & 12 & -14 & 10 \end{array}$$ ← signs alternate

-1 is a lower bound.

Review exercises

1. $\left\{-\dfrac{1}{2}, \dfrac{5}{2}\right\}$ **2.** $\left\{\dfrac{-5 - \sqrt{5}}{2}, \dfrac{-5 + \sqrt{5}}{2}\right\}$

3. $\left\{\dfrac{1 - i\sqrt{31}}{4}, \dfrac{1 + i\sqrt{31}}{4}\right\}$ **4.** $P(-2) = -27$

5. $P(4) = 247$ **6.** $f^{-1}(x) = \sqrt[3]{x + 4}$

Exercise 10-2

Answers to odd-numbered problems

1. $\pm 1, \pm 2$ **3.** $\pm 1, \pm 2, \pm 3, \pm 4, \pm 6, \pm 8, \pm 12, \pm 24$

5. $\pm 1, \pm\dfrac{1}{4}, \pm\dfrac{1}{2}, \pm 2, \pm 4, \pm 8$ **7.** $-3, -2, 4$ **9.** $1,2,3,4$

11. $-1, \dfrac{3}{4}, -i, i$ **13.** $\dfrac{1}{2}, -i, i,$ 3 of multiplicity 2

15. $\{1, -2, 3\}$ **17.** $\{-1\}$ **19.** $\{-1, 1, 2\}$

21. $\left\{2, -3, \dfrac{5}{2}\right\}$ **23.** $\left\{-\dfrac{3}{2}, \dfrac{1}{2}, 3\right\}$

25. $\left\{-1, -\dfrac{1}{3}, 2, 3\right\}$ **27.** $\left\{1, -\dfrac{3}{4}, i\sqrt{2}, -i\sqrt{2}\right\}$

29. $\left\{3, \dfrac{1}{2}, -\dfrac{1}{2}, -4, -2\right\}$ **31.** $\{-1\}$ **33.** Given $\sqrt{2}$ is a solution, then $x^2 - 2 = 0$, and, the solution set is $\{-\sqrt{2}, \sqrt{2}\}$. However, by the theorem on rational zeros, the possible rational solutions are ± 1 and ± 2. Since $\sqrt{2}$ is a solution but is not one of the possible rational solutions, $\sqrt{2}$ is not rational.   **35.** Given $1 + \sqrt{2}$ is a solution, then $1 - \sqrt{2}$ is a solution and $[x - (1 + \sqrt{2})][x - (1 - \sqrt{2})] = x^2 - 2x - 1 = 0$. By the rational zeros theorem, the possible rational solutions are ± 1. Since $(1 + \sqrt{2})$ is not one of these possible rational solutions, $1 + \sqrt{2}$ is not rational.   **37.** Given $x^n + c = 0$, n is even, $n > 0$ and $c > 0$, then $x^n = -c$ $(-c < 0)$ and $x = \pm\sqrt[n]{-c}$. But $\sqrt[n]{-c}$ does not exist in R when n is even and $n > 0$. Thus, $x^n + c = 0$, has no real solutions when $n > 0$, $c > 0$, and n is even.

39. Given $P(x) = x^4 + 3x^2 + 2$, since there are no changes in signs, then there are no positive real solutions. $P(-x) = x^4 + 3x^2 + 2$ does not change signs, so there are no negative real solutions. Thus, there are no real solutions.

Solutions to trial exercise problems

4. Given $P(x) = 2x^4 - 3x^3 + 2x^2 - 3x - 3$, then $a_0 = -3$ and $a_n = 2$. Possible values of a_0 are ± 1 and ± 3 and possible values of a_n are $\pm 1, \pm 2$. The possible rational zeros are $\pm 1, \pm 3, \pm\dfrac{3}{2}, \pm\dfrac{1}{2}$.

11. Given $P(x) = 4x^4 + x^3 + x^2 + x - 3$, the possible rational zeros are $\pm 1, \pm\dfrac{1}{4}, \pm\dfrac{1}{2}, \pm 3, \pm\dfrac{3}{4}; \pm\dfrac{3}{2}$. By Descarte's rule of signs, there is 1 positive real zero and 3 or 1 negative real zeros. Consider 1 as a zero.

$$\begin{array}{r|rrrrr} 1 & 4 & 1 & 1 & 1 & -3 \\ & & 4 & 5 & 6 & 7 \\ \hline & 4 & 5 & 6 & 7 & 4 \end{array}$$ ← 1 is not a zero, 1 is an upper bound

Try $\dfrac{3}{4}$ as a zero.

$$\begin{array}{r|rrrrr} \dfrac{3}{4} & 4 & 1 & 1 & 1 & -3 \\ & & 3 & 3 & 3 & 3 \\ \hline & 4 & 4 & 4 & 4 & 0 \end{array}$$ ← $\dfrac{3}{4}$ is a zero

Next, try -1 as a zero using $4x^3 + 4x^2 + 4x + 4$.

$$\begin{array}{r|rrrr} -1 & 4 & 4 & 4 & 4 \\ & & -4 & 0 & -4 \\ \hline & 4 & 0 & 4 & 0 \end{array}$$ ← -1 is a zero

We now have $4x^2 + 4 = 4(x^2 + 1) = 0$.
Given $x^2 + 1 = 0$,
$x^2 = -1$
$x = \pm\sqrt{-1} = \pm i$
The zeros are $-1, \dfrac{3}{4}, -i, i$.

17. $x^4 + 4x^3 + 6x^2 + 4x + 1 = 0$. Consider -1 as a possible solution.

$$\begin{array}{r|rrrrr} -1 & 1 & 4 & 6 & 4 & 1 \\ & & -1 & -3 & -3 & -1 \\ \hline -1 & 1 & 3 & 3 & 1 & 0 \\ & & -1 & -2 & -1 & \\ \hline -1 & 1 & 2 & 1 & 0 & \\ & & -1 & -1 & & \\ \hline & 1 & 1 & 0 & & \end{array}$$

← -1 is a solution
← -1 is a solution
← -1 is a solution

We now have $x + 1 = 0$, so $x = -1$. Thus, -1 is a solution of multiplicity 4 and the solution set is $\{-1\}$.   **22.** $2x^3 - 5x^2 + 8x - 3 = 0$. The possible rational solutions are $\pm 1, \pm\dfrac{1}{2}, \pm 3, \pm\dfrac{3}{2}$ and there are 3 or 1 positive real solutions and no negative real zeros. Using 1 as a possible solution,

$$\begin{array}{r|rrrr} 1 & 2 & -5 & 8 & -3 \\ & & 2 & -3 & 5 \\ \hline & 2 & -3 & 5 & 2 \end{array}$$ ← 1 is not a solution

Using possible solution $\dfrac{1}{2}$,

$$
\begin{array}{r|rrrr}
\dfrac{1}{2} & 2 & -5 & 8 & -3 \\
 & & 1 & -2 & 3 \\
\hline
 & 2 & -4 & 6 & \boxed{0}
\end{array}
$$
$\leftarrow \dfrac{1}{2}$ is a solution

we now have $2x^2 - 4x + 6 = 0$
$2(x^2 - 2x + 3) = 0$.

Using the quadratic formula,
$$x = \dfrac{2 \pm \sqrt{4 - 12}}{2} = \dfrac{2 \pm 2i\sqrt{2}}{2} = 1 \pm i\sqrt{2}$$
The solution set is $\left\{\dfrac{1}{2}, 1 + i\sqrt{2}, 1 - i\sqrt{2}\right\}$.

27. $4x^4 - x^3 + 5x^2 - 2x - 6 = 0$. The possible rational solutions
are $\pm 1, \pm \dfrac{1}{4}, \pm \dfrac{1}{2}, \pm 2, \pm 3, \pm \dfrac{3}{4}, \pm \dfrac{3}{2}, \pm 6$. There are 3 or 1 positive
real solutions and 1 negative real solution. Using possible solution 1,
trial and error leads to $-\dfrac{3}{4}$.

$$
\begin{array}{r|rrrrr}
1 & 4 & -1 & 5 & -2 & -6 \\
 & & 4 & 3 & 8 & 6 \\
\hline
-\dfrac{3}{4} & 4 & 3 & 8 & 6 & \boxed{0} \\
 & & -3 & 0 & -6 & \\
\hline
 & 4 & 0 & 8 & \boxed{0}
\end{array}
$$
$\leftarrow 1$ is a solution

$\leftarrow -\dfrac{3}{4}$ is a solution

We now have $4x^2 + 8 = 0$
$4(x^2 + 2) = 0$
$x^2 + 2 = 0$, so $x^2 = -2$ and $x = \pm\sqrt{-2} = \pm i\sqrt{2}$.
The solution set is $\left\{1, -\dfrac{3}{4}, -i\sqrt{2}, i\sqrt{2}\right\}$.

Review exercises

1. $\{x | -2 \le x \le 3\}$ **2.** $\{x | x \le 4\}$ **3.** $\{x | x > -6\}$ **4.** 16 **5.** 2
6. 4

7.

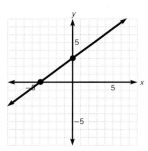

8.

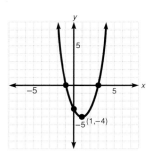

Exercise 10–3

Answers to odd-numbered problems

1. $P(-2) = 3$ and $P(-1) = -2$ so there is an r such that
$r \in [-2, -1]$. **3.** $P(1) = -4$ and $P(2) = 5$ so there is an r such
that $r \in [1, 2]$. **5.** $P(0) = -5$ and $P(1) = 1$ so there is an r such
that $r \in [0, 1]$. **7.** $P(-2) = -1$ and $P(-1) = 3$ so there is an r
such that $r \in [-2, -1]$. **9.** $P(2.1) = -0.73$ and $P(2.2) = 1.65$ so
there is an r such that $r \in [2.1, 2.2]$. **11.** $P(-1) = -31$ and $P(0)$
$= 16$ so there is an r such that $x \in [-1, 0]$. **13.** -1.13 **15.** 0.88
17. -2.00 **19.** -0.63 **21.** 2.63 **23.** 2.88 **25. a.** 3 or 1
positive real zeros, 0 negative real zeros **b.** 0.875 is the positive real
zero.

Solutions to trial exercise problems

2.
$$
\begin{array}{r|rrrr}
-1 & 2 & 17 & 31 & -20 \\
 & & -2 & -15 & -16 \\
\hline
 & 2 & 15 & 16 & -36
\end{array}
$$
$P(-1) = -36$

$$
\begin{array}{r|rrrr}
2 & 2 & 17 & 31 & -20 \\
 & & 4 & 42 & 146 \\
\hline
 & 2 & 21 & 73 & 126
\end{array}
$$
$P(2) = 126$

Since $P(-1) = -36$ and $P(2) = 126$ there exists r such that $r \in [-1, 2]$.
9. $P(2.1) = (2.1)^5 - 6(2.1)^3 + 14 = 40.84 - 55.57 + 14 = -0.73$
$P(2.2) = (2.2)^5 - 6(2.2)^3 + 14 = 51.84 - 63.89 + 14 = 1.65$
Since $P(2.1) = -0.73$ and $P(2.2) = 1.65$, then there exists r such that $r \in [2.1, 2.2]$.
15. $P(0) = -5$ $P(0.5) = -3.625$ $P(0.75) = -1.86$
$ P(0.5) = -3.625$ $P(0.75) = -1.86$ $P(0.875) = -0.59$
$ P(1) = 1$ $P(1) = 1$ $P(1) = 1$
The zero is approximately 0.88 after three subdivisions since -0.59 is closer to 0 than 1.
22. We know that $\sqrt[3]{10}$ is a zero of $P(x) = x^3 - 10$. Now $P(2) = (2)^3 - 10 = -2$
and $P(3) = (3)^3 - 10 = 17$, thus there is a zero in the interval $[2, 3]$
$P(2) = -2$ $P(2) = -2$ $P(2) = -2$
$P(2.5) = 5.625$ $P(2.25) = 1.39$ $P(2.125) = -0.40$
$P(3) = 17$ $P(2.5) = 5.625$ $P(2.25) = 1.39$
Thus, $\sqrt[3]{10} \approx 2.125$ after three subdivisions since -0.40 is closer to 0 than 1.39.

Review exercises

1.

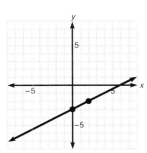

2.

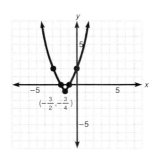

$(-\frac{3}{2},-\frac{3}{4})$

3. x-intercept, $(-4,0)$; y-intercept, $(0,10)$ **4.** x-intercepts, $\left(\frac{1-\sqrt{37}}{6},0\right)$, $\left(\frac{1+\sqrt{37}}{6},0\right)$; y-intercept, $(0,-3)$
5. x-intercepts, $(-2\sqrt{3},0)$, $(2\sqrt{3},0)$; y-intercepts, $(0,-2)(0,2)$
6. $x + 3y = 7$ **7.** $\{2\}$; -1 is an extraneous solution.

Exercise 10–4

Answers to odd-numbered problems

1.

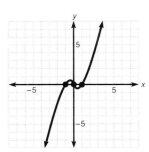

3.

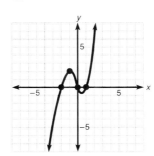

5.

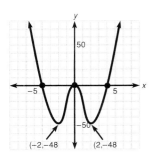

$(-2,-48)$ $(2,-48)$

7.

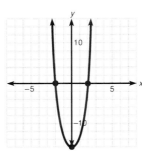

9.

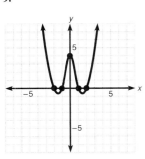

11.

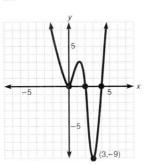

$(3,-9)$

13.

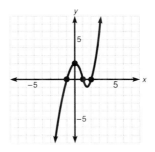

15.

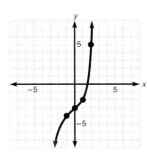

17. y-intercept, $(0,24)$; x-intercepts, $(-2,0)$, $(3,0)$, $(4,0)$

$x < -2$	$y < 0$
$-2 < x < 3$	$y > 0$
$3 < x < 4$	$y < 0$
$x > 4$	$y > 0$

The function is negative on the intervals $(-\infty,0)$ and $(3,4)$ and positive on the intervals $(-2,3)$ and $(4,\infty)$.
19. y-intercept, $(0,0)$; x-intercepts, $(0,0)$, $(2,0)$, $(-6,0)$

$x < -6$	$y > 0$
$-6 < x < 0$	$y < 0$
$0 < x < 2$	$y < 0$
$x > 2$	$y > 0$

The function is negative on the intervals $(-6,0)$ and $(0,2)$ and positive on the intervals $(-\infty,-6)$ and $(2,\infty)$.
21. y-intercept, $(0,36)$; x-intercepts, $(-2,0)$, $(3,0)$

$x < -2$	$y > 0$
$-2 < x < 3$	$y > 0$
$x > 3$	$y > 0$

The function is positive on the intervals $(-\infty,-2)$, $(2,3)$ and $(3,\infty)$.
23. y-intercept, $(0,0)$; x-intercepts, $(-5,0)$, $(2,0)$, $(0,0)$

$x < -5$	$y > 0$
$-5 < x < 0$	$y > 0$
$0 < x < 2$	$y < 0$
$x > 2$	$y > 0$

The function is negative on the interval $(0,2)$ and positive on the intervals $(-\infty,-5)$, $(-5,0)$, and $(2,0)$.

Solutions to trial exercise problems

5. $y = f(x) = x^4 - 16x^2 = x^2(x^2 - 16) = x^2(x + 4)(x - 4)$
The y-intercept $(0,0)$; the x-intercepts are $(-4,0)$, $(0,0)$, $(4,0)$

	x^2	$x + 4$	$x - 4$	Product	
$x < -4$	+	−	−	+	$y > 0$
$-4 < x < 0$	+	+	−	−	$y < 0$
$0 < x < 4$	+	+	−	−	$y < 0$
$x > 4$	+	+	+	+	$y > 0$

x	$y = f(x)$	
-5	225	$(-5,225)$
-2	-48	$(-2,-48)$
2	-48	$(2,-48)$
5	225	$(5,225)$

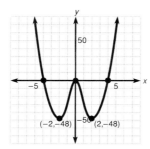

14. $y = f(x) = x^3 + 3x^2 - x - 3 = x^2(x + 3) - 1(x + 3)$
$= (x^2 - 1)(x + 3) = (x + 1)(x - 1)(x + 3)$
The y-intercept is $(0,-3)$.
The x-intercepts are $(-3,0)$, $(-1,0)$, and $(1,0)$.

	$x + 1$	$x - 1$	$x + 3$	Product	
$x < -3$	−	−	−	−	$y < 0$
$-3 < x < -1$	−	−	+	+	$y > 0$
$-1 < x < 1$	+	−	+	−	$y < 0$
$x > 1$	+	+	+	+	$y > 0$

x	$y = f(x)$	
-4	-15	$(-4,-15)$
-2	3	$(-2,3)$
0	-3	$(0,-3)$
2	15	$(2,15)$

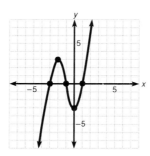

20. If we multiply constants $0 \cdot 4 \cdot (-8) = 0$, the y-intercept is $(0,0)$. Set each factor equal to zero and solve. The x-intercepts are $(-4,0)$, $(0,0)$, and $(8,0)$.

	x^2	$x + 4$	$x - 8$	Product	
$x < -4$	+	−	−	+	$y > 0$
$-4 < x < 0$	+	+	−	−	$y < 0$
$0 < x < 8$	+	+	−	−	$y < 0$
$x > 8$	+	+	+	+	$y > 0$

The function is positive on $(-\infty, -4)$ and $(8, +\infty)$ and negative on $(-4,0)$ and $(0,8)$.

Review exercises

1. $\{x \mid x \in R, x \neq -3\}$ **2.** $\{y \mid y \in R, y \neq -5,7\}$
3. $\{x \mid x \in R, x \neq -2,2\}$ **4.** $\dfrac{2}{3}$
5. $\dfrac{9}{0}$ undefined   **6.** $d = 2\sqrt{10}, m = \dfrac{1}{3}$
7. $\{25\}$; 4 is extraneous

Exercise 10–5

Answers to odd-numbered problems

1. $f(x) = \dfrac{3}{x - 2}$;   **a.** Since $x = 2$ when $x - 2 = 0$, the line $x = 2$ is a vertical asymptote.   **b.** The line $y = 0$ is a horizontal asymptote.   **3.** $f(x) = \dfrac{4}{3x - 2}$ **a.** Since $x = \dfrac{2}{3}$ when $3x - 2 = 0$, the line $x = \dfrac{2}{3}$ is a vertical asymptote.   **b.** The line $y = 0$ is a horizontal asymptote.   **5.** $f(x) = \dfrac{x - 2}{x + 3}$; $x = -3$ is a vertical asymptote. $y = 1$ is a horizontal asymptote.   **7.** $f(x) = \dfrac{3x - 1}{2x + 9}$; $x = -\dfrac{9}{2}$ is a vertical asymptote. $y = \dfrac{3}{2}$ is a horizontal asymptote.   **9.** $f(x) = \dfrac{5}{x^2 - 3x - 10}$; $x = 5$ and $x = -2$ are vertical asymptotes. $y = 0$ is a horizontal asymptote.   **11.** $f(x) = \dfrac{x^2 + 2}{x - 3}$; $x = 3$ is a vertical asymptote. $y = x + 3$ is an oblique asymptote.   **13.** $f(x) = \dfrac{x - 2}{2x^2 - 5x - 3}$; $x = -\dfrac{1}{2}$ and $x = 3$ are vertical asymptotes. $y = 0$ is a horizontal asymptote.   **15.** $f(x) = \dfrac{x^2 - 5}{x^2 + 2}$; there are no vertical asymptotes. $y = 1$ is a horizontal asymptote.   **17.** $f(x) = \dfrac{3}{x - 4}$ **19.** $f(x) = \dfrac{-1}{(x + 1)(x - 3)}$

17. $f(x) = \dfrac{3}{x-4}$

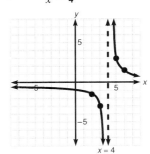

19. $f(x) = \dfrac{-1}{(x+1)(x-3)}$

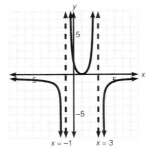

21. $f(x) = \dfrac{3x}{x^2-1}$

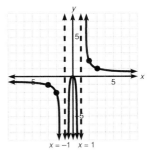

23. $f(x) = \dfrac{x+5}{x-3}$

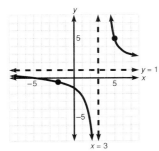

25. $f(x) = \dfrac{4-x}{x+2}$

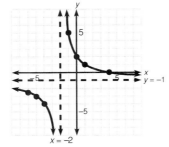

27. $f(x) = \dfrac{x^2-1}{x^2-2x+1}$

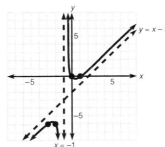

29. $f(x) = \dfrac{x-4}{(x+2)(x-3)}$

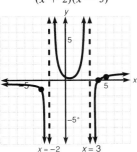

31. $f(x) = \dfrac{x^2-x}{x+1}$

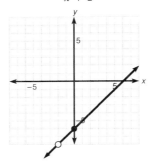

33. $f(x) = \dfrac{3x^2+2}{3x-1}$

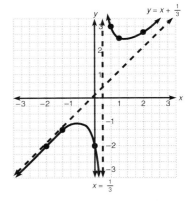

35. $f(x) = \dfrac{3}{x^2+2}$

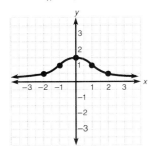

37. $f(x) = \dfrac{x^2-4x-12}{x+2}$

39. a. x-intercept, $\left(-\dfrac{b}{a},0\right)$ **b.** y-intercept, $\left(0,\dfrac{b}{d}\right)$

c. Vertical asymptote, $x = -\dfrac{d}{c}$ **d.** Horizontal asymptote, $y = \dfrac{a}{c}$

Solutions to trial exercise problems

6. a. Since $x = 5$ when $x - 5 = 0$, $x = 5$ is the vertical asymptote.

b. Since the degree of the numerator is the same as the degree of the denominator, $y = \dfrac{\text{coefficient of } -x}{\text{coefficient of } x} = \dfrac{-1}{1} = -1$ is the horizontal asymptote.

11. a. Since $x = 3$ when $x - 3 = 0$, $x = 3$ is the vertical asymptote.

b. Since the degree of the numerator is one more than the degree of the denominator, divide $(x^2 + 2) \div (x - 3) = x + 3 + \dfrac{11}{x - 3}$. Then $y = x + 3$ is the oblique asymptote.

15. a. Since $x^2 + 2 \neq 0$ for any real number x, there is no vertical asymptote.

b. Since the degree of the numerator equals the degree of the denominator, $y = \dfrac{1}{1} = 1$ is the horizontal asymptote.

19. a. Since $x = -1$ when $x + 1 = 0$ and $x = 3$ when $x - 3 = 0$, $x = -1$ and $x = 3$ are vertical asymptotes.

b. Since the degree of the numerator is less than the degree of the denominator, $y = 0$ (x-axis) is the horizontal asymptote.

x	y
-3	$-\dfrac{1}{12}$
-2	$-\dfrac{1}{4}$
0	$\dfrac{1}{3}$ (y-intercept)
1	$\dfrac{1}{4}$
2	$\dfrac{1}{3}$
4	$-\dfrac{1}{5}$
5	$-\dfrac{1}{12}$

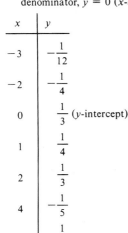

Plot the points, draw the asymptotes in dashed lines, and connect the points in a smooth curve.

28. 1. Since the denominator $x^2 + 4x + 4 = (x + 2)^2$ and $x = -2$ when $x + 2 = 0$, $x = -2$ is the vertical asymptote.

2. Since the degree of the numerator equals the degree of the denominator, the coefficients of x^2, 1, determines the horizontal asymptote to be $y - \dfrac{1}{1} = 1$.

Note: $f(x) = \dfrac{x^2 - 4}{x^2 + 4x + 4} = \dfrac{(x + 2)(x - 2)}{(x + 2)^2} = \dfrac{x - 2}{x + 2}$ $(x \neq -2)$, so we need only graph $y = \dfrac{x - 2}{x + 2}$.

x	y
0	-1 (y-intercept)
2	0 (x-intercept)
-1	-3
4	$\dfrac{1}{3}$
-3	5
-5	$\dfrac{7}{3}$
-7	$\dfrac{9}{5}$

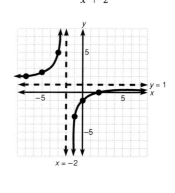

Plot the points, draw the asymptotes in dashed lines, and connect the points in a smooth curve.

31. 1. Since $x = -1$ when $x + 1 = 0$, $x = -1$ is the vertical asymptote.

2. Since the degree of the numerator is greater than the degree of the denominator, divide $(x^2 - x) \div (x + 1) = x - 2 + \dfrac{2}{x + 1}$, the oblique line $y = x - 2$ is an asymptote.

x	y
0	0 (y-intercept)
1	0 (x-intercept)
2	$\dfrac{2}{3}$
$-\dfrac{1}{2}$	$\dfrac{3}{2}$
-2	-6
-3	-6
-4	$-\dfrac{20}{3}$

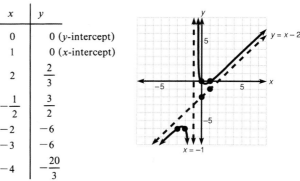

Plot the points, draw the asymptotes in dashed lines, and connect the points in a smooth curve.

35. 1. Since $x^2 + 2 \neq 0$ for any real number x, there is no vertical asymptote.
2. Since the degree of the numerator is less than the degree of the denominator, $y = 0$ is a horizontal asymptote.

x	y
0	$\dfrac{3}{2}$
-1	1
1	1
-2	$\dfrac{1}{2}$
2	$\dfrac{1}{2}$
-3	$\dfrac{3}{11}$
3	$\dfrac{3}{11}$

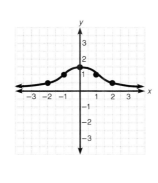

Plot the points, draw the asymptotes in dashed lines, and connect the points in a smooth curve.

Review exercises

1. x^2 **2.** $\dfrac{y^3}{x^4}$ **3.** $\dfrac{1}{16x^6y^2}$ **4.** $x = \dfrac{4}{3-y}$ **5.** $\{-6\}$
6. $\{x \mid -3 \leq x \leq 5\}$

Chapter 10 review

1. $P(x) = x^3 - 6x^2 + x - 6$ **2.** $P(x) = x^4 - 10x^3 + 39x^2 - 70x + 50$ **3.** $P(x) = x^5 - 3x^4 - 14x^3 + 122x^2 - 15x + 125$ **4.** $1, 2 + 3i, 2 - 3i$ **5.** $1 - i, 1 + i, \dfrac{1}{2}$ **6.** $-i, i, 2i, -2i$

7. either 2 positive reals and 1 negative real or no positive real and 1 negative real zeros **8.** either 1 positive real and 2 negative reals or 1 positive real and no negative real zeros **9.** no positive reals and 4, 2, or 0 negative real zeros **10.** 2 is an upper bound; -5 is a lower bound **11.** 1 is an upper bound; -1 is a lower bound **12.** 2 is an upper bound; -1 is a lower bound **13.** $\dfrac{1}{3}, \dfrac{-1 + i\sqrt{3}}{2}, \dfrac{-1 - i\sqrt{3}}{2}$

14. $-4, -\dfrac{1}{2}, \dfrac{1}{3}$ **15.** $-\dfrac{3}{2}, 2, 1$ **16.** $\{-3, 2\}$; 2 of multiplicity 2

17. $\{-2, 1 - \sqrt{7}, 1 + \sqrt{7}\}$ **18.** $\{1, -\dfrac{2}{3}, -\sqrt{2}, \sqrt{2}\}$

19.

```
2 | 3  -2  -6
  |     6   8
  --------------
    3   4   2
```

```
1 | 3  -2  -6
  |     3   1
  --------------
    3   1  -5
```

$P(2) = 2; P(1) = -5$
Since $P(2) > 0$ and $P(1) < 0$, there is a zero in $[1, 2]$.

20.

```
1/2 | 3   7     0    -4
    |     3/2  17/4  17/8
    ------------------------
      3  17/2  17/4  -15/8
```

$P\left(\dfrac{1}{2}\right) = -\dfrac{15}{8}$

```
1 | 3   7    0   -4
  |     3   10   10
  --------------------
    3  10   10    6
```

$P(1) = 6$

Since $P\left(\dfrac{1}{2}\right) < 0$ and $P(1) > 0$, there is a zero in $\left[\dfrac{1}{2}, 1\right]$.

21.

```
-1 | 5  -9   -4    9
   |    -5   14  -10
   --------------------
     5 -14   10   -1
```

$P(-1) = -1$

```
2 | 5  -9  -4   9
  |    10   2  -4
  ------------------
    5   1  -2   5
```

$P(2) = 5$

Since $P(-1) < 0$ and $P(2) > 0$, there is a zero in $[-1, 2]$.
22. $\sqrt[3]{53} \approx 3.76$
23.   **24.**

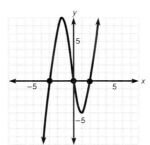

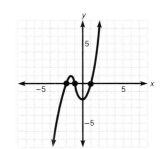

25.   **26.**

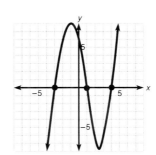

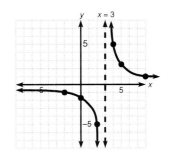

27.

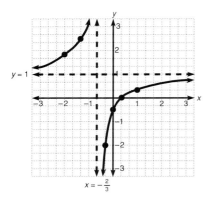

$x = -\frac{2}{3}$

28.

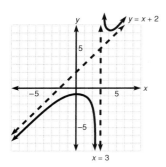

$x = 3$

Chapter 10 cumulative test

1. a. $A \cup B = \{-7,-3,-2,0,1,7,8,9\}$; $A \cap B = \{-3,0,8\}$

2. distributive property   **3.** $\frac{44}{5}$ **4.** $S = 256$ **5.** $\{-28\}$

6. $a = \frac{2S - n^2d + nd}{n}$ **7.** \$8,000   **8.** $\left\{x \mid 1 < x \le \frac{8}{3}\right\}$

9. $\left\{2, -\frac{1}{2}\right\}$ **10.** $\{x \mid -2 < x < 3\}$ **11.** $\left\{-\frac{4}{3}, 2\right\}$ **12.** $\left\{\frac{19}{2}\right\}$

13. $\left\{-\frac{7}{5}\right\}$ **14.** $x^{1/4}$ **15.** $x^{3/8}$ **16.** $2\sqrt{6}$ **17.** $\frac{2\sqrt{3}}{9}$

18. $6\sqrt{6} - 6$ **19.** $23 - 2i$ **20.** $t = 2$ seconds

21. $\{x \mid x \le -1\} \cup \{x \mid x \ge 3\}$

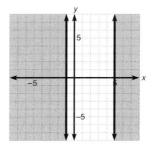

22. parallel   **23.** $d = \sqrt{34}$; midpoint $\left(-\frac{1}{2}, \frac{11}{2}\right)$

24. a. $y = x + 3$ **b.** $2x + 3y = 12$

25.

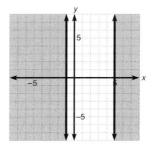

26. a. function   **b.** function   **c.** not a function

27.

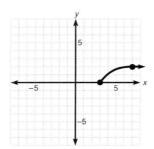

28. a. 6   **b.** -17 **c.** 4   **29. a.** $k = 32$ **b.** $x = 32$

30.   **31.**

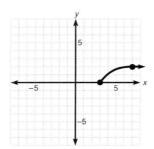

$\left(-\frac{1}{2}, -\frac{9}{4}\right)$

32. a. circle   **b.** parabola   **c.** ellipse   **33.** $P(-2) = -35$

34. -1 is a solution.   **35.** $\frac{2}{3}$

36.   **37.**

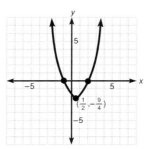

$x = 4$

Chapter 11

Exercise 11–1

Answers to odd-numbered problems

1.

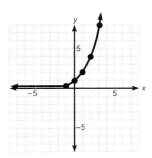

3.

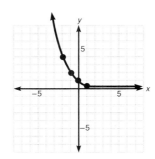

5.

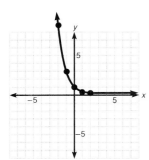

7.

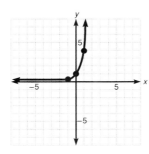

9.

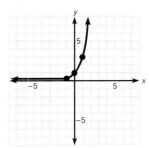

11.

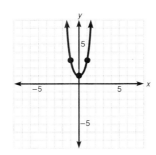

13. $\{5\}$ **15.** $\left\{\dfrac{5}{2}\right\}$ **17.** $\left\{\dfrac{3}{2}\right\}$ **19.** $\{-2\}$

21. $\{-2\}$ **23.** $\left\{\dfrac{-5}{4}\right\}$ **25.** $\left\{\dfrac{2}{9}\right\}$ **27.** $\{-3\}$

29. $\{3\}$ **31.** $\{3\}$ **33.** $\{-3\}$ **35.** $\left\{-\dfrac{1}{6}\right\}$ **37. a.** 4,500
b. 40,500   **c.** 23,383   **d.** $1,500(3)^{t/12}$ **39. a.** \$20,000
b. \$40,000   **c.** $5,000(2)^{t/9}$ **41. a.** 1,000   **b.** 250

Solutions to trial exercise problems

5. Using $g(x) = 3^{-x}$ when
 (1) $x = -2$, $g(-2) = 3^{-(-2)} = 3^2 = 9$ $(-2,9)$
 (2) $x = -1$, $g(-1) = 3^{-(-1)} = 3$ $(-1,3)$
 (3) $x = 0$, $g(0) = 3^{-0} = 1$ $(0,1)$
 (4) $x = 1$, $g(1) = 3^{-1} = \dfrac{1}{3}$ $\left(1,\dfrac{1}{3}\right)$

 (5) $x = 2$, $g(2) = 3^{-2} = \dfrac{1}{3^2} = \dfrac{1}{9}$ $\left(2,\dfrac{1}{9}\right)$

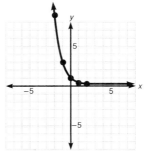

10. (1) $f(-1) = 2^{-1-1} = 2^{-2} = \dfrac{1}{4}$ $\left(-1,\dfrac{1}{4}\right)$

 (2) $f(0) = 2^{-0-1} = 2^{-1} = \dfrac{1}{2}$ $\left(0,\dfrac{1}{2}\right)$

 (3) $f(1) = 2^{1-1} = 2^0 = 1$ $(1,1)$
 (4) $f(2) = 2^{2-1} = 2^1 = 2$ $(2,2)$
 (5) $f(3) = 2^{3-1} = 2^2 = 4$ $(3,4)$

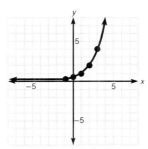

15. $4^x = 32$. Since $4 = 2^2$ and $32 = 2^5$, then $(2^2)^x = 2^5$ and
$2^{2x} = 2^5$. Then $2x = 5$ and $x = \dfrac{5}{2}$. The solution set is $\left\{\dfrac{5}{2}\right\}$.
24. $9^{2x} = 27$. Since $9 = 3^2$ and $27 = 3^3$, then $(3^2)^{2x} = 3^3$ and
$3^{4x} = 3^3$. Thus $4x = 3$ and $x = \dfrac{3}{4}$. The solution set is $\left\{\dfrac{3}{4}\right\}$.
35. $27^{2x+1} = 9$. Since $27 = 3^3$ and $9 = 3^2$, then $(3^3)^{2x+1} = 3^2$ and
$3^{6x+3} = 3^2$. Thus $6x + 3 = 2$, $6x = -1$, and $x = -\dfrac{1}{6}$. The

solution set is $\left\{-\dfrac{1}{6}\right\}$.   **40. a.** Let $t = 2$. Then $A = 200(3)^{-0.5(2)} =$

$200(3)^{-1} = \dfrac{200}{3} = 66\dfrac{2}{3}$ grams.   **b.** Let $t = 12$. Then

$A = 200(3)^{-0.5(12)} = 200(3)^{-6} = \dfrac{200}{3^6} = \dfrac{200}{729}$ grams.

Review exercises

1. $f^{-1}(x) = \dfrac{x + 3}{4}$

2. domain $= \left\{ x \mid x \in R, x \neq \dfrac{5}{3} \right\}$

3. $f[g(x)] = 2x^2 + 3$ 4. $x^{1/6}$

5. xy^3 6. $x^{2/3}$ 7. $\{-1,5\}$

Exercise 11–2

Answers to odd-numbered problems

1.

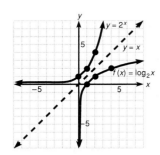

3.

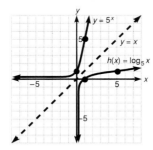

5.

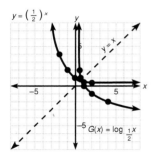

7. $\log_3 81 = 4$ 9. $\log_2 64 = 6$ 11. $\log_{1/2}\left(\dfrac{1}{8}\right) = 3$

13. $\log_{3/2}\left(\dfrac{8}{27}\right) = -3$ 15. $2^4 = 16$ 17. $10^3 = 1,000$

19. $4^{-2} = \dfrac{1}{16}$ 21. $10^{-4} = 0.0001$ 23. $\left(\dfrac{1}{2}\right)^{-3} = 8$

25. $\log_2 32 = 5$ 27. $\log_5 25 = 2$ 29. $\log_2\left(\dfrac{1}{4}\right) = -2$

31. $\log_6\left(\dfrac{1}{216}\right) = -3$ 33. $\log_8 8 = 1$ 35. $\log_{1/3}\left(\dfrac{1}{81}\right) = 4$

37. $\log_{5/4}\left(\dfrac{125}{64}\right) = 3$ 39. $\log_7 \sqrt[3]{7} = \dfrac{1}{3}$ 41. $\{2\}$

43. $\{3\}$ 45. $\{7\}$ 47. $\{b \mid b > 0, b \neq 1\}$ 49. $\{81\}$ 51. $\left\{\dfrac{1}{36}\right\}$

53. $\left\{\dfrac{1}{256}\right\}$ 55. $\{x \mid x > 2\}$ 57. $\{x \mid x < -1 \text{ or } x > 1\}$

Solutions to trial exercise problems

5. Graph the equation $y = \left(\dfrac{1}{2}\right)^x$ and reflect this curve about the line $y = x$ to obtain the graph of $G(x) = \log_{1/2} x$.

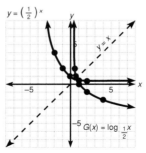

11. $\left(\dfrac{1}{2}\right)^3 = \dfrac{1}{8}$ is equivalent to $\log_{1/2}\left(\dfrac{1}{8}\right) = 3$.   21. $\log_{10} 0.0001 = -4$ is equivalent to $10^{-4} = 0.0001$.

29. $\log_2\left(\dfrac{1}{4}\right) = x$ is equivalent to $2^x = \dfrac{1}{4}$. Since $\dfrac{1}{4} = \left(\dfrac{1}{2}\right)^2 = 2^{-2}$, then $2^x = 2^{-2}$ and $x = -2$. $\log_2\left(\dfrac{1}{4}\right) = -2$.

34. $\log_{1/2}\left(\dfrac{1}{8}\right) = x$ is equivalent to $\left(\dfrac{1}{2}\right)^x = \dfrac{1}{8}$. Since $\dfrac{1}{8} = \left(\dfrac{1}{2}\right)^3$, then $\left(\dfrac{1}{2}\right)^x = \left(\dfrac{1}{2}\right)^3$ and $x = 3$. $\log_{1/2}\left(\dfrac{1}{8}\right) = 3$.

43. $\log_x\left(\dfrac{1}{27}\right) = -3$ is equivalent to $x^{-3} = \dfrac{1}{27}$. Since $\dfrac{1}{27} = \dfrac{1}{3^3} = 3^{-3}$, then $x^{-3} = 3^{-3}$ and $x = 3$. The solution set is $\{3\}$.

48. $\log_{10} x = -2$ is equivalent to $10^{-2} = x$. Then $x = \dfrac{1}{100}$ and the solution set is $\left\{\dfrac{1}{100}\right\}$.   55. Since $x - 2 > 0$ for $x > 2$, then $\log_6(x - 2)$ is defined for $\{x \mid x > 2\} = (2, +\infty)$.

Review exercises

1. $C = 25$ 2. $\dfrac{5 - y}{2y + 3}$ 3. $\dfrac{3z^2 - 5z - 8}{z - 2}$ 4. $\dfrac{1}{b^4 c^{12}}$ 5. $\dfrac{2}{3}$

6. $\{0\}$

Exercise 11–3

Answers to odd-numbered problems

1. $\log_b 3 + \log_b 5$ 3. $\log_3 7 - \log_3 13$ 5. $3 \log_b 5$ 7. $4 \log_{10} 5$

9. $2 \log_{1/2} 2 + 2 \log_{1/2} 3$ 11. $3 \log_5 2 + \log_5 3 - 2$

13. $3(\log_b x + \log_b y)$ 15. $\dfrac{1}{2}\log_b x + \dfrac{1}{3}\log_b y$ 17. $\log_b(2x + 3y)$

19. $\log_3 85$ 21. $\log_e \dfrac{11}{16}$ 23. $\log_6 64$ 25. $\log_{10} 432$ 27. $\log_4 6$

29. $\log_{10}\dfrac{200}{3}$ 31. $\log_b x^{2/3}$ 33. $\log_b \sqrt[4]{\dfrac{x^3}{y}}$

35. $\log_5\left(\dfrac{x^2 + 2x - 24}{9x^2}\right)$ 37. $\log_b(14) = 1.1461$; $b^{1.1461} = 14$

39. $\log_b(49) = 1.6902$; $b^{1.6902} = 49$ **41.** $\log_b(42) = 1.6272$; $b^{1.6232} = 42$ **43.** $\log_b\left(\dfrac{7}{10}\right) = -0.1549$; $b^{-.1549} = \dfrac{7}{10}$
45. $\log_b\sqrt{7} = 0.4226$; $b^{.4226} = \sqrt{7}$ **47.** $\log_b\sqrt[4]{9} = 0.2386$; $b^{.2386} = \sqrt[4]{9}$ **49.** $\left\{\dfrac{64}{5}\right\}$ **51.** {20}   **53.** {67}   **55.** {3}   **57.** {4}; -25 is extraneous   **59.** {3}; -9 is extraneous   **61.** $\left\{\dfrac{-1 + \sqrt{33}}{4}\right\}$; $\dfrac{-1 - \sqrt{33}}{4}$ is extraneous   **63.** $\left\{\dfrac{-1 + \sqrt{5}}{2}\right\}$; $\dfrac{-1 - \sqrt{5}}{2}$ is extraneous   **65.** {1,5}   **67.** *Proof:* Let $\log_b u = m$ and $\log_b v = n$. Then $b^m = u$ and $b^n = v$. $\dfrac{u}{v} = \dfrac{b^m}{b^n} = b^{m-n}$. So $\log_b \dfrac{u}{v} = m - n = \log_b u - \log_b v$.
69. Let $u = 1$, $v = 1$, and $b = 2$. Then
(1) $\log_b(u + v) = \log_2(1 + 1) = \log_2 2 = 1$;
(2) $\log_b u = \log_2 1 = 0$;
$\log_b v = \log_2 1 = 0$; thus,
$\log_2 1 + \log_2 1 = 0$ and
$\log_b(u + v) \neq \log_b u + \log_b v$.

Solutions to trial exercise problems
8. Since $24 = 2^3 \cdot 3$, then $\log_b(24) = \log_b(2^3 \cdot 3) = \log_b 2^3 + \log_b 3 = 3\log_b 2 + \log_b 3$.   **10.** Since $\dfrac{15}{16} = \dfrac{3 \cdot 5}{2^4}$, then $\log_{10}\left(\dfrac{15}{16}\right) = \log_{10}\left(\dfrac{3 \cdot 5}{2^4}\right) = \log_{10} 3 + \log_{10} 5 - \log_{10} 2^4 = \log_{10} 3 + \log_{10} 5 - 4\log_{10} 2$.   **12.** $\log_b\left(\dfrac{x^3}{y^2}\right) = \log_b x^3 - \log_b y^2 = 3\log_b x - 2\log_b y$
13. $\log_b(xy)^3 = 3\log_b(xy) = 3[\log_b x + \log_b y]$ **17.** $\log_b(2x + 3y) = \log_b(2x + 3y)$, since none of the properties apply to addition.
24. $2\log_2 7 + \log_2 5 = \log_2 7^2 + \log_2 5 = \log_2 49 + \log_2 5 = \log_2(49 \cdot 5) = \log_2(245)$ **28.** $\log_e 4 + 2\log_e 3 - \log_e 2 = \log_e 4 + \log_e 3^2 - \log_e 2^3 = \log_e 4 + \log_e 9 - \log_e 8 = \log_e\left(\dfrac{4 \cdot 9}{8}\right) = \log_e\left(\dfrac{36}{8}\right) = \log_e\left(\dfrac{9}{2}\right)$ **31.** $\dfrac{1}{3}\log_b(x^2) = \log_b(x^2)^{1/3} = \log_b x^{2/3}$
41. $\log_b(42) = \log_b(2 \cdot 3 \cdot 7) = \log_b 2 + \log_b 3 + \log_b 7 = (0.3010) + (0.4771) + (0.8451) = 1.6232$. Then $b^{1.6232} = 42$.
42. $\log_b\left(\dfrac{27}{4}\right) = \log_b 27 - \log_b 4 = \log_b 3^3 - \log_b 2^2 = 3\log_b 3 - 2\log_b 2 = 3(0.4771) - 2(0.3010) = 1.4313 - 0.6020 = 0.8293$. Then $b^{0.8293} = \dfrac{27}{4}$.   **54.** $\log_4(x - 5) - \log_4 3 = 2$ is equivalent to $\log_4\left(\dfrac{x - 5}{3}\right) = 2$, which is equivalent to $\dfrac{x - 5}{3} = 4^2$. Then $\dfrac{x - 5}{3} = 16$, $x - 5 = 48$, and $x = 53$. Thus the solution set is {53}.
57. $\log_{10}(x + 21) + \log_{10} x = 2$ is equivalent to $\log_{10} x(x + 21) = 2$. Then $x(x + 21) = 10^2$, $x^2 + 21x = 100$, and $x^2 + 21x - 100 = 0$. Factoring, $(x + 25)(x - 4) = 0$, so $x = -25$ or $x = 4$. But if $x = -25$, then $\log_{10}(x + 21) = \log_{10}(-4)$, which does not exist. The solution set is {4}; -25 is extraneous.   **62.** $\log_5(x - 2) = \log_5 5 - \log_5(x + 3)$; $\log_5(x - 2) + \log_5(x + 3) = 1(x - 2)(x + 3) = 5^1$; $x^2 + x - 6 = 5$; $x^2 + x - 11 = 0$; $x = \dfrac{-1 \pm \sqrt{1 + 44}}{2(1)} = \dfrac{-1 \pm \sqrt{45}}{2} = \dfrac{-1 \pm 3\sqrt{5}}{2}$; solution set is $\left\{\dfrac{-1 + 3\sqrt{5}}{2}\right\}$. Since $\dfrac{-1 - 3\sqrt{5}}{2}$ is negative, it is extraneous.

Review exercises
1. ellipse; y-intercepts, $(0, -\sqrt{3})$, $(0, \sqrt{3})$; x-intercepts, $(-3,0)$ $(3,0)$
2. parabola; y-intercept, $(0, -20)$; x-intercepts, $(-5,0)$, $(4,0)$
3. hyperbola; y-intercepts, $(0, -3)$, $(0,3)$; no x-intercepts   **4.** $\left\{\dfrac{4}{3}\right\}$
5. $-\dfrac{4\sqrt{6}}{3}$ or $\dfrac{4\sqrt{6}}{3}$ **6.** 9   **7.** $\left\{x \mid x \leq \dfrac{2}{5} \text{ or } x \geq \dfrac{6}{5}\right\} = \left(-\infty, \dfrac{2}{5}\right] \cup \left[\dfrac{6}{5}, \infty\right)$

Exercise 11–4

Answers to odd-numbered problems
1. $\log 8 = 0.9031$; $10^{0.9031} = 8$ **3.** $\log(53) = 1.7243$; $10^{1.7243} = 53$ **5.** $\log(80,200) = 4.9042$; $10^{4.9042} = 80,200$
7. $\log(794,000,000) = 8.8998$; $10^{8.8998} = 794,000,000$
9. $\log 0.00863 = -2.0640$; $10^{-2.0640} = .00863$
11. $\log 0.000000107 = -6.9706$; $10^{-6.9706} = 0.000000107$
13. 5.7024   **15. a.** -4.1367 **b.** 37.9912   **17. a.** 281.19   **b.** 10.99
19. 2.7773
21. 14.85   **23.** 34.7712
25. 6,990   **27.** 82,000,000
29. .0000165   **31.** 337.5   **33.** .005887   **35.** .0000003368
37. 48.37
39. 18.97   **41.** 3.3010   **43.** 1.585×10^{-6}

Solutions to trial exercise problems
1. $\log 8 = 0.9031$. Then $10^{0.9031} = 8$.   **28.** Antilog $8.7528 - 10 =$ antilog -1.2472. Press $\boxed{1}$ $\boxed{0}$ $\boxed{\cdot}$ $\boxed{2}$ $\boxed{4}$ $\boxed{7}$ $\boxed{2}$ $\boxed{+/-}$ $\boxed{=}$ and read "0.0565979 on the display. Antilog $8.7528 - 10 = 0.0566$.
30. Antilog 8.8340 is found by pressing $\boxed{1}$ $\boxed{0}$ $\boxed{y^x}$ $\boxed{8}$ $\boxed{\cdot}$ $\boxed{8}$ $\boxed{3}$ $\boxed{4}$ $\boxed{0}$ $\boxed{=}$. Read 6.823×10^8. Then antilog $8.8340 = 682,300,000$.   **43.** pH $= -\log(\text{H}^+)$. Then $5.8 = -\log(\text{H}^+)$; $-5.8 = \log(\text{H}^+)$. Find antilog -5.8 by pressing $\boxed{1}$ $\boxed{0}$ $\boxed{y^x}$ $\boxed{5}$ $\boxed{\cdot}$ $\boxed{8}$ $\boxed{+/-}$ $\boxed{=}$. Read 0.000001585. Then $\text{H}^+ = 0.000001585$.

Review exercises
1. 14   **2.** $4ab - 3a - 5b$ **3.** $\sqrt{2}$ **4.** $-1 - 13i$
5. $13\dfrac{1}{3}$ min   **6.** $66\dfrac{2}{3}$ cm³ of 10%; $33\dfrac{1}{3}$ cm³ of 4%

Exercise 11–5

Answers to odd-numbered problems
1. 1.2920   **3.** 2.3219   **5.** -2.6804 **7.** 1.2183   **9.** 1.6094
11. 1.0043   **13.** 1.9066   **15.** -2.9878 **17.** 5.8493
19. 1.1332   **21.** 2.9459   **23.** 4.0986   **25.** 13.6   **27.** 0.884
29. 0.0542   **31.** 18.3 years   **33.** 55.9 years   **35.** 12,979.3 years
37. 269.9 ft   **39.** 13.5 days

Solutions to trial exercise problems
7. $\log_{23} 45.6 = \dfrac{\log 45.6}{\log 23} = \dfrac{1.6590}{1.3617} = 1.2183$ **10.** $\log_e 3 = \ln 3$. Press $\boxed{3}$ $\boxed{\ln}$ and read "1.0986" on the display. Thus, $\log_e 3 = 1.0986$.   **12.** $\log_e(107) = \ln(107)$. Press $\boxed{1}$ $\boxed{0}$ $\boxed{7}$ $\boxed{\ln}$ and read "4.6728" on the display. Thus, $\log_e(107) = 4.6728$.   **21.** $\ln 7e = \ln 7 + \ln e = 2.3026 \log 7 + 1 = 1.9459 + 1 = 2.9459$
37. $I = I_0 e^{-kd}$, given $k = 0.00853$, $I = 0.10 I_0$, and we want d. Then $0.10 I_0 = I_0 e^{-0.00853d}$; $0.10 = e^{-0.00853d}$; $-0.00853d = \ln 0.10$; $d = \dfrac{\ln 0.10}{-0.00853} = 269.9$. The light is reduced to 10% at a depth of 269.9 feet.

Review exercises

1. domain = $\{-4,2,0,1\}$ **2.** domain = $\left\{x\middle|x \geq \dfrac{3}{2}\right\}$ **3.** no, since $(-4,3)$ and $(2,3)$ have the same second component

4. 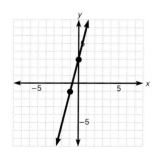   **5.**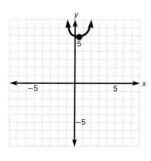

Exercise 11–6

Answers to odd-numbered problems

1. $\{3.17\}$ **3.** $\{0.29\}$ **5.** $\{-2.10\}$ **7.** $\{1.19\}$ **9.** $\{0.26\}$ **11.** $\{0.07\}$

13. $\{1.71\}$ **15.** $\{2.46\}$ **17.** $n = \dfrac{\log y}{3 \log x}$ **19.** 4.2%   **21.** 3.9 years   **23.** 2.14 hours   **25.** .396 hours or 24 minutes   **27.** 11.5 centuries   **29.** 2.05 years   **31.** 3.7%

Solutions to trial exercise problems

7. $6^{2x-1} = 12$; $\log 6^{2x-1} = \log 12$; $(2x-1)\log 6 = \log 12$; $2x \log 6 - \log 6 = \log 12$; $2x \log 6 = \log 12 + \log 6$; $x = \dfrac{\log 12 + \log 6}{2 \log 6}$

$= \dfrac{1.0792 + 0.7782}{1.5564} = 1.19$. The solution set is $\{1.19\}$.

14. $2^{x^2} = 7$; $\log 2^{x^2} = \log 7$; $x^2 \log 2 = \log 7$; $x^2 = \dfrac{\log 7}{\log 2} = \dfrac{0.8451}{0.3010}$

$= 2.81$. Then $x = \pm 1.68$. The solution set is $\{1.68, -1.68\}$.

17. $y = x^{3n}$; $\log y = \log x^{3n}$; $\log y = 3n \log x$; $n = \dfrac{\log y}{3 \log x}$.

26. $q = q_0(0.96)^t$. Now $q = \dfrac{1}{2}q_0 = 0.5q_0$; so $0.5q_0 = q_0(0.96)^t$;

$0.5 = (0.96)^t$; $\log 0.5 = \log(0.96)^t$; $\log(0.5) = t \log(0.96)$;

$t = \dfrac{\log(0.5)}{\log(0.96)} = \dfrac{-0.3010}{-0.0177} = 17.0$. The half-life of radium is 17.0 centuries.

Review exercises

1. $(5x + 4)(x - 1)$ **2.** $(5z + 6)(5z - 6)$ **3.** $x(5x^2 - 2x + 1)$
4. $(3z - 2)^2$ **5.** $\{-\sqrt{7}, \sqrt{7}, -i\sqrt{2}, i\sqrt{2}\}$ **6.** $\{x|-5 \leq x < 2\}$
7. $\sqrt{101}$; $\left(1, \dfrac{3}{2}\right)$

Chapter 11 review

1. a. 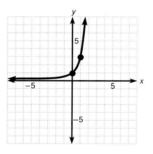   **b.**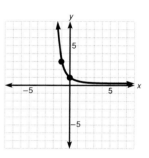

c.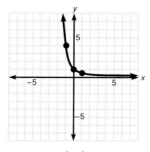

2. a. $\{4\}$ **b.** $\left\{\dfrac{7}{8}\right\}$

3.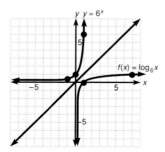

4. $\log_4 \dfrac{1}{64} = -3$ **5.** $\left(\dfrac{1}{3}\right)^{-3} = 27$ **6.** 3   **7.** -4 **8.** $\dfrac{1}{3}$

9. $\{5\}$ **10.** $\left\{\dfrac{1}{243}\right\}$ **11.** $\{4\sqrt{2}\}$

12. $x > 3$ or $x < -5$ **13.** $\log_b 7 + 3 \log_b 2$
14. $\log_4 5 - \log_4 2 - \log_4 3$ **15.** $2 \log_5 3 - 2 \log_5 2$ **16.** $\log_5 28$

17. $\log_6 5$ **18.** $\log_b \dfrac{x^4 y^2}{z}$

19. $\log_4\left(\dfrac{x + 3}{x - 4}\right)$ **20.** $\log_b\left(\dfrac{x^3(2x - 1)}{(x + 1)^3}\right)$

21. 1.3423; $b^{1.3423} = 22$ **22.** 2.0826; $b^{2.0826} = 121$
23. 1.9821; $b^{1.9821} = 96$ **24.** 0.3471; $b^{3471} = \sqrt[3]{11}$

25. -0.8652; $b^{-0.8652} = \dfrac{3}{22}$

26. 1.1710; $b^{1.1710} = \dfrac{27}{\sqrt[4]{11}}$ **27.** $\{48\}$

28. $\left\{\dfrac{2}{9}\right\}$ **29.** $\left\{\dfrac{23}{4}\right\}$ **30.** $\left\{\dfrac{83}{26}\right\}$

31. 2.5340 **32.** 5.7050 **33.** -2.1331 **34.** -0.7784
35. 15.2 yr **36.** 1.09 **37.** 0.45 **38.** 3.86
39. 1.2 days **40.** 2.0 days **41.** $\{1.76\}$ **42.** $\{-0.79\}$
43. $\{2.81\}$ **44.** $\{0.40\}$ **45.** $\{-1.04\}$ **46.** 2.20 hours

Chapter 11 cumulative test

1. $\{-3,-2,-1,0,1,2,3,4,5\}$ **2.** -8 **3.** -81
4. $4a^2 - 4ab + b^2$ **5.** $25x^2 - 9y^2$ **6.** $12y^2 - 14y - 10$

7. $-18a^5b^3$ **8.** $-\dfrac{27b^6}{a^3c^9}$ **9.** $\dfrac{a^{12}}{9b^{10}}$

10. $\{1\}$ **11.** $\{y|y \le 2\}$ **12.** $\left\{x\Big|-2 \le x < \dfrac{5}{2}\right\}$

13. $\left\{\dfrac{5}{4},\dfrac{1}{4}\right\}$ **14.** $\left\{x\Big|-\dfrac{2}{3} \le x \le 2\right\}$

15. $\{x|x < -11 \text{ or } x > 1\}$ **16.** $\left\{\dfrac{3}{2},-1\right\}$

17. $\left\{\dfrac{5}{4}\right\}$ **18.** $\dfrac{10a + 6b}{8a^2b^2}$

19. $\dfrac{15y}{y - 7}$ **20.** $\dfrac{x^2 - 9x + 20}{x^2 + 2x - 15}$
21. 1 **22.** $13 - 4\sqrt{3}$ **23.** 25 **24.** $-i$
25. $7\sqrt{2}$ **26.** $-(6 + 2\sqrt{2} + 3\sqrt{3} + \sqrt{6})$
27. $\left\{\dfrac{3 + \sqrt{37}}{2},\dfrac{3 - \sqrt{37}}{2}\right\}$ **28.** $\{8\}$; -1 is extraneous

29. $\{z|-5 \le z \le 8\}$ **30. a.** $2x - y = -6$ **b.** $2x - 3y = -42$

c. $4x - 3y = 29$ **31. a.** 26 **b.** -25 **c.** -44 **d.** 5

32.   **33.**

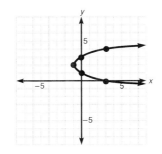
(0,4)
(-6,0)

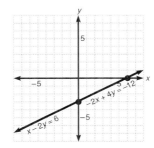

34. a. circle **b.** parabola **c.** hyperbola **d.** ellipse

35. $-4, -2, \dfrac{1}{2}, 2$

36.

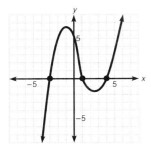

37. $\{2\}$ **38. a.** 2 **b.** $\dfrac{1}{36}$ **39.** 1.36 **40.** $\left\{\dfrac{17}{8}\right\}$ **41.** $\log_b\left(\dfrac{8}{125}\right)$

Chapter 12

Exercise 12–1

Answers to odd-numbered problems

1. yes **3.** yes **5.** no

7. $\{(3,2)\}$ **9.** $\{(-2,1)\}$

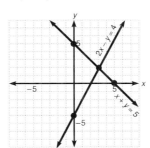

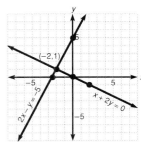
(-2,1)

11. $\{(x,y)|x - 2y = 6\}$; dependent

13. $\{(-2,-5)\}$ **15.** $\{(-2,3)\}$ **17.** $\{(2,3)\}$ **19.** $\{(3,1)\}$
21. $\left\{\left(\dfrac{3}{2},1\right)\right\}$ **23.** $\{(-1,-4)\}$ **25.** $\{(x,y)|3x + y = 2\}$;

dependent **27.** $\emptyset$; inconsistent **29.** $\left\{\left(\dfrac{5}{9},\dfrac{10}{9}\right)\right\}$

31. $\left\{\left(\dfrac{5}{3},\dfrac{1}{2}\right)\right\}$ **33.** $\left\{\left(\dfrac{3}{5},0\right)\right\}$ **35.** $\left\{\left(-2,\dfrac{7}{2}\right)\right\}$

37. $\{(3,-2)\}$ **39.** $\{(7,-4)\}$ **41.** $\left\{\left(\dfrac{7}{2},-\dfrac{3}{2}\right)\right\}$ **43.** $\{(1,-3)\}$

45. $\{(-1,-4)\}$ **47.** $\{(4,12)\}$ **49.** $\{(-2,-5)\}$ **51.** $\{(6,6)\}$
53. $\{(-1,0)\}$ **55.** $\emptyset$; inconsistent

57. $\{(x,y)|2x - y = 7\}$; dependent **59.** $\left\{\left(\dfrac{5}{2},2\right)\right\}$

61. $\left\{\left(-\dfrac{18}{7},\dfrac{22}{7}\right)\right\}$ **63.** $\left\{\left(-\dfrac{12}{11},-\dfrac{63}{11}\right)\right\}$ **65.** $\left\{\left(-1,\dfrac{1}{4}\right)\right\}$

67. $\left\{\dfrac{7}{5},-\dfrac{7}{4}\right\}$ **69.** $x + y = 502$

71. $x = y + 6$ **73.** $\ell = 3w + 4$ **75.** $x - y = 33$

Solutions to trial exercise problems

19. $3x + 2y = 11$ $3x + 2y = 11$ $3 - y = 2$ The solution set is $\{(3,1)\}$.

$\underline{x - y = 2}$ (times 2) $\underline{2x - 2y = 4}$ $-y = -1$

                        $5x = 15$ $y = 1$

                        $x = 3$

24. $3x - y = 10$ (times -2) $-6x + 2y = -20$

$\underline{6x - 2y = 5}$ $\underline{6x - 2y = \;\;5}$ The system is inconsistent. There are no common

                           $0 = -15$ solutions. The solution set is $\emptyset$.

31. $\dfrac{1}{2}x + \dfrac{1}{3}y = 1$ (times 6) $3x + 2y = 6$ $3x + 2y = 6$ $\dfrac{1}{2}x + \dfrac{1}{3}\left(\dfrac{1}{2}\right) = 1$

$\dfrac{1}{4}x - \dfrac{2}{3}y = \dfrac{1}{12}$ (times 12) $\underline{3x - 8y = 1}$ (times -1) $\underline{-3x + 8y = -1}$ $\dfrac{1}{2}x + \dfrac{1}{6} = 1$

                                         $10y = 5$ $3x + 1 = 6$

                                         $y = \dfrac{1}{2}$ $3x = 5$

                                                     $x = \dfrac{5}{3}$

The solution set is $\left\{\left(\dfrac{5}{3}, \dfrac{1}{2}\right)\right\}$.

36. $(0.3)x - (0.8)y = 1.6$ (times 10) $3x - 8y = 16$ $3x - 8y = 16$ $8 + 4y = 12$

$\underline{(0.1)x + (0.4)y = 1.2}$ (times 10) $\underline{x + 4y = 12}$ (times 2) $\underline{2x + 8y = 24}$ $4y = 12 - 8 = 4$

                                             $5x = 40$ $y = 1$

                                             $x = 8$

The solution set is $\{(8,1)\}$.

39. $2x + y = 10$ $2x + (-x + 3) = 10$ $y = -(7) + 3$

$\underline{y = -x + 3}$ $x + 3 = 10$ $y = -4$

                                   $x = 7$ The solution set is $\{(7,-4)\}$.

44. $3x - 5y = 4$ $x = -2y - 2$, substituting $3(-2y - 2) - 5y = 4$

$\underline{x + 2y = -2}$ $-6y - 6 - 5y = 4$

                                       $-11y = 10$

                                           $y = -\dfrac{10}{11}$

Then $x = -2\left(-\dfrac{10}{11}\right) - 2 = \dfrac{20}{11} - 2 = -\dfrac{2}{11}$

The solution set is $\left\{\left(-\dfrac{2}{11}, -\dfrac{10}{11}\right)\right\}$.

49. $4x - 3y = 7$ Substituting,

$\underline{y = -5}$ $4x - 3(-5) = 7$

                  $4x + 15 = 7$ The solution set is $\{(-2,-5)\}$.

                      $4x = -8$

                       $x = -2$

61. $-\dfrac{1}{3}x + y = 4$ (3) $-x + 3y = 12$

$x = -\dfrac{1}{2}y$ (3) $2x = -y - 2$,   then $y = -2x - 2$

Substituting in $-x + 3y = 12$, $-x + 3(-2x - 2) = 12$

                                   $-x - 6x - 6 = 12$

                                       $-7x = 18$

                                          $x = -\dfrac{18}{7}$

$y = -2\left(-\dfrac{18}{7}\right) - 2$

$= \dfrac{36}{7} - 2 = \dfrac{22}{7}$

The solution set is $\left\{\left(-\dfrac{18}{7}, \dfrac{22}{7}\right)\right\}$.

66. Let $p = \dfrac{1}{x}$ and $q = \dfrac{1}{y}$.

$$\dfrac{2}{x} - \dfrac{3}{y} = 1$$

$$\dfrac{1}{x} + \dfrac{2}{y} = 2$$

$$2p - 3y = 1$$
$$p + 2q = 2 \text{ Multiply by } -2 \rightarrow$$

$$2p - 3q = 1$$
$$\underline{-2p - 4q = -4}$$
$$-7q = -3$$
$$q = \dfrac{3}{7}$$

Using $p + 2q = 2$,

$$p + 2\left(\dfrac{3}{7}\right) = 2$$

$$p + \dfrac{6}{7} = 2$$

$$p = 2 - \dfrac{6}{7} = \dfrac{8}{7}$$

$p = \dfrac{8}{7}$ and $q = \dfrac{3}{7}$.

Then $\dfrac{1}{x} = \dfrac{8}{7}$ and $x = \dfrac{7}{8}$

$\dfrac{1}{y} = \dfrac{3}{7}$ and $y = \dfrac{7}{3}$

The solution set is $\left\{\left(\dfrac{7}{8}, \dfrac{7}{3}\right)\right\}$.

Review exercises

1. $-x^3 - 5x^2 - x + 5$ **2.** $x^3 + 8$ **3.** $25y^2 - 20yz + 4z^2$
4. $\{1\}$; -9 is extraneous   **5.** 9,36   **6.** 12

Exercise 12–2

Answers to odd-numbered problems

1. 8 ft by 12 ft   **3.** $w = 6$ m; $\ell = 9$ m   **5.** 6 ft and 15 ft   **7.** 11 volts and 36 volts   **9.** \$6,000 at $6\dfrac{1}{2}$%; \$14,000 at 7%

11. \$15,000 at 6%; \$15,000 at 8%   **13.** \$16,800 at 8%; \$19,200 at 7%   **15.** \$9,000 at 12%; \$16,000 at 15%   **17.** 15 suits at \$205; 17 suits at \$152   **19.** 23 laborers (5 cat operators)   **21.** 403 children tickets were sold (100 adult tickets were sold).   **23.** 19 quarters; 24 dimes   **25.** 1,600 kg of 85% pure copper; 400 kg of 60% pure copper

27. 320 liters of 6% acid; 480 liters of 3.5% acid   **29.** $28\dfrac{4}{7}$ ml

31. 2 ℓ   **33.** The jogger runs for $\dfrac{6}{7}$ of an hour or 51 min 26 sec

35. speed of boat is 20 mph; speed of stream is 4 mph
37. The mother jogs at 2 mph; the daughter jogs at 4 mph.

39. cyclist's rate $= 12\dfrac{8}{9}$ mph; pedestrian's rate $= 4\dfrac{8}{9}$ mph

41. 25 mph; 75 mph   **43.** $3x - 2y = 1$; $5x - 2y = 18$; the point of intersection is $\left(\dfrac{17}{2}, \dfrac{49}{4}\right)$.

45. $\left(\dfrac{11}{5}, -\dfrac{8}{5}\right)$

Solutions to trial exercise problems

1. Let $w =$ width of the rectangle and $\ell =$ length of the rectangle.
(1)   Then $2\ell + 3w = 48$
$\underline{2\ell + 2w = 40}$ (times -1)

(2) $2\ell + 3w = 48$
$\underline{-2\ell - 2w = -40}$
$w = 8$

(3) $2\ell + 2(8) = 40$
$2\ell + 16 = 40$
$2\ell = 24$
$\ell = 12$

The room is 8 feet wide and 12 feet long.
12. Let $x =$ amount invested at 7%. Let $y =$ amount invested at 9%.
Then $x + y = 16,000$
$0.07x = 0.09y$
(1) $x + y = 16,000$ (times 9)
$7x - 9y = 0$

(2) $9x + 9y = 144,000$
$\underline{7x - 9y = 0}$
$16x = 144,000$
$x = 9,000$
Then $y = 7,000$

Jamie invested \$9,000 at 7% and \$7,000 at 9%.
17. Let $x =$ number of suits sold at \$152 each. Let $y =$ number of suits sold at \$205 each.
Then $x + y = 32$ (times -152)
$152x + 205y = 5,659$
(1) $-152x - 152y = -4,864$
$\underline{152x + 205y = 5,659}$
$53y = 795$
$y = 15$

(2) $x + 15 = 32$
$x = 17$

15 suits were sold at \$205 each and 17 were sold at \$152 each.
21. Let $x =$ the number of children's tickets sold. Let $y =$ the number of adult tickets sold.
$x + y = 503$
$1.25x + 3.50y = 853.75$
(1) $x + y = 503$ (times -125)
$125x + 350y = 85,375$

$-125x - 125y = -62,875$
$\underline{125x + 350y = 85,375}$
$225y = 22,500$
$y = 100$

(2)   Then $x + 100 = 503$
$x = 403$
There were 403 children's tickets sold.

30. Let x = the amount of pure salt and
y = the amount of final solution.
Then $\quad x + 9 = y \qquad x + 9 = y$
$\quad \underline{x + 0.2(9) = 0.4y} \rightarrow \underline{10x + 18 = 4y}$
Substituting $x + 9$ for y in $10x + 18 = 4y$,
$10x + 18 = 4(x + 9)$
$10x + 18 = 4x + 36$
$\qquad 6x = 18$
$\qquad x = 3$
Thus, 3 cubic centimeters of pure salt must be mixed with 9 cubic centimeters of 20% salt solution to obtain a 40% salt solution.

38. Let x = number of hours at 4.5 mph. Let y = number of hours at 4 mph.
Then $x + y = 7$
$4.5x + 4y = 30$
(1) $\quad -4x - 4y = -28 \qquad$ (2) $\quad 4 + y = 7$
$\quad \underline{4.5x + 4y = 30} \qquad\qquad\qquad y = 3$
$\quad 0.5x \qquad\quad = 2$
$\qquad x = 4$
They rowed 4 hours at 4.5 mph and 3 hours at 4 mph.

39. Let x = rate of the cyclist. Let y = rate of the pedestrian.
Then $2\frac{1}{4}x + 2\frac{1}{4}y = 40$
$5x = 5y + 40$ or $x = y + 8$
(1) $\quad \frac{9}{4}x + \frac{9}{4}y = 40 \quad$ (times 4)
$\qquad\quad x - y = 8 \quad$ (times 9)

(2) $\quad 9x + 9y = 160$
$\quad \underline{9x - 9y = 72}$
$\quad 18x \qquad = 232$
$\qquad x = \frac{232}{18} = \frac{116}{9}$

(3) $\quad \left(\frac{116}{9}\right) = y + 8$
$\qquad \frac{116}{9} = y + 8$
$\qquad y = \frac{116}{9} - 8 = \frac{44}{9}$
$\qquad y = \frac{44}{9}$

The cyclist travels at $\frac{116}{9} = 12\frac{8}{9}$ mph and the pedestrian walks at $\frac{44}{9} = 4\frac{8}{9}$ mph.

43. (1) Using $(-1,-2)$ and $(3,4)$, $m = \dfrac{4 - (-2)}{3 - (-1)} = \dfrac{6}{4} = \dfrac{3}{2}$.
Then $y - 4 = \frac{3}{2}(x - 3)$
$\quad 2y - 8 = 3x - 9$
$\quad -3x + 2y = -1$
(2) Using $(4,1)$ and $(2,-4)$, $m = \dfrac{1 - (-4)}{4 - 2} = \dfrac{5}{2}$.
Then $y - 1 = \frac{5}{2}(x - 4)$
$\quad 2y - 2 = 5x - 20$
$\quad -5x + 2y = -18$
(3) Solving $\begin{array}{l} -3x + 2y = -1 \\ \underline{-5x + 2y = -18} \end{array}$ (-1) $\qquad \begin{array}{l} -3x + 2y = -1 \\ \underline{5x - 2y = 18} \\ 2x \qquad = 17 \\ \qquad x = \frac{17}{2} \end{array}$

Then $-3\left(\dfrac{17}{2}\right) + 2y = -1$
$\qquad -\dfrac{51}{2} + 2y = -1$
$\qquad\qquad 2y = \dfrac{49}{2}$
$\qquad\qquad y = \dfrac{49}{4}$
$\left(\dfrac{17}{2}, \dfrac{49}{4}\right)$

Review exercises
1. $2x - 3y = -7$ **2.** $x - 2y = 6$ **3.** $2x + y = 3$ **4.** -8
5. -3 **6.** $2i$ **7.** $2xy\sqrt{2y}$

Exercise 12–3

Answers to odd-numbered problems
1. $\{(3,1,2)\}$ **3.** $\{(3,1,2)\}$ **5.** $\{(2,3,1)\}$
7. $\{(3,-1,2)\}$ **9.** $\{(5,-5,7)\}$ **11.** inconsistent; $\emptyset$
13. dependent; $\{(x,y,z)|x - 4y + z = -5\}$

15. $\{(1,-2,1)\}$ **17.** $\left\{\left(\dfrac{43}{3}, -3, \dfrac{122}{3}\right)\right\}$

19. $\{(8,1,-5)\}$ **21.** $\{(-1,-3,2)\}$ **23.** $\left\{\left(\dfrac{2}{5}, -\dfrac{23}{5}, -2\right)\right\}$

25. $\{(1,0,3)\}$ **27.** $\left\{\left(8, \dfrac{5}{2}, \dfrac{5}{2}\right)\right\}$

29. dependent; $\{(x,y,z)|2x + 8y - 2z = 6\}$ **31.** $33°, 47°, 100°$
33. 25 m, 36 m, 61 m **35.** p_1 (expensive) $= \$21.00$; p_2 (middle-priced) $= \$15.00$; p_3 (cheapest) $= \$11.00$ **37.** 16—$750 stamps;
26—$1,500 stamps; 4—$25,000 stamps **39.** $a = \dfrac{1}{2}$,
$b = -\dfrac{9}{2}$, $c = 2$ **41.** 10—$5 bills, 15—$10 bills, 8—$20 bills

Solutions to trial exercise problems

1. $x + y + z = 6$
$x - 2y - z = -1$
$x + y - z = 2$

(1) Using $x + y + z = 6$
$\underline{x - 2y - z = -1}$
$2x - y \quad\quad = 5$

(3) Solving $2x - y = 5$ (times 2)
$2x + 2y = 8$

(5) Substituting
$2(3) - y = 5$
$6 - y = 5$
$-y = -1$
$y = 1$

The solution set is $\{(3,1,2)\}$.

(2) Using $x + y + z = 6$
$\underline{x + y - z = 2}$
$2x + 2y \quad\quad = 8$

(4) Then $4x - 2y = 10$
$\underline{2x + 2y = 8}$
$6x \quad\quad = 18$
$x = 3$

(6) Substituting in $x + y + z = 6$
$3 + 1 + z = 6$
$4 + z = 6$
$z = 2$

8. $x + 2y + 3z = 5$
$-x + y - z = -6$
$2x + y + 4z = 4$

(1) Using
$x + 2y + 3z = 5$
$\underline{-x + y - z = -6}$
$3y + 2z = -1$

(2) Using
$-x + y - z = -6$ (times 2)
$2x + y + 4z = 4$

$\underline{\begin{array}{l}-2x + 2y - 2z = -12\\ 2x + y + 4z = 4\end{array}}$
$3y + 2z = -8$

(3) Solving
$3y + 2z = -1$ (times -1)
$\underline{3y + 2z = -8}$

$\begin{array}{l}-3y - 2z = 1\\ \underline{3y + 2z = -8}\\ 0 = -7\end{array}$

The system is *inconsistent*. There are no common solutions. The solution set is $\emptyset$.

13. $x - 4y + z = -5$
$3x - 12y + 3z = -15$
$-2x + 8y - 2z = 10$

Using $x - 4y + z = -5$ (times -3)
$3x - 12y + 3z = -15$

$\begin{array}{l}-3x + 12y - 3z = 15\\ \underline{3x - 12y + 3z = -15}\\ 0 = 0\end{array}$

The system is *dependent*. The solution set is $\{(x,y,z)|x - 4y + z = -5\}$ (or either of the other two equations).

18. $x - y = -1$
$x + z = -2$
$y - z = 2$

(1) Using $x + z = 2$
$\underline{y - z = -2}$
$x + y = 0$

(2) Using $x - y = -1$
$\underline{x + y = 0}$
$2x \quad\quad = -1$
$x = -\dfrac{1}{2}$

(3) Using $x - y = -1$ and substituting $-\dfrac{1}{2}$ for x

$-\dfrac{1}{2} - y = -1$

$-y = -\dfrac{1}{2}$

$y = \dfrac{1}{2}$

(4) Using $y - z = 2$, $\dfrac{1}{2} - z = 2$,

$-z = \dfrac{3}{2}$

$z = -\dfrac{3}{2}$

The solution set is $\left\{\left(-\dfrac{1}{2}, \dfrac{1}{2}, -\dfrac{3}{2}\right)\right\}$.

33. Let $x =$ the length of the shortest side
$y =$ the length of the middle side
$z =$ the length of the longest side.

Then $x + y + z = 122$
$z = x + y$ and
$2x = z - 11$

(1) Add $x + y + z = 122$
$\underline{-x - y + z = 0}$
$2z = 122$
$z = 61$

$x + y + z = 122$
$-x - y + z = 0$
$2x - z = -11$

(2) Substitute in $2x - z = -11$
$2x - 61 = -11$
$2x = 50$
$x = 25$

(3) Substitute in $x + y + z = 122$
$25 + y + 61 = 122$
$y + 86 = 122$
$y = 36$

The sides have length 25 meters, 36 meters, and 61 meters.

38. (1) Using $(0,5)$,
$5 = a(0)^2 + b(0) + c$
$c = 5$

(2) Using $(-1,2)$,
$2 = a - b + c$

(3) Using $(2,17)$,
$17 = a(2)^2 + b(2) + c$
$17 = 4a + 2b + c$

(4) Solve the system.
$a - b + c = 2$
$4a + 2b + c = 17$
$c = 5$

(5) Substitute 5 for c.
$a - b + 5 = 2$
$4a + 2b + 5 = 17$

then $a - b = -3$
$4a + 2b = 12$

(times 2) and $2a - 2b = -6$
$\underline{4a + 2b = 12}$
$6a = 6$
$a = 1$

(6) Substitute 1 for a in
$a - b = -3$
$1 - b = -3$
$-b = -4$
$b = 4$

Therefore $a = 1$, $b = 4$, $c = 5$.

Review exercises

1. 2.47×10^{-4} **2.** $\dfrac{2y^2 - 6y + 18}{(y - 3)(y + 1)(y - 1)}$ **3.** 16

4.

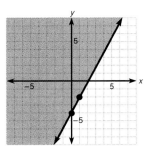

5. $d = \sqrt{65}; \left(-\dfrac{5}{2}, -1\right)$ **6.** $\left\{-\dfrac{2\sqrt{3}}{3}, \dfrac{2\sqrt{3}}{3}, -i, i\right\}$

Exercise 12–4

Answers to odd-numbered problems

1. $\{(2,-1), (-1,2)\}$ **3.** $\{(7,-5), (-5,1)\}$

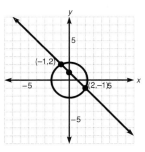   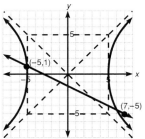

5. $\{(2,1), (-1,4)\}$

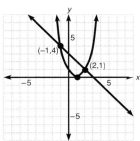

7. $\left\{\left(-\dfrac{2}{9}, -\dfrac{22}{9}\right), (2,2)\right\}$ **9.** $\{(4,1), (-4,-1), (i,-4i), (-i,4i)\}$

11. $\{(2,2), (-2,-2)\}$ **13.** $\left\{\left(\dfrac{\sqrt{6}}{2}, -1\right), \left(-\dfrac{\sqrt{6}}{2}, -1\right)\right\}$

15. $\{(3,0), (-3,0)\}$ **17.** $\{(2\sqrt{3}, \sqrt{13}), (2\sqrt{3}, -\sqrt{13}),$
$(-2\sqrt{3}, \sqrt{13}), (-2\sqrt{3}, -\sqrt{13})\}$ **19.** $\{(3,3), (3,-3), (-3,3),$
$(-3,-3)\}$ **21.** $\{(3,0), (-3,0)\}$ **23.** $\{(-\sqrt{7}, -i\sqrt{3}),$
$(\sqrt{7}, -i\sqrt{3}), (-\sqrt{7}, i\sqrt{3}), (-\sqrt{7}, i\sqrt{3})\}$ **25.** $\{(\sqrt{3}, \sqrt{3}),$
$(-\sqrt{3}, -\sqrt{3})\}$ **27.** 15 in. by 8 in.   **29.** 9 and 7   **31.** $x = 2$
when $p = 3; \left(x = \dfrac{15}{4}\text{ when }p = \dfrac{8}{5}; \text{ not logical}\right)$

Solutions to trial exercise problems

9. $x^2 - y^2 = 15$
$xy = 4$

(1) $y = \dfrac{4}{x}$ and substituting, $x^2 - \left(\dfrac{4}{x}\right)^2 = 15$

$$x^2 - \frac{16}{x^2} = 15$$
$$x^4 - 15x^2 - 16 = 0$$
$$(x^2 - 16)(x^2 + 1) = 0$$
$$(x + 4)(x - 4)(x^2 + 1) = 0$$

Then $x = 4$ or $x = -4$; $x = \sqrt{-1} = i$ or $x = -\sqrt{-1} = -i$

(2) $y = \dfrac{4}{x}$, so $y = \dfrac{4}{4} = 1$ or $y = \dfrac{4}{-4} = -1$; or $y = \dfrac{4}{i} = -4i$ or $y = \dfrac{4}{-i} = 4i$

The solution set is $\{(4,1),\ (-4,-1),\ (i,-4i),\ (-i,4i)\}$.

21. $2x^2 - 9y^2 = 18$

$\dfrac{4x^2 + 9y^2 = 36}{6x^2 \qquad = 54}$

$x^2 = 9$
$x = \pm 3$

(1) When $x = 3$
$2(3)^2 - 9y^2 = 18$
$18 - 9y^2 = 18$
$-9y^2 = 0$
$y = 0$

(2) When $x = -3$
$2(-3)^2 - 9y^2 = 18$
$18 - 9y^2 = 18$
$-9y^2 = 0$
$y = 0$

The solution set is $\{(3,0),\ (-3,0)\}$.

24. $x^2 + y^2 = 10$ (times -1)
$x^2 - 3xy + y^2 = 1$

$-x^2 \qquad - y^2 = -10$
$\dfrac{x^2 - 3xy + y^2 = 1}{\qquad -3xy = -9}$
$xy = 3$
$y = \dfrac{3}{x}$

(2) Substituting in $x^2 + y^2 = 10$

$$x^2 + \left(\frac{3}{x}\right)^2 = 10$$
$$x^2 + \frac{9}{x^2} = 10$$
$$x^4 + 9 = 10x^2$$
$$x^4 - 10x^2 + 9 = 0$$
$$(x^2 - 9)(x^2 - 1) = 0$$

So $x^2 - 9 = 0$; $x^2 = 9$; $x = \pm 3$
$x^2 - 1 = 0$; $x^2 = 1$; $x = \pm 1$

When $x = 3$, $y = \dfrac{3}{3} = 1$

$x = -3$, $y = \dfrac{3}{-3} = -1$

$x = 1$, $y = \dfrac{3}{1} = 3$

$x = -1$, $y = \dfrac{3}{-1} = -3$

The solution set is $\{(3,1),\ (-3,-1),\ (1,3),\ (-1,-3)\}$.

29. Let $x =$ the larger number.
Let $y =$ the smaller number.
Then $x + y = 16$
$x^2 - y^2 = 32$

(1) Now $y = 16 - x$, and substituting, $x^2 - (16 - x)^2 = 32$
$x^2 - 256 + 32x - x^2 = 32$
$32x = 288$
$x = 9$

(2) Substituting, $9 + y = 16$, so $y = 7$.
The numbers are 9 and 7.

32. Let ℓ = the length of the rectangle.
Let w = the width of the rectangle.
Then $\ell w = 260$
Since there is a 2-inch square cut out of each corner, the length and width of the box is 4 inches shorter than the length and width of the piece of cardboard. Thus, using ℓwh = volume,
$$2(\ell - 4)(w - 4) = 288$$
So, (1) $\ell = \dfrac{260}{w}$ and
$$\ell w - 4\ell - 4w + 16 = 144$$
Then $\ell w - 4\ell - 4w = 128$
(2) Substituting $\dfrac{260}{w}$ for ℓ,
$$w\left(\dfrac{260}{w}\right) - 4\left(\dfrac{260}{w}\right) - 4w = 128$$
$$260 - \dfrac{1,040}{w} - 4w = 128$$
$$-\dfrac{1,040}{w} - 4w = -132 \text{ (times } -w)$$
$$1,040 + 4w^2 = 132w$$
So $4w^2 - 132w + 1,040 = 0$
$$4(w^2 - 33w + 260) = 0$$
$$4(w - 13)(w - 20) = 0$$
If $w = 13$, then $\ell = 20$ and if $w = 20$, then $\ell = 13$
(3) Then $\ell = \dfrac{260}{13} = 20$
or $\ell = \dfrac{260}{20} = 13$
The rectangle has dimensions 13 inches by 20 inches.

Review exercises

1. -6 **2.** 12 **3.** $\left\{y \mid y \geq \dfrac{2}{3}\right\}$ **4.** -17 **5.** 3 **6.** 0
7. $\{x \mid 1 \leq x \leq 4\}$

Exercise 12–5

Answers to odd-numbered problems

1.

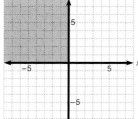

3.

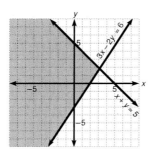

5.

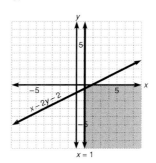

7.

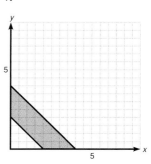

9.

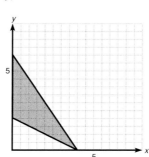

11.

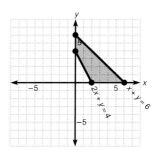

13. maximum value of 15 at $(3,3)$; minimum value of 2 at $(0,1)$
15. maximum value of 2,500 at $(50,40)$; minimum value of 750 at $(0,30)$ **17.** maximum value of 190 at $(30,40)$ and minimum value of 20 at $(20,0)$

19.

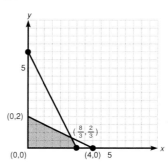

maximum value at $\left(\dfrac{8}{3}, \dfrac{2}{3}\right)$

21.

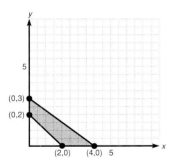

minimum value of 8 at (2,0)

23. profit function $P = 100x + 125y$

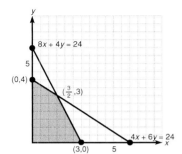

use $4x + 6y \leq 24$
$8x + 4y \leq 24$
$x \geq 0$
$y \geq 0$

To maximize profit, must manufacture $1\frac{1}{2}$ cases of chocolate chip and 3 cases of peanut butter cookies

25. $P = 40x + 60y$

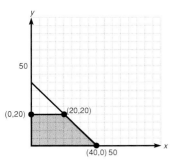

$0 \leq x \leq 50$
$0 \leq y \leq 20$
$x + y \leq 40$
maximum profit at (20,20) with maximum profit of $2,000

27. $C = 12x + 10y$

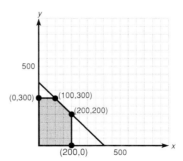

$0 \leq x \leq 200$
$0 \leq y \leq 300$
$x + y \leq 400$
To minimize cost, plant no corn and 200 acres of wheat.

29. $C = 35x + 50y$

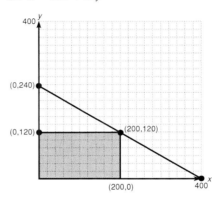

$0 \leq x \leq 200$
$0 \leq y \leq 120$
$8x + 12y \leq 3,200$
To minimize cost, manufacture no 14-inch and 120 16-inch saws.

Solution to trial exercise problems

8.

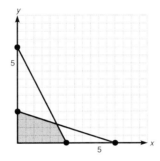

14.

vertex	(0,6)	(3,5)	(6,2)	(6,0)	(0,0)
$f = 6x + y$	6	23	38	36	0

maximum value of 38 at (6,2) and minimum value of 0 at (0,0)
25. Let x = sets of model A and y = sets of model B. Then $x \leq 50$ and $y \leq 20$ and $x + y \leq 40$. The profit function $P = 40x + 60y$.

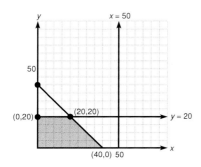

vertex	(0,20)	(20,20)	(40,0)
$P = 40x + 60y$	1,200	2,000	1,600

maximum profit of $2,000 when 20 sets of each type are manufactured

Review exercises

1. -1 **2.** -42 **3.** 7 in.   **4.** $\emptyset$; $-\dfrac{3}{5}$ is extraneous   **5. a.** 1

b. $9x^2 + 6x - 3$ **c.** -8

Chapter 12 review

1. $\{(3,1)\}$ **2.** $\left\{\left(\dfrac{1}{2},\dfrac{1}{3}\right)\right\}$ **3.** $\{(x,y)|2x + 3y = 4\}$; dependent

4. $\emptyset$; inconsistent   **5.** $\left\{\left(\dfrac{5}{4},-\dfrac{1}{4}\right)\right\}$ **6.** $\left\{\left(-\dfrac{12}{5},\dfrac{26}{5}\right)\right\}$

7. $F_1 = \dfrac{39}{5}$; $F_2 = \dfrac{3}{5}$ **8.** $\{(3,-4)\}$ **9.** $\{(-1,-10\}$

10. $\left\{\left(-\dfrac{2}{7},\dfrac{1}{7}\right)\right\}$ **11.** $\left\{\left(-\dfrac{11}{4},\dfrac{27}{4}\right)\right\}$ **12.** $\left\{\left(-\dfrac{7}{3},-\dfrac{5}{6}\right)\right\}$

13. $\left(\dfrac{11}{5},-\dfrac{3}{5}\right)$ **14.** $\ell = 60$ ft; $w = 30$ ft   **15.** 4 false alarms; 26 real alarms   **16.** $1,800 at 6½%; $1,200 at 7%   **17.** 16⅔ g of 20% tin; 33⅓ g of 5% tin   **18.** $72 for topcoat; $105 for suit   **19.** 5.2 hr

20. $\{(1,2,3)\}$ **21.** $\left\{\left(\dfrac{1}{10},\dfrac{1}{2},-\dfrac{3}{10}\right)\right\}$ **22.** $\{(3,-5,-1)\}$

23. $\{(3,2,4)\}$ **24.** $a = 3, b = -8, c = 2$

25. $F_1 = \dfrac{35}{2}$, $F_2 = -5$, $F_3 = 25$ **26.** $41°, 57°, 82°$

27. 135 nickels; 120 dimes; 30 quarters   **28.** $\{(1,1)\}$ **29.** $\{(4,3),$ $(3,4)\}$ **30.** $\{(3,2i), (3,-2i)\}$ **31.** $\{(2\sqrt{2},-2), (-2\sqrt{2},-2)\}$

32. $\left\{\left(\dfrac{4}{3},\dfrac{5}{3}\right)\right\}$ **33.** $\{(2,2i), (2,-2i), (-2,2i), (-2,-2i)\}$

34. $\{(\sqrt{19},\sqrt{3}), (\sqrt{19},-\sqrt{3}), (-\sqrt{19},\sqrt{3}), (-\sqrt{19},-\sqrt{3})\}$
35. 9 in. by 17 in.

36.

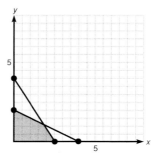

37. maximum value of 13 at (5,2); minimum value of 0 at (0,0)
38.

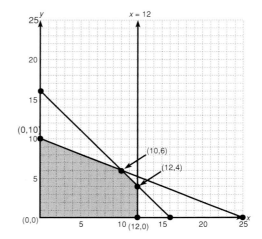

Maximum profit is $112 when raising 12 geese and 4 pigs.

Chapter 12 cumulative test

1. -258 **2.** $V = 110$ **3.** $3a^2 - 2a + 6 + \dfrac{15}{a - 3}$

4. $6a^5 - 5a^3 + 3a$ **5.** $4x^2 - 56x - 45$ **6.** $\dfrac{2x^4}{y^5}$ **7.** $\left\{\dfrac{14}{13}\right\}$

8. $\left\{\dfrac{1 + \sqrt{13}}{3}, \dfrac{1 - \sqrt{13}}{3}\right\}$ **9.** $\{-3 + \sqrt{14}, -3 - \sqrt{14}\}$

10. $\left\{z|z \geq -\dfrac{4}{3}\right\}$ **11.** $\{x|-3 \leq x \leq 1\}$ **12.** $\left\{\dfrac{5}{2},-\dfrac{3}{2}\right\}$

13. $\left\{x|x < \dfrac{1}{2} \text{ or } x > \dfrac{7}{6}\right\}$ **14.** $\dfrac{4}{2y - 3}$

15. $\dfrac{4x^2 - 13x - 12}{3}$ **16.** $\dfrac{4x + 2}{(x - 5)(x + 4)(x - 3)}$ **17.** $\dfrac{2y}{y - 7}$

18. $\dfrac{3p^2ab}{14q}$ **19.** 9   **20.** $\dfrac{1}{y^3}$ **21.** $a^{5/4}$ **22.** $4\sqrt{5}$

23. $2\sqrt{3} + 3\sqrt{2} - 4\sqrt{6} - 12$ **24.** $-9 + 40i$ **25.** $-2y\sqrt[3]{x}$
26. $\emptyset$; 36 and 1 are extraneous

27.

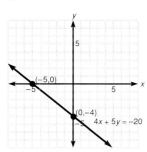

28.

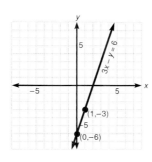

29. $5x + 7y = 28$ **30.** $x + 4y = 13$

31.

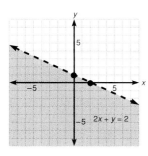

32. a. $f(-3) = -9$ **b.** $(f - g)(4) = -16$ **c.** $(f \circ g)(-1)$
$= 9$ **33.** neither even nor odd   **34.** $f^{-1}(x) = \dfrac{x + 3}{4}$

35. a. circle   **b.** hyperbola   **c.** parabola   **36.** $\{-1, 5, -i\sqrt{3}, i\sqrt{3}\}$

37. $x = -\dfrac{3}{4}$ is horizontal asymptote; $y = \dfrac{3}{4}$ is vertical asymptote;

y-intercept, $-\dfrac{2}{3}$; x-intercept, $\dfrac{2}{3}$ **38.** $\left\{\dfrac{3}{4}\right\}$ **39.** $\left\{\dfrac{1}{8}\right\}$

40. $\left\{\left(\dfrac{3}{17}, -\dfrac{4}{17}\right)\right\}$ **41.** $\left\{\left(-\dfrac{1}{6}, -\dfrac{2}{3}, \dfrac{1}{6}\right)\right\}$

42. $\{(-i\sqrt{10}, -3), (-i\sqrt{10}, 3), (i\sqrt{10}, -3), (i\sqrt{10}, 3)\}$

43. maximum value of $37\dfrac{1}{2}$ at $\left(\dfrac{15}{2}, 0\right)$

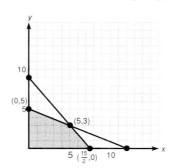

Chapter 13

Exercise 13–1

Answers to odd-numbered problems

1. 11   **3.** -24 **5.** 5   **7.** -35 **9.** -12 **11.** -14 **13.** 0
15. -4 **17.** 29   **19.** -14 **21.** -24 **23.** a^3 **25.** $y^3 - x^2y$
27. -19 **29.** 37

Solutions to trial exercise problems

3. $\begin{vmatrix} 4 & -2 \\ -6 & -3 \end{vmatrix} = -12 - (12) = -24$ **9.** $\begin{vmatrix} 1 & 2 & 3 \\ 3 & 2 & 1 \\ 2 & 1 & 3 \end{vmatrix} = 1\begin{vmatrix} 2 & 1 \\ 1 & 3 \end{vmatrix} - 3\begin{vmatrix} 2 & 3 \\ 1 & 3 \end{vmatrix} + 2\begin{vmatrix} 2 & 3 \\ 2 & 1 \end{vmatrix}$ (Use first column.)

$= 1(6 - 1) - 3(6 - 3) + 2(2 - 6)$
$= 1(5) - 3(3) + 2(-4) = 5 - 9 - 8 = -12$

16. $\begin{vmatrix} -1 & -1 & -1 \\ 2 & 2 & 2 \\ 3 & -3 & 3 \end{vmatrix} = -2\begin{vmatrix} -1 & -1 \\ -3 & 3 \end{vmatrix} + 2\begin{vmatrix} -1 & -1 \\ 3 & 3 \end{vmatrix} - 2\begin{vmatrix} -1 & -1 \\ 3 & -3 \end{vmatrix}$ (Use second row.)

$= -2(-3 - 3) + 2(-3 + 3) - 2(3 + 3)$
$= -2(-6) + 2(0) - 2(6) = 12 + 0 - 12 = 0$

$\begin{vmatrix} a & 0 & a \\ 0 & a & 0 \\ 0 & 0 & a \end{vmatrix} = a\begin{vmatrix} a & 0 \\ 0 & a \end{vmatrix} - 0\begin{vmatrix} 0 & a \\ 0 & a \end{vmatrix} + 0\begin{vmatrix} 0 & a \\ a & 0 \end{vmatrix}$ (Use first column.)

$= a(a^2 - 0) - 0(0 - 0) + 0(0 - a^2)$
$= a^3 - 0 + 0 = a^3$

27. $\begin{vmatrix} 1 & 2 & 3 & -1 \\ 2 & 0 & 1 & 3 \\ -2 & 1 & 0 & -1 \\ 0 & 3 & 2 & 0 \end{vmatrix} = 1\begin{vmatrix} 0 & 1 & 3 \\ 1 & 0 & -1 \\ 3 & 2 & 0 \end{vmatrix} - 2\begin{vmatrix} 2 & 1 & 3 \\ -2 & 0 & -1 \\ 0 & 2 & 0 \end{vmatrix} + 3\begin{vmatrix} 2 & 0 & 3 \\ -2 & 1 & -1 \\ 0 & 3 & 0 \end{vmatrix} - (-1)\begin{vmatrix} 2 & 0 & 1 \\ -2 & 1 & 0 \\ 0 & 3 & 2 \end{vmatrix}$ (Use first row.)

$= 1\left[0\begin{vmatrix} 0 & -1 \\ 2 & 0 \end{vmatrix} - 1\begin{vmatrix} 1 & 3 \\ 2 & 0 \end{vmatrix} + 3\begin{vmatrix} 1 & 3 \\ 0 & -1 \end{vmatrix} \right] - 2\left[2\begin{vmatrix} 0 & -1 \\ 2 & 0 \end{vmatrix} + 2\begin{vmatrix} 1 & 3 \\ 2 & 0 \end{vmatrix} + 0\begin{vmatrix} 1 & 3 \\ 0 & -1 \end{vmatrix} \right]$ (Use first column.)

$+ 3\left[2\begin{vmatrix} 1 & -1 \\ 3 & 0 \end{vmatrix} + 2\begin{vmatrix} 0 & 3 \\ 3 & 0 \end{vmatrix} + 0\begin{vmatrix} 0 & 3 \\ 1 & -1 \end{vmatrix} \right] + 1\left[2\begin{vmatrix} 1 & 0 \\ 3 & 2 \end{vmatrix} + 2\begin{vmatrix} 0 & 1 \\ 3 & 2 \end{vmatrix} + 0\begin{vmatrix} 0 & 1 \\ 1 & 0 \end{vmatrix} \right]$

$= 1[0 - 1(-6) + 3(-1)] - 2[2(2) + 2(-6) + 0] + 3[2(3) + 2(-9) + 0] + 1[2(2) + 2(-3) + 0]$
$= 3 + 16 - 36 - 2 = 19 - 38 = -19$

Review exercises

1. $\{(-2,2)\}$

2. $y = 3x - 2$

3. $x = -2$

4. $2x + 3y = -6$

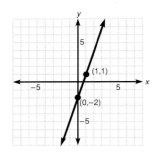

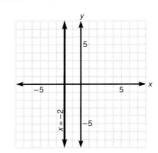

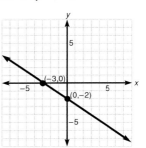

5. 16   **6.** $\{1,4\}$

Exercise 13–2

Answers to odd-numbered problems

1. $\left\{\left(\dfrac{5}{3}, -\dfrac{1}{3}\right)\right\}$ **3.** $\left\{\left(\dfrac{-2}{19}, \dfrac{24}{19}\right)\right\}$ **5.** $\emptyset$; inconsistent

7. $\{(x,y)|2x + 3y = -1\}$; dependent   **9.** $\{(0,-1)\}$

11. $\left\{\left(\dfrac{5}{2}, 4\right)\right\}$ **13.** $\{(-2,5)\}$ **15.** $\{(9,-2)\}$ **17.** $\{(-1,2,3)\}$ **19.** $\{(0,0,0)\}$

21. $\{(x,y,z)|3x - y - 6z = 5\}$; dependent   **23.** $\emptyset$; inconsistent
25. $\{(x,y,z)|3x + 2y + 4z = 3\}$; dependent   **27.** $\emptyset$; inconsistent

29. $\left\{\left(1, -6, -\dfrac{1}{2}\right)\right\}$ **31.** $\{(2,2,2)\}$ **33.** $\left\{\left(\dfrac{4}{3}, 3, \dfrac{8}{3}\right)\right\}$ **35.** $\{(-5,-2,1)\}$

37. $w = 10\dfrac{1}{2}$ ft; $\ell = 12\dfrac{1}{2}$ ft   **39.** cream-filled, $0.28; jelly-filled,

$0.32   **41.** 13 and 21   **43.** 60 oz of 15%; 40 oz of 8%
45. pop, $2; candy, $1.50   **47.** $A = 73°$, $B = 73°$, $C = 34°$
49. 101—$3; 121—$4; 103—$5   **51.** 10, 11, 15
53. 11 ft, 27 ft, 36 ft

Solutions to trial exercise problems

1. $x - y = 2$
$2x + y = 3$ $D = \begin{vmatrix} 1 & -1 \\ 2 & 1 \end{vmatrix} = 1 - (-2) = 3$

$$D_x = \begin{vmatrix} 2 & -1 \\ 3 & 1 \end{vmatrix} = 2 - (-3) = 5$$

$$D_y = \begin{vmatrix} 1 & 2 \\ 2 & 3 \end{vmatrix} = 3 - 4 = -1$$

Then $x = \dfrac{D_x}{D} = \dfrac{5}{3}$; $y = \dfrac{D_y}{D} = -\dfrac{1}{3}$

The solution set is $\left\{ \left(\dfrac{5}{3}, -\dfrac{1}{3} \right) \right\}$.

4. $4x - y = 3$
$8x - 2y = 1$ $D = \begin{vmatrix} 4 & -1 \\ 8 & -2 \end{vmatrix} = -8 + 8 = 0$

$$D_x = \begin{vmatrix} 3 & -1 \\ 1 & -2 \end{vmatrix} = -6 + 1 = -5$$

$$D_y = \begin{vmatrix} 4 & 3 \\ 8 & 1 \end{vmatrix} = 4 - 24 = -20$$

Since $D = 0$, and at least one of D_x and D_y is nonzero, the system is *inconsistent*. The solution set is $\emptyset$.

11. $6x - 2y = 7$
$2y = 8$ $D = \begin{vmatrix} 6 & -2 \\ 0 & 2 \end{vmatrix} = 12 - 0 = 12$

$$D_x = \begin{vmatrix} 7 & -2 \\ 8 & 2 \end{vmatrix} = 14 - (-16) = 30$$

$$D_y = \begin{vmatrix} 6 & 7 \\ 0 & 8 \end{vmatrix} = 48 - 0 = 48$$

$$x = \frac{D_x}{D} = \frac{30}{12} = \frac{5}{2}; \quad y = \frac{D_y}{D} = \frac{48}{12} = 4$$

The solution set is $\left\{ \left(\dfrac{5}{2}, 4 \right) \right\}$.

17. $4x - y + 2z = 0$
$2x + y + z = 3$
$3x - y + z = -2$

$$D = \begin{vmatrix} 4 & -1 & 2 \\ 2 & 1 & 1 \\ 3 & -1 & 1 \end{vmatrix} = 4 \begin{vmatrix} 1 & 1 \\ -1 & 1 \end{vmatrix} - 2 \begin{vmatrix} -1 & 2 \\ -1 & 1 \end{vmatrix} + 3 \begin{vmatrix} -1 & 2 \\ 1 & 1 \end{vmatrix} \text{ (Use first column.)}$$

$$= 4(2) - 2(1) + 3(-3)$$
$$= -3$$

$$D_x = \begin{vmatrix} 0 & -1 & 2 \\ 3 & 1 & 1 \\ -2 & -1 & 1 \end{vmatrix} = 0 \begin{vmatrix} 1 & 1 \\ -1 & 1 \end{vmatrix} - 3 \begin{vmatrix} -1 & 2 \\ -1 & 1 \end{vmatrix} + (-2) \begin{vmatrix} -1 & 2 \\ 1 & 1 \end{vmatrix} \qquad x = \frac{D_x}{D} = \frac{3}{-3} = -1$$

$$= 0 - 3(1) - 2(-3) = 3$$

$$D_y = \begin{vmatrix} 4 & 0 & 2 \\ 2 & 3 & 1 \\ 3 & -2 & 1 \end{vmatrix} = 4 \begin{vmatrix} 3 & 1 \\ -2 & 1 \end{vmatrix} - 2 \begin{vmatrix} 0 & 2 \\ -2 & 1 \end{vmatrix} + 3 \begin{vmatrix} 0 & 2 \\ 3 & 1 \end{vmatrix} \qquad y = \frac{D_y}{D} = \frac{-6}{-3} = 2$$

$$= 4(5) - 2(4) + 3(-6) = -6$$

$$D_z = \begin{vmatrix} 4 & -1 & 0 \\ 2 & 1 & 3 \\ 3 & -1 & -2 \end{vmatrix} = 4 \begin{vmatrix} 1 & 3 \\ -1 & -2 \end{vmatrix} - 2 \begin{vmatrix} -1 & 0 \\ -1 & -2 \end{vmatrix} + 3 \begin{vmatrix} -1 & 0 \\ 1 & 3 \end{vmatrix} \qquad z = \frac{D_z}{D} = \frac{-9}{-3} = 3$$

$$= 4(1) - 2(2) + 3(-3) = -9$$

The solution set is $\{(-1, 2, 3)\}$.

30. $x + y = 1$

$\quad y + 2z = -2$

$\quad 2x - z = 0$

$$D = \begin{vmatrix} 1 & 1 & 0 \\ 0 & 1 & 2 \\ 2 & 0 & -1 \end{vmatrix} = 1\begin{vmatrix} 1 & 2 \\ 0 & -1 \end{vmatrix} - 0\begin{vmatrix} 1 & 0 \\ 0 & -1 \end{vmatrix} + 2\begin{vmatrix} 1 & 0 \\ 1 & 2 \end{vmatrix} \text{(Use first column.)}$$

$$= 1(-1) - 0 + 2(2) = 3$$

$$D_x = \begin{vmatrix} 1 & 1 & 0 \\ -2 & 1 & 2 \\ 0 & 0 & -1 \end{vmatrix} = 1\begin{vmatrix} 1 & 2 \\ 0 & -1 \end{vmatrix} - (-2)\begin{vmatrix} 1 & 0 \\ 0 & -1 \end{vmatrix} + 0\begin{vmatrix} 1 & 0 \\ 1 & 2 \end{vmatrix}$$

$$= 1(-1) + 2(-1) + 0 = -3$$

$$x = \frac{D_x}{D} = \frac{-3}{3} = -1$$

$$D_y = \begin{vmatrix} 1 & 1 & 0 \\ 0 & -2 & 2 \\ 2 & 0 & -1 \end{vmatrix} = 1\begin{vmatrix} -2 & 2 \\ 0 & -1 \end{vmatrix} - 0\begin{vmatrix} 1 & 0 \\ 0 & -1 \end{vmatrix} + 2\begin{vmatrix} 1 & 0 \\ -2 & 2 \end{vmatrix}$$

$$= 1(2) - 0 + 2(2) = 6$$

$$y = \frac{D_y}{D} = \frac{6}{3} = 2$$

$$D_z = \begin{vmatrix} 1 & 1 & 1 \\ 0 & 1 & -2 \\ 2 & 0 & 0 \end{vmatrix} = 1\begin{vmatrix} 1 & -2 \\ 0 & 0 \end{vmatrix} - 0\begin{vmatrix} 1 & 1 \\ 0 & 0 \end{vmatrix} + 2\begin{vmatrix} 1 & 1 \\ 1 & -2 \end{vmatrix}$$

$$= 1(0) - 0 + 2(-3) = -6$$

$$z = \frac{D_z}{D} = \frac{-6}{3} = -2$$

The solution set is $\{(-1, 2, -2)\}$.

39. Let x = cost of cream-filled; y = cost of jelly-filled.

$\quad 5x + 7y = 3.64$

$\quad \underline{3x + 9y = 3.72}$

$$D = \begin{vmatrix} 5 & 7 \\ 3 & 9 \end{vmatrix} = (5)(9) - (3)(7) = 24$$

$$D_x = \begin{vmatrix} 3.64 & 7 \\ 3.72 & 9 \end{vmatrix} = 9(3.64) - 7(3.72) = 6.72$$

$$D_y = \begin{vmatrix} 5 & 3.64 \\ 3 & 3.72 \end{vmatrix} = 5(3.72) - 3(3.64) = 7.68$$

$$x = \frac{D_x}{D} = \frac{6.72}{24} = 0.28; \quad y = \frac{D_y}{D} = \frac{7.68}{24} = 0.32$$

cream-filled, \$0.28 and jelly-filled, \$0.32

48. Let $x, y,$ and z be the three numbers.

(1) $x + y + z = 47$

(2) $2x + y + z = 56$

(3) $x + 3y + z = 81$

(2) $\quad 2x + y + z = 56$ (2) $2x + \ y + z = 56 \rightarrow$ $-6x - 3y - 3z = -168$

(1) $\underline{(-)x + y + z = 47}$ (3) $\underline{\ x + 3y + z = 81}$ $\underline{\ \ x + 3y + \ \ z = 81}$

$\qquad x \qquad\quad = 9$ $\qquad\qquad\qquad\qquad\qquad\qquad$ $-5x \qquad - 2z = -87$

$-5(9) - 2z = -87$ $\qquad x + y + z = 47$

$-45 - 2z = -87$ $\qquad 9 + y + 21 = 47$

$\qquad -2z = -42$ $\qquad\quad y + 30 = 47$

$\qquad\quad z = 21$ $\qquad\qquad\quad y = 17$

The three numbers are 9, 17, and 21.

Review exercises

1. $(-2,12), (-1,9), (0,6), (1,3), (2,0)$ **2.** $\left\{ x \mid x < -3 \text{ or } x > \dfrac{1}{2} \right\}$

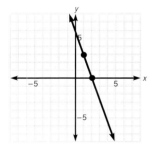

3. $\left\{ \dfrac{1}{3}, 5 \right\}$ **4.** $\{-3, 1\}$ **5.** $37 - 20\sqrt{3}$ **6.** 23 **7.** i

Exercise 13-3

Answers to odd-numbered problems

1. $\{(2,3)\}$ **3.** $\{(-2,1)\}$ **5.** $\{(3,-1)\}$

7. $\left\{ \left(-\dfrac{1}{17}, -\dfrac{5}{17} \right) \right\}$ **9.** $\left\{ \left(-5, \dfrac{1}{2} \right) \right\}$

11. $\{(x,y) \mid 4x - 2y = 1\}$; dependent   **13.** $\{(0,-1)\}$

15. $\left\{ \left(\dfrac{1}{5}, \dfrac{4}{5}, -\dfrac{16}{5} \right) \right\}$ **17.** $\{(2,0,-1)\}$

19. $\{(4,1,0)\}$ **21.** $\{(x,y,z) \mid x - y = 1\}$; dependent

23. $a = -\dfrac{8}{7}, \; b = \dfrac{12}{7}$ **25.** 10 small, 7 intermediate, 7 large

Solutions to trial exercise problems

3. $x - 4y = -6$ augmented matrix $\begin{bmatrix} 1 & -4 & -6 \\ 3 & 1 & -5 \end{bmatrix}$
  $3x + y = -5$

We want 0 in the second row, first column. Multiply row one by -3 and add to row two. We get $\begin{bmatrix} 1 & -4 & -6 \\ 0 & 13 & 13 \end{bmatrix}$

We now have the system $x - 4y = -6$

$$13y = 13$$

Then $y = 1$ and replace y by 1 in the first equation $x - 4(1) = -6$
$$x - 4 = -6$$
$$x = -2$$

The solution set is $\{(-2,1)\}$.

10. $-x - y = 4$ augmented matrix $\begin{bmatrix} -1 & -1 & 4 \\ 2 & 2 & -1 \end{bmatrix}$
  $2x + 2y = -1$

Multiply row one by 2 and add to row two.

$\begin{bmatrix} -1 & -1 & 4 \\ 0 & 0 & 7 \end{bmatrix}$ We obtain the second row statement $0 = 7$, which is false.

The system is inconsistent and the solution set is $\emptyset$.

14. $x + 3y - z = 5$ augmented matrix
$3x - y + 2z = 5$
$x + y + 2z = 7$

$$\begin{bmatrix} 1 & 3 & -1 & | & 5 \\ 3 & -1 & 2 & | & 5 \\ 1 & 1 & 2 & | & 7 \end{bmatrix}$$

Multiply row one by -3 and add to row two.

$$\begin{bmatrix} 1 & 3 & -1 & | & 5 \\ 0 & -10 & 5 & | & -10 \\ 1 & 1 & 2 & | & 7 \end{bmatrix}$$

Multiply row one by -1 and add to row two.

$$\begin{bmatrix} 1 & 3 & -1 & | & 5 \\ 0 & -10 & 5 & | & -10 \\ 0 & -2 & 3 & | & 2 \end{bmatrix}$$

Multiply row two by $-\frac{1}{10}$.

$$\begin{bmatrix} 1 & 3 & -1 & | & 5 \\ 0 & 1 & -\frac{1}{2} & | & 1 \\ 0 & -2 & 3 & | & 2 \end{bmatrix}$$

Multiply row two by 2 and add to row three.

$$\begin{bmatrix} 1 & 3 & -1 & | & 5 \\ 0 & 1 & -\frac{1}{2} & | & 1 \\ 0 & 0 & 2 & | & 4 \end{bmatrix}$$

$x + 3y - z = 5$
$y - \frac{1}{2}z = 1$
$2z = 4$
$z = 2$

Replace z by 2 in $y - \frac{1}{2}z = 1$
$y - 1 = 1$
$y = 2$
Replace z by 2 and y by 2 in $x + 3y - z = 5$.
$x + 3(2) - 2 = 5$
$x + 6 - 2 = 5$
$x + 4 = 5$
$x = 1$
The solution set is $\{(1,2,2)\}$.

Review exercises

1. x-intercept, $(-4,0)$; y-intercept, $(0,8)$ **2.** x-intercept, $(-3,0)$;
y-intercept, $\left(0,\frac{9}{2}\right)$ **3.** no x-intercept; y-intercept, $(0,6)$

4. $x^2 + 8x + 16 = (x + 4)^2$ **5.** $y^2 - 5y + \frac{25}{4} = \left(y - \frac{5}{2}\right)^2$

6. $z^2 - \frac{1}{2}z + \frac{1}{16} = \left(z - \frac{1}{4}\right)^2$ **7.** $(y - 7)^2$ **8.** $(x + 5)^2$

Chapter 13 review

1. 7 **2.** 24 **3.** 41 **4.** 108 **5.** 39 **6.** 88 **7.** $\{(1,-1)\}$

8. $\left\{\left(-\frac{5}{26},\frac{2}{13}\right)\right\}$ **9.** $\left\{\left(\frac{15}{4},9\right)\right\}$ **10.** $\left\{\left(\frac{8}{3},6\right)\right\}$ **11.** $\{(0,0,0)\}$

12. $\left\{\left(\frac{7}{6},\frac{-5}{6},\frac{-7}{6}\right)\right\}$ **13.** $\left\{\left(\frac{4}{7},-\frac{2}{7}\right)\right\}$ **14.** $\left\{\left(-3,\frac{14}{5},\frac{22}{5}\right)\right\}$

15. 24 pennies; 5 nickels; 15 dimes **16.** $2,000 at 8% and
$1,400 at 7%

Chapter 13 cumulative test

1. a. $-9,0,4$ **b.** $-9,-\frac{3}{2},-\frac{1}{9},0,4,\frac{9}{2}$ **c.** $-\sqrt{2},\sqrt{3}$ **2.** -9

3. $x = \frac{5}{3 - y}$ **4. a.** $6y^2 + 19y - 7$ **b.** $36x^2 - 60x + 25$
c. $9x^2 - 4y^2$ **d.** $y^3 - 1$ **5.** $3xy(2x + y - 4x^2y^2)$
6. $(7x + 2)(x - 2)$ **7.** $(3y - 5x)^2$ **8.** $3(x + 5y)(x - 5y)$
9. $(2a + b)(4a^2 - 2ab + b^2)$ **10.** $(2a - b)(2b - c)$

11. $\left\{\frac{23}{3}\right\}$ **12.** $\left\{y | y \le \frac{11}{3}\right\}$ **13.** $\{-1,4\}$ **14.** $\left\{x | -\frac{1}{2} \le x \le 3\right\}$

15. $\left\{y | y < -2 \text{ or } y > \frac{2}{3}\right\}$ **16.** $\left\{-\frac{2}{5},2\right\}$ **17.** $\left\{-\frac{13}{3}\right\}$

18. $\{81\}$; 1 is extraneous **19.** $\frac{32 - 11x}{(3x - 1)(2x + 5)}$

20. $\frac{2x^2 - 13x + 9}{(x - 6)(x + 5)(x - 5)}$ **21.** $\frac{(a - b)(a + 5)}{a - 1}$ **22.** $\frac{y - 2x}{3x - 2y}$

23. $-3\sqrt{3} - 3$ **24.** $9\sqrt{2}$ **25.** 9 **26.** $8 - i$ **27.** $\frac{9}{25} - \frac{13}{25}i$

28. $\{2\}$; -9 is extraneous **29.** $\left\{x | x \le -\frac{5}{2} \text{ or } x > -1\right\}$

$= \left(-\infty, -\frac{5}{2}\right] \cup (-1,\infty)$ **30.** $x - 3y = -5$

31. $3x - y = 18$ **32.** $3x - 4y = 31$

33.

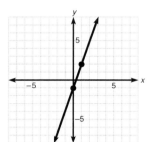

34.

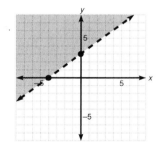

35. a. $f(2) = 3$ **b.** $f(-3) = 18$ **c.** $\frac{f(x + h) - f(x)}{h} = 2x +$
$h - 2$ **36.** $g^{-1}(x) = \sqrt[3]{x + 3}$ **37.** $(f + g)(-4) = 8$;
$\left(\frac{f}{g}\right)(x) = \frac{2x - 1}{x^2 + 1}$; $f[g(3)] = 19$ **38. a.** parabola **b.** ellipse
c. hyperbola **d.** circle

39.

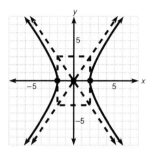

40. $-3,2,4$

41.

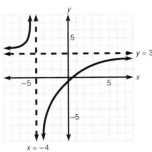

42. $\left\{\dfrac{12}{17}\right\}$ **43. a.** $\left\{\dfrac{3}{2}\right\}$ **b.** $\{1.3\}$ **44.** $\left\{\left(\dfrac{17}{7}, -\dfrac{8}{7}\right)\right\}$

45. $\left\{\left(5, \dfrac{15}{2}, \dfrac{3}{5}\right)\right\}$

Chapter 14

Exercise 14–1

Answers to odd-numbered problems

1. $7, 11, 15, 19, 23$ **3.** $\dfrac{2}{3}, \dfrac{1}{3}, \dfrac{2}{9}, \dfrac{1}{6}, \dfrac{2}{15}$ **5.** $-6, \dfrac{7}{2}, \dfrac{8}{5}, \dfrac{9}{8}, \dfrac{10}{11}$

7. $\dfrac{2}{5}, \dfrac{2}{5}, \dfrac{8}{15}, \dfrac{4}{5}, \dfrac{32}{25}$ **9.** $-1, 7, -13, 19, -25$ **11.** $1, -\dfrac{9}{5}, 3,$

$-\dfrac{81}{17}, \dfrac{81}{11}$ **13.** $1, 1, 1, 1, 1$ **15.** 32 **17.** $\dfrac{1}{111}$ **19.** $\dfrac{65}{41}$ **21.** 89

23. -247 **25.** $a_n = 2n + 4$ **27.** $a_n = 5n - 3$

29. $a_n = (n + 1)^2$ **31.** $a_n = \dfrac{1}{2^{n-1}}$ **33.** $a_n = \dfrac{n + 2}{2n + 3}$

35. $a_n = (-1)^n(4n + 2)$ **37. a.** $27{,}000$ **b.** $243{,}000$

c. $1{,}000(3^n)$ **39.** $\dfrac{5}{2}$ ft; $\dfrac{5}{32}$ ft; $10\left(\dfrac{1}{2}\right)^n$ ft   **41. a.** $\$16{,}000$;

$\$17{,}500; \$19{,}000; \$20{,}500; \$22{,}000; \$23{,}500$ **b.** $\$16{,}000 + 1500$
$(n - 1)$ **c.** $\$44{,}500$

Solutions to trial exercise problems

11. $a_n = (-1)^{n-1} \cdot \dfrac{3^n}{2^n + 1}$; $a_1 = (-1)^{1-1} \cdot \dfrac{3}{2 + 1} = 1 \cdot \dfrac{3}{3} = 1$;

$a_2 = (-1)^{2-1} \cdot \dfrac{3^2}{2^2 + 1} = (-1) \cdot \dfrac{9}{4 + 1} = -\dfrac{9}{5}$; $a_3 = (-1)^{3-1} \cdot$

$\dfrac{3^3}{2^3 + 1} = 1 \cdot \dfrac{27}{8 + 1} = 3$; $a_4 = (-1)^{4-1} \cdot \dfrac{3^4}{2^4 + 1} = (-1) \cdot \dfrac{81}{16 + 1}$

$= -\dfrac{81}{17}$; $a_5 = (-1)^{5-1} \cdot \dfrac{3^5}{2^5 + 1} = 1 \cdot \dfrac{243}{32 + 1} = \dfrac{243}{33} = \dfrac{81}{11}$.

The first five terms are $1, -\dfrac{9}{5}, 3, -\dfrac{81}{17}, \dfrac{81}{11}$.   **21.** $a_{14} =$

$(-1)^{14}[6(14) + 5] = 1 \cdot (84 + 5) = 89$ **24.** $a_8 = 2(8)^2[3(8) - 1]$

$= 2(64)(23) = 2{,}944$ **30.** $\dfrac{1}{3}, \dfrac{1}{9}, \dfrac{1}{27}, \dfrac{1}{81}, \dfrac{1}{243}, \cdots$. The

numerators are all 1 and the denominators are powers of 3. So

$a_n = \dfrac{1}{3^n}$.   **35.** $-6, 10, -14, 18, -22, \cdots$. The signs alternate so we

have factor $(-1)^n$. Since $10 - 6 = 4$, $14 - 10 = 4$, $18 - 14 = 4$,
there is a common difference of 4 so a term of the general term is $4n$.
Since $4(1) + 2 = 6$ and $4(2) + 2 = 10$, we find the general term is
$a_n = (-1)^n(4n + 2)$.   **37.** Since the culture triples every hour, we
want (a) $a_3 = 1{,}000(3^3) = 27{,}000$; (b) $a_5 = 1{,}000(3^5) = 243{,}000$;
(c) $a_n = 1{,}000(3^n)$.

Review exercises

1. a. -6 **b.** 4 **c.** 14 **2.** -10 **3.** $\left\{\left(\dfrac{8}{7}, -\dfrac{4}{7}\right)\right\}$

4. $x = -3$ **5.** $x + 3y = -4$ **6.** $\left\{-\dfrac{1}{2}\right\}$

Exercise 14–2

Answers to odd-numbered problems

1. 75 **3.** $\dfrac{47}{20}$ **5.** -33 **7.** $\dfrac{41}{6}$ **9.** 55 **11.** 60

13. 10 **15.** 50 **17.** 10 **19.** $\dfrac{743}{840}$

21. $\dfrac{937}{168}$ **23.** $\dfrac{5269}{900}$ **25.** $\dfrac{7}{36}$ **27.** 24

29. 72 **31.** 45 **33.** $\dfrac{77}{60}$ **35.** $\dfrac{573}{60}$ **37.** -13 **39.** $\sum\limits_{i=1}^{5} i$

41. $\sum\limits_{i=1}^{6} i^3$ **43.** $\sum\limits_{i=1}^{4} \left(\dfrac{i + 1}{i + 2}\right)$ **45.** $\sum\limits_{i=1}^{4} \left(\dfrac{2i}{3^{i-1}}\right)$ **47.** $\sum\limits_{i=1}^{4} \left(\dfrac{2i + 3}{3i + 1}\right)$

49. $\sum\limits_{i=1}^{6} (-1)^i 2^i$

Solutions to trial exercise problems

2. $a_n = 5n - 1$
$a_1 = 5(1) - 1 = 4$
$a_2 = 5(2) - 1 = 9$
$a_3 = 5(3) - 1 = 14$
$a_4 = 5(4) - 1 = 19$
$S_4 = 4 + 9 + 14 + 19 = 46$

13. $\sum\limits_{k=1}^{5} k(k-3) = 1(1-3) + 2(2-3) + 3(3-3) + 4(4-3) + 5(5-3) = -2 + (-2) + 0 + 4 + 10 = 10$

19. $\sum\limits_{k=1}^{5} \dfrac{1}{k+3} = \dfrac{1}{1+3} + \dfrac{1}{2+3} + \dfrac{1}{3+3} + \dfrac{1}{4+3} + \dfrac{1}{5+3}$

$= \dfrac{1}{4} + \dfrac{1}{5} + \dfrac{1}{6} + \dfrac{1}{7} + \dfrac{1}{8} = \dfrac{210 + 168 + 140 + 120 + 105}{840} = \dfrac{743}{840}$

24. $\sum\limits_{j=1}^{5} (-1)^j \cdot \dfrac{3}{2j} = (-1)^1 \cdot \dfrac{3}{2(1)} + (-1)^2 \cdot \dfrac{3}{2(2)} + (-1)^3 \cdot \dfrac{3}{2(3)} + (-1)^4 \cdot \dfrac{3}{2(4)} + (-1)^5 \cdot \dfrac{3}{2(5)}$

$= \left(-\dfrac{3}{2}\right) + \dfrac{3}{4} + \left(-\dfrac{1}{2}\right) + \dfrac{3}{8} + \left(-\dfrac{3}{10}\right)$

$= -\dfrac{23}{10} + \dfrac{9}{8} = \dfrac{-92 + 45}{40} = -\dfrac{47}{40}$

35. $\sum\limits_{j=0}^{5} \dfrac{2j+1}{j+1} = \dfrac{2(0)+1}{0+1} + \dfrac{2(1)+1}{1+1} + \dfrac{2(2)+1}{2+1} + \dfrac{2(3)+1}{3+1}$

$+ \dfrac{2(4)+1}{4+1} + \dfrac{2(5)+1}{5+1} = 1 + \left(\dfrac{3}{2}\right) + \left(\dfrac{5}{3}\right) + \left(\dfrac{7}{4}\right)$

$+ \left(\dfrac{9}{5}\right) + \left(\dfrac{11}{6}\right) = \dfrac{573}{60}$

43. $\dfrac{2}{3} + \dfrac{3}{4} + \dfrac{4}{5} + \dfrac{5}{6}$; The numerator is the number of the term plus 1, that is, $(n+1)$, and the denominator is 2 plus the number of the term. Then $a_n = \dfrac{n+1}{n+2}$ and we have 4 terms. $\sum\limits_{i=1}^{4} \left(\dfrac{i+1}{i+2}\right)$

48. $2 - 5 + 8 - 11 + 14$; The signs alternate starting with the first term positive, $(-1)^{n+1}$. Each term differs by 3 and the first term is $3(1) - 1$, second term is $3(2) - 1$. So $a_n = (-1)^{n+1}(3n-1)$ and we have $\sum\limits_{j=1}^{5} (-1)^{j+1}(3j-1)$.

Review exercises

1. 11 **2.** 10 **3.** 22 **4.** 54 **5.** $-\dfrac{1}{18}$ **6.** $\{2.262\}$

Exercise 14–3

Answers to odd-numbered problems

1. arithmetic; $d = 1$ **3.** arithmetic; $d = 2$ **5.** not arithmetic

7. arithmetic; $d = \dfrac{1}{2}$ **9.** arithmetic; $d = \dfrac{5}{3}$ **11.** $a_{16} = 79$

13. $a_{17} = -58$ **15.** $a_{12} = \dfrac{17}{3}$ **17.** $a_{14} = 9$ **19.** $a_{16} = 65$

21. $a_{25} = 105$ **23.** $n = 6$ terms **25.** $n = 15$ terms **27.** $n =$ 29 terms **29.** $S_{16} = 288$ **31.** $S_{14} = -175$ **33.** $S_{18} = 85\dfrac{1}{2}$

35. $S_{14} = 280$ **37.** $S_{19} = -95$ **39.** 575 **41.** 165

43. -440 **45.** 51 **47.** 13 **49.** $S_{13} = 494$ **51.** $S_{15} = -165$

53. $S_{11} = -\dfrac{319}{6}$ **55.** 121 cans **57.** 165 cans **59.** 3,422

61. 240 ft; 1,600 ft **63.** \$1,010 per month **65.** \$82,080
67. \$590

Solutions to trial exercise problems

4. Since $6 - 4 = 2, 8 - 6 = 2, 10 - 8 = 2$, the sequence is arithmetic and $d = 2$. **7.** Since $2 - \dfrac{3}{2} = \dfrac{1}{2}, \dfrac{5}{2} - 2 = \dfrac{1}{2},$

$3 - \dfrac{5}{2} = \dfrac{1}{2}$, the sequence is arithmetic and $d = \dfrac{1}{2}$.

15. Using $a_n = a_1 + (n-1)d, a_{12} = 2 + (12-1)\dfrac{1}{3} = 2 +$

$(11)\dfrac{1}{3} = 2 + \dfrac{11}{3} = \dfrac{17}{3}$. **28.** Using $a_n = a_1 + (n-1)d$, we want

n when $a_n = -2, a_1 = \dfrac{5}{3}$, and $d = \dfrac{4}{3} - \dfrac{5}{3} = -\dfrac{1}{3}$. Then $-2 =$

$\dfrac{5}{3} + (n-1)\left(-\dfrac{1}{3}\right); -2 = \dfrac{5}{3} - \dfrac{1}{3}n + \dfrac{1}{3}; -2 = 2 - \dfrac{1}{3}n;$

$-4 = -\dfrac{1}{3}n; n = 12$. **33.** Using $S_n = \dfrac{n}{2}(a_1 + a_n)$, we want

S_{18} when $n = 18, a_1 = \dfrac{1}{2}$, and $a_{18} = 9$. So $S_{18} = \dfrac{18}{2}\left(\dfrac{1}{2} + 9\right)$

$= 9\left(\dfrac{19}{2}\right) = \dfrac{171}{2}$ or $85\dfrac{1}{2}$. **37.** Using $S_n = \dfrac{n}{2}(a_1 + a_n)$, we want

S_{19} when $n = 19, a_1 = 5 - 1 = 4; a_{19} = 5 - 19 = -14$.

$S_{19} = \dfrac{19}{2}[4 + (-14)] = \dfrac{19}{2}(-10) = -\dfrac{190}{2} = -95$.

46. $\sum_{k=1}^{14} \frac{1}{2}k = \frac{14}{2}\left(\frac{1}{2}+7\right) = 7\left(\frac{15}{2}\right) = \frac{105}{2}$ or $52\frac{1}{2}$.

53. We want S_{11} using $S_n = \frac{n}{2}(a_1 + a_n)$. When $n = 11$, $a_1 = \frac{1}{6}$,

and $a_{11} = \frac{1}{6}+(11-1)(-1) = \frac{1}{6}+(10)(-1) = \frac{1}{6}+(-10)$

$= -\frac{59}{6}$. So $S_{11} = \frac{11}{2}\left[\frac{1}{6}+\left(-\frac{59}{6}\right)\right] = \frac{11}{2}\left(-\frac{58}{6}\right) = \frac{11}{2}\left(-\frac{29}{3}\right)$

$= -\frac{319}{6}$ or $-53\frac{1}{6}$. **57.** We want S_n when $a_1 = 30$, $d = 27 - 30$

$= -3$, and $a_n = 3$. Now, using $a_n = a_1 + (n-1)d$, we have
$3 = 30 + (n-1)(-3)$; $-27 = -3n+3$; $-30 = -3n$; $n = 10$.

So $S_{10} = \frac{10}{2}(30+3) = 5(33) = 165$ boxes.

62. We want n when $S_n = 3,600$, $a_1 = 16$, and $d = 32$. Using

$S_n = \frac{n}{2}[2a_1 + (n-1)d]$, $3,600 = \frac{n}{2}[2(16) + (n-1) \cdot 32]$;

$3,600 = \frac{n}{2}[32 + 32n - 32]$; $3,600 = \frac{n}{2}(32n)$; $3,600 = 16n^2$;

$n^2 = 225$; $n = 15$ sec. **67.** We want a_{19} when $a_1 = 50$, $n = 19$,
and $d = 30$. Using $a_n = a_1 + (n-1)d$, $a_{19} = 50 + (19-1)30 =$
$50 + 18(30) = 50 + 540 = 590$. Thus \$590 was deposited on her
eighteenth birthday.

Review exercises

1. $\frac{5}{y-6}$ **2.** 5 **3.** $\{4\}$ **4. a.** parabola **b.** hyperbola **c.** ellipse

5. $-2 - \sqrt{6}$ **6.** $\frac{3(a+1)}{3a+5}$

Exercise 14–4

Answers to odd-numbered problems

1. geometric; 27,81,243; $r = 3$ **3.** geometric; $\frac{1}{54},\frac{1}{162},\frac{1}{486}$; $r = \frac{1}{3}$

5. geometric; $-\frac{1}{2},\frac{1}{4},-\frac{1}{8}$; $r = -\frac{1}{2}$ **7.** geometric; $\frac{-2}{9},\frac{2}{27},\frac{-2}{81}$;

$r = -\frac{1}{3}$ **9.** not geometric **11.** $a_n = 3(2)^{n-1}$

13. $a_n = 27\left(\frac{-2}{3}\right)^{n-1}$ **15.** $a_n = (\sqrt{3})^{n-1}$

17. $a_n = \frac{-1}{15}(-3)^{n-1}$ **19.** $a_5 = 162$ **21.** $a_4 = 3$

23. $a_6 = -160$ **25.** $a_5 = -\frac{1}{8}$ **27.** $a_7 = 576$

29. $a_9 = 1,792$ **31.** $S_5 = 1,694$ **33.** $S_4 = -\frac{312}{25}$

35. $S_6 = 27,993$ **37.** $S_9 = \frac{511}{512}$ **39.** $S_5 = -305$

41. 87,380 **43.** -183 **45.** $\frac{4,118}{2,187}$

47. $\frac{31,122}{15,625}$ **49.** $-\frac{22,344}{3,125}$ **51.** $40\frac{7}{27}$ ft **53.** \$512

55. \$10,737,418.23 **57.** $\frac{2,101}{3,125} \approx 0.67$ of the tank

Solutions to trial exercise problems

5. Since $\frac{-2}{4} = -\frac{1}{2}$ and $\frac{1}{-2} = -\frac{1}{2}$ the sequence is geometric with

$r = -\frac{1}{2}$ and the next three terms are $-\frac{1}{2},\frac{1}{4},-\frac{1}{8}$. **15.** Using

$a_n = a_1 r^{n-1}$, since $\frac{\sqrt{3}}{1} = \sqrt{3}$ and $\frac{3}{\sqrt{3}} = \sqrt{3}$, then $r = \sqrt{3}$ and

$a_1 = 1$. Thus $a_n = 1(\sqrt{3})^{n-1} = (\sqrt{3})^{n-1}$. **25.** Since $a_1 = -32$

and $r = -\frac{1}{4}$, using $a_n = a_1 r^{n-1}$, $a_5 = (-32)\left(-\frac{1}{4}\right)^{5-1}$

$= (-32)\left(-\frac{1}{4}\right)^4 = (-32)\left(\frac{1}{256}\right) = -\frac{1}{8}$. **26.** Now $r = \frac{18}{3} = 6$

and $a_1 = 3$, then $a_6 = 3(6)^{6-1} = 3(6)^5 = 3(7,776) = 23,328$.

33. Using $S_n = \frac{a_1 - a_1 r^n}{1-r}$, given $a_1 = -10$ and $r = \frac{1}{5}$,

$S_4 = \frac{-10-(-10)\left(\frac{1}{5}\right)^4}{1-\frac{1}{5}} = \frac{-10+10\left(\frac{1}{625}\right)}{\frac{4}{5}} = \frac{-10+\frac{2}{125}}{\frac{4}{5}}$

$= \frac{-1,250+2}{100} = \frac{-1,248}{100} = -12\frac{12}{25}$. **34.** Now $r = \frac{18}{9} = 2$

and $a_1 = 9$, so $S_7 = \frac{9-9(2)^7}{1-2} = \frac{9-9(128)}{-1} = \frac{9-1,152}{-1}$

$= 1,143$. **42.** Now $r = -2$ and $a_1 = -2$. We want S_7.

$S_7 = \frac{-2-(-2)(-2)^7}{1-(-2)} = \frac{-2+2(-2)^7}{3} = \frac{-2+2(-128)}{3}$

$= \frac{-2-256}{3} = \frac{-258}{3} = -86$ **47.** Now $r = \frac{2}{5}$ and

$a_1 = 3\left(\frac{2}{5}\right) = \frac{6}{5}$. We want S_6. $S_6 = \frac{\frac{6}{5} - \frac{6}{5}\left(\frac{2}{5}\right)^6}{1-\frac{2}{5}}$

$= \frac{\frac{6}{5}-\frac{6}{5}\left(\frac{64}{15,625}\right)}{\frac{3}{5}} = \frac{6-\frac{384}{15,625}}{3} = \left(6-\frac{384}{15,625}\right)\frac{1}{3} =$

$2 - \frac{128}{15,625} = \frac{31,122}{15,625}$.

51. Since the ball will travel each height *twice*, after the initial drop of 9 ft, then we want $9 + 2 \sum_{i=1}^{5} 9\left(\frac{2}{3}\right)^i$. Now

$$2\sum_{i=1}^{5} 9\left(\frac{2}{3}\right)^i = 2\left[\frac{6 - 6\left(\frac{2}{3}\right)^5}{1 - \frac{2}{3}}\right] = 2\left[\frac{6 - 6\left(\frac{2}{3}\right)^5}{\frac{1}{3}}\right] = 2\left[\frac{6 - 6\left(\frac{32}{243}\right)}{\frac{1}{3}}\right]$$

$$= 2\left[\left(6 - \frac{192}{243}\right) \cdot 3\right] = 2\left[18 - \frac{192}{81}\right] = 2\left[\frac{1,458 - 192}{81}\right] = 2\left(\frac{1,266}{81}\right)$$

$$= 2\left(\frac{422}{27}\right) = \frac{844}{27} \text{ or } 31\frac{7}{27}. \text{ The ball has traveled } 9 + 31\frac{7}{27} = 40\frac{7}{27} \text{ ft at the sixth strike.}$$

Review exercises

1. $16x^2 + 8xy + y^2$ **2.** $27x^3 - 54x^2y + 36xy^2 - 8y^3$

3. $x^2 - 12x + 36 = (x - 6)^2$ **4.** $\frac{4}{3}$ **5.** $\frac{6}{5} + \frac{3}{5}i$

6. $\left\{\frac{1 - i\sqrt{31}}{4}, \frac{1 + i\sqrt{31}}{4}\right\}$

Exercise 14–5

Answers to odd-numbered problems

1. 3   **3.** -6 **5.** $\frac{9}{10}$ **7.** $-\frac{1}{2}$ **9.** 18   **11.** 5   **13.** no sum

15. 4   **17.** $-\frac{2}{5}$ **19.** $\frac{3}{4}$ **21.** $\frac{1}{3}$ **23.** $\frac{31}{110}$ **25.** $\frac{2}{55}$

27. 570 in.   **29.** 84 in.   **31.** 25 mg

Solutions to trial exercise problems

3. Using $S_\infty = \frac{a_1}{1 - r}$, $S_\infty = \frac{-3}{1 - \frac{1}{2}} = \frac{-3}{\frac{1}{2}} = -6$.

6. Using $S_\infty = \frac{a_1}{1 - r}$, $S_\infty = \frac{4}{1 - \left(-\frac{1}{2}\right)} = \frac{4}{\frac{3}{2}} = 4 \cdot \frac{2}{3} = \frac{8}{3}$.

9. Now $a_1 = 12$ and $r = \frac{4}{12} = \frac{1}{3}$, so $S_\infty = \frac{12}{1 - \frac{1}{3}} = \frac{12}{\frac{2}{3}} = 12 \cdot \frac{3}{2} = 18$.

13. $a_1 = 6$ and $r = \frac{-8}{6} = -\frac{4}{3}$. So S_∞ does not exist since $\left|-\frac{4}{3}\right| > 1$.

16. Now $a_1 = \left(\frac{7}{8}\right)^{1+1} = \left(\frac{7}{8}\right)^2 = \frac{49}{64}$ and $r = \frac{7}{8}$, so $\sum_{k=1}^{\infty} \left(\frac{7}{8}\right)^{k+1} = \frac{\frac{49}{64}}{1 - \frac{7}{8}} = \frac{\frac{49}{64}}{\frac{1}{8}} = \frac{49}{64} \cdot \frac{8}{1} = \frac{49}{8}$ or $6\frac{1}{8}$.

23. $0.28181\overline{81} = 0.2 + 0.08181\overline{81}$. Now for $0.08181\overline{81}$ $a_1 = 0.081$ and $r = \frac{0.00081}{0.081} = 0.01$. Then $0.08181\overline{81} = \frac{0.081}{1 - 0.01} = \frac{0.081}{0.99}$

$= \frac{81}{990} = \frac{9}{110}$. Thus $0.28181\overline{81} = \frac{2}{10} + \frac{9}{110} = \frac{22 + 9}{110} = \frac{31}{110}$.

28. Now $a_1 = 16$ cm and $r = \frac{7}{8}$. We want $S_\infty = \frac{a_1}{1 - r} = \frac{16}{1 - \frac{7}{8}} = \frac{16}{\frac{1}{8}} = 16 \cdot 8 = 128$.

The bob travels 128 cm before coming to rest.

Review exercises

1. $x^3 + 3x^2y + 3xy^2 + y^3$ **2.** $\dfrac{1}{a^2}$ **3.** a^3b^2

4.

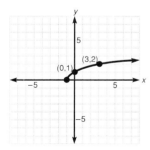

5. $f[g(x)] = 4x^2$ **6.** $(a^2 + b^2)(a + b)(a - b)$
7. $3(x + 3y)(x - 3y)$ **8.** $3(x + 2y)(x^2 - 2xy + 4y^2)$

Exercise 14–6

Answers to odd-numbered problems

1. P_1: $1 = \dfrac{1(1 + 1)}{2} = \dfrac{2}{2} = 1$

P_2: $1 + 2 = \dfrac{2(2 + 1)}{2}$

$3 = \dfrac{2(3)}{2} = 3$

P_3: $1 + 2 + 3 = \dfrac{3(3 + 1)}{2}$

$6 = \dfrac{3(4)}{2}$

$6 = 6$

P_4: $1 + 2 + 3 + 4 = \dfrac{4(4 + 1)}{2}$

$10 = \dfrac{4(5)}{2}$

$10 = 10$

3. P_1: $1^3 = \dfrac{1^2(1 + 1)^2}{4}$

$1 = \dfrac{1(2^2)}{4}$

$1 = 1$

P_2: $1^3 + 2^3 = \dfrac{2^2(2 + 1)^2}{4}$

$1 + 8 = \dfrac{4(3)^2}{4}$

$9 = 9$

P_3: $1^3 + 2^3 + 3^3 = \dfrac{3^2(3 + 1)^2}{4}$

$1 + 8 + 27 = \dfrac{9(4)^2}{4}$

$36 = 9(4)$

$36 = 36$

P_4: $1^3 + 2^3 + 3^3 + 4^3 = \dfrac{4^2(4 + 1)^2}{4}$

$1 + 8 + 27 + 64 = \dfrac{16(5)^2}{4}$

$100 = 4(25)$

$100 = 100$

5. P_1: $1 < 2^1$ (true)
P_2: $2 < 2^2$
$2 < 4$ (true)
P_3: $3 < 2^3$
$3 < 8$ (true)
P_4: $4 < 2^4$
$4 < 16$ (true)

7. P_1: $2 = 1(1 + 1)$
$2 = 2$ so P_1 is true
Assume P_k: $2 + 4 + 6 + \cdots + 2k = k(k + 1)$ for some k
Prove: P_{k+1}: $2 + 4 + 6 + \cdots + 2(k + 1) = (k + 1)[(k + 1) + 1]$
Proof: By induction hypothesis
$2 + 4 + 6 + \cdots + 2k = k(k + 1)$
$2 + 4 + 6 + \cdots + 2k + 2(k + 1) = k(k + 1) + 2(k + 1)$
$= (k + 1)(k + 2)$
$= (k + 1)[(k + 1) + 1]$
Thus, P_k implies P_{k+1} is true and P_n is true for all positive integers n.

9. P_1: $3 = 1(2 \cdot 1 + 1)$
$3 = 1(3)$
$3 = 3$ so P_1 is true
Assume P_k: $3 + 7 + 11 + \cdots + (4k - 1) = k(2k + 1)$ is true for some k.
Prove: P_{k+1}: $3 + 7 + 11 + \cdots + [4(k + 1) - 1]$
$= (k + 1)[2(k + 1) + 1]$
Proof: By induction hypothesis, $3 + 7 + 11 + \cdots + (4k - 1)$
$= k(2k + 1)$
$3 + 7 + 11 + \cdots + (4k - 1) + [4(k + 1) - 1]$
$= k(2k + 1) + [4(k + 1) - 1]$
$= 2k^2 + k + 4k + 4 - 1$
$= 2k^2 + 5k + 3$
$= (k + 1)(2k + 3)$
$= (k + 1)[2(k + 1) + 1]$
Thus, P_k implies P_{k+1} is true and P_n is true for all positive integers n.

11. P_1: $3 = 3(1)^2$

$3 = 3$ so P_1 is true

Assume P_k: $3 + 9 + 15 + \cdots + (6k - 3) = 3k^2$

Prove P_{k+1}: $3 + 9 + 15 + \cdots + [6(k + 1) - 3]$
$= 3(k + 1)^2$

Proof: By induction hypothesis, $3 + 9 + 15 + \cdots + (6k - 3)$
$= 3k^2$

$3 + 9 + 15 + \cdots + (6k - 3) + [6(k + 1) - 3]$
$= 3k^2 + [6(k + 1) - 3]$
$= 3k^2 + 6k + 6 - 3$
$= 3k^2 + 6k + 3$
$= 3(k^2 + 2k + 1)$
$= 3(k + 1)^2$

Thus, P_k implies P_{k+1} is true and P_n is true for all positive integers n.

13. P_1: $\dfrac{1}{1 \cdot 2} = \dfrac{1}{1 + 1}$

$\dfrac{1}{2} = \dfrac{1}{2}$ so P_1 is true

Assume P_k: $\dfrac{1}{1 \cdot 2} + \dfrac{1}{2 \cdot 3} + \dfrac{1}{3 \cdot 4} + \cdots + \dfrac{1}{k(k + 1)} = \dfrac{k}{k + 1}$

Prove P_{k+1}: $\dfrac{1}{1 \cdot 2} + \dfrac{1}{2 \cdot 3} + \dfrac{1}{3 \cdot 4} + \cdots + \dfrac{1}{(k + 1)[(k + 1) + 1]} = \dfrac{k + 1}{(k + 1) + 1}$

Proof: By induction hypothesis,

$$\dfrac{1}{1 \cdot 2} + \dfrac{1}{2 \cdot 3} + \dfrac{1}{3 \cdot 4} + \cdots + \dfrac{1}{k(k + 1)} = \dfrac{k}{k + 1}$$

$$\dfrac{1}{1 \cdot 2} + \dfrac{1}{2 \cdot 3} + \dfrac{1}{3 \cdot 4} + \cdots + \dfrac{1}{k(k + 1)} + \dfrac{1}{(k + 1)[(k + 1) + 1]} = \dfrac{k}{k + 1} + \dfrac{1}{(k + 1)[(k + 1) + 1]}$$

$$= \dfrac{k}{k + 1} + \dfrac{1}{(k + 1)(k + 2)}$$

$$= \dfrac{k(k + 2)}{(k + 1)(k + 2)} + \dfrac{1}{(k + 1)(k + 2)}$$

$$= \dfrac{k^2 + 2k + 1}{(k + 1)(k + 2)}$$

$$= \dfrac{(k + 1)^2}{(k + 1)(k + 2)}$$

$$= \dfrac{k + 1}{k + 2}$$

$$= \dfrac{k + 1}{(k + 1) + 1}$$

Thus, P_k implies P_{k+1} is true and P_n is true for all positive integers n.

15. P_1: $1^3 = \dfrac{1^2(1 + 1)^2}{4}$

$\quad\quad 1 = \dfrac{1(2)^2}{4}$

$\quad\quad 1 = 1 \quad\quad P_1$ is true

Assume P_k: $1^3 + 2^3 + 3^3 + \cdots + k^3 = \dfrac{k^2(k + 1)^2}{4}$

Prove P_{k+1}: $1^3 + 2^3 + 3^3 + \cdots + (k + 1)^3 = \dfrac{(k + 1)^2[(k + 1) + 1]^2}{4}$

Proof: By induction hypothesis,

$1^3 + 2^3 + 3^3 + \cdots + k^3 = \dfrac{k^2(k + 1)^2}{4}$

$1^3 + 2^3 + 3^3 + \cdots + k^3 + (k + 1)^3 = \dfrac{k^2(k + 1)^2}{4} + (k + 1)^3$

$\quad\quad\quad\quad\quad\quad\quad\quad\quad\quad\quad = \dfrac{k^2(k + 1)^2}{4} + \dfrac{4(k + 1)^3}{4}$

$\quad\quad\quad\quad\quad\quad\quad\quad\quad\quad\quad = \dfrac{k^2(k + 1)^2 + 4(k + 1)^3}{4}$

$\quad\quad\quad\quad\quad\quad\quad\quad\quad\quad\quad = \dfrac{(k + 1)^2[k^2 + 4(k + 1)]}{4}$

$\quad\quad\quad\quad\quad\quad\quad\quad\quad\quad\quad = \dfrac{(k + 1)^2(k^2 + 4k + 4)}{4}$

$\quad\quad\quad\quad\quad\quad\quad\quad\quad\quad\quad = \dfrac{(k + 1)^2(k + 2)^2}{4}$

$\quad\quad\quad\quad\quad\quad\quad\quad\quad\quad\quad = \dfrac{(k + 1)^2[(k + 1) + 1]^2}{4}$

Thus, P_k implies P_{k+1} is true and P_n is true for all positive integers n.

17. P_1: $1 < 2^1$ (true)

Assume P_k: $k < 2^k$

Prove P_{k+1}: $(k + 1) < 2^{k+1}$

Proof: By induction hypothesis, $k < 2^k$. Multiply each member by 2.

$2 \cdot k < 2^k \cdot 2$

$2k < 2^{k+1}$ \quad\quad property of exponents

Since $2k = k + k$, consider $k + 1 < k + k$. Subtracting k from each member, $1 < k$ (true). Thus, $k + 1 < 2k < 2^{k+1}$ and, by transitive property, $k + 1 < 2^{k+1}$. Thus, P_k implies P_{k+1} is true and P_n is true for all positive integers n.

19. P_3: $2^3 > 2(3)$

$\quad\quad 8 > 6 \quad\quad$ so P_3 is true

Assume P_k: $2^k > 2k$

Prove P_{k+1}: $2^{k+1} > 2(k + 1)$

Proof: By induction hypothesis, $2^k > 2k$.

Multiply each member by 2. Then

$2 \cdot 2^k > 2 \cdot 2k$

$2^{k+1} > 4k$

Since $2(k + 1) = 2k + 2$, we must show that $4k > 2k + 2$. Subtract $2k$ from each member. Then $2k > 2$, which is true for all $k \geq 3$. Therefore,

$\quad\quad\quad 2^{k+1} > 4k > 2(k + 1)$

and $\quad\quad 2^{k+1} > 2(k + 1) \quad\quad$ by transitive property

P_k implies P_{k+1} is true so P_n is true for all positive integers $n \geq 3$.

21. P_1: When $n = 1$, $n^2 + n = (1)^2 + 1 = 2$ and 2 *is* a factor of 2.

Assume P_k: 2 is a factor of $k^2 + k$ for some k.

Prove P_{k+1}: 2 is a factor of $(k + 1)^2 + (k + 1)$.

Proof: $(k + 1)^2 + (k + 1) = k^2 + 2k + 1 + k + 1$

$$= k^2 + 3k + 2$$
$$= (k^2 + k) + (2k + 2)$$
$$= (k^2 + k) + 2(k + 1)$$

By induction hypothesis, 2 is a factor of $k^2 + k$ and 2 is a factor of $2(k + 1)$. Thus, 2 is a factor of $(k + 1)^2 + (k + 1)$. P_k implies P_{k+1} is true so P_n is true for all positive integers n.

23. P_1: $7^1 - 1 = 6$ and 6 is divisible by 6, so P_1 is true.

Assume P_k: $7^k - 1$ is divisible by 6 for some positive integer k.

Prove P_{k+1}: $7^{k+1} - 1$ is divisible by 6.

Proof: $7^{k+1} - 1 = 7^k \cdot 7 - 1$

$$= 7^k(6 + 1) - 1$$
$$= (7^k \cdot 6) + (7^k - 1)$$

By induction hypothesis, $7^k - 1$ is divisible by 6 and $7^k \cdot 6$ is also divisible by 6 since 6 is a factor. Thus, $7^{k+1} - 1$ is divisible by 6 and we can say $7^n - 1$ is divisible by 6 for every positive integer n.

25. P_1: $(ab)^1 = a^1 b^1$ is true by property of exponents.

Assume P_k: $(ab)^k = a^k b^k$

Prove P_{k+1}: $(ab)^{k+1} = a^{k+1} b^{k+1}$

Proof: By induction hypothesis, $(ab)^k = a^k b^k$ and multiply each term by ab.

$$(ab)^k \cdot (ab) = (a^k b^k)(ab)$$
$$(ab)^{k+1} = (a^k \cdot a) \cdot (b^k \cdot b)$$
$$(ab)^{k+1} = a^{k+1} \cdot b^{k+1}$$

P_k implies P_{k+1} is true so P_n is true for every positive integer n.

Solutions to trial exercise problems

2. P_1: $1^2 = \dfrac{1(1 + 1)[2(1) + 1]}{6}$

$$1 = \frac{1 \cdot 2 \cdot 3}{6}$$
$$1 = 1$$

P_2: $1^2 + 2^2 = \dfrac{2(2 + 1)[2(2) + 1]}{6}$

$$1 + 4 = \frac{2(3)(5)}{6}$$
$$5 = 5$$

P_3: $1^2 + 2^2 + 3^2 = \dfrac{3(3 + 1)[2(3) + 1]}{6}$

$$1 + 4 + 9 = \frac{3(4)(7)}{6}$$
$$14 = 2 \cdot 7$$
$$14 = 14$$

P_4: $1^2 + 2^2 + 3^2 + 4^2 = \dfrac{4(4 + 1)[2(4) + 1]}{6}$

$$1 + 4 + 9 + 16 = \frac{4(5)(9)}{6}$$
$$30 = 2 \cdot 5 \cdot 3$$
$$30 = 30$$

16. P_1: $1^3 = (1)^2[2(1)^2 - 1]$

$$1 = 1(2 - 1)$$
$$1 = 1 \qquad P_1 \text{ is true}$$

Assume P_k: $1^3 + 3^3 + 5^3 + \cdots + (2k - 1)^3 = k^2(2k^2 - 1)$

Prove P_{k+1}: $1^3 + 3^3 + 5^3 + \cdots + [2(k + 1) - 1]^3 = (k + 1)^2[2(k + 1)^2 - 1]$

Note: $(k + 1)^2[2(k + 1)^2 - 1] = (k + 1)^2(2k^2 + 4k + 1)$

$$= 2k^4 + 8k^3 + 11k^2 + 6k + 1$$

Proof: By induction hypothesis,

$1^3 + 3^3 + 5^3 + \cdots + (2k - 1)^3 = k^2(2k^2 - 1)$

$1^3 + 3^3 + 5^3 + \cdots + (2k - 1)^3 + [2(k + 1) - 1]^3$

$= k^2(2k^2 - 1) + [2(k + 1) - 1]^3$

$= k^2(2k^2 - 1) + (2k + 1)^3$

$= 2k^4 - k^2 + 8k^3 + 12k^2 + 6k + 1$

$= 2k^4 + 8k^3 + 11k^2 + 6k + 1$

$= (k + 1)^2[2(k + 1)^2 - 1]$

Thus, P_k implies P_{k+1} is true and P_n is true for all positive integers n.

Review exercises

1. $A \cup B = \{1,2,3,4,5,7,8,9\}$; $A \cap B = \{5,7\}$ **2.** 24   **3.** $\dfrac{xy}{x + y}$

4. $\{2,3\}$ **5.** $\{2\}$; -4 is extraneous

Chapter 14 review

1. 7,11,15,19,23   **2.** $5, \frac{10}{3}, 3, \frac{20}{7}, \frac{25}{9}$

3. $\frac{-4}{7}, \frac{4}{9}, \frac{-4}{11}, \frac{4}{13}, \frac{-4}{15}$

4. $2, -4, 8, -16, 32$ **5.** $a_6 = -14$ **6.** $a_7 = -17$

7. $a_9 = \frac{6561}{2}$ **8.** $a_{11} = \frac{683}{59,049}$

9. $a_n = 2n + 3$ **10.** $a_n = 5n - 2$

11. $a_n = \frac{n+1}{4n-1}$ **12.** $a_n = (-1)^n(5n - 1)$

13. $a_n = 0.25n + 2.75$; $a_8 = \$4.75$ **14.** 36

15. 196   **16.** $\frac{229}{20}$

17. $-\frac{37}{75}$ **18.** $\sum\limits_{i=1}^{4} (3i + 2)$

19. $\sum\limits_{k=1}^{5} (-1)^{k+1}\left(\frac{k+3}{k+4}\right)$

20. $a_{15} = 61$ **21.** $a_{17} = 77$ **22.** $a_{21} = 83$ **23.** $a_{19} = -42$
24. $n = 15$ **25.** $n = 21$ **26.** $S_{15} = 585$ **27.** $S_{21} = -420$

28. $S_{29} = 290$ **29.** $S_{25} = \frac{375}{2}$ **30.** $\$16,500$ **31.** $a_n = 3(2)^{n-1}$

32. $a_n = (-1)^n\left(\frac{3}{4}\right)^n$ **33.** $a_5 = 405$ **34.** $a_4 = -\frac{8}{9}$

35. $a_3 = 16$ **36.** $a_7 = -576$ **37.** $S_5 = 363$

38. $S_6 = -\frac{189}{4}$ **39.** $S_5 = -\frac{543}{1,024}$ **40.** $S_7 = \frac{4,372}{2,187}$

41. $S_\infty = 12$ **42.** $S_\infty = -\frac{5}{2}$ **43.** $S_\infty = \frac{-3}{5}$ **44.** $S_\infty = -\frac{1}{45}$

45. $\frac{35}{99}$ **46.** $\frac{214}{495}$ **47.** 24 meters

48. P_1: $2 = 3^1 - 1$
$\qquad 2 = 2$
$\qquad P_2$: $2 + 6 = 3^2 - 1$
$\qquad\qquad 8 = 9 - 1$
$\qquad\qquad 8 = 8$
$\qquad P_3$: $2 + 6 + 18 = 3^3 - 1$
$\qquad\qquad 26 = 27 - 1$
$\qquad\qquad 26 = 26$

49. P_1: $2 = \dfrac{1[5(1) - 1]}{2}$

$\qquad 2 = \dfrac{1(4)}{2}$

$\qquad 2 = 2 \qquad P_1$ is true

Assume P_k: $2 + 7 + 12 + \cdots + (5k - 3) = \dfrac{k(5k - 1)}{2}$

for some k
Prove P_{k+1}: $2 + 7 + 12 + \cdots + [5(k + 1) - 3] = \dfrac{(k + 1)[5(k + 1) - 1]}{2}$

Proof: By induction hypothesis,

$2 + 7 + 12 + \cdots + (5k - 3) = \dfrac{k(5k - 1)}{2}$

$2 + 7 + 12 + \cdots + (5k - 3) + [5(k + 1) - 3]$

$= \dfrac{k(5k - 1)}{2} + [5(k + 1) - 3]$

$= \dfrac{k(5k - 1)}{2} + (5k + 2)$

$= \dfrac{k(5k - 1) + 2(5k + 2)}{2}$

$= \dfrac{5k^2 - k + 10k + 4}{2}$

$= \dfrac{5k^2 + 9k + 4}{2}$

$= \dfrac{(k + 1)(5k + 4)}{2}$

$= \dfrac{(k + 1)[5(k + 1) - 1]}{2}$

P_k implies P_{k+1} is true and P_n is true for all positive integers n.

50. P_1: $1 + 2(1) \le 3^1$
$\qquad\qquad 3 \le 3 \qquad$ true for $n = 1$
Assume P_k: $1 + 2k \le 3^k$ for some k
Prove P_{k+1}: $1 + 2(k + 1) \le 3^{k+1}$
Proof: By induction hypothesis, $1 + 2k \le 3^k$. Multiply each member by 3 to get
$\qquad 3(1 + 2k) \le 3 \cdot 3^k$
$\qquad\quad 3 + 6k \le 3^{k+1}$
Now $1 + 2(k + 1) = 2k + 3$ and
$\qquad\qquad 6k + 3 \ge 2k + 3$
$\qquad\qquad\quad 6k \ge 2k$
Thus, $1 + 2(k + 1) \le 3 + 6k \le 3^{k+1}$ implies
$\qquad 1 + 2(k + 1) \le 3^{k+1}$
P_k implies P_{k+1} is true and P_n is true for all positive integers n.

Chapter 14 cumulative test

1. $\{-7, -6, -5, -4, -3, -2, -1, 0, 1, 2, 3\}$ **2. a.** $\{2, 7\}$
b. $\{-4, -1, 0, 2, 4, 7, 9\}$ **c.** $\emptyset$ **3.** $\{-4, -3, -2, -1, 0, 1, 2, 3, 4, 5, 6, 7, 8\}$
4. 13   **5. a.** $\dfrac{-2a^7}{b^5}$ **b.** $\dfrac{a^6}{4b^4}$ **c.** $\dfrac{b^4}{a^6}$

6. a. $4y^2 + 28y + 49$
b. $16x^2 - 4y^2$ **c.** $x^3 - 8$ **7.** $4xy(3 - xy^2 + 2x^2y)$
8. $(7y + 1)(y - 5)$ **9.** $(2a - 5b)^2$ **10.** $2(2x + 5y)(2x - 5y)$
11. $2(2a - b)(4a^2 + 2ab + b^2)$ **12.** $(3a - b)(2x - y)$
13. $\left\{\dfrac{21}{25}\right\}$ **14.** $\left\{x \mid x < -\dfrac{29}{2}\right\}$ **15.** $\left\{x \mid -\dfrac{5}{2} < x \le 1\right\}$

16. $\{4, 1\}$ **17.** $\left\{x \mid \dfrac{1}{3} < x < 3\right\}$ **18.** $\left\{c \mid c \le -\dfrac{2}{5} \text{ or } c \ge 2\right\}$

19. $\{8,-3\}$ **20.** $\left\{-\dfrac{5}{4}\right\}$ **21.** $\dfrac{32a + 9}{(2a - 1)(4a + 3)}$

22. $\dfrac{y^2 - 5y + 1}{(y - 7)(y + 6)(y - 6)}$ **23.** $\dfrac{3a^2 - 4a + 1}{a^2 - 3a - 4}$ **24.** -2

25. $3x^2 + 13x + 37 + \dfrac{112}{x - 3}$ **26.** $17\sqrt{3}$ **27.** $\dfrac{4 + 7i}{13}$

or $\dfrac{4}{13} + \dfrac{7}{13}i$ **28.** $12 - 8\sqrt{5} + 9\sqrt{2} - 6\sqrt{10}$

29. $\left\{\dfrac{7 + \sqrt{73}}{6}, \dfrac{7 - \sqrt{73}}{6}\right\}$ **30.** $\{5\}$; 2 is extraneous

31. $\{y | 1 \le y \le 3\}$ **32. a.** $3x - 2y = -1$ **b.** $x + 4y = -8$

c. $2x + 3y = 0$ **33.** $m = \dfrac{3}{5}, b = \dfrac{9}{5}$ **34. a.** -18 **b.** 13

c. 5

35.

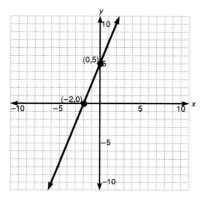

36.

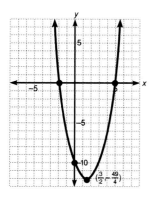

37.

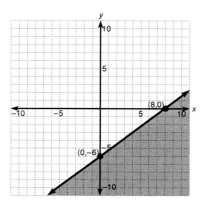

38.

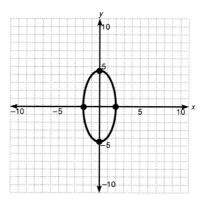

39. a. circle   **b.** hyperbola   **c.** parabola   **d.** ellipse

40. $\left\{\left(\dfrac{11}{19}, -\dfrac{12}{19}\right)\right\}$ **41.** $\{(-2,3),(2,-3)\}$ **42.** 149

43. $\left\{\left(\dfrac{29}{11}, -\dfrac{5}{11}\right)\right\}$ **44.** $\log_b\left(\dfrac{30}{8}\right)$ or $\log_b\left(\dfrac{15}{4}\right)$ **45.** 1.77   **46.** 0.7

47. 3.58   **48.** $\displaystyle\sum_{i=1}^{5} (6i - 3)$ **49.** 49   **50.** -44 **51.** $\dfrac{351}{2}$

52. $-\dfrac{1}{54}$ **53.** $-\dfrac{2}{3}$ **54.** $\dfrac{26}{111}$

55. P_1: $2^2 = \dfrac{2(1)(1 + 1)[2(1) + 1]}{3}$

$4 = \dfrac{2(2)(3)}{3}$

$4 = 4$ P_1 is true

Assume P_k: $2^2 + 4^2 + 6^2 + \cdots (2k)^2 = \dfrac{2k(k + 1)(2k + 1)}{3}$ for some k

Prove P_{k+1}: $2^2 + 4^2 + 6^2 + \cdots + [2(k + 1)]^2 =$
$\dfrac{2(k + 1)[(k + 1) + 1][2(k + 1) + 1]}{3}$

Proof: By induction hypothesis,

$2^2 + 4^2 + 6^2 + \cdots + (2k)^2 = \dfrac{2k(k + 1)(2k + 1)}{3}$

$$2^2 + 4^2 + 6^2 + \cdots + (2k)^2 + [2(k + 1)]^2 = \frac{2k(k + 1)(2k + 1)}{3} + [2(k + 1)]^2$$

$$= \frac{2k(k + 1)(2k + 1)}{3} + 4(k + 1)^2$$

$$= \frac{2k(k + 1)(2k + 1)}{3} + \frac{12(k + 1)^2}{3}$$

$$= \frac{(k + 1)[2k(2k + 1) + 12(k + 1)]}{3}$$

$$= \frac{(k + 1)(4k^2 + 2k + 12k + 12)}{3}$$

$$= \frac{2(k + 1)[2k^2 + 7k + 6]}{3}$$

$$= \frac{2(k + 1)(k + 2)(2k + 3)}{3}$$

$$= \frac{2(k + 1)[(k + 1) + 1][2(k + 1) + 1]}{3}$$

P_k implies P_{k+1} is true and P_n is true for all positive integers n.

Exercise 15–1

Answers to odd-numbered problems

1.

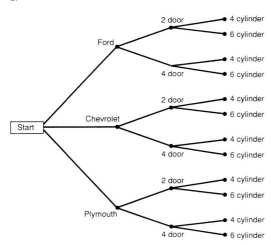

3. H for Heads, T for Tails

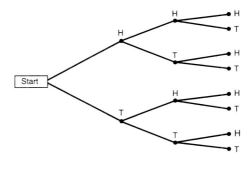

5.

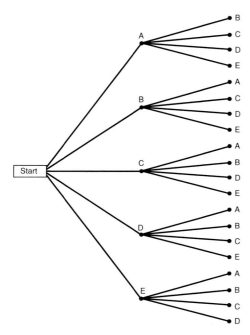

7.

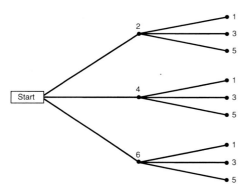

9. 24   **11.** 132   **13.** 3,600   **15.** 256   **17.** 256   **19.** 432
21. 2,880   **23.** 16   **25. a.** 625   **b.** 120   **c.** 375
d. 72   **e.** 250   **f.** 48   **g.** 36   **h.** 12   **i.** 75
j. 12   **27.** 24   **29.** 5,040   **31.** 5,040
33. 60,480   **35.** 362,880   **37.** 35   **39.** 210
41. 184,756   **43.** 3,838,380   **45.** 7,059,052

Solutions to trial exercise problems

15. Since each question could be answered two ways (true or false), the number of ways that all the questions could be answered is:
$2 \cdot 2 \cdot 2 \cdot 2 \cdot 2 \cdot 2 \cdot 2 \cdot 2 = 2^8 = 256$

20. Since either a boy or a girl can sit in the first position, we have 8 people to choose from. Then there are the 4 of the other sex for the second position, 3 from each sex for positions 3 and 4, 2 of each sex left for positions 5 and 6, and 1 of each sex left to fill the last two seats.
$8 \cdot 4 \cdot 3 \cdot 3 \cdot 2 \cdot 2 \cdot 1 \cdot 1 \cdot = 1,152$

Review exercises

1. $(x + 3)(x + 5)$ **2.** $(x + y^2)(x^2 - xy^2 + y^4)$ **3.** $9(x + 2)$
$(x - 2)$ **4.** $(4x - 1)(x + 3)$ **5.** $(2x - y)(a + 3b)$
6. $(2a - 3)(4a^2 + 6a + 9)$ **7.** $(2x - 3)^2$ **8.** $(6x - 1)(x + 3)$

Exercise 15–2

Answers to odd-numbered problems

1. 360   **3.** 120   **5.** 10   **7.** 32,760
9. 665,280   **11.** 504   **13.** 13,800   **15.** 40,320
17. 110   **19.** 40,320   **21.** 1,816,214,400   **23. a.** 360   **b.** 720
c. 720   **d.** 60   **e.** 120   **25.** 120   **27.** 719   **29.** 1,680
31. 9,979,200   **33.** 83,160   **35.** 30   **37.** 3,780   **39.** 554,400

Solutions to trial exercise problems

19. The contractor has 8 lots on which he is going to build 8 different houses. The fact that 5 lots are on one side of the street and 3 lots are on the opposite side of the street is unnecessary information. This is a permutation of the 8 houses.
$8! = 40,320$

26. The four blocks can be arranged in $_4P_4 = 4! = 24$ different ways. Since only 1 way is correct, there are $24 - 1 = 23$ different ways in which she can spell her name wrong.

33. We can write $3a^2b^4c^5$ without exponents as $3aabbbbccccc$. Since either are two a's, four b's, and five c's, the possible distinguishable permutations would be $_8P_8 = \dfrac{8!}{0!} =$

$\dfrac{12!}{1!2!4!5!} = 83,160$

Review exercises

1. $\{-2\}$ **2.** $\left\{x \middle| x < \dfrac{13}{2}\right\}$ **3.** $\{33\}$ **4.** $\left\{x \middle| x \geq -\dfrac{7}{11}\right\}$ **5.** $\{0\}$

6. $\left\{x \middle| -\dfrac{9}{2} \leq x \leq \dfrac{3}{2}\right\}$ **7.** $\emptyset$ **8.** $\{x \mid 1 < x < 6\}$

Exercise 15–3

Answers to odd-numbered problems

1. 3,003   **3.** 1   **5.** 18   **7.** 5,245,786   **9.** 184,756   **11.** 4,845
13. 2,598,960   **15.** 495   **17.** 27,132   **19.** 1,287   **21.** 70
23. 91,390   **25.** 30,260,340   **27.** 105   **29.** 1,330   **31.** 1,001;
3,632,428,800   **33. a.** 24,310   **b.** 272   **c.** 8,821,612,800
d. 6,188   **35.** 5,040; 144   **37.** 120   **39. a.** 125   **b.** 75
c. 20   **d.** 27   **41.** 1,400   **43. a.** 179,520   **b.** 924   **c.** 593,775
45. 21,378,500   **47.** 18,643,560   **49.** 1,716   **51.** 380
53. 725,760   **55.** 4,838,400

Solutions to trial exercise problems

19. There are 13 diamonds among the 52 cards. This is a combination of the 13 diamonds taken 5 at a time.
$$_{13}C_5 = \frac{13!}{(13 - 5)!5!} = \frac{13!}{8!5!} = 1,287$$

Review exercises

1. $\left\{-1, \dfrac{5}{3}\right\}$ **2.** $\left\{-3, \dfrac{1}{3}\right\}$ **3.** $\{x \mid x < -6 \text{ or } x > 1\}$

4. $\left\{x \middle| -\dfrac{1}{3} \leq x \leq 3\right\}$ **5.** $\left\{x \middle| -\dfrac{5}{3} < x < \dfrac{7}{3}\right\}$ **6.** $\{0,5\}$

7. $\left\{x \middle| x \leq -1 \text{ or } x \geq \dfrac{5}{3}\right\}$ **8.** $\left\{x \middle| x \leq -\dfrac{6}{5} \text{ or } x \geq 2\right\}$

Exercise 15–4

Answers to odd-numbered problems

1. $\dfrac{1}{2}$ 3. $\dfrac{3}{4}$ 5. $\dfrac{1}{8}$ 7. $\dfrac{1}{2}$ 9. $\dfrac{3}{8}$ 11. $\dfrac{5}{16}$ 13. $\dfrac{1}{13}$

15. $\dfrac{1}{4}$ 17. $\dfrac{6}{13}$ 19. $\dfrac{6}{13}$ 21. $\dfrac{4}{13}$ 23. $\dfrac{7}{13}$ 25. $\dfrac{12}{13}$ 27. $\dfrac{6}{13}$

29. $\dfrac{3}{4}$ 31. $\dfrac{1}{2}$ 33. $\dfrac{1}{38}$ 35. $\dfrac{18}{19}$ 37. 0   39. $\dfrac{7}{12}$ 41. 0

43. $\dfrac{1}{663}$ 45. $\dfrac{80}{221}$ 47. $\dfrac{20}{663}$ 49. $\dfrac{33}{66,640} \approx 0.0005$

51. $\dfrac{253}{9,996} \approx 0.0253$ 53. $\dfrac{2109}{8330} \approx 0.253$ 55. $\dfrac{1,625}{4,998} \approx 0.3251$

57. $\dfrac{143}{16,660} \approx 0.0086$ 59. $\dfrac{1}{7,059,052}$ 61. $\dfrac{1}{9,366,819}$

63. $\dfrac{1}{12,271,512}$ 65. $\dfrac{1}{4}$ 67. $\dfrac{1}{6}$

Solutions to trial exercise problems

7. At least two tails. The outcomes that make up this event are *HTT, THT, TTH,* and *TTT.* The sample space has 8 outcomes. Then
$$P(A) = \frac{n(A)}{n(S)} = \frac{4}{8} = \frac{1}{2}$$

19. A card between 3 and 10. The endpoints are not included. The statement is the same as the cards from 4 to 9. The probability is
$$P(4 \cup 5 \cup 6 \cup 7 \cup 8 \cup 9)$$
$$= P(4) + P(5) + P(6) + P(7) + P(8) + P(9)$$
$$= \frac{4}{52} + \frac{4}{52} + \frac{4}{52} + \frac{4}{52} + \frac{4}{52} + \frac{4}{52}$$
$$= \frac{4+4+4+4+4+4}{52} = \frac{24}{52} = \frac{6}{13}$$

23. The card is from 2 to 6 or a spade. These two events are not mutually exclusive. If A represents the event of a card from 2 to 6, B represents the event of a spade and $A \cap B$ represent the 2, 3, 4, 5, or 6 of spades, the probability is found as follows:
$$P(A \cup B) = P(A) + P(B) - P(A \cap B)$$
$$= \frac{n(A)}{n(S)} + \frac{n(B)}{n(S)} - \frac{n(A \cap B)}{n(S)}$$
$$= \frac{20}{52} + \frac{13}{52} - \frac{5}{52}$$
$$= \frac{28}{52} = \frac{7}{13}$$

37. The number will be white. Since there are no white numbers, this is an impossible event and the probability is 0.

45. Three men and three women. Let A represent the event of selecting 3 men and 3 women. Since we are making two separate selections and there is no order in the selection, we have
$$n(A) = {}_{10}C_3 \cdot {}_8C_3$$
$$= \frac{10!}{7!3!} \cdot \frac{8!}{5!3!}$$
$$= 120 \cdot 56$$
$$= 6,720$$

The sample space is combinations of 18 people taken 6 at a time.
$$n(S) = {}_{18}C_6 = \frac{18!}{12!6!} = 18,564$$

The probability is found as follows:
$$P(A) = \frac{n(A)}{n(S)} = \frac{6,720}{18,564} = \frac{80}{221}$$

Review exercises

1. $\{-2,4\}$ 2. $\{x \mid -2 < x < 5\}$ 3. $\left\{ \dfrac{-3 \pm i\sqrt{11}}{2} \right\}$

4. $\left\{ -3, \dfrac{1}{4} \right\}$ 5. $\{0,3\}$ 6. $\left\{ \dfrac{-1 \pm \sqrt{33}}{4} \right\}$

7. $\{x \mid x \le -4 \text{ or } x \ge -2\}$ 8. $\{x \mid -1 < x < 0\}$

Exercise 15–5

Answers to odd-numbered problems

1. $a^4 - 12a^3 + 54a^2 - 108a + 81$ 3. $p^6 + 6p^5q + 15p^4q^2 + 20p^3q^3 + 15p^2q^4 + 6pq^5 + q^6$ 5. $16a^4 + 96a^3 + 216a^2 + 216a + 81$ 7. $\dfrac{p^6}{64} - \dfrac{3}{16}p^5q + \dfrac{15}{16}p^4q^2 - \dfrac{5}{2}p^3q^3 + \dfrac{15}{4}p^2q^4 - 3pq^5 + q^6$
9. $a^{10} + 5a^8b^2 + 10a^6b^4 + 10a^4b^6 + 5a^2b^8 + b^{10}$ 11. $x^8 - 4x^6y^3 + 6x^4y^6 - 4x^2y^9 + y^{12}$ 13. $32a^{15} + 80a^{12}b + 80a^9b^2 + 40a^6b^3 + 10a^3b^4 + b^5$ 15. $84a^6b^3$ 17. $-2,002a^9b^5$ 19. $7,920q^8$
21. $18,144k^6$ 23. $56a^{10}b^3$ 25. $165a^6b^{16}$ 27. 1.0263 29. 0.8587
31. $924a^9 (a > 0)$

Solutions to trial exercise problems

1. $(a - 3)^4 = a^4 + \dfrac{4}{1!}a^3(-3)^1 + \dfrac{4 \cdot 3}{2!}a^2(-3)^2 + \dfrac{4 \cdot 3 \cdot 2}{3!}a(-3)^3 +$
$\dfrac{4 \cdot 3 \cdot 2 \cdot 1}{4!}(-3)^4 = a^4 - 12a^3 + 54a^2 - 108a + 81$

9. $(a^2 + b^2)^5 = (a^2)^5 + \dfrac{5}{1!}(a^2)^4(b^2)^1 + \dfrac{5 \cdot 4}{2!}(a^2)^3(b^2)^2$
$+ \dfrac{5 \cdot 4 \cdot 3}{3!}(a^2)^2(b^2)^3 + \dfrac{5 \cdot 4 \cdot 3 \cdot 2}{4!}(a^2)^1(b^2)^4 + \dfrac{5 \cdot 4 \cdot 3 \cdot 2 \cdot 1}{5!}(b^2)^5$
$= a^{10} + 5a^8b^2 + 10a^6b^4 + 10a^4b^6 + 5a^2b^8 + b^{10}$

21. Given $(6 - k)^9$, we want the seventh term, where $n = 9$, $r = 7$, $x = 6$, and $y = -k$. Using $\dfrac{n!}{[n - (r-1)]!(r-1)!}x^{n-(r-1)}y^{r-1}$
$= \dfrac{9!}{3!6!}(6)^3(-k)^6 = \dfrac{9 \cdot 8 \cdot 7}{3!}(216)(k)^6 = 84(216)k^6 = 18,144k^6$.

27. We want $(1.002)^{13} = (1 + 0.002)^{13} \approx 1^{13} + \dfrac{13}{1!}(1)^{12}(0.002) +$
$\dfrac{13 \cdot 12}{2!}(1)^{11}(0.002)^2 + \dfrac{13 \cdot 12 \cdot 11}{3!}(1)^{10}(0.002)^3 = 1 + 0.026 +$
$0.000312 + 0.000002288 = 1.026314288 \approx 1.0263$.

31. Since there are 13 terms, we want the seventh term of $(a + \sqrt{a})^{12}$. Now $n = 12$, $r = 7$, $x = a$, and $y = \sqrt{a}$.
So $\dfrac{n!}{[n-(r-1)]!(r-1)!}x^{n-(r-1)}y^{(r-1)} = \dfrac{12!}{6!6!}a^6(\sqrt{a})^6$
$= \dfrac{12 \cdot 11 \cdot 10 \cdot 9 \cdot 8 \cdot 7}{6 \cdot 5 \cdot 4 \cdot 3 \cdot 2 \cdot 1}a^6a^3 = 11 \cdot 3 \cdot 4 \cdot 7\, a^9 = 924a^9$. The middle term of $(a + \sqrt{a})^{12}$ is $924a^9$. $(a > 0)$

Review exercises

1. $x^{5/4}$ 2. $2x^2y\sqrt[3]{4y}$ 3. $13\sqrt{2}$ 4. $\dfrac{\sqrt[3]{2a^2b}}{2ab}$ 5. $\dfrac{3\sqrt{10} + 5\sqrt{6}}{30}$

6. $a + 2\sqrt{ab} + b$ 7. $\dfrac{\sqrt{x} + 2}{x - 4}$ 8. $\sqrt{7} - 2$

Chapter 15 review

1.

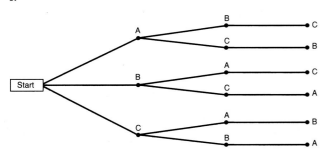

2.

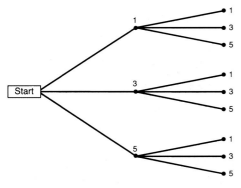

3. 1,512   **4.** 1,024   **5.** 1,024   **6.** 120   **7.** 1,320   **8. a.** 24
b. 24   **c.** 12   **d.** 12   **e.** 6   **9.** 840   **10.** 5,040
11. 53,130   **12.** 133,784,560   **13.** 38,760   **14.** 220

15. a. $\dfrac{3}{8}$ **b.** $\dfrac{7}{8}$ **16. a.** $\dfrac{1}{13}$ **b.** $\dfrac{1}{4}$ **c.** $\dfrac{5}{13}$ **d.** $\dfrac{4}{13}$ **e.** $\dfrac{4}{13}$

f. $\dfrac{3}{4}$ **17.** $x^7 + 35x^6 + 525x^5 + 4{,}375x^4 + 21{,}875x^3 + 65{,}625x^2$
$+ 109{,}375x + 78{,}125$ **18.** $32a^5 - 240a^4b + 720a^3b^2 - 1{,}080a^2b^3$
$+ 810ab^4 - 243b^5$ **19.** $\dfrac{a^4}{16} - \dfrac{3a^3b}{2} + \dfrac{27a^2b^2}{2} - 54ab^3 + 81b^4$

20. $84{,}480a^7$
21. $19{,}702{,}683a^8b^6$ **22.** $608{,}256x^{10}y^2$

Final examination

1. $\{-4,-3,-2,-1,0,1,2,3,4,5\}$ **2. a.** $\{-6,-4,-2,0,1,6,9\}$
b. $\{-4,6,9\}$ **c.** $\emptyset$ **3.** -46
4. distributive and commutative   **5. a.** $9y^2 - 4$ **b.** $16x^2 - 8x + 1$

c. $y^3 + 8$ **6. a.** $\dfrac{a^6}{b^9}$ **b.** $-8a^4b^3c^3$ **c.** $\dfrac{y^6}{x^4}$

7. $\{-11\}$ **8.** $\{1\}$ **9.** $\left\{-\dfrac{10}{3}\right\}$ **10.** $(5y - 1)(y + 3)$

11. $3xy(x - 3y + 5x^2y^2)$ **12.** $(3y - z)(x + 2y)$
13. $5(x - 3)(x + 1)$ **14.** $2(y + 7)(y - 7)$ **15.** $(3x - 7z)^2$
16. $\{x|x \le 2\}$ **17.** $\{x|1 \le x \le 2\}$ **18.** $\{x|2 \le x \le 8\}$
19. $\{x|x < 2 \text{ or } x > 5\}$ **20.** $\{-1,2\}$ **21.** $\left\{\dfrac{1 - \sqrt{33}}{8}, \dfrac{1 + \sqrt{33}}{8}\right\}$

22. $\{1\}$; -6 is extraneous   **23.** $\{-2\sqrt{2}, 2\sqrt{2}, -i\sqrt{2}, i\sqrt{2}\}$
24. $\{y|y \le -4 \text{ or } y \ge 8\}$ **25.** $\{x|x < -3 \text{ or } x > -2\}$

26. $\dfrac{10y + 17}{(2y - 1)(y + 5)}$ **27.** $\dfrac{-x^2 + 5x}{(x - 3)(x - 1)(x + 3)}$

28. $\dfrac{x^2 - x + 1}{2x + 1}$ **29.** $\dfrac{3y - 2x}{y + 5x}$

30. $9\sqrt{2}$ **31. a.** $30 - 5\sqrt{3} + 18\sqrt{2} - 3\sqrt{6}$ **b.** $7 - 24i$

32. $\dfrac{4}{5} - \dfrac{3}{5}i$ **33.** $m = \dfrac{5}{2}$; x-intercept, $(2,0)$; y-intercept, $(0,-5)$

34. a. $x + 3y = 11$ **b.** $2x + y = -3$ **35. a.** -13 **b.** 16

c. $2x + h$ **36.** $f^{-1}(x) = \dfrac{x + 6}{9}$ **37. a.** x-intercepts, $(7,0)$ and

$(-1,0)$; y-intercept, $(0,-7)$ **b.** vertex, $(3,-16)$ **c.** $x = 3$
38. x-intercepts, $(-\sqrt{2},0)$, $(\sqrt{2},0)$; y-intercepts, $(0,-\sqrt{5})$, $(0,\sqrt{5})$;
ellipse   **39. a.** hyperbola   **b.** parabola   **c.** circle   **40.** $4x^2 -$

$15x + 47 - \dfrac{142}{x + 3}$ **41.** $\{-1,2,5\}$

42.

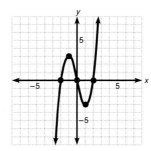

43. $\log_3\left(\dfrac{45}{4}\right)$ **44.** $\left\{\dfrac{25}{7}\right\}$ **45.** $\{1.68\}$ **46.** ≈ 3.96

47. $\{(1,-2)\}$ **48.** 34   **49.** $\left\{\left(-\dfrac{11}{7}, \dfrac{9}{7}\right)\right\}$

50. $\displaystyle\sum_{i=1}^{5} (-1)^{i + 1}(6i - 1)$ **51.** 133   **52.** 96   **53.** 5,053

54. $-\dfrac{6}{5}$ **55.** $\dfrac{1}{48}$ **56.** 86,400   **57.** 8,821,612,800   **58.** 720

59. 142,506   **60. a.** 286,650   **b.** 45   **c.** 1,081,575   **61.** 462

62. $\dfrac{4}{429}$ **63.** $\dfrac{160}{429}$ **64.** $\dfrac{33}{66{,}640}$ **65.** $\dfrac{253}{9{,}996}$ **66.** $924a^6$

67. $81x^4 - 216x^3y + 216x^2y^2 - 96xy^3 + 16y^4$

Index

Symbols

Symbol	Means or Is Read	Section Number
$\{a,b,c\}$	The set whose elements are a, b, and c	1–1
ϵ	Is an element of	1–1
$\subseteq$	Is a subset of	1–1
$\emptyset$	Empty set or null set	1–1
$\cup$	Union	1–1
$\cap$	Intersection	1–1
$N = \{1,2,3,\cdots\}$	The set of natural numbers	1–1
$W = \{0,1,2,3,\cdots\}$	The set of whole numbers	1–1
$J = \{\cdots,-3,-2,-1,0,1,2,3,\cdots\}$	The set of integers	1–1
$\{x \mid x \text{ satisfies some conditions}\}$	Set-builder notation	1–1
$Q = \left\{\dfrac{p}{q} \middle\vert p \text{ and } q \epsilon J, \text{ and } q \neq 0\right\}$	The set of rational numbers	1–1
H	The set of irrational numbers	1–1
R	The set of real numbers	1–1
$<$	Is less than	1–1
$>$	Is greater than	1–1
$\leq$	Is less than or equal to	1–1
$\geq$	Is greater than or equal to	1–1
$\|x\|$	Absolute value of x	1–1
$[a,b]$	The interval from a to b	2–5
(a,b)	The interval between a and b	2–5
∞	Infinity	2–5
a^n	The nth power of a	3–1
a^{-n}	a to the negative n power $\left(a^{-n} = \dfrac{1}{a^n}, a \neq 0\right)$	3–3
a^0	a to the zero power $(a^0 = 1, a \neq 0)$	3–3
$\sqrt[n]{a}$	The principal nth root of a	5–1
$a^{1/n}$	The principal nth root of a	5–1
i	Imaginary number $(i = \sqrt{-1})$	5–7
$a + bi$	Complex number	5–7
$\pm$	Plus or minus	6–1
$\dfrac{-b \pm \sqrt{b^2 - 4ac}}{2a}$	Quadratic formula	6–3
(x,y)	Ordered pair of numbers	7–1
$d = \sqrt{(x_2 - x_1)^2 + (y_2 - y_1)^2}$	The distance between any two points (x_1,y_1) and (x_2,y_2)	7–2
$\left(\dfrac{x_1 + x_2}{2}, \dfrac{y_1 + y_2}{2}\right)$	Midpoint of a line segment	7–2